Proceedings of the International Symposium for Production Research 2018

Numan M. Durakbasa · M. Günes Gencyilmaz
Editors

Proceedings of the International Symposium for Production Research 2018

Springer

Editors
Numan M. Durakbasa
Faculty of Mechanical and Industrial Engineering
Technische Universität Wien
Vienna, Austria

M. Günes Gencyilmaz
Faculty of Engineering
Istanbul Aydın University
Istanbul, Turkey

ISBN 978-3-030-06396-2 ISBN 978-3-319-92267-6 (eBook)
https://doi.org/10.1007/978-3-319-92267-6

Softcover re-print of the Hardcover 1st edition 2019

This Springer imprint is published by the registered company Springer Nature Switzerland AG
The registered company address is: Gewerbestrasse 11, 6330 Cham, Switzerland

Foreword

Dear Colleagues and Friends,

It is a great pleasure to welcome you to the "International Symposium for Production Research—ISPR 2018" at the TU Wien, between 28 and 31 August, 2018, with the theme of "**Impacts of Industry 4.0 on Production Systems**". We are particularly pleased with our collaboration with Society for Production Research, İstanbul, Turkey, in hosting this year's Symposium in Vienna.

The purpose of the symposium is to bring together researchers, scientists and experts at universities, companies, institutions, communities, associations and societies to share ideas and vision on production and operations management and technology and to provide a forum for experts and professionals to discuss the actual theme "Impacts of Industry 4.0 on Production Systems" and the further relevant developments in this area.

We hope that this gathering will offer a venue where the participants will enrich their professional knowledge base and will present an opportunity to all the participants to build and deepen friendships with experts, industry and corporate professionals and authorities.

August 2018

Kurt Matyas
Vice-rector for Academic Affairs

Preface

The 18th International Symposium for Production Research "ISPR2018" was held in Vienna, Austria, from 28 to 31 August 2018, with the theme of "Impact of Industry 4.0 on Production Systems". This symposium was organized in Vienna, for the second year in a row, by TU Wien and Society for Production Research, Istanbul, Turkey.

The generic theme of "Industry 4.0" was first adopted in the symposium held in 2016 and maintained in the following two symposia in 2017 and 2018 but with an emphasis on the relevant developments and progress on the various aspects of this "fourth industrial revolution", thus symposium adopting the same main theme but taking various aspects of it into account, with the purpose of drawing the attention of the researchers to the influences of this new industrial era on the production systems and production management.

The world of science and technology is under increasing influence of the requirements of Industry 4.0. Transition to a new era seems inevitable for every sector of industry. Due to the importance of this theme, ISPR2018 hosted numerous distinguished speakers from both the academia and industry to hear their views on the impacts of Industry 4.0 on the various components of production systems.

This volume of proceedings from the symposium contains 76 refereed papers in 18 categories shown in the contents of the proceedings. Participants from 11 countries attended the symposium. 12 invited talks and 76 papers were presented in 19 sessions.

We are very grateful to our host institution, the Vienna University of Technology, for its invaluable support and hospitality and for enabling this symposium to be organized, for a second time, in their premises. In particular, we would like to express our gratitude to Vice-Rector Prof. Kurt Matyas, also the Honorary Chairman of the Scientific Committee of this symposium for his leadership and generous support. We also thank to Prof. Detlef Gerhard, Dean of the Faculty of Mechanical and Industrial Engineering, Prof. Friedrich Bleicher, Head of the Institute for Production Engineering and Laser Technology for their interest and support for this symposium.

We would like to thank all the keynote and invited speakers whose contributions enhanced the success of the symposium.

In organizing this event, our colleagues in Vienna and Istanbul contributed endless hours of hard work, energy and wisdom to make this event the success it was. On the Vienna side, our sincere thanks go to the staff of the Department for Interchangeable Manufacturing and Industrial Metrology of the Institute for Production Engineering and Laser Technology, in particular, Mr. Erol Güclü.

On the Istanbul side, we are very grateful to Prof. Zaim and Prof. Çebi who made available the resources of their departments, and to Ms. Hatice Camgöz Akdağ who was involved in every aspect of the organization from the very beginning. Our special thanks go to Ms. Tuğçe Beldek, and Mr. Kemal Konyalıoğlu, as well as our research assistants, for their hard and dedicated work.

We would like to express our gratitude to the board members of the Society for Production Research in Istanbul for their strong support and dedication to make the symposium a success.

Our very special thanks go to our colleagues, the participants of this symposium. Undoubtedly, they are the core component of this organization.

We would like to recognize and thank our dear colleagues who graciously accepted to join the honorary and scientific committees or who served as peers in this event.

Finally, no such event is possible without the generous support of patrons and sponsors. In this regard, we would like to thank Dr. Michael Ludwig, the Mayor of Vienna, for hosting a reception for the participants of this symposium, and to all the corporations and individuals who provided invaluable financial and intellectual contributions.

And last, but not least, we are grateful to Ms. Silvia Schilgerius, Senior Editor Applied Sciences from Springer Nature, for her able guidance, professionalism and patience.

M. Güneş Gençyılmaz
Numan M. Durakbaşa

Production of the Future

The Industry 4.0 is an innovative concept and model for future enterprises which is initiated with the aim to provide cost-effective, efficient, agile and optimal ways to customer-driven design and production. Companies operating in various production areas in the future will be adapting more and more Industry 4.0 technologies that were originally modelled in smart manufacturing and multi-functional integrated factories (MFIF). The transition to the Industry 4.0 requires models to be integrated by utilizing advanced information analytics, artificial intelligence and interconnected Industrial Internet of things (IIoT) as a part of automated and robotic applications, networked intelligent machines and instruments.

Intelligence is an essential feature of future development and production systems and intelligent production is a major component of future businesses, also according to the technological developments especially in the field of precision engineering at micro/nano and pico scale production. Modern production engineering and production metrology and its industrial application started on the basis of the scientific, technical and organisational work of E. Abbe, William Taylor and F. W. Taylor and at the end of the twentieth century—the "Quality Control Century"—the development has reached nanotechnology and is proceeding to pico- and phytotechnology. To meet market demands in a global industrial world, manufacturing enterprises must be flexible and agile enough to respond quickly to product demand changes also according technological developments especially in the field of precision engineering at micro-, nano- and pico-scale production with support of artificial intelligence (AI) and modern Information technology (IT), modern cost-effective customer-driven design and manufacturing.

Adequate knowledge in the areas of intelligent production and especially intelligent metrology are important presuppositions to achieve waste-free production and low costs of manufacturing and accuracy at the same time within the sophisticated production systems. This is of extreme importance in the current and future worldwide competition in industry and production engineering and at the same time in the face of increasingly higher costs of energy and raw material. Learning with self-improving ability makes possible the way to "Zero Error" production. Fuzzy logic will be applied for quality function deployment (QFD) and for the monitoring and forecasting of maintenance of measuring instruments, and further on for intelligent design and for a high-level knowledge-based expert system for tolerancing and quality planning. The supervision of quality in the process chain as well as its optimization by means of knowledge and neuronal-network-based learning management system, and a self-learning system with neuronal networks can be used in the factory of the future according to Industry 4.0 concept to learn stepwise from deviations and to improve the processes continuously.

On the basis of the technological developments, the manufacturing sector faces radical structural changes, with the digital transformation offering enormous opportunities but also presenting high-level challenges. The challenge today is implementing the concepts of the automation creatively with the possibilities of new sensors and instruments that allow metrological technologies and integrate telepresence production processes towards a complete platform industry 4.0 not only in large production complexes but mainly in the SMEs.

Amid the challenges of competitive and digital technologies in the advanced manufacturing industry, the universities and industry cooperation are imperative. The new aspects of cooperation between the research institutions and the manufacturing industry will provide a development capacity for high-quality and innovative products in the frame of Industry 4.0 concept.

August 2018

Günes Gencyilmaz
Numan Durakbasa

Organization

ISPR2018 was organized by TU Wien, Austria, and the Society for Production Research, Turkey. The symposium took place on the TUtheSky Campus of the TU Wien.

Editors

Numan M. Durakbaşa
M. Güneş Gençyılmaz

Co-editors

Peter Kopacek
Ayhan Toraman
Selim Zaim
Jorge Martin Bauer
Serpil Erol
Semra Birgün
Kemal Güven Gülen
Andreas Otto
Alptekin Erkollar
Mahmut Tekin

Honorary Chair

Kurt Matyas	TU Wien, Austria

Symposium Chairs

Numan M. Durakbaşa	TU Wien, Austria
M. Güneş Gençyılmaz	Istanbul Aydın University, Turkey

International Honorary Committee

Andrew Kusiak	The University of Iowa, USA
Duc Truong Pham	University of Birmingham, UK
Albert Weckenman	University of Erlangen-Nürnberg Germany
Daniela Popescu	Technical University of Cluj-Napoca, Romania
Detlef Gerhard	TU Wien
Sorin Popescu	Technical University of Cluj-Napoca, Romania
Vidosav Majstorovic	University of Belgrade, Serbia
Stanislaw Adamczak	Kielce University of Technology, Poland
Friedrich Bleicher	TU Wien
Lubomir Šooš	Slovak University of Technology in Bratislava, Slovakia
Jozef Peterka	Slovak University of Technology, Slovakia

Organizing Committee

Ayhan Toraman, Turkey
Selim Zaim, Turkey
Güneş Gençyılmaz, Turkey
Erol Güçlü, Austria
Semra Birgün, Turkey
Serpil Erol, Turkey
Kemal Güven Gülen, Turkey
Alptekin Erkollar, Turkey
Eva Walcher, Austria
Ferhan Çebi, Turkey
Günther Poszvek, Austria
Alp Baray, Turkey
Hatice Camgöz Akdağ, Turkey
Haluk Soyuer, Turkey
Hür Bersam Bolat, Turkey
Ece Soyuer, Austria
Ezequiel Simon, Austria
Jorge Bauer, Austria
Tuğçe Beldek, Turkey
Kemal Konyalıoğlu, Turkey
Ignacio Dorna, Austria
Gökçen Baş, Austria
Gamze Uğur Tuncer, Austria
Oskar Mayer, Austria
David Riepl, Austria
Liane Höller, Austria

Prepared for Publishing by

Tuğçe Beldek, Turkey
A. Kemal Konyalıoğlu, Turkey

Scientific Committee

E. Hekelová	City University Bratislava, Slovakia
F. Holesovsky	J. E. Purkyně University in Ústí nad Labem, Czech Republic
Z. Irani	University of Bradford, UK
V. Işler	Hasan Kalyoncu University, Turkey
C. Kahraman	Istanbul Technical University, Turkey
İ. Kara	Başkent University, Turkey
T. Katoka	Kindai University, Japan
M. Klumpp	Duisburg University, Germany
D. Kocaoğlu	Portland State University, USA
T. Koç	Istanbul Technical University, Turkey
P. Kopacek	Vienna University of Technology, Austria
M. Králik	Slovak University of Technology in Bratislava, Slovakia
L. Kräuter	Vienna University of Technology, Austria
C. Kubat	Sakarya University, Turkey
U. Kula	American University of the Middle East, Kuwait
A. Kulakli	American University of the Middle East, Kuwait
J. Kundrak	University of Miskolc, Hungary
A. Kusiak	University of Iowa, USA
V. Majstorovic	University of Belgrade, Serbia
I. Mankova	Technical University of Kosice, Slovakia
A. Markopoulos	National Technical University of Athens, Greece
K. Matyas	Vienna University of Technology, Austria
M. Novák	J. E. Purkyně University in Ústí nad Labem, Czech Republic
Y. Omurtag	Robert Morris University, USA
N. Özçakar	Istanbul University, Turkey
D. Özdemir	Bilgi University, Turkey
A. Özok	Okan University, Turkey
E. Öztemel	Marmara University, Turkey
J. Peterka	Slovak University of Technology, Slovakia
D. Pham	University of Birmingham, UK
M. Pokusová	Slovak University of Technology in Bratislava, Slovakia
S. Popescu	Technical University of Cluj-Napoca, Romania
D. Popescu	Technical University of Cluj-Napoca, Romania
E. Pucher	Vienna University of Technology, Austria
A. Sağbaş	Namık Kemal University, Turkey
J. Sanz Juan	Polytechnic University of Valencia, Spain
A. Sharif	University of Bradford, UK
A. Smith	University of Auburn, USA
Ľ. Šooš	Slovakia University of Technology, Slovakia
A. Sorguç	Middle East Technical University, Turkey
A. Soysal	Doğuş University, Turkey
H. Soyuer	Ege University, Turkey
K. Stepien	Kielce University of Technology, Poland
B. Tan	Koç University, Turkey
M. Tanyaş	Maltepe University, Turkey

F. Taşgetiren	Yaşar University, Turkey
H. Taşkin	Sakarya University, Turkey
M. Tekin	Selçuk University, Turkey
A. Toraman	Istanbul Technical University, Turkey
J. Torgersen	Norwegian University of Science and Technology, Norway
O. Torkul	Sakarya University, Turkey
G. Ulusoy	Sabancı University, Turkey
G. Varga	University of Miskolc, Hungary
K. Velíšek	Slovak University of Technology, Slovakia
J. Wang	National University of Singapore, USA
A. Weckenmann	University of Erlangen-Nürnberg, Germany
M. Yalçintaş	Istanbul Commerce University, Turkey
M. Yenisey	Istanbul University, Turkey
M. Yurci	Yıldız Technical University, Turkey
S. Zaim	Istanbul Şehir University, Turkey
W. Zębala	Cracow University of Technology, Poland
M. Zerenler	Selçuk University, Turkey

Reviewers

Baray, Ş.
Bayraktar, E.
Bayyurt, N.
Bereketli-Z., İ.
Berto, F.
Beyca, Ö.
Blecha, P.
Bolat, H.
Budak, I.
Bulut, Ö.
Camgöz Akdağ, H.
Crisan, L.
Çebi, F.
Dragomir, M.
Dregelyi-Kiss, Á.
Ekinci, E.
Erkollar, A.
Erol, S.
Esnaf, Ş.
Gergin, Z.
Gülen, K.
Kazançoğlu, Y.
Kesikburun, D.
Kızılay, D.
Kopacek, P.
Krauter, L.
Küçükdeniz, T.
Mankova, I.
Mullaoğlu, G.
Novak, M.
Öner, A.
Öner, E.
Özcan, S.
Özdemir, D.
Öztürkoğlu, Y.
Pokusova, M.
Soyuer, H.
Staiou, E.
Stepien, K.
Şahingöz, Ö.
Tekin Temur, G.
Torgersen, J.
Üney Yüksektepe, F.
Üstündağ, A.
Varga, G.
Yurtseven, C.
Zaim, S.
Zębala, W.

Contents

Process Management

Production Planning and Control

Productivity and Performance Management

Quality Management

Artificial Intelligent Applications

Multi-objective Optimization Study in Face Milling of Steel

János Kundrák[1], Angelos P. Markopoulos[2(✉)], Tamás Makkai[1], Nikolaos E. Karkalos[2], and Antal Nagy[1]

[1] Institute of Manufacturing Science, University of Miskolc, Miskolc-Egyetemváros, Miskolc 3515, Hungary
{kundrak, tamas.makkai, antal.nagy}@uni-miskolc.hu
[2] School of Mechanical Engineering, Section of Manufacturing Technology, National Technical University of Athens, Heroon Politechniou 9, 15780 Athens, Greece
{amark, nkark}@mail.ntua.gr

Abstract. High productivity of parts manufacturing in industrial practice is closely related, not only to time efficiency, but also to the production of parts with high surface quality and considerable lifespan. Face milling is widely used for the efficient creation of accurately flat surfaces, for a large variety of part sizes and materials. However, determining the process parameters, which can lead to the achievement of all required conditions can be considered as a multi-objective problem. This problem can be sufficiently solved using suitable optimization techniques. In the present work, it is attempted to determine the optimum parameters for face milling of steel parts, in order to achieve minimum cutting forces and surface roughness, as well as maximum possible material removal rate. For that reason, after regression models are derived to correlate process parameters with cutting forces and surface roughness, an optimization process is carried out with two different optimization methods, namely Genetic Algorithm and Fireworks Algorithm and after the determination of the optimum process parameters, results concerning the efficiency of optimization algorithms are discussed as well.

Keywords: Face milling · Design of experiments · Optimization Genetic algorithm · Fireworks algorithm

1 Introduction

Machining processes are essential in metalworking industries, as they can accurately and efficiently produce a large variety of part features. Especially, in the automotive and aerospace industries, turning and milling are widely used for the fabrication of a considerable part of the required components. When it is required to create high quality flat surfaces, face milling process is usually selected due to its simplicity and ability to achieve high material removal rates.

For the study of machining processes, it is required to derive the correlation between process parameters and the outcome of the process, e.g. cutting forces, temperatures, surface quality and residual stresses, among others. In the relevant literature

N. M. Durakbasa and M. G. Gencyilmaz (Eds.): ISPR 2018, *Proceedings of the International Symposium for Production Research 2018*, pp. 3–15, 2019.
https://doi.org/10.1007/978-3-319-92267-6_1

of face milling, there exist many studies on cutting forces, surface quality and workpiece integrity [1–4]. Regarding surface quality, Mikó [5] developed a surface roughness predictive model for milling of a complex free-form surface using statistical methods. Mikó and Farkas [6] presented a study where both flatness and surface roughness were evaluated after face turning and face milling of various workpieces, with a view to determine strategies for the control of these deviations. An interesting approach concerning the theoretical prediction of surface roughness parameters was presented by Kundrák and Felhő [7], who used a CAD software to model the machined surface for face milling cases with square and dodecagonal inserts. In a later work, Felhő and Kundrák [8] developed also a theoretical model for the prediction of surface roughness in the case of face milling with octagonal and circular inserts. Finally, Mikó, Tóth and Varga [9] conducted an analysis regarding surface roughness for the case of ball end milling under various process conditions.

Apart from cutting forces and surface quality, it is considerably important to obtain high material removal rates in industrial practice. However, as the goals of obtaining high surface quality, low power consumption, low tool wear and workpiece damage but also high material removal rate are contradictory, it is required to conduct an optimization process for the determination of suitable process conditions, which will enable the achievement of each goal at a sufficient level. Wang, Liu and Wang [10] conducted a multi-objective optimization for turning using NSGA-II algorithm, emphasizing on energy consumption, machining cost and surface quality. Their approach was able to perform sufficiently for energy and cost objectives but improvement on surface quality was limited. Yan and Li [11] conducted multi-objective optimization for milling using grey relational analysis (GRA) and response surface methodology (RSM), focusing on surface quality, production rate and energy consumption. Anand et al. [12] also performed optimization on turning regarding energy consumption during machining of various metals such as steel, aluminum and brass. Sharma and Pandey [13] derived the optimum parameters regarding residual stress minimization during ultrasonic turning process by RSM. Moreover, Fu, Zhao and Liu [14] were able to determine the optimum cutting parameters in high-speed milling with a model combining GRA with principal component analysis.

In the present work, a multi-objective optimization approach is presented, with the aim of obtaining process parameters, which will maximize material removal rate during face milling, under the constraints of maintaining lowest possible cutting forces and surface roughness. For that reason, the three components of cutting forces, two surface roughness indicators and material removal rate are included in the objective function. The methodology presented comprises of an efficient DOE method which reduces considerably the experimental work, the derivation of regression models correlating input and output quantities of the face milling process and finally the optimization process using Genetic Algorithm (GA) and Fireworks Algorithm (FA) techniques. After the results are presented, discussion on the efficiency of the proposed method is conducted and useful conclusions are drawn.

2 Methodology

In this work, results from face milling experiments on steel workpieces with various process conditions are employed in order to determine optimum process parameters according to the desired goals. In the flowchart of Fig. 1, the steps followed in the present work are presented in detail. The desired goal for this work is the maximization of material removal rate under the constraint of maintaining cutting forces and surface roughness as low as possible. More specifically, the requirement for low cutting forces leads to minimum tool wear and power consumption of the machine tool, the requirement for minimum surface roughness leads to better surface quality and maximization of material removal rate leads to increase of production rate.

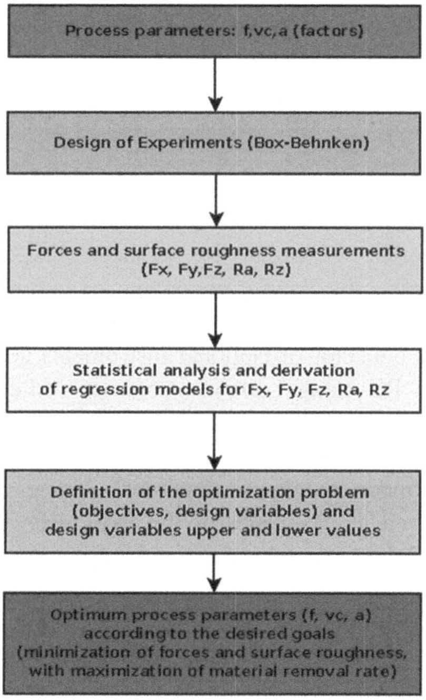

Fig. 1. Flowchart of the methodology followed for the determination of optimum process parameters

The 15 experiments were designed according to Box-Behnken method for 3 factors at 3 levels each, as can be seen in Table 1. For these experiments, a milling head with a single cutting insert was employed in a Perfect Jet MCV-M8 vertical machining center. The milling head diameter was 80 mm, the width of cut 58 mm and the workpiece material was normalized C45 (1.0503) carbon steel. Typical chemical composition for C45 is: 0.45% C, <0.40% Si, 0.65% Mn, <0.40% Cr, <0.10% Mo, <0.40% Ni and the rest is Fe.

Table 1. Process parameters values for each experiment

Number of experiment	Feed ($mm \cdot tooth^{-1}$)	Cutting speed ($m \cdot min^{-1}$)	Depth of cut (mm)
1	0.2	50	0.25
2	1.6	50	0.25
3	0.2	400	0.25
4	1.6	400	0.25
5	0.2	225	0.1
6	1.6	225	0.1
7	0.2	225	0.4
8	1.6	225	0.4
9	0.9	50	0.1
10	0.9	400	0.1
11	0.9	50	0.4
12	0.9	400	0.4
13	0.9	225	0.25
14	0.9	225	0.25
15	0.9	225	0.25

During the experiments, the three components of cutting forces were recorded using Kistler 9257A dynamometer, three Kistler 5011A charge amplifiers and CompactDAQ-9171 data collector of National Instruments company. These measurements were stored on a PC using LabView software. After the experiments were conducted, surface roughness measurements were performed; for the analysis presented hereafter, surface roughness measurements concerning Ra (arithmetic mean deviation of the profile) and Rz (mean roughness depth or average maximum height of the profile) are used, measured at the center of the workpiece width. Ra and Rz were chosen as they are considered among the most important roughness indicators in industrial practice.

For the optimization process, two different optimization techniques, namely the well-established Genetic Algorithm technique and the relatively new, but promising, Fireworks Algorithm technique are employed in this work. In the GA models, the candidate solutions of the optimization problem are represented by chromosomes, composed by a number of genes equal to the number of design variables. The optimization process consists of applying various operators such as parent selection, crossover and mutation to the candidate solutions in order to produce better offspring solutions and eventually reach the global optimum of the desired objective function.

In the FA [15], fireworks are placed in several locations within the search area and after exploding they produce sparks, which are dispersed in the search area. Each position of spark is evaluated according to the objective function; fireworks which are located in favorable positions are considered "good" and produce more sparks, uniformly distributed around the explosion location, whereas "bad" fireworks produce fewer sparks, dispersed in larger area in order to avoid getting trapped in unfavorable regions. After the evaluation of sparks, new firework positions are chosen and the process continues until a termination criterion is met [16].

3 Results and Discussion

A. *Experimental results*

After the face milling experiments were conducted with the process parameters values described in Table 1, results regarding cutting forces and surface roughness are obtained and presented in Table 2.

Table 2. Results of the face milling experiments

No of experiment	Feed (mm · tooth−1)	Cutting speed (m · min−1)	Depth of cut (mm)	F_x (N)	F_y (N)	F_z (N)	Ra (μm)	Rz (μm)
1	0.2	50	0.25	93.56	121.25	132.95	3.68	19.78
2	1.6	50	0.25	356.49	602.21	466.05	16.66	74.49
3	0.2	400	0.25	70.89	115.03	181.31	0.50	8.98
4	1.6	400	0.25	236.15	533.3	521.11	17.64	58.73
5	0.2	225	0.1	33.92	50.99	100.88	1.00	11.15
6	1.6	225	0.1	113.5	192.45	325.5	17.58	58.93
7	0.2	225	0.4	113.34	182.87	212.63	0.91	4.22
8	1.6	225	0.4	243.35	963.18	547.31	16.24	59.95
9	0.9	50	0.1	102.86	160.2	184.49	10.63	48.90
10	0.9	400	0.1	63.43	136.34	253.3	8.89	36.05
11	0.9	50	0.4	353.62	586.61	421	7.29	37.60
12	0.9	400	0.4	245.38	508.03	382.68	7.27	29.33
13	0.9	225	0.25	194.08	342.46	337.91	9.26	38.43
14	0.9	225	0.25	194.39	341.4	335.61	9.18	37.76
15	0.9	225	0.25	193.57	343.73	340.2	9.13	37.71

B. *Derivation of regression models*

As presented in the flowchart of Fig. 1, before the optimization process is started, it is necessary to derive the correlations between process parameters, cutting forces and surface roughness. For that reason, regression models are derived, for which statistical analysis is conducted to determine the more appropriate regression model type. Moreover, ANOVA allows for the determination of the importance of each term of the regression model, something that is important for the optimization process as well, as variations of important terms have very significant impact on the results.

Regarding the forces components, the equations which relate F_x, F_y and F_z with the process parameters were derived in a previous work [17], as follows:

$$F_x^* = 0.22534 - 0.12708 f_z + 1.45348 \cdot 10^{-4} v_c - 0.62854 a - 6.68472 \cdot 10^{-6} f_z v_c + 0.11431 \cdot f_z a - 1.5524 \cdot 10^{-4} v_c a + 0.034551 f_z^2 - 1.20023 \cdot 10^{-7} v_c^2 + 0.76339 a^2 \quad (1)$$

where $F_x^* = \frac{1}{\sqrt{F_x}}$

$$F_y = 13.79216 - 26.3030f_z + 0.11952v_c + 165.94405a - 0.12794f_z v_c + 1521.07143f_z a - 0.524v_c a \quad (2)$$

$$F_z = -48.84095 + 220.01786f_z + 0.095164v_c + 582.225a \quad (3)$$

Regarding the surface roughness parameters Ra and Rz, the regression equations were derived using Minitab software and are presented afterwards:

$$Ra = 1.98 + 8.36f_z - 0.01469v_c + 8.75a + 0.00849f_z v_c - 2.98f_z a + 0.0164v_c a + 0.86f_z^2 - 30.2a^2 \quad (4)$$

$$\mathrm{Rz} = 16.41 + 38.13f_z - 0.0868v_c + 30.2a - 0.01012f_z v_c + 18.9f_z a + 0.0436v_c a - 1.92f_z^2 + 0.000113v_c^2 - 154a^2 \quad (5)$$

It can be seen from Eqs. (4) and (5), that for both Ra and Rz, second order regression models are the most appropriate. For the Ra model, R^2, R^2-adjusted and R^2-predicted values were found to be 99.6%, 98.88% and 93.63%, respectively and for the Rz 99.62%, 98.94% and 93.99%, respectively. In order to assess the validity of the regression models for surface roughness, both the experimental and predicted values for each case are presented in Table 3. Furthermore, in Fig. 2, experimental and predicted results for the roughness parameters for each experiment are portrayed. From both Table 3 and Fig. 2, it can be asserted that most of the predicted values are sufficiently close to the experimental ones and so, the regression models can accurately describe the correlation between the milling process parameters and surface roughness.

After the derivation of regression models for Ra and Rz, it is possible to determine the effects of each process parameter to Ra and Rz. In Figs. 3 and 4, the response surfaces for the surface roughness indicators are presented. Furthermore, in order to observe more directly the effect of each process parameter to Ra and Rz, the main effects plots are also presented in Figs. 5 and 6. From Figs. 3 and 5, it can be observed that Ra is mostly affected from feed and that an increase in feed results to a considerable increase of Ra, whereas an increase of cutting speed leads to a slight decrease of Ra and an increase of the depth of cut leads to a slight increase and then decrease of Ra. Moreover, almost similar trends can be observed for Rz from Figs. 4 and 6. ANOVA results also confirm that, in both cases feed was the most important factor, so it is expected to affect considerably the optimization process.

Finally, material removal rate is defined as follows:

$$Q = aBv_f \quad (6)$$

where B represents the width of cut and v_f represents the feed in $[\mathrm{mm} \cdot \mathrm{min}^{-1}]$ units. Thus Q is expressed in $[\mathrm{mm}^3 \cdot \mathrm{min}^{-1}]$ units.

Table 3. Experimental and predicted results for Ra and Rz

Number of experiment	$Ra_{experim}$ (μm)	Ra_{pred} (μm)	$Rz_{experim}$ (μm)	Rz_{pred} (μm)
1	3.68	3.39	19.78	19.22
2	16.66	16.82	74.49	73.67
3	0.50	0.28	8.98	9.74
4	17.64	17.86	58.73	59.23
5	1.00	1.65	11.15	12.53
6	17.58	17.77	58.93	60.54
7	0.91	0.67	4.22	2.57
8	16.24	15.55	59.95	58.51
9	10.63	10.24	48.90	48.06
10	8.89	8.34	36.05	33.81
11	7.29	7.77	37.60	39.78
12	7.27	7.60	29.33	30.11
13	9.26	9.17	38.43	37.94
14	9.18	9.17	37.76	37.94
15	9.13	9.17	37.71	37.94

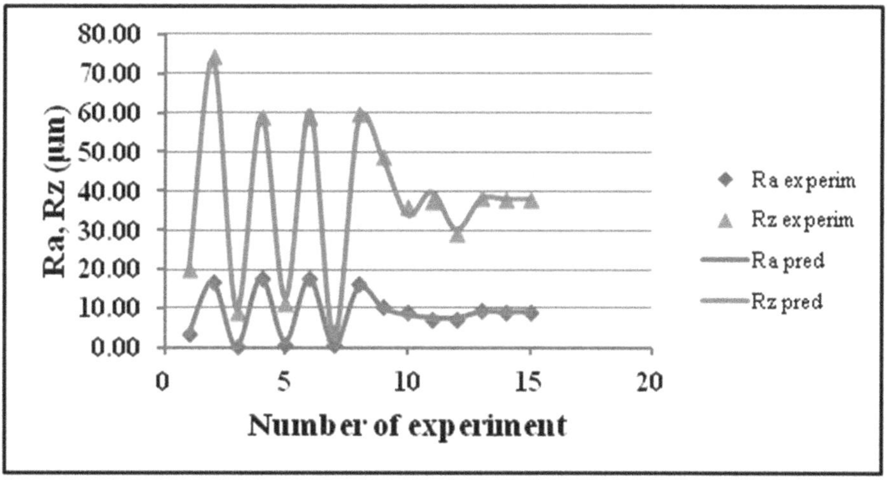

Fig. 2. Comparison of experimental and predicted values of Ra and Rz for each experiment

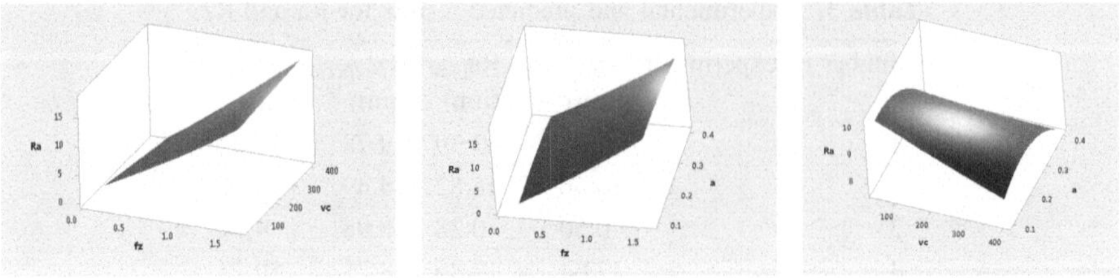

(a) (b) (c)

Fig. 3. Response surfaces of Ra in respect to: (a) fz and vc, (b) fz and a, (c) vc and a

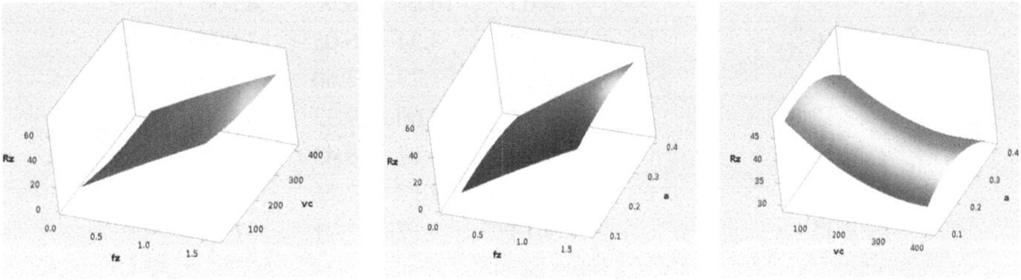

(a) (b) (c)

Fig. 4. Response surfaces of Rz in respect to: (a) fz and vc, (b) fz and a, (c) vc and a

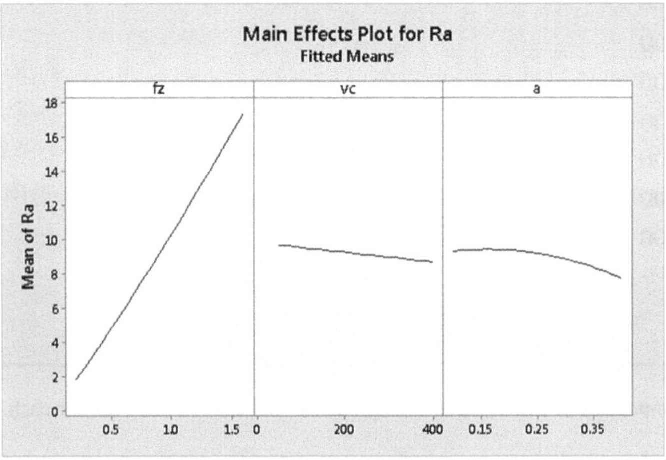

Fig. 5. Main effects plot for Ra

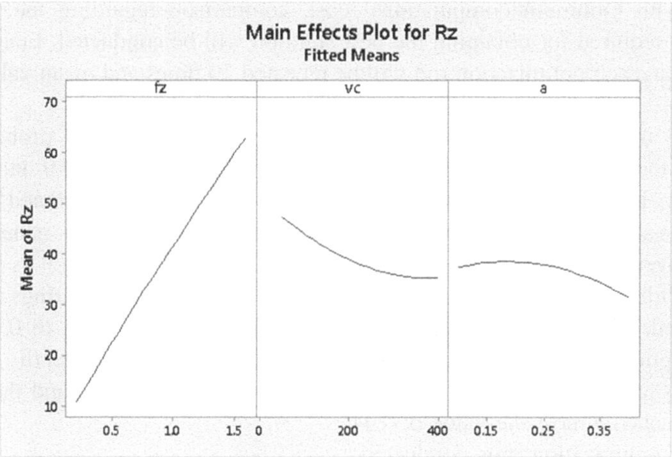

Fig. 6. Main effects plot for Rz

C. *Definition of optimization problem and determination of optimum parameters*

As it was aforementioned, the optimization problem will involve cutting forces, surface roughness indicators as well as the volumetric material removal rate. This is reflected on the objective function of the problem which is defined as follows:

$$minF = w_1 F_x + w_2 F_y + w_3 F_z + w_4 \mathrm{Ra} + w_5 \mathrm{Rz} + \frac{w_6}{Q} \tag{7}$$

where w_i are weighting factors. Although this problem is a multi-objective one, it is transformed in a single-objective problem with a more complex objective function in order to be more convenient and to simplify the processing of results. As every goal is considered of equal importance, the value of every weighting factor is 0.167. Furthermore, as the value of each constituent of the objective function is of different order of magnitude, it is essential to normalize each term by its maximum value so as no term has greater importance, if it has a relatively greater value than the other quantities. For the optimization problem, three design variables are defined, namely depth of cut, cutting speed and feed and their lower and upper boundaries are the same as the lower and upper values of the face milling experiments. The range of these limits, especially for cutting speed and feed, is broad enough in order to be able to take into account a wide range of process responses, as well as respect the technological limitations, which lead to restrictions in the limit values for these parameters.

For the optimization process using GA and FA, MATLAB software is used. As these algorithms are considerably different, it is not possible to conduct a direct comparison between them. However, for each algorithm two parameters will be varied at three levels in order to determine the parameters leading to the best solution and then, comparison between the best solutions from the runs using the two techniques will be compared. The settings that will be not varied are the default settings of the MATLAB codes for the two techniques. Furthermore, as it is desired to find good solutions with minimum computational cost, comparison regarding the number of evaluations required for obtaining the best solution will be conducted. Finally, it is to be noted that, each optimization run will be repeated 20 times and mean values will be reported.

For the optimization runs conducted with GA, the crossover probability and number of individuals for each generation are varied, from 0.7 to 0.9 and 15 to 45, respectively. In Fig. 7, the results concerning the best solution determined by the GA with various settings are presented. From this figure it can be seen that, generally, cases with crossover probability equal to 0.8 have better performance, as well as some cases with probability equal to 0.7. It was determined that the optimum settings for the GA were: 45 individuals per generation and crossover probability equal to 0.7. For that case, the optimum process parameters for feed, cutting speed and depth of cut were 0.766 mm · tooth^{-1}, 399.93 m · min^{-1} and 0.399 mm, respectively and the objective function reached almost the value 0.7194.

For the optimization runs conducted with FA, the amplitude of the explosions

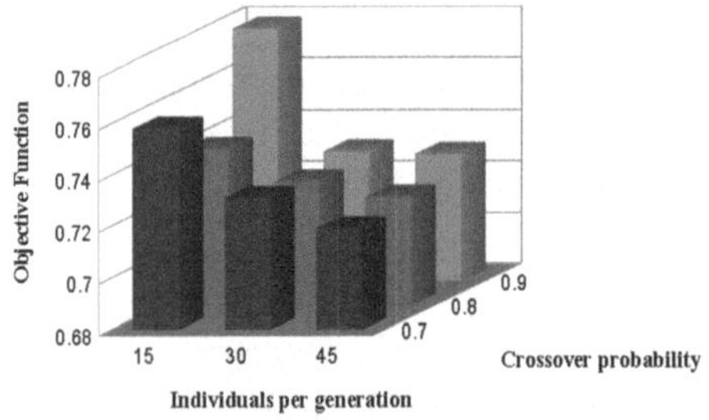

Fig. 7. Performance of GA according to various settings values

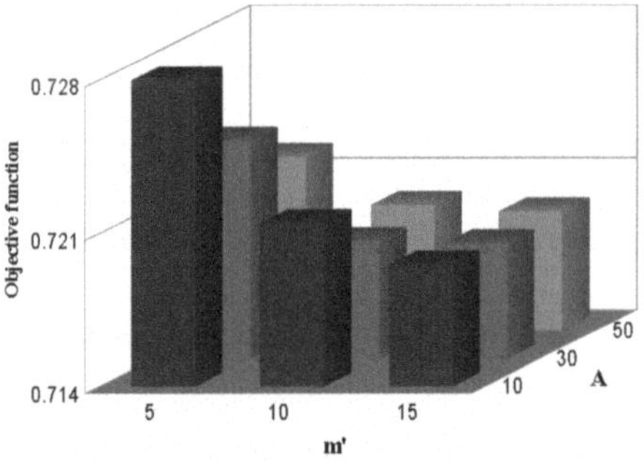

Fig. 8. Performance of FA according to various settings values

(A) and number of mutated solutions (m’) are varied, from 10 to 50 and 5 to 15, respectively. In Fig. 8, the results concerning the best solution determined by FA in each case are presented. It can be seen that the variation between these values is much lower than in the case of GA, indicating that FA produced solutions very close to the optimum for this range of settings. It was determined that the optimum settings for the FA were: m’ = 15 and A = 30. For the specific case, the optimum process parameters for feed, cutting speed and depth of cut were 0.764 mm · tooth^{-1}, 399.95 m · min^{-1} and 0.399 mm, respectively and the objective function value was almost 0.7193.

By comparing the optimum cases between GA and FA, it is observed that both algorithms reached almost the same solution. However, the GA with the best settings produced the optimum solution on average after 7560 evaluations of the objective function, whereas the FA after 6050 evaluations and also the convergence of FA was much quicker, as can be seen in Fig. 9. Thus, it can be concluded that the FA was more time-efficient than the GA.

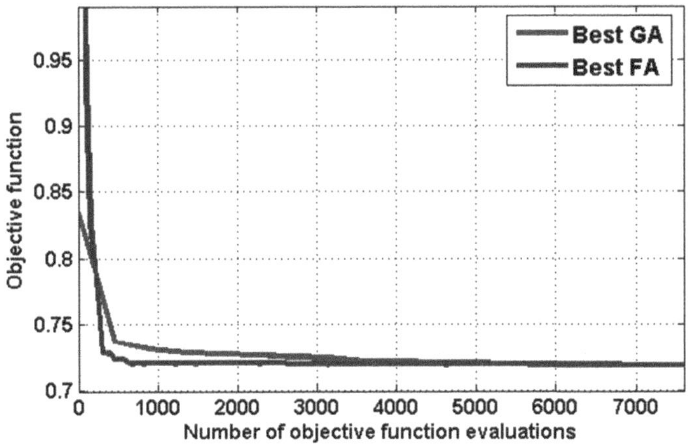

Fig. 9. Convergence plot for the best GA and FA runs

As for the optimum result presented in Table 4, it can be seen that it is reliable as it indicates that the required goals can be achieved by using a moderate value for feed and the maximum values of cutting speed and depth of cut. The values of cutting forces, surface roughness indicators and material removal rate for these process parameters are presented in Table 5. These results show that the optimization approach was able to keep the forces at a low to moderate level, as well as maintain moderate surface roughness values and achieve a moderate material removal rate, something that can be attributed also to the use of equal weights for the various goals. Thus, it can be concluded that the proposed experimental methodology of using DOE, statistical and optimization techniques was successful, as it enabled a quick determination of the maximum allowable material removal rate under minimum cutting forces and surface roughness constraints, allowing for an efficient face milling process with low energy consumption and acceptable surface quality.

Table 4. Optimum values of process parameters

Parameters	Feed ($mm \cdot tooth^{-1}$)	Cutting speed ($m \cdot min^{-1}$)	Depth of cut (mm)	Objective function value
Optimum value	0.764	399.95	0.399	0.7193

Table 5. Values of forces, surface roughness indicators and material removal rate for the optimum process parameters

	F_x (N)	F_y (N)	F_z (N)	Ra (N)	Rz (N)	Q ($mm^3 \cdot min^{-1}$)
Max	356.49	963.18	547.31	17.64	74.49	59076
Min	33.92	50.99	100.88	0.50	4.22	230.77
Optimum	198.98	367.44	345.12	5.43	22.81	20784

4 Conclusions

In the present work, multi-objective optimization was performed for cases of face milling of steel workpieces. The desired goals of the optimization process were the maximization of production rate, represented by material removal rate under the constraint of low energy consumption and tool wear, achieved by minimization of cutting force components and the constraint of minimum surface roughness. For the achievement of these goals, two different optimization techniques, namely Genetic Algorithm and the novel and promising Fireworks Algorithm were employed and their performance was compared. Several conclusions of this work are discussed afterwards.

At first, the correlation between process parameters and surface roughness indicators, namely Ra and Rz was established by regression models. It was found that for both Ra and Rz second order models are the most appropriate. In both cases, feed was identified as the most important factor.

Then, the optimization problem was defined by employing an objective function that incorporated all desired goals, weighted by weighting factors. More specifically, the minimization of the three components of cutting force and surface quality indicators Ra and Rz, as well as the maximization of material removal rate was desired. Both optimization techniques were able to determine sufficiently low objective function values in a short time period and it was shown that FA technique performed better, as it enabled the achievement of the best solution with a lower number of objective function evaluations.

The optimization algorithms, based on the aforementioned goals, provided the maximum material removal rate value possible under the imposed restrictions, maintaining also low cutting forces values and acceptable surface roughness. As these results are proven reliable, it is proposed that a more comprehensive study, containing also other important requirements for machining processes such as minimum production cost or minimum residual stresses should be conducted, by incorporating these goals to the model, in order to further improve the performance of face milling.

Acknowledgment. The authors greatly appreciate the support of the National Research, Development and Innovation Office – NKFIH (No. of Agreement: OTKA K 116876).

This study was carried out as part of the EFOP-3.6.1-16-00011 "Younger and Renewing University – Innovative Knowledge City – institutional development of the University of Miskolc aiming at intelligent specialization" project implemented in the framework of the Szechenyi 2020 program. The realization of this project is supported by the European Union, co-financed by the European Social Fund.

References

1. Masmiati N, Sarhan AAD, Hassan MAN, Hamdi M (2016) Optimization of cutting conditions for minimum residual stress, cutting forces and surface roughness in end milling of S50c medium carbon steel. Measurement 86:253–265
2. Pimenov DYu, Guzeev VI, Mikolajczyk T, Patra K (2017) A study of the influence of processing parameters and tool wear on elastic displacements of the technological system under face milling. Int J Adv Manuf Technol 92:4473–4486

3. Hadad M, Ramezani M (2016) Modeling and analysis of a novel approach in machining and structuring of flat surfaces using face milling process. Int J Mach Tools Manuf 105:32–44
4. Hricova J, Naprstkova N (2015) Surface roughness optimization in milling aluminium alloy by using the Taguchi's design of experiment. Manuf Technol 15(4):541–546
5. Mikó B (2016) Surface quality prediction in case of steep free form surface milling. Key Eng Mater 686:119–124
6. Mikó B, Farkas G (2017) Comparison of flatness and surface roughness parameters when face milling and turning. Dev Mach Technol 7:18–27
7. Kundrák J, Felhő C (2016) 3D roughness parameters of surfaces face milled by special tools. Manuf Technol 16(3):532–538
8. Felhő C, Kundrák J (2014) Comparison of theoretical and real surface roughness in face milling with octagonal and circular inserts. Key Eng Mater 581:360–365
9. Mikó B, Tóth B, Varga B (2017) Comparison of theoretical and real surface roughness in case of ball-end milling. Solid State Phenom 261:299–304
10. Wang Q, Liu F, Wang X (2014) Multi-objective optimization of machining parameters considering energy consumption. Int J Adv Manuf Technol 71:1133–1142
11. Yan J, Li L (2013) Multi-objective optimization of milling parameters- the trade-offs between energy, production rate and cutting quality. J Clean Prod 52:462–471
12. Anand Y, Gupta A, Abrol A, Gupta A, Kumar V, Tyagi SK, Anand S (2016) Optimization of machining parameters for green manufacturing. Cogent Eng 3:1153292
13. Sharma V, Pandey PM (2016) Optimization of machining and vibration parameters for residual stresses minimization in ultrasonic assisted turning of 4340 hardened steel. Ultrasonics 70:172–182
14. Fu T, Zhao J, Liu W (2012) Multi-objective optimization of cutting parameters in high-speed milling based on grey relational analysis coupled with principal component analysis. Front Mech Eng 7(4):445–452
15. Tan Y, Yu C, Zheng S, Ding K (2015) Introduction to fireworks algorithm. Int J Swarm Intell Res 4:39–70
16. Karkalos NE, Markopoulos AP (2018) Determination of Johnson-Cook material model parameters by an optimization approach using the fireworks algorithm. Procedia Manuf 22:107–113
17. Kundrák J, Markopoulos AP, Makkai T, Karkalos NE (2018) Correlation between process parameters and cutting forces in the face milling of steel. In: Jármai K, Bolló B (eds) Vehicle and Automotive Engineering 2. VAE 2018, ser. Lecture Notes in Mechanical Engineering, Springer, Cham, pp 255–267

RSM and Neural Network Modeling of Surface Roughness During Turning Hard Steel

Pavel Kovač[1(✉)], Mirfad Tarić[2], Dragan Rodić[1], Bogdan Nedić[3], Borislav Savković[1], and Dušan Ješić[4]

[1] Faculty of Technical Sciences, University of Novi Sad, Novi Sad, Serbia
{pkovac, rodicdr, savkovic}@uns.ac.rs
[2] Technical School, Sarajevo, Sarajevo, Bosnia and Herzegovina
mirfad.taric@hotmail.com
[3] Faculty of Engineering, University of Kragujevac, Kragujevac, Serbia
nedic@kg.ac.rs
[4] International Technology Management Academy, Novi Sad, Serbia
jesic@hotmail.com

Abstract. In the paper examined was the influence of the cutting regime parameters on surface roughness parameters during turning of hard steel with cubic boron nitrite cutting insert. In this study for modeling of surface finish parameters was used central compositional design of experiment and artificial neural network. The values of surface roughness parameters R_a and R_t were predicted by this two-modeling methodology and determined models were then compared. The results showed that the proposed systems can significantly increase the accuracy of the product profile when compared to the conventional approaches. The results indicate that the design of experiments with central composition plan modeling technique and artificial neural network can be effectively used for the prediction of the surface roughness for hard steel and determined significand cutting regime parameters.

Keywords: RSM · Neural network · Surface roughness · Hard steel

1 Introduction

Increasingly, research in manufacturing processes and systems is evaluating processes to improve their efficiency, productivity and quality. The quality of finished products is defined by how closely the finished product adheres to certain specifications, including dimensions and surface quality. Surface quality is defined and identified by the combination of surface finish, surface texture, and surface roughness. Surface roughness (R_a, R_{max}) is the commonest index for determining surface quality [1, 2].

Manufacturing processes do not allow achieving the theoretical surface roughness due to defects appearing on machined surfaces and mainly generated by deficiencies and imbalances in the process. Due to these aspects, measuring procedures are necessary; as it permits one to establish the real state of surfaces to manufacture parts with higher accuracy. To know the surface quality, it is necessary to employ theoretical models making it feasible to do predictions in function of response parameters [3–5].

N. M. Durakbasa and M. G. Gencyilmaz (Eds.): ISPR 2018, *Proceedings of the International Symposium for Production Research 2018*, pp. 16–25, 2019.
https://doi.org/10.1007/978-3-319-92267-6_2

A lot of analytically methods were also developed and used for predicting surface roughness. An empirical model for prediction of surface roughness in finish turning [6]. Nonlinear regression analysis, with logarithmic data transformation is applied in developing the empirical model. Metal cutting experiments and statistical tests demonstrate that the model developed in this research produces smaller errors and has a satisfactory result. The mathematical models for modeling and analyzing the vibration and surface roughness in the precision turning with a diamond cutting tool [7].

Recently, some initial investigations in applying the basic artificial intelligence approach to model of machining processes, have appeared in the literature, concludes that the modeling of surface roughness in machining processes has mainly used Artificial Neural Networks and fuzzy set theory [8, 9]. Average mean arithmetic surface roughness, Ra using artificial neural network was predicted in [10]. Surface roughness and surface finish have been considered in [11–14]. Research of the influence of machining parameters combination to obtain a good surface finish in turning and to predict the surface roughness values using fuzzy modeling is presented in [15]. Also, may notice that the neural network used in the study, where the enabling resolution of the problem that is difficult to define and mathematically model. This can be seen in the work where the neural network was based on the face milling machining processes, where is aimed to produce the relationship of cutting force versus instantaneous angle φ [16]. Use of coolants and lubricants in hard machining were presented in [17, 18].

In this paper, cutting speed, feed and depth of cut as machining regime were selected. Response surface methodology and artificial neural network models of surface roughness parameters R_a and R_{max} were developed for modeling.

2 Experimental Procedure and Material

Machining tests was done on the universal lathe - Prvomajska DK480. In the study was used interchangeable insert of CBN (cubic boron nitrite) CNMA 120404 ABC 25/F producer ATRON Germany. Used was appropriate insert holder for external processing PCLNR 25 25 M16.

The markings of the cutting tips according to DIN 4983 more closely define the geometry, as follows: the shape of the plate C → rhomb; the rake angle N → = 0, C → = 7; tolerance class M; Type of tile → with opening A, W and G; length of cutting blade → 12.7 mm (12); cutting edge thickness → 4.76 mm (04); radius of tool tip → 0.4 mm (04). All inserts have a rake angle (–6°) (Table 1).

Table 1. Experimental input factor levels

Factor levels	Cutting speed v (m/min)	Feed f (mm/rev)	Dept of cut a (mm)
Highest +1.41	180	0,25	0,7
High +1	160	0,2	0,5
Middle 0	120	0,1	0,22
Small –1	90	0,05	0,1
Smallest –1.41	80	0,045	0,07

The machining regime was:

- Cutting speed (m/min),
- Feed f (mm/rev), and
- depth of cut, a (mm).

Preparation of the workpiece was carried out before the experimental performance. The workpiece was thermally treated steel Č3840 (90MnCrV8) whose hardness after heat treatment was 55 HRC, cross-section Ø34 mm and length 500 mm. Before machining start It was necessary to remove a certain layer of material in order to avoid throwing-ovality and the results were more reliable. The length of the bar of 500 mm, was divided into 24 fields with a length of 10 mm on which the longitudinal cutting was performed. Each field on workpiece was planned for the measurement of one experimental point. Cutting without the presence of cooling and lubricating agents was provided

Measuring the surface roughness parameters with the Talysurf-6 measuring device was done. After processing by a computer, the results, was printing or writing on screen. The personal computer was connected to the Talysurf-6 measuring device using a serial connection COM-3. Instead of the printer, a computer was connected with a special adapter with a measuring machine Talysurf-6. The basic parts of the measuring device Talysurf-6 are shown in Fig. 1.

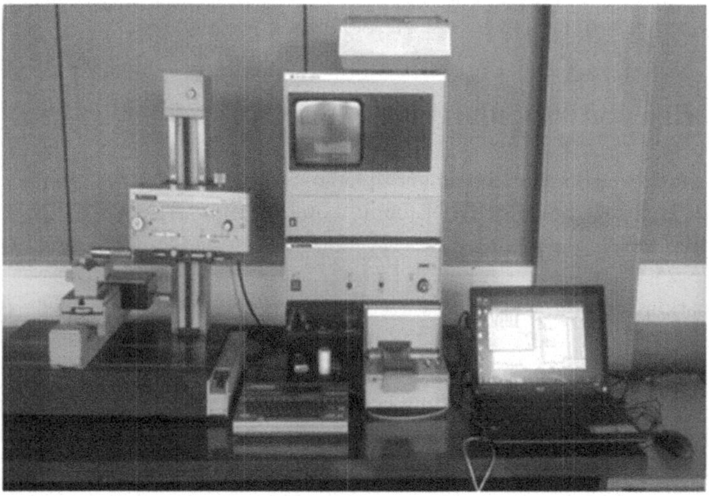

Fig. 1. Surface roughness measurement system Talysurf-6 connected with computer

The measured was values of surface roughness parameters: R_a, R_{max}. The measurement results of these parameters and estimated values by central compositional three factorial models are given in Table 2.

Implementation of factorial experimental plan: in the Table 3 are given results of dispersion analyses, adequacy of models and significance of parameters.

Table 2. The measurement and modeled results - input parameters

No.	Factor			R_i measured		R_i RSM model		R_i neural network	
	v [m/min]	f [mm/rev]	a [mm]	R_a [μm]	R_{max} [μm]	R_a [μm]	R_{max} [μm]	R_a [μm]	R_{max} [μm]
1	90	0,05	0,10	0.61	3.7	0.39	2.86	0.6592	3.7574
2	160	0,05	0,10	0.36	2.1	0.38	2.80	0.3557	1.9229
3	90	0,20	0,10	0.81	4.5	0.88	4.81	0.8331	4.6422
4	160	0,20	0,10	0.62	3.46	0.86	4.71	0.5893	2.9841
5	90	0,05	0,50	0.71	5.1	0.39	2.59	0.7777	5.0833
6	160	0,05	0,50	0.47	4.8	0.38	2.54	0.6832	5.3483
7	90	0,20	0,50	0.8	4.2	0.89	4.36	0.8527	4.0420
8	160	0,20	0,50	0.73	4.1	0.87	4.26	0.8288	4.4420
9	120	0,10	0,22	0.53	2.8	0.58	3.50	0.3581	1.6534
10	120	0,10	0,22	0.48	2.9	0.58	3.50	0.3588	1.6588
11	120	0,10	0,22	0.33	1.8	0.58	3.50	0.3582	1.6539
12	120	0,10	0,22	0.34	1.8	0.58	3.50	0.3558	1.6365
13	80	0,10	0,22	0.42	2.2	0.59	3.55	0.2565	1.5225
14	180	0,10	0,22	0.66	3.1	0.57	3.45	0.59537	2.7949
15	120	0,045	0,22	0.27	2.1	0.36	2.59	0.3057	1.7402
16	120	0,25	0,22	1.80	8.1	1.00	4.93	1.8051	8.2975
17	120	0,10	0,07	0.60	6.6	0.58	3.75	0.7843	6.6161
18	120	0,10	0,70	0.70	3.9	0.59	3.26	0.7456	3.9992
19	80	0,10	0,22	0.47	2.9	0.59	3.55	0.4533	2.5444
20	180	0,10	0,22	0.652	3.48	0.57	3.45	0.6655	3.2542
21	120	0,045	0,22	0.33	2.15	0.36	2.59	0.2790	1.8978
22	120	0,25	0,22	1.9	8.2	1.00	4.93	1.8304	8.3473
23	120	0,10	0,07	0.73	6.2	0.58	3.75	0.8895	6.6167
24	120	0,10	0,70	0.445	2.48	0.59	3.26	0.5301	2.2659

Table 3. Adequacy of models and significance of parameters

Model adequacy		R_a	R_{max}
		Fa = 3,2288	Fa = 4,0491
Significance of parameters	F_{ro}	143,19	655,65
	Fr1 (v)	0,05 (*)	0,03 (*)
	Fr2 (f)	53,95	18,75
	Fr3 (a)	0,01(*)	0,682 (*)

Analyze of adequacy of models shows that both models are adequate because coefficients are smaller than $F_t = 6.61$. Cutting speed and depth of cut are not significant because values are smaller than $F_t = 4.47$.

2.1 Artificial Neural Network Modelling

Artificial neural network (ANN) method is becoming useful as the alternative approach to conventional techniques, or as the component of integrated systems. It is an attempt to predict, within a specialized software, the multiple layers of a number of elementary units called neurons [14]. The MATLAB software, Neural Network Toolbox function, was used to create, train, validate, and predict the different ANNs reported in this research.

In this work, one of the most popular feed-forward networks was selected. This network is a multi-layer architecture proving to be an excellent universal approximation of nonlinear functions. The feed-forward neural network was trained by TRAINLM algorithms. The TRAINLM is a network training function that updates weight and bias values to Levenberg-Marquardt optimization.

Learning is a process by which the free parameters of the neural network are adapted through a continuous process of simulation by the environment in which the network is embedded. The learning function can be applied to individual weights and biases within the network. The LEARNGDM learning algorithms in feed-forward networks are used to adapt networks. Gradient descent method (GDM) was used to minimize the mean squared error between the network output and the actual error rate. It trains the network with gradient descent with the momentum back-propagation method. The back-propagation learning in feed-forward networks belongs to the real of supervised learning, in which the pairs of input and output values are fed into the network for many cycles, so that the network 'learns' the relationship between the input and the output.

For this study, feed-forward network was selected since this architecture interactively creates one neuron at a time. This is an optimization procedure based on the gradient descent rule which adjusts the weights of the network to reduce the system error is hierarchical. The network always consists of at least three layers of neurons: the input, output, and middle hidden layer neurons. The input layer has inputs, which are: v, the cutting speed (m/min); f, the feed (mm/rev); and a, [mm] the depth of cut. The outputs are the values of surface roughness parameters: arithmetic mean roughness R_a and maximal roughness hight R_{max}. These parameters were set to optimize by the neural network performance: the number of hidden layers is 12, the number of iterations is 100 and the number of neurons in the hidden layer is 20.

In this study, a part of the experimental data was used for training and the remaining data was used for testing the network. Each input has an associated weight that determines its intensity. The neural network can be trained to perform certain tasks where the data is fed into the network through an input layer.

This is processed through one or more intermediate hidden layers and finally it is fed out to the network through an output layer as shown in Fig. 2. It must be highlighted that the best network architecture is reached by trial and error after considering different combinations of the number of neurons in the hidden layer, the number of hidden layers, spread parameter, and learning rate, depending on the type of neural network being used.

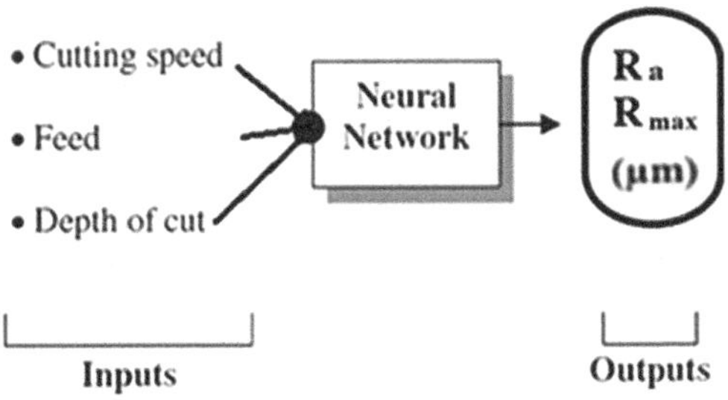

Fig. 2. Network input and output layer

3 Results and Discussions

Equations for surface roughness modeling by design of experiment determined by central compositional plan.

$$R_a = 2,8264 \cdot v^{-0,04471} \cdot s^{0,58975} \cdot a^{0,00716}$$

$$R_{max} = 9,0036 \cdot v^{-0,3717} \cdot s^{0,37380} \cdot a^{-0,6148}$$

As mentioned before, neural network modeling was used for analysis and optimization of surface roughness in turning process. The obtained results of neural network model are given in the Table 4, side by side with the obtained experimental results. For reduction of a deviation, is needed to increase the number of inputs.

Table 4. Experimental values and values obtained by neural network with percentage deviation for 6 testing points

No.	Factor			R_i – experimental roughness		R_i – modeled roughness	
	v [m/s]	s [mm/rev]	a [mm]	R_a [µm]	R_{max} [µm]	R_a [µm]	R_{max} [µm]
1	81	0.1	0.22	0.47	2.9	0.4533	2.5444
2	182	0.1	0.22	0.652	3.48	0.6655	3.2542
3	121	0.045	0.22	0.33	2.15	0.2790	1.8978
4	122	0.25	0.22	1.9	8.2	1.8304	8.3473
5	123	0.1	0.07	0.73	6.2	0.8895	6.6167
6	119	0.1	0.7	0.445	2.48	0.5301	2.2659
	Average deviation %					**10.30**	**8.62**

Calculation of percental deviation for measured and model surface roughness values was performed according next formula:

$$E = \frac{\left|Ri_{\exp} - Ri_m\right|}{Ri_{\exp}} \cdot 100\%$$

Where are: Ri_{exp}- experimental value, Ri_m- model value.

Calculated percental deviation for first 18 experimental points are for Ra is 8.94 and for R_{max} is 9.94. Experimental values and values obtained by neural network with percentage deviation for 6 testing points for neural network are in Table 4.

Any change in the cutting speed leads to a slowly corresponding change in the value of surface roughness. The cutting speed has a small and decreasing effect, Fig. 3. Influence of feed on value surface roughness is higher than the cutting speed effect.

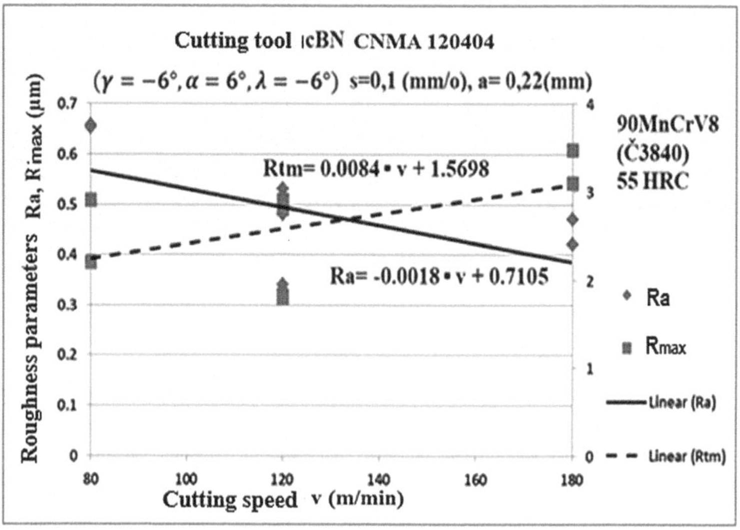

Fig. 3. The surface roughness (R_a, R_{max}) versus cutting speed

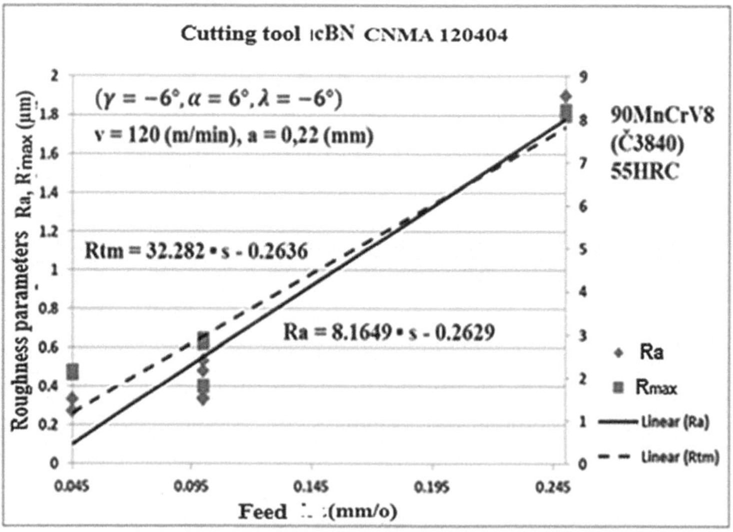

Fig. 4. The surface roughness (R_a, R_{max}) versus feed

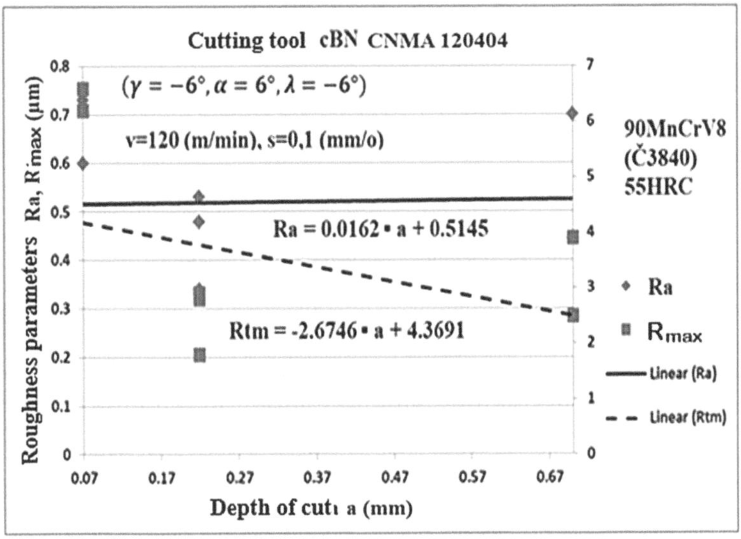

Fig. 5. The surface roughness (R_a, R_{max}) versus the cutting depth

Increasing feed increase surface roughness, Fig. 4. Depth of cut at least influences the wear on the flank surface and surface roughness values slightly, Fig. 5.

Any change in the cutting speed leads to a slowly corresponding change in the value of surface roughness. The cutting speed has a small and decreasing effect, Fig. 4. Influence of feed on value surface roughness is higher than the cutting speed effect. Increasing feed increase surface roughness, Fig. 5. Depth of cut at least influences the wear on the flank surface and surface roughness values slightly.

4 Conclusion

Intelligent optimization techniques give the influence of cutting conditions on machining surface quality during turning hard material, are investigated through experimental verification. The investigation results confirm the highly consent of experimental research and intelligent techniques modeling. The intelligent optimization techniques and experimental results show some good information which could be used by future researches for optimal control of machining conditions. This paper has successfully established neural network model, for predicting the workpiece surface roughness parameters. Figures 4 and 5 shows the compared predicted values obtained by experiment and estimated by neural network shows a good comparison with those obtained experimentally. The average deviations of models are checked and are found to be adequate. The model adequacy can be further improved by considering more variables and ranges of parameters.

Acknowledgment. The paper is the result of the research within the project TR 35015 financed by the ministry of science and technological development of the Republic of Serbia SRB/SK bilateral project.

References

1. Chen JC, Savage M (2001) A fuzzy-net-based multilevel in-process surface roughness recognition system in milling operations. Int J Adv Manuf Technol 17:670–676
2. Quintana G, Garcia-Romeu ML, Ciurana J (2009) Surface roughness monitoring application based on artificial neural networks for ball-end milling operations. J Intell Manuf 22: 607–617
3. Sivarao IR, Castillo WJG, Taufik (2000) Machining quality predictions: comparative analysis of neural network and fuzzy logic. Int J Electr Comput Sci IJECS 9:451–456
4. Drégelyi-Kiss Á, Horváth R, Mikó B (2013) Design of experiments (DOE) in investigation of cutting technologies. In: Development in machining technology (DIM 2013), Cracow, pp 20–34
5. Maňková I, Vrabeľ M, Beňo J, Kovač P, Gostimirovic M (2013) Application of Taguchi method and surface response methodology to evaluate of mathematical models for chip deformation when drilling with coated and uncoated twist drills. Manuf Technol 13(4): 492–499
6. Hadi SG, Ahmed SG (2006) Assessment of surface roughness model for turning process. In: Knowledge enterprise: intelligent strategies in product design, manufacturing, and management. International federation for information processing (IFIP), vol 207, pp 152–158
7. Chen CC, Chiang KT, Chou CC, Liao YC (2011) The use of D-optimal design for modeling and analyzing the vibration and surface roughness in the precision turning with a diamond cutting tool. Int J Adv Manuf Technol 54:465–478
8. Choudhary A, Harding J, Tiwari M (2009) Data mining in manufacturing: a review based on the kind of knowledge. J Intell Manuf 20(5):501–521
9. Grzenda M, Bustillo A, Zawistowski P (2012) A soft computing system using intelligent imputation strategies for roughness prediction in deep drilling. J Intell Manuf 23:1733–1743
10. Balic J, Korosec M (2002) Intelligent tool path generation for milling of free surfaces using neural networks. Int J Mach Tools Manuf 42:1171–1179
11. Pérez CJL (2002) Surface roughness modeling considering uncertainty in measurements. Int J Prod Res 40(10):2245–2268
12. Azouzi R, Gullot M (1997) On-line prediction of surface finish and dimensional deviation in turning using neural network-based sensor fusion. Int J Mach Tools Manuf 37(9):1201–1217
13. Ho SY, Lee KC, Chen SS, Ho SJ (2002) Accurate modeling and prediction of surface roughness by computer vision in turning operations using an adaptive neuro-fuzzy inference system. Int J Mach Tools Manuf 42(13):1441–1446
14. Zębala W, Gawlik J, Matras A, Struzikiewicz G, Ślusarczyk Ł (2014) Research of surface finish during titanium alloy turning. Key Eng Mater 581:409–414
15. Rajasekaran T, Palanikumar K, Vinayagam BK (2011) Application of fuzzy logic for modeling surface roughness in turning CFRP composites using CBN tool. Prod Process 5 (2):191–199

16. Savković B, Kovač P, Gerić K, Sekulić M, Rokosz K (2013) Application of neural network for determination of cutting force changes versus instantaneous angle in face milling. J Prod Eng 16(2):25–28
17. Kundrák J, Varga G (2013) Use of coolants and lubricants in hard machining. Tech Gaz 20 (6):1081–1086
18. Kovac P, Rodic D, Pucovsky V, Savkovic B, Gostimirovic M (2013) Application of fuzzy logic and regression analysis for modeling surface roughness in face milling. J Intell Manuf 24(4):755–762

Big Data and Analytics

A Data Mining Approach to Predict E-Commerce Customer Behaviour

Büşra Altunan, Ebru D. Arslan, Merve Seyis, Merve Birer, and Fadime Üney-Yüksektepe(✉)

Department of Industrial Engineering, İstanbul Kültür University, İstanbul, Turkey
altunanbsr@gmail.com, ebrudarslan@gmail.com, mrvsys95@hotmail.com, birer.merve@hotmail.com, f.yuksektepe@iku.edu.tr

Abstract. Watsons, established in 1841, is one of the world's leading beauty and personal care industry with more than 6300 stores in 11 different markets. In addition to 280 Watsons's stores, online shopping is also an alternative for Turkish customers.

Usage of online tools and technologies for many different purposes has increased by 21st-century companies thanks to the great accessibility of the Internet. Therefore, due to current trends, many customers prefer online shopping. Among them, some of them are adding the products to their market basket but unfortunately; they are leaving the website without a purchase. This causes an important problem for most of the e-commerce retailers. In this context, research in e-commerce needs to determine the reasons that encourage them to buy as much as they need to understand the behaviours of consumers. However, it is difficult and complex to identify customer behaviour and the factors that encourage purchasing. Data mining techniques have become the focus of such analyses. This project mainly focuses on data classification approach to predict customer's behaviour during the website visits of Watsons e-commerce channel.

Based on the problem to predict e-commerce customers' behaviour at the customers' first entrance to the Watsons's website, data classification approach is studied. The collected data from Watsons' e-commerce site are analysed in the WEKA by applying different data classification algorithms. Two distinct data sets are determined; while dataset 1 includes all attributes, dataset 2 is created by considering the results of attribute selection algorithms. Both dataset 1 and dataset 2 are examined separately in WEKA software by using seven classifier. The data classification algorithm with the highest accuracy is selected. Some rules and trees are suggested to the company to predict future customer behaviours. To encourage the customers who leave the website of Watsons, rules are interpreted for suggestion.

Keywords: E-commerce · Data mining · Classification · Customer behaviour Online shopping

N. M. Durakbasa and M. G. Gencyilmaz (Eds.): ISPR 2018, *Proceedings of the International Symposium for Production Research 2018*, pp. 29–43, 2019.
https://doi.org/10.1007/978-3-319-92267-6_3

1 Introduction

In today's changing technology, the approach of people about shopping has altered. Shopping over the Internet is become more and more preferable in place of going traditional shopping [1]. Usage of online tools and technologies for many different purposes has increased by 21st century companies thanks to great accessibility of the Internet. Currently, huge data in data repository is owned and e-commerce is accepted by lots of corporation [2]. In addition, recognizing the customers has become more difficult based on the increasing customers, products, competitors and decreasing reacting time in comparison with earlier. Also, some various influences make the customer relationships more complicated [3]. Consequently, companies notice that realizing their customers enhanced and rapidly satisfying their desires and requirements are definitely necessary. Moreover, responding process for that is shortening. Taking action after customer dissatisfaction's symptoms are observed is not more possible anymore [3].

In this context, research in e-commerce needs to determine the reasons that encourage them to buy as much as they need to understand the behaviours of consumers. However, it is difficult and complex to identify customer behaviour and the factors that encourage purchasing. Data mining techniques have become the focus of such analyses [1].

Data mining is the technique by which specific algorithms are used to reach meaningful and necessary information from large data sets. The data mining technique has been widespread in various fields where large data sets are effectively examined and analysed [4].

In e-commerce, the main goal of the data mining approach is to figure out forms of utilization, trying to clarify the customers' behaviour and interest. Thus, different types of data mining techniques such as classification, clustering, association rules, or sequential patterns are used successfully [1].

This project mainly focuses on data classification approach to predict customer's behaviour during the website visits of Watsons e-commerce channel. Due to current trends, many customers prefer online shopping. Among the online channel customers, some of them purchase the products in their market basket. On the other hand, some of them leave the website without a purchase. Aim of the study is to predict e-commerce customers' behaviour at the customers' first entrance to the Watsons's website. The past data related to customer web site visits and their demographical information is used for the data classification problem of the company. By comparing well-known data classification methods, the best one who has the highest prediction accuracy is determined. Finally, results are shared with the company and a decision support system is recommended to predict the customer's behaviour at her first entrance to the website.

The paper starts with the literature review on data mining in Sect. 2. Section 3 introduces the company information and problem definition according to current application. Section 4 involves the detailed steps of the methodology. By using the data obtained from the company, computational results are provided in Sect. 5. Section 6 gives the conclusion of this study.

2 Literature Review

Due to big data and its implementations, size of the information becomes immensely significant nowadays. In order to detect default risks, banking and insurance industries benefit from big data and data mining analysis. In master thesis by Reference [5], 22745 samples and 14 attributes have been used which were taken from Turkish Statistical Institution. The target of that study is to choose the best classification algorithm which can help to identify default risk of individuals depending on their characteristics. Several classification algorithms such as Bayes Network, Naive Bayes, Logistic Regression, J48, Random Forest, MultiLayer Perceptron had been selected and compared by using accuracy, root mean squared error, Roc curve, recall, precision statistics and the most suitable classification of algorithm was detected. Data mining software, named WEKA was used to do analysis [5].

One of the leading medical challenges all over the world is Diabetes Mellitus. While deteriorating human, economic and social fabric, the penetration of diabetes is boosting fast. Regarding numerous clinical searches, advisement is to use some of incorporate data mining techniques for diabetics and its improvement. All of these traditional systems are typically based on a single classifier or plain combinations. In recent time, for developing the accuracy rate of such systems applying set of classifiers, comprehensive efforts are being done. Reference [6] classified individual's diabetes mellitus status by using disease's risk factors and applied the Adaboost and Bagging techniques in accordance with J48 (c4.5) decision tree as a base of learner. This classification is performed in the system of Canadian Primary Care Sentinel Surveillance with three different ordinal adult's team. The empirical results indicated entire performance of Adaboost ensemble method is much better than Bagging as well as independent J48 decision tree [6].

Anaklı [7] briefly gives instances of data mining applications, at the present time, KDD databases and datamining are being used in many fields in order to gather extensive information. The definition of datamining is subtraction of fascinating, implied and helpful models or information gathered through massive quantity of data [8]. The most effective data mining methods are classification and prediction [8, 9]. Quality Improvement (QI) and control are two separate areas for applying data mining findings. Astuldill et al. [10] examined data mining methods used for quality improvement in literature. For instance, Huang [11] proposed a decision tree method to define crucial factors that affect the ratio of defectives. Integrated machine learning technics are recommended to explain specific quality problems by Kang [12]. Fan et al. [13] combined the idea of data mining, quality control and process knowledge. While predicting the future values of categorical data, classification technic has been used. Statistical based algorithms, decision tree based algorithms, artificial Neural Network-based systems and other classification algorithms are the main classification methods used in literature [7].

Data mining can be organized according to the different family of problems that it solves. These problems include classifying items into previously known categories, grouping items according to their similarities, discovering association rules from transactions, identifying atypical data, and predicting a continuous (dependent) variable. This section gives a brief overview of these types of problem. There are several core techniques in data mining that are used to build data mining. Data mining algorithms are divided into two groups such as predictive (supervised) learning and descriptive (unsupervised) learning. Supervised learning is a known example of creating a model using output and inputs and estimating the output variable when a new input data is given. Output variables are also referred to as dependent variables because they are influenced by one or more input variables also the input variable is independent variable. Classification and regression is supervised learning methods. Output is unavailable when learning is unsupervised. The goal of unsupervised learning is to estimate input variables and output variables. There is no control mechanism. The user builds and evaluates the model on its own. It is used for forecasting future or current situation. Clustering and association analysis is unsupervised learning.

Recently, online shopping has become increasingly common in daily lives. Understanding users' interests and behaviours is crucial for adapting e-commerce web sites to the needs of customers. The information about the behaviour and interests of users is collected in web server logs. The analysis of such data has focused on applying data mining techniques where a highly static characterization is used to model the behaviour of users, and the order of their actions is often ignored. Therefore, combining a view of the process followed by a user during a session can help to define more complex behaviour patterns. To handle this issue, this article suggests a linear-temporal logic model control approach for analysing structured e-commerce Web logs. Web logs can be easily converted into event records that capture the behaviour of users when by defining a common way of matching log records according to e-commerce structure. Then, different predefined queries can be made to identify different behavioural patterns taking into account different actions performed by a user during a session. Finally, the usefulness of the suggested approach has been tested by applying it to a real case study of a Spanish e-commerce web site. As a result of this study have identified findings that have made possible to suggest some improvements in the web site design with the aim of increasing its efficiency [1].

Data mining is as a field of basic and applied research in computer science. The purpose of this study is to evaluate, propose and improve the use of some of the recent approaches, architectures and web mining techniques (collecting personal information from customers) are the means of utilizing data mining methods to induce and extract useful information from web information (that means User's behaviour). Data mining has been applied in the fields of e-commerce and e-business for service to customer. In web mining, clustering method could be used to identify customer behaviours in e-commerce based on similar click-streams and general site access behaviours. In the literature, most of the algorithms offered to deal with clustering web sessions treat sessions as sets of visited pages within a time period and that the sequence of click-stream visitation do not consider is a significant feature when comparing similarities between web sessions [3].

Data mining and specifically web data mining techniques plays an important role in electronic commerce that can be used for many purposes. Electronic commerce is the use of information and communication technologies through the internet platform to share business information, keep business relationships and conduct business transactions. Data mining techniques are becoming more useful to discover and understand unknown customer models thanks to fast growth of electronic commerce and the large amounts of data collected through operational transactions in recent years. Reference [8] briefly describes some examples of data mining application in electronic commerce. Clustering or grouping customers with similar browsing behaviour allows the identification of common features of customers, in this way providing a more appropriate and personalized service to customers and provides a better understanding to customer behaviour. Some basic problems are described that can be solved using data mining techniques in electronic commerce. As an example, the technique of association rules is described in more detail and some application related methods are described [8].

Nowadays, wide data volume, structured and unstructured data, is big data that collected from customer's internal processes, supplier, markets and business environment. Especially, opportunities are created for companies that use electronic commerce (e-commerce). Reference [2] presents three common algorithms that are association, clustering and prediction from data mining methods for e-commerce. Furthermore, obtainable by three data mining algorithm in terms of market segmentation highlights some benefits which are sale forecasting, basket analysis, customer relationship management to e-commerce companies. The main objective of this study is to review the implementation of data mining in e-commerce thus focusing on structured and unstructured data gathered from a variety of sources and cloud computing services thanks to proves the importance for data mining. Moreover, this study evaluates some of the challenges of data mining, such as spider identification, data transformations, making understandable data models for business users, supporting the slow changing dimensions of data, making the data transformation and model building accessible to business users. As a result of this study, the competitiveness of e-commerce companies with large amounts of data is increasing.

3 Problem Definition

3.1 Company Information

Watsons, established in 1841, is one of the world's leading beauty and personal care industry with more than 6300 stores in 11 different markets including China, Hong Kong, Taiwan, Macau, Singapore, Thailand, Malaysia, Philippines, Indonesia, Turkey and Ukraine. Watson Group is the largest international health and beauty retailer in Asia and Europe with over 14,000 stores and 13 retail brands in 24 markets. The Watson Group has one of the world's largest retail portfolios, with retail formats, retail brands and geographic reach. In addition to 280 Watsons's stores, online shopping is also an alternative for Turkish customers.

3.2 Problem Definition

Due to current trends online shopping is more preferable instead of traditional shopping in recent times. E-commerce satisfies customers with endless product catalogues viewings, comparing prices and offering better service quality. Therefore, today many companies use e-commerce channel to buy and sell quickly. Also, companies notice that realizing their customers enhanced and rapidly satisfying their desires and requirements are definitely necessary. However, recognizing the customers' behaviour about purchasing has become more difficult based on the increase in the number of customers, products and competitors.

Watsons complains about customers who leave the website without purchasing as lots of company and needs to determine the reasons that encourage them to buy as much as they need to understand the behaviours of consumers. This project mainly focuses on data classification approach to predict customer's behaviour during the website visits of Watsons e-commerce channel.

4 Methodology

4.1 Data Mining Software: WEKA 3.6

WEKA (Waikato Environment for Knowledge Analysis) is a data mining and machine learning software developed in New Zealand. It is also improved with programming language of Java. WEKA is open source software issued under the GNU General Public License. WEKA includes various type of classification techniques and allows user to use many algorithms such as pre-processing, classification and visualization tools.

Once the data set is prepared, it needs to be uploaded to WEKA to process it. WEKA only accepts files with an extension of ARFF (Attribute-Relation File Format), CSV, C4.5, binary, libsvm, XRFF; also read files from the SQL database via the url. The WEKA accepts numeric or nominal attributes, and reads the data from a simple file. At the same time, data can also be retrieved from the database, but in this case it is expected that the data will be in the form of a file.

WEKA Software has 7 different classifier methods such as Bayes, Functions, Lazy, Meta, Misc, Rules and Trees.

4.2 Proposed Solution Methodology

The flowchart of the methodology is given in Fig. 1. All steps are explained in detail as follows:

- Obtaining old data from the company: Old data is taken from Watsons which consist of customers' information about age, sex, city, average visit duration, browser type, purchasing a product or not and the shopping day. Due to the size of the given data, some part of the data is eliminated.
- Controlling missing value in Excel: After sample data selection, it is transferred to Excel. Missing values and duplicate data are removed from the list.
- Selecting attributes: According to given data, all attributes are selected to estimate customers' behaviour about purchasing a product or not.
- Determining data sets: Two datasets are generated. Data set 1 includes all of attributes and Data set 2 is created by considering the results of attribute selection methods.

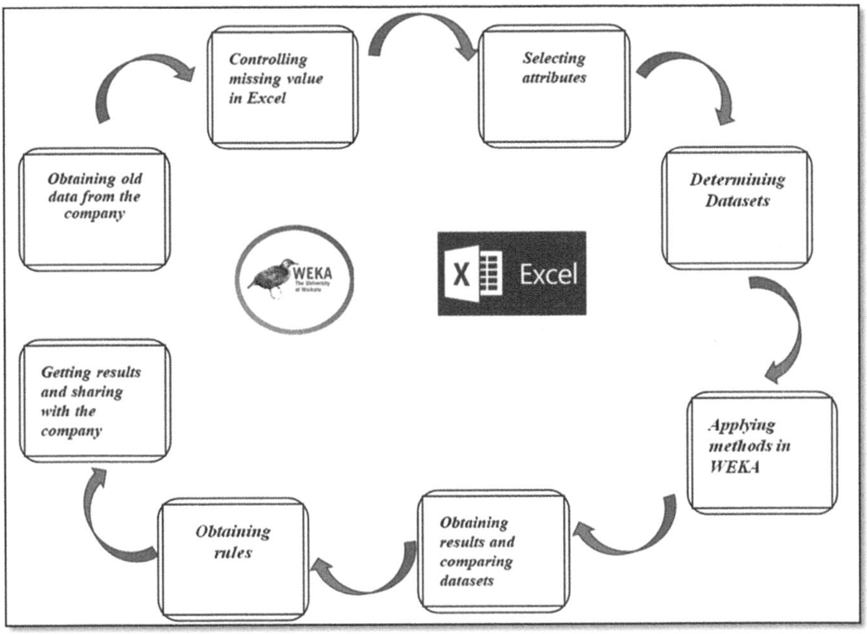

Fig. 1. Flowchart of the methodology

- Applying methods in WEKA: WEKA is used to run the selected methods by one by for all input data.
- Obtaining results and comparing data sets: After applying methods in WEKA, the best ones of each group are selected according to their accuracy. This illustrates overall results for comparing Data set 1 and Data set 2.
- Obtaining rules: Rules are obtained separately for Data set 1 and Data set 2 in WEKA.
- Getting results and sharing with the company: Based on the results and findings, significant suggestions are made for the company.

5 Implementation and Results

Proposed solution methodology given in Sect. 4.2 is implemented by using WEKA software.

5.1 Data Collection

In the first step, the period to be analysed is determined as 19–25 March 2018. Then, demographic and behavioural data of e-commerce customers are taken from Watsons for this period.

5.2 Controlling Missing Values

After eliminated some of the data, it is transferred to Excel. Then, missing values and duplicate data removed from the list. After all these steps, 1317 data is selected to be analysed in WEKA software.

5.3 Data Set Description

According to the data, 7 attributes are identified such as day, age, sex, city, browser type, average visit duration, purchase or not. In addition, classes are defined as purchase or not purchase.

Data set description is given in Table 1. For each attribute, their name, types and properties are listed. For categorical attributes, all possible attribute values are listed and the numbers in the parenthesis are the number of instances with that attribute value. For numerical attributes, max, min, mean and standard deviation values are given.

Table 1. Data set description

No	Attribute name	Characteristics	
		Types	Properties
1	Sex	Categorical	Female (1051) Male (266)
2	Age	Categorical	18–24 (374) 25–34 (572) 35–44 (211) 45–54 (70) 55–64 (53) 65+ (37)
3	Day	Categorical	Monday (277) Tuesday (240) Wednesday (160) Thursday (160) Friday (160) Saturday (160) Sunday (160)
4	Average visit duration (seconds)	Numerical	Min: 0 Max: 9440 Mean: 838.741 Std. Dev.: 1135.717
5	Browser type	Categorical	Chrome (334) Firefox (36) Safari (576) Internet explorer (92) Other (279)

(*continued*)

Table 1. (*continued*)

No	Attribute name	Characteristics	
		Types	Properties
6	City	Categorical	İstanbul (367) Ankara (159) İzmir (105) Adana (50) . . . Antalya (43)
7	Purchase or not	Categorical	Purchase (658) Not purchase (659)

5.4 Preprocessing

The 1317 data are transferred to WEKA and 8 selection methods are used in preprocessing part of the study. After the methods are studied one by one for all input data, "selected attributes" for all method are obtained. The results are obtained from WEKA. Then, suggested datasets are determined and demonstrated by Table 2. While data set 1 include all attributes, data set 2 is created by considering same attributes which are included in all sequence. First 3 attributes of each method's sequence are chosen. Finally, data set 2 attributes are determined as 5, 2, 3, 6, 1, 7. In this study, both dataset 1 and dataset 2 are examined separately.

Table 2. Suggested data sets

Data sets	Attributes
Data set 1	1, 2, 3, 4, 5, 6, 7
Data set 2	5, 2, 3, 6, 1, 7

1: Sex
2: City
3: Browser type
4: Age
5: Average visit duration (second)
6: Day
7: Purchase or not

5.5 Results

In this study, both dataset 1 and dataset 2 are examined separately in WEKA software. 7 defined methods in Sect. 4.1 such as Bayes, Functions, Lazy, Meta, Misc, Rules and Trees are implemented to determine best classification method for both dataset 1 and dataset 2. Results of accuracy rate and confusion matrixes are obtained.

5.5.1 Results of Data Set 1

Totally, 41 classification methods are studied for dataset 1.

For Bayes method, there is one highest accuracy, BayesNet with a value 86.8641%. Simple Logistic is the highest accuracy with a value 85.6492% for Functions method and KStar is the highest accuracy with a value 88.2308% for Lazy method. For Meta, the highest accuracy with a value 86.7122% is for Attribute Selected Classifier and Multi Class Classifier Updateable. There is a unique accuracy which is Input Mapped Classifier with a value 49.8861% for Misc method and JRip is the highest accuracy with a value 86.8641% for Rules method. Finally, J48 is the highest accuracy with a value 87.6693% for Trees method. As a result, KStar has the highest accuracy rate with a value 88.2308% in the dataset 1 (Tables 3 and 4).

Table 3. Result of lazy for data set 1

Type	Method	Accuracy
Lazy	IBK	77.3728%
	KStar	**88.2308%**
	LWL	86.5604%

Table 4. Confusion matrix of KStar for data set 1

		Predicted classes	
		Purchase	Not purchase
Actual classes	Purchase	586	72
	Not purchase	83	576

Accuracy: (586 + 576)/1317 = 0.8823
Error rate: 1−0.8823 = 0.1177

5.5.2 Rules of Data Set 1

As Jrip has a considerably high accuracy, obtained JRip rules are given below:

JRip Rules

(Average Visit Duration $\geq$ 470 seconds) $\rightarrow$ Class =Purchase (626.0/74.0)

If customer's average visit duration is more than 470 s, it is classified as "purchase".

(Average Visit Duration $\geq$ 220.67 seconds) and (Sex = Male) and (Browser Type = Safari) $\rightarrow$ Class = Purchase (24.0/2.0)

If customer is male and using Safari with an average visit duration more than 220.67 s, it is classified as "purchase".

(Average Visit Duration $\geq$ 120 seconds) and (Sex = Male) and (Day = Monday) $\rightarrow$ Class = Purchase (11.0/0.0)

If customer is male and selected shopping day is Monday also average visit duration is more than 120 s, it is classified as "purchase".

(Average Visit Duration $\geq$ 102 seconds) and (Sex = Male) and (Day = Tuesday) $\rightarrow$ Class = Purchase (7.0/0.0)

If customer is male and selected shopping day is Tuesday also average visit duration is more than 102 s, it is classified as "purchase".

(Average Visit Duration $\geq$ 225.4 seconds) and (Age = 55-64) $\rightarrow$ Class = Purchase (5.0/0.0)

Else $\rightarrow$ Class = Not Purchase (644.0/61.0)

If customer's age between 55 to 64 and has average visit duration more than 225.4 s, it is classified as "not purchase".

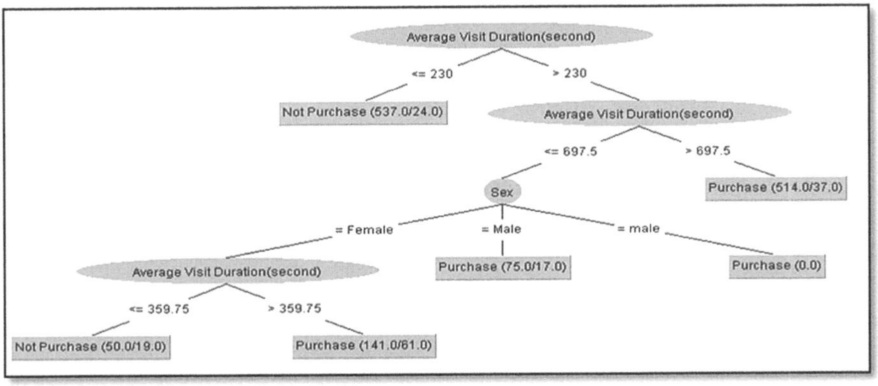

Fig. 2. J48 Tree of data set 1

Figure 2 shows the decision tree obtained from J48 method. If consumers' average visit duration is less than 230 s, it is classified as "not purchase". Secondly, if consumers have average visit duration more than 697.5 s, so it is classified as "purchase".

5.5.3 Results of Data Set 2

Totally, 41 classification methods are studied for data set 2.

For Bayes method the best is Bayes Net with accuracy value of 86.8641%. Simple Logistic is the highest accuracy with a value 85.1177% for Functions method and KStar is the highest accuracy with a value 87.4715% for Lazy method. For Meta, the highest accuracy with a value 87.0919% is Classification Via Regression. There is a unique accuracy which is Input Mapped Classifier with a value 49.8861% for Misc method and JRip is the highest accuracy with a value 86.6363% for Rules method. Lastly, J48 is the highest accuracy with a value 87.6234% for Trees method. As a result, J48 has the highest accuracy rate with a value 87.6234% in the dataset 2 (Tables 5 and 6).

Table 5. Results of trees for data set 2

Type	Method	Accuracy
Trees	Decision stump	86.7122%
	Hoeffding tree	87.0159%
	J48	**87.6234%**
	LMT	85.4214%
	Random forest	84.9658%
	Random tree	79.2711%
	REP tree	85.4973%

Table 6. Confusion matrix J48 for data set 2

		Predicted classes	
		Purchase	Not purchase
Actual classes	Purchase	625	33
	Not purchase	130	529

Accuracy: (625 + 529)/1317 = 0.8762
Error rate: 1−0.8762 = 0.1238

Figure 3 shows the decision tree obtained from J48 method. If consumers who are male, have average visit duration is between 230 s and 697.5 s, it is classified as "purchase". Secondly, if customer who are female and have visit duration between 230 s and 359.75 s, so it is classified as "not purchase".

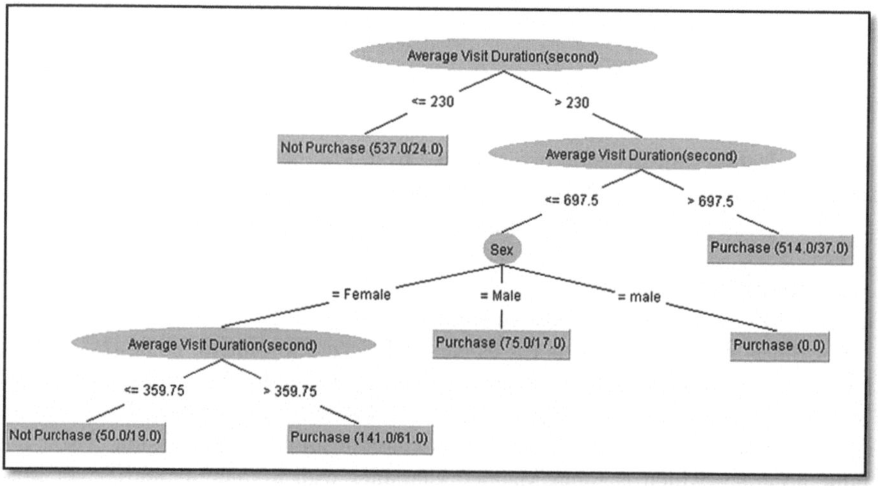

Fig. 3. J48 Tree of data set 2

5.5.4 Rules of Data Set 2

As Jrip has a considerably high accuracy, obtained JRip rules are given below:

JRip Rules:

(Average Visit Duration $\geq$ 248.5 seconds) $\rightarrow$ Class =Purchase (763.0/136.0)

If customer's average visit duration is more than 248.5 s, it is classified as "purchase".

(Average Visit Duration $\geq$ 166.67 seconds) and (Sex = Male) $\rightarrow$ Class =Purchase (18.0/7.0)

Else $\rightarrow$ Class = Not Purchase (536.0/20.0)

If customer is a male and average visit duration is more than 166.67 s, it is classified as "not purchase".

5.5.5 Comparison of Results and Suggestions

When accuracy rates are compared, KStar has the highest accuracy rate with a value 88.2308% in the data set1 and J48 has the highest accuracy rate with a value 87.6234% in the data set 2. However, there is no big differences in highest accuracy rates among KStar and JRip, dataset 1 has the highest accuracy rate with 88.2308% for KStar method.

Since Jrip method in Rules has the highest accuracy for both dataset 1 and dataset 2. In this manner, JRip rules are interpreted to estimate class correctly. According to obtained rules some significant criteria are defined. For instance, if customer's average visit duration is more than 470 s, it is classified as "purchase" but, if customer's age between 55 to 64 and has average visit duration more than 225.4 s, it is classified as "not purchase". Moreover, if customer is a male and average visit duration is more than 166.67 s, it is classified as "not purchase". In addition, J48 method in Trees has the highest accuracy for both data set 1 and data set 2. Decision tree of J48 which is shown in Fig. 3 is accepted as another rule.

When average visit duration is recognized more than 470 s purchasing will be done. Therefore, after 470 s free shipping option should be offered for over price of the market basket 100 TL. In addition to this, When average visit duration is considered with age, if customer's age between 55 to 64 and has average visit duration more than 225.4 s, the customer leave the website without purchasing. For this type of customer, promotion or discount could be suggested before 225.4 s duration to encourage them for purchasing.

6 Conclusion

Watsons complains about customers who leave the website without purchasing as lots of company and needs to determine the reasons that encourage them to buy as much as they need to understand the behaviors of consumers. This project mainly focuses on data classification approach to predict customer's behavior during the website visits of Watsons e-commerce channel. Aim of the study is to predict e-commerce customers' behavior at the customers' first entrance to the Watsons's website.

According to proposed solution defined in Fig. 1 in Sect. 4, after data are taken from Watsons between 19–25 March 2018, missing values and duplicate data are eliminated in Excel. Before data set1 and data set 2 are determined thanks to WEKA, 7 attributes are determined such as day, age, sex, city, browser type, average visit duration, purchase or not, also there are two classes as purchase or not purchase. By comparing well-known data classification methods, the best one who has the highest prediction accuracy is determined for both data set 1 and data set 2, separately. As a result of this, KStar has the highest accuracy rate with a value 88.2308% and J48 has the highest accuracy rate with a value 87.6234% respectively in dataset1 and dataset 2, respectively. Data set 1 is the best one who has the highest prediction accuracy for KStar method when compared with data set 2. In addition to this part of study, according to highest accuracy rate of rules method; some rules are obtained from WEKA. To encourage the customers who leave the website of Watsons, rules are interpreted for suggestion.

When average visit duration is more than 470 s purchasing will be done. Therefore, after 470 s free shipping option should be offered for over price of the market basket 100 TL. By the help of that, the customers who will make a purchase will increase their market basket over 100 TL. Consequently, the profit of the company will increase. In addition to this, when average visit duration is considered with age, if customer's age between 55 to 64 and has average visit duration more than 225.4 s, the customer leave the website without purchasing. For this type of customer, promotion or discount could be suggested before 225.4 s duration to encourage them for purchasing.

For future use, more attributes could be added to the input data of the data mining classification approach for obtaining better accuracy rate of class prediction. A decision support system which is recommended could be developed to estimate the tendency of customers who visit the website.

Acknowledgment. We would like to acknowledge Burak Kurtulmuş (E-Commerce Operations Unit Manager of Watsons Turkey) for giving us permission to use departmental data that is necessary to complete this study.

References

1. Hernández S, Álvarez P, Fabra J, Ezpeleta J (2017) 'Analysis of users' behavior in structured e-commerce websites. IEEE 5:11941–11958
2. Ismail M, Ibrahim MM, Sanusi ZM, Nat M (2015) Data mining in electronic commerce: benefits and challenges. Int J Commun Netw Syst Sci 8:501–509
3. Satish B, Sunil P (2012) Study and evaluation of user's behavior in e-commerce using data mining. Res J Recent Sci 1:375–387
4. Jia W, Zhou H, Shi Y (2017) Optimization of data-mining-based e-commerce enterprise marketing strategy. Revista de la Facultad de Ingenieraí U.C.V 32(13):273–278
5. Çığşar B (2017) Kredi risklerinde veri madenciliği sınıflandırma algoritmaları. Unpublished MS thesis, Çukurova Üniversitesi, Adana
6. Perveen S, Shahbaza M, Guergachib A, Keshavjee K (2016) Performance analysis of data mining classification techniques to predict diabetes. Procedia Comput Sci 82:115–121
7. Anaklı Z (2009) A comparison of data mining methods for prediction and classification types of quality problems. Unpublished. MS thesis, Middle East Technical University, Ankara

8. Han J, Kamber M (2001) Data mining: concepts and techniques. Morgan Kaufmann Publishers, Burlington
9. Rokach L, Maimon O (2006) Data mining of improving the quality of manufacturing: a feature set decomposition approach. J Intell Manuf 17:285–299
10. Astudill C, Bardeen M, Cerpa N (2014) Editorial: data mining in electronic commerce - support vs. confidence. J Theor Appl Electron Commer Res 9. https://doi.org/10.4067/s0718-18762014000100001
11. Huang H, Wu D (2005) Product quality improvement analysis using data mining: a case study in ultra-precision manufacturing industry. In: Wang L, Jin Y (eds) FSKD 2005. LNAI, vol 3614, pp 577–580
12. Kang B, Park S (2000) Integrated machine learning approaches for complementing statistical process control procedures. Decis Support Syst 29:59–72
13. Fan C, Guo R, Chen A, Hsu K, Wei C (2001) Data mining and fault diagnosis based on wafer acceptance test data and in-line manufacturing data. In: IEEE, pp 171–174

Data Mining in Digital Marketing

Mahmut Tekin[1(✉)], Mehmet Etlioğlu[1], Özdal Koyuncuoğlu[2], and Ertuğrul Tekin[1]

[1] Selçuk University, Konya, Turkey
mahtekins@gmail.com, mehmetetlioglu@gmail.com, ertugrultekin42@gmail.com
[2] Necmettin Erbakan University, Konya, Turkey
okoyuncuoglu@konya.edu.tr

Abstract. With Industry 4.0, the Internet of objects, Internet services and cyber-physical systems have led to radical changes in every aspect of society. Artificial intelligence and intelligent systems that emerge together with technological developments are rapidly advancing towards becoming technologies that we use in almost every field of our lives and that are convenient for us. Thanks to these developments, computer systems, processor speeds and storage capacities have also increased cheaper computer systems and increasing processor speed and storage capacities have caused to be collected huge amounts of data. We have produce huge of data by the log files of WEB servers formed by the web sites we visit, blogs, photos, videos, texts etc. that we share through social media tools. Analysing increased diversity and volume of data and the result of this analysis is that more meaningful information and interpretation of the acquired knowledge is beyond what human competence and relational databases can do. At this point, data mining which allows large quantities of data to be transformed into meaningful and useful information, offers many advantages and facilities.

Data mining enables the use of computer programs to find correlations and rules that provide meaningful, potentially useful future predictions from large amounts of available data. Nowadays, data mining is successfully applied in medicine, banking and insurance, telecommunication, marketing and customer service sectors. In the field of marketing, data mining techniques enable businesses to understand hidden patterns in their past history. Thus, it is possible to plan and realize new marketing campaigns in a fast and cost-effective manner, develop product and promotion activities for specific customer segments, price determination, customer preferences and product positioning, effect on sales, customer satisfaction, point-of-sale data analysis, supply and store placement optimizations as well as profits. This study is a review of literature to emphasize the importance of data mining and to identify applications related to data mining in digital marketing and customer relationship management. This work will enable data mining techniques to be used effectively and efficiently by businesses and to enable more ARGE activities in this regard.

Keywords: Data mining · Marketing

N. M. Durakbasa and M. G. Gencyilmaz (Eds.): ISPR 2018, *Proceedings of the International Symposium for Production Research 2018*, pp. 44–61, 2019.
https://doi.org/10.1007/978-3-319-92267-6_4

1 Introduction

It has become very important to reach, analyze, use and store information that will help to make strategic decisions in an environment where information is power today. The rapid development in information communication technologies and the cheapening of hardware have led to the formation and storage of large quantities of data. The data express a value after transformation into information. Significant information is obtained by analyzing large quantities of data with various statistical methods. Innovations in emerging information technologies and database management systems as well as meaningful information derived from data contribute to the effectiveness of the strategic decision-making process and the development of new strategies.

Together with technological improvements, successful works are being carried out in many areas such as artificial intelligence, robots, virtual reality, autonomous devices, machine learning, big data and data processing subjects. Along with these developments, traditional marketing has also evolved to carry a customer focused personalized digital lane. The ability to use information communication technologies and to meet customer demands and needs that are increasing day by day with the right time, place and minimum cost has become the building block of sustainable competition. This is only possible with technology. Businesses need to know a lot of information such as demographic information, interests, personalities, habits, purchasing behaviors to reach the target audience, understand the customers and establish sustainable communication with them. Organizations must make strategic decisions about the future by making inferences and analyzes from this information. Significant personalization of meaningful information by processing these data has recently attracted the interest of all businesses. Machine learning and data mining offers more opportunities for businesses in terms of giving accurate and consistent strategic decisions and sustainability. Machine learning allows data to be processed by programming them in a manner similar to data mining. With the automation of machine learning offers the use of algorithms to decompose the data, the ability to expand software knowledge without human intervention, the ability of computers to learn from the data without needing additional programming and the ability of a machine to learn input tasks and use them for automation.

2 Data Mining

Data mining is the search and analysis of meaningful and useful links and rules through computer programs that help predict the future from large amounts of data. In addition, data mining is a data analysis technique that helps to find the link between them by examining the relationships within a very large amount of data and enables the retrieval of information that is hidden in database systems [1]. According to [2], data mining is defined as a secondary analysis discipline that interacts with statistics, database technology, pattern recognition, machine learning, and secondary analysis of unpredictable relationships in large databases. In addition data mining is a field with many disciplines. This set of disciplines; database technologies, statistics, machine learning, visualization and other disciplines (Fig. 1).

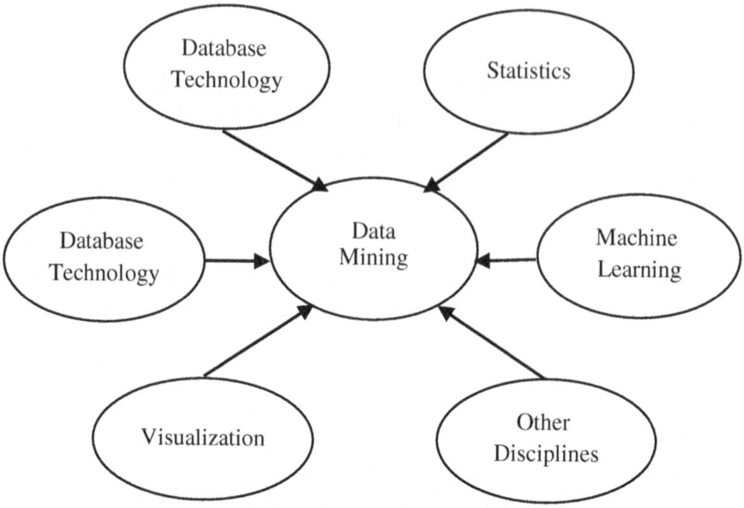

Fig. 1. Data mining relationship

Technological developments have made it easier for raw data to transform knowledge to respond to management and market needs to generate new opportunities and have forced organizations to work on data mining [3].

- As a result of the diversification of measuring instruments and the development of automatic data collection tools, the number and types of data collected have increased.
- As a result of the development of databases and database technology, a large amount of data is stored in the data repository.
- As a result of the development of computer and data processing technology, collected data can be analyzed quickly.

Data mining is a key issue in the field of marketing where intense competition is taking place, especially in order to achieve profit and market share. The answers to the questions such as which customer, which product, when to buy, who gives up the supplier and what kind of variables can be done to give up or acquisition such customers and what causes the loss of product value are in the data. Data mining solutions are needed to find these answers. With Data Mining, companies improve decision-making processes by revealing previously unknown information. Using data mining techniques; it is possible to reduce costs, increase revenues, productivity, uncover new opportunities, make new discoveries, automate labor intensive activities, identify frauds and improve customer experience. In sum, we can say that the data mining is due to the need to sort out the amount of collected data and increase the competence to make the right decisions in an increasingly competitive manner. Data mining is a process at the same time. In addition to revealing data by abstract excavations in the data stacks, it is also part of this process of separating the patterns during information discovery process and preparing them for the next step. The steps followed in the data mining process are generally as follows [4];

1. Identification of the problem,
2. Preparation of data,
3. Establishment and evaluation of the model,
4. Using the model,
5. Monitoring the model

Models used in data mining are examined under two main headings as predictive and descriptive [5]. In predictive models, it is aimed to develop a model from the known data and use this model to estimate the outcome values for unknown data sets. In descriptive models, it is possible to identify patterns in existing data that can be used to guide decision making [6]. According to functions of data mining models are examined in three parts as classification and regression, clustering and association rules. The most popular data mining programs which used in nowadays are SAS and SPSS statistical programs.

2.1 Classification and Regression

Classification and regression are models of data analysis that reveal important data classes or predict future data trends. While categorization predicts categorical values, regression is used to predict continuity values [7]. The main techniques used in classification and regression models are [8];

- Decision Trees,
- Artificial Neural Networks,
- Genetic Algorithms,
- K-Nearest Neighbour,
- Memory Based Reasoning,
- Saf Bayes.

2.2 Clustering

Clustering is the process of dividing into data classes or clusters. While the elements in the same cluster are similar to each other, they are different from the elements of other clusters. Clustering is used in many areas such as data mining, statistics, biology and machine learning. In the clustering model, there are no classes of data in the classification model [9]. Mainly clustering methods [7];

- Partitioning methods,
- Hierarchical methods,
- Density-based methods,
- Grid-based methods,
- Model-based methods.

2.3 Association Rules

Association rules reveals the relationships among the data. Depending on these rules, later it is predicted whether or not the data are co-existing [10]. It is the process of finding associations between different data in large data sets. The analysis is the discovery of relational rules for feature values that co-occur at a high frequency in a given set of data. The following two process steps are followed while there are coherence rules in large databases [11].

- There are often repeated items. Each of these items is repeated at least as often as the predetermined minimum number of supports.
- Establish strong association rules from frequently repeated items. These rules should meet minimum support and minimum confidence values.

3 Advanced Data Mining Techniques

Text, web and multimedia mining are interrelated areas that have been working much in recent years. Text mining is the analysis of very large documents and the creation of hidden patterns in text-based data. Web mining includes analysis of the web-related data, including web content, page structures, and web link statistics. Multimedia mining is also used for extracting interesting information for multimedia data sets, such as audio, video, images, graphics, speech, text and combination of several types of data set which are all converted from different formats into digital media [12]. These are:

3.1 Web Mining

Web Mining can be described as data mining on Internet data. Web mining is a sub-area of data mining; methods used in data mining are used in web mining. Analyzing and extracting information from web related documents and other data obtained automatically [13]. Web mining surveys can be classified as web content mining, web structure mining and web usage mining [14]. Web content mining is concerned with the discovery of useful information from the content of websites. Web structure mining is focused on the link between web structure and web sites. Web usage mining is working on the input patterns of web users.

Many topics can be exploited with web mining such as making detailed inferences about users, editing content according to users' tendencies, making improvements to increase the usability of the website, taking various security precautions after detecting various anomalies. In recent years, with the proliferation of e-commerce and online shopping services, work in this area has resulted in competition, revealing the importance of web mining [15].

3.2 Text Mining

Text mining is a combination of natural language processing (NLP) and data mining. Data mining is a search for data using large quantities of data that can be used to predict the future. In text mining, data mining methods are used to classify, group, or relate data and relational, statistical results are created between models to create models. The generated model allows you to make a prediction about this record when a new record is not in the dataset you created [16].

In its simplest meaning, text mining is the practice of data mining that considers work as a text data source. Objectives to obtain structured data via text. For example, classification of texts, clustering, concept/entity extraction, production of granular taxonomy, sentimental analysis, document summarization, entity relationship modeling. In order to reach these targets, information mining activities such as information retrieval, lexical analysis, word frequency distribution, pattern recognition, tagging, information extraction, data mining, and even visualization [17].

3.3 Multimedia Mining

Rapid developments in multimedia and storage technology has led to enormous growth in databases. In general, multimedia database systems store and manage a large number of multimedia objects such as images, video, audio, and hypertext data [18]. Information discovery in multimedia documents deals with unstructured information. For this reason, we need tools such as discovering relationships between objects or sections within multimedia document components, classifying images based on content, casting templates, categorizing speech and music and recognizing and monitoring objects in video streams. In this context, data mining techniques are used to reach the desired information from images, video, audio and text data. Multimedia data mining can be defined as "the process of finding interesting patterns of media data such as audio, video, images and text that are not normally accessed by basic queries and aggregated results" [19].

4 Data Mining Applications Areas

Data mining has become a strategic analytical method that is used today in the decision making process and in the achievement of organizational goals. Data mining applications have frequently used finding hidden relationships with financial indicators, determining of customer buying patterns in marketing and in insurance determining of risky customer. Today, data mining is primarily used by companies in the financial, communication, and marketing sectors. Data mining allows firms to determine internal factors such as price, production planning, and personnel skills. In addition, economic indicators allow to determine external factors such as competition and market structure. Thus, the positive and negative effects on the sales of the companies, the satisfaction of the customers and the profits of the companies can be determined. In this framework, data mining can be applied to many areas [20]. Data mining has a wide range of uses, such as medicine, finance and banking, insurance and health services, marketing, astronomy, biology and telecommunications.

5 Data Mining in Marketing

One of the most important actors of electronic marketing activities is creating technological infrastructure and internet. It is essential that businesses update their technological structure outside of their web sites, detailed content, products, employees and customer relationships in order to be successful in electronic marketing activities, respond to expectations and integrate with other systems [21]. In this context, it is very important to store and evaluate information about customers in the databases so that they can make decisions about the future.

The importance of data mining and usage rate in marketing decision have increased. Data mining which helps businesses demonstrate the evolution of customer relationships is the process of exploring the many dimensions of customer relationships, market trends, and behavioral models. The information held in the marketing database is crucial for the business to make strategic decisions. For this reason, data must be organized just like a marketing function. Businesses need consistent information to sustain their existence in today's ever-changing, complex market environment. In this context, data mining emerges as an important tool providing consistent information to businesses [22]. In this sense, data mining is the process of scanning database for the analysis of big data, revealing invisible patterns and associations and finding meanings of data by revealing trends in the market [23]. For this reason, in the marketing approach, data mining is most commonly used in database-based marketing and customer relationship management.

Data Mining is currently used by companies with strong consumer-focused retail, financial, communications and marketing organizations. Data mining allows businesses to understand hidden patterns in historical transaction data thus helping to plan and implement new marketing campaigns quickly and cost-effectively. Businesses use data mining methods to develop product and promotion activities for specific customer segments, such as pricing, customer preferences and product positioning, impact on sales, customer satisfaction, point of sale data analysis, supply and store placement optimizations, and profitability [24]. Market segmentation, competitiveness analysis, customer valuation and cross-selling analyzes are conducted with data mining [25]. Successful applications can be made in many different areas such as individualized campaigns, sales policies, new products, cross and additional sales and market arrangement for information obtained by data mining in marketing. According to [3]. these activities are;

Organizing Campaign Conditions: Selecting existing customer segments in various campaigns to be organized and developing approaches to behavioral characteristics of these customers. This is aimed at increasing the response rate to marketing or retail campaigns and reducing the cost of managing customer relationship management.

Organizing Special Campaigns: Applying special campaigns to potential customers which determining they can be most profitable. Transforming the most costly customers into less costly customers. For example, the most banking transactions are revealed to those who are directed to internet banking instead of branch banking.

Organizing Sales Policies for Customers: Finding "model" customer groups that share the same characteristics (income level, interests, spending habits, etc.) and determine sales terms and prices based on purchase profiles.

Developing New Product: By designating the features that different customer groups need, developing different products that meet the customer's expectations by extracting features they do not need.

Development of Market and Shelf Layout: The most typical example of the use of association rules is the application of a market basket. This process solves the purchasing habits of customers by finding associations between the products in the purchases they make. The discovery of these types of associations reveals the knowledge of what products customers receive together, and market managers can increase sales ratios and develop effective sales strategies by identifying market and shelf orders in this information [26].

Cross Sales: Finding connections and associations between product sales and models that can understand customer groups according to credit card expenditures, hidden correlations between different financial indicators, what customer profile means what, when and why, and offering additional products by developing these linkage-based estimates.

Additional Sales: To find the best customers or customer groups, to develop personalized products and services by identifying the needs of these customer groups, and to create product offerings that customers can not give up in this way.

5.1 Customer Relationship Management-CRM

It has observed that the differences among the products have decreased, the products have got shorter time to market, the profit margins have decreased and the customers' lifestyles and purchasing habits have changed in the globalizing world. This environment also obliges business to change their approach to customers. One of the tools used by businesses in this context is customer relationship management. The aim of CRM is to find customers, to reach customers more efficiently, to provide customers with appropriate goods and services by understanding their demands, to create customer loyalty by providing customer satisfaction and to make more profit by making more sales to customers [27]. CRM consists of a set of processes and systems that support business strategy to build long-term and profitable relationships with specific customers [28]. Customer relationship management aims to optimize the customer value of an organization through data analysis and communication. An important tool or method of businesses which using in their CRM applications is data warehouse and data mining. Data mining is the selection and evaluation of data of a particular feature from a very large number of data with computer programs, the recognition of consumers and the development of marketing applications accordingly. In the early stages of CRM, applications such as customer related data, call center, sales campaigns, relationship management were given importance. In later CRM applications, the customer's life-long value is analyzed and the applications for the customer's re-purchasing are taken as basis. Data mining is the process of revealing meaningful new relationships, patterns, and trends by examining a wide range of data stored in the data warehouse using pattern recognition techniques and statistical and mathematical techniques. In the process of data mining, hidden information is extracted in the data by using statistical analysis methods and artificial intelligence algorithms and it is the basis for the decisions of the managers [27].

Customer segmentation is the process of separating customers into different, meaningful and homogeneous subgroups based on various qualities and characteristics. It is used as a differentiated marketing tool. It enables organizations to understand customers and create different strategies [29]. Customer segmentation is a popular application of data mining with existing customers. A segmentation project begins with identifying business objectives and ends with the presentation of differentiated marketing strategies for segments [30]. There are many different types of partitioning based on certain criteria or qualities used for partitioning. The customer segmentation categories obtained by the apriori algorithm from the association rules are given below [31];

Buyer-Readiness Segmentation: The division of prospects and customers into groups reflecting the different stages which consumers normally pass through during the purchase process. These usually comprise ignorance, awareness, knowledge, preference and conviction.

Benefits Segmentation: Dividing the market into groups according to the different benefits that consumers seek from the product.

Behaviour Segmentation: The division of customers into groups based on attitude, usage or response to a product or promotion.

Occasion Segmentation: The division of customers into groups which consume a product or service at particular times, in certain situations, in response to particular events or according to seasonal or cyclical times.

Psychographic/Lifestyle Segmentation: The division of customers into groups based on lifestyle, social behaviour, values, sensitivities and personality characteristics.

Demographic Segmentation: The division of customers into different groups based on demographic variables such as age, gender, family size, income, occupation, education, language, religion, race and nationality.

Life-Cycle Segmentation: The division of customers into different groups that recognise the different needs of consumers at different stages in their life.

Geographic Segmentation: The division of customers into different groups based on countries, regions, climate and population density.

Loyalty Segmentation: The division of customers into different groups based on different degrees of loyalty to supplier or brand.

Product Segmentation: The division of customers into different groups based on levels and type of usage of the product or service.

Profitability Segmentation: The division of customers into different groups based on the different levels of value or profitability of the customers.

Interaction Segmentation: The division of customers into different groups based on their preferences regarding channels, payment method, promotions and communications.

Satisfaction Segmentation: The division of customers into different groups based on their recorded satisfaction levels, complaint history, fault history and upgrade history.

The firm's potential benefit to each client is called lifetime customer value [32]. Customer lifetime value is a measure of the profit generating potential or value of the increasingly important customer in managing customer relationships. Predicting the customer's lifetime value has become very important for companies competing in highly rich media environments [33]. Determining firm's lifetime value of customer accurately will determine the areas where the company will operate in the future and will provide innumerable benefits, especially in offering appropriate benefits to its customers. The customer profile defines the customers' characteristics such as age, income and lifestyle. It is a basic tool for marketers to create customer profiles, to provide better service to existing customers and to communicate with customers to keep them. Profiling is achieved by gathering the demographic and behavioral information from the collected data related to the customers. Creating a customer profile is also used to target new customers who use external resources, such as demographic data purchased from various sources. This data is used for customer segmentation by customers' purchasing characteristics [34].

5.2 Customer Value Analysis

Customer value is one of the most basic marketing strategies today. It refers to the emotional process that occurs in the mind as a result of the monetary value comparison with the customer's purchased product/ service. In terms of businesses, it is the difference between customer value, total customer value and total customer cost [35]. According to [36], the concept of customer value summarized the following characteristics;

- Customer value is perceived value by the customer.
- Customer value is total benefit and total costs are perceived by the customers proportionally.
- It is important to reduce costs as well as to increase benefits in increasing customer value.
- Customers' sense of value is subjective.
- Product and service characteristics as well as presentation and customer relationships play an important role in the formation of customer value.

Value analysis enables organizations to direct their product development activities to areas of greatest customer value. More specifically, value analysis helps to establish a relationship between the expenditure and functional requirements of the product as defined by the customer. Then A value sequence can be created mathematically to define the relationship between the cost of a product system and the customer value placed in those systems. Value analysis is also the key to the successful implementation of target costing, the process of profit planning and cost management used to control costs in the design phase of new product development. Cost targets are derived by subtracting the necessary profit margins from market driven product prices. In practice, once a cost target for a product has been identified, a team tries to design a product that meets customer needs more than the target cost. However, the target cost is not just about cost reduction. More importantly, organizations use target costing to increase product value for customers [37].

5.3 Determining Customers' Purchasing Patterns

Businesses have stored much data about customer, product and service characteristics and customer interactions. Although log analysis programs are important in determining customer behavior, they can provide relatively simple analysis of customer behavior. In contrast, data mining can provide highly developed customer research analysis by finding hidden patterns in databases [38]. Data mining requires the creation and recording of customer and product/service characteristics and customer interactions. Such as understanding and forecasting of customer purchasing actions and tendencies, customer profiles, industry analysis are increasingly being used in business.

5.4 Increasing the Rate of Answering in Mail Campaigns

Mail marketing is a popular marketing communication tool that brings the highest investment in direct marketing. But getting a high response rate, as well as having excellent cost effectiveness of email campaigns, is a big challenge for marketers [39]. An important element of ecommerce applications is web advertising. Companies can send ads directly to their customers using different channels, such as email campaigns and content-based ads [40]. E-mail communication with customers provides cost effectiveness and time benefits. However, if companies want to use their e-mail as a direct communication channel with their customers, they have to understand how e-mail campaigns affect their attitudes and behaviors [41]. So businesses can turn their e-mail campaigns into competitive advantage by optimizing them.

In e-mail campaigns, segmentation is done to target the email list to interests, purchasing behavior, demographics, and specific email campaigns to a message that is likely to respond to your message or to your offer. Segmentation in direct marketing has become more productive in recent years due to the development of database marketing techniques. Data mining approaches emerge as the best way to develop existing marketing strategies by segmenting existing customers or directing them to the market. In recent years, database marketing techniques have evolved into statistical techniques such as chi square automatic interaction detection and logistic regression from simple

RFM models (renewing customer purchases, purchasing frequency and amount of money spent) [39]. Recently, neural network models have been used in the field of database marketing [42].

5.5 Retention of Existing Customers and Acquisition of New Customers

Customer life cycle refers to the stages of a relationship between a customer and a business. The customer life cycle is very important because it is directly related to customer revenue and profitability. There are three ways to increase customer value. These; Increasing purchases of products or services by existing customers, selling products with higher margins and retaining customers for longer [43]. The customer life cycle analyzes the change over time in the relationship between the customer and the business. These customer groups are examined in four stages;

Prospects: The potential that is not yet a customer in the target market.

Responders: Responding members. They are candidates who have performed an activity with the business through web or mobile devices with registration or application.

Active Customers: Customers who use or continue to use the product or service.

Former Customers: They may be "bad" customers who do not pay their bills or are exposed to high costs.

According to [44], the customer life cycle consists of four dimensions: customer identification, customer acquisition, customer retention, and customer development. These four dimensions can be seen as the closed loop of a customer management system. These four dimensions are trying to maximize customer value in the long run. For this reason, data mining techniques help achieve such a goal by extracting or identifying hidden customer attributes and behaviors from large databases. Models can be created by extracting meaningful information from these data belonging to the customers with the data mining techniques mentioned below. Each data mining technique can perform one or more of the following data models [45];

- Association Rules,
- Classification,
- Clustering,
- Forecasting,
- Regression,
- Sequence discovery,
- Visualization.

With the analysis of data mining and business intelligence information, operators give customer ratings by analyzing billing information, customer service relationships, website visits and other information to prevent customer losses. It is aimed to retain customers by offering different incentives and offers to high-risk customers who are considering leaving as a result of this rating [46].

5.6 Market Basket Analysis

Data mining technique is used in Market Basket Analysis. With the development of barcode applications, all information related to the products that transferred to the electronic medium when the sales process is realized. These data which are usually collected at the sales points of the markets called market basket data. In a record the market basket data, information such as the transaction number, the date and the product code, quantity and price for the purchased products are included [47]. The purpose of market basket analysis is to find the relationships between sales and to draw up rules related to them. Knowing these relationships can be used to increase the company's profits [26]. This technique helps to understand the relationship between the different items that the customer puts into the shopping chart and the purchasing behavior of the customers [48]. If it is known that buyers of product X are very likely to buy product Y with X and if a customer is buying a product X but not a product Y, then we say that he is a potential Y customer. By doing data analyzes like this, we can reach sales estimates for the next month for each product, promotional applications and lineups can be made for the products purchased together, customers can be grouped according to the products they buy, potential customers for a new product can be identified [49].

5.7 Sales Forecasting

Retailers use sales forecasts specifically for inventory control. Answering the question of when a customer will shop again after shopping. Data mining is used to determine the purchasing habits of customers with varying price increases in the field of data mining marketing to determine the relationships between cross sales analysis and product sales and their relationships and to determine which products customers are purchasing according to customer profile determination studies [50].

Sales forecast refers to sales expectations for a certain future time period. it is important to understand consumer trends and influences new product designs because statistical data in the past also show preference for products. The purpose of making sales forecasts; the managers need to realize planning, execution, organizing and control activities. Managers take advantage of these estimates in many factors such as production planning, labor and financial resource needs, inventory levels and raw material purchases. Sales forecasts are an integral part of marketing planning because forecasting is necessary to ensure that marketing decisions are made effectively and the planning function is successful. The knowledge of how market conditions can give an image which sales forecasts for business in the future. Sales forecasts form the basis of marketing programs, production unit requests, business programs, budgets, production schedules, personnel expenses, expansion programs and procurement plans. Stages of sales forecasting period [51];

- Determining the intended use of sales forecasts
- Division of business goods into heterogeneous groups,
- Determining the factors affecting the sale of each good,
- Selection of sales forecasting technique,
- Collection of data,

- Analysis of data,
- Control of results from data,
- Application of forecasting to operating activities
- Periodic review of forecast results.

Quantitative methods include qualitative methods based on subjective evaluation and techniques based on technological developments, while quantitative methods are divided into time series techniques and mixed (based on economic and causal relations) techniques [51].

Qualitative Forecasting Techniques: Qualitative estimation methods can be divided into market research, expert opinion and Delphi technique. Qualitative, in other words, qualitative estimation methods can also be called subjective or judicial methods. If no information is available on the subject or if the collection is costly, the entity should proceed to a forecasting system that does not require numerical data.

Quantitative Forecasting Techniques: The use of quantitative techniques in the decision-making process is based on numerical data, as in quantitative methods, as well as statistical methods of analyzing past numerical data related to the variables or circumstances to be estimated, rather than experts' opinions and opinions. Calculation of sales forecasts using numerical methods is time consuming and expensive. Examples of quantitative estimation techniques include causal techniques and time series analysis.

6 Data Mining in Other Areas

Data mining is widely used in scientific and engineering, banking and finance, customer relationship management, fraud detection, security and intelligence, education, healthcare and biomedical. These areas:

6.1 Data Mining in Finance and Banking

A lot of data is generated in online transactions, so the ability to determine the right information at the right time becomes financially important. Nowadays many banks and financial institutions offers a wide range of banking services such as invest, credit, credit card and so on. The data collected by these organizations is generally reliable, complete and of high quality, providing reliable information.

Many field enabled methods such as data mining, risk analysis, fraud detection, credit payment forecasting, customer acquisition and retention analysis, cross selling, customer credit policy analysis, customer segmentation for target market, classification and determination of the amount of money to be distributed over ATM during the day it is used. Reducing risk is used to assess risks in the banking and insurance sectors, which have a high potential for loss after product or service delivery. Banks predict a financial risk when they give credit to customers, predicted risk models and the possibility that the borrowers can not repay their loans. Models are set up by the way of consumer behaviors of past periods, and it is determined that no one will fulfill the payments

The risk of dishonesty is also an important issue for banks. When the credit card is lost, the banks undertake some of the losses incurred during the loss. Fraud detection systems have been devised to reduce damages in this period. It is one of the methods used to pre-define typical spending patterns of customers, to detect sudden changes in expenditure trends, and to stop approving purchases in this direction [52].

Successful results are obtained in stock exchange transactions such as stock market price estimation, general market analysis and optimization of trading strategies, determination of customer loss causes in insurance activities, prevention of irregularities, reduction of main expenses and determination of policy prices [25]. Use of data mining is beneficial in financial activities [53];

- Collecting and analyzing customer behavior data and taking strategic decisions increases customer loyalty,
- Helps to find hidden relationships between various financial indicators to detect suspicious activity in order to identify high-potential risk activities,
- Supports real-time decision making,
- Helps define fraudulent actions by gathering past data and transforming it into valid and useful information.
- Data mining helps to predict the life value of each customer in the bank and to serve each segment appropriately by offering special opportunities and discounts [43].

6.2 Data Mining in Health Services

Data mining practices in healthcare can have enormous potential and usefulness. However, the success of health care data mining depends on the availability of healthy data. From this point of view, the healthcare sector has a critical focus on how to better acquire, store, prepare, and deduce data. Standardization of clinical vocabulary knowledge and data sharing among institutions are included in the health sector to increase the effectiveness of data mining practices [48]. Data mining applications in health sector can be classified as medical diagnosis and determination of appropriate treatment period, classification, estimation of test results and medical product development applications. In terms of estimation and decision making, data mining techniques have a significant expansion in the medical sector [54] with regard to diverse diseases such as diabetes, heart disease, liver diseases, cancer and others. It is also frequently used in gene studies in the biomedical field. It is very difficult to find the genes that lead to diseases among thousands of genes. Sequence analysis and similarity search methods developed in data mining have made it easier to analyze on DNA data.

6.3 Telecommunication

Data mining has been successfully used in many issues such as quality improvement, fraud detection, error density estimation, customer acquisition and retention analysis in the field of telecommunications. One of the most important problems of the telecommunication sector is loss of customers. If businesses can predict which customers will lose, they can develop strategies to hold these customers and organize low cost effective campaigns [50]. Telecommunication companies are faced with increasing competition to force existing pricing schemes to be aggressively marketed to protect existing customers and attract new innovations. Telecommunication companies carry out information discovery from the data they have acquired in order to retain customers and make strategic decisions. This information discovery [43];

Call Detail Record Analysis: Telecommunication companies accumulate detailed call records. Companies can develop attractive pricing and feature promotions by identifying customer segments with similar usage models.

Customer Loyalty: Some customers may want to change their service providers to take advantage of competing firms' attractive incentives. Companies can use data mining to determine the characteristics of customers who are likely to be loyal so companies focus on customers who will maximize their profits.

7 Conclusion

Technology is progressing rapidly and the power is increasing day by day. With the increase of computers' storage capacities, the number of information recording areas are also increasing. Hence the analysis of the available data and the consequence of these methods of data deduction are becoming increasingly important for decision makers. Data mining is now widely used in service sectors such as marketing, banking and insurance, as well as many field applications where decision making is required. Many applications can be made especially in the field of marketing such as determining customer buying habits, determining the relationships between customers' demographic characteristics, increasing campaign response rate, retaining existing customers and acquiring new customers, increasing customer satisfaction and reducing complaint numbers, facilitating market basket analysis, facilitating customer relationship management practices, making customer assessments, helping to make accurate sales forecasts, and conducting various customer analyzes. In this context, the use of data mining in the marketing field has a strategic importance. Using data, web, text and multimedia mining methods and techniques for marketers to make strategic decisions and forecasts for the future will help them to compete sustainably.

References

1. Kalıkov A (2006) Veri Madenciliği ve Bir E-Ticaret Uygulaması (Yüksek Lisans Tezi). Gazi Üniversitesi, Ankara
2. Hand DJ (1998) Data mining: statistics and more? The American Statistician 52:112–118
3. Argüden Y, Erşahin B (2008) Veri Madenciliği. Alkim, İstanbul
4. Shearer C (2000) The crisp-DM model: the new blueprint for data mining. J Data Warehous 5(4):13–23

5. Zhong N, Zhou L (1999) Methodologies for knowledge discovery and data mining. In: Third Pacific-Asia Conference, Pakdd 1999, Beijing, China, 26–28 April 1999
6. Özekes S (2000) Veri Madenciliği Modelleri ve Uygulama Alanları. İstanbul Ticaret Üniversitesi Dergisi
7. Han J, Kamber M (2000) Data Mining. Multiscience Press, San Francisso
8. Akpınar H (2000) Veri Tabanlarında Bilgi Keşfi ve Veri Madenciliği. İstanbul Üniversitesi İşletme Fakültesi Dergisi
9. Rajkumar GD, Swami A (1998) Clustering Data Without Distance Functions. IEEE Bull Tech Committee Data Eng 21(1):9–14
10. Alaeddinoğlu MF, Aydın T, Deniz D (2012) Birliktelik Kuralları ile Mekansal-Zamansal Veri Madenciliği. EÜFBED - Fen Bilimleri Enstitüsü Dergisi 5(2):191–212
11. Zaki MJ (1999) Parallel and distributed association mining: a survey. IEEE Concurr, special issue on Parallel Mechanisms for Data Mining 7(5):14–25
12. Waris M, Azam F, Muzaffar AW (2012) A survey of issues in multimedia databases. Int J Comput Appl 46(7):887–895
13. Mobasher B, Cooley R, Srivastava J (2000) Automatic personalization based on web usage mining. Commun ACM 43(8):142–151
14. Kosala R, Blockeel H (2000) Web mining research: a survey. ACM SIGKDD Explor 2(1): 1–15
15. Çınar I, Bilge HŞ (2006) Web Madenciliği Yöntemleri ile Web Loglarının İstatistiksel Analizi ve Saldırı Tespiti. Bilişim Teknolojileri Dergisi 9(2):115–127
16. Kaşıkçı T, Gökçen H (2014) Metin Madenciliği ile E-Ticaret Sitelerinin Belirlenmesi. Bilişim Teknolojileri Dergisi 7(1):25–32
17. Şeker ŞE (2014) Metin Madenciliği (Text Mining). http://bilgisayarkavramlari.sadievrenseker.com/2014/06/15/metin-madenciligi-text-mining/
18. Yoshitaka A, Ichikawa T (1999) A survey on content-based retrieval for multimedia databases. IEEE Trans Knowl Data Eng 11:81–93
19. Wasnik C (2012) Tools, techniques and models for multimedia database mining. Int J Netw Parallel Comput 1(2):1–5
20. Tüzüntürk S (2010) Veri Madenciliği ve İstatistik. İktisadi ve İdari Bilimler Fakültesi Dergisi, Cilt XXIX: 55–60
21. Tekin M, Zerenler, M (2016) Pazarlama. (2.Baskı). Günay Ofset, Konya
22. Tandoğan GK, Tetik D (2010) Otel İşletmelerinde Pazarlama Aracı Olarak Veri Madenciliğinin Kullanımı. 11. Ulusal Turizm Kongresi 1(64): 784–794
23. Arnold E, Price L, Zinkhan G (2004) Consumers. McGraw-Hill, Boston
24. Rajkumar P (2017) 14 useful applications of data mining. http://bigdata-madesimple.com/14-useful-applications-of-data-mining/
25. Köktürk F, Ankaralı H, Sümbüloğlu V (2009) Veri Madenciliği Yöntemlerine Genel Bakış. Türkiye Klinikleri J Biostat 1(1):20–25
26. Döşlü A (2008) Veri Madenciliğinde Market Sepet Analizi ve Birlikteik Kurallarının Belirlenmesi. (Yüksek Lisans Tezi), Yıldız Teknik üniversitesi, İstanbul
27. Aydın K (2012) CRM ve veri madenciliği. https://www.perakende.org/crm-ve-veri-madenciligi-1340026634h.html
28. Ling R, Yen DC (2001) Customer relationship management: An analysis framework and implementation strategies. J Comput Inf Syst 41:82–97
29. Kotler P, Keller KL (2005) Marketing management, 12th edn. Prentice Hall, Upper Saddle River
30. Tsiptsis K, Chorianopoulos A (2009) Data mining techniques in CRM: Inside Customer Segmentation. Wiley Hoboken

31. Kelly S (2003) Interactive Marketing, vol. 4, Henry Stewart Publications, USA
32. Drew J, Mani J, Betz D, Datta P (1999) Statistics and data mining techniques for lifetime value modeling. ACM, USA
33. Aeron H, Kumar A, Moorthy J (2012) Data mining framework for customer lifetime value-based segmentation. Database Mark Cust Strategy Manag 19(1):17–30
34. Ziafat H, Shakeri M (2014) Using data mining techniques in customer segmentation. J Eng Res Appl 4(9):70–79
35. Çetintürk İ (2017) Müşteri Değeri, Müşteri Tatmini ve Marka Sadakati: Üniversite Sosyal Tesisleri Üzerine Bir Araştırma. Seyahat ve Otel İşletmeciliği Dergisi 14(2): 93–109
36. Uzkurt C (2007) Müşteri Değeri ve Tatmininin Satın Alım Sonrası Gelecek Eğilimlere Etkisi Üzerine Ampirik Bir Çalışma. Dumlupınar Üniversitesi Sosyal Bilimler Dergisi 17:25–43
37. Dummer W, Masters M, Swenson D (2015) Delivering Customer Value Through Value Analysis. Thomson Reuters, USA
38. Gupta AK, Gupta C (2010) Analyzing customer behavior using data mining techniques: optimizing relationship with customer. 6(1): 92–98
39. Theerthaana P, Sharad S (2014) A study to improve the response in email campaigning by comparing data mining segmentation approaches in aditi technologies. Int J Manag Bus Res 4(4):273–293
40. Xuerui WL, Ying C, Ruofei Z, Mao J (2010) Click-through rate estimation for rare events in online advertising. In: Online multimedia advertising Techniques and Technologies
41. Shan L et al (2016) Predicting ad click-through rates via feature-based fully coupled interaction tensor factorization. Electron Commer Res Appl 16:30–41
42. Yang AX (2004) How to develop new approaches to RFM segmentation. J Target Measur Anal Mark 13(1):50–60
43. Rygielski C, Wang JC, Yen DC (2002) Data mining techniques for customer relationship management. Technol Soc 24:483–502
44. Swift RS (2001) Accelarating customer relationships: using CRM and relationship technologies. Prentice Hall PTR, Upper saddle river, N.J
45. Ngai EWT, Xiu L, Chau DCK (2009) Application of data mining techniques in customer relationship management: a literature review and classification. Expert Syst Appl 36:2592–2602
46. Matillion (2018) 5 real life applications of Data Mining and Business Intelligence. https://www.matillion.com/insights/5-real-life-applications-of-data-mining-and-business-intelligence/
47. Han J, Kamber M (2000) Data mining: concepts and techniques. Morgan Kaufmann Publishers, Burnaby
48. Padhy N, Panigrahi R (2012) The survey of data mining applications and feature scope. Int J Comput Sci Eng Inf Technol 2(3):16
49. Alpaydın E (2000) Zeki Veri Madenciliği: Ham Veriden Altın Bilgiye Ulaşma Yöntemleri. In: Bilişim 2000 Eğitim Semineri, vol. 1(10)
50. Timor M, Şimşek UT (2008) Veri Madenciliğinde Sepet Analizi ile Tüketici Davranışı Modellemesi Yönetim, vol. 19
51. Ukuş SG (2014) Veri Madenciliğinin Satış Tahminleri Açısından Önemi ve Bir Araştırma. (Yüksek Lisans Tezi), Galatasaray Üniversitesi, İstanbul
52. Uyumaz Ö (2017) Bankacılık Sektöründe Pazarlama Stratejilerinin Belirlenmesinde Sınıflandırma ve Veri Madenciliği. (Yüksek Lisans Tezi), Beykent Üniversitesi, Ankara
53. Varone M, Mayer D (2016) 5 Data mining applications. http://www.expertsystem.com/5-data-mining-applications/
54. Patel S, Patel H (2016) Survey of data mining techniques used in healthcare domain. Int J Inf Sci Tech 6(1/2):53–60

Computer Aided Manufacturing

Investigation of Flatness and Angularity in Case of Ball-End Milling

Balázs Mikó(✉) and Gábor Rácz

Óbuda University, Budapest, Hungary
miko.balazs@bgk.uni-obuda.hu, racz.gabor.a@gmail.com

Abstract. Free-form surfaces have an important role in the mould and die industry. The accurate manufacturing of these surfaces are unimaginable without ball-end milling and application of CAM systems. The quality of the surfaces means not only the surface roughness, but the dimensional and geometric accuracy. The principles of GPS standards have increasing importance. The macro and micro accuracy depends on many parameters and circumstances, like tool geometry, cutting parameters, tool path properties and geometric parameters of the machined surface. The article presents the study of flatness and angularity error in case of 3D plane surface machined by ball-end milling.

Keywords: GPS · Geometric tolerances · Flatness · Angularity · Ball-end milling Surface roughness

1 Introduction

The CAM systems ensure effective tools in order to generate CNC programs for complex part geometry, and the simulation of the machining increases of the reliability and efficiency of the cutting process. In case of application of CAM systems, the cutting tools, cutting parameters and cutting strategies are selected by the manufacturing engineer. These parameters determine the micro and macro accuracy of the machined surface.

Beside the dimensional accuracy and surface quality the application of geometric tolerances have growing importance in part manufacturing thanks to the advantages of GPS/GD&T system. The application of geometric tolerances has several aspects in the life cycle of a product. Standards [1, 2] define the interpretation and indication of the different symbols, but in the field of measuring and production technologies there are lot of open questions.

The geometric tolerances have increasing importance in the engineering practice, as Plowucha et al. [3] presents. The article presents the related standards, and suggests European cooperation in education. The evaluation of the different geometric error has several mathematical methods, which provide different results. Radlovacki et al. [4] presents a new method for evaluating minimal zone (MZ) based flatness error and compares with least square method (LSM). The MZ gives smallest error than the LSM.

The flatness means that the produced surface shall be contained between two parallel planes, and the tolerance's value gives the maximum distance of planes [1]. In case of

N. M. Durakbasa and M. G. Gencyilmaz (Eds.): ISPR 2018, *Proceedings of the International Symposium for Production Research 2018*, pp. 65–72, 2019.
https://doi.org/10.1007/978-3-319-92267-6_5

angularity, the definition is the same, but the parallel planes must be inclined at the theoretically exact angle to a datum plane (Fig. 1). The flatness tolerance zone has three degree of freedom: two rotations ("pitch" and "roll") and one linear ("altitude"). The angularity has only one degrees of freedom, because the definition of datum plane eliminates the rotation. Therefore the value of the flatness is always smaller than the angularity. Generally, the larger degree of freedom means smaller geometric tolerance value in case of same toleranced geometric element.

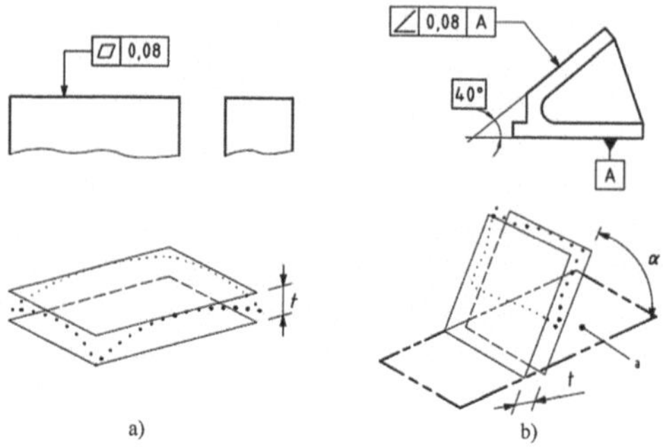

Fig. 1. Indication and definition of flatness and angularity [1]

Based on measured points' coordinates and considering the possible degree of freedom of the actual tolerance type, the evaluation algorithms can calculate the distance between the boundary planes.

The surface roughness can be predicted based on tool and cutting parameters, not only in case of 2D roughness parameters, but in 3D topography. In case of face milling Felhő et al. [5] presents an experiment based modified geometric model. The flatness error can interpret not only in maco level, but micro level too. Nalodny and Kaplonek [6] uses the flatness to characterize the 3D surface topology, 4 × 4 mm surfaces were investigated, and the ISO 12781 based flatness parameters were compared as one of a surface roughness parameters. These parameters showed a little bit different nature as conventional surface roughness parameters.

Kumar and Rajamohan [7] investigate the flatness and the surface roughness in case of end milling on 100 × 20 mm flat surface AL6063 aluminium alloy. The effect of the spindle speed, feed rate, axial and radial depth of cut were studied, and found that the feed (v_f) is the significant parameter from the viewpoint of surface roughness, but the axial and radial depth of cut (a_p, a_e) are the significant parameters from the viewpoint of flatness. Sheth and George [8] investigates face milling of cast iron workpieces from viewpoint of surface roughness and flatness error. The surface roughness and the flatness change parallely, feed increases them, spindle speed decreases, but the depth of cut has just little effect.

The ball-end milling technology has lot of investigated aspects, like the surface roughness, cutting forces, chip formation, 3D and 3D strategies, tool geometry optimization etc. Beno et al. [9] suggest a fragmentation of the surface in order to create a better surface roughness evaluation. The parameters of ball-and milling have effect to the surface macro and micro geometry, but Huang et al. [10] presents the effect on mechanical properties of the test part: residual stress and fatigue life. The higher feed increases the residual stress and decreases the fatigue life. The different tool path angles give different results: the 45° tool path gives the highest residual stress, but in case of fatigue life the effect of the tool path direction depends on the feed. Vyboishchik [11] presents the mathematical model of the periodical changes of the mechanical load of the ball-end milling cutter, as the origin of the form deviation of the free form surface.

The current article focuses on the CAM assisted CNC milling technology of angular plane surfaces by ball-end milling cutter (Fig. 2), and investigates two selected geometric tolerances: flatness and angularity. The effect of the surface properties (position and inclination), and the cutting parameters (width of cut and tool path direction) were studied. The aim of the research is to discover the manufacturing background of the geometrical deviance of surfaces in field of cutting technologies, and support the manufacturing process planning.

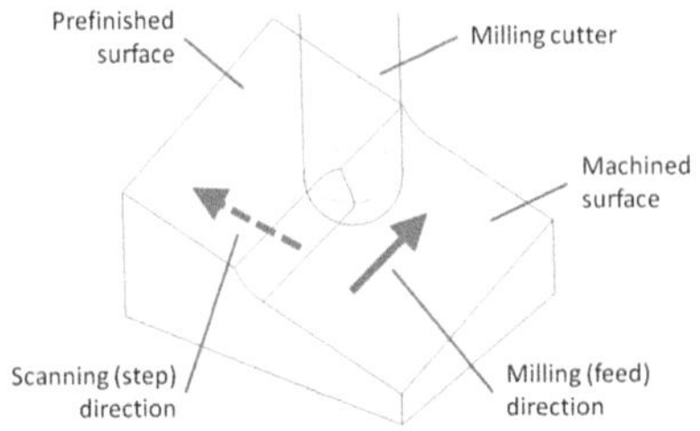

Fig. 2. Ball-end milling (theoretic figure)

2 Methods and Equipment

The test parts were made of C60 steel (1.1221; C: 0.60–0.70%; Mn: 0.50–0.80%; P: 0.04% (max); S: 0.05% (max); Rm 650 MPa), and the geometry of them was designed in consideration of the aim of the research: what is the effect of the surface properties and tool path orientation to the micro and macro accuracy of a surface. In this project the micro geometry means the surface roughness (Ra and Rz parameters) and the macro geometry is the flatness and the angularity. The position of the plane can be defined by two angles: A1 is the steepness of the plane and the A2 is the orientation based on X axis (Fig. 3). The three levels were defined in case of A1 and A2 parameters: A1 = 10° – 20° – 30° and A2 = 30° – 45° – 60°.

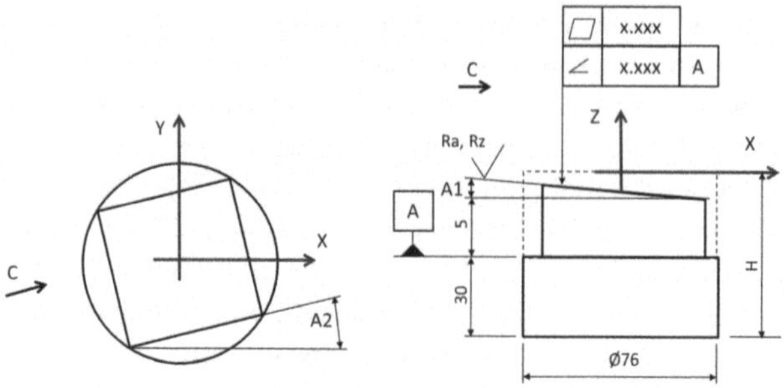

Fig. 3. Test part geometry

The test parts were milled by Mazak A410-II CNC machining centre, and the CNC programs were generated by CATIA v5 integrated CAD/CAM system. The plane surface was milled by a ball-end milling cutter (ID: Fraisa X7420.450; diameter: D = 10 mm; radius: R = 5 mm; teeth number: z = 2). The cutting speed ($v_c = 140$ m/min, n = 4450 1/min), the feed ($f_z = 0.06$ mm, $v_f = 534$ mm/min) and the depth of cut (a_p = 0.3 mm) was constant during the test. The value of width of cut was changed in three level: $a_e = 0.1$ mm – 0.3 mm – 0.5 mm. The direction of milling tool path was angled with X axis and it is marked by A3. Three levels were defined: A3 = 0° – 45° – 90°. In case of A3 = 45°, the feed speed components (v_{fx}, v_{fy}) are the same, and in other cases one of these components is larger, and the other is smaller.

The test settings were designed based on Taguchi L9 DoE method (Table 1), so 9 sets were performed.

Table 1. Test settings (Taguchi L9)

No.	A_1	A_2	A_3	a_e	A_1 [°]	A_2 [°]	A_3 [°]	a_e [mm]
1	1	1	1	1	10	30	0	0,1
2	1	2	2	2	10	45	45	0,3
3	1	3	3	3	10	60	90	0,5
4	2	1	2	3	20	30	45	0,5
5	2	2	3	1	20	45	90	0,1
6	2	3	1	2	20	60	0	0,3
7	3	1	3	2	30	30	90	0,3
8	3	2	1	3	30	45	0	0,5
9	3	3	2	1	30	60	45	0,1

The surface roughness was measured by Mitutoyo SJ-301 instrument. The results were calculated at an average of 5 measures, where the measuring direction was perpendicular to the milling direction.

The geometric deviances were measured by Mitutoyo Crysta-Plus C544 CNC coordinate measuring machine with Renishaw PH1 touch probe, and the geometric deviations were calculated by Geopak Mcosmos-1 v2.3 software tool. The flatness (FL) and the angularity (ANG) were calculated based on 16 points (Fig. 4) and the reference (base) plane (A – B_FL) was defined with 7 points, and the values were calculated at an average of three measures. The same position of test parts was ensured by a modular fixture (Fig. 4). The statistical analysis was performed by Minitab v14.

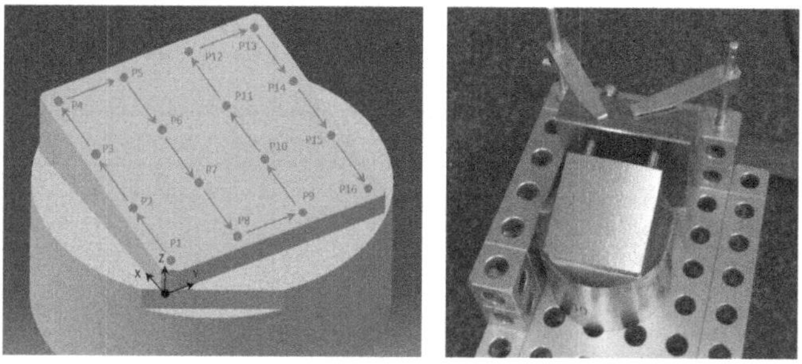

Fig. 4. Measuring strategy and the fixture on CMM

3 Results

The results of the measures are collected in Table 2. The angularity values are larger than the flatness, which meet the expectation. Therefore the angularity is stricter requirement. Because of the applied Taguchi DoE method, the main effects analysis can show the relationship between the data.

Table 2. Measured value of flatness, angularity, flatness of the datum plane and surface roughness

No	FL [mm]	ANG [mm]	FL_B [mm]	Ra [μm]	Rz [μm]
1	0.004	0.006	0.006	0.24	1.59
2	0.004	0.007	0.006	0.58	2.60
3	0.006	0.007	0.002	1.32	6.24
4	0.005	0.006	0.007	0.68	3.67
5	0.002	0.006	0.007	0.25	1.95
6	0.009	0.012	0.010	2.30	12.88
7	0.005	0.006	0.007	0.90	5.93
8	0.011	0.013	0.006	2.29	12.20
9	0.004	0.008	0.007	1.35	7.28

The main effects plot for Rz surface roughness shows (Fig. 5), that the steepness (inclination) of the surface (A1) and the width of cut (ae) have a notable effect, as based on the general geometric model predicts. But the rotational position of the surface (A2) and the direction of the tool path (A3) have a really large importance, moreover the difference of these two parameters show large effect. The difference of A2 and A3 is the relative tool path direction. The larger value means steeper tool path, the 0 value means horizontal tool paths. The steeper surface and tool path mean worst surface quality, but this combined parameter requires more investigation and attention. The width of cut has same effect, but the in case of large value other parameters moderates the effect.

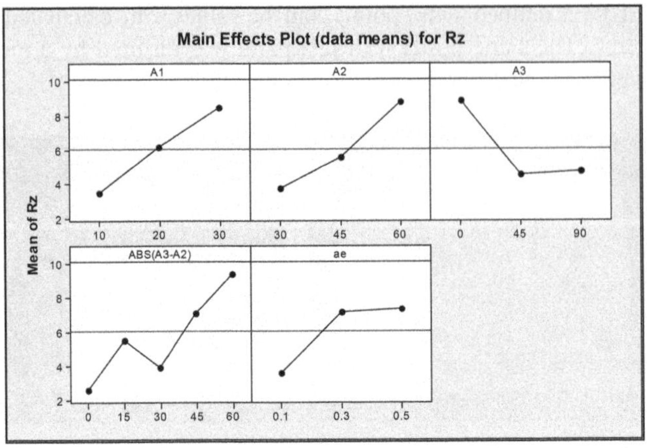

Fig. 5. Main effects plot for Rz surface roughness

Based on the main effects plot (Fig. 6), the flatness of the inclined surface is defined by the surface roughness, the width of cut, the relative tool path direction and the inclination of the surface. The surface roughness has the largest effect, so the rough surface has worst flatness deviation, the surface roughness and the flatness error change parallely.

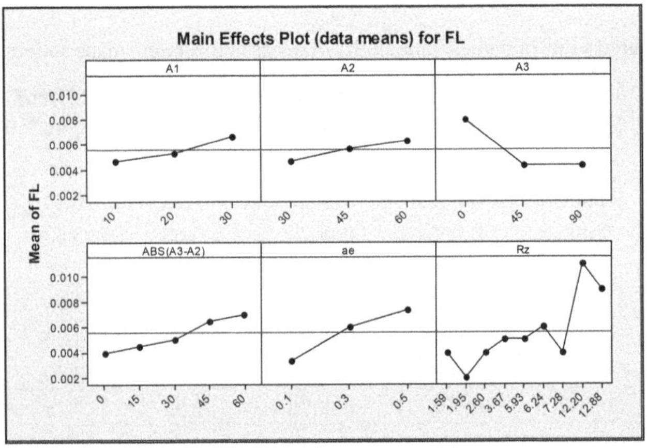

Fig. 6. Main effects plot for flatness

In case of angularity (Fig. 7), the nature of the effect of the parameters is same like in case of flatness. The definitions of the flatness and the angularity are very close to each other, the only difference is the definition of the datum surface. The value of the angularity is always larger than the flatness (Table 2).

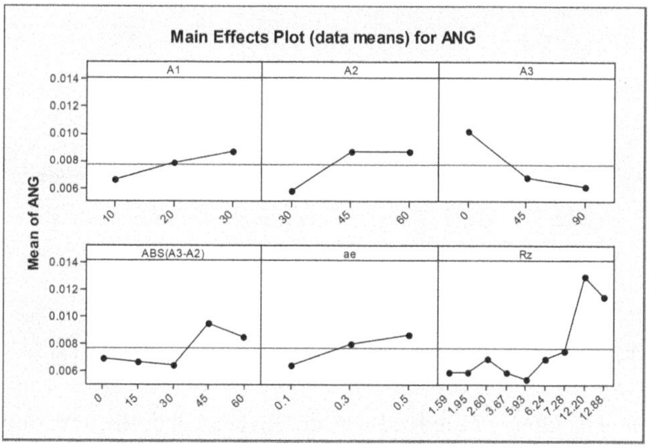

Fig. 7. Main effects plot for angularity

Theoretically, in case of investigated test parts the flatness value of the datum planes (FL_B) must be equal, because they were manufactured with the same tool and parameters. But the Table 2 shows different values, however the standard deviation in case of all measured datum plane is only 2.2 μm.

4 Conclusions

The selection of the appropriate geometric tolerance and tolerance zone is a complex engineering task, where the manufacturing and measuring aspects are more serious, than the design aspect. The definition of flatness and the angularity are close to each other, but they have a different degree of freedom. The surface roughness describes the micro geometry accuracy, but it cannot be defined independently from dimensional and geometric accuracy, considering the working, manufacturing and assembly circumstances.

Based on measured data the investigated geometric deviations and the surface roughness are not independent properties, the larger surface roughness means worst flatness and angularity parameters (Fig. 8). This perception matches with the data of Sheth and George [8], but in case of Kumar and Rajamohan [7] this trend is not so clear. The ranges of the Ra and the flatness are very different: in the first case the Ra values are larger than 1.5 μm, in the second case the Ra values are smaller than 0.5 μm.

The possible cause of this experience in case of ball-end milling is the same effect of the parameters. The larger tool load causes larger deformation and vibration (see [11]), which have the same effect on the geometrical and surface roughness deviations.

The inclination and the position of the surface have effect on the micro and macro quality of the surface, which makes the manufacturing planning (cutting parameters, tool path etc.) difficult. The relative tool path direction has an important role. The steep tool path is disadvantageous from the viewpoint of surface roughness and flatness/

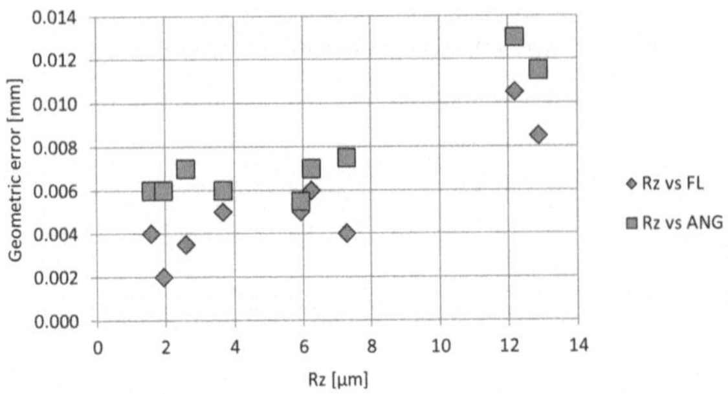

Fig. 8. Flatness and angularity in function of surface roughness

angularity error. The effect of the feed was not investigated; the next step is the deeper analysis of the tool path and the feed.

References

1. ISO 1101-2017 (2017) Geometrical product specifications (GPS) — Geometrical tolerancing — tolerances of form, orientation, location and run-out
2. ASME Y14.5 (2009) Dimensioning and tolerancing: engineering drawing and related documentation practices. American Society of Mechanical Engineers
3. Plowucha W, Jakubiec W, Humienny Z, Hausotte T, Savio E, Dragomir M, Bills P, Marxer M, Wisla N, Mathieu L (2014) Geometrical product specification and verification as toolbox to meet up-to-date technical requirements. In: Proceedings of the 11th international scientific conference "coordinate measuring technique" CMT pp 131–139
4. Radlovacki V, Hadzistevic M, Strbac B, Delic M, Kamberovic B (2016) Evaluation minimum zone flatness error using new method – bundle of plains through one point. Precis Eng 43:554–562
5. Felhő Cs, Karpuschewski B, Kundrák J (2015) Surface roughness modelling in face milling. Procedia CIRP 31:136–141
6. Nadolny K, Kaplonek W (2014) Analysis of flatness deviation for austenitic stainless steel workpieces after efficient surface machining. Meas Sci Rev 14(4):204–351
7. Kumar D, Rajamohan G (2015) Optimization of surface roughness and flatness in end milling of aluminium alloy AL 606-íT6. Int J Adv Eng Technol 8(6):937–949
8. Sheth S, George PM (2016) Experimental investigation and prediction of flatness and surface roughness during face milling operation of WCB material. Procedia Technol 23:344–351
9. Beno J, Manková I, Izol P, Vrabel' M (2016) An approach to the evaluation of multivariate data during ball end milling free-form surface fragments. Measurement 84:7–20
10. Huang W, Zhao J, Xing A, Wang G, Tao H (2018) Influence of tool path strategies on fatigue performance of high-speed ball-end-milled AISI H13 steel. Int J Adv Manufact Technol 94:371–380
11. Vyboishchik A (2017) The influence of dynamical parameters on surface quality during ball-end milling. Procedia Eng 206:1029–1033

Decision Making

A DBN Based Prognosis Model for a Complex Dynamic System: A Case Study in a Thermal Power Plant

Demet Özgür-Ünlüakın[1(✉)], İpek Kıvanç[2], Busenur Türkali[1], and Çağlar Aksezer[1]

[1] Department of Industrial Engineering, Işık University, Istanbul, Turkey
{demet.unluakin, caglar.aksezer}@isikun.edu.tr, busenur.turkali@isik.edu.tr

[2] Department of Industrial Engineering, Boğaziçi University, Istanbul, Turkey
ipek.kivanc@boun.edu.tr

Abstract. With the development of industry, complexity of systems and equipment has increased extensively. This results in the introduction of many interdependencies (stochastic, structural and economic) among the components of systems. Neglecting these interdependencies, when planning maintenance actions, leads to undesirable outcomes such as prolonged downtime and higher costs. That is why a multi-component system approach needs to be taken into account in maintenance planning models. However, maintenance planning is a difficult task in multi-component systems because of their complexities.

Energy production systems are notable examples of such complex structures consisting of many interacting components. Maintenance planning is extremely crucial for this sector since any unexpected malfunction leads to very serious costs. Therefore, the aim of this study is to formulate the maintenance problem of a multi-component dynamic system in thermal power plants focusing on system reliability prognosis. Bayesian networks (BN) are probabilistic graphical models that have been extensively used to represent and model the causal relations. A dynamic Bayesian network (DBN) is an extended BN which has a temporal dimension. We propose to use DBNs to prognose the reliabilities of components and processes of a dynamic system in a thermal power plant and show that this representation is efficient to model the interdependencies and degradations in such a system.

Keywords: Power plants · Multi-component systems · Prognosis Reliability · Maintenance · DBNs

1 Introduction

Maintenance is the periodic repairs to extend the useful life of a system [1]. In recent decades, complexity of systems increases rapidly with the industrialization. This causes an unignorable rise in maintenance costs. Maintenance management, which is especially more important for manufacturing companies that have multi-component systems, is a complex process. Taking into consideration that there are different

N. M. Durakbasa and M. G. Gencyilmaz (Eds.): ISPR 2018, *Proceedings of the International Symposium for Production Research 2018*, pp. 75–84, 2019.
https://doi.org/10.1007/978-3-319-92267-6_6

maintenance procedures for each component, it is hard to select a maintenance policy for the whole system. Furthermore, dependencies among the components of the systems should be defined. These dependencies can be classified as: stochastic, structural and economic [2]. Stochastic dependency can be defined as the effect of the degradation of one component on the lifetime distribution of the other components. On the other hand, when there is a structural dependency among components, it is required to replace the still operational components simultaneously with the failed ones. Economic dependency is another type of dependency that causes either cost saving or higher costs while grouping the maintenance of components compared to individual maintenances.

Since energy production systems are complex structures with many components and subsystems, maintenance modeling of such systems is highly critical. Progress in industrialization and mechanization, and also the growing population of the world have led to a serious increase in the need for energy. Developing countries with a high ratio of young population like Turkey import fossil fuels such as oil and natural gas in order to meet this increasing energy demand, which caused economic dependency. In recent years, a new energy policy has been proposed, aimed at utilization of coal reserves of the country so as to break free from its economic dependence on the imports. Since 2005, in order to meet the increasing energy demand in Turkey, minister of energy, Turkey announced that they would support domestic resources more. Parallel to this, the discovery of new coalfields and development of known lands have been aimed. 7.38 billion tons of new lignite reserves have been identified by the end of 2014, in addition to the existing reserves of 8.3 billion tons, as a result of these efforts [3]. Some necessary information regarding the energy sources and electricity production in Turkey is illustrated in Table 1. The second column of the Table 1 gives the distribution of electricity production according to the given production sources in the last one year. Natural gas, coal and lignite thermal power plants provide 71% of electricity production. On the other hand, the source with the largest number of power plants is hydraulic, whereas natural gas power plants are the second. If considered in terms of installed power, hydroelectric and natural gas power plants are in the first places followed by coal and lignite power plants. In the light of this information, it is obvious that thermal power plants are crucial in Turkey in terms of both electricity generation and number of stations. Because of the long setup time, coal or lignite fired thermal power plants should operate continually. For this reason, they are considered as primary energy production resources. As can be seen in Table 1, the number of natural gas-fired thermal power plants is 221 and the production percentage is 36.71%. However, the number of coal or lignite fired thermal power plants is only 40, but still around 34.29% of the total gross electricity are generated by these resources in the last one year. Hence, utilization of coal or lignite fired thermal plants is greater than natural gas fired thermal plants in terms of production. This is the reason why effective prognosis and maintenance is crucial in especially in coal or lignite fired thermal power plants.

Since the role of complex systems, such as the ones in thermal power plants, is undeniable in our lives, their maintenance is inevitable. In recent years, a great deal of attention has been paid to the maintenance and troubleshooting of the complex systems. This is because the importance of maintenance and reliability analyses has increased due to their role in increasing performance, product quality, on-time delivery, environmental and safety requirements, and overall plant cost efficiency [4]. These types of

Table 1. Information about electricity production sources in Turkey

Production source	Annual production (GWh)	Production percentage	Number of active power plants	Installed capacity (MW)
Natural gas	108,206	36.71%	221	26,750
Coal, Lignite	101,064	34.29%	40	17,907
Hydraulic	57,612	19.55%	623	27,212
Wind	18,965	6.43%	178	6,504
Geothermal	5,893	2.00%	40	1,028
Other Thermic	1,451	0.49%	56	970
Biogas	2,226	0.76%	100	530

problems can be modelled efficiently in a finite time horizon with Dynamic Bayesian Networks (DBNs). DBNs are extended form of standard Bayesian networks by adding the time dimension [5].

Bayesian networks have advantages of modeling and analyzing the complex systems, predicting the future condition of the systems, making diagnostics, inferring failure probabilities, and also updating the probabilities according to the observed evidences in comparison with the other dependability analysis methods such as Markov Chains and Fault Trees. In recent years, BNs found considerable usage in reliability, safety and maintenance as well [6]. Bayesian networks have especially been used to define the dependency between the components in decision-theoretic troubleshooting problems (DTTP) [7], which became widespread in the late 1980s to diagnose the cause of the fault [8]. These studies, and many others, handle the problem in a static manner. When dynamic systems are of interest, temporal dimension should be added to BNs directing the usage of DBNs, which have been used recently in the literature in the areas of dependence, risk analysis, reliability and maintenance [9].

There has been an increasing number of research on degradation prediction using DBNs in recent years. In a multi-component dynamic system, a novel opportunistic predictive maintenance method with DBNs is proposed to determine the degradation trends of each component separately and system as a whole where group maintenance is possible to find the optimal group maintenance times resulting in the minimum expected total maintenance cost in a given planning horizon [10]. In another study [11], a DBN based degradation model is used to predict the condition of technical systems. DBN has been preferred because of its success in modeling stochastic processes sufficiently in this study. Condition monitoring systems, expert knowledge and observation data are used to identify the potential failures early. The impact of time or exogenous variables on the failure (degradation) modes of the system is analyzed in [12] where DBN gives a straightforward way to specify the dependencies and to supply a compact representation of the problem.

DBNs are used to characterize, represent and propagate uncertainties in reliability analysis of complex systems [6]. A DBN based maintenance model is provided for high-speed train control systems where the uncertainty problem is high [13]. In this case, DBN is used to solve the fault diagnosis problem. An integrated safety prognosis model is proposed in [14] to tackle the problem of complexity and uncertainty of fault propagation. The proposed model consists of several submodels such as hazard and operability, degradation, dynamic Bayesian network, monitoring, assessment, risk evaluation and prediction models, which are integrated to each other to reduce fault occurrence probability. Degradation trend of the system is predicted to allow corrections before faults occur. In another study [15], a DBN based failure prognosis method is provided to model the response of the protection layers and effects of the external environment quantitatively on complex industrial systems. Furthermore, DBNs are used to define the conditional probabilities of hazardous events and their consequences in order to prevent serious accidents during offshore drilling [16].

The rest of the paper is organized as follows: Sect. 2 defines the problem. The proposed Dynamic Bayesian Network based model is presented in Sect. 3. Section 4 gives the analyses and finally Sect. 5 concludes and points future work.

2 Problem Definition

In this study, we propose a DBN based maintenance model to a real system, i.e. an air-gas system within a thermal power plant to predict the future degradation of the system. Thermal power plants are facilities that transform the chemical energy of solid, liquid and gas fuels to thermal energy and then transform this thermal energy to electrical energy. In general, thermal power plants have eight systems, which are shown in Fig. 1, and all these systems consist of different subsystems. Fuel supply system supplies the required fuel for combustion. Air-gas system provides the air required for combustion and emits combustion gases. Ash and slag system provides that ash; slag and other solid wastes, resulting from the burning in the boiler, are disposed of under appropriate conditions. In water purification system, clean water required for other systems is provided by separating particles. Water steam cycle system transforms water into steam and converts dead steam back into water. The dead steam is condensed in the condenser and converted to the water phase. In cooling water cycle, necessary water is supplied and hot water is also cooled in the condenser. Via the flue gas purification system, release of the harmful gases, generated during production such as CO, NO, NO_2, SO_2 and VOC, are prevented to be released to atmosphere. Electrical system consists of all other electrical components needed in the energy production process.

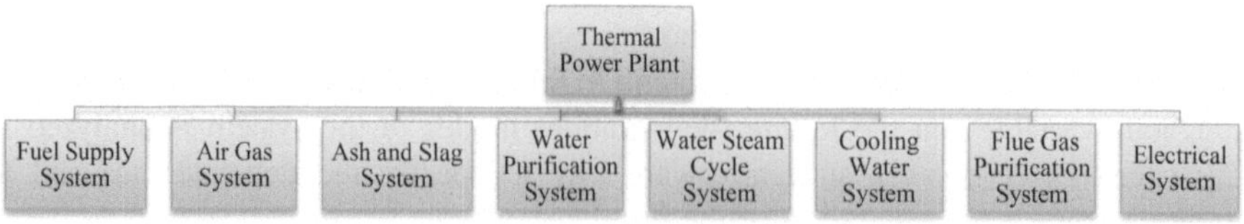

Fig. 1. Main systems of a thermal power plant

The air-gas system in a thermal power plant is selected to work on since it has many interacting components. It provides the air for the combustion system and disposes off the gas after combustion. The air-gas system consists of five main subsystems of which the general operating principle is as follows: The fresh air required for combustion is absorbed from the atmosphere by the fresh air fan and incorporated into the system. The temperature of fresh air is increased by steam in steam preheaters. The purpose of this process is to protect the regenerative air heater from corrosion. The heated air is further heated by the hot gas leaving the economizer tubes in the regenerative air heaters, which are rotating on their own axes. The gas that gives off heat passes through the electro filter, while the heated air is used for dehumidifying and also burning the coal.

In this paper, we propose a DBN formulation for the regenerative air heater, which is an important subsystem of the air-gas system in thermal power plants, and perform a degradation prediction analysis for the components and processes of this subsystem using the constructed DBN model.

3 DBN Based Model

A Bayesian network (BN) is a probabilistic graphical notation that represents the probabilistic relationship among a set of variables and the dependencies among them which is used to characterize the uncertainty [5]. A Bayesian Network consists of two main parts: The qualitative part consists of a directed acyclic graph (DAG) whose nodes represent random variables and arrows represent probabilistic dependencies between the variables. On the other side, the quantitative part defines the conditional probabilities of the variables. A dynamic Bayesian network (DBN) is an extended form of a BN by adding the time dimension to a BN to handle the dynamic behavior of random variables. A DBN consists of a set of time slots in which each slice contains a BN. In different time slices, temporal probabilistic dependence between the variables is defined by temporal arrows.

A two-slice temporal BN (2TBN) can be defined as a product of the conditional probabilities in the 2TBN as given in (1) and a DBN can be defined by using the 2TBNs as in (2).

$$P(X_t|X_{t-1}) = \prod_{i=1}^{N} P(X_t^i|Pa(X_t^i)) \tag{1}$$

$$P(X_{1:T}) = \prod_{t=1}^{T}\prod_{i=1}^{N} P(X_t^i|Pa(X_t^i)) \tag{2}$$

where X_t^i is the ith node in a time-slice t, $Pa(X_t^i)$ is the parent of X_t^i including all parents in both the same and the previous time slices, N is the number of random variables in a time slice, T is the number of time slots and $P(X_{1:T})$ represents the joint probability $P(X_1, X_2, \ldots, X_T)$.

The regenerative air heater (RAH) subsystem mainly consists of two motor groups, two hub reduction gears, a honeycomb and a RAH isolation material. The honeycomb provides a rotation on rotating parts; the isolation is located in the middle of this rotating fixture. Each motor group consists of a ball bearing, windings, an isolation material and an integrated shaft and rotor. The RAH has no separate shaft since it has a small motor structure. When one of the motor groups fails, RAH continues to operate with the single motor group. If both motor groups fail, then the RAH is disabled. Rotor rotation is affected by the states of ball bearings, windings, isolation and shaft-and-rotor. The function of a ball bearing is to connect two machine elements that move relative to one another in such a manner that the frictional resistance to motion is minimal. Problems with bearing affect windings and shaft-and-rotor alignment. Ball bearing lock-ups also cause heating and short-circuit in motor windings because of over-current. The ball bearing can be replaced directly due to aging. Motor windings may burn out when the temperature of motor windings is increased more than the rated conditions. In the case of isolation loss; the windings, which affect the rotation of rotor, may cause short circuit, so isolation loss can burn motor windings over time. The shaft-and-rotor is the component in which the rotation is provided. When the ball bearing is locked or misaligned, the shaft-and-rotor may not rotate or has axial misalignment. The shaft directly affects the rotation of the rotor. Hub reduction gear is an arrangement by which an input speed can be lowered for a requirement of slower output speed, with the same or more output torque. Therefore, the hub reduction gear rotation is directly affected by the rotation of the rotor and the state of the mechanical structure of the gear. For RAH rotation, at least one of the two gears must rotate. The amount of leakage in the isolation of RAH directly affects the performance of RAH. It is assumed that a small amount of leakage is available in the isolation. However, an increase in the amount of leakage causes performance problems in RAH. On the other side, coal quality also affects RAH exit temperature since low quality coal increases first the gas temperature and then slagging which causes the RAH exit temperature to decrease.

The constructed DBN model is given in Fig. 2 where the arcs with "1" on them show the temporal relations and the other arcs are the causal relations. We use Genie Modeler [17] to construct the DBN model.

Nodes to which temporal arcs are pointing are referred as dynamic nodes which are components and are also degrading over time. Dynamic nodes used in the DBN model are represented in Table 2 with their state definitions.

There is also another type of node used in the model: Process nodes are the ones showing the rotation processes and RAH exit temperature. The states of these processes are given in Table 3.

Other than the dynamic and process nodes, there are also exogenous variables resulting from the environment but affecting the processes. Coal rank and slagging are of exogenous types of nodes available in the model. Their states are shown in Table 4.

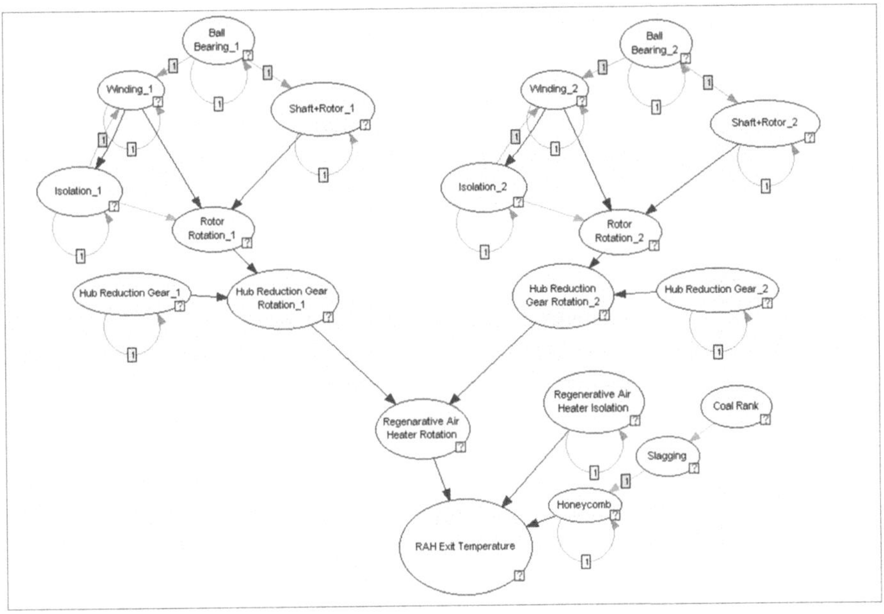

Fig. 2. The DBN model of the RAH subsystem

Table 2. Dynamic nodes in the DBN model

Dynamics nodes	State space
Isolation	No Leakage, Medium leakage, High leakage
Winding	Original, Wound, Burned
Ball Bearing	Normal, Bedding is not doing well, Locked
Shaft + Rotor	Normal, Unaligned, Unable to rotate
Hub Reduction Gear	Normal, Broken
Honeycomb	New, Cleaned, Dirty
Regenerative Air Heater Isolation	No Leakage, Medium leakage, High leakage

Table 3. Process nodes in the DBN model

Process nodes	State space
Rotor Rotation	Rotate, Do not rotate
Hub Reduction Gear Rotation	Rotate, Do not rotate
Regenarative Air Heater Rotation	Rotate, Do not rotate
Regenarative Air Heater Exit Temperature	Normal, Low, Super low

Table 4. Exogenous nodes in the DBN model

Exogenous nodes	State space
Slagging	Exist, Do not exist
Coal Rank	Conforming, Nonconforming

4 Analyses

We run the DBN model for a planning horizon of 300 days (one year excluding the days of routine annual maintenance) using Matlab and BNT [18] and prognose the reliabilities of components and processes, which are shown in Fig. 3a and b respectively. Windings and motor isolation are interdependent to each other. So, their reliabilities behave very similar as seen in Fig. 3.a (the two lowest reliability curves). Since the condition of the ball bearing affects the condition of the shaft-and-rotor directly, reliability of the shaft and rotor fall aside below the reliability of the ball bearing although the mean time to failure of the shaft-and-rotor is higher. Besides, reliability of the hub reduction gear rotation is always lower than reliability of the rotor rotation because rotor rotation is a prerequisite of hub reduction gear rotation. A similar relation exists between the reliabilities of RAH exit temperature and RAH rotation. Since there are two parallel-connected motor groups and in order to enable the rotation of RAH, it is sufficient that one of the parallel-connected motor groups works. That's why, reliabilities of RAH rotation and RAH exit temperature are higher than the reliabilities of rotor rotation of two motor groups which can be seen in Fig. 3.b.

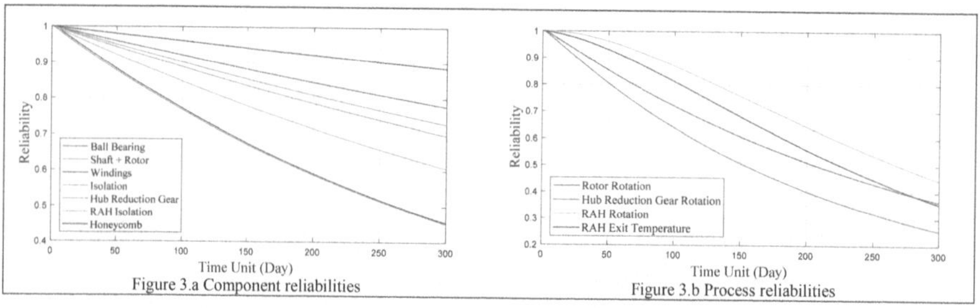

Fig. 3. Reliabilities without maintenance

In order to increase the process reliabilities, we plan maintenance activities at time period t = 151 consisting of ball bearing and hub reduction gear replacements and shaft-and-rotor adjustments. We assume that these maintenance activities are perfect, so that the mentioned components return to their best states at that time period. Then we monitor the effects of these on process reliabilities. Figure 4.a and b indicate the effects of the aforementioned maintenance activities respectively. With the help of the replacements and adjustments, reliabilities of ball bearing, shaft-and-rotor, hub reduction gear become one at t = 151 which can be seen obviously in Fig. 4.a. Furthermore, as it is shown in Fig. 4.b, process reliabilities are also affected by these maintenance activities. Replacement of ball bearing and shaft-and-rotor adjustment improve the reliability of rotor rotation. Replacement of hub reduction gear increases the reliability of hub reduction gear rotation. That's why, reliabilities of RAH rotation and RAH exit temperature also increase because of the reliability improvements obtained in their predecessors.

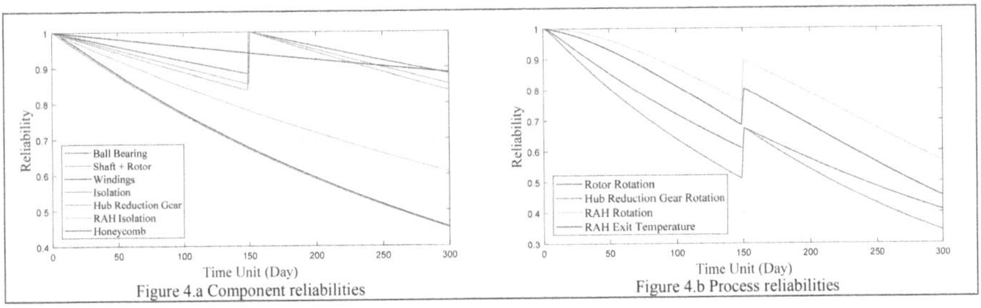

Fig. 4. Reliabilities with maintenance

5 Conclusions

We study the maintenance problem of a complex dynamic system consisting of many interacting components in thermal power plants. We propose to use DBNs to represent the causal dependencies and temporal relations for this maintenance. We perform system health monitoring via prognosis of reliabilities of components and processes. We compare the reliabilities resulting from two scenarios; one is without any maintenance and the other is with the maintenance of some critical components. As a result of these analyses, we can say that DBNs are efficient modelling tools in defining the dependencies among components and the degradations of components and also modelling the maintenance activities.

In this work, we focus on the effects of restricted size of maintenance, which is scheduled at a single time point. As a future research, one can improve the model by considering more maintenance activities which can be performed at any time point.

Acknowledgment. This work was supported by the Scientific and Technological Research Council of Turkey (TÜBİTAK) under grant no: 117M587.

References

1. Sharma GY, Deshmuk S (2011) A literature review and future perspectives on maintenance optimization. J Qual Maint Eng 17:5–25
2. Nicolai RP, Dekker R (2008) Optimal maintenance of multi-component systems: a review. In: Complex system maintenance handbook. Springer Series in Reliability Engineering. Springer, pp 263–286
3. (2018) The Ministry of Energy and Natural Resources website. http://www.enerji.gov.tr

4. Kutucuoglu K, Hamali J, Irani Z, Sharp J (2001) A framework for managing maintenance using performance measurement systems. Int J Oper Prod Manag 21:173–194
5. Murphy KP (2002) Dynamic Bayesian networks: representation, inference and learning. Ph. D. thesis, University of California, Berkeley
6. Weber P, Medina-Oliva G, Simon C, Iung B (2012) Overview on Bayesian networks applications for dependability, risk analysis and maintenance areas. Eng Appl Artif Intell 25 (4):671–682
7. Kalagnanam J, Henrion M (1990) A comparison of decision analysis and expert rules for sequential diagnosis. Mach Intell Pattern Recogn 9:271–281
8. Heckerman D, Breese JS, Rommelse K (1995) Troubleshooting under uncertainty
9. Weber P, Jouffe L (2006) Complex system reliability modeling with Dynamic Object Oriented Bayesian Networks (DOOBN). Reliab Eng Syst Saf. 91:149–162
10. Hu J, Zhang L, Liang W (2012) Opportunistic predictive maintenance for complex multi-component systems based on DBN-HAZOP model. Process Saf Env Prot 90:376–388
11. Lorenzi A, Kempf M (2015) Degradation processes modelled with dynamic bayesian networks. In: IEEE 13th international conference on industrial informatics, pp 1694–1699
12. Weber P, Munteanu P, Jouffe L (2004) Dynamic Bayesian Networks modelling the dependability of systems with degradations and exogenous constraints. In: 11th IFAC symposium on information control problems in manufacturing (INCOM 2004), vol 4, pp 671–682
13. Cheng Y, Xu T, Yang L (2013) Bayesian network based fault diagnosis and maintenance for high-speed train control systems. In: 2013 international conference on quality, reliability, risk, maintenance, and safety engineering
14. Hu J, Zhang L, Ma L, Liang W (2011) An integrated safety prognosis model for complex system based on dynamic Bayesian network and ant colony algorithm. Expert Syst Appl 38:1431–1446
15. Hu J, Zhang L, Tian W, Zhou S (2017) DBN based failure prognosis method considering the response of protective layers for the complex industrial systems. Eng Fail Anal 79:504–519
16. Wu S, Wenpei Z, Zhang L, Lungteigen MA (2016) A DBN-based risk assessment model for prediction and diagnosis of offshore drilling incidents. J Natural Gas Sci Eng 34:139–158
17. (2018) Genie Modeler website. https://www.bayesfusion.com/genie-modeler
18. (2014) BNT Toolbox. https://github.com/bayesnet/bnt

An Assessment Model for Organizational Adoption of Industry 4.0 Based on Multi-criteria Decision Techniques

Fatma Demircan Keskin(✉), İnanç Kabasakal, Yunus Kaymaz, and Haluk Soyuer

Department of Business Administration, Ege University, Izmir, Turkey
{fatma.demircan.keskin, inanc.kabasakal, yunus.kaymaz, haluk.soyuer}@ege.edu.tr

Abstract. Through the last decade, Industry 4.0 has been a transformative wave of change that affects organizations in several aspects. One of the most crucial elements in this context is the organizational adoption for upcoming changes. In particular, recent literature involves several maturity models where various challenges were discussed in several dimensions/pillars. Such models provide assessments that identify critical aspects that require more intensive efforts for development. However, a more holistic approach that signifies the overall organizational readiness has been scarce in previous research. This study presents an analytical model that provides an overall estimation of organizational readiness to Industry 4.0. This model is a three-stage model. The first stage involves the selection of relevant criteria by high/mid-level decision makers in organizations to be assessed from the criteria pool based on the KPIs discussed in the literature. In the second stage, the model uses decision makers' pairwise comparisons for relevant criteria and determines the importance weights of these criteria using the AHP (Analytic Hierarchy Process) method. In the last stage, decision-makers evaluate the level of organizations with the group decision under relevant criteria. The TOPSIS (Technique for Order Preference by Similarity to Ideal Solution) method with predetermined ideal and negative ideal solutions is applied using these evaluations to reveal the readiness level of organizations to Industry 4.0. The model is applied to a company operating in the apparel industry in Turkey, and organizational readiness level of the company to Industry 4.0 is evaluated.

Keywords: Industry 4.0 · Industry 4.0 adoption
Multi-criteria decision-making · AHP · TOPSIS

1 Introduction

With the integration of Internet & communication technologies into manufacturing processes, businesses have been in an inevitable transformation, namely Industry 4.0, through the last decade. Internets of Things (IoT) and Cyber-Physical Systems (CPS) have been emphasized as the primary drivers of the technological change in prior studies (e.g. [1, 2]). Accordingly, numerous studies from a diversity of disciplines have

N. M. Durakbasa and M. G. Gencyilmaz (Eds.): ISPR 2018, *Proceedings of the International Symposium for Production Research 2018*, pp. 85–100, 2019.
https://doi.org/10.1007/978-3-319-92267-6_7

been introduced, focusing on the benefits of technological progress. However, it can also be emphasized that such a leap in technology might cause major organizational issues, such as adoption of technology in organizations. For instance, it has been estimated that some of the jobs will become obsolete, thus will be replaced by others soon. Intuitively, it can be expected that the skill requirements of future jobs will require complex and non-routine skills. Likewise, from an organizational perspective, it can be speculated that capabilities of businesses will drastically change to meet the market demand and to survive in a harsh competition.

The Industrial Revolution, which started with the introduction of mechanical production systems using water and steam power, has moved into a second phase by mass production with the help of electric power. With the development of information technology in the third stage, the digital revolution and the industrial revolution that passed through are now in the fourth stage; namely, Industry 4.0. It can be argued that Industry 4.0 is a set of opportunities originated from Internet & Communication Technologies (ICT), especially IoT and CPS. The everlasting Industry 4.0 revolution also transforms the production systems. In such context, businesses face organizational challenges as well as operational challenges. Some significant factors that obstruct adoption of technology can be listed as follows: the uncertainty about the financial efforts that small and medium businesses need to concider in that case costs of the implication of the process [3], the necessity of high skilled knowledge [4], the inconsistency between the existing IT structures and technologies within the businesses [5]. The need of the state of the art ICT components and equipment are inevitable for the operations of such a digital transformation. The perception of Industry 4.0 as a complex structure by businesses and the lack of a systematic approach to the implementation of Industry 4.0 raise uncertainty about Industry 4.0.

The incorporation of Industry 4.0 technologies into the manufacturing process and adoption of Industry 4.0 is a long, multi-step process. Thus, in this study with the help of a multi-criteria decision technique, an assessment score can be derived. Based on the generated assessment score businesses are going to determine their positions at which stage they are in Industry 4.0 applications.

This study aims to identify a methodology and a tool to help to clarify the requirements and thus, helps to adapt ever-increasing competitive environment.

2 Literature Review: Maturity Dimensions and Corresponding Levels

The relevant literature about the readiness levels or maturity models of businesses in the context of Industry 4.0 has been addressed in our study. Moreover, white papers presented by industry leaders/consulting companies have been reviewed.

Reference [3] introduces a readiness and maturity model for manufacturing businesses for assessing the Industry 4.0. With additional models discovered in our literature review, the major studies that present a maturity model with dimensions have been listed in Table 1.

Among the models listed in Table 2, the first objective of our study was to generate a collection of criteria that might be addressed to assess the maturity of an organization

Table 1. Existing maturity and readiness models (Adapted from [3])

EXISTING MATURITY AND READINESS MODELS (ADAPTED FROM [3])

Model Name	Source/ Institution	Assessment Approach
The Connected Enterprise Maturity Model (2014)	Rockwell Automation [6]	A five-stage approach was introduced for achieving Industry 4.0. The model introduces four dimensions with a technology assessment focus.
IMPULS – Industrie 4.0 Readiness (2015)	VDMA, RWTH Aachen, IW Consult [7]	A five-level readiness level model which includes six dimensions and eighteen items were introduced. Challenges through progressing and overcome methods were highlighted.
Empowered and Implementation Strategy for Industry 4.0 (2016)	Lanza et al. [8]	Assessment of Industry 4.0 maturity as a quick check and part of a process model for realization; gap-analyses and toolbox for overcoming maturity-barriers are intended; no details about items and development process presented.
Industry 4.0 / Digital Operations Self Assessment (2016)	PricewaterHouse Coopers (PwC) [9]	The work has six dimensions with four levels; also it provides an online self-assessment application.
A maturity model for assessing Industry 4.0 readiness and maturity of manufacturing enterprises (2016)	Schumacher et al. [3]	A model developed using a multi-methodological approach with a literature review conceptual modeling and qualitative and quantitative methods for empirical validation.
An Industry 4 readiness assessment tool (2016)	University of Warwick, Crimson&Co, Pinsent Mason (WMG) [10]	A readiness assessment tool for businesses with six dimensions and related thirty-seven sub-dimensions which provides a benchmark among researched organizations
Development of an Assessment Model for Industry 4.0: Industry 4.0-MM (2017)	Gökalp et al. [11]	The proposed model compares seven maturity models in literature and provides a holistic approach to guide companies toward achieving a higher maturity stage to maximize economic benefits.

for Industry 4.0. However, as noticed from the prior research, those criteria were grouped into separate groups/dimensions. Supposedly, to assess an organization with such criteria would inherently require selection of relevant criteria for the organization. In this manner, the division of criteria might ease the relevant criteria selection process for the organization. For instance, for an organization with virtually no supply chain activities, a manager would not involve a criteria group, named "logistics and supply chain" in assessment.

Among the previous criteria and dimensions presented in Table 2, the dimensions listed in the assessment tool developed by [10] has been chosen. Accordingly, the dimensions have been chosen as the following:

- Products and Services
- Manufacturing and Operations
- Strategy and Organization
- Supply Chain
- Business Model
- Legal Considerations

In the following subsections, the assessment criteria derived from various studies have been presented according to six dimensions mentioned.

2.1 Products and Services

With the diffusion of digital technologies on various sectors/industries in the last decade, the use of digital features has become more common in products. In particular, various digital services have been augmented onto physical products. More generally, such applications can be interpreted within the context of product/service integration. Moreover, businesses that incorporate digital features into their products also utilize such products for data collection, feedback, post-sales support, and customization. As in mobile phones or TVs, products with digital capabilities also host data-driven and value-added services. Within the context of digitalization/augmentation of products/services, the maturity criteria derived from prior research are listed as the following (Adapted from [7]):

- Penetration of ICT in product lifetime
- ICT Add-on Functionalities On Products For Usage Data Collection
- Data Usage for Product Development and Support for Pre/Post Sales
- Digital features of products augmented with smart services
- Degree of product customization
- Degree of product personalization
- Share of revenue for data-driven services

2.2 Manufacturing and Operations

The manufacturing and operations dimensions pose a critical and a core area of the Industry 4.0. Although it can be argued that there are various dimensions, it can be easily interpreted that the modernization of the production units in an organization needs both close monitoring and a significant amount of investment. Arguably, the integration of production and the operations to the other aspects requires more effort.

Within the context of manufacturing and operations, the maturity criteria derived from prior research are listed as the following table (Adapted from [7]):

- Equipment Infrastructure - Current Situation
- Equipment Infrastructure – Vision
- Use of Digital Modelling
- Level of Data Collection
- Level of Data Usage For Process Optimization
- IT Systems & Process Integration
- Share of System Integrated Data
- Availability of Autonomous Embedded Tasks
- Implementation of Self Reacting Processes
- Degree of IT Security
- Cloud Usage

2.3 Strategy and Organization

Industry 4.0 leads businesses to improve their strategies with an organizational perspective. The context of Industry 4.0 differs among organizations with both in contextual and operational level. In previous literature, it was mentioned that Industry 4.0 involves various activities beyond the manufacturing or operational activities. Workforce analysis, leadership, and management commitment and other related long term and short term strategies need to be structured, and if necessary they should be changed according to organizations' Industry 4.0 strategy. The strategy and organization dimension includes the adoption level of the Industry 4.0 through an organizational and departmental level. The maturity criteria derived from prior research for this dimension are listed as the following table (Adapted from [7]):

- Level of Strategy Implementation
- Definition of Indicators
- Investments on Industry 4.0
- Level of Innovation Management
- Measurement of KPI Metrics for Industry 4.0
- Workforce Competence
- Departmental Collaboration For Improvement
- Leadership & Degree of Commitment
- Analysis of Industry 4.0 Investments (Financial, Cost/Benefit)

2.4 Supply Chain and Industry 4.0 Integration

Supply chain integration is a crucial structure in Industry 4.0 applications, and this essential structure directly affects other participants in the chain. The most significant challenges in passing Industry 4.0 implementations are not primarily in the singular sense of enterprises trying to incorporate these technological transformations into their structures, but rather at the stage of integrating suppliers and customers within the chain while trying to apply those changes to the whole structure [10].

Reference [10] has introduced five subheadings for four levels (Stock Control with Real-Time Data, Supply Chain Integration, Supply Chain Visibility, Supply Chain Flexibility, Lead Time) within the integration of supply chain and Industry 4.0.

In the same way, [9] stated that the supply chain structures in the digitalization have changed and it has defined as a network of raw material suppliers, carriers carrying these raw materials and finally a customer waiting for these requests and this network is shaped according to several key technologies. These technologies are integrated planning and application systems, logistics visibility, autonomous logistics, smart warehousing and logistics, smart spare part management, advanced analytics.

Within supply chain dimension the criteria used in the model are listed in the table below (Adapted from [7, 10]):

- Use of Intelligent Inventory Control
- Degree of Supply Chain Integration and Communication
- Visibility through channels
- Supply Chain Flexibility
- Customer Focus Level
- Lead Time Improvement
- Supply Chain Security

2.5 Business Model

The digitalization of processes enables various opportunities for businesses for competitive advantage. In particular, such transformation forms a fertile basis to introduce new business models. It can be argued that the usage of data and use/implementation of data-driven services is an important measure that indicates the degree of such transformation. In particular, analysis of customer data (analytical CRM) facilitates the alignment of business strategies to existing/potential customers. Modern business practices that depend on multiple marketing channels also require the integration of such channels.

In the 'Business Model' dimension of the maturity model adapted for our study; data-driven services, collection/analysis of data, integration of channels are all essential aspects that facilitate implementation of business models aligned for customer needs. Moreover, new business models based on automated processes help to decrease costs and increase efficiency. For such reason, real-time tracking of goods and automated task scheduling are good indicators of organizational maturity for the Industry 4.0 era.

The maturity criteria derived from prior research for this dimension are listed as follows (Adapted from [7, 10]):

- Use of data-driven services offered with (physical) products
- Data collection and analysis to expand existing portfolio and business model
- Aligning strategies to customer needs
- Real-time and automated scheduling
- Integrated marketing channels
- Real-time tracking

2.6 Legal Considerations

Considering various facets of the Industry 4.0, legal factors are crucial for businesses in some circumstances. Industry 4.0 is a newly nourishing area for many organizations. Due to this fact, it can be expected that some organizations have limited capability of understanding the Industry 4.0 and its facets. The maturity model adapted in our study involves criteria such as contract evaluation, risk assessment, data sharing and protection, and protection of intellectual property. Such dimensions mostly relate to the organizational issues and legal procedures within the business to business context as well as business to consumer context. The dimensions are listed as follows (Adapted from [10]):

- Contract Evaluation
- Risk Assessment
- Data Sharing and Protection
- Protection of Intellectual Property

As described briefly above, the dimensions and related criteria can be seen in Table 2.

Table 2. The dimension and criteria table according to the previous literature

Dimension	Criteria
Products and Services	▪ Penetration of ICT in product lifetime ▪ ICT Add-on Functionalities On Products For Usage Data Collection ▪ Data Usage for Product Development and Support for Pre/Post Sales ▪ Digital features of products augmented with smart services ▪ Degree of product customization ▪ Degree of product personalization ▪ Share of revenue for data-driven services
Manufacturing and Operations	▪ Equipment Infrastructure - Current Situation ▪ Equipment Infrastructure – Vision ▪ Use of Digital Modelling ▪ Level of Data Collection ▪ Level of Data Usage For Process Optimization ▪ IT Systems & Process Integration ▪ Share of System Integrated Data ▪ Availability of Autonomous Embedded Tasks ▪ Implementation of Self Reacting Processes ▪ Degree of IT Security ▪ Cloud Usage
Strategy and Organization	▪ Level of Strategy Implementation ▪ Definition of Indicators ▪ Investments on Industry 4.0 ▪ Level of Innovation Management ▪ Measurement of KPI Metrics for Industry 4.0 ▪ Workforce Competence ▪ Departmental Collaboration For Improvement ▪ Leadership & Degree of Commitment ▪ Analysis of Industry 4.0 Investments (Financial, Cost/Benefit)
Supply Chain	▪ Use of Intelligent Inventory Control ▪ Degree of Supply Chain Integration and Communication ▪ Visibility through channels ▪ Supply Chain Flexibility ▪ Customer Focus Level ▪ Lead Time Improvement ▪ Supply Chain Security
Business Model	▪ Use of data-driven services offered with (physical) products ▪ Data collection and analysis to expand existing portfolio and business model ▪ Aligning strategies to customer needs ▪ Real-time and automated scheduling ▪ Integrated marketing channels ▪ Real-time tracking
Legal Considerations	▪ Contract Evaluation ▪ Risk Assessment ▪ Data Sharing and Protection ▪ Protection of Intellectual Property

3 Methodology

This study proposes a three-stage model using multi-criteria decision-making techniques based on the evaluation dimensions presented in the literature to determine the organizations' readiness levels for Industry 4.0.

In the proposed model, the first stage involves selection of relevant criteria by decision makers. In particular, main and sub-criteria are chosen from a criteria pool for self-assessment of organizational readiness level to Industry 4.0. On the criteria selection screen, explanations about all criteria are provided, and organizations are guided to select the most relevant criteria with their characteristics. Subsequently, main/sub criteria chosen are structured hierarchically.

In the second stage, weights of the relevant main and sub-criteria are determined using Analytic Hierarchy Process (AHP), a method widely addressed in prior research on complex decision making. This stage involves pairwise comparisons of the criteria and approval the comparison matrices, conducted by the decision-making group.

In the third stage, decision-making group evaluates the level of the organization under the chosen relevant criteria and approve the evaluations. The readiness level of the organization is determined at this stage using the logic of distance from ideal and negative ideal solutions of the TOPSIS method. The ideal and negative ideal solution sets are predefined as the best and worst achievable levels and are weighted by the related criteria weights.

The steps of AHP and TOPSIS methods used in the proposed approach are briefly explained below.

3.1 The AHP Method

The process of determining the criteria weights in the AHP method involves the following steps, respectively [12, 13]).

1. Define the problem.
2. Determine the goal, main and sub-criteria and arrange them in a hierarchical structure.
3. Perform pairwise comparisons between main and sub-criteria separately and arrange each of these comparisons as a pairwise comparison matrix given in Eq. (1).

$$\mathrm{A} = \begin{bmatrix} a_{11} & a_{12} & \ldots & a_{1n} \\ a_{21} & a_{22} & \ldots & a_{2n} \\ \vdots & & & \vdots \\ a_{n1} & a_{n2} & \ldots & a_{nn} \end{bmatrix} \quad (1)$$

In pairwise comparison matrix of criteria, the fundamental scale in AHP shown in Table 3 is used.

Table 3. The fundamental scale in AHP [14]

Intensity of Importance	Definition
1	Equal importance
3	Moderate importance of one over another
5	Strong importance
7	Very strong importance
9	Extreme importance
2,4,6,8	Intermediate values between the two adjacent judgments
Reciprocals of above	If activity i has one of the above non-zero numbers assigned to it when compared with activity j, then j has the reciprocal value when compared with i

4. Calculate the global priority weight of each criterion.

The importance weights of main and sub-criteria (local weight) are found using the pairwise comparison matrices. Global weights of sub-criteria are obtained by multiplying the local weights of the sub-criteria by the weight of the corresponding main criteria.

To compute the criterion weights, pairwise comparison matrix A is normalized using Eq. (2). Then weights are calculated by Eq. (3) [15].

$$a_{ij}^{*} = \frac{a_{ij}}{\sum_{i=1}^{n} a_{ij}} \quad j = 1, 2, \ldots, n \tag{2}$$

$$w_i = \frac{\sum_{j=1}^{n} a_{ij}^{*}}{n} \quad i = 1, 2, \ldots, n \tag{3}$$

5. Checking the consistency of all pairwise comparison matrices.

3.2 The TOPSIS Method

The fundamental principle of the TOPSIS method, firstly introduced by [16], is that the selected alternative is the closest to the positive ideal solution and the farthest away from the negative ideal solution [17]. In this study, the calculation logic of relative closeness to the ideal solution of the TOPSIS method is used to assess organizations' readiness level.

Positive and negative ideal solutions are predefined taking into consideration of relevant criteria chosen by the organization to be evaluated and the criteria weights calculated using AHP method. The organization's relative closeness of the ideal solution is calculated using these predefined ideal solutions.

The procedure of TOPSIS method for benchmarking is as follows [18, 19]):

1. Construction of the decision matrix and normalized decision matrix.
   a_{ij}, elements in decision matrix A represent the value of the alternative i with respect to the criterion j. The normalized decision matrix's elements are calculated by using Eq. (4).

$$r_{ij} = \frac{a_{ij}}{\sqrt{\sum_{i=1}^{m} a_{ij}^2}} \quad i = 1, \ldots, m;\ j = 1, \ldots, n \tag{4}$$

2. Construction of the weighted normalized decision matrix. The elements of normalized decision matrix are weighted using the formula given in Eq. (5).

$$v_{ij} = w_i r_{ij} \quad i = 1, \ldots, m;\ j = 1, \ldots, n \tag{5}$$

where w_i is the weight of the ith criterion, and $\sum_{i=1}^{n} w_i = 1$

3. Determine the ideal (A*) and negative ideal (A^-) solution.

$$A^* = \left\{ \left(max_i v_{ij} | j \in J' \right), \right\} \left(min_i v_{ij} | j \in J'' \right) \qquad A^* = \left\{ v_1^*, v_2^*, \ldots, v_n^* \right\} \tag{6}$$

$$A^- = \left\{ \left(min_i v_{ij} | j \in J' \right), \right\} \left(max_i v_{ij} | j \in J'' \right) \qquad A^- = \left\{ v_1^-, v_2^-, \ldots, v_n^- \right\} \tag{7}$$

In Eqs. (6) and (7), J′ represents benefit and J″ represents cost criteria.

4. Calculation of the separation measures. For each alternative, separation from ideal and negative ideal solutions are calculated using Eqs. (8) and (9), respectively.

$$S_i^* = \sqrt{\sum_{j=1}^{n} \left(v_{ij} - v_j^* \right)^2} \quad i = 1, \ldots, m \tag{8}$$

$$S_i^- = \sqrt{\sum_{j=1}^{n} \left(v_{ij} - v_j^- \right)^2} \quad i = 1, \ldots, m \tag{9}$$

5. Calculation of the relative closeness to the ideal solution for each alternative using Eq. (10).

$$C_i^* = \frac{S_i^-}{S_i^- + S_i^*} \quad i = 1, \ldots, m \tag{10}$$

4 An Application of the Proposed Model

The proposed model in this study is applied to assess the level of Industry 4.0 readiness of a company which is one of the leading companies operating in the apparel industry. This company has a vision for Industry 4.0, making significant investments in this area and covering a significant distance in the implementation of Industry 4.0. In this section, step by step implementation of the developed 3-stage assessment approach is presented.

Stage 1- Criteria Selection by a Group Working
In this study, a criteria pool is formed based on the KPIs introduced in the literature. In the criteria pool, the dimensions described in the maturity dimensions' section are considered as the main criteria, and the criteria of these dimensions are regarded as the sub-criteria. This pool contains 6 main criteria and 44 sub-criteria (number of sub-criteria that the main criteria have are 7, 11, 9, 7, 6, 4 respectively). The decision-making group comprises two mid-level managers. One of them works in the technology department focusing on Industry 4.0 implementation, and the other works in the production department. The decision-making group came together to evaluate all the main and sub-criteria and selected relevant ones by reaching the final consensus of their discussions. The hierarchical structure of the chosen criteria is given in Fig. 1.

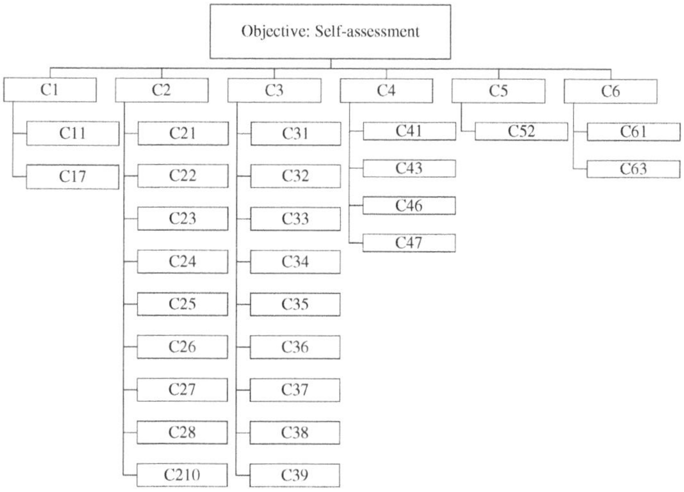

Fig. 1. Hierarchical structure of the chosen relevant main and sub-criteria

The decision-making group has chosen the majority of sub-criteria from "Manufacturing and Operations" and "Strategy and Organization" dimensions. On the other hand, only one (out of six) sub-criteria was chosen within the dimension: "Business Model."

Table 4. Pairwise comparison matrix for main criteria

	C1	C2	C3	C4	C5	C6
C1	1	0.17	0.33	6	4	5
C2	4	1	2	5	5	6
C3	3	0.50	1	6	6	9
C4	0.17	0.20	0.17	1	2	4
C5	0.25	0.20	0.17	0.50	1	3
C6	0.20	0.17	0.11	0.25	0.33	1

Stage 2- Determination of Weights of Relevant Main and Sub-criteria

The decision-making group has made pairwise comparisons among the main criteria. Pairwise comparison matrix for the main criteria based on the consensus of the decision-making group is given in Table 4. The consistency of this matrix has been checked for validation.

For each main criterion, decision-making group has made pairwise comparisons between sub-criteria that are under that main criterion. They have considered all sub-criteria are of equal importance. The global weights of all criteria resulting from the implementation of the AHP method are presented in Table 5:

Table 5. Global weights of criteria

Criterion	Global Weight
C1	**0.1659**
C11	0.0830
C17	0.0830
C2	**0.3658**
C21	0.0406
C22	0.0406
C23	0.0406
C24	0.0406
C25	0.0406
C26	0.0406
C27	0.0406
C28	0.0406
C210	0.0406

Criterion	Global Weight
C3	**0.3009**
C31	0.0334
C32	0.0334
C33	0.0334
C34	0.0334
C35	0.0334
C36	0.0334
C37	0.0334
C38	0.0334
C39	0.0334

Criterion	Global Weight
C4	**0.0764**
C41	0.0191
C43	0.0191
C46	0.0191
C47	0.0191
C5	**0.0585**
C52	0.0585
C6	**0.0324**
C61	0.0162
C63	0.0162

According to the company's assessment, the most important dimension affecting the level of readiness to Industry 4.0 is "Manufacturing and Operations" with a weight of $\sim$37%. Subsequently, the dimension "Strategy and Organization" comes next, with a weight of 30%. The descending order of other dimensions according to their importance level is: 'Products and Services', 'Supply Chain and Industry 4.0 Integration', 'Business Model', and 'Legal Considerations'.

Stage 3- Evaluation the Level of Organization and Calculation the Assessment Score

Decision-making group has evaluated the level of the organization under all chosen criteria using the scale 1 (immature) to 5 (mature). Evaluation of the organization with respect to all relevant criteria based on the consensus of the decision-making group is given in Table 6.

Table 6. Evaluation of the organization with respect to all relevant criteria

Criterion	C11	C17	C21	C22	C23	C24	C25	C26	C27	C28	C210
Organization Level	4	3	3	2	2	4	4	5	4	3	5
Criterion	C31	C32	C33	C34	C35	C36	C37	C38	C39		
Organization Level	5	4	5	4	3	4	4	4	5		
Criterion	C41	C43	C46	C47	C52	C61	C63				
Organization Level	4	3	3	4	3	3	4				

Decision-making group has scored the company's level as the highest in "strategy and organization" dimension among all evaluation dimensions. The average score of this dimension is 4.22. In all other dimensions, the company score is around 3.50.

In this study, organizations are assessed individually. For this reason, the decision matrix presented in Table 7 is also the normalize decision matrix.

Table 7. Weighted evaluation of the organization with respect to all relevant criteria

Criterion	C11	C17	C21	C22	C23	C24	C25	C26	C27	C28	C210
Organization	0.332	0.249	0.122	0.081	0.081	0.163	0.163	0.203	0.163	0.122	0.203
Criterion	C31	C32	C33	C34	C35	C36	C37	C38	C39		
Organization	0.167	0.134	0.167	0.134	0.100	0.134	0.134	0.134	0.167		
Criterion	C41	C43	C46	C47	C52	C61	C63				
Organization	0.076	0.057	0.057	0.076	0.176	0.049	0.065				

Predetermined ideal and negative ideal solutions are as follows:

$$A^* = \{5,5,5,5,5,5,5,5,5,5,5,5,5,5,5,5,5,5,5,5,5,5,5,5,5,5,5\}$$

$$A^- = \{1,1,1,1,1,1,1,1,1,1,1,1,1,1,1,1,1,1,1,1,1,1,1,1,1,1,1\}$$

Calculated separation measures and relative closeness of the organization to the ideal solution are given in Table 8.

Table 8. Separation measures and relative closeness of the organization to the ideal solution

S^+	S^-	C^*
0.333	0.582	0.636

As shown in Table 8, the relative closeness of the organization to the ideal solution is 0,636. When assessed holistically with all maturity dimensions, the maturity level of the organization is between 3 and 4 on the scale of 1 (immature) to 5 (mature). The organization needs to decrease the S^+ value and increase the S^- value to improve its maturity level. For this, it should primarily focus on the sub-criteria with the highest level of importance and the lowest evaluation score. The greatest improvement in the assessment score of the company will be achieved if the level in C17 (share of revenue for data driven services) increases from 3 to 4. Subsequently, raising C11 (penetration of ICT in product lifetime) or C17 level from 4 to 5 will provide the greatest assessment score contribution.

5 Conclusions

Industry 4.0 transformation has various aspects relating to competitive, technological, and organizational issues. In prior studies, Industry 4.0 was described as a 'disruptive' wave of innovations that deeply affect businesses [20]. It can be argued that such comments highlight the substantial amount of effort required to conform to the upcoming level of competition. In this point of view, organizational issues that will emerge due to such transformation will also be crucial and challenging. In our literature review, several studies were encountered that pioneered in addressing the Industry 4.0 requirements for organizations and developing corresponding maturity models.

Our study aims to develop a self-assessment model that measures the maturity of an organization against the Industry 4.0 environment. For such purpose, a maturity model (with six dimensions and various criteria) was adopted based on prior models, and an assessment form was designed. In the form, the level of maturity for each criterion was declared with descriptions of two extreme states (the least mature vs. the most mature).

The dimensions involved in our study might correspond to different levels of importance across various industries. For instance, Industry 4.0 maturity criteria included in the 'Supply Chain' dimension might have higher priority for a textile manufacturer, than a hosting company. The assessment approach proposed in this study has a wide range of potential uses across various industries since the importance levels of evaluation dimensions are determined by pairwise comparisons.

An international apparel industry leader currently active in Turkey was chosen for data collection. The form was used to receive the response from mid-level management. Relative importance levels of criteria were calculated with the application of

AHP method on pairwise comparison data. The TOPSIS method was implemented to the evaluation score of the organization under all relevant criteria, and the relative closeness of the organization to the ideal maturity level was determined.

The model proposed in this study enables organizations from different industries, sizes, and characteristics to select relevant main and sub-criteria from criteria pool based on the group decision of high/mid-level managers and make self-assessment based on these chosen relevant criteria. Thus, organizations can assess themselves on a number of relevant dimensions, and they can determine the dimensions that require more attention to improve their readiness level. In further studies, the model can be developed to provide benchmarks among organizations with similar characteristics.

Acknowledgment. This study was conducted at Department of Business Administration/Ege University; and has been funded by Ege University, Turkey (Scientific Research Project, Code: 2016-İKT-008).

References

1. Almada-Lobo F (2016) The Industry 4.0 revolution and the future of manufacturing execution systems (MES). J Innov Manag 3(4):16–21
2. Kagermann H, Wahlster W, Helbig J (2013) Recommendations for implementing the strategic initiative INDUSTRIE 4.0
3. Schumacher A, Erol S, Sihn W (2016) A maturity model for assessing Industry 4.0 readiness and maturity of manufacturing enterprises. Procedia CIRP 52:161–166
4. Schuh G, Potente T, Wesch-Potente C, Weber AR, Prote JP (2014) Collaboration mechanisms to increase productivity in the context of industrie 4.0. Procedia CIRP 19:51–56
5. Niesen T, Houy C, Fettke P, Loos P (January 2016) Towards an integrative big data analysis framework for data-driven risk management in industry 4.0. In: 2016 49th hawaii international conference on system sciences (HICSS). IEEE, pp 5065–5074
6. Rockwell Automation (2014) The Connected Enterprise Maturity Model, Rockwell Automation
7. Lichtblau K, Stich V, Bertenrath R, Blum M, Bleider M, Millack A, Schmitt K, Schmitz E, M.S.: IMPULS - Industrie 4.0-Readiness, VDMA's IMPULS-Stiftung, Aachen, Cologne (2015)
8. Lanza G, Nyhuis P, Ansari SM, Kuprat T, Liebrecht C (01–02/2016) Befähigungs- und Einführungsstrategien für Industrie 4.0, Karlsruhe/ Hannover
9. PricewaterhouseCoopers (2016) The Industry 4.0/Digital Operations Self Assessment
10. Agca O, Gibson J, Godsell J, Ignatius J, Davies CW, Xu O (2017) An industry 4 readiness assessment tool. University of Warwick, Crimson & Co., Pinsent Mason
11. Gökalp E, Şener U, Eren PE (October 2017) Development of an assessment model for industry 4.0: industry 4.0-MM. In: International conference on software process improvement and capability determination. Springer, Cham, pp 128–142
12. Saaty TL (2008) Decision making with the analytic hierarchy process. Int J Serv Sci 1 (1):83–98
13. Saaty TL, Kearns TL (2014) Analytical planning: the organization of system, vol. 7. Elsevier
14. Saaty TL (1990) How to make a decision: the analytic hierarchy process. Eur J Oper Res 48 (1):9–26

15. Chen CF (2006) Applying the analytical hierarchy process (AHP) approach to convention site selection. J Travel Res 45(2):167–174
16. Hwang CL, Yoon K (1981) Multiple attribute decision making: methods and applications, a state-of-the-art survey. Lecture Notes in Economics and Mathematics Systems, No. 186. Springer-Verlag, New York
17. Lai YJ, Liu TY, Hwang CL (1994) Topsis for MODM. Eur J Oper Res 76(3):486–500
18. Opricovic S, Tzeng GH (2004) Compromise solution by MCDM methods: a comparative analysis of VIKOR and TOPSIS. Eur J Oper Res 156(2):445–455
19. Yang T, Chou P (2005) Solving a multiresponse simulation-optimization problem with discrete variables using a multiple-attribute decision-making method. Math Comput Simul 68(1):9–21
20. Schmidt R, Möhring M, Härting RC, Reichstein C, Neumaier P, Jozinović P (2015) Industry 4.0-Potentials for creating smart products: empirical research results. In: International conference on business information systems. Springer, Cham, pp 16–27

Analysis of Reactive Maintenance Strategies on a Multi-component System Using Dynamic Bayesian Networks

Demet Özgür-Ünlüakın(✉) and Ayşe Karacaörenli

Department of Industrial Engineering, Işık University, Istanbul, Turkey
demet.unluakin@isikun.edu.tr,
ayse.karacaorenli@isik.edu.tr

Abstract. Recently effective planning and management of maintenance activities have gained great importance. Although proactive maintenance approach is preferred against reactive maintenance approach in some highly critical systems, it is inevitable to give up the latter one because of the probabilistic nature of faults. Hence, reactive maintenance is still widely applied. Therefore the aim of this study is to develop an effective reactive maintenance strategy for a dynamic system consisting of four components. Components are hidden and degrading over time. However it is possible to receive partial observations about the condition of the system. One can replace components at any time period. Our aim is to minimize the total number of replacements in a given planning horizon. We propose several approaches within the framework of reactive maintenance for this system and compare their performances by simulating these using Dynamic Bayesian Networks.

Keywords: Multi-component systems · Maintenance · Reliability
Decision analysis · DBNs

1 Introduction

Systems have become more complex with the advancement of technology. Parallel to this, the interaction between the components has also increased and this makes it difficult to make maintenance decisions. Therefore, it is important to diagnose effectively in order to make proper maintenance decisions. In multi-component systems, developing an effective maintenance policy for the system plays a more important role for the health of the system compared to single-component systems. Maintenance actions can be examined under two main subjects. In general, maintenance actions can be categorized as either reactive maintenance or proactive maintenance. In reactive maintenance, repairs are realized after the component has already broken down or has given a failure. Proactive maintenance actions, on the other hand, consist of preventive and predictive maintenance activities which are performed to avoid failures or identify defects that could lead to failure. Proactive maintenance is important to increase system reliability by performing early diagnosis-based maintenance activities without waiting for a problem. As the availability of new maintenance techniques arises and the economic

N. M. Durakbasa and M. G. Gencyilmaz (Eds.): ISPR 2018, *Proceedings of the International Symposium for Production Research 2018*, pp. 101–110, 2019.
https://doi.org/10.1007/978-3-319-92267-6_8

consequences of the maintenance actions are understood, a direct impact on the maintenance policies is expected. In one kind or another, various maintenance policies may be considered to trigger proactive or reactive maintenance interventions [1].

The concept of effective maintenance is based on the prognostic process which provides the selection of best maintenance action. Prognosis includes the process monitoring, diagnosis, prediction and decision making. The prognostic process depends on combination of a probabilistic approach in the modeling of degradation mechanism. Detailed information about the prognosis is quite applicable for a dynamic model of an industrial system [2]. In another study, a prognostic model using dynamic Bayesian networks is proposed for a system [3]. Furthermore, prognosis is used for improving reliability, performance and ensuring safety of complex systems. An assessment model for a gas turbine compressor system is developed using Dynamic Bayesian Networks (DBNs) [4]. Ant colony algorithm is also used to obtain the propagation paths of faults.

A survey is done to study the maintenance policies in deteriorating systems [5]. Effective maintenance strategy is developed by creating various maintenance policies which are compared with respect to system performance, maintenance costs, maintenance times and system reliability. The aim of maintenance policies is to improve system reliability, availability and mean time between failures for reducing downtime and frequency of failures incurring a maintenance cost. In this paper, the authors examine the stochastic behavior of the system under several maintenance policies and determine the optimal maintenance policies. Generally, optimal maintenance policy is determined according to achieving the minimum maintenance costs, maximum system reliability and minimum maintenance time. An opportunistic predictive maintenance based DBN-HAZOP model is applied for a real life multi-component system [6]. Two approaches are offered, which are local predictive maintenance approach and global opportunistic predictive maintenance approach. Then, an effective maintenance policy integrating these approaches provides the optimal maintenance time, reliability and maintenance cost.

There are many studies in the literature using DBNs for reliability prediction of dynamic systems, but a limited number exists for developing effective maintenance methodologies for complex dynamic systems. A study [7] offers two approaches which are myopic and look-ahead for a predictive maintenance plan. The aim is to minimize the total number of replacements of components in a predetermined planning horizon. Group maintenance [8] is another alternative policy for multi-component systems with economic dependencies. Positive addictions are often attributed to installation costs, while negative dependencies are related to shut-down costs. Group maintenance policy can reduce some of the installation costs. However, it can also increase shut-down costs. In another study, the conditional-based maintenance is applied and a maintenance interval is used to reduce high institutional costs. This study evaluates their potential for cost savings and compares their policies with a failure and an age-based policy [9].

The rest of the paper is organized as follows: Sect. 2 presents DBNs as our methodology. The empirical DBN model is given in Sect. 3. Section 4 explains the details of the proposed reactive maintenance strategies. We present the computational analyses in Sect. 5 and finally Sect. 6 concludes the study and points future work.

2 Dynamic Bayesian Networks

Bayesian networks (BN) are a member of the family of probabilistic graphical models which provide a framework to represent and model complex domains, large number of interacting random variables, using probability distributions. In these graphs, nodes represent random variables and edges between the nodes represent the probabilistic dependency between the random variables. Bayesian networks have a graphical model structure known as directed acyclic graph. These networks provide an efficient representation and implementation of the multivariate probability distribution of a set of random variables using conditional independence. With this feature, the number of parameters used to calculate the multivariate probability distribution of variables is reduced.

A dynamic Bayesian network (DBN) is an extended form of a BN by adding the time dimension to a BN to handle the dynamic behavior of random variables. A DBN consists of a set of time slots in which each slice contains a BN. In different time slices, temporal probabilistic dependence between the variables is defined by temporal arrows.

A two-slice temporal BN (2TBN) can be defined as a product of the conditional probabilities in the 2TBN as given in (1) and a DBN can be defined by using the 2TBNs as in (2).

$$P(X_t|X_{t-1}) = \prod_{i=1}^{N} P(X_t^i \mid Pa(X_t^i)) \tag{1}$$

$$P(X_{1:T}) = \prod_{t=1}^{T} \prod_{i=1}^{N} P(X_t^i \mid Pa(X_t^i)) \tag{2}$$

where X_t^i is the ith node in a time-slice t, $Pa(X_t^i)$ is the parent of X_t^i including all parents in both the same and the previous time slices, N is the number of random variables in a time slice, T is the number of time slots and $P(X_{1:T})$ represents the joint probability $P(X_1, X_2, \ldots, X_T)$.

3 The Empirical Model

Due to developing technology and industry, relation among the components of multi-component systems has increased and thus this makes it difficult to make maintenance decisions. An empirical dynamic model consisting of four components was chosen to reflect this complex relationship. In the empirical model, there exist three processes and one observable node. The relation between the components and the processes are shown in Fig. 1 where the model is built with Genie Modeler [10]. It is not possible to observe the components and the processes directly. But the states of the processes and the components can be inferred by the observable node. It is assumed that all components, i.e., C1, C2, C3, C4, can be replaced at any time period.

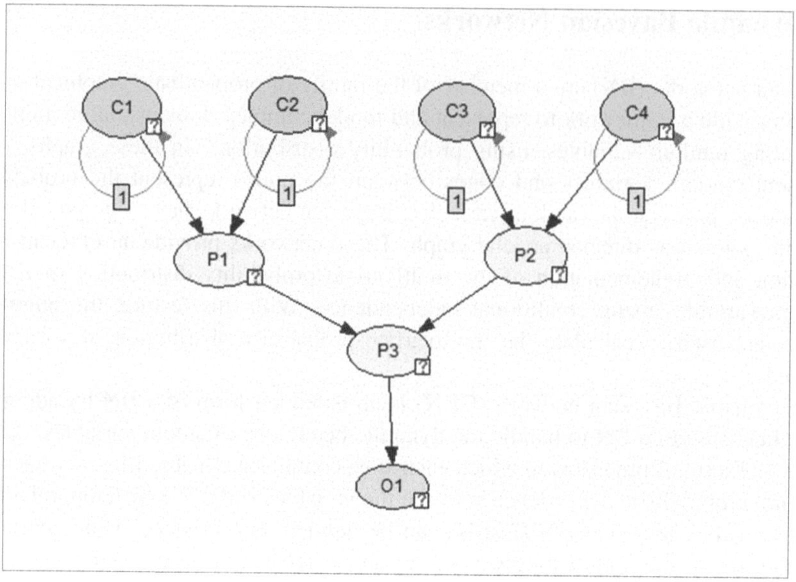

Fig. 1. Empirical DBN model (Color figure online)

All components and the observable node have three states, and all processes have two states which are given in Table 1 where states W, D, F, G, Y, R stands for "working", "degradation", "failure", "green", "yellow" and "red" respectively. All components are subject to degradation and hence they are modeled with temporal nodes. These temporal relations are shown with circular arrows in Fig. 1. The processes are defined as the result of the interaction between their predecessor components. Main process node is P3, directly connected to the observable node O1 which is used to collect information from the process node.

Table 1. Types of nodes in the DBN model

Node type	Nodes	State space
Replaceable nodes	C1, C2, C3, C4	{W, D, F}
Process nodes	P1, P2, P3	{W, F}
Observable node	O1	{G, Y, R}

Temporal property of component nodes are defined in the transition probabilities of components. Transition probabilities of C1 and C3 are given in Table 2, and transition probabilities of C2 and C4 are given in Table 3. Components C1 and C3 degrade faster than components C2 and C4.

Finally, Table 4 gives the causal relation between the main process node and the observable node, in other words, the conditional probability of the observable node O1 given the states of the main process node P3.

Table 2. Transition probabilities of C1 and C3

(self)[t]	(self)[t-1]		
	W	D	F
W	0.994	0	0
D	0.004	0.996	0
F	0.002	0.004	1

Table 3. Transition probabilities of C2 and C4

(self)[t]	(self)[t-1]		
	W	D	F
W	0.997	0	0
D	0.002	0.998	0
F	0.001	0.002	1

Table 4. Conditional probability of O1

O1	P3	
	W	F
G	0.997	0
Y	0.002	0
R	0.001	1

4 Proposed Solution

We propose four approaches within the framework of reactive maintenance for a multi-component dynamic system and apply these on the empirical model given in Sect. 3. Our aim is to minimize the total number of replacements in a given discrete time planning horizon. We simulate the proposed maintenance strategies using the empirical DBN model. The general simulation framework is given in Fig. 2.

In each time period, observable node is inferred and an observation is sampled according to its probability distribution at that time period. If it is undesirable, meaning "R" in the empirical model, this indicates mostly a faulty system state. Thus, reactive maintenance action(s) should be performed so that the system gets rids of from its faulty state. When a reactive maintenance decision is made, one of the proposed methods is applied to select the component to replace, and evidence is updated accordingly as replacements are performed. Then another observation is sampled for the observable node according to its probability distribution, taking into account the accumulated evidence so far. If this observation is undesirable again, another reactive maintenance action is performed, searching for the components that have not replaced in that period, until we get a desirable observation. In this case, we increase the time period by one and continue with the next period similarly until we reach the end of the planning horizon.

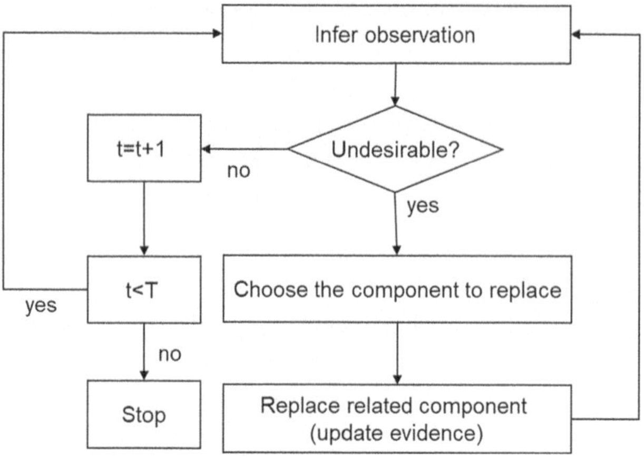

Fig. 2. General framework of the simulation

We use the following notations in the definition of the proposed reactive maintenance strategies:

O_t: State of O1 node in time t.
P: State of P3 node in time t.
ε: Accumulated evidence consisting of the replacement history.
C_{it}: State of component node C_i in time t.
ef_{it}: Efficiency measure of component i in period t.

We propose four methods to decide which component to replace which are inspired from [11]. Each method selects the component to replace using the efficiency measure ef_{it}. Methods FEM and FEL are fault oriented and select the component to replace as given in (3) whereas methods REM and REL are replacement oriented and select the component to replace as given in (4).

$$i^* = argmax\{ef_{it}\} \quad (3)$$

$$i^* = argmin\{ef_{it}\} \quad (4)$$

A. *Fault Effect Myopic Method (FEM)*
This method considers the failure probability of replaceable components in a time period given that we get an undesirable observation in that time period. The method selects the component with the highest conditional failure probability at the time a reactive maintenance decision is made. We evaluate the failure probability of each component i at time t as in (5) and replace the component with the highest failure probability.

$$ef_{it}^{FEM} = P(C_{it} = "F" | \, \varepsilon \cup O_t = "R") \quad (5)$$

B. *Fault Effect Look-ahead Method (FEL)*
A similar failure probability calculation with the FEM method is performed, but this time we consider the failure probability of replaceable components in the next time period given that we get an undesirable observation in a period. Equation (6) evaluates the failure probability of each component *i* at time *t* + *1* and we replace the component with the highest failure probability.

$$ef_{it}^{FEL} = P(C_{i,t+1} = "F" | \varepsilon \cup O_t = "R") \tag{6}$$

C. *Replacement Effect Myopic Method (REM)*
We evaluate the impact of replacement of each component on receiving an undesirable observation in a time period. The efficiency measure is calculated as in (7) and the component that leads to the lowest failure probability is replaced.

$$ef_{it}^{REM} = P(O_t = "\mathrm{R}" | \varepsilon \cup C_{it} = "W") \tag{7}$$

D. *Replacement Effect Look-ahead Method (REL)*
A similar failure probability calculation with the REM method is performed, but this time we consider the impact of replacement of each component on receiving an undesirable observation in the next time period. Equation (8) evaluates the efficiency measure of each component and we replace the component with the lowest efficiency score.

$$ef_{it}^{REL} = P(O_{t+1} = "\mathrm{R}" | \varepsilon \cup C_{it} = "W") \tag{8}$$

5 Computational Analyses

The proposed approaches are simulated in Matlab using BNT toolbox [12]. We run the simulation for a planning horizon of 100 time periods. Each method is replicated 30 times. We evaluate the performances of the methods with respect to total number of replacements performed in the given planning horizon. In addition to the four methods, we also design a random method to select the components to replace randomly so that we can compare the performances of the four methods with also this random method.

In order to understand how the system behaves in time, we infer the failure probabilities of the main process node, the observable node and the components, firstly without performing any maintenance. Figure 3 shows this prognosis results without any maintenance. As expected, failure probabilities increase in time when no maintenance is planned.

We further infer the failure probabilities of the main process node, the observable node and the components under a reactive maintenance strategy using the REM method. Figure 4 shows the prognosis results. When a component is replaced at a time period, the failure probability of that component decreases to zero at that time as seen in Fig. 4a since replacement actions are assumed to be perfect maintenance. REM method plans four replacements for component C1, one replacement for components C2 and C3, and no replacement for component C4. Effect of these replacements can be also seen on the failure probability of the main process P3 and the observable node O1 in Fig. 4b.

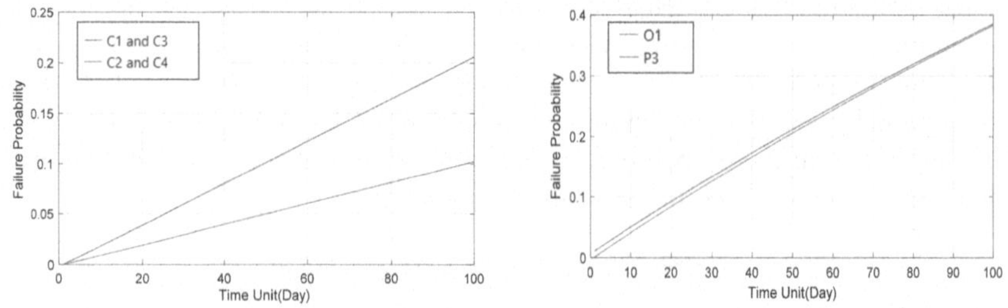

Figure 4.a Component failure trends without maintenance

Figure 4.b Main process failure trends without maintenance

Fig. 3. Future failure trends of the empirical model without maintenance

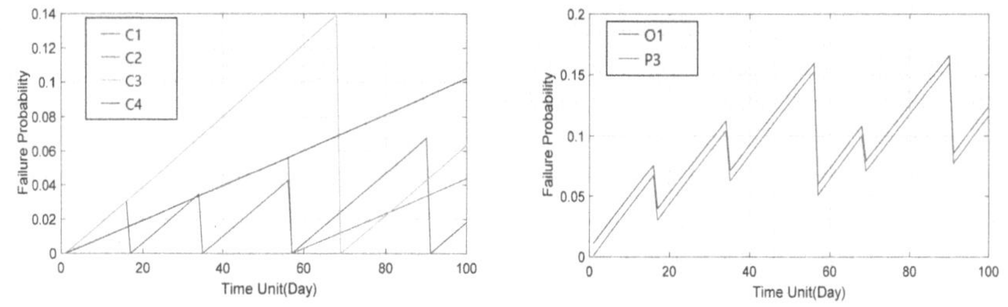

Figure 5.a Component failure trends with REM method

Figure 5.b Main process failure trends with REM method

Fig. 4. Future failure trends of the empirical model with REM

Replication results for all five methods are presented in Table 5. Total number of replacements performed in the given horizon is indicated inside of the table for 30 replications. Replication average and standard deviation for each method can be seen in the right side of the table.

Table 5. Replication results

	R1	R2	R3	R4	R5	R6	R7	R8	R9	R10	R11	R12	R13	R14	R15	Average
	R16	**R17**	**R18**	**R19**	**R20**	**R21**	**R22**	**R23**	**R24**	**R25**	**R26**	**R27**	**R28**	**R29**	**R30**	**Std.Dev**
FEM	8	7	6	11	8	10	9	9	9	14	10	9	6	6	9	**8.833**
	11	9	10	11	12	7	8	10	9	9	8	7	8	7	8	**1.809**
FEL	9	7	10	7	10	9	14	9	6	6	7	9	10	9	9	**8.900**
	8	6	8	9	12	10	9	9	8	10	11	11	8	10	7	**1.777**
REM	12	9	9	12	12	11	10	8	10	10	8	7	9	13	9	**9.433**
	8	9	9	7	11	12	10	10	9	8	7	12	8	7	7	**1.764**
REL	11	9	9	9	11	9	10	8	7	7	10	10	7	10	8	**8.900**
	7	8	10	6	12	9	6	10	9	9	10	8	10	10	8	**1.469**
Random	15	10	13	12	12	13	11	14	19	11	8	10	10	14	12	**11.867**
	13	7	9	13	8	14	13	14	10	13	14	11	9	9	15	**2.553**

All five methods, including the random method, are compared using one-way ANOVA. The test result gives a P-value of 0.000. This indicates that at least one of the methods is significantly different than the others. So we further apply post-ANOVA and use Tukey's test [13] to obtain all pairwise mean comparisons. The results are given in Fig. 5. Pairwise comparisons are done by constructing 95% confidence intervals for the difference of pairwise means. If an interval does not contain zero, the corresponding means are significantly different. Random method, as expected, is significantly worse than all the proposed methods. However, one cannot say that performances of the four methods are significantly different.

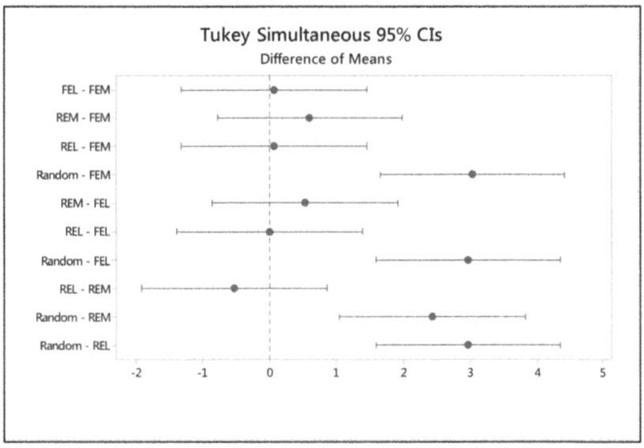

Fig. 5. Post-ANOVA test results

6 Conclusions

Since multi-component dynamic systems involve many interacting components and sub-systems, they are referred as complex systems. Developing a maintenance policy is also a complex task for such systems. We study the reactive maintenance of a four-component dynamic system. We model the system using DBNs. DBNs provide flexibility in modelling and efficient computation.

We propose four methods, inspired from [11], to develop a reactive maintenance plan for a four-component dynamic system. Our aim is to minimize the total number of replacements. We also develop a random selection method to be compared with the four methods. We compare the performances of these methods using the results of 30 replications. Although all proposed methods perform significantly better than the random method, there exists no significant difference among them. As a future study, one can work on a model having more number of components and different probability settings to analyse the performance of the methods.

In this study, the objective is to minimize the total number of replacements, and the cost of maintenance is not considered. Therefore, this study can be extended so that it aims to reduce the total cost of maintenance including cost of actions and other break down related costs for future work. The maintenance algorithm can also be improved for proactive policies as a future study.

Acknowledgment. This work was supported by the Scientific and Technological Research Council of Turkey (TÜBİTAK) under grant no: 117M587.

References

1. Kobbacy K, Murthy P (2008) Complex System Maintenance Handbook. Springer, London Limited
2. Muller A, Suhner MC, Iung B (2008) Formalisation of a new prognosis model for supporting proactive maintenance implementation on industrial system. Reliab Eng Syst Saf. 93:234–253
3. McNaught KR, Zagorecki A (2009)Using dynamic bayesian networks for prognostic modelling to inform maintenance decision making. In: Proceedings of IEEE international conference on industrial engineering and engineering management (IEEM 2009), Hong Kong, pp 1155–1159
4. Hu J, Zhang L, Ma L, Liang W (2011) An integrated safety prognosis model for complex system based on dynamic Bayesian network and ant colony algorithm. Expert Syst Appl 38:1431–1446
5. Wang H (2002) A survey of maintenance policies of deteriorating systems. Eur J Oper Res 139:469–489
6. Hu J, Zhang L, Liang W (2012) Opportunistic predictive maintenance for complex multi-component systems based on DBN-HAZOP model. Process Saf Environ Prot 90:376–388
7. Özgür-Ünlüakın D, Bilgiç T (2006) Predictive maintenance using dynamic probabilistic networks. In: Proceedings of 3th european workshop on probabilistic graphical models (PGM 2006), Czech Republic, pp 239–246
8. Vu HC, Barros A, Berenguer C (2014) Maintenance grouping strategy for multi-component systems with dynamic contexts. Reliab Eng Syst Saf 132:233–249
9. Zhu Q, Peng H, Houtum G-J (2015) A condition-based maintenance policy for multi-component systems with a high maintenance setup Cost. OR Spect 37:1007–1035
10. Genie Modeler website (2018). https://www.bayesfusion.com/genie-modeler
11. Özgür-Ünlüakin D, Bilgiç T (2014) Replacement policies for a complex system with unobservable components using dynamic Bayesian networks. Int J Comput Intell Syst 7 (Suppl 1):68–83
12. BNT Toolbox (2014). https://github.com/bayesnet/bnt
13. Ghosh MN, Sharma D (1963) Power of Tukey's test for non-additivity. J Roy Stat Soc Ser B (Methodological) 25:213–219

Comparison of Hydraulic Bending Machines for Profile, Pipe and Beams in Manufacturing Companies with Electre Method

Elif Çirkin(✉), Aşkın Özdağoğlu, and Kevser Yılmaz

Division of Production Management and Marketing, Faculty of Business, Dokuz Eylul University, İzmir, Turkey
{elif.yuce,askin.ozdagoglu,kevser.yilmaz}@deu.edu.tr

Abstract. There have been different evaluation criteria which effect the decision making process. The methods that help the decision making process by taking into these criteria account together are called multi criteria decision making methods. Electre (Elimination Et Choix Traduisant la Realite) method is one of them. Within this study, hydraulic bending machines for profile, pipe and beams will be compared according to shaft diameters, rolls diameters, working speed and motor power evaluation criteria with the help of Electre method.

Keywords: Production management · Multi-criteria decision making Electre

1 Introduction

There are many alternatives that businesses have to examine while trying to solve the problems they confront. If there was only one alternative, decision making would not have been an issue. However, within the business environment when choosing between various alternatives, it is inevitable that there are many evaluation criteria to be examined together. When many alternatives are examined according to a large number of evaluation criteria, the methods utilized are referred to as multi-criteria decision making methods.

Within the scope of this study, firstly a literature review regarding the applications in which the Electre method is used will be given and then the processes of the Electre method will be explained. In the scope of application, the hydraulic profile and pipe bending machines will be compared and evaluated with the help of the Electre method according to the diameter of the shafts, the diameter of the rolls, the working speed and the motor power for the manufacturing enterprises.

2 Literature Review

Electre method which was first developed by Benayoun et al. is used to compare the alternatives pairwise for each criterion and aims to provide the strength of preferring one over the other [1]. Electre method can be defined as a family of outranking methods

N. M. Durakbasa and M. G. Gencyilmaz (Eds.): ISPR 2018, *Proceedings of the International Symposium for Production Research 2018*, pp. 111–123, 2019.
https://doi.org/10.1007/978-3-319-92267-6_9

consisting of seven different models (I, II, III, IV, A, IS and TRI) derived from the original Electre I [2]. Electre I can be benefitted for selection problems, Electre TRI for assignment problems, and for ranking problems Electre II-IV have been developed [3].

When the literature is examined, it can be understood that enterprises using production activities choose the Electre method for selection and evaluation among various alternatives. This section will refer to some of the applications used in the Electre method. The Electre method is based on the binary superiority comparison between the alternatives for the criteria. With the help of method, decision makers and researchers can incorporate a large number of quantitative and qualitative criteria into the decision making process, and they can weigh the criteria for their purposes and choose the most appropriate alternative by gathering their weight. This method can be successfully applied to real-world problems in many areas such as environmental management, energy, agriculture and forestry, finance, media and marketing, transportation, and project selection [4].

Within the study of some authors based on the surveys, quantitative criteria are determined. Through the analyses and with the help of Electre method, the location of a new store for a cargo company is attempted to specified [5]. Another study benefitted Electre method is for a layout design problem. Both quantitative (handling cost) and qualitative (adjacency and distance requests between departments) objectives are considered in the model and best alternative has been chosen [6].

In order to assess an action plan for the diffusion of renewable energy technologies at regional scale, the multi-criteria decision making methodology (Electre) is applied. 12 evaluation criteria and three possible decisional scenarios have been defined and the best scenario is found [7]. Accordingly, another study is done for the decision making of sustainable energy systems based on the criteria including technical, economic, environmental and social aspects [8].

This method is also used for laptop purchasing decision problem. Based on the evaluations of purchasing manager, IT manager and a senior manager such criteria as processor speed, display card, system memory, hard drive capacity, battery life, weight, brand reliability, and price, five alternatives are determined. Out of the five alternatives, the best one is found for managers [9]. In another study, Electre method is benefitted in order to find the optimum facility location for a textile company located in Uşak, Turkey [10].

Furthermore, the basic functions of a machine in the industrial field have been determined and Electre I and Electre II methods have been used to evaluate the alternatives according to these eight basic functions. Some of the functions can be expressed as steel feed, bending according to specified properties, material transfer activity, material and part type [11]. Sixteen risk factors are determined to evaluate the side effects of illegal drugs and evaluated by Electre method. The Delphi method is used to score according to these risk factors and eleven experts working on this field are consulted. According to the Delphi method of operation, some of these experts are asked to make self-reliant evaluations in order not to influence the decisions of other experts and the results are gathered. These sixteen risk factors are summarized in four main categories. These include individual health, community health, violations of civil rights and criminal behavior. Some of the risk factors can be defined as physical dependence, psychological dependence, sudden poisoning, chronic poisoning, injury to the user, availability of medication, access to information on possible effects of medication, reliability of drug quality, reliability of distribution and sales network, disturbance of society [12]. Analytical Hierarchy Process, Promethee and Electre methods have been used in multi-criteria decision making methods in the evaluation of three alternatives related to renewable energy resources in Southern Cyprus. Five criteria have been identified for evaluating renewable energy source alternatives. These criteria are the maturity of the technology, the initial investment cost, the efficiency, the potential of this technology to be implemented in the country, and the adoption of this technology by the society. Analytical Hierarchy Process method is used to determine the importance levels of the criteria. In the analysis of alternatives according to these evaluation criteria, both Promethee and Electre methods are used and alternatives are listed for both cases, and then the results are found to be the same. In addition to these evaluation criteria discussed in the study, it is stated that future studies might also consider the political will of the country's government on renewable energy sources and carbon emissions [13]. In examining outsourcing contract alternatives for businesses, a highly qualified utility theory is used to determine the weights of the evaluation criteria, and Electre method is integrated in an effort to make comparisons of the alternatives according to these criteria [14].

The Simos process along with the SRF computer software are applied in determining the weights of the evaluation criteria required as inputs in the multi-criteria decision making problems using Electre method [15]. Since the Electre III method which is among the versions of the Electre method is not easily understood by decision makers and is considered to be very complicated, an arrangement on the Electre I version has been offered to assess the continuity of transport activities between countries. Some of the evaluation criteria considered in this review are the gross domestic product (GDP), the contribution of transport activities to GDP, total cargo carried in ton-kilometer, total passenger transport in passenger-kilometer, the impact of transport activities on employment growth, ratio of energy consumption to GDP, emissions of greenhouse gases, airborne particles, the quality of public transport, and the ratio of population killed in traffic accidents. In this context, Bulgaria, the Czech Republic, Latvia, Lithuania, Hungary, Poland, Slovakia and Romania which are entered the European Union after 2004, has been compared with the countries like Croatia and Turkey which are still the candidate countries by the year 2010 [16]. Analytical Hierarchy Process is used for prioritization of the required general rule alternatives for business activities as well as determination of importance ratings of criteria and seven different general rule structures are ranked by Electre method in group decision making process [17]. Different management options related to the management of national parks located on the Pacific coast of western Canada are examined according to the main criteria of ecosystem protection, public service and effective management of funds. In this context, Analytical Hierarchy Process and Electre method are used together [18].

As can be seen from the studies in the literature, there are many different areas using Electre method such as evaluation of different method strategies, comparison of transportation sector alternatives, comparison of business outsourcing alternatives, examination of production technology in energy sector, assessment of risk factors in pharmaceutical industry and examination of the functions of machines in manufacturing enterprises. Owing to today`s competitive business environment, companies are forced to make right decisions on time as well as at minimum cost in order to survive. Machine selection is one of the multi-criteria decision-making processes, and of strategic importance for most companies. According to the requirements including technical specifications, the right decision should be made as quick as possible. This study deals with the actual application of machine selection using the Electre method. By shedding light on the results, managers and/or engineers who are responsible for machine selection are expected to be able to device better strategies and make appropriate decisions. This study, further can set an example for other selection problems having a small set of data. Moreover, this study also makes contribution to literature by defining outranking relations by pairwise comparisons to decide which alternative is preferable and/or similar in the fuzzy environment.

3 Electre Method

The Electre (Elimination Et Choix Traduisant la Reality) method is a multi-criteria decision-making method first proposed by Benayoun in 1966. The method is based on binary superiority comparisons between alternative decision points for each rating factor. The method goes through the solution in eight steps [19]. The steps of the Electre method are described below ([13, 20–22]).

Step 1. First, the Decision Matrix (D) is created. In the row of the decision matrix, i represents $(i = 1, 2, \ldots, m)$ the alternatives and in the columns j represents the criteria $(j = 1, 2, \ldots, n)$. D matrix is the initial matrix generated by the decision maker. The decision matrix is shown as Eq. 1:

$$D_{ij} = \begin{bmatrix} x_{11} & x_{12} & . & x_{1n} \\ x_{21} & x_{22} & . & x_{2n} \\ . & . & . & . \\ x_{m1} & x_{m2} & . & x_{mn} \end{bmatrix} \quad (1)$$

Step 2. The Normalized decision matrix (R) needs to be generated by applying one of the normalization types to the decision matrix (D). There are different methods of performing the normalization process.

- Vector normalization

$$r_{ij} = \frac{x_{ij}}{\sqrt{\sum_{i=1}^{m} x_{ij}^2}}, i = 1, 2, \ldots, m; \;\; j = 1, 2, \ldots, n$$

- Linear normalization (1)

$r_{ij} = \frac{x_{ij}}{x_j^*} i = 1, 2, \ldots, m; \;\; j = 1, 2, \ldots, n; \;\; x_j^* = max_i(x_{ij})$ when the best case for criterion is maximization

$r_{ij} = \frac{x_j^-}{x_{ij}} i = 1, 2, \ldots, m;\; j = 1, 2, \ldots, n;\; x_j^- = min_i(x_{ij})$ when the best case for criterion is maximization

$r_{ij} = 1 - \frac{x_{ij}}{x_j^*} i = 1, 2, \ldots, m;\; j = 1, 2, \ldots, n;\; x_j^* = max_i(x_{ij})$ when the best case for criterion is minimization

- Linear normalization (2)

$r_{ij} = \frac{x_{ij} - x_j^-}{x_j^* - x_j^-} i = 1, 2, \ldots, m;\; j = 1, 2, \ldots, n;\; x_j^* = max_i(x_{ij}),\quad x_j^- = min_i(x_{ij})$ when the best case for criterion is maximization

$r_{ij} = \frac{x_j^* - x_{ij}}{x_j^* - x_j^-} i = 1, 2, \ldots, m;\; j = 1, 2, \ldots, n;\; x_j^* = max_i(x_{ij}),\quad x_j^- = min_i(x_{ij})$ when the best case for criterion is minimization

- Linear normalization (3)

$$r_{ij} = \frac{x_{ij}}{\sum_{i=1}^{m} x_{ij}}, i = 1, 2, \ldots, m;\; j = 1, 2, \ldots, n$$

- Non-monotone normalization

$e^{-\frac{z^2}{2}}, z = \frac{x_{ij} - x_j^0}{\sigma_j}$, x_j^0 the most appropriate value for j and the standard deviation of σ_j j values. Non-monotone normalization is very rarely used in the literature. For normalized decision matrix (R), vector normalization is a commonly used method. Here, the vector normalization formula for the normalized decision matrix (R) is given as Eq. 2.

$$r_{ij} = \frac{x_{ij}}{\sqrt{\sum_{i=1}^{m} x_{ij}^2}}, i = 1, 2, \ldots, m;\; j = 1, 2, \ldots, n \tag{2}$$

For example, to compute the r11 element of the R matrix, the element x11 of the matrix D is obtained by dividing the square of the sum of the squares of the first column elements of the matrix. At the end of the calculations, the R matrix is obtained as in Eq. 3:

$$R_{ij} = \begin{bmatrix} r_{11} & r_{12} & . & r_{1n} \\ r_{21} & r_{22} & . & r_{2n} \\ . & . & . & . \\ r_{m1} & r_{m2} & . & r_{mn} \end{bmatrix} \tag{3}$$

Step 3. Weighted Normalize Decision Matrix (Y) is created using the weight values of each measure.

Criteria may differ in importance from the perspective of the decision maker. The Y matrix is calculated to reflect these significance differences to the Electre solution. First, weight values (wi) related to evaluation factors are determined $\left(\sum_{i=1}^{n} w_i = 1\right)$. In the next step, the elements in each column of the R matrix are multiplied by the corresponding wi value to form the Y matrix. Y matrix is shown in Eq. 4:

$$Y_{ij} = \begin{bmatrix} w_1r_{11} & w_2r_{12} & . & w_nr_{1n} \\ w_1r_{21} & w_2r_{22} & . & w_nr_{2n} \\ . & . & . & . \\ w_1r_{m1} & w_2r_{m2} & . & w_nr_{mn} \end{bmatrix} \tag{4}$$

Step 4. Using the Weighted Normalize Decision Matrix (Y), concordance set (Q (a, b)) elements must be specified for each of the binary alternative comparisons.

There is no precise measurement method for the determination of concordance sets and the calculation of concordance indexes depending on these elements. The frequently used method can be expressed as follows. To determine the concordance sets, the Y matrix is used, the decision points are compared with each other in terms of evaluation factors, and the sets are determined by the relation shown in Eq. 5:

$$Q(a,b) = \{c, y_{ac} \geq y_{bc}\} \tag{5}$$

Equation 5 is based on comparing the size of the row elements relative to each other. For example, for the C (a, b) concordance set for a = 1 and b = 2, the first and second row elements of the Y matrix are compared to each other and if there are five evaluation criteria, the concordance set C (a, b) shall consist of maximum five members. In the given example, comparing rows 1 and 2,

$$y_{11} < y_{21}$$
$$y_{12} = y_{22}$$
$$y_{13} < y_{23}$$
$$y_{14} > y_{24}$$
$$y_{15} > y_{25}$$

if these results are encountered, c = 2, c = 4 and c = 5 according to Eq. 5 will fit and C (a, b) concordance set will be $C(1,2) = \{2,4,5\}$.

Attention should be paid to the meanings of the evaluation factors when forming the concordance sets in the Electre method. For example, if the relevant evaluation factor is profit, Eq. 5 will be used for the concordance set. However, if the evaluation factor is cost, then the condition for the concordance set is $y_{ac} < y_{bc}$ inequality.

Step 5. The indexes of concordance (C (a, b)) and discordance (D (a, b)) are established. Concordance sets are used to generate concordance indexes (C (a, b)). The C matrix is m * m in dimension. The elements of the matrix C (a, b) can be calculated by Eq. 6.

$$C(a,b) = \sum\nolimits_{j \in Q(a,b)} w_j \tag{6}$$

If we take the example given above, the value of the C (1,2) element of C matrix for C (1,2) = {2,4,5} will be $C(1,2) = w_2 + w_4 + w_5$. The C (a, b) matrix is shown in Eq. 7:

$$C(a,b) = \begin{bmatrix} 1 & C(1,2) & C(1,3) & . & C(1,m) \\ C(2,1) & 1 & C(2,3) & . & C(2,m) \\ C(3,1) & C(3,2) & 1 & . & C(3,m) \\ . & . & . & 1 & . \\ C(m,1) & C(m,2) & C(m,3) & . & 1 \end{bmatrix} \tag{7}$$

Discordance indexes $(D(a,b))$ can be calculated as in Eq. 8:

$$D(a,b) = \frac{max_c(y_{bc} - y_{ac})}{max|y_{ac} - y_{bc}|} \tag{8}$$

All binary comparison values will be considered as the largest absolute difference value to be found in the denominator of Eq. 8. For example, $D(1,2)(a = 1\, ve\, b = 2)$ elements are obtained by comparing the first and second row elements of Y matrix. For D (1,2), the reciprocal of all the elements in the first and second rows of the Y matrix in the fraction of Eq. 8 is found and the largest one is selected. For the denominator of Eq. 8, all the differences between the two comparisons in the Y matrix are examined and the largest one is selected. This value obtained for the denominator is used to calculate all of the discordance index values.

Like the C (a, b) matrix, the matrix D (a, b) is also dimensioned as m x m. The matrix D (a, b) is shown in Eq. 9:

$$D(a,b) = \begin{bmatrix} 0 & D(1,2) & D(1,3) & . & D(1,m) \\ D(2,1) & 0 & D(2,3) & . & D(2,m) \\ D(3,1) & D(3,2) & 0 & . & D(3,m) \\ . & . & . & 0 & . \\ D(m,1) & D(m,2) & D(m,3) & . & 0 \end{bmatrix} \tag{9}$$

Step 6. The concordance threshold (C *) and discordance threshold (D*) values are determined. These values can be determined by the decision maker considering the criteria. It is even possible to use different threshold values in the examination for different evaluation criteria. In this case, concordance thresholds such as C_1^*, C_2^* will be found according to the evaluation scale. If two different evaluations such as strong superiority and weak superiority in comparison between alternatives are desired, different values can be used for threshold values in accordance with $C^* > C^-$ - and $D^* < D^-$ conditions. If a standard concordance threshold value is desired, the arithmetic mean of the concordance index values can be obtained. The determination of the concordance threshold value using the concordance index values is shown in Eq. 10.

$$C^* = \frac{\sum_{a=1}^{m} \sum_{b=1}^{m} C(a,b)}{m^2} \tag{10}$$

The m that is in the equation shows the total number of alternatives evaluated. Similarly, if a standard discordance threshold value is desired, the arithmetic average of the discordance index values can be obtained. The determination of the discordance threshold value by using the discordance index values is shown in Eq. 11

$$D^* = \frac{\sum_{a=1}^{m} \sum_{b=1}^{m} D(a,b)}{m^2} \tag{11}$$

The concordance threshold and discordance threshold values can be compared with the concordance index values and discordance index values to determine which alternative has a strong superiority to which alternative. If an alternative has a stronger superiority than the other, the intersection cell of the row of the alternative which is superior to the superiority matrix structure and the column of the weak alternative is denoted by S^F. In some cases, the comparative thresholds for concordance thresholds and discordance thresholds are not favoured because there should be distinctive advantages among the alternatives. In this case, the symbol S^f is shown to symbolize weak superiority by using two different thresholds including concordance and discordance represented by C^*, C^-, D^*, D^-. If there is one threshold of concordance threshold and discordance threshold, the method is Electre I, and if two thresholds of concordance and discordance are found, the method is expressed as Electre II. Binary comparisons of alternatives for superiority are made as in Eq. 12.

$$\begin{cases} C(a,b) \geq C^* ve\, D(a,b) \leq D^* \Rightarrow a\, is\, stronger\, than\, b \\ C(a,b) \geq C^- ve\, D(a,b) \leq D^- \Rightarrow a\, is\, weaker\, than\, b \end{cases} \tag{12}$$

Step 7. The most suitable alternative is chosen according to the superiority matrix. The alternatives can be sorted according to the priorities found in Eq. 12. and the highest one is selected. Evaluating the rows of the matrix S shows the alternatives. For example, if the matrix S is computed as follows, it is understood from the size of the matrix (5 × 5) that five different alternatives are compared.

$$S = \begin{matrix} A_1 \\ A_2 \\ A_3 \\ A_4 \\ A_5 \end{matrix} \begin{bmatrix} & S^F & & & \\ & & & & \\ S^F & S^F & & S^F & S^F \\ & & & & S^F \\ S^F & & & S^F & \end{bmatrix}$$

In the S matrix; $S(1,2), S(3,1), S(3,2), S(3,4), S(3,5), S(4,5), S(5,1)$ and $S(5,4)$ show strong superiority. S(1,2) denotes the value at the intersection of row 1 and column 3. Accordingly, the alternative in row 1 has a strong superiority when all the evaluation criteria are taken into account, according to alternative 2 in column 1. Since the third row has strong superiority over all other columns, it would be the right decision to choose the 3rd alternative.

4 Application

Within the scope of the application technical information has been collected for three different hydraulic profiles and pipe bending machines. Technical specifications for the evaluation of hydraulic profile and pipe bending machines are the diameter of the shaft, the diameter of the rolls, the working speed and the engine power. In all of these evaluation criteria, the highest value indicates the best result. However, the measurement units of these evaluation criteria are different. Evaluation criteria and measurement units related to them are presented in Table 1.

Table 1. Evaluation criteria and measurement units

Evaluation Criteria	Diameter of the shaft	Diameter of the rolls	Working Speed	Engine Power
Measurement unit	milimeter	milimeter	$\frac{\text{meter}}{\text{second}}$	kilowatt
Unit symbol	mm	mm	$\frac{\text{m}}{\text{sec}}$	kw

The values of three different hydraulic profiles and pipe bending machines according to these evaluation criteria will form the decision matrix calculated in Eq. 1. The decision matrix is presented in Table 2 [23].

Table 2. Hydraulic profile and pipe bending machines decision matrix [21]

	Evaluation criteria 1	Evaluation criteria 2	Evaluation criteria 3	Evaluation criteria 4
	Diameter of the shaft	Diameter of the rolls	Working speed	Engine power
Unit symbol	mm	mm	$\frac{m}{sec}$	kw
Alternative 1	150	470	6	22
Alternative 2	180	580	6	30
Alternative 3	600	650	5	30

By applying the vector normalization shown in Eq. 2 to the decision matrix, the normalized decision matrix is prepared. The normalized decision matrix is as shown in Table 3.

Using the weight values of each measure, the Weighted Normalized Decision Matrix (Y), shown in Eq. 4, is constructed. The Weighted Normalize Decision Matrix (Y), which is formed by considering that all of these evaluation criteria have the same importance level, appears in Table 4.

By using the Weighted Normalize Decision Matrix (Y), the Concordance Set elements for each of the binary alternative comparisons are determined by Eq. 5.

Table 3. Normalized decision matrix

	Evaluation criteria 1	Evaluation criteria 2	Evaluation criteria 3	Evaluation criteria 4
	Diameter of the shaft	Diameter of the rolls	Working speed	Engine power
Alternative 1	0,232873	0,474820	0,609208	0,460336
Alternative 2	0,279448	0,585948	0,609208	0,627730
Alternative 3	0,931493	0,656666	0,507673	0,627730

Table 4. Weighted normalization decision matrix

	Evaluation criteria 1	Evaluation criteria 2	Evaluation criteria 3	Evaluation criteria 4
	Diameter of the shaft	Diameter of the rolls	Working speed	Engine power
Alternative 1	0,058218	0,118705	0,152302	0,115084
Alternative 2	0,069862	0,146487	0,152302	0,156933
Alternative 3	0,232873	0,164167	0,126918	0,156933

Compared to the first and second rows according to the values in Table 4, the following results were found.

$$y_{11} < y_{21}$$
$$y_{12} < y_{22}$$
$$y_{13} = y_{23}$$
$$y_{14} < y_{24}$$

Accordingly, it will be formed as $C(1,2) = \{3\}$. The concordance set elements for all binary alternative comparisons are shown below.

$$C(1,2) = \{3\}$$
$$C(1,3) = \{3\}$$
$$C(2,1) = \{1,2,3,4\}$$
$$C(2,3) = \{3,4\}$$
$$C(3,1) = \{1,2,4\}$$
$$C(3,2) = \{1,2,4\}$$

With the help of Eq. 6, the concordance index values are determined. The concordance matrix containing the concordance index values is as follows.

$$C(a,b) = \begin{matrix} Alternative\ 1 \\ Alternative\ 2 \\ Alternative\ 3 \end{matrix} \begin{bmatrix} 1 & 0,25 & 0,25 \\ 1 & 1 & 0,5 \\ 0,75 & 0,75 & 1 \end{bmatrix}$$

Discordance index values obtained using Eq. 8 are presented in the following matrix.

$$D(a,b) = \begin{matrix} Alternative\ 1 \\ Alternative\ 2 \\ Alternative\ 3 \end{matrix} \begin{bmatrix} 0 & 0,239608 & 1 \\ 0 & 0 & 0,933333 \\ 0,145336 & 0,145336 & 0 \end{bmatrix}$$

Concordance threshold (C^*) is calculated by using Eq. 11 and discordance threshold (D^*) is determined by using Eq. 12. Concordance threshold (C^*) and discordance threshold (D^*) are demonstrated below.

$$C^* = 0,722222$$
$$D^* = 0,273735$$

As seen in Eq. 12, this concordance threshold and discordance threshold values are compared with the concordance index values and discordance index values to specify which alternative has a strong superiority to which alternative. The matrix of the superiority (S) resulting from these comparisons is given below.

$$S = \begin{matrix} Alternative\ 1 \\ Alternative\ 2 \\ Alternative\ 3 \end{matrix} \begin{bmatrix} & & \\ S^F & & \\ S^F & S^F & \end{bmatrix}$$

To explain the expressions in the superiority matrix, the symbol S^F, which is written at the intersection of row 2 and column 1, shows that Alternative 2 has a strong superiority to Alternative 1. The symbol S^F, which is written at the intersection of row 3 and column 1, shows that Alternative 3 has a strong superiority to Alternative 1. The symbol S^F, which is written at the intersection of row 3 and column 2, shows that Alternative 3 has a strong superiority to Alternative 2. Accordingly, Alternative 3, which has a strong superiority over all other alternatives, becomes the best alternative and follows Alternative 2.

5 Conclusions

Businesses may face various alternatives in the problems they encounter. As matter of fact, it is an inevitable requirement to have multiple alternatives in order to be able to mention a decision-making process. As the problem faced in the businesses becomes more complex and complicated, the number of factors that need to be taken into consideration are equally increasing. The methods benefitted to solve the problems in this structure are pointed out in the general name of multi-criteria decision making methods.

In this study, primarily a literature review covering the applications used in the Electre method is given, then the operation of the Electre method is explained and the method is applied to an actual machine selection problem. The scope of the application regarding finding the best hydraulic profile and pipe bending machine alternatives based upon the evaluation criteria including diameter of the shaft, the diameter of rolls, the working speed and the motor power is expected to contribute to the field of decision-making process of manufacturing management especially for metal industries in suggesting the use of Electre method and comparing three alternatives with regard to four evaluation criteria. From a theoretical point of view, it provides a framework to researchers and academicians to explore the best alternatives in such fields as personnel selection, supplier selection, examination of production technology and project selection when there is a small set of data with a few alternatives and evaluation criteria.

The primary advantage of the Electre method is that the comparison of the alternatives can be obtained although there is not an obvious preference for one of those. Therefore, when compared to other multi-criteria decision making methods which are sensitive to the decision maker`s thoughts and beliefs, Electre method is much more reliable. Furthermore, this method has the ability to handle both quantitative and qualitative assessments. That is to say, the vector normalization process does not hinder that the evaluation criteria have different units of measurement. The application of this method concerns environmental impact assessment, environmental as well as energy planning, renewable energy sources' problems, energy systems technology and waste management. Based on a study regarding the overview and selection of multi-criteria evaluation methods, some properties such as mathematical background and using weight coefficients, parameters, thresholds, and indexes for Electre Method is quite strong. Moreover, understanding the problem, sensitivity analysis, selecting criteria/sub-criteria and determination of their weight coefficients, measurement scales and assessment of the performance are positively moderate in comparison with other techniques consisting of AHP, MAUT, SMART which is derived from MAUT (Multiple Attribute Utility Theory), AMS and PROMETHEE. Even though some properties are similar with this method, Electre offers additional options for thresholds when defining criteria models and it also provides with ease of use and low requirements on time and money [24].

The constraints on the study can be defined as follows. First, the weight of each assessment measure is considered equal. However, these weight values may not be equal in all of the operating problems. These priorities may change depending on the nature of the problem encountered. In addition, some of the evaluation criteria used in this study may not be needed, but some other factors may need to be examined. The presence of binary comparisons in the calculation process of the Electre method constitutes a limitation of the method. If the number of alternatives increases, binary comparisons will increase exponentially, making the calculation process more difficult, which may lead to delays. In situations where it is necessary to make quick decisions in rapidly changing business conditions, this creates a weakness of the method against some other multi-criteria decision making methods such as Moora and Topsis.

References

1. Zeleney M, Cochrane JL (1973) Compromise programming. In: Multi-criteria decision making. University of South Carolina, Columbia, pp 262–301
2. Benayoun R, Roy B, Sussman N (1966) Manual de Reference du Programme Electre. SEMA, Paris

3. Tscheikner-Gratl F, Egger P, Rauch W, Kleidorfer M (2017) Comparison of multi-criteria decision support methods for integrated rehabilitation prioritization. Water 9:1–28
4. Şahin S (2015) Electre. In: Çok Kriterli Karar Verme Yöntemleri. Dora Yayıncılık, Bursa
5. Yücel M, Ulutaş A (2009) Çok Kriterli Karar Yöntemlerinden Electre Yöntemiyle Malatya'da Bir Kargo Firması için Yer Seçimi. Sosyal Ekonomik Araştırmalar Dergisi 327–344
6. Aiello G, Enea M, Galante G (2005) A multi-objective approach to facility layout problem by genetic search algorithm and Electre method. Robot Comput Integr Manuf 447–455
7. Beccali M, Cellura M, Mistretta M (2003) Decision-making in energy planning. Application of the Electre method at regional level for the diffusion of renewable energy technology. Renew Energy 2063–2087
8. Wang J, Jing Y, Zhangand C, Zhao J (2009) Review on multi-criteria decision analysis aid in sustainable energy decision-making. Renew Sustain Energy Rev 13(9):2263–2278
9. Ertuğrul İ, Karakaşoğlu N (2010) Electre ve Bulanık AHP Yöntemleri ile Bir İşletme İçin Bilgisayar Seçimi. Dokuz Eylül Üniversitesi İktisadi ve İdari Bilimler Fakültesi Dergisi 25 (2):23–41
10. Akyüz Y, Soba M (2013) Electre Yöntemiyle Tekstil Sektöründe Optimal Kuruluş Yeri Seçimi: Uşak İli Örneği. Uluslararası Yönetim İktisat ve İşletme Dergisi 9(19)
11. Armaghan N, Renaud J (2012) An application of multi-criteria decision aids models for case-based reasoning. Inf Sci 210:55–66
12. Amsterdam JGCV, Best W, Opperhuizen A, Wolff FAD (2004) Evaluation of a procedure to assess the adverse effects of illicit drugs. Regul Toxicol Pharmacol 39:1–4
13. Theodorou S, Florides G, Tassou S (2010) The use of multiple criteria decision making methodologies for the promotion of RES through funding schemes in Cyprus, a review. Energy Policy 38:7783–7792
14. Almeida ATD (2007) Multicriteria decision model for outsourcing contracts selection based on utility function and ELECTRE method. Comput Oper Res 34:3569–3574
15. Figueira J, Roy B (2002) Determining the weights of criteria in the ELECTRE type methods with a revised Simos' procedure. Eur J Oper Res 139:317–326
16. Bojkovic N, Anic I, Pejcic-Tarle S (2010) One solution for cross-country transport-sustainability evaluation using a modified ELECTRE method. Ecol Econ 69:1176–1186
17. Choi DH, Ahn BS, Kim SH (2005) Prioritization of association rules in data mining: multiple criteria decision approach. Expert Syst Appl 29:867–878
18. Rudolphi W, Haider W (2003) Visitor management and ecological integrity: one example of an integrated management approach using decision analysis. J Nat Conserv 11:346–354
19. Triantaphyllou E (2000) Multi-criteria decision making methodologies: a comparative study. In: Applied optimization, vol 44. Kluwer Academic, Dordrecht
20. Yaralıoğlu K (2010) Karar Verme Yöntemleri. Detay Yayıncılık, Ankara
21. Wang X, Triantaphyllou E (2008) Ranking irregularities when evaluating alternatives by using some ELECTRE methods. Omega 36:45–63
22. Shih H-S, Shyur H-J, Lee ES (2007) An extension of TOPSIS for group decision making. Math Comput Model 45:801–813
23. Şahinler Metal Makina Endüstri A.Ş. (2017). http://www.sahinlermetal.com/kataloglar
24. Papadopoulos AM, Konidari P (2011) Overview and selection of multi-criteria evaluation methods for mitigation/adaptation policy instruments. PROMITHEAS-4 Report

Evaluation of Metal Forming Machine Alternatives Used for Production Activities with MOORA Method

Aşkın Özdağoğlu, Kevser Yılmaz, and Elif Çirkin(✉)

Division of Production Management and Marketing, Faculty of Business, Dokuz Eylul University, İzmir, Turkey
{askin.ozdagoglu,kevser.yilmaz,elif.yuce}@deu.edu.tr

Abstract. This study aims at evaluating of machine alternatives that the companies need for metal forming with MOORA method. Evaluation criteria have been determined as engine power, approach speed, working speed and return speed. Preference ranking of 61 different models have been analysed by taking into account all of these criteria together from the viewpoint of the companies.

Keywords: Multi-criteria decision making · MOORA method
Metal forming machines · Production management

1 Introduction

Businesses must make decisions about the problems they face regularly. Decision-making is actually an election process which means that there are many alternatives available for companies when make decisions. To evaluate and decide on these alternatives, it is certain that businesses should consider many factors together. There are various multi-criteria decision-making methods that can be used for companies to evaluate alternatives and decide one of them. Moora method is one of the multi-criteria decision method. In the scope of this study, firstly literature review will be given about the fields where Moora method is used in operation activities and then the method will be explained and later, in application stage, the order of preference of hydraulic braking press alternatives for sheet metal processing, which is a type of machinery that businesses need during manufacturing activities, will be evaluated by Moora method. For this purpose, motor power, descent speed, press speed and take off speed were determined as evaluation criteria from the specifications of hydraulic press braking machines and these criteria were sorted by considering together.

2 Literature Review

When the literature is examined, it is seen that the manufacturing enterprises use the Moora method to select and evaluate among various alternatives. In this section, some applications used in the Moora method will be mentioned.

N. M. Durakbasa and M. G. Gencyilmaz (Eds.): ISPR 2018, *Proceedings of the International Symposium for Production Research 2018*, pp. 124–135, 2019.
https://doi.org/10.1007/978-3-319-92267-6_10

The Moora method has been used to maximize profit and minimize the product cost, improve performance and reduce fuel consumption, reduce the weight, increase the strength, and choose the material required for the production of an engine flywheel. For this purpose, various measurements were made by determining metal fatigue value, metal hardness value per unit weight and fragility criteria and compared with ten different raw material alternatives whose main substance is carbon-epoxy, kevlar-epoxy or glass fiber. The Moora method was used to select the most suitable raw material to be used in the production of cold storage tanks for liquid nitrogen transport. There are seven raw material alternatives that are candidates for this raw material selection problem. Seven evaluation criteria were used to compare these alternatives. These evaluation criteria were determined as endurance index, yield, stress multiplier, density, thermal expansion coefficient, heat permeability and heat value. The Moora method has been used to select the appropriate work material for a product that needs to be processed in an environment rich in oxygen at high temperatures. The selection problem of this work material includes six alternative and four evaluation criteria. The evaluation criteria are hardness grade, abrasion resistance, workability rate and cost. Within the scope of the same study, Moora was used for the selection of the last raw material to be used for construction of the sailboat ship. The sailboat requires a large sail area in order to produce a sailboat capable of sufficient speed, which increases the mechanical impact of the sailboat. In addition, the effects of the saltwater environment should not be forgotten. Fifteen different raw material alternatives have been examined for use in the construction of sailboat posts and four criteria have been set to use the assessments. These evaluation criteria are strength, strain, wear resistance and cost. The basic types of alternatives are epoxy-glass, epoxy-carbon and heat-resistant species.

In another study, Moora and standard deviation method were used together to optimize welding processes. It is known that the welding process of the materials affects the quality of weldment. For this study, five large weld deposits, each consisting of three parts, were created and analyzed. The standard deviation method is used to determine the weights of the mechanical test results, and the Moora method is used to optimize the parameters of the welding operation. As a result of the study, optimum values of the welding current, welding voltage, electrode diameter and welding speed were determined for welding with the best mechanical properties and it was confirmed that the optimization of the processes used in the welding process affects the welding qualities [2].

Moreover, to solve the selection problem of Non-Conventional Machining Processes, it was aimed to find the best among alternatives using Moora and AHP methods. In the study, firstly different performance criteria were defined, and 4 different non-traditional machining processes were evaluated within the framework of nine criteria, then relative importance was determined by comparing the criteria with the AHP method and different non-traditional machining processes were graded by Moora method and the best machining process was determined [3]. Moreover, Kodrat and his colleagues used the Moora method to help marketers decide on the ideal marketing location choice [4].

Yazdani et al., conducted a sensitivity analysis using different variant decision-making methods. They did an application in material selection. First, expert opinions were taken for the evaluation methods and weights of the methods were determined and then the best material was selected by using different multivariate decision-making methods. Waspas and Moora methods were compared for consistency of results [5].

In another study, intuitive fuzzy sets and Moora method were used together to evaluate the suppliers and determine the best supplier by Pérez-Domínguez and his colleagues. Firstly, the criteria and alternatives were evaluated with intuitive fuzzy sets and then the best supplier was selected by Moora method [6]. In addition, the Moora method was also used in an article which written in 2016 for evaluating the 9 alternatives based on 3 criteria, results play an important role for users to select best one within alternatives [7].

Furthermore, Jagadish made a study was carried out for accurate evaluation of cutting fluid for green production. Choosing the right cutting fluid has a vital proposition for companies to reduce environmental impact, improve quality and reduce costs. This fluid is used to increase the life cycle of the product in cutting and lubrication processes for the producers. Three alternatives were evaluated in this study. The main criteria were determined as cost, quality and impact on the environment and then sub-criteria related to them were established. A total of 10 criterions were taken as basis for the evaluation of alternatives. The AHP and Moora methods were used together when assessments were made. The AHP method was used in the formulation of decision matrices and weights, and the alternatives were ranked using Moora method [8].

Additionally, to integrate the subjective evaluations of different decision makers in the group decision making process, the Moora method is combined with the multi-modal and fuzzy logic structure containing the multiplicative form. The mentioned method has been used by the human resources department to select workers [9]. Topsis, Moora and AHS methods used in the study to solve the selection problems of the CNC lathe. Twenty-two different CNC lathes have been compared for this purpose [10]. The results of the Multimoora method and the Topsis method, which were developed based on the Moora method, were compared in the evaluation of alternative electricity generation technologies [11].

The Moora method was also used in the selection of personnel. Firstly, the main and sub criteria that the candidates have been identified, weights of these criteria have been determined according to the SWARA method, and then the most suitable candidates have been listed by using the Moora method [12]. Moreover, some studies are also available in the literature using Moora method for calculating group performance data using individual performance scores [13].

The mechanical arms of industrial robots, which are general purpose, reprogrammable machines in accordance with certain anthropometric properties, are the most important and vital anthropometric component. Also, the ability to make decisions, the response capacity to various sensor inputs, communication with other machines such as material transfer, assembly, loading, painting, welding, control resolution, accuracy, reproducibility, loading capacity, human- machine coordination ability, programming flexibility, maximum speed, memory capacity, and after-sales service quality that the seller offers are the most important factors in the selection of an industrial robot. These factors include load capacity, repeatability, maximum speed, memory capacity, and criteria to be used for robot selection for robot arm reach and seven different industrial robots have been investigated using the Moora method. For the selection of the flexible manufacturing system, preferences for eight different flexible manufacturing technologies were ranked according to evaluation criteria which are labor costs, machine preparation costs, response to market expectations, quality improvement opportunities, capital cost and maintenance cost. Considering various factors such as cost of capital, speed of rotation, 21 different CNC lathe machines have been evaluated. Automatic control systems have been investigated considering factors such as initial investment cost, output ratio, operating cost, environmental impact [14].

3 MOORA Method

MOORA (Multi-Objective Optimization on the basis of Ratio Analysis) is a simultaneous optimization process under two or more overlapping qualities or objective constraints. The MOORA method starts with a decision matrix showing the performance of different alternatives for different qualifications or objectives [15].

$$X = \begin{bmatrix} x_{11} & x_{12} & \ldots & x_{1n} \\ x_{21} & x_{22} & \ldots & x_{2n} \\ \ldots & \ldots & \ldots & \ldots \\ x_{m1} & x_{m2} & \ldots & x_{mn} \end{bmatrix} \quad (1)$$

In this decision matrix,
i = alternative
j = qualifications or criteria
m = total number of alternatives
n = total number of qualifications or criteria
x_{ij} = i. alternative j. it is expressed as performance measurement value in terms of criteria.

Then, a denominator value representing all alternatives is calculated. This value is the square root of the sum of the squares of the performance measures relative to a measure of all the most preferred alternatives. This ratio can be expressed as given in Eq. (2).

$$x_{ij}^{*} = \frac{x_{ij}}{\left[\sum_{i=1}^{m} x_{ij}^{2}\right]^{1/2}} (j = 1, 2, \ldots, n) \quad (2)$$

In this equation, x_{ij} value j. criteria or qualifications i. is the number without the unit in the range [0,1] representing the normalized performance of the alternative. For a multi-purpose optimization, these normalized performance values (for useful attributes) are added in the case of maximization and subtracted from the minimization case (for useless attributes) to find a single value for each alternative. In this case, the optimization problem occurs as in Eq. 3.

$$y_i = \sum_{j=1}^{g} x_{ij}^{*} - \sum_{j=g+1}^{n} x_{ij}^{*} \quad (3)$$

In this equation,
g = the number of attributes or criteria to be maximized
$n - g$ = the number of qualifications or criteria to be minimized
y_i = all qualifications or criteria i. The normalized value for the alternative

In many cases it appears that certain qualities or criteria are more important than others. To give more importance to a quality or measure, the relevant ratio can be multiplied by the weight value (significance coefficient) of that attribute or measure. Considering the weights of the qualities, Eq. 3 transforms Eq. 4.

$$y_i = \sum\nolimits_{j=1}^{g} w_j x_{ij}^* - \sum\nolimits_{j=g+1}^{n} w_j x_{ij}^* \; (j = 1, 2, \ldots, n) \tag{4}$$

$w_j = j$. Attribute or measure weight (significance) coefficient

y_i may take a positive or negative value depending on the sum of the largest (useful attributes or criteria) and the smallest (useless attribute or criteria) sum in the decision matrix. The order of the values of y_i indicates the final priorities. The best alternative is the highest value of y_i, while the worst alternative is the lowest value of y_i.

4 Application

Moora method converts to different measurement units to standardized one unit that give ability to researchers to make evaluation the different measurement units in same study. Moreover, when there are very much number of alternatives in decision data set, this method is one of the suitable evaluation methods to compare the alternatives and to find best one. So that, in this study Moora method used to assess the various alternatives of hydraulic bending press models of a firms which operate in machining industry. Hydraulic bending press models are often used to form molds or angles by pressing steel sheet. Hydraulic bending press models produced by companies producing sheet metal working machines were searched and their values were determined as technical features of engine power, descent speed, press speed and take off speed. The measurement units of these specifications are presented in Table 1.

Table 1. Measurement units of evaluation criteria

Evaluation criteria	Approach speed	Pressing speed	Return speed	Motor power
Measurement unit	Millimeter/second	Millimeter/second	Millimeter/second	Kilowatt

When the units of measurement shown in Table 1 are examined, it is seen that not all of them are the same. The normalization operations performed during the operation process of the Moora method allow examination of the evaluation criteria with different measurement units.

The values of the engine power, descent speed, press speed and take off speed of the hydraulic braking press models produced by the companies that produce sheet metal working machines will form the decision matrix in Eq. 1. Technical data of the decision matrix are shown in Table 2.

Table 2. Technical specifications of alternative models

	Evaluation criteria 1	Evaluation criteria 2	Evaluation criteria 3	Evaluation criteria 4
	Approach speed	Pressing speed	Return speed	Motor power
Measurement unit	Mm/sec	Mm/sec	Mm/sec	Kw
Alternative 1	300	25	300	7,5
Alternative 2	250	20	250	11
Alternative 3	250	20	250	22
Alternative 4	70	100	10	4
Alternative 5	70	90	10	4
Alternative 6	65	90	9	5,5
Alternative 7	80	80	10	7,5
Alternative 8	80	95	10	7,5
Alternative 9	90	100	10	11
Alternative 10	75	90	8	11
Alternative 11	75	100	8	15
Alternative 12	80	85	8	18,5
Alternative 13	60	80	9	22
Alternative 14	75	90	8	11
Alternative 15	75	100	8	15
Alternative 16	80	85	8	18,5
Alternative 17	75	90	8	11
Alternative 18	75	100	8	15
Alternative 19	80	100	8	18,5
Alternative 20	60	80	9	22
Alternative 21	140	17	170	6
Alternative 22	200	14	165	8
Alternative 23	140	17	170	6
Alternative 24	200	14	165	8
Alternative 25	200	14	165	8
Alternative 26	200	12	190	11
Alternative 27	160	12	190	15
Alternative 28	200	12	190	11
Alternative 29	200	12	190	15
Alternative 30	180	12	190	15
Alternative 31	180	10	185	19
Alternative 32	140	11	135	22
Alternative 33	140	11	150	30
Alternative 34	110	8	130	30
Alternative 35	80	7	65	30
Alternative 36	180	12	190	15

(*continued*)

Table 2. (*continued*)

	Evaluation criteria 1	Evaluation criteria 2	Evaluation criteria 3	Evaluation criteria 4
	Approach speed	Pressing speed	Return speed	Motor power
Measurement unit	Mm/sec	Mm/sec	Mm/sec	Kw
Alternative 37	180	10	185	19
Alternative 38	140	11	150	30
Alternative 39	200	12	190	15
Alternative 40	180	12	190	15
Alternative 41	180	10	185	19
Alternative 42	140	11	135	22
Alternative 43	140	11	150	30
Alternative 44	110	8	130	30
Alternative 45	80	7	65	30
Alternative 46	80	8	75	37
Alternative 47	70	6	80	55
Alternative 48	70	6	80	75
Alternative 49	200	12	190	15
Alternative 50	180	10	185	19
Alternative 51	110	8	130	30
Alternative 52	80	8	75	37
Alternative 53	130	11	125	19
Alternative 54	80	11	75	30
Alternative 55	80	8	65	30
Alternative 56	80	7	65	30
Alternative 57	80	8	75	37
Alternative 58	80	6	65	37
Alternative 59	70	6	50	45
Alternative 60	70	6	55	55
Alternative 61	70	6	55	75

References: http://www.ermaksan.com.tr/tr-TR/katalog/SpeedBend-TR.pdf, 18.04.2017; http://dirinler.com.tr/tr/urunler/hidrolik-presler/c-tipi-hidrolik-presler/cdhc-serisi; 18.04.2017

In this step, using Eq. 2, j. in terms of criteria or qualifications i. a number without a unit in the range [0,1] representing the normalized performance of the alternative is calculated according to all the alternatives and all evaluation criteria. Table 3 shows the results of applying the Eq. 2 to all alternatives and all evaluation criteria.

Importance level of each criterion is multiplied by these normalized performance values. It is thought here that all the evaluation criteria have equal importance. These normalized performance values (for useful qualities) multiplied by the significance level of the evaluation measure are then added in the case of maximization and

Table 3. Normalized performance values

	For evaluation criteria 1	For evaluation criteria 2	For evaluation criteria 3	For evaluation criteria 4
Alternative 1	0,279148	0,064740	0,292374	0,035941
Alternative 2	0,232624	0,051792	0,243645	0,052714
Alternative 3	0,232624	0,051792	0,243645	0,105428
Alternative 4	0,065135	0,258960	0,009746	0,019169
Alternative 5	0,065135	0,233064	0,009746	0,019169
Alternative 6	0,060482	0,233064	0,008771	0,026357
Alternative 7	0,074440	0,207168	0,009746	0,035941
Alternative 8	0,074440	0,246012	0,009746	0,035941
Alternative 9	0,083744	0,258960	0,009746	0,052714
Alternative 10	0,069787	0,233064	0,007797	0,052714
Alternative 11	0,069787	0,258960	0,007797	0,071883
Alternative 12	0,074440	0,220116	0,007797	0,088655
Alternative 13	0,055830	0,207168	0,008771	0,105428
Alternative 14	0,069787	0,233064	0,007797	0,052714
Alternative 15	0,069787	0,258960	0,007797	0,071883
Alternative 16	0,074440	0,220116	0,007797	0,088655
Alternative 17	0,069787	0,233064	0,007797	0,052714
Alternative 18	0,069787	0,258960	0,007797	0,071883
Alternative 19	0,074440	0,258960	0,007797	0,088655
Alternative 20	0,055830	0,207168	0,008771	0,105428
Alternative 21	0,130269	0,044023	0,165679	0,028753
Alternative 22	0,186099	0,036254	0,160806	0,038337
Alternative 23	0,130269	0,044023	0,165679	0,028753
Alternative 24	0,186099	0,036254	0,160806	0,038337
Alternative 25	0,186099	0,036254	0,160806	0,038337
Alternative 26	0,186099	0,031075	0,185170	0,052714
Alternative 27	0,148879	0,031075	0,185170	0,071883
Alternative 28	0,186099	0,031075	0,185170	0,052714
Alternative 29	0,186099	0,031075	0,185170	0,071883
Alternative 30	0,167489	0,031075	0,185170	0,071883
Alternative 31	0,167489	0,025896	0,180298	0,091051
Alternative 32	0,130269	0,028486	0,131568	0,105428
Alternative 33	0,130269	0,028486	0,146187	0,143765
Alternative 34	0,102354	0,020717	0,126696	0,143765
Alternative 35	0,074440	0,018127	0,063348	0,143765
Alternative 36	0,167489	0,031075	0,185170	0,071883
Alternative 37	0,167489	0,025896	0,180298	0,091051
Alternative 38	0,130269	0,028486	0,146187	0,143765
Alternative 39	0,186099	0,031075	0,185170	0,071883

(*continued*)

Table 3. (*continued*)

	For evaluation criteria 1	For evaluation criteria 2	For evaluation criteria 3	For evaluation criteria 4
Alternative 40	0,167489	0,031075	0,185170	0,071883
Alternative 41	0,167489	0,025896	0,180298	0,091051
Alternative 42	0,130269	0,028486	0,131568	0,105428
Alternative 43	0,130269	0,028486	0,146187	0,143765
Alternative 44	0,102354	0,020717	0,126696	0,143765
Alternative 45	0,074440	0,018127	0,063348	0,143765
Alternative 46	0,074440	0,020717	0,073094	0,177310
Alternative 47	0,065135	0,015538	0,077966	0,263569
Alternative 48	0,065135	0,015538	0,077966	0,359413
Alternative 49	0,186099	0,031075	0,185170	0,071883
Alternative 50	0,167489	0,025896	0,180298	0,091051
Alternative 51	0,102354	0,020717	0,126696	0,143765
Alternative 52	0,074440	0,020717	0,073094	0,177310
Alternative 53	0,120964	0,028486	0,121823	0,091051
Alternative 54	0,074440	0,028486	0,073094	0,143765
Alternative 55	0,074440	0,020717	0,063348	0,143765
Alternative 56	0,074440	0,018127	0,063348	0,143765
Alternative 57	0,074440	0,020717	0,073094	0,177310
Alternative 58	0,074440	0,015538	0,063348	0,177310
Alternative 59	0,065135	0,015538	0,048729	0,215648
Alternative 60	0,065135	0,015538	0,053602	0,263569
Alternative 61	0,065135	0,015538	0,053602	0,359413

subtracted from the minimization case (for useless qualities) to find a single value for each alternative. In this application, the motor power, the descent speed, the press speed and the take-off speed are useful criteria since all the technical specifications are desirable. Using Eq. 4, the value of y_i is obtained for each alternative. The values of y_i obtained for each alternative using Eq. 4 are given in Table 4.

The best alternative is the highest value of y_i, while the worst alternative is the lowest value of y_i. The order of the alternatives according to the obtained values of y_i is shown in Table 5.

When Table 5 is analyzed, it is seen that the most suitable first three models are alternatives 1, 3 and 2 respectively. The three worst alternatives are Alternatives 35, 45 and 56.

Table 4. y_i values of alternative models

	y_i value		y_i value		y_i value
Alternative 1	0,168051	Alternative 22	0,105374	Alternative 42	0,098938
Alternative 2	0,145194	Alternative 23	0,092181	Alternative 43	0,112177
Alternative 3	0,158372	Alternative 24	0,105374	Alternative 44	0,098383
Alternative 4	0,088252	Alternative 25	0,105374	Alternative 45	0,074920
Alternative 5	0,081778	Alternative 26	0,113765	Alternative 46	0,086390
Alternative 6	0,082168	Alternative 27	0,109252	Alternative 47	0,105552
Alternative 7	0,081824	Alternative 28	0,113765	Alternative 48	0,129513
Alternative 8	0,091535	Alternative 29	0,118557	Alternative 49	0,118557
Alternative 9	0,101291	Alternative 30	0,113904	Alternative 50	0,116183
Alternative 10	0,090840	Alternative 31	0,116183	Alternative 51	0,098383
Alternative 11	0,102106	Alternative 32	0,098938	Alternative 52	0,086390
Alternative 12	0,097752	Alternative 33	0,112177	Alternative 53	0,090581
Alternative 13	0,094299	Alternative 34	0,098383	Alternative 54	0,079946
Alternative 14	0,090840	Alternative 35	0,074920	Alternative 55	0,075567
Alternative 15	0,102106	Alternative 36	0,113904	Alternative 56	0,074920
Alternative 16	0,097752	Alternative 37	0,116183	Alternative 57	0,086390
Alternative 17	0,090840	Alternative 38	0,112177	Alternative 58	0,082659
Alternative 18	0,102106	Alternative 39	0,118557	Alternative 59	0,086262
Alternative 19	0,107463	Alternative 40	0,113904	Alternative 60	0,099461
Alternative 20	0,094299	Alternative 41	0,116183	Alternative 61	0,123422
Alternative 21	0,092181				

Table 5. Alternatives ranked according to y_i values

Order of preference	Alternative	y_i value	Order of preference	Alternative	y_i value
1	Alternative 1	0,168051	32	Alternative 32	0,098938
2	Alternative 3	0,158372	33	Alternative 42	0,098938
3	Alternative 2	0,145194	34	Alternative 34	0,098383
4	Alternative 48	0,129513	35	Alternative 44	0,098383
5	Alternative 61	0,123422	36	Alternative 51	0,098383
6	Alternative 29	0,118557	37	Alternative 12	0,097752
7	Alternative 39	0,118557	38	Alternative 16	0,097752
8	Alternative 49	0,118557	39	Alternative 13	0,094299
9	Alternative 31	0,116183	40	Alternative 20	0,094299
10	Alternative 37	0,116183	41	Alternative 21	0,092181
11	Alternative 41	0,116183	42	Alternative 23	0,092181
12	Alternative 50	0,116183	43	Alternative 8	0,091535
13	Alternative 30	0,113904	44	Alternative 10	0,090840
14	Alternative 36	0,113904	45	Alternative 14	0,090840

(*continued*)

Table 5. (*continued*)

Order of preference	Alternative	y_i value	Order of preference	Alternative	y_i value
15	Alternative 40	0,113904	46	Alternative 17	0,090840
16	Alternative 26	0,113765	47	Alternative 53	0,090581
17	Alternative 28	0,113765	48	Alternative 4	0,088252
18	Alternative 33	0,112177	49	Alternative 46	0,086390
19	Alternative 38	0,112177	50	Alternative 52	0,086390
20	Alternative 43	0,112177	51	Alternative 57	0,086390
21	Alternative 27	0,109252	52	Alternative 59	0,086262
22	Alternative 19	0,107463	53	Alternative 58	0,082659
23	Alternative 47	0,105552	54	Alternative 6	0,082168
24	Alternative 22	0,105374	55	Alternative 7	0,081824
25	Alternative 24	0,105374	56	Alternative 5	0,081778
26	Alternative 25	0,105374	57	Alternative 54	0,079946
27	Alternative 11	0,102106	58	Alternative 55	0,075567
28	Alternative 15	0,102106	59	Alternative 35	0,074920
29	Alternative 18	0,102106	60	Alternative 45	0,074920
30	Alternative 9	0,101291	61	Alternative 56	0,074920
31	Alternative 60	0,099461			

5 Conclusion

Businesses are constantly faced with making a choice between many alternatives. There are a variety of multi-criteria decision-making methods used to choose between alternatives. The Moora method is one of these methods. Within the scope of this study, preference order of the hydraulic braking press alternatives for the sheet metal processing, which is a type of machinery that may be needed by the enterprises during the production activities, has been prepared for the 61 different models in consideration of motor power, descent speed, discharge speed and departure speed evaluation criteria. In this study, it was thought that the evaluation criteria had equal importance. However, it is also possible that the evaluation criteria considered in the process of producing solutions to the unique problems of the enterprises have different levels of importance. There may also be other technical features that need to be considered according to the business expectations. These two points are the constraints of this application. Another issue is that, to speed up the decision-making process, enterprises can consider the cost factors, so that instead of looking at all of the 61 models examined here, some models may be left out of the assessment directly due to investment constraints. When examined these constraints, it will be possible to produce solutions to the different problems of the enterprises. Moreover, this method helps managers in decision making process, form a briefing the compare the alternatives and find best one. Finally, Moora method can be applied different sectors to solve the decision-making problems o firms and compare various criteria/alternative to find best criteria/alternative.

References

1. Karande P, Chakraborty S (2012) Application of multi-objective optimization on the basis of ratio analysis (MOORA) method for materials selection. Mater Des 37:317–324
2. Achebo J, Odinikuku WE (2015) Optimization of gas metal arc welding process parameters using standard deviation (SDV) and multi-objective optimization on the basis of ratio analysis (MOORA). J Miner Mater Charact Eng 3(4):298–308
3. Madić M, Radovanović M, Petković D (2015) Non-conventional machining processes selection using multi-objective optimization on the basis of ratio analysis method. J Eng Sci Technol 10(11):1441–1452
4. Kodrat KF, Supiyandi S, Mesran M (2018) Application of multi-objective optimization on the basis of ratio analysis (MOORA) in strategic location marketing. Int J Sci Res Sci Technol 4(2):681–686
5. Yazdani M, Zavadskas EK, Ignatius J, Abad MD (2016) Sensitivity analysis in MADM methods: application of material selection. Eng Econ 27(4):382–391
6. Pérez-Domínguez L, Alvarado-Iniesta A, Rodríguez-Borbón I, Vergara-Villegas O (2015) Intuitionistic fuzzy MOORA for supplier selection. Dyna 82(191):34–41
7. Modanloo V, Doniavi A, Hasanzadeh R (2016) Application of multi criteria decision making methods to select sheet hydroforming process parameters. Decis Sci Lett 5(3):349–360
8. Ray A (2015) Green cutting fluid selection using multi-attribute decision making approach. J Inst Eng (India) Ser C 96(1):35–39
9. Balezentis A, Balezentis T, Brauers WKM (2012) Personnel selection based on computing with words and fuzzy MULTIMOORA. Expert Syst Appl 39:7961–7967
10. İç YT (2012) An experimental design approach using TOPSIS method for the selection of computer-integrated manufacturing technologies. Robot Comput Integr Manuf 28:245–256
11. Streimikiene D, Balezentis T, Krisciukaitien I, Balezentis A (2012) Prioritizing sustainable electricity production technologies: MCDM approach. Renew Sustain Energy Rev 16: 3302–3311
12. Cetin EI, Icigen ET (2017) Personnel selection based on step-wise weight assessment ratio analysis and multi-objective optimization on the basis of ratio analysis methods. World Acad Sci Eng Technol Int J Soc Behav Educ Econ Bus Ind Eng 11(11):2538–2542
13. Stanujkic D (2016) An extension of the ratio system approach of MOORA method for group decision-making based on interval-valued triangular fuzzy numbers. Technol Econ Dev Econ 22(1):122–141
14. Chakraborty S (2011) Applications of the MOORA method for decision making in manufacturing environment. Int J Adv Manuf Technol 54:1155–1166
15. Brauers WKM, Zavadskas EK (2009) Robustness of the multi-objective MOORA method with a test for the facilities sector. Technol Econ Dev Econ Baltic J Sustain 15:352–375

Prediction of Industry 4.0's Impact on Total Productive Maintenance Using a Real Manufacturing Case

Ebru Turanoglu Bekar[1(✉)], Anders Skoogh[1], Nihan Cetin[2], and Osman Siray[2]

[1] Department of Industrial and Materials Science, Chalmers University of Technology, Gothenburg, Sweden
{ebrut, anders.skoogh}@chalmers.se
[2] Delphi Technologies, Izmir, Turkey
{nihan.cetin, osman.siray}@delphi.com

Abstract. With rapid advancements in industry, technology and applications, many concepts have emerged in the manufacturing environment. It is generally known that the term 'Industry 4.0' was published to highlight a new industrial revolution. Maintenance is a key operation function since it is related to all the manufacturing processes and focuses not only on avoiding the equipment breakdown but also on improving business performance. Monitoring of manufacturing systems for maintenance helps to identify equipment condition and failures before equipment breakdowns. Total Productive Maintenance (TPM) is widely used maintenance strategy to gain a competitive advantage in the industry. In this context, this paper focuses on the incompletely perceived link between the Industry 4.0's key technologies and TPM. Conjoint analysis, which is a multi-attribute decision making method based on experimental design, is implemented to quantify the impacts of Industry 4.0's key technologies on pillars of TPM. Moreover, interaction between key technologies of Industry 4.0 and pillars of TPM demonstrates several opportunities for achieving synergies thus leading to a successful implementation of future interconnected smart factories. The major contribution of is to provide a guideline and technology roadmap for investment decision for industries that are under the transformation phase towards future smart factory and offers a space for further scientific discussion.

Keywords: Industry 4.0 · Total Productive Maintenance (TPM)
Key technologies of Industry 4.0 · Impact of Industry 4.0 · Conjoint analysis
Prediction

1 Introduction

Information and communication technology (ICT) is undergoing rapid development, and many recent technologies have emerged in manufacturing and services [1]. Through these technologies, real-time integrating and analysing massive malicious data can be obtained and shared to facilitate rapid and accurate decision making. The connection of physical manufacturing equipment and devices over the Internet together

N. M. Durakbasa and M. G. Gencyilmaz (Eds.): ISPR 2018, *Proceedings of the International Symposium for Production Research 2018*, pp. 136–149, 2019.
https://doi.org/10.1007/978-3-319-92267-6_11

with big data analytics in the digital world (e.g., the cloud) has resulted in the emergence of a revolutionary means of production, namely, cyber-physical production systems (CPPSs). CPPSs are a materialization of the general concept cyber physical systems (CPS) in the manufacturing environment. The interconnection and interoperability of CPS entities in manufacturing shop floors together with analytics and knowledge learning methodology provide an intelligent decision support system [2]. The wide spread application of CPS (or CPPS) has accompanied by the fourth stage of industrial production, namely, Industry 4.0 [3].

Industry 4.0 is set to drive changes in production processes, engineering and the global competitive landscape through the development of digital networks. This presents new challenges to the managers of process equipment, making reliability and Total Productive Maintenance (TPM) goal of Zero Breakdowns even more crucial. In this study, we aim to provide clarity by predicting the impacts of Industry 4.0's key technologies on TPM. Hence, this work addresses the following research questions:

- Which key technologies of Industry 4.0 have the highest statistically significant impacts on TPM?
- How will Industry 4.0 depend on reliable equipment and how this can be managed through TPM?
- How will Industry 4.0 change the way TPM is implemented?
- How TPM programs are both adapting to this changing environment and providing critical support to the manufacturing process?

To answer these research questions, the remainder of this study is structured as follows. First, related literature is identified and relevant concepts are defined. Second, the research methodology is described. Third, results with regard to the research questions are presented based on the findings from real manufacturing case study. Finally, this paper concludes with a critical discussion and an outlook on future research in this context.

2 Literature Review

2.1 Industry 4.0 and Its Key Technologies

Smart factories are the central object of Industry 4.0. This concept figures out the usage of modern and advanced information and communication systems and future-oriented technologies to enable human begins, machines, products and resources to exchange information with each other [1, 4]. Many disruptive technologies, such as cloud computing, Internet of Things (IoT), big data analytics, and artificial intelligence and cognitive technologies and so on have emerged. These technologies are permeating the manufacturing industry and make it smart and capable of addressing current challenges, such as increasingly customized requirements, improved quality, and reduced time to market [5]. Industry 4.0 has elicited much interest from the industry and academe [6]. A recent literature survey identified the basic concept, perspectives, key technologies, and industrial applications of Industry 4.0 and examined its challenges and future research trends [1, 7].

Different technologies have been classified as pillars for Industry 4.0 [8]. For example, nine key technologies that can support the smart transformation of an industrial production system were identified in a report from the Boston Consulting Group [9]. A first class of technologies refers to autonomous robots and additive manufacturing, advanced "hardware" technologies based on innovative devices and equipment that support the automation of a production process [10]. One of the main aspects regarding automation solutions is the shift towards collaboration, adaptability, and autonomy to react to external changes [11], resulting in the implementation of CPS that provide cognitive capabilities to physical assets [12].

A second class of technologies encompasses solutions for data management, acquisition, transformation, visualization, and integration. In this class, the technologies such as the (Industrial) Internet of Things (IoT) (the core technology for connecting objects and devices in manufacturing systems, allowing communication and cooperation), big data analytics, cloud technology (along with cybersecurity solutions), simulation and augmented reality were identified [10]. These two classes of technologies allow the realization of horizontal and vertical integration [13].

Many researchers have been working on the key technologies of Industry 4.0 [1]. In this study, some of these technologies are used to measure the potentially impacting on the pillars of TPM [4, 9, 14]. Table 1 summarizes these technologies and their abbreviations used in this study.

Table 1. A template to list the key technologies potentially impacting the pillars of TPM

Key technologies of Industry 4.0	Abbreviations
Additive Manufacturing (3D Printing)	AM
Augmented or Virtual Reality (Human-Machine Interaction)	HMI
Autonomous Robots	AR
Big data Analytics	BDA
Cyber-security	CS
Horizontal and Vertical System Integration (Machine to Machine Communication)	M2M
Simulation	SIM
The Internet of Things (Smart objects such as sensors, actuators and cloud computing)	IoT

With Industry 4.0, *AM* methods such as 3-D printing, which are used mostly to produce small batches of customized products that offer construction advantages like lightweight designs and reduce transport distances and stock on hand [9]. The approach of *HMI* is based on the information sharing and collaboration between production machines and employees by interfaces like virtual reality or augmented reality [4]. This system supports a variety of services such as selecting parts in a warehouse and sending repair instructions over mobile devices [9]. Another application is virtual training (Siemens has developed a virtual plant operator planning). In this virtual world, operators can learn to interact with machines by clicking on a cyber-representation. They also can change parameters and retrieve operational data and maintenance instructions [9]. *AR* are interconnected so that they can work together and automatically adjust their actions to fit the next unfinished product in line. High-end sensors and control units enable close collaboration with humans. For example, industrial-robot supplier ABB is launching a two-armed robot called YuMi that is specifically designed to assemble products (such as consumer electronics) alongside human [9]. *BDA* are depending directly on sensor and actuator generated data of smart devices and smart machines [17]. This data enables the research on a huge amount of statistic process data directly out of the machines to identify instable process parameters or avoid quality issues inside defined tolerance ranges. A predictive maintenance is one possible result of this application [4]. With the increased connectivity and use of standard communications protocols that come with Industry 4.0, the need to protect critical industrial systems and manufacturing lines from *CR* threats increases dramatically. As a result, secure, reliable communications as well as sophisticated identity and access management of machines and users are essential [9]. One of the main concepts of the vision of Industry 4.0 is the *M2M* based on CPS. This concept enables intelligent applications like auto-adaptive controlling of interconnected machines and equipment without human interaction and combines the approach of vertical and horizontal integration [4]. *Horizontal integration* refers to the creation of a network enabling real time communication among suppliers, manufacturers and final customers [10]. *Vertical integration* connects machines and data on different levels. This means for example a gapless data connection of machine processes with manufacturing executive system (MES) and enterprise resources planning (ERP) [4]. It involves the information technology structure inside an enterprise enabling data exchange from the sensors/actuators level to the ERP level, in order to provide real-time decision-making and corrective actions from the business to the shopfloor [9]. *SIM* allows operators to test and optimize the machine settings for the next product in line in the virtual world before the physical changeover, thereby driving down machine setup times and increasing quality [14]. *IoT* and services creates the environment to connect such smart objects (sensors and actuators to interact with the physical world and additional middleware enables services like cloud computing) with the global internet [4, 15].

Manufacturing leaders have the opportunity to develop improved operations strategies and to realize key business objectives based on these technologies they may choose to employ at various points in the manufacturing processes [3]. For example, in a smart factory with machines interconnected with information and communication systems, when a machine breaks down, it sends error notifications to respective shopfloor and maintenance personnel. The maintenance worker then checks the error code for solutions and gets necessary tools and parts for repairing. Meanwhile the manufacturing execution system can reschedule the jobs to mitigate the impact of breakdown [16]. With more advanced analytics and big data environment, machines are equipped to be self-aware and self-maintained. Such machines assess their own health and degradation and utilise data from other machines to avoid potential maintenance issues [17]. The ability to anticipate potential breakdown and identify root cause needs to be developed in the control systems. For example, ERP systems have included comprehensive frameworks for predictive maintenance. It integrates between machine data,

ERP data, sensory data and predictive algorithms [18]. So *machine-worker communication*, *self-maintenance assessment* and *predictive maintenance control system* notably improve TPM in the factory [19].

2.2 Concept and Pillars of TPM

It has been recognized by many researchers and practitioners that maintenance is a crucial support function for the manufacturing processes with significant investment in physical assets and a main contributor to the performance and profitability of a manufacturing system [20]. TPM is a strategic tool that is widely used to make manufacturing industries competitive and effective in the field of maintenance [21]. The main practices of TPM are generally called "the pillars or elements of TPM" shown in Fig. 1. These pillars are explained as the following paragraphs [22]:

1. Autonomous Maintenance – (Pillar 1) is mother pillar. It brings the ownership approach in the operator. Cleaning and maintenance is taken care by the operator himself. It places responsibility for routine maintenance, such as cleaning, lubricating, and inspection, in the hands of operators.
2. Focused Improvement – (Pillar 2) aims to identify and minimize waste, quality and manufacturing losses. Elimination of losses helps improving OEE. Small groups of employees work together pro-actively to achieve regular, incremental improvements in the operation of equipment.
3. Planned Maintenance – (Pillar 3) schedules maintenance tasks based on predicted and/or measured failure rates; deals with Preventive break down (Time Base Maintenance). It establishes Preventative and Predictive Maintenance systems for equipment and tooling. Effectiveness of maintenance department is increased to the point where 8 big losses are not generated.
4. Quality Maintenance – (Pillar 4) detects design errors and prevents them from entering into production processes. It applies root cause analysis to eliminate recurring sources of quality defects; monitors the factors affecting variability in product quality. It focuses to accomplish zero quality defects.
5. Education and Training – (Pillar 5) fill in gaps in knowledge which is required to achieve TPM goals. Training and educational opportunities are given to operators, maintenance personnel and managers; boost the moral and expertise of the operators and persons involved by to providing soft skill training and technical training.
6. Early Equipment Management – (Pillar 6) directs practical knowledge and understanding of manufacturing equipment gained through TPM towards improving the design of new equipment; establish systems to shorten new product or equipment development prototyping lead time. Achieve stable commissioning of new product and equipment vertical start up.
7. Safety, Health and Environment – (Pillar 7) these are about maintaining a safe and healthy working environment; Assures safety and prevents adverse environmental impacts which are important priorities in any TPM effort. It achieves and sustains zero accidents, creates healthy, rewarding and pleasant workplace.

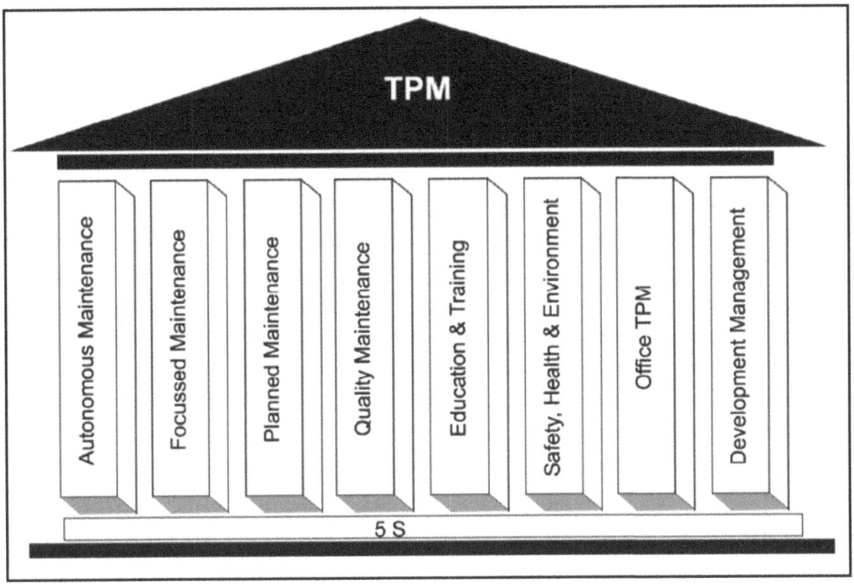

Fig. 1. Eight pillars approach for TPM implementation [23]

8. TPM in Administration – (Pillar 8) Applying TPM techniques to administrative functions. The goal is to achieve zero functional losses, create highly efficient offices and provide effective service and support to other departments. Administrative and support departments can be seen as process plants whose principal tasks are to collect, process, and distribute information. Process analysis should be applied to streamline information flow.

2.3 Review on the Impact of Industry 4.0's Key Technologies

As these waves of change continue to shape the competitive landscape, manufacturers must decide how and where to invest in new technologies, and identify which ones will drive the most benefit for their organizations. In this context, a lot of studies have been proposed in literature in order to understand the overall impact of Industry 4.0 and its key technologies. For instance, Sangameshwran [23] have discussed the impact, challenges and opportunities of digitization and concluded with examples of recommended policy action for German Industry. Kagermann [24] have addressed the research question of how new and sustainable business models and structures for Industry 4.0 might look like and in which direction existing traditional business concepts have to be developed to deploy a strong business impact of Industry 4.0. Arnold et al. [25] have analysed the influence of the IoT on business models of established manufacturing companies within several industries. Demartini et al. [26] have demonstrated technological concepts of Industry 4.0 and related enabling technologies (such as CPPS and IoT) that could support decentralization and manufacturing flexibility. Their application allows to orchestrate and execute manufacturing and production processes with the aim of supporting individualized production, small lot sizes, and small batches and provide defining of digitalized opportunities for specific type of flexibility that has an impact on the manufacturing system starting from an analysis of potential improvements of current MES. Wagner et al. [4] have presented an Industry 4.0 impact matrix on lean production systems. It considers elements of lean production systems with Industry 4.0 technologies and gives a first estimation of impact. In the described development process of a cyber-physical Just-in-Time delivery the matrix is used to find a stabilizing application for a Just-in-Time material supply process. Tjahjono et al. [14] have presented a preliminary analysis of the impact of Industry 4.0 technologies on four functions of supply chain management such as procurement, transport logistics, warehouse and order fulfilment. They have showed that the opportunities and threats of these technologies according to the functions of supply chain. Bokrantz et al. [27] have described a probable future of maintenance organizations in digitalized manufacturing in the year 2030 for the largest companies in the Swedish manufacturing industry. They have illustrated wildcard scenarios for future maintenance based on the Delphi method including e.g. advancement of data analytics, increased emphasis on education and training, novel principles for maintenance planning with a systems perspective, and stronger environmental legislation and standards. Lin et al. [28] have examined the strategic response to Industry 4.0 for Chinese automotive industry and to identify the critical factors for its successful implementation. They have used a comprehensive framework to build the structural models, for the validation of these models they also have used statistical tools and interpretive structural modeling method to further analyze these derived factors for depicting the relationship. Thus, they have illustrated which factors have impact on the strategic response and whether their relationships are positive or negative. According to the result, it can be shown that company size and nature do not increase the use of advanced production technologies, while other factors have positive impacts on improving the technology adoption among the companies surveyed. Müller et al. [29] have explained the relevance of Industry 4.0-related opportunities and challenges as drivers for Industry 4.0 implementation in the context of sustainability, taking a differentiated perspective on varying company sizes, industry sectors, and the company's role as an Industry 4.0 provider or user using partial least square structural equation modeling. This model have been applied for a sample of 746 German manufacturing companies from five industry sectors. The results have illustrated that strategic, operational, environmental and social opportunities are positive drivers of Industry 4.0 implementation, whereas challenges with regard to competitiveness and future viability as well as organizational change depending on different company characteristics.

It can be seen from literature review, in order to understand the impacts of Industry 4.0 and its key enabling technologies on manufacturing processes have been analyzing intensively. However, no work has analyzed the impact of Industry 4.0's key technologies on TPM, and has investigated which key technologies of Industry 4.0 has the highest impact on TPM. Therefore this is the first study to quantify the impacts of Industry 4.0's key technologies on TPM based on experimental design using a real manufacturing case study.

3 Research Methodology

3.1 Research Method: Conjoint Analysis

In this study, after a general list is determined for the key technologies of Industry 4.0 (shown in Table 1), conjoint analysis, which is a multi-attribute decision making method based on experimental design, is implemented to quantify the impacts of Industry 4.0's key technologies on TPM pillars.

Conjoint analysis is used as a way to map the strategic thinking of respondents because it is one of the most widely used methodologies for analysing personal preferences [30]. Conjoint analysis requires respondents to rate different scenarios with varying combinations of attribute levels.

Conjoint analysis can measure preferences at the individual level and reveal hidden motivations which may not even be apparent to the respondents themselves, as well as providing realistic choices and scenarios for the respondents to consider [31]. A further advantage of conjoint analysis is that it gives a psychological profile of respondents' preferences and corresponding decision-making processes, because it uses algebraic theory to study cognitive processes and to develop statistical estimations [32, 33].

3.2 Collection of Data

Each of TPM pillar is analyzed by the decision makers (TPM, Production, and Quality managers) to quantify the impact of key technologies of Industry 4.0. These decision makers work at strategic levels in a manufacturing company operating in the automotive industry in Aegean Free Zone since 2002. 860 Direct and 130 indirect employees work in the company. Along with the core operating departments, there are support functions including the TPM department. In the TPM department, there are one lead engineer and four supervisors and 28 maintenance technicians. The overall TPM activities of this company are managed by the TPM-office.

The relevant data for measuring the impact of each key technologies of Industry 4.0 have been collected according to evaluations of each decision maker for each pillar of TPM. Table 2 gives the experimental design using the Taguchi OA L18 table and the relevant data for pillar 1 of TPM. According to this table, experiment numbers of Taguchi OA L18 table, the identified key technologies of Industry 4.0 (shown in Table 1) and their possible levels of impact such as low, medium and high and impact evaluations of decision makers are listed in the columns. For instance, in experiment number 1, the total impact on pillar 1 of TPM have been ranked in the 18th, 18th, and 17th by the decision makers (TPM Manager, Production Manager and Quality Manager), respectively when the all factors have the low level of impact. In experiment number 5, when the AM, SIM and IoT have the low level of impact, the HIM, AR and BDA have the medium level of impact, and the CS and M2M have the high level of impact, the total impact on pillar 1 of TPM have been ranked in the 16th by all decision makers. In this study, the relevant data is only given for pillar 1 of TPM as an example.

Table 2. The relevant data for Pillar 1 of TPM

Taguchi OA L18 Table	Key technologies of Industry 4.0 (Factors)								Ranking of total impact		
Experiment number	AM	HMI	AR	BDA	CS	M2M	SIM	IoT	Production manager	Quality manager	TPM manager
1	1	1	1	1	1	1	1	1	18	18	17
2	1	1	2	2	2	2	2	2	17	14	14
3	1	1	3	3	3	3	3	3	6	3	1
4	1	2	1	1	2	2	3	3	5	1	8
5	1	2	2	2	3	3	1	1	16	16	16
6	1	2	3	3	1	1	2	2	13	10	6
7	1	3	1	2	1	3	2	3	2	4	7
8	1	3	2	3	2	1	3	1	4	6	5
9	1	3	3	1	3	2	1	2	8	11	4
10	2	1	1	3	3	2	2	1	15	17	18
11	2	1	2	1	1	3	3	2	14	9	9
12	2	1	3	2	2	1	1	3	11	13	12
13	2	2	1	2	3	1	3	2	9	7	13
14	2	2	2	3	1	2	1	3	10	12	11
15	2	2	3	1	2	3	2	1	1	15	10
16	2	3	1	3	2	3	1	2	7	8	15
17	2	3	2	1	3	1	2	3	1	5	2
18	2	3	3	2	1	2	3	1	3	2	3

Level of Factors, 1-LOW, 2-MEDIUM, 3-HIGH

4 A Case Study in the Automotive Industry

4.1 Evaluation of Industry 4.0's Impact on TPM Pillars

In the conjoint analysis, the full factorial design is generally used to obtain the statistical results. However, in this case, if the full factorial design is used for conjoint analysis, there would be too much combinations (3^8) because each Industry 4.0's key technologies have three impact levels such as Low, Medium and High. Hence, the numbers of design should be reduced. In this context, the Taguchi design methodology is used for reducing the number of the designs. Then the conjoint analysis is performed by using of the Taguchi OA L18 table.

4.2 Results

Experimental designs for each TPM pillars are solved by using IBM SPSS 24.0 statistics program. In Table 3, the IBM SPSS 24.0 output of pillar 1 of TPM is presented. This table gives the model summary, ANOVA results and coefficients of predictors (factors or key technologies of Industry 4.0). For the other pillars of TPM, only significance values of coefficients of predictors are given in Table 4.

Table 3. SPSS 24.0 output of pillar 1 of TPM

Model	R	R Square	Adjusted R Square	Std. Error of the Estimate	Change Statistics				
					R Square Change	F Change	df1	df2	Sig. F Change
1	.891	.793	.756	2.585	.793	21.572	8	45	.000

ANOVA

Model		Sum of Squares	df	Mean Square	F	Sig.
1	Regression	1152.880	8	144.110	21.572	.000
	Residual	300.620	45	6.680		
	Total	1453.500	53			

Coefficients

Model		Unstandardized Coefficients		Standardized Coefficients	t	Sig.	95,0% Confidence Interval for B		Correlations			Collinearity Statistics	
		B	Std. Error	Beta			Lower Bound	Upper Bound	Zero-order	Partial	Part	Tolerance	VIF
1	(Constant)	30.278	2.536		11.938	.000	25.169	35.386					
	AM	.481	.703	.046	.684	.497	−.935	1.898	.046	.102	.046	1.000	1.000
	HMI	−3.583	.431	−.564	−8.318	.000	−4.451	−2.716	−.564	−.778	−.564	1.000	1.000
	AR	−1.278	.431	−.201	−2.966	.005	−2.145	−.410	−.201	−.404	−.201	1.000	1.000
	BDA	.000	.431	.000	.000	1.000	−.868	.868	.000	.000	.000	1.000	1.000
	CS	.000	.431	.000	.000	1.000	−.868	.868	.000	.000	.000	1.000	1.000
	M2M	.000	.431	.000	.000	1.000	−.868	.868	.000	.000	.000	1.000	1.000
	SIM	−3.194	.431	−.503	−7.416	.000	−4.062	−2.327	−.503	−.742	−.503	1.000	1.000
	IoT	−2.694	.431	−.424	−6.255	.000	−3.562	−1.827	−.424	−.682	−.424	1.000	1.000

According to ANOVA results given in Table 3, the value of test statistic F is found 21.572. Also significance level α (Sig.) is found 0.000. Then, it is concluded that the proposed multiple regression model for the impact of Industry 4.0's key technologies on pillar 1 of TPM (autonomous maintenance) are statistically significant. The correlation coefficient of this model is also found 0.891. This means that pillar 1 of TPM and the key technologies of Industry 4.0 are highly correlated with each other. If we look at the significance level α (Sig.) of the predictors in the coefficients part in Table 3, we can see that the technologies "HMI", "AR", "SIM", and "IoT" have very high statistically impacts on pillar 1 of TPM with the 0.000 significance values (<0.05). The other technologies "AM", "BDA", "CS", and "M2M" have not statistically significance impacts on pillar 1 of TPM (>0.05).

Table 4. The significance values of coefficients of predictors for the other pillars of TPM

Key technologies of Industry 4.0 (Predictors)	Significance (Sig.) value							
	Pillar 1	Pillar 2	Pillar 3	Pillar 4	Pillar 5	Pillar 6	Pillar 7	Pillar 8
AM	.497	.069	.016	.029	.102	.017	.010	.018
HMI	.000	.833	.002	.002	.001	.358	.000	.000
AR	.005	.000	.392	.110	.868	.447	.001	.356
BDA	1.000	.000	.000	.010	.001	.097	.069	.644
CS	1.000	.000	.193	1.000	.836	.016	.807	.000
M2M	1.000	.011	.008	.376	.740	.282	.696	.293
SIM	.000	.000	.004	.042	.005	.000	.000	.190
IoT	.000	.000	.014	.081	.005	.177	.016	.000

We can see the significance values of the key technologies of Industry 4.0 with respect to other pillars of TPM in Table 4. According to this table, it can be seen that highlighted cells are statistically significant for the corresponding pillars of TPM (<0.05). For instance, all of the technologies are statistically significant except the technologies "AM" and "HMI" for pillar 2 of TPM (focused improvement) and only the technologies "AM", "CS, and "SIM" have highly statistically significance impacts on pillar 6 of TPM (early equipment management).

In the next step, we have prioritized the impacts of these technologies by estimating part-worth utilities or relative weights for each pillar of TPM. They are calculated based on the standardized coefficients of predictors and presented in Table 5. Additionally, it can be concluded that the technologies "SIM", "IoT", and "HMI" have highly statistically significance impacts on almost all pillars of TPM.

Table 5. Relative weights of Industry 4.0's key technologies for all pillars of TPM

Key technologies of Industry 4.0 (Predictors)	Part-worth utilities (Relative weights-%)							
	Pillar 1	Pillar 2	Pillar 3	Pillar 4	Pillar 5	Pillar 6	Pillar 7	Pillar 8
AM	4.68	15.96	22.48	22.94	16.96	26.25	24.25	16.00
HMI	34.89	1.11	18.63	20.35	22.32	6.04	25.25	34.40
AR	12.45	21.43	4.78	10.18	1.04	4.99	20.63	3.73
BDA	0.00	25.88	24.66	16.72	22.32	11.03	10.32	1.86
CS	0.00	20.32	7.30	0.00	1.30	16.27	1.36	22.14
M2M	0.00	13.92	15.35	5.57	2.07	7.09	2.17	4.26
SIM	31.11	20.87	16.86	13.08	18.16	29.40	29.87	5.34
IoT	26.24	23.37	14.10	11.15	18.42	8.92	13.85	34.14

According to Table 5, the technologies that are "HMI", "SIM", "IoT", and "AR" have the highest relative weights with the values 34.89%, 31.11%, 26.24%, and 12.45% respectively. Furthermore, the technologies that are "CS", "BDA" "M2M", and "AM" have the lowest relative weights with the value 0%, 0%, 0%, and 4.68% respectively. That means the technologies "HMI" and "SIM" have the highest impact on autonomous maintenance (pillar 1 of TPM). For example, the technology "BDA" has the highest importance for focused improvement (pillar 2), planned maintenance (pillar 3) and education and training (pillar 5) since it includes advance data analytics methods and supports predictive maintenance (especially in keeping with pillar 3). Consequently, the technology "SIM" has the highest importance while the technology "CS" has the lowest importance almost all pillars of TPM.

4.3 Discussions

With the results of this study, the research questions are generally answered as following. For the first question "Which key technologies of Industry 4.0 have the highest statistically significant impacts on TPM", the answer is that "from the conjoint analysis performed, it can be seen that the implementation of certain technologies, such as simulation (SIM), internet of things (IoT), human machine interaction (HMI), big data analytics (BDA), and additive manufacturing (AM) have the higher impact than the others like cybersecurity (CS), machine to machine interaction (M2M) and autonomous robots (AR) on all pillars of TPM". For the second question "How Industry 4.0 will depend on reliable equipment and how this can be managed through TPM", the answer is that "according to correlation coefficients and ANOVA results, it can be concluded that the proposed models for predicting the impact of Industry 4.0's key technologies on all pillars of TPM are highly statistically significant and also these models illustrates high correlations between the key technologies of Industry 4.0 and TPM. Since we can see the highest technological impact on TPM (SIM, IoT, HMI, BDA and AM), it provides a useful framework from the management perspective especially for investment decision on key technologies that need to be integrated with TPM urgently." For the third question, "How Industry 4.0 will change the way TPM is implemented", the answer is that SIM and HMI technologies (especially virtual reality) change the way of autonomous maintenance (pillar 1) of TPM. They provide a lot benefits namely, increased flexibility, quality standards, efficiency and productivity. Additionally, the BDA technology provide highly valuable information for planned maintenance including preventive and predictive maintenance decisions." For the last questions "How TPM programs are both adapting to this changing environment and providing critical support to the manufacturing process", the answer is that we can see the statistically significant key technologies of Industry 4.0 for all practices of TPM, according to these results more simulation and visualization tools should be integrated into TPM implementation program to help machine-operator communication, smart devices like sensors and actuators (IoT) should be used to collect more data for focused improvement (forecasting overall equipment effectiveness-OEE), planned maintenance, and quality maintenance, machine-worker communication, and BDA technology should be adapted to TPM program to predictive maintenance control system. As a result, this study is expected to provide a roadmap for the manufacturing industry which Industry 4.0's key technologies are important and how they can integrate them with TPM in the near future.

5 Conclusions and Future Directions

New possibilities from ICT are keeping with TPM. Improvements or adaptations of evolution of these new technologies need to be analyzed over their influence on TPM. This paper shows that the key technologies of Industry 4.0 can stabilize and support pillars of TPM statistically. The experimental design for measuring the impacts of Industry 4.0's key technologies on TPM gives a statistical framework to start design and development of Industry 4.0 technological solutions for industries.

Despite the fact that the impacts of Industry 4.0's key technologies on TPM pillars have been discussed in this paper, due to its theoretical nature, the generalizability of the paper is limited to only one industrial case. For future research, detailed empirical research should be performed using different manufacturing cases from the different industries.

References

1. Lu W (2017) Industry 4.0: a survey on technologies, applications and open research issues. J Ind Inf Integr 6:1–10
2. Liu C, Jiang P (2016) A cyber-physical system architecture in shop floor for intelligent manufacturing. Procedia CIRP 56:372–377
3. Zheng P, Wang H, Sang Z, Zhong RY, Liu Y, Liu C, Mubarok K, Yu S, Xu X (2018) Smart manufacturing systems for Industry 4.0: conceptual framework, scenarios, and future perspectives. Front Mech Eng 13(2):137–150
4. Wagner T, Herrmann C, Thiede S (2017) Industry 4.0 impacts on lean production systems. Procedia CIRP 63:125–131
5. Rittinghouse JW, Ransome JF (2016) Cloud computing: implementation, management, and security. CRC Press, Boca Raton
6. Liu Y, Xu X (2016) Industry 4.0 and cloud manufacturing: a comparative analysis. J Manuf Sci Eng 139(3)
7. Thames L, Schaefer D (2017) Industry 4.0: an overview of key benefits, technologies, and challenges. In: Thames L, Schaefer D (eds) Cybersecurity for Industry 4.0. Springer, Cham, pp 1–33
8. Mittal S, Khan M, Wuest T (2016) Smart manufacturing: characteristics and technologies. In: Harik R, Rivest L, Bernard A, Eynard B, Bouras A, Bouras A (eds) IFIPAICT, vol 492. Springer, Cham, pp 539–548
9. Rüssmann M, Lorenz M, Gerbert P, Waldner M, Justus J, Engel P, Harnish M (2015) Industry 4.0: the future of productivity and growth in manufacturing industries. Technical report, The Boston Consulting Group
10. Cimini C, Pinto R, Pezzotta G, Gaiardelli P (2017) The transition towards Industry 4.0: business opportunities and expected impacts for suppliers and manufacturers. In: Lödding H, et al (eds) APMS 2017, Part I, IFIP AICT, vol 513. Springer, Cham, pp 119–126
11. Park HS (2013) From automation to autonomy-a new trend for smart manufacturing. In: DAAAM international scientific book, vol 3, pp 75–110
12. Lee J, Ardakani HD, Yang S, Bagheri B (2015) Industrial big data analytics and cyber-physical systems for future maintenance & service innovation. Procedia CIRP 38:3–7
13. Kagermann H, Wahlster W, Johannes H, Gerbert P, Waldner M, Justus J, Engel P, Harnish M (2013) Recommendations for implementing the strategic initiative Industrie 4.0. Technical report, Industry-Science Research Alliance
14. Tjahjono B, Espluguesb C, Ares E, Pelaez G (2017) What does Industry 4.0 mean to supply chain? Procedia Manuf 13:1175–1182
15. Lee I, Lee K (2015) The Internet of Things (IoT): applications, investments, and challenges for enterprises. In: McMullen JS (ed) Business horizons. Elsevier, pp 431–440
16. Lucke D, Constantinescu C, Westkämper E (2008) Smart factory-a step towards the next generation of manufacturing. In: Manufacturing systems and technologies for the new frontier. Springer, London, pp 115–118
17. Lee J, Kao HA, Yang S (2014) Service innovation and smart analytics for industry 4.0 and big data environment. Procedia CIRP 16:3–8
18. Haddara M, Elragal A (2015) The readiness of ERP systems for the factory of the future. Procedia Comput Sci 64:721–728
19. Sanders A, Elangeswaran C, Wulfsberg J (2016) Industry 4.0 implies lean manufacturing: research activities in Industry 4.0 Function as enablers for lean manufacturing. J Ind Eng Manag 9(3):811–833

20. Bartz T, Siluk CJM, Bartz APB (2014) Improvement of industrial performance with TPM implementation. J Qual Maint Eng 20(1):2–19
21. Piechnicki AS, Sola AVH, Trojan F (2015) Decision-making towards achieving world-class total productive maintenance. Int J Oper Prod Manag 35(12):1594–1621
22. Shinde DD, Prasad R (2018) Application of AHP for ranking of total productive maintenance pillars. Wireless Pers Commun 100(2):449–462
23. Sangameshwran P, Jagannathan R (2002) HLL's manufacturing renaissance. Indian management, pp 30–35
24. Kagermann H (2015) Change through digitization value creation in the age of Industry 4.0. In: Management of permanent change, pp 23–45
25. Arnold C, Kiel D, Voigt K-I (2016) How the industrial Internet of Things changes business models in different manufacturing industries. Int J Innov Manag 20(8):1–25
26. Demartini M, Tonelli F, Damiani L, Revetria R, Cassettari L (2017) Digitalization of manufacturing execution systems: the core technology for realizing future smart factories. In Proceedings of the 22nd Summer School "Francesco Turco" - Industrial Systems Engineering, Italy, pp 326–333
27. Bokrantz J, Skoogh A, Berlin C et al (2017) Maintenance in digitalized manufacturing: Delphi-based scenarios for 2030. Int J Prod Econ 191:154–169
28. Lin D, Lee CKM, Lau H, Yang Y (2018) Strategic response to Industry 4.0: an empirical investigation on the Chinese automotive. Ind Manag Data Syst 118(3):589–605
29. Müller JM, Kiel D, Voigt K-I (2018) What drives the implementation of Industry 4.0? The role of opportunities and challenges in the context of sustainability. Sustainability 10(1): 1–24
30. Carroll JD, Green PE (1995) Psychometric methods in marketing research: part 1. Conjoint analysis. J Mark Res 32(4):385–391
31. Kim G, Kim A, Sohn SY (2009) Conjoint analysis for luxury brand outlet malls in Korea with consideration of customer lifetime value. Expert Syst Appl 36(1):922–932
32. Bronn PS, Olson EL (1999) Mapping the strategic of public relations managers in a crisis situation: an illustrative example using conjoint analysis. Public Relat Rev 25(3):351–368
33. Kuhfeld FW (2006) Marketing research method in SAS. SAS Institute, North Carolina

The Selection of a Process Management Software with Fuzzy Topsis Multiple Criteria Decision Making Method

Arda Yiğit Şen[1(✉)], Neslihan Semiz[1], Buse Güneş[1], Duygu Algül[1], Zeynep Gergin[1], and Nurcan Demirok Dönmez[2]

[1] Industrial Engineering Department, İstanbul Kültür Üniversity, Istanbul, Turkey
senyigitarda@gmail.com, neslihan.semiz.53@gmail.com, buseegunees@gmail.com, duygualgul.da@gmail.com, z.gergin@iku.edu.tr

[2] Process and Project Management Department, Defacto, Istanbul, Turkey
nurcan.donmez@defacto.com.tr

Abstract. Today's retail companies are facing with many different criteria in choosing the software they use to manage their processes. The choice of appropriate software is a multi-criteria decision-making problem that requires process managers to come together with multiple factors as well as with different decision makers. In this study, the problem of choosing process management software for a retailer operating in the textile sector is discussed. The main problem in existing software used in the company's process management department is that it does not cover all process flows causing lack of communication with other softwares. With this study, it is aimed to reduce time loss in operations by selecting a software that combines processes in one system. Thus, in all departments, there will be a unified and single reporting system for real-time analysis of statistics and better cooperation. In this study, the Analytic Hierarchy Process (AHP) and Fuzzy TOPSIS approaches are presented for this multi criteria decision making problem. In the first step, the AHP implementation and the pairwise comparisons taken from the seven decision makers determined the importance weights of the integrated decision criteria. In the second step, the Fuzzy TOPSIS method is carried out for the selection of the best software with the quantitative and qualitative evaluations of the decision makers.

Keywords: Multi-criteria decision making · AHP · Fuzzy TOPSIS Software selection

1 Introduction

Today's evolving technology directly influences the software selection process of companies with the various innovations that it brings. Selection of the best software appropriate for their operations is a difficult process that almost every company faces. The main purpose of software selection in companies is to choose the software that meets the expectations of the company and realizes at optimum cost. While this choice is

N. M. Durakbasa and M. G. Gencyilmaz (Eds.): ISPR 2018, *Proceedings of the International Symposium for Production Research 2018*, pp. 150–167, 2019.
https://doi.org/10.1007/978-3-319-92267-6_12

made, the expectations of the companies are taken into account and these expectations are considered as decision criteria. When there are more than one criteria, the choices are made with the help of multiple criteria decision making methods. Having more than one of the selected criteria is an important reason as many departments use this software at the same time, and the expectations of each department are different. If there are more than one criteria to decide, methods such as AHP, ANP, ELECTRE, TOPSIS and PROMETHEE can be used by the decision maker. When all these methods are used, fuzzy numbers can be used to translate verbal values into numerical values if the decision criteria are evaluated by a scale composed of other than numerical values.

In this study, the problem of choosing process management software for a retailer operating in the textile sector is discussed. The Project is completed in cooperation with the Process and Project Management Department of a retailer company operating in textile sector. The Analytic Hierarchy Process (AHP) and Fuzzy TOPSIS approaches are implemented for decision making. It is aimed to reduce time loss in operations by selecting software that combines processes in one system. Thus, in all departments, there will be a unified and single reporting system for real-time analysis of statistics and better communication and cooperation will be achieved in all departments. This paper is organised as follows; Literature review on the topic is given in the following section. This part includes brief description of the decision-making tools, the AHP method and the Fuzzy TOPSIS methods together with the related research done on the topic. The succeeding section explains the methodology of the study. The implementation and the results are given in Sect. 4. Finally the study is concluded in Sect. 5.

2 Literature Review

2.1 Decision Making Methods

Multiple criteria decision making is the process that aims to reach the most appropriate result in the selection processes that require evaluation of two or more criteria. The most widely used method is Analytic Hierarchy Process (AHP), others are TOPSIS, ELECTRE, VIKOR, ANP, PROMETHEE. All multiple criteria decision making methods are shown in Fig. 1. AHP and Fuzzy TOPSIS are explained in detail in the succeeding sections as they are used fort his study.

2.2 Analytical Hierarchy Process

AHP is first created by Myers and Alpert, later developed by Thomas Lorie Safety used for decision making. The AHP method clearly shows the relationship between the goal, the criteria and the decision alternatives. In AHP method, the decision maker can be objective and subjective. AHP is an approach that can simplify and express the system in a hierarchical structure associated with subsystems in decision problems of systems with complex relationships between its components, and to allow them to be modeled and analyzed with an intuitive and logical idea [1]. This method can be used to determine the weights of each criteria when there are more than one criteria. Criteria can not be evaluated on the same scale because they represent different concepts. The

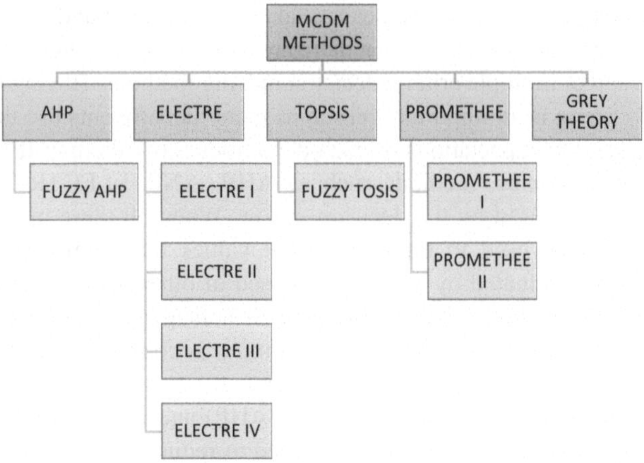

Fig. 1. Multiple criteria decision-making methods (Aruldoss, Lakshmi, & Venkatesan, 2013)

only way to evaluate these criteria is to make pairwise comparisons. Therefore, AHP, which is a binary comparison method, is used in this study. The analytical hierarchy method is also a method of achieving the result by reducing the complex decision problem (multiple alternatives and multiple criteria) by pairwise comparisons. In the AHP method, a questionnaire is often used to make pairwise comparisons of criteria [2]. The survey is evaluated by people who are knowledgeable about the content of the survey. The answers given are used to determine the matrices to be used in the method. The weights of the criteria are calculated with matrices [3].

2.3 Fuzzy TOPSIS Method

Another widely used method used by decision makers for multiple criteria decision making is TOPSIS. The use of fuzzy values in the TOPSIS method began in 1992 when Chen and Hwang found a reference to the work on the classic TOPSIS method [4]. After that, this method has become a method used in decision making where there are non numerical evaluations for decision criteria. The fuzzy TOPSIS method is a method that helps to evaluate and rank alternatives under uncertainty according to a certain criteria and make the most accurate choice. The TOPSIS approach uses a fuzzy theory with a linguistic value instead of a crisp value. Using linguistic values instead of numerical values is often a better approach. Evaluations made in this way are evaluated by linguistic variables. The method allows linguistic evaluation of criteria evaluations that can not be expressed by numerical values [5]. The fuzzy TOPSIS method uses triangular fuzzy numbers for linguistic variables, to determine the order of the alternatives and the weights of the criteria. Using the positive and negative ideal solution, the Fuzzy TOPSIS method finds the most appropriate alternative. The most appropriate alternative chosen as a result of the method should be the closest to the positive ideal point and the farthest away from the negative ideal point. Where there is no ideal solution, the closest optimal solution is chosen [6]. The TOPSIS method is a very simple method that does not involve complex algorithms and complex mathematical models. Hence, when the company wants to repeat the selection with additional alternatives in the future, the same method can easily be repeated by different decision makers of the company.

3 Other Research Related on the Topic

A number of different studies of Analytic Hierarchical Process (AHP) and Fuzzy TOPSIS are included in the literature review. In addition to these, there are also studies carried out when these two and other methods are used together. Reference [7] selected the most proper tour operator and used AHP and Fuzzy Topsis methods together in tourism business. AHP and Fuzzy TOPSIS method are applied together to determine the best tour operator alternative that the hotels deal with. The criteria for selecting tour operators for a five-star hotel operating in the province of Isparta are weighted by the AHP. The obtained weights are evaluated by using Fuzzy Topsis method and the suitable tour operator for determined hotel. Tunca research [8] used AHP-based TOPSIS and ELECTRE methods on accounting software selection. In this study, the selection process of the accounting package program is considered. In order to select the most suitable accounting package program in study, the most suitable solution is searched by selecting three main and fifteen sub criteria. Criteria with the highest importance are the ability to report and ease of use of menus, respectively. As a result of the questionnaire that is made to accounting professionals, weights of criteria are determined by AHP method. The weights are then evaluated using TOPSIS and ELECTRE methods. The ranking and superiority of the alternatives in the results of the ELECTRE and TOPSIS methods are found to be the same. For this reason, it is concluded that the two methods support each other. In the study of selection a supplier for a telecommunication company AHP and TOPSIS methods are used together [9]. Firstly, criteria are determined by taking the opinions of the managers and a decision making hierarchy is established. Criteria's wights are obtained from AHP method. After AHP, the order of importance of the criteria has been determined. Product quality is determined as the most important criterion at the end of the work. Then the most suitable supplier is selected from four alternatives by using TOPSIS method. After the TOPSIS method is applied, the ranking of the supplier companies is made, and Company1 is chosen as the best supplier. Selecting a program to be used by the IT department of a university to develop Enterprise Project Management Software is aimed in [1]. For this purpose, AHP and TOPSIS methods are used together. The weights of the criteria are determined by using the AHP method, and the tools of various companies are evaluated by the TOPSIS method. As a result, the most useful EPM tool for the institution is determined within the alternatives. It can be seen that the selection of the software that can be used within other public institutions and private sector enterprises can be done by going out of this study. Reference [10] aimed to simplify group decision-making by using both using fuzzy extension of AHP and Fuzzy TOPSIS and its implementation for selecting a software in an electronic firm. Precedence values of criteria is calculated by using fuzzy extension of AHP method as the first step of the Project. The researcher reports little computation time and simplicity as the reason of the using this method. As the second step, the fuzzy TOPSIS is applied to rank the software selection process according to four software alternative criteria to be completed and identify the most suitable alternative for the firms's objectives. As a result, the weights of criteria are determined for purchasing fee, software reliability and references, respectively in descending order. Alternative 1 is chosen as the most appropriate alternative. Reference [11] aimed to decide the most suitable production alternative that can be produced from raw fabric, dyed fabric and curtain fabric alternatives with AHP method. The criterion and the sub-criteria, which effect the decision in the study, are determined with the business owners and managers with the help of pairwise comparison matrix. As a result of the study, it is seen that the most suitable production alternative for the operation is the production of curtain fabric. As the highest preference should be selected from defined criteria and sub criteria for enterprise production of curtain fabric had the highest priority in terms of salability, profitability, labor productivity and machine productivity.

4 Problem Definition

This study is conducted with the second largest retail company in Turkey's textile sector. Founded in 2003 and opened its first store in 2004, the company has taken its place among the leading brands in apparel and fashion sector in Turkey in a short time, such as 11 years. Today, with over 305 stores abroad has been operating in the retail textile sector in Turkey. In the short term, the experience and rapid growth acceleration have encouraged the company to open up to foreign markets also. In addition to its design products, the company, which has been deceived by innovation and R&D investments, offers its customers differentiated products in the area of innovation and recycling.

Developments in communication and information technologies have provided the company to transfer works in many areas to electronic environment. This leads to the emerge of different software that provides business integrity in companies with many features for different departments. However, administrators should select a software meeting all the requirements most for their business. There are many problems in the current software used in the process management department that does not cover entire process flows, which leads to difficulty in monitoring and tracking the processes. The other problem is current software does not have capability to communicate with other softwares. This causes lack of communication and inconvenient flow for all processes between departments. Moreover, the idle time spent in communication is redundant. There is a need for a software which is fully integrated with other programs. It is aimed that, with such a software, faults are reduced, tasks are getting faster, problems in the business process can be noticed in a short time and result in reduction of the operational costs by integrated programs. The new information system is needed to increase the productivity of users and to enable them to access the data or transaction pages they need more quickly. In addition, it needs to reduce overall bottlenecks and slowdowns in business processes, thereby increasing overall business efficiency. Process-oriented solutions are needed to work in a fast and consistent manner between executives and employees.

5 Methodology and Implementation

In this study, the problem of process management software selection is solved by using AHP and Fuzzy TOPSIS methods respectively. The methodology part of this study consists of 3 stages as shown in Fig. 2.

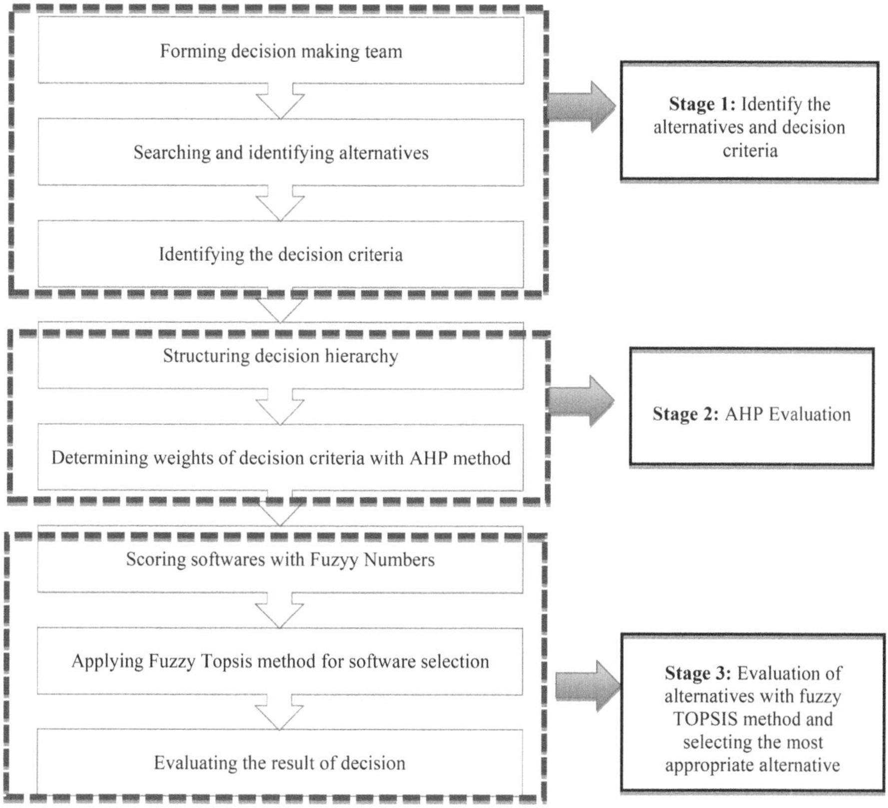

Fig. 2. Process map of the methodology

In the first stage, a decision-making team is formed from the employees working in the department to be used in the software to be selected. With the help of the decision making team, the criteria and softwares to be used in the methods have been determined. In the second stage, the weights of predetermined criteria are calculated using the AHP method. A hierarchical structure is established that included criteria and alternatives for use in the AHP method. In the last stage, linguistic expressions are used when evaluating alternatives. Alternative softwares are sorted and the most appropriate alternative is selected by the fuzzy TOPSIS method.

5.1 Identifying the Alternatives and Decision Criteria

In the first step, a decision-making team is formed from the employees working in the department to be used for decision making. With the help of the decision making team, the decision criteria and alternative softwares to be used in the methods are determined. Within the scope of the study, a team of decision makers within the organization is established primarily to determine the selection criteria for process management software. This team, is a group of 7 people consisting of employees of project management and process development department of the company. In this study, the four criteria that are used to determine the criteria that can be used in process management software selection according to the needs of the business are introduced to the decision makers for their evaluation. As a result of the negotiations, five process management software and four decision criteria such as 'Cloud', 'Price', 'User friendly', and 'Platform' are decided for the implementation. Cloud refers to the ability to access an online site with a certain amount of memory, even if the software is installed in the device, via the Internet with devices such as phones, computers and tablets. Platform criteria refers to operating systems such as IOS and Android, where the software to be evaluated is supported. With the price criterion, the installation fee of the software to be selected is expressed. The user friendly feature is a general notion of functionality and consistency of the icons, the software informing the user, the software being intuitive, error warning and effective error reduction.

5.2 AHP Evaluation

AHP is based on a decision hierarchy using a predefined scale of pairwise comparison of the criteria and affect importance value of point decision in terms of these factors. With the proposed approach the aggregated preferences assist the group members to understand each other and reach consensus [12]. In the AHP method, a hierarchical structure is first established to solve the problem. Then, the consistency of the values in the matrix is checked by determining the consistency ratio. If the consistency rate is acceptable, the order of priorities of the alternatives is accepted.

The first step of the AHP is the hierarchical organization of the decision problem. The choice of a hierarchical structure for the solution of problems means that the problem is divided into several levels. This structure is called modeling and forms a hierarchy to analyze the decision.

The second is the creation of a pairwise comparative matrix and the relative importance level of each criteria is calculated by pairwise comparisons, that is, comparing two criteria with each other. The decision maker sets the importance level between the criteria based on the scale 1–9, as given in Table 1. The results of the comparison are defined as integers ranging from 1 to 9. Here, the criteria having a higher integer value means that it is more important than the criteria having a smaller integer value than itself.

The calculation of the weights of the criteria should be made after the comparison matrix is created and the criteria are given numerical values. The comparison matrix is used to calculate the significance weights among the criteria after being used in the numerical values reported in Table 1.

Table 1. Nine-point intensity important scale [5]

Intensely of importance	Definition
1	Equal importance
3	Moderate importance of one factor over another
5	Strong or essential importance
7	Very strong importance
9	Extreme importance
2, 4, 6, 8	Values for inverse comparison

Let C_j be the set of criteria which is $j = 1,2,...,n$. The comparison matrix C gives the pairwise comparison of the criteria in the hierarchy. Matrix A consists of elements a_{ij} which are the results of binary comparisons of criteria where $i = 1,2,....,n;\ j = 1,2,....,n$. The matrix A is as shown below:

$$A = \begin{bmatrix} a_{11} & \cdots & a_{12} & \cdots & a_{1n} \\ \vdots & & \vdots & & \vdots \\ a_{21} & \cdots & a_{22} & \cdots & a_{2n} \\ \vdots & & \vdots & & \vdots \\ a_{n1} & \cdots & a_{n2} & \cdots & a_{nn} \end{bmatrix} a_{ii=1},\ a_{ji} = 1/a_{ij},\ a_{ij} \neq 0 \quad (1)$$

As the weight is based on bi-directional comparisons, the criteria can be weighted more dependably because comparison of the two features are easier than overall assignment. Before computing the vector of priorities, normalization of the comparison matrix must be in the range [0, 1]. The normalized matrix R is shown below;

$$R = \begin{bmatrix} r_{11} & r_{12} & \cdots & r_{1n} \\ r_{21} & r_{22} & \cdots & r_{2n} \\ \vdots & \vdots & & \vdots \\ r_{n1} & r_{n2} & \cdots & r_{nn} \end{bmatrix} \quad (2)$$

The comparison matrix includes pairwise comparisons of the criteria in the hierarchical structure. The purpose is to detect their relative priorities relating t o every single of the elements at the further higher step.

Step 3: Determination of Relative Weight of Criteria and Calculation of the Ratio

Once the eigenvector (W) has been computed and the relative significance ratings of the criteria have been determined, the consistency (CR) of the comparison matrix must be calculated. The eigenvector is calculated using the following formula:

$$w_i = \frac{1}{n}\sum_{j=1}^{n} \frac{a_{ij}}{\sum_{ji=1}^{n} a_{ij}} \quad (3)$$

The goal is to determine whether the decision maker is consistent when making comparisons between the criteria. If the CR is greater than 0.10, the decision maker has to repeat the values that it has entered due to the inconsistency. The closer the CR is to the zero, the higher the consistency of the decision matrix. The following formula is used to calculate the consistency of the comparison matrix.

$$CR = \frac{CI}{RI} \tag{4}$$

$$CI = \frac{\lambda_{max} - n}{n - 1} \tag{5}$$

The maximum eigenvalue (λ_{max}) in the formula used in calculating the CI is calculated using the below given formula:

$$\lambda_{max} = \frac{1}{n}\sum_{i=1}^{n} \frac{(AW)_i}{W_i} \tag{6}$$

If A is consistent the λ_{max} and the rank of matrix A is equal to n. In this instance, the relative criteria can be discussed. Any row or column of matrix A must be normalized to calculate the weights of each criterion. The consistency is provided when the following condition is satisfied.

$a_{ij} \,.\, a_{jk} = a_{ik}$ ($\forall$ i, j, k),
$\lambda_{max} = n$ and
CI = 0.

The columns of the comparison matrix are then multiplied by the relative priorities, and a weighted sum vector is constructed. After the elements of the weighted total vector are divided into their corresponding relative priority, the result gives the arithmetic mean λ_{max}. Consistency may be used to assess the consistency of decision makers or hierarchies. If $\lambda_{max} > n$, level of inconsistency needs to be measured. A matrix is accepted as consistent if CR < 0.1. RI ratios according to matrix size are shown in Table 2.

Table 2. The relationship between RI value and count of criteria [13]

n	1	2	3	4	5	6	7
RI	0.00	0.00	0.58	0.90	1.12	1.24	1.32
n	**9**	**10**	**11**	**12**	**13**	**14**	**15**
RI	1.45	1.49	1.51	1.48	1.56	1.57	1.59

There is a difficulty in group decision making about how to construct a group decision from individual choices. There are seven decision makers in this study and, the aggregation of individual preferences is done at the last stage. Initially, from each given comparison matrix the priority weights are obtained as the decision makers' individual preferences. A survey is conducted to compare each criteria in order to determine the expectations of the decision-making team. The results of the questionnaires are entered into the program called "Super Decision" for pairwise comparison of criteria and the weight and inconsistency ratios of each criteria are obtained for each participant. Then, they are aggregated based on the average with the importance weights of the group members as given in Table 3. The consistency ratio of the binary comparison matrix is calculated as $0.0844 < 0.1$. The weights are consistent since the reliability ratio is smaller than 0.1.

Table 3. Aggregated results obtained with AHP

Criteria	W	λmax	CI	RI	CR
Cloud	0.14186	4.22788	0.07596	0.90	0.0844
Platform	0.15236				
Price	0.16988				
User Friendly	0.53589				

5.3 Evaluation of Alternatives with Fuzzy TOPSIS Method and Selecting the Most Appropriate Alternative

In many cases, while people are evaluating criteria, they may not want or are not able to give clear or numeric values. In uncertain situations where net values can not be used, people express themselves with linguistic variables. Then, they can be converted to numerical values with fuzzy numbers [14]. In cases where linguistic variables are preferred to numerical variables, the classical TOPSIS method can not used, but the fuzzy TOPSIS method can used. The method developed by Hwang and Yoon [15], is a fuzzy TOPSIS method which is one of the multi criteria decision making methods. The values represented by the linguistic variables with the fuzzy TOPSIS method can be evaluated by those who know to evaluate these variables.

A fuzzy set is a set of elements that have no definite bounds and that have certain membership grades. In the definition of membership functions, the neighbors of numbers are used, and they are generally represented by triangular membership functions and trapezoidal membership functions, generally according to the condition of this neighborhood. Triangular membership functions are preferred in terms of ease of calculation. A fuzzy triangular number can be expressed as $\tilde{\alpha} = (a_1, a_2, a_3)$ as shown in Fig. 3.

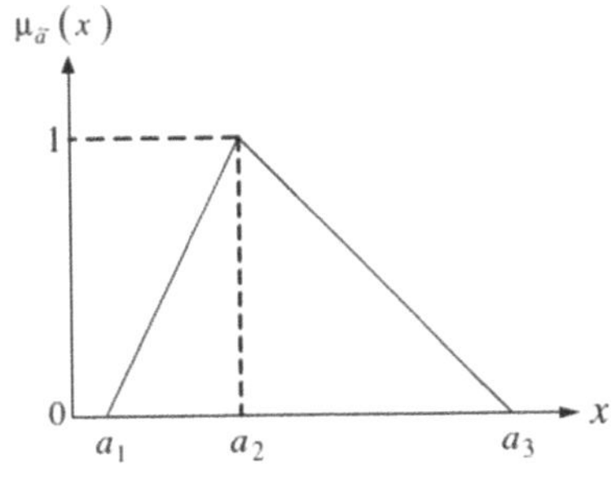

Fig. 3. A triangular fuzzy number $\tilde{\alpha}$. (Amiri, 2010)

The steps of Fuzzy TOPSIS Method are as follows;

Step 1: Creation of the decision matrix ($\tilde{A}_{ij}$)

$$\tilde{A}_{ij} = \begin{bmatrix} \tilde{a}_{11} & \cdots & \tilde{a}_{12} & \cdots & \tilde{a}_{1n} \\ \vdots & & \vdots & & \vdots \\ \tilde{a}_{21} & \cdots & \tilde{a}_{22} & \cdots & \tilde{a}_{2n} \\ \vdots & & \vdots & & \vdots \\ \tilde{a}_{m1} & \cdots & \tilde{a}_{m2} & \cdots & \tilde{a}_{mn} \end{bmatrix} \quad (7)$$

M and n represent the evaluation criteria and the number of alternatives in the matrix A_{ij}, respectively. The matrix elements show the performance levels of the software alternatives to the jth criteria. Determine the linguistic values (i = 1, 2… m; j = 1, 2… n) for the alternatives according to the criteria. (i is the criteria, j is to produce alternatives.)

Step 2: Normalization of Fuzzy decision matrix

When the A_{ij} decision matrix is normalized, a normalized decision matrix called as R_{ij}.

$$\tilde{R}_{ij} = \begin{bmatrix} \tilde{r}_{11} & \cdots & \tilde{r}_{12} & \cdots & \tilde{r}_{1n} \\ \vdots & & \vdots & & \vdots \\ \tilde{r}_{21} & \cdots & \tilde{r}_{22} & \cdots & \tilde{r}_{2n} \\ \vdots & & \vdots & & \vdots \\ \tilde{r}_{m1} & \cdots & \tilde{r}_{m2} & \cdots & \tilde{r}_{mn} \end{bmatrix} \quad (8)$$

where

$$\tilde{r}_{ij} = \left(\frac{a_{ij}}{c_j^+}, \frac{b_{ij}}{c_j^+}, \frac{c_{ij}}{c_j^+}\right) \quad \text{and} \quad c_j^+ = maxc_{ij} \text{ (Benefit criteria)} \quad (9)$$

$$\tilde{r}_{ij} = \left(\frac{a_j^-}{c_{ij}}, \frac{a_j^-}{c_{ij}}, \frac{a_j^-}{c_{ij}}\right) \quad \text{and} \quad a_j^- = maxc_{ij} \quad \text{(Cost criteria)} \quad (10)$$

The characterization of normalized triangular fuzzy numbers by the above-mentioned normalization method is preserved between [0, 1].

Step 3: Calculation of the weighted normalized matrix ($\tilde{v}$)

After the normalized decision matrix Rij is constructed, weights of each criteria are processed to form a weighted normalized decision matrix. The weighted normalized fuzzy decision matrix is constructed as follows:

$$\tilde{v}_{ij} = \begin{bmatrix} w_1\tilde{r}_{11} & \cdots & w_2\tilde{r}_{12} & \cdots & w_n\tilde{r}_{1n} \\ \vdots & & \vdots & & \vdots \\ w_1\tilde{r}_{21} & \cdots & w_2\tilde{r}_{22} & \cdots & w_n\tilde{r}_{2n} \\ \vdots & & \vdots & & \vdots \\ w_1\tilde{r}_{m1} & \cdots & w_2\tilde{r}_{m2} & \cdots & w_n\tilde{r}_{mn} \end{bmatrix} \quad (11)$$

$i = 1, 2, \ldots, m$; $j = 1, 2, \ldots, n$ where $\tilde{v}_{ij} = \tilde{r}_{ij}(x)\, w_j$

Step 4: Describing Ideal (A^+) and Negative Ideal (A^-) Solutions

V_{ij}, is a normalized positive triangular fuzzy numbers according to the weighted normalization decision matrix, which are in the [0, 1] closed interval.

Equations (12) and (13) describe the positive ideal solution and the negative ideal solution points of fuzzy TOPSIS.

$$A^+ = \{\tilde{v}_1^+, \tilde{v}_2^+, \ldots, \tilde{v}_n^+\} \quad \text{where} \quad \tilde{v}_j^+ = (1, 1, 1) \quad (12)$$

$$A^- = \{\tilde{v}_1^-, \tilde{v}_2^-, \ldots, \tilde{v}_n^-\} \quad \text{where} \quad \tilde{v}_j^- = (0, 0, 0) \quad (13)$$

Step 5: Calculating the Distance Size

In Eqs. (14) and (15), the distance (d_i^+, d_i^-) of each weighted alternatives from FPIS and FNIS are calculated.

$$d_i^+ = \sum\nolimits_{j=1}^{n} dv((\tilde{v}_{ij}, \tilde{v}_j^+) \quad (14)$$

$$d_j^+ = \sum\nolimits_{j=1}^{n} dv((\tilde{v}_{ij}, \tilde{v}_j^-) \quad (15)$$

$$i = 1, 2, \ldots, m$$

If $\tilde{v}_{ij} = \left(\tilde{a}_{ij}, \tilde{b}_{ij}, \tilde{c}_{ij}\right)$ and $\tilde{v}_j^+ = (1, 1, 1)$ and $\tilde{v}_j^- = (0, 0, 0)$

The distance between two fuzzy numbers $dv\left(\tilde{a}, \tilde{b}\right)$, $a\,and\,b$ given here by the following equation.

$$dv\left(\tilde{v}_{ij}, \tilde{v}_j^+\right) = \sqrt{\frac{1}{3}\left[(\tilde{a}_{ij} - 1)^2 + (\tilde{b}_{ij} - 1)^2 + (\tilde{c}_{ij} - 1)^2\right]}, \quad (16)$$

$$dv\left(\widetilde{v}_{ij}, \widetilde{v}_j^-\right) = \sqrt{\frac{1}{3}\left[(\widetilde{a}_{ij} - 0)^2 + (\widetilde{b}_{ij} - 0)^2 + (\widetilde{c}_{ij} - 0)^2\right]}. \tag{17}$$

Step 6: Calculate Similarities to Ideal Solution

The approximation coefficients of each alternative are calculated to specify the order of the alternatives as follows.

$$CC_i = \frac{d_i^-}{d_i^- + d_i^+} \qquad \mathrm{i} = 1, 2, \ldots, \mathrm{m} \tag{18}$$

Step 7: Sorting Alternatives

The order of the alternatives is determined on the basis of calculated coefficient and the most suitable one is selected.

In this study, decision makers evaluated each software separately according to each criterion with a questionnaire supplied with the scale given in Table 4.

Table 4. Fuzzy numbers used in the implementation

Linguistic variables	Fuzzy numbers
Very low (VL)	(0, 0, 0.25)
Low (L)	(0, 0.25, 0.5)
Medium (M)	(0.25, 0.5, 0.75)
Good (G)	(0.5, 0.75, 1)
Very good (VG)	(0.75, 1, 1)

The decision-making team assessed the criteria set using linguistic variables. The conversion of linguistic variables to fuzzy numbers is shown in Table 5. And the results of the Fuzzy TOPSIS implementation of the study is summarized in the Tables 6, 7, 8, 9, 10 and 11.

The A_{ij} fuzzy decision matrix is normalized using Eq. (11). Each alternative is considered a benefit criterion. The normalization performed is shown in Table 7.

The normalized fuzzy decision matrix is multiplied by the weight values of the criteria. The weight values of the criterion are shown in Table 3. As a result of the multiplication result, a weighted normalized decision matrix is obtained. The weighted normalized decision matrix obtained is given below, in Table 8.

After the weighted normalized decision matrix is constructed, the Eqs. (12) and (13) are used to determine the positive and negative ideal solutions. The positive ideal solution by the fuzzy method $\widetilde{v}_j^+ = (1, 1, 1)$ while the negative ideal solution is defined by the formula $\widetilde{v}_j^- = (0, 0, 0)$.

Table 5. All ratings of the decision makers in triangular fuzzy numbers

Criteria	Softwares	DM1	DM2	DM3	DM4	DM5	DM6	DM7
Price	Process street	(0, 0, 0.25)	(0.25, 0.5, 0.75)	(0.25, 0.5, 0.75)	(0, 0, 0.25)	(0, 0, 0.25)	(0, 0, 0.25)	(0.25, 0.5, 0.75)
	Bizagi	(0, 0.25, 0.5)	(0, 0.25, 0.5)	(0.25, 0.5, 0.75)	(0.25, 0.5, 0.75)	(0, 0, 0.25)	(0, 0.25, 0.5)	(0, 0, 0.25)
	Comidor	(0.5, 0.75, 1)	(0.5, 0.75, 1)	(0.5, 0.75, 1)	(0.25, 0.5, 0.75)	(0, 0, 0.25)	(0.5, 0.75, 1)	(0, 0.25, 0.5)
	Ms Visio	(0.5, 0.75, 1)	(0.5, 0.75, 1)	(0.25, 0.5, 0.75)	(0.25, 0.5, 0.75)	(0, 0.25, 0.5)	(0.5, 0.75, 1)	(0, 0.25, 0.5)
	Simul8	(0.2S, 0.5, 0.75)	(0, 0, 0.25)	(0, 0.25, 0.5)	(0, 0.25, 0.5)	(0.75, 1, 1)	(0.5, 0.75, 1)	(0, 0, 0.25)
Platform	Process street	(0.5, 0.75, 1)	(0.5, 0.75, 1)	(0.25, 0.5, 0.75)	(0.25, 0.5, 0.75)	(0.25, 0.5, 0.75)	(0.5, 0.75, 1)	(0, 0, 0.25)
	Bizagi	(0.5, 0.75, 1)	(0.25, 0.5, 0.75)	(0.25, 0.5, 0.75)	(0.25, 0.5, 0.75)	(0.25, 0.5, 0.75)	(0.5, 0.75, 1)	(0.5, 0.75, 1)
	Comidor	(0.5, 0.75, 1)	(0.25, 0.5, 0.75)	(0.5, 0.75, 1)	(0.5, 0.75, 1)	(0.75, 1, 1)	(0.5, 0.75, 1)	(0.5, 0.75, 1)
	Ms Visio	(0, 0.25, 0.5)	(0, 0.25, 0.5)	(0.5, 0.75, 1)	(0.5, 0.75, 1)	(0, 0.25, 0.5)	(0.5, 0.75, 1)	(0, 0.25, 0.5)
	Simul8	(0.5, 0.75, 1)	(0.5, 0.75, 1)	(0.25, 0.5, 0.75)	(0.25, 0.5, 0.751	(0, 0.25, 0.5)	(0.25, 0.5, 0.75)	(0.5, 0.75, 1)
User Friendly	Process street	(0.06, 0.31, 0, 56)	(0.44, 0.69, 0.94)	(0.5, 0.75, 1)	(0.25, 0.50, 0.75)	(0, 0.13, 0.38)	(0.06, 0.25, 0.50)	(0.25, 0.50, 0.75)
	Bizagi	(0.25, 0.5, 0, 75)	(0.38, 0.63, 0.88)	(0.31, 0.56, 0.81)	(0.19, 0.44, 0.69)	(0.25, 0.50, 0.69)	(0.31, 0.56, 0.81)	(0.25, 0.50, 0.75)
	Comidor	(0.25, 0.5, 0.75)	(0.44, 0.69, 0.81)	(0.38, 0, 63, 0.88)	(0.19, 0.44, 0.69)	(0.25, 0.44, 0.63)	(0.25, 0.50, 0.75)	(0.13, 0.38, 0.63)
	Ms Visio	(0.56, 0.81, 1)	(0.38, 0.63, 0.81)	(0.56, 0.81, 0.94)	(0.13, 0.38, 0.63)	(0.50, 0.75, 0.88)	(0.44, 0.69, 0.94)	(0.50, 0.75, 1)
	Simul8	(0.38, 0.63, 0.88)	(0.63, 0, 88, 1)	(0.31, 0.56, 0.81)	(0.19, 0.44, 0.69)	(0.38, 0.56, 0.75)	(0.56, 0.81, 1)	(0.50, 0.75, 1)
CLOUD	Process street	(0.25, 0.5, 0.75)	(0.5, 0.75, 1)	(0.25, 0.5, 0.75)	(0.25, 0.5, 0.75)	(0.5, 0.75, 1)	(0.25, 0.5, 0.75)	(0.5, 0.75, 1)
	Bizagi	(0.5, 0.75, 1)	(0.5,0.75,1)	(0.5, 0.75, 1)	(0.5, 0.75, 1)	(0.5, 0.75, 1)	(0.5, 0.75, 1)	(0.5, 0.75, 1)
	Comidor	(0.5, 0.75, 1)	(0.5,0.75,1)	(0.5, 0.75, 1)	(0.25, 0.5, 0.75)	(0.75, 1, 1)	(0.75, 1, 1)	(0.5, 0.75, 1)
	Ms Visio	(0.75, 1, 1)	(0.5,0.75,1)	(0.25, 0.5, 0.75)	(0.25, 0.5, 0.75)	(0, 0.25, 0.5)	(0.5, 0.75, 1)	(0.5, 0.75, 1)
	Simul8	(0.5, 0.75, 1)	(0.25,0.5,0.75)	(0.25, 0.5, 0.75)	(0, 0.25, 0.5)	(0.25, 0.5, 0.751	(0.5, 0.75, 1)	(0.5, 0.75, 1)

Table 6. Fuzzy decision matrix

Fuzzy decision matrix					
	Process street	Bizagi	Comidor	MS Visio	Simul8
Price	(0.107, 0.214, 0.464)	(0.071, 0.250, 0.500)	(0.321, 0.536, 0.786)	(0.286, 0.536, 0.786)	(0.214, 0.393, 0.607)
Platform	(0.321, 0.536, 0.786)	(0.357, 0.607, 0.857)	(0.500, 0.750, 0.964)	(0.214, 0.464,0.714)	(0.321, 0.571,0.821)
Cloud	(0.357, 0.607, 0.857)	(0.500, 0.750, 1)	(0.536, 0.786, 0.964)	(0.393, 0.643, 0.857)	(0.321, 0.571, 0.821)
User Friendly	(0.223, 0.447, 0.697)	(0.277, 0.527, 0.769)	(0.270, 0.511, 0.734)	(0.439, 0.689, 0.886)	(0.421, 0.661, 0.876)

$$A^* = ((1,1,1),\ (1,1,1),\ (1,1,1),\ (1,1,1))$$

$$A^- = ((0,0,0),\ (0,0,0),\ (0,0,0),\ (0,0,0))$$

Fuzzy positive ideal solution and fuzzy negative ideal solution's distances are calculated with using Eqs. (14) and (15). The results of each alternative are summed up to include distances between the positive ideal and negative ideal solutions of the alternatives. An example for Process Street is displayed below, and all results are listed in Table 9.

$$\text{d(Process Street.}\,A^*) = \sqrt{\frac{1}{3}\left[(1-0.023)^2 + (1-0.046)^2 + (1-0.100)^2\right]} = 0.944$$

$$\text{d(Process Street.}\,A^-) = \sqrt{\frac{1}{3}\left[(0-0.023)^2 + (0-0.046)^2 + (0-0.100)^2\right]} = 0.065$$

After determining the distances, closeness coefficients are calculated for each alternative with the help of Eq. (18) as the final step. Table 11 shows the distances from FPIS and FNIS and the closeness coefficients obtained using these distances. An example for Process Street is displayed below, and the results of all calculations are given in Table 11.

$$d^-(\text{Process Street}) = 0.065 + 0.092 + 0.091 + 0.300 = 0.547$$

$$d^*(\text{Process Street}) = 0.944 + 0.914 + 0.914 + 0.734 = 3.506$$

$$CC_i(\text{Process Street}) = \frac{0.547}{(0.547 + 3.506)} = 0.135$$

The results showed that MS Visio is selected the best alternative with a closeness coefficient of 0.178 value. Simul 8 and Comidor Follows it with very close values as 0.170 and 0.169 respectively. Hence, it can be interpreted that the results are very sensitive to the decision makers preferences. Small changes in the weights or an additional decision maker can alternate the decision. The decision rank of the softwares according to the results is also given in Table 11.

Table 7. Normalized fuzzy-decison matrix

Normalized fuzzy-decison matrix					
	Process street	Bizagi	Comidor	MS Visio	Simul8
Price	(0.136, 0.272, 0.590)	(0.09, 0.318, 0.636)	(0.408, 0.682, 1)	(0.364, 0.682, 1)	(0.272, 0.500, 0.772)
Platform	(0.333, 0.556, 0.815)	(0.37, 0.63, 0.889)	(0.519, 0.778, 1)	(0.222, 0.481, 0.741)	(0.333, 0.592, 0.852)
Cloud	(0.357, 0.607, 0.857)	(0.500, 0.750, 1)	(0.536, 0.786, 0.964)	(0.393, 0.643, 0.857)	(0.321, 0.571,0.821)
User Friendly	(0.252, 0.505, 0.787)	(0.313, 0.595, 0.868)	(0.305, 0.577, 0.828)	(0.495, 0.778, 1)	(0.475, 0.746, 0.989)

Table 8. Weighted normalized fuzzy decision matrix

Weighted normalized fuzzy decision matrix					
	Process street	Bizagi	Comidor	MS Visio	Simul8
Price	(0.023, 0.046, 0.100)	(0.015, 0.054, 0.108)	(0.069, 0.116, 0.170)	(0.062, 0.116, 0.170)	(0.046, 0.085, 0.131)
Platform	(0.051, 0.085, 0.124)	(0.056, 0.096, 0.135)	(0.079, 0.119, 0.152)	(0.034, 0.073, 0.113)	(0.051, 0.090, 0.130)
Cloud	(0.051, 0.086, 0.122)	(0.071, 0.106, 0.142)	(0.076, 0.112, 0.137)	(0.056, 0.091, 0.122)	(0.046, 0.081, 0.116)
User Friendly	(0.135, 0.271, 0.422)	(0.168, 0.319, 0.465)	(0.163, 0.309, 0.444)	(0.265, 0.417, 0.536)	(0.255, 0.400, 0.530)

Table 9. Distances between each Alternatives and A^* with respect to each criterion

	Price	Platform	Cloud	User friendly
d(Process Street. A^*)	0.944	0.914	0.914	0.734
d(Bizagi. A^*)	0.942	0.905	0.894	0.693
d(Comidor. A^*)	0.883	0.884	0.892	0.704
d(Ms Visio. A^*)	0.885	0.927	0.911	0.604
d(Simul8. A^*)	0.913	0.910	0.919	0.616

Table 10. Distances between each Alternatives and A^- with Respect to each Criterion

	Price	Platform	Cloud	User friendly
d(Process Street. A^-)	0.065	0.092	0.091	0.300
d(Bizagi. A^-)	0.070	0.101	0.110	0.340
d(Comidor. A^-)	0.125	0.120	0.111	0.326
d(Ms Visio. A^-)	0.124	0.080	0.093	0.421
d(Simul8. A^-)	0.094	0.096	0.086	0.410

Table 11. d_i values and closeness coefficients

	d^-	d^*	CC_i	Rank
Process Street	0.547	3.506	0.135	5
Bizagi	0.621	3.434	0.153	4
Comidor	0.683	3.363	0.169	3
Ms Visio	0.718	3.328	0.178	1
Simul8	0.686	3.359	0.170	2

6 Conclusion

Today, decision makers face many complex situations. In such complex situations, the evaluations of decision makers are often uncertain. Hence various decision making approaches are used to reach the best alternative. In this study, a software selection problem for a textile retailer is studied is using AHP and fuzzy TOPSIS methods. While the weights of the software criteria are determined by the AHP method, the ranking of the selected alternatives is done by the fuzzy TOPSIS. In AHP stage, decision makers make pairwise comparisons to give weights for the identified decision criteria, which are price, cloud, user friendly and platform. According to AHP results, the highest weight is given to 'user friendly' criteria. The order of importance of the other criteria are as follows; price, platform and cloud respectively. In the following stage, fuzzy TOPSIS method is applied with the fuzzy evaluations of the decision-making team. The linguistic expressions such as very good, good, medium, low and very low are used to evaluate the five software alternatives with respect to four criteria, and then the evaluations are translated to the fuzzy numbers. The normalized fuzzy decision matrix and the weighted normalized decision matrix are constructed. The fuzzy positive ideal solution and the fuzzy negative ideal solution points are defined and the distance and proximity coefficient of each alternative is calculated. Consequently, the Fuzzy Topsis method shows Ms Visio as the best option among the alternatives due to its highest closeness coefficient calculated. The remaining alternatives Simul8, Comidor, Bizagi and Process Street are ranked as the second, third, fourth and fifth alternatives respectively.

The proposed two-stage fuzzy-TOPSIS approach needs to be applied systematically in accordance with the structure of firms. In addition to being convenient to implement the proposed method, there are some limitations. Because the method can evaluate both qualitative and quantitative factors, the effectiveness of the model is dependent on the ability to obtain clear and accurate information from decision makers. For this reason, the weight of decision makers, and the criteria must be determined objectively. In addition, the specified criteria and their weight must reflect and be consistent with the decision of the firms. On the other hand, decision makers may want to evaluate more criteria and alternatives for process software selection. In this case, the solution of the problem will require more processing and effort. The study provides a guided approach that researchers and practitioners can use in strategic decision making processes. At the same time, the two-stage fuzzy-TOPSIS approach presented can be used as a resource for managers to handle complex decision problems such as process software selection.

As a final remark, it should be noted that, MS Visio is very close in the ranking to two alternatives, with a closeness coefficient of 0.178 value. Simul 8 and Comidor follow it with very close values, such as 0.170 and 0.169 respectively. If more than seven decision makers were used in the study, the methodology could lead to more robust results. Hence, it can be interpreted that the results are very sensitive to the decision makers preferences. Small changes in the weights or an additional decision maker can alternate the decision. According to the results of AHP method, only the userfriendly criteria has a high weight, and the others have close values. Consequently, future studies are recommended for repeating the decision making with different methods and/or more decision makers. It seems that the choice of software that can be used in other special retail businesses can be done by going out of this study. This study presents a guiding approach, and can be used by institutions developing software. The presented two-stage AHP-fuzzy TOPSIS approach can also be used as a resource for similar businesses for similar decision problems.

Acknowledgment. We want to thank to DeFacto Turkey's Process and Project Management Department, all the personnel that act as decison maker, and especially Nurcan DÖNMEZ for the support at the implementation of this study.

References

1. Ömürbek N, Makas Y, Ömürbek V (2015) AHP ve Topsis yöntemleri ile kurumsal proje yönetim yazılımı seçimi. Süleyman Demirel Üniversitesi Sosyal Bilimler Enstitüsü Dergisi, pp 59–80
2. Dağdeviren M, Akay D, Kurt M (2004) İş Değerlendirme Sürecinde Analitik Hiyerarşi Prosesi ve Uygulaması. Gazi Üniversitesi Mühendislik Mimarlık Fakültesi Dergisi 19 (2):131–138
3. Aslan N (2005) Analitik Network Prosesi. Yüksek Lisans Tezi, Yıldız Teknik Üniversitesi, Fen Bilimleri Enstitüsü, İstanbul
4. Chen SJ, Hwang CL (1992) Fuzzy Multiple Attribute Decision Making Methods and Applications. Springer, Berlin
5. Amiri MP (2010) Project selection for oil-fields development by using the AHP and Fuzzy TOPSIS methods. Expert Syst Appl 37:6219. https://doi.org/10.1016/j.eswa.2010.02.103
6. Moayeri M, Shahvarani A, Behzadi M, Hosseinzadeh-Lotfi F (2015) Comparison of Fuzzy AHP and Fuzzy TOPSIS methods for math teachers selection. Indian J Sci Technol 8:1- 10
7. Karaatlı M, Nuri Ö, Aksoy E, Karakuzu H (2014) Turizm işletmeleri için AHP temelli bulanık TOPSIS yönetimi ile tur operatörü seçimi. Anadolu Univ J Soc Sci 14(2): 53–68
8. Tunca MZ, Aksoy E, Bülbül H, Ömürbek N (2015) Use of AHP-based TOPSIS and ELECTRE methods on accounting software selection. Niğde Üniversitesi İktisadi ve İdari Bilimler Fakültesi Dergisi 8:53–71
9. Günay Z, Ünal ÖF (2016) Selection of supplier with AHP and Topsis (sample of a telecommunication firm in Turkey). PESA Uluslararası Sosyal Araştırmalar Dergisi 2:37–53. http://dergipark.gov.tr/pesausad/issue/26079/274881
10. Efe B (2015) An integrated fuzzy multi criteria group decision making approach for ERP system selection. Appl Soft Comput: 106–117. https://doi.org/10.1016/j.asoc.2015.09.037
11. Başkaya Z, Akar C (2005) Ürün alternatifi seçiminde Analitik Hiyerarşi Süreci: Tekstil işletmesi örneği. Anadolu Üniversitesi Sosyal Bilimler Dergisi 5(1):273–286
12. Entani T (2009) Interval AHP for a group of decision makers. IFSA-EUSFLAT, pp 155–160
13. Saaty TL, Tran LT (2007) On the invalidity of fuzzifying numerical judgments in the Analytic Hierarchy Process. Mathematical and Computer Modelling, p 966
14. Zadeh LA (1975) The concept of a linguistic variable and its application to approximate reasoning. Inf Sci 8:199–249(I), 301–357(II)
15. Yoon K, Hwang CL (1981) Methods and Applications A State-of-the-Art Survey. Springer, Heidelberg

Fuzzy Logic

A Multimoora Method Application with Einstein Interval Valued Fuzzy Numbers' Operator

Hatice Camgöz-Akdağ, Gökhan Aldemir, and Aziz Kemal Konyalıoğlu(✉)

Management Engineering Department, Istanbul Technical University, Istanbul, Turkey
{camgozakdag, aldemirg, konyalioglua}@itu.edu.tr

Abstract. Multi Criteria Decision Making (MCDM) is a process to decide the best possible ranking between alternatives. In fuzzy sense, fuzzy MCDM methods aim to get closer the best alternative. Interval valued intuitionistic fuzzy sets can be applied in MCDM and they form many fuzzy methods. One of them is MULTIMOORA which tries to deal with lessening incertitude options. MULTIMOORA method becomes important because of its usage compared to other MCDM methods. In this paper, the data in which Baležentis and Zeng (2013) studied is used and GIFTNOWGA operator which Baležentis and Zeng studied is replaced by Einstein operator to investigate how choosing different operators affect the final ranking of alternatives.

Keywords: MCDM · Decision making · Fuzzy · Multimoora Operator

1 Introduction

Decision making has been an interesting topic throughout centuries. This interesting topic mainly comes from the change of point of view by comparison in order to decide which alternative or situation fits in the best way [1].

MCDM is an important topic in the 21st century because of evaluating alternatives. Alternatives can have different proprieties to grab them and they might also have different features. In decision making process, it is essential that an operative process should be implemented in order to have a qualified decision making [2]. As another perspective, MCDM can be defined as an approach containing subjective rankings and evaluation in order to select the best [3].

Apart from crisp valued- MCDM methods, many studies about fuzzy MCDM methods have been developed to optimize all given criteria [4]. To give an example, Yang et al. have used MCDM to decide the best alternative of vendors by using integrated fuzzy MCDM techniques and they assumed that all criteria are independent via subjective preferences of experts [5]. As another example, Liang et al. used fuzzy linguistic assessment to evaluate personals by putting subjective weightings and ratings of decision makers into effect by implementing not only subjective but also objective

N. M. Durakbasa and M. G. Gencyilmaz (Eds.): ISPR 2018, *Proceedings of the International Symposium for Production Research 2018*, pp. 171–178, 2019.
https://doi.org/10.1007/978-3-319-92267-6_13

assessments [6]. In the same area, Dursun and Karsak have used another fuzzy MCDM approach of 2 tuple linguistic model to evaluate personnel by the aim of dealing with operational heterogeneity [7].

In the fuzzy MCDM approaches, the operator which Is used is also important. For example, Chiclana et al. have used the Ordered Weighted Geometric Operator (OWG) to obtain an incorporation of fuzzy problems [8]. Bordogna and Pasi evaluated linguistic aggregation operators to designate of the aggregation criteria [9].

MULTIMOORA method is an essential method by the way of MCDM. Baležentis et al. have implemented fuzzy MULTIMOORA method to assign indicators and situation of Lithuania in the European Union [10].

According to the studies explained above, it is aimed to use fuzzy MULTIMOORA method which is generally used in fuzzy MCDM and change the operator to observe final rankings of using Einstein operator compared to GIFTNOWGA operator in which Baležentis and Zeng used.

2 Background of the Multimoora Method

The MULTIMOORA (Multi-Objective Optimization by Ratio Analysis plus Full Multiplicative Form) method is an alternative for multi-criteria decision-making techniques based on weighting criteria, which was developed by Brauers and Zavadskas in 2010 [11]. MULTIMOORA has been applied in various areas where alternatives need be compared. MULTIMOORA method originated MOORA method and Full Multiplicative Form method for Multi-Criteria Decision Making [12]. Therefore, MULTIMOORA method consists of MOORA-Ratio System, MOORA-Reference Point System and Full Multiplicative Form method. In this context, the steps of MOORA-Ratio System, MOORA-Reference Point, Full Multiplication methods take place in the following section.

The MOORA method has three important assumptions. These assumptions include the use of metrics, which imply counting number, the availability of alternatives that take a discrete value, and the well-defined objectives [13]. After meeting assumptions, a decision matrix of alternatives and criteria are established. The decision matrix is in Eq. (1), with n-criteria and m-alternatives. The vector normalization method requires normalizing the decision matrix in the MOORA-Ratio method. Equation (2) normalizes the values of each alternative based on the criterion to obtain x_{ij^*} values.

$$\mathrm{X} = \begin{pmatrix} x_{11} & \cdots & x_{1n} \\ \vdots & \ddots & \vdots \\ x_{m1} & \cdots & x_{mn} \end{pmatrix} \tag{1}$$

$$x_{ij^*} = \frac{x_{ij}}{\sqrt{\sum_{j=1}^{j=m} x_{ij}^2}} \tag{2}$$

These indicators are added if the desirable value of indicator is maximum and subtracted if desirable value is minimum. The summarizing index of each alternative is given in Eq. 3 [13].

$$y_j^* = \sum\nolimits_{i=1}^{g} x_{ij^*} - \sum\nolimits_{i=g+1}^{n} x_{ij^*} \tag{3}$$

The performance values y_j^* are ordered from small to large to rank the alternatives. For preference problems, the alternative with the highest performance value should be preferred. The MOORA-Reference Point method uses the normalized decision matrix obtained by the Eq. (2) in the MOORA-Ratio method. The best values of each criterion are the reference values of the alternatives. In other words, if the criterion is benefit-oriented, it is the greatest value and if it is cost-oriented, it is the lowest value reference point. Tchebycheff min-max metric method is used to measure the difference between the reference points of the alternatives. The distance value d_{ij} is obtained by Eq. (4), where r_i is the reference point. Equation (4) and the reference series have the distances from the reference point of the alternatives by taking the absolute value differences of the normalization decision matrix values obtained from Eq. (2). Then the maximum distance values of the alternatives are determined. These values are ordered from small to large for alternative sorting. If the problem is a preference problem, the alternative with the smallest distance value should be preferred [13].

$$d_{ij} = min_j(max_i(\left|r_i - x_{ij^*}\right|)) \tag{4}$$

The Full Multiplication method is used in nonlinear, non-additive, preference or sorting problems where weighting and normalization cannot be used. The utility and cost performance values of the alternatives included in the decision matrix are obtained by multiplying the benefit side criterion values by Eq. (5) and the cost side criterion values by the Eq. (6). In these equations g = 1..., i are cost-oriented criteria, and k = i + 1,..., n are cost-oriented criteria. The utility performance value is compared to the cost performance value and the general performance value of the alternate is obtained as in Eq. (7). The alternatives are sorted from small to large according to their general performance values [11].

$$A_j = \prod\nolimits_{g=1}^{i} x_{gi} \tag{5}$$

$$B_j = \prod\nolimits_{k=i+1}^{n} x_{ki} \tag{6}$$

$$U_j = \frac{A_J}{B_J} \tag{7}$$

3 Methodology

Zavadskas et al. (2015) give two numerical examples of real-world civil engineering problems and rank the alternatives based on the suggested method. Then, they compare the results to the rankings yielded by some other methods of decision making with IVIF information. The comparison shows the conformity of the proposed IVIF-MULTIMOORA method with other approaches. The proposed algorithm is favourable because of the abilities of IVIFS to be used for imagination of uncertainty and the MULTIMOORA method to consider three different viewpoints in analysing engineering decision alternatives [11].

Wu et al. (2018) propose a strongly robust method to solve multi-experts multi-criteria decision making problems with linguistic evaluations. To enrich the computation and to improve the measures of probabilistic linguistic term set, they firstly define an expectation function of it. In addition, they advance three kinds of probabilistic linguistic distance measures reflecting on the difference of linguistic terms and probabilities at the same time to make up for the defects of the existing distance measures, and then propose the similarity and correlation measures. Integrating the subjective opinions with the correlation coefficients between criteria, they put forward a combined weight determining method. The robustness of the ranking method, MULTIMOORA, is enhanced by the improved Borda rule. Based on these research findings, a probabilistic linguistic MULTIMOORA method is proposed. Finally, the developed method is applied to an empirical example concerning the selection of shared karaoke television brands. The effectiveness of the proposed method is verified by some comparative analysis [16].

Aytekin (2016) identifies the importance weights of the patients, which are effective in preference of the hospital and the hospitals located in the city center of Eskişehir were listed with MULTIMOORA as the most criterion decision making technique in these factors. As a result, while it is determined that the most effective criteria for selecting hospitals is the availability of all kinds of services and specialists, it is seen that the competition levels of public hospitals were at a level that could compete with private hospitals [17].

There are lots of operators used in fuzzy decision making such as GITFNWGA (Generalized Interval-Valued Trapezoidal Fuzzy Numbers Weighted Geometric Aggregation), GITFNOWGA (The Generalized Interval-Valued Trapezoidal Fuzzy Numbers Ordered Weighted Geometric Aggregation) and GITFNHGA (The Generalized Interval-Valued Trapezoidal Fuzzy Numbers Hybrid Geometric Aggregation). Although there are lots of studies about fuzzy decision-making methods in the literature, there can be found limited studies about MULTIMOORA Method, to that end studies relating to Einstein Operator aggregation with MULTIMOORA. Balezentis and Zeng (2012) extend the MULTIMOORA Method with type 2 fuzzy sets with GITFNOWGA operator to select best candidate for a manager position in an R&D department [14].

This study aggregates type 2 fuzzy numbers with Einstein Operator to search whether the ranking of candidates will remain same or different and compares results with the study of Balezentis and Zeng (2012) [14]. There are four candidates named

A_1, A_2, A_3, A_4 and three decision makers labeled as DM_1, DM_2, DM_3. Decision makers assess the four candidates based on the five benefit criteria, which are proficiency in identifying research ideas (C_1), proficiency in administration (C_2), personality (C_3), experience (C_4), and self-confidence (C_5) [14]. Table 1 shows linguistic variables and respective generalized interval-valued trapezoidal fuzzy numbers [15].

Table 1. Linguistic term generalized interval-valued trapezoidal fuzzy number (Wei and Chen (2009))

Absolutely poor (AP)	[(0.0, 0.0, 0.0, 0.0; 0.8), (0.0, 0.0, 0.0, 0.0; 1.0)]
Very poor (VP)	[(0.00, 0.00, 0.02, 0.07; 0.8), (0.0, 0.0, 0.02, 0.07; 1.0)]
Poor (P)	[(0.04, 0.10, 0.18, 0.23; 0.8), (0.04, 0.10, 0.18, 0.23; 1.0)]
Medium poor (MP)	[(0.17, 0.22, 0.36, 0.42; 0.8), (0.17, 0.22, 0.36, 0.42; 1.0)]
Medium (F)	[(0.32, 0.41, 0.58, 0.65; 0.8), (0.32, 0.41, 0.58, 0.65; 1.0)]
Medium good (MG)	[(0.58, 0.63, 0.80, 0.86; 0.8), (0.58, 0.63, 0.80, 0.86; 1.0)]
Good (G)	[(0.72, 0.78, 0.92,0.97;0.8), (0.72, 0.78, 0.92, 0.97; 1.0)]
Very good (VH)	[(0.93, 0.98, 1.0, 1.0;0.8), (0.93, 0.98, 1.0, 1.0; 1.0)
Absolutely good (AG)	[(1.0, 1.0, 1.0, 1.0; 1.0), (1.0, 1.0, 1.0, 1.0; 1.0)]

Table 2 shows the evaluation of each decision makers for four candidates. Then, matrix A converts the linguistic terms into the generalized interval valued trapezoidal fuzzy numbers by Einstein Operator aggregation. Since all the criteria are benefit criteria, so we do not need to normalize them.

Table 2. Linguistics variables

		C_1	C_2	C_3	C_4	C_5
DM_1	A1	VG	VG	VG	VG	VG
	A2	G	VG	VG	VG	MG
	A3	VG	MG	G	G	G
	A4	G	F	F	G	MG
DM_2	A1	G	MG	G	G	VG
	A2	G	VG	VG	VG	MG
	A3	G	G	MG	VG	G
	A4	VG	F	MG	F	G
DM_3	A1	MG	F	G	VG	VG
	A2	MG	MF	G	MG	G
	A3	VG	VG	VG	VG	MG
	A4	MG	VG	MG	VG	F

[A]=	[(0,997	0,999	1	1	0,8);	(0,997	0,999	1	1	1)]
	[(0,986	0,993	1	1	0,8);	(0,986	0,993	0,9996	1	1)]
	[(1	1	1	1	0,8);	(1	1	1	1	1)]
	[(0,997	0,999	1	1	0,8);	(0,997	0,999	1	1	1)]
	[(0,99	0,998	1	1	0,8);	(0,99	0,998	1	1	1)]
	[(0,999	1	1	1	0,8);	(0,999	1	1	1	1)]
	[(0,997	0,999	1	1	0,8);	(0,997	0,999	1	1	1)]
	[(0,981	0,996	1	1	0,8);	(0,981	0,996	1	1	1)]
	[(0,998	1	1	1	0,8);	(0,999	1	1	1	1)]
	[(1	1	1	1	0,8);	(0,993	1	1	1	1)]
	[(0,997	0,999	1	1	0,8);	(0,999	0,999	1	1	1)]
	[(0,93	0,958	0,993	0,998	0,8);	(0,999	0,958	0,9935	0,9976	1)]
	[(1	1	1	1	0,8);	(0,976	1	1	1	1)]
	[(0,999	1	1	1	0,8);	(0,976	1	1	1	1)]
	[(1	1	1	1	0,8);	(0,976	1	1	1	1)]
	[(0,994	0,999	1	1	0,8);	(0,976	0,999	1	1	1)]
	[(1	1	1	1	0,8);	(1	1	1	1	1)]
	[(0,977	0,987	0,999	1	0,8);	(0,977	0,987	0,999	0,9998	1)]
	[(0,986	0,993	1	1	0,8);	(0,986	0,993	0,9996	1	1)]
	[(0,956	0,977	0,998	1	0,8);	(0,956	0,977	0,9975	0,9995	1)]

The four candidates are ranked according to Ratio System, cf. Eq. (4) which can be seen in Table 3.

Table 3. The ratio system

	RS_i	Distance	Rank
A_1	[(4.985,4.997,5,5,0.8), (4.962,4.997,5,5,1)]	4,533	4
A_2	[(4.961,4.98,4.999,5,0.8), (4.932,4.98,4.999,5,1)]	4,627	2
A_3	[(4.979,4.992,5,5,0.8), (4.957,4.992,5,5,1)]	4,586	3
A_4	[(4.858,4.929,4.991,4.997,0.8), (4.909,4.929,4.991,4.997,1)]	4,704	1

Table 4 shows the results of Reference Point System, four candidates are ranked based on their distance by Eq. (2) and Table 5 indicates the results of Full Multiplicative Form.

Table 4. Reference point

	$Max_j\{d(\beta_{ij}, \beta_j)\}$	Rank
A_1	0.103	1
A_2	0.108	3
A_3	0.105	2
A_4	0.119	4

Table 5. Full multiplicative form

	RS_i	Distance	Rank
A_1	[(0.838,0.920,1,1,0.8), (0.763,0.921,1,1,1)]	2.326995	4
A_2	[(0.811,0.895,1,1,0.8), (0.712,0.895,1,1,1)]	2.364969	2
A_3	[(0.832,0.913,1,1,0.8), (0.761,0.913,1,1,1)]	2.352266	3
A_4	[(0.681,0.784,1,1,0.8), (0.679,0.784,1,1,1)]	2.537938	1

Table 6 includes the results of Ratio System, Reference Point and Full Multiplicative Form; the MULTIMOORA ends with final ranking by dominance theory.

Table 6. Multimoora ranking

	Ratio system	Reference point	Full multiplicative form	MULTIMOORA (Final rank)
A_1	4	1	4	4
A_2	2	3	2	2
A_3	3	2	3	3
A_4	1	4	1	1

The results show that the final ranking of MULTIMOORA will remain differ with the use of Einstein operator. The results show that final ranking of Einstein operator is differ from the final ranking of GITFNOWGA operator conducted by Balezentis and Zeng (2012). It indicates managers should give importance to selection of the operator in multi criteria decision making methods. They should support the conclusion of the MULTIMOORA method with other experience and methods.

4 Conclusions

In today's world decision making problems have a considerable importance and this importance will continue to grow in the future. To that end, studies relating to decision making area will increase especially in fuzzy decision-making studies which consider uncertainty. Even though the literature consists of numerous studies about fuzzy decision-making methods, there can be found limited numbers of the MULTIMOORA method related studies. MULTIMOORA is rooted from MOORA method and Full Multiplicative Form method for Multi-Criteria Decision Making. Thus, MULTIMOORA method consists of MOORA-Ratio System, MOORA-Reference Point System and Full Multiplicative Form method. MULTIMOORA method is widely used to solve problems of preferencing and/or sequencing which facilitates jobs of decision makers.

In the context of this research, the data from the study of the MULTIMOORA Method with type 2 fuzzy sets and GITFNOWGA operator was conducted from Balezentis and Zeng (2012) is used to evaluate whether the final ranking of MULTIMOORA will remain be same/differ with the use of Einstein operator. The results show that final ranking of Einstein operator is differ from the final ranking of GITFNOWGA operator conducted by Balezentis and Zeng (2012). As result, the conclusions from this study illustrates that the selection of an aggregation operator is extremely crucial due to it may change the results of rankings.

References

1. Greco S, Figueira J, Ehrgott M (2005) Multiple criteria decision analysis. Springer's International series
2. Hsieh TY, Lu ST, Tzeng GH (2004) Fuzzy MCDM approach for planning and design tenders selection in public office buildings. Int J Proj Manag 22(7):573–584
3. Shyur HJ, Shih HS (2006) A hybrid MCDM model for strategic vendor selection. Math Comput Model 44(7–8):749–761
4. Liang GS (1999) Fuzzy MCDM based on ideal and anti-ideal concepts. Eur J Oper Res 112 (3):682–691
5. Yang JL, Chiu HN, Tzeng GH, Yeh RH (2008) Vendor selection by integrated fuzzy MCDM techniques with independent and interdependent relationships. Inf Sci 178 (21):4166–4183
6. Liang GS, Wang MJJ (1994) Personnel selection using fuzzy MCDM algorithm. Eur J Oper Res 78(1):22–33
7. Dursun M, Karsak EE (2010) A fuzzy MCDM approach for personnel selection. Expert Syst Appl 37(6):4324–4330
8. Chiclana F, Herrera F, Herrera-Viedma E (2000) The ordered weighted geometric operator: properties and application in MCDM problems. In: Proceedings of the 8th conference information processing and management of uncertainty in knowledgebased systems (IPMU)
9. Bordogna G, Pasi G (1995) Linguistic aggregation operators of selection criteria in fuzzy information retrieval. Int J Intell Syst 10(2):233–248
10. Baležentis A, Baležentis T, Valkauskas R (2010) Evaluating situation of Lithuania in the European Union: structural indicators and MULTIMOORA method. Technol Econ Dev Econ 16(4):578–602
11. Brauers WKM, Zavadskas EK (2010) Project management by MULTIMOORA as an instrument for transition economies. Technol Econ Dev Econ 16(1):5–24
12. Baležentis T, Zeng S (2013) Group multi-criteria decision making based upon interval-valued fuzzy numbers: an extension of the MULTIMOORA method. Expert Syst Appl 40 (2):543–550
13. Brauers WK, Zavadskas EK (2006) The MOORA method and its application to privatization in a transition economy. Control Cybern 35:445–469
14. Baležentis A, Baležentis T, Brauers WK (2012) Personnel selection based on computing with words and fuzzy MULTIMOORA. Expert Syst Appl 39(9):7961–7967
15. Wei SH, Chen SM (2009) Fuzzy risk analysis based on interval-valued fuzzy numbers. Expert Syst Appl 36(2):2285–2299
16. Brauers WKM, Zavadskas EK (2011) MULTIMOORA optimization used to decide on a bank loan to buy property. Technol Econ Dev Econ 17(1):174–188
17. Aytekin A (2016) Hastaların Hastane Tercihinde Etkili Kriterler ve Hastanelerin MULTIMOORA ile Sıralanması: Eskişehir Örneği. İşletme ve İktisat Çalışmaları Dergisi 4(4):134–143

Healthcare Systems and Management

The Comparison and Similarity Study Between Green Buildings and Green Hospitals: A General View

Hatice Camgöz-Akdağ(✉), Aziz Kemal Konyalıoğlu, and Tuğçe Beldek

Management Engineering Department, Istanbul Technical University, Istanbul, Turkey
{camgozakdag,konyalioglua,beldek}@itu.edu.tr

Abstract. Nowadays sustainability is one of the most important topics in the developing world. Sustainability might include three main aspects in which one of the dimensions is environmental sustainability. Due to changing conditions of the world, decreasing natural resources and increasing needs of public conscience towards environment, not only environmental sustainability conscience, but also the word "green" gain importance. Furthermore, in healthcare and construction areas, green constructions, which mainly are green buildings and green hospitals, are outstandingly taken into consideration. Even if green hospitals are sub-buildings of green buildings, they have a lot of similarities and differences. The aim of the study is to give a general view by comparing green hospitals and green buildings and have a perspective about the main features of green buildings and green hospitals. In this study, two different analytic hierarchy processes are applied to compare green hospitals and green buildings and to have a perspective about which features of green buildings and hospitals are the most crucial. Besides, as a different spectacle, this study will investigate of which factors should be encouraged to have a better quality of green buildings and hospitals.

Keywords: Sustainability · Green hospitals · Green buildings · Construction Analytic hierarchy process

1 Introduction

Global warming causes so many different problems such as pollution, climate changing and lack of natural resources. These topics lead green production in every area to leave sustainable environment. Natural products and services are very popular and preferable today. Product and process design is very critical when life-cycle is considered for sustainability. For long time periods, especially for constructions, reusable and recyclable materials are preferred in the design phase. The design of a building with green products is not enough to make the whole system sustainable. Thinking energy and water usage, easy access for customers, clean and fresh in-door environment and eco-friendly devices have different importance weights while constructing a green building. To classify different levels of green buildings some standardization systems are being used such as LEED, BREEAM and Green Star. These kind of systems give certifications to green

N. M. Durakbasa and M. G. Gencyilmaz (Eds.): ISPR 2018, *Proceedings of the International Symposium for Production Research 2018*, pp. 181–190, 2019.
https://doi.org/10.1007/978-3-319-92267-6_14

buildings, according to their properties in terms of environmental aspects. Gaining points from those criteria lead to different levels of certification. Although the green standards are specifically defined, if those buildings will be used for different kind of sectors, some special criteria also have been considered.

For healthcare sector, human life is the one and only focus point to orient the medical system on. Patient focused healthcare is very important to reach high quality of service with high rate of recovery. Using healthy and hygienic equipment, leaving clean inside environment, providing patients' satisfaction with the process and also with the hardware of the whole building are very important. These vital topics are similar to green building design requirements. In this study it will be discussed if green building and green hospital design criteria are similar or have relation each other. Analytical Hierarchy Process is used for finding out similarities or differences of green buildings and green hospitals features.

After introduction a literature review is given to understand green building design and green hospital design in detail. In third part, methodology which is AHP will be defined and the analysis will be shown respectively in the application part of the study. At last conclusion will give the results which explains the comparison of two different design systems.

2 Literature Review

2.1 Green Building Design

Like any other industries, green production is very popular today at construction sites. At the design phase, architectures are considering different types of shapes and materials to provide energy saving. Green building de-sign provides savings in different scopes: nearly 30% energy, 35% carbon, 30–50% water usage, 50–90% waste cost savings will be reached [1]. There are different definitions of green building in literature. A green building is better designed than a traditional building in case of its effect to the environment. Another definition is the building that provides an important development and in-novation within its environment. Green building is not only the consumer but also a manufacturer of energy and water. During its life cycle, it presents the healthiest environment while using water, energy and land sources efficiently [2]. Green design definition is being used for years as the effect of buildings' in terms of environmental issues and to show the difference of green and regular buildings. The requirements include topics such as health, waste, comfort and emissions:

- Avoiding environmental damage
- Avoiding a new infrastructural system
- Avoiding environmental damage during construction
- Reducing emissions
- Reducing need of energy, water and material resources
- Providing comfort area for residents
- Avoiding harmful material indoor usage

These standardization of green building requirements puts reducing environmental damage and using less material while increasing comfort and health level in the first place [3].

2.2 Green Hospital Design

Green building operations is vital for both environment and people. The growing trend of environmental awareness and practice in management (especially supply chain) systems has had its effect on many sectors, including healthcare and hospitals. Having touched upon Green supply chain and Green management in general, turning towards Green practices in hospitals, starting with the, again US focused, —Hospitals 2020 initiative, which aims —to accelerate the development, use, and diffusion of environmentally preferable products, practices, and construction of Green buildings in hospitals and medical practices worldwide ||. The seven key elements include hospitals 'food, water, and (alternative) energy consumption, waste production, and related factors of building de-sign, energy efficiency, and transportation in and around the hospital. World Health Organizations seven key elements for green hospital are as follows [4]:

1. With efficiency measures, reducing cost and energy consumption
2. Building to reduce resource and energy demand and being sensitive to climate conditions
3. Producing/consuming clean, renewable energy
4. Make personnel and people coming to the hospital prefer walking and cycling
5. Sustainability of producing/consuming green food for personnel and patients
6. Reducing waste and using alternative disposal techniques
7. Finding safe alternatives to save water instead of bottled ones

For green hospital design standards, LEED (The Leadership in Energy and Environmental Design), BREEAM (Building Research Establishment's Environmental Assessment Method) and Green Star certifications systems can be considered. Table 1 shows the details of those standards within the grading systems and differences. Being environmental assessment is the common issue for these systems [5].

United States developed LEED and Australia uses Green Star system for green building assessment. BREEAM is being used widely and can be said that it is more preferable than the others [5].

German Healthcare Partnership made a study on green hospital design. The hospital selected was in China for the case study. The aim was to help decision makers for the construction design and drawing a road map for other countries. The study is based on again LEED and another system DGNB which is German Sustainable Building Council. Figure 1 shows basic elements related to green hospitals [6].

Table 1. Rating systems of green hospitals [5]

Country-System	Year	Elements and points	Ratings and level of certification
UK-BREEAM	2008	Management (12), Health and Wellbeing (15), Energy (19), Transport (8), Water (6), Materials (12.5), Waste (7.5), Land Use & Ecology (10), Pollution (10), Innovation (10). Total points = 110	Unclassified < 30 Pass ≥ 30 Good ≥ 45 V Good ≥ 55 Excellent ≥ 70 Outstanding ≥ 85
US-LEED	2009	Sustainable Sites (18), Water Efficiency (9), Energy and Atmosphere (39), Materials and Resources (16), Indoor Environmental Quality (18), Innovation In Design (6), Regional Priority Credits (4). Total points = 110	Certified 40–49 Silver 50–59 Gold 60–79 Platinum 80 and above
AUSTRALIA-Green Star	2009	Management (17), Indoor Environment Quality (32), Energy (29), Transport (12), Water (14), Materials (35), Land use & Ecology (8), Emissions (20), Innovation (5). Total points = 172	Best Practice (4 star) 45–59 Australian Excellence (5 star) 60–74 World Leadership (6 star) 75–100

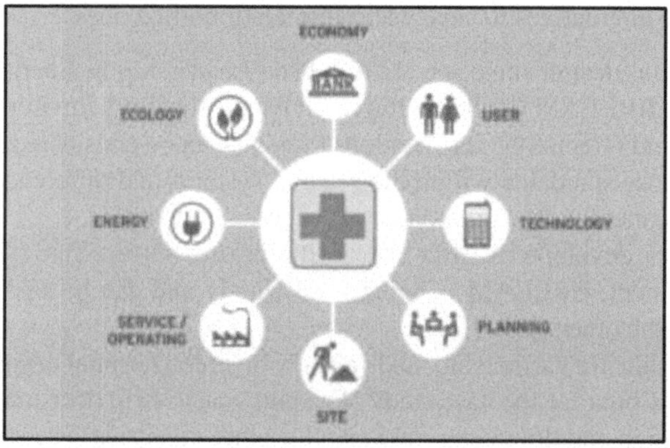

Fig. 1. Basic elements for green hospitals [6] (Color figure online)

3 Methodology

AHP for which stands Analytic Hierarchy Process was developed by Saaty in 1980. AHP can be taken into consideration in order to make a decision under multi-criteria. AHP is a useful and powerful hierarchical process which mainly aims to solve complex problems by using simple comparisons. Also, AHP can meet various objectives based on multiple criteria and alternatives.

AHP firstly commence the process by calculating the weights. Actually, AHP make a pairwise comparison to calculate weights. According to Saaty, the matrix A is a real-valued matrix whose entry a_{ij} implies the relative importance of ith criterion compared to kth criterion [7].

After deciding the decision problem, the matrix between factors named after pairwise comparison matrix should be built to decide importance scale. The matrix A whose diagonal entries are 1, is given below. Hereby, it should be noticed that A is a matrix of dimension n x n.

$$A = \begin{bmatrix} a_{11} & \cdots & a_{1n} \\ \vdots & \ddots & \vdots \\ a_{n1} & \cdots & a_{nn} \end{bmatrix}$$

Also, the entry a_{ij} has an equality as given below

$$a_{ij} = \frac{1}{a_{ji}}$$

Besides, one may need to calculate the matrix C to normalize the entries. In another word, the column vectors of matrix A are used to decide the normalized matrix with row averages. The entries of matrix B is found as follow:

$$c_{ij} = \frac{a_{ij}}{\sum_{i=1}^{N} a_{ij}}$$

The scale of importance is given as below (Table 2).

Table 2. The scale of importance [7]

Intensity of importance	Definition	Explanation
1	*Equal importance*	Two activities contribute equally to the objective
2	Weak or slight	
3	*Moderate importance*	Experience and judgment slightly favor one activity over another
4	Moderate plus	
5	*Strong importance*	Experience and judgment strongly favor one activity over another
6	Strong plus	
7	*Very strong or demo - strated importance*	An activity is favored very strongly over another; its dominance demonstrated in practice
8	*Very, very strong*	
9	*Extreme importance*	The evidence favoring one activity over another is of the highest possible order of affirmation

According to this scale and normalized matrix, the weights of each criteria should be decided. The weights are calculated as given below based on the normalized matrix. Here, by noting that the row average is taken to find weights as well.

$$w_i = \frac{\sum_{j=1}^{n} c_{ij}}{n}$$

After finding weights, the consistency ration (CR) is calculated. CR shows whether the values of matrices are consistent or not.

4 Application

The application will be for choosing the most important features of green buildings and green hospitals. Analytic Hierarchy Process (AHP) will facilitate to choose the most important features of each criteria (Fig. 2).

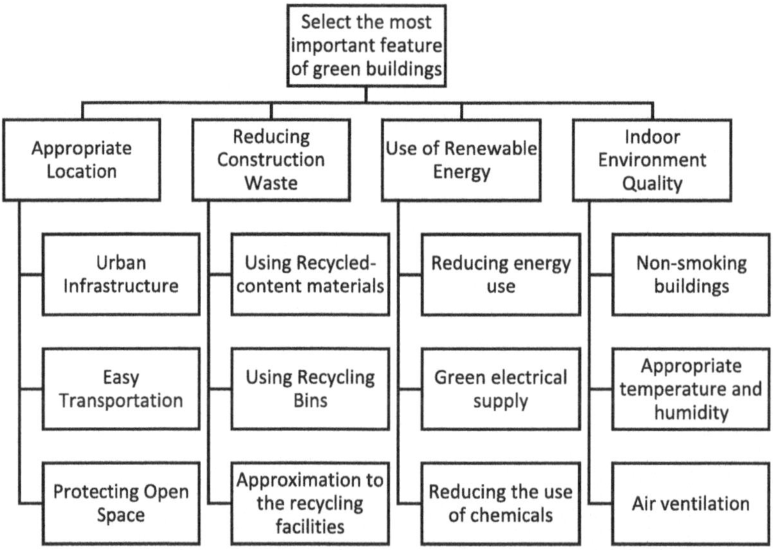

Fig. 2. Hierarchy list of green buildings

According to the hierarchy list, the importance values have been asked to the experts. For the first criterion "Appropriate Location", the experts answered and the following matrix has been built.

In this case, the criteria are "urban infrastructure, easy transportation and protecting open space". Based on the answers, the matrix A_1 is as follow:

$$A_1 = \begin{bmatrix} 1 & 3 & 8 \\ 1/3 & 1 & 5 \\ 1/8 & 1/5 & 1 \end{bmatrix}$$

Based on the matrix A_1, the normalized matrix C_1 and vector W_1 are found as follows

$$C_1 = \begin{bmatrix} 0,68571 & 0,71428 & 0,57142 \\ 0,22857 & 0,23809 & 0,35714 \\ 0,08571 & 0,04761 & 0,07142 \end{bmatrix}$$

$$W_1 = \begin{bmatrix} 0,657 \\ 0,275 \\ 0,068 \end{bmatrix}$$

While investigating, CR (consistency ratio) is %7 which means the system is consistent provided that CR is smaller than %10. In the calculation, it is seen that the most important criterion under "appropriate location" is *urban infrastructure.*

The second criterion is "reducing construction waste". According to the importance values taken by experts, matrix A_2 is as follow:

$$A_2 = \begin{bmatrix} 1 & 2 & 5 \\ 1/2 & 1 & 6 \\ 1/5 & 1/6 & 1 \end{bmatrix}$$

Based on the matrix A_2, the normalized matrix C_2 and vector W_2 are found as follows

$$C_2 = \begin{bmatrix} 0,58823 & 0,63157 & 0,41666 \\ 0,29411 & 0,31578 & 0,5 \\ 0,11764 & 0,05263 & 0,83333 \end{bmatrix}$$

$$W_2 = \begin{bmatrix} 0,545 \\ 0,370 \\ 0,085 \end{bmatrix}$$

CR (consistency ratio) in this case has been found as %9 which means the system is consistent provided that CR is smaller than %10. The most important criterion under "reducing construction waste" is "Using Recycled-content materials" based on weights.

Thirdly, the calculation is based on the criterion "use of renewable energy". Based on the answers taken from experts, the matrix A_3 has been built as given below:

$$A_3 = \begin{bmatrix} 1 & 6 & 3 \\ 1/6 & 1 & 1/4 \\ 1/3 & 4 & 1 \end{bmatrix}$$

Based on the matrix A_3, the normalized matrix C_3 and vector W_3 are found as follows

$$C_3 = \begin{bmatrix} 0,6666 & 0,54545 & 0,70588 \\ 0,1111 & 0,09090 & 0,05882 \\ 0,2222 & 0,36363 & 0,23529 \end{bmatrix}$$

$$W_3 = \begin{bmatrix} 0,639 \\ 0,087 \\ 0,274 \end{bmatrix}$$

Referring to the weights found above, it is clear that the most important criterion is "*reducing energy use*". CR is 7% which enables to decide that the values are consistent.

Finally, for green buildings, it is tended to evaluate the last criterion "indoor environment quality". For this criterion, the calculation based on experts' answers is given as below:

$$A_4 = \begin{bmatrix} 1 & 1/8 & 1/2 \\ 8 & 1 & 7 \\ 2 & 1/7 & 1 \end{bmatrix}$$

Based on the matrix A_4, the normalized matrix C_4 and vector W_4 are found as follows

$$C_1 = \begin{bmatrix} 0,0909 & 0,09859 & 0,05888 \\ 0,72727 & 0,78873 & 0,82352 \\ 0,18181 & 0,11267 & 0,11764 \end{bmatrix}$$

$$W_4 = \begin{bmatrix} 0,083 \\ 0,780 \\ 0,137 \end{bmatrix}$$

Hereby, the "appropriate temperature and humidity" criterion is the most important feature of indoor environment quality. After evaluating the criteria of green buildings, green hospitals evaluation criteria are as follow (Fig. 3).

While evaluating, the importance values of each criteria, the values have been asked to the experts. Firstly, based on the answers taken from experts, the matrix B_1 which explains importance values of "sustainability factors in hospital buildings", is as follow

$$B_1 = \begin{bmatrix} 1 & 1/3 & 4 \\ 3 & 1 & 8 \\ 1/4 & 1/8 & 1 \end{bmatrix}$$

Based on the matrix B_1, the normalized matrix C_1 and vector W_1 are found as follows

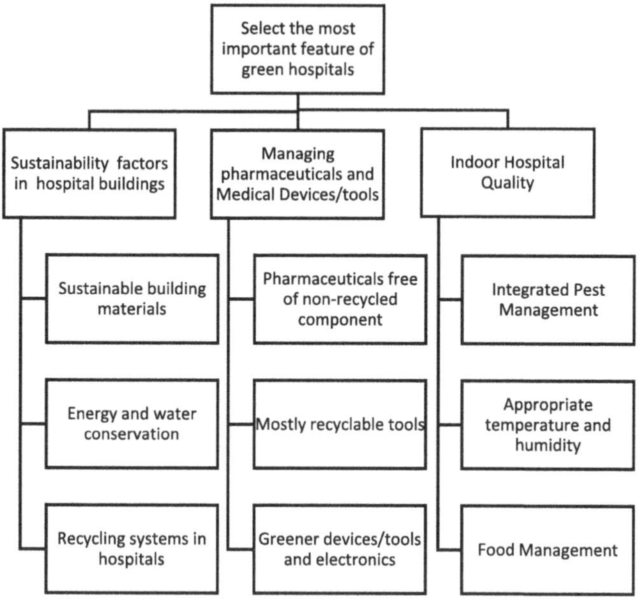

Fig. 3. Hierarchy list of green hospitals

$$C_1 = \begin{bmatrix} 0,23529 & 0,22857 & 0,30769 \\ 0,70588 & 0,68571 & 0,61528 \\ 0,05882 & 0,08571 & 0,07692 \end{bmatrix}$$

$$W_1 = \begin{bmatrix} 0,257 \\ 0,669 \\ 0,074 \end{bmatrix}$$

Under sustainability factors in hospital buildings, the most important feature is "energy and water conservation" with regard to weights found.

Secondly, under "Managing pharmaceuticals and Medical Devices" criteria, the matrix B_2 which explains importance values of "sustainability factors in hospital buildings", is as follow

$$B_2 = \begin{bmatrix} 1 & 1/4 & 1/5 \\ 4 & 1 & 1/2 \\ 5 & 2 & 1 \end{bmatrix}$$

Based on the matrix B_1, the normalized matrix C_1 and vector W_1 are found as follows

$$C_2 = \begin{bmatrix} 0,1 & 0,07692 & 0,11764 \\ 0,4 & 0,30769 & 0,29411 \\ 0,5 & 0,61538 & 0,58823 \end{bmatrix}$$

$$W_2 = \begin{bmatrix} 0,098 \\ 0,334 \\ 0,568 \end{bmatrix}$$

In this case, the most important feature is "greener devices/tools and electronics".

Finally, the last criterion is indoor hospital quality. By following the same procedure, one finds that the most important feature is *appropriate temperature and humidity* by the weight of 0,069.

5 Conclusion

When two different analytic hierarchy processes were applied to given hierarchy list, we found that even if green hospitals are subsets of green buildings, they have different features and the most important features change with regard to their application areas. Furthermore, they have different hierarchy structures. Green buildings are the crucial features of Using Recycled-content materials, urban infrastructure and reducing energy use but for green hospitals, they are not the same. It is obvious by calculations that the most important features of green hospitals are greener devices/tools and electronics and energy-water conservation.

By contrast, it is found that they have similarities also. For green buildings, in the same category "indoor environment quality", the most important feature is appropriate temperature and humidity while it is the same for green hospitals. For further researches, these evaluations can be done under fuzziness and more recommendations from experts can be included in the study.

References

1. Council UGB (2001) Leadership in energy and environmental design (LEED)
2. Terekli G, Özkan O, Bayın G (2013) Çevre dostu hastaneler: Hastaneden yeşil hastaneye
3. Cole RJ (2012) Transitioning from green to regenerative design. Build Res Inf 40(1):39–53
4. World Health Organization and Health Care Without Harm (2016) Healthy Hospitals-Healthy Planet Healthy People: Addressing Climate Change in Health Care Setting. World Health Organization, England. http://www.who.int/globalchange/publications/climatefootprint_report.pdf?ua=1. Accessed 01 Oct 2016
5. Sahamir SR, Zakaria R (2014) Green assessment criteria for public hospital building development in Malaysia. Procedia Environ Sci 20:106–115
6. GHP Green Hospital. http://ghp-initiative.de/key-areas/green-hospital/. Accessed 01 Apr 2018
7. Saaty TL (1980) The Analytic Hierarchy Process, McGraw Hill, New York. International, Translated to Russian, Portuguese, and Chinese, Revised editions, Paperback (1996, 2000), RWS Publications, Pittsburgh

Industrial Applications

Accuracy of Ducts Made with Various Processing Strategies

L. Nowakowski(✉), M. Skrzyniarz, and E. Miko

Department of Mechanical Engineering and Metrology,
Kielce University of Technology, Kielce, Poland
{lukasn,mskrzyniarz,emiko}@tu.kielce.pl

Abstract. This article has presented the results of the analysis regarding the size and shape accuracy also selected spatial parameters of the surface of ducts that were made with the use of various processing strategies. The milling process of passage ducts was performed at a DMU-50 vertical processing center with the use of a Guhring 3309 carbide cutter with diameter of 8 mm. The cutting tests involved making seven ducts with the size of 12 × 60 × 10 mm in a semi-finished product made from aluminum alloy type 2017. The analysis of the size and shape accuracy of the made ducts was performed with the use of the ZEISS Prismo Navigator coordinate measuring machine. The analysis of the surface was conducted on a TalSurf CCI optical profilometer.

Keywords: CNC milling · Duct · Groove · Processing strategy
Size and shape accuracy · Surface parameters

1 Introduction

The process of shell end milling is used for making short, shallow withdrawals, especially in the case of ducts, pockets, and keyways. Milling of ducts with shell end mills results in generating large cutting force, especially in the first travel in the processing, when there are three surfaces formed at the same time, which causes vibrations and bending of the tool during the processing. Vibrations and the bending of the tool are factors that affect the size and shape accuracy [1, 6], of the made ducts and are the cause for lower efficiency of the milling process. Errors regarding the size and shape that occurred as a result of the chosen processing strategy during the milling process of ducts can lead to lowered practical properties of machine elements [5, 10, 12]. Errors in the execution of precision parts [2] might lead to improper operation of the device, e.g. through faster wear of mating components, which can generate vibrations, frictional resistance and causes the temperature increase of working elements [3, 8]. The processing strategy is directly connected with the accuracy of the movements of the tool, which are numerically controlled during the performance of the technological process [4, 9, 11]. It is a significant factor from the point of view of forming precise machine parts.

N. M. Durakbasa and M. G. Gencyilmaz (Eds.): ISPR 2018, *Proceedings of the International Symposium for Production Research 2018*, pp. 193–200, 2019.
https://doi.org/10.1007/978-3-319-92267-6_15

2 Methods

The subject of research was to analyze the size and shape accuracy of 7 ducts that were made with the use of various processing strategies. The milling process of passage ducts was performed at a DMU-50 vertical processing center with the use of a Guhring 3309 carbide cutter with diameter of 8 mm and number of blades z = 2 pcs. The cutting tests involved the same cutting parameters: cutting speed v_c = 250 m/min, feed per blade f_z = 0.02 mm/blade for seven ducts with the size of 12 × 60 × 10 mm in a semi-finished product made from aluminum alloy type 2017 (Fig. 1). Table 1 presents the tool tracks for each processing strategy, including the time it lasted for where a_p - depth of cut, $a_{e1..3}$ - width of cut for subsequent travels.

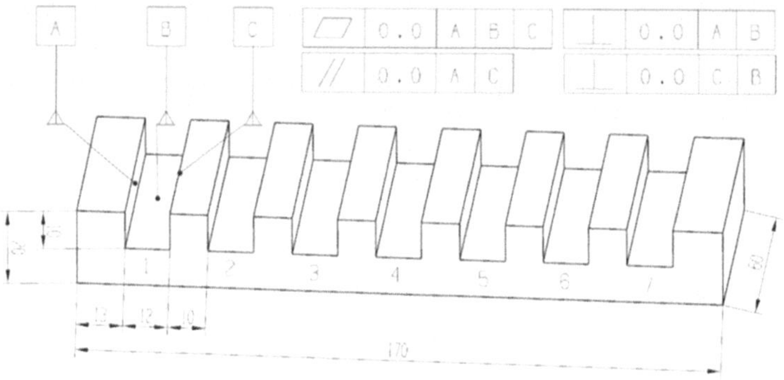

Fig. 1. The workpiece view with GD&T

Strategy 1 of processing involved making a duct with width of 12 mm and depth of 10 mm with two travels of the tool. The width of cut for the first travel was 8 mm, for the second one −4 mm. The depth of cut was 10 mm. The travel direction of the tool resulted in forming the walls through climb milling, while the processing time of the duct was 24 s. Strategy 2 involved a different depth of cut (a_p = 2.5 mm), which extended the time of processing the duct four times, i.e. 96 (four penetrations of the tool), but decreased the load on the tool. Strategy 3 applied a zig-zag processing tract with parameters a_p = 10 mm, a_e = 0.8 mm, and it lasted for 65 s. The processing of duct 4 involved the application of trochoidal milling, in which the cutter removed material by moving on a track formed by a sequence of constant spirals, with maximum radial engagement, a_e = 0.8 mm and the depth of cut of a_p = 10 mm. The processing lasted for 169 s. In strategy 5, the shape of the tool track was planned in a way that allowed processing of the side walls through climb milling with the following parameters a_p = 10 mm, a_{e1} = 8 mm, a_{e2} = 2 mm, a_{e3} = 2 mm. The processing lasted for 48 s. Strategies 6 and 7 were conducted with the use of processing cycle 253 (duct milling) implemented in the HEIDENHAIN iTNC 530 controller. The difference between both strategies was that in the first strategy, the cutter would go into the material up to the full depth a_p = 10 mm, while in the second one, the cutter would

Table 1. Comparison of tool tracks with the parameters and processing time, AP - depth of cut, AE1..3 - width of cut for subsequent travels

No. strategy	Path tool		Cutting parameters	Time
1			a_p =10 mm a_{e1}=8 mm a_{e2}=4 mm	24 s
2		x4	a_p =2,5 mm a_{e1}=8 mm a_{e2}=4 mm	96 s
3	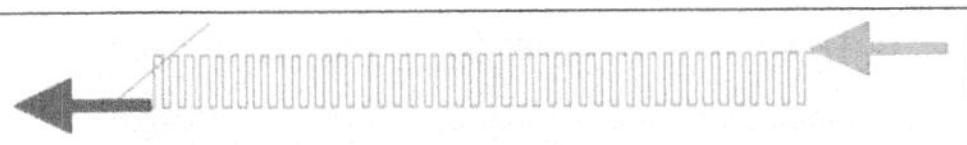		a_p =10 mm a_e=0.8 mm	65 s
4	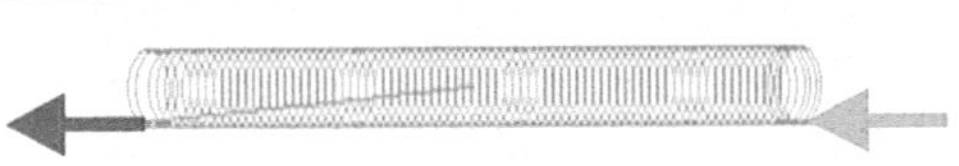		a_p =10 mm a_e=0.8 mm	169 s
5			a_p =10 mm a_{e1}=8 mm a_{e2}=2 mm a_{e3}=2 mm	48 s
6	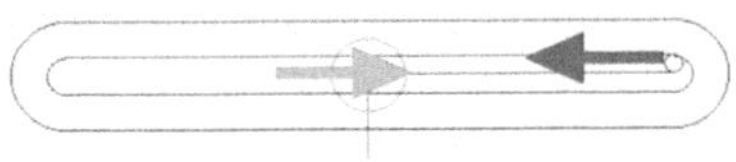	x4	a_p =2,5 mm a_{e1}=8 mm a_{e2}= 2 mm a_{e3}= 2 mm	259 s
7	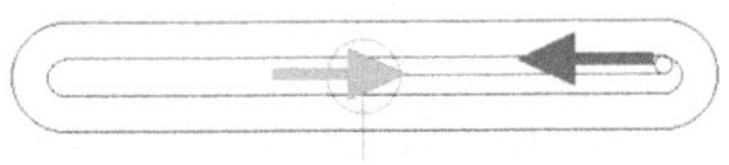		a_p=10 mm a_{e1}=8 mm a_{e2}= 2 mm a_{e3}= 2 mm	85 s
			START	END

perform a four-stage processing, each time going deeper by 2.5 mm. Processing of the duct with width of 12 mm, depth of 10 mm, and length of 60 mm while using the processing cycle no. 253 at feed-in depth of 10 mm lasted 85 s. In the case of feed-in depth of 2.5 mm, it would extend to 259 s.

3 Results

The analysis of the size and shape accuracy of the made ducts was performed with the use of the ZEISS Prismo Navigator coordinate measuring machine to measure: the width of the duct, flatness of the bottom, flatness of the walls, perpendicularity of the walls against the bottom, and parallelism of the walls (Fig. 1). The detailed list of measurement results has been presented in Table 2.

During the analysis of measurement results presented in Table 2, it was determined that two out of seven processing strategies (strategy 1 and 3) resulted in making ducts that were wider than the programmed width of 12 mm. In the case of the other five processing strategies, the obtained width of the groove was lower than the programmed one and was within the tolerance of −0.018 µm to −0.042 µm, which reflects the IT9 accuracy class. The lowest size and shape accuracy from all of the analyzed properties

Table 2. Results of measurement of size and shape accuracy for the made ducts

	Channel number						
Measured feature	**1**	**2**	**3**	**4**	**5**	**6**	**7**
Width S, mm	12,004	11,982	12,050	11,972	11,958	11,987	11,981
⏥ 0.0 A	0,0318	0,0052	0,0668	0,0112	0,0069	0,0047	0,0071
⏥ 0.0 B	0,0104	0,0095	0,0136	0,0084	0,0121	0,0135	0,098
⏥ 0.0 C	0,0345	0,0100	0,0681	0,0092	0,0069	0,0070	0,0073
⊥ 0.0 A B	0,0754	0,0176	0,1143	0,0222	0,0424	0,0131	0,0117
⊥ 0.0 C B	0,0765	0,0228	0,1198	0,0168	0,092	0,0095	0,0111
// 0.0 A C	0,0613	0,0075	0,1056	0,0236	0,0403	0,0126	0,0090

was obtained in strategy no. 3, which applied a zig-zag track for the tool, for which the working conditions of the tool were unstable due to the constant changes of the direction the tool worked in and changes of the milling process from out-cut milling to climb milling. The lowest value of flatness deviation, at the level of 8 μm, was observed in trochoidal processing. What is more, it was determined that the flatness deviation for all the other processing strategies were similar and within the range of 10–13 μm. The measurement results for side walls of the ducts proved that the load on the tool had a significant impact on the flatness of surfaces [7]. A reduction of the load on the tool by reducing the value of width of cut to $a_e = 4$ mm or depth of cut to $a_p = 2.5$ mm would make the flatness of the side walls of the duct be within the range of 5–10 μm, while in the case of travels during which more load was imposed on the tool ($a_e = 8$ mm and $a_p = 10$ mm), the measured flatness of the surfaces of side walls was 32 μm on average. The load on the tool during duct milling has also a significant impact on the perpendicularity of side walls to the bottom of ducts, in the cases where the load on the tool was the biggest during the processing, strategy no. 1 ($a_p = 10$ mm, $a_{e1} = 8$ mm, $a_{e2} = 4$ mm) or the working conditions of the tool were unstable, strategy no. 2, the measured average errors in perpendicularity were, respectively, about 75 μm for strategy 1 and about 110 μm for strategy 2. The lowest values of errors concerning perpendicularity were measured in strategies 6 and 7, where the ducts were performed with the use of processing cycle no. 253 (duct milling) of the HEIDENHAIN iTNC 530 controller. The last analyzed property was the error in parallelism of the side surfaces of the ducts. The lowest error value of the side surfaces of the ducts was measured in strategy 2 (7.5 μm), strategy 7 (9 μm), and 6 (12 μm). The reason behind such results was that in all those strategies, the selected parameters caused the lowest load on the tool during the processing of side walls.

As a result of the measurements, spatial parameters of formed surfaces were obtained. The quality of the surface has been described with parameters S_F, i.e. without filtering out the wavy finish. In table no. 1, the most important practical parameters of the formed surface of grooves have been presented along with their isometric images.

Table 3. Isometric images representing the bottom of formed grooves with regard to the applied processing strategy, along with the SF practical parameters of the surface

Method I	Method II
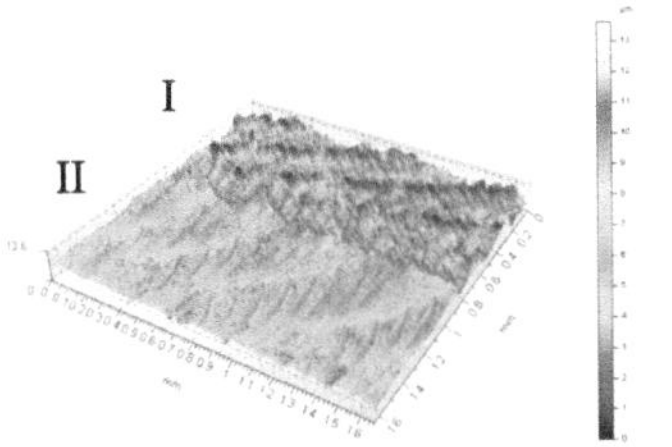	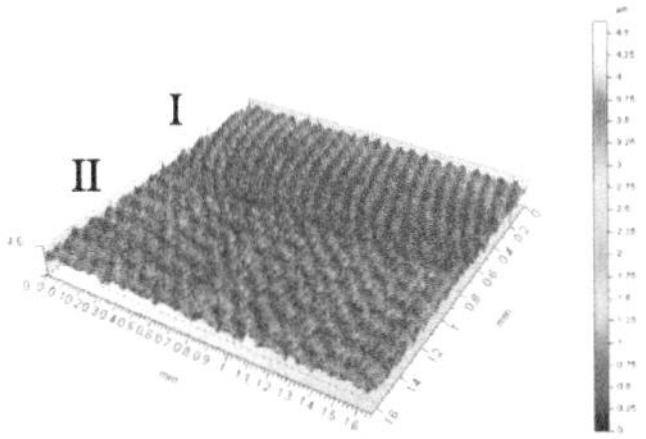
I - Sa = 1.665 μm, Sku = 2.716, Ssk = 0.295, Sal = 0.066 mm, Str = 0.17, Std = 120.8° II- Sa = 0.592 μm, Sku = 2.883, Ssk = 0.026, Sal = 0.053 mm, Str = 0.158, Std = 97°	I - Sa = 0.39 μm, Sku = 2.479, Ssk = -0.044, Sal = 0.018 mm, Str = 0.127, Std = 102.7° II - Sa = 0.352 μm, Sku = 2.932, Ssk = -0.16, Sal = 0.013 mm, Str = 0.245, Std = 86.7°
Method III	Method IV
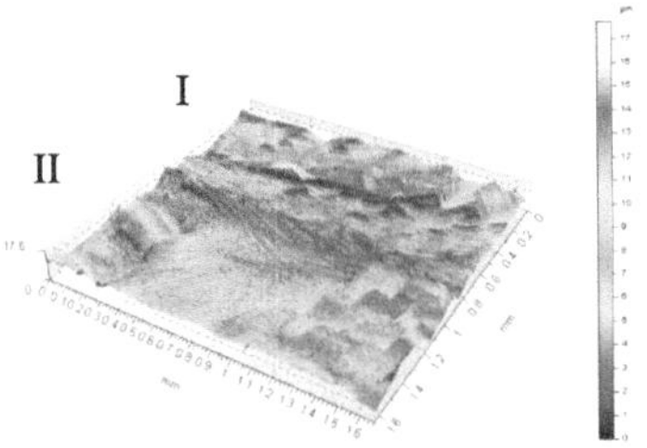	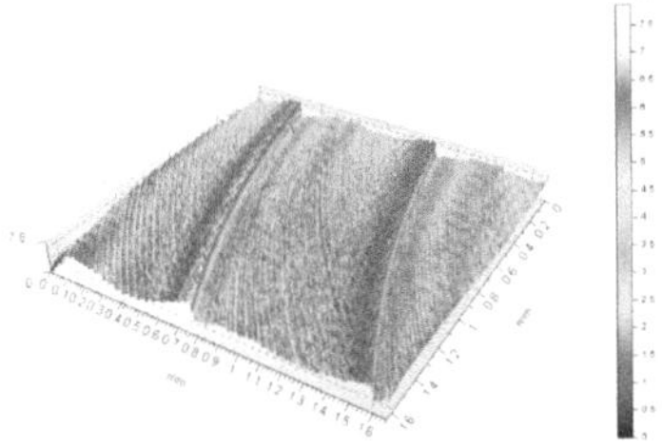
I - Sa = 2.94 μm, Sku = 2.45, Ssk = 0.529, Sal = 0.121 mm, Str = 0.136, Std = 7.97° II - Sa = 3.18 μm, Sku = 1.81, Ssk = -0,4, Sal = 0.341 mm, Str = 0.699, Std = 1.47°	Sa = 0.68 μm, Sku = 3.379, Ssk = 0.238, Sal = 0.058 mm, Str = 0.067, Std = 108.5°
Method V	Method VI
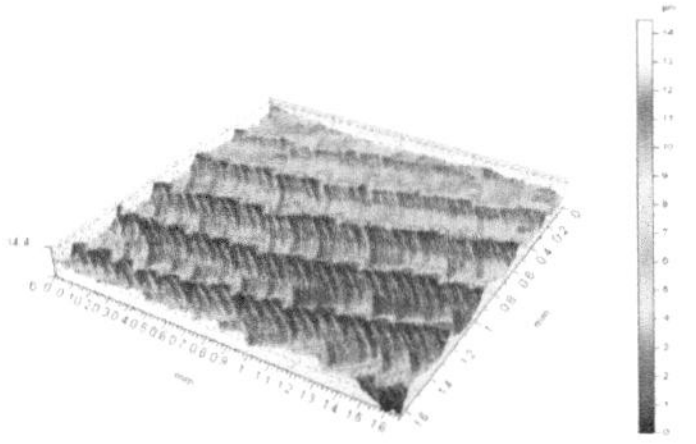	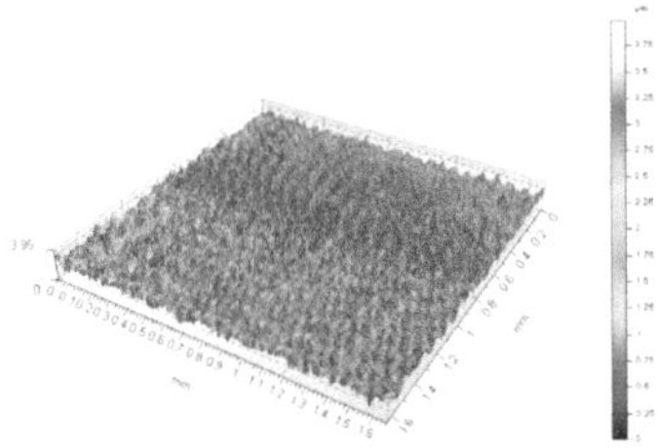
Sa = 1.75 μm, Sku = 2.85, Ssk = 0.557, Sal = 0.067 mm, Str = 0.078, Std = 101.9°	Sa = 0.32 μm, Sku = 2.95, Ssk = -0.147, Sal = 0.015 mm, Str = 0.29, Std = 87.21°
Method VII	
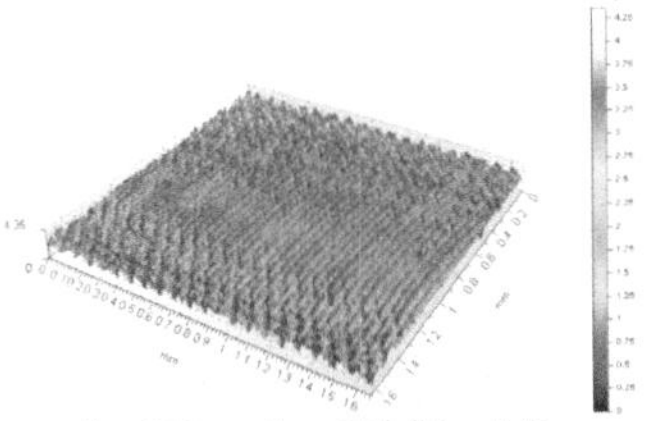	
Sa = 0.33 μm, Sku = 2.76, Ssk = -0.01, Sal = 0.01 mm, Str = 0.059, Std = 87.27°	

Table 4. Isometric images representing the side walls of formed grooves with regard to the applied processing strategy, along with the SF practical parameters.

Method I 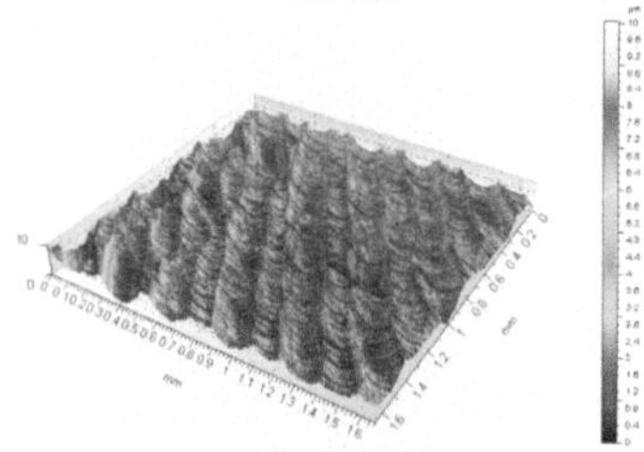 Sa = 1.333 μm, Sku = 2.339, Ssk = -0.063, Sal = 0.05 mm, Str = 0.052, Std = 0.736°	Method II 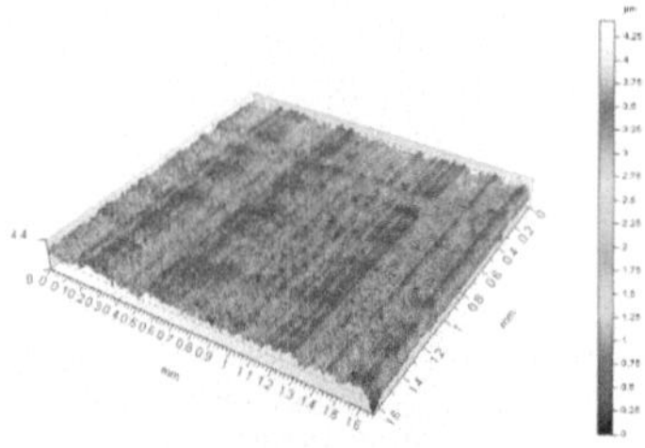 Sa = 0.37 μm, Sku = 2.848, Ssk = -0.064, Sal = 0.047 mm, Str = 0.057, Std = 89.7°
Method III 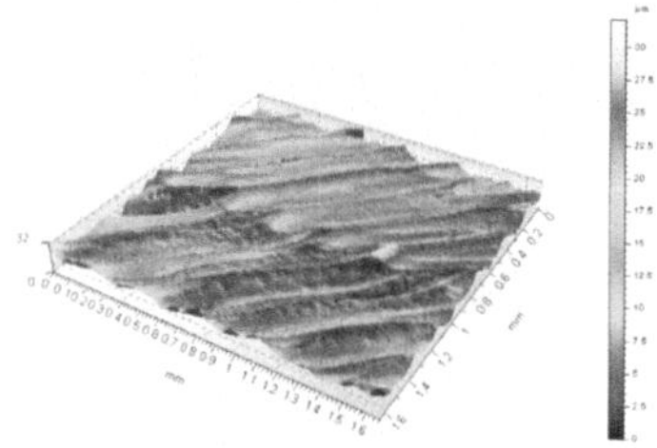 Sa = 3.92 μm, Sku = 2.71, Ssk = 0.255, Sal = 0.045 mm, Str = 0.119, Std = 45.04°	Method IV 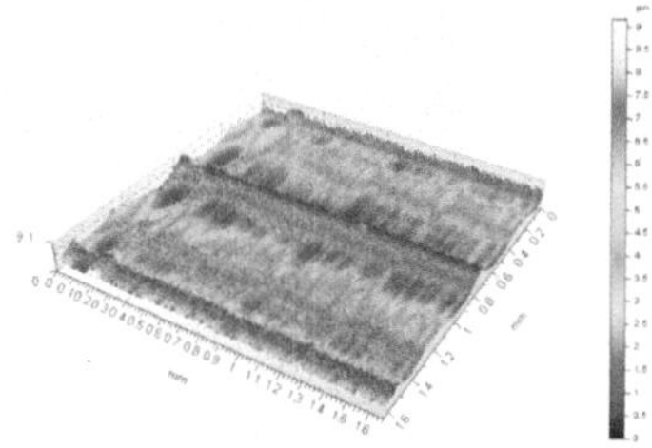 Sa = 0.911 μm, Sku = 3.916, Ssk = 0.761, Sal = 0.091 mm, Str = 0.11, Std = 89.7°
Method V  Sa = 0.372 μm, Sku = 3.235, Ssk = 0.286, Sal = 0.05 mm, Str = 0.06, Std = 91.2°	Method VI 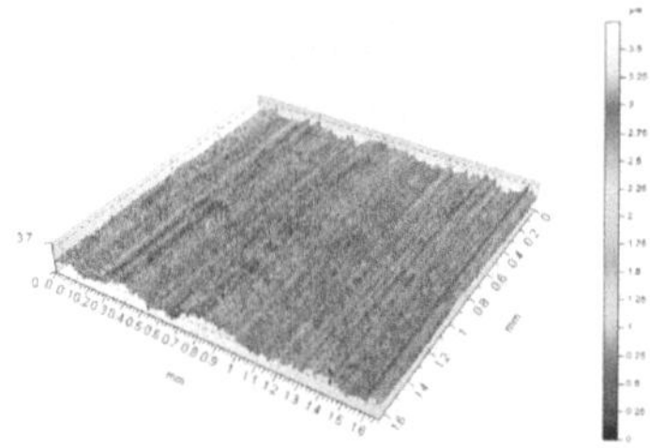 Sa = 0.397 μm, Sku = 2.63, Ssk = -0.061, Sal = 0.076 mm, Str = 0.091, Std = 90.14°
Method VII 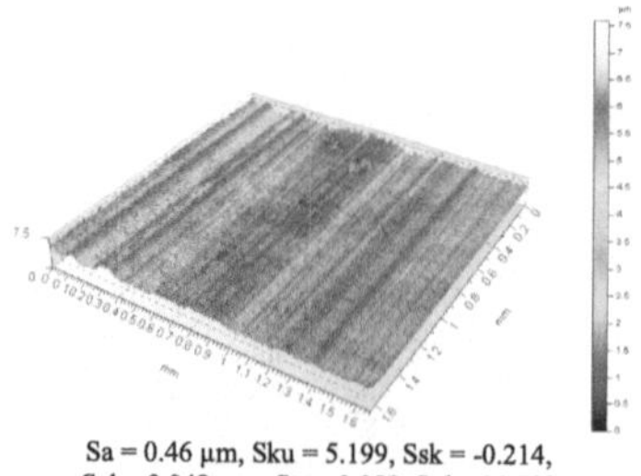 Sa = 0.46 μm, Sku = 5.199, Ssk = -0.214, Sal = 0.049 mm, Str = 0.059, Std = 90.78°	

The bottoms of grooves no. 1 and 2 were analyzed as two distinct surfaces, since the differences between their parameters and structure lay varied significantly.

The substantial difference in the geometrical structure of the surface of the formed bottom was affected by the direction of movements of the cutting edge and its load during the processing, which results from the wrapping angle. In method 1, during the first travel of the tool with a full wrapping angle (milling width of a_e = 8 mm), a geometrical structure of the surface is formed, in the case of which the parameter Sa = 1.665 μm. In that case, the arithmetic mean height of the surface is three times bigger than in the second travel with width of a_e = 4 mm, in the case of which the parameter Sa = 0.592 μm. A similar value describing the practical values of the surface were obtained for method 5, in the case of which the arithmetic mean height of the surface was Sa = 1.75 μm. The lowest value of the Sa parameter was achieved for method 6 and it amounts 0.39 μm. Method 4 of trochoidal milling results in formation of grooves on the surface that are the effect of the bending of the tool during the processing, while the distance between those grooves is represented through the value of the lead step of the trochoid of 0.8 mm (Tables 3 and 4).

Apart from analyzing the surface properties of the bottom being formed, an analysis of side surfaces was also conducted. The results of that analysis have been presented in Table no. 2. The analysis of measurement results proved that the highest load of the tool during milling with method I causes formation of the largest unevenness on the side surface of milled grooves. In the case of a wall formed as a result of machining with width a_e = 8 mm, Sa = 4.761 μm, while in the case of the second wall, this parameter amounted Sa = 1.333 μm. Methods II and V feature the same spatial parameters and are similar to each other in terms of the value, i.e. Sa ≈ 0.37 μm. In the strategy of trochoidal milling, the parameter of arithmetic mean height of the surface Sa amounted 0.911 μm. In this method, the isometric surfaces show the maximum heights of the surfaces and the distance between them equals the width of machining in the applied strategy.

4 Conclusions

1. When analyzing the measurements, it was decided that the type of the processing strategy affected the size and shape accuracy of the milled ducts.
2. The lowest size and shape accuracy was proved for the processing strategy that involved a zig-zag track of the tool. In the case of the track defined in such a way, the working conditions of the tool are unstable due to the changes in the direction the tool works in and the changes in the milling process, from out-cut milling to climb milling.
3. The most efficient processing strategy for the duct in the aluminum alloy 2017, when taking into consideration the size and shape accuracy and the time of processing, is strategy 7. This strategy applies the processing cycle no. 253 of duct milling in the HEIDENHAIN iTNC 530 controller, conducted with the following parameters a_p = 10 mm, a_{e1} = 8 mm, a_{e2} = 2 mm, a_{e3} = 2 mm.
4. The impact of the processing strategy is a significant factor that influences the quality of the formed surface, for both the bottom and the side walls.

5. Among the analyzed strategies, the lowest value of the Sa parameter of the formed bottom and side wall of the groove was measured for strategy no. 6.
6. Thanks to the processing strategy, we may influence the parameters of surface roughness, as well as their expansion and lay

Acknowledgment. The publication was created as a result of research and development carried out by the Polish Bearing Factory, Kraśnik S.A. together with the Kielce University of Technology in the project entitled "The Establishment of R & D Centre in FŁT-Kraśnik S. A." under the Smart Growth Operational Programme 2014-2020, co-financed from the European Regional Development Fund No. CBR/1/50-52/2017 from 07/04/2017.

References

1. Adamczak S, Zmarzly P, Kozior T, Gogolewski D (2017) Analysis of the dimensional accuracy of casting models manufactured by fused deposition modeling technology. In: Proceedings of the 23rd international conference engineering mechanics 2017, Svratka, pp 66–69
2. Adamczak S, Zmarzły P (2017) Influence of raceway waviness on the level of vibration in rolling-element bearings. Bull Pol Acad Sci: Tech Sci 64(4):541–551
3. Bartoszuk M, Grzesik W (2011) Numerical prediction of the interface temperature using updated finite difference approach. In: Modelling of machining operations book series: advanced materials, vol 223, pp 231–239
4. Kuczmaszewski J, Pieśko P, Zawada-Michałowska M (2017) Influence of milling strategies of thin-walled elements on effectiveness of their manufacturing. Procedia Eng 182:381–386
5. Lazoglu I, Manav C, Murtezaoglu Y (2009) Tool path optimization for free form surface machining. CIRP Ann Man Technol 58:101–104
6. Nowakowski L., Skrzyniarz M, Miko E (2017) The analysis of relative oscillation during face milling. In Proceedings of the 23rd international conference engineering mechanics 2017, Svratka, pp 730–733
7. Nowakowski L, Skrzyniarz M, Miko E (2017) The assessment of the impact of the installation of cutting plates in the body of the cutter on the size of generated vibrations and the geometrical structure of the surface. In: Proceedings of the 23rd international conference engineering mechanics 2017, Svratka, pp 734–737
8. Nowakowski L, Skrzyniarz M, Miko E, Takosoglu J, Blasiak S, Laski P, Bracha G, Pietrala D, Zwierzchowski J, Blasiak M (2016) Influence of the cutting parameters on the workpiece temperature during face milling. In: Proceedings of the international conference experimental fluid mechanics 2016, Mariánské Lázně, pp. 523–527
9. Perez H, Diez E, Perez J, Vizan A (2013) Analysis of machining strategies for peripheral milling. Procedia Eng 63:573–581
10. Ramos AM, Relvas C, Simoes JA (2003) The influence of finishing milling strategies on texture roughness and dimensional deviations on the machining of complex surfaces. J Mater Process Technol 136:209–216
11. Toh CK (2004) A study of the effects of cutter path strategies and orientations in milling. J Mater Process Technol 152:346–356
12. Zhang XF, Xie J, Xie HF, Li LH (2012) Experimental investigation on various tool path strategies influencing surface quality and form accuracy of CNC milled complex freeform surface. Int J Adv Manuf Technol 59:647–654

Acquisition of Measurement Data on a Stand for Durability Tests of Rolling Bearings

Jaroslaw Zwierzchowski[1(✉)], Dawid Pietrala[1], Pawel Andrzej Laski[1], and Henryk Lomza[2]

[1] Kielce University of Technology, Kielce, Poland
{j.zwierzchowski,dpietrala,pawell}@tu.kielce.pl
[2] PBF, Kraśnik, Poland
hlomza@flt.krasnik.pl

Abstract. A stand for durability testing of rolling bearings is presented in this paper. The project assumptions of the stand, mechanical construction and architecture of data acquisition and analysis system (hardware and software) are described in this paper. Finally, there are conclusions about stand's capabilities included. The device allows is capable of testing the running characteristics of bearings under varying test conditions and loads, speeds, lubrication conditions and temperatures. This stand provides precise measurements and autonomic work which is necessary in long-term researches. Compact construction of the stand containing overload protection system and automatic turn off system, activated after exceeding limit values of operating parameters or diagnostic signals, allows autonomic work without frequent human supervision. The data obtained from the stand will be used in future researches. In addition, they allow to test the service life under the set operating conditions.

Keywords: Rolling bearings · Durability test

1 Introduction

Measurements of bearings durability are very important in the engineering industry. Through durability measurements of bearings, we can evaluate the bearing's operating time in various devices, which contributes to the right moment of their replacement and, as a result, we can avoid damage to the machine or vehicles. Correctly designed strength tests help the bearing company determine their lifetime, warranty, maximum loads and service recommendations. In this article, we will focus on presenting a measuring stand designed for the needs of a rolling bearing factory. We will describe the method of data acquisition, we will describe the software dedicated to stand, and in particular, we will focus on vibration measurement.

N. M. Durakbasa and M. G. Gencyilmaz (Eds.): ISPR 2018, *Proceedings of the International Symposium for Production Research 2018*, pp. 201–207, 2019.
https://doi.org/10.1007/978-3-319-92267-6_16

2 Durability Test Stand

The article presents a test stand used to conduct durability tests of journal bearings and roller bearings. The device is adapted for simultaneous testing of four bearings with an inside diameter from 60 mm to 120 mm. The stand consists of base on which the body is mounted, horizontal pressure unit, vertical pressure unit and drive unit.

Figure 1a shows general construction of measurement stand for bearings tests. Whereas Fig. 1b shows the cross-section of the central part of the measuring machine, where the bearing measuring units (Fig. 1b-1) are located. The general construction of this machine part can be described as follows. Test head is mounted in the body and contains set of four bearings (Fig. 1b-1), drive shaft, set of vibration test sensors (Fig. 1b-3), temperature sensors and lubrication system components. Each of the pressure clamping sets consists of a hydraulic cylinder with a piston diameter of 200 mm and a nominal pressure of 25 MPa, two actuator inductive end position sensors and a ZEPWN CL16 m force sensor measuring between 0–400 kN for vertical clamping pressure and 0–300 kN for horizontal one. The spindle drive assembly consists of a 5.5 kW AC electric motor, belt transmission, torque sensor and rotary speed sensor KTR Dataflex 32. The stand is equipped with four temperature sensors designed for bearing temperature measurement and four acceleration sensors measuring the vibrations of the IFM VSA004 bearings. The drive assembly and pressure control assembly are shown in Fig. 2.

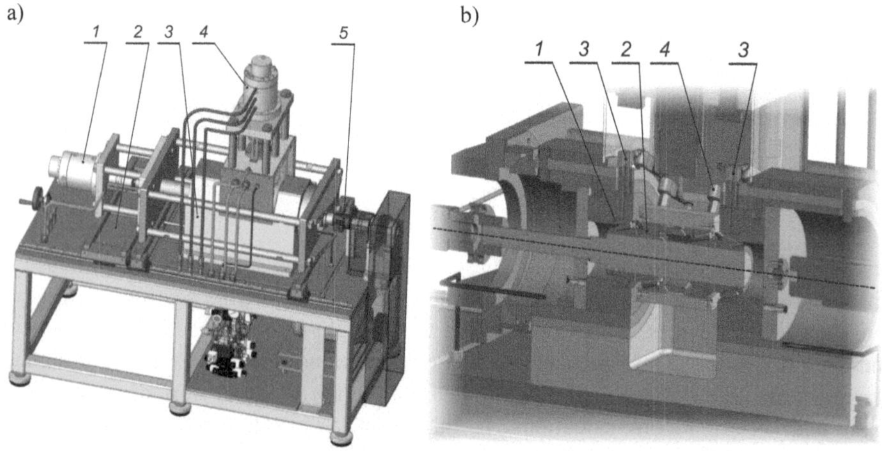

Fig. 1. Stand for measuring the durability of rolling bearings. (a) general view of the measurement stand 1 - longitudinal load cylinder, torque sensor, 2 - base, 3 - body, 4 - radial load actuator, 5 - torque sensor (b) 1 - support bearings 2 - test bearings, 3 - vibration sensors

A. *Test procedure*

The test procedure consists in mounting in the test head a set of test bearings and possibly support bearings. For each series of tested bearings dedicated components of

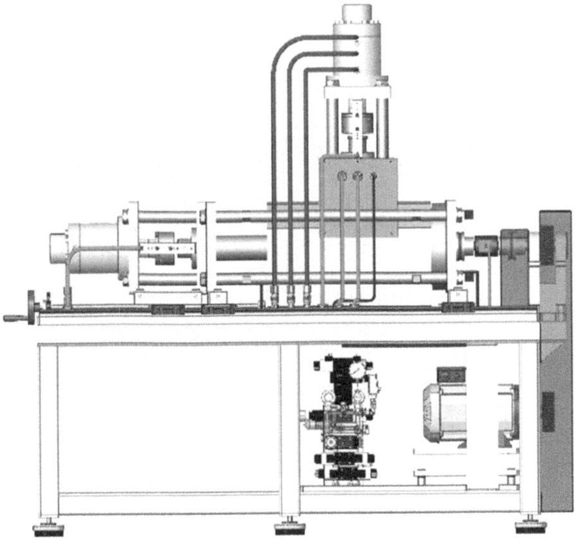

Fig. 2. General view of the measuring station showing the pressure control unit and the drive unit

the measuring head are intended. Next, acceleration and temperature sensors as well as accessories for the lubrication and cooling system are mounted in the head. The head prepared in this way is mounted in the body of the station. In the next step, the drive unit is started.

The rotational speed of the shaft in head is increased to the nominal value, depending on the size of the bearings from 1200 rpm to 4000 rpm. Then, axial and radial loads with the value determined by the test program are exerted on the bearing set using hydraulic cylinders. What's more, the station allows us to programmatically change the value of the load at any time during the test. A simple language for setting test parameters is shown in Fig. 3a. The order and time of the test execution is controlled by a PC computer which sends specific instructions to the measuring machine controller via serial RS232 communication. During the test, the temperature of the measuring node is controlled by means of flow lubrication. During the station's operation, the vibrations generated by each of the bearings, the temperature of each bearing, the temperature of the lubricant, the torque and the rotational speed of the head and the forces generated by the pressure units are measured. All operating parameters of the device and the tested bearings are systematically archived by the control and measurement system. An example of a graph is shown in Fig. 3b.

B. *Bearing vibration measurement*

One of the ways to determine if the bearing works correctly is to measure the vibration of the bearing. In the proposed measuring machine, there is the possibility of simultaneously testing 4 bearings. For vibration measurement, we selected IFM VSA002 sensors that are connected to the VSA004 control module. Communication

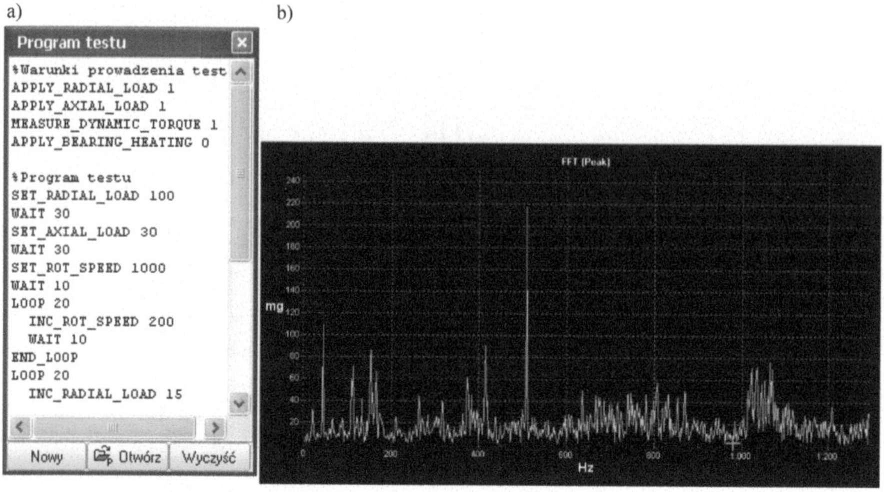

Fig. 3. (a) Sample test program (b) Peak monitoring chart illustrated on the prepared software

between the device and sensors is carried out using the TCP/IP protocol. Sensor manufacturer has joined the library with functions to control their sensors. In developed stand program we use this library. The prepared software was written in a high-level C++ language. When writing the application, it should be noted that the measurement process on the sensor controller side takes place in an infinite loop, therefore the entire processing of acquired measurement data should be carried out in a separate program thread. Contact with the controller is done using the so-called callback function, it means that the user of the driver writes the body of this function and it is responsible for the speed of the sensor driver. In the proposed software, work with the controller is limited only to cyclical copying of memory, setting of operating parameters and logical replacement of sensors (they cannot be read at the same time). All work related to data interpretation is done in a separate thread on a PC.

This programming approach allows you to quickly receive the following data from the sensor:

(a) monitoring of raw data,
(b) spectrum monitoring,
(c) peak monitoring.

Monitoring sensors raw data allows us to collect measurement data without processing on the sensor controller, this way allows us to conduct your own analysis of the signal. Before the measurement, the sensors should be calibrated according to the procedure specified in the device's specification and the sensor's scale should be determined. On the basis of further research, we additionally checked the sensors at the laboratory stand at the Kielce University of Technology. For this purpose, we used the inductor of nominal frequencies. Figure 4 is a chart with sample data illustrated with the software included with the sensors. From the controller you can read the signal with sampling rate 20kS/s or 100kS/s. Each received raw sample from the controller is a 16-bit integer.

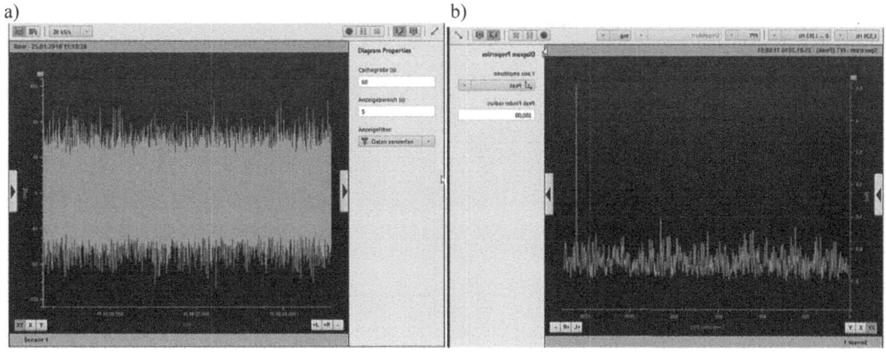

Fig. 4. Sample graphs obtained during vibration measurement (a) graph of data from the measuring sensor without processing (raw data) (b) graph of measurement data from the sensor showing the spectrum

Spectrum monitoring is the basic function of the controller. After selecting this option, the controller sends 850 samples that are full spectrum. During the measurement, we can choose the FFT or HFFT standard. Before starting the measurement, the measurement resolution should be determined, which ranges from 24.414 Hz to 0.192 Hz. As mentioned above, the control system only sends 850 samples, so the frequency range that can be displayed on the screen is the product of 850 and resolution. The abscissa (y) axis can be scaled in mm, mm/s, mg, where mg is the basic and default unit. If you choose HFTT analysis, the software user must choose the type of filter for HFFT.

Rotating elements in bearings are a very important part of every machine. In case of progressive wear, single damage frequencies are generated for the rolling bearing. The frequencies of rolling bearing damage depend on the geometry of the bearing (determined by the type of bearing and manufacturer) and is unique to each bearing. The frequency of damage must be calculated and taken into account as the rotational speed of the shaft with a rotational speed of 1 Hz.

An important measurement parameter is the calculation of RMS (root mean square), when using IFM sensors we can get the result as a-RMS (time domain) monitor the RMS of the acceleration and v-RMS (time domain) monitor the RMS of the vibration velocity in a frequency range adjustable via filters. Figure 5 shows a schematic representation of the RMS calculation method, where the selected *dt* interval is a measuring divergence.

In the developed software dedicated to the measuring device, the important measurement data is the peak in the time domain, as shown in Fig. 5b. In this case, the distance dt is also a measurement section. Peak measures the maximum amplitude on a dynamic input within the set measurement time. Due to the very short measurement time, this type of measurement is particularly suitable for machine protection, for example in emergency situations. The measuring range can be set within the limits (0.64 to 1.3 s).

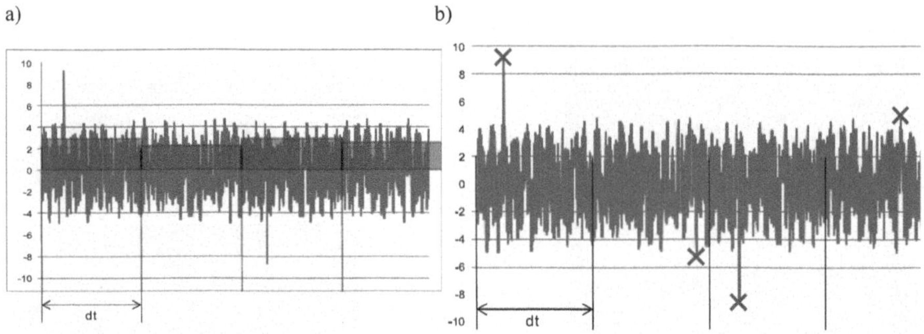

Fig. 5. Graph showing (a) RMS (root mean square) on the measuring section dt (b) peak marked in red also on the measuring part (Color figure online)

3 Conclusions

The article describes the concepts of a measuring system that tests the durability of rolling bearings designed for the needs of the PBF bearing factory in Kraśnik. A computer program was designed to monitor the most important parameters of the bearing's operation. The software allows the analysis of data sent from vibration sensors, temperature sensors. The user of the measuring station can, to a large extent, set the bearing load parameters through a user-friendly interface. The work describes the operation of sensors investigating vibrations and their implementation in the software. The possibility of using information pre-calculated by the sensor controller was investigated. It can be assumed that the data received from the sensors can be used to monitor the bearing durability test process. However, a detailed analysis and employee alerting system on bearing damage should be made on the basis of raw data from the sensor. It should be noted that four bearings are subjected to a test at a time.

Acknowledgment. The publication was created as a result of research and development carried out by Fabryka Łożysk Tocznych Kraśnik SA together with the Kielce University of Technology in the project entitled "Establishment of a R & D Center in FŁT Kraśnik SA" under the Intelligent Development Operational Program 2014-2020, co-financed from the funds of European Regional Development No. CBR/1/50-52/2017 from 07/04/2017

References

1. Simrit: Simmerings and Rotary Seals (2007) Freundenberg Simrit GmbH & Co. KG/Technical Manual
2. SKF rolling bearings catalogue (2013) ©SKF Group 2013. www.skf.com/binary/77-121486/SKF-rolling-bearings-catalogue.pdf
3. SKF Group (2012) Bearing designs – bearing testing. In: Railway technical handbook, vol 1, pp 99–105
4. Randall RB, Antoni J (2010) Rolling element bearing diagnostics – a tutorial. Mech Syst Signal Process 25:485–520

5. Wieczorek AN (2015) Designing machinery and equipment in accordance with the principle of sustainable development. Manag Syst Prod Eng 17(1):28–34. https://doi.org/10.12914/MSPE-05-01-2015
6. Jurecki R, Pokropiński E, Więckowski D, Żołądek Ł (2017) Design of a test rig for the examination of mechanical properties of rolling bearings. Manag Syst Prod Eng 25(1):22–28
7. Mehdigholi H, Rafsanjani H, Behzad M (2011) Estimation of rolling bearing life with damage curve approach. Pol Marit Res 18:66–70

Amplitude Surface Texture Parameters of Models Manufactured by FDM Technology

Tomasz Kozior(✉) and Stanisław Adamczak

Faculty of Mechatronics and Mechanical Engineering,
Kielce University of Technology, Kielce, Poland
{tkozior,adamczak}@tu.kielce.pl

Abstract. The paper present test results of a measurement surface texture of models manufactured by FDM (fused deposition modeling) technology. The material used for constructing sample models was a structural material - ABS P430. A Dimension 1200es machine by Stratasys was used for the construction of the models. The process parameter, of which the impact on the quality of the surface texture was analysed, was the model orientation on a virtual building platform. A Form Talysurf PGI 1230, Taylor Hobson contact profiler was used for the measurements, and the test results were compared with the results obtained during the measurements with an optical profilometer.

Keywords: Additive technology · FDM · ABS · Roughness surface texture
Process parameters

1 Introduction

The dynamic technological development observed over the last several decades, regarding mechatronic systems in particular, significantly affects the development of all sorts of manufacturing technologies. It applies to both the development of conventional manufacturing methods, such as founding or machining, plastic forming, as well as unconventional manufacturing technologies (additive, water-jet, laser or electrochemical technologies). In the case of additive manufacturing technologies, the development of precision mechatronic systems improved the input material feeding process (powder, liquid resin, rod) and increase production speed. Moreover, the development of chemistry resulted in a dynamic growth in the manufacturing of new materials, which enabled the accommodation of generative (additive) technologies to new industry types. All of the aforementioned improvements resulted in a particularly extensive development of the FDM (Fused Deposition Modeling) technology. This technology, known since the 1980s [1] due to the low price of materials used and a relatively simple building nature, is one of the most commonly used forms of additive production of prototypes.

Unconventional manufacturing methods, thanks to the layered nature of construction from 3D CAD models, have numerous advantages, among others: rapid construction time, low manufacturing cost, the possibility to create models with complex shapes (outer, in particular) and a wide range of used materials. Metal-based, as well as plastic-base materials are widely used. In the case of technologies utilizing

N. M. Durakbasa and M. G. Gencyilmaz (Eds.): ISPR 2018, *Proceedings of the International Symposium for Production Research 2018*, pp. 208–217, 2019.
https://doi.org/10.1007/978-3-319-92267-6_17

plastics, it is possible to use materials such as: polyamids (e.g. with a glass fibre addition) - SLS (selective laser sintering), liquid polymer resins - PJM (polyjet matrix), Stereolitography and ABS in the FDM technology. Moreover, when using the technologies based on metallic powders, e.g., SLM (selective laser melting), it is available to use materials such as, for example, corrosion-resistant 316L steels. Additive technologies are mainly used for the construction of prototypes, short production runs, as well as single final models. They fit perfectly, where the process instrumentation manufacturing cost is high and, in many cases, exceeds the cost of the end product. In the case of the SLM technology, there is a possibility to construct durable prototypes used for functionality tests, as well as in the dental industry, where permanent prosthetic dentures are made. Moreover, the wide application range of additive technologies based on plastics includes the construction of injection inserts, casting dies and moulds, particularly in the case of lost wax (investment moulding) [2, 3], e.g., in the medical industry [4]. In the medical industry, due to the high requirements regarding mechanical properties and corrosion resistance of the materials used, SLM technology is widely used [5].

In addition, the selected materials have international ISO certificates, which confirm meeting the basic requirements in regard to biocompatibility of the used materials. Plastics used in layered technologies, thanks to a wide range of used materials, are widely used in, e.g., construction models of technical seals [6]. In addition, due to the possibility of making so-called overprints, FDM technology is widely used in textile industry [7–10].

Regardless of the manufacturing process (conventional, unconventional), each technology is characterized by process parameters, crucially affecting the mechanical properties of the constructed models, their dimensional-shape accuracy and wear process, e.g. abrasive. Studies describing the impact of individual parameters of fused deposition technology on the aforementioned properties are shown, among others, in the papers [11–14]. In the case of the FDM technology, described in the research paper, several process parameters are also taken into account. The main parameters impacting a number of the aforementioned properties are: the thickness of the constructed layer (Lt, mm), model orientation on a virtual construction platform (Pd, °), process chamber temperature (T, °C), model cleaning process, etc. In the case of the thickness of a constructed layer, the selection of a nozzle through which the model, as well as supporting material is forced, is crucial. The model orientation on a construction platform in many cases depends on the geometry of the constructed model and the dimensions of the machine's (printer) operating chamber. The operating chamber temperature is set arbitrarily, depending on the used material. Moreover, there is an option to control such parameters as the nozzle (printhead) movement speed.

The execution accuracy of models constructed with the use of generative technologies was described in several research papers [5, 12, 13]. Initial literature analysis enables to determine that the value of surface geometric structure parameters, the amplitude parameters in particular, greatly impacts the wear of the generated models [15], therefore, it is important to determine, by way of experiments, the impact of the manufacturing process parameters on the SGP parameters. Experimental research demonstrated a clear impact of the model orientation on a number of mechanical properties, with the results presented in the paper [11].

2 FDM Technology

The paper discusses the tests covering samples made of ABS P430 construction material, with the use of a Dimension 1200es machine, Stratasys company. This technology utilizes two types of materials. The model and supporting material are forced through a printhead nozzle, where they are heated up to a temperature slightly lower than the material melting temperature (200–220 °C). In the case of the FDM technology, there is a possibility to build physical model on already existing objects, which, as demonstrated by publication [7–10] is used in, for example in textile industry. Due to the fact that this technology requires the construction of supporting structure, the construction of complicated models with a complex internal structure is limited. Due to the dynamic development of chemistry and the possibility to construct new materials, the efforts in the field of all additive technologies are concentrated on constructing soluble supporting materials. Examples of mechanical properties declared by the material manufacturer are shown in Table 1. Figure 1 presents the manufacturing process in FDM technology

Table 1. Mechanical properties of ABS P430 [16]

Mechanical properties	Young's modulus	Tensile strength	Charpy notched impact strength	Hardness Shore'a	Density
ABS P430	2320	37	106	75	1040
Unit	MPa	MPa	kJ/m2	Scale D	kg/m^3
Standard	EN ISO 527		ISO 180/1A	ISO 868	ASTM D792

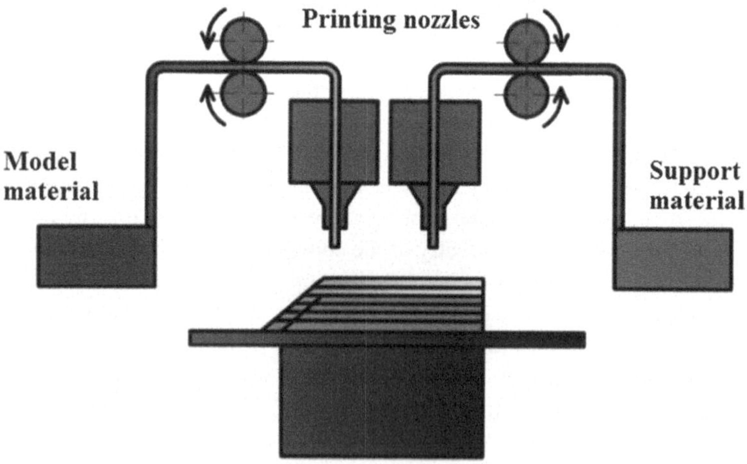

Fig. 1. The principle of FDM technology

3 Research

The sample shape (outer diameter 31.75 mm, inner diameter 19.1 mm and height - 8 mm) were designed in the Solidwork software. The analysed process parameter was the model orientation on the construction platform (set at three levels of variation) and its impact on the surface geometric structure parameters. Three characteristic orientations of sample models relative to the machine were selected - 0°, 45° and 90° (Fig. 2). A contact profiler Form Talysurf PGI 1230, Taylor Hobson company was used for the tests. The measuring instrument is equipped with a tip with a 2 µm blade radius, and the test parameters used are: measuring range - 4 mm, 0.8 mm cut-off filter. The samples models on a tester are shown in Fig. 4. The test involved the execution of statistical calculations with the use of the Statistica software. The test results were compared with the results obtained during an optical measurement using a Talysurf CCI Lite profiler by Taylor Hobson [18]. The test results indicated that there were significant differences in the values of sample surface roughness parameters for different location of the models on the virtual platform. Samples before cleaning, on the building platform are shown in Fig. 3.

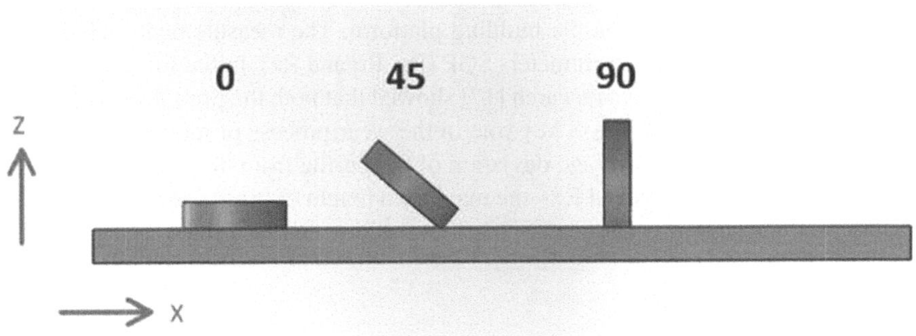

Fig. 2. Model placement on the wirtual platform

Fig. 3. Psychical model of the samples

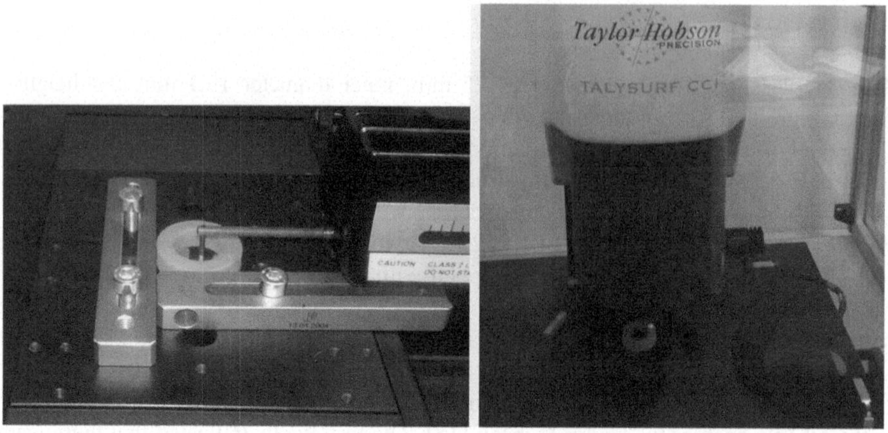

Fig. 4. Samples on the tested stands (a) contact measurement, (b) optical measurement

All types of samples (three variants of location) were made in the amount of five pieces, where for each sample the measurement was made three times. This means that for each model orientation, fifteen measurements of geometric surface structure parameters were carried out on the building platform. The measurement include three main so-called the amplitude parameters SGP (Ra, Rq and Rz). Initial literature analysis and tribological results of own research [17] showed that both the print direction and the SGP amplitude parameters play a key role in the wear process of models produced by additive technologies. The average deviation of the profile from the mean line Ra, Rq - root mean square roughness and Rz - the maximum height of profile are one of the most commonly used roughness parameters used in machine construction.

4 Results

The results of measurements of the amplitude parameters of the geometric structure of the surface are presented in Table 2 and Fig. 5. The purpose of the comprehensive analysis of the measurement results was the calculation based on the determination of the Pearson coefficient (Table 3).

Based on the results of the tests presented in Table 2, it can be concluded that the highest values of surface roughness parameters were measured on samples made with a given angle Pd - 45°. The average value of the Ra parameter in this case is equal 16.5 μm, Rq - 19.048 μm, and Rz - 66.183 μm. The most advantageous variant characterized by the smallest SGP parameters is the location of the sample models at an angle of Pd - 0°. In this case, the average values are: Ra - 6,023 μm, Rq - 8,014 μm and Rz - 36,396 μm respectively. The views of the surface profile and with the spatial view of the geometrical structure of the surface obtained directly from measuring machines Form Talysurf PGI 1230 and Talysurf CCI Lite respectively for three orientations of the samples relative to the building platform are shown in Figs. 6 and 7. The tribological tests presented in [17] clearly indicate that the highest wear was in the samples made with a given angle of

Table 2. Research results of measurement roughness surface parameters

Sample number	Roughness parameters		
	Ra	Rq	Rz
$\bar{x}$ 1-15	6,023	8,014	36,396
S	1,863	2,371	10,504
$\bar{x} \pm Up$	3,97	5,054	22,385
$\bar{x}$ 16-30	16,5	19,048	66,186
S	12,112	14,106	49,895
$\bar{x} \pm Up$	25,811	30,05	106,325
$\bar{x}$ 31-45	8,73	10,529	38,472
S	8,242	9,653	34,78
$\bar{x} \pm Up$	17,564	20,571	74,116

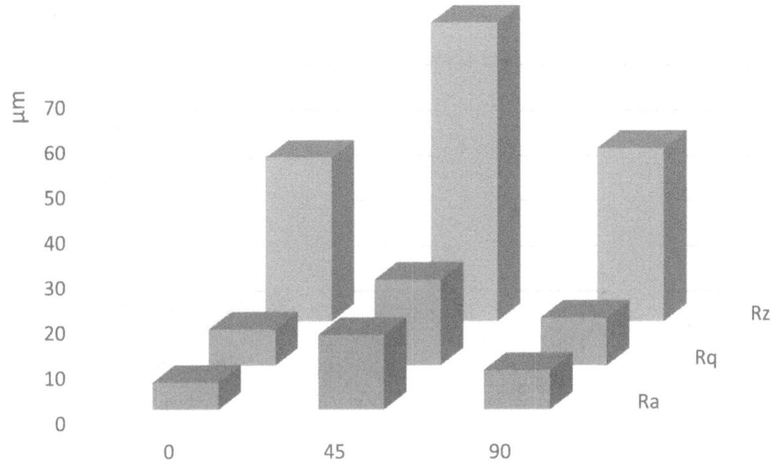

Fig. 5. Research results (a) 0°, (b) 45° and (c) 90°

Table 3. Calculated Pearson coefficient

Sample number	Pearson coefficient		
	Ra	Rq	Rz
$\bar{x}$ 1-15	30,928	29,59	28,861
$\bar{x}$ 16-30	73,408	74,025	75,385
$\bar{x}$ 31-45	94,418	91,686	90,404

Pd - 45°, where the SGP's amplitude parameters had the highest values. Comparing the test results with those presented in the publication [18], where the measurement of sample surfaces also related to SGP amplitude parameters determined by the optical method, it should be stated that in both cases the Pd - 45° is the least advantageous variant of placing the models on the building platform (the highest SGP parameters). In addition, the results of studies [18] have shown that also spatial parameters (measurement of 3D parameters) have a similar tendency for the angle Pd - 45°.

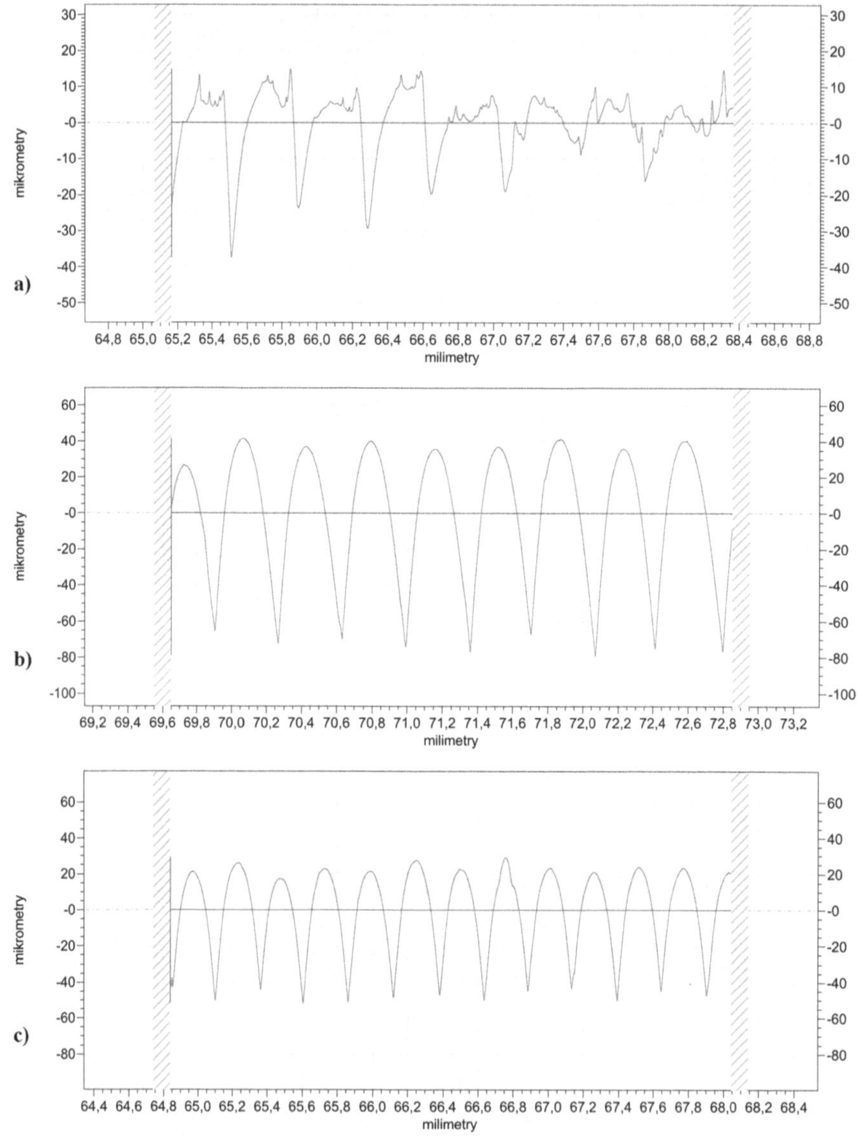

Fig. 6. Surface texture profile (a) 0°, (b) 45° and (c) 90°

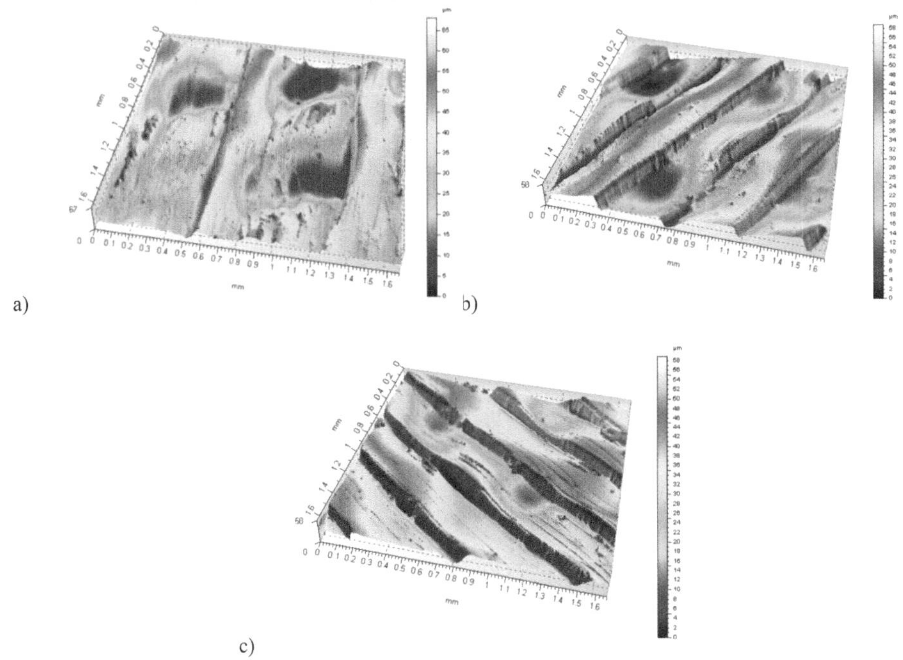

Fig. 7. 3D view of samples surface texture (a) 0°, (b) 45° and (c) 90°

A more in-depth analysis of the results, especially taking into account the standard deviation and the Pearson coefficient, indicates that there are very big discrepancies in the results, especially for samples made with a set angle of Pd - 45°. In this case, the deviation value is 12.112 and the Pearson coefficient – 73.408, which is a relative reference of the standard deviation against the mean value. When comparing the test results with the previous measurements using an optical profiler, it should be noted that the spread of the results presented with the use of the Pearson coefficient, also in the case of the previous studies, adopted the highest value for the sample model orientation angle of Pd - 45°.

5 Conclusions

Based on the presented results of own studies and a review of the subject literature, the following general conclusions can be drawn:

The orientation of the models on the construction platform is the crucial process parameter impacting the quality of the surface geometric structure and its parameters. The least favourable model positioning variant during construction, exhibiting the highest values of SGP amplitude parameters is the inclination of the samples relative to the construction platform at an angle of Pd - 45°.

The spread of the results calculated for contact measurements indicates the highest value for the samples prepared with a set angle of Pd - 45°. This is also confirmed by the analyses of previous tests conducted with the use of an optical profiler. In addition, tests have shown that printing samples with a pre-set angle of 45° adversely affects the technical capabilities of the measurement (low number of measured measuring points).

The specimens of the 45° and 90° are clearly visible residues of supporting material that could not be removed without disturbing the microgeometry of the structure.

In the future, it is planned to extend the scope of the research with a larger range of inclination of the sample models, and to study the impact of SGP parameters on the abrasive wear of sample models.

Acknowledgment. The publication was created as a result of executing a research-development project with the Polish Bearing Factory, Kraśnik S.A. and Kielce University of Technology in the project entitled "*Establishing a R&D Centre at FŁT Kraśnik S.A.*" within the framework of the smart growth 2014–2020 Operational Programme co-financed by the European Regional Development fund no. CBR/1/50-52/2017 of /7/4/2017.

References

1. Crump S (1989) Apparatus and method for creating three-dimensional objects, US 5121329 A
2. Adamczak S, Zmarzły P, Kozior, T, Gogolewski D (2017) Analysis of the dimensional accuracy of casting models manufactured by fused deposition modeling technology. Eng Mech 66–69
3. Coniglio N, Sivarupan N, El Mansori M (2018) Investigation of process parameter effect on anisotropic properties of 3D printed sand molds. Int J Adv Manuf Technol 94:2175–2185
4. Dikova T, Vasilev T, Dzhendov D, Ivanova E (2017) Investigation the fitting accuracy of cast and SLM Co-Cr dental bridges using CAD software. J IMAB 23:1688–1696
5. Chlebus E, Gruber K, Kurzac T, Kurzynowski T, Stopyra W (2016) Influence of laser power on the penetration depth and geometry of scanning tracks in selective laser melting. Laser Technology 2016: Progress and Applications of Lasers
6. Kundera Cz, Bochnia J (2014) Investigating the stress relaxation of photopolymer O-ring seal models. Rapid Prototyp J 20:533–540
7. Fafenrot S, Grimmelsmann N, Wortmann M, Ehrmann A (2017) Three-Dimensional (3D) printing of polymer-metal hybrid materials by fused deposition modeling. Materials (Basel) 10:1–14
8. Sabantina L, Kinzel F, Ehrmann A, Finsterbusch K (2015) Combining 3D printed forms with textile structures - mechanical and geometrical properties of multi-material systems. Mater Sci Eng, 1–5
9. Martens Y, Ehrmann A (2017) Composites of 3D-printed polymers and textile fabrics. Proceedings Paper
10. Grimmelsmann N, Martens Y, Schal P, Meissner, H, Ehrmann, A (2016) Mechanical and electrical contacting of electronic components on textiles by 3D printing. Proceedings Paper, pp 66–71
11. Kozior T, Kundera Cz (2017) Evaluation of the influence of parameters of FDM technology on the selected mechanical properties of models. Procedia Eng 192:463–468
12. Budzik G, Burek J, Bazan A, Turek P (2016) Analysis of the accuracy of reconstructed two teeth models manufactured using the 3DP and FDM technologies. Strojniski Vestnik – J Mech Eng 62:11–20

13. Polák R, Sedláček F, Raz K (2017) Determination of FDM printer settings with regard to geometrical accuracy. In: Katalinic B (ed) Proceedings of the 28th DAAAM International Symposium-2017. Published by DAAAM International, Vienna, Austria
14. Salazar AG, Martín MAP, Garcia-Granada AA, Reyes G, Puigoriol JM (2018) A study of creep in polycarbonate fused deposition modelling parts. Mater Design 141:414–425
15. Grzesik W (2015) Effect of the machine parts surface topography features on the machine service. Mechanik :587–593
16. www.stratasys.com. 14 June 2018
17. Kozior T, Kundera Cz (2016) Assessment of tribological properties of polymers used in additive technologies SLS and PJM. Tribologia, pp 73–84
18. Kozior T, Kundera Cz (2018) Surface texture of models manufactured by FDM technology. International conference – Electromachining 2018, Bydgoszcz,– Web of Science base - in review

Analysis of the Impact of Ball Bearing Track Waviness on Their Frictional Moment

Stanisław Adamczak, Łukasz Gorycki(✉), and Włodzimierz Makieła

Department of Mechanical Technology and Metrology,
Kielce University of Technology, Kielce, Poland
{adamczak, lgorycki, wmakiela}@tu.kielce.pl

Abstract. Frictional moment is one of the parameters that determine the time and quality of working bearings. Therefore, it is important that it is possible to accurately predict the value of the frictional moment. This article analyses the impact of the track waviness on the frictional moment. In the research, a correlation analysis of selected wavelength parameters and frictional moment was performed and frictional moment. Those parameters wavelength are not included in the models used to determine the theoretical value of the frictional moment.

Keywords: Frictional moment · Ball bearing · Track waviness

1 Introduction

The frictional moment is one of the most important operational parameters in the bearings. On its basis, it is possible to predict bearing wear and its maximum working time. Therefore, it is very important that there is a possibility to determine with the highest accuracy the theoretical value of the frictional moment. The currently used models that allow to determine the theoretical value of the frictional moment do not take into account all the factors that affect it, which undermines their accuracy [1]. However, the formulas that allow a more accurate determination of the frictional moment can only be used under very specific conditions [2]. For this reason, an analysis of the impact of selected factors on the resistive torque was carried out. The way the balls move on the treadmill is very important in relation to the frictional moment. It is known that the surface roughness will affect the friction and thus the moment of rolling friction (frictional moment) [3]. However, it is not known what effect track waviness has on the frictional moment.

This article presents the analysis of the influence of selected wavelength parameters on the resistive torque. This is part of the research analysing the impact of selected factors that are not included in the main models designed to determine the theoretical value of the resistance model [4].

N. M. Durakbasa and M. G. Gencyilmaz (Eds.): ISPR 2018, *Proceedings of the International Symposium for Production Research 2018*, pp. 218–223, 2019.
https://doi.org/10.1007/978-3-319-92267-6_18

2 Methodology of Measurements

The measurements were carried out on open bearings 6203. These bearings were divided into 5 groups according to their subtype and were marked by subsequent Roman numbers from I to V. All bearings were made in the same accuracy class. In addition, each group consists of bearings taken successively from a given production series, which minimizes the impact of tool wear on the differences in the track waviness. The bearing sub-type itself should also have no effect on the track waviness.

Frictional moment tests were carried out on the STPM Momentometer (Fig. 1) designed and made at the Kielce University of Technology [5]. This device allows you to measure the frictional moment of deep-groove and angular contact bearings with given parameters: rotational speed from 1 to 26,000 rpm, radial load up to 200 N and axial load up to 400 N. The measurement time of each bearing was six minutes, with one minute is the time required to distribute a small amount of Droser 10 oil with was introduced to secure the bearing track. The remaining 5 min is the correct measurement during which the value of the frictional moment was recorded. The measurements were performed at a radial load of 100 N and a rotational speed of 8000 rpm.

Fig. 1. Momentometer STPM

The waviness measurements were carried out on the measuring system for the shape outline of Talyrond 73. located in the Laboratory of Computer Measurements Geometrical Size in Kielce University of Technology. On this device, the measurements of the waviness of the outer and the inner rings tracks were carried out. Tracks measurements were made in cross-sections determined by operation angle of these bearings [1]. The ROFORM software was used to analyze the measured profile, in which a Gauss filter of 15–500 was used to filter the results.

3 Measurement Results and Calculations

The results of the frictional moment and waviness measurement for the external (er) and inner (ir) ring are presented in Table 1, in the form of the entire range of variation.

Table 1. Results of the frictional moment and selected wavelength measurements

Groups		I	II	III	IV	V
Frictional moment	N mm	0,86–1,73	0,79–1,46	0,96–2,13	0,68–1,84	0,77–1,30
W_t ir	µm	0,023–0,126	0,025–0,083	0,029–0,192	0,035–0,291	0,025–0,193
W_p ir	µm	0,011–0,052	0,012–0,068	0,013–0,140	0,017–0,068	0,011–0,178
W_v ir	µm	0,012–0,097	0,012–0,160	0,012–0,116	0,014–0,223	0,011–0,046
W_5 ir	µm	0,017–0,050	0,016–0,036	0,019–0,131	0,023–0,225	0,016–0,053
W_a ir	µm	0,003–0,006	0,003–0,004	0,003–0,020	0,003–0,029	0,003–0,005
W_t er	µm	0,032–0,071	0,039–0,170	0,035–0,250	0,035–0,216	0,047–0,480
W_p er	µm	0,015–0,032	0,021–0,117	0,017–0,188	0,017–0,137	0,024–0,073
W_v er	µm	0,016–0,250	0,018–0,053	0,017–0,063	0,018–0,096	0,021–0,081
W_5 er	µm	0,022–0,044	0,023–0,509	0,025–0,189	0,023–0,171	0,029–0,145
W_a er	µm	0,004–0,055	0,005–0,015	0,005–0,028	0,004–0,027	0,004–0,032

Measured wavelength parameters:
W_t - Waviness
W_p - Maximum peak
W_v - Maximum valley
W_5 - Surface macrowaviness
W_a - Arithmetic mean deviation of the profile

Using the Statistica software, a correlation analysis of selected wavelength parameters and a frictional moment was carried out [6]. On the basis of these results, it was determined whether there is a need for more accurate analysis, which allowed to determine the quantitative dependence of the frictional moment on the selected wavelength parameters. The correlation results are presented in Table 2, the results marked in bold were considered significant by Statistica.

Table 2. Analysis of the correlation between the frictional moment and the selected parameters of the waviness

	W_t ir	W_p ir	W_v ir	W_5 ir	W_a ir	W_t er	W_p er	W_v er	W_5 er	W_a er
Groups I–V	0,137	−0,006	0,163	0,187	0,186	−0,030	0,119	−0,083	0,132	−0,074
Group I	−0,046	−0,350	0,122	−0,116	−0,184	0,062	−0,289	−0,471	0,090	−0,477
Group II	0,259	0,394	0,174	0,129	0,038	0,261	0,411	−0,125	0,495	0,359
Group III	−0,104	−0,080	−0,103	−0,120	−0,143	−0,057	−0,067	−0,013	−0,039	−0,038
Group IV	**0,504**	**0,506**	0,390	**0,543**	**0,524**	**0,620**	**0,584**	**0,629**	**0,609**	**0,607**
Group V	−0,273	−0,236	−0,229	−0,264	−0,342	−0,435	**−0,617**	**−0,592**	**−0,592**	**−0,579**

The most important for the analysis are the results from the first row (groups I-V). The level of frictional moment correlation with the analyzed parameters does not exceed ± 0.2. Based on these results, it can be assumed that there is no correlation between the selected waveform parameters and the frictional moment. For this also there is no basis for conducting an additional analysis defining the relationship between these parameters, e.g. creating a mathematical model.

The correlation analysis carried out showed that there is a significant interdependence only between the waviness of the bearing rings from group IV. The results from this group stand out in a significant way compared to other groups. In order to confirm whether this is really just a deviation from the norm, a graph of the dependence of the absolute value of the correlation coefficient on the maximum value of the W_t ir parameter was made (Fig. 2). Based on this chart, it can be concluded that the higher value of the W_t ir parameter generates a higher value of the correlation coefficient. It should be concluded that the influence of waviness on the frictional moment may be significant for surface irregularities, which is characterized by large values of the W_t ir parameter.

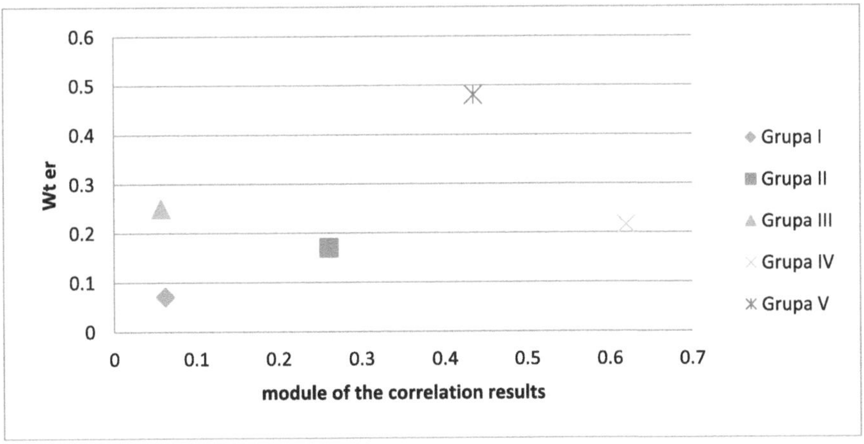

Fig. 2. A graph of the dependence of the absolute value of the correlation coefficient on the maximum value of the Wt ir parameter

Table 3. Analysis of the correlation between the rolling bearing's wave parameters. Parameters with high significance of influence on other factors were marked in bold.

	W_t ir	W_p ir	W_v ir	W_5 ir	W_a ir	W_t er	W_p er	W_v er	W_5 er	W_a er
W_t ir	1,000	**0,690**	**0,796**	**0,942**	**0,869**	**0,478**	**0,712**	**0,389**	**0,401**	**0,477**
W_p ir	**0,690**	1,000	**0,304**	**0,453**	**0,344**	0,198	**0,319**	**0,266**	**0,317**	**0,271**
W_v ir	**0,796**	**0,304**	1,000	**0,821**	**0,778**	**0,420**	**0,629**	**0,289**	**0,282**	**0,380**
W_5 ir	**0,942**	**0,453**	**0,821**	1,000	**0,967**	**0,533**	**0,791**	**0,398**	**0,398**	**0,508**
W_a ir	**0,869**	**0,344**	**0,778**	**0,967**	1,000	**0,546**	**0,819**	**0,351**	**0,392**	**0,528**
W_t er	**0,478**	**0,198**	**0,420**	**0,533**	**0,546**	1,000	**0,674**	**0,305**	**0,368**	**0,429**
W_p er	**0,712**	**0,319**	**0,629**	**0,791**	**0,819**	**0,674**	1,000	**0,416**	**0,558**	**0,624**
W_v er	**0,389**	**0,266**	**0,289**	**0,398**	**0,351**	**0,305**	**0,416**	1,000	**0,239**	**0,332**
W_5 er	**0,401**	**0,317**	**0,282**	**0,398**	**0,392**	**0,368**	**0,558**	**0,239**	1,000	**0,363**
W_a er	**0,477**	**0,271**	**0,380**	**0,508**	**0,528**	**0,429**	**0,624**	**0,332**	**0,363**	1,000

In order to confirm the legitimacy of using the selected wavelength parameter in the quantitative assessment, calculations of mutual dependence between particular parameters characterizing the treadmill waveform were carried out. The results of the calculations in Table 3 confirm the strong dependence between the tested parameters and allow to state that in further studies, the dependence of the resistive torque on the surface waviness, it is enough to use one selected parameter, e.g. W_t ir, which will be used in determining the mathematical model of the dependence of the frictional moment on the selected bearing parameters [7].

4 Conclusions

1. The results of the conducted tests have shown that the rolling surface of the bearing race can have an effect on the frictional moment at high values of this surface irregularity.
2. Further testing should by performed on bearings from group IV in order to demonstrate the causes of a significant correlation between the frictional moment, and their waviness
3. Correlation results between individual waviness parameters showed a strong dependence. This justifies the use of one selected waviness parameter in further research.

Acknowledgment. The publication was created as a result of research and development carried out by the Polish Bearing Factory, Kraśnik S.A. together with the Kielce University of Technology in the project entitled "The Establishment of R & D Centre in FŁT-Kraśnik S. A." under the Smart Growth Operational Programme 2014-2020, co-financed from the European Regional Development Fund No. CBR/1/50-52/2017 from 07/04/2017.

References

1. Rolling bearings (2012) SKF Group
2. Kusznierewicz Z Metoda obliczania momentu tarcia w łożyskach tocznych kulko-wych zwykłych niedociążonych, PAK 57/211
3. Adamczak S (2008) Pomiary geometryczne powierzchni. Zarysy kształtu, falistość i chropowatość, WNT 2008
4. Adamczak S, Gorycki Ł, Makieła W (2016) The analysis of the impact of the design parameters on the friction torque in ball bearings, Tribologia, pp 11–19, May 2016
5. Patent: PL 214217 z 2012r
6. Copyright StatSoft, Inc. (2018). https://www.statsoft.pl/. Accessed 07 June 2018
7. Muciek A (2012) Wyznaczanie modeli matematycznych z danych eksperymentalnych. Oficyna Wydawnicza Politechniki Wrocławskiej 2012

Application of Wavelet Transform to Determine Surface Texture Constituents

Damian Gogolewski(✉) and Włodzimierz Makieła

Department of Manufacturing Engineering and Metrology, Kielce University of Technology, Kielce, Poland
{dgogolewski,wmakiela}@tu.kielce.pl

Abstract. The paper studies the possibility of application of a two-dimensional wavelet transform to analyse surface texture signals. The tests were directed at determining the impact of wavelet transformation parameters in the scope of using the wavelet analysis to separate the constituents of measured surface irregularities. The surface roughness signals obtained as a result of the Gaussian filter application and after wavelet decomposition were compared. Based on the obtained results, it was concluded that a wavelet transform may be used to separate 3D surface texture constituents.

Keywords: Filtration · Gaussian filter · Wavelet analysis · Surface roughness Statistical analysis

1 Introduction

A number of techniques may be used to successfully divide the constituents of surface texture formed in the course of the machining process. However, these methods apply to, especially, surface profiles. Expanded selected procedures may be used to execute the filtration process covering a 3D surface geometrical structure. However, there are certain limitations associated with the measurement method and the characteristics of signals obtained from measuring machine. Gaussian filters are most commonly used in such an analysis, with their application methodology specified in the standard PN-EN ISO 4287:1999/A1:2010. The filter type is defined based on obtained, selected *Ra, RSm* etc. surface roughness parameter values. Individual filters have defined lengths of measuring sections, over which a profile measurement should be conducted.

The size of the obtained measurement surface becomes an issue when measuring the surface texture of machine part elements using optical method. The use of lenses with a significant magnification, results in decreasing the studied area. A Talysurf CCI optical profilometer was used in the course of the studies. The measurements were taken using a $\times 10$ magnification, obtaining a signal corresponding to a 1.6 mm $\times$ 1.6 mm area. Therefore, for a significantly irregular surface, the filtration of the obtained signal may not be performed, according to the above standard, using Gaussian filters. Currently, there is no standard stipulating the methodology of filtration for such surfaces. Filtration with the use of wavelet function may be used as an alternative method for separating surface topography constituents.

N. M. Durakbasa and M. G. Gencyilmaz (Eds.): ISPR 2018, *Proceedings of the International Symposium for Production Research 2018*, pp. 224–231, 2019.
https://doi.org/10.1007/978-3-319-92267-6_19

When analysing the current state of the art, it must be noted that the wavelet transform is increasingly used in many fields of science. Many researchers used wavelet transform for the diagnostic analysis of signals [1, 2], including the analysis of 2D and 3D surface irregularities signals [3–6].

2 Test Methodology

The concept of the multi-resolution analysis developed by Mallat, which divides the analysed base signal into an approximated signal and a detail signal, utilized the idea of discrete signal processing. When analysing a signal with the use of a two-dimensional wavelet transform, the signal is filtered at each consecutive approximation level using two inter-complementing filters: high-pass and low-pass. The obtained approximated signal corresponds to low-frequency information, while the details present high-frequency information regarding the signal. Therefore, it can be concluded that the decomposition of measured surface on consecutive levels of an analysis using wavelet transform leads to separating the surface texture constituents. The surface roughness signal will be described as the sum of details formed at consecutive analysis levels to a determined, selected decomposition level. The studies were aimed at verifying the concept of adapting two-dimensional wavelet analysis to decompose surface topography signals into constituents with various frequencies (roughness, waviness and form) and to estimate the analysis level, for which a signal formed as a result of detail summing can be considered as a surface roughness signal. The analysis was conducted for selected mother wavelets.

Studies with the use of specific statistical tests were executed for this purpose. Filtration utilizing a 0.8 mm × 0.8 m Gaussian filter, as well as utilizing selected mother wavelets was conducted for individual signals of the measured surface. The obtained surface roughness signals were compared using selected statistical tests. The T2 Hotelling test at a significance level of $\alpha = 0.05$, Pearson correlations of a roughness signal for a surface formed as a result of the Gaussian filtration and a signal decomposed at a selected analysis level were used for the comparative studies. It was also checked whether there was a difference between selected surface roughness parameters [7].

The tests were conducted on C45 steel elements. The face milling process was conducted with set cutting parameters on a VMC 800 vertical machining centre by AVIA. Cutting tests were executed at defined process parameters. The input parameters of the milling process were: cutting depth over a range of $a_p = 0.1 - 2$ mm, feed per tooth $f_z = 0.1 - 0.22$ mm/tooth and cutting speed $v_c = 200$–350 m/min. Sixty-four test samples were obtained after the cutting tests.

3 Test Results

The first stage of the study was to determine the signal correlation coefficient value, in order to define the decomposition level, to which the filtration was to be conducted. The indicated analysis level has the highest probability of an obtained signal after a conducted wavelet analysis and Gaussian filtration. The Pearson correlation coefficient value at subsequent wavelet decomposition levels was determined for individual mother wavelets, for sample no. 8 (a_p = 2 mm, f_z = 0.14 mm/tooth, v_c= 200 m/min), for sample no. 29 (a_p = 0.1 mm, f_z = 0.22 mm/tooth, v_c= 250 m/min), for sample no. 34 (a_p = 0.5 mm, f_z = 0,1 mm/tooth, v_c= 300 m/min), for sample no. 59 (a_p = 1 mm, f_z = 0.18 mm/tooth, v_c= 350 m/min). Figure 1 was developed with the purpose of better visualization of the obtained results.

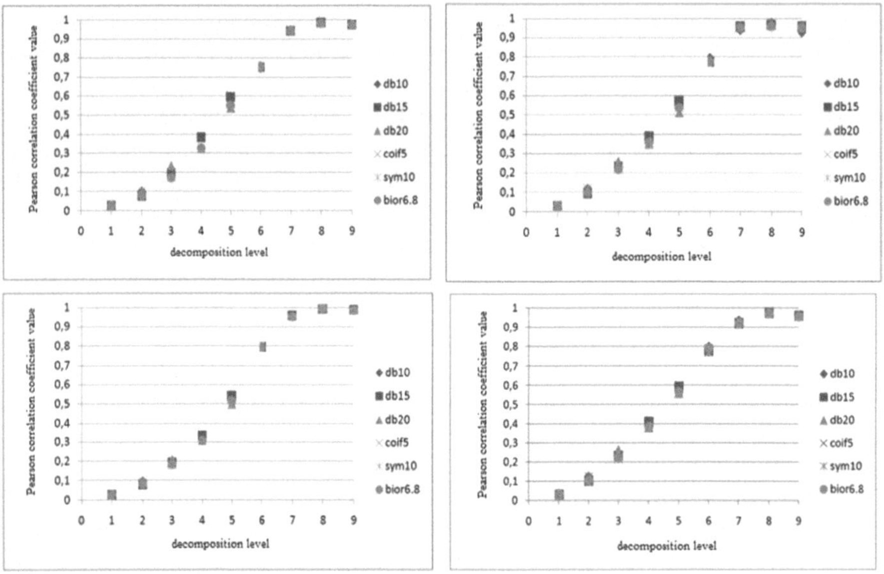

Fig. 1. Pearson correlation coefficient values – (a) sample no. 8 (b) sample no. 29 (c) sample no. 34 (d) sample no. 59

The values obtained at the initial analysis stages indicated no correlation between the studied signals. However, the coefficient value increases at subsequent decomposition levels. In the case of the selected samples, an almost complete correlation between signal coefficients was reached at the seventh, eighth and ninth levels for all mother wavelets used in the analysis. When analysing the obtained data in the form of the figures above, it should be noted that the signal obtained as a result of summing the details formed in the course of the measured surface analysis process, with the use of a two-dimensional wavelet transform, reaches the highest coefficient of conformity with the signal formed after the Gaussian filtration, at the eighth analysis level. Smaller values of the Pearson correlation coefficient were obtained for the selected mother wavelets at the next, ninth decomposition level.

Similar calculations were also conducted for the other surfaces of prepared samples. Depending on the mother wavelet used in the analysis and the studied sample, an almost complete correlation was reached at the seventh or eighth decomposition level.

The obtained values of the correlation coefficient for all studied surfaces showed that both signals exhibited the greatest coefficient correlation at the eighth analysis level.

For a selected level it was purposeful to determine the mother wavelets, for which it could be stated with a defined probability that the matrices of coefficients describing both surfaces did not significantly differ from each other. The T2 Hotellinga test at a significance level of $\alpha = 0.05$ was conducted for particular mother wavelets. Taking into account the size of the analysed matrices, the critical test value was determined at F = 1.1084. Table 1 contains the obtained statistical values for the selected samples, calculated for a surface formed as a result of summing the details up to the eighth decomposition level and as a result of Gaussian filtration.

Table 1. Calculation results for the eighth decomposition level - Hotelling T2 statistics

Mother wavelet	Sample no. 8	Sample no. 29	Sample no. 34	Sample no. 59
db10	0.00106	0.00169	0.00183	0.00252
db15	0.00107	0.00165	0.00183	0.00254
db20	0.00108	0.0016	0.00182	0.00271
coif5	0.00106	0.00185	0.00179	0.00272
sym10	0.00106	0.00185	0.00178	0.00262
bior6.8	0.00107	0.00185	0.00182	0.00267

Pursuant to the information contained in the table, it can be concluded that the analysis with the use of the T2 Hotelling test showed that the signals at the assumed significance level did not differ from each other. The study conducted on a sample of sixty-four showed that the signal resulting from a wavelet filtration can be treated as a 3D surface roughness signal.

The next study stage was the evaluation of selected roughness parameters for a surface formed with the use of wavelet filters. A comparative analysis involving the values of individual parameters obtained with the use of Gaussian filters (Table 2), as well as a result of the wavelet-transform-based decomposition, was conducted. The following 3D roughness parameters were selected for this purpose: *Sq, Sku, Sz, Sa, Sal, Str, Sdq, Vmc*.

Table 2. Surface roughness parameter values – Gaussian filter

Roughness parameters		Sample no. 8	Sample no. 29	Sample no. 34	Sample no. 59
Sq	μm	1.5527	1.1204	1.5467	0.9771
Sku		3.9334	6.0472	3.4655	2.7482
Sz	μm	8.7778	7.8959	11.3638	6.5753
Sa	μm	1.1777	0.7904	1.2495	0.8031
Sal	mm	0.0352	0.0364	0.0343	0.0383
Str		0.0417	0.044	0.0407	0.0887
Sdq		0.1502	0.1267	0.1538	0.1182
Vmc	mm^3/mm^2	0.0014	0.0008	0.0015	0.0009

Table 3 contains the calculated values of selected surface roughness height parameters for the selected samples, determined at the eighth level of a two-dimensional wavelet decomposition. Moreover, the relative difference "δ" of parameters was calculated, assuming the parameter values obtained by Gaussian filtration as the reference value. The applied, different filtration method determines the separation of other, subtle signal information from the input signal, and obtaining slightly different parameter values. According to [8], the characteristics of wavelet filters allow to obtain better filtration results due to more sharper transmission characteristics of the filter. Therefore, it was decided that two surfaces did not differ significantly, when the value of an index determining the relative change of roughness parameter is not higher than 20%.

Table 3. Calculation results -height parameter – eighth decomposition level

Mother wavelet	Sample no. 8		Sample no. 29		Sample no. 34		Sample no. 59	
	Parameter value	δ,%	Parameter value	δ,%	Parameter value	δ,%	Parameter value	δ,%
Sq, μm								
db10	1.62	4.36	1.14	1.88	1.58	2.26	1.01	3.04
db15	1.6	3.3	1.12	0.2	1.58	2.46	1.01	2.94
db20	1.6	2.78	1.11	0.91	1.58	2.44	1.01	2.89
coif5	1.59	2.33	1.12	0.46	1.58	2.34	1.01	3.24
sym10	1.59	2.32	1.12	0.29	1.58	2.44	1.01	3.15
bior6.8	1.59	2.37	1.12	0.07	1.58	2.43	1.01	3.3
Sku								
db10	3.75	4.71	5.51	8.81	3.26	6.05	2.74	0.28
db15	3.73	5.09	5.61	7.26	3.29	4.95	2.8	1.81
db20	3.74	4.96	5.96	1.49	3.37	2.82	2.8	2.01
coif5	3.8	3.48	5.89	2.56	3.37	2.68	2.89	5.08
sym10	3.78	3.98	5.64	6.69	3.35	3.44	2.85	3.79
bior6.8	3.78	3.99	5.69	5.89	3.34	3.55	2.84	3.32
Sz, μm								
db10	8.78	0.02	8.35	5.75	11.58	1.87	6.85	4.18
db15	8.76	0.22	8	1.32	11.47	0.98	7.04	7.08
db20	8.54	2.7	8.09	2.44	11.65	2.52	7	6.42
coif5	8.8	0.29	8.22	4.15	11.65	2.5	7.12	8.3
sym10	8.91	1.48	8.29	4.98	11.65	2.53	7.13	8.38
bior6.8	8.89	1.3	8.34	5.59	11.62	2.3	7.14	8.56
Sa, μm								
db10	1.25	5.88	0.84	5.66	1.28	2.55	0.82	2.59
db15	1.23	4.32	0.82	3.24	1.28	2.72	0.82	2.42
db20	1.22	3.39	0.79	0.01	1.28	2.51	0.82	2.23
coif5	1.2	1.91	0.79	0.35	1.28	2.42	0.82	2.4
sym10	1.2	1.92	0.8	1.39	1.28	2.57	0.82	2.45
bior6.8	1.2	1.92	0.8	1.7	1.28	2.46	0.82	2.7

When analysing the tabulated results for selected roughness height parameters, it can be concluded that the surface roughness values obtained for signals formed with the use of a two-dimensional wavelet transform only slightly differ from the value determined using Gaussian filters.

A similar list of the test results was prepared also for the other parameter groups. Table 4 includes test results for the Sdq hybrid parameter, Table 5 shows the analysis results for spatial parameters, while Table 6 shows the results for the Vmc volumetric parameter.

Table 4. Calculation results - Sdq hybrid parameter –eighth decomposition level

Mother wavelet	Sample no. 8		Sample no. 29		Sample no. 34		Sample no. 59	
	Parameter value	δ,%	Parameter value	δ,%	Parameter value	δ,%	Parameter value	δ,%
db10	0.17	12.8	0.14	14.2	0.17	13.19	0.14	14.88
db15	0.17	12.79	0.14	14.19	0.17	13.2	0.14	14.89
db20	0.17	12.79	0.14	14.19	0.17	13.2	0.14	14.89
coif5	0.17	12.79	0.14	14.19	0.17	13.2	0.14	14.89
sym10	0.17	12.79	0.14	14.19	0.17	13.2	0.14	14.89
bior6.8	0.17	12.79	0.14	14.19	0.17	13.2	0.14	14.89

Table 5. Calculation results - spatial parameters – eighth decomposition level

Mother wavelet	Sample no. 8		Sample no. 29		Sample no. 34		Sample no. 59	
	Parameter value, 10^{-2}	δ,%	Parameter value, 10^{-2}	δ,%	Parameter value, 10^{-2}	δ,%	Parameter value, 10^{-2}	δ,%
Sal, mm								
db10	3.93	11.87	3.73	2.58	3.56	3.61	4.14	2.59
db15	3.82	8.74	3.6	0.94	3.57	3.83	4.13	7.99
db20	3.77	7.13	3.51	3.52	3.56	3.72	4.14	8.15
coif5	3.72	5.96	3.55	2.53	3.56	3.56	4.19	9.48
sym10	3.72	5.95	3.55	2.34	3.56	3.77	4.17	8.92
bior6.8	3.73	6.07	3.59	1.28	3.56	3.78	4.18	9.09
Str								
db10	4.66	11.87	4.48	1.95	4.22	3.61	9.12	2.83
db15	4.53	8.74	4.59	4.25	4.23	3.83	9.27	4.49
db20	4.46	7.13	4.63	5.23	4.22	3.72	9.6	8.19
coif5	4.41	5.96	4.63	5.23	4.21	3.56	9.73	9.65
sym10	4.41	5.95	4.62	5.02	4.22	3.77	9.67	9.06
bior6.8	4.42	6.07	4.61	4.87	4.22	3.78	9.7	9.34

Table 6. Calculation results -Vmc volumetric parameter, mm^3/mm^2 – eighth decomposition level

Mother wavelet	Sample no. 8		Sample no. 29		Sample no. 34		Sample no. 59	
	Parameter value, 10^{-3}	δ,%	Parameter value, 10^{-4}	δ,%	Parameter value, 10^{-3}	δ,%	Parameter value, 10^{-4}	δ,%
db10	1.52	10.4	7.93	5.15	1.54	2.47	9.7	2.28
db15	1.47	6.87	7.82	3.68	1.54	2.49	9.7	2.31
db20	1.44	4.68	7.38	2.18	1.53	1.89	9.65	1.76
coif5	1.4	1.49	7.4	1.92	1.53	1.97	9.66	1.87
sym10	1.41	1.9	7.59	0.61	1.53	1.95	9.69	2.18
bior6.8	1.4	1.81	7.58	0.52	1.53	1.8	9.72	2.52

The calculation results listed in the tables above indicate that wavelet analysis can be used to separate surface texture constituents. For all selected surface roughness parameters at the eighth decomposition level, the obtained values of the δ coefficient were smaller than the critical value (20%).

In the course of analysing the obtained results, it is crucial to get a bigger picture of not only the parameter relative difference index value, but also make sure that the characteristics of the signal are significantly different. The surface layer of elements is described using certain elements, in relation to which, to a certain extent, although the obtained values are quantitatively different, it cannot be stated that the signal characteristics are significantly different. Such examples include the Str parameter. The literature indicates that the values of parameter Str > 0.5 characterize strongly isotopic surfaces, while Str < 0.3 strongly anisotropic. Obtaining a relative difference between parameter values does not always mean that a signal changed its nature.

When considering the obtained values of individual parameters and the values of the δ index, it was demonstrated that satisfactory results were obtained for the selected mother wavelets. It was decided that the relative difference criterion for parameter value was met, when at most the value of one of the selected parameters resulted in exceeding the threshold of the adopted critical value. In the case of almost all surfaces, it was demonstrated that the signals obtained through two different filtration methods did not vary significantly. The Table 7 below shows the number of samples, for which the criteria assumed in the mother wavelet function were satisfied.

Table 7. Number of samples satisfying the filtration criteria

Mother wavelet	db10	db15	db20	coif5	sym10	bior6.8
Number of samples	61	62	61	61	61	61

In the case of the samples, for which the adopted test assumptions were not satisfied, a significant difference in the values of the parameters determining the length of the fastest autocorrelation function disappearance section was mainly recorded (Sal). Nonetheless, in most cases the values of the δ index varied in the range up to 40%.

The conducted studies prove that a two-dimensional wavelet transform may be used to separate the surface texture constituents.

4 Conclusions

The paper verified the concepts of applying a two-dimensional wavelet transform to divide a measured surface into surface texture constituents, i.e., form, waviness and surface roughness. The conducted comparative analysis of the surface roughness signal obtained as a result of Gaussian filtration and a signal formed as a result of summing details created up to the eighth decomposition level with the use of a two-dimensional wavelet transform confirmed that in the case of selected mother wavelets, the obtained signals did not differ significantly. The applied, different filtration method determines the separation of other, subtle signal information from the input signal. Therefore, it can be concluded that a two-dimensional wavelet transform may be used to separate the surface texture constituents.

The use of an alternative filtration method based on wavelet analysis may be adopted mainly for analysing three-dimensional surface signals, for which it is troublesome, or in more than one case even impossible, to perform an analysis utilizing Gaussian filters. For very rough elements, the area where the 3D surface measurements should be executed determines that in the case of optical instruments, it is necessary to use surface stitching, which greatly extends the execution time of one measurement. Therefore, the possibility of using an alternative method, which will enable executing the filtration process, and consequently, obtain a surface roughness signal is crucial.

Acknowledgment. The paper has been elaborated within the framework of the research project entitled "Theoretical and experimental problems of integrated 3D measurements of elements' surfaces", reg. no.: 2015/19/B/ST8/02643, ID: 317012, financed by National Science Centre, Poland.

References

1. Djebala A, Ouelaa N, Hamzaoui N (2008) Detection of rolling bearing defects using discrete wavelet analysis. Meccanica 43(3):339–348
2. Wang LN, Wang HB, Cai YH, Wang SL (2012) Fault diagnosis system of rolling bearing based on wavelet analysis. Appl Mech Mater 166–169:951–955
3. Abdul-Rahman HS, Jiang X, Scott P (2013) Freeform surface filtering using the lifting wavelet transform. Precis Eng 37:187–202
4. Jiang X, Blunt L (2004) Third generation wavelet for the extraction of morphological features from micro and nano scalar surface. Wear 257:1235–1240
5. Stępień K (2014) Research on a surface texture analysis by digital signal processing method. Tehnicki Vjesnik – Tech Gaz 21(3):485–493
6. Zahouani H, Mezghani S, Vargiolu R, Dursapt M (2008) Identification of manufacturing signature by 2D wavelet decomposition. Wear 264:480–485
7. Makieła W, Gogolewski D (2016) Metoda rozdzielania składowych nierówności powierzchni za pomocą dwuwymiarowej transformaty falkowej. Mechanik 11:1716–1717 (in Polish)
8. Lingadurai K, Shunmugan MS (2006) Metrological characteristics of wavelet filter used for engineering surfaces. Measurement 39:575–584

Assessment of the Accuracy of Laser Vibrometer for Measurement of Bearing Vibrations

Stanisław Adamczak and Mateusz Wrzochal(✉)

Department of Manufacturing Engineering and Metrology,
Kielce University of Technology, Kielce, Poland
{adamczak,mwrzochal}@tu.kielce.pl

Abstract. The advantage form no interference in the measurement system causes that Doppler laser vibrometers could be used in combination with other contact sensors, and also very helpful for tests determining the metrological properties of rolling bearing vibration measurement systems. Therefore, it is necessary to study problems related to the impact of the laser beam on the external surfaces of the tested bearings, as well as to determine the scope of use of laser vibrometers in the study of bearing vibration measurement systems.

Keywords: Bearings · Vibrations · Vibrometer · Measurements Accuracy

1 Introduction

In the 80s of the twentieth century, the first laser vibrometers were created, since then very quickly they began to find newer applications. A very extensive overview through the basic use of vibrometers can be found in [1]. Among the available publications, you can also find items that describe in detail specific applications, e.g. in paper [2], the use of a laser vibrometer to assess the quality and safety of building materials is presented.

The wide range of these devices results in the fact that we are able to use them to study relatively large structures (buildings [3]) and small components of machine parts (bearing balls [4]). Recently, the use of vibrometers for the diagnosis of machines and mechanical devices is also very noticeable. For example, the authors in [5, 6] present the use of a Doppler laser vibrometer to detect damage in working electric motors. The article [7] describes the use of a vibrometer for monitoring high-speed milling, while the study [8] describes the practical use of a vibrometer in the quality control of household appliances. In [9], the use of the vibrometer for RPM measurements is described, whereas in [10] the same authors take up the subject of developing a model for illustrating the interactions of the beam with rotating elements (rotors). A very important element of machines and mechanical devices containing rotating units is a rolling bearing. Laser vibrometry is also widely used in measurements related to rolling bearings. It is currently difficult to find research on the comparison of electrodynamic velocity sensors with a laser vibrometer in relation to vibration measurements of rolling bearings.

N. M. Durakbasa and M. G. Gencyilmaz (Eds.): ISPR 2018, *Proceedings of the International Symposium for Production Research 2018*, pp. 232–238, 2019.
https://doi.org/10.1007/978-3-319-92267-6_20

In electrodynamic sensors, a moving plunger on which the coil is wound is in contact with the operating bearing. The sensor housing is located in a fixed mounting system, which is a reference system. Inside the sensor there is a permanent magnet, in whose field the winding moves under the effect of the vibrations of the bearing, thus creating an electromotive force. If the length of the coil and the induction generated by the permanent magnet are immutable, the resulting force is proportional only to the speed of the oscillating object. Measurement of bearing vibrations with electrodynamic sensor is shown in Fig. 1a.

Fig. 1. Measurement with electrodynamic sensor (a) and laser vibrometer (b)

Measurement with a laser vibrometer (Fig. 1b) consists in a precision laser interferometer being compared by a device with a beam reflected from a vibrating object. In the case where the object being studied moves, the frequency of the reflected object's light is changed as a result of the Doppler effect. The frequency of the wave increases as the subject approaches the source, but decreases when it moves away. Knowing the frequency difference, you can directly determine the vibration speed of the object.

2 Methods

The STPPD anderometer [11] located in the Wheel Bearing Research Laboratory at the Kielce University of Technology in Kielce, equipped with the SG4.3 electrodynamic sensor, was used for the study. Vibration measurements were carried out on a group of fifty 6204 bearings, five different manufacturers marked with the letters F, N, T, R and S. Each manufacturer represented 10 bearings. The rolling bearings tested had nitrile seals, individually marked by each manufacturer. The bearings were mounted on a shaft rotating at 1800 rpm/min and a three-point tightening of the outer ring in the axial direction with a force of 6 kg was used. The sensor SG 4.3, which contacts the external bearing surface, registers the vibration velocity in the radial direction. The vibrations were measured three times. Then the same bearing was applied to the shaft in the reverse manner and three measurements were also made. Between successive measurements, the outer ring of the bearing was turned by a third of turn. In addition to the

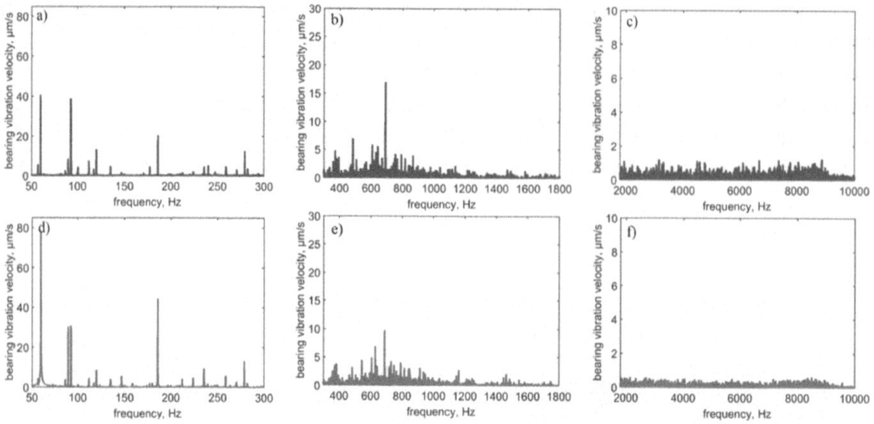

Fig. 2. Vibration Speed Signal in the frequency domain. Measurement with the vibrometer low band (a), Medium (b) and high (c). Electrodynamic sensor measurement - low band (d), Medium (e) and high (f)

anderometer, a PSV-500 vibrometer was placed, the beam of which recorded the oscillation of the outer ring also in the axial direction, at the point shifted by 90o relative to the sensor SG 4.3. Signals from both devices were recorded in parallel for the same period of 6 s. The described methodology allowed to obtain six measurements of each bearing for each device, resulting in 60 pairs of results for each of the five series. The frequency of data sampling by both systems was the same. Due to the fact that the devices collected data from two perpendicular directions, the average amplitude values from three measurements for one side were calculated (Fig. 3).

We can use measuring accuracy (MA) to compare measurement results from two devices [12]. To determine it, the experimental relative measurement error for comparable instruments is calculated, assuming that the electrodynamic sensor as a typical sensor for measuring vibrations of rolling bearings is a standard sensor.

$$MA = \left|\bar{w}_{\Delta z} \pm k_p s\right|_{max} * 100\% \quad (1)$$

where: $w_{\Delta z}$- average value of measurement error, k_p-extension factor (P = 0.95) determined from normal distribution tables, s- deviation of mean error of measurement.

3 Measurement of the Root Mean Square Value in Three Frequency Bands

The measurement accuracy was compared on the basis of the root mean square values of the vibration velocity signal in three frequency ranges: low: 50–300 Hz, medium: 300–1800, and high: 1800–10000 Hz. The effective value tested in these bands is typical for the control of newly manufactured rolling bearings. These bands are of course closely related to the rotational speed [11, 13, 14]. Each of the three bands is

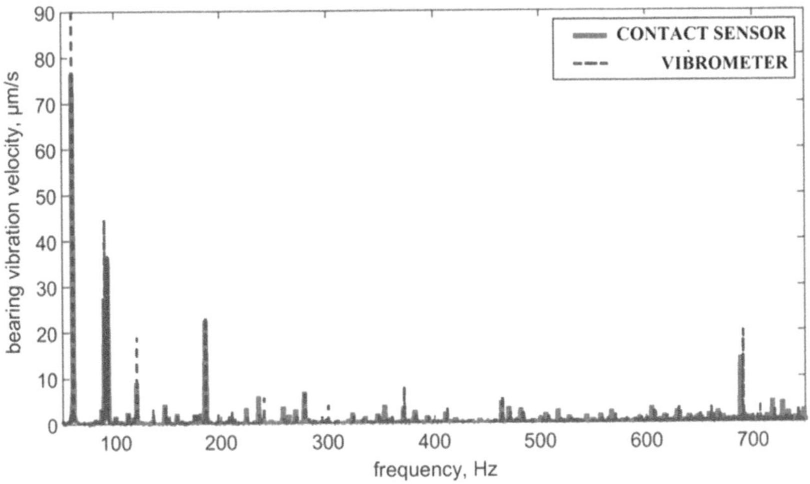

Fig. 3. An example of comparison of vibration spectrum of type 6204 rolling bearing in the Range 50–700 Hz, obtained from the sensor SG4.3 (Red) and the head PSV-500 (Blue), at the same time (Color figure online)

assigned other suspected faults. For example, a high RMS value in the low band may indicate, for example, a difference in ball diameter or roundness of the rings, in the medium band with ball failure or treadmill errors, in the high band with too high surface roughness of the ball or treadmill, or impurities. Obtained signals in the frequency domain can be found in Fig. 2.

The results of the measurement accuracy depending on the frequency band and the bearing manufacturer can be found in Table 1.

Table 1. Measurement accuracy of the PSV-500 vibrometer, depending on the bearing manufacturer and the frequency band of the measured effective value of the vibration velocity

Frequency band, Hz		50–300	300–1800	1800–10000
Measurement accuracy of the RMS value in a band, %	F	70.87	19.84	170.25
	N	73.89	23.24	152.43
	T	65.66	42.37	186.9
	R	100.26	49.53	292.48
	S	66.79	39.42	187.56

4 Measurement of Selected Amplitudes in the Low and Medium Frequency Band

Six harmonics with the highest amplitudes were selected for comparative analysis. The first one (for most measurements the highest) corresponds to 60 Hz. This frequency coincides with the second harmonic of the rotational frequency as well as the ball rotation frequency. For both measurements with electrodynamic sensor and laser, in the vicinity of 90 Hz you can see two components next to each other. The first corresponds to the frequency of 90 Hz, this is the third harmonic of the rotational frequency. The second one can be identified by the frequency of 92.5 Hz, corresponding to the basic frequency of the outer ring. For a frequency of 120 Hz (the fourth harmonic of the rotational frequency coincides with the frequency of the second harmonic of the rotation of the ball) also appears the component that has the lowest amplitude of the six selected for the comparative calculations. The next peak adopted for analysis is assigned to the frequency calculated at the basis of the formula, 185 Hz which corresponds to the second harmonic frequency of the outer ring. In the spectrum there is also a component with a frequency of 690 Hz. The components between the discussed and those larger than 690 Hz do not always appear, which makes it difficult to combine them. Due to the fact that the devices collected data from two perpendicular directions, the average amplitude values from three measurements for one side were calculated.

The results of the accuracy of the measurement depending on the specific harmonics of the signal and the bearing manufacturer can be found in Table 2.

Table 2. Measurement accuracy of the PSV-500 vibrometer, depending on the bearing manufacturer and the frequency of the measured vibration velocity amplitude.

Harmonic frequency, Hz		60	90	92,5	120	185	690
Measurement accuracy of the specified harmonic,%	F	88.45	92.56	56.54	76.13	69.87	40.99
	N	87.87	90.71	51.94	74.13	66.1	41.92
	T	53.57	73.86	63.7	76.82	93.71	42.08
	R	189.54	137.63	181.24	111.12	178.74	42.44
	S	53.98	69.92	52.06	79.39	90.76	42.01

5 Conclusions

Analysis of the measurement accuracy results clearly shows that it depends on the frequency range in which we compare. For the medium frequency range, we obtain the best measurement accuracy of 20–50% depending on the manufacturer. In the high frequency band there is a clear problem with the accuracy of the measurement. The results are very unfavorably high. This may be due to the fact that the surface of the outer ring on which the laser falls is rounded. The beam of light falling on the ring may not reflect completely and some of the photons do not return to the vibrometer. The outer surface of the bearing also has a metallic gloss, which in some measurements caused a problem with the detection of the measuring point on the bearing by the

PSV-500 device. The analysis of the results shows that the measurement accuracy also depends on the manufacturer supplying the bearing. The outer surface of the R bearing is the worst suited for laser measurements. We can also see the similarity of results obtained for groups F and N and T and S.

The measurement accuracy of the amplitude measurement for the PSV-500 vibrometer ranges from 41% up to 190%. In the case of such a comparison of results, we can also confirm the similarity of the measurement accuracy for groups F and N and T and S. Similarly as in the case of effective value comparison here also there is a clear problem with the measurement accuracy by the group bearing vibrometer R. It is difficult to find a clear relationship between measurement accuracy and low-frequency harmonics (up to 300 Hz). On the other hand, the harmonic accuracy corresponding to 690 Hz is clearly better than the others.

Replacement of typical electrodynamic sensors with laser heads brings additional advantages such as: lack of sensor wear by contact with bearings, or the possibility of measuring in more than one direction, as well as the possibility of using it in combination with other sensors.

The direction of further tests should be comparative tests using a compact head with the tested station with a single-point laser. The surfaces on which the laser falls in terms of the structure, condition, geometry or absorption of the external surfaces of the rolling bearings produced, and hence an attempt to find such a surface that is most suitable for the Doppler beam of the laser vibrometer should also be examined.

Acknowledgment. The publication was created as a result of research and development carried out by the Polish Bearing Factory, Kraśnik S.A. together with the Kielce University of Technology in the project entitled "The Establishment of R&D Centre in FŁT-Kraśnik S.A." under the Smart Growth Operational Programme 2014-2020, co-financed from the European Regional Development Fund No. CBR/1/50-52/2017 from 07/04/2017.

References

1. Castellini P, Martarelli M, Tomasini EP (2006) Laser Doppler Vibrometry: development of advanced solutions answering to technology's needs. Mech Syst Signal Process 20(6):1265–1285
2. Longo R, Vanlanduit S, Vanherzeele J, Guillaume P (2010) A method for crack sizing using Laser Doppler Vibrometer measurements of Surface Acoustic Waves. Ultrasonics 50(1):76–80
3. Gioffré M, Cavalagli N, Pepi C, Trequattrini M (2017) Laser doppler and radar interferometer for contactless measurements on unaccessible tie-rods on monumental buildings: Santa Maria dellaConsolazione Temple in Todi. J Phys Conference Series, vol 778, Conference 1
4. Valliapan R, Lieu DK (1992) Defect characterization of roller bearing surfaces with laser Doppler Vibrometry. Precis Eng 14(1):35–42
5. Dwojak J, Rzepiela M, Struzik I (2011) Wykorzystanie wibrometru laserowego do diagnostyki eksploatacyjnej silników elektrycznych na podstawie własnych doświadczeń. Zeszyty Problemowe – Maszyny Elektryczne Nr 89/2011, 57–63

6. Cristalli C, Paone N, Rodríguez RM (2006) Mechanical fault detection of electric motors by laser Vibrometer and accelerometer measurements. Mech Syst Signal Process 20(6):1350–1361
7. Tatar K, Gren P (2008) Measurement of milling tool vibrations during cutting using laser Vibrometry. Int J Mach Tools Manuf 48(3–4):380–387
8. Paone N, Scalise L, Stavrakakis G, Pouliezos A (1999) Fault detection for quality control of household appliances by non-invasive laser Doppler technique and likelihood classifier. Measurement 25(4):237–247
9. Halkon BJ, Rothberg S (2016) A practical guide to laser Doppler vibrometry measurements directly from rotating surfaces. Presented at the IMechE Vibrations in Rotating Machinery (VIRM11), Manchester, 13–15th
10. Halkon BJ, Rothberg S (2006) Vibration measurements using continuous scanning laser Vibrometry: advanced aspects in rotor applications. Mech Syst Signal Process 20(6):1286–1299
11. Adamczak S, Zmarzły P (2017) Influence of raceway waviness on the level of vibration in rolling-element bearings. Bull Polish Acad Sci Tech Sci 65(4):541–551
12. Adamczak S, Zmarzły P, Stępień K (2016) Identification and analysis of optimal method parameters of the V-block waviness measurements. Bull Polish Acad Sci Tech Sci 64 (2):325–332
13. Standard: ISO 15242-1:2015 (2015) Rolling bearings - Measuring methods for vibration - Part 1: Fundamentals
14. Lacey SJ (2008) An overview of Bearing Vibration Analysis. Maint Asset Manag 23(6):32–42

Challenges of Miniaturizing a Precision Gear

Numan M. Durakbasa, Jorge M. Bauer, Osman Bodur, and Günther Poszvek(✉)

The Department of Interchangeable Manufacturing and Industrial Metrology/Production Metrology and Quality, Vienna, Austria
durakbasa@ift.tuwien.ac.at, doktorjbauer@yahoo.com.ar, bodur.osman@hotmail.com, guenther.poszvek@tuwien.ac.at

Abstract. Miniaturization offers two options to come closer to the micro and nano scale: bottom-up or top-down. Each method has its own and unique benefits, although in practice the top down approach is more common, as most companies find the top down approach to be the simpler method. So can similar processes use the top-down method to reduce the size of a precision gear? In our first attempt, a plant gear similar to a harmonic drive gearbox is to be made it smaller, lighter and cheaper using a 3D printing method. Then industry 4.0 lab is used to see how the production process meets I 4.0 requirements. Working on RP4 gears, which have been patented by TU Vienna, nearly no backlash can be expected due to the use of radial ball bearings. If the RP4 gear is miniaturized to one tenth of the original size, functionality issues can be expected because of the smaller auxiliary parts, and no matching radial ball bearings are available in the official market. Thus, our focus is on redesigning the existing gear which will have the same functionality. Furthermore, the geometric tolerances of the existing are compared with the new gears produced by additive manufacturing to understand the functional and metrological problems. Possible solutions may lie in modifying those bearings or in creating a new generative design instead.

Keywords: Backlash · Gearbox · Micro ball bearings · Miniaturization 3D printing · Metrology · Industry 4.0. SPC · Geometrical tolerances GPS

1 Introduction

In times when everyone is thinking and writing about smart factories (Industry 4.0), it is worth giving an example of the development of production in the field of mechanical engineering. This paper presents an example of how to reduce the size of a precision gear. The possibilities of downsizing are studied using an RP4 gearbox which was created during the development of a robot for automated measurement in an international cooperation project. The characteristics of the gearbox are very similar to those of a "harmonic-drive" gear, but without relying on a flexible body. While downsizing has taken on more and more concrete forms, other problems have emerged, opening up an intensive study of the possibilities of small ball bearings. With all the challenges that have arisen, the geometric product specification (GPS) has always been of help and support.

N. M. Durakbasa and M. G. Gencyilmaz (Eds.): ISPR 2018, *Proceedings of the International Symposium for Production Research 2018*, pp. 239–253, 2019.
https://doi.org/10.1007/978-3-319-92267-6_21

After miniaturizing existing gear, it was not possible to find in ISO 15:2017 and in suppliers' catalogues any standard radial ball bearings with a minimum 2.5 mm outside diameter and 1 mm width. Therefore, new micro ball bearings were created, but they had durability problems that not only arose from the geometrical design, but also from the material and tribology that affects the friction coefficient between raceway and ball bearings surfaces. Accordingly, tribological research for micro ball bearings was required before producing the new radial ball bearings by additive manufacturing. In that way, it was possible to identify the right material to withstand the forces of the new gear. Another consideration is the friction coefficient between surfaces due to the design parameters of micro ball bearings. Hence, the basic static load rating, which determines the permanent deformation of the ball bearings, had to be determined. The first micro ball bearing prototype was a starting point for finding the most suitable bearing for RP4 gear considering the design circumstances.

2 Method and Theoretical Part

This chapter focuses on the description of the different methods of downsizing RP4. Our main consideration was the GPS system during the whole design stage. The dimensions and definition of the properties relate in part to the entire gearbox and in particular to the bearings used.

2.1 Problem of Miniaturization

When determining the technical drawings, strict compliance with normative regulations is of great importance. It is no challenge for today's CAD programs to display a work object in an enlarged or reduced scale, but production shows the limits of manufacturing methods. So generally, there are two approaches to the manufacture of products: top-down and bottom-up. The top-down approach often uses the traditional workshop or micro-fabrication methods where externally controlled tools are used to cut, mill, and shape materials into the desired shape and order. Different patterning techniques, such as lithography are in this category. In this small-size manufacturing processes many new technologies have to be considered.

2.2 GPS - General Geometrical Product Specification

To create a model, many examples given in standard ISO 1101 [11]. The decision to create parts with some functions of the GPS system means using position specification of a centreline as it can be seen in Fig. 1, which also shows the tolerancing. A united geometry element is shown in Fig. 2, with its tolerancing symmetrical to the theoretically exact location of the median plane to the reference axis a. A position specification of a midplane is seen in Fig. 3 with tolerancing. The extracted centreline of each hole must be within a cylindrical zone of diameter 0.1 whose axis coincides with the theoretically exact location of the hole relative to the reference planes C, A and B. These three samples of the standard are not representative of all possibilities of the GPS but only for a first attempt to understand and apply the technology. In doing so, there are limits in the CAD programs: drawing as it is used in the construction field is not the same when drawing in the micro dimension. There are limitations given by the program itself as SolidWorks™ is limited to draw features under 100 nanometres. If a given dimension is required, it is no problem when the calculation can be done by the power of ten. All other calculations result in unexpected rounding errors which do not comply with the standards.

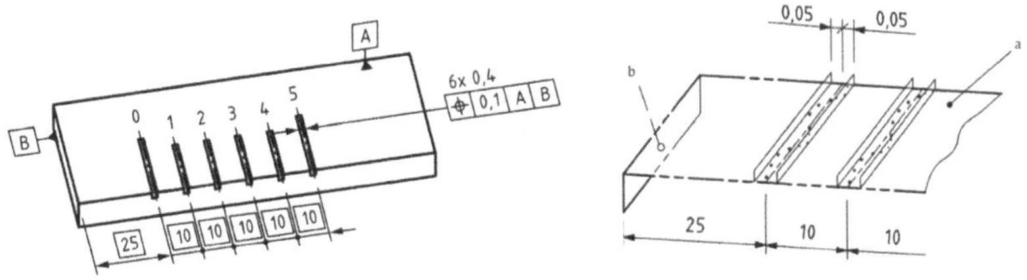

Fig. 1. Position specification of a centerline 3D - tolerance zones Source: [11]

Practical engineers should use the series of preferred numbers according to DIN 323 (German Standards), a universal number system agreed by international standards (ISO 3, ISO 17, ISO 497). It serves as a comprehensive order to simplify the technical and economic creation as grades sizes of any kind (e.g. lengths, areas, volumes, forces, pressures, moments, voltages, speeds, powers) with the aim of keeping the required amount of numbers to a necessary minimum.

2.3 Tribological Characteristics and Material Applications of Micro Ball Bearings

Dae-San Oh and Kyeong Hee Kang examined some tribological characteristics of borosilicate glass micro ball bearings of sub-60-µm diameter. In that study, the friction coefficient was limited in the V shaped groove (Fig. 4a) When the experiment of micro ball without thin film, the surface had intensive forces that totally damaged the micro balls over 2000 revolutions of testing [6].

When the case with the 33, 67 and 126 nm thick Ag films were applied on the grooves under 400 mN over 100 revolutions, the friction coefficient increased depending on the thickness of film and those films were divided between the surface and the sliding path of the micro balls, also transferred on to the micro ball surface. The observation of the micro ball in electron microscope images and X-ray spectroscopy showed no damage, which confirmed that the organic and metallic thin films are able to protect the micro ball bearing by reducing the friction coefficient. Further development for practical applications might be the effects on friction of temperature, humidity, speed and the type of wear. OTS-SAM (octadecyltrichlorosilane) exhibited remarkable friction and wear characteristics as compared to Ag thin film (Fig. 4b). With optimized thin film application, the development of durable micro-ball bearings is to be anticipated [6].

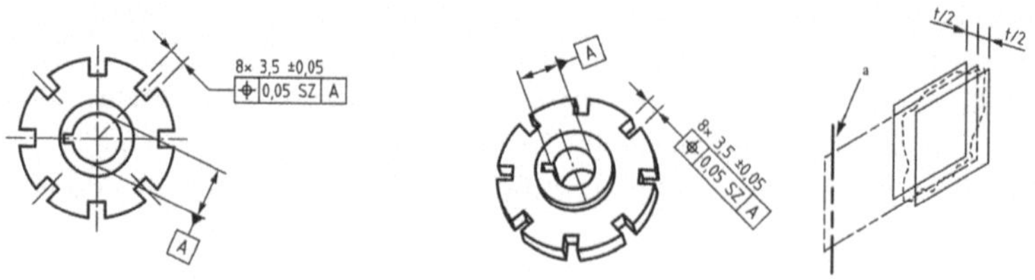

Fig. 2. United geometry element 2D and 3D - tolerance zones Source: [11]

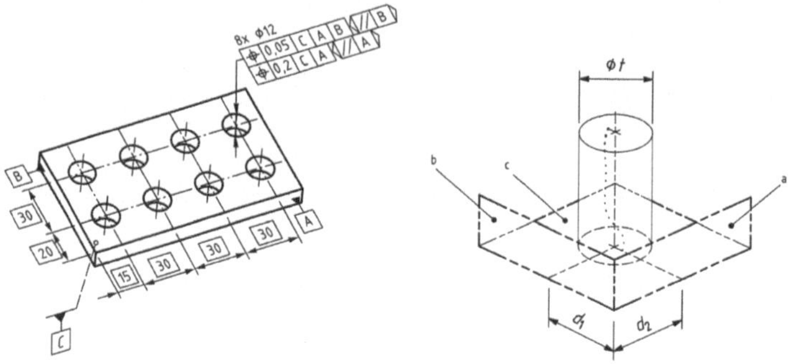

Fig. 3. Position specification of a midplane 3D - tolerance zones Source: [11]

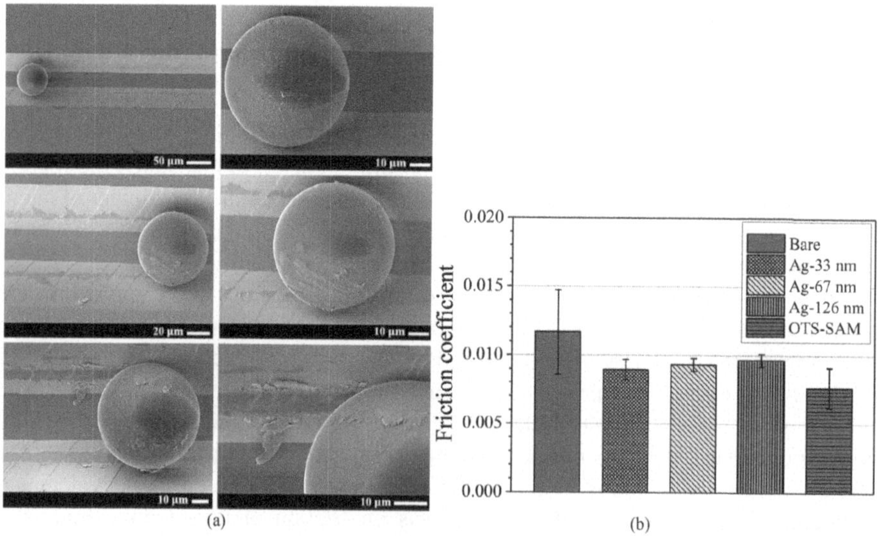

Fig. 4. (a) Friction coefficient with respect to coating under 400 mN normal force Source: [6] (b) SEM images of micro-ball and V-shaped groove with Ag film of thickness (1) 33 nm, (2) 67 nm, and (3) 126 nm Source: [6]

Mike Waits and Bruce Geil had an application on the encapsulated ball bearings for rotary micro machines. In the studies mentioned, stainless steel ball bearings of 285 μm diameter were located in the groove which had silicon race, and customized by using Silicon Dioxide and Cr/Au/AuSn/Au adhesion layer as shown in Fig. 5a [7].

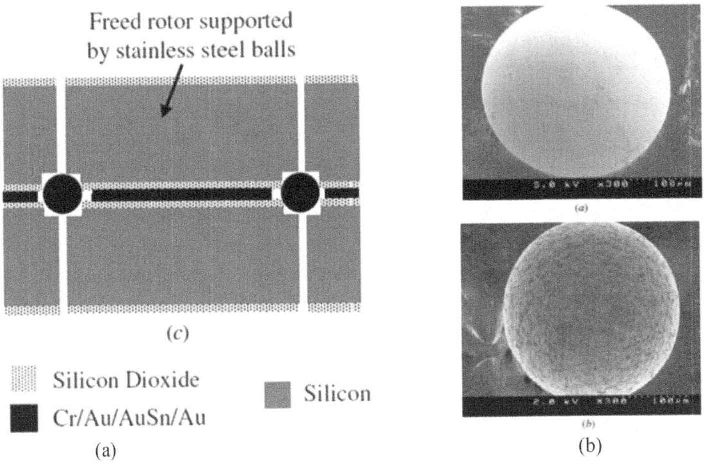

Fig. 5. (a) Fabrication flow showing encapsulation using a Cr/Au/AuSn/Au adhesion layer and release using deep reactive ion etching (b) SEM images showing (a) Fresh stainless-steel ball and (b) Stainless steel after 39 min of an average 3.2 krpm (kilo revolutions per minute rotation rate) Source: [7]

After 39 min of operation at 400 °C, the results demonstrated that the stainless-steel balls affect the load distribution and the friction coefficient (Fig. 5b). Also the geometry of the silicon race has a large influence on loading and wear. Another result was about the ball size and material with the lubrication style such as SiC or diamond, which impressed the frictional features [7].

Sujeet K. Sinha and Robin Pang tested 53 μm ± 3.7 μm micro ball bearings between 15 mm diameter Si plates with and without a channel, which is shown in Fig. 6a. Because of the silicon plates, friction decreased (Fig. 6b) on the surface bearings in a 100 rpm process. Besides, the rolling life cycle of the glass ball bearing was more than 1 million rotations(?). That means micro spheres would have been candidates due to their low expenses and attractive tribological features for reducing wear in micro-mechanical devices [8].

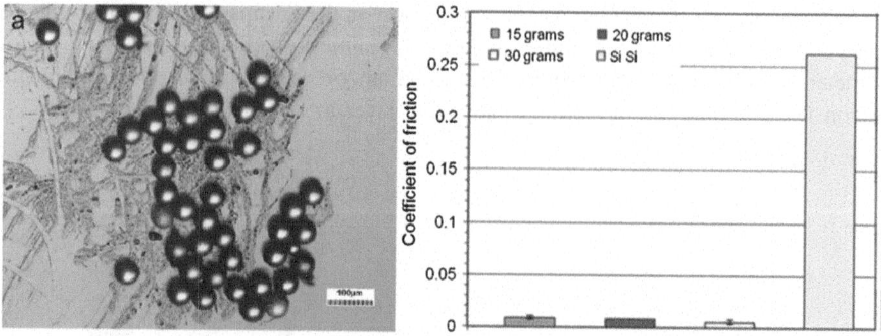

(a) (b)

Fig. 6. (a) Micrograph image showing low viscous-glass-melt stains on silicon surface when pressure is applied to the glass balls and rolled gently. The melt layer separates the ball from the silicon surface and is believed to have some lubricating effect Source: [8] (b) Coefficient of friction in rolling test for glass balls between two silicon plates Source: [8]

2.4 Design Parameters of Micro Ball Bearings

The maximum stress at the centre of the contact area between micro ball bearings and raceway is determined by:

$$\sigma_{max} = \frac{3Q}{2\pi ab} \tag{1}$$

in which Q denotes the load exerted normal to the race from the ball, a is the semi-major axis radius of an elliptical contact area, and b is the semi-minor axis radius for an elliptical contact [5].

The following equations use bearing selection for the number of balls in the single-row radial ball bearings, as they have been our system,

$$D_p = (D_o + D_i)/2 \tag{2}$$

$$D = 0.3(D_o - D_i) \tag{3}$$

$$N = \mathrm{Int}\left(0.6\,\pi\,D_p/D\right) \tag{4}$$

In normal ball-race contact geometries, contact stress, contact area and normal approach are of basic significance for the design of bearings [5]. The separation of ball and raceway surfaces (Fig. 7) is:

$$h = Ax^2 + By^2 = x^2/2R' + y^2/2R'' \tag{5}$$

$$R' = R_1' + R_2' \tag{6}$$

$$R'' = R_1'' + R_2'' \tag{7}$$

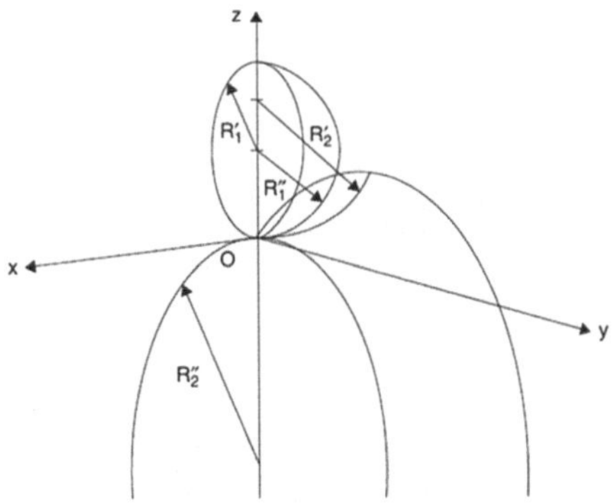

Fig. 7. Ball and raceway contact geometries Source: [5]

In the precision bearing applications, the lubrication is minimum and axial preload is effective bearing load. So, friction torque is:

$$F_s = X_s F_r + Y_s F_a \tag{8}$$

$$M = 0.9 \cdot z \cdot [Y_s \cdot F_s/C_s]^y \cdot \text{ctan}(\alpha) \cdot d_m \tag{9}$$

Cs is the basic static load rating of the bearing, dm is pitch diameter. Ys is 0.5 and Xs is 0.6 for the single row ball bearings. Besides, y = 0.55 and z = 0.0007 [5].

The above equation shows how important the basic static load rating is, so it is necessary to calculate Cs parameters for new bearing design [3, 4]:

$$Q_{max} = \frac{5F_r}{iZ\cos\alpha} \tag{10}$$

Fr = Cs will happen because of stress criterion and it can be demonstrated [3, 4]:

$$C_s = 0.2\, i\, Z\, Q_{max} \cos\alpha \tag{11}$$

It is also explained by ISO 76 [9];

$$C_s = 44(1-\gamma)i\, l\, Z\, D\cos\alpha \tag{12}$$

2.5 Miniaturization of RP4 Gear and New Micro Ball Bearing Design

Figure 8 shows our RP4 as a CAD file in the beginning, before starting to miniaturize. Many parts of the RP4 were tried smaller, lighter and cheaper by additive manufacturing. But the deep grove ball bearings [12] might be hard to make if they are downsized to a 1:10 scale, because balls will be in micro scale and need to be redesigned under 2 mm.

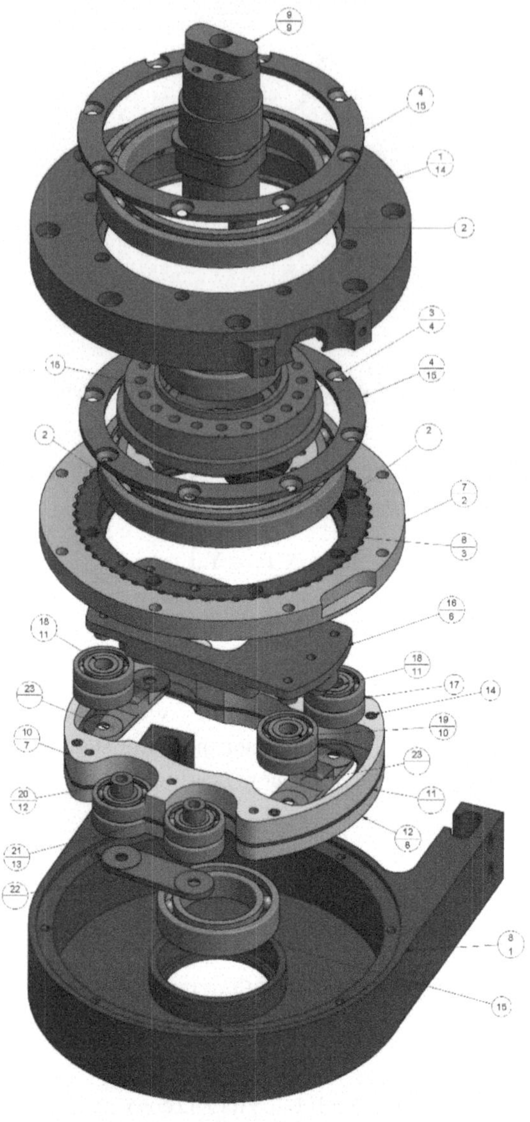

Fig. 8. 3D parts of RP4 gear for additive manufacturing

The dimensions and calculation data of the new ball bearings were renewed by the equations in the chapter "Design Parameters of Micro Ball Bearings". Also, the first micro ball bearing prototype is a starting point for finding the most suitable bearing for RP4 gear considering the design circumstances, is shown in Fig. 9.

$$d = 0.8\,\text{mm}, \quad D = 1.6\,\text{mm}, \quad B = 0.4\,\text{mm}, \quad r_{1,2} = 0.05\,\text{mm}.$$

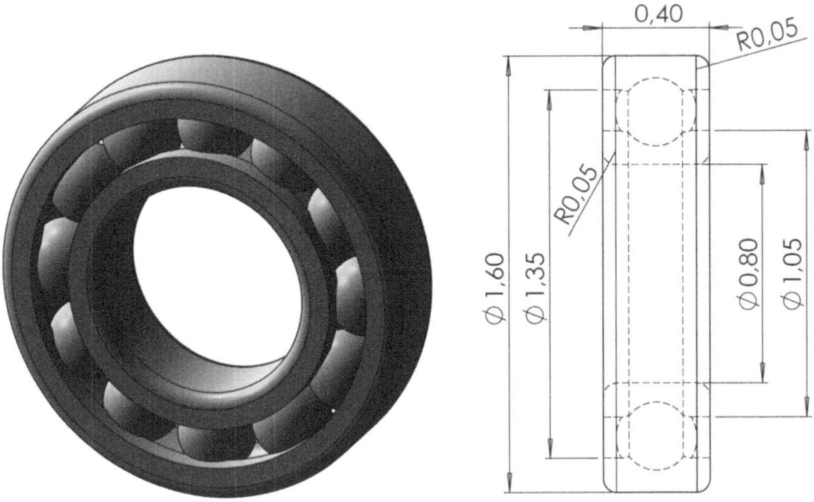

Fig. 9. New deep grove ball bearing technical drawings in CAD program

In Fig. 8, radial ball bearings are shown by number 18. In RP4 gear there are 16 pieces ball bearings which help other mechanical parts to create elliptical movement. In the next chapter, the geometrical tolerances of existing fundamental parts are compared with new parts of RP4 gear after additive manufacturing.

2.6 Comparing the Geometrical Tolerances of the Existing and New Gear Parts

There are some important parts which affect the elliptical movement in RP4 gear. In Fig. 8, these parts are numbers 2, 7 and 20. The geometrical tolerances of these parts are presented in the technical drawings below.

Figures 10 and 11 show some of the most important geometrical tolerances in existing RP4 gear according to ISO 1101:2017 [11]. The outer diameter of part number 7 is ∅ 96h6 mm with a hole tolerances and width of 6 ± 0.05 mm. The outer diameter of part number 2 is ∅ 65H68 mm with the shaft tolerances. Other important dimensions are R 34.25 ± 0.05 mm, and the distance between centre points of the bars number 20 is 22 ± 0.05 mm.

Application of SPC involves three main phases of activity: First, understanding the process and the specification limits. Second, eliminating assignable (special) sources of variation so that the process is stable. Third, monitoring the ongoing production process, assisted by the use of control charts to detect significant changes of mean or variation. Thus the changes of geometrical tolerances can be easily detected in additive manufacturing. Before SPC analyses, almost all parts except ball bearings and bolts have been manufactured using 3D printer, as seen in Fig. 12.

After manufacturing these parts, the geometrical tolerances of these important dimensions of parts numbers 2, 7, and 20 are measured by CMM [13], which has

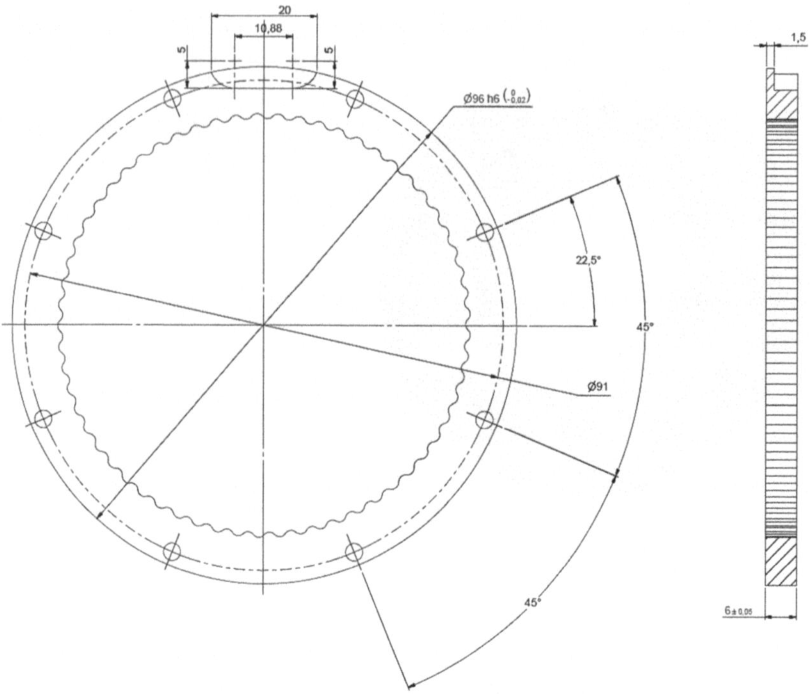

Fig. 10. Part number 7 according to Fig. 8

0.1 μm scale resolution, at the operating temperature range of 18°–22 °C, and 4 bar air pressure (Fig. 13).

Some important dimensions, which are shown in Figs. 10 and 11, have been measured 10 times and SPC application charts have been created for part number 2 in Fig. 14.

As seen in Fig. 14, the measuring data for part number 2 at Ø 65H68 mm show upper and lower deviations of 65.046 mm and 65.000 mm respectively in ISO 286. Furthermore, the upper and lower tolerances for R 34.25 ± 0.05 mm result in 35.30 mm and 34.20 mm. Unfortunately, these measuring results exceed the upper and lower deviation for the required shaft parameter (Ø65H68 mm) according to ISO 286 after producing them by additive manufacturing [14]. But the measuring results are good for the geometrical tolerances for R 34.25 ± 0.05 mm because measuring results are within tolerances.

SPC application for part number 7 in Fig. 15;

For Ø 96h6 mm, the upper and lower deviations are 96.000 mm and 95.978 mm respectively. Likewise, after calculating the upper and lower tolerances for 6 ± 0.05 mm which are respectively 6.05 mm and 5.95 mm, measuring results are within the tolerance range. However, the hole measuring results are not in the tolerance range according to ISO 216.

SPC application for part number 20 in Fig. 16;

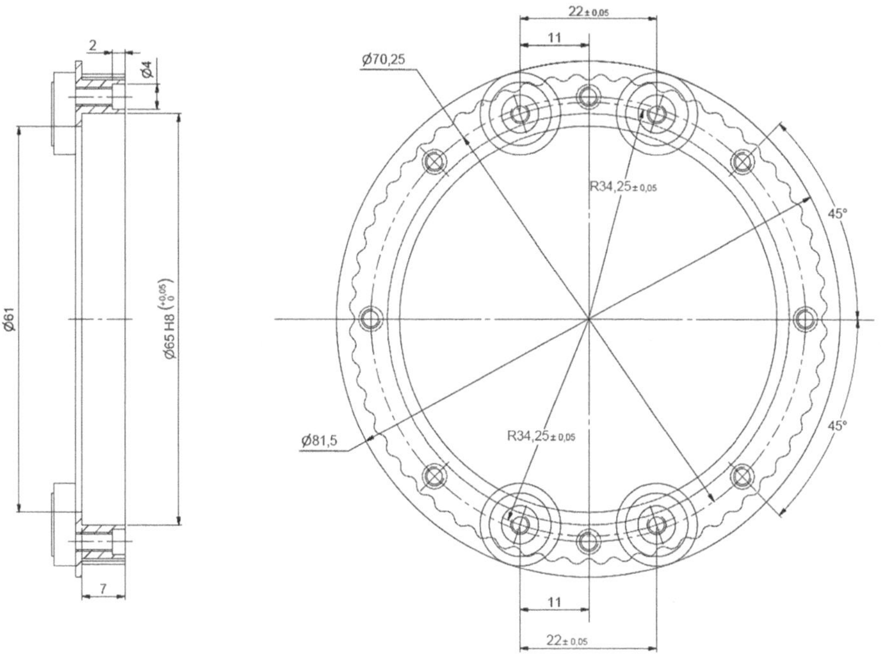

Fig. 11. Parts numbers 2 and 20 according to Fig. 8

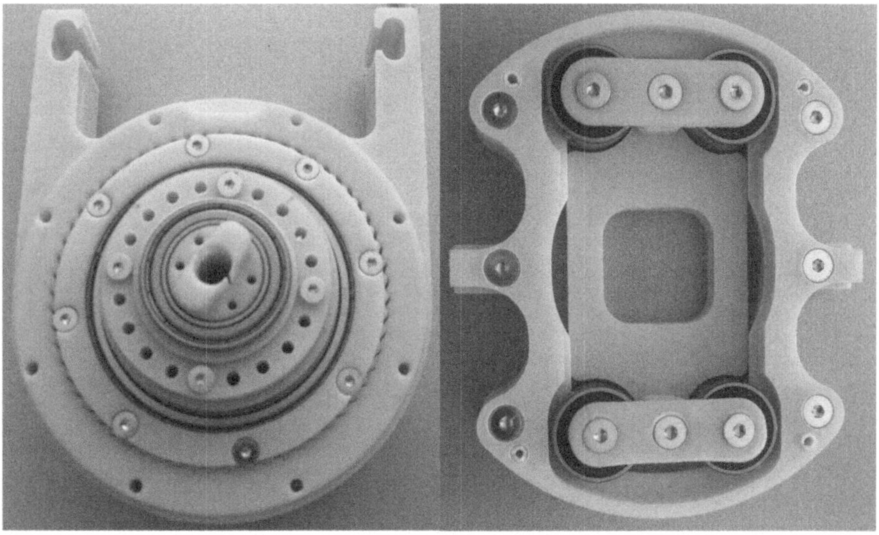

Fig. 12. Manufactured parts and assembling by 3D printer

Fig. 13. CMM at laboratories of TU Vienna

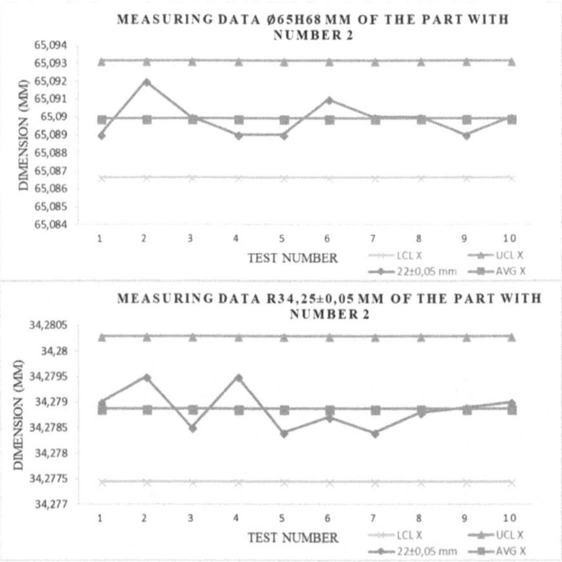

Fig. 14. Measuring data for the dimensions ∅65H68 mm and R34.25 ± 0.05 mm of part number 2

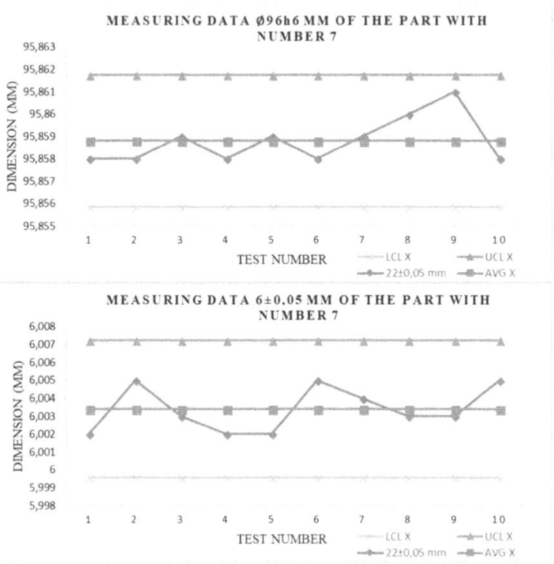

Fig. 15. Measuring data for the dimensions Ø96H6 mm and 6 ± 0.05 mm of part number 7

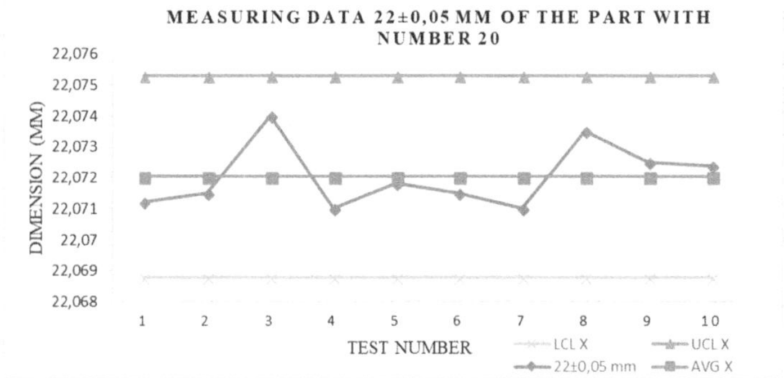

Fig. 16. Measuring data for the dimension 22 ± 0.05 mm of part number 20

After the calculation of the upper and lower deviations of dimension 22 ± 0.05 mm, which are respectively 22.05 mm and 21.95 mm, the measuring results of that additively manufactured part are out of the tolerance range, so SPC applications show that the parts produced out of the tolerance ranges by 3D printer have increased backlash. Thus, the printed RP4 gear cannot be defined as a harmonic drive with zero-backlash, because all geometrical errors create functional and metrological problems and affects the backlash of the harmonic drive geometry.

3 Conclusions

The production of the miniaturized RP4 gearbox has shown that additive manufacturing requires strict control of data from production to embedded technology. This means meeting geometric tolerances, optimizing the generative design on a microscale, and using material with suitable tribological properties. Considering all these steps, this experiment gives:

- a great start to design micro ball bearings with desirable materials, which is borosilicate glass for balls, octadecyltrichlorosilane film for the silicon raceway groove;
- generative design parameters for the micro radial ball bearings and precision bearings because of the lack of radial ball bearings under 2 mm in the official markets.
- metrological problems of the 3D printing model and as a solution, these components of RP4 gear have to be manufactured in a high-precision printer. Afterwards, the 3D models on a 1:10 scale can also be produced in high precision printers.

As a future work, new generative prototypes for the micro ball bearings will be designed and produced with high accuracy, and will then be compared with the geometrical tolerances of CMM.

Nomenclature

σ_{max}	*maximum stress at the central point of the contact area*
Q	*load exerted normal to the race from the ball*
a	*semi-major axis radius of an elliptical contact area*
b	*semi-minor axis radius for an elliptical contact*
D_p	*pitch circle diameter of bearing*
D_o	*outside diameter of bearing*
D_i	*bore diameter of bearing*
D	*ball diameter*
Int	*integral*
N	*number of balls*
h	*separation of the surfaces*
A, B	*curvatures*
x, y	*coordinates*
R', R''	*principal radii of curvature*
R_1, R_2	*raceway radius in rolling direction*
F_s	*static equivalent applied load*
X_s, Y_s	*equivalent load factors*
F_r	*radial applied loads*
F_a	*axial applied loads*
M	*load dependent torque for precision bearings*
y, z	*load factors for the precision bearings*
$ctan$	*cotangents*
d_m	*pitch diameter*
α	*contact angle*

i	*number of rows*
Z	*number of rolling elements per row*
Q_{max}	*max Rolling element load*
γ	*D cos(a)/dm*
l	*Roller effective length*
krpm	*rotation rate - kilo revolutions per minute*
LCL	*Lower Control Limits*
UCL	*Upper Control Limits*
AVG	*Average Measuring*

References

1. Durakbasa N, Bas G, Kräuter L, Poszvek G (2013) The assessment of industrial manufacturing systems towards advanced operations by means of integrated modeling approach. In: 21st IBIMA International Business Information Management Association Conference, Vienna
2. Durakbasa N, Poszvek G, Bas G, Bauer J (2015) Developments in precision engineering: high precision metrology applications to improve efficiency and quality. In: XXI IMEKO world congress, Prague
3. Harris TA, Kotzalas MN (2007) Essential concepts of bearing technology, 5th edn. Taylor & Francis Group, Florida
4. Harris TA, Kotzalas MN (2007) Advanced concepts of bearing technology, 5th edn. Taylor & Francis Group, Florida
5. Wardle F (2015) Ultra-precision bearings. Woodhead Publishing, pp 37–146
6. Oh D-S, Kang K-H (2018) Tribological characteristics of micro-ball bearing with V-shaped grooves coated with ultra-thin protective layers. Tribol Int 119:481–490
7. Waits CM, Geil B, Ghodssi R (2007) Encapsulated ball bearings for rotary micro machines. J Micromech Microeng 17:S224–S229
8. Sinha SK, Pang R, Tang X (2010) Application of micro-ball bearing on Si for high rolling life-cycle. Tribol Int 43:S178–S187
9. ISO 76 (2006) Rolling bearings—static load ratings
10. ISO 15. Rolling bearings — Radial bearings — Boundary dimensions, general plan 27
11. EN ISO 1101. Geometrische Produktspezifikation (GPS) – Geometrische Tolerierung – Tolerierung von Form, Richtung, Ort und Lauf. Deutsche Fassung EN ISO 1101:2017
12. PUB BU/P1 10000/3 EN (2016) SKF rolling bearings catalogue, license from Shutterstock.com
13. Aberlink innovative metrology (2018). https://www.aberlink.com/products/cmm/axiom-too-hs/
14. ISO 286-1:2010 (2010) Geometrical product specifications (GPS) — ISO code system for tolerances on linear sizes — Part 1: Basis of tolerances, deviations and fits

Cross Mark Coordinate Determination and Automatic Registration for Offset Printing

N. Gökhan Kasapoğlu[1(✉)], Zeynep Gergin[2], M. Güneş Gençyılmaz[3], Fadime Üney-Yüksektepe[2], Fatma Kutlu-Gündoğdu[2], and Ayşe Bilge Torbalı[2]

[1] Electrical and Electronics Engineering Department, Istanbul Aydın University, Istanbul, Turkey
gokhankasapoglu@aydin.edu.tr

[2] Industrial Engineering Department, Istanbul Kültür University, Istanbul, Turkey
{z.gergin, f.yuksektepe, f.kutlu, a.torbali}@iku.edu.tr

[3] Industrial Engineering Department, Istanbul Aydın University, Istanbul, Turkey
gunesgencyilmaz@aydin.edu.tr

Abstract. Offset printing is the method for producing commercialized printed media as newspapers and magazines. There are various factors contributing to the overall print quality such as; paper, ink pigment penetration, ink water balance, hygiene of environment, air temperature and humidity. Moreover, quality control is still operator dependent process where the operator is responsible for the visual checks of the printed material. One of the major checks applied is the correct registration of the plates in order to have a sharp image. The misregistration check may require more than one iterations creating setup time variation. Consequently, this highly time consuming setup takes considerable amount of time, especially in low circulation amounts. In this study a novel method called Cross Mark Coordinate Determination and Automatic Registration (CMCDR) is proposed for setup time reduction. CMCDR is based on x and y intensity profiles of registration marks and applied in a printing shop for automatic registration. For this purpose a traditional registration cross for overlapped color components (CMYK) as well as one registration cross for each individual color component is used to determine misclassification error. Using x and y intensity profiles of individual color component to determine misregistration errors and corrections of misregistrations are reduced with only one iteration.

Keywords: Offset printing · Printing quality · Setup time · Misregistration Automatic registration · Image processing

1 Introduction

Offset printing is a circuitous printing process, which is the generally utilized in the creation of daily papers and magazines printed in colorful forms. Printing four main colors, cyan (C), magenta (M), yellow (Y), and black (K) upon each other contribute to the multicolor pictures [1]. Figure 1 display the colors generated from the four main colors.

N. M. Durakbasa and M. G. Gencyilmaz (Eds.): ISPR 2018, *Proceedings of the International Symposium for Production Research 2018*, pp. 254–264, 2019.
https://doi.org/10.1007/978-3-319-92267-6_22

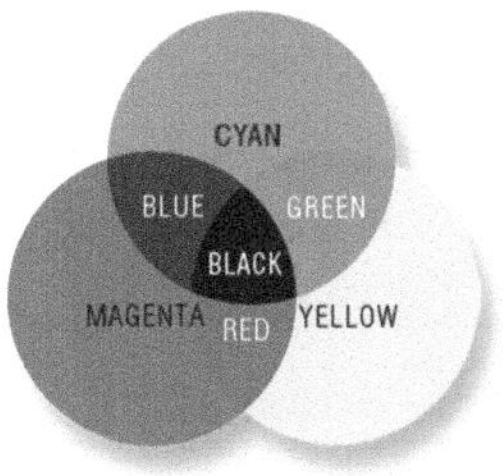

Fig. 1. Color wheel [2] (Color figure online)

There are two types of offset presses [3];

(1) Sheet-fed offset press: In a sheet-fed press the paper is sustained to the press in single sheets and each sheet is printed separately. The printed paper is then gathered in a tray and stacked flawlessly. Sheet-fed prints frequently give preferred quality over web-fed prints. However, web-fed can be more efficient for bigger runs.
(2) Web-offset press: In a web-offset the paper is fed to the printing machine from a continuous roll. In the wake of printing the roll is either rewound onto a spool or cut into sheets relying upon the prerequisites. Web-offset print is realized at a considerably higher speed than sheet-fed printing and thereby is more economical for bigger runs.

A run of the offset has three barrels – the printing plate chamber, the elastic cover barrel and the substrate or impression barrel. Every one of these parts cooperates to create the printed images on the paper. The paper is fed to the system and meets the printing plates having different surface properties that transfer the image on the plate onto the paper with the help of the fountain solution. The ink supplied is transferred onto the paper as the paper passes through the plates and the rubber blankets. This process is repeated as much as the number of colors to be printed. Hence, in the four colored printing machine the paper is processed in four chambers for four main colors (CMYK). These key elements are shown in Fig. 2 below;

The paper passing through the plates carries different color parts of the image and the final image is aimed to be a clear output to mention a quality print. There are various quality factors contributing to the overall print quality such as; paper, ink pigment penetration, ink water balance, hygiene of environment, air temperature and humidity. A large variety of nonconformities could appear during printing such as; miss-registration of printing plates, wrinkle formation [5], poorly formed paper surface [6] piling, and mottling. Verikas et al. [7] lists the print quality attributes to reach a clear image under ten factors such as; color deviation, misregistration of plates, dot deformation, dot gain, ink density deviation, edge sharpness, the number of missing dots, mottling, trapping and print through level. Although technologically advanced printing machines are better in controlling the print quality attributes, still printing quality is mostly operator dependent. An experienced operator does first the misregistration and then the color controls on the initial copies, and then the full run is

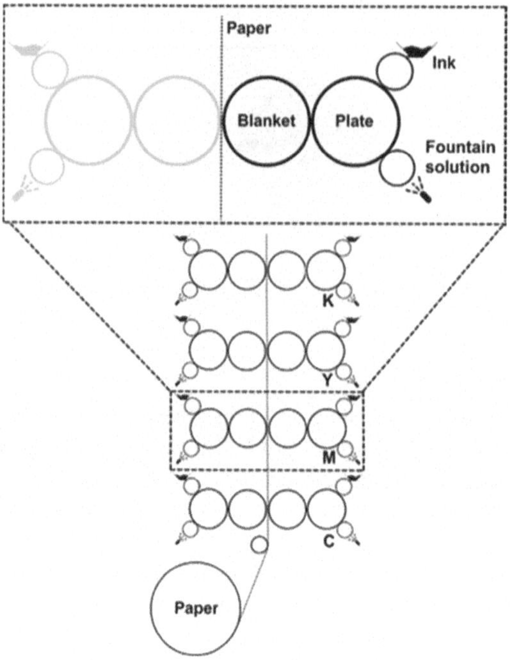

Fig. 2. Key elements of offset printing [4]

released after his approval. The sequence of controls for overall printing quality is given in the flowchart in Fig. 3.

The machine operator takes samples from the printed copies, checks the cross marks visually. The registration crosses are used in various shapes and put on every color plate and transferred to the paper. The difference between a perfect registered and a misregistered cross mark is given in Fig. 4. In an optimum registered situation, all crosses are on the same coordinates. The operator estimates the misregistration degree of printing plates in X, Y and Z directions, and makes the corrections on the machine. This process is highly dependent on the operator perception, and may require more than one iterations creating setup time variation. This highly time consuming setup takes considerable amount of time, especially in low circulation amounts. After this iterative process is completed, color densities are controlled with respect to the proofs by using a densitometer. Other attributes listed are also controlled periodically throughout the full run of the circulated amount. However, if the circulation is in low amounts mostly none of the quality problems listed are experienced.

As the manual procedures are time consuming and the results are not objective as they rely on operator abilities, automated printing quality inspection systems are highly admired. Some researchers suggested the inspection of one or several print quality attributes, such as the actual size and quality of printed dots, mottling, CCD camera-based estimation of ink density, mottling automatic detection and classification of various printing defects [8–13]. Printing quality control expert system, which consists

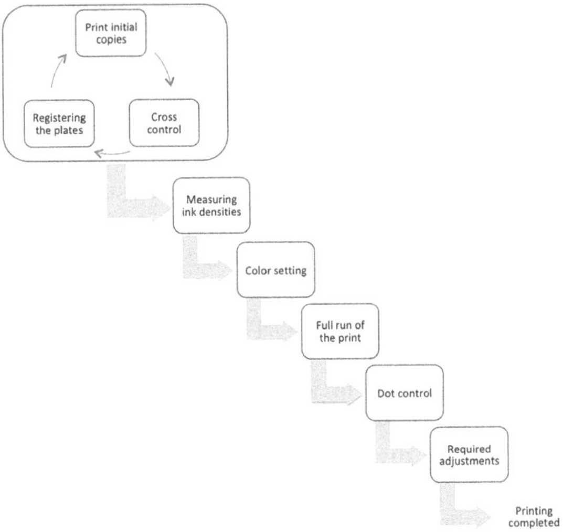

Fig. 3. The sequence of the controls for quality attributes in offset printing

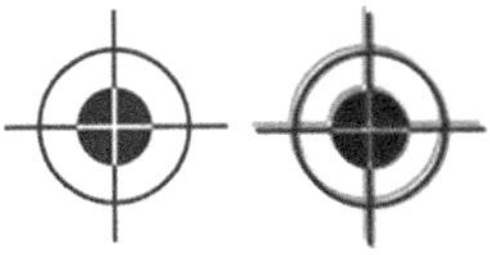

Fig. 4. A perfect registered (left), and a misregistered (right) cross mark

of knowledge base, user interface and inference engine for controlling offset printing dot variations were proposed by Chuang and Lai [14]. Asikainen [15] developed a linear model to predict an overall image quality index using four quality attributes called colorfulness, contrast, noise, and colour differences measured on a sample image taken from a picture printed. Eerola et al. [16] utilized Bayesian systems to display the general print quality by a gathering of individuals in electrophotography printing. The prepared system structure mirrored the connection between instrumental estimations, subjective print quality properties, and the general quality. Guan et al. [17] built up a case based thinking framework for offset print quality control. An assortment of print quality cases are put away in a learning base and abused for decision making in the printing procedure. Perner [13] built up a learning-based image inspection framework for programmed imperfection discovery, arrangement, and misprint analysis in offset printing. Previous studies with regard to misregistration have almost exclusively focused on medical image analysis however; on the basis of offset printing, misregistration has never been studied. Therefore, this paper has a significant importance to improve offset printing quality regarding decrease on set up time of printing.

2 Cross Mark Coordinate Determination and Automatic Registration

In offset printing, colors of traditional registration marks overprint and form a new color. If there is a registration error then printed image will blur and color of the image changed. Manual registration checking is observing registration mark visually under the magnifying glass. If any misregistration detected then positions of the cylinders where color plates are installed on will be arranged in order to correct registration.

Automatic methods based on image processing can be used to detect misregistration errors [18]. However checking the displacement of different color registration marks by separating RGB to CMYK may be problematic. Overprinted CMY generates black color and it is not possible to identify the printed black color comes from overprinted CMY or K alone. Cross mark as a registration mark is very thin and it is very difficult to detect it using digital image processing. High resolution cameras can measure its intensity values properly. For moderate resolution cameras pixels contain more than one color (mixture of colors) may occur in captured images. In this study Cross Mark Coordinate Determination and Automatic Registration (CMCDR) is propose to overcome these difficulties. CMCDR is performed with individual color registration marks for each color C, M, Y and K are used in addition to traditional registration mark for overprinting colors of CMYK. The layout of the registration marks for proposed CMCDR is illustrated in Fig. 5. In this layout cross marks are used as registration marks. Processing steps of proposed CMCDR can be seen as follows:

Cross **M**ark **C**oordinate **D**etermination and Automatic **R**egistration (CMCDR)
• Image capturing from region of interest where cross marks are printed
• RGB to gray conversion of the image
• Determine border pixels of the captured image of the printed paper
• Calculate coordinates of the individual cross marks using border pixels and x and y profiles
• Find the difference between current coordinates of cross marks and their desired coordinates in pixel unit using cross mark "K" as a reference
• Convert the pixel units to mm for fine tuning of the appropriate registration

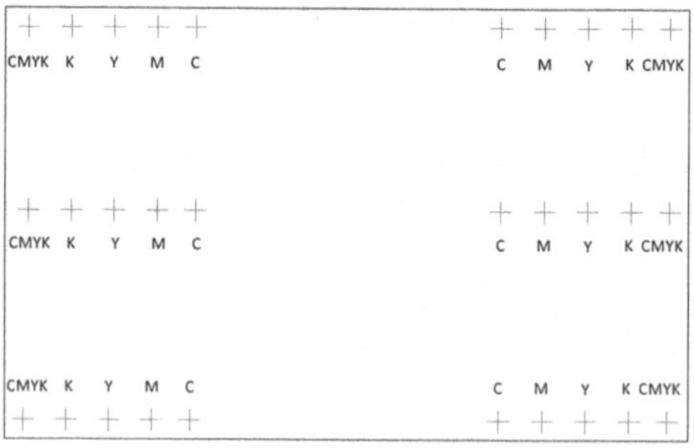

Fig. 5. The layout of the proposed registration marks

The width and height of the cross marks are 1 cm and distance between centers of the two perfect registered neighbor cross marks is 2 cm. The upper left corner cross mark K is used as a reference mark and other marks positions can be calculated relative to the reference mark.

In this study x and y profiles are used to specify positions of the reference marks. x profile (P_x) is a one dimensional array and each element of this array shows sum of the intensity values of a cross mark image over y direction. Similar way y profile (P_y) is a one-dimensional array of sum of the intensity values of a cross mark image over x direction. Let assume that I is a gray scale image of a cross mark with size of nxm. x profile (P_x) can be calculated as follows:

$$P_x(i) = \sum_{j=1}^{m} I(i,j), i = 1\ldots n \tag{1}$$

Similar way y profile (P_y) can be calculated as follows:

$$P_y(j) = \sum_{i=1}^{m} I(i,j), j = 1\ldots m \tag{2}$$

A captured image of a registered cross mark is shown in Fig. 6 with reference coordinate system. For this image P_x and P_y are plotted in Fig. 7(a) and (b), respectively.

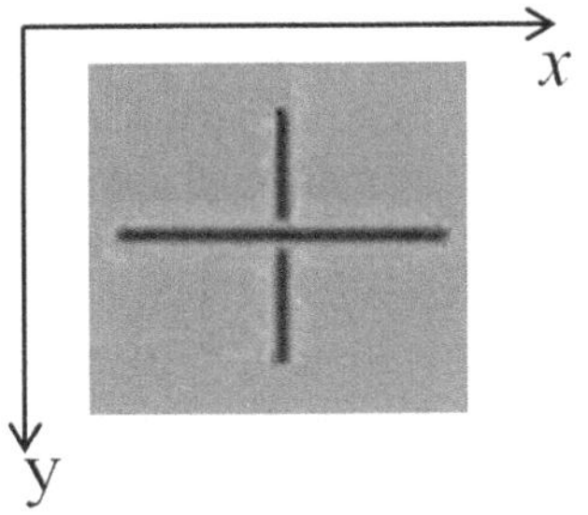

Fig. 6. Registered image of a cross mark

Position of the cross mark (x, y) is the index of the minimum value of x profile (P_x) and y profile (P_y) and calculated as follows:

$$x = \min_i \{P_x(i)\} \tag{3}$$

$$y = \min_j \{P_y(j)\} \tag{4}$$

For the cross mark shown in Fig. 6, $x = 48$ and $y = 42$ are found. This is the position of the cross mark in the captured image. Captured images are color images and

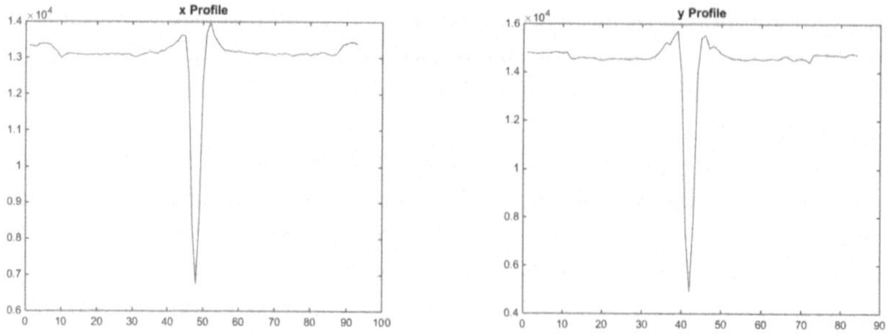

(a) (b)

Fig. 7. (a) x profile (P_x) (b) y profile (P_y) of a registered cross mark

they needed to be converted into gray scale before the extraction of x profile (P_x) and y profile (P_y). RGB to gray scale conversion can be done using the following function

$$I_{gray-scale}(i,j) = \alpha.I(i,j,1) + \beta.I(i,j,2) + \gamma.I(i,j,3) \tag{5}$$

where $\alpha = 0.2989, \beta = 0.5870\, and\, \gamma = 0.1140$ and $I(i,j,1), I(i,j,2)\, and\, I\,(i,j,3)$ are red, green and blue components of RGB image, respectively.

In Fig. 8, a misregistered cross mark is shown. x profile (P_x) and y profile (P_y) plots of the misregistered cross mark are depicted in Fig. 9(a) and (b) respectively. When we compare the profiles of misregistered and registered cross marks, then the difference can be seen for the minimum value of the profiles and its neighbors. As seen in Fig. 9(a) and (b), neighbor pixels to the minimum pixel show the overlapping colors. However it is very difficult to separate individual color cross marks from CYMK cross mark. Overprinted CMY generates black color and it is not possible to identify the printed black color comes from overprinted CMY or K alone. RGB to CMYK conversion can be applied to perform this separation. After the separation x profile (P_x) and y profile (P_y) can plot for each color component as seen in Fig. 10(a), (b), (c) and (d) for C, M, Y and K colors, respectively.

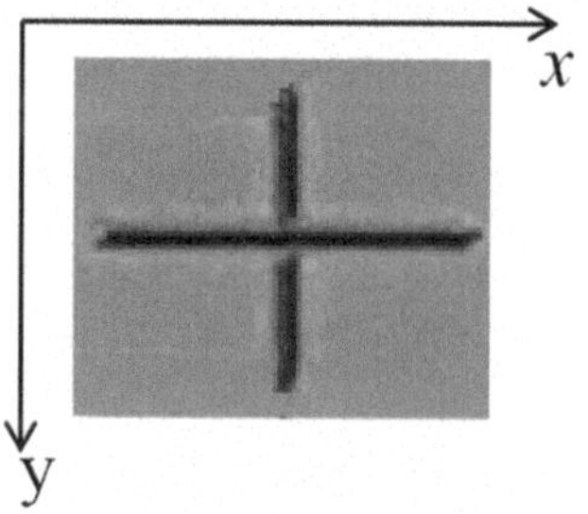

Fig. 8. Misregistered image of the cross mark

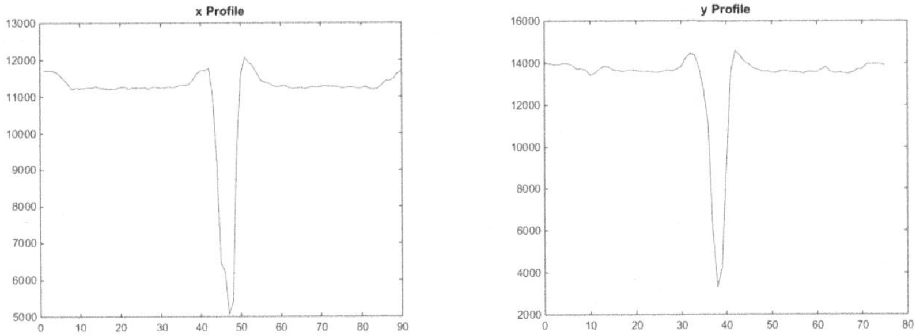

(a) (b)

Fig. 9. (a) x profile (P_x) (b) y profile (P_y) of a misregistered cross mark

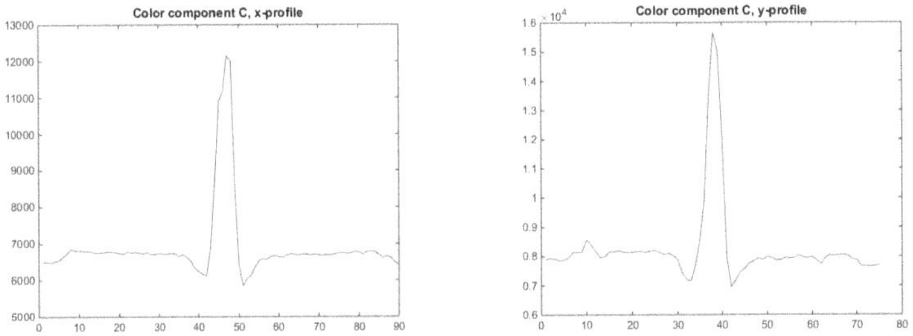

(a) (b)

Fig. 10. x profile (P_x) and y profile (P_y) of CMYK color components (Color figure online)

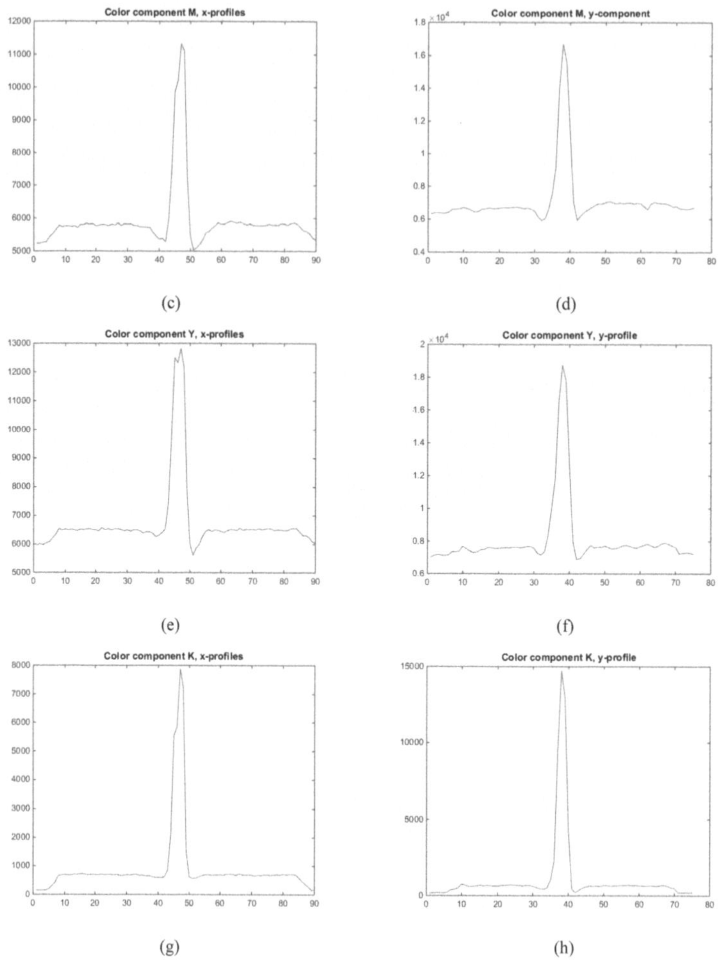

Fig. 10. (*continued*)

3 Experiments and Results

This study is conducted with a four colored sheet feed offset machine. 4 test samples are collected together with the initial X and Y coordinates. A moderate resolution camera with 12 Mega pixels was used for image capturing. 5 mm length is depicted with approximately 100 pixels. Proposed registration marks for four images are depicted in Fig. 11(a), (b), (c), and (d), respectively.

In this study registration were performed based on K color component and positions of other color components were arranged based on that component. To find proper positions of the registration marks x and y profiles were used. Table 1 shows the positions of the registration marks and fine tuning values to proper registration are tabulated in Table 2.

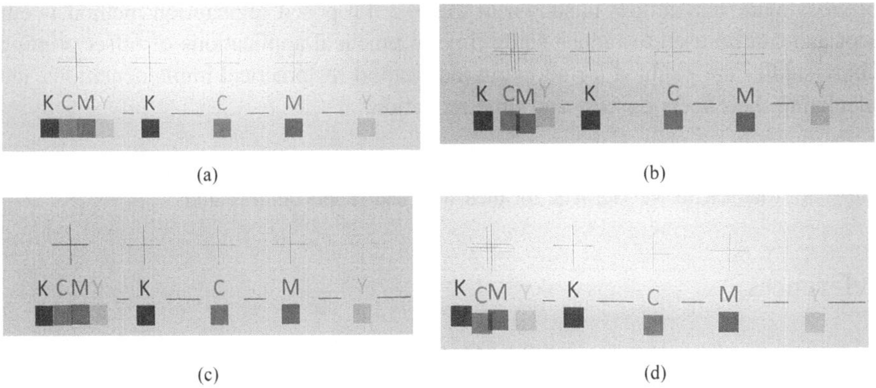

Fig. 11. Images for automatic registration experiments

Table 1. Positions of registration marks in pixel unit

	K		C		M		Y	
	Line	Sample	Line	Sample	Line	Sample	Line	Sample
Image (a)	316	1541	317	1944	319	2344	319	2750
Image (b)	284	1426	285	1866	298	2247	267	2652
Image (c)	289	1218	289	1629	288	2029	292	2428
Image (d)	268	1451	302	1854	284	2233	285	2676

Table 2. Fine tuning values for cross mark registration in mm unit

	Image (a)		Image (b)		Image (c)		Image (d)	
	X	Y	X	Y	X	Y	X	Y
K	–	–	–	–	–	–	–	–
C	0.15	0.05	2	0.05	0.55	–	0.15	1.7
M	0.15	0.15	1.05	0.7	0.55	−0.05	−0.9	0.8
Y	0.45	0.15	1.3	−0.85	0.5	0.15	1.25	0.85

4 Conclusion

Several factors contribute to the print quality, and the machine operator is majorly responsible for the visual checks of the printed material. One of the checks applied is the correct registration of the plates in order to have a sharp image. This is called misregistration check or cross mark check, and may require more than one iteration. Increased iterations result in higher and varying setup durations.

In this study a method CMCDR with a new layout for cross mark registration are proposed for automatic registration of offset printing. Image processing based registration method is introduced to find proper positions of the cross marks. X and Y profiles were used to specify the positions of the cross marks in X and Y directions, respectively. Experiments were conducted with 4 different offset printed images. Positions of the registration marks are found in pixel unit in Table 1. Fine tuning values for cross mark registration is shown in Table 2. Proposed registration method is efficient and can be used to reduce setup time in practical applications of offset printing. Future studies are planned for adopting the method for practical implementations, and calculating the effects on cost and time reduction.

Acknowledgment. The researchers express their gratitude to Uğur Gergin and Levent Gergin from G.M. Matbaacılık ve Tic. A.Ş. for their help and support in this study.

References

1. Kipphan H (2001) Handbook of Print Media. Springer, Heidelberg
2. (2018) Offset Lithography website. http://www.offsetprintingtechnology.com/wp-content/uploads/2011/09/color_wheel.png
3. (2018) Offset Lithography website. http://www.offsetprintingtechnology.com/wp-content/uploads/2011/09/offset-printing-control.jpg
4. Lundström J, Verikas A, Tullander E, Larsson B (2013) Assessing, exploring, and monitoring quality of offset colour prints. Measurement 46:1427–1441
5. Gottlebe R, Hubler A (2001) Wrinkle formation during the web motion in offset presses. Technical association of the graphic arts, San Diago, pp 186–210
6. Wood JR, McDonald D, Ferry P, Short C, Cronin DC (1998) The effect of paper machine forming and pressing on offset linting – Forming and consolidation in the presses strongly influence sheet linting. Pulp and Paper Canada, Ontario
7. Verikas A, Lundström J, Bacauskiene M, Gelzinis A (2011) Advances in computational intelligence-based print quality assessment. Expert Syst Appl 38:13441–13447
8. Antoine C, Lloyd MD, Antoine J (2001) A robust thresholding algorithm for halftone dots. J Pulp Paper Sci 27:268–272
9. Sadovnikov A, Lensu L, Kalviainen H (2007) Automated mottling assessment of colored printed areas. Computer Science, Aalborg 2007
10. Verikas A, Bacauskiene M (2008) Estimating ink density from colour camera RGB values by the local kernel ridge regression. Eng Appl Artif Intell 21:35–42
11. Kowalczyk GE, Trksak RM (1998) Image analysis of ink-jet quality for multi-use office paper. TAPPI J 81:181–190
12. Jang W, Allebach JP (2005) Simulation of print quality defects. J Imaging Sci Technol 49:1–18
13. Perner P (1994) A knowledge-based image-inspection system for automatic defect recognition, classification, and process diagnosis. Mach Vis Appl 7:135–147
14. Chuang CP, Lai FP (1997) Developing a prototype of quality control expert system for offset printing dot variations troubleshooting. TAGA
15. Asikainen R (2010) Quality analysis of a printed natural reference image. Aalto University Press, Espoo
16. Eerola T, Lensu L, Kamarainen J, Leisti T, Ritala R, Nyman G, Kalviainen H (2011) Bayesian network model of overall print quality: construction and structural optimisation. Pattern Recogn Lett 32:1558–1566
17. Guan LM, Lin J, Chen GJ, Chen M (2006) Study for the offset printing quality control expert system based on case reasoning. In: International conference on mechatronic and embedded systems and applications, Beijing
18. Haoxue L, Wenjie Y, Min H, Bing W, Yanfang X (2009) Detection and control algorithm of multi-color printing registration based on computer vision. In: 2nd international congress on image and signal processing (CISP 2009), Tianjin

Design and Development of Human Machine Interface for Unmanned Aerial Vehicle Control

Fatih Akkoyun, İsmail Böğrekci(✉), Pınar Demircioğlu, and Salih Vardin

Faculty of Engineering, Mechanical Engineering Department, Adnan Menderes University, Aydın, Turkey
{fatih.akkoyun,ibogrekci,pinar.demircioglu,svardin}@adu.edu.tr

Abstract. Unmanned Aerial Vehicle (UAV) also known as drone implies to a pilotless aircraft, a flying machine without a human pilot onboard. Most recently, UAV is widely used in civil and military based applications such as remote sensing, scientific research and aerial surveillance. In this study, a special drone with control mechanism was designed and developed. The main purpose was to build an efficient and user friendly remote control system to fulfill the deficiency of advanced control units. A graphical user interface (GUI) with touch screen was designed and implemented for motion control. The control unit had capability to reach each motor on UAV separately via duplex communication and control their speed precisely with pulse width modulation method (PWM). As a result of this study, a simple and efficient UAV system was developed that has ability to display and analyze the collected field data. An agricultural robot developed will be integrated with the proposed UAV in order to control the agricultural field operations.

Keywords: Graphical User Interface · Human Machine Interface
Pulse Width Modulation · Unmanned Aerial Vehicles · Remote control
Wireless communication

1 Introduction

Unmanned Aerial Vehicles controlled from ground are sort of an aircraft with remote control, which was once referred as "drone" in 1950s, and as UAV in today's World [1, 2]. As these vehicles were referred as "Remote Controlled Vehicles (RPV)" in the period between 1960 and 1970, they are also known as "Unmanned Aircraft Systems (UAS)" [1].

Whereas UAVs can be utilized in both military and civilian purposes, their significance increases further [3]. The general interest to these low-altitude, small-sized, light-weight unmanned aircrafts has elevated substantially [4]. Manufacturing costs of these aircrafts that can be controlled from long distance through a control system or computer screens have comparably decreased and thus, they are now at affordable levels [3]. Owing to advanced technology in UAVs, these multi-purpose aircrafts are incited in industrial and academic domains since it is possible employ UAVs in numerous applications such as military tasks, agricultural applications, and observation conveniently [5–7].

N. M. Durakbasa and M. G. Gencyilmaz (Eds.): ISPR 2018, *Proceedings of the International Symposium for Production Research 2018*, pp. 265–274, 2019.
https://doi.org/10.1007/978-3-319-92267-6_23

In real World, enhancement of process and equipment, and minimization of relevant costs for UAVs are primarily important as much as the same with other products [9]. In UAVs, necessity for multi-purpose systems with advanced mobility is evident. For instance, UAV system must fit in backpack of a soldier; noise level generated by the motors of the vehicle is required to be minimized; and take-off and landing should be completed in minimum distance as much as possible.

Besides, the aircraft, as a whole, should be light weight and ensure convenient operation. These characteristics are required by the UAV systems utilized in non-military duties such as urban monitoring and audit of power lines [9].

The design of the GUI that provides convenience in operation for UAVs is essentially important because manual control of an aircraft on long distance and at high altitude may not be stable and reliable. A GUI system can be helpful for a user in terms of course of direction, position, system and monitoring environment and facilitates the operation process [7].

As these vehicles necessitate wireless long distance control [1], their operation and data analysis require remote control and a computer respectively. In majority of studies on unmanned aircrafts, a control arm or laptop-control arm combination was utilized as remote control unit. Albeit the control arm can be carried by hand, operation without screen limits efficient usage. When control arm used together with laptop computers, the data received from the field can be displayed on screen; nevertheless, this control unit can only be carried on back. Review of studies on this issue reveals that convenient and agile control of UAVs requires advanced interface units. Accordingly, a mobile control unit with advanced interface has turned into an essential requirement of these systems.

The core purpose of this study conducted by taking control needs of unmanned aircraft vehicles is to ensure convenient and efficient remote control of UAVs and to display field data on graphic screen functionally during control operation. As the present study conducted by considering lack of advanced control unit in unmanned aircraft vehicles, a hand held interface control unit with graphic screen and touchscreen panel, which can analyse and display field data was designed so that an UAV system that can be operated conveniently and efficiently was implemented. Figure 1 shows the implemented UAV and HMI systems together.

In the structured system, an advanced interface with touchscreen panel and graphic screen was used as remote control unit. In the remote control unit, the Radio Frequency (RF) transmitter and receiver, Liquid Crystal Display (LCD) screen, touchscreen panel and the Field Programmable Gate Array (FPGA) were used. On the other hand, in the UAV unit, small, light-weight and low-cost micro-controller was employed. Revolution control of electric motors in this unit was maintained by the Electronic Speed Controllers (ESC) operated by the micro-controller. Wireless communication between remote control unit and UAV unit is maintained by means of the RF transceiver. Moreover, the field data received by the FPGA board from the UAV unit can optionally be transferred to the computer environment for analysis; and this can be used separately from the computer of the control unit.

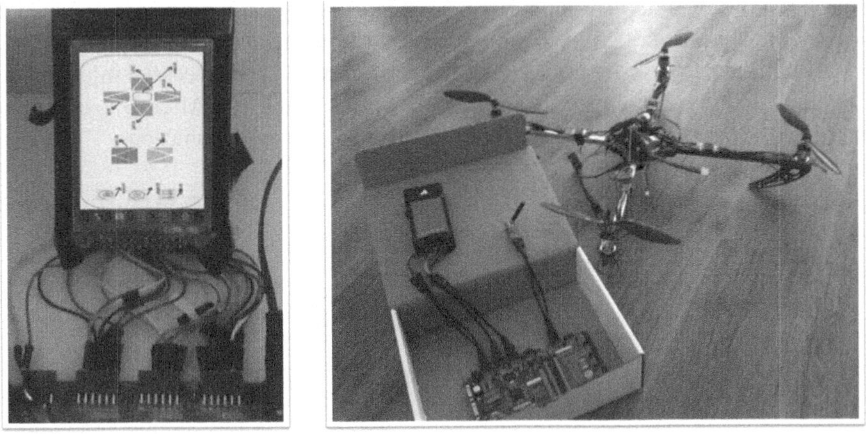

Fig. 1. Implemented UAV and HMI systems

2 Material and Method

2.1 Material

The structured UAV in the present study was planned to be controlled by means of an advanced wireless unit remotely and received field data to be displayed on graphic-based screen. Along with this, a system capable of wireless-based control of the UAV conveniently and efficiently through touchscreen interface; and to display field data received from the UAV on the control screen were structured.

In the implemented system, control data received from the touchscreen screen control unit was processed in the FPGA board; and it is transmitted to the controller in UAV over wireless network. Microcontroller in the UAV unit controls motors driven by the ESCs based on the data received from the control unit. On the other hand, two-way communication was established so as to display field data received from the UAV unit on the graphic-based screen on the remote control unit.

2.2 Control Unit

In order to ensure convenient and efficient usage of the control unit, one of the basic elements of the study, graphic screen with touchscreen panel was utilized as human-machinery interface. This unit, considering need for processing collected field data, was designed with the FPGA board for advanced characteristics such as high sampling rate, stable structure, and capability of parallel processing. Besides, control unit was designed by putting mobility into forefront.

In regard to internal communication of the control unit, environmental units are connected to the touchscreen panel by means of serial communication; graphic-based LCD screen was connected by means of parallel communication. Touchscreen panel and LCD Screen IP (Intellectual Property) were added onto the FPGA board. The design of the control unit is presented in the block diagram in Fig. 2. In the design of the remote control unit, 32-bit MicroBlaze microprocessor with RISC architecture, developed by Xilinx Company was preferred. The data collected from the field by means of the FPGA with MicroBlaze Processor at the remote control interface are processed in the format appropriate to the LCD Screen. After these data processed by the MicroBlaze, they are sent to the IL9230 driver on the LCD Screen over parallel communication.

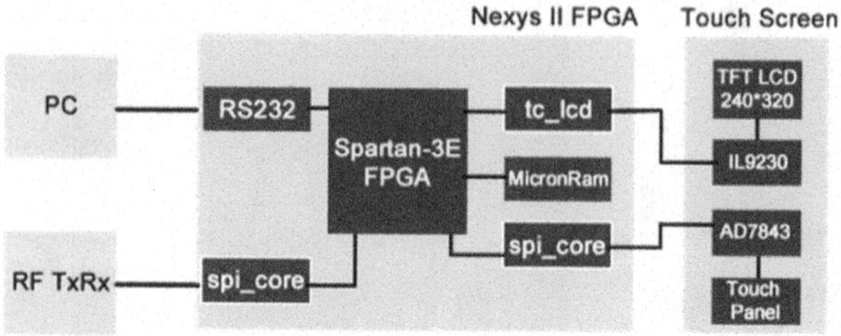

Fig. 2. Block diagram of the remote control unit

For remote control interface, LCD Screen, touchscreen panel with graphic base referred as ELT240320TP, was used. IL9230 in the LCD Screen with touchscreen panel, processes captured color data and commands and displays them on the TFT LCD screen. Touchscreen panel section receives (X, Y) coordinates of the point that the operator touches through AD7843 analog-digital transformer; and transmits them to communication IP core. Serial communication IP core processes incoming data; and thus, determines the point touched.

In order to provide wireless communication in the remote control unit, RF transceiver called as PMODRF1 was connected to the FPGA board as a peripheral unit. By means of this unit, data received from the touchscreen panel can be transmitted to the microcontroller in the UAV unit; and data received from the UAV unit can be transmitted to the control screen. While the FPGA board is connected to the PC through serial port in this unit, the data can be saved after they are downloaded into PC.

2.3 The UAV Unit

The UAV unit was designed in small, light-weight and mobile form so that it can fully perform guidance commands received from the control unit. This unit was structured on the ground of "quadrotor" [8, 10] consisting of 4 motors, and considered in small UAV class.

Small-sized UAVs, same as the ones used in the present study, are electric-driven systems. Usage of brushless DC motors with high torque-weight rate in these UAVs provides saving from weight and space. Due to number of advantages of brushless DC motors compared to the induction motors and brush motors in terms of torque-speed, high efficiency, high torque, low vibration, and noise free operation, the application of brushless DC motors in small UAVs turns out to be appropriate [11]. Considering advantages such as high torque, high efficiency, and larger space on vehicle, brushless DC motors were preferred in the UAV structured in this study.

Materials used in implementation of the UAV presented in Fig. 3 were microcontroller, RF transceiver, ESC, motor and battery, which were selected in the light of studies in this field [7–11]. CerebotNano brand microcontroller which contains ATmega168 integrated circuit manufactured by the ATMEL was used in the UAV. PmodRF1 – Wireless Radio Transceiver manufactured by Digilent was preferred. The UAV unit includes four externally revolving brushless DC electric motor, four units of ESC, and four units of HK401b digital gyroscopes. As power feed, one unit of lightweight ZIPPY Flightmax 2200 mAh 3S1P 20C LiPo battery capable of providing high current feed was used.

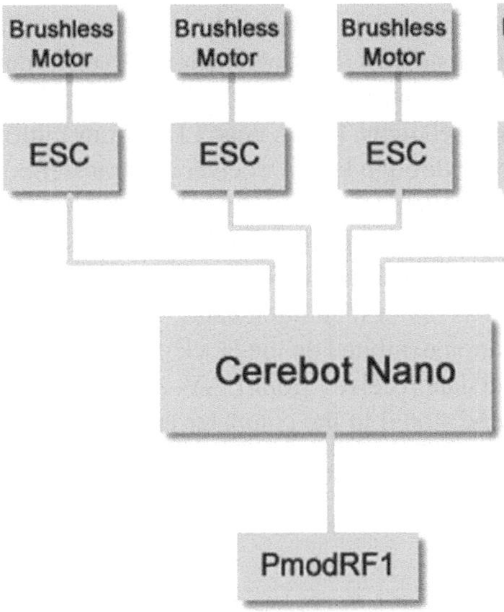

Fig. 3. UAV block diagram

In the UAV unit, motors driven by ESCs, RF transceiver maintaining communication with the control unit were added on the controller of the UAV unit as peripheral unit. Control of the motor revolutions was ensured by microcontrollers through ESCs. Stable and balanced cruise of the UAV in the air and its stable motion were established by the digital gyroscopes. By means of the function of the RF transceiver in the UAV unit, data exchange was maintained with the remote control unit. In regard to maintenance of the control of the motor revolution, while RF transceiver receives motor revolution data from the control unit, they are sent to the microcontroller in the UAV unit; then, they are transformed into PWM signals before transmitted to the ESCs.

In general, control data is received through touchscreen panel in the remote control unit in the form of data entry; then, they are transmitted to the UAV by means of RF transceiver after processed in the FPGA. These data sent are received by the RF transceiver in the UAV unit and transferred to the microcontroller. After process of these received data, they are sent to the ESCs as an operation signal. Internal and external communication in units, desired revolution level can be acquired on the UAV motors based on the data entered in the touchscreen.

2.4 Wireless Communication Between Units

During wireless communication process between human-machine interface (HMI) and UAV, RF transceivers maintaining wireless communication between HMI and UAV were connected to both the controller at the UAV unit and in the FPGA board at the control unit as peripheral unit. Serial commutation in these units was coded in C language. After testing of the consistency of the data communication between these two units, wireless communication process is completed.

In inter-unit communication process, control data is obtained as data entry from the touchscreen panel in the remote control unit; and these data is transmitted to the UAV by RF transceivers after processed in the FPGA board. Data transmitted to the UAV are received by the microcontroller through RF transceiver and transferred to the ESCs. On the other way of the communication, data processed by the microcontroller in the UAV are transmitted back by the RF transceiver in the UAV to the RF transceiver in the remote control unit. Data received by the RF transceiver are processed in the FPGA and displayed on the touchscreen as motor revolution data.

Based on the communication exhibited on the block diagram in Fig. 4, UAV can be controlled both remotely and data received from UAV. Additionally, data received by control unit from UAV can be stored in the computer environment through the serial port of the FPGA board when necessary.

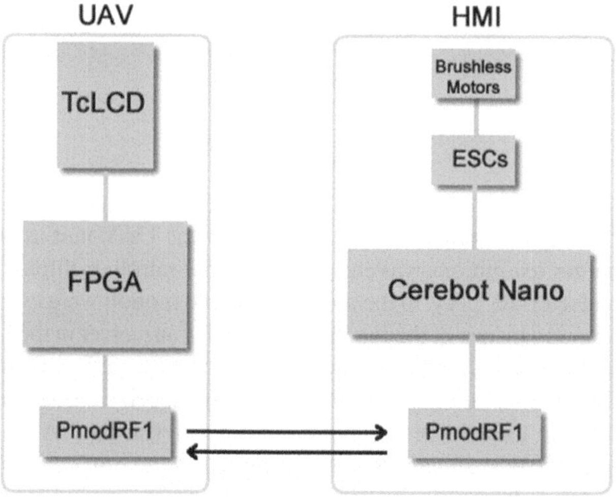

Fig. 4. Communication between UAV - HMI units

2.5 Method

In order to ensure long distance control of UAVs conveniently and efficiently, there is need for developed intermediate units. Since these vehicles are required to be operated remotely, whereas management of the vehicle necessitates remote controller, data analysis necessitates a computer. Additionally, in order to ensure convenient and fast control of the UAV from the long distance, an advanced intermediary unit is required. However, target weight to accomplish flight does not permit an advanced computer to be placed on the UAV. In the study conducted by taking these requirements into account, it was aimed to control the UAV conveniently and efficiently and to display field data functionally on the interface screen during operation of the vehicle.

In the present study, the need for convenient mobility and advanced interface were considered in the remote control unit; and a touchscreen interface with graphic screen was used as a control panel. In the remote control unit, radio transceiver, touchscreen interface and FPGA (Field Programmable Gate Array) which have parallel computing process to ensure simultaneous data transmission with the computer parallel were used. Wireless communication between remote control and the UAV is maintained by radio frequency. In the UAV unit, Cerebot microcontroller, a low-cost, small and light-weight microcontroller, was preferred by placing cost and weight measures in prominence. It was also provided that revolutions of the brushless motors in the UAV unit were controlled by the ESCs (Electronic Speed Controllers) subject to microcontrollers. Besides, capacity of the FPGA board to transfer data received by the UAV to the computer allows analysis of the field data by means of computer. Moreover, the system was structured in a way to allow using the control unit separately from the computer.

2.6 The UAV Flight Theory and the Control Technique

In order to ensure convenient remote control the UAV through touchscreen interface, the vehicle was designed as quadrotor; and four units of brushless direct current motors were used; four propellers were assembled on the UAV each of which can make right angle with another. When it is viewed from Fig. 5, F1 motor is located on the front side of the vehicle; and F2 and F3 denote left and right sides of the vehicle respectively. F4 motor is located on the rear side of the vehicle. During movement, signals regarding motor speed are transmitted from the FPGA to the controller by considering these locations so that upward, downward, right, left, forward and backward movements can be accomplished by the UAV, respectively. Hence, following guidance functions were structured to guide the UAV unit;

Forward Movement: While revolutions of F2, F3 and F4 motors are increased in the same rate, F1 is held at the same rate.

Backward Movement: While revolutions of F1, F2 and F3 motors are increased in the same rate, F4 is held unchanged.

Right Movement: While revolutions of F1, F2 and F4 motors are increased in the same rate, F3 is held unchanged.

Left Movement: While revolutions of F1, F3 and F4 motors are increased in the same rate, F2 is held unchanged.

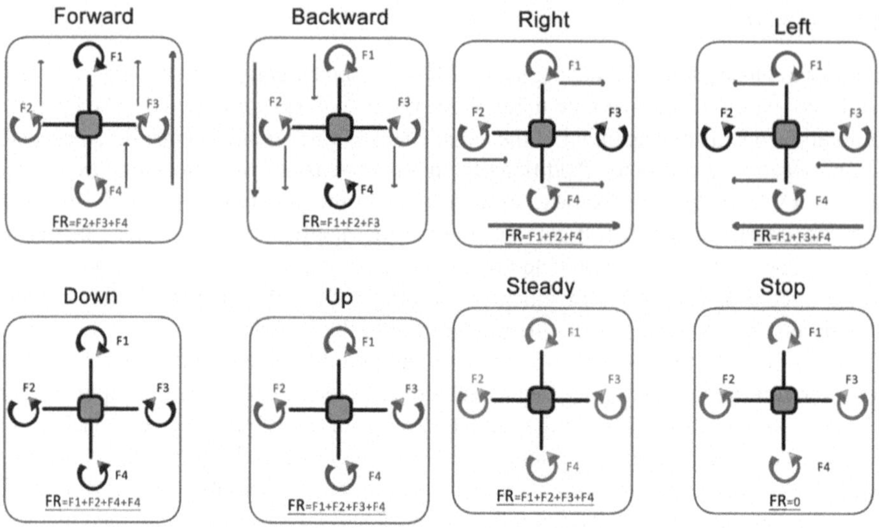

Fig. 5. Direction of the rotation of UAV motors

Downward Movement: Revolutions of F1, F2, F3, and F4 motors are reduced in the same rate.

Upward Movement: Revolutions of F1, F2, F3, and F4 motors are increased in the same rate.

Thus, the UAV can be controlled remotely by only sending downward, upward, right, left, forward and backward movement commands. That is, vehicle movement becomes rather convenient for the operator. Owing to interface used as control unit screen presented in Fig. 6, user can operate the UAV remotely by only selecting direction of the movement.

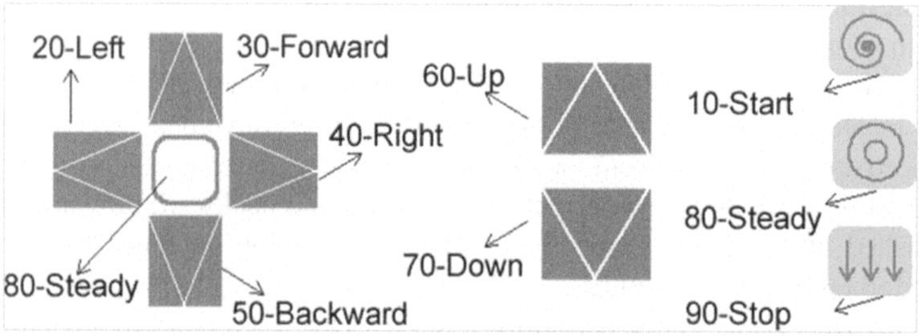

Fig. 6. Remote control unit screen

In order to minimize data exchange between the UAV unit and the control unit, following functions were formed. Along with the remote control of the UAV, following function in Table 1 were created in C programming language.

Table 1. The UAV guidance data

Value sent to UAV	Function	Description
100	Start	Start all motors at the beginning level
200–290	Left	It will be increased along the movement direction – 0–90 units of revolution value
300–390	Forward	It will be increased along the movement direction – 0–90 units of revolution value
400–490	Right	It will be increased along the movement direction – 0–90 units of revolution value
500–590	Backward	It will be increased along the movement direction – 0–90 units of revolution value
600–690	Upward	It will be increased along the movement direction – 0–90 units of revolution value
700–790	Downward	It will be increased along the movement direction – 0–90 units of revolution value
800	Stabilize	Stabilize position of the vehicle
900	Stop	Stop all motors

3 Conclusions

The control unit and data analysis screen are combined on a single unit by accomplishing control of the UAV through an interface with touchscreen. Thus, operation of unmanned aircraft vehicle was facilitated and field data was displayed on graphic screen to develop its functionality. Therefore, convenient and efficient operation of unmanned aircraft vehicle through touchscreen interface with graphic screen and wireless communication is achieved. The analysis of field data are received from the UAV and displayed on graphic screen through a wireless connection successfully. Fast and effectiveness of the UAV operator are provided. So, immediate intervention capability to the incidents to

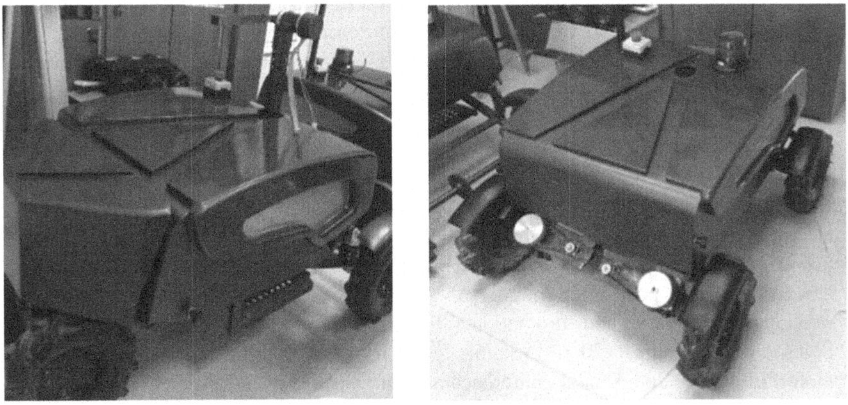

Fig. 7. Agricultural ground vehicles Alp1 (left) Alp2 (right)

control desired area is improved. The designed system offers operator a convenient and mobile control unit with graphic screen. An aircraft vehicle can be controlled from long distance. Owing to the advantage of control interface with graphic screen, this low-cost and convenient UAV offering capability to be controlled from long distance contributes efficiency of the designed system.

The present study created an environment to transmit sound and video from the UAV to the control interface; a concrete ground for further studies establishing control over UAVs from long distance was set. Here a quadcopter which have ability to integrate with a ground control system was developed. With the support of the on-going FP7-Eranet-TÜBİTAK project (UserPA-112O465), the agricultural robots for surveying vineyards were developed. In this international project two agricultural robots were developed for observing vineyards autonomously shown in the following Fig. 7 (left) and (right). At present, this aerial vehicle is able to support the developed agricultural ground vehicles for remote controlling and communicating with each other.

Acknowledgment. This project is funded by FP7-ERANET-ICTAGRI and TUBITAK and BAP-KRM at Adnan Menderes University in Turkey. The authors would like to acknowledge the support provided by funding agencies. Also, we are thankful to our partners in UserPa consortium.

References

1. Sullivan JM (2005) Revolution or evolution? The rise of the UAVs. IEEE Technol Soc Mag 25(3):43–49
2. ODTÜ Saham, İnsansız Hava Aracı (2013). http://www.saham.metu.edu.tr/tr/library.php?docid=90. Accessed 22 Mar 2013
3. Cook KLB (2007) The Silent Force Multiplier: The History and Role of UAVs in Warfare. In: Aerospace Conference, IEEE, pp 1–7
4. Chiu CC, Lo CT (2011) Vision-Only Automatic Flight Control for Small UAVs. IEEE Trans Veh Technol 60(6):2425–2437
5. Feng L, Ben MC, Yew LK (2007). Integration and Implementation of a Low-costand Vision-based UAV Tracking System, Control Conference, Chinese
6. Zhou G, Zang D (2007) Civil UAV System for Earth Observation, Geoscience and Remote Sensing Symposium. In: IEEE international conference on IGARSS 2007
7. Azfar AZ, Hazry D (2011) Simple GUI Design for monitoring of a remotely operated quadrotor Unmanned Aerial Vehicle(UAV). In: IEEE 7th international colloquium on signal processing and its applications (CSPA)
8. Santos M, López V, Morata F (2010) Intelligent fuzzy controller of a quadrotor. In: International conference on intelligent systems and knowledge engineering (ISKE)
9. Rangel RK, Kienitz KH, Brandão MP (2011) Development of a multi-purpose portable electrical UAV System, fixed & rotative wing. In: Aerospace Conference 2011. IEEE
10. Fowers SG, Lee DJ, Tippetts BJ, Lillywhite KD (2007) Vision aided stabilization and the development of a quad-rotor micro UAV. In: International symposium on computational intelligence in robotics and automation. CIRA 2007
11. Solomon O, Famouri P (2006) Dynamic performance of a permanent magnet brushless DC motor. In: IEEE 32nd Annual conference on industrial electronics, IECON 2006

Design of a Test Stand for Rolling Bearing Durability Testing

Pawel Andrzej Laski[1,2(✉)], Dawid Sebastian Pietrala[1,2], Jarosław Zwierzchowski[1,2], and Henryk Łomża[1,2]

[1] Kielce University of Technology, Kielce, Poland
{pawell,dpietrala,j.zwierzchowski}@tu.kielce.pl, hlomza@flt.krasnik.pl

[2] Fabyka Łożysk Toczych - Kraśnik Joint-Stock Company, Kraśnik, Poland

Abstract. The paper presents a design basis adopted for the construction of a test bench for rolling bearing durability testing. This article presents preliminary engineering requirements and the design of a special test rig for the examination of rolling bearings without the necessity of mounting the bearings in a specific machine component unit. The rig testing is widely used in consideration of numerous good points of such a method. The on test rigs consists in reproducing as accurately as possible the real conditions of operation of the bearings when mounted in the device for which they are intended. The device allows is capable of testing the running characteristics of bearings under varying test conditions and loads, speeds, lubrication conditions and temperatures. In addition they allow to test the service life under the set operating conditions.

Keywords: Rolling bearings · Rig tests · Designing of test rigs

1 Introduction

Rolling bearings are elements of machine parts that support axles mounted in them and shafts or movable elements. They are designed to transfer transverse and longitudinal forces acting on machine components, fixed and supported by bearings on the body of the device (housing, base frame). However, they mainly serve to reduce the frictional resistance when the journals move relative to the bearing housings. This article discusses the stand for durability tests of transversal and tapered rolling bearings. One of the methods of bearings durability testing is mounting them as parts in real device. However, this action entails significant costs and can lead to permanent damage to the device in which they were used. Therefore, it is expedient to make test stand where the bearings can be subjected to series of loads without the risk of permanent damage to the target device for which they were intended.

2 Model Stand for Durability Tests

Based on the literature [1–3] and own studies of the Department of Mechanical Technology and Metrology of the Kielce University of Technology, it has been determined

N. M. Durakbasa and M. G. Gencyilmaz (Eds.): ISPR 2018, *Proceedings of the International Symposium for Production Research 2018*, pp. 275–279, 2019.
https://doi.org/10.1007/978-3-319-92267-6_24

that one of the parameters that has the greatest impact on bearing durability is the power load. It was assumed that in the durability tests, the series of tests will cover assemblies consisting of main pair bearings and an additional two bearings performing a support function. If four bearings of the same type are used, a simultaneous test of all of them

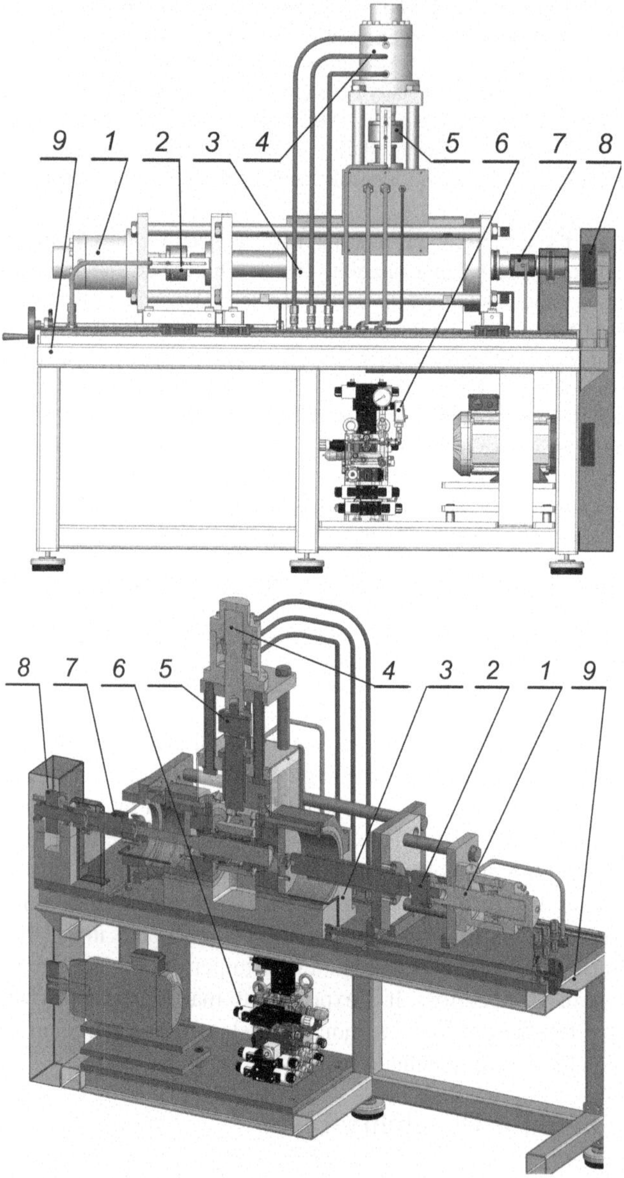

Fig. 1. Stand for durability tests. 1 - hydraulic cylinder for heavy load, 2 - longitudinal force sensor, 3 - body, 4 - radial load hydraulic cylinder, 5 - radial load sensor, 6 - hydraulic control unit,7 - torque gauge, 8 - drive assembly

is possible. The designed station enables conducting durability tests for several types of bearings. Stand for durability tests consists of a frame made of closed profiles, on which the supporting plate constituting the base is attached. The main teams of the post are attached to the plate, among others body, power unit, power load assemblies, hydraulic and electrical connections. An electric drive assembly is attached to the support frame. The proposed station uses an asynchronous AC motor. When carrying out durability tests, it is often necessary to change the rotational speed of the main shaft on which the tested bearings are mounted. A frequency converter was used to change the speed of the drive. The overall dimensions of the body and a set of sleeves allow easy adaptation of the station for subsequent tests using several types of bearings [4].

Figure 1 presents a view of a 3D solid model of a test stand for rolling bearings. In order to induce longitudinal and transverse loading of the tested bearings, hydraulic cylinders were used. Because of durability tests are carried out at long time intervals, they require maintaining the load at a given level throughout the entire test. To enable the introduction of loading forces over a large range of spans, special cylinders with two piston surfaces were used [5].

When carrying out durability tests, the load is controlled in closed loop feedback with using a force sensor. The system is therefore hydraulic force load servo. This solution was used both for the radial load unit of Fig. 2 and for the longitudinal load unit of Fig. 3.

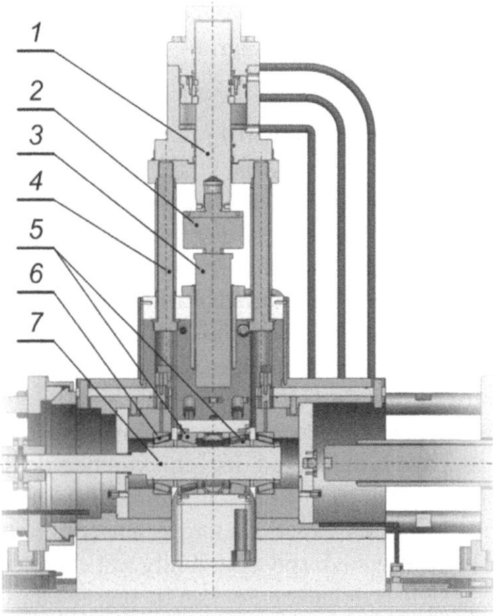

Fig. 2. Cross-section of the radial load unit. 1 - radial load hydraulic cylinder, 2 - radial load force sensor, 3 - intermediate load bolt, 4 - guide columns, 5 - pair of tested bearings, 6 - a pair of supporting bearings, 7 - main shaft

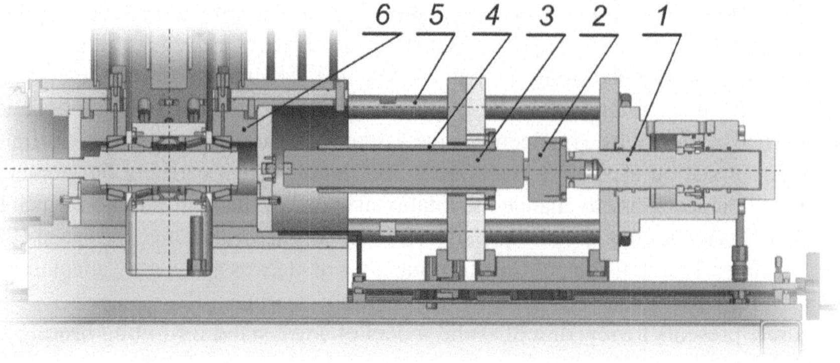

Fig. 3. Cross section of the longitudinal load unit. 1 - hydraulic cylinder for longitudinal load, 2 - longitudinal load force sensor, 3 - medium load bolt, 4 - guide bushing, 5 - guide rollers, 6 - longitudinal loading sleeve assembly

One of the most important parameters when conducting long-term durability tests is to maintain bearing units in given temperature limits. During the tests, constant temperature control with the thermocouples used for each of the bearings, both main and support bearings is carried out. The temperature is maintained in the given range using an external hydraulic power station. The external hydraulic system enables both cooling and heating of the lubricating oil. This is especially important because durability tests take place in rooms without temperature control.

3 Method of Conducting Durability Tests

The test stand is equipped with devices measuring diagnostic signals: vibrations (accelerations), temperature of tested bearings, torque. In addition, operating parameters are measured, such as: instantaneous rotational speed of the shaft, power frequency and longitudinal and radial load forces of the tested bearings. Signals from the torque sensor, temperature sensors are conditioned in both the PLC and the proprietary software. The signal from the vibration sensors are pre-treated in the data acquisition module and sent to the proprietary software for further analysis. The station provides the option of manually setting parameters as well as in the programmed test cycle. In addition, the asynchronous motor slippage is monitored. Its value can be calculated using the measured rotor speed and power frequency. Slip is an indirect measure of the load moments of the tested rolling bearing. In the event of a torque sensor failure based on the engine slip value, an alarm can be triggered, and a durability test completed safely.

Vibration and speed signals, temperatures can be presented as diagrams or graphs. It is also possible to set alarm thresholds for signals from vibration sensors, torque, temperature and speed. After exceeding one of the thresholds, the system will automatically complete the measurements, the load actuators will be withdrawn, and the main engine drive will be safely switched off. Stand for conducting durability tests of rolling bearings can work without direct human supervision and gives the possibility of conducting long-term tests.

4 Conclusions

Designed test stand fulfills the assumed functions and allows to conduct durability tests to the full extent. The station enables testing at radial, axial and radial-axial loads. The versatility of the stand consists in the fact that when using a set of interchangeable sleeves, it is possible to carry out tests of ball bearings as well as inclined roller bearings in a large range of diameters. The durability tests are carried out even for a period of several months. The length of such a long-lasting experiment allows you to follow the development of damages based on vibroacoustic signals and create curves of durability of rolling bearings. The test stand for rolling bearings is a professional device for industrial applications in long-term bearing life tests. The stand ensures precise measurements and autonomous work, which is necessary in long-term research. The compact design together with the switch-off system after exceeding the limits of operating parameters or diagnostic signals, enables autonomous operation without frequent human supervision.

Acknowledgment. The publication was created as a result of research and development carried out by Fabryka Łożysk Tocznych Kraśnik SA together with the Kielce University of Technology in the project entitled "Establishment of a R & D Center in FŁT Kraśnik SA" under the Intelligent Development Operational Program 2014–2020, co-financed from the funds of European Regional Development No. CBR/1/50-52/2017 from 07/04/2017.

References

1. Wieczorek A (2015) Designing machinery and equipment in accordance with the principle of sustainable development. Manag Syst Prod Eng 17:28–34
2. Jurecki R, Pokropiński E, Więckowski D, Żołądek Ł (2017) Design of a test rig for the examination of mechanical properties of rolling bearings. Manag Syst Prod Eng 25(1):22–28
3. Janecki D, Zwierzchowski J (2015) A method for determining the median line of measured cylindrical and conical surfaces. Meas Sci Technol 26(8):085001
4. Randall RB, Antoni J (2011) Rolling element bearing diagnostics—a tutorial. Mech Syst Signal Process 25(2):485–520
5. Janecki D, Zwierzchowski J (2009) The bird-cage method used for measuring cylindricity. A problem of optimal profile matching. In: Proceedings of the XIX IMEKO world congress: fundamental and applied metrology, pp 1784–1789
6. Simrit (2007) Simmerings and rotary seals. Freundenberg Simrit GmbH & Co. KG/Technical Manual
7. SKF rolling bearings catalogue. ©SKF Group 2013. www.skf.com/binary/77-121486/SKF-rolling-bearings-catalogue.pdf

Implementation of a Real-Time Data Acquisition System with Wireless Sensor Network for Temperature Measurement

İsmail Böğrekci(✉), Fatih Akkoyun, Pinar Demircioğlu, and Salih Vardin

Faculty of Engineering, Mechanical Engineering Department, Adnan Menderes University, Aydın, Turkey
{ibogrekci,fatih.akkoyun,pinar.demircioglu,svardin}@adu.edu.tr

Abstract. Remote data acquisition and monitoring have an important role for industry and home appliances to provide real time status of monitored constituent. Nowadays, Wireless Sensor Technology (WST) is one of the common topics for research and development. Advanced technology continuously increases abilities of components related to these types of systems while reducing their sizes and costs. In this study a real-time wireless temperature data monitoring system including four nodes to acquire temperature based data and one base station to monitor these collected data was developed. Communication between remote terminal units (RTU) and base station was implemented using Radio Frequency (RF) transceivers. Time-division multiple access (TDMA) method was established to provide sequential access to each RTU. Developed system could be adequate for continuous communication and also it was able to support additional RTUs. Laboratory tests were conducted to demonstrate developed system's functionality. During these tests, RTUs placed in standard refrigerators were used to observe temperature differences using wireless sensor network (WSN).

Keywords: Wireless Sensor Network · RF communication
Real-time data acquisition · Temperature measurement · Remote monitoring

1 Introduction

A wireless sensor network (WSN) contains radio frequency (RF) transmitter and receiver, microcontrollers, power units and remote sensors. Wireless sensor networks with many capabilities have been designed to find solutions for specific problems or to make applications possible that were not available with traditional technologies. New applications considered to be impossible before will be within reach with WSN technologies. With the introduction of Industry 4.0, WSN systems would be sought out eventually in the near future. Figure 1 illustrates the importance of wireless data systems, a sub group of Internet of Things (IoT).

N. M. Durakbasa and M. G. Gencyilmaz (Eds.): ISPR 2018, *Proceedings of the International Symposium for Production Research 2018*, pp. 280–291, 2019.
https://doi.org/10.1007/978-3-319-92267-6_25

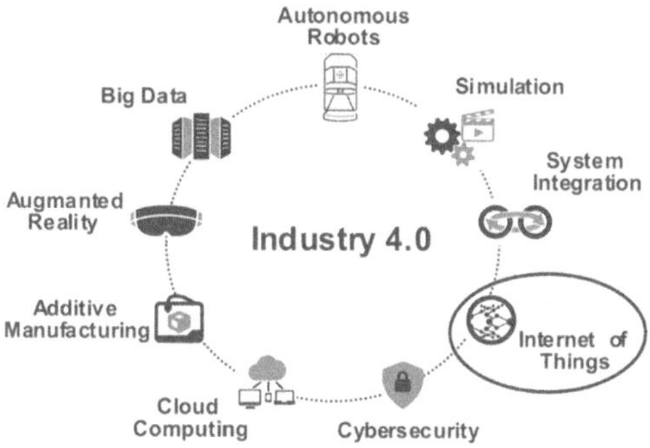

Fig. 1. Internet of Things in Industry 4.0

Continuous measurement and control of certain environmental variables are of great importance in various sectors and areas. In a previous study, continuous measurement of the temperature data was transmitted wirelessly to a computer screen [1]. In the study, authors claimed that sensor networks had important roles in order to prevent catastrophic failures with reliable and continuous monitoring. In this study, a WSN was developed to acquire and monitor field data from different nodes in real time. These nodes were named as remote terminal units (RTUs) in this paper, each RTU was placed in a different ambient.

Temperature values are kept under control in areas such as the health sector (drug stores, pharmacies, hospitals, etc.), food storage areas (cold chain supply vehicles etc.) and greenhouses. It is important to control the temperature closely to be within a certain range.

Today, many systems are utilized to provide the necessary conditions for electronic equipment rooms. Relevant to our study, in case of a possible cooling system failure, the failure could cause problems in mechanical and electronic parts due to operation at higher or intolerable temperatures. It is necessary to be able to sort out the reasons of the fault, and to fix the system remotely by possible controlling of the ambient conditions. If remote conditioning would not be enough, the faulty part or parts should be pre-identified before determination of appropriate solution. Choice of right equipment could also be counted as an important factor for system performance for identification, monitoring and data transfer. Sensors and other equipment could determine the problem

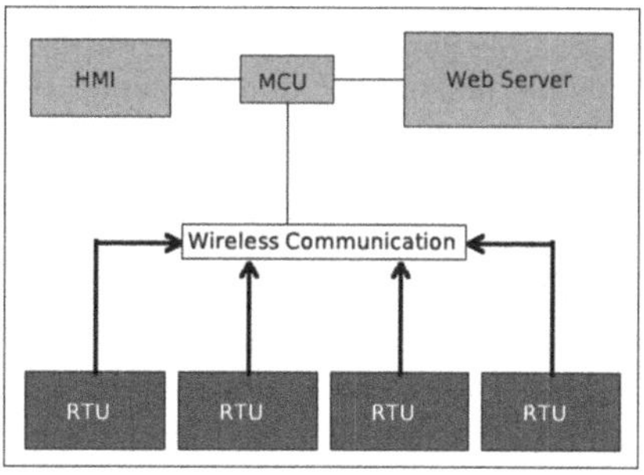

Fig. 2. Block diagram of the real-time wireless data monitoring station: HMI (Human Machine Interface), MCU (Main Control Unit), RTU (Remote Terminal Unit)

though they are mostly not able to solve the problem covered by themselves. The real time wireless temperature data monitoring station block diagram is displayed in Fig. 2.

2 Literature Review

Categorization of applications in industry of wireless application areas were studied in a previous research by Rashid and Rehmani, (2016) [2]. A wireless prototype system was developed by Vivoni and Camilli (2003) to acquire geo-referenced environmental real-time data. The system could also transmit, display and store the data between remote locations and multiple field teams [3]. Perkins et al. (2002) introduced a self-organizing sensor network with low power and low cost. The network was developed by Motorola Labs [4]. The system could be utilized to acquire environmental, agricultural and different process parameters. Rusko et al. (1999) developed a sensor and remote identification system using acoustic wave resonators. The system was activated in the European 433.92 MHz ISM-band [5]. Developed system claimed to have environmental compatibility and long durability for applications such as identification and environmental monitoring.

Some of agricultural studies consist of wireless networks as well. Gomide et al. (2001) developed a mobile field data acquisition system in order to gather data for spatial- variability. The system could also collect data about crop management [6]. Data acquisition and control systems were mounted on farm machines and additional vehicles like a data collection vehicle and a manager vehicle were utilized in the study. A silage yield mapping system consisting of a Bluetooth wireless communication module, a GPS, a moisture sensor and load cells were developed by Lee et al. (2002) [7]. An infrared thermometer system operating wirelessly was developed so as to collect in-field data by Mahan and Wanjura (2004) with cooperation of a private company [8]. Damas et al. (2001) developed a distributed automatic irrigation system which could be controlled remotely in order to control 1500 ha irrigated area. The area contained by Spain, was divided into sub-regions and hydrants were installed. The total number of hydrants in seven sections was 1850 [9]. An automated fertilizer applicator was designed by Cugati et al. (2003) especially for tree crops [10]. Jensen et al. (2000) developed a web server in order to provide information on disease infestation and pests. Weather forecasts were also contained in the information provided by the server [11].

Machine control, process control and machine guidance were also among popular topics for implementing wireless networks. Guo and Zhang (2002) established a WLAN based system enabling real time vehicle to vehicle data communication. Exchanging information amongst the vehicles was the main aim of the system established. The information could contain vehicle states and operation control variables [12]. An autonomous tractor driving fully autonomously at least 90% of the time was developed to carry out spray operations in fields by Charles and Stenz (2003). Through a radio link, a precise control of the autonomous tractor was also possible by a supervisor [13]. An autonomous guidance tractor was established by Ribeiro et al. (2003) for spray operations. The tractor was designed to be utilized in olive trees fields and citric fields in Spain [14]. Stentz et al. (2002) developed a visualization agent for human operators to supervise and to remotely control a fleet of semi-autonomous tractors utilizing WLAN. Consequently, Stentz et al. established a wireless link between human supervisor and tractors [15]. In order to prevent collisions between vehicles and humans Chung et al. established a wireless personal safety radio device (WPSRD) [16].

For robotic control and facility automation, a distributed wireless robot control system was designed by Hirakawa et al. (2002). Local link controllers were used to control the robot manipulator [17]. Moisture content of wood was monitored in real time by Heimerdinger (2000). A wireless probe system (WPS) was utilized during the drying process [18].

Due to gradual safety and security demand increments, complete documentations from field to customer for food products have gained major importance [19]. Traceability system on product, animal and human tracking could be addressed as well structured when implemented with newly accepted RFID technology. Data collecting could also be possible with this technology [20]. A remote health monitoring system was designed for cattle containing many units such as a respiration transducer, a core body temperature sensor, a GPS unit, an electronic belt, ambient temperature transducer and a pulse oximeter. Communication of the system was implemented wirelessly with a base station via Bluetooth telemetry [21]. Taylor and Mayer (2004) inspected a "smart" system related to comprehensive animal management [22].

There were studies conducted for food packaging, disposable and inexpensive RFID biosensor tags were inspected by Wentworth (2003). His study claimed that these biosensor tags could be utilized for checking contamination, inventory and history of food products. In order to detect bacteria, the biosensor used antigen-antibody reaction and the biosensor was based on an acoustic wave platform [23]. For numerous topics like promotional opportunities, automatic checkout, smart appliances, smart recycling, smart packaging and marketing opportunities, Chandler (2003) investigated the RFID tags for their possible potential [24].

Wireless sensor systems are also used for transportation and food inspection. For animal transportation, a monitoring system was designed by Geers et al. (1998) and Gebresenbet et al. (2003). The on-the-road system consisted of sensors in the animal compartment. The sensors were utilized to monitor the air quality, animal behaviours and vibration. It was also possible to identify the animals with the sensors [25, 26]. For food processing plants, a handheld PC was designed and developed by Najjar et al. (1997). Quality inspectors of the plant could create and modify a form using the handheld PC. This data then could be sent to plant manager's computer through a wireless data communication having specification of 16 bit, 2 Mb per second and full duplex audio [27].

3 Materials and Method

A wireless temperature measurement system was designed and produced in this study. The data was transferred wirelessly via RF signals. The system was able to display the received data on PC screen. The system was intended to collect the temperature data of the devices or indoor environment for the end users. Real life events could be saved or transferred as data. The data can be later processed or used for many different purposes in WSN by sensor nodes [2].

Design goals and design criteria should be indicated carefully before the manufacturing process. The design goals and criteria were listed as the following;

1. The first design criterion was to reduce to cost. And one of the most important criteria was to produce products with the readily available materials in the market.
2. Dimensions of the system casings and sensor casings should be flexible and portable to easily move.
3. Energy requirement of the system was also an important parameter. Energy consumption could be affected by RF transceiver distance, data exchange rate, communication distance and number of remote terminal units.
4. The interface of the PC monitoring should be basic as possible so that following data on the screens should be easy and improvements could be implemented when needed.

Micro Controller Unit (MCU), temperature sensor and RF transceiver module were the main components for tests carried out in this study. Table 1 contains the electronic components and corresponding descriptions.

Table 1. Electronic components of the remote data monitoring system

Electronic Unit	Model	Figure	Description
Wireless Communication	DRF4432D20I		20dBm ISM RF Transceiver Module.
Microprocessor	At32MDB01		Development board was used as Data Acquiring Unit and Main Control Unit.
Web Monitoring	PHP Software		Web Interface for monitoring of the field data
Local Monitoring	C++ Software		PC Desktop Interface for monitoring of the field data
USB to serial UART interface	FT232RL FTDI		USB to serial UART interface

Microcontrollers are special purpose computers. They are embedded inside some other device in order to control the actions or features of the device. Specially prepared programs are run by the microcontrollers, and microcontrollers are usually dedicated to one task. The special program for each separate application could be stored in read-only memory (ROM) or Flash memory. Microcontrollers are often low-power devices. In this study, Atmega32A microcontroller was chosen as the microcontroller.

LM35-DZ was used as the temperature sensor for the real-time wireless temperature data collecting station. Integrated-circuit temperature sensors are applied in a general way, in this study the LM35 is no exception. Mounting of the temperature sensor could be implemented as gluing or cementing to surface, and the sensor's temperature is generally assumed to be within about 0.01 °C of the surface temperature. The ambient air and the surface temperature were presumed to be almost the same. If the surface temperature was lower than the ambient temperature, the actual temperature of the sensor would be expected to be an intermediate temperature between the surface temperature and the ambient temperature values. Also the dimensions of the utilized sensor were suitably small for remote applications. At the water level, calibration and trimming could be carried out to ensure low cost.

DRF4432D20I RF Transceiver-Receiver Module was used for the study. The module could operate in 40 channels. The channels had 1 MHz channel space reducing the interface from adjacent channels. NetID and NodeID were contained by the module. For the modules to communicate, modules should only have the same NetID. The NodeID of modules however could differ and this would not affect the modules. Network module ID could be reserved for future uses. The task of the component was mainly to send the collected temperature data to the Node Point on the microprocessor. The data viewing distance limit was about 1 km. And the mini gauges of RF transceiver could provide flexible sizing. A virtual COM port was created using USB on a computer by USB-TTL UART Module. Various standard Baud Rates were supported for serial communication by the module. Battery selection for many applications is an undeniably important parameter to get stable and continuous results. Dimensions, voltage, weight, and charging method of the battery are important factors to consider. A Li-Po type of battery unit with two cells was utilized in the system to supply required energy for electronic circuits. The wireless data transfer was implemented between a single RTU and the Remote Data Monitoring System as shown in the Fig. 3.

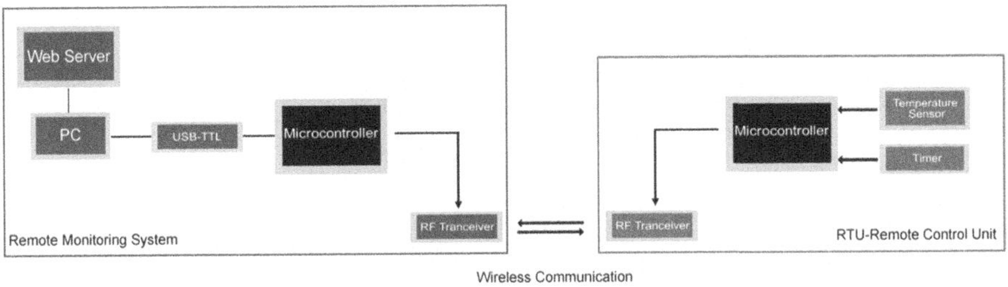

Fig. 3. Block diagram of the remote data monitoring system and remote terminal unit

The system measured the temperature data of the environment of interest and the data could be followed at the collection point at certain time intervals. As shown in Fig. 4, the block diagram of the remote terminal unit (RTU) was established.

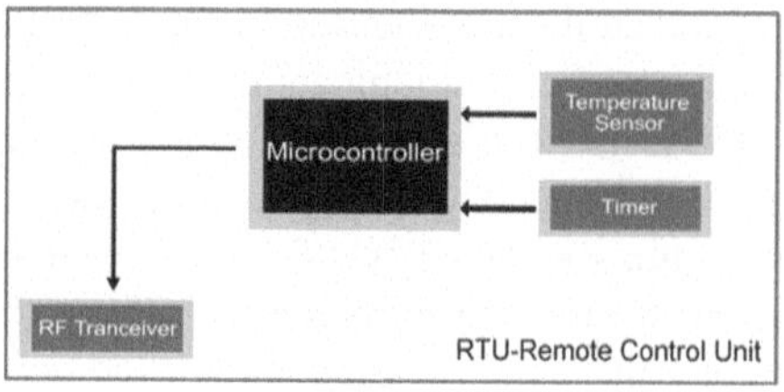

Fig. 4. Remote terminal unit block diagram

Temperature data collectors (RTUs) are presented in Fig. 5. The remote units consisted of a MCU circuit board, RF transceiver and temperature sensor, powered by Li-Po batteries and these batteries could function for several months continuously for this application.

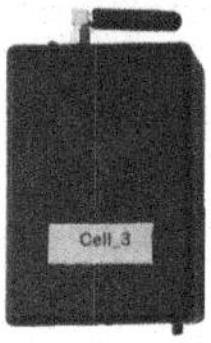

Fig. 5. Remote Terminal Unit (RTU)

The block diagram of the central unit is as shown in Fig. 6. The central unit displayed the data received at the collection point in graphical user interfaces (GUI) on the PC screen.

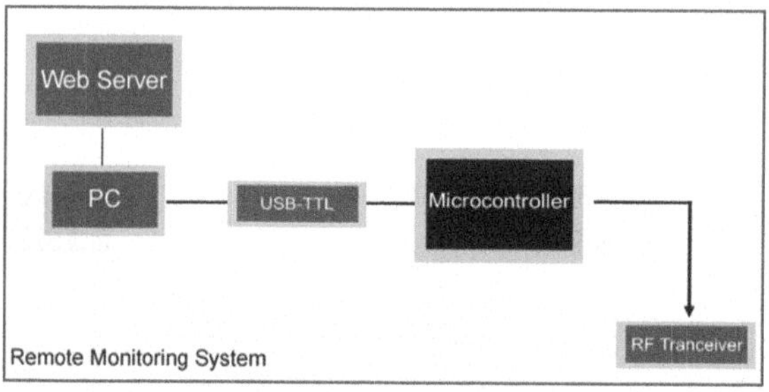

Fig. 6. Block diagram of remote monitoring system

User interface was designed to interact with electronic devices through graphical icons and visual indicators. Instead of giving text-based data to user, typed command labels or text navigations were created. Users not familiar with a web server or desktop user interface might evaluate the pros and cons of each to help them determine which choice is best for them.

Desktop applications are used to perform certain duties in both laptops and desktops. In a networked environment, some desktop applications could be used by multiple people. Rapid development of web applications is responsible for replacing desktop applications for the reasons of better functioning, better usability and mostly portability. Server-client architecture is usually preferred while a web application is being created. This could be counted as one of the many reasons why web applications are spreading widely nowadays. Even though web applications are slightly better than desktop applications, a small chance of becoming outdated could be possible.

After data acquisition, the data file was transferred to the web environment. Saved files from the desktop application were updated at certain intervals. So the web server depended on the desktop application during the prototype phase. In Fig. 7 the web interface is shown.

Fig. 7. WEB interface

4 Results and Discussion

Typical results were obtained during testing phase for the sensors. Remote thermal unit 1 (RTU-1) was placed in freezer section and the system was able to monitor the temperature changes in the freezer. Figure 8 shows the temperature data recorded at a specific time interval.

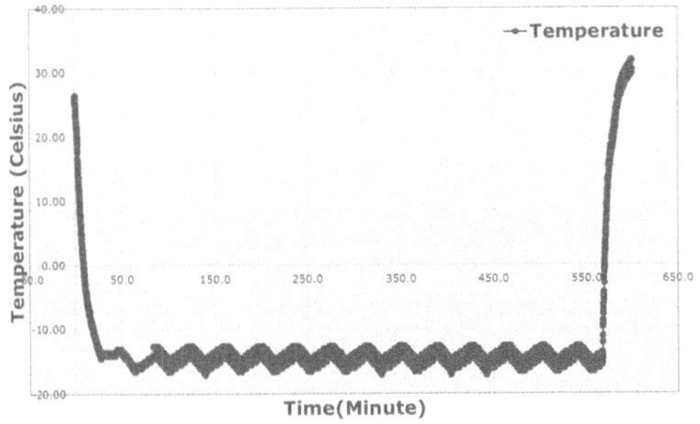

Fig. 8. Freezer temperature data recorded by RTU-1

When the RF transceiver power was set to 27 dBm, the system update time varied between 0.5–3 s depending on the sensor location. However, a few error bits were also discovered in the measured temperature data because of electromagnetic interface (EMI) from other RF transceiver devices. Operation of the RF transceiver devices at the same ISM frequency band could be the reason for the error bits.

The RTUs reported initial ambient temperatures as they were activated at room temperature prior to locating them in refrigerators. There was a brief time for RTU units to settle and transfer the refrigerator temperature values. This time was approximately 10 min as the graphs indicated. During the period at refrigerator, RTU-1 and RTU-2 units transmitted temperatures under threshold of 10 °C consistently. The measurement of RTU-1 and RTU-2 are given in Figs. 9 and 10 as graphs.

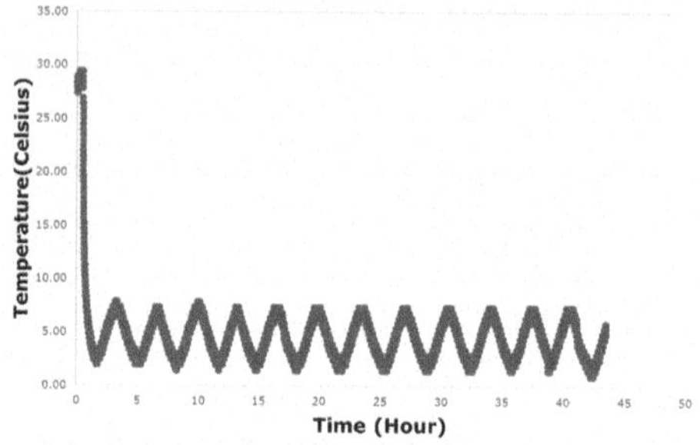

Fig. 9. Refrigerator temperature data recorded by RTU-1

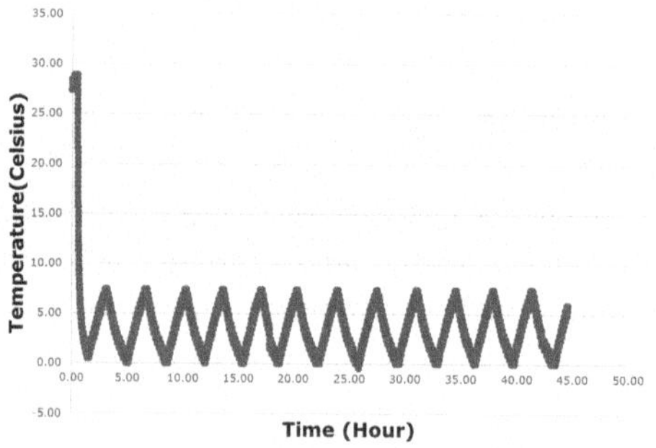

Fig. 10. Refrigerator temperature data recorded by RTU-2

Comparison graphics are shown in Fig. 11 for ambient air at the place where the refrigerators are located. RTU-3 and RTU-4 were employed for recording and transferring ambient temperature data. All the tests for RTUs were executed at the same interval. Test conditions were evaluated according to environment temperature and refrigerator temperature to measure the reaction of the sensors.

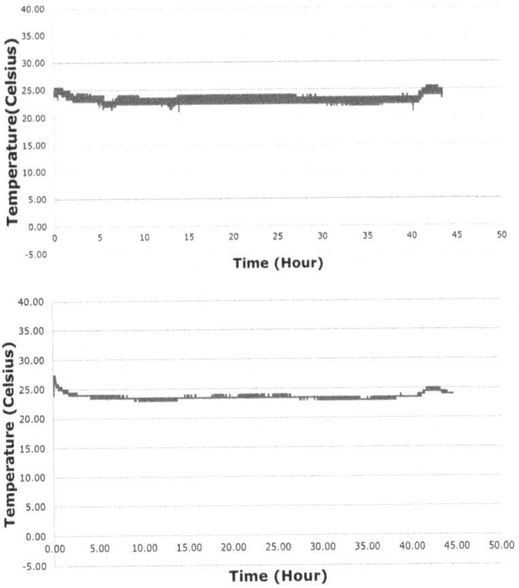

Fig. 11. Ambient air temperature data recorded by RTU-3 (top) and RTU-4 (bottom)

It was observed from the tests that temperature sensors could detect the temperature in the refrigerator correctly. The tests were carried out on two different types of refrigerators, and as a result refrigerator temperature set values were confirmed to be −18 °C for the freezer sections (see Fig. 12) and 4 °C for refrigerator sections.

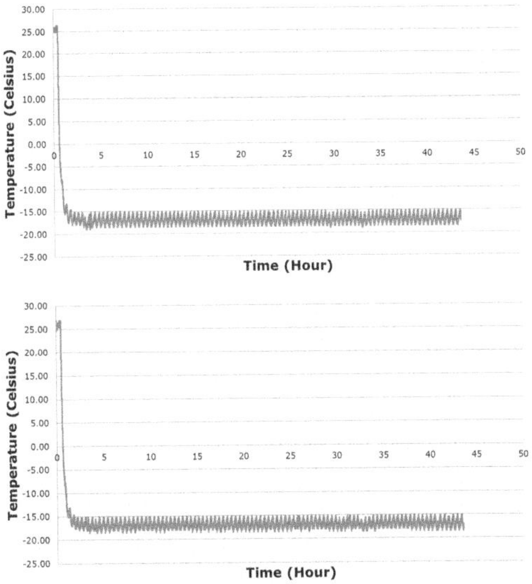

Fig. 12. Freezer temperature data recorded by RTU-1 (top) and RTU-2 (bottom)

5 Conclusion

The study was based on the monitoring of the data to be controlled in specific areas and the implemented system could be further modified to be utilized in various industrial fields. Microprocessor read the data from the sensor and transferred the data created through wireless network successfully.

Certain circuit boards were prepared in our theory of infrastructure and the necessary software was installed to carry out the test phase in order to achieve the goals of the study. Prior to the test phase, design and manufacturing processes were explained, and throughout the conducted experiments, functions of the RTU and MCU were tested, then results were discussed. The tests were conducted in two refrigerators. The measurement and transfer tasks were executed at room temperature and at refrigerator temperatures. Controlling the status and information of the station environment were carried out simultaneously. Desktop application was implemented as unit interface and web publishing was also prepared to share or view measured temperatures in an attempt to provide the data continuously to users.

The study could also be further improved to send an information message to a desired mobile phone in order to inform the user instead of watching the data on the computer screen for a certain threshold value. It was also possible to remotely interfere with the cooling system to activate so as to compensate for a possible temperature discontinuity.

References

1. Dargie W, Poellabauer C (2010) Fundamentals of wireless sensor networks. JohnWiley&Sons
2. Rashid B, Rehmani MH (2016) Applications of wireless sensor networks for urban areas: a survey. J Netw Comput Appl 60:192–219
3. Vivoni ER, Camilli R (2002) Real-time streaming of environmental field data. Comput Geosci 29:457–468
4. Perkins M, Correal N, O'Dea B (2002) Emergent wireless sensor network limitations: a plea for advancement in core technologies. In: 1st IEEE international conference on sensors, pp 1505–1509, June 2002
5. Rusko M, Buff W, Binhack M, Goroll M, Ehrenpfordt J, Klett S (1999) Passive resonator identification TAG for narrow-band wireless telemetry. In: Proceedings of the IEEE ultrasonics symposium, vol 1, pp 377–380, October 1999
6. Gomide RL, Inamasu RY, Queiroz DM, Mantovani EC, Santos WF (2001) An automatic data acquisition and control mobile laboratory network for crop production systems data management and spatial variability studies in the Brazilian center-west region. The American Society of Agriculture Engineers, ASAE Paper No.:01-1046
7. Lee WS, Burks TF, Schueller JK (2002) Silage yield monitoring system. The American Society of Agriculture Engineers, ASAE Paper No.: 02-1165
8. Mahan J, Wanjura D (2004) Design and construction of a wireless infrared thermometry system. The USDA annual report, Project number: 6208-21000-012-03, September 2004
9. Damas M, Prados AM, Gómez F, Olivares G (2001) HidroBus® system: fieldbus for integrated management of extensive areas of irrigated land. Microprocess Microsyst 25:177–184

10. Cugati S, Miller W, Schueller J (2003) Automation concepts for the variable rate fertilizer applicator for tree farming. In: The Proceedings of the 4th european conference in precision agriculture, June 2003
11. Jensen AL, Boll PS, Thysen I, Pathak BK (2000) Pl@nteInfo: a web-based system for personalized decision support in crop management. Comput Elect Agric 25:271–293
12. Guo LS, Zhang Q (2002) A wireless LAN for collaborative off-road vehicle automation. In: Proceedings of automation technology for off-road equipment conference, pp 51–58, July 2002
13. Charles K, Stenz A (2003) Automatic spraying for nurseries. USDA annual report, Project number: 3607-21620-006-03, August 2003
14. Ribeiro A, Garcia–Perez L, Garcia-Alegre, Guinea MC (2003) A friendly man-machine visualization agent for remote control of an autonomous tractor GPS guided. In: The proceedings of the 4th European conference in precision agriculture, June 2003
15. Stentz A, Dima C, Wellington C, Herman H, Stager D (2002) A system for semi-autonomous tractor operations. Auton Robots 13:87–104
16. Chung YC, Olsen SL, Wojcik L, Song Z, He C, Adamson S (2001) Wireless safety personnel radio device for collision avoidance system of autonomous vehicles. In: Digest of 2001 IEEE antennas and propagation society international symposium, pp 121–124, July 2001
17. Hirakawa AR, Saraiva AM, Cugnasca CE (2002) Wireless robust robot for agricultural applications. In: Proceedings of the world congress of computers in agriculture and natural resources, pp 414–420, March 2002
18. Heimerdinger U (2000) Wireless probes revolutionize moisture measurement when drying wood. In: Proceedings of the 51st western dry kiln association meeting, pp 63–66, May 2000
19. Thysen I (2000) Agriculture in the information society. J Agric Eng Res 76:297–303
20. Sahin E, Dallery Y, Gershwin S (2002) Performance evaluation of a traceability system: an application to the radio frequency identification technology. In: Proceedings of the 2002 IEEE international conference on systems, man and cybernetics, vol 3, pp 647–650, October 2002
21. Nagl L, Schmitz R, Warren S, Hildreth TS, Erickson H, Andresen D (2003) Wearable sensor system for wireless state-of-health determination in cattle. In: Proceedings of the 25th annual international conference of the IEEE engineering in medicine and biology society, pp 3012–3015, September 2003
22. Taylor K, Mayer K (2004) TinyDB by remote. In: Presentation in Australian mote users' workshop, February 2004
23. Wentworth SM (2003) Microbial sensor tags. In: The 2003 IFT (the institute of food engineering) annual meeting book of abstracts, July 2003
24. Chandler S (2003) Vision of the future for smart packaging for brand owners. In: Proceedings of the international conference on smart and intelligent packaging, pp 253–269, October 2003
25. Gebresenbet G, Ljungberg D, Van de Water G, Geers R (2003) Information monitoring system for surveillance of animal welfare during transport. In: Proceedings of the 4th European conference in precision agriculture, June 2003
26. Geers R, Saatkamp HW, Goossens K, Van Camp B, Gorssen J, Rombouts G, Vanthemsche P (1998) TETRAD: an on-line telematic surveillance system for animal transports. Comput Electron Agric 21:107–116
27. Najjar LJ, Thompson JC, Ockerman J (1997) Wearable computer for quality assurance inspectors in a food processing plant. In: Proceedings of the 1st IEEE international symposium on wearable computers, pp 163–164, October 1997

Integrations Management and Product Development for New Product

Semih Dönmezer(✉)

Eskişehir Vocational School, Anadolu University, Eskişehir, Turkey
semihdonmeze@anadolu.edu.tr

Abstract. In recent years, many industries have designed their production in accordance with the principles of lean production, increased flexibility and achieved great success in productivity. Currently, production is facing on a new breakthrough.

Virtual Reality is a promising new medium that's changing the way we approach 3D Modeling. It allows artists to immerse themselves in a virtual workspace where they can construct, carve, and paint their visions like never before. Also, VRP is definitely a requirement of market, because the Comprehensive design by VRP carry with usability, salability, maintainability, acceptability, manufacturability and availability. Efficiency contribution of the VRP reduces the time to time market, reduces innovation risks and increases the innovation head start. What is virtual prototyping. Virtual prototyping is the first step for the technology and unified technology acceptance. Extended Technology Acceptance model, Technology life cycle and competitive position, technologies portfolios are important parts of the product developer. The best examples of virtualization include the digitization of product development, for example through the use of 3-D models and the development of data-based product models for the various value-added processes. The digitization of process designing is such as CAD-3D simulation, augmented reality, 3-D user interface, stereoscopic desk top VRP, modelling. In this study, these topics are investigated comprehensively and the future suggestions on the industry 4.0 is presented for the complex product development.

Keywords: Virtual reality prototype · Design for Manufacturing
Dynamic modeling · Product technology acceptance model · Complexity

1 Introduction

Successful New Product Development is a critical mainspring for business success. Many companies are striving to offer new products to the market. A study sponsored by the Institute of Marketing Science in 1990 found that on average 25% of recent sales of successful companies came from new products introduced in the last three years to gain a reference. New product development can also be a potential source of significant in the economies for the firm. New products can use most of the same raw material inputs as the existing products of the company and can be sold by the existing sales force of the company - which can lead to significantly lower unit costs and higher profit for the firm. In addition, new product development can be an important leverage for the firm to

N. M. Durakbasa and M. G. Gencyilmaz (Eds.): ISPR 2018, *Proceedings of the International Symposium for Production Research 2018*, pp. 292–308, 2019.
https://doi.org/10.1007/978-3-319-92267-6_26

use in relation to its distribution channel partners. In view of marketing perspective, companies that develop the organizational structures and processes and needed to produce new products effectively and efficiently are more likely to be in line with the needs and desires of their customer [1].

Regrettably, new product development is an extremely difficult and complex process. Innovation is inherently risky and companies can spend a significant amount of time and money on new product ideas, without giving any guarantees that they will be able to live commercially. Many new products fail and new product development scenes are filled with samples.

In this paper, we discuss a dispersed and integrated new product development process that has proven to enhance success and mitigate failure in product developmentdocument.

2 Manufacturing Management Strategies

A manufacturing strategy must deal with a variety of issues from operational to tactical to strategic levels. Todays, big industries are divided into business units for effective and stream lined decision making for successful products. for successful production, it is needed for flexible manufacturing to cope with environmental uncertainties. The most common production problems are machine failures and process-quality. Such as supplier-deliver problems, competitors' behavior and customer volatility problems should be cope of with. Because of rapid technological shifts. Every company should choose between two alternatives "buy or make". Company discusses in detail connected integration low cost choice and saving product design among the different functions of their financial, marketing and logistics matters. Another argument is reduction of uncertainties via better control over the product, quality, lead times and pricing stratifies. It can be concentrated on minimize in process inventory via precise scheduling carried out on computers [2].

Engineering Design
Engineering design is always directly influenced by social needs. The basic idea of the design is to create high quality products at a reduced cost. However, design paradigms change over time, depending on market forces. Some of the known paradigms are DFA (Design for Assembly), DFM (Design for Manufacturing), DFX (Design for Anything), and CE (Concurrent Engineering). DFA seeks to streamline product design for easy assembly. Engineering Design is now more environmentally conscious by integrating the concept of sustainability into new product development (NPD). Incorporating sustainability leads the designer to examine the environmental impact of the product, from development to commercialization and disposal.

Sustainable design goes beyond ensuring that a design meets the functional requirements of a product. It looks at the entire lifecycle of the product, from "birth" to "death," also known as "cradle to grave" (McDonough and Braungart 2002). Design can mean setting parameters of system. Also, design would be the detailing of the materials, shapes and tolerances of each part of product. It is an activity that starts with sketches of parts and assemblies. Then it goes to the CAD workstation, where assembly drawings are to be passed to manufacturing and assembly. Design for assembly (DFE) is concerned with reducing product assembly cost, design for manufacturing (DFM) is concerned with reducing overall part production cost. In this case, it may be needed to chance the design and this situation will be cause serious cost problem and besides considerably delay in final products. Reducing costs, shortens time to market will become more important, but it is possible that such problems can be solved by using DFMA. Using DFMA reduced development time from two years to one years in the car industry.

As a terminology; The term "DFMA" comes with the combination of DFA (Design for Assembly) and DFM (Design for Manufacturing). The basic concept of it is that the design engineers apply the DFMA paradigm or software to analyze the manufacturing and assembly problems at the early design stage [3].

- technology has problem solving potential,
- product is troubleshooting,
- technology is knowledge of an enterprise,
- knowledge is as good and technology as good for asset of company,
- a good quality and well-crafted design pulls you up,

The Consequences of Inadequate DFM

The consequences of inadequate DFM **for delivery are**; line stoppers, parts do not assembly correctly much pruning, grooming, and tuning to get products out the door, poor yield, invariably results in late delivery. **The behavioral hurdles** to good DFM are; the lack of DFM training, lack of DFM knowledge, parts designed with no consideration of how it is to be built, over the wall syndrome, after release, no longer engineering's problem, never enough time to design parts right the first time, always enough time do it over. **The attitude hurdles** to good DFM are; tradition designer limited knowledge of manufacturing process reluctance to accept suggestion from supplier regarding design issues, we don't seem to allot time to design system properly up front. **The bottom line consequences**; unhappy customers, sometimes problems get shipped to the field, customers losing confident in our products, problems increase overall costs resulting in loss of ability to complete, low product quality leads to poor customers' satisfaction, proper performance, and eventually to high costs, post-launch redesigns [5].

Knowledge management (**KM**) is the process of creating, sharing, using and managing the knowledge and information of an organization. It refers to a multidisciplinary approach to achieving organizational objectives by making the best use of knowledge [6].

Technology is knowledge of an enterprise has problem solving potential and product is troubleshooting. Knowledge is as good and technology as good for asset of company and a good quality and well-crafted design pulls you up. Good product is not ill, service is not necessary. You acquire knowledge through research, through knowledge will acquire technology, so technology will be your most important asset. Quality Function Deployment, increases market acceptance and reduces design risk needs analysis. Technological developments bring some advantages to the companies and affect them positively or negatively in the market. These changes can be briefly described as follows, technological change and effect on companies, shorter product life cycles stronger segment splitting, new forms of the competitor, decentralized decisions, deregulation especially in the energy market and finally increasing globalization. So, you always have to count on the market developments, chances and risks.

Effective Product Development

Effectiveness and efficiency are only enduring existence basis of a company and Effective product development phase drive the company's performance. Based on this background, every company should ask itself again the question. *why should the customer just buy from us? what can and should we do, so he does it.* and *what can we do to fulfil our customers' wishes as possible as in the most economical way?* so that "he principle of survival of the fittest".

Effectivity

Significant; A competitively priced position can be achieved with regard to the customer's subjectively significant selling points. The performance, which is superior in comparison to the competition, therefore it has to start with performance features that are particularly important for the customer.

Perception; Performance advantages must be recognized by the customer as such, the consumer must be subjectively convinced that the producer has a performance advantage in terms of significant features. It is crucial to clearly differentiate between technical objective circumstances and the subjective perception of customers.

Efficiency

Defence capability; in order to be considered as a strategic advantage, the advantage must have a certain temporal consistency. it must not be short-term competitiveness and thus recoverable. Aaker speaks of a *sustainable competitive advantage* in this context (see Aaker 1988 and 2007) For example, a price reduction that does not rely on a much lower cost structure than the competition is not a permanent competitive advantage, as competitors are in a position to react very quickly and strongly. Competitor benefits can only be achieved with measures that cannot be copied immediately.

Economics; a company can better fulfil existing needs of customers than the competition. If it is based solely on the condition of a customer advantage, it will be inevitably result in misleading consumer orientation which aligns all researches for customer "wish not fulfilment" is another efficiency condition to be satisfied in the form an economic advantage [7].

Stressors in marketing

Time to market stress – Break Even $\frac{\text{Required time for BE}^2}{\text{Available time BE}} < 1$

Solution Stress $\frac{\text{R\&D time for required innovation lead}}{\text{R\&D time for achievable innovation lead}} < 1$

Profitability Stress $\frac{\text{Required ROI}}{\text{Achievable ROI}}$

If > 1, then stress caused by short resources [8]

Economic growth and new technologies will only occur at changing markets, and this is a reality for technological innovation. Well-designed product will promote growth, when commodities and service markets are receptions to buy by customers. When A purchase request to the customer is given, company decide to get risk/benefit relationship for the potential customer and leads to technological innovation. For a long time in the market, should be necessary research-oriented, innovative (high profit/high risk), fragmented with numerous small chance window, dynamic with high speed, short product-life cycle for using net-effect in market. A purchase request of market, new designed goods encourages market acceptance for the innovation. To capitalize on the growth drivers of technological innovation, innovative goods markets need to be identified, analyzed and processed. Recognizing and stimulating new needs promotes to get receptiveness, finally it must be demand-driven instead of supply-driven. Good design is going to open the window big change for the company to enter market and window of opportunity [9]. See Fig. 1.

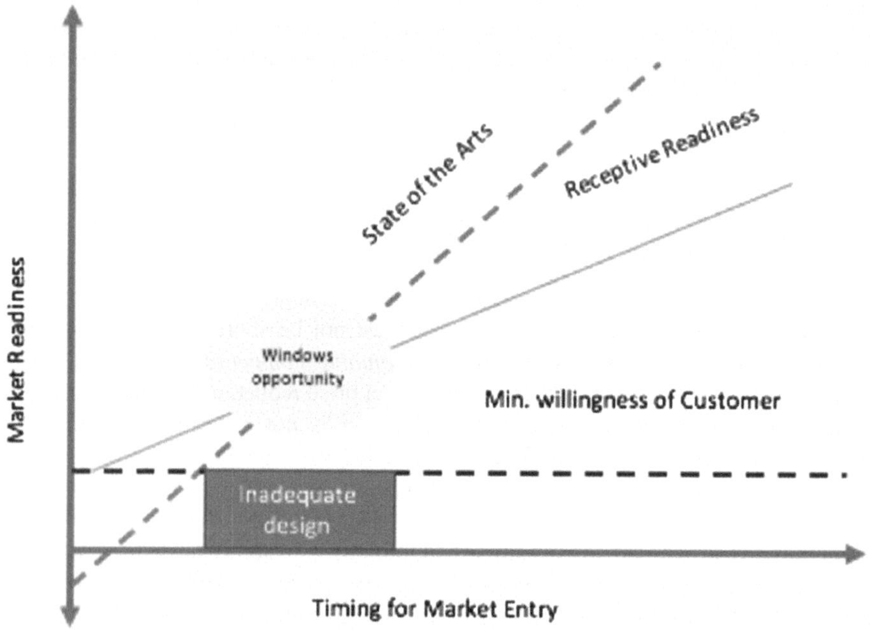

Fig. 1. Chance window for company - Day and Freeman [10]

Pitfalls of innovative behavior [9]: The following are the five main pitfalls that people and organizations inclined to become more innovative. Read how many of them apply to you or your organization:

1. Identify the wrong problem.
2. To judge ideas too fast.
3. Stop with the first good idea (never the best).

4. Do not get "the bandits on the train" (get support from those who could derail your train)
5. Follow rules that do not exist.

Best practice study for Development of new products, according to 189 reports, ca. 1/3 high technologic goods, 80% products dominated product technology.

In view of these results, cross functional teams ca. 76%, new product development process 55%. Development time of innovative product is 2,95 years. Most frequent obstacle for successful product lead to lack of resources, besides, dropping out rate for the successful design is only 1 product concept of 11 successful. Companies launch their new product average 37 new products in 5 years and the new products of companies that introduced to the market, is successful of 58%.

Business models/business models; What are business models? Definition: A business model sets the roles & Connections to customers, partners & Suppliers of a company [11].

Core Strategy; Business strategy, business mission lead to reduce risk in an emergency, provide position related information.

Dynamic Modeling of Product Development Process

Successful design and development projects are critical success for ever industries. Project performance must be understood that is dynamic concurrence relationship. Developing products faster, better and cheaper than competitors has become to success in many markets. This will provide competitive advantages to the business. Increasing concurrence and cross-functional development also dramatically increases the dynamic complexity of product development (Smith and Eppinger 1997; Ford 1996; Wetherbe 1995; Osborne 1993). However, the mental models used by developers and managers to value, estimate, and manage projects have generally not been improved to include dynamic performance effects. The resulting lack of understanding (Diehl and Sterman 1995, Sterman 1994, Paich and Sterman 1993, Rechtin 1991) and inadequate decision heuristics (Kleinmuntz 1993) have contributed to the oft-quoted poor management of development projects (Construction Industry Institute 1990, Womack, Jones and Roos Dertouzos, Lester and Solow 1989; Davis and Ledbetter 1987; Pugh 1984; Abdel-Hamid 1984; Brooks 1975).

Product Complexity

The study of product complexity has been hampered by the lack of consensus around a precise definition. Product complexity according to Baldwin & Clark, proportional to total member of design decision, Griffin (1997a) The number of functional designed into a product, Keski and Heikkia (2002) Represented by number of physical modules and also by the degree of dependency, Gupt and Krishnan (1999) Number of components, Tatikonda and Stock (2003) Proportional to the interdependence of technologies. Complexity is the state of possessing a multiplicity of elements manifesting relatedness. Product complexity is a design state resulting from processing a multiplicity of elements manifesting relatedness among, product architectural elements [12].

Wrong Design, Two Sample of Dramatic Results

New product failure rates are substantial; the cost of failure can be enormous. Various studies routinely report that 30–35% of products introduced to the market end up failing, even when the product is simply a line extension of an existing brand, or a new brand introduced in a category where the firm already has a successful product [1].

Imagination is more important than knowledge, albert Einstein, why is product development important, because product development is the key element the success of company, long term sustainability and material condition and best available tools of our lives. A small mistake in product development phase will cause to injure the reputation of a company.

Mercedes Benz introduces the A-Class car in 1997 after a $1.5 Bn development cost. In Sweden, during Mosse-test the A-Class car flipped over at 60 km/h, immediately after the test, 2500 newly-sold cars were recalled and sales almost stopped. Mercedes-Benz added stability control (ESP) and redesigned the car's suspension. Cost of this change was $250,000,000. A serious mistake in product development can cost the company and repeated mistakes in product development will also cost company. Likewise, Firestone disaster was more than 250 deaths and 700 injures in the US were as a result of Ford explorers rolling over after the treat separated on firestone tyres. Ford replaced 13 M tyres and take a $ 3 Bn charge afterwards, repot of Fords was $ 551 M quarterly loss and Ford' market share falls by 22%, cut 10% of its white-collar works and finally Ford CEO Jacques Nasser resigned [13].

When Costs are Determined

When a product is designed, 80% of the cost has already been determined. and by the time a product goes into the production 95% of its costs will be determined, thus it will be very difficult to remove cost at that late a date. The most profound implication for product development is that 60% of a products cumulative lifetime costs is committed by the concept architecture phase. This explains why the design phase is so important. The Toyota philosophy confirms that the cost of product is largely determinate at the planning and design stage. not much in the way of cost improvements can be expected once full-scale productions begins. Skillful improvements at the planning and design stage are then ten times more effective that at manufacturing stage [14].

Using RT technology, which designing the prototype or production tooling based on RP parts, complements RP when large quantities of similar parts containing complex features, made economically utilizing materials close to or identical to end production materials and with normal production processes are required [15].

Rapid production permits producing the prototype at a cost advantage and favorable lead-time manufacturing. RP parts function as design prototypes to iron out flaws in casting or tooling design and functional prototypes to address the positioning of gates, vents and runners. Product optimization of molding parameters and evaluation of molded patterns can be conducted effectively [16].

Sustainable Production, Life Cycle Engineering, Sustainable Dynamics

Value creation is a key element for ensuring societal prosperity. At the same time, sustainable development the future of global human wellbeing and based on profound environmental, social and economic mechanisms, so described the direct and indirect effects of value creation. Both aspect is closely linked.

"*The sustainability dynamics model*" for the first time ever enables the visual and qualitative capabilities and as a major system of the tree sustainability dimension of environment, economy (enterprises) and society (individual). The principal of this dynamic model allows high level and principle trade off discussion and qualitative reasoning. Principles of model theory system dynamics dimension.

- The primary effects on environment,
- The primary effect on society and living standards,
- The primary effect on the economy and manufacturing process, factories and logistic.

The major element to transform manufacturing towards "higher sustainability" with a compatible environment.

1. Casual relations,
2. Magnitude and scale drivers and
3. Latency and timely duration decency [17].

High Performance Product Development at Toyota

Initial Quality Study (IQS) by Make (2017)

Speed to Market; Product development lead times that were half of their competitors,

Most efficient Process; best in industry engineering efficiencies and fraction of the engineering changes than competitors. Design drive the world's most efficient plants T. P.S. and smoothest launches.

The **Toyota Production System (TPS)** is an integrated socio-technical system developed by Toyota, that comprises its management philosophy and practices. The TPS organizes manufacturing and logistics for the automobile manufacturer, including interaction with suppliers and customers. The system is a major precursor of the more generic "Lean manufacturing". Taiichi Ohno, was a Japanese industrial engineer and business man. He is considered to be the father of the Toyota Productions System, which become Lean Manufacturing and Eiji Toyoda, he was largely responsible for bringing Toyota to profitability and worldwide prominence during his tenure. Japanese industrial engineers, developed the system between 1948 and 1975.

Toyota's product development system is as powerful and profound as their manufacturing system and may be even more important [18].

Engineering Design

Organizations now fail or succeed based upon their ability to respond quickly to changing customer demands and to utilize new technological innovations.

Engineering design is always directly impacted by societal needs. The basic premise of engineering design is to create products of high quality at reduced cost. However, design paradigms change with time, depending on market forces. Sustainable design goes beyond ensuring that a design meets the functional requirements of a product. It looks at the full life cycle of the product, from "birth" to "death," also known as from "cradle to grave" (McDonough and Braungart 2002).

Product life cycle is the product to be ready for production. design for sustainability, life cycle assessment is a method for measuring the environmental impacts with all stages of the product life cycle.

- Identify all interactions between a given activity in the PLC and the environment.
- Quantify each of the interactions in terms of the negative impacts on the environment using the standard impact metrics.
- Evaluate the total load on the environment using the quantification that covers all interactions of the PLC with the environment [19].

Standardization Efforts and Road Mapping During Design Process

Collaborative road-mapping efforts have helped to identify and define the important technical hurdles that prevent the vision of additive manufacturing from being realized today. (Bourell et al. 2009.) Most of these challenges require considerable metrological effort in order to be resolved successfully.

1. *Material Properties*; The material properties of parts manufactured by additive manufacturing are not well understood and lack the high-quality data required for the use of Additive Manufacturing parts in high stress applications such as turbine blades or jet engine components.
2. *Process Understanding*; Many metal-based processes involve the heating, melting and cooling of metal, which at the fundamental level is understood as the difficulty in measuring these processes. This leads to a lack of process understanding and difficulties in optimizing the processes.
3. *Qualification and certification*; Currently there are no standardized methods for qualification and certification of Additive Manufacture input materials, processes and parts. Current empirical methods for qualifying parts are cumbersome and impractical because of the large number of test samples required and the large number of process parameters for AM.
4. *Lack of AM standards*; recently, there are very few additive manufacturing-specific standards developed in a consensus method by a standards development organization (SDO)
5. *Data formats*; there are only two general data formats, one (STL), which is widely used but has some limitations, and another (AMF), which overcomes some of the STL limitations, but is not widely used because of the limited number of systems compatible.
6. *Standardization efforts*; AM standards are essential if AM technologies are to be used more widely. Standards provide technically correct and consistent methods that, if followed correctly, give confidence everyone in the industry is following the same, right paths [20].

Real Options to New Product Development by Modeling

Design theory and product development support generating multiple alternative solutions to overcome the design problems [21]. According to aim, tools and strategies should be developed to help design teams to offer them best alternative. This tool are decision trees and comparison matrices. These alternatives allow them useful analyses and effective performance, cost and to predict unexpected problems.

Concept design; the product development program starts with the chief engineer, developing concept, written document describing vision of company and product specifications.

System level design; after concepts continue to develop ideas based on selected model offer. Body engineering design studies on these ideas. These efforts include all planning required to realize the design of product, cross sections of structural parts, joint efforts, design layout, usability tests.

Detailed design, includes all body, engineering drawings without tolerance. The prototype are assembled as a "pattern built. Fit and function are adjusted through modifications to the part and then full prototype are built and testing. This process includes concept approval, styling approval and final approval. If option is necessary, get to PD strategy [22],

IF ($S \geq C$) THEN (Change to New Strategy) ELSE (Retain Current Strategy)

Where, ***S*** **vector** of signaling parameter values used to describe performance, ***C*** **vector** of parameter values describing conditions that must be met to justify a change in PD Strategy.

Quality Management System and Tools for Quality During Design Phase

A main concern of an organization should be the quality of its products, thereby, satisfy customer's expectations, meet a well-defined need, comply with applicable standard and specification, comply with requirement of society, reflect environmental needs, are made available at competitive price and provided economically.

One main concern of quality system is to meet the customer's need and expectation. Costumers need a confidence for a good and want to deliver the desired quality and keep it upright consistently.

The process model; the term "process" is defined in ISO 9000 as "set of interrelated or interacting activities which transforms inputs to outputs". Inputs and outputs are defined both process. A process in an organization are generally planned and carried out und under controlled conditions to add value. Inputs and outputs may be tangible or intangible.

Process approach; a desired result is achieved more efficiency when activities and related resources are managed as a process.

ISO 9001 stresses the importance for an organization to identify. Implement, mange and continually improve the effectiveness of the process. Efficiency and effectiveness based on a process-based approach, performance improvements. Process effectiveness and efficiency can be assessed through internal or external review process and to be evaluated on a maturity scale.

The most commonly used quality control method in Turkey is Deming cycle which is PDCA-plan-do, study-ac, Six sigma and FMEA Failure mode and effect analysis [23].

TAM (Technology Acceptance Model)

This theory is based on theory of reasoned action and the theory of planned behavior under the theories in social phycology. (Paul et al. 2015) The Technology Acceptance Model was developed by Davis (1989) as an adaptation of the Theory of Reasoned Action and was originally used to model the user acceptance of information technologies (Davis 1989). See Fig. 6.

Davis's Technology Acceptance Model (TAM) seems to be the most promising. TAM is an intentional model developed specifically to explain and/or predict the user acceptance of computer technologies. User acceptance is a critical success factor for IT adoption and can be sufficiently explained, accurately predicted, and effectively managed by a variety of relevant factors [24].

TAM Is an information systems theory and the models explain how users come to accept and use the technology. Model suggest that when users are presented with a new technology a number of factors influence their decision about how and when they will use TAM.

(https://en.wikipedia.org/wiki/Technology_acceptance_model)

The model assumes two dimensions are critical to the acceptance of technologies

1. Perceived Usefulness and
2. Perceived Ease of Use [25].

Innovation is critical to business to be succeeded in today's global marketplace. A competitive advantage can be achieved by effectively applying new technologies and processes to the challenges of current design practice. Opportunities surround all aspects of product development (ergonomics, manufacturing, maintenance, product life cycle, etc.) with the greatest potential impact in the early stages of the product design process. Prototyping and evaluation are necessary steps in the current product development process. Although computer modeling and analysis practices are currently being applied at various levels, the construction of unique physical prototypes makes the current typical process very expensive and time consuming.

New technologies are needed to help the industry make a faster and more efficient decision-making process. VR technology has become a new level over the past two decades. VR has changed the way scientists and engineers consider computers to perform mathematical simulations, data visualization and decision making. VR technology combines multiple human-computer interfaces to provide different sensations (visual, haptic, auditory, etc.) that give the user a sense of presence in the virtual world [26].

The planning of virtual assembly processes is a critical step in product development. In this process, details of the assembly operations that describe how different parts are assembled are formalized. It has been found that assembly processes often account for most of the cost of a product [27].

What does VRP do for Business to Business Marketing?
Market requirements

- Shortened product life cycles,
- Shortened market entry time for new product,
- Increasing demand on the shelf-life of the innovation advantage,
- Increasing variety of product variants,
- Increasing fragmentation of the markets does not mean more retail business but niche Market,
- Opportunity window and short-lived,
- Business Conflict: lower volumes with increasing variety of product variants,
- Which product characteristics are relevant to the purchase decision.

What does the customer really want?

- The customer does not always know which solution really meets his needs,
- The customer knows his needs, but does not have sufficient ability/ knowledge how to meet the needs.
- A problem-solving product is always a compromise between the product itself and how it is used.

VRP helps to reduce the customer's understanding distance.
The product developer does not exactly cover the customer needs

- The developer has too little knowledge about the concrete use of the product.

Conway Law: "The structure of a system reflects the structure of the organization."

- Extension: "The product reflects the character of the organization that built it"

VRP helps reduce the gap between customers and developers.
Newly developed products are usually unsatisfactory

- Learning curve effect solves only part of the problem,
- Quality improvement comes mostly from experience gained through repeated use of the product,
- The initial level of quality of newly developed products is weak.

VRP reduces design risks
The Weak aspects

- Customers can not formulate their needs,
- Developers cannot implement what the customer wants,
- Newly developed products are always unsatisfactory,
- Product must start from the end purpose of the customer,

Efficiency contribution of the VRP

- Disclose tacit acceptances of customers
- To disclose/ understand the preferences of customers and/ or product characteristics
- To focus the product developer on the needs of the customer

Contribution of the VRP

1. Lower Time to Market;
   - Shorter prototyping times,
   - Improved communication in the cross-functional area,
   - Chance for improved amortization.
2. Reduction of innovation risk;
   - Early involvement of customers/ suppliers via VRP,
   - Better understanding of the use by the customer,

3. Increasing the innovation advantage;
   - Better insight into the preferences of the customer,
   - Better articulation of unconscious wishes [28].

Design concept of Valliant
See Tables 1 and 2.

Table 1. Valliant strategy - The new product development decision process

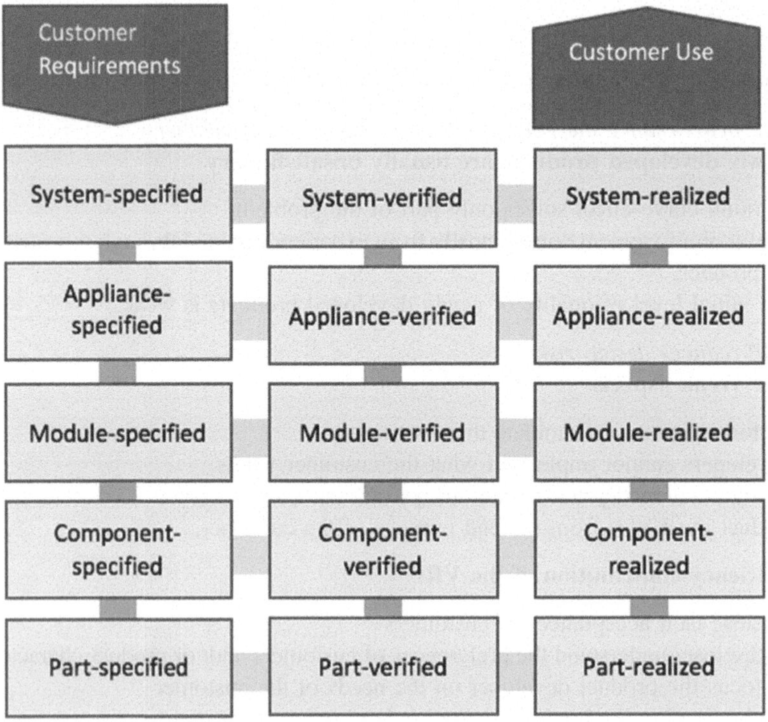

Discussing on Valliant
It seems to me that Valliant products are based on fundamental inventions rather than innovative design. They make design the product fit design for market. A common production strategy advocates assigning responsibility for a product to a design-team. This team designs the product, schedules its production and stays responsible until the product reaches maturity, and provides customer support.

Product marketability is an important factor in the design and subsequent manufacturing of consumer products. Marketing efforts frequently concentrate on highlighting the non-functional, eye-pleasing design features of products in their promotion.

Table 2. Valliant strategy and conception

Development functions of valliant			
Marketing	Research & Development	Purchasing	Production
• Marketing research • System and product strategy • Customer review • Customer complaints	• Competitor and technology analyzer • Component & part definition • Functional & process control • Simulating using, digital prototypes • System definition • Definition of main component items • Product safety review • Design verification • Electric/electronic verification • Quality strategy • Project FMEA • Analysis of product quality • Project charter & synchro planning • Project performance • Make or buy decision	• Purchasing • Supplier selection	• Production & process verification

With more than 13,000 employees worldwide, Valliant, headquartered in Remscheid, is one of the market and technology leaders in the field of heating technology. We offer energy-saving and environmentally friendly systems for heating, cooling and hot water - especially on the basis of renewable energies. These systems ensure maximum living comfort - just like our service offering that accompanies the entire product life cycle [29].

Arçelik Product Development
See Fig. 2.

Discussing on Arçelik
Arçelik design team rightfully argues that products/systems must be designed based on two fundamental axioms for efficient manufacturing and effective utilization by the customer. The functional requirements of the product/system should be addressed as independently as possible, and simplicity of design should be an important objective. Review several products/systems that have been, or could have been, designed and manufactured in accordance with these axioms (e.g., light dimmers that are also used as on/off switches, etc.). Quality improvement is a manufacturing strategy that should be adopted by all enterprises. Although quality control is a primary concern for any manufacturing company, engineers should attempt to improve quality. In statistical terms, all variances should be minimized, and furthermore, where applicable, the mean values should be increased (e.g., product life, strength, etc.) [30].

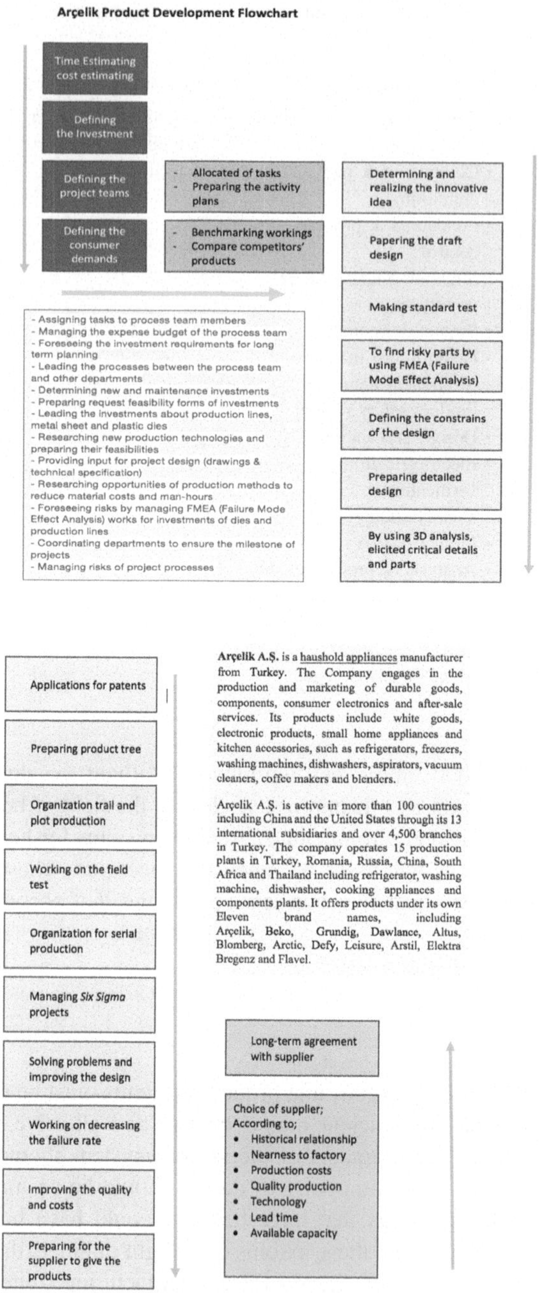

Fig. 2. Arçelik production process

Design of experiments (DOE) is a statistical approach that can be used in the design of physical or simulation-based experiments for the determination of optimal variable values. Such factorial-based experiments help engineers in the narrowing of the field of search to those parameters that have the greatest impact on the performance of the product as well as limiting the combinatory number of variations of these variables

3 Conclusions

All successful designs begin with customer requests. Designing a product is a very difficult process for engineers. Successful design and development project are critical success for every industry. Project performance must be understood that is dynamic concurrence relationship. Developing products faster, better and cheaper than competitors has become to success in many markets.

For successful product development, it is necessary Marketing as pre-design, identification of market opportunities, costumer needs, target pricing and promotion of product. design, product quality, product cost, includes development costs, development time, development capacity, manufacturing, production system, supply chain. Why is good product development difficult, because product of trade off, dynamics, details, time pressure and economics. During designing process, first, company is to consider market opportunity, consider existing product platform, new technologies, to identify production and corporate constrains. Company needs also standardization efforts and road mapping during design process.

References

1. Hauser JR, Dahan E (2008) New product development. McGraw Hill, Inc., Columbus Ohio. Draft corrections by John Hauser, 22 July 2008
2. Benhabib B (2003) Manufacturing Design, Production, Automation, and Integration. University of Toronto
3. Stienstra D Introduction to design assembly and manufacturing
4. https://de.scribd.com/document/19345183/DFMA
5. Anderson DM Design for manufacturability- How to use concurrent engineering to rapidly develop low-cost, high quality products for lean production. CRC Press, S.6
6. http://www.wikizero.org/index.php?q=aHR0cHM6Ly9lbi53aWtpcGVkaWEub3JnL3dpa2kvS25vd2xlZGdlX21hbmFnZW1lbnQ
7. Bachhaus K, Voeth M (2010) Industriegüter Marketing, Vahlen, 9. Auflage 2010, ss. 23–24
8. N00092 ETC Project SR-AT, Hi-Tech Zentrum in der grenzüberschreitenden Region-TU, Wien
9. Hasenauer R (2007) High Tech Marketing, WU-TU Wien -Lecture note
10. Day G, Freeman J N00092 ETC Project SR-AT, High-Tech Zentrum in der grenzüberschreitenden Region-TU Wein
11. See Weill and Vitale (2001)
12. Jacobs MA (2007) Product complexity: a definition and impacts on operations. Department of Management Information Systems, Operations Management, and Decision Sciences, University of Dayton

13. Yazdani B (2008) All you ever needed to know about product development. Nottingham Business School
14. Design for manufacturability- How to use concurrent engineering to rapidly develop low-cost, high quality products for lean production, David M. Anderson, CRC Press, S.9
15. Radstok E (1999) Rapid Prototyping. J 5:164–169
16. Cheah CM, Chua CK, Lee CW, Feng C, Totong K (2005) Rapid prototyping and tooling techniques: a review of applications for rapid investment casting. Int J Adv Manuf Technol
17. Sustainability Manufacturing, Sustainable Production, Life Cycle Engineering and Management, Sustainable Production, Life Cycle Engineering and Management, Rainer Stark and Kai Lindow, Springer, Switzerland, 2017 S. 21
18. https://de.wikipedia.org/wiki/Toyota-Produktionssystem
19. Parsaei HR, Kamrani AK (2013) Engineering and management innovation. CRC press, Boca Raton
20. Slotwinski JP, Moylan SP (2014) Metals-Based Additive Manufacturing: Metrology Needs and Standardization Efforts, American society for precision engineering, National Institute of Standards and Technology Gaithersburg, Maryland, USA, spring topical meeting volume 57
21. Pugh S (1991) Total design: integrated methods for successful product en- gineering. Addison-Wesley, Wokingham, U.K
22. Ford DN, Sobek DK (2005) Adapting real options to new product development by modeling the second toyota paradox. IEEE Trans Eng Manag 52(2):175–185
23. Durakpasa MN, Osanna PH Quality in Industrie, Austauschbau und Messtechnik. Technische Universität Wien, p 91
24. Hu PJ, Chau PYK, Liu Sheng OR, Tam KY (1999) Examining the technology acceptance model using physician acceptance of telemedicine technology. J Manag Inf Syst Fall 16 (2):91–112
25. Rehder E, Karla J Adaption des Technology Acceptance Model für den Onlinevertrieb von Versicherungsprodukten, RWTH Aachen, Lehrstuhl für Wirtschaftsinformatik und Operation Research
26. Bryson 1996, Eddy and Lewis 2002, Zorriassatine et al 2003, Xianglong et al 2001
27. Boothroyd and Dewhurst (1989)
28. Hasenauer R (2007) Business to Business marketing and High-Tech Marketing, Lecture Notes. WU-Wien
29. https://www.vaillant.de/ueber-uns/
30. Nelson P (1996) Repenning, reducing cycle time and development time at ford electronics. Department of Operation Management, System Dynamics, Sloan School of Management, Massachusetts Institute of Technology

Measurement Error on the Reconstruction Step in Case of Industrial Computed Tomograph

Ágota Drégelyi-Kiss[1(✉)] and Numan M. Durakbasa[2]

[1] Bánki Donát Faculty of Mechanical and Safety Engineering, Óbuda University, Budapest, Hungary
dregelyi.agota@bgk.uni-obuda.hu
[2] Department Interchangeable Manufacturing and Industrial Metrology, Institute of Production Engineering and Laser Technology, TU Wien, Vienna, Austria
numan.durakbasa@tuwien.ac.at

Abstract. More recently, the industrial CT equipment is used not only for non-destructive analysis but to perform geometrical evaluations. The three dimensional, optical dimensional measurements made by CT are popular because the measurement time is much shorter than in case of traditional 3D measurement machines, furthermore the inner geometries can be determined by non-destructive manner. In this article the design and measurement plan of an aluminium test cube by industrial CT are described and the evaluation of the measurement data are presented.

Keywords: Measurement error · Measurement uncertainty · Industrial CT Reconstruction process

1 Introduction

Dimensional metrology has an important role in the quality control of manufactured (cut, milled or turned) parts. The tolerances of the characteristics (i.e. diameter or length) of these parts are usually within the range of 0.003 mm to 0.01 mm. There are several dimensional measurement equipments which are good for the control of these specifications such as simple devices (e.g. micrometer, dial calliper) or CMM machines with tactile or optical probes. Nowadays new technologies have been developed for dimensional measurements. One of them is industrial computed tomography (CT) metrology using x-rays [1, 2]. The advantages of CT technique are as follows: non-contact, capability of measuring both internal and external geometries, fast scanning time [3]. Due to the automation of the measurements and the requirements towards faster measurement processes the dimensional measurements are preferred to perform by CT equipments in industrial practice.

The basis of the verified dimensional measurements is the traceability of the measurement scale values of a certain device to the metre definition internationally accepted [4]. In case of industrial CT equipment the traceability and the calibration are not solved totally because of the complexity of the CT measurement processes [5, 6].

N. M. Durakbasa and M. G. Gencyilmaz (Eds.): ISPR 2018, *Proceedings of the International Symposium for Production Research 2018*, pp. 309–323, 2019.
https://doi.org/10.1007/978-3-319-92267-6_27

The measurement process of industrial CT has several important steps (Fig. 1). In order to make verified dimensional data a calibration is used with a ball bar. This calibration should be repeated when the distance between the detector and the object changes. The length calibrated is as long as the size of the object examined.

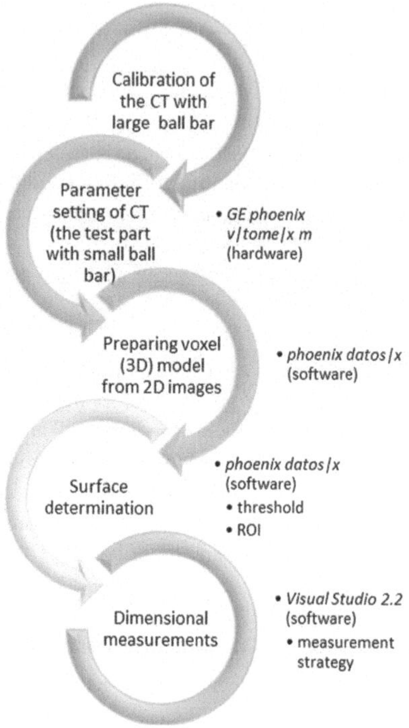

Fig. 1. Flowchart of the CT measurement process (in the investigated case)

The dimensional measurement of a certain part by industrial CT has several difficulties. The accuracy and the precision of this procedure is lower than in case of tactile CMM. The quality of the dimensional measurement is better if an etalon characteristic is measured at the same time with the test part [7]. Therefore a small ball bar (appr. 16 mm) was chosen. The test piece with this ball bar is scanned simultaneously. This is important because dimensional measurement values have become more precise, the values are corrected with this calibrated ball bar value during the dimensional measurements phase. The selection of the threshold is essential in case of CT measurements. It has large emphasis on the measurement results [8, 9].

The accuracy of the measurements performed by CT can be performed in two ways. In the first case the error is calculated as the difference between the measured value and value from the CAD data. There are problems with this approach. The CAD data are theoretical and do not contain the errors of the manufacturing process. In the second case the values measured by CT can be compared with the true value of the characteristic which can be determined by a more precise measurement (e.g. high-precision tactile CMM) [10].

The difficulties of the comparability of dimensional measurements are the eventual differences in the measurement strategies. To perform the comparison between CT and CMM measurement it has to be declared the measurement strategies of the dimensional measurements [11–13].

During this research a milled aluminium test part was designed and manufactured. Some geometry was measured by tactile CMM machine to get the reference values of the characteristics, and measurements were performed by industrial computer tomography where the parameters of the reconstruction process were varied systematically. The effects of these threshold and other parameters were investigated on the dimensional measurement error.

2 Materials and Methods

2.1 Aluminium Test Work Piece

A test work piece was designed and produced from aluminium. The design of this test part can be seen in Fig. 2. The overall size of this part is 90 mm × 90 mm × 90 mm. The material is AlMgSi1. The manufacturing was prepared in MSN-500 type milling machine. The measured values were the diameters of the 14 holes and the distances between the holes (1-2, 1-3, 1-4, 1-5, 1-6, 1-7; 8-9, 8-10, 8-11, 8-12, 8-13, 8-14). The ball bar which is used for the correction in the dimensional measurement phase was connected to the test piece in the middle of the upper part of the cube (Fig. 3).

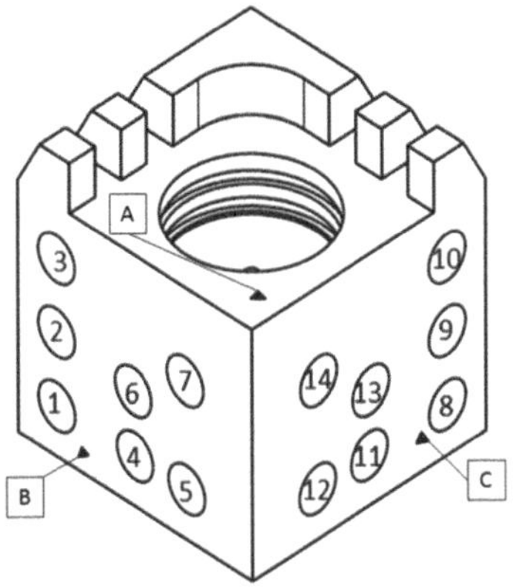

Fig. 2. Test part with numbered holes (size: 90 × 90 × 90 mm)

2.2 The Devices/Equipment Used

The precise and accurate measurement of the test piece was performed at TU WIEN, Department for Interchangeable Manufacturing and Industrial Metrology, High

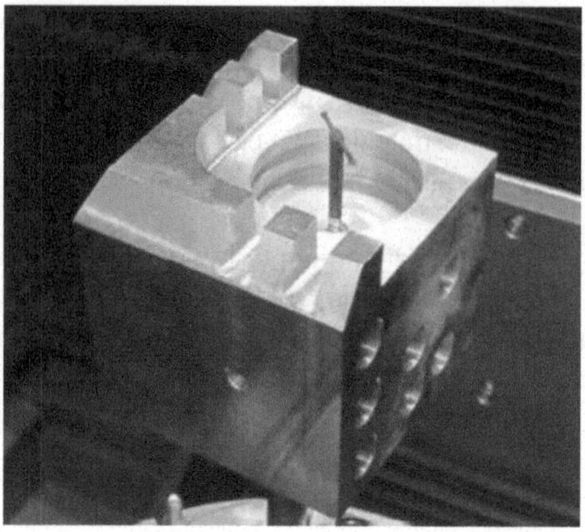

Fig. 3. The aluminium test piece with ball bar

Precision Measurement Room – Nanometrology Laboratory on a CMM ZEISS UMM500, whose main technical characteristics are: (a) the number of axis – 3 (X, Y, Z); (b) the measuring range – 500 × 200 × 300 mm; (c) MPE – 0.4+L/600 μm; (d) resolution – 0.1 μm.

The measurement strategy of the CMM measurement is as follows:

- Base A, Base B and Base C: 12 measurement points equally spreaded (see Fig. 2)
- Numbered (1 to 14) holes: 4-4 measurement points in different circle (see Fig. 2)
  - cylinders were fitted, the *diameter*s are given
  - the intercept of the axis of the cylinders and the Base B or C are calculated, the distances between these calculated points means the *distances* between the cylinders.
- All the features were fitted using Least-Square method.
- The measurements were performed 5 times, and the average value has become the reference value of a certain parameter (diameter or distance).

The 3D micro-CT system used was GE phoenix v|tome|x m. The specification of the CT system components can be found in Table 1. The setting parameters are in Table 2.

Measurements of the CT data were carried out using software VGStudio Max 2.2 from Volume Graphics. Measurements were performed on voxel data.

2.3 Standards

Two ball bars were used during the investigations. One ball bar with 99.9276 mm calibrated ruby sphere distance was used for the correction of the CT equipment. The other ball bar with 15.9329 mm calibrated ruby sphere distance was used for the

Table 1. The specification of the CT system used according to the manufacturer

System component	Value/attribute/type
Max. voltage/power	300 kV/500 W
Min. voxel size	Down to 1 micron (microfocus tube)
Detector type	Temperature stabilized digital GE DXR detector array, 200 μm pixel size, 2000 × 2000 pixels, 400 × 400 mm, extremely high dynamic range > 10000:1
Measurement accuracy	4+L/100 μm referring to VDI 2630-1.3 guideline
Manipulation	Granite based precision 5-axes manipulator (6-axes with detector shift)
Software	phoenix datos|x 3D computed tomography acquisition and reconstruction software

Table 2. An overview of scanning parameter used

Parameter	Value
Voltage	250 kV
Current	360 μA
X-ray pre-filter	0.5 mm Sn + 0.5 mm Cu
Number of views	1440
Number of image averaging	5
Number of skipped images	2
Uncorrected voxel size	73.44 μm

correction after the reconstruction and surface determination phase. The basis of this correction is to refine the new voxel size (s) in the following way [3]:

$$CF = \text{Reference value}/\text{measured value}$$

$$s = s_0 \cdot CF, \text{ where } s_0 \text{ is the original voxel size}$$

2.4 Experimental Design

There are several parameters which can be changed during the reconstruction and the surface determination process. The values of the following reconstruction parameters were modified during the experiments:

- correction of the software (phoenix datos|x) for the consideration of the background radiation during the rotation (so-called ROI setting no or yes status)
- threshold automatic or manual in case of ruby sphere (Ruby auto thr. or man. thr.)
- threshold automatic or manual in case of aluminium (Al auto thr. or man. thr.)

Table 3. The experimental design

Run	ROI	Ruby	Al
1	no	auto	auto
2	yes	auto	auto
3	no	auto	manual
4	yes	auto	manual
5	no	manual	auto
6	yes	manual	auto
7	no	manual	manual
8	yes	manual	manual
9	yes	manual	manual
10	yes	manual	manual

The settings and the experimental design can be found in Table 3. As it is seen there are repeated measurements (Run 8-10) to determine the uncertainty of the fitting procedure.

Therefore one scanning was performed, and the reconstructions were replicated 10 times related to the experimental design settings. The diameters of the holes and the distances between the holes were calculated.

3 Results

3.1 Reference Values of the Test Workpiece

The aluminium test workpiece were measured five times in CNC mode, and the means as reference values and the standard deviations of the base data can be found in Table 4. It is seen that the measurement uncertainty of the CMM is quite low and the measured values of the diameters and the distances largely differ from the nominal values. This is because the errors of the manufacturing process. The bases of the measurement error and uncertainty of CT are the true values of the characteristics measured by the tactile CMM.

3.2 Measurement Data by Industrial CT

The aluminium test piece with a ball bar was scanned once during our investigations. Before the scanning calibration of the CT equipment was performed by a large ball bar (with 99.9276 mm calibrated value). After this setting and correction was made and the CT scanning process was done for aluminium test piece with a small ball bar (with 15.9329 mm calibrated value).

The setting parameters of the CT scanning process can be seen in Table 2. During the dimensional measurements the parameters of the reconstruction and surface determination were varied as it can be found in Table 3.

Table 4. Precise measurements of the characteristics of the test part

No. of cylinders	Nominal value [mm]	Reference values [mm]	Std. dev. [mm]	Distance codes	Nominal value [mm]	Reference values [mm]	Std. dev. [mm]
cyl 01	14	13.9865	0.0006	1-2	22	22.0184	0.0001
cyl 02	14	13.9725	0.0005	1-3	44	44.0474	0.0006
cyl 03	14	13.9836	0.0005	1-4	30	29.9960	0.0012
cyl 04	14	13.9930	0.0006	1-5	50	49.9828	0.0004
cyl 05	14	13.9945	0.0002	1-6	35	35.0181	0.0004
cyl 06	14	13.9880	0.0005	1-7	59	59.0161	0.0012
cyl 07	14	13.9668	0.0050	8-9	22	22.0155	0.0004
cyl 08	14	13.9927	0.0002	8-10	44	44.0112	0.0005
cyl 09	14	13.9891	0.0003	8-11	30	30.0280	0.0006
cyl 10	14	13.9782	0.0006	8-12	50	49.9972	0.0045
cyl 11	14	13.9904	0.0003	8-13	35	34.9882	0.0006
cyl 12	14	13.9799	0.0018	8-14	59	58.9562	0.0004
cyl 13	14	13.9609	0.0006				
cyl 14	14	13.9771	0.0000				

Measurement of Small Ball Bar

The first step of the dimensional evaluation was the measurement of the distance between the two ruby balls. The results are in Table 3. It can be stated that the setting parameters have large impact on the distance. There are large bias values depending on the settings (Table 5).

Table 5. Determination of the distance between two ruby balls

	ROI on		ROI off	
	auto thr.	*man. thr.*	*auto thr.*	*man. thr.*
Measured value [mm]	15.922	15.908	15.910	15.907
Calibrated value	15.9329 mm			

It has to be emphasized that only one CT scan has prepared, and the calculation method of the data has given the various measured values for the determination of the distance between the two ruby spheres. If the ROI setting is in off status, i.e. the correction is not used for the background radiation during the rotation, then there are small differences between the results calculated by the different threshold methods, the difference is 3 μm. To take into consideration the ROI setting the differences in threshold methods have large impact on the measured size.

Measurements of Aluminium Test Piece

The measurement errors are calculated as the measured values by CT minus the measured values by the CMM. The measurement error values are in Fig. 4 for the diameters of the numbered cylinders. The measurement errors related to a certain cylinder differ from each other, the differences are almost 0.05 – 0.07 mm. If the data are grouped by the setting factors (Fig. 5) it can be seen that the largest negative measurement errors are in case of ROI – on setting, and automatic threshold for the ruby sphere. It can be stated if the background radiation is neglected during the calculation (i.e. ROI has off setting), the deviations become smaller (in case of red square or blue circle).

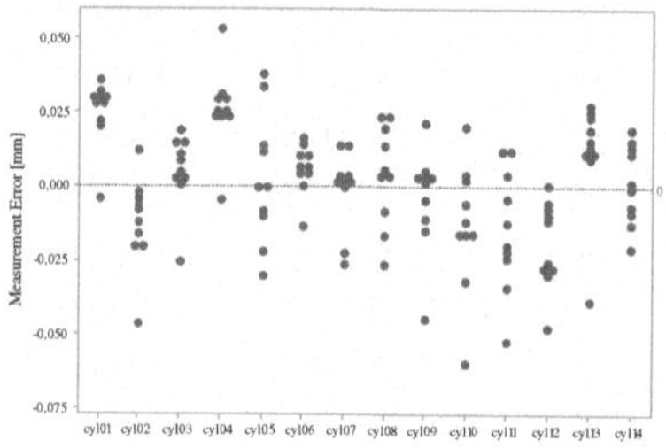

Fig. 4. Measurement errors of the measured cylinders

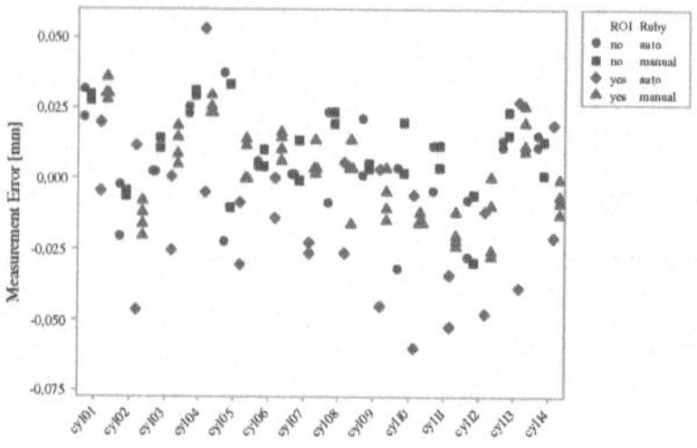

Fig. 5. Measurement error of diameters grouped by the setting parameters (Color figure online)

The measurement errors of the measured distances between the holes are in Fig. 6. The measurement error becomes larger with the distance in absolute value. In case of manual threshold of the ruby sphere the measurement errors are positive, in case of consideration of background radiation and automatic threshold related to ruby sphere the errors are negative values (Fig. 7). Figures 8 and 9 visualize the measurement error values grouped by the nominal distances, i.e. distance 1–2 equals in nominal value to 8–9 because of the symmetry.

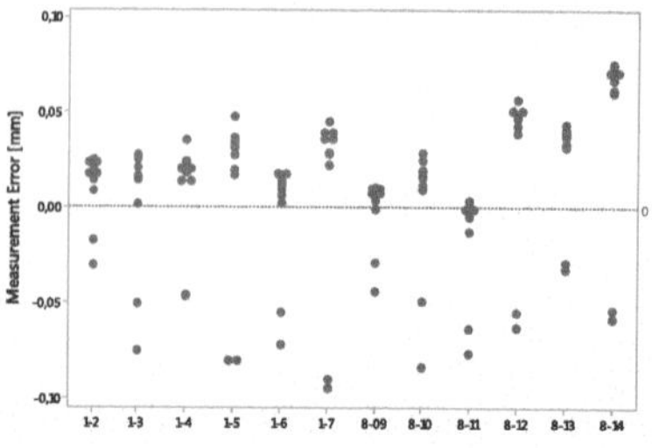

Fig. 6. Measurement errors of the measured distances

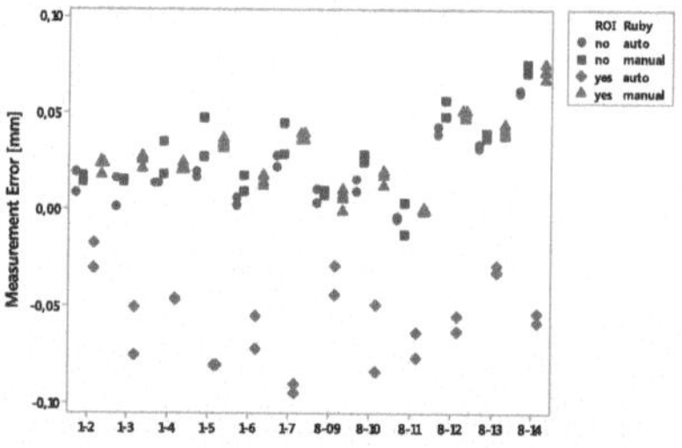

Fig. 7. Measurement error of distances grouped by the setting parameters

It can be seen that the ROI setting yes level has larger standard deviation than no level for all investigated cases.

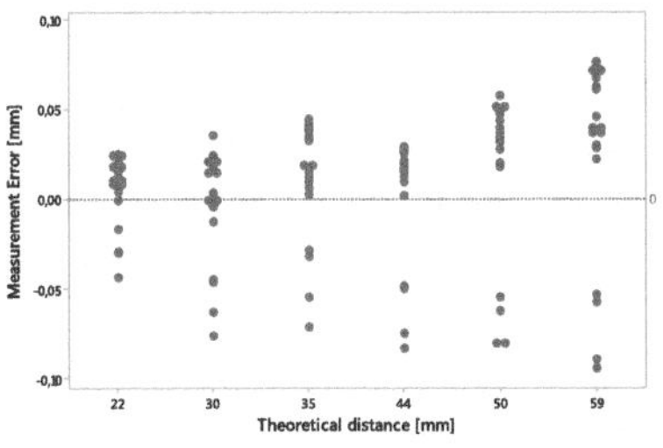

Fig. 8. Measurement errors of the distances with nominal values by length

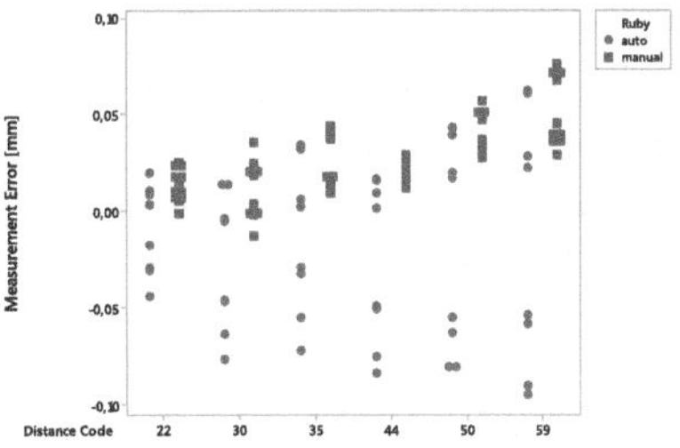

Fig. 9. Measurement errors of the distances with nominal values by length grouped by the setting parameters

4 Statistical Analysis (the Significance of Setting Parameters)

4.1 Repeated Measurements

In order to estimate the uncertainty of dimensional measurement capability of the VGStudio Max 2.2 software one evaluation setting (ROI: on; Ruby: manual; Al: manual; see row 8-10 in Table 2) was estimated three times. One-way ANOVA was calculated for the diameters and distances, respectively (Tables 6 and 7). It can be stated that there are significant differences in measurement errors among the measured cylinders, and among the various distances as well. The pooled standard deviation (uncertainty) for the measurement error of the diameters is 7.44 μm, and this uncertainty in case of distance measurements is 2.85 μm. These uncertainties are originated only from the VGStudio Max 2.2 software calculation method for the determination of diameters and distances.

Table 6. Anova for the diameters, repeated measurements

Source	DF	Adj SS	Adj MS	F-Value	P-Value
No.of cyls	13	0,010311	0,000793	14,33	0,000
Error	28	0,001549	0,000055		
Total	41	0,011861			
Model Summary					
S	R-sq	R-sq(adj)	R-sq (pred)		
0,0074386	86,94%	80,87%	70,61%		

Table 7. Anova for the distances, repeated measurements

Source	DF	Adj SS	Adj MS	F-Value	P-Value
Distance sign	11	0,013325	0,001211	148,84	0,000
Error	24	0,000195	0,000008		
Total	35	0,013521			
Model Summary					
S	R-sq	R-sq(adj)	R-sq (pred)		
0,0028529	98,56%	97,89%	96,75%		

4.2 Anova

Statistical analysis is prepared for the estimation of the various factors related to the measurement error. Four-way crossed ANOVA is calculated for fixed factors with interactions up to second order.

Diameters

There are differences in measurement errors related to the various cylinders as can be seen in the main effects plot (Fig. 10). The other three factors (ROI, Ruby and Al thresholds) have significant effect on the measurement results. The largest effect is related to the ROI factor.

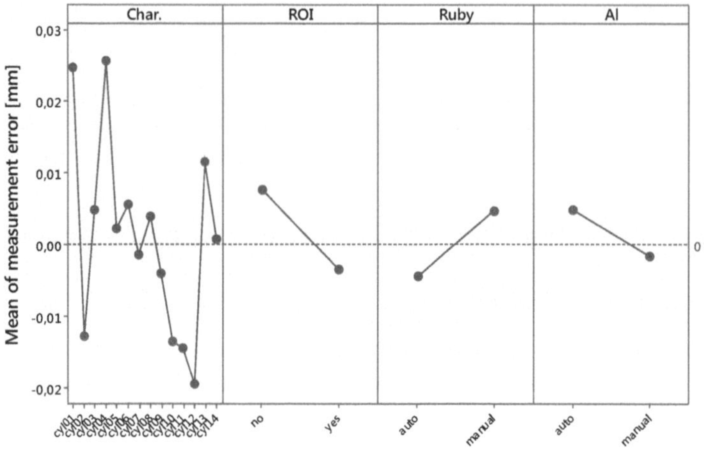

Fig. 10. Main effect plot for the diameters

It can be seen in the interaction plot of the measurement error of the diameters (Fig. 11) that the threshold setting (i.e. automatic or manual) in case of ruby spheres has the same effects for the values for each, individual diameters.

In case of manual threshold setting of ruby sphere the measurement errors are smaller (on average 5 μm) in absolute values than in case of automata threshold setting.

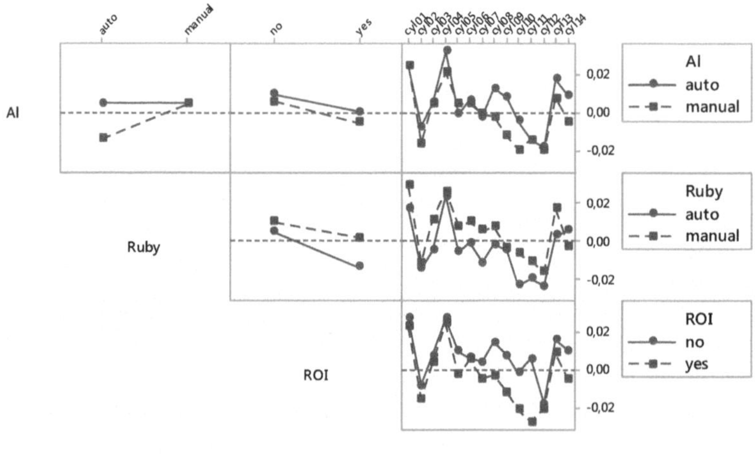

Fig. 11. Interaction plot for measurement errors of diameters (in [mm])

Table 8. Anova for the diameters

Source	DF	Adj SS	Adj MS	F-Value	P-Value	Model Summary			
Char.	**13**	**0,021076**	**0,001621**	**10,45**	**0,000**				
ROI	**1**	**0,005538**	**0,005538**	**35,72**	**0,000**	S	R-sq	R-sq (adj)	R-sq (pred)
Ruby	**1**	**0,003494**	**0,003494**	**22,53**	**0,000**	0,0124528	78,05%	62,34%	19,27%
Al	**1**	**0,002003**	**0,002003**	**12,91**	**0,001**				
Char.*ROI	13	0,002586	0,000199	1,28	0,240				
Char. *Ruby	13	0,001561	0,000120	0,77	0,684				
Char.*Al	13	0,001678	0,000129	0,83	0,625				
ROI*Ruby	**1**	**0,000934**	**0,000934**	**6,03**	**0,016**				
ROI*Al	1	0,000509	0,000509	3,28	0,074				
Ruby*Al	**1**	**0,003548**	**0,003548**	**22,88**	**0,000**				
Error	81	0,012561	0,000155						
Total	139	0,057231							

The difference between the two levels of Al threshold is 1.6 μm, and in case of ROI is 3.2 μm on average related to the absolute value of measurement errors.

The results of ANOVA analysis (Table 8) show that all of the main factors (numbered diameters, ROI, Ruby and Al) have significant effects on the measurement error value at 95% significance level. There are interactions between ROI and Ruby; and between Ruby and Al settings. The residual analysis shows (Fig. 9) that the residuals are independent and normally distributed with 0 expected value therefore fulfill the requirements of using ANOVA and the results are adequate. The pooled standard deviation of this diameter measurement error is 12.5 μm (Fig. 12).

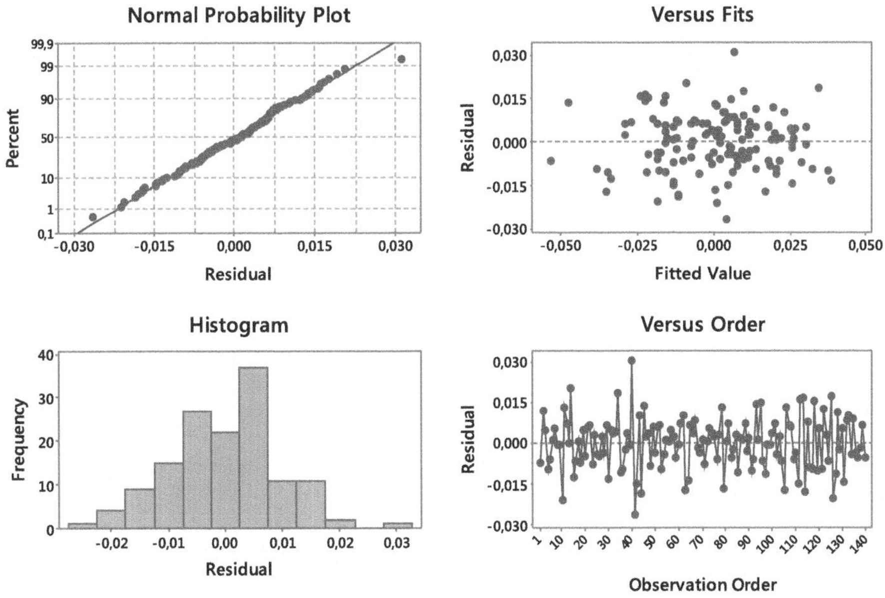

Fig. 12. Residuals plot in case of ANOVA analysis of diameters

Distances

The main effects plot for the distance measurement errors is in Fig. 13. There are differences in measurement error based on the parameters. The largest effect is caused by the Ruby threshold setting (auto or manual), in average the difference is about 50 µm. The lowest effect can be found in case of Al threshold setting. The last part of Fig. 13 shows the effect of the nominal distance value for measurement error. It can be stated that the measurement error increase with the length of the distance.

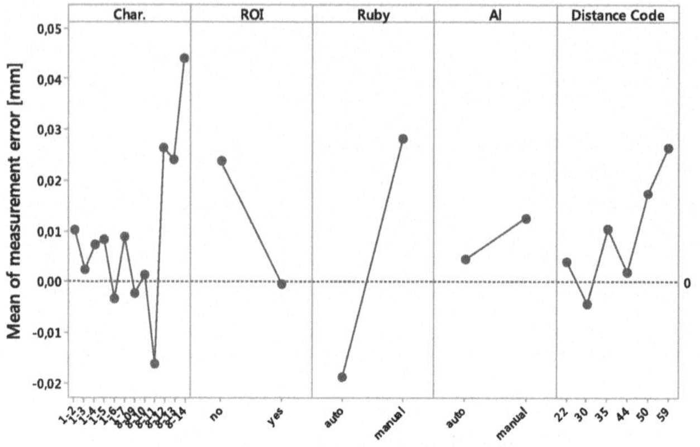

Fig. 13. Main effect plot for the distances

The interaction plot (Fig. 14) shows that in case of Ruby manual threshold setting there are positive measurement error values; whilst in case of Ruby automatic threshold setting there errors are negative for all investigated cases.

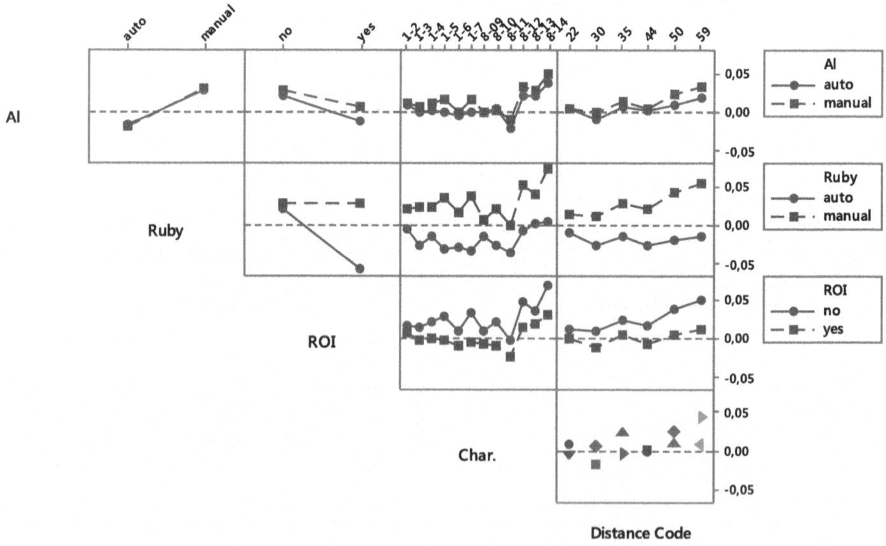

Fig. 14. Interaction plot for measurement errors of distances (in [mm])

The results of the ANOVA analysis for the distances (Table 9) show that Al threshold setting has not significant effect on the measurement error. The other main factors have significant impact on the results at 95% level. The residuals are independent and normally distributed with 0 expected value therefore fulfill the requirements of using ANOVA and the results are adequate. The pooled standard deviation of this diameter measurement error is 8.6 μm.

Table 9. Anova for the distances

Source	DF	Adj SS	Adj MS	F-Value	P-Value	Model Summary			
Char.	**11**	**0,024520**	**0,002229**	**29,95**	**0,000**				
ROI	**1**	**0,038860**	**0,038860**	**522,04**	**0,000**	S	R-sq	R-sq (adj)	R-sq (pred)
Ruby	**1**	**0,059477**	**0,059477**	**799,01**	**0,000**	0,0086278	97,17%	95,11%	89,46%
Al	1	0,000003	0,000003	0,04	0,840				
Char. *ROI	**11**	**0,004165**	**0,000379**	**5,09**	**0,000**				
Char. *Ruby	**11**	**0,008215**	**0,000747**	**10,03**	**0,000**				
Char.*Al	11	0,000720	0,000065	0,88	0,564				
ROI*Ruby	**1**	**0,039488**	**0,039488**	**530,48**	**0,000**				
ROI*Al	**1**	**0,000714**	**0,000714**	**9,59**	**0,003**				
Ruby*Al	1	0,000207	0,000207	2,78	0,100				
Error	69	0,005136	0,000074						
Total	119	0,181213							

4.3 Comparison Between the Measurement of Diameters and Distances

There are differences between the significance of the main factors for the two investigated cases (Table 10). In case of the diameter measurements the Al threshold setting is a significant factor at 95% level, in case of distances it is not so relevant. It is because the calculation and determination of the distance are more complicated mathematically and structurally. The standard deviation of only repetition for the diameter measurements is about 7 μm, for the whole diameter measurements 12 μm (Tables 6 and 8, respectively); whilst the standard deviation of the only repetition is 3 μm, and for the whole measurements is about 9 μm in case of distance measurements (Tables 7 and 9, respectively). The uncertainties have lower values in case of distance measurements than in case of diameter measurements.

Based on the results, the effects of the main and the interaction factors show that with this equipment and method the following factor levels should be fixed: the Al threshold is manual, the Ruby threshold is manual, ROI setting is off.

Table 10. The results of the anova significant factors for both investigated cases at 95% level

Factors	Diameter measurement	Distance measurement
No. of Characteristics (Char.)	significant	significant
ROI	significant	significant
Ruby threshold	significant	significant
Al threshold	significant	non-significant
Char. * ROI	non-significant	significant
Char. * Ruby	non-significant	significant
Char. * Al	non-significant	non-significant
ROI * Ruby	significant	significant
ROI * Al	non-significant	significant
Ruby * Al	significant	non-significant

5 Conclusion

In determining the uncertainty of dimensional measurements with industrial CT, a number of factors need to be considered. Factors influencing the measurement process not only result from adjusting the parameters of the equipment, but the reconstruction of the X-ray scans and the subsequent evaluation also result in significant errors and uncertainties. Within this research it was highlighted how much the geometric dimensions of a prepared test piece can be influenced by reconstruction adjustment parameters and calculation methods.

The levels of the parameters within the reconstruction step and the uncertainties have been determined within this research. In the future research this levels will be fixed and the parameters of the CT equipment (such as current, voltage, histogram) will be varied.

Acknowledgment. The research was supported by the ÚNKP-17-IV-6 New National Excellence Program of the Ministry of Human Capacities. The authors would like to thank for the measurements for Continental Ltd.

References

1. Kruth J-P, Bartscher M, Carmignato S, Schmitt R, De Chiffre L, Weckenmann A (2011) Computed tomography for dimensional metrology. CIRP Ann Manuf Technol 60(2):821–842. https://doi.org/10.1016/j.cirp.2011.05.006
2. Bartscher M, Hilpert U, Goebbels J, Weidemann G (2007) Enhancement and proof of accuracy of industrial computed tomography (CT) measurements. CIRP Ann Manuf Technol 56(1):495–498. https://doi.org/10.1016/j.cirp.2007.05.118
3. Cantatore A, Müller P (2011) Introduction to computed tomography, DTU Mechanical Engineering. Kgs. Lyngby, Denmark
4. JCGM 200:2012 International vocabulary of metrology – Basic and general concepts and associated terms (VIM)

5. Claverley JD, Leach RK (2015) A review of the existing performance verification infrastructure for micro-CMMs. Prec Eng 39:1–15
6. Kowaluk T, Woźniak A (2016) Methods for verifying the dimensional and material properties on industrial CT scanners according to VDI/VDE 2630 Blatt 1.3. In: Advanced Mechatronics Solutions. Springer, Cham, pp 341–346
7. Bartscher M, Neuschaefer-Rube U, Wäldele F (1860) Computed tomography - A highly potential tool for industrial quality control and production near measurements. NUMB 2004:477–482
8. Kowaluk T, Woźniak A (September 2017) Influence of measurement parameters settings on the results of the CT measurement. In: International conference mechatronics. Springer, Cham, pp 607–612
9. Kowaluk T, Wozniak A (2017) A new threshold selection method for X-ray computed tomography for dimensional metrology. Prec Eng 50:449–454
10. Kasperl S, Hiller J, Krumm M (2009) Computed tomography metrology in industrial research and development. MP Mater Test 51(6):405–411
11. Carmignato S, Pierobon S, Rampazzo P, Parisatto P, Savio E (2012) CT for industrial metrology, accuracy and structural resolution of CT dimensional measurements. In: Proceedings of the conference on industrial computed tomography (ICT), pp 161–172
12. Stojadinovic SM, Majstorovic VD, Durakbasa NM, Sibalija TV (2016) Ants colony optimisation of a measuring path of prismatic parts on a CMM. Metrol Meas Syst 23(1): 119–132
13. Weckenmann A, Krämer P (2013) Predetermination of measurement uncertainty in the application of computed tomography. Geometric Variations, Product Lifecycle Management, pp 317–330

Precision Metrology for Additive Manufacturing

Binnur Sagbas[1(✉)], Tahir Hakan Boyacı[1], and Numan M. Durakbasa[2]

[1] Department of Mechanical Engineering, Yildiz Technical University, Istanbul, Turkey
bsagbas@gmail.com, thr.hkn@gmail.com

[2] Department of Interchangeable Manufacturing and Industrial Metrology, Institute for Production Engineering and Laser Technology, TU Wien, Vienna, Austria
numan.durakbasa@tuwien.ac.at

Abstract. Additive manufacturing (AM) is a layer-by-layer manufacturing method from 3D digital model, without using molds, fixtures and tools. This rapidly developing technology provides opportunity to obtain customized products with complex geometries, freeform shapes, internal cavities and lattice. However, lack of quality insurance procedures and policies are the gaps waiting to be filled. Precision metrology is the key factor for inspection of dimensional tolerances and surface quality of AM manufactured parts. For filling these gaps precision measurement techniques have to be developed parallel with additive manufacturing technology. With the 4th industrial revolution it would be possible to inspect the produced parts during manufacturing process by using in-line measurement techniques. Coordinate metrology (CMM), optical metrology and industrial computed tomography (x-CT) are primary precision measurement techniques used for quality control of the AM parts. In this study an application of the optical metrology for AM parts are given and measurement steps are investigated in terms of controlling dimensional deviations.

Keywords: Optical metrology · Precision metrology · Dimensional metrology
Fused Deposition Modeling (FDM) · 3D printing

1 Introduction

Additive Manufacturing (AM) is emerging technology that has growing interest by manufacturing community [1]. AM technology provides opportunity for manufacturing complex geometries which are difficult to manufacture by conventional techniques. Therefore increasing number of studies focused on the AM processes. Moreover it is concerned in future manufacturing plans of the governments [2–4]. However there are many gaps about the quality assurance and inspection of AM parts in terms of dimensional and geometrical properties especially for complex geometries, hollow parts and lattice structures [5–8].

Metrology is the key factor for development of quality assurance and inspection procedure. Lord Kelvin said that "If you can't measure it, you can't improve it". So, for

N. M. Durakbasa and M. G. Gencyilmaz (Eds.): ISPR 2018, *Proceedings of the International Symposium for Production Research 2018*, pp. 324–332, 2019.
https://doi.org/10.1007/978-3-319-92267-6_28

improving AM technology suitable metrological systems should be involved in the AM manufacturing processes. In current application three basic metrological systems such as coordinate metrology (CMM), optical metrology and industrial computed tomography (x-CT) have been used for quality control of the AM parts. These metrological systems provide feedback to improve manufacturing processes [6, 9].

Mechanical probe based, tactile CMMs have been used for inspection of manufactured parts. These systems determine spatial coordinates of any points on the surface of the measured part. CMMs use Cartesian coordinate system and generally incorporate three linear axis but with a fourth rotary axis they are available with four axis [10]. CMMs are able to measure form deviations with high accuracy. However, they can measure just limited number of points on the part surface and they are relatively slow. For complex geometries and hollow surfaces it is difficult to reach these surfaces by a CMM probe. With the Fourth Industrial revolution metrological systems are tended to place in manufacturing system to make inspection at the same time with manufacturing. CMMs are not suitable for this type of in-line metrology [6].

In-situ and post-process dimensional and form measurements are critical issues for quality control of AM parts. For developing optimal process conditions it is important to characterize shape deformation which occurs due to the relaxation of thermal stresses [5]. For improving metrological ability, non-contact type optical measurement systems which provide opportunity to collect number of measurement points have been used. These systems are faster and more practical for complex geometries. Passive and active systems are the two groups of optical form measurement system [11]. Passive systems mostly based on photography and they don't need spatial modulated lightening to operate while they require specific level of static ambient illumination. Active optical form measurement systems are categorized in to five groups. Such as; Interferometry and Confocal, Conoscopic Holography, Laser and White Light Triangulation, Time of Flight (phase-based) and Time of Flight (pulse-based) [11]. Maximum volume of measurement area (distance), Dimensional tolerance, Resolution and accuracy and Measurement time are the industrial requirements that must be considered for selecting the appropriate form measurement systems [6].

Industrial X-ray computed tomography (XCT) is one of the new generation metrological systems which promise a nondestructive method to dimensional characterization of inner geometries on AM manufactured part. Although standardization studies are still in draft stage, XCT has begun to give accurate measurement results for inspection of inner geometries in AM parts. The system is relatively expensive, slow and difficult to integrate manufacturing system for in-line metrology but with continuing research and recent developments it is becoming an appealing alternative both for geometrical and surface topography [8, 9, 11–13]

In this study, inspection of dimensional properties of an FDM manufactured part was applied by optical measurement system and deviations from nominal dimensions are evaluated.

2 Materials and Methods

In this study, PLA filament has been used for manufacturing test part by FDM technique. The part is designed to have geometrical features such as diameter, angle, inner surface and distance of two points. The dimensions of the test part can be seen in Fig. 1 below.

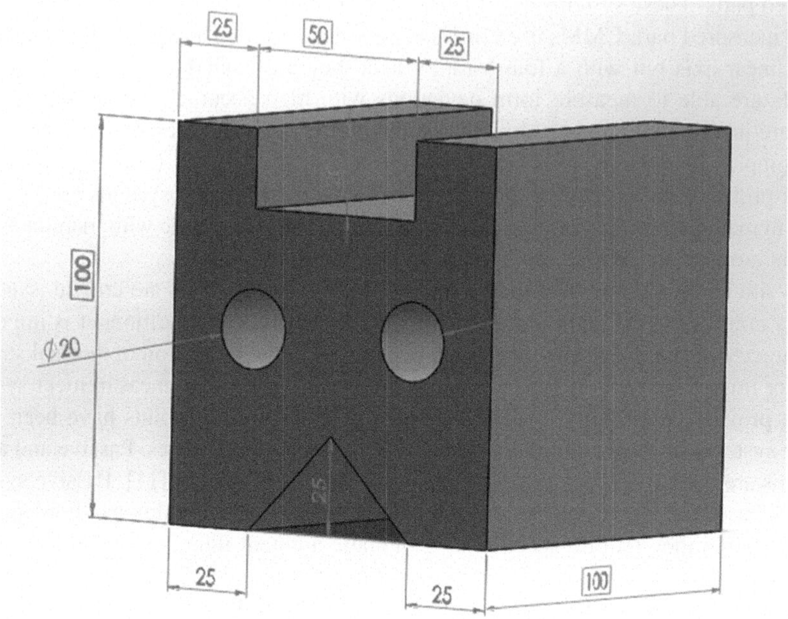

Fig. 1. The dimensions of the test part from general perspective

After modelling, the CAD model of the part was converted to the STL file format to be able to transfer the part's information to the FDM machine. The FDM machine and its properties can be seen in Fig. 2 and Table 1.

The FDM process parameters used for manufacturing of test part and definition step of these parameters can be seen in Table 2 and Fig. 3 respectively.

For dimensional control of the test part a professional 3D optical scanner, Solutionix Rexcan was used. The system uses twin CCD camera and phase-shifting optical triangulation that provide more precise measurement data. A laser spot or line is projected on to the measured part by laser triangulation system. Then by using a mirror for deflecting the beam, the spot or line is scanned across measured part. The height of the scanned points is calculated by the performed triangulation at each position of the mirror. Resolution of the system is defined by the triangulation angle. If the angle is small the resolution is low [6]. Resolution of the system, used in this study, is 0.003 mm while the precision 0,001 mm.

Before starting to measurement, the sample part coated with a special spray and target points placed on necessary region of the part. Then calibration of the scanner was applied for minimizing the errors which are aroused from different laser beam intensity,

Table 1. Technical properties of the FDM machine

Specification	Value
Printing volume	200 mm × 200 mm × 200 mm
Resolution/Detail	80 micron – 300 micron
Position precision	XY 11 micron (0.011 mm), Z 2.5 micron (0.0025 mm)
Printer size	451 mm × 471 mm × 418 mm
Software	Matter control
File formats	STL, OBJ, AMF
Raw materials	PLA and ABS

Fig. 2. A photo of FDM machine

Table 2. FDM process parameters

Layer height (coarse)	0.3 mm
Fill density	0.15
Infill type	GRID
Extruder temperature	210 °C
Bed temperature	70 °C
Infill speed	40 mm/s
Top solid infill speed	30 mm/s

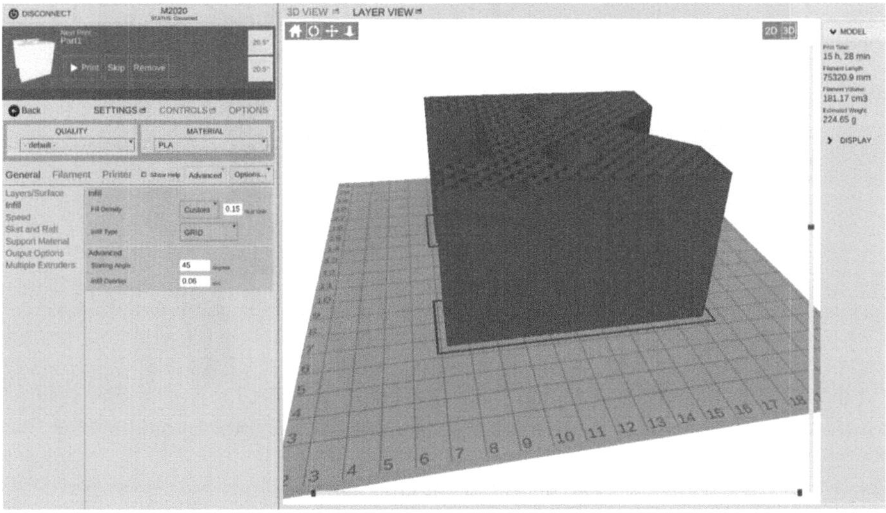

Fig. 3. Deciding fill density parameter on FDM machine's software

variation in projected angle, height of the reflected beam and so on. For calibrating a measurement device traceable reference objects are used with 10° projection angle. Calibration and measurement steps can be seen in Figs. 4 and 5 respectively.

Fig. 4. Calibration step of 3D optical scanner

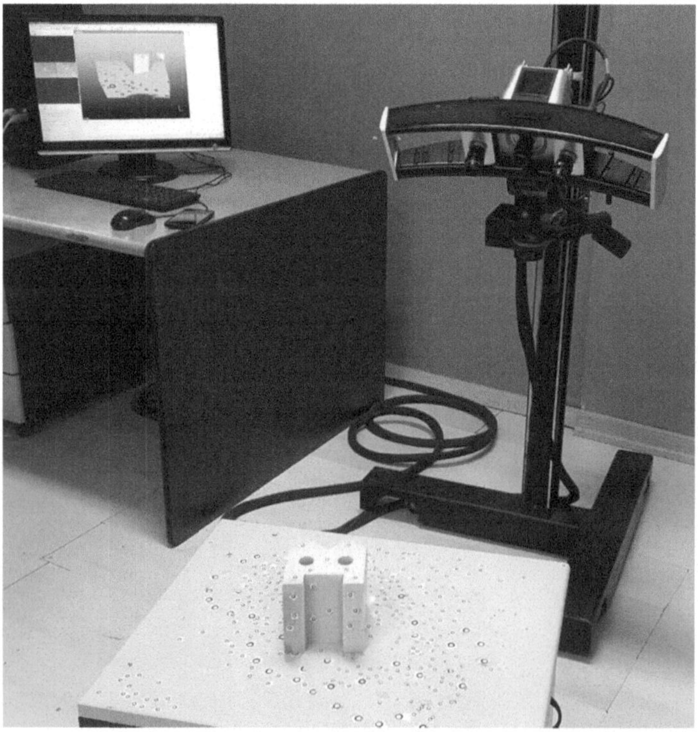

Fig. 5. Measurement of test part

3 Results and Discussion

After scanning the test part, the collected measurement data are converted into a new CAD model. This CAD model had new dimensions depending on the results from scanning step. During the 3D optical scanning process 10.369.983 triangle obtained. For meshing the solid model of the part, 978086 sample points was used. Geometrical dimensions of the part obtained from these data with 0,2448 mean deviation and −1,9803 to 0,9719 error range. Figures 6 and 7 shows both the designed cad model and the new CAD model of the part built up by using the measurement results, obtained by 3D scanner on the manufactured part.

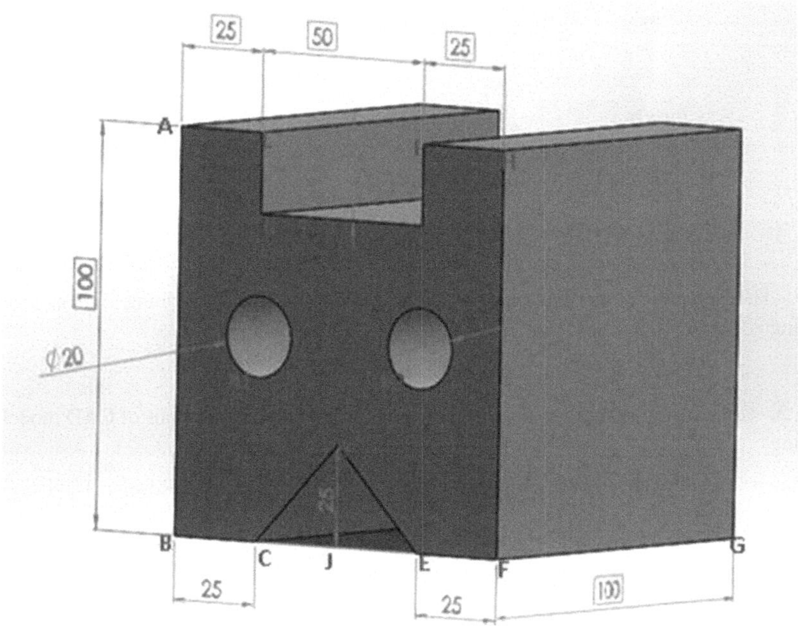

Fig. 6. The specified points on the general perspective

For defining the measured distance, the measurement points were named by letters. These letters and the measurement points can be seen in Fig. 7. Measurement results and nominal results reported in Table 3.

Two solid models, that the designed one and created one with measurement data are matched with each other. Deviations of the manufactured part's dimensions from nominal dimensions were calculated. Graphical representation of the deviations from nominal dimensions can be seen in Fig. 8.

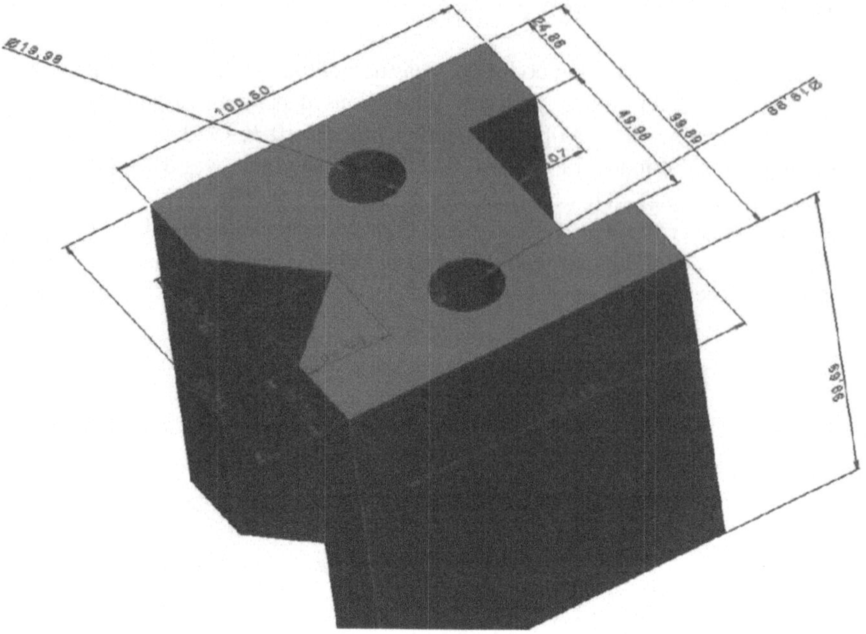

Fig. 7. The general perspective of new CAD model which is describing dimensions after scanning

Table 3. Comprasion designing (nominal) dimensions and new dimensions of CAD model after scanning

Dimensions (length between points and diameters)	Design model (mm)	Results (mm)	Deviation (mm)
A-L	25	24.86	0,14
L-K	20	20.07	−0,07
I-L	50	49.98	0,02
H-I	25	25.05	−0,05
F-H	100	100.56	−0,56
E-F	25	25.10	−0,10
D-J	25	24.98	0,02
E-J	25	24.98	0,02
J-C	25	24.98	0,02
C-B	25	24.83	0,17
B-A	100	100.50	−0,50
R1	20	19.98	0,02
R2	20	19.99	0,01

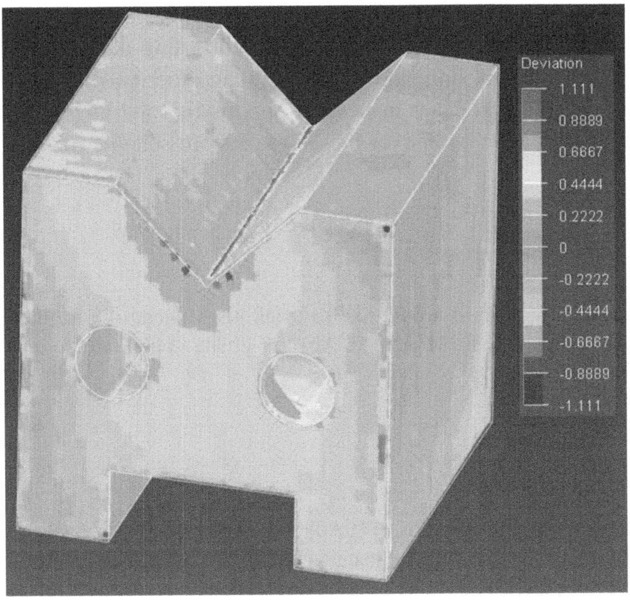

Fig. 8. Graphical representation of matched models which shows deviations from nominal values

It is clear from the Fig. 8 that general form of the manufactured part is conformed with nominal model. The deviations are frequently at the sharp edged of the part. The most intense deviation can be seen at the center of the V channel. For decreasing deviations from nominal geometry, work with lower layer thickness may a solution but while the production time would increase at the same time. For obtaining the best results, processing conditions mus be optimized according to desired specifications of the manufactured part.

4 Conclusions and Future Perspectives

In this study general information about metrological systems such as CMM, optical metrology and industrial CT for AM technology are given. Also dimensional properties of a test part manufactured by FDM are inspected by optical measurement system and deviations from nominal dimensions are evaluated.

It can be concluded that;

- Coordinate metrology, optical metrology and industrial computed tomography are the three basic measurement system used for metrological evaluation of AM parts.
- Coordinate measuring machines are mechanical probe based contact systems which provide more accurate data than current non-contact systems but they are relatively slow and not suitable for in-line inspection of the AM process. Moreover CMMs measures only limited number of the points on a part's topography.

- For inspection of internal features CMMs and optical metrology are inadequate where industrial CT provides opportunity for collecting data about these features. However, this system is slow and expensive than other two alternatives.
- Optical systems provide faster inspection opportunity and large set of point cloud data rather than CMMs. Moreover these systems are suitable for in-line inspection of AM parts.

For making more comprehensive decisions, in future studies the systems must be compared for same AM parts.

Acknowledgement. The authors would like to thank 4B Mühendislik for their support about PLA sample manufactured by FDM method. Also for giving opportunity to inspect the product by their optical scanner.

References

1. Gibson I, Rosen DW, Stucker B (2014) Additive manufacturing technologies: 3D printing, rapid prototyping, and direct digital manufacturing. Springer, New York
2. Foresight Report (2013) The future of manufacturing: a new era of opportunity and challenge for the UK
3. Additive Manufacturing Technology, A Roadmap for unlocking future growth opportunities for Australia CSIRO 2016
4. Ford SLN (2014) Additive manufacturing technology: potential implications for U.S. manufacturing competitiveness. J Int Commer Econ. http://www.usitc.gov/journals. Published electronically September 2014
5. Leach RK (2016) Metrology for additive manufacturing. Meas Control 49(4):132–135 London- Institute of Measurement and Control
6. Stavroulakis PI, Leach RK (2016) Review of post-process optical form metrology for industrial-grade metal additive manufactured parts. Rev Sci Instrum 87:041101
7. Townsend A, Senin N, Blunt L, Leach RK, Taylor JS (2016) Surface texture metrology for metal additive manufacturing: a review. Precis Eng 46:34–47
8. Thompson A, Körner L, Senin N, Lawes S, Maskery I, Leach R (2017) Measurement of internal surfaces of additively manufactured parts by X-ray computed tomography. In: 7th conference on industrial computed tomography, Leuven, Belgium (iCT 2017)
9. Villarraga H, Lee CB, Charney SP, Tarbutton JA, Smith ST (2015) Dimensional metrology of complex inner geometries built by additive manufacturing In: Spring topical meeting, vol 60
10. Leach R (2014) Fundamental principles of engineering nanometrology, 2nd edn. Elsevier Inc., San Diego, pp 252–253 ISBN: 978-1-4557-7753-2
11. Se S, Pears N (2012) 3D Imaging, Analysis and Applications. Springer, Chap 3, p 96
12. Thompson A, Maskery I, Leach RK (2016) X-ray computed tomography for additive manufacturing: a review. Meas Sci Technol 27:072001
13. Villarraga H, Morse EP, Hocken RJ, Smith ST (2014) Dimensional metrology of internal features with X-ray computed tomography. In: Proceedings of the ASPE annual meeting, vol 59

The Evaluation of a Contact Profilometer Measuring Tip Movement on the Surface Texture of the Sample

Paweł Zmarzły(✉) and Stanisław Adamczak

Faculty of Mechatronics and Mechanical Engineering, Kielce University of Technology, Kielce, Poland
{pzmarzly,adamczak}@tu.kielce.pl

Abstract. Measuring instruments based on contact methods are commonly used in industrial, as well as laboratory conditions for the evaluation of geometric structure of the surface elements of machine parts. One of the flaws of such measuring systems is the risk of damaging a measured surface due to the contact with the measuring tip. The paper discusses research aimed at evaluating the impact of a stylus profiler measuring tip passage on the condition of the surface layer of samples made from three different materials, i.e. C45 steel, EN-AW-2017A aluminium and Vero White liquid resin. The tests involved thirty measurements of surface roughness of the selected samples and evaluating the changes in the basic roughness parameters due to the passage of a profiler measuring probe. The condition of the analysed samples was evaluated after the roughness measurements using a Talysurf CCI non-contact profilometer. The tests demonstrated that a multiple passage of a stylus profiler measuring tip influences the change in the measured surface roughness parameters. Moreover, the roughness measurement on the surface of a sample made from EN-AW-2017A aluminium caused a clear scratch.

Keywords: Contact instruments · Surface roughness · Stylus profiler
Measuring tip · Rapid technologies

1 Introduction

There are two basic groups of methods used for analysing the surface geometry: stylus (contact) and optical (non-contact) methods [1, 2]. Despite the intensive development of optical methods, they find their application mainly in laboratories [3]. Whereas, the most popular and available are the instruments based on contact methods. Instruments of this type can be used to measure roundness or waviness deviations of cylindrical machine parts [4], as well as to evaluate the roughness of flat surfaces [5].

Among contact instruments, thanks to their compact design, portable stylus profilers are increasingly more often used to measure roughness directly in industrial conditions [6, 7]. Both in stationary, as well as portable devices, the stylus tip moves in contact with the measured surface, modelling the measured profile. In order to obtain a real model of a profile, stylus tip with a small rounding radius in the range of 1.5–12.5 μm

N. M. Durakbasa and M. G. Gencyilmaz (Eds.): ISPR 2018, *Proceedings of the International Symposium for Production Research 2018*, pp. 333–342, 2019.
https://doi.org/10.1007/978-3-319-92267-6_29

and a minimum measuring force of 0.004–0.06 N are used. Despite such small measuring tip force values, in the case of measuring elements with low hardness, there is a risk of scratching the measured surface, which can significantly impact the measurement result [8, 9]. It is particularly visualized for elements made of plastics or aluminium alloys. Moreover, most portable instruments used for roughness measurements, which are commonly used in industrial practice are equipped with a skid. Forsyth and Scott [10] in their paper indicate that using a skid method significantly contributes to scratching a measured surface, compared to a non-skid method.

An intense development of rapid technologies makes elements made from plastics become more and more popular [11], especially in foundry industry [12]. Unfortunately, due to high reflexivity and anisotropies of certain plastics, surface roughness measurement with optical methods is hindered or almost impossible. Whereas using contact methods utilizing stylus profilers, there is a risk of scratching a surface. There are research papers aimed at analysing the impact of force and geometry of a measuring tip on the condition of a measured surface [13]. However, the analysis is conducted only for samples made from one material, in a conventional manner. Whereas, there is a deficiency of research aimed at evaluating the state of a surface texture of elements manufactured with rapid technologies (3D printers).

As a result, tests should be conducted, aimed at evaluating the way the movement of a stylus profiler measuring tip impacts the condition of a measured surface of elements made of several different materials. The evaluation was visual and by analysing the changes in the values of basic roughness parameters, i.e., Ra, Rt, RSm, Rsk. Three samples, made of different materials, i.e., C45 steel, EN-AW-2017A aluminium alloy and liquid polymer resin were used for the tests. In order to analyse the impact of measuring tip passage on the changes in selected roughness parameters, thirty roughness measurements at the same location in the sample were taken. Moreover, for the purpose of detailed evaluation of the condition of sample surface layer, the evaluation of the surface topography was conducted with the use of a Talysurf CCI non-contact system by Taylor Hobson was performed after the roughness measurements.

The research presented in this paper is particularly important for the modern industries facilities that implement the concept of the "Industry 4.0". Most industrial plants are used Computer Aided Quality Control, where measuring systems based on contact method. However, these methods are not fully universal and cannot be used to measure certain materials. The research results indicate some risk of application contact method to analyse surface quality of selected product. This will be guidance for industry to applying appropriate measuring devices for specific applications.

2 Experimental Tests

The experimental tests conducted for the purposes of the article were entirely executed at the laboratories of the Kielce University of Technology, i.e., in the Laboratory of Computer-Based Measurement of Geometrical Quantities and the Laboratory of Unconventional Manufacturing Technology. The first stage of the tests was the preparation of the samples for roughness measurements. Next, the roughness was measured using an

SJ-210 portable measuring instrument by Mitutoyo. The last stage of the experimental tests was the evaluation of the condition of the tested surfaces using a Talysurf CCI measuring system.

2.1 Test Samples

Three different samples were used for the tests. Two samples were made of C45 steel and an EN-AW-2017A aluminium alloy. The measured surface of the steel and aluminium samples was polished. Whereas one sample was made of plastic - a Vero White liquid polymer resin with the use of the PJM (PolyJet Matrix) rapid technology, executed through a Connex 350 machine. The sample was created with a set thickness of the constructed layer of 0.03 mm in the "Digital Material" print mode. The sample was set at an angle of 90° in relation to the operating platform of the machine. Such a printing angle ensures satisfactory mechanical properties and high dimensional-shape accuracy of the manufactured sample [14]. The sample material was selected, so that they would be characterised by different mechanical properties. The Table 1 shows selected mechanical properties of the tested materials.

Table 1. Selected mechanical properties of C45 steel [15], EN-AW-2017A aluminium alloy [16] and Vero White polymer resin [17]

Material	Modulus of elasticity, E	Tensile strength, Rm	Yield point, Re	Hardness
C45 Steel	198–207 GPa	560–850 MPa	275–490 MPa	229 HB
Aluminium alloy EN-AW-2017A	72.5 GPa	350–390 MPa	240–260 MPa	101–110 HB
Vero White	2.495 GPa	50 MPa	-	85D (Shore Hardness - D scale)

2.2 Test Instruments

An SJ-210 portable stylus profiler by Mitutoyo was used for measuring roughness parameters. The instrument is equipped with a battery; therefore, it can be used for non-stationary measurements. However, in order to maintain constant measurement conditions and stabilize the measurement procedure, a granite stand was used. In addition, using a stand enables eliminating accidental changes of the profiler measuring tip position during roughness measurements. Figure 1 shows an SJ-210 measuring instrument mounted in a stand bracket together with the tested sample. Whereas, Table 2 shows the technical specification of the SJ-210 instrument.

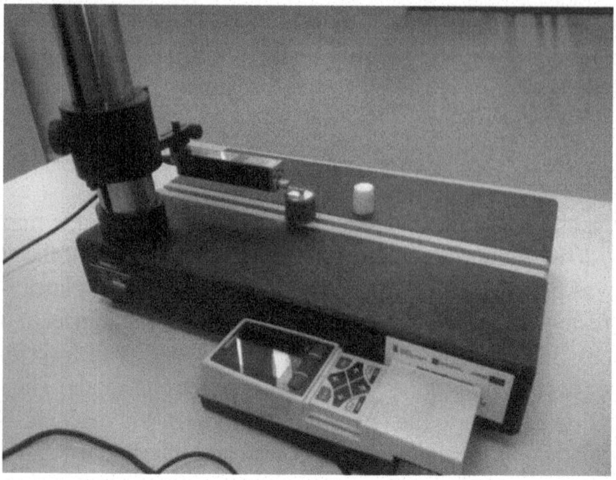

Fig. 1. Test stand used for roughness measurements

Table 2. Technical specification of an SJ-210 stylus profiler [18]

Drive unit	
Measurement range	16 mm
Passage	17.5 mm
Measuring speed	0.25 mm/s; 0.5 mm/s; 0.75 mm/s
Detector	
Measurement method	Differential - inductive
Range	360 μm
Measuring head	
Tip	Diamond
Measuring tip radius	2 μm
Measuring tip angle	60°
Measuring tip measuring force	0.75 mN
Skid radius	40 mm

The parameters presented in the table are taken from the technical specification of the measuring instrument [18]. Due to the operation of the measuring system, certain parameters may deviate from the values presented by the manufacturer for a new system. Due to the fact that the force of a stylus profiler measuring tip may impact the condition of the measured surface, hence, the values of roughness parameters, therefore real measuring force, of the measuring tip was additionally determined. The measuring force was determined using a precision balance with a resolution of 0.001 g. The true measuring force of 0.49 mN was then determined. It is a value lower than given by the manufacturer (see Table 2).

In order to evaluate the impact of a contact instrument measuring tip flight on the measured surface condition of steel, aluminium and plastic samples, the authors analysed the changes of four roughness parameters, i.e., Ra – arithmetical mean roughness value, Rt – total height of the roughness profile, Rsk – skewness of the roughness profile, RSm – Mean width of the roughness profile elements. For the C45 steel and EN-AW-2017A aluminium samples, a sampling length of lr = 0.25 mm, while for a liquid polymer resin samples of lr = 2.5 mm was selected. Moreover, 3D surface topography analysis, with the use of a CCI non-contact profilographometer was conducted for each measured sample. The tests were aimed at detecting potential scratches, which could have occurred as a result of a contact profiler measuring tip passage.

3 Analysis of Measured Results

The results of the conducted tests are shown in Table 3 and in the form of graphs depicting the change of individual roughness parameters, depending on the number of measurements. Due to a large number of measurement results, Table 3 shows the following, for each group of roughness parameters: $\bar{x}$ – mean value, x_{max} – maximum value, $x_{\min}$ – minimum value, $R = x_{max} - x_{min}$ – range, s – mean standard deviation.

Table 3. Summary of roughness measurement results

	C45				EN-AW-2017A				Vero White			
	Ra, μm	Rt, μm	RSm, μm	Rsk	Ra, μm	Rt, μm	RSm, μm	Rsk	Ra, μm	Rt, μm	RSm, μm	Rsk
$\bar{x}$	0.019	0.330	41.547	−1.300	0.156	0.895	92.406	−0.256	9.412	68.005	643.873	−1.075
x_{max}	0.024	0.486	110	−0.663	0.535	2.564	212.50	1.284	9.527	68.800	724.500	−1.062
$x_{\min}$	0.018	0.250	27.6	−2.045	0.015	0.105	2.185	−2.964	9.269	66.560	558.000	−1.096
R	0.006	0.236	82.4	1.382	0.520	2.459	210.315	4.248	0.258	2.240	166.500	0.034
s	0.001	0.065	19.737	0.417	0.154	0.671	56.539	0.986	0.044	0.622	50.467	0.008

Based on the results shown in Table 3, it can be concluded that the C45 steel samples are characterised by the lowest roughness value. While the Vero White liquid polymer resin samples, prepared with the PJM rapid technology, are characterised by the highest roughness values. This may result from the fact that the surfaces of steel and aluminium samples were subject to grinding and polishing, while the sample made with the PJM technology was not subject to any finishing. This is confirmed by the mean value of the Rt parameter, which specifies the maximum height of a roughness profile. For the samples made in the PJM technology, the value of this parameter is almost two hundred times higher than for C45 steel samples. Whereas, when analysis a parameter specifying the skewness of a roughness profile, i.e. Rsk, it can be concluded that its mean values are negative for all analysed samples. This proves that the material concentrates near the vertices of the roughness profile, which means surfaces with a shape close to a plateau. A negative value of a roughness profile asymmetry also informs about the presence of "deep valleys" on the surface of a studied material.

In order to better depict the impact of a contact profiler measuring tip passage on the values of roughness parameters of the measured surface, graphs showing the change of individual parameters depending on the number of measurements were presented. Figure 2 shows the impact of a measuring tip passage on the values of roughness parameters for a C45 steel sample, Fig. 3 for the EN-AW-2017A sample, while Fig. 4 shows the results obtained for a sample made of Vero White liquid resin.

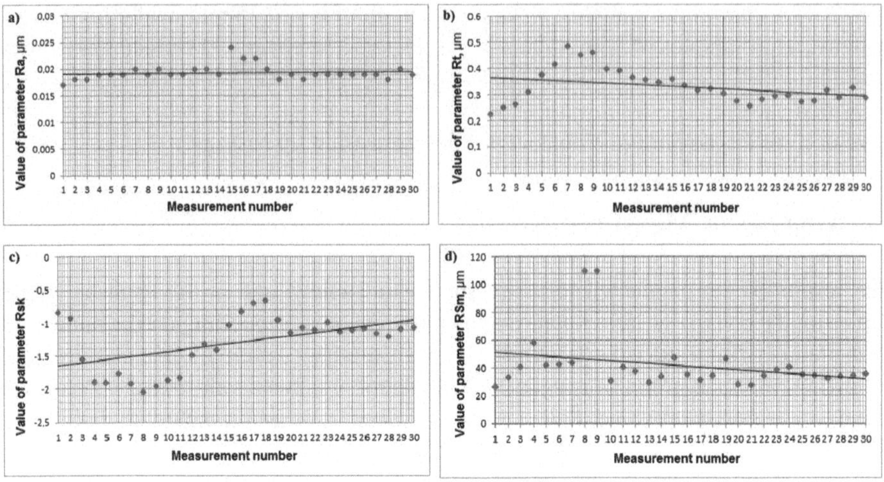

Fig. 2. Analysis of the impact of an SJ-210 profiler measuring tip movement on the values of roughness parameter in the C45 steel sample

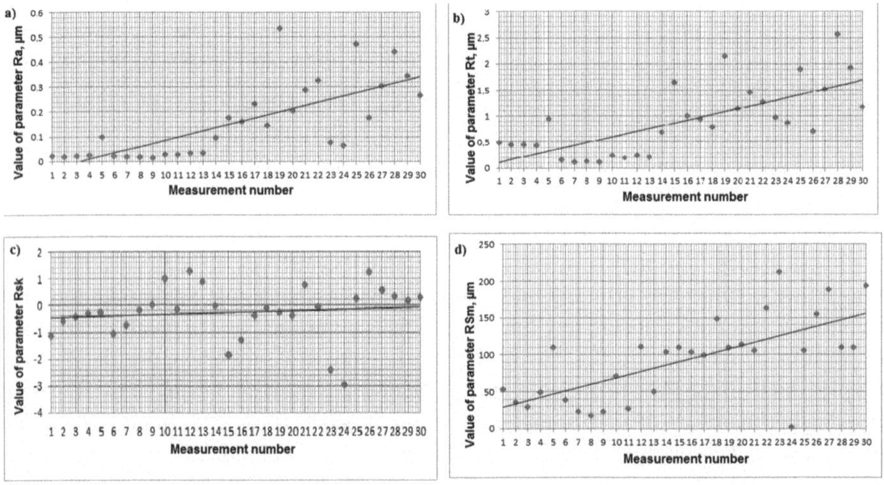

Fig. 3. Analysis of the impact of an SJ-210 profiler measuring tip movement on the values of roughness parameter in the EN-AW-2017A aluminium sample

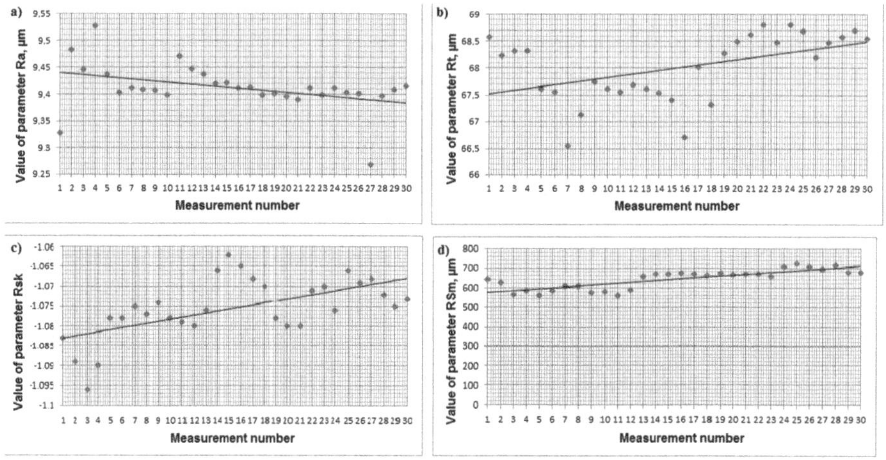

Fig. 4. Analysis of the impact of an SJ-210 profiler measuring movement on the values of roughness parameter in the Vero White liquid polymer resin sample

When analysing the results shown in Fig. 2a, we can see that together with an increase of roughness measurements, the value of the Ra parameter stays at the same level. Whereas the values of the Rt parameter, after an initial increase (up to the seventh roughness measurement level), exhibit a decreasing tendency. Whereas, while considering the Rsk parameter, which describes the roughness profile skewness, after the initial decrease of this parameter, we can observe its sudden increase, followed by stabilization. However, negative values of this parameter point to unevenness vertices similar to a plateau. When analysing the RSm parameter, which defines the mean distance between unevenness, it can be concluded that together with an increase in the number of roughness measurements, the mean gap between roughness vertices decreases.

When considering the roughness measurement results for the aluminium sample, it can be concluded that together with an increase in the number of a contact profiler measuring tip passages, the value of the analysed roughness parameters also increases. While for the first thirteen measurements, the roughness parameters stay at the same level. Whereas after passing the thirteenth measurement, the roughness measurement values exhibit a tendency to increase. This may result from the measured surface being damaged (scratched) by the measuring tip.

When analysing the liquid polymer resin sample surface roughness measurement results, it is impossible to draw unequivocal conclusions. As can be seen in Fig. 4a, together with an increase in the number of roughness measurements, the value of the Ra parameter decreases, while the total height of the roughness profile, described by the Rt parameter, increases (Fig. 4b). This may result from the fact that after initial scratching of the surface by the measuring tip, consecutive move of the tip might smooth the measured surface, hence, the decreasing roughness values. Negative values of the Rsk parameter (Fig. 4c) mean that surface unevenness of a sample prepared in the PJM technology are of a flatter character. Whereas, when interpreting the trend line shown in Fig. 4d, it can be concluded that together with an increase in the number of measurements, the distance between unevenness vertices increases.

Due to the fact that a passage of a contact profiler tip might have caused the scratches on the measured surface, the surface topography of the studied samples was analysed. The test showed that only in the case of the EN-AW-2017A aluminium sample a clear, 0.105 mm wide and 1.87 μm deep scratch was identified (Fig. 5). Whereas the other samples were not scratched. Figure 5 shows the view of a scratch on the surface of an aluminium sample, which appeared after a series of roughness measurements.

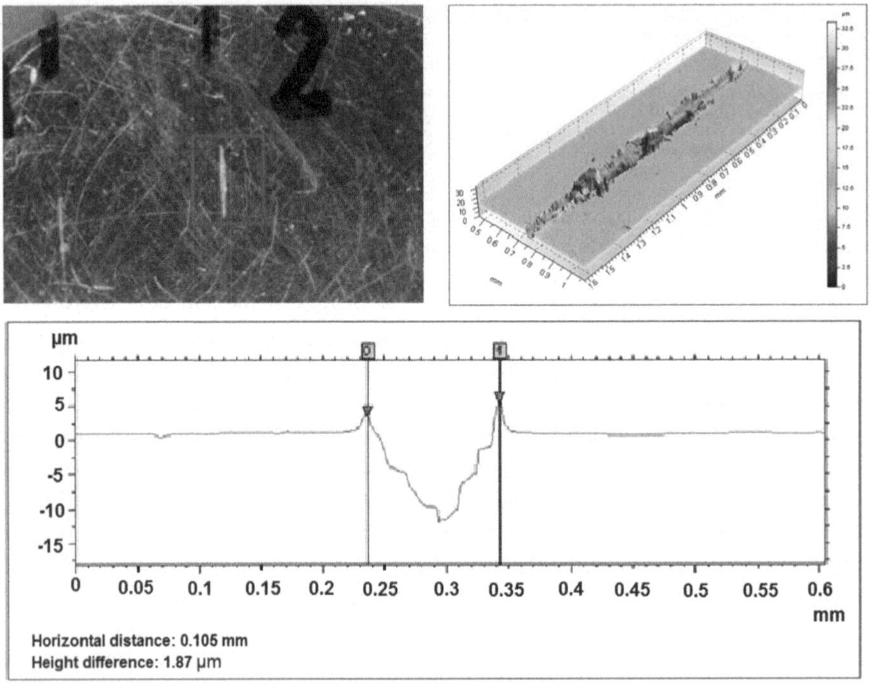

Fig. 5. View of a scratch on the surface of a sample made from EN-AW-2017A aluminium alloy

4 Summary and Conclusions

The main objective of the studies presented in this paper was the evaluation of the impact of a stylus profiler measuring tip movement on the measured surface condition. Three samples, prepared from different materials, i.e., C45 steel, EN-AW-2017 aluminium alloy and Vero White liquid polymer resin were tested. The tested materials exhibited various mechanical properties.

General analysis of obtained roughness values showed that the smallest roughness was obtained for a C45 steel sample, while the greatest for a Vero White liquid polymer resin sample made with the PJM rapid technology. Whereas the differences in the roughness values were significant. For example, when analysing the Ra parameter obtained for the tested samples, it can be concluded that an average Ra value from a C45 steel sample was Ra = 0.019 μm, while for a Vero White sample was Ra = 9.412 μm. This value is almost 500 times higher than for steel. It stems from the fact that the steel sample surface was polished, while the surface of the sample made with the incremental technology was not subject to any other finishing processing.

When analysing the impact of multiple passages of a measuring tip of a stylus profiler on the values of selected roughness parameters, it can be concluded that for a steel sample and a sample made from liquid polymer resin, after a moderate increase in the value of roughness parameters after several initial measurements, the roughness values stabilize or exhibit a decreasing tendency. It can result from the fact that a measuring tip super-finishes the measured surface, hence, the roughness decreases. Whereas in the case of an aluminium sample, the roughness value increases together with the number of measurements.

When analysing the topography of a surface measured with a Talysurf CCI instrument, it can be concluded that only the surface of the aluminium sample was scratched. No scratches were recorded on other surfaces.

While summarizing the test results presented in the paper, one can conclude that a multiple flights of a measuring tip impact roughness parameter values, hence, the condition of the measured surface of all analysed materials. Whereas the shape of the measuring tip, in most cases, does not impact the measured surface in a damaging manner. Materials with a very smooth surface are an exception.

The study presented in this paper is preliminary. The authors will thoroughly analyse the measuring tip geometry and analyse the elastic deflections due to tip passage in consecutive papers. The impact of a movement of a measuring tip without a skid will also be analysed. Moreover, the range of the tested materials will also be expanded.

Acknowledgments. The paper has been elaborated within the framework of the research project entitled "Theoretical and experimental problems of integrated 3D measurements of elements' surfaces", reg. no.: 2015/19/B/ST8/02643, ID: 317012, financed by the National Science Centre, Poland.

References

1. Whitehouse DJ (2011) Handbook of surface and nanometrology, 2nd edn. University of Warwick, Coventry
2. Demircioglu P, Durakbasa MN (2011) Investigations on machined metal surfaces through the stylus type and optical 3D instruments and their mathematical modeling with the help of statistical techniques. Measurement 44(4):611–619
3. Makieła W, Gogolewski D (2017) Variability evaluation of signal in two-dimensional wavelet decomposition using fractal dimension. Procedia Eng 192:243–248
4. Adamczak S, Zmarzły P (2017) Influence of raceway waviness on the level of vibration in rolling-element bearings. Bull Polish Acad Sci Tech Sci 65(4):541–551
5. Miller T, Adamczak S, Świderski J, Wieczorowski M, Łętocha A, Gapiński B (2017) Influence of temperature gradient on surface texture measurements with the use of profilometry. Bull Polish Acad Sci Tech Sci 65(1):53–61

6. Xia W, Ni Ch, Xie G (2016) The influence of surface roughness on wettability of natural/gold-coated ultra-low ash coal particles. Powder Technol 288:286–290
7. Stępień K (2015) Testing the accuracy of surface roughness measurements carried out with a portable profilometer. Key Eng Mater 637:69–73
8. Brown CA, Savary G (1991) Describing ground surface texture using contact profilometry and fractal analysis. Wear 141(2):211–226
9. Pawlus P, Śmieszek M (2005) The influence of stylus flight on change of surface topography parameters. Precis Eng 29:272–280
10. Forsyth I, Scott D (1982) Characterization of micromachined mirror surfaces from high speed diamond fly cutting. Wear 83(2):251–263
11. Kundera Cz, Bochnia J (2014) Investigating the stress relaxation of photopolymer O-ring seal models. Rapid Prototyp J 20(6):533–540
12. Adamczak S, Zmarzły P, Kozior T, Gogolewski D (2017) Analysis of the dimensional accuracy of casting models manufactured by fused deposition modeling technology. Eng Mech 2017:66–69
13. Chetwynd DG, Liu X, Smith ST (1992) Signal fidelity and tracking force in stylus profilometry. Int J Mach Tools Manuf 32:239–245
14. Adamczak S, Bochnia J, Kaczmarska B (2015) An analysis of tensile test results to assess the innovation risk for an additive manufacturing technology. Metrol Measur Syst 22(1):127–138
15. Standard EN 10083-2:2006: Quenched and tempered steels. Technical delivery conditions for unalloyed quality steels
16. Standard EN 485-2:2016-10: Aluminium and aluminium alloys - sheet, strip and plate. Mechanical properties
17. https://www.stratasysdirect.com. Accessed 14 June 2018
18. https://www.mitutoyo.com/. Accessed 14 June 2018

Industry 4.0 Applications

Approach of Medıum-Sized Industry Enterprıses to Industry 4.0 a Research in Konya

Leyla Polat(✉) and Güzide Karakuş

Necmettin Erbakan University, Konya, Turkey
{lpolat, gkarakus}@konya.edu.tr

Abstract. The Industrial Revolution, which began with the invention of steam-powered machines in the 18th century, caused a great resonance in the world and greatly affected the fate of mankind. Thanks to day-to-day technology, the rapidly developing manufacturing sector has overtaken the third generation and is now entering into its the 4. decade. As mentioned above, the industrial revolution, which started with steam power, then took advantage of the electrical power, and later, with the development of the information industry, the electronic digital revolution took its first step. Today, the 4th Industrial Revolution is called Industry 4.0.

Industry 4.0 consists of many sub-headings. When this technology is used in the manufacturing industry, it takes the name of the Internet of Things (IoT), similarly when it is used in the service sector it is also called the Internet of Services (IoS). The fact that objects, services are in a single network, and that they can be managed from a centralized system, contributes positively and brings a number of threats as well. A number of precautions are taken against the cyber attack that may occur in this direction and studies are being conducted on this subject.

Generally speaking, the factories that switch to Industry 4.0 technology will enter into a modular system, which means that the machines that are used in the factory will communicate on the internet and work without human power. In other words, the objects will be moved to the virtual platform, and the business will be made smart by being able to be controlled from the virtual platform. This will lead to a reduction in time loss, a reduction in costs, an increase in productivity, and an improvement in business performance. The aim of this study is to understand how Industry 4.0 that is spreading around the world rapidly yet mainly started by large businesses in Turkey is perceived in Konya which has intensive industry in Turkey. In this context, the exploratory research method was used. During the implementation phase, a questionnaire was prepared to evaluate the approaches of enterprises to Industry 4.0, and a face-to-face interview was conducted with two business managers. Within the scope of study, it is aimed to encourage enterprises to adapt the technological age introducing the 4. Industry Revolution and also to contribute to the academic literature about Industry 4.0 which is newly being researched in Turkey.

Keywords: Industry 4.0 · Internet of things · Industrialization in konya
Industry 4.0 in turkey

N. M. Durakbasa and M. G. Gencyilmaz (Eds.): ISPR 2018, *Proceedings of the International Symposium for Production Research 2018*, pp. 345–354, 2019.
https://doi.org/10.1007/978-3-319-92267-6_30

1 Introduction

Technology, which is one of the biggest factors of the development and recognition of a country, is constantly seen in every field today. As it is difficult to keep pace with technology which is constantly improving, the societies that catch up with this era are one step forward. The fact that technology is so effective and each object has a technological dimension is an innovation brought by the concept of internet of things called IoT. According to the European Technology Platform, the definition of IOT, which is one of the building blocks of the scientific age we live in, "Creation a smart network by the objects which are concrete and abstract and have predetermined duties and also their interaction of objects with eachother and with the other networks" (Turak 2015: 3). These interactions are made possible by the presence of sensors and wireless communication networks on the objects. The sensors on the objects allow them to be perceived, controlled and shared on the wireless network (Oral and Cakir 2017: 747). According to the definition made by the RFID Group, which makes the most use of IoT and makes full use of RFID technology, IoT is to be able to position objects in a common network by means of certain standard communication protocols and to communicate with each other depending on each other (Gubbi et al. 2012: 4).

The emergence of the IOT, which is a blessing in the information industry that develops at an unstoppable pace and deeply affects human life, dates back to the early 90 s. In these years, the display of an object by academicians and the distribution of these images through a network such as the internet is regarded as the first application of the Internet of Things (IOT) (López-De-Armentia, Casado-Mansilla and López-De-Ipiña 2012: 1). The concept of IoT at the same time of application was initially confronted by the use of the Radio Frequency Identification (RFID) system, one of the subtitles of the IoT by a private company at the end of the 90 s and by their making many gains from this system (Ashton 2009: 1). Understanding the importance of IoT has taken a short while, however, it has not reached to the desired level (Cankat Bati et al. 2017: 2). The concept of IoT has been rapidly spreading and has actively played a role in all aspects of life with the improvements at technologies, hardware, decrease in cost and the biggest factor in use of IPv6 protocol a little in 2012 about Internet of Things (World IPV6 Launch 2018).

The fact that objects interact with each other on a wireless network without human factor provides many benefits in terms of time and performance in the tasks of that object. For this reason, the concept of IoT is readily available in many areas of our daily lives, and accelerates its development with great acceleration (Xu, He and Li 2014: 2233). Since the object is a general expression, the use areas are considerably widened because the IoT item covers all the objects. Below are listed areas where IoT is more or less active (Xu et al. 2014: 2234) (Bandyopadhyay and Sen 2011: 60–65) (Yigitbasi 2011: 105).

- Aviation Sector
- Production Sector
- Communication Sector
- Health Sector
- Logistics and Supply Chain Management

- Retail Sector
- Environmental Protection
- Transport Sector
- Agriculture Sector
- Entertainment, media Sector
- Insurance Sector
- Recycling
- Security Sector
- Information Sector
- Food Sector

The concept of IOT, as seen in the above list, is confronted in many different areas. In this study, the manufacturing sector which is experiencing a new period thanks to IoT handled. The traditional manpower and labor-intensive production system that took place in the factories was replaced by the machines communication each other thanks to required protocols and manufacturing system which has autonomous control system. In Part 1 of this report, this new production management system, which is called Industry 4., is introduced. In the 2. Part, the present situation about Industry 4.0 in Turkey is investigated. In the 3. Part, a questionnaire is conducted on the awareness and applications of Industry 4.0 on two enterprises operating in Konya, which is an industrial city consisting of many SMEs. In the last part, based on the results obtained, suggestions are presented for both businesses and academic world.

2 General Description of Industry 4.0

The industrial revolution, which started with the use of steam-powered machines, is in a period when dreams have turned into reality. This period, which was revealed by Germany that proved to be a developed country for industrialization and technology in 2011, is called Industry 4.0 (Rojko 2017: 80). Industry 4.0 offers many possibilities for businesses and brings useful applications. But for that, it is expected that company executives will improve their operational capabilities by identifying an end-to-end strategy. Thus, production diversity, speed and customer satisfaction is promised (Ganzarain and Errasti 2016: 1119). The term "smart" come into life thanks to IoT technology is met at manufacturing as "Smart Factories". According to the report on Industry 4.0 by German National Academy of Sciences and Engineering (Acatech) which have Professor Kagermann and other leading scientists of the world, the following are the possibilities and and the solutions of existing problems at smart factories switched into fourth industrial revolution (Kagermann, Wahlster and Helbig 2013: 15–16):

- To meet individual customer needs,
- The flexibility structure established by cyber-physical systems and some network structures,
- Taking the most appropriate decisions in a short time as sensitive to the outside world,
- Using a small number of resources to maximize product extraction, improve efficiency,

- Establishing new business models and offering smart services by processing factory data,
- To evaluate the social rights and working conditions of factory employees individually and to ensure the working order accordingly,
- Establishing work-life balance by ensuring satisfaction of employees and customers.

In order to move towards the above-mentioned objectives, the physical world in smart factories needs to be digitized through cyberspace through IoT technology. These systems are called Siber-Physical Systems (SFS). In these systems, the information about these objects is switch to the cyber area by using sensors, warnings, barcodes embedded in the objects in the factory using wireless internet network. By this means, 2 different areas are met at one point (Broy, Cengarle and Geisberger 2012: 2). SFS, one of the greatest technology blessings of the 20th century, is used at many areas that will enhance the quality of life within the scope of robotic, smart housing, structures, defense industry, aviation and smart city, especially in medical fields (Lee 2008: 363). In this study, how this system works in factories, especially the relation between medium sized companies and this technology is investigated.

The overall technological structure in the industry, which is called the smart factory that switched into the Industry 4.0 era by using Ciber - Physical Systems, is given in Fig. 1. As shown in the figure, if smart factory should be defined in general, smart factory is a structure which establish control mechanizm in other words automation that id managed exclusively integrating the software and hardware within which it is built and provides a quick and efficient solution to the customers without any waste of work power and resources at the same time (Radziwon et al. 2014: 1187).

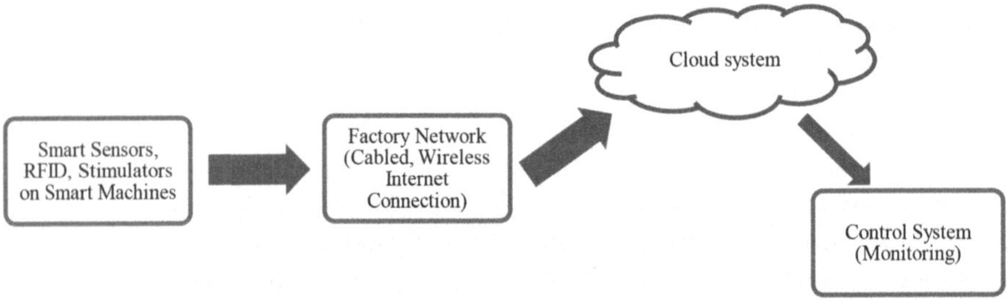

Fig. 1. Smart factory operation structure Resource: Wang et al. 2016: 3

Artificial intelligence applications are being used in the 4. Industry era thanks to internet of things. The smart machines mentioned in Fig. 1 are designed using Machine Learning algorithms and in this way the machines learn their own tasks in production without any employee support and know in advance what kind of response or result they will give. At the same time, these smart machines can decide for themselves the maintenance and repair requirements (Lee, Kao and Yang 2014: 5). At the same time, they can communicate with each other by using a common network of the machines through the sensors located on them, and they can observe what operations are performed by the control system. Data should be collected and processed using different technologies such as sensors, RFID, WSN from machines and other devices in order to control all operations performed through automation and to display existing conditions. The place where these things happen is the step called Cloud System. Creating applications to be used in accordance with different aims thanks to storing all the data, their management and data mining and presenting them to the management is actualized at this step (Ma et al. 2013: 1144–1149). The concept of Industry 4.0, which incorporates many technologies and ultimately provides solutions to businesses and customers in a fast and efficient manner, combines many different fields of work and requires multidisciplinary work. The rapid development of the information sector is leading to this and many other sectors are concerned.

3 Industry 4.0 in Turkey

Industry 4.0, at which Germany is a pioneer and which the whole world is following, not only improves country economies but also increases the social welfare levels of countries when they are considered as a whole (Kotynkova 2017: 249). Before examining the studies about this topic if we focus on the journey of our country between industrial revolutions, it is seen that industrialization was first started in the Tanzimat Period and related factories were established in many different sectors (Ertin 1998: 165). Afterwards, thanks to unstoppable development of technology and the expansion of our country to the outside world, our country transitioned to the 2. and 3. Industrial Eras. The small and medium sized enterprises except for the developed large-scale enterprises located in our country live in the 3rd industry era in their factories. For businesses manufacturing in large-sized factories, it is now possible to mention Industry 4.0 (Electric 2016: 51).

According to the joint report by TUSIAD and Boston Consulting Group on Industry 4.0, when Turkey catches the train, it will achieve a high amount of value-added production and will show significant improvements in the categories determined in the report (TUSIAD and Group Boston Consulting 2016: 14).

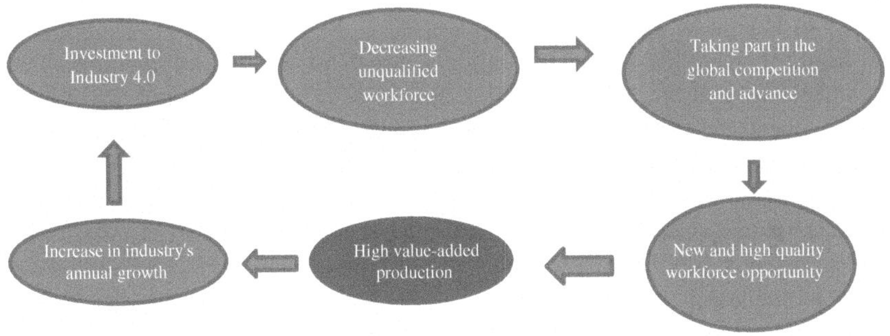

Fig. 2. Business model with Industry 4.0 in Turkey Resource: TUSIAD and Group Boston Consulting 2016: 37

Many domestic and foreign large businesses which take the Fig. 2, modeling the achievements the applications of Industry 4.0 in Turkey, start to required researches about this issue and pilot scheme. For example, Bosch, one of the leading companies in technology, has collaborated with Bogazici University to evaluate the employment in Industry 4.0 and to develop innovative and dynamic projects to cover up the lack of projects in this area. In this pilot scheme, 40 university students designated by consultants from both sides are studying in the field of cyber safety, one of the new generation of technology, and internet of things in Bosch Manisa factory and observe the major projects which are integrated with Industry 4.0 that doesn't remain a theory (Turkey's Industry 4.0 Platform 2017). Also, Siemens, one of the leading companies in the technology and has a pioneering role in the emergence of the 4. industrial era, has initiated a joint study with the leading information companies, Bogazici Yazılım and Atos, in Turkey. These companies are producing various solutions to respond to all kinds of problems in areas such as hardware, consulting, human resources, etc. (Siemens, Bogazici Yazılım and Atos 2016). If the Industry 4.0 in our country is looked from a general point of view, it is seen that the companies which have a voice about technology put their projects into practice and they are collaborating with many businesses carrying on multidisciplinary work.

4 The Industry 4.0. Awareness and Practices of Medium-Sized Enterprises in Konya

Konya, known as agriculture city of Turkey because of its fertile land within its boundaries and concentrated on its development in this direction, has made a move in industry as well as agriculture. The number of industrial establishments has increased from 47 in 1950 to 91 in 1960 and the industrial activities in the following years showed a steady development. Organized Industrial Zones built in the city center and small industrial sites in the center and districts are the main factors that accelerate this development. As a consequence of the developments in manufacturing industry in Konya after 1965, there was a serious tendency towards the creation of Organized Industrial Zones within the framework of the 3. Development Plan (KSO, E.T.: 2018). As of 2018, there are 9 organized industrial zones, 38 small industrial sites and 3 private OIZs in Konya along with the counties.

99 percent of Konya's industrial sector consists of small and medium-sized enterprises (SMEs). It is the " Konya SME Capital" with its 32 thousand SMEs.

In Konya, one of the cities which provides the most employment and added-value in Turkey, automotive supply industry, casting and shoemaking sectors are the forefront also there are important initiatives towards machinery, food and defense industry (https://www.dunya.com/kose-yazisi/konya039da-hizli-bir-gelisme-var/13647, E.T.: 2018).

As in large enterprises, there is a range of activities in medium-sized enterprises ranging from purchasing to production, storage to logistics and maintenance. The Industry 4.0 promises a 10–30% reduction in costs in these areas for factories (Bauernhansl et al. 2016: 9). Thus, a cost reduction will ensure the enterprise gain competitive advantage by making a significant impact on productivity and profitability.

In this context, it is the main objective of this study to determine the awareness and practices of the enterprises operating in Konya, an SME capital city, about this issue.

Within the scope of study, firstly, the studies on Industry 4.0 has been investigated and a questionnaire is created. Questionnaires were filled and evaluated by conducting interviews with two companies operating in Konya using depth interview technique.

The first company is an enterprise which produce cylinder for milling machines along with casting and machining activities and also it is one of the top five enterprises in the international market (Enterprise A). The enterprise has a high investment cost and strong range of products with technology.

The second company is a large enterprise which is one of the pioneers of the food sector in Konya and it follows all the technological developments in the sector and obtains the competitive advantage by R & D activities (Enterprise B).

In the first question, a question to determine whether they have information about Industry 4.0 is asked to enterprises. According to responses, According to the responses, Enterprise A does not have detailed information about Industry 4.0, whereas Enterprise B has various information in the technical and literature context.

In the second question, it is asked to enterprises whether their factories are ready for a end-to-end change. Enterprise A says that hey have implemented innovation activities for product and process improvement and that they use robots in certain parts of the plant to reduce human errors. However, it is understood that the entire manufacturing process is not ready for Industry 4.0 transformation. They point out that in Konya region there are no human resources and necessary technical equipment for this. Enterprise B indicates that they contacted to the related companies for the necessary equipments, they have invested for these and they are getting ready for this radical change.

When they are both asked what benefits they will receive and what damages they will face if they apply this change, both enterprises say that the infrastructure and qualified workforce costs needed in the adaptation process will affect the economic structure of the factories. But on the other hand, they argue that when they shift to this production management system, there will be some benefits that human errors will be minimized, waste rates will decrease, production speed and productivity will increase.

When asked about whether they will survive the transition problem and whether their current employees will be affected by this situation, Enterprise A predicts that this will not affect the blue-collar workers, or even increase the number of technical personnel. Enterprise B also assumes that there will be no employment problems.

Lastly, when the enterprises have been asked whether Industry 4.0 is an opportunity for the customers, both enterprises stated that the increase in quality and productivity and decrease in cost will be a great opportunity for the customers also the satisfactions rates will increase.

5 Conclusions and Recommendations

Technological developments play an important role in underlying an enterprise's production system and shaping its product and process development structure (Richardson 1972: 888). The new technological era in industrialization, which is called Industry 4.0, aims to provide structural changes in production management and enable the machines and equipments to communicate with one another in a faster and more efficient way with better quality and more economical products and processes. In this study, the awareness and applications of Industry 4.0 are evaluated in Konya, which is increasingly getting stronger in manufacturing industry but mostly composed of small and medium sized enterprises.

A questionnaire has been created based on the studies in literature and the interviews made with the enterprises and it has been filled out by interviewing with the persons in the top management of the two enterprises that are strong in the sector and developing with R&D activities. In accordance with the information obtained from interviews, the possible problems that enterprises might encounter during the application of 4. Industrial technology and shown in Table 1. When Table 1 is examined, it is seen that the Enterprise B, which operates in food sector, is ready for this innovation and change also it is carrying out the necessary preparatory work. On the other hand, Enterprise A, which operates at milling machines and is a casting company, does not have enough information about this and therefore it is seen that the factory is not ready for this change.

Table 1. Current situation of enterprises in Industry 4.0 applications

The requirements for Industry 4.0	Current Situation of enterprises	
	Enterprise A	Enterprise B
Need for qualified personnel	Insufficient	Insufficient
Technical equipment	Insufficient	Partially available
Additional investment requirement	Available	Partially available
Orientation of employees to Industry 4.0	Insufficient	Partially available

Obviously, it is seen in the studies and applications of Industry 4.0 in Turkey that medium-sized enterprises do not have enough information about the subject and they think that this transformation is monopolised by the large enterprises. Particularly, not having required human resources in accordance with the development level of regions and the need for large investments make enterprises prejudge about Industry 4.0. While the companies which have proven their existence worldwide such as Siemens, Bosch, Arzum have started pilot scheme at their factories in Turkey, the dissemination of the system will be achieved again with the support of large enterprises and governmental incentives.

In the scope of the study conducted, it is seen that the awareness of middle sized enterprises which are located in our country and have an important role in the development of this country, about Industry 4.0 is insufficient. In this context, it is important that the technology leaders in our country should disseminate their studies and announce the sample of good practices. Universities need to improve their curricula within the context of Industry 4.0 and raise qualified human resources in line with current requirements. It is suggested that the importance of this subject should be discussed in academic studies and theoretical and practical projects in this field should be created.

References

Ashton K (2009) That "Internet of Things" Thing, RFID. http://www.rfidjournal.com/articles/pdf?4986. Accessed 19 Jan 2018

Bandyopadhyay D, Sen J (2011) Internet of Things -Applications and Challenges in Technology and Standardization. https://arxiv.org/pdf/1105.1693.pdf. Accessed 19 Jan 2018

Bauernhansl T et al (2016) Wgp-Standpunkt Industrie 4.0, Wissenschaftliche Gesellschaft für Produktionstechnik

Broy M, Cengarle MV, Geisberger E (2012) Cyber-physical systems: imminent challenges. In: Proceedings of the 17th Monterey conference on large-scale complex IT systems: development, operation and management, pp 1–28. https://doi.org/10.1007/978-3-642-34059-8

Cankat Batı A et al (2017) Gelişmiş Çalışma Ortamı ve Enerji Verimliliği için Nesnelerin İnterneti Bazlı Akıllı Ofis Uygulaması IoT based smart office application for advanced indoor working environment and energy efficiency. In: 2017 25th Signal processing and communications applications conference (SIU), Antalya. https://doi.org/10.1109/siu.2017.7960664

Electric S (2016) Endüstri 4.0 Devrim Değil Evrim, Elektrik Mühendisliği, 459, pp 51–52

Ertin G (1998) Türkiye'de Sanayi, Türkiye Coğrafyası, p 20

Ganzarain J, Errasti N (2016) Three stage maturity model in SME' s towards industry 4.0. J Ind Eng Manag 9(5):1119–1128. https://doi.org/10.3926/jiem.2073

Gubbi J et al (2012) Internet of Things (IoT): a vision, architectural elements, and future directions. Future Gener Comput Syst. https://doi.org/10.1016/j.future.2013.01.010

http://www.kso.org.tr/sayfa/tr/tarihce-1. Erişim Tarihi: 07 May 2018

Kagermann H, Wahlster W, Helbig J (2013) Recommendations for implementing the strategic initiative INDUSTRIE 4.0. Final report of the Industrie 4.0 WG, p 82. https://doi.org/10.13140/rg.2.1.1205.8966

Kotynkova M (2017) Re-industrialization of Europe: Industry 4.0 and the future of work. Eur Sci J 7881:249–256

Lee EA (2008) Cyber physical systems: design challenges. In: 2008 11th IEEE International symposium on object and component-oriented real-time distributed computing (ISORC), pp 363–369. https://doi.org/10.1109/isorc.2008.25

Lee J, Kao HA, Yang S (2014) Service innovation and smart analytics for Industry 4.0 and big data environment. Procedia CIRP 16:3–8. https://doi.org/10.1016/j.procir.2014.02.001

López-de-Armentia J, Casado-Mansilla D, López-de-Ipina D (2012). Fighting against vampire appliances through eco-aware things. In: 2012 Sixth international conference on innovative mobile and internet services in ubiquitous computing (IMIS). IEEE, pp 868–873. https://doi.org/10.1109/imis.2012.112

Ma M, Wang P, Chu C-H (2013) Data management for Internet of Things: challenges, approaches and opportunities. In: 2013 IEEE International conference on green computing and communications and IEEE Internet of Things and IEEE cyber, physical and social computing, pp 1144–1151. https://doi.org/10.1109/greencom-ithings-cpscom.2013.199

Oral O, Çakır M (2017) Nesnelerin İnterneti Kavramı ve Örnek bir Prototipin Oluşturulması', Mehmet Akif Ersoy Üniversitesi Fen Bilimleri Enstitüsü Dergisi, vol 1, pp 172–177, 1309–2243). https://mestek.mehmetakif.edu.tr/eBildiriKitabi/bildiriler/61-564.pdf Accessed 19 Jan 2018

Radziwon A et al (2014) The smart factory: exploring adaptive and flexible manufacturing solutions. Procedia Eng 69:1184–1190. https://doi.org/10.1016/j.proeng.2014.03.108

Richardson GB (1972) The organisation of industry. Econ J 82(327):883–896

Rojko A (2017) Industry 4.0 concept: background and overview. Int J Interact Mobile Technol (iJIM) 11(5):77. https://doi.org/10.3991/ijim.v11i5.7072

Siemens, Boğaziçi Yazılım and Atos (2016) Dijital Fabrikalar, Siemens e-dergi

Turak Y (2015) Nesnelerin İnterneti Ve Güvenliği. http://www.yigitturak.com/wp-content/uploads/IoTGuvenligi.pdf. Accessed 19 Jan 2018

Türkiye'nin Endüstri 4.0 Platformu (2017) Bosch Türkiye ve Boğaziçi Üniversitesi Sanayi 4.0 için Ortak Program Hazırladı, Türkiye'nin Endüstri 4.0 Platformu. http://www.endustri40.com/bosch-turkiye-ve-bogazici-universitesi-sanayi-4-0-icin-ortak-program-hazirladi/. Accessed 6 May 2018

TÜSİAD and Group Boston Consulting (2016) Türkiye'nin Küresel Rekabetçiliği için Bir Gereklilik Olarak Sanayi 4.0

Wang S et al (2016) Implementing smart factory of Industrie 4.0: an outlook. Int J Distrib Sens Netw 2016:1–10. https://doi.org/10.1155/2016/3159805

World IPV6 Launch (no date) This Time It is For Real. http://www.worldipv6launch.org/. Accessed 28 Mar 2018

Xu LD, He W, Li S (2014) Internet of Things in industries: a survey. IEEE Trans Ind Inf 10 (4):2233–2243. https://doi.org/10.1109/TII.2014.2300753

Yiğitbaşı ZH (2011) Nesnelerin İnterneti ve Makineden Makineye Kavramları için Kilit Öncül, IPv6', in Ulusal IPv6 Konferansı, pp 103–108. http://www.ipv6.net.tr/docs/ipv6konf/pdf/15.pdf. Accessed 19 Jan 2018

https://www.dunya.com/kose-yazisi/konya039da-hizli-bir-gelisme-var/13647. E.T.: 07 May 2018

Cultural Aspects in the Adoption of ERP Systems: A Holistic View

Gülay Ekren[1(✉)], Tuğba Koç[2], Alptekin Erkollar[2], and Birgit Oberer[2]

[1] Sinop University, Sinop, Turkey
gekren@sinop.edu.tr
[2] Sakarya University, Sakarya, Turkey
{tcekici, erkollar, oberer}@sakarya.edu.tr

Abstract. Over the past years, Enterprise Resource Planning (ERP) systems have become an essential business driver with its main features such as modularity, real-time data accessibility and sharing an integrated database on the way to Industry 4.0 Many companies have utilized the advantages of integrated solutions of ERP systems offered by multinational corporations. Nowadays, ERP systems have been widely adopted in specially developed countries as well as developing countries. In developing countries, companies usually use national ERP systems instead of foreign-based ERP systems. However, ERP vendors, consultants have offered different global ERP templates containing their multi-language and high qualified customization options, but also have some cultural adoption problems which may exist during the implementation of ERP systems. Thus, ERP systems face many challenges in, especially developing countries including cultural issues. This study focuses on cultural issues in ERP adoption which might be different across cultural contexts. It is aimed to assess the impact of cultural aspects in the implementation of ERP systems via holistic view by using Hofstede's cultural dimensions (power distance, individualist/collectivist, uncertainty avoidance, etc.). Studying such aspects may provide guidelines for the success of ERP system adoption in a global perspective.

Keywords: Cultural aspects · National cultural dimensions · ERP adoption ERP implementation · Hofstede

1 Introduction

Developing countries are faced with many difficulties when applying and using the management processes and information systems (IS) techniques of western technologies imported from developed countries. IS is a kind of socio-technical system as the economy, politics, social factors, education, skill levels and culture are critical factors that influence the adoption and use of these systems. Therefore, global organizations need to become aware of cultural differences in primarily developing countries to deploy their information technology (IT) successfully. Otherwise, these organizations can be failed if the IT designs are not sufficiently tailored to those countries' cultural and industrial norms [1, 2]. Moreover, multinational corporations (MNCs) which conduct their activities across national borders come across with national culture as a

N. M. Durakbasa and M. G. Gencyilmaz (Eds.): ISPR 2018, *Proceedings of the International Symposium for Production Research 2018*, pp. 355–366, 2019.
https://doi.org/10.1007/978-3-319-92267-6_31

significant matter that inhibits innovation through the subsidiaries of MNCs [35]. Some of the main challenges need to overcome cultural issues related to IS, such as (1) current studies ignoring indefinite culture and the culture of the plural structure by executives, (2) the importance of the dynamics and how it affects the adoption of the IS, (3) culture researchers mostly using less sophisticated methods to obtain more profound insights into the cultural environment through thick descriptions [3].

Enterprise Resource Planning (ERP) systems are one of the IS that is consist of "*a set of business applications or modules, which links various business units of an organisation such as financial, accounting, manufacturing, and human resources into a tightly integrated single system with a common platform for flow of information across the entire business* (p. 184)" [4]. ERP systems are increasingly used by many organizations worldwide to improve competitiveness as well as to support their managerial decision-making processes. By developing an integrated enterprise-wide system through an ERP system, many companies replace their existing legacy systems and software applications with the internal business systems which are connected to the systems of customers and suppliers [4, 5]. Most of the ERP vendors provide flexibility for small and medium-sized enterprises (SMEs) to minimize costs as well as future upgrades [6]. However, this effort is not enough for the standardization of ERP systems in global perspective. The standardization of ERP systems should be studied within the context of business processes, suppliers and customers (horizontal integration) as well as within the standardization of ERP software packages (vertical integration) [7].

European headquarters of ERP systems have many problems when implementing an ERP system developed in Western countries into a company of Eastern countries, as it is known that there is no single global ERP template that compatible with all subsidiaries in European, American, African or Asian countries. Avison and Malaurent (2007) provided a case study describing a failed implementation of an ERP project carried out by a French firm in its Chinese subsidiary [8]. This study revealed a set of difficulties that negatively have an impact on the implementation of ERP systems in a cultural context as follows: (1) limited employee involvement (need more time as well as effective communication with local employees), (2) language and communication difficulties (the global ERP template is in English, everyone can't understand the ERP instructions in English, the ERP system usually not support national characters; e.g., SAP R/3 4.5 did not support Unicode system), (3) consistency of local laws and regulations (the unexpected locally reporting obligations related to accounting and bidding processes), (4) strong hierarchy (losing face of local managers; the necessity to respect the company hierarchy), (5) national characteristics (patience required for national culture as local staff can be acted in a fairly passive attitude to expected problems). Moreover, a set of cultural factors that affect the implementation of ERP systems are identified; such as mismatch of globally used technologies with local culture, lack of ownership culture, management of culture, reluctant to change in cultural views, considering cultural fragmentation in the marketplace, regarding the readiness of culture, the existence of multiple subcultures, diversity of information flows, communication culture, sectoral differences, discrimination of gender, and impatience of culture [9].

Boersma and Kingma (2005) developed a cultural perspective on ERP systems and organizations as the implementation of ERP systems should devote attention not only adapting to the software packages but also to the "artefacts" including scanners, PCs, cables, mainframes, interfaces, reports as well as "people" including ERP consultants, programmers and operators [7]. Besides, the implementation of the ERP systems needs several country-specific requirements especially in Asian organizations in terms of data (format, relationship), process (access, control, operational), and output (presentation format, information content) as the business models underlying most ERP packages reflect European or U.S. industry practices [10]. On the other hand, the cultural dynamics and their fit with ERP systems also require investigation in the concept of cultural and linguistic concepts as Gramscian approach can be used to understand ERP system deployment practices regarding language and culture. Since Gramscian approach provides a rich theoretical language to explore language channels and how it limits organizational approaches to ERP implementation [5]. Besides, Rasmy et al. (2005) developed a conceptual model in Egypt to understand the application of large IS such as ERP within the challenging organizational culture. They stated that ERP implementation problems become acuter because of the challenging Egyptian culture which is entirely different from cultures where these systems are developed [11]. This study offers a self-assessment tool for ERP implementers such as MNCs, vendors, and consultants to consider cultural issues when planning for ERP implementation. Additionally, national culture has an impressive effect on ERP implementation for developing countries [12]. A range of social science theories of culture are revealed such as Schein (1992), Hofstede (1994), Trompenaars (1994) to predict the impact of corporate and national culture on ERP implementation. These theories can be applied to explain culture-related problems which arise during ERP implementations [13, 37–39]. Although, some of researchers [14] found Hofstede's cultural framework rather crude and ordinary for studying cultural differences in IS research, some of the others [15, 16] especially suggest this framework to evaluate national culture in terms of globalisation which is placing higher demands on IS design, development, and management.

As culture is one of the critical success factors relevant to ERP systems implementations [12, 17, 18], the research related to cultural issues is still immature. Therefore, there is a need for further research to find out the reasons for the unsuccessful adoption and implementation of ERP systems globally.

The present study examines cultural issues in ERP adoption which might be different across cultural contexts. For the better understanding of ERP systems using across the non-Western cultures, this study aimed to assess the impact of cultural aspects in the implementation of ERP systems via holistic view by using Hofstede's national cultural dimensions, especially for the developing countries context. The rest of this study is organized as; a literature review of ERP systems from the point of Hofstede's national cultural dimensions, followed by research methodology, discussions on cultural differences from European to Asian subsidiaries with regards to ERP implementation and adoption, and finally conclusions.

2 Hofstede's Cultural Dimensions and ERP Systems

Extra skills are required to overcome culturally related difficulties in the adoption of IS globally. Three main cultural dimensions, as organizational culture, national culture and individual culture, must be considered to understand the adoption and the use of IT products in a global perspective [1]. National culture refers the differences in values between groups of nations or regions. On the other hand, organizational culture is based on strategic practices between organizations or parts/subsidiaries within the same organization (https://www.hofstede-insights.com). Hofstede's national cultural dimensions are commonly used to analyze management practices in a cultural context. These dimensions are recognized as follows:

Hofstede's National Culture Dimensions.
Hofstede's national culture dimensions are frequently used to understand cross-cultural differences between countries. These dimensions are such as power distance (PD), individualism/collectivism (IC), uncertainty avoidance (UA), masculinity/femininity (MF), long term/short term orientation (LS), and indulgence/restraint (IR) [19]. According to Hofstede, PD has been defined as the distribution of power in organizations and institutions unequally. Secondly, IC can be described as the integration of individuals into primary groups. Thirdly, UA can be defined as the level of stress in a society in the face of an unknown future. Fourthly, MF can be defined as the division of emotional roles between women and men. Then, LS can be defined as the choice of focus for efforts of people towards past, present or future. Lastly, IR can be defined as the gratification versus control of basic human desires about enjoying life [19]. Index scores of these dimensions get changed according to the various countries as shown in Table 1.

Alshare, El-Masri and Lane (2015) conducted a study to assess the effect of students' culture on learning ERP systems based on Hofstede's national cultural dimensions. The results show that the Hofstede's three cultural dimensions as uncertainty avoidance, masculinity/femininity, and power distance play an important moderating role in explaining students' efforts to learn ERP. The findings also indicated that the efforts of students towards learning ERP software from masculine cultures have stronger than the students from feminine cultures [20]. Besides, the ERP systems' perceived usefulness and the ease of use has stronger effects on attitudes if the students come from high power distance cultures. Additionally, the perceptions of students' with low uncertainty avoidance cultures have more effort expectancy from ERP systems than the students with high uncertainty avoidance cultures. Consequently, the students who come from cultures with high masculinity, high power distance, and high uncertainty avoidance need more emphasis on creating a positive attitude to learning ERP systems as well as the usefulness of learning ERP systems and making easier the use of ERP systems. Additionally, the effect of cultural factors was measured in a study to contribute the usage of off-the-shelf/ERP packages in India by using Hofstede's cultural dimensions power distance, uncertainty avoidance, individualism, and masculinity. As individualism and masculinity are positively correlated with growth in the usage of off-the-shelf/ERP packages [21].

The national culture of the country that the projects take place also affects the selection process of ERP systems. Livermore and Ragowsky (2002) emphasized the

Table 1. Levels of Hofstede's national culture dimensions according to countries [19]

Dimensions	Levels	Countries
Power distance	Small power distance	German- and English-speaking Western countries
	Large power distance	East European, Latin, Asian and African countries
Individualism/collectivism	Low collectivism, high individualism	Developed countries, Western countries
	Neither collectivism nor individualism	Japan
	High collectivism, low individualism	Developing countries, Eastern countries
Uncertainty avoidance	Weak uncertainty avoidance	English-speaking Western countries, Nordic and Chinese culture countries
	Strong uncertainty avoidance	Eastern- and Central Europe, Latin countries, Japan and German-speaking countries
Masculinity/femininity	Low masculinity, high femininity	Nordic countries, Netherlands, some of the Latin and Asian countries (e.g., France, Spain, Portugal, Chile, Korea, Thailand)
	High masculinity, low femininity	Japan, German-speaking countries, and some Latin countries (e.g., Italy and Mexico), some of English-speaking Western countries
Long term/short term orientation	Long-term oriented	East Asian countries as well as Eastern- and Central Europe
	Medium oriented	South- and North-European, South Asian countries
	Short-term oriented	U.S.A., Australia, Latin American, African and Muslim countries
Indulgence/restraint	High indulgence, low restraint	South and North America, Western Europe and some of the Sub-Sahara African countries
	Neither indulgence nor restraint	Mediterranean Europe
	Low indulgence, high restraint	Eastern Europe, Asia, and Muslim countries

cultural aspects of the ERP selection process based on Hofstede's national dimensions and stated that culture makes a difference in ways that are predicted by the Hofstede's model as if one country's values are high on a particular dimension, the decisions of managers on the selection and implementation of ERP systems is consistent with expected culture profile [22].

Besides his national culture framework, Hofstede also defined a framework for organizational cultures. According to a research conducted by Hofstede (2011), the organizational culture of companies can be explained by his multi-focus model. This model provides insights into the organization's actual culture (actual way of working) and organization's strategic direction towards change. Six cultural dimensions of this model are; (1) the effectiveness of the organization (means-oriented vs. goal-oriented), (2) business ethics (internally-driven vs. externally-driven), (3) internal control (easy-going work discipline vs. strict work discipline), (4) social control (local vs. professional), (5) the accessibility to the organization (open system vs. closed system), (6) management philosophy (employee-oriented vs. work-oriented) [19].

3 Research Methodology

In this study, an extensive literature review on the implementation of IS, particularly ERP systems in different cultural contexts was conducted on relevant journal articles, conference proceedings, books from online repositories as well as in case examples provided by companies focusing on key themes as "culture" and "ERP systems." The review first examined the literature on the adoption of IS in various cultures. However, the primary focus of this study aimed to discover the cultural factors that affect the ERP implementation. The research steps are as follows: (a) an extensive literature review on ERP implementation in different cultural contexts, (b) identifying the Hofstede's national cultural framework according to countries, (c) examining the relationship between Hofstede's cultural framework and ERP implementation over the case studies based on literature review, and (d) identifying the similar themes between success and failure cases. The case studies are chosen according to the following reasons: (1) all cases must reflect a range of misfits over ERP implementation practices, and (2) all cases must lay emphasis on the difficulties when applying and using ERP systems in their cultural context based on Hofstede's culture dimensions.

Business and management practitioners, trainers, and researchers widely used the Hofstede as well as GLOBE (the expanding model of Hofstede's national culture dimensions) national culture dimensions to understand cultural differences between countries, however, Hofstede's cultural framework is the most popular one providing a standardized template for facilitating analysis of cultural related management behaviour [23]. Besides, Hofstede's national cultural framework is the most widely cited in existence and also seen as the most significant cross-cultural study of work-related values. Hence, the observations and analysis provided by this framework support the dynamics of cross-cultural relationships [16, 24]. For this reason, Hofstede's national culture is the dominant cultural framework to explain cross-cultural differences in the adoption of ERP systems.

4 Cultural Adoption to ERP Systems

The success and failure of western-designed ERP systems not only depend on technical issues of hardware and software but also cultural and social issues such as language and communication specificities, economic and structural reasons [8]. According to Hawking (2007), the global ERP template includes standardized definitions of organizational structures, business processes, and master data. Although they have been increasingly using large companies, the use of these templates locally has lack of flexibility to take advantages of regional opportunities due to the cultural differences [25]. Thus, European headquarters of ERP systems aspire to standardize a global template conform with possible local settings as well as cultural differences from European to Asian subsidiaries (Fig. 1).

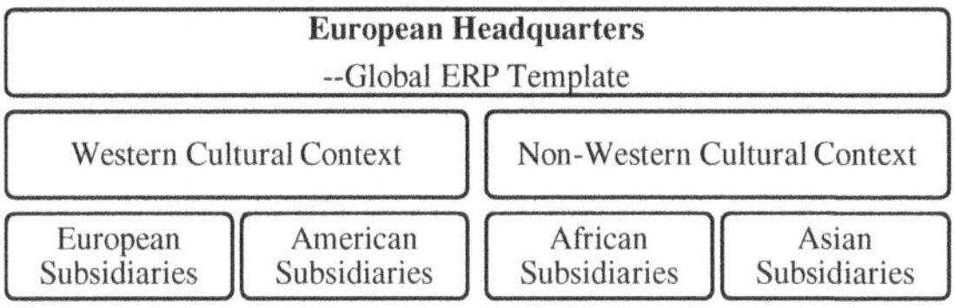

Fig. 1. Cultural aspects of a global ERP template [8]

4.1 ERP Adoption Within the Western Cultural Context

Hofstede (2010) characterized most of the European nations as being large in power distance, weak in uncertainty avoidance, low in masculinity, high in individualism, and long-term oriented. Van Everdingen and Waarts (2003) investigated the effects of Hofstede's national culture dimensions (PD, UA, IC, MF, LS) to explain cross-country differences in innovation adoption rates based on a survey covering 2647 medium-sized firms from 10 European countries (Belgium, Denmark, Finland, France, Italy, Netherlands, Norway, Spain, Sweden, and the United Kingdom). Results indicate that national culture directly influenced the ERP adoption rates. Although individualism positively influenced the early adoption of ERP systems, the other dimensions influenced both early and later adoptions. On the other hand, higher levels of the UA, PD and masculinity dimensions in a country negatively affected ERP adoption, while higher levels of long-term orientation have a positive influence [29]. Similarly, Waarts and Van Everdingen (2005) indicated that European countries are highly individualistic, accepting medium levels of power, rather feminine and uncertainty avoiding as well as focusing short-term orientation [30]. However, it is essential to building a long-term partnership with the company to understand the company's culture and industrial norms by the vendors and implementation partners [31].

4.2 ERP Adoption Within the Non-Western Cultural Context

Hofstede (2010) characterised most of the Asian nations as being high in uncertainty avoidance, power distance, and masculinity, but low in individualism. The implementation issues of ERP systems in Asian countries context have been trying to be defined for years. Hawking (2007) identified global implementation issues into Asian countries through content analysis. The data was collected from 2000 industry presentations focused on global implementations of SAP user group conferences, events and industry forums from 2000 to 2006 conducted of both locally and globally. The findings show that the major issues are language, culture, and currency [25]. On the other hand, eight differences between the Chinese ERP implementation of four state-owned enterprises and four private ventures as shown in Table 2:

Table 2. ERP implementation differences in China [26]

ERP implementation differences	State-owned enterprises	Private ventures
#1 Primary project aims	To increase consistency and reduce costs like overhead	Continuous effort to enhance their supply chain management and adopt the best business practices
#2 Role of top management	Reluctance to involve the ERP project	Hands-on leadership to demonstrate a commitment
#3 Role of the cross-functional steering committee	The more centralized management structure	More frequent meetings and a sharper focus on problem resolution
#4 Role of consultants	Rarely engaged outside help	more likely to hire consultants
#5 Scope of implementation	The application of ERP moderates the functional conflicts between departments	Broader and more cross-functional ERP application
#6 Pace of implementation	Resistant to change, incremental and continuous implementation	Faster and less gradual implementation
#7 Implementation problems	more data maintenance problems after adopting the ERP system	Less frequent and less severe problems
#8 Evaluation and outcomes	Less systematic evaluation and control, cost reductions are common, but quality or supply chain enhancements are rare	More systematic evaluation and control as well as achieve more substantial quality and supply chain improvements

Chinese organizational culture is a significant determinant of ERP implementation success. Thus, China's enterprises and Western-based ERP vendors have to adapt business processes and business models of their ERP systems in conformable with Chinese national culture which is embedded in their organizational culture [27]. Based on Hofstede's (2001) organizational culture dimensions including parochial vs.

professional, open vs. closed system, and loose vs. tight control, Zhang et al. (2005) stated that "Chinese people are more tolerant to unclear information, relying more on personal experience, keeping more information among themselves than their Western counterparts. However, the ERP system deployment requires clear and accurate data/information focuses on business processes and inter-department cooperation, which is incompatible with Chinese organizational culture. Thus, to obtain ERP systems implementation success, Chinese enterprises should take their organizational culture into account and try to change their culture to the modern management requirements regarding the three dimensions particularly (p. 71)" [27]. On the other hand, Ge and Voß (2009) indicated that Chinese domestic ERP vendors have a comparative advantage on the ERP implementation against MNCs because of some reasons such as; (1) The Chinese language uses pictographic symbols as each character represents an object or concept, so it is difficult to translate an ERP system from English to Chinese accurately and thoroughly. Foreign ERP systems usually have a mixture of English and Chinese interfaces as well as completely in English help files that confused Chinese users, (2) China's accounting standards are dissimilar to international accounting standards so foreign vendors have to modify their financial accounting modules to generate the correct formats to meet local requirements, (3) foreign vendors can be forced to provide adequate and comprehensive customer support on time [28].

On the other hand, Hofstede (2010) characterized African nations as being large in power distance, low in individualism, high in restraint, and short-term oriented. El Sawah et al. (2008) present a self-assessment tool for ERP implementers in Egypt to improve the understanding of the Egyptian organizations on how to implement ERP systems within the challenging organizational culture. The results show that Egyptian organizational culture has a negative impact on ERP implementation success [36]. Additionally, the impact of organizational culture are examined on the adoption of IS (e.g., ERP systems) in Libya. According to findings, organizations in Libya are not ready to accept and adopt IS [32].

Rajapakse and Seddon (2005) defined the factors affecting the implementation of ERP systems in Asian countries (e.g., Sri Lanka) such as low adoption rates particularly in national and organizational cultural issues, high costs for organizations, and limited national infrastructure (e.g., telecommunication standards). According to this study, the adoption problems of ERP systems can be explained by cultural differences between Western countries and Asian countries through the two national cultural dimensions of Hofstede as "power distance" and "individualism/collectivism" [15]. This study concentrated only two dimensions of Hofstede (PD, IC) which are thought to be leading the cultural problems in the implementation of western-based ERP systems in Asian countries. Supportive evidence is summarized in Table 3.

Table 3. The impact of power distance and individualism on the adoption of ERP systems [15]

Cultural dimensions	Supportive evidence for Asian Countries
Power distance	Centralised decision making as subordinates highly depends on superiors, less open in discussions, operational-level staff needs high supervision as they are not disciplined and must be followed up, expecting others to do a better job while having poor performance, lack of accountability, absorbing only incremental changes, not drastic ones
Individualism	Poor performance as a result of high numbers of staff as well as long service histories, the staff is reluctant to take additional responsibilities, lack of self-learning, unwilling to accept new technologies, changes are not welcomed

5 Conclusions

Most of ERP systems are being used in the developing countries provided not only by domestic companies but also by MNCs. From management and information systems' perspective, because of the complexity of ERP systems and the process of adoption to these systems are precluded in concern with the lack of long-term strategies, process management, and project experience as well as low IT maturity and small firm size [33]. Moreover, ERP systems are not only a software package but also consist of business models or practices that reflect the Western (European or American) business processes. Therefore, ERP developers, consultants, vendors and MNCs primarily need to adapt their products and services to different national contexts as well as to different types of organizations within a single country [26].

Many researchers have emphasized Hofstede's cultural dimensions which impact on the successful implementation of ERP systems. However, there is still a need in IS field for an understanding of how national culture affects the development, usage, and management of the ERP systems [34]. Enterprises have to adjust their strategies on the adoption of new products and systems according to their countries' traits [29]. Besides, ERP implementers such as MNCs, vendors, and consultants have to consider cultural issues when planning for ERP implementation.

This study revealed that the ERP system adoption or implementation problems within the context of the national culture in western countries (as the birthplace of ERP) and in non-western countries are not similar. Hofstede's national cultural framework may be the appropriate vehicle to explain or to solve these problems. This study may provide guidelines for the success of ERP system adoption in a global perspective.

References

1. Talet N, Al-Wahaishi S (2011) The relevance cultural dimensions on the success Adoption and Use of IT. In: 3rd international conference on advanced management science. IPEDR, vol 19
2. Kummer TF, Leimeister JM, Bick M (2012) On the importance of national culture for the design of information systems. Bus Inf Syst Eng 4(6):317–330
3. Jackson S (2011) Organizational culture and information systems adoption: a three-perspective approach. Inf Organ 21(2):57–83
4. Beheshti HM (2006) What managers should know about ERP/ERP II. Manag Res News 29 (4):184–193
5. Willis R, Chiasson M (2007) Do the ends justify the means? A Gramscian critique of the processes of consent during an ERP implementation. Inf Technol People 20(3):212–234
6. Ruivo P, Johansson B, Oliveira T, Neto M (2013) Commercial ERP systems and user productivity: a study across European SMEs. Procedia Technol 9:84–93
7. Boersma K, Kingma S (2005) Developing a cultural perspective on ERP. Bus Process Manag J 11(2):123–136
8. Avison D, Malaurent J (2007) Impact of cultural differences: a case study of ERP introduction in China. Int J Inf Manag 27(5):368–374
9. Zaglago L, Apulu I, Chapman C, Shah H (2013) The impact of culture in enterprise resource planning system implementation. In: Proceedings of the world congress on engineering, vol 1, pp 516–521
10. Soh C, Kien SS, Tay-Yap J (2000) Enterprise resource planning: cultural fits and misfits: is ERP a universal solution? Commun ACM 43(4):47–51
11. Rasmy MH, Tharwat A, Ashraf S (2005) Enterprise resource planning (ERP) implementation in the Egyptian organizational context. In: Proceedings of the EMCIS international conference
12. Asemi A, Jazi MD (2010) A comparative study of critical success factors (CSFs) in the implementation of ERP in developed and developing countries. Int J 2(5):99–110
13. Krumbholz M, Maiden NAM (2000). How culture might impact on the implementation of enterprise resource planning packages. In: International conference on advanced information systems engineering. Springer, Heidelberg, pp 279–293
14. Gurung A, Prater E (2017) A research framework for the impact of cultural differences on IT outsourcing. In: Global sourcing of services: strategies, issues and challenges, pp 49–82
15. Rajapakse J, Seddon P (2005) ERP adoption in developing countries in Asia: a cultural misfit. In: 28th information systems seminar in Scandinavia, Kirstiansand, pp 6–9
16. Jones ML, Alony I (2007) The cultural impact of information systems–through the eyes of Hofstede–a critical journey
17. Shah SIH, Bokhari RH, Hassan S, Shah MH, Shah MA (2011) Socio-technical factors affecting ERP implementation success in Pakistan: an empirical study. Aust J Basic Appl Sci 5(3):742–749
18. Annamalai C, Ramayah T (2013) Does the organizational culture act as a moderator in Indian enterprise resource planning (ERP) projects? An empirical study. J Manuf Technol Manag 24(4):555–587
19. Hofstede G (2011) Dimensionalizing cultures: the Hofstede model in context. Online Read Psychol Cult 2(1):8
20. Alshare KA, El-Masri M, Lane PL (2015) The determinants of student effort at learning ERP: a cultural perspective. J Inf Syst Educ 26(2):117

21. Agrawal VK, Haleem A (2005) Environmental pressures, culture, and factors contributing to the usage of various categories of application software. Glob J Flex Syst Manag 6(2):31
22. Livermore C, Ragowsky A (2002) ERP systems selection and implementation: a cross-cultural approach. In: AMCIS 2002 Proceedings, p 185
23. Venaik S, Brewer P (2016) National culture dimensions: the perpetuation of cultural ignorance. Manag Learn 47(5):563–589
24. Erumban AA, De Jong SB (2006) Cross-country differences in ICT adoption: a consequence of culture? J world Bus 41(4):302–314
25. Hawking P (2007) Implementing ERP systems globally: challenges and lessons learned for Asian countries. J Bus Syst Gov Eth 2(1):21–32
26. Martinsons MG (2004) ERP in China: one package, two profiles. Commun ACM 47(7): 65–68
27. Zhang L, Lee MK, Zhang Z, Banerjee P (2003) Critical success factors of enterprise resource planning systems implementation success in China. In: Proceedings of the 36th annual Hawaii international conference on system sciences, 2003. IEEE, pp 10-pp
28. Ge L, Voß S (2009) ERP application in China: an overview. Int J Prod Econ 122(1):501–507
29. Van Everdingen YM, Waarts E (2003) The effect of national culture on the adoption of innovations. Mark Lett 14(3):217–232
30. Waarts E, Van Everdingen Y (2005) The influence of national culture on the adoption status of innovations: an empirical study of firms across Europe. Eur Manag J 23(6):601–610
31. Sun H, Ni W, Lam R (2015) A step-by-step performance assessment and improvement method for ERP implementation: action case studies in Chinese companies. Comput Ind 68:40–52
32. Twati JM, Gammack JG (2006) The impact of organisational culture innovation on the adoption of IS/IT: the case of Libya. J Enterp Inf Manag 19(2):175–191
33. Huang Z, Palvia P (2001) ERP implementation issues in advanced and developing countries. Bus Process Manag J 7(3):276–284
34. Ford DP, Connelly CE, Meister D (2009) Hofstede's dimensions of national culture in IS research. In: Handbook of research on contemporary theoretical models in information systems, pp 455–481
35. Efrat K (2014) The direct and indirect impact of culture on innovation. Technovation 34 (1):12–20
36. El Sawah S, Abd El Fattah Tharwat A, Hassan Rasmy M (2008) A quantitative model to predict the Egyptian ERP implementation success index. Bus Process Manag J 14(3): 288–306
37. Schein EH (1992) Organisational culture and leadership. Jossey-Bass Publishers, San Francisco
38. Hofstede G (1994) Cultures and organisations, intercultural co-operation and its importance for survival. Software of the mind, author of culture's consequences. Harper Collins, London
39. Trompenaars F (1994) Riding the waves of culture: understanding cultural diversity in business. Nicholas Brealey Publishing, London

Defining the Pros and Cons of Cloud ERP Systems: A Turkish Case

Tuğba Koç[1(✉)], Gülay Ekren[2], Birgit Oberer[3], and Alptekin Erkollar[1]

[1] Management Information System, Sakarya University, Sakarya, Turkey
{tcekici,erkollar}@sakarya.edu.tr
[2] Sinop University, Sinop, Turkey
gekren@sinop.edu.tr
[3] Business Administration, Sakarya University, Sakarya, Turkey
oberer@sakarya.edu.tr

Abstract. In recent years, ERP systems have provided a competitive advantage as well as their increasing significance for companies in a rapidly globalizing world. Accordingly, changing technologies and trends in the world are closely followed in Turkey, too. Traditional products have started to give their ways to cost-effective products, mainly provided by rental models via cloud technologies. This transformation process is first being developed and popularized by big players, and then a range of local solutions have arisen. Mobile technologies and IoT solutions via cloud technologies also positively drive this change. There are different types of ERP systems such as traditional, cloud-based, hosted or on premise. Cloud-based ERP solutions are becoming more attractive with increasing features. Currently, the overall resources of companies can be integrated through cloud-based ERP. Therefore, many companies have adopted these systems. There is a range of vendors offering ERP solutions in Turkey. Most of these vendors introduce themselves as successful players on the progress of the ERP sector in Turkey. The present study attempts to analyze the cloud-based ERP solutions of these vendors and to find out the pros and cons of cloud-based ERP systems in Turkey. For this purpose, interviews, as common qualitative tools, will be conducted on convenient samples to collect data and to gather and to explore insights as well as perceptions of cloud consultants and project managers of ERP vendors. Then, obtained data will be identified and clarified for SMEs which are concerned with ERP systems. The findings of this study are aimed to provide meaningful insights of consultants and project managers of cloud-based ERP systems. The present study is expected to share experiences on the progress of the ERP sector as well as the cloud-based ones in Turkey towards the way to Industry 4.0. Finally, some recommendations will be given to SMEs to support them on their journey towards cloud-based ERP.

Keywords: Cloud-based ERP · ERP · SMEs · Turkey · Qualitative study

N. M. Durakbasa and M. G. Gencyilmaz (Eds.): ISPR 2018, *Proceedings of the International Symposium for Production Research 2018*, pp. 367–378, 2019.
https://doi.org/10.1007/978-3-319-92267-6_32

1 Introduction

Enterprise Resource Planning (ERP) systems, also known as integrated management systems, are a disruptive innovation which aims to map all the processes and data within the companies. Although there exist some debates among the scholars of management information systems about "What ERP exactly refers?" [1, 2], as reference [3] advocated that ERP systems are the most critical deployment of information technology (IT) among firms, and considered as the top list of IT-enabled business innovations [4]. Additionally, the primary aim of ERP systems is to integrate main functions of the firms such as accounting management, financial management, customer relationship management, human resource management, supply chain management, and also they supports the other activities like risk management, collaborative working, IT governance. As reference [5] stated that such systems could integrate both resources and information systems by the execution of internal transactions and advanced planning via the real-time data access opportunity.

These systems are usually expected to bring significant opportunities by offering rapid business solutions to firms of all sizes [6–8]. In [9], a case study was conducted in one of the largest Dutch publisher, named as SDU Company, to observe their stages throughout the ERP implementation process. At the end of the project, SDU achieved to adapt the ERP system, and the findings show that, as time progressed, paralleling with the decreasing resistance, operational efficiency started to improve and business process improvement became clear. [10] claimed that to get the most significant benefits from any ERP system, it must be planned a high level of organizational changing, before. In other words, if a manager wants to have an ERP only can be able to control his/her employees easily, then it is hard to expected to get increasing rate of return on investments (ROI) or attaining competitive advantage or bettering the reorganization of internal resources.

Even ERP offers so many advantages to the firms, selecting and implementing the right one is not only crucial but also a complex process. Some researches [10, 11] demonstrated how selection and then adoption of ERP systems could be a compelling task in which various parameters and managerial competencies are required to reach the best fit between the business priorities and the system itself. [12] reviewed 154 articles and conducted a meta-analysis and find out that as well as the literature agrees upon that; without successful implementation, potential benefits of the system would not be forthcoming. [13] emphasized that many companies dedicated themselves to select and to implement an ERP system without evaluating the effectiveness of the adopted ERP system. There are plenty of factors affecting the implementation process which are called as Critical Success Factors (CSFs) in the literature, however the main reason for the failure of the system is the lack of organizations being able to properly align existing business processes with the processes of packaged ERP systems [6, 14]. In [15], a survey was conducted to reveal the basis of successful ERP adoption among the European mid-size companies. It is indicated that four Scandinavian countries and The Netherlands, which are recognized as having high-level individualism, a small power distance, low level of masculinity and low level of uncertainty avoidance led to the greatest level of ERP adoption [16]. The results show that the most important factor for selecting the

ERP systems is the best fit with current information system as well as the flexibility of the offered system. Cost, user-friendliness, scalability and the supplier support are the other essential selection criteria.

As the customization and scalability features of a system increase, the possibility of fitting with the existing information systems will also have an increase. Ever since the beginning of ERP systems in the early 1990s, companies have struggled to balance high costs because of the customized features of their systems. Early on, the sector leaders like SAP and ORACLE only offer an on-premises model which has fully integrated modules with an integrated database but suffers from other important issues. As a wide variety of IT services move to online, on-premises model has started losing its reputation, but it still offers a best-practice for some companies. In our study, we have tried to give information about a third model called "cloud-based ERP solutions" and its differences from "on-premises" and "hosted" models. After a literature-based discussion on cloud ERP systems for SMEs, a range of suggestions will be provided according to the interviews conducted with the managers of cloud-based companies in Turkey.

2 Traditional ERPs vs. Cloud-Based ERPs

In spite of the benefits potentially offered by ERP systems [9, 17, 18], the evaluation issue plays an essential role regardless the company size; during the planning phase it is critical for companies to figure out whether a specific ERP system fits their business practices. In the same line, one of the most misleading expectations of traditional software systems is that the company hopes to gain instant value from the use of the software application as soon as it is installed [19]. When it comes to the first wave of the development of ERP systems, it may be a legacy to get benefit from on-premises ERP just after it is installed, however some researchers [20, 21] stated that companies have still more tendency for implementing on premise ERP systems rather than hosted or cloud-based ones.

The roots of the cloud-based ERP systems are originated from the term cloud computing [22], and has three main layers as Software as a Service (SaaS), Platform as a Service (PaaS), and Infrastructure as a Service (IaaS). Even though there is no universal definition of cloud computing [23], some common characters such as virtualization, agility, scalability, flexibility, need-based configurability, pay-as-you-go can be located in cloud-based systems [24]. According to [25], the most significant feature that distinguishes cloud-based ERP services from others is; intending not to buy the whole system but having the only relevant components and using the pay-and-go basis for fulfilling the companys' needs. With the flexibility and scalability of cloud-based ERP, the system can be accessed via the Internet for a wide variety of users in much lower costs. In this sense, it becomes also easier to customize the new system according to companies' requirements as well as the implementation times which take as far as a few months. Table 1 shows a brief comparison for different types of ERP systems:

Table 1. ERP systems deployment [26]

	On-premises	Hosted	Cloud-based
Implementation size	Large	Medium	Small to medium
Solution complexity	High	Medium	Low
Capital costs	High	Medium	Low
Operating costs	Low to medium	Medium	Medium
Implementation time	12–36 months	9–18 months	4–8 months

According to Table 1, choosing the right deployment method would enable the company to fully take the advantage for their IT investment. As [15] also analyzed as well as [20] claimed that the selection process of ERP systems consist of a combination of functionality, cost, implementation speed, integration capabilities with the existing systems, as well as company size, individual needs, support of managers or/and suppliers and so other sectoral/company-specific factors. Choosing the most suitable ERP system is a radically complex and uncertain process [27], and also a challenging process. However, there is one more option by blending the systems. For instance, [7] suggested that the hybrid model is expected to dominate the IT strategy of firms' in the future. In general, companies are willing to remove their non-critical components into the cloud whereas the primary ones are located in on premise ERP database. Besides, cloud-based ERP systems generally preferred to support business activities [28]. Even if the cloud is still regarded as supporting tool with some security and data-lock problems [29, 30]; however low initial costs is the greatest advantage of cloud. With the limited IT resources, the cloud supports firms by focusing their specialization area and by creating value-based services. Thanks to the benefits of cloud computing such as flexibility and standardization, any firm can easily implement cloud-based ERP system and integrate it into their existing system with minimum effort and complexity. Cloud technologies may be an opportunity for enterprises before deciding to implement and adopt ERP systems, in case their advantages and disadvantages are evaluated correctly.

3 Cloud-based ERPs and SMEs

ERP systems have become inevitable not for only enterprises which have complex processes but also for small and medium sized enterprises (SMEs), since all of them have to manage their data and have to provide solutions for their customers as soon as possible. Although the effective use of business information is a strategic goal for companies of any size, nowadays most of the ERP systems available on the market are too expensive for the financial capabilities of smaller companies [10]. [15] emphasized that, besides the large companies have already adopted themselves with the ERP solutions, the midsize market is also catching up.

Most companies have distributed workforce and it is important to provide them one common platform that all employees can reach and manage their work. After having the opportunities for low-cost implementation, customer-based integrated services become most popular for SMEs as they select SaaS-based ERP systems instead of traditional ones [31, 32]. [31] also stated that meeting specific business needs and providing reliable, high-speed connectivity via Internet are also other factors that have driven SMEs toward SaaS-based ERP models. [33] conducted a survey to investigate different companies from different countries to discover their step to modern ERP solutions. Their results revealed that on premise ERP models are dominantly located, however, majority of the companies tend to move hybrid and cloud ERP models by 2025. Findings also show that the main constraint for moving to any cloud ERP system is the lack of clarity on the roadmap of vendor/solution partner. Figure 1 shows a comparison of determinants for cloud-based ERP systems according to implementation size and systems complexity of companies:

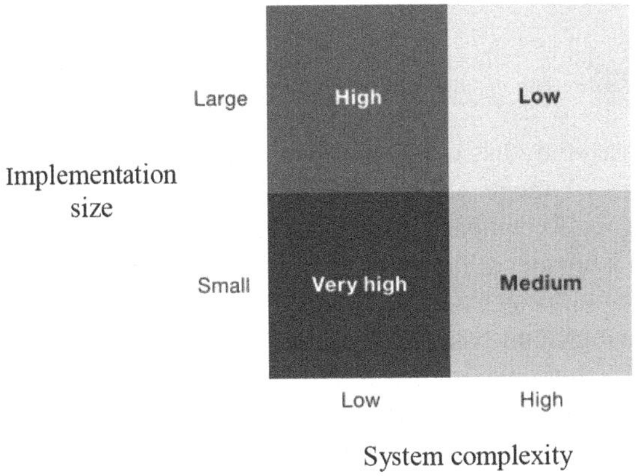

Fig. 1. Determinants of successful cloud-based ERP system [26]

In Fig. 1, implementation size refers to the application area of the selected system which means that how many processes are expected to be improved via the existing system; system complexity indicates the amount of functionality and the extent of integration. Companies with low system complexity and small implementation size are considered potentially the most profitable ones to select the cloud-based ERP systems. According to [10], there is a relationship between the size and the organizational complexity of companies during the selection of ERP systems. Therefore, it can be said that cloud-ERP solutions are highly suitable for SMEs which are fragile during economic crises, and the trepidation destructively, and trying to avoid themselves from making initial IT investments before feeling sure that their investments will return a positive financial value [34]. In contrast with the large-sized companies, SMEs operate with limited IT budget so that, cloud-based ERP systems are an opportunity for them by providing low costs for implementation and maintenance [35]. As [36] draw attention to the fact that there is still a lack of investigation on the impact of SaaS business application adoption on the firm performance of SMEs. They argued that the cloud-based solutions are highly recommended for SMEs, since the ROI is a more critical issue for them in comparison with larger enterprises.

Besides the above-mentioned benefits, it has to be stated that selecting and adoption of cloud–based ERP system is a challenging process for SMEs as well as for larger ones, since the cloud services are still young and consists of immature complexity in its own

[21, 37]. For instance, [29] stated that although it is less costly to own cloud-based services, it is so much challenging to remove errors in these very large-scale distributed systems. Another study's results showed that customization in cloud potentially brings a more significant challenge than in on premise ERP system [38]. Another drawback in cloud ERP system is; not having total control over the system and being dependent on the Internet for running the business processes [39].

Cloud technologies are thought to carry the companies' one step forward; but also they may impair the ability of the companies to innovate [40]. Therefore, the companies should know '*what they need*' and must tailor their business processes regarding their unique requirements.

4 Methodology

In this study, a case study has conducted in a company in Turkey which uses cloud-based ERP systems. A range of semi-structured interviews are carried out with four personnel working in this company as a general manager, an employee, a consultant and an intern. To gain a broader perspective, it is aimed to reach different operating levels. An aforementioned company named as Company 1 to provide anonymity. This company is a single case driving cloud-based ERP sector in Turkey, and keeps living for 26 years. According to [41], interviews in a case study are used to generate facts opinions and insights; since it is preferred to use interviews as a qualitative measure in our study.

Company 1 was established in 1992, as the first Microsoft distributor in Turkey. In addition to Microsoft products, they began to work with the different kind of software as well as hardware products in different brands in addition to their accessories and add-ons. They are still working across three different channels as Retailers, Distributors, and Corporations by the year of 2004. In 2014, they decided to invest data centers to catch the cloud technology trends all over the world. They also signed a contract with SAP in SAP Business One segment in 2016, as a first 'distributor model' in the world which means that they can train and host their technical employees to serve their own customers. Thus customers have the opportunity for 7/24 technical support instead of buying a new one. Then, Company 1 has become the only official place which train cloud or on premise consultants. Company 1 is a unique business in Turkey, not only in ERP training sector, but also in the industry of cloud-based ERP services. Besides SAP Business One, Company 1 also supports Microsoft Dynamics NAV and Microsoft Dynamics 365 solutions of more than 7000 business partners.

During the semi-structured interviews, the questions categorized in three main titles as (1) Cloud-based ERP readiness, (2) Organizational challenges for cloud-based ERP systems, (3) Future of cloud-based ERP systems. These titles are inspired from a research conducted by [30]. All interviews are digitally recorded. Table 2 shows the demographics of the participants, including the duration of each interview.

Table 2. Overview of participants

	Age	Sex	Position	Experience in sector	Duration of interview
Participant 1 (P1)	53	Male	General manager	More than 25 years	73 min
Participant 2 (P2)	41	Male	Project coordinator	More than 10 years	30 min
Participant 3 (P3)	25	Male	Consultant	2 years	13 min
Participant 4 (P4)	23	Male	Intern	3 months	18 min

P1 is an senior general manager in Company 1 who is worked in another company with several positions such as the manager of sales teams, creating and organizing reseller channels, and the presidency of sales group. He has been in various sectors before such as software development, and media sales marketing. Recently, he is the head of the cloud platform of Company 1. He is dealing with deciding the cloud strategies of the company, organizing the employees including consultants and making decisions for future investments. Then, P2 is an experienced project coordinator who responsible for all the possible project collaborations with SAP. The greatest advantage of having an interview with him is to get information about the technical issues of cloud platform deeply, since he is working for technical side. P3 and P4 are relatively young and inexperienced employees. P3 has not too much experience in ERP sector; however he is an SAP consultant. P4 is a long-term intern mostly performs in business intelligence side of the company which is one of the critical dimensions of cloud-based ERP systems.

5 Findings

Once the company has decided to have an ERP system or maintain it, the most crucial question is about **ERP readiness** level of employees and technical systems. Since implementing an ERP system is expensive and time-consuming process. Thus it is important to be sure that company is ready to accept and adapt it, before making any ERP investment [42]. P1 mentioned that he has already thought about doing something like cloud 15 years ago. *"I intended to integrate all the work files and customer information which are used by one of my financial advisor friend in his office. Unfortunately, since the internet connection provided by modem and any other infrastructure problems made it impossible."* In general, all participants agreed that the cloud is not a novel technology, but also it is still immature one. E-Mail services that we have used for a long time can also be evaluated as cloud technology. However, a few believe cloud-based ERP is closely related to this phenomenon. P2 draw attention to a misunderstanding about the security concerns. *"People believe that if they choose a cloud-based ERP system, their critical data may hand over to someone else's hand. When they feel like that, I always ask them the same question: 'Are you sure about your emails are only seen by you? If you easily say yes, then the cloud will be the same for you. Go ahead.'"* P1 also agreed to this idea and expressed *"In fact, the cloud is the commercial version of mailing. If anybody doesn't hesitate to you use mail services, it has to be same for the cloud. Since the cloud idea is new for Turkey adaptation may take time and persuading the people who will use this technology is challenging."* Also, P3 similarly expressed

"Some of our customers even do not know what cloud means. Think about how is difficult to explain the potential opportunities of cloud offers to them and convince to transform their current system. According to their side, works are already on their way, so there is no need to change anything." All interviewers believe that cloud-based ERP systems are beneficial for all sectors and address every size of business (IT democracy). However, P1 admitted *"In comparison with other sectors, the manufacturing sector is the most abstainer. The manufacturer produces and wants to see his product. He wants to touch it and wants to feel belonging. For instance, he feels ineligible if he does not have servers in his company. Thus, it becomes more difficult to adapt manufacturing sector for cloud-based systems than other ones."*

We asked our participants about what do they think about the **challenging factors of cloud-based ERP systems**. In Gartner 2013 report it has stated that the cloud vendors moved slowly to accept the cloud technologies [43]. The same situation is still same for Turkey nowadays. All of our interviewers complained about the unwillingness technology companies. They all agreed that firstly the distributors have to aware of the importance of the topic and try to inform their customers for using cloud-based ERP. When P3 talked about his customer experiences: *"When a company does not have an IT department on its own, they usually tend to outsource it. Unless any technology firm mentions about cloud technologies to these non-IT firms, how they will ensure to get more benefit having a cloud-based ERP? Unluckily, in Turkey, most of the technology firms avoid recommending cloud-based systems to their customers, since they want to earn lump sum at one time instead of earning by installments."* About this issue, P1 commented: *"Actually, once you sell your cloud-based ERP to one customer then it becomes your member and company's network begins to take shape. As your customers use the system, you also continue to earn. What it means that the profit of the technology firms (distributors) exponentially increase depending on the number of their customers. In this point, our company's mission is leading the technology firms about their profit will increase more over a long period if they choose cloud-based ERP and recommend it to their customers."* According to P2, there are two other main problems that have to be dealing with. One of them is infrastructure problems. There exists no successful cloud-based ERP implementation without high speed and continuous internet connection. The other one is legal regulations; however, this issue is not a big problem for SMEs. Contrary to general belief [44], P1 and P2 insisted that security and data migration problems do not frequently encounter during the cloud-based ERP adaptation process. Both of them believes data-lock problems can be prevented if the technical employee is capable enough in his field and also security concerns do not worth to worry if you trust your vendor. P1 expressed that convincing their customers and gaining their trusts are challenging. He gave an example and said that: *"Most of our customers asked me how they can trust their data are in safe? I immediately answered them: Do you think your data are inaccessible to your employees and how do you trust any of them?"* P1 continued to explain: *"Data problems that occur within the company like stolen data hijacked data, duplicated data or non-integrated data causes from the employees. Also, any company can face to face with unexpected natural disasters, and protected system suffers from assuring the security of data. However, we support our customers 7/24* *online and if needed via our consultants. Their all data are in our responsibility. All they can do that, doing their own business."*

Our last title is about the **future of cloud-based ERP systems**. *"Although cloud technology does not meet the compulsory requirement for all companies, it gives a chance to a company having an extended partner which organizes all their data via high sophisticated systems and offers to them focusing their target,"* P1 said. P4 who is relatively young and inexperienced employee suggested business intelligence applications will be the complementary factors for cloud-based ERP in the future. Real-time reports and key performance indicators become more critical, and managers want to integrate these add-ons their current cloud systems. P3 supported this idea and added *"This is another reason to choose cloud ERP. Managing data correctly needs expertise."* Recently, 99% of SMEs in Turkey use domestic ERPs, and none of them supports cloud services. P3 insisted that this trend will evolve to another side within 1 or 2 years and most of the SMEs will want to transform their current system into the cloud. In the future, SMEs can prefer to use partner-managed private cloud which is expected to increase its popularity, and they will tend to use it instead of the public cloud. P1 believed that: *"Once your competitors start to use any version of cloud ERP (public/private/partner-managed private) and your current ERP begin to suffer from meeting your requirements, it will be inevitable to choose cloud."* P3 emphasized that *"Thanks to the cloud operational expenditures (OpEx) have the critical importance; however there is no need for capital expenditures (CapEx) as it used to be."*

Table 3. Summary of findings

Cloud-based ERP readiness	Challenges for cloud ERP	Future of cloud ERP
• Relatively new for Turkey • Not ready but promising • Lack of consciousness • IT democracy	• Infrastructure • Data migration • Unwilling buyers/customers • Money grubber distributors • Intrinsic factors (Company culture, habits, etc.)	• Rapid adaptation • Compulsory use • Business Intelligence • Extrinsic factors (Imitating the competitors, competitive advantage, etc.) • Partner-managed private cloud • CapEx to OpEx

6 Conclusions

Table 3 provides an overview of our results. Considering interviews lasted nearly 2 h, it has to be stated that cloud ERP is still a new technology for Turkey. However, this technology becomes inevitable for all kind of sectors and every size of companies. Although domestic ERPs are dominant for SMEs in Turkey nowadays, it is not considered as a permanent situation. Especially for SMEs, cloud ERP systems pay-as-you-go logic offer several advantages for them. As they aware of the benefits of cloud and contribute to adding value to their business, they start to move toward this way. As the cloud technologies spread around SMEs, the term of 'IT democracy' which guarantees reachable technology for everyone, for instance, SAP full packet for a company which has only 5 employees, will be occurred. Capital investment will lose its importance, and operational value of a company will increase, instead. Besides there are some potential challenges such as lack of standards, infrastructure problems and other intrinsic factors throughout this modification can be occurred; all of our interviewers believe this adaptation process will happen very quickly. On the contrary to the literature, our experienced informants believe that the security and the data migration problems do not frequently encounter during the cloud-based ERP adaptation process. Both of them believes data related problems such as data lock-in, data confidentiality, data auditability, other API problems can be prevented and controlled if the technical employee of the distributor is capable enough in his field. Another issue about distributors is their hesitation towards cloud ERP. They have to be convinced to understand that the opportunities of the cloud are more than their familiar on-premise ERP packages; so that their customers are also informed about.

In this study, we aimed to present general views about cloud-based ERP technologies in Turkey via conducting a case study with semi-structured interviews. This study can be an effort to answer these questions: "Is Turkey ready for cloud ERP?, What are the main challenges?, What is expected for the future?". Cloud-based ERP can be seen as a promising sector for Turkey; however it is still immature. Cloud solutions offer advantages for many sectors with its high turnover costs, especially for seasonal ones such as tourism and restaurant management. For further studies, there is need more research on the readiness of cloud ERP systems in Turkey by including more other ERP consultant firms.

References

1. Pawlowski S, Boudreau M-C (1999) Constraints and flexibility in enterprise systems: a dialectic of system and job. In: Americas conference on information systems, Milwaukee, pp 13–15
2. Laudon K, Laudon J (2000) Management information systems: organization and technology in the networked enterprise. Prentice Hall, Upper Saddle River
3. Davenport T (1998) Putting the enterprise into the enterprise system. Harv Bus Rev 76:121–131
4. Beatty R, Williams C (2006) ERP II: best practices for successfully implementing an ERP upgrade. Commun ACM 49(3):105–109
5. Klaus H, Rosemann M, Guy G (2000) What is ERP? Inf Syst Front 2:141–162
6. May J, Dhillon G, Caldeira M (2013) Defining value-based objectives for ERP systems planning. Decis Support Syst 55(1):98–109
7. Ruivo P, Oliveira T, Neto M (2012) ERP use and value: Portuguese and Spanish SMEs. Ind Manag Data Syst 112(7):1008–1025
8. Galy E, Sauceda M (2014) Post-implementation practices of ERP systems and their relationship to financial performance. Inf Manag 51(3):310–319
9. Benders J, Batenburg R, van der Blonk H (2006) Sticking to standards; technical and other isomorphic pressures in deploying ERP-systems. Inf Manag 43:194–203
10. Buonanno G, Faverio P, Pigni F, Ravarini A, Sciuto D, Tagliavini M (2005) Factors affecting ERP system adoption: a comparative analysis between SMEs and large companies. J Enterp Inf Manag 18(4):384–426

11. Rajan C, Baral R (2015) Adoption of ERP system: an empirical study of factors influencing the usage of ERP and its impact on the end user. IIMB Manag Rev 27:105–117
12. Addo-Tenkorang R, Helo P (2014) ERP SaaS value chain: a proposed SaaS model for manufacturing SCM networked activities. Int J Bus Inf Syst 17(3):355–372
13. Wei CC (2008) Evaluating the performance of an ERP system based on the knowledge of ERP implementation objectives. Int J Adv Manuf Technol 39(1–2):168–181. https://doi.org/10.1007/s00170-007-1189-3
14. Wood T, Caldas M (2001) Reductionism and complex during ERP implementations. Bus Process Manag J 7(5):387–393
15. Van Everdingen Y, Hillegersberg J, Waarts E (2000) ERP adoption by European mid-size companies. Commun ACM 43(4):27–31
16. Hofstede G (1997) Culture's consequences: international differences in work related values. Sage, Beverly Hills
17. Beard J, Summer M (2004) Seeking strategic advantage in the post-net era: viewing ERP systems from a resource-based perspective. J Strateg Inf Syst 13:129–150
18. Häkkinen L, Hilmola O (2008) ERP evaluation during the shakedown phase: lessons from an after-sales division. Inf Syst J 18(1):73–100
19. Al-Mashari M, Al-Mudimigh A, Zairi M (2003) Enterprise resource planning: a taxonomy of critical factors. Eur J Oper Res 146:352–364
20. Castellina N (2011) SaaS and cloud ERP trends, observations, and performance. Aberdeen Group. http://www.meritsolutions.com/resources/whitepapers/Aberdeen-Research-SaaS-Cloud-ERP-Trends-2011.pdf
21. Martens C, Hamerman P, Moore C, Magarie A (2011) The state of ERP in 2011: customers have more options in spite of market consolidation
22. Sharma R, Keswani B (2014) Study of cloud-based ERP services for small and medium enterprises. Rev Sist Inf FSMA 13:2–10
23. Mell P, Grance T (2011) The NIST definition of cloud computing. Recommendations of the National Institute of Standards and Technology
24. Wu M (2011) Cloud computing: hype or vision. In: Zhang J (eds) Applied informatics and communication, part IV, pp 346–353. Springer, Heidelberg
25. Sharif A (2010) It's written in the cloud: the hype and promise of cloud computing. J Enterp Inf Manag 33(2):131–134
26. Utzig C, Holland D, Horvath M, Manohar M (2013) ERP in the cloud: is it ready? are you? PwC
27. Vetschera R, Chen Y, Hipel K, Marc Kilgour D (2010) Robustness and information levels in case-based multiple criteria sorting. Eur J Oper Res 202:841–852
28. Salleh S, Teoh S, Chan C (2012) Cloud enterprise systems: a review of literature and its adoption. In: Pacific Asia Conference on Information Systems (PACIS), Paper 76
29. Armbrust M, Fox A, Griffith R, Joseph A, Katz R, Konwinski A, Zaharia M (2010) A view of cloud computing. Commun ACM 53(4):50–58
30. Haddara M, Elragal A (2015) The readiness of ERP systems for the factory of the future. In: Conference on Health and Social Care Information Systems and Technologies, vol 64, pp 721–728. Procedia
31. Shukla S, Agarwal S, Shukla A (2012) Trends in cloud-ERP for SMB's: a review. Int J New Innov Eng Technol 1(1):7–11
32. Navaneethakrishnan C (2013) A comparative study of cloud-based ERP systems with traditional ERP and analysis of cloud ERP implementation. Int J Eng Comput Sci 2(9):2866–2869

33. Ruivo P, Rodrigues J, Oliveira T (2015) The ERP surge of hybrid models - exploratory research into five and ten years forecast. In: International conference on project management/ conference on health and social care information systems and technologies, pp 594–600. Procedia
34. Johansson B, Karlsson L, Laine E, Wiksell V (2016) After a successful business case of ERP – what happens then? In: Conference on ENTERprise information systems/international conference on project, pp 383–392. Procedia
35. Beaubouef B (2011) Changing the game for ERP cloud implementations: cloud can bring out the best of ERP, 23 November 2011. https://gbeaubouef.wordpress.com/2011/11/23/cloud-erp-advantage/. Accessed 23 Mar 2018
36. Rodrigues J, Ruivo P, Oliveira T (2014) Software as a service value and firm performance - a literature review synthesis in small and medium enterprises. In: International conference on health and social care information systems and technologies, pp 206–211. Procedia
37. López C, Ishizaka A (2017) GAHPSort: a new group multi-criteria decision method for sorting a large number of the cloud-based ERP solutions. Comput Ind J 92–93:12–24
38. Mijač M, Picek R, Stapić Z (2013) Cloud ERP system customization challenges. In: Central European Conference on Information and Intelligent Systems, pp 132–140. https://doi.org/10.13140/2.1.2258.0488
39. Weng F, Hung M-C (2014) Competition and challenge on adopting cloud ERP. Int J Innov Manag Technol 5(4):309–313
40. Hoffman P (2010) Cloud computing: the limits of public clouds for business applications. IEEE Internet Comput 14(6):90–93
41. Yin R (1984) Case study research: design and methods. Sage, Beverly Hills
42. Abdinnour-Helm S, Lengnick-Hall M, Lengnick-Hall C (2003) Pre-implementation attitudes and organizational readiness for implementing an Enterprise Resource Planning system. Eur J Oper Res 146:258–273
43. Gartner (2013) Survey Analysis: Adoption of Cloud ERP, 2013 Through 2023
44. Abd Elmonem M, Nasr E, Geith M (2016) Benefits and challenges of cloud ERP systems – a systematic literature review. Future Comput Inform J 1:1–9

Design of Car Seat Mechanism for Disabled

Michael Pasteka(✉) and Marian Králik

Institute of Manufacturing Systems, Environmental Technology and Quality Management, Slovak University of Technology in Bratislava, Bratislava, Slovakia
{michael.pasteka,marian.kralik}@stuba.sk

Abstract. This article is focused on development of mechanism for easy accessibility seat for disabled people, especially for easy getting in and out of the vehicle (Volkswagen UP 3 door version). Mechanism is easily adaptable for any other vehicle of similar type. Kinematics of mechanism is especially designed for seniors and people with lighter moving problems driving around town in a small city car. Project was released.

Keywords: Power seat · Seniors · Disabled · Car seat · Vehicle Volkswagen Up!

JEL Classification: Economic development · Innovation Technological change and growth

1 Introduction

The idea of designing a car seat mechanism for the disabled and elderly came up as a result of an analysis of holes in the market. Nowadays, the car seats for the disabled exist, even whole cars adjusted to the needs of the immobile drivers have been constructed. The main aims of this project were a design proposal, dimensioning and realization of a prototype mechanism for the car seat based on the fact that the basic functional principles, after verifying of functionality, could be applied into mass production. Such component would be available as an option on purchase of a new vehicle. These aims resulted in many factors that need to be followed in terms of versatility, minimum installation requirements, weight, safety, durability, reliability etc. Thanks to the cooperation of the Slovak University of Technology in Bratislava and Volkswagen Slovakia a.s., three-door version of Volkswagen Up! has been chosen. This automobile is the ideal representative of a small European city vehicle and the three-door version offers enough space for a mechanism such as this one.

2 Technical Requirements

2.1 Lateral Guidance

Some of the fundamental properties of a safe and comfortable car seat are good body posture, support in the lumbar region of spine, lateral guidance in the back and thighs

N. M. Durakbasa and M. G. Gencyilmaz (Eds.): ISPR 2018, *Proceedings of the International Symposium for Production Research 2018*, pp. 379–390, 2019.
https://doi.org/10.1007/978-3-319-92267-6_33

and support of the seat up to the knee. It must be emphasized that the possibilities of entering and exiting the vehicle are considerably worsened by the prominent profiling and lateral guidance.

2.2 The Effects of Vibration

In addition to the body support, the seat serves as an absorber of vibration as well. Spine, as well as the individual organs, are sensitive to excitation at different frequencies. The effects of this excitation come from, for instance, the unevenness of the road, vibration of the car engine or the drive mechanism. The car seat will make use of the different combination of various materials for the padding, so that the impact of harmful frequencies on the human body of any kind could be eliminated to the maximum extent possible. It mainly concerns frequencies around 8 Hz.

2.3 The Seat Ejecting Mechanism

The ejecting mechanism is the ideal solution of how to enable the usage of such seat in a small car and how to simplify the entering of a vehicle to an even simpler level than the standard seat allows. So we get the technical solution for a quality, safe and comfortable seat with all the aspects of simple entry and exit without compromise.

3 Concept Design of Mechanism

3.1 Definition of Movements

Basic movements of the mechanism are aligning the seat base to the horizontal position and turning the seat out of the vehicle (Fig. 1). The alignment process is complex because the seat leans and lifts up, both in one movement of the actuation mechanism. The second movement is the rotation of the seat by 90° around the vertical axis in the direction outside of the automobile. All of these movements are electrically operated using standard one-way seat actuators. These actuators consist of one-way electric motor and worm gearbox. After the specification of movements, the design of geometry and the kinematics of movements was needed. Here the mechanism was divided into various construction units.

- mounting frame,
- alignment frame,
- vertical movement mechanism,
- rotary movement mechanism,
- rotary holder of the seat,
- safety mechanism.

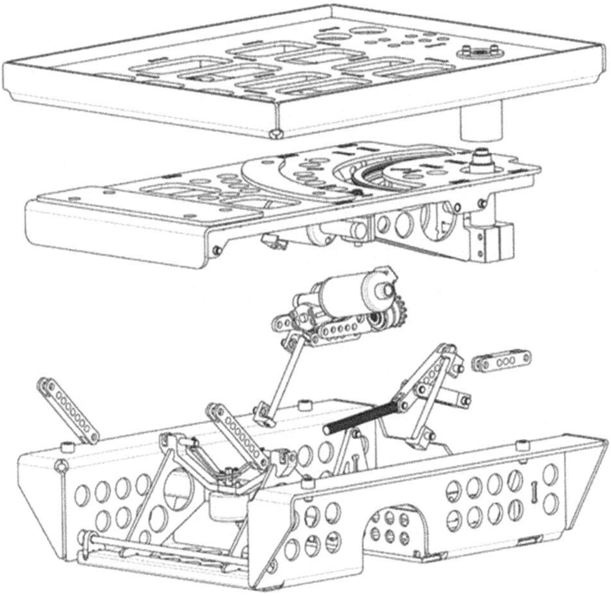

Fig. 1. Structure of the mechanism

3.2 Kinematic a Dynamic Analysis

In this part of the project, two basic movements of the mechanism are analysed: rotary and alignment movement. Multi-body model simulating the movement of the mechanism in the design process was used for this analysis. The results were the kinematic and dynamic parameters of the model.

The rotary movement is realized with the one-way electric motor, at the output of which the worm gearbox is placed. The gearbox reduces the speed of rotation of the output shaft and multiplies torque at the same time.

The geometry of movements and individual components, as well as its physical properties were acquired by creating models in CATIA software.

The position of the rotation axis of the seat had to be defined with a model of the vehicle's body in such way that the required effect would be achieved, meaning the maximal possible ejection of the seat out of the automobile while eliminating any collision with other parts of the vehicle.

4 Design Solution of the Mechanism

All the parts needed for the functioning of the mechanism are described in the design proposal.

4.1 Mounting Frame

Mounting frame (Fig. 2) must guarantee firmness, since one of its parts is a security equipment that needs to transfer all forces and shocks to the vehicle body without damage in case of an accident. Firm handles for the arms are to be placed on the frame to ensure the hold and movement of the alignment frame. Structure of the frame consists of steel sheets, 3, 5 and 8 mm thickness, cut with laser or water beam by the CAD data, bent on a bending machine and thin-walled steel profiles with a rectangle cross-section. The frame consists of a welded part, sliding sleeve and assembly material.

Mounting frame was constructed according to the requirements of regulations EHK 17 (1).

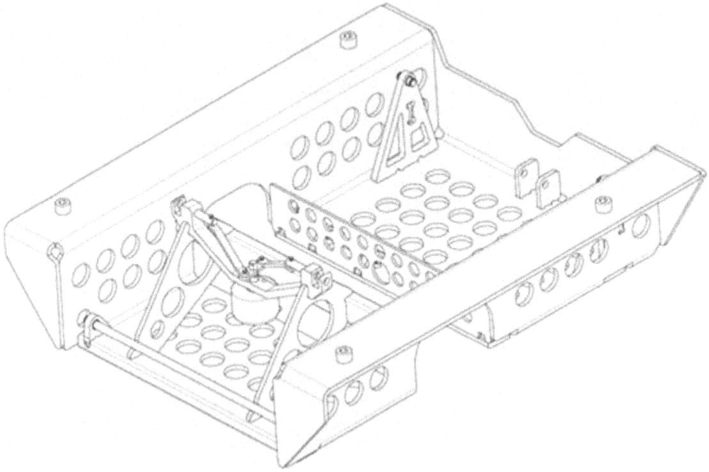

Fig. 2. Mounting frame

4.2 Alignment Frame

This component is secured to the mounting frame with arms and lever stroke mechanism. It is the most complex part of this assembly because it serves as a carrier of the force from the holder of the seat to the mounting frame, to the vehicle body. In such small installation space, it is necessary to achieve high strength against deformation and to place various components so that the frame would fullfill its function. The basis is made of 5 mm thick sheet with relief openings cut out, bent on a bending machine. A number of holders and stiffeners cut out by laser are attached to it. The transfer of forces to the mounting frame is conducted by the arms and mounting pins. The forces from the holder of the seat are carried through so called 'thompson shaft' and an arc guide from THK company (HCR15A 150R). The support of the holder is provided by delrin sliders mounted to the alignment frame (Fig. 3).

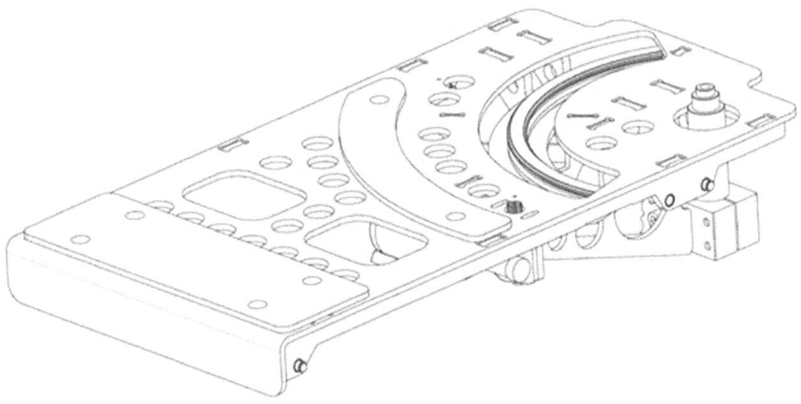

Fig. 3. Alignment frame

4.3 Vertical Movement Mechanism

The alignment and the lift of the alignment frame to the horizontal position is provided by geared actuator through lever mechanism. The actuator is secured to a simple welded part made of laser cut sheet metal through dural block similar to the one in the alignment frame. The output of the actuator is connected to a custom made sprocket. The setting of the actuator with a smaller sprocket is on slots thanks to which it is possible to adjust the chain tension. The rotating movement is carried through a 05B chain to a second sprocket. Besides the actual actuator, compression springs assist to overcoming the applied load during the lifting of the mechanism as well. These are set on linear spring guides (Fig. 4).

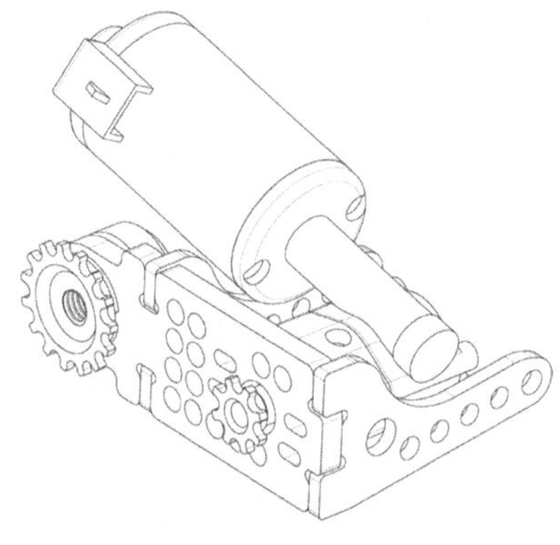

Fig. 4. Vertical movement mechanism

4.4 Rotary Movement Mechanism

The rotary movement of the holder of the car seat going outwards is executed by a pair of gearing. The outlet from the other actuator is attached to a custom made sprocket. The sprocket is made of hardened tool steel made by an EDM wire cut. Embedding is eroded with submersible electrodischarge machining. After the wheel machining is completed, its surface is nitridated for firmness.

The second gearing, quarter circle with internal gear, made of tool steel as well, is placed at the holder of the seat (Fig. 5). This gearing is not hardened because with such small thickness, there is a risk of rupture. The component is cheaper to produce as well. The manufacturing process consists of cutting out the gearing with an EDM wire cut and then regrinding it from 4 mm thickness to 4.7 mm thickness. The gearing is nitridated as well.

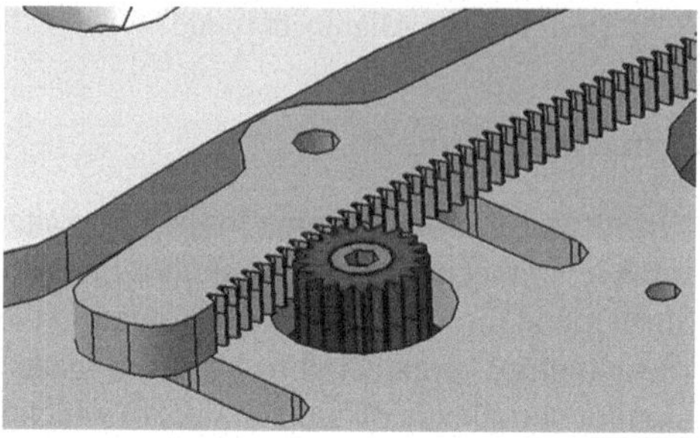

Fig. 5. Rotary movement mechanism

4.5 Rotary Holder of the Seat

This part of the mechanism serves as a carrier of the forces from the seat to other parts of the mechanism. The car seat is placed on the upper part of the holder with its original rails and height-adjustable mechanism. This component is especially important in terms of firmness. The requirement is to achieve the smallest deformations possible while reaching the greatest strength. The seat mechanism is designed to the requested carrying capacity of 150 kg. Besides this mounting to the alignment frame, the transfer of the forces through the THK arc guide is used as well. Two linear guides are attached to the seat holder (Fig. 6). The pair of linear guides reduces torque straining of the guides and uses the strong side of these guides in transmitting the direct forces (pressure and pull).

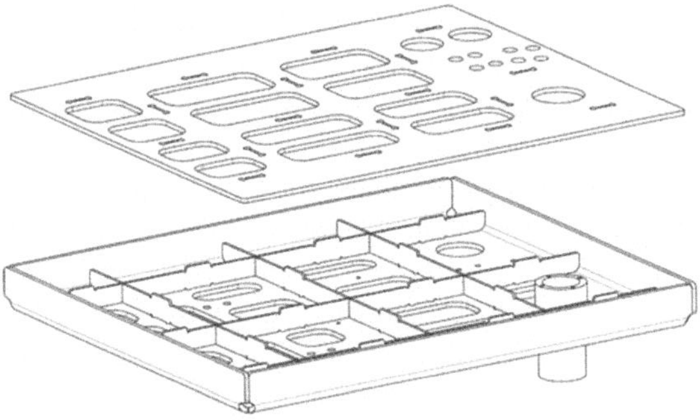

Fig. 6. Rotary holder of the seat

4.6 Safety Mechanism

The last core component of the mechanism is the safety mechanism. The most part of this mechanism is firmly connected with the mounting frame. The safety mechanism is automatically controlled by the inputs of the automobile. The fixed part of the mechanism forms conducts for the retractable safety pins. In case of a collision of the vehicle, this mechanism is able to transfer the forces from the seat to the body of the vehicle. The function of the mechanism is as follows: The seat retracts into the automobile to the driving position. The driver closes the door. Sensor of the door notices the closed door and sends a signal to the actuator of the safety mechanism. The mechanism then, with the help of the rods, retracts the pins through the openings in safety points in the holder of the seat directly to the opposite opening of the lock. This lock will then unlock only after opening the door of the vehicle.

In Fig. 7, the mechanism is seen in the unlocked position, prepared for ejecting the car seat outwards.

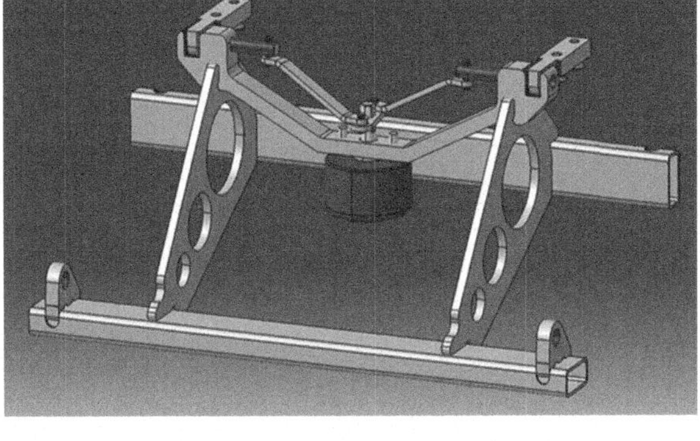

Fig. 7. Safety mechanism

5 Designing of the Cad Model and the Manufacturing Technology

In this chapter, due to the limited space, we will only describe the main parts and solutions that led to the overall and final solution.

To design the geometry and kinematics of the mechanism, obtaining the space possibilities of the given automobile was needed. Such data can be acquired by measuring but in our case, we were able to work directly with the model of the vehicle. This guarantees that the manufactured prototype will not collide with the body of the vehicle and that there is a possibility to install it into the vehicle without any additional adjustments (Fig. 8).

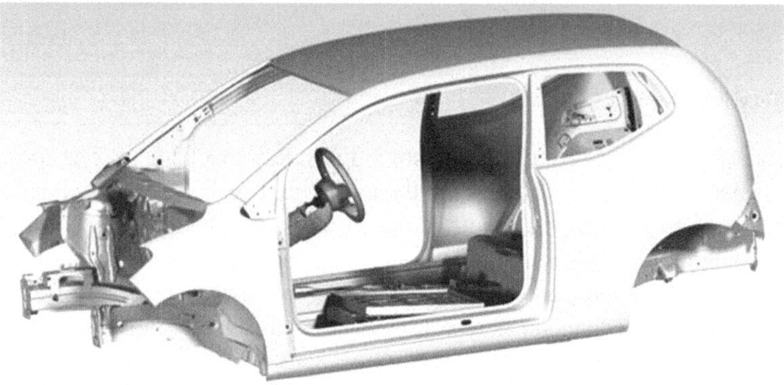

Fig. 8. Mechanism in the interior of the vehicle

The first step in creating the reference points was spacial definition of mounting screws. Then, in the CAD software environment, the angles and distances in which the original seat had to be moved were measured. After obtaining this information, a 2D line model in Sketcher was created. As a result of this trial were the necessary lengths of the arms, levering positions of the rotary axis etc.

Available technologies for which the model is designed are:

- laser cutting
- water beam cutting
- bending
- milling
- turning
- EDM wire cutting
- EDM die sinking
- grinding
- heat treatment: hardening, induction hardening, cementation, nitridation
- surface treatment: coloring, burnishing.

The final mechanism is compiled of approximately 250 components and other assembly material. Outer measurements of the whole mechanism are l/w/h in 610/470/180.

The whole mechanism can be directly installed into the Volkswagen Up! automobile without any additional adjustments. The original seat needs to be removed from the vehicle body before installing the mechanism. After unscrewing the four M10 screws that hold the rails of the original seat to the body of the vehicle, it is possible to remove the car seat from the automobile. Then the mechanism is placed in the same position and attached with the original M10 screws in the place of the original seat.

The next step is to insert the original car seat back into the vehicle. The rails for the movement are screwed down to the holder of the seat through the original mounting openings. The attached M10 screws are used. Besides strength analyses, kinematic trials had been administered. The mechanism was placed into the vehicle body, all its parts were set in motion and the seat was attached to the holder. During this test, it was observed whether or not a collision would happen, for instance with the roof of the vehicle, B-column of the door or the steering wheel (Fig. 9).

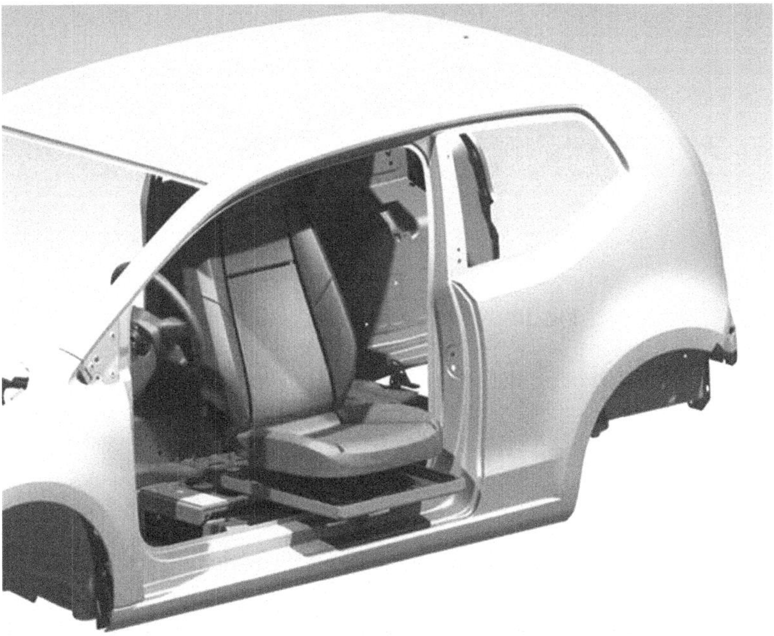

Fig. 9. Extended position of the seat and the collision test

6 Stress Analysis

The results of the administered simulations were used as a basis for the stress analysis. The forces and moments applied are identical to the ones that were used in calculations.

Linear, so called static analysis was done in the environment of CATIA V5 program, in Generative Structural Analysis module.

Figure 10 from von Mises analysis of tensions shows the concentration of tension in the surrounding of the bearing part. It also shows the block which an onerous force affects in the same way as in the analysis of the alignment frame- in a situation when a person sits at the edge of the car seat. The place of maximum tension is located on the inner spur. The value of the maximum tension is 109 MPa. The welded part of the rotary holder is made of the same material as the alignment frame. All parts of the mechanism had been gradually analyzed and submitted to the stress analysis. We present only one of the parts, because of the limited space.

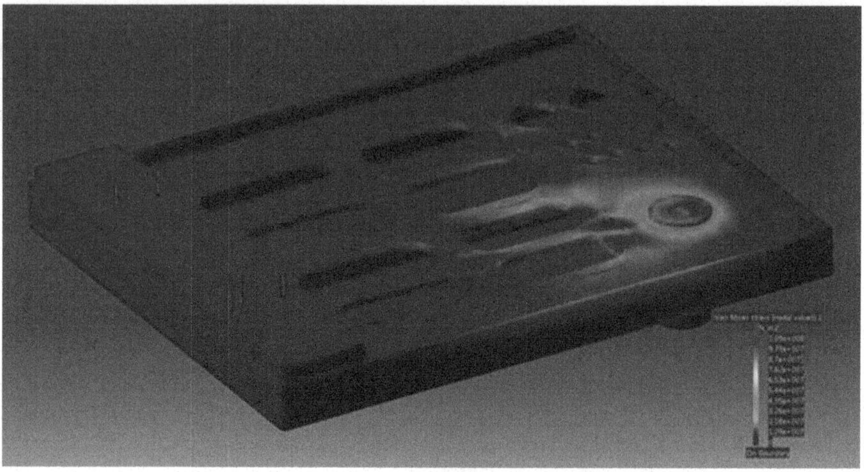

Fig. 10. Von Mises analysis of tensions

After the realization of all the stress analyses followed the making of the technical documentation itself.

7 Manufacturing

The manufacture process began with cutting of all the parts intended for welding and machining. All the needed parts, such as bearings, linear guides and bushings, were ordered from suppliers. Some of the parts had to be thermically treated by hardening, tempering, cementation or nitridation. Components, such as 'Lift gear worm' and a few others, needed the manufacture on the CNC machining center (Fig. 11). In the second picture there is a toothed ridge and a sprocket made of hardened tool steel with the help of EDM wire cutting and die skinking, tolerance of 0.003 mm.

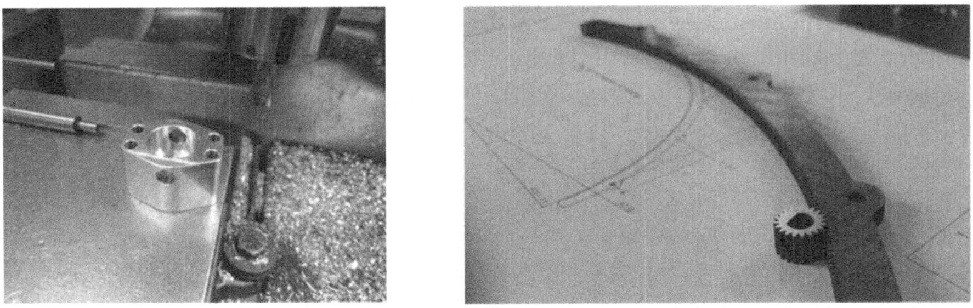

Fig. 11. Milling of a dural component and the gearing for the rotary movement

The difficulty of the whole manufacturing process of the car seat for the disabled, from the design to the installation itself, can be documented with the number of manufactured parts. The final product in the extended position is in Fig. 12.

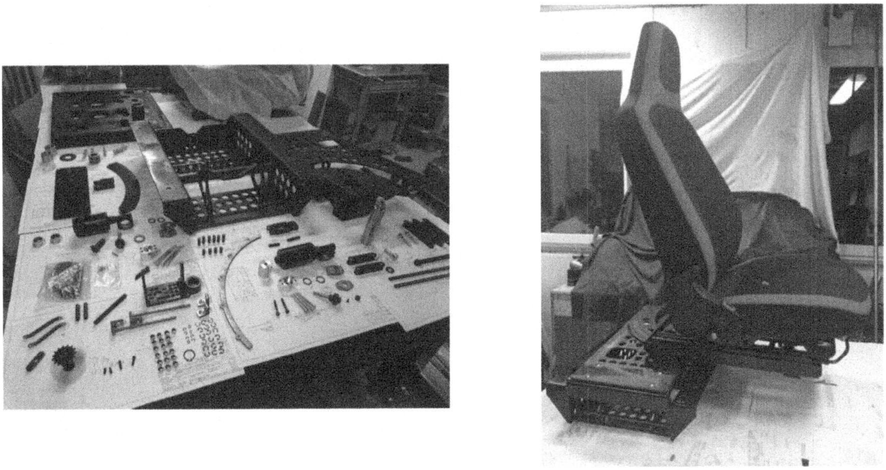

Fig. 12. The parts before assembling and the mechanism with the seat in the extended position

8 Conclusion

The aim of this project was to design, manufacture and confirm the functionality of the rotary mechanism of the car seat for a small city vehicle. In the project were realized: virtual model and design, construction, stress analysis and the manufacture of the mechanism itself, along with the documentation.

The function of the mechanism was attested by a CAD system model as well as a prototype. The practicality of its use will be shown in time, meanwhile a maximum effort will be given into the implementation of the mechanism in practice, directly in the automobiles.

The work mentioned was realized with a financial help of Volkswagen Slovakia Foundation, technical support of Technouniversum, s.r.o. And the Center of Innovations of the Faculty of Mechanical Engineering, Slovak University of Technology in Bratislava.

Acknowledgment. The article was written with the help of projects: VEGA 1/0652/16 and KEGA 035STU-4/2017.

References

1. E/ECE/324, E/ECE/TRANS/505: 1995, Revízia 5:2014: EHK 17 (Predpis č. 17), Dohoda o prijatí jednotných technických predpisov pre kolesové vozidlá, vybavenie a časti, ktoré sa môžu montovať a/alebo používať na kolesových vozidlách a o podmienkach vzájomného uznávania homologizácii udelených na základe týchto predpisov
2. Paštéka M (2016) Návrh mechanizmu sedadla automobilu pre imobilných pasažierov. Diplomová práca. Vedúci práce: Magdolen, L. STU Strojnícka fakulta v Bratislave
3. Králik M, Morávek I (2008) Programovanie CNC výrobných strojov. Edícia učebníc. Slovenská technická univerzita v Bratislave, FX, s.r.o. ISBN 978-80-89313-36-5
4. Schaeffler (2014) Technical pocket guide. Herzogenaurach 695 s.: MATNR 084713119-0000/ STT/GB-D/20140370
5. Julina M, Kovář J, Venclík V: Mechanika II. Kinematika, Praha
6. Vávra P (2006) Strojnícke tabuľky. 781 s. ISBN 80-89223-07-09

Developments in Biomedical Techniques Through Digital Transformation of Industry 4.0 – Applications in Tissue and Dental Implants

Gökcen Bas[1(✉)], Wolfgang Pirker[2], Lukas Kräuter[1], Erol Güclü[1], and Lara Durakbasa[3]

[1] AuM, Vienna University of Technology, Vienna, Austria
{goekcen.bas, lukas.kraeuter, erol.gueclue}@tuwien.ac.at
[2] Face Your Face HandelsGmbH, Vienna, Austria
info@face.at
[3] Comenius University, Bratislava, Slovakia
lara.durakbasa@gmx.at

Abstract. The biomedical techniques have been developed over the last decade rapidly through digitalisation, modern devices and sophisticated measurement equipment. In order to meet the demand in functionality of the non-technical structures, combined challenges of biological chemical and mechanical reactions need to be studied in detail. This study focuses on the latest developments with applications in modern production techniques and sophisticated measuring devices in strategic directions of tissue and dental implants. 3D printing of non-technical structures such as dental implants and scaffolds for tissue engineering are being researched. Computer tomography and techniques are applied to create and evaluate the shape of these structures targeting high accuracy and functionality. The digitalisation and modelling of the structures through advanced equipment are defined as key points in transition to the Industry 4.0 concept in biomedical technologies.

Keywords: Biomedical techniques · Tissue · Dental implant · 3D-printing Measurement technology · High precision · Industry 4.0

1 Introduction

Biomedical sciences and research are taking an important role with digital transformation in technology [1]. The clinical implantation techniques are an indispensable part of the health and medicine field. In order to increase the functionality and quality of tissue and prosthesis implantation, modern production and precision metrology have to work together. The measurement science forms the fundamental basis of checking requirements as well as the quality of the structures associated during biomedical applications. There is a continuously growing demand in the capability of biomedical

N. M. Durakbasa and M. G. Gencyilmaz (Eds.): ISPR 2018, *Proceedings of the International Symposium for Production Research 2018*, pp. 391–401, 2019.
https://doi.org/10.1007/978-3-319-92267-6_34

techniques. From that, modern metrology techniques also arise with a general trend towards higher performance and accuracy.

The term, "Industry 4.0" was introduced to the technical field, emerged in 2011 to refer a new era in industry, following a research agenda and recommendations for implementation in 2013 by Acatech – the German Academy of Engineering Sciences [2]. Industry 4.0 includes a higher priority on individualized products, which requires transformations in the designing, manufacturing, and service of industrial and health systems and products as well as their business models [3]. For meeting the demands of these developments, a sustainable strategy and continuous improvement model is needed. This study focuses on research in developments to better explore and understand certain technologies of biomedical techniques.

2 Measurement Technology and Biomedical Techniques

Metrology is the science of measurement, which is devoted to the practice of measurement tasks. The know-how obtained from measurements is developed on the basis of metrology and products are evaluated accordingly. Thus, metrology and inspection serve as the control function that ties together quality of design with quality of conformance. The benefits of inspection are that it helps to evaluate the degree of conformance or nonconformance to specifications, provides information for reporting deficiencies early in the production process, and helps to assure that desired quality requirements have been met. Since quality decisions are based on inspection and measurement, undesirable consequences may result if these tasks are not performed properly [4].

The main research in non-technical structures such as tissue and implants focus on determining the capability of implants with its quality, biocompatibility and functionality. Biomedical techniques of metrology are used to measure the quality and quantity of these parameters.

Optoelectronic measurement techniques have been developed over the years for practical applications such as computer tomography, interferometry, laser scanners and 3D scanners. The development of optical measurement techniques are further continuing for the future. Figure 1 represents the development of dimensional increase in optical measurement techniques.

To the other class of metrology techniques belong contact mode measurement methods such as tactile Coordinate Measurement Machine and Stylus Surface Profiler. In contradiction to the optical measurement techniques, the surface and material is measured by contact. The characteristics and properties of the structures such as surface roughness and material corrosion resistance can be discussed by measuring the size, geometrical form, size of roughness and surface topography.

The production of very precise components and biomedical implants require the development of the necessary metrology, and a wide range of metrology techniques. The products can differ for the evaluation of surfaces and structures down to the 0.1 nm level.

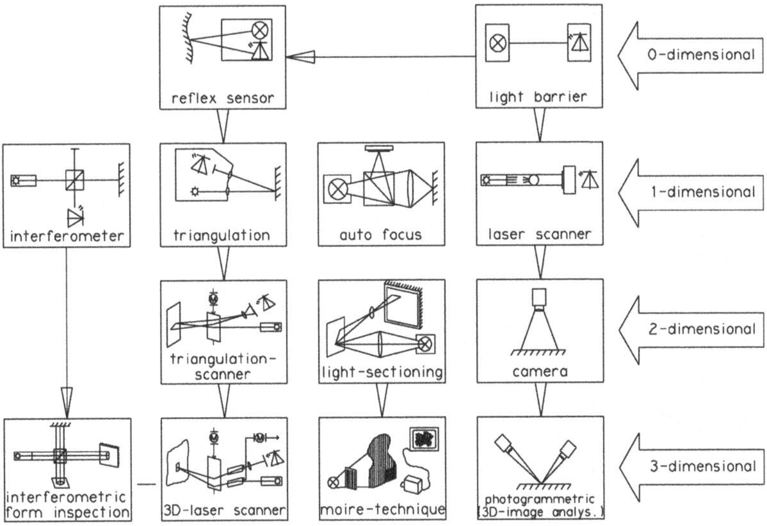

Fig. 1. Overview of different dimensional optical measurement techniques

Micro- and nano-metrology measurement techniques provide this powerful array of instruments with required capability and accuracy with three major scientific attributes:

- Reliability: Measurement results have to mirror the real surface structure and statistic and systematic errors may be reduced to an absolute minimum.
- Comparability: Measurement results must be comparable when they are measured with different measurement systems of the same kind. Ideally measurement results taken with different systems should be comparable too.
- Reproducibility: Several measurements of the same sample under the same conditions must result in the same outcome. Changes in measurement conditions must result in comprehensible changes in the measured parameters.

To meet the increasing demand in different fields for measurement instrumentation while combining research and applications, the High Precision Measurement and Nano Metrology Laboratory at the AuM-TUWien (The Department of Interchangeable Manufacturing and Industrial Metrology/Production Metrology and Quality, Vienna University of Technology) has become an international node. The transition to Industry 4.0 applications in both technical and biomedical applications is possible at this Laboratory with adequate competency, knowledge and experience.

The vision and planning of digitalisation and technical developments at the Laboratory of AuM is represented in the Figure on a basis of continuous development cycle (Plant-Do-Check-Act) between the equipment such as Computer Tomography, Coordinate Measurement Machine and Digital Applications (Fig. 2).

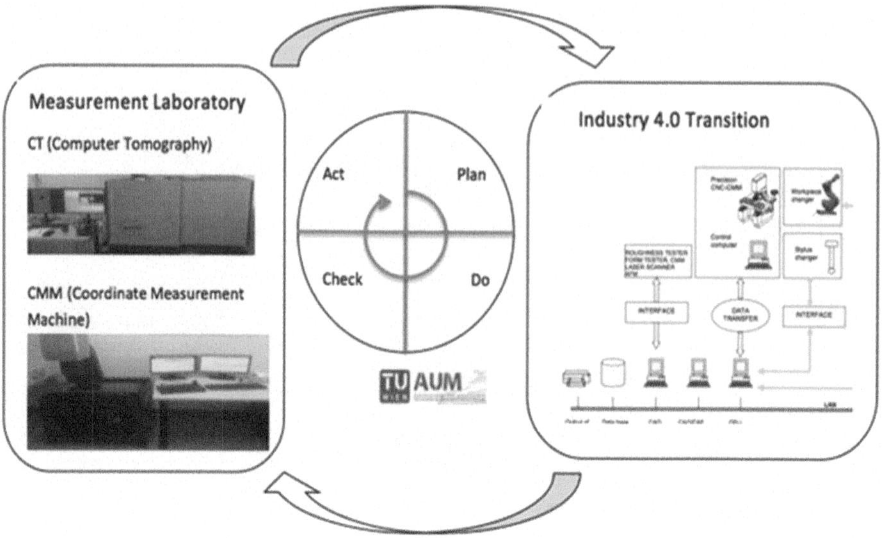

Fig. 2. Applications at the high precision measurement and nano metrology laboratory at the AuM-TUWien

3 Metrology and Production Techniques for Biomedical Applications

There exists rapid development in biomedical technique and this demands the application of modern production technique, intelligent design, metrology and sophisticated devices. These can be applied to create and evaluate the shape of nontechnical structures in biomedical applications with high accuracy. Figure 3 illustrates the measurement and evaluation of the shape of an artificial tooth as a 3D model used for implantation in human jaws. The results of such measurements form the basis for the improvement and optimisation of future work for biomedical applications.

Nowadays implant surfaces are created by a large variety of manufacturing processes and techniques such as: CNC turning, milling, broaching casting, grinding, polishing, honing, electro chemical etching, welding, brazing, stamping, bending, sandblasting, etc. as well as new methods of additive manufacturing technology. 3D printing as a part of additive manufacturing is a process to create objects from a 3D digital model. Figure 4 shows the steps involved in 3D printing of an object from a 3D CAD model.

The manufacturing accuracy range and allowable tolerances for implants exist between 10 nm–100 µm. If the implant require natural surface, then it's important to compare to the feature of natural bone surface that contains about 100 nm. The surface features also affect the interaction of bone tissue and body parts in terms of wear and friction creating special patterns. The utilization of new techniques in bioengineering enables the production of three-dimensional objects with complex geometries and fine structures.

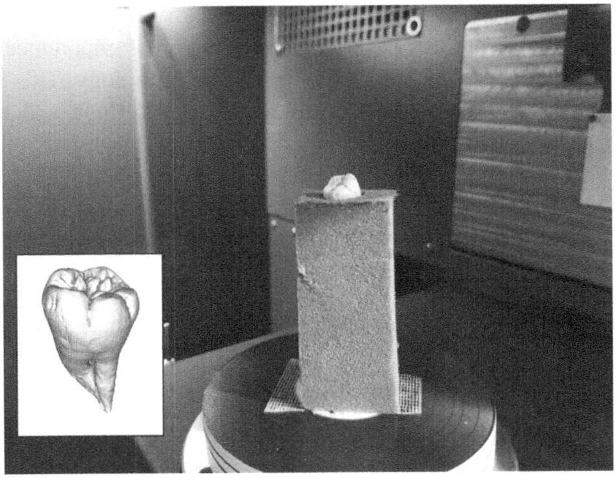

Fig. 3. Measurement and evaluation of a tooth using computed tomography (Werth [5])

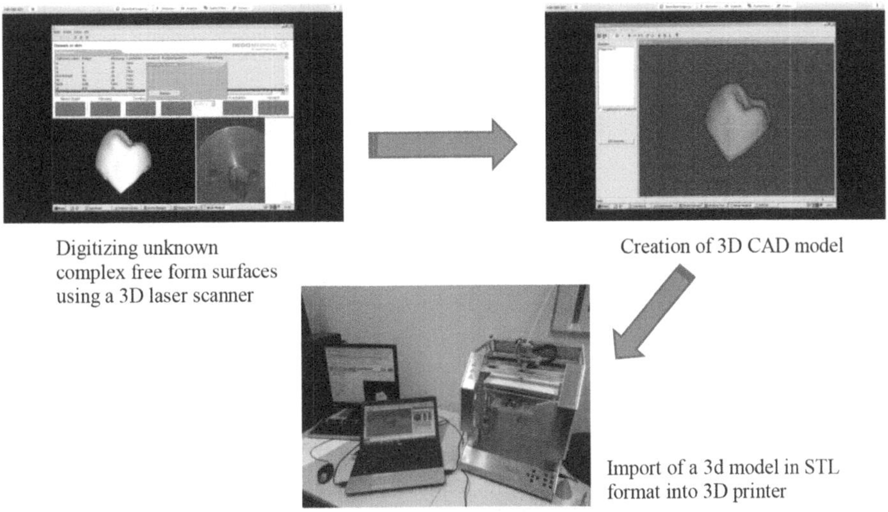

Fig. 4. Steps in a 3D printing process of an unknown complex free form surface

The new techniques enabling new devices have made it possible to establish metrology applications, which can measure the surface of a structure in 3-dimension with high quality and accuracy to fulfill the requirements for use in biomedical implants and tissue engineering. This presents the implementation of modern technology and digitalisation in line with the micro-/nano-metrology functionality, to provide high-quality and cost-effective health care.

The presented overview about the research area metrology in the fields of biomedicine and bioengineering demonstrates clearly the great importance of this development and it goes directly into the direction of nanotechnology. This is fully in line with the general ideas of production metrology, nanometrology and engineering of technical surfaces [6]. Figure 5 represents today's possible measured atomic structure topography with a Scanning Probe Microscopy.

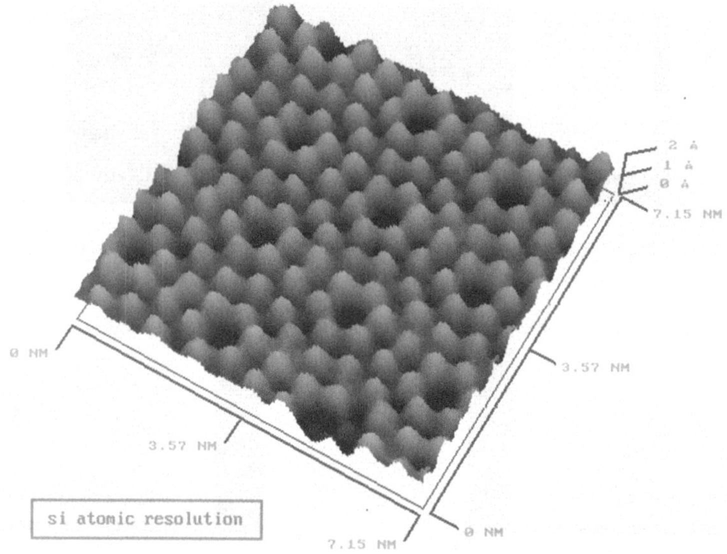

Fig. 5. The measurement result of an atomic structure topography with Scanning Probe Microscopy

4 Biomedical Techniques in Tissue Implants

Application in tissue and dental implants are showing great promises in the biomedical techniques with rapid developments in measurement and printing technologies. The printing of real-tissue and real-bone materials are still in research with problems of structural integrity and functionality for wound repair [7].

3D Bioprinting as an attractive method is one of the most actual research areas also in the context of Industry 4.0 production concept using the 3D Printing technology. This technology shows potential in tissue engineering general for the creation of cells, tissues and organs. Human skin produced by 3D printing for skin grafting may be of vital importance to victims of accidents or burn wounds. Current engineered skin constructs have been used, but they lack many of the characteristics of natural skin. In this 3D Bioprinting technology, skin is modeled as a 3D structured, consisting of multiple 2D constructs. Different cell types (keratinocytes, fibroblasts) are precisely put in a special spatial configuration. The primary purpose of the usage of 3D skin printing is for the in vitro generation of skin substitutes, in situ small cosmetic and in situ large wound reparations. Two systems are working together to achieve a three-dimensional print, the movement and the delivery system. There are some parameters important for the delivery system, which are dependent on the type of ink, printer size and the size of the output orifice. To these parameters belong the spatial resolution, throughput, bio-ink, viability, single cell precision and delivery matrix [8]. Bio-inks are substances made of living cells that can be used for 3D printing of complex tissue models.

Bioprinting methods:

- Ink-jet bioprinting: In this method the bio-ink filled in a cartridge is dropped onto a substrate. The inkjet-bioprinting method is based on thermal, piezoelectric and electromagnetic technics.
- Laser based bioprinting: Bio-ink is vaporized by a laser pulse and then transferred to a receiving substrate.
- Extrusion based bioprinting system: In this system, mechanical or pneumatic forces dispense bio-ink through a nozzle.
- Stereolithography printing: This method is based on the polymerization of photo-sensitive polymers by precisely controlled light reflected from digital micromirrors.

Movement Systems: The parameters important for this system are the movement range, spatial precision, spatial accuracy and movement speed.

The printing technologies for implants are based on metrology models of 2-dimensional and 3-dimensional data. As an example, the technology for printing skin models use in vitro generation of skin substitutes precisely put in a special spatial configuration with delivery systems of ink-jet bioprinting, laser-assisted bioprinting, or extrusion delivery system. The important parameters for these systems are the movement range, spatial precision, spatial accuracy and movement speed [9–11]. Thus, the transition of Industry 4.0 applications with developed techniques will provide the future of biomedicine with better results.

5 Biomedical Applications in Dental Clinics

Moreover, the biomedical applications in dental clinics have been one of the most common topic as complicated procedure. Developing micro- and nano-measurement techniques have enabled an investigation strategy increase the quality of operations [12]. The correlation between the root surface profile and surface roughness variation is an important feature to evaluate the state of the prosthesis integration process that is possible through micro- and nano-metrology. Figure 6 presents a real tooth to be evaluated for producing the exact model of implant and Fig. 7 presents the measurement of the real tooth in 3-dimension by using the Computer Tomography. The concept of tooth replacement with custom-made root analog implants has proved to introduce successful results used by a patented application in dental clinics [13].

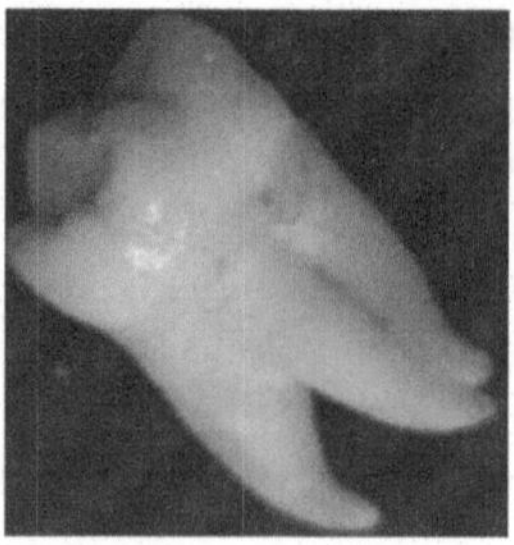

Fig. 6. A real tooth taken from the patient for producing the implant

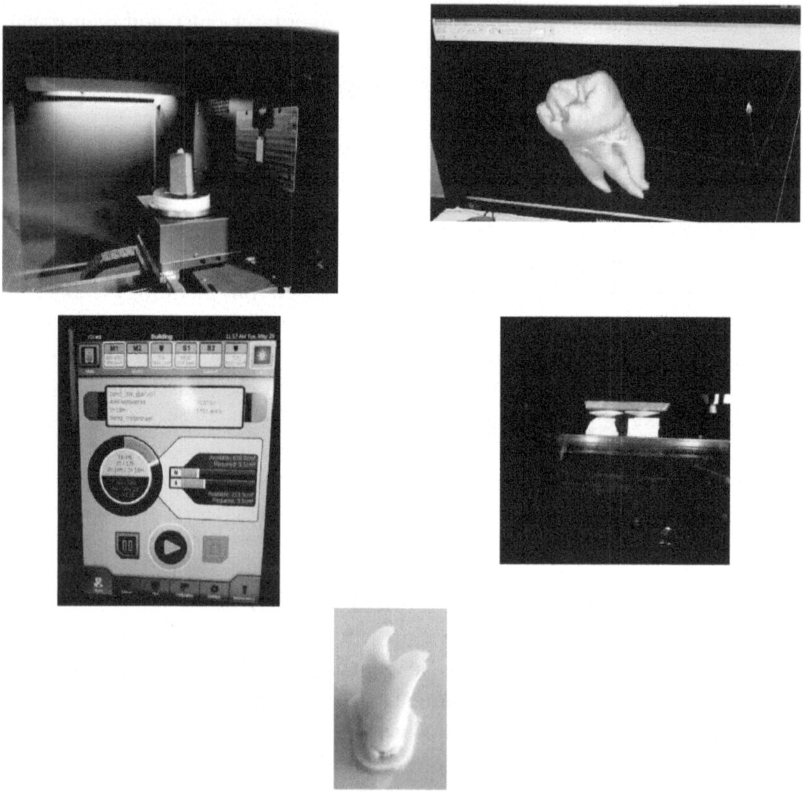

Fig. 7. The computer tomography evaluation of a real tooth (Alphacam [14])

The target in dental implants is evaluating the real tooth surface and manufacturing the implants accordingly. The actual results represent difference in surface roughness values. Thus, high precision metrology techniques and also clinical expertise need to work together for further better implants than traditional conventional ones.

The biomedical implant processes need to evaluate the quality, efficiency for better health of patients as well as below stated guidelines:

- Quality Assurance
- Health and Safety Check
- Service Quality
- Patient Feedback
- Waste/Environmental Monitoring
- Efficiency
- Cost Analyse
- mass customization
- cost efficience

Preparations of the dental implants are carried out based on the model established as presented in Fig. 8. A proposal of the process management system integration in a dental clinic is represented in the Fig. 9 by using a professional process management software [15].

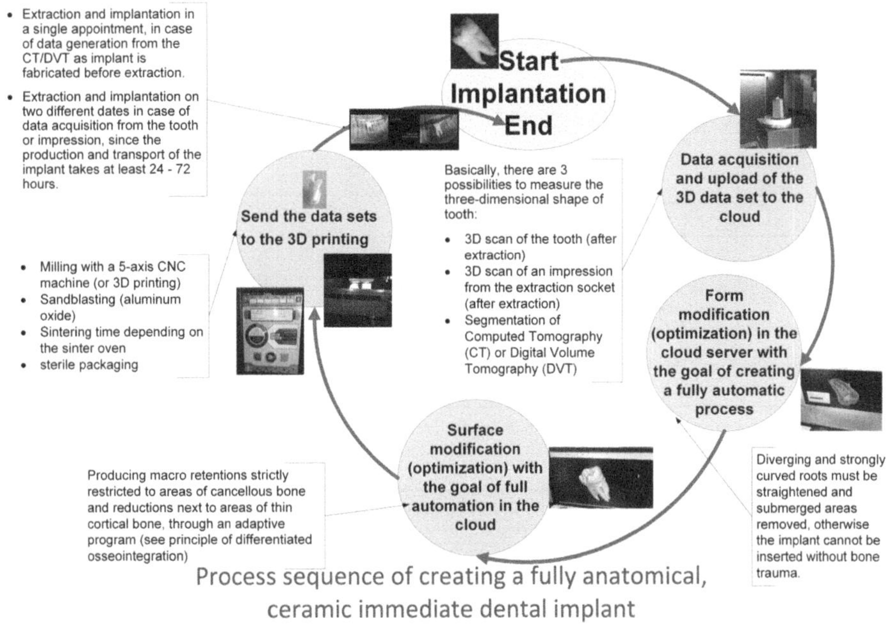

Fig. 8. The process flow of the dental implants

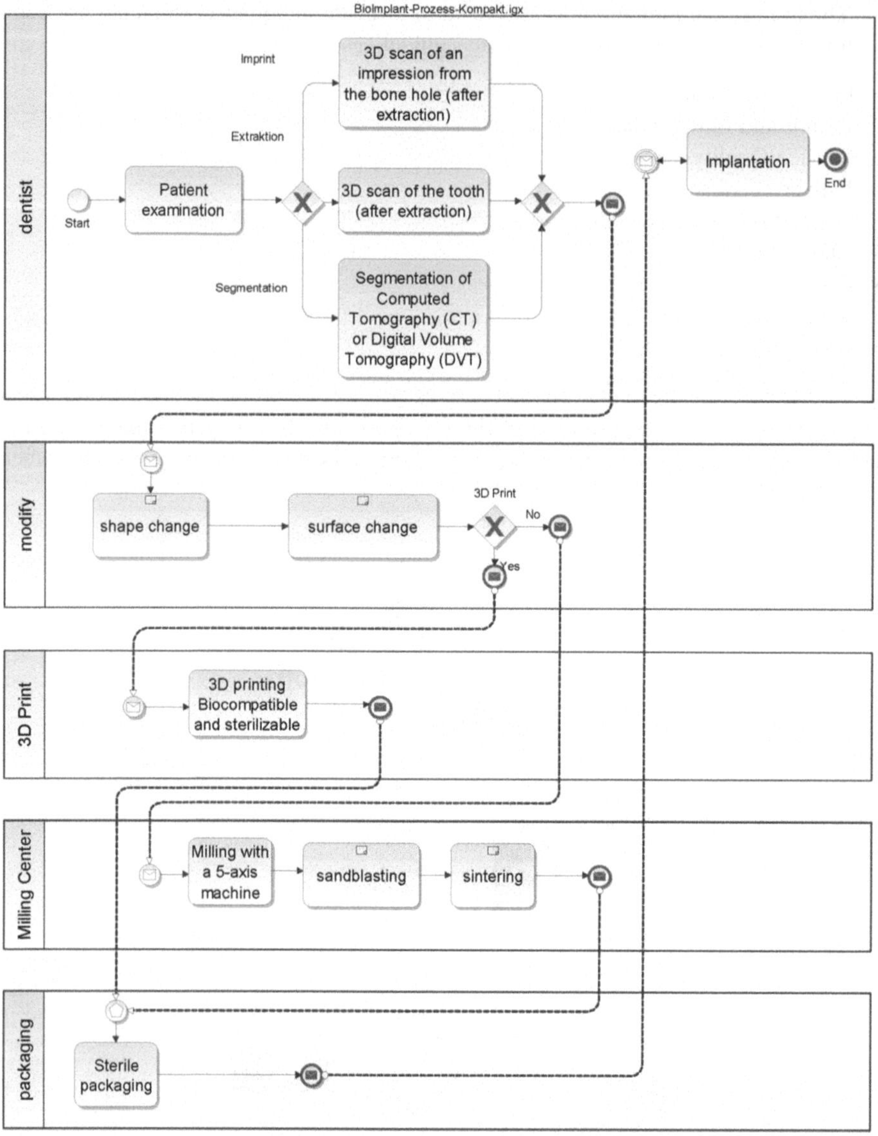

Fig. 9. The model of processes in dental implantation

6 Conclusions

The developments in the fields of metrology, biomedicine and digital technologies indicate a great promise in the research, development and implementation of biomedical applications. This study researches the applications of tissue and dental implants based on measurement and printing technologies. With the fundamental basis of measurement science evaluation, the nontechnical structures associated with biomedical applications show a general trend towards higher accuracy and quality, which will improve the health of patients.

References

1. Herrmann M, Boehme P, Mondritzki T, Ehlers JP, Kavadias S, Truebel H (2018) Digital transformation and disruption of the health care sector: internet-based observational study. J Med Internet Res 20(3):e104
2. BMBF. https://www.bmbf.de/de/zukunftsprojekt-industrie-4-0-848.html
3. Schuh G, Potente T, Wesch-Potente C, Weber A, Prote J (2014) Collaboration mechanisms to increase productivity in the context of Industrie 4.0. Procedia CIRP 19:51–56
4. Durakbasa NM, Osanna PH (2017) Quality in Industry. TU-Wien Vienna University of Technology, Vienna
5. Werth TomoScope. http://werthinc.com/products/tomoscope-xs/
6. Afjehi-Sadat A, Durakbasa NM, Osanna PH (2005) Quality control in bio-medicine and bio-engineering through application of sophisticated production metrology. Elektronika XLVI (10):46–50 ISSN 0033-2089
7. Kamolz LP, Lumenta DB (2013) Dermal replacements in general, burn, and plastic surgery: tissue engineering in clinical practice. Springer, Wien ISBN 9783709115855
8. Binder K, Skardal A (2016) Human skin bioprinting: trajectory and advances. In: Skin tissue engineering and regenerative medicine, pp 401–420. ISSN 978-0-12-801654-1
9. Tarassoli SP, Jessop ZM, Al-Sabah A, Gao N, Whitaker S, Doak S, Whitaker IS (2017) Skin tissue engineering using 3D bioprinting: an evolving research field. J Plast Reconstr Aesthetic Surg. 71:1–9
10. Svecova D, Danilla T (2017) Textbook of Dermatology. Comenius University Bratislava, Bratislava. ISBN 9788022342773
11. Koller J (2014). Burns. Comenius University Bratislava, Bratislava. ISBN 9788022335768
12. Osanna PH, Durakbasa MN, Yaghmei K, Kräuter L (2009). Quality control and nanometrology for micro/nano surface modification of orthopaedic/dental implants. Measurement. M. Tysler et al Institute of Measurement Science, pp 167–172. SAS; Komprint s.r.o., Bratislava. ISBN 978-80-969672-1-6
13. Pirker W, Kocher A (2011) Root analog Zirconia implants: true anatomical design for molar replacement – a case report. Int J Periodontics Restor Dent 31(6):662–668
14. Alphacam. http://www.alphacam.com/
15. iGrafx Process 2015 for Six Sigma. https://www.igrafx.com

Digitized Production – Its Potentials and Hazards

Petr Blecha[1(✉)], Numan Durakbasa[2], and Michal Holub[1]

[1] Institute of Production Machines, Systems and Robotics, Department of Production Machines, Faculty of Mechanical Engineering, Brno University of Technology, Brno, Czech Republic
blecha@fme.vutbr.cz, Michal.Holub@vutbr.cz

[2] Department for Interchangeable Manufacturing and Industrial Metrology and Nanometrology Laboratory, Institute for Production Engineering and Laser Technology, Vienna University of Technology, Vienna, Austria
durakbasa@ift.at

Abstract. The current possibilities of production automation allow the manufacturers of machine tools to consider new possibilities of acquiring the necessary information on the ongoing production process and the suitable ways of their use. This paper presents the potential of the applications for data acquisition from the control systems, implementation of the sensor systems into the production process, big data processing and development of a knowledge base allowing for efficient decision-making of the operator and planning of production tasks or maintenance of the machine, as well as the risks associated with remote access to machines within production digitalization in connection with the elements of Industry 4.0.

Keywords: Industry 4.0 · Production accuracy · Risk analysis
Smart machine tool

1 Introduction

Long-term development of automation of the industrial processes has been accelerated in the recent years thanks to development of information technologies and high-speed computers. This evolution of industry is nowadays focused on digitization of the whole life cycle of a product, which provides a new potential for increasing competitiveness of the companies (higher effectiveness of production, reduction of costs) as well as for realization of long-term goals associated with environmental protection and vision of the knowledge-based economy (Fig. 1) [1, 2].

Thanks to this potential, it is now possible to set up excellent starting conditions for development activities connected with digitization of a company´s infrastructure, i.e. production and support processes in a digitized company, that would allow for early prediction of the monitored machinery behaviour on the basis of the input data, advanced simulations, modelling and the developed knowledge systems. The subsequent visualization of the state of the machinery in the virtual or augmented reality environment may contribute to the effectiveness of the decision-making processes in all of the involved persons within the digitized company or its digitized part.

N. M. Durakbasa and M. G. Gencyilmaz (Eds.): ISPR 2018, *Proceedings of the International Symposium for Production Research 2018*, pp. 402–411, 2019.
https://doi.org/10.1007/978-3-319-92267-6_35

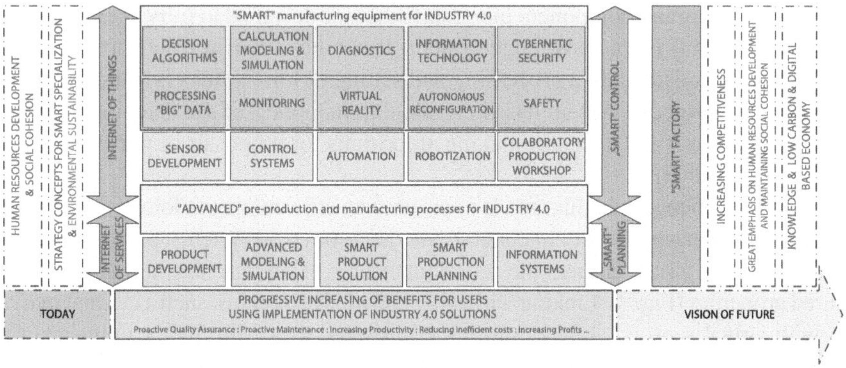

Fig. 1. Evolution of automation in industrial companies [1]

2 Digitized Production Potential

In order to accomplish the vision of "smart factories" that would realize "smart planning" of production tasks and "smart control" of production, it is necessary to begin with development of digitized machines based on employment of a virtual copy of their kinematic-dynamic model (the so-called virtual twin). Virtual twin aims to influence the expected implications in manufacturing when introducing INDUSTRY 4.0 elements published by the European Parliamentary Research Service [3].

- Increased flexibility due to automation, data system management.
- Bulk customization will allow for economically efficient production of small series, even individual pieces, thanks to the ability to quickly reconfigure the machines and add-on production.
- Production speed increases due to digital models and simulations, data managed supply chains will accelerate the production process by 120% by shortening the time after receipt of the order to the start of production, and by 70% by shortening the time from finishing the production to delivery to the customer.
- High potential is identified in the increase in quality by reducing the error rate due to built-in sensors and machines that gain intelligence by accessing a large amount of data. The EU estimates savings of 160 billion euros in total for the hundreds of the most important European producers.
- Productivity increases due to a number of INDUSTRY 4.0 elements. For example, productivity growth after predictive maintenance integration is defined by 20% due to a 50% reduction in downtime. Another type of productivity gain is optimizing the number and utilization of employees.
- The concept also involves a higher level of customer engagement in the product development process, thanks to the process of return of European production from low-income countries.
- Variation in business models with the aim to diverge them from a purely cost-based competition to cost-competitiveness, including, apart from costs, also the innovation speed, flexibility in customization, quality.

The most important component of a smart machine designed to fully use the potential of INDUSTRY 4.0 is its knowledge system. It can be either in the form of a simple database using statistical data processing and setting of limit values (virtual twin), or in the form of an advanced system based on calculation models of the machine for prediction of its behaviour and decision-making algorithms for proposals of preventive measures (smart machine).

The most challenging in this area is to propose an efficient methodology for the development of smart machines with integrated cybernetic-physical systems, based on multidisciplinary models, which ensures the self-maintenance of the machine at a given level of required properties (Fig. 2). Linking with other knowledge systems, such as virtual machine twin with virtual sensors, allows to fuse the measured data with the data obtained from simulations. The aim is to support the decision-making processes that affect the self-configurability of the machine, autonomous machine parameter correction and self-optimization and/or repair of overloaded and damaged main functional parts of the machine.

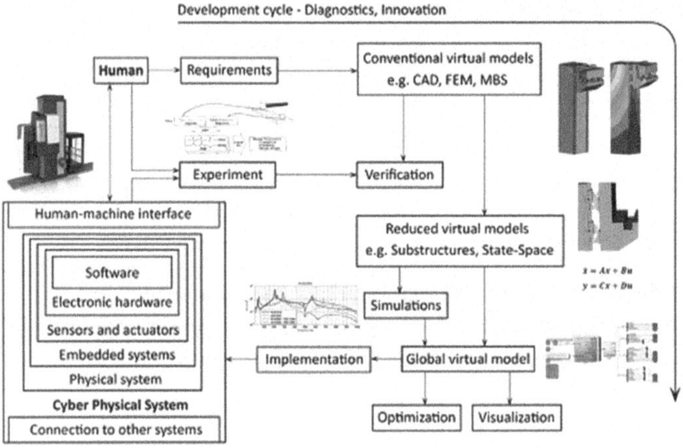

Fig. 2. Implementation of virtual model into CPS [4]

During development of the virtual twins of the machines and production processes, individual approach to every machine or process is necessary. Digitization of the whole machines and production processes is very time and cost demanding and therefore it is not always an efficient solution. From the aspect of effectiveness, it is therefore logical to develop a virtual twin only for the node of the machine that provides the data necessary for further analyses and simulations of the machine or the production process. Effective employment of the I4.0 concept, however, requires achievement of a benefit for its users. Such a benefit may be:

(a) competitive advantage (e.g. higher production accuracy);
(b) more efficient use of the production means (e.g. predictive maintenance, elimination of bottlenecks in production, higher number of realized orders);
(c) support of decision-making processes (e.g. reduction of costs for pre-production and production phases of the product´s life cycle);

(d) fulfilment of legislative requirements (e.g. reduction of energy consumption, reduction of greenhouse gasses production);
(e) prevention of not good products, possibility of virtual audits via internet connection);
(f) increased comfort of the workplace environment (e.g. elimination of the effects of weather on the production means, fulfilment of hygiene standards, elimination of tiring and monotonous work, e.g. by cooperation of persons with a machine/robot);
(g) development of the image of a hi-tech company etc.

3 Hazards Associated with Digitized Production

In order to achieve higher productivity of production, improved dimensional and geometric accuracy of a workpiece or higher safety and reliability of the machinery, it is necessary to analyse the risks of the digitized production, with its significant part being the identification of the relevant hazards. The following scheme shows our perception of the smart machine tool structure (Fig. 3).

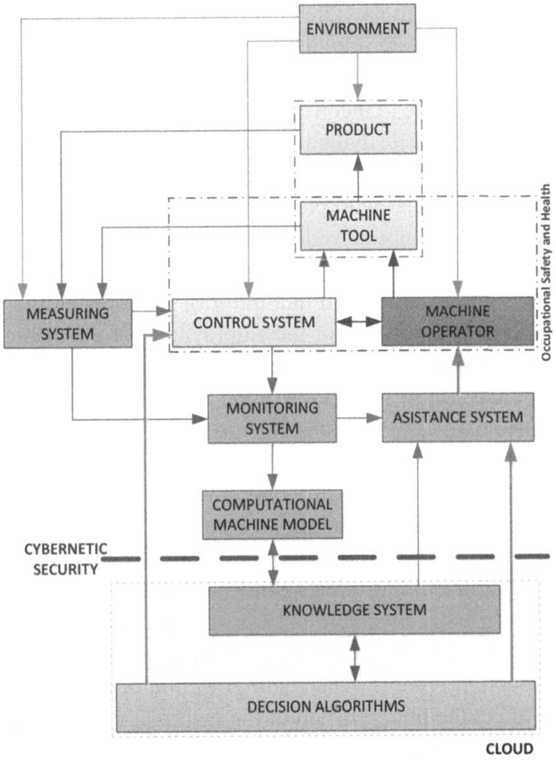

Fig. 3. Smart machine tool structure

Our approach to identifying hazards associated with digitized production is based on the above-described smart machine tool structure (Fig. 3). We have identified the following eleven areas that are exposed to different hazards. These hazards, in our view, are associated with significant risks from the perspective of production digitalization:

- **The ambient environment** may have a negative influence on the behaviour of the machine in the sense of geometric accuracy (quasi-static error). Changes in the ambient temperature affect the length dilatation of the workpieces. Furthermore, the vibrations generated by the neighbouring machinery also change the quality of the workpiece surface. Last, but not least, the changes of temperature, pressure and humidity may lead to increased uncertainty of the measurement by the multisensory systems implemented in the construction of the machine and of the sensors employed in workpiece check-ups. Besides, the operator´s concentration may be disturbed by increased level of noise, temperature or vibrations.
- **A product** must meet the requirements of customers expressed as the required dimensional and geometric accuracy of the product. Accuracy of the machine should correspond to the requirements put on the product portfolio. The product itself may have a negative effect on the accuracy of production as its weight may cause elastic deformations of the machine and thus decrease its geometric accuracy. In thin-walled workpieces, it is necessary to pay attention to the fact that incorrect clamping may lead to deformation of the semi-product and thus to deformation of the machined surfaces once the workpiece is released. Such a workpiece may be deformed also by the cutting process itself due to incorrect setting of the cutting conditions.
- **A machine tool** is a man-made dynamic system used for transformations of a workpiece with the required dimensional and shape accuracy. It is directly linked to the ambient environment and the workpiece, which affects its final properties. If the production engineer would not consider the characteristics of the machine and the tool, it could lead to production of a not good part. Such characteristics are, for example, chatter of the machine and durability of the tool. Similarly, the designer who would neglect the conditions for the employment of the machine, could negatively affect its properties such as static and dynamic compliance or thermal deformations. This would result in a deviation of the real tool trajectory against the workpiece from the ideal one.
- **Control system** of a machine may have an effect on the final production accuracy of the machine and thus also on the quality of the workpiece. Modern systems allow us to monitor the stability of the machining, temperature drifts of TCP, application of volumetric accuracy, to compensate the effects of workpiece imbalances and to warn the operator that the machining process has been incorrectly prepared. However, these functions are usually available only in the machines with integrated sensor system. On the other hand, incorrect setting of the limit values may have a negative effect on the production process. Unsuitable selection of the measuring system and its position may affect the final position of the workpiece and the tool.
- **Operator of the machine** is nowadays usually an integral part of the production process. His/her behaviour may cause either intentional or unintentional violation of the expected technological process. Such violation can be in the form of incorrect adjustment of the tool, use of excess clamping force, poorly cleaned surface of the clamping area or unsuitable positioning of the workpiece on the machine. The operator may also neglect the maintenance of the machine and thus cause its unplanned

stand off. Insufficient attention of the operator may lead to selection of a wrong production program, collision of the tool with the clamping elements or with the workpiece itself.

- **Measuring system** serves as a support for evaluation of the momentary state of a monitored parameter and it can be used as an information for the operator and for the control system of the machine; it can also be used for further processing with the aim to control the work process. A measuring system may be oriented at monitoring of the workpieces, tools or other characteristics such as the temperature at the bearings or the load of the table surface. A correctly chosen measuring system with adequate uncertainty of the measurement is able to provide the monitoring or the control system with relevant data indicating the level of interference from the ambient environment as well as from the operator of the machine.
- **Monitoring system** of a smart machine serves for collection and preparation of the data before their further processing. The total amount of the raw data provided by the individual sensors may cause system overload or slow down. For this reason, it is necessary to process such Big Data in a suitable way and to use only that information that are needed as input values into the relevant calculation models, the so-called virtual twins.
- **Computational model** of a smart machine requires selection of a suitable mathematical model of the monitored process (static and dynamic) and determination of the physical parameters within this model. Their determination usually requires performance of time-demanding experiments in every single smart machine. An efficient way of solving the demanding applications is a combination of a numeric and experimental approach. From the long-term perspective it represents collection of data from the monitoring system, their storage and processing within the creation of the knowledge module. Use of cloud services comes with the risk of cybernetic attacks, alteration or staling of the computational model.
- **Knowledge base** gathers the experiences with the interactions between the causes (i.e. machine behaviour) and their consequences (i.e. product accuracy). Based on the obtained knowledge it is then possible to adjust the calculation model of a smart machine and to visualize the processed data from the calculation model as a support for decision-making of the machine´s operator or other competent persons. The knowledge base may include the limit values of the monitored parameters that would guarantee continuous production within the required tolerances. The character and the nature of the stored data as well as the access to the internet pose the hazard of targeted cybernetic attacks with the aim to alter or steal the data from the knowledge database. Such attack could be performed from the outside as well as inside the company.
- **Decision-making algorithms** of the smart machines are able to perform autonomous decisions regarding the setting of the machine constants of the machine in order to ensure the required accuracy of the production. In general, it is the consideration of the effects of the ambient environment on the machine and of the dead-weight of the workpiece clamped at the machine. Another task of the decision-making algorithm is planning of the maintenance in order to prevent unplanned stand offs due to failures of some of its components. Such decisions would require high amount of knowledge

as well as experience of the machine´s operators. The missing knowledge of the designers on the ways of use of the individual machines in various customers could lead to proposal of unsuitable preventive measures. This is the reason for employment of artificial intelligence that would continuously improve its algorithms and, based on the possible damage associated with application of incorrect decision, it would choose the most suitable form of reaction to the unfavourable development of the monitored parameters (autonomous interference, warning to the operator, recommendations for operators, error messages). Due to the necessity of a great amount of experimental data and case studies there is a presumption of development of the decision-making algorithms on the part of the machine´s producer and providing of this know-how in the form of "Internet of services". However, here is also the possibility of cybernetic attacks.

- **Occupational Safety and Health (OSH)** during the work with a machine is associated both with the safe design of the machine and with the suitable design and application of the safety parts of its control system. Safe design of a machine can be supported by the information on the real work load of the machine during the production process in various industrial areas and safe operation of the machine by attention to the expected durability of its safety components (e.g. number of activation cycles of the individual components). A cybernetic attack aimed at a smart machine could interfere with the machine constants or even safety-related elements of software, which could finally result in an injury of the operator. However, the potential of digitization offers other possibilities how to monitor the operator´s behaviour to identify the dangerous procedures within the machine´s operation in time.

In order to achieve sufficient reduction of risks of the hazards described above, it is necessary to design smart machines, not only equipped with appropriate sensor system, but also with their structure itself corresponding to the expected use (operational load) and with a control system having sufficient performance level of both functional and cybernetic safety. First of all, it is therefore necessary to analyse the data from real operation of the machines and to build up a suitable knowledge base.

4 Discussion

Effective implementation of the I4.0 concept in companies requires focus on research and development activities in the field of basic as well as applied research, industrial experience of the key research and development specialists and innovation of study programmes so that their graduates would be able to effectively use all of the benefits of the I4.0 concept in real industrial practice (knowledge of mechanical engineering, electrotechnics and informatics). As regards education, it is necessary to update the current curricula to provide the fourteen skills listed below. The references shown in this list indicate the corresponding science and research activities of our department whose results have already been integrated into the study plans of the study branches "Design of Product Machines and Equipment" and "Quality, reliability and safety" guaranteed by our department.

(1) modelling of real objects and industrial processes in the virtual environment (simulation models, computational models) [5];
(2) prediction of real-object behaviour and industrial processes behaviour followed by experimental verification of the simulated data on a real object [6];
(3) acquisition and analysis of operational data for correction of computational models [7];
(4) visualization of virtual models in the immersive and/or augmented virtual reality environment including the possibility of active interaction with the virtual objects and monitoring of the effect of such interaction on their behaviour [8];
(5) linking of the real and virtual objects, visualization of the state and condition of the real objects in the virtual environment [9];
(6) development of smart systems for detection of the state and condition of the production and operational equipment including testing of the unconventional solutions (wireless sensors, virtual sensors, autonomous sensors using energy harvesting etc.) [10];
(7) testing of hi-tech measurement systems for diagnostics of production and operational equipment and development of procedures for their efficient employment [11];
(8) application of adaptive local linear models for the tasks aimed at identification, filtration and control of the advanced mechatronic systems, also with the use of distributed sensor systems based on IoT [12];
(9) development of the concept of machine and robot self-calibration with the application of advanced methods of measurement and monitoring (adaptability) [13];
(10) testing of heterogeneous materials for advanced applications and use of additive technology in the field of ultrafast actuators [14];
(11) development of the procedures of virtual prototyping of production systems with regard to the length of the production cycle, optimization of energy consumption and greenhouse gasses emissions [15];
(12) development of predictive and proactive multiparametric procedures of maintenance assurance with the aim to ensure planned operability of the machinery during its whole life cycle [16];
(13) testing of alternative sources of electricity and heat to ensure autonomous supply and energy self-reliance [17];
(14) last, but not least, the procedures from risk management and design of safe machines [18].

5 Conclusion

Industry 4.0 and the associated digitization of products, machines, production processes as well as of the whole factories bears a great potential for increase of production efficiency and thus also for better competitiveness of the companies. However, the new technologies such as the Internet of Things and Internet of Services require the remote acquisition of data and distribution of know-how via the internet or at least intranet connection. This is connected with the appearance of new threats that may stop or significantly slow down the production. In further development of industry digitization, it will be necessary to pay adequate attention mainly to cybernetic safety of the smart machines. Besides, we will have to overcome other barriers preventing effective employment of digitized production, especially the lack of qualified specialists with knowledge covering both the field of engineering (including metrology, automation and risk management) and electrotechnics and informatics.

Acknowledgment. The presented work was supported by AKTION (Grant No. 79, p. 4) and by Brno University of Technology, Faculty of Mechanical Engineering, Czech Republic (Grant No. FSI-S-17-4477).

References

1. Blecha P (2017) Industry 4.0 R&D area of possible cooperation at institute of production machines, systems and robotics - research and reference. Brno University of Technology
2. Marek J(May 2018) Industry 4.0 - a comprehensive solution. MM Průmyslové spektrum. https://www.mmspektrum.com/clanek/prumysl-4-0-komplexni-reseni.html
3. Ron D (2015) Industry 4.0: digitalisation for industry and growth: what will Industry 4.0 change? research service European Parliamentary. http://www.europarl.europa.eu/RegData/etudes/BRIE/2015/568337/EPRS_BRI(2015)568337_EN.pdf
4. Kovar J, Mouralova K, Ksica F, Kroupa J, Andrs O, Hadas Z (2016) Virtual reality in context of Industry 4.0 proposed projects at Brno University of Technology. In: proceedings of the 17th international conference on mechatronics - mechatronika, ME 2016
5. Augste J, Holub M, Knoflicek R, Novotny T, Vyroubal J (2013) Monitoring of energy flows in the production machines. In: mechatronics 2013: recent technological and scientific advances
6. Tuma J, Tuma Z, Synek M (2016) Verification of prediction method for energy consumption of machine tool feed axes. MM Sci J, 1634–1638. https://doi.org/10.17973/mmsj.2016_12_2016201
7. Ksica F, Hadas Z (2017) Position-dependent response simulation of machine tool using state-space models. MM Sci J, 2120–2127. https://doi.org/10.17973/mmsj.2017_12_201799
8. Tuma Z, Kotek L, Tuma J, Bradac F (2016) Application of augmented reality for verification of real workplace state. MM Sci J 2016:1487–1490. https://doi.org/10.17973/MMSJ.2016_11_2016166
9. Tuma Z, Tuma J, Knoflicek R, Blecha P, Bradac F (2014) The process simulation using by virtual reality. In: Procedia engineering, pp 1015–1020
10. Hadas Z, Janak L, Smilek J (2018) Virtual prototypes of energy harvesting systems for industrial applications. Mech Syst Signal Process 110:152–164. https://doi.org/10.1016/j.ymssp.2018.03.036
11. Holub M, Jankovych R, Andrs O, Kolibal Z (2018) Capability assessment of CNC machining centres as measuring devices. Measurement 118:52–60. https://doi.org/10.1016/J.MEASUREMENT.2018.01.007
12. Andrs O, Hadas Z, Kovar J, Vetiška J, Singule V (2013) Model-based design of mobile platform with integrated actuator - Design with respect to mechatronic education. Mechatronics: recent technological and scientific advances. Springer, Cham, pp 891–898

13. Kubela T, Pochyly A, Singule V (2015) Investigation of position accuracy of industrial robots and online methods for accuracy improvement in machining processes. In: 2015 international conference on electrical drives and power electronics, EDPE 2015 – Proceedings, pp 385–388
14. Krbalova M, Blecha P (2017) Environmental management in design process of machinery. MM Sci J, 1762–1768. https://doi.org/10.17973/mmsj.2017_02_2016209
15. Iskandirova M, Blecha P, Holub M, Dudarev I (2014) Assessing the impact of mechatronic systems on the environment. In: proceedings of the 16th international conference on mechatronics, mechatronika, pp 706–710
16. Opocenska H, Nahodil P, Hammer M (2017) Use of multiparametric diagnostics in predictive maintenance. MM Sci J, 2090–2093. https://doi.org/10.17973/mmsj.2017_12_201792
17. Hadas Z, Holub M, Blecha P, Vetiska J, Singule V (2014) Energy analysis of energy harvesting from machine tool vibrations. MM Sci J, 4 p. ISSN 18031269
18. Blecha P, Blecha R, Bradáč F (2011) Integration of risk management into the machinery design process. Mechatronics: recent technological and scientific advances. Springer, Heidelberg, pp 473–482

Evaluation of Industry 4.0 Readiness Level: Cases from Turkey

Gül T. Temur[1], Hür Bersam Bolat[2(✉)], and Sıtkı Gözlü[2]

[1] Department of Engineering Management, Bahcesehir University, Istanbul, Turkey
gul.temur@eng.bau.edu.tr
[2] Department of Management Engineering, Istanbul Technical University, Istanbul, Turkey
{bolat,gozlus}@itu.edu.tr

Abstract. The Fourth Industrial Revolution or Industry 4.0 is the name of recent digital transformation era. This transformation is conducted to improve digital operational abilities, perform collaboration and integration in the ecosystem, manage data, and develop cyber security. Industry 4.0 has become a very popular issue recently, but there is lack of understanding about its impact on real practices and the barriers expected to be encountered in the near future. Under the uncertain environment of digitalization, the companies need an evaluation method to define their strengths and weaknesses in the adaptation process. In this study, the methodologies for evaluating Industry 4.0 readiness level of companies are reviewed, and one of the most applicable methods is utilized for evaluation of three leading Turkish companies from different industries. In addition, the awareness of companies is also evaluated in terms of operational and socio-economical perspectives. Little research has been conducted on (1) reviewing the opinions of the companies on the socio-economic effects of Industry 4.0 and (2) evaluating Industry 4.0 readiness in developing countries. This study contributes by evaluating three cases dealing with the perceptions, awareness, and readiness of the Turkish companies in Industry 4.0 adaptation process. The result of the research shows that companies failing to develop road maps and new workforce planning strategies in the adaptation process under the uncertain environment of Industry 4.0 prefer to be the follower rather than the pioneer.

Keywords: Digitalization · Industry 4.0 · Socio-economic effects of industry 4.0 Readiness level

1 Introduction

The new age of digital transformation is called the Fourth Industrial Revolution or Industry 4.0. The purpose of this transformation to structure digital abilities, perform collaboration in the ecosystem, manage data, to manage cyber security, and implement two-speed systems/data architecture to differentiate quick-release cycles [1]. It directly affects design, production, distribution, and exploitation processes [2]. The strategies of product life cycle have become more global, multidisciplinary, innovation oriented, and customer focus [3]. Industry 4.0 applications provide many advantages for the

N. M. Durakbasa and M. G. Gencyilmaz (Eds.): ISPR 2018, *Proceedings of the International Symposium for Production Research 2018*, pp. 412–425, 2019.
https://doi.org/10.1007/978-3-319-92267-6_36

management of new product life cycles by (1) increasing flexibility, (2) decreasing lead times, (3) increasing customer specific production with small batches, and (4) proposing new offerings by the help of big data analytics [4].

Popular technologies in Industry 4.0 are simulation, augmented reality, autonomous robots, Internet of Things (IoT), cloud, cybersecurity, additive manufacturing, horizontal and vertical system integration, and Big Data analytics [5]. Utilization of computerized, intelligent, flexible and highly efficient systems, synchronization of material flow, and integrating customers, suppliers and companies have been increased by the effect of the popular technologies of Industry 4.0 [6]. The main elements of Industry 4.0 are connection between machines, equipment, systems and organizations within the value chain and to control the process autonomously. Such developments will directly change the production processes and the structure of factories. Factories become smarter, and include autonomous equipment. It becomes automatic to plan, organize and control production process [6]. Besides that, utilization of internet technology has been increasing the intelligence of factories, improving ergonomic oriented issues, and increasing the efficiency of resources [7]. Enhancement of interaction between production machines can be achieved by the help of IoT and augmented reality techniques. IoT can let machines obtain and analyze the information via computer network. Furthermore, by establishing effective prediction models, unexpected negative events can be detected earlier [6]. It is remarkable to mention that technological effect has not only originated from the developments made by the effect of internet. There are more technological fields [8]:

- Information and communication technology
- Technologies for keeping the security of basic sources (food, water and energy)
- New health technologies
- New manufacturing and automation technologies.

The potential of these technologies can differ according to the industry. For instance, in logistics industry, Physical Internet (PI), which consists of global logistics web on physical, digital and operational interconnectivity, is very popular. By utilization of such kind of IT oriented tools, analysing, communicating, designing, understanding, and optimizing operations have been improved in logistics. Also, smart containers which have identifiers and IoT elements are connected to internet [9].

Although Industry 4.0 has become a popular issue recently, most of the managers are not aware of its effects on real practices. One of the most leading surveys on Industry 4.0 was conducted by the Institute for Industrial Management at the University of Aachen in Germany by gathering the responses of more than 400 developed countries. The results indicate that 85% of the production companies realize the importance of the topic; and the most important developments will be performed within the next five years in the fields of information interoperability, data standardization, and advanced analytics [10]. The studies that evaluate Industry 4.0 adaptation process especially in developing countries are quite few. In this study, the perceptions and awareness of leading companies in Turkey are evaluated in terms of operational and socio-economic factors, and the readiness level scores are computed. The strengths and weaknesses of the companies are revealed by checking the results of readiness evaluation and the barriers they can face with are discussed. In the cases, three leading companies from construction, textile, and wire production industries are taken into consideration. In this pursuit, a detailed questionnaire is conducted with their Industry 4.0 coordinators.

The rest of the study is organized as follows: Sect. 2 gives a brief information and literature review for readiness level and Industry 4.0 readiness evaluation. Section 3 includes the Industry 4.0 awareness and readiness level evaluation methodology. Section 4 illustrates conducting the readiness evaluation procedure for companies from construction, textile and wire production sectors. Section 5 discusses the results briefly. Section 6 presents the research results and further plan of the study.

2 Industry 4.0 Readiness Level

2.1 Readiness Level

In a system, the success of improvement of a new process is directly affected from the prior performance of research and development (R&D) studies. The level of prior efforts give foresight about the readiness of the system for new capabilities. In the Cambridge dictionary, readiness is defined as "willingness or a state of being prepared for something" [11]. The preparation for something is related with the prior effects. This term is mostly used for technology adaptation and utilization process. Readiness evaluations for technology are utilized to clarify the maturity of new capability requirements [12]. Evaluation consists of determination of performance goals, readiness level and hardness degree of R&D studies. Mankins categorizes the technology readiness into 9 levels and proposes a descriptive analysis of each level that begins from reporting basic principles and ends with proving actual system [12]. Especially in high-tech environments, such as Industry 4.0 development or adaptation processes, high importance is attached to technology readiness level.

Technology readiness levels propose a measurement system, but it is also required to develop a metric for determination of integration level. By the help of technology readiness and integration readiness levels, the system maturity can be evaluated comprehensively [13].

2.2 Industry 4.0 Readiness Level

There are different readiness level measurements specified for different operations or processes, such as Design Readiness Level, Manufacturing Readiness Level, Software Readiness Level, and Operational Readiness Level [14]. After "Industry 4.0" is announced as a new concept by German government in 2012, researchers have begun to attach high importance to develop new metrics to measure the "Industry 4.0 Readiness Level". Industry 4.0 Readiness Level evaluations are a metric development and measurement process to check the maturity of related system's technologies and infrastructure before full adaptation is performed. In order to be ready for adapting Industry 4.0 necessities, some pre-conditions should be supplied [15] such as (1) standardization (of systems, platforms, protocols, connections, etc.), (2) work organization, (3) availability of products, new business models, know-how protections, and (4) availability of skilled workers, research, professional development, legal framework. The readiness level of

Industry 4.0 within companies can be measured by evaluation of these 4 items, but the evaluation of readiness level differs regarding the perspective of the academics, practitioners and international consortia in the relevant domains. Most of the readiness and maturity models are proposed for manufacturing industry, which is directly and dramatically affected by Industry 4.0 revolution. IMPULS model [16] consists of 6 evaluation factors including strategy and organization, smart factory, smart operations, smart products, data-driven services, and employees. Tonelli et al. (2016) propose another value modelling and mapping for Industry 4.0 in which "Gartner Maturity Model" takes place [17]. In this model, there are five stages including (1) reacting the firms focusing on operations, (2) forecasting supply chain plans, (3) integrating the firms to the maturity process, (4) collaborating the companies that catch the maturity, and (5) coordinating the barriers between demand and supply. Schumacher et al. (2016) provide a maturity level for evaluating the readiness of Industry 4.0 in terms of nine dimensions; strategy, leadership, customers, products, operations, culture, people, governance and technology [18]. They test the proposed model in many firms. Akdil et al. (2018) present other maturity and readiness model that are consisted of three important dimensions such as smart products and services; smart business processes; strategy and organization [19]. The dimensions are measured conducting surveys, which are graded by scores between 0 and 3 and by checking the scores. The maturity of firms are categorized into 4 parts, such as absence, existence, survival, and maturity.

3 Methodology

This study mainly has two goals:

1. Evaluation of the perceptions and awareness of companies in terms of operational and socio-economic factors,
2. Evaluation and computation of the readiness level by applying the IMPULS model.

Recognizing these goals, the methodology is designed by conducting questionnaires. It is important to state that questionnaires are responded by Industry 4.0 coordinators of each firm whose ideas can represent the perceptions of the related company.

In the first part of the questionnaire, following questions are asked for evaluation of the perceptions and awareness of companies in terms of operational and socio-economic factors:

- Current strategy of the company on concerning Industry 4.0 applications
- The capability of the company on application of Industry 4.0 processes
- The most important motivation source and reason for interest in Industry 4.0 processes
- The existence of a working group/department has the title of digital transformation or Industry 4.0
- The role of foreign capital partnership on Industry 4.0 adaptation
- The existence of an Industry 4.0 project directed by collaborating with a consultant company
- The existence of an Industry 4.0 project directed by collaborating with a university

- The awareness on the patents taken out by universities and techno parks on development of Industry 4.0
- The number of patents taken out by the company on Industry 4.0
- The percentage of Industry 4.0 investments in total annual investment
- The participation of the company to educational activities for development of Industry 4.0
- The most important operations in which Industry 4.0 tools are utilized
- The opinion of the company on Industry 4.0's effect on ethnic discrimination
- The opinion of the company on Industry 4.0's effect on unemployment
- The opinion of the company on Industry 4.0's effect on inequality of women and men
- The opinion of the company on Industry 4.0's effect on income inequality
- The opinion of the company on Industry 4.0's effect on social interaction

In the second part of the questionnaire, following questions are asked for evaluation and computation of the readiness level by applying the IMPULS model:

Strategy and organization:

- The existence of a road map for Industry 4.0 adaptation process
- The existence of a new workforce planning for Industry 4.0 adaptation process
- The strategy development capability for Industry 4.0 adaptation process
- The attitude to improve process compliance in Industry 4.0
- The increase of investments on different operations
- The existence of corporate technology and innovation management capability

Smart factory:

- The capability on collecting the data
- The capability on using real time data for automatic production planning and control
- The reasons for using data

Smart operations:

- The capability on integration of the systems and machines
- The operations in which internal integration has been conducted
- The operations in which external integration has been conducted
- The existence of autonomous control
- The existence of internal data storage
- The security of internal data storage
- The security of cloud systems
- The security of data sharing with business partners
- The operations in which cloud systems are utilized.

Smart products:

- The Information and Communication Technologies add-on functionalities of products

Data-driven services:

- The position of the company on data driven services
- The share of data used in the company

- The share of data used outside the company
- The phases in which the data is utilized

Employees:

- The employee skill sets for Industry 4.0

Three cases in three leading companies in the construction, textile and wire production sectors are conducted. After descriptive analysis is performed, readiness levels are computed. The visualization is presented by using radar chart in order to indicate the strong and weak dimensions of the companies.

4 Cases from Turkey

4.1 Industry 4.0 Adaptation Process of a Construction Company

First of all, the proposed questionnaire is conducted in a leading Turkish construction company. The company was established thirty-one years ago and it has 342 employees. It does not have any foreign capital partnership and not export any of its products.

The results of the questionnaire show that;

- The company is moderately interested in Industry 4.0 processes.
- The company defines itself as a good follower of developments for Industry 4.0.
- The most important motivation sources are "to maintain an innovative spirit, need to gain experience, the necessity of being a technology leader, and market requirements and competitive pressure".
- In this company, the reasons for interest in Industry 4.0 processes are "to increase productivity in production and management systems".
- The company does not have a working group/department, which has the title of "digital transformation for Industry 4.0". However, they aim to establish such a department in the next five years.
- The company states that they would like to improve an Industry 4.0 adaptation process in accordance with the national requirements.
- The company mentions that they can work with a national and/or international consulting group.
- The company states that they are not in collaboration with any national or international university on Industry 4.0 projects.
- The company has high awareness on the patents generated by the universities and techno-parks on development of Industry 4.0, but they have not employed any patent related to Industry 4.0 in the last five years.
- In the company, the percentage of Industry 4.0 investments in the total annual investment is 20%.
- The company does not have any contribution to educational activities for the development of Industry 4.0.
- In the company, the most important operations are R&D and production, where Industry 4.0 tools are used.
- The company highly believes that Industry 4.0 will increase ethnic discrimination.

- The company does not believe that Industry 4.0 will increase unemployment.
- The company does not believe that Industry 4.0 will increase the inequality between women and men.
- The company highly believes that Industry 4.0 will increase unequal distribution of income.
- The company highly believes that Industry 4.0 will decrease social interaction.

The results show that the company believes that they are good in following the new challenges, but they are not sufficient to act proactively. They are conservative in being more competitive and innovative. The competitive pressure has direct impact on their own Industry 4.0 adaptation process. Furthermore, the company has an optimistic idea about the effect of Industry 4.0 on unemployment, but they believe that equal distribution of income will be affected negatively.

For the evaluation of the readiness through the items of IMPULS model, it is assumed that each of the evaluation factors has the same importance weight. In our case, there are 24 factors ($sa_1, sa_2, \ldots, sa_{24}$); therefore each of them has the importance value 0.0416 ($w_1 = w_2 = \ldots = w_{24} = 0.0416$). The total of importance weights is "1". Under these conditions, the readiness of the company (R_C) is evaluated by computing the weighted average using the following formula:

$$R_C = \frac{w_1 * sa_1 + w_2 * sa_2 + \ldots + w_{24} * sa_{24}}{w_1 + w_2 + \ldots + w_{24}} \tag{1}$$

The readiness score of the company is computed as follows:

$$R_{C(construction)} = \frac{\begin{array}{c} 0.0416 * 0 + 0.0416 * 0 + 0.0416 * 2 + 0.0416 * 3 + 0.0416 * 4 + 0.0416 * 1 + 0.0416 * 5+ \\ 0.0416 * 5 + 0.0416 * 5 + 0.0416 * 5 + 0.0416 * 4.14 + 0.0416 * 4.42 + 0.0416 * 3 + 0.0416 * 5+ \\ 0.0416 * 5 + 0.0416 * 5 + 0.0416 * 5 + 0,0416 * 5 + 0.0416 * 3.125 + .0416 * 3 + 0.0416 * 5+ \\ 0.0416 * 4 + 0.0416 * 4.33 + 0.0416 * 4.28: \end{array}}{1}$$

$$R_{C(construction)} = 3.75$$

With the score of 3.75, the company can be accepted as an "experienced" company. In order to demonstrate the readiness level of the company in terms of the six main evaluation factors, a radar chart is prepared. As shown in Fig. 1, the readiness score in terms of strategy and organization is very low as compared to others because the company fails in improving a road map and new workforce planning. The second important reason is that the company has not applied new strategies for Industry 4.0 yet.

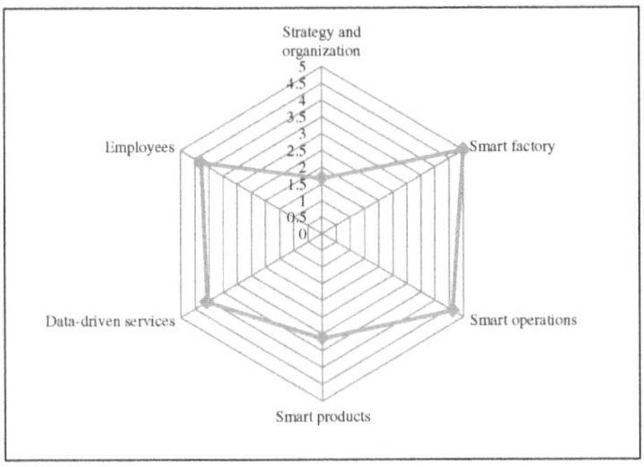

Fig. 1. Radar chart demonstrating the readiness of the construction company in terms of 6 main factors

4.2 Industry 4.0 Adaptation Process of a Textile Company

Secondly, the proposed questionnaire is conducted in a leading Turkish textile company. The company was established forty-five years ago and it has 1140 employees. It does not have any foreign capital partnership and exports its products to eight countries.

The results of the questionnaire show that;

- The company is aware of Industry 4.0 processes and aims to act in this direction.
- The company defines itself as a good follower of developments for Industry 4.0.
- The most important motivation sources are "to maintain an innovative spirit, obtain an opportunity to differentiate its products, and market requirements and competitive pressure".
- In this company, the reason for interest in Industry 4.0 processes is "to increase productivity in production and management systems".
- The company does not have a working group/department, which has the title of "digital transformation for Industry 4.0". However, they aim to establish such a department in the next five years.
- The company states that they would like to improve an Industry 4.0 adaptation process in accordance with the criteria proposed by leading countries.
- The company mentions that they can collaborate with a national and/or international consulting group.
- The company states that they are not in collaboration with any national or international university on Industry 4.0 projects.
- The company has no awareness on the patents generated by the universities and techno-parks on development of Industry 4.0, and they have not employed any patent related to Industry 4.0 in the last five years.
- In the company, the percentage of Industry 4.0 investments in the total annual investment is 10%.
- The company does not have any contribution to educational activities for the development of Industry 4.0.

- In the company, the most important operation is production where Industry 4.0 tools are used.
- The company believes that Industry 4.0 will not increase ethnic discrimination.
- The company believes that Industry 4.0 will not increase unemployment.
- The company believes that Industry 4.0 will not increase inequality of women and men.
- The company believes that Industry 4.0 will not increase unequal distribution of income.
- The company believes that Industry 4.0 will not decrease social interaction.

The results show that the company believes that they are aware of Industry 4.0 processes, but they do not have enough capability to act for implementing Industry 4.0 applications in all aspects. They seem that they are satisfied for being a good follower, and by the effect of the market requirements and competitive pressure. They have potential to act effectively on this issue in the near future. It is also interesting to notice that the company is very optimistic about the socio-economic effects of Industry 4.0. Although textile industry is labour intensive and the digitalization may have direct impact on the workforce, the company prefers to ignore such negative outputs. This is probably because that the company does not have enough experience in implementing real processes for Industry 4.0. Also, the company may believe that new job opportunities can be increased, which might prevent dramatic negative results on unemployment.

For the evaluation of the readiness through the items of IMPULS model, the same assumptions provided in the previous case are accepted. By using Eq. (1), the readiness score of the company is computed as follows:

$$R_{C(textile)} = \frac{\begin{array}{c}0.0416 * 0 + 0.0416 * 0 + 0.0416 * 0 + 0.0416 * 0 + 0.0416 * 2.16 + 0.0416 * 1 + 0.0416 * 4+\\ 0.0416 * 4 + 0.0416 * 3.57 + 0.0416 * 1.66 + 0.0416 * 1.57 + 0.0416 * 1.57 + 0.0416 * 0 + 0.0416 * 3+\\ 0.0416 * 4 + 0.0416 * 3 + 0.0416 * 3 + 0,0416 * 1.66 + 0.0416 * 3.25 + .0416 * 0 + 0.0416 * 4+\\ 0.0416 * 0 + 0.0416 * 2.33 + 0.0416 * 3.57:\end{array}}{1}$$

$$R_{C(textile)} = 1.96$$

With the score of 1.96, the company can be accepted as a "beginner". In order to demonstrate the readiness level of the company in terms of six main evaluation factors, a radar chart is constructed. As shown in Fig. 2, the readiness score in terms of strategy and organization is very low as compared to others because the company fails in improving a road map and new workforce planning. Besides that, it is stated that Industry 4.0 process does not affect the firm strategy; and the adaptability of the company to Industry 4.0 has not been evaluated yet. Interestingly, there is no any new strategic improvement in this company.

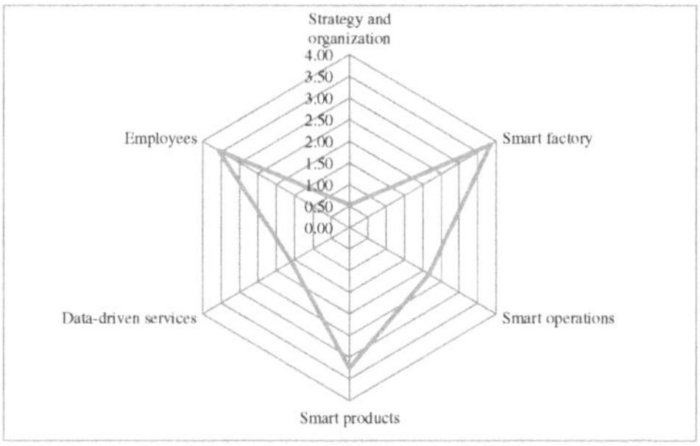

Fig. 2. Radar chart demonstrating the readiness of the textile company in terms of 6 main factors

The second factor that has low score is data driven services. They do not share any data with external stakeholders and not use any data driven system.

4.3 Industry 4.0 Adaptation Process of A Wire Production Company

Thirdly, the proposed questionnaire is conducted in a leading Turkish wire production company. The company was established thirty-six years ago and it has 779 employees. It does not have any foreign capital partnership and exports its products to 39 countries.

The results of the questionnaire show that;

- The company is moderately interested in Industry 4.0 processes.
- The company defines itself as a good follower of developments for Industry 4.0.
- The most important motivation sources are "to maintain an innovative spirit, obtain an opportunity to differentiate its products, and market requirements and competitive pressure".
- In this company, the reason for interest in Industry 4.0 processes is "to increase productivity in production systems and management systems".
- The company has recently established a working group/department, which has the title of "digital transformation for Industry 4.0".
- The company states that they would like to improve an Industry 4.0 adaptation process in accordance with the criteria developed by leading countries.
- The company mentions that they can collaborate with an international consulting group.
- The company states that they are not in collaboration with any national or international university on Industry 4.0 projects.
- The company has low awareness on the patents generated by the universities and techno-parks on development of Industry 4.0, and they have not employed any patent related to Industry 4.0 in the last five years.
- In the company, the percentage of Industry 4.0 investments in the total annual investment is 5%.
- The company does not have any contribution to educational activities for the development of Industry 4.0.

- In the company, the most important operations are logistics and production where Industry 4.0 tools are used.
- The company believes that Industry 4.0 will not increase ethnic discrimination.
- The company believes that Industry 4.0 will not increase unemployment.
- The company believes that Industry 4.0 will not increase inequality of women and men.
- The company believes that Industry 4.0 will moderately increase unequal distribution of income.
- The company believes that Industry 4.0 will not decrease social interaction.

The results show that the company believes that they are interested in Industry 4.0 processes, but their capability is limited to only "being a follower". The company is optimistic about the socio-economic effects of Industry 4.0 except its effect on unequal distribution of income. Contrasting to other companies, although it is a production company, the investment on technology is low. In order to become the pioneer and leader in Industry 4.0 applications, the percentage of Industry 4.0 investments in the total annual investment should be increased.

For the evaluation of the readiness through the items of IMPULS model, the same assumptions given in the previous case are accepted. By using Eq. (1), the readiness score of the company is computed as follows:

$$R_{C(production)} = \frac{\begin{array}{c} 0.0416*0+0.0416*0+0.0416*2+0.0416*3+0.0416*1.66+0.0416*2+0.0416*4+ \\ 0.0416*4+0.0416*3.57+0.0416*1+0.0416*2.14+0.0416*0+0.0416*0+0.0416*3+ \\ 0.0416*3+0.0416*3+0.0416*3+0,0416*0+0.0416*0+.0416*0+0.0416*5+ \\ 0.0416*3+0.0416*3+0.0416*4.14: \end{array}}{1}$$

$$R_{C(production)} = 2.1$$

With the score of 2.1, the company can be accepted as a "beginner" company. In order to demonstrate the readiness level of the company in terms of six main evaluation ctors, a radar chart is used. As shown in Fig. 3, the readiness score in terms of smart products is very low compared to others. Because there are not any information and communication technologies add-on functionalities of products. Strategy and organization and smart operations have also very low scores. The reasons are the same as the other companies in terms of strategy and organization. However, for smart operations, the reasons for low scores are (1) there is not any external integration among any operations, (2) there is not any autonomous control, and (3) there is not any cloud system.

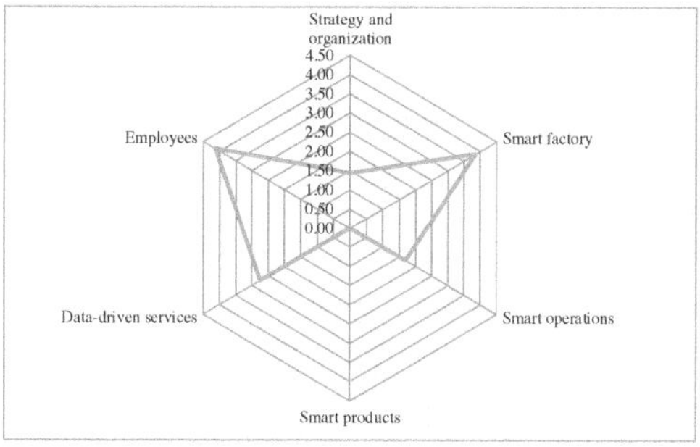

Fig. 3. Radar chart demonstrating the readiness of the wire production company in terms of 6 main factors

5 Findings and Discussion

The most important findings of the study are:

- Companies fail to develop new road maps and new workforce planning strategies in the adaptation process. This is the most important impact in decreasing the score of readiness level.
- Data-driven services are not used in the companies efficiently. This output has the secondly most important impact in decreasing the score of readiness level.
- Under the uncertain environment of Industry 4.0, the companies tend to be the follower rather than the pioneer.
- Interestingly; integration, being one of the vital necessities of Industry 4.0, does not affect the wire production company although it exports its products to many developed countries.
- It is interesting to observe that there is lack of awareness, but they have highly optimistic ideas on the socio-economic factors of Industry 4.0. It makes adaptation easy to accept and implement, but it can also prevent employing necessary actions before some unsatisfactory results for the society occur. Therefore, the firms should change their mentality and point of views to protect the rights of society for sustainability.
- None of companies have any idea about the patents taken by universities and they are not collaborating in any project with any university. It clearly shows that their interests on the capabilities and opportunities of universities are really poor. In order to follow the new challenges in high-tech processes such as Industry 4.0 revolution, they have to collaborate with R&D study groups, universities, consortiums besides the support of consulting groups.

In addition to these important results, it may be stated that the readiness level scores are calculated by considering the perceptions of the company. It is not the result of a real and detailed observation on the practices of Industry 4.0 adaptation processes. Therefore, this score is based on perceptions and possible to have high subjectivity.

6 Conclusion

This study aims to evaluate the perceptions and readiness levels of leading companies on Industry 4.0 adaptation processes in Turkey. As a result of the evaluation process, it is revealed that there is lack of motivation to be the pioneer and leader in Industry 4.0. Hence, the companies tend to continue to be a "follower" in order to preserve their current conditions under control. Although, they are conservative to apply new developments in a short time, they believe that Industry 4.0 will have positive effects on socio-economic conditions. Moreover, there is lack of consciousness and effort on new strategy development. They have not prepared any road map and new workforce plans yet.

This study contributes by dealing with some cases in Turkey as an emerging economy, considering the points of view and awareness on Industry 4.0 adaptation process in such environment. This study also introduces readiness level measurement applications for a developing country. In the future, the survey can be conducted with many other companies to perform advanced statistical analysis. Furthermore, because the scores are originated from the perceptions, which have subjectivity, fuzzy set theory can be utilized for scoring. Moreover, rather than assuming that all weights are equal, the importance of the evaluation factors can be computed by the help of decision making tools such as analytical hierarchy process.

References

1. McKinsey&Company (2014) www.mckinsey.com
2. Schlaepfer RC, Koch M, Merkofer P (2015) Industry 4.0: challenges and solutions for the digital transformation and use of exponential technologies. Deloitte AG. http://deloittecom/content/dam/Deloitte/ch/Documents/manufacturing/chen-manufacturing-industry-4-0-24102014.pdf, January 2018
3. Mitra S (2015) Digital manufacturing POV. Hewlett-Packard Development Company. http://www.petrochemconclave.com/presentation/2015/Dr.SMitra.pdf. December 2017
4. Heng S (23 April 2014) Industry 4.0 upgrading of Germany's industrial capabilities on the horizon. Deutsche Bank research, Germany. https://www.dbresearch.com/PROD/RPS_ENPROD/PROD0000000000451959/Industry_4_0%3A_Upgrading_of_Germany%E2%80%99s_industrial_ca.PDF, January 2018
5. Audio-Tech Business Book Summaries, Inc. (2015) Industry 4.0 and the U.S. manufacturing renaissance. Trends E-Magazine, 4–10 June
6. Zawadzki P, Zywicki K (2016) Smart product design and production control for effective mass customization In the industry 4.0 concept. Manag Prod Eng Rev 7(3):105–112
7. Stork A (2015) Visual computing challenges of advanced manufacturing and industrie 4.0. IEEE Computer Graphics and Applications, March–April 2015
8. Global Trends 2030 (2012) Alternative Worlds, National Intelligence Council. https://globaltrends2030.files.wordpress.com/2012/11/globaltrends-2030-november2012.pdf. January 2018
9. Maslaric M, Nikolicic S, Mircetic D (2016) Logistics response to the industry 4.0: the physical internet. Open Eng 6:511–517
10. Industry 4.0 – The State of the Nations, INFOSYS. http://images.experienceinfosys.com/Web/Infosys/%7Bf0e3bb53-176a-4b5a-991b-0708c00fc0a9%7DResearchReport.pdf. January 2018

11. Cambridge Business English Dictionary (2011)
12. Mankins JC (2009) Technology readiness evaluations: a retrospective. Acta Astronaut 65:1216–1223
13. Sauser BJ, Forbes E, Long M, McGrory SE: Defining an integration readiness level for defense acquisition. In: International symposium of the international council on systems engineering
14. Sauser B, Ramirez-Marquez JE, Magnaye R, Tan W (2009) A systems approach to expanding the technology readiness level within defense acquisition. 2009, report for the Acquisition Chair of the Graduate School of Business & Public Policy at the Naval Postgraduate School
15. Smit J, Kreutzer S, Moeller C, Carlberg M (2016) Industry 4.0, study for the ITRE committee, policy department a: economic and scientific policy. Eur Parliament, Brussels
16. Lichtblau K, Stich V, Bertenrath R, Blum M, Bleider M, Millack A, Schmitt K, Schmitz E, Schröter M (2015) IMPULS - industrie 4.0- readiness. Impuls-Stiftung des VDMA, Aachen-Köln
17. Tonelli F, Demartini M, Loleo A, Testa C (2016) A novel methodology for manufacturing firms value modeling and mapping to improve operational performance in the industry 4.0 era. Procedia CIRP 57:122–127
18. Schumacher SE, Sihn W (2016) A maturity model for assessing Industry 4.0 readiness and maturity of manufacturing enterprises. Procedia CIRP 52:161–166
19. Akdil KY, Ustundag A, Cevikcan E (2018) Maturity and readiness model for industry 4.0 strategy. In: Ustundag A, Cevikcan E (eds) Industry 4.0: managing the digital transformation. Springer

Industry 4.0 Scorecard of Turkish SMEs

Zeynep Gergin[1(✉)], Fadime Üney-Yüksektepe[1], M. Güneş Gençyılmaz[2], A. Tülin Aktin[1], Kemal Güven Gülen[3], Doğan Aybars İlhan[1], Uğurcan Dündar[1], Özay Cebeci[2], and Ali İhsan Çavdarlı[4]

[1] Industrial Engineering Department, Istanbul Kültür University, Istanbul, Turkey
{z.gergin, f.yuksektepe, t.aktin, a.ilhan, u.dundar}@iku.edu.tr

[2] Industrial Engineering Department, Istanbul Aydın University, Istanbul, Turkey
gunes.gencyilmaz@aydin.edu.tr, ozaycebeci@stu.aydin.edu.tr

[3] Industrial Engineering Department, Tekirdag Namık Kemal University, Tekirdağ, Turkey
kggulen@nku.edu.tr

[4] KOSGEB Small and Medium Sized Enterprises Development Organization, Ankara, Turkey
ali.cavdarli@kosgeb.gov.tr

Abstract. Today the industry is revolutionizing into a new era with the total integration of all elements of the production through digitalization and establishing communication between each other. This new revolution is named as "Industry 4.0". Turkey is categorized as a developing country in terms of economy and Turkish economy is ranked at 17th place in world Gross Domestic Product (GDP) ranking. The country's economy met the expectations and reached 7.4% GDP growth in 2017 while worldwide average growth was 3.6%, and developing countries had the average of 4.6%. Turkish government encourages enterprises to keep up with industry 4.0, and KOSGEB (Small and Medium Sized Enterprises Development Organization) is supporting the SMEs with various programs. This can be seen as an opportunity for Turkish industry to catch the trend and increase its growth rapidly. And at this point the exact knowledge of the current situation is crucial. Small and Medium Sized Enterprises (SME) has the 99.8% of the total according to Turkish Statistical Institute (TÜİK). Hence, this project is initiated to identify the current situation of SMEs of Turkish industry. A comprehensive questionnaire is prepared to find out the readiness and requirements of the SMEs for Industry 4.0 transformation. This questionnaire is applied to SMEs operating in Marmara Region of Turkey. The outcome of the research shows a scorecard for the SMEs in terms of technology usage, readiness to Industry 4.0, and also increased their awareness on the concept and the importance of the new era. The future studies are continuing as a second phase of this research in order widen the research to all regions of the country, and to develop the transformation capabilities of the selected enterprises.

Keywords: Economy · Industry 4.0 · Turkish economy · Turkish industry

N. M. Durakbasa and M. G. Gencyilmaz (Eds.): ISPR 2018, *Proceedings of the International Symposium for Production Research 2018*, pp. 426–437, 2019.
https://doi.org/10.1007/978-3-319-92267-6_37

1 Introduction

The industry is one of the biggest elements of the economy that converts raw materials into products. So far, industry had three revolutions. The first revolution took place at the end of 18th century with the start of steam-powered mechanization. The second revolution was initiated with the usage of electricity in mechanization and mass production concept at the beginning of 20th century. At the end of 1970s, the computer integrated manufacturing started, and it was called the third revolution of the industry. Today the industry is revolutionizing into a new era with the integration of machines, humans, internet, products and each element of the production with each other which is named as "Industry 4.0" [1]. This revolution is not just coming or at the edge of the corner, it has already stepped inside. It's not only the first revolution in the history of industry that is pre-named, it is also the first one that can be shaped in purpose [2].

In the literature there are several definitions of Industry 4.0 [2–8, 9]. Most commonly it aims to form a system that has communication between humans, machines and all resources. And unlike previous designs, in Industry 4.0 decentralized production processes, through smart products that know their destinations, history etc., are playing a key role rather than centrally controlled processes [9]. On the contrary of the variety of Industry 4.0 definitions, the researchers almost fully achieved a consensus about what it includes. It is more appropriate to use definitions for these constituents that are used by TÜSİAD (Turkish Industry & Business Association) in the report that they published with the association of BCG (Boston Consulting Group) in 2016 [7]. In the report these constituents are called technology drivers of Industry 4.0 which are listed below;

Big Data and Analytics: It is the analytics of large data sets that are coming from various sources to assist real-time decision-making.

Autonomous Robots: It is the concept of robots that are more flexible, cooperative and autonomous with lower prices.

Simulation: It is recreating the production environment in a virtual world that mimics the real world with real-time data.

Horizontal and Vertical System Integration: It is the total integration of the functions of the companies, in the company and cross company as well, in the global basis.

The Industrial Internet of Things: It is the virtual platform of everything in the process. It allows parts of the production to communicate with each other and a central controller.

Cyber Security: With the increase in connected devices in virtual world, cyber security's importance increases.

Cloud: It is a system where data stored and shared. With the increased integration of computer embedded and human systems, the more data should be shared and stored globally.

Additive Manufacturing. It is a concept that generally used to build a prototype or produce some individual parts.

Augmented Reality: It is an augmentation of real world with a virtual world where computer-generated images are imposed to real world vision.

Literature research shows that, most of the studies are focused on defining Industry 4.0 and its constituent parts. However, there are only a few numbers of research that focus on scoring the readiness of companies [2–4, 7]. In addition, the lack of studies focusing on Small and Medium Sized Enterprises (SME) is seen especially in Turkey.

PWC published a report in 2016 that investigates the readiness of companies in 26 + countries, 9 major industrial sectors and 2000+ attendants [6]. They declared that Industry 4.0 is not a future trend anymore, it is now in companies' agendas. Furthermore, findings show that core businesses are attuning themselves to the trend much faster than other businesses. Also, attendees of the survey believe that digitization will be doubled until 2020. Based on a survey applied for Turkish companies, TÜSİAD's report highlights that the Turkish Industry is aware of what Industry 4.0 will bring to the table with pros and cons, and is willing to attune itself to Industry 4.0.

Turkey is categorized as a developing country in terms of economy [10] and Turkish economy is ranked at 17th place in world GDP ranking [11]. The country's economy met the expectations and reached 5.1% GDP growth in 2017 while worldwide average growth is 3.6% and developing countries have the average of 4.6% [12]. The more detailed look on economic situation of Turkey shows that Turkish industry has 16.6% of the GDP in the country. [13] Small and Medium Sized Enterprises (SME) has the 99.8% of the enterprises share of Turkey according to Turkish Statistical Institute (TÜİK) [14]. SMEs are defined as the enterprises that has less 250 employees and less than 40 million Turkish Liras (TL) endorsement or balance [15].

As being one of the major drivers in Turkish economy, SMEs' current situation and potential for future are needed to be well understood in terms of Industry 4.0 perspective. It is better for SMEs policy makers and institutions to develop proper strategies for near and far future to be able to stay competitive. This can be possible by;

- defining current industrialization level and awareness on new trends and technology as the first stage, (Academicians, SMEs, Institutions)
- then informing policy makers and SME's about defined topics for them to discuss possible actions to be taken, (Academicians, Policy makers, SMEs, Institutions, NGOs)
- arranging corresponding incentives, funds, campaigns etc. to keep up with the new technology and trends (Policy makers, institutions)
- developing and/or adopting efficient methods, high technology equipment/systems etc. satisfying Industry 4.0 requirements (SMEs).

Turkish industry fell behind at following the trends and industrial revolutions in the history. This can be seen as an opportunity for Turkish industry to catch the trend and increase its growth rapidly. Turkish government also encourages enterprises with various incentives to keep up with industry 4.0. When advancement is being made, the exact knowledge of the current situation is crucial. Adopting Industry 4.0 implementation requires a four stage approach as given in Fig. 1.

This study focuses on the first stage of the proposed course of actions. In this project we aim to see the current situation of Turkish industry, and measure the level of awareness and readiness of SMEs for Industry 4.0. Based on the concept, a

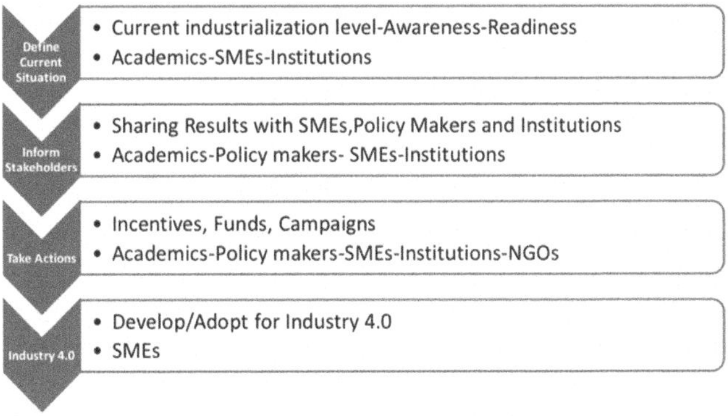

Fig. 1. Stages of Industry 4.0 adoption

questionnaire is prepared and applied to SME's in Marmara Region of Turkey. The outcome is giving a score for each SME and showing the situation of SME's in Turkey in terms of technology usage and readiness for Industry 4.0 adoption.

2 Methodology

2.1 Questionnaire Design

A questionnaire with 28 questions are deigned specifically for the research. Questions are prepared by academicians considering the relevant literature including papers, technical reports, and comments of KOSGEB and SME representatives. Questionnaire is designed both on an e-survey platform and in Microsoft Word document to be distributed to targeted SME's. The aim of the questionnaire is mainly to take information on SMEs' profile, current industrialization level of SMEs, and awareness and readiness of SME's on Industry 4.0. Hence it consists three main Sections.

(1) *Company Information*

Company information is crucial when analyzing the collected data to make various types of comparisons and conclusions based on sector, location, size etc. Therefore, the first section has six questions with regard to the general information about the company like its sector, location, number of employees of, and the position of the respondent that answers the questions.

(2) *Industrialization Level*

In the second section of the questionnaire, the aim is to find the score showing the industrialization level of the SMEs. There are 12 questions collecting data from SMEs on their Industry 4.0 adoptions on various processes such as, order taking, product design/development, work order management, purchasing, manufacturing technology & hardware usage, software/automation & software usage, quality-control, inventory & warehouse management, maintenance management, and mobile application usage.

Considering the definitions and requirements of all industrial revolutions, each answer option of a question corresponds to an industrialization score between 1.0 and 4.0. A question from this part is given in Table 1 as an example.

Table 1. A sample question for scoring industrialization level

Which one(s) is/are most suitable for work order system in your production environment?	Score
Verbal communication is used to give an order and to get feedback about the order	1.0
Written communication is used to give an order without a prepared form	1.3
Order is sent to corresponding units via PC using a prepared form	2.8
A work order management system/software is used	3.3
An e-system/software, consisting of e-tickets attached to the products, is used	4.0

(3) *Barriers and Enablers for Adopting Industry 4.0 Implementations*

A crucial factor affecting survival of SMEs is being well prepared to stay competitive in the market. Because being aware of new technology and trends plays an important role for SMEs to be ready for future. Hence in the third section 10 questions asked to get relevant data on awareness and readiness including some questions on challenges and enablers for SMEs to invest for Industry 4.0 technology, or adopt requirements of Industry 4.0.

2.2 Validity and Reliability of the Questionnaire

The validity and reliability of the questionnaire is tested with two groups of people. Initially, a Delphi study is applied with a group that consists of academicians and industry representatives dealing with Industry 4.0 concepts. The clarity and representativeness of the questionnaire with regard to Industry 4.0 concepts, and the scoring of the options are evaluated for validity with this group of experts. For the reliability test, a small sample of respondents is selected, and their answers are collected from the same set of respondents twice. The data obtained from the two test implementations are compared in order to find the differences. The responses are found to be 97% consistent, and the reliability of the questions is concluded.

3 Implementation and Results

In the first stage of the project, the survey is implemented to the SME's location in Marmara Region. In total of 193 SMEs participated in the survey. Approximately, 38% of the companies are located in İstanbul whereas remaining ones are in Bursa, Tekirdağ, Çorlu and Kocaeli. In the sector analysis part, majority sectors are Rubber & Plastics Manufacturers, Metal Manufacturers and Machine & Equipment Manufacturers.

To determine the industrialization levels, the scores analyzed with the answers to the 12 questions in the second section of the questionnaire is given in Table 2. First column represents the question numbers and second column gives the measured process. For each process, minimum, maximum and average scores are calculated. Among the processes, the lowest average score is obtained for warehouse management processes. On the other hand, mobile application usage has the highest average score. SMEs have also a high score for traceability, production technology and software implementations. The remaining processes have a moderate average score. Among all processes, there exists a SME with an average score close to 4.0 or equal to 4.0. However, minimum scores are relatively low for each process except traceability, purchasing, production technology, software and mobile application usage. Those companies should be supported to improve their current situation for those processes.

Table 2. Overall results for industrialization level

Question	Process	Minimum score	Maximum score	Average score
6	Order receiving	1.00	3.80	2.26
7	R&D	1.00	4.00	2.92
8	Work order	1.00	4.00	2.76
9	Traceability	3.30	4.00	3.62
10	Purchasing	2.50	4.00	2.64
11	Production technology	2.80	4.00	3.47
12	Software	2.60	4.00	3.12
13	Quality control	2.20	3.50	2.73
14	Inventory management	2.00	4.00	2.63
15	Warehouse management	1.00	4.00	1.82
16	Maintenance management	1.50	4.00	2.27
19	Mobile application usage	3.30	3.80	3.68
Overall scores		1.73	3.55	2.69

Overall industrialization score of the 193 companies in Marmara Region is 2.69. Figure 2 shows the overall scores of the companies. To obtain that evaluation, the scores are rounded to the nearest integer. For instance, if the score is in between 1.50 to 2.49, the company's score is assumed as 2.00. As a result, 66 companies are categorized in the group with industrialization score of 2. Majority of the companies exist in the group with a score 3. On the contrary, only two of the companies are currently has a score of 4. Consequently, the overall score 2.69 is a representative overall score for SMEs. The SME with the lowest overall score (1.73) manufactures paper-based products in Bursa. In contrast, the highest overall scores (3.55) and (3.52) are achieved by a metal manufacturer and a textile producer, respectively.

When industries are analyzed separately, it is observed that they have different but close average scores. As displayed in Table 3, the average industrialization score for 'Printing and Copying Recorded Media Materials' industry has the maximum with a value of 3.10. There are no other sector that passed the 3.0 level, however they are all above 2.5 except one. Companies manufacturing outfits and computers, electronics &optic materials are listed as the second and third runners with 2.95 and 2.88 scores respectively. The only sector below 2.5 is the 'manufacture of other transport vehicles' sector with a result of 2.30 on average.

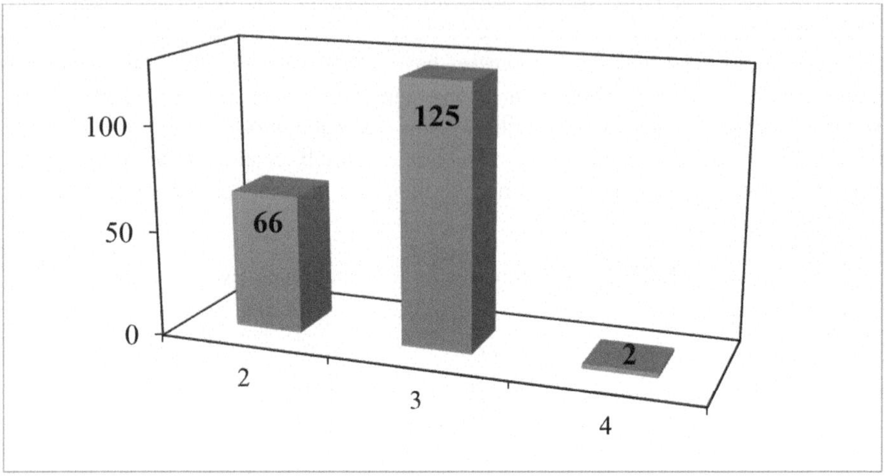

Fig. 2. Overall industrialization scores

Table 3. Average scores of industries

Sector	Score
Printing and copying recorded media materials	3.10
Outfit manufacturing	2.95
Manufacturing computers, electronics and optic materials	2.88
Manufacturing rubber and plastic materials	2.84
Base metal industry	2.77
Manufacturing chemicals and chemical products	2.75
Manufacturing basic pharmaceutics and pharmaceutics materials	2.71
Manufacturing electronic equipments	2.69
Furniture manufacturing	2.69
Manufacturing non-metallic minerals	2.68
Manufacturing leather and its bi-products	2.68
Manufacture of motor vehicles, trailers and semi-trailers	2.67
Manufacture of textile products	2.65
Installation and repair of machinery and equipment	2.64
Manufacture of wood, wood and cork products (except furniture); Manufacture of articles of straw, straw and similar materials	2.63

(*continued*)

Table 3. (*continued*)

Sector	Score
Manufacture of fabricated metal products, except machinery and equipment	2.62
Manufacture of paper and paper products	2.62
Other productions	2.62
Manufacture of food products	2.60
Manufacture of machinery and equipment not elsewhere classified	2.54
Manufacture of other transport vehicles	2.30

Remaining 10 questions are to gather some information on the awareness, readiness and interest of SMEs for Industry 4.0, and the summary of the results are given in Table 4. Most of the SMEs under consideration are currently using sensor systems, automation and preventive maintenance. On the other hand, almost majority of the companies are not aware of the Internet of Things, Additive (3D) Manufacturing, Augmented Reality and Big Data. As most of the SMEs know or currently using Automation, Improved Robot Technologies, Sensor Systems, Smart Warehouses and Mobile Systems, they want to invest money mostly on those technologies.

Table 4. Awareness of technologies

Technologies	Not known	Known but not in use	Currently in use	Planned to invest and use
3D Printing	14.69%	60.45%	16.38%	8.47%
Sensor systems	5.00%	41.67%	42.22%	11.11%
Virtual reality	37.21%	56.98%	1.74%	4.07%
Cloud technologies	23.30%	46.59%	23.30%	6.82%
Internet of things	43.20%	43.79%	4.73%	8.28%
Additive (3D) manufacturing	43.86%	42.11%	8.77%	5.26%
Improved robot technologies	23.12%	52.60%	12.72%	11.56%
Automation	8.65%	35.14%	42.70%	13.51%
Modeling and simulation	13.48%	48.31%	31.46%	6.74%
Online education	24.71%	57.06%	10.00%	8.24%
Electronic performance reports	21.84%	47.70%	20.69%	9.77%
Big data	42.26%	46.43%	7.14%	4.17%
Augmented reality	49.70%	44.97%	1.78%	3.55%
Cyber security	20.59%	44.12%	28.24%	7.06%
Integrated product development	31.58%	41.52%	17.54%	9.36%
Predictive maintenance	37.65%	37.65%	21.18%	3.53%
Preemptive maintenance	18.97%	30.46%	45.40%	5.17%
Smart warehouses	23.81%	61.31%	4.76%	10.12%
Mobile systems	21.05%	50.29%	16.96%	11.70%

Benefits of Industry 4.0 from the perspective of SMEs are shown in Fig. 3. With 32.05%, increase in productivity gets the highest value. Decrease in costs stand in the second position with 28.41%. Improve demand forecast and increase in demand has considerably high occurrence with 20% and 17.73%, respectively.

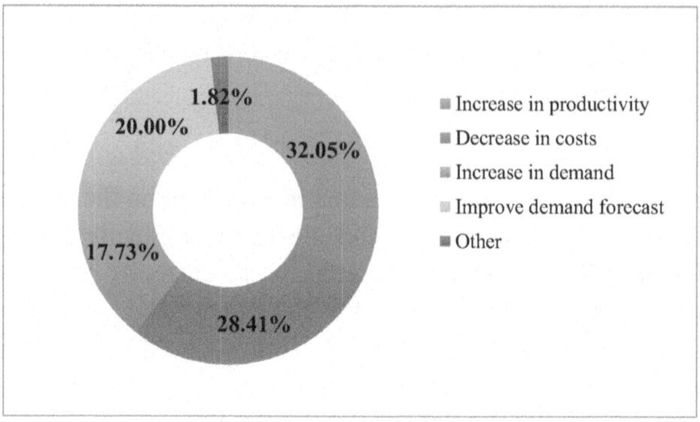

Fig. 3. Benefits of industry 4.0

Table 5 summarizes the comments of SMEs on the barriers of Industry 4.0. Majority of them think that, high cost of new technologies, financial resource shortage and high cost of or lack of consultancy are among the important barriers with a high effect. Thus, rather than the other affects, cost related issues constitutes the highest

Table 5. Barriers of industry 4.0

Barrier	No effect	Low effect	High effect
Current hardware is not suitable	23.62%	39.37%	37.01%
Current software is not suitable	25.41%	33.61%	40.98%
Financial resource shortage	7.79%	31.17%	61.04%
High cost of the new technologies	12.79%	27.91%	59.30%
Weak technical support from suppliers	20.71%	37.86%	41.43%
Complexity of new technologies	31.88%	41.25%	26.88%
Mismatch of work flow in the facility	26.53%	44.22%	29.25%
Lack of knowledge on new technologies	21.58%	43.17%	35.25%
Lack of education of workers in production	16.94%	37.90%	45.16%
Technical personnel's insufficient knowledge/capability	16.28%	48.06%	35.66%
Resistance of employees to change	26.47%	39.71%	33.82%
Resistance of top management to change	50.64%	34.62%	14.74%
Lack of support from universities/government	15.69%	40.20%	44.12%
High cost or lack of consultancy	10.64%	30.85%	58.51%
Low demand	22.70%	40.43%	36.88%
Lack of support of the big clients	26.62%	41.01%	32.37%
Uncertainty of return on investment	16.22%	40.54%	43.24%
Hesitation on data security and privacy	34.04%	43.97%	21.99%

problem that should be improved. On the other hand, it is observed that neither employees nor top managements resistance to change has an effect on the transition. This can be interpreted as no resistance to higher industrialization, hence an interest on new technologies.

Answers to the question "Which information sources are important for your company while you are gathering information on Industry 4.0?" are summarized in Table 6. All of the listed sources are commonly considered as more important. SMEs think that websites are more important among others information sources. Participating to trade fairs and memberships in trade unions are the second and third information sources after the web sites. SMEs also declared their interest in attending academic symposiums for collaboration with universities.

Table 6. Information sources for Industry 4.0

Information sources	Not important	Less important	More important
Websites	8.33%	20.00%	71.67%
Articles on search engines	11.57%	31.40%	57.02%
Trade fairs	10.71%	26.79%	62.50%
Direct mails	17.65%	36.76%	45.59%
Papers in journals	16.22%	39.86%	43.92%
Trade unions	9.45%	29.92%	60.63%
Presentations in congress	10.85%	34.11%	55.04%
Consumers	17.19%	33.59%	49.22%
Holding company	15.45%	30.08%	54.47%
Equipment vendors	10.29%	36.76%	52.94%
Consultants	9.76%	34.96%	55.28%
Universities	16.24%	31.62%	52.14%
Government foundations	19.79%	32.29%	47.92%

4 Conclusions

The industry is revolutionizing into a new era with the integration of machines, humans, internet, products and each element of the production with each other which is named as "Industry 4.0". This study is initiated due to the fact that the number of researches focusing on the awareness and readiness of Turkish SMEs for Industry 4.0 is scarce.

SMEs in Turkey have the 99.8% share of the total economy according to Turkish Statistical Institute (TÜİK), and Turkish government encourages enterprises to keep up with the new technological advances of this new era. This can be seen as an opportunity for Turkish industry to catch the trend and increase its growth rapidly. Hence, this project is initiated to identify the current situation of SMEs that are operating in various industries with regard to Industry 4.0. Consequently, a questionnaire is deigned to give an industrialization score between 1.0 and 4.0 to the companies. The responses to the questionnaire are collected via an e-survey platform together with sent e-mails.

The results show that overall industrialization score of the 193 companies analyzed in Marmara Region is 2.69. When the scores are rounded to integers, it is found that 66 companies have the industrialization score of 2.0, and the majority has a score of 3.0. On the contrary, only two of the companies currently have a score of 4.0. The SME with the lowest overall score (1.73) manufactures paper-based products in Bursa. In contrast, the highest overall scores (3.55) and (3.52) are achieved by a metal manufacturer and a textile producer, respectively. When the average scores of the different sectors are analyzed 'Printing and Copying Recorded Media Materials' industry has the maximum industrialization score with a value of 3.10. There are no other sector that passed the 3.0 level, however they are all above 2.5 except 'manufacture of other transport vehicles' sector with a result of 2.30 on average.

The findings on the awareness, readiness and interest of SMEs for Industry 4.0 show that most of the SMEs are currently using sensor systems, automation and preventive maintenance. On the other hand, many companies are not aware of the internet of things, additive (3D) manufacturing, augmented reality and big data. Benefits of Industry 4.0 from the perspective of SMEs are perceived as increase in productivity and demands, and decrease in costs. Whereas majority claims the high costs as the main barrier for implementations of Industry 4.0 technologies.

Based on the findings of this research it can be concluded that the SMEs require more information on Industry 4.0 adoption. Hence, KOSGEB should organize more educational seminars on these topics for training SMEs and increasing awareness. What's more the governments should develop new financial support programs to overcome the cost barriers of the SMEs. The future studies are continuing as a second phase of this research in order widen the research to all regions of the country, and to develop the transformation capabilities of the selected enterprises.

Acknowledgment. This research is made possible with the support of KOSGEB. The researchers express their gratitude to KOSGEB Management for their support on this study and for the ongoing research to cover all regions of Turkey.

References

1. Turkish Industry & Business Association (2017) Türkiye'nin Sanayide Dijital Dönüşüm Yetkinliği, TÜSİAD-T/2017, 12–589
2. Drath R, Horch A (2014) Industrie 4.0: Hit or Hype? IEEE Industr Electron Mag 8(2):56–58
3. Hermann H, Pentek T, Otto B (2016) Design principles for industrie 4.0. In: 49th Hawaii international conference on system sciences
4. Siemens (2016) "Endüstri 4.0 Yolunda"
5. Impuls (2015) "Industrie 4.0- Readiness", Aachen, Köln
6. PriceWaterhouseCoopers (PWC) Industry 4.0: Building the digital enterprise. https://www.pwc.com/gx/en/industries/industries-4.0/landing-page/industry-4.0-building-your-digital-enterprise-april-2016.pdf
7. Turkish Industry & Business Association (TÜSİAD) (2016) Industry 4.0 in Turkey as an Imperative for Global Competitiveness: An Emerging Market Perspective, TÜSİAD-T/2016-03/576

8. McKinsey Industry 4.0: How to Navigate Digitization of the Manufacture Sector. https://www.mckinsey.com/business-functions/operations/our-insights/industry-four-point-o-how-to-navigae-the-digitization-of-the-manufacturing-sector
9. Kagermann H, Lukas W, Wahlster, W (2011) Industrie 4.0: Mit dem Internet der Dinge auf dem Weg zur 4. industriellen Revolution. VDI nachrichten 13
10. United Nations, State of Commodity Dependence 2016, United Nations Conference on Trade and Development 2017
11. GDP ranking (n.d.). https://data.worldbank.org/data-catalog/GDP-ranking-table. Accessed 14 Jan 2018
12. GDP Map (n.d.). http://www.imf.org/external/datamapper/NGDP_RPCH@WEO/OEMDC/ADVEC/WEOWORLD/TUR. Accessed 14 Jan 2018
13. Turkish Statistical Institute (TÜİK) (September 2017) Annual Gross Domestic Product, 2016, Press release, No: 27817
15. Turkish Statistical Institute (TÜİK) (November 2017) Small and Medium Sized Enterprises Statistics, 2016, Press release, No: 21540
15. Ministry of Science (2015–2018) Industry and Technology, Kosgeb, Kobi Stratejisi ve Eylem Planı (KSEP)

Measurement Technology & Quality & Justicia in Industry 4.0

Andreas Bauer, Gökcen Bas, Numan M. Durakbasa, Lukas Kräuter, and Gamze Ugur–Tuncer(✉)

Vienna University of Technology, AuM, Vienna, Austria
{goekcen.bas,durakbasa,lukas.kraeuter}@tuwien.ac.at, e1576078@student.tuwien.ac.at

Abstract. Development and integration of advanced manufacturing technologies in the so labelled Industry 4.0 hold the promise of benefits as well as challenges. The need for addressing challenges lay out the requirement for the measurement & law & quality fundamentals. Industry of the future defined with guidelines will not only provide the added value in industry but moreover for humanity by means of a systems approach in the new era of industry. In this study, we establish the concepts and activities required as pillars of "Justicia/Justice" by means of measurement & law & quality in advanced manufacturing operations. A universal basic model will highlight main factors and increase insight in key success players integrating future plans in a meaningful way.

Keywords: Measurement · Quality · Justicia · Industry 4.0
Advanced manufacturing

1 Introduction

The manufacturing industry transforms with the value added by digitization, information and communication applications in the design, manufacturing and service. This transformation to the next generation advanced operation label a new era "Industry 4.0" [1]. Industry 4.0 requires meeting the increasing demand in industry more efficiently with high-tech features. In order to integrate high-tech features, comprehensive and appropriate strategies and models are required to overcome the challenges of continuous advancement [2].

The challenge of maintaining a sustainable development towards the goals of advancement, value creation in social, environmental and economic dimensions play vital role [3]. Therefore, research focus on both opportunities and challenges in strategic, operational, environmental, social as well as economical characteristics of the advanced technology implementations [4–6].

European Commission defined six Key enabling technologies (KETs) in 2009, which consist of micro and nano-electronics, nanotechnology, industrial biotechnology, advanced materials, photonics, and advanced manufacturing technologies contributing to growth and job creation while achieving re-industrialisation, energy, and climate

N. M. Durakbasa and M. G. Gencyilmaz (Eds.): ISPR 2018, *Proceedings of the International Symposium for Production Research 2018*, pp. 438–450, 2019.
https://doi.org/10.1007/978-3-319-92267-6_38

change targets simultaneously [7]. The Key enabling technologies (KETs) identified as priority for EU industrial policy receive support under EU research and innovation programmes with an allocated budget of EUR 6.6 billion [8]. Together with the Industry 4.0 transition and the KETs, the manufacturing sector has a higher priority on products that requires transformations in the industrial systems and their business models [9]. In this study, we highlight the main factors of measurement & law & quality in advanced manufacturing operations in order to establish "Justicia/Justice" enabling organizations to integrate future plans in a meaningful and sustainable way.

2 Metrology and Measurement-Technology

The defined KETs and individual product developments make it fundamental to cooperate the advanced manufacturers with technology developers for precision in micro-scale and nanoscale production in the re-industrialised facilities. The concept of micro-components, nanostructures, micro and nano surface characteristics have brought the concepts of robust and repeatable fabrication of micro-/nanostructures as strict requirements [10]. This future needs of advanced manufacturing industry in high precision engineering and workpieces call for measurement instrumentation that can be applied reliably in modern production processes, together with international standards defining parameters and tolerances in the nanometer scale, using in a high accurate environment such as high precision metrology laboratory.

The operating manufacturing plants are developing to keep up with the digitisation and automation applications for advanced industrial processes in every re-industrialisation transition [11]. The simulation of the control systems for development of computer aided, automated production are demand driven aspects that states the future of the manufacturing industry [12]. Moreover, further optimization and quality parameters can be fulfilled by means of testing through virtual process models [13].

Advanced technologies in precision engineering, machining, biotechnology, optics, electronics, materials, will increasingly require high-accuracy qualification of mechanical and electrical properties in addition to physical dimensions. The need for new high precision and innovative measurement techniques are indispensible to provide reliable measurement results with small uncertainties at the micro and nano-scale production concurrent with the latest manufacturing developments.

Increasing demands on precision and accuracy of measurement results in manufacturing and metrology is a particular important challenge in industry. Adequate competencies, knowledge and experience in metrology and measurement techniques are required for sustainable development.

The essential step in the direction of overcoming challenges of the re-industrialised products can be derived from a statement of Lord Kelvin around 135 years ago: "I often say that when you can measure what you are speaking about, and express it in numbers, you know something about it; but when you cannot measure it, when you cannot express it in numbers, your knowledge is of a meager and unsatisfactory kind" [14]. Besides, the measurement and measurement technology are key success players leading new inventions when high technology meets the metrology.

Measurement technology has the following main tasks:

- Registration of quantity and quality of substance, energy and/or information where these are procured, processed and/or sold in all industrial processes, automatisation, to increase safety, reliability, and up time.

For all analytical purposes and for a better understanding, a basic model of the activities is needed that isn't restricted to one field of application alone. A basic model consists of three main items: test object, measurement equipment and an operator. Additionally we have information-flow and/or energy-flow. To identify the flow in more detail we define a main direction leading from the test-object to the measurement-equipment and again from the measurement-equipment to the operator. In the other direction there's the command from operator to test equipment and additionally a feedback from the test-equipment to the test-object. In some very basic descriptions, this feedback is defined to be zero. But each interaction follows the basic principle of "action et reaction". That means it's a must to have the feedback included into the model. Recording the approach, calculation result and range of measurement-value ensures traceability of the results later on and enables us to easily get back to the complete basic-model when we are simplifying the measurement by skipping the feedback values. Additionally, the supporting energy for the measurement, test equipment and the environment of measurement are required to be modelled to complete the basic measurement-model.

3 Quality

Quality is degree to which a set of inherent characteristics fulfils requirements [15]. The origins of quality management and quality assurance began in manufacturing operations, and many of the tools for quality analysis and improvement were developed for manufacturing problems. In the late 1980s, and into the 1990s, business began to recognize the importance of quality service in achieving customer satisfaction and competing in the global marketplace. The modern philosophy of production was created by the work of Taylor who decomposed a job into individual work tasks. The inspection tasks were separated from production tasks, which led to the creation of a separate Quality Department in the manufacturing organisations [16]. In a very important sense, this recognition has expanded the definition and concept of quality to include nearly in any organizational improvement such as the reduction of manufacturing cycle time and improved worker skills. In addition to industrial organizations and the manufacturing industry, also service organizations build up quality systems. Ancillary services in manufacturing companies as well as "stand-alone" service organisations such as hospitals and banks realised the benefits of a focus on quality. The general experiences both in the industrial practice and in the everyday life show a general trend towards higher expectations concerning the quality of products and services [17].

The overall total quality management model is represented in the Fig. 1 as a basis for developing the management system required for high quality and efficiency, which is applicable for all organizations.

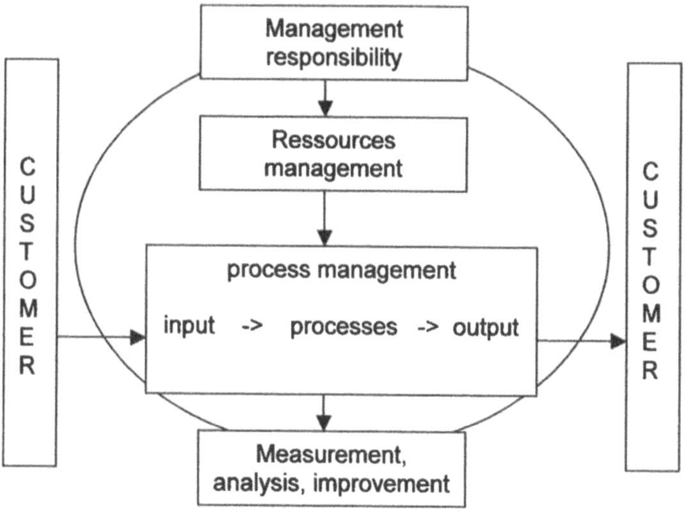

Fig. 1. The total quality management model applicable for all organizations

The future of industrial developments in transition to Industry 4.0 puts emphasis on the relationship between economical evaluation and increase in the quality of products in the manufacturing sector. The quality of products affect the competition and continuity of production, production costs, the production extent, the job productivity and the efficiency of these products at their application or their use as well as adding value in the needs of population. The extraction of high-quality information is a task of the measurement technology. The characteristic of quality can be qualitative or quantitative and there are various classes of characteristics, such as physical, mechanical, electrical, chemical or biological, sensory, behavioural, temporal, ergonomic and functional. Therefore, the quality management and measurement technology are indispensible to each other.

With advancing technologies, new products will be developed. The research and development process model can be used as the basic to fulfil the conformance of compulsory high quality product requirements more efficiently. The compulsory requirements are derived from the contractual terms and this methodology enables to implement achievement. The waterfall-model in a research and development process allows to track every phase and to enter the next development phase only when the critical test (or sum of all measurement-, inspection- and test-steps). If in one phase, there's no chance to go ahead because of heavy problems, there's the path back to the phase before to correct and improve the situation. Figure 2 represents the waterfall-model of a research and development process [18].

The development phase of a product can be separated with the analytical phase strictly by implementing the V-model in quality approach. Figure 3 represents a V-model for the development of a product, which is divided between left and right side indicating that the measurement and test conditions are derived from the requirements. Additionally, it indicates that if the test object fails, the implementation gets back to the

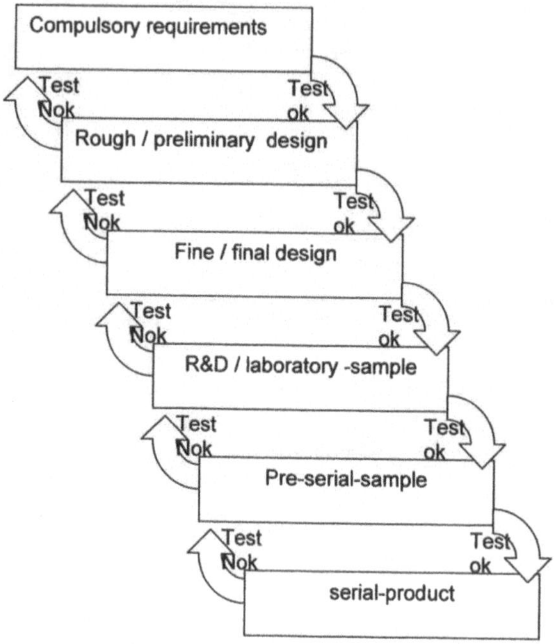

Fig. 2. The waterfall-model of a research and development process

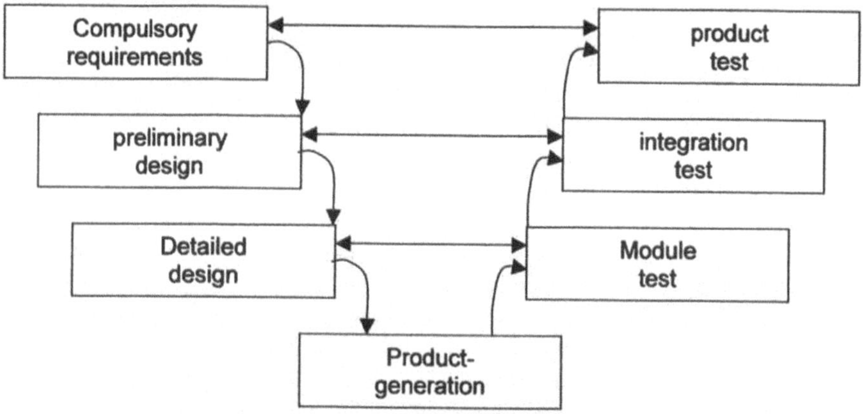

Fig. 3. The V-model for the development of a product

corresponding phase. The whole tight branch is only dedicated to measurement and test activities [18].

The basic problem with the systematic approach is the fact that we deal with errors. So at the first glance, it seems that we have no systematic access available, which leads to the bottom up error. All scientific disciplines can be affected by measurement errors. Therefore, analyse of the possibility of measurement-errors is a part of the quality assurance. Figure 4 represents an example of measurement error analyses from a measurement model [18].

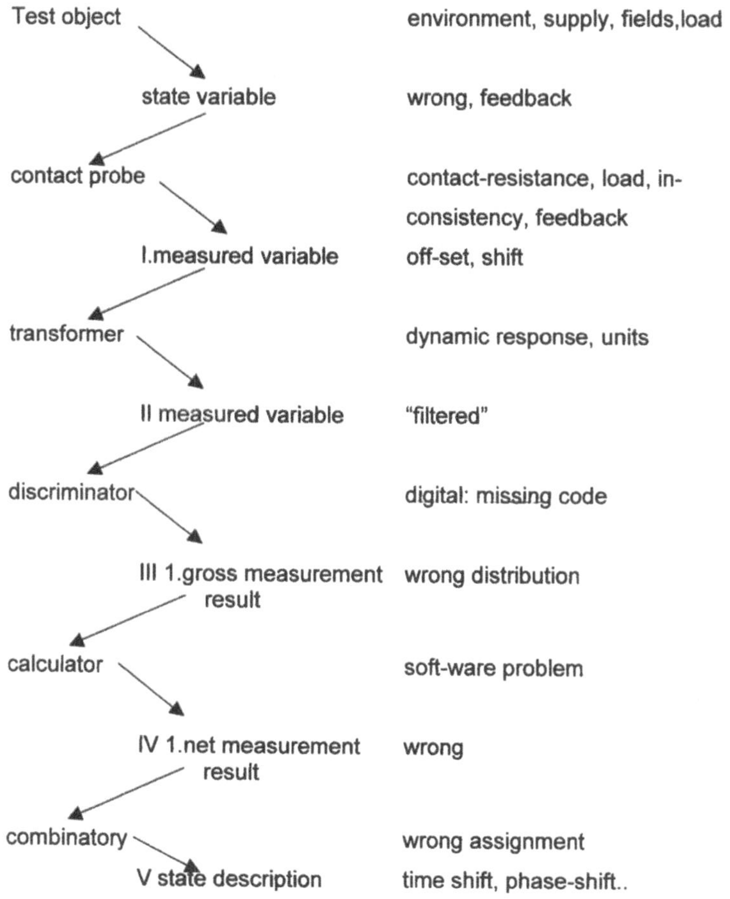

Fig. 4. An example of measurement error analyses from a measurement model

The theory of error formation distinguishes between three kinds of error:

- Gross errors
- Systematic errors
- Errors by chance

Using this definition, we can assess the opportunities we have to deal with (Fig. 5).

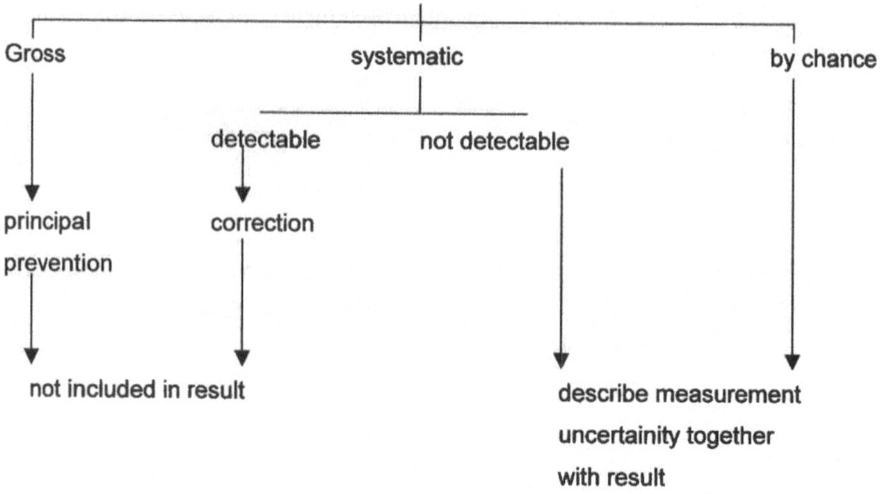

Fig. 5. The measurement error formation kinds

4 Risk Management and Industry 4.0

Risk is the effect of "uncertainty" on objectives, of an organization which is an inevitable part of all decision making processes in doing business (ISO guide 73) [19]. Therefore it should be carefully assessed and accordingly treated in order to achieve business objectives.

There exist two main types of risks, risks having positive impacts (opportunities) and risks having negative impacts. The risks having negative impacts should be eliminated and those having positive should be managed appropriately. Furthermore, there are three groups of risk occurrence: [20]

- operational: those arising from operational activities of an organization, only have negative impacts and they should be eliminated.
- strategic: those might have positive or negative impacts and can be managed/fine-tuned in order to generate superior returns.
- external: those arising out of organization's control, where there's no way to avoid or manage but the impact can be reduced through increasing the resilience of the organization.

For all types of organizations, there is a need to understand the *overall level of risk present within their processes and activities*. According to the type of business, it is important for them to recognize and prioritize significant risks and identify the weakest critical controls, in order to enhance the likelihood of achieving their objectives and increase the protection of their assets.

The global financial crisis in 2008 highlighted the importance of adequate risk management. Since then, new risk management standards have been published, including the international standard, ISO 31000 'Risk management – Principles and guidelines' which was first released in 2009, offering a structured approach for implementing enterprise risk management (ERM) [21]. It provides direction and standards on how companies can integrate risk-based decision making into their governance, planning, management, reporting, policies, values and culture. It proposes an organization-wide approach to risk management which allows to consider the potential impact of all types of risks on all processes, activities, stakeholders, products and services.

One of the important distinguishing properties from other frameworks is that, at the same time with ISO 31000, ISO also produced Guide 73:2009 'Risk management – Vocabulary – Guidelines for use in standards' which creates a common language, supporting common understanding in the organization. A revised version of ISO 31000 was published in 2018 to take into account the evolution of the market and new challenges faced by business and organizations. ISO 31000:2018 places more emphasis on both the involvement of senior management and the integration of risk management into the organization.

The above mentioned general concepts are addressed under the "*Principles*" of ISO risk management standard (Fig. 6). ISO 31000 describes the components of a risk management implementation in "*Framework*" (Fig. 6). Framework includes the essential steps in the implementation and ongoing support of the risk management process. The initial component here is "mandate and commitment" by the top management and this is followed by design of framework for managing risk, implementing risk management, monitoring and reviewing the framework and improving the framework. This framework of continuous improvement is in concordant with Deming's Plan-Do-Check-Act (PDCA) cycle.

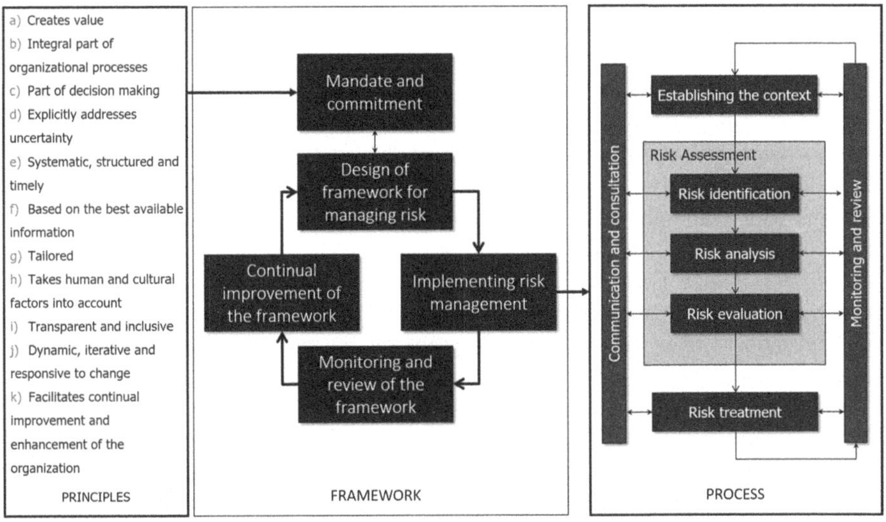

Fig. 6. ISO risk management standard, integrating RM to the overall management system: relationships between the risk management principles, framework and process

The “Do” part is the implementation of risk management process, which is elaborated within the “*Process*” box of Fig. 6. Risk assessment is the core of risk management process which comprises of identification, analysis and evaluation of risk. Risk identification establishes the exposure of the organization to risk and uncertainty. This requires a thorough knowledge of the organization, from the market conditions to its specific environment, as well as an understanding of strategic and operational objectives. In risk analysis activity, the risks which need the attention of the management are identified. Risk evaluation is used to compare the estimated risk against the given risk criteria to determine the significance of the risk. Risk treatment is the activity of selecting and implementing appropriate control measures to change the risk. Risk treatment includes risk control (or mitigation), and also covers risk avoidance, risk transfer and risk financing. Any system of risk treatment should provide efficient and effective internal controls. An organization has to understand the applicable laws and accordingly implement a system of controls.

Feedback mechanisms of ISO 31000 recognize the importance of feedback by way of two mechanisms. These are “monitoring and review” and “communication and consultation”. Monitoring and review makes sure that the organization monitors its risk performance and takes lessons from experience. Communication and consultation is also a part of the risk management process, which also supports risk framework [21].

4.1 Risk Management Framework in Industry 4.0

In many cases, the implementation of Industry 4.0 has shown that the relations between humans, systems and objects have become more complicated and real-time optimized network. The volume and availability of real time generated data causes new necessities of the infrastructure, technologies and management. Therefore, Industry 4.0 creates new categories of risks, mainly on information security.

The new categories of operational risks in manufacturing area are defined to be; “*process management*” which involves information risk stemming from possible data losses, loss of integrity and available information; “*maintenance*” which is basically problems with availability and integrity of data for maintenance; “*operation methods and tools used*” which indicates errors in data processing; “*machines and manufacturing technologies*” which is sensitivity and vulnerability of data related to cyber-attacks; “*human sources*” indicates not enough qualified workers and “*machine environments*” represents attacks from Internet network, problems related to electromagnetic compatibility and electromagnetic emissions affecting manufacturing machines [22].

It is suggested by Tupa et al. [22] that information security management system (ISMS) can be a solution for these new risks and also can effectively be integrated to ERM within the framework of Deming’s PDCA cycle. This framework enables the requirements for ISMS and quality management system to be achieved at the same time.

The PLAN phase is “organizational vision and objectives” where the objectives, processes, procedures and policies -including the ISMS policy- relevant to RM and to organization’s overall policies and objectives are established. The DO phase is the process itself where the policy, controls, processes and procedures are implemented and operated. CHECK phase is the assessment of performance compared to policies, objectives and reporting results to the management. ACT phase is the improvement phase where the necessary corrections and preventive actions are taken based on audits and management review.

The security policy also proposed to be extended by risk management aspects to an integrated corporate policy. Thereby, the requirements of all stakeholders, as well as legal and regulatory requirements, would be considered and appropriate corporate risk objectives and strategies would be established.

5 Target of Justicia

During the history in law, in context to the change of technology, it was always the case that, first the change in the technology appeared and after a period of time new laws regarding the change were established, based on the existing laws. Historic examples of the "technology leads – laws follow" –principle are: the steam engine as in German "Dampfkesselverordnung – DampfkV (Ordinance on SteamBoilers)" to nowadays "Betriebssicherheitsverordnung (BetrSV) (Workplace Safety Ordinance)".

It can be presumed that, for the next technological evolution in the industrial field – which has at present various names, like Industry 4.0 (more common in (Central) Europe) or Internet of Things (IoT – worldwide use) – the main aim is to connect vertical and horizontal processes in various levels over the border and influences of companies to make machines able to decide autonomic for a faster delivered product or service for the costumer by minimizing the costs of the producers and maximize the profit for each party, which is involved. This needs a high developed supply chain management, where existing data will be updated (for example: trends in the price of raw materials, ongoing tracing of the stocks in the store, trends in the orders of the costumers, traffic survey for optimized logistics in time and price, etc.). A most common used example for the IoT in household consumption is, that an "intelligent" refrigerator will order a new preferred product for a person, if the expiration date will be exceed or if a customer plans to make a party and invites friends over a social media platform, the data will be linked and the number of preferred drinks of the invited persons will be ordered according to previous events – as far as their data profile is available.

The question in this field is the human rights aspect, especially the European Convention on Human Rights – ECHR, article 8: The right of privacy. Since the collection of personaldata, which is (partly or fully) crosslinked over many internet site and also the crosslink data analysis reached in the past years a very high developed stage, where for example the pregnancy of a woman can be detected by the change of her pattern of communication by the algorithms running in the background of social media sites and/or internet stores, this topic becomes more into the focus of data rights activists.

The question according to the ECHR article 8 in this case is – even if the person has agreed to store her data of her activities during her communication – that the result of a statistical analysis and/or pattern matching of the behaviour of the person and the impact of the reached result belongs also to the person itself and cannot be sold or shared without her distinctive agreement.

That data's being stored is a well-known fact by a majority of users of the web, the circumstance, that this data might be analysed and used without the distinctive permission of the users and that it will be used in a way that, the vast majority of them would never had agreed to, was not obvious. This knowledge spread out to the public quit recently, mainly since the case of «Cambridge Analytics» and their activities on social media site "Facebook" widespreadly took place in the media.

How this case will influence the future behaviour of internet users in the www is not known right now – maybe the impact of changing one's habit or awareness for this problem will be for the majority of the users neglectable – for sure is: only a minority of users were and will be sensitive on one's personal data and their protection against unauthorized use and the misuse or influence of the results/analysis of this data on the data creators itself.

And if the power of those data protection sensitive users will be big enough to influence the legislative body in their nations or economic area (European Union) can be an interesting task, which would be worth to follow the previous introduced main historic principle "technology leads – laws will follow" – but only if the pressure of the voters on the politicians in the legislative body will be intensive, strong and long enough for their attention-getting-values-filters.

5.1 Industrial Field of Metrology

The area of legal consequences in the industrial field of metrology is divided into the categories of "deficiencies, defects, non-conformities" under two main domains:

- Warranty (independent of guilt/fault)
- Compensation of damage (dependent on guilt/fault)

According to the Austrian law of ABGB §932 Rechte aus Gewährleistung (Rights associated with warranties) has primary and secondary approaches:

- Verbesserung (Improvement)
  - after treatment
  - supplement
- Austausch der Sache (Replacement)
- Angemessene Preisminderung (Appropriate Price-Reduction)
- Aufhebung des Vertrages (Dissolve the Contract)

To discuss the primary legal consequences of warranty, following analysis is performed:

- Product oriented view: Product value and significance of deficiency
- Supplier oriented view: Demand to improve (after treatment, supplement)
- Buyer oriented: Inconvenience of customer

This gives a total overview shown below in the Fig. 7 [18].

Warranty		Analysis
Primary	improvement	product value
	Replacement	significance of defect/deficiency
		Demand for supplier
		Inconvenience of customer

Fig. 7. The overview analysis the warranty

6 Conclusions

Designing, producing, and/or selling products as the core of industry, keeping track of new technologies and emerging customer needs require to master many competencies. Understanding the necessity of combining three individual disciplines' (metrology & quality & law) knowledge is a key driver and even more importantly value adding fundamental for growth and development. The studies of system-approach and transparency of working documents present the next great potential in transition to the Industry 4.0.

Acknowledgment. This study is a hommage to Dr. DI. A. Bauer DWTI MBA for his academic contributions.

References

1. Acatech (2013) Umsetzungsempfehlungen für das Zukunftsprojekt Industrie 4.0 – Abschlussbericht des Arbeitskreises Industrie 4.0. BMBF homepage. https://www.bmbf.de/files/Umsetzungsempfehlungen_Industrie4_0.pdf
2. Osanna PH, Durakbasa MN, Crisan L, Bauer JM (2009) The management and exchange of knowledge and innovation in environments of collaborating small and medium sized enterprises. Commun IBIMA (CIBIMA), Off J Int Bus Inf Manag Assoc, I, 7, pp 130 – 136
3. Stock T, Seliger G Opportunities of sustainable manufacturing in industry 4.0. Procedia CIRP, 40:536–541. ISSN: 2212-8271
4. Müller J, Kiel D, Voigt K (2018) What drives the implementation of industry 4.0? The role of opportunities and challenges in the context of sustainability. Sustainability 10(1):247
5. Buhr D (2015) Soziale Innovationspolitik für die Industrie 4.0. WISO Diskurs. Friedrich-Ebert-Stiftung, Bonn
6. Daniel K, Müller J, Arnold C, Voigt KI (2017) Sustainable industrial value creation: benefits and challenges of industry 4.0. Int J Innov Manag 21(8):1740015
7. Communication from the commission to the european parliament, the council, the european economic and social committee and the committee of the regions. Preparing for our future: developing a common strategy for key enabling technologies in the EU. SEC(2009) 1257
8. European Commission (February 2016). KETs (Key Enabling Technologies). http://s3platform.jrc.ec.europa.eu/documents/20182/117542/S2E_Fiche_KETs.pdf/72f2e425-9fbc-4e99-a26a-10122b53ce31

9. Schuh G, Potente T, Wesch-Potente C, Weber A, Prote J (2014) Collaboration mechanisms to increase productivity in the context of industrie 4.0. Procedia CIRP 19:51–56
10. Durakbasa MN, Bauer J, Posvek G (2017) Advanced metrology and intelligent quality automation for industry 4.0-based precision manufacturing systems. In: Solid state phenomena, 1, vol 261. Trans Tech Publications, pp. 432–438
11. Kopacek P (1999) Intelligent manufacturing: present state and future trends. J Intell Rob Syst 26:217–229
12. Durakbasa MN (2014) Geometrical product specifications and verification for the analytical description of technical and non- technical structures. Department of Interchangeable Manufacturing and Industrial Metrology, TU-Wien, Austria
13. Bookjans M, Weckenmann A (2010) Virtual quality management - validation of measurement systems by the use of simulation technologies. Physics Procedia 5:745–752
14. Kelvin L PLA, vol 1. Electrical units of measurement, 03 May 1883
15. EN ISO 9000:2015 Quality management systems – Fundamentals and vocabulary
16. Taylor FW (1911) The principles of scientific management. Harper & Brothers
17. Durakbasa MN, Osanna PH (2017) Quality in industry. TU-Wien Vienna University of Technology, Vienna
18. Bauer A (2013) Measurement-technology & Law & Q, Abteilung Austauschbau und Messtechnik. TU-Wien Vienna University of Technology, Vienna
19. ISO Guide 73:2009, Risk Management – Vocabulary. https://www.iso.org/standard/44651.html
20. Göstl P, Schwaiger W (2016) Risk-based planning and control systems: integration of different risk types into strategic, financial and operational management systems. In: Management control association conference, university of antwerp, 7–9 September 2016
21. A Risk Management Standard: irm, airmic, ALARM 2010. https://www.theirm.org/media/886062/ISO3100_doc.pdf
22. Tupa J, Simota J, Steiner F Aspects of risk management implementation for industry 4.0. In: 27th international conference on flexible automation and intelligent manufacturing, FAIM2017, Modena, Italy, 27–30 June 2017

Performance Analysis of Vehicle-Specific Methods and Sensors for Autonomous Vehicles

Ernst Pucher(✉), Andreas Gruber, Mathias Innerkofler, and Marco Buhmann

Institute for Powertrains and Automotive Technology, Vienna University of Technology, Vienna, Austria
{ernst.pucher,andreas.gruber,mathias.innerkofler, marco.buhmann}@tuwien.ac.at

Abstract. This article deals with the performance of modern sensor systems for autonomous vehicles. The examined automobile was equipped with state-of-the-art sensor technology and provides a solid basis for the further close-to-production development of the increasing requirements for environmental recognition. With regard to Industry 4.0, too, parallels can be deduced, since progressive networking between cars and infrastructure is assumed by the fact that the IoT (Internet of Things) is being expanded in both areas.

The following publication evaluates the performance of vehicle-specific methods and sensors in modern vehicle assistance systems under real driving conditions. In addition, a high-resolution vehicle dynamics measurement was performed on a partially automated test vehicle. The test results show that for good weather conditions and good road infrastructure, there is no discernible difference in driving behavior between manual and partly autonomous operation. Finally, a route model was created based on simulations using MATLAB SIMULINK. The deviation of the simulated from the measured lateral acceleration is discussed.

Keywords: Autonomous vehicle · Internet of Things (IoT)
Vehicle-to-Infrastructure (V2I) communication · Sensor technology
Simulation

1 Introduction

The future road traffic on public roads, if one believes the traffic planners and car manufacturers should be carried out with autonomous vehicles. The slow introduction of automation levels began several years ago. The purpose of these measures is to increase the efficiency of the flow of traffic, thereby averting the threat of collapse in conurbations. Furthermore, the additional electronic monitoring of the "human system" as a driver and operator to reduce the reaction times in dangerous situations, since the evaluation algorithms on the powerful computers of the system electronics can obtain much faster response times of assistance systems. This aspect pursues the goal of reducing the lethal consequences of traffic accidents, or preventing them altogether and thus drastically reducing the number of road deaths.

N. M. Durakbasa and M. G. Gencyilmaz (Eds.): ISPR 2018, *Proceedings of the International Symposium for Production Research 2018*, pp. 451–463, 2019.
https://doi.org/10.1007/978-3-319-92267-6_39

Many of the new cars on the market today already have systems of automation levels 1 and 2, and various prototypes that can drive completely autonomously are already or are being tested on public roads in Europe as well as in the USA. These technologies have evolved very rapidly and a very strong development progress is expected in the future as well. The motivation of the following report was to provide a brief insight into the state of the art in currently used assistance systems in modern vehicles. Furthermore, the possibilities for driving dynamics measurement or simulation of vehicle movements in the surrounding room will be shown and discussed.

The sensors used today for tracking and other assistance systems are already good performing on well-marked lanes and easily detectable road markings. In order to be able to determine differences between a partially autonomous and manual operation of the test vehicle, driving dynamics measurements were carried out on rural roads and motorway sections and an evaluation of the measurement results obtained thereby.

2 The Partially Autonomous Test Vehicle

The examined SAE Level 2 automated vehicle was a Volvo V90 D4 built in 2017, shown in Fig. 1. In this level of automation, the vehicle can already perform several tasks normally performed by the driver at the same time. Thus, for example, by means of the so-called Pilot Assist mode of the manufacturer Volvo the lane boundary detected and the speed to the car in front are respected independently. In addition, this Pilot Assist mode allows semi-autonomous driving as standard up to a speed of 130 km/h.

Fig. 1. Test vehicle - Volvo V90 with Pilot Assist

3 Sensor Technology of the Test Vehicle

The sensor system for detecting the driving dynamics of a vehicle should provide important information for the driver assistance systems and must meet the following requirements:

- System Requirements
- Installation requirements/geometry
- Environmental requirements
- Legal requirements and standards

The modern lane departure warning systems are based on information obtained from roadway estimation algorithms. The key technology for determining the course of the lane lies in the correct and robust evaluation of combined measurement results from radar sensors and cameras, which are based on the different reflection signals from lane markings and from dark or grey asphalt.

3.1 Radar Sensors

Radar sensors are used in driver assistance systems or (partially) autonomous vehicles for distance control and object recognition. The principle of a radar sensor is based on the generation of modulated intense light sources to illuminate the field of view and to detect distant objects. Of the detected objects, a certain part of the radiation power is reflected back to the sensor and detected by an optical receiver. In automotive applications, the transmitter and receiver of the sensor are placed close to each other, or use the same optics in an equiaxed radiation configuration. To ensure successful object detection, the received signal must be greater than the sensitivity limit of the optical sensor. The radar systems in use today are among the most important sensors in semi-autonomous vehicles, since they are also relatively insensitive to weather and light conditions. Radar sensors with four different frequency bands are used in road traffic applications, with the 76.5 GHz range currently dominating in automotive applications. These so-called millimeter-wave radars can detect objects very accurately and are also less sensitive to obstructions caused by fog, snow and rain. Due to their robustness against environmental influences, these radar systems are used in various assistance systems. Although the millimeter-wave radars are considered to be quite robust, various researches have shown that snow-covered radar sensors can reduce their performance by as much as 30%. Even with heavy rain, there is a non-negligible reduction in the maximum detectable range of objects [1, 3].

3.2 Cameras

Today, almost all elements of traffic environments, such as traffic signs and road markings, are designed to be perceived by the human eye through different shapes and colors. Therefore, it is also of great importance for machine perception by autonomous vehicles to detect and properly interpret such signs and other road users in real operation. This can best be done with camera-based systems. The determination of the road structure and of objects takes place in camera systems by an exact analysis of digital images. In order to be able to realize a real-time-capable system in this case, only the information which is important for a safe continuation of travel is extracted from the images obtained. This is done by comparing the different gray levels of each individual pixel. The different grey levels indicate different structures, which subsequently leads to a representation of the contour of objects. The resulting structure must be imaged on a plurality of pixels, wherein the number of pixels can be increased by raising the silicon area at the same pixel pitch, or by reducing the pixel pitch with the same number of pixels. Another important criterion of cameras for driver assistance systems is the exact and correct contrast resolution of objects. This is especially important when detecting dark objects when driving at night. Realization of the correct contrast resolution is achieved by A/D converters, which resolve the signal with bit stages of eight to twelve bits. In addition to the aforementioned spatial resolution and the contrast resolution but also the temporal resolution plays a crucial role. Today's modern camera systems use refresh rates of approximately 30 frames per second [2].

4 Driving Dynamics of Autonomous Vehicles

The driving dynamics of motor vehicles is considered separately according to the three translational and the three rotational degrees of freedom of movement of the vehicle body, with above all the degrees of freedom of movement transverse to the vehicle longitudinal axis in autonomously operated vehicles are of great importance. The term "lateral dynamics" summarizes all processes that influence the driving stability, the cornering behavior and the tracking or course attitude. The movement processes in the longitudinal direction of the vehicle and the resulting power and energy requirements are investigated in the longitudinal dynamics. The effects of the vibration behavior in the vehicle's vertical axis on driving comfort or driving safety are dealt with in the vertical dynamics of the vehicle.

5 Results of a Vehicle Dynamics Measurement of a Semi-autonomous Vehicle

5.1 Real-World Vehicle Dynamics Measurement on a Semi-autonomous Test Vehicle

On 25.04.2017, in the Wiener Neustadt area, a real-world vehicle dynamics measurement with high-resolution acceleration sensors was carried out on the Volvo V90 D4 test vehicle. The acceleration and deceleration processes were recorded in real time in all three main vehicle axes. In addition, a gyroscope was used to measure the lateral accelerations and rotation rates in curves. Together with a GPS signal so the exact position of the vehicle could be determined.

5.2 Measurement Route

The investigated vehicle was driven in city traffic of Baden near Vienna with an average speed of about 20 km/h. The motorway was completed on the A2 between Bad Vöslau and Leobersdorf. Here the average speed was 110 km/h. The entire driven measuring section is shown in Fig. 2. Furthermore, the long right-hand bend of a motorway exit of the A22 in Vienna as shown in Fig. 3 was examined for driving dynamic characteristics.

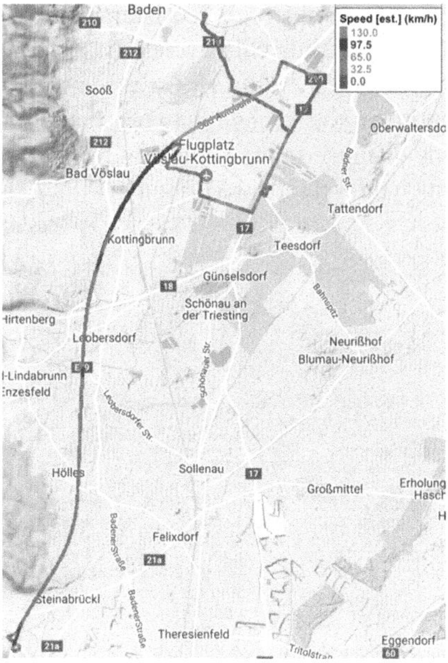

Fig. 2. Entire recorded test section

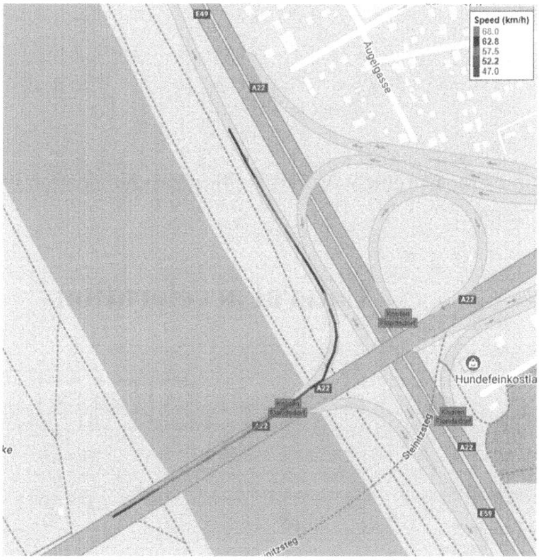

Fig. 3. Recorded long right turn from an urban motorway

5.3 Results from the Driving Dynamics Measurement

It should be noted that in the subsequent evaluation of the measurement results, only a partial section in which part was driven autonomously and manually is treated. The course shown in Fig. 4 represents the recorded lateral acceleration of a country road section with the associated marked driving events. Based on this figure, the dynamic driving behavior of the vehicle in the Pilot Assist mode is displayed. The Pilot Assist mode helps the driver to move the vehicle between the side lanes of the lane while maintaining a pre-set time distance to the vehicle in front. The marker 1 indicates the section of highway on which the vehicle was driven in Pilot Assist mode. The following

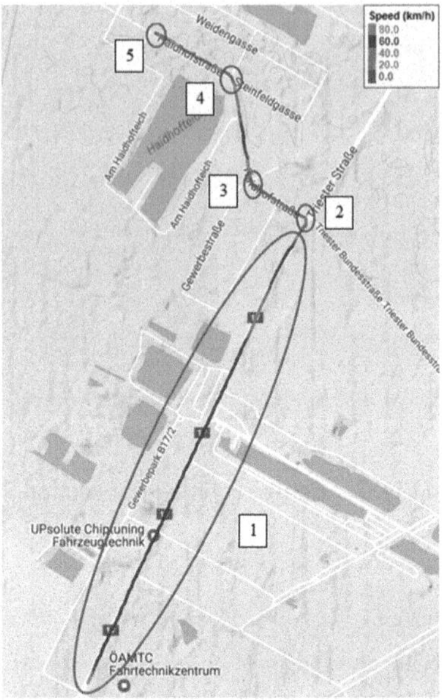

Fig. 4. Vehicle dynamics measurement Volvo V 90, measured section of track in Pilot Assist mode and manual operation

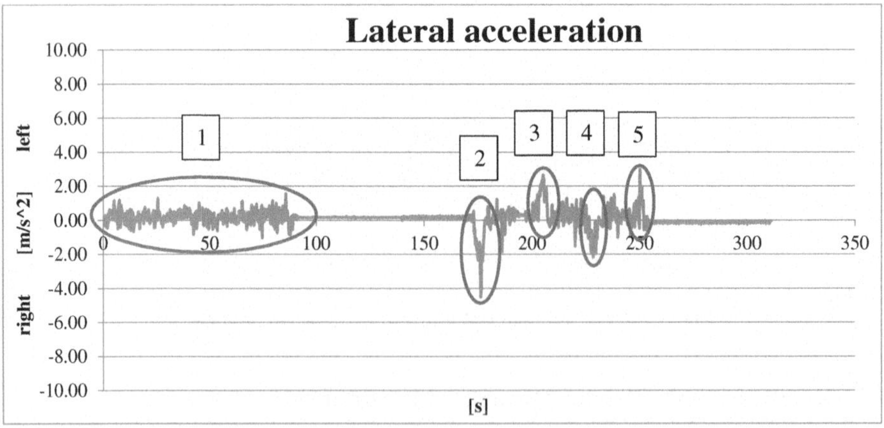

Fig. 5. Vehicle dynamics measurement Volvo V 90, measured lateral accelerations in Pilot Assist mode (Marker 1) and manual operation (Marker 2 to 5).

Figs. 5, 6 and 7 show the measured section, the speed and the associated acceleration values in two main vehicle axes. It could be observed during the test drive in Pilot Assist mode that with good road infrastructure a smooth regulation of the speed and slight corrective steering angles occurred and on the basis of the recorded data no distinction could be made to a manual operation. The rides on the highway also showed that in oncoming vehicles in certain situations came to an unexpected deceleration of the vehicle. It has been assumed that these delays can be attributed to the limitations of camera and radar units. Furthermore, it could be observed that insufficient road markings led to jerky steering interventions and to an orientation of the vehicle in the direction of the center of the road. The marks 2 to 5 indicate manual turning operations and are intended to show the possibilities of driving dynamics measurement.

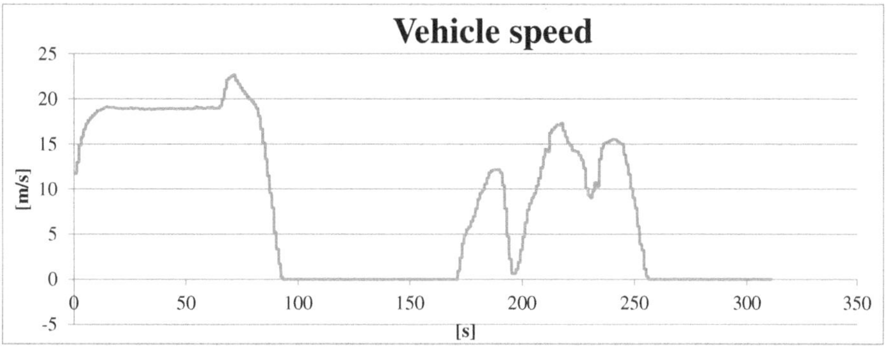

Fig. 6. Driving dynamics measurement Volvo V 90, speed measurements in Pilot Assist mode and manual operation

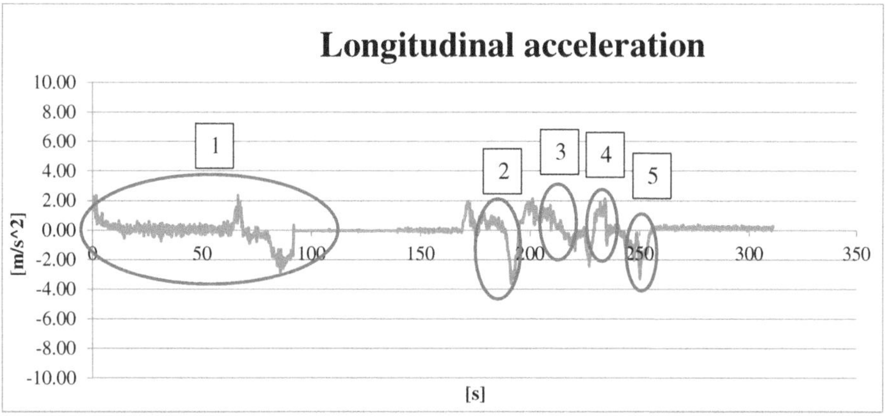

Fig. 7. Vehicle dynamics measurement Volvo V 90, measured longitudinal accelerations in Pilot Assist mode and manual operation

6 Simulation of Driving Dynamics on Recorded Streets

6.1 Vehicle Model

In addition to the driving dynamic measurement already described during a real road trip, a procedure has been developed with which the driving dynamics can be calculated simulative on GPS-recorded streets and evaluated [4]. It proved to be useful to model the vehicle by a single-track approach. This is characterized by the fact that the vehicle's center of gravity is at the height of the road, whereby no rolling and pitching movements occur [5]. The kinematic relationships and the balance of forces are explained in Fig. 8 and the corresponding Table 1.

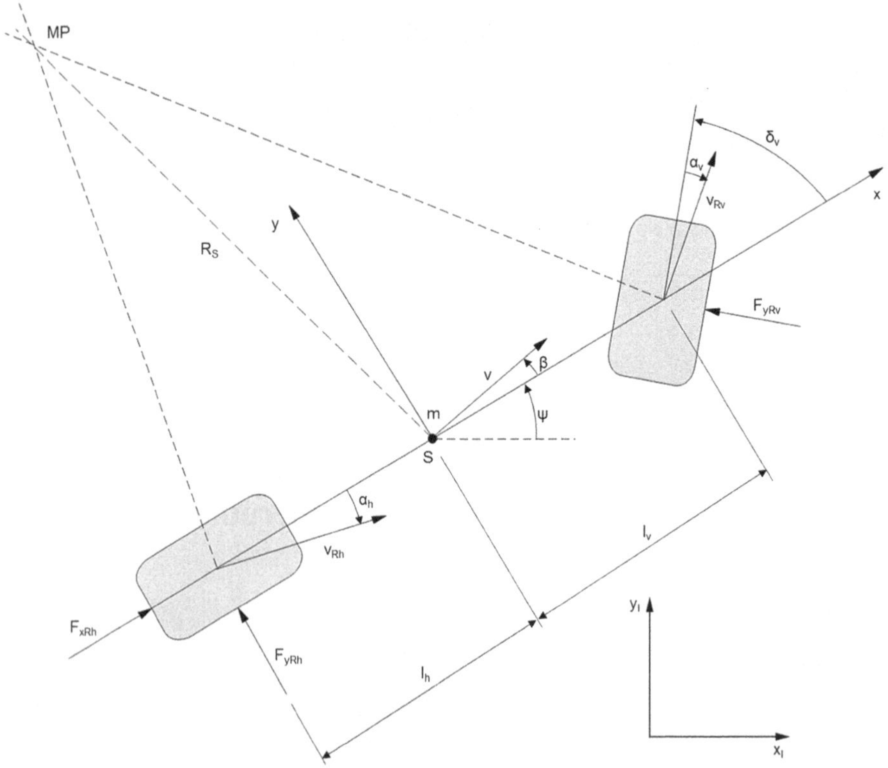

Fig. 8. Kinematics and force ratios of the single-track model

Table 1. Formula symbol of the one-track model

MP … instantaneous centre of the vehicle
S … center of gravity from the vehicle
R_S … arc radius of the center of gravity
x, y … vehicle − fixed coordinate axes
x_I, y_I … inertial coordinate system of the route
ψ … yaw angle of the vehicle's longitudinal axis with respect to the x_I − axis
β … sideslip angle
v … velocity vector of the center of gravity
v_{Rv} … velocity vector of the front wheel
v_{Rh} … speed vector of the rear wheel
$\alpha_{v,h}$ … slip angle of the front and rear tires
δ_v … steering angle of the front wheel
m … vehicle mass
$F_{yRv,yRh}$ … lateral force of the front or rear wheel
F_{xRh} … longitudinal force of the rear wheel
$l_{v,h}$ … distance from the center of gravity to the front or rear axis in x − direction

With a linearization for small angles, the vehicle model can finally be described using the equations of motion (1) in the x-direction, (2) in the y-direction and (3) around the z-axis.

$$m\dot{v} = F_{xRh} \tag{1}$$

$$\dot{\beta} = \dot{\psi}\left[\frac{c_{Rh}l_h - c_{Rv}l_v}{mv^2} - 1\right] - \beta\left[\frac{c_{Rh} + c_{Rv} + m\dot{v}}{mv}\right] + \frac{c_{Rv}}{mv}\delta_v \tag{2}$$

$$\ddot{\psi} = -\dot{\psi}\left[\frac{1}{J_z v}\left(c_{Rh}l_h^2 + c_{Rv}l_v^2\right)\right] + \beta\left[\frac{c_{Rh}l_h - c_{Rv}l_v}{J_z}\right] + \frac{c_{Rv}l_v}{J_z}\delta_v \tag{3}$$

6.2 Route Model

In order to be able to replace the driving dynamics measurement on the road with a simulation, a mathematical model of the driving route is necessary in addition to a vehicle model. The raw data of the route model are initially available as GPS data points. In a first step of pre-processing, these are converted into a planar x-y coordinate system. In further steps, the route is smoothed and the data points are interconnected by a cubic spline interpolation to ensure jerk-free acceleration changes across the direction of travel during steering maneuvers.

Figure 9 explains how pre-processing works after the GPS raw data has been transferred to the x-y plane. Shown is an exemplary offset in the y-direction or transverse to the direction of travel, as it occurs due to inaccuracies in the recording with the GPS technology used. The data points are first offset by an algorithm, resulting in a smoothed course. In a final step, a spline interpolation is performed, where the points are connected by cubic splines.

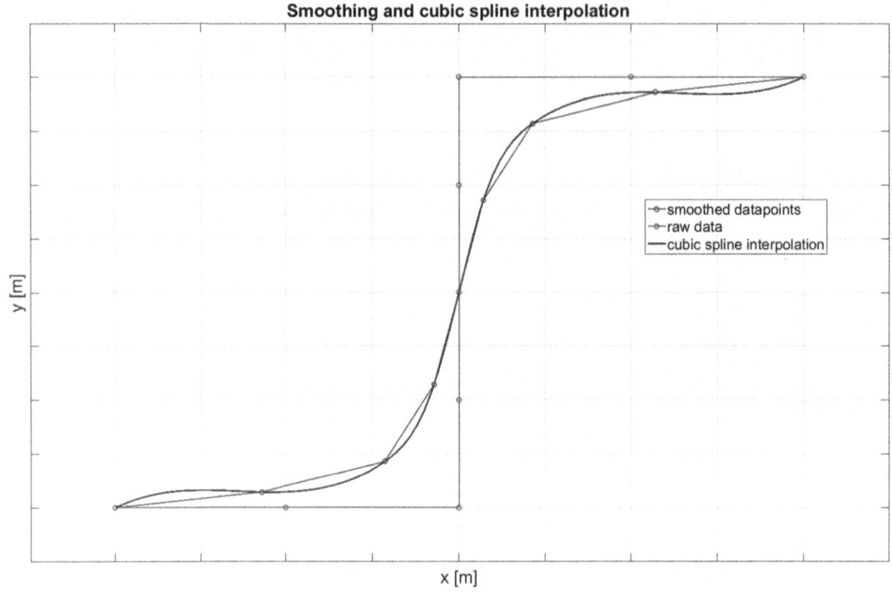

Fig. 9. Illustration of how the preprocessing of GPS raw data from a recorded route works

6.3 Comparison of the Measured and Simulated Values of the Lateral Acceleration

In order to make a comparison of measured and simulated driving dynamics values, the road shown in Fig. 3 was used in the form of a long right-hand bend. During a journey in which the test vehicle was controlled by a real driver, the driving distance with a powerful GPS receiver at a recording rate of ten hertz, and the thereby occurring on the vehicle lateral acceleration were recorded simultaneously.

The recorded GPS raw data was subjected to the pre-processing described in the previous section and then fed to the simulation realized in MATLAB SIMULINK. The schematic structure of the simulation structure is shown in Fig. 10. Particularly noteworthy is the feedforward control in the transverse regulation system, which specifies the current required steering angle resulting from the given driving route. As a controlled variable in the transverse control, the transverse distance between the actual position of the vehicle and the desired position that is located on the trajectory of the route model, is defined. In the longitudinal control, the vehicle speed acts as a controlled variable and is adjusted from the GPS data during the simulation of the recorded speed.

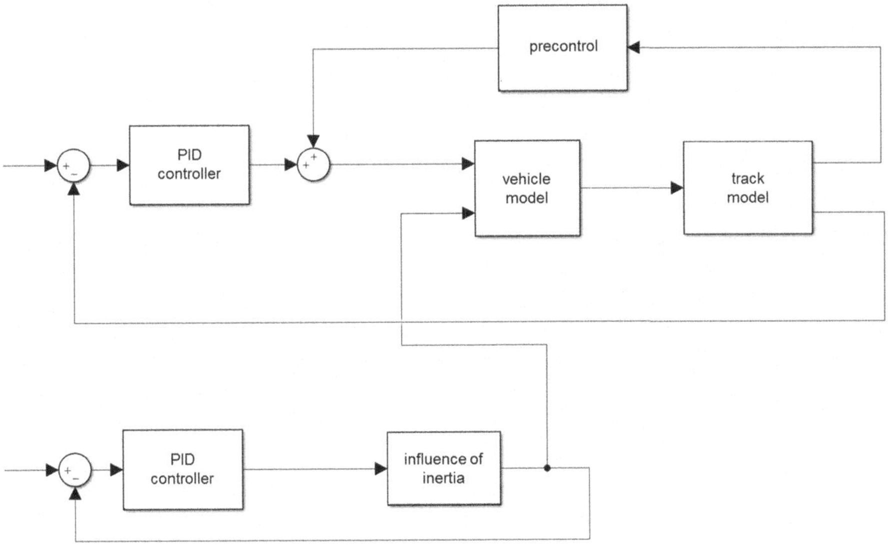

Fig. 10. Schematic structure of the applied transverse (top) and longitudinal control (bottom); the speed represents the link size

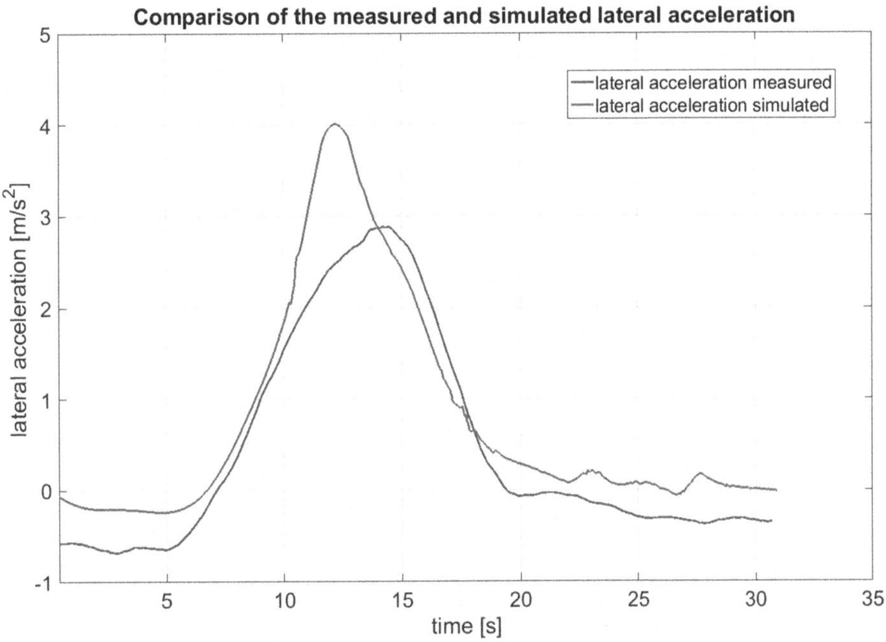

Fig. 11. Comparison of the measured and simulated lateral acceleration

The comparison of the measured [6] and the simulated lateral acceleration is shown in Fig. 11. Since no exact data on mass, inertia and tire behavior of the test vehicle were known, the vehicle model could not be modelled accordingly. Despite the resulting inaccuracies in the simulation, the simulative determination of the lateral acceleration allowed good coverage with the data from the real measurement.

7 Conclusions and Outlook

The lane keeping and automatic distance assistants with the entire expansion stages still have a substantial potential for improvement, especially in terms of use in adverse weather conditions. Although the cameras and radar sensors used in this case achieve considerable performance in good environmental conditions, the human vehicle driver cannot yet be replaced by the assistance systems used today. The measurements carried out in this publication show that the conditions of the ground markings and the road infrastructure have a significant influence on the performance of assistance systems with cameras and radar sensors. Leaving the carriageway was compensated by the vehicle independently with smooth steering interventions on well-marked country sections. The evaluations of the recorded acceleration values also show that the steering interventions in the Pilot Assist mode on road sections with good infrastructure can hardly be distinguished from manual operation, which underlines the good function of the system in principle. The biggest problems showed the lane departure warning however with bleached road markings and/or with bad street infrastructure.

If the environment can be recognized without problems by the car sensor system and additional support of high-tech infrastructure with sufficient certainty, and subsequently a mathematically described driving trajectory can be formed, vehicles can move relatively easily along the trajectory autonomously, which is illustrated by the presented simulation.

Looking at the current state of technology, it becomes apparent that in the near future, no fully autonomous vehicle will be available as a commercial product. On roads with well-equipped infrastructure, however, is already an automatic distance, tracking and longitudinal control with additional adaptive cruise control prior art. These vehicles can only be classified in automation levels 0 and 1. The systems of these vehicles, however, are classified only as driver assistance systems and one cannot yet speak of an autonomous operation. With today's sensors, the systems are able to take over driving functions in favorable weather conditions and thereby assist the driver in certain driving situations. However, since especially in adverse weather conditions, such as in fog, snow or heavy rain, a higher reliability is expected from autonomous vehicles, than by people. The development has to be still in progress to deliver better sensor technologies and support from the infrastructure to get faster decisions from the artificial intelligence. In the near future, it can be assumed that autonomous systems with higher automation levels will initially be used, above all, in known operating environments with calculable boundary conditions, such as on construction sites and in factory halls.

The main problem with not fully automated cars is that as a supervisor, the human should be in investigative mode at all times to intervene in danger situations and take control of the vehicle. The autonomous vehicles should support the driver and protect him and the surroundings from accidents in the event of carelessness or visually imperceptible hazards. Therefore, legislators must give people new rules and laws to make traffic safer. In summary, it can be said that the automotive industry and the research institutes will develop on- and off-board sensors and more powerful electronics, bringing the announced automation levels of the vehicles onto the road in the foreseeable future.

Acknowledgment. We want to thank our partners from industry and Austrian Department of Transport for their cooperation and support.

References

1. Rasshofer R, Spies M, Spies H (2011) Influences of weather phenomena on automotive laser radar systems. Adv Radio Sci 9:49–60
2. Winner H, Hakuli S, Lotz F, Singer C (2009) Handbuch Fahrerassistenzsysteme - Grundlagen, Komponenten und Systeme für aktive Sicherheit und Komfort. Vieweg+Teubner Verlag, Wiesbaden ISBN: 978-3-8348-9977-4
3. Yamawaki T, Yamano S (1998) 60 - GHz Millimeter - Wave Automotive Radar, Seite 3–14
4. Kampfer A, Pucher E (2005) Principles of systems that enable autonomous driving of vehicles. In: Proceedings of the 8th International IEEE Conference on Intelligent Transportation Systems, pp 930–935
5. Kritayakirana K, Gerdes JC (2012) Autonomous vehicle control at the limits of handling. Int J Veh Auton Syst 10(4):271–296
6. ISO 3888-2:2011-03 (2011) Passenger cars - Test track for a severe lane-change maneuvers - Part 2: Obstacle avoidance

Smart Factories: A Review of Situation, and Recommendations to Accelerate the Evolution Process

K. Öncü Şen[1,2(✉)], M. Numan Durakbasa[1], Merve Vildan Baysal[2], Gizem Şen[3], and Gökçen Baş[1]

[1] TU Wien - Vienna University of Technology, Vienna, Austria
{numan.durakbasa, goekcen.bas}@tuwien.ac.at, e0626662@student.tuwien.ac.at

[2] Information Technologies, TUBITAK, Ankara, Turkey
vildan.simsek@tubitak.gov.tr

[3] Budget and Performance Division, TUBITAK, Ankara, Turkey
gizem.sen@tubitak.gov.tr

Abstract. Smart factory has become increasingly important in the last years as an indispensable manufacturing model of the future. In this context, studies on the development of technologies in the field of smart factory systems intensively continue in the world. Turkey is also trying to contribute through national grant mechanisms to encourage technology development efforts in this area. In this study, the situation of smart factory system technologies in Turkey has been explained. By examining the academic studies and by carrying out a questionnaire study to companies and academicians studying in this field, suggestions are made about which of the smart factory system technologies should be given priority to gain momentum in this field.

Keywords: Smart factory · Industry 4.0 · Priority areas

1 Introduction

Current production systems are generally set up as unintelligent systems. Recent trends such as the rise of the fourth industrial revolution (Industry 4.0) enable the overlapping of digital and physical worlds with the help of information technology (IT). Eight (Autonomous Robots, Simulations, Internet of Things (IoT), Cyber-security, Cloud computing, Additive Management, Augmented reality, Big data and analytics) today's and close future's technologies are helping to transform production processes. Innovative information and communication technologies (ICT) have already been in use in production processes. But with Industry 4.0 the momentum of digitalization gained speed in production processes [1]. Industry 4.0 may affect to change traditional production processes to obtain higher efficiencies. Some of the contexts in the Industry 4.0 are;

- Big data and analytics, expresses to the storing and complete evaluation for the making better system management and assist real-time decision making.

N. M. Durakbasa and M. G. Gencyilmaz (Eds.): ISPR 2018, *Proceedings of the International Symposium for Production Research 2018*, pp. 464–479, 2019.
https://doi.org/10.1007/978-3-319-92267-6_40

- The expected robotics based production will finally interact with each another and work alongside with employees' and learn from them. This robotic based production will help to reduce the manufacturing cost and production time.
- Simulations will be used widely for the line based production processes to get better real-time production outcomes and represent the physical production processes in a virtual model, which can include production equipment, outcomes, and employees. This will allow the decision makers to evaluate and performance tuning the production equipment settings to obtain the best production process, reduce machine setup times and increase total quality.
- Internet of Things (IoT) refers that more devices will be communicated and interact both with each other and with more centralized controllers when needed. IoT will help to decentralize analytics and decision making in local for the real-time reactions.
- Growing production-related needs increased the communication across main components. Therefore these needs, all collected data, and functionality will increasingly be controlled into the cloud, enabling more data-driven services for production monitoring and control systems and cloud based system help to obtain higher efficiencies. Day by day the performance of cloud technologies in progress and nowadays it is performing reaction times of just several milliseconds.
- Additive manufacturing methods is popularly used in nowadays to produce samples of personalized goods that offer production advantages, such as 3-D printing. Production has kicked off the adoption to the additive manufacturing to reduce the complexity and expensive designs [2]. With using the today's image processing technologies augmented reality based systems have been started to support the variety of services, such as parts in storage and sending repair instructions via mobile devices. These support systems are currently in beginner level, but in the near future, production and storing processes are going to use the augmented reality to support the employees' for real time informing to improve decision making and operations.

Industry 4.0 enables connectivity and communications to increase dramatically. Additionally, critical industrial systems are needed to protect and production processes from cyber security threats. Consequently, secure, complex identity and reliable communications access are necessary.

Today, production systems depend on information systems more than ever. Therefore, intelligent systems are evolving to smart systems. In order a factory to be smart, all processes should be managed [3]. These features enabled the systems to be more efficient and qualified. The Industry 4.0 and its components can be used in the production process to boost the transformation to digitalization. This digitalization process is formed by 3 groups with interaction. The first group is academicians who lead the technology. The second group consists of Research and Development (R&D) driven Small Medium Enterprise's (SMEs) who can produce the hi-tech product. And the last one is the manufacturer (factories, large industry) who can make investment on hi-tech product. All these groups designate the technology maturity level of Turkey. So we focused on these three groups and carried out questionnaires on them.

2 Related Work

The New Industrial Revolution represents the digital transformation of production systems as advances in information and communication technologies. The digitalisation of every stage of the production chain, the provision of machine-human-infrastructure interaction and the development of 'Intelligent Manufacturing Systems' have created a paradigm shift in the industry [4]. The countries that make this change will have an advantage in the global market. "New Industrial Revolution: Intelligent Manufacturing Systems Technology Road Map" study of TUBITAK dated 03.01.2017, evaluated the digital maturity level of the Turkish companies as between Industry 2.0 and Industry 3.0 [5]. It is stated that making value-added production with this level of industry is difficult. Because of this situation, It is possible that Turkey's living in comparative weakening global competitiveness, will lead to reduce of global market share, rise unemployment and decline labour quality. In this case, Turkey has remained low investment, it will slide into an economic vicious circle where low value-added production. On the other hand, if you try to create ground-breaking changes in global competitiveness with stable Industry 4.0 investments, you will get more shares in the global value chain [6]. To break this chain, it is necessary to determine the policies and goals of the country and apply it decisively.

3 National Goals and Policies

National policies underscore the need for strengthening competition and industrial productivity to enhance medium-high technology exports.

A. The Place of Manufacturing Industry Transformation in Policy Documents

(1) *Supreme Council for Science and Technology (BTYK) Decisions:* According to decree 2016/01, Smart Manufacturing Systems Technology Roadmap is established transition of Turkish industry for increasing international competitiveness in technology production:

- Developing an implementation and monitoring model for smart manufacturing in coordination with all stakeholders
- Increasing goal-oriented R&D efforts in critical and pioneering technology areas (cyber-physical systems, AI/sensor/robotics, Internet of Things, big data, cyber security, cloud techs, etc.)
- Designing support mechanisms for manufacturing infrastructures to develop critical and pioneering technologies

(2) *Goals for 2023:* The Government's 2023 Goals, which cover the 63-item goals, set for the 100th anniversary of the Republic, and which are road maps, play a leading role in the transformation of the manufacturing industry. Most of the identified objectives have the ability to directly and indirectly influence the manufacturing industry.

(3) *65th Government Program:* Within the framework of the 65th Government Program, it was aimed to give more support to high-tech investments, to invest

in advanced technology class, to be included in priority investments and to benefit from 5th region support.

(4) *10th Development Plan:* It is targeted that the innovation production capacity be raised by closely pursuing the developments in the field of science and technology and that transformation in the production structure be reached by integrating these innovations into production structure. Realization of Transformation of Production Concentration of R&D is one of the components of Development Plan, which contributes innovation and design activities in the medium-high and high-tech products

(5) *Program for 2018:* Among the main aims and objectives of the Program, Turkey's international competitiveness and in the manufacturing industry to increase its share of world exports by performing the conversion switch to high value added structure and is located to increase the share of high-tech industries. Many policies and measures have been detailed in order to achieve this purpose.

(6) *Turkey Industrial Strategy Document (2015–2018):* In order to accelerate the transformation of the Turkish industry into an industry structure in which high value-added and high-tech products are produced predominantly, the policy of "realizing technological transformation in the industry" was included. For this purpose, 15 action plans have been included and it has been planned to carry out actions at the determined time intervals together with the responsible and cooperating organizations.

(7) *Medium Term Program (2018–2020):* In the Medium Term Programme, the digital transformation (effective use and domestic production of additive manufacturing, robotics, the Internet of Things, big data, augmented reality technologies) roadmap for manufacturing industry will be analysed. Industry and technology development zones will be analysed and restructured with respect to their technical equipment, technology development capacities and utilization levels.

4 Methodology

Smart factory technologies are the production methods of the future. In this study, a research and analysis on smart factory technologies has been carried out. Questionnaires were conducted on three different groups consisting of academicians, R&D companies and manufacturers (factories) to investigate which of the smart factory system technologies should be given priority and suggestions were made based on the questionnaire results.

A. Questionnaire Construction

Three draft questionnaires were prepared as the first step of the data collection process. The method of obtaining expert opinion was followed during the validation phase of the questionnaires. The questions of the draft questionnaires that emerged as a result of the literature research were evaluated by two academicians specialized in the field and ten Information Technology (IT) specialists working in TUBITAK. In accordance with the views of these two groups, the draft forms of the questionnaires have been finalized. During the process of sending questionnaires to academicians, R&D companies and manufacturers, careful attention has been paid to inform the purpose of the questionnaire, the scope, and for what purpose the results are to be used. Questionnaires were applied online using the "Online Surveys" [7] application. An information message containing the purpose of the questionnaire, the scope, for what purpose the results are to be used and the link of the questionnaire was prepared and sent to the designated target group via e-mail. There were 4 questions in the questionnaires: 1 question was about smart factory technology awareness assessment, 1 question was about exploring the sources of smart factory technology awareness, 1 question was about detecting the firm or person's current situation about infrastructure/R&D/P&D/works in this area, 1 question was about detecting the firm or person's investment/R&D/work plans to this area. The 5-point Likert Scale was used in the question intending to determinate prioritization of the smart factory technologies.

B. Selection of the Research Sample

Research sample contains R&D firms that have carried out at least 1 project in the smart factory area, manufacturers (factories) and academicians studying in this field. Questionnaires were submitted to 448 companies and 200 academics who fulfilled these conditions. A total of 150 people, including 76 manufacturers, 36 R&D companies and 38 academicians, participated to the questionnaires.

C. Reliability Analysis of Questionnaire Results

Reliability analysis measures the consistency of responses to a questionnaire according to a pre-determined scale. The results of questionnaires were analyzed by obtaining "Cronbach's alpha" [8] values. IBM SPSS package program was used for this evaluation. The Cronbach Alpha (α) reliability value of the questionnaire for the academicians was determined as 0.884, the Cronbach Alpha (α) reliability value of the questionnaire for the R&D companies was determined as 0.848 and the Cronbach Alpha (α) reliability value of the questionnaire for the manufacturer companies was determined as 0.805. When Table 1 was examined, it was determined that the reliability of the results of the questionnaires are at "good" level.

Table 1. Cronbach's alpha reliability table

Cronbach's alpha	Reliability
$\alpha \geq .9$	Excellent
$.9 > \alpha \geq .8$	Good
$.8 > \alpha \geq .7$	Acceptable
$.7 > \alpha \geq .6$	Questionable
$.6 > \alpha \geq .5$	Poor
$.5 > \alpha$	Unacceptable

D. Limitations and Challenges

A number of difficulties has experienced in obtaining data. TUBITAK business application software has been used in the process of determining the firms and academicians forming the target group. This process took more time than planned due to the high number in target group. The questionnaires were submitted to the obtained contacts through e-mail channel. Because respondent numbers were low, reminder e-mails were sent to increase attendance to the questionnaires. Hence collecting the results of the questionnaires took more time than planned.

E. Results

The questions of questionnaires and answers to these questions are provided below. The analysis of these answers will be discussed in the "Conclusion" section.

Q1. *What is your level of knowledge in Smart Factory Systems and related advanced technologies (Industry 4.0)?* (Fig. 1)

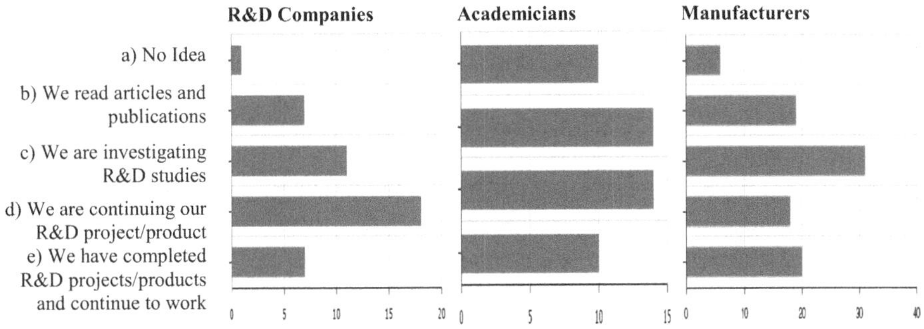

Fig. 1. Answers of first question

Q2. *From which source (s) did you obtain your knowledge of Smart Factory Systems and related advanced technologies (Industry 4.0)?*

Q3. ***(R&D Companies)*** *Would you rate your R&D activities in Smart Factory Systems and related advanced technologies (Industry 4.0)?* (Fig. 2)

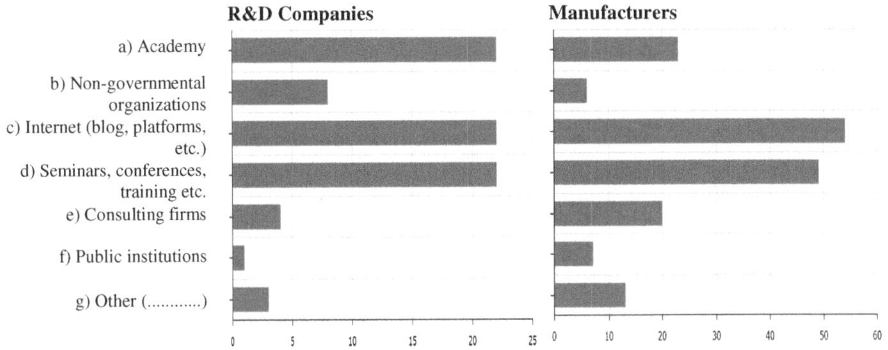

Fig. 2. Answers of second question

(a) *R&D activities in real-time data collection for the production/service or environmental areas using digital sensor networks (Internet of Things (IOT), Wireless/wired Sensor Networks (WSN), Geographic Information Systems (GIS), etc.)*
(b) *Innovative sensor technologies (Industry related physical, chemical, biological, optical, micro-nano sensors; smart actuators; industrial, wireless, digital sensors; artificial vision, innovative sensor applications; development of sensors resistant to extreme conditions)*
(c) *R&D activities in storing and analyzing data related to production/service processes, accessing data independent of time and space (cloud computing and big data)*
(d) *R&D activities in cyber security infrastructure*
(e) *R&D activities in analysis tools, optimization, simulation, etc. technologies*
(f) *R&D activities in collecting and analyzing customer data in design, engineering and production/service processes*
(g) *R&D activities in software supported planning, production/service, supply chain areas*
(h) *R&D activities in the areas of early warning systems related to performance of all production/service processes (intelligent systems, machine learning, virtual reality, etc.)*
(i) *R&D activities in the field of robots (smart manufacturing robots, equipment and software/management systems)*
(j) *R&D activities in M2X infrastructure (Machine-Machine, Machine-Human, Machine-Infrastructure)*
(k) *R&D activities in the area of autonomous production*

(1): No Idea (2): We read articles and publications on related topics (3): We are investigating R & D studies on related topics (4): We are continuing our R & D project/product work on related topics (5): We have completed R & D projects/products in related topics and continue to work (Fig. 3).

Q4. (R&D Companies) *Would you prioritize your investment/research/development plans for Smart Factory Systems and related advanced technologies (Industry 4.0)?*

(1): Not a priority, (2): Long term investment/research/development planned (3): Technical analysis is carried out (4): Short-term investment/research/development will be done (5): Investment/research/development studies started/completed (Fig. 4).

Q3. (Academicians) *Would you rate your article/publication/project work in Smart Factory Systems and related advanced technologies (Industry 4.0)?*

(a) *R&D activities in real-time data collection for the production/service or environmental areas using digital sensor networks (Internet of Things (IOT), Wireless/wired Sensor Networks (WSN), Geographic Information Systems (GIS), etc.)*
(b) *Data processing technologies (big data analysis, processing, correlation, etc.)*
(c) *Cyber security*
(d) *Innovative sensor technologies (Industry related physical, chemical, biological, optical, micro-nano sensors; smart actuators; industrial, wireless, digital sensors; artificial vision, innovative sensor applications; development of sensors resistant to extreme conditions)*

(1): No Idea (2): We read articles and publications on related topics (3): We are investigating R & D studies on related topics (4): We are continuing our R & D project/product work on related topics (5): We have completed R & D projects / products in related topics and continue to work.

	(1)		(2)		(3)		(4)		(5)			
	Σ	%	Σ	%	Σ	%	Σ	%	Σ	%	Ø	±
a)	3x	8,33	5x	13,89	9x	25,00	12x	33,33	7x	19,44	3,42	1,20
b)	7x	19,44	10x	27,78	8x	22,22	7x	19,44	4x	11,11	2,75	1,30
c)	6x	16,67	6x	16,67	8x	22,22	10x	27,78	6x	16,67	3,11	1,35
d)	7x	19,44	12x	33,33	9x	25,00	5x	13,89	3x	8,33	2,58	1,20
e)	5x	13,89	7x	19,44	9x	25,00	10x	27,78	5x	13,89	3,08	1,27
f)	1x	2,78	6x	16,67	10x	27,78	12x	33,33	7x	19,44	3,50	1,08
g)	5x	13,89	6x	16,67	9x	25,00	10x	27,78	6x	16,67	3,17	1,30
h)	6x	16,67	7x	19,44	9x	25,00	8x	22,22	6x	16,67	3,03	1,34
i)	6x	16,67	6x	16,67	11x	30,56	9x	25,00	4x	11,11	2,97	1,25
j)	6x	16,67	10x	27,78	10x	27,78	4x	11,11	6x	16,67	2,83	1,32

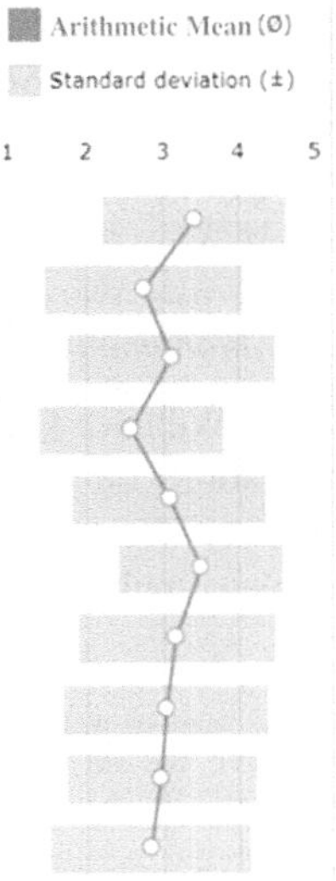

Fig. 3. Answers of R&D companies to third question

(1): Not a priority, (2): Long term investment/research/development planned (3): Technical analysis is carried out (4): Short-term investment/research/development will be done (5): Investment/research/development studies started /completed.

	(1)		(2)		(3)		(4)		(5)			
	Σ	%	Σ	%	Σ	%	Σ	%	Σ	%	Ø	±
a)	5x	13,89	7x	19,44	3x	8,33	6x	16,67	15x	41,67	3,53	1,54
b)	13x	36,11	7x	19,44	2x	5,56	6x	16,67	8x	22,22	2,69	1,64
c)	7x	19,44	6x	16,67	6x	16,67	10x	27,78	7x	19,44	3,11	1,43
d)	7x	19,44	11x	30.56	9x	25,00	2x	5.56	7x	19,44	2,75	1,38
e)	5x	13,89	4x	11,11	7x	19,44	9x	25,00	11x	30,56	3,47	1,40
f)	4x	11,11	8x	22,22	7x	19,44	10x	27,78	7x	19,44	3,22	1,31
g)	6x	16,67	6x	16,67	5x	13,89	10x	27,78	9x	25,00	3,28	1,45
h)	5x	13,89	7x	19,44	5x	13,89	8x	22,22	11x	30,56	3,36	1,46
i)	7x	19,44	8x	22,22	8x	22,22	7x	19,44	6x	16,67	2,92	1,38
j)	7x	19,44	11x	30,56	6x	16,67	6x	16,67	6x	16,67	2,81	1,39
k)	8x	22,22	8x	22,22	7x	19,44	7x	19,44	6x	16,67	2,86	1,42

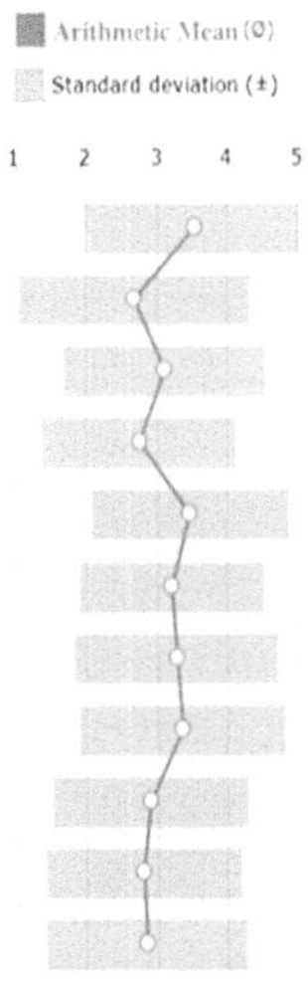

Fig. 4. Answers of R&D companies to fourth question

(e) *Automation (Simulation, Modelling, etc.)*
(f) *Embedded manufacturing materials (equipment, software, decision support systems)*
(g) *Early warning systems (intelligent systems, machine learning, virtual reality, prediction models, etc.)*
(h) *Robots (smart manufacturing robots, equipment and software/management systems)*
(i) *M2X infrastructure (Machine-Machine, Machine-Human, Machine-Infrastructure)*
(j) *Autonomous Systems*

(1): We have knowledge. We don't have studies yet (2): We have read articles and publications on related topics (3): We are investigating academic studies on related topics (4): We are working on academic projects/articles on related topics (5): We have completed/published articles and continue to study (Fig. 5).

(1): We have knowledge. We don't have studies yet (2): We have read articles and publications on related topics (3): We are investigating academic studies on related topics (4): We are working on academic projects/articles on related topics (5): We have completed/published articles and continue to study.

	(1)		(2)		(3)		(4)		(5)			
	Σ	%	Σ	%	Σ	%	Σ	%	Σ	%	Ø	±
a)	11x	28,95	6x	15,79	9x	23,68	6x	15,79	6x	15,79	2,74	1,45
b)	8x	21,05	5x	13,16	7x	18,42	10x	26,32	8x	21,05	3,13	1,46
c)	16x	42,11	9x	23,68	5x	13,16	4x	10,53	4x	10,53	2,24	1,38
d)	19x	50,00	6x	15,79	5x	13,16	7x	18,42	1x	2,63	2,08	1,28
e)	11x	28,95	5x	13,16	7x	18,42	7x	18,42	8x	21,05	2,89	1,54
f)	20x	52,63	8x	21,05	6x	15,79	2x	5,26	2x	5,26	1,89	1,18
g)	12x	31,58	9x	23,68	4x	10,53	6x	15,79	7x	18,42	2,66	1,53
h)	16x	42,11	8x	21,05	6x	15,79	7x	18,42	1x	2,63	2,18	1,25
i)	18x	47,37	5x	13,16	7x	18,42	7x	18,42	1x	2,63	2,16	1,28
j)	13x	34,21	9x	23,68	9x	23,68	3x	7,89	4x	10,53	2,37	1,32

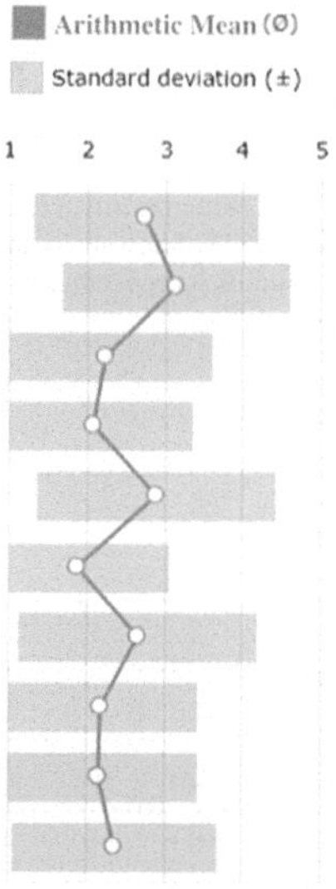

Fig. 5. Answers of academicians to third question

***Q4.* (Academicians)** *Would you prioritize your near-term study plans for Smart Factory Systems and related advanced technologies (Industry 4.0)?*

(1) Not Priority (2) Low Priority (3) Moderate Priority (4) Priority (Fig. 6)

***Q3.* (Manufacturers)** *Would you specify the level of use of Smart Factory Systems and related advanced technologies (Industry 4.0) in your production line processes?*

(1) Not Priority (2) Low Priority (3) Moderate Priority (4) Priority

	(1)		(2)		(3)		(4)			
	Σ	%	Σ	%	Σ	%	Σ	%	Ø	±
a)	7x	18,42	8x	21,05	15x	39,47	8x	21,05	2,63	1,02
b)	3x	7,89	5x	13,16	7x	18,42	23x	60,53	3,32	0,99
c)	10x	26,32	10x	26,32	7x	18,42	11x	28,95	2,50	1,18
d)	15x	39,47	7x	18,42	10x	26,32	6x	15,79	2,18	1,14
e)	9x	23,68	4x	10,53	11x	28,95	14x	36,84	2,79	1,19
f)	19x	50,00	6x	15,79	11x	28,95	2x	5,26	1,89	1,01
g)	7x	18,42	6x	15,79	17x	44,74	8x	21,05	2,68	1,02
h)	14x	36,84	9x	23,68	5x	13,16	10x	26,32	2,29	1,23
i)	13x	34,21	7x	18,42	12x	31,58	6x	15,79	2,29	1,11
j)	9x	23,68	8x	21,05	12x	31,58	9x	23,68	2,55	1,11

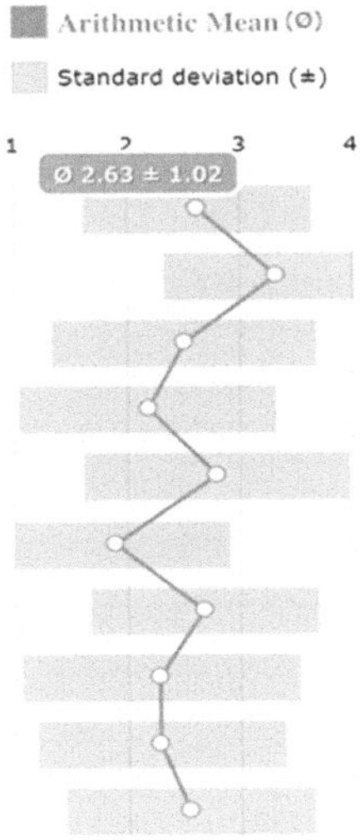

Fig. 6. Answers of academicians to fourth question

(a) *R&D activities in real-time data collection for the production/service or environmental areas using digital sensor networks (Internet of Things (IOT), Wireless/wired Sensor Networks (WSN), Geographic Information Systems (GIS), etc.)*
(b) *Storing and analyzing data related to production/service processes, accessing data independent of time and space (cloud computing and big data)*
(c) *Cyber security Infrastructure*
(d) *Analysis tools, use of optimization, simulation, etc. technologies*
(e) *Collecting and analyzing customer data in design, engineering and production/service processes*
(f) *Software supported planning, production/service, supply chain areas*
(g) *Use of early warning systems on the performance of all your processes (intelligent systems, machine learning, virtual reality, etc.)*
(h) *Robots (smart manufacturing robots, equipment and software/management systems)*
(i) *M2X infrastructure (Machine-Machine, Machine-Human, Machine-Infrastructure)*
(j) *Autonomous Production*

(1): It is not used (2): Little use (3): Moderately use (4): Heavily use (Fig. 7)

***Q4.* (Manufacturers)** *Would you ou prioritize your investment plans for the implementation/integration of Smart Factory Systems and related advanced technologies (Industry 4.0) into your existing processes?*

(1): Not a priority, (2): Long term investment planned (3): Technical analysis is carried out (4): Short-term investment will be done (5): Investment studies started/completed (Fig. 8).

(1): It is not used (2): Little use (3): Moderately use (4): Heavily use

	(1)		(2)		(3)		(4)			
	Σ	%	Σ	%	Σ	%	Σ	%	Ø	±
a)	16x	21,92	29x	39,73	13x	17,81	15x	20,55	2,37	1,05
b)	14x	19,18	28x	38,36	17x	23,29	14x	19,18	2,42	1,01
c)	9x	12,33	25x	34,25	25x	34,25	14x	19,18	2,60	0,94
d)	16x	21,92	19x	26,03	25x	34,25	13x	17,81	2,48	1,03
e)	12x	16,44	20x	27,40	29x	39,73	12x	16,44	2,56	0,96
f)	10x	13,70	21x	28,77	23x	31,51	19x	26,03	2,70	1,01
g)	30x	41,10	25x	34,25	12x	16,44	6x	8,22	1,92	0,95
h)	24x	32,88	22x	30,14	17x	23,29	10x	13,70	2,18	1,05
i)	23x	31,94	24x	33,33	16x	22,22	9x	12,50	2,15	1,02
j)	20x	27,40	33x	45,21	17x	23,29	3x	4,11	2,04	0,82

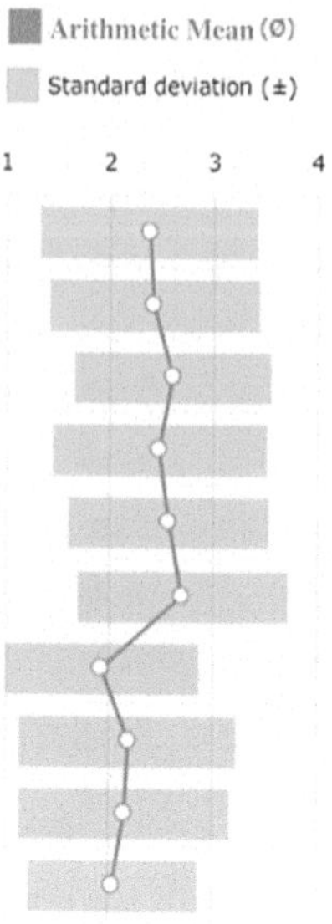

Fig. 7. Answers of manufacturers to third question

(1): Not a priority, (2): Long term investment planned (3): Technical analysis is carried out (4): Short-term investment will be done (5): Investment studies started /completed.

	(1)		(2)		(3)		(4)		(5)			
	Σ	%	Σ	%	Σ	%	Σ	%	Σ	%	Ø	±
a)	8x	10,81	20x	27,03	14x	18,92	12x	16,22	20x	27,03	3,22	1,39
b)	5x	6,76	14x	18,92	18x	24,32	15x	20,27	22x	29,73	3,47	1,28
c)	10x	13,51	8x	10,81	22x	29,73	9x	12,16	25x	33,78	3,42	1,40
d)	6x	8,11	14x	18,92	15x	20,27	19x	25,68	20x	27,03	3,45	1,29
e)	9x	12,16	14x	18,92	20x	27,03	10x	13,51	21x	28,38	3,27	1,38
f)	6x	8,11	13x	17,57	16x	21,62	14x	18,92	25x	33,78	3,53	1,34
g)	10x	13,51	20x	27,03	19x	25,68	13x	17,57	12x	16,22	2,96	1,29
h)	15x	20,27	9x	12,16	19x	25,68	17x	22,97	14x	18,92	3,08	1,39
i)	17x	22,97	13x	17,57	17x	22,97	12x	16,22	15x	20,27	2,93	1,45
j)	13x	17,57	16x	21,62	20x	27,03	18x	24,32	7x	9,46	2,86	1,24

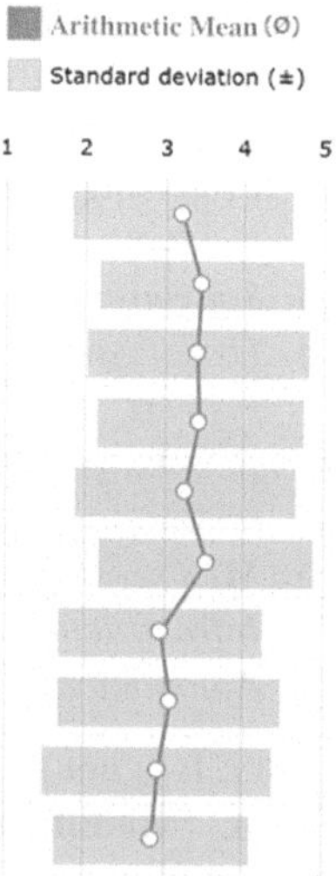

Fig. 8. Answers of manufacturers to fourth question

5 Conclusion

5.1 Level of Knowledge in Smart Factory Systems and Related Advanced Technologies

According to "New Industrial Revolution: Intelligent Manufacturing Systems Technology Road Map" study of TUBITAK dated 03.01.2017, only 22% of the companies that participated in the questionnaire had extensive knowledge on smart factory systems and related advanced technologies. In this study, 78.4% of the R&D companies and 73.97% of the manufacturers have comprehensive knowledge. Over the past 2 years, it has been seen that acquiring comprehensive knowledge has increased 3.5 times and a positive development has been observed. It is thought that 79.2% awareness rate of academicians contributed to this increase in awareness rate. It would be beneficial for the universities to communicate more closely with the industry and transfer their academic knowledge and culture to the industry and to create a similar academic discipline in the industry.

5.2 Information Sources in Smart Factory Systems and Related Advanced Technologies

It was determined that R & D and manufacturer companies participating in the questionnaire obtained 78% of the acquisitions in Smart Factory Systems and related advanced technologies from the following sources:

- Information on the internet
- Seminars, conferences, training, etc. activities
- Academy
- Civil Society Organizations (NGOs)

If more effective and focused work on these 4 sources can be achieved, the level of awareness can be increased and trends can be directed. The effectiveness of these 4 sources can be enhanced by the following suggestions:

- It is recommended to transfer the domestic vision and international goals to all stakeholders and make the correct orientations by using TUBITAK's, Universities' and Ministries' web pages efficiently.
- Companies and universities can be guided by informing them about the conversion to the national smart factory with face-to-face interaction in the seminars that will be organized.
- It is believed that the establishment of support programs that strengthen the communication between the academy and the companies will enable the academicians to be closer to the industry and to be more effective in their R&D work.
- It is evaluated that the source (capital) problem in communication to R&D and manufacturer companies can be solved by preparing special support programs for Civil Society Organizations (NGOs). It is also believed that the NGOs can be brought together with academicians to ensure that the information about smart factory technologies is more academic, up-to-date and international.

5.3 The Level of Use of Smart Factory Systems and Related Advanced Technologies

The following aspects emerged within the scope of the study. These aspects are provided separately for R&D companies and Manufacturers.

R&D Companies:

- The activities in the area of "cyber security" infrastructure are at minimum level
- They cannot perform sufficient R&D work in the areas of "innovative sensor technologies", "early warning systems for the performance of all production processes", "robots", "M2X technologies". These technologies contain costly hardware requirements.
- The activities in "real-time data collection for the production/service or environmental areas using digital sensor networks", "collecting and analyzing customer data in design, engineering and production/service processes", "analysis tools, use of optimization, simulation, etc. technologies", "storing and analyzing data related to production/service processes, accessing data independent of time and space", "software supported planning, production/service, supply chain areas", "autonomous production" has been found to be in relatively better condition.

Manufacturers:

- The activities in the area of "using early warning systems on the performance of all their processes" are at minimum level.
- "Software supported planning, production/service, supply chain technology" used at a relatively high level.
- The use levels of "real-time data collection for the production/service or environmental areas using digital sensor networks", "storing and analyzing data related to production/service processes", "accessing data independent of time and space", "cyber security infrastructure", "use of optimization, simulation, etc. technologies", "collecting and analyzing customer data in design, engineering and production/service processes" are promising, even if not at the expected level.
- The use levels of "Early warning systems for the performance of all production processes", "robots", "M2X technologies, "autonomous production" are below the expected level. It is necessary to take measures to improve the integration and use of these technologies.

The results of this study and the "New Industrial Revolution: Intelligent Manufacturing Systems Technology Road Map" study, appeared to be similar. "New Industrial Revolution: Intelligent Manufacturing Systems Technology Road Map" study evaluated the digital maturity level of the companies as between Industry 2.0 and Industry 3.0. The increase in the level of awareness in the related technology fields of R&D companies and manufacturers in our country has shown that the conversion of smart factories can take place in the medium term. The studies of academicians also accelerated in related fields.

5.4 Investment/Research/Development Plans for Smart Factory Systems and Related Advanced Technologies

The comparison table containing the plans of R&D companies, manufacturers and academicians is presented in Table 2. 1-value is considered priority and 11-value is considered non-priority.

Table 2. Comparison of the plans

Areas	R&D companies	Manufacturers	Academicians
Real-time data collection for the production/service or environmental areas using digital sensor networks (Internet of Things (IOT), Wireless/wired Sensor Networks (WSN), Geographic Information Systems (GIS), etc.)	2	1	4
Innovative sensor technologies (Industry related physical, chemical, biological, optical, micro-nano sensors; smart actuators; industrial, wireless, digital sensors; artificial vision, innovative sensor applications; development of sensors resistant to extreme conditions)	11	10	10
Storing and analyzing data related to production/service processes, accessing data independent of time and space (cloud computing and big data)	4	6	3
Cyber security infrastructure	10	11	6
Analysis tools, optimization, simulation, etc. technologies	5	2	2
Collecting and analyzing customer data in design, engineering and production/service processes	1	5	1
Software supported planning, production/service, supply chain areas	3	4	9
Early warning systems related to performance of all production/service processes (intelligent systems, machine learning, virtual reality, etc.)	6	3	8
Robots (smart manufacturing robots, equipment and software/management systems)	7	7	7
M2X infrastructure (Machine-Machine, Machine-Human, Machine-Infrastructure)	9	9	7
Autonomous production	8	8	5

The R&D study plans of the R&D companies are similar to the investment plans of the manufacturers. Robot, M2X, autonomous production, cyber security infrastructure and innovative sensor technologies are seen to have the lowest level of R&D work and investment plans. These results are believed to be due to the high technology requirements of the related technologies, high investment costs and inadequate domestic product. As the cost of investment increases, it is considered that R&D investments and smart factory transformation investment extend the recycling period and therefore all investment plans are adversely affected. Plans are focused on the following areas: "Real-time data collection for the production/service or environmental areas using digital sensor networks", "Storing and analyzing data related to production/service processes, accessing data independent of time and space", "Analysis tools, optimization, simulation, etc. technologies", "Collecting and analyzing customer data in design, engineering and production/service processes", "Software supported planning, production/service, supply chain areas".

5.5 Suggestions

The reason for the tendency of focusing on the "Real-time data collection for the production/service or environmental areas using digital sensor networks", "Storing and analyzing data related to production/service processes, accessing data independent of time and space", "Analysis tools, optimization, simulation, etc. technologies", "Collecting and analyzing customer data in design, engineering and production/service processes", "Software supported planning, production/service, supply chain areas" is that these areas are software based, costs are lower and it is relatively easy to integrate them into production processes. It is thought that this tendency go round in circles to the demand for supply or vice versa. Investments in smart factory technology with hardware content, incentives for areas that are needed nationally through attractive support programs for R&D activities should be intensified to break this turnaround. It was also observed that the academicians who participated in the survey had almost the same tendency of their plans as the manufacturers and R&D companies. This tendency is due to the fact that software based technology studies can be done more individually, relatively lab independent and work cost is lower. In order to change this tendency of academicians, laboratory support can be given to encourage study in related fields by reducing the cost of laboratory setup/modernization. Thus, the staff-related problems of R&D companies will be reduced as much as possible by triggering national studies in related technologies. In conclusion, it is suggested to construct grant programs with 100% grants in areas where planning of companies is lower. Grant programs can lead organizations to work in these areas.

Acknowledgment. We thank The Scientific and Technological Research Council of Turkey (TUBITAK) that provided permission for the research.

References

1. Durakbasa MN et al (2017) Advanced metrology and intelligent quality automation for Industry 4.0-based precision manufacturing systems. Solid State Phenom 261:432–438
2. Durakbasa MN, Bas G, Bauer JM, Poszvek G (2014) Trends in precision manufacturing based on intelligent design and advanced metrology. Key Eng Mater 581:417–422
3. Kopacek P (1999) Intelligent manufacturing: present state and future trends. J Intell Robot Syst 26(3):217–229
4. Durakbasa MN, Bauer JM, Bas G, Kräuter L (2016) Novel trends in the development of autonomation and robotics in metrology for production in the future. IFAC-PapersOnLine. 49 (29):6–11
5. TÜBİTAK (2016) New Industrial Revolution: Intelligent Manufacturing Systems Technology Road Map Report. http://www.tubitak.gov.tr/sites/default/files/akilli_uretim_sistemleri_tyh_v27aralik2016.pdf
6. Turkey's global competitiveness as a necessity for Industry 4.0 TÜSİAD-T/2016-03/576. ISBN 978-605-165-016-6
7. Online Surveys (2018). https://www.onlineanketler.com/
8. Reynaldo J, Santos A (1999) Cronbach's Alpha: a tool for assessing the reliability of scales. J Extension (JOE), **37**(2)

Technology Selection for Digital Transformation: A Mixed Decision Making Model of AHP and QFD

Hasan Erbay and Nihan Yıldırım(✉)

Management Engineering, Istanbul Technical University, Istanbul, Turkey
hasan.erbay@tr.bosch.com, yildirimni@itu.edu.tr

Abstract. By the introduction of Industry 4.0 technologies and tools, manufacturing companies and especially SME (Small and medium sized companies) from supplier tiers of main manufacturing value chains faced the challenge of uncertain environment and lack of methods in making effective technology identification, prioritization and selection in accordance with their productivity and effectiveness improvement needs. Especially from a developing country perspective, being decision makers in the user side of Industry 4.0 value chain had been in a quest for decision making models that can be used in technology investments for Industry 4.0 technologies. However, due to the fact that, balancing potential benefits, barriers/challenges of Industry 4.0 tools with current or potential improvement needs in manufacturing systems require high level of knowledge and expertise both in manufacturing and Industry 4.0 technologies, most of the researches which focus on few pillars of the problem remained insufficient as they lack interdisciplinary and qualitative approaches. Especially for technology-dependent and follower countries such as Turkey that has limited resources and strict constraints, previous researches with proposed decision making models on digital transformation are very limited. In this context, this study proposes a multi-dimensional and hybrid technology evaluation methodology for technology selection on Industry 4.0 technologies that covers technological issues, competitive competencies and managerial pillars together. By combining the pillars of technological tools, benefits and challenges of their usage in manufacturing industry, the proposed model utilizes multi -criteria decision support model, namely AHP, a Needs Analysis framework based on Quality Function Deployment approach. Data collected via interviews and survey, the qualified participant having manufacturing excellence are preferred. Outputs from the model is expected to serve decision makers in SMEs of manufacturing industry for technology selection and use case scenario generation for the best and appropriate strategy.

Keywords: Industry 4.0 · Digital transformation
Quality function deployment · Analytical hierarchy process

N. M. Durakbasa and M. G. Gencyilmaz (Eds.): ISPR 2018, *Proceedings of the International Symposium for Production Research 2018*, pp. 480–493, 2019.
https://doi.org/10.1007/978-3-319-92267-6_41

1 Introduction

In recent years, digital transformation in manufacturing industry that is conceptualized as industry 4.0 and underlined as the Fourth Industrial Revolution have gained importance and popularity. Industry 4.0 namely digital transformation is defined as the integrating of complex machineries with smart sensors and software to predict, manipulate and plan for more sustainable business and societal outcomes. This transformation referred to the trends towards digitization of the manufacturing environments which takes the automation of processes and business systems to a next level by introducing intelligent usage of inter-communicating machines through the usage of technologies such as smart sensors to get reliable data for analysis, robotics technology to replace the labor for dangerous tasks as well as precise operations, data analytics to analyze what is happening and what will probably happen, cloud computing for a faster and reliable cooperation among companies with a performance, image processing to execute the operations with higher level of quality in a digitized environment, VR/AR to support labor and make robots more autonomous, 3D printing to shorten time to market and various state-of-art technologies with far too numerous benefits.

New technologies arise uncertainty and unclarity for those organizations planning to transform their business models into digital. As the use of these technologies are newly emerging in manufacturing environments and stands as an uncertainty environment, identification and selection of the Industry 4.0 technologies and tools occurred as a challenge that is faced by manufacturing industry decision makers in the user side. Benefits and the barriers of the digital transformation is unclear, and selection of the best strategy is difficult requiring high level of knowledge and expertise competency.

Theoretical and practical frameworks of roadmaps and decision making models that can be utilized for guiding manufacturing industry still remains insufficient or immature. Automotive supplier industry is chosen as the representing sector of manufacturing industries. Approaches on digital transformation are described and analyzed by the industry 4.0 technology providers, and the numbers of academic studies for users are limited in the literature. The literature is lack of studies especially for the dependent and follower countries such as Turkey that has limited resources and strict constraints. Although there are some reports and research studies on the digital transformation proposing an overall evaluation with generic expressions, it is needed to analyze manufacturing processes based on a detailed perspective for Turkish automotive industry.

In this context, based on this emerging need, this study aims to present a hybrid method of decision making on prioritizing Industry 4.0 technologies and tools for technology selection and planning in manufacturing industry.

This study proposes a multi-dimensional and hybrid methodology that covers technological issues, competitive competencies and managerial pillars. By combining the pillars of technological tools, benefits and challenges of their usage in manufacturing industry, the proposed model utilizes multi-criteria decision support model, namely AHP, a Needs Analysis framework based on Quality Function Deployment approach. Data collected via interviews and survey, the qualified participant having manufacturing excellence are preferred. AHP (Analytical Hierarchy Process) is selected the best tool to assure relativity and prioritization as the technologies are pushed to be used all together, but the resources are limited. QFD (Quality Function Deployment) is selected as it is needed to choose the preferable strategy among the multiple dimensions and their interactions proposing an intervention matrix. Outputs from AHP and QFD were utilized for technology selection and use case scenario generation for the best strategy.

The proposed decision making model is expected to represent a solid practical and unique example that can guide the entrepreneur and the government in terms of the potential benefits, technologies and the barriers of industry 4.0 and enable them to better and systematically understand the potential impacts and benefits of digital transformation, and also to allocate resources more effectively.

2 Literature Review

Originally initiated in Germany, Industry 4.0, the fourth industrial revolution, has attracted much attention in recent literatures. It is closely related with the Internet of Things (IoT), Cyber Physical System (CPS), information and communications technology (ICT), Enterprise Architecture (EA), and Enterprise Integration (EI) (Lu 2017). Table 1 summarizes the academic and research studies on the concept of I4.0.

Schuh and Gartzen (2015) proposed the connection between digital transformation and the required competence. Due to increasing complexity and dynamics of products and processes, employees in today have to be qualified for more than just repetitive operations. Individualization of products in industrial mass production (also known as mass customization) poses new and major challenges on the qualification of production workers. I4.0, IoT and CPS do not only enable real-time communication, transparency, but also bring major changes on the shop floor level. CPS can support production line workers by offering new ways of gathering, processing and visualization of process data. Providing this information offers advanced opt ions for full automation but also supports the efficient error-free task execution by the production personnel (Fig. 1).

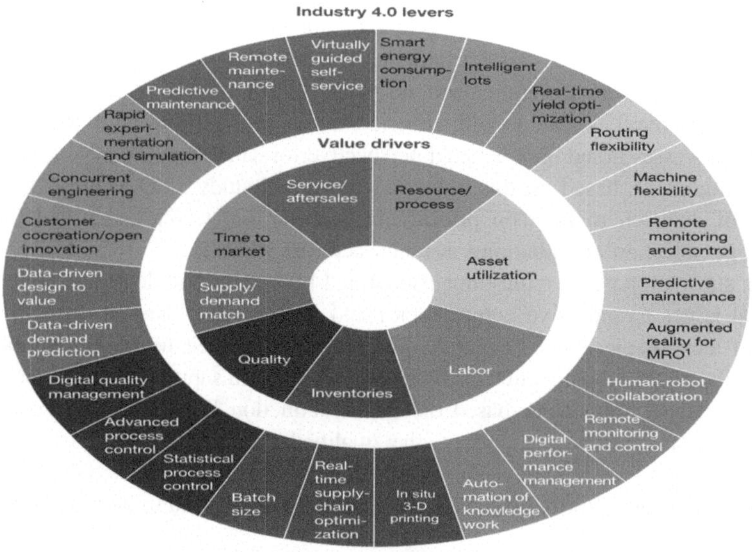

Fig. 1. Digital drivers compass (McKinsey and Company 2017)

Shafiq et al. (2016) defines Industry 4.0 in the context of conversion of the product and service within the entire value chain; the goals of Industry 4.0 are to provide IT-enabled mass customization of manufactured products; to make automatic and flexible adaptation of the production chain; to track parts and products; to facilitate communication among parts, products, and machines; to apply human-machine interaction (HMI) paradigms; to achieve IoT-enabled production optimization in smart factories; and to provide new types of services and business models of interaction in the value chain. Aßmann and Resenhoeft (2016) discusses the new industrial revolution in the basis of creation completely novel services: constant quality control, efficient production for batch sizes of one, and improved energy management. Sensors collect information from their surroundings, software draws new conclusions from the data, and uses it to create new business models with surprising customer benefits. Keywords here include big data, data mining, and data analytics – all of which are brimming with commercial potential. It is no coincidence that i4.0 is arriving precisely at this point in time. People first needed to make great strides in sensors, robotics, microelectronics, and data processing. WiFi, fast databases, sensors, RFID tags, or cloud computing have by and large become relatively inexpensive over the past few years. The increasing connectivity of all process steps in i4.0 also leads to a situation where skills such as self-organization, flexibility, self-learning, the willingness and ability to learn, innovative thinking, and communication skills become more vital. This applies to all jobs and careers. Workers enjoy new freedom to choose where and – to a certain extent – when to work. Industry 4.0 creates 400,000 new jobs.

Roblek et al. (2016) analyze the importance and the effect of the digital transformation and eventually the interconnected technologies for the creation of value for society and organizations.

Table 1 summarizes the academic and research studies on the concept of I4.0. These resources were utilized in our study for defining the pillars of sub-technologies and application areas of our research.

Table 1. Academic and research studies and corresponding subjects

Concept approach	Author and subject
Work-based learning through I4.0	Schuh and Gartzen (2015) proposed the connection between **digital transformation and the required competence**. Due to increasing complexity and dynamics of products and processes, **employees in today have to be qualified** for more **than just repetitive operations**
Value drivers compass map	McKinsey and Company (2015) presented a useful and compact digitalization map so called **digital drivers' compass**
Lean through I4.0	Sanders et al. (2016) analyses the **similarity** between the **lean manufacturing and Industry 4.0**. He investigated the relevance of the lean and Industry 4.0 that is capable of establishing lean. **Tagging each entity** in the production system guarantees supplying **correct products to the right destinations and improvement of lead times**

(*continued*)

Table 1. (*continued*)

Concept approach	Author and subject
Decision support systems by I4.0	In Industry 4.0, **machines, parts, systems and human beings** will be **highly connected** and highly **integrated**. Every physical object will formulate a Cyber-Physical System (CPS) and it will always be linked to its digital footprint
Mass customization through smart production	Zawadzki and Żywicki (2016 proposes general concept of **smart design** and production control as key elements for efficient operation of a smart factory. **Design process of individualized products** and organization **of their production** in the context of **realization** of the **mass customization** strategy, which allows **a shortened time of development** for a new product
Smart maintenance through mach. learning	Isaacs et al. (2017) proposed a roadmap of making factories **smarter through machine learning**. Machines and systems have become more intelligent and more connected. Using **sophisticated data analytics**, combining the **wealth of sensor** data and other asset-related lifecycle parameters, companies are shifting their focus on **"Preventative" maintenance with greater emphasis on "Predictive" maintenance**
OEE opt. via data analytics	Garbie (2016) proposed a practical way to get started in manufacturing IIoT by means of **data analytics to monitor OEE** and to **optimize the processes**
Rise of AI	Mongo DB White Paper (2017) More Data, Cheaper Computation, More Sophisticated Algorithms, Broader Investment

Considering all these application areas and technologies, Oesterreich and Teuteberg (2016) explore the state of the art as well as the state of practice of Industry 4.0 relating technologies by pointing out the PESTEL (political, economic, social, technological, environmental and legal) implications of its adoption illustrated in Table 2. This approach provided the benefit/challenge pillar as a framework of our research and source for interview and Delphi survey design.

Table 2. Benefits and challenges of industry 4.0 in the context of pestel (Oesterreich and Teuteberg 2016)

Implications	Perspective
Benefits	PESTEL
Cost savings: Elimination of labour-intensive task via automation	Economic
Time savings: Shorter duration of design by innovative technologies	Economic
On-time and on-budget delivery: Reducing delivery time and keep projects under budget	Economic

(*continued*)

Table 2. (*continued*)

Implications	Perspective
Improving quality: Elimination of errors via predictive models	Economic
Improving collaboration and communication: Simultaneous engineering among the shareholders via cloud platforms	Economic
Improving customer relationship: Use of simulation technologies to get better customer needs for customization	Economic
Enhancing safety: Reducing risks via wearables on dangerous tasks	Social
Improving the image of the industry: The digital transformation of the whole industry can help to improve its image	Social
Improving sustainability: Controlling high carbon dioxide emissions, energy consumption and wastes suggests sustainable conversion processes	Environmental
Challenges	PESTEL
Hesitation to adopt: Resistance against the transformation within an organization	Political
High implementation cost: High cost of technology, training and skilled workforce	Economical
Organizational and process changes: Redesign of the current business models	Economical
Need for enhanced skills: Need for competent employees	Social
Knowledge management: Difficulty against large volume of data along with the entire value chain	Social
Acceptance: Conservatism and inability to adapt by staff members of its companies	Social
Lack of standards and reference architectures: Need for complete and international standard, software incompatibility	Technological
Data security and data protection: The concern for large amount company data against unauthorized access	Technological
Enhancement of existing communication networks: Requirement for a fast and reliable internet access, broadband connectivity	Technological
Regulatory compliance: Ethical and legal concerns about the tracking and monitoring of employees	Legal

3 Methodology

The main objective of this study to describe, analyze and understand the I4.0 benefits, tools and challenges for an organization targeting to implement smart factory solutions in their strategic decision making. A new methodology has been developed which is a combination of the Quality Function Deployment approach (QFD) and Analytical Hierarchy Process (AHP). This study proposes a multi-dimensional and hybrid methodology that covers technological issued, competitive competencies and managerial pillars. By combining the pillars of technological tools, benefits and challenges of their usage in manufacturing industry, the proposed model utilizes multi-criteria decision support model, namely AHP, a Needs Analysis framework based on Quality

Function Deployment approach. The proposed methodology suggests a flexible and adaptable structure which can be employed for various cases. After the weights (global importance) and relevance values (local importance) are set by QFD and AHP, is structured to identify which benefits (WHAT) can be gained by means of which tools and technologies (HOW) under what kind of challenges (DESPITE -AGAINST). Data collected via interviews and survey from the qualified participants from manufacturing industry, who has Industry 4.0 expertise, knowledge and manufacturing excellence.

In this study the main I4.0 benefits are defined and are associated with the appropriate tools and challenges in a needs analysis method which is derived from QFD as adapted from Herron and Braiden (2006). By means of combined structured survey, the experts rated the relevancies among the I4.0 elements based on ***QFD approach in a scale 1 to 4 to imply*** the overall scores of the benefits, challenges and tools respectively based on QFD. Those values are normalized to obtain the relevance weights.

Construction of the research is given in Fig. 2. After identifying the I4.0 pillars through content analysis and expert views, we collected data via survey for QFD relevance and AHP pairwise comparisons and set the global importance of each pillar. Finally, we could be able to rank the QFD relevancies in a matrix among Benefits - Tools, and Tools – Challenges dimensions. Process flow of the data collection and analysis phases is also presented in Fig. 3.

At the very first step, the main pillars are identified as listed in Table 3 with the participation of the experts working in the related fields such as I4.0 project management, IoT management, data science through Delphi Technique as shown step-wise in Figs. 2 and 3.

AHP (Analytical Hierarchy Process) is selected the best tool to assure relativity and prioritization as the technologies are pushed to be used all together, but the resources are limited. AHP is a method in which a complex, disintegration to smaller sections then put in a hierarchy structure. Saaty in 1994 express that AHP help to disintegrate the unstructured program to hierarchy logical decision (Pakizehkara et al. 2016). The experts prioritized the identified I4.0 elements in the structured survey sessions using a 1 to 9 scale as shown in Fig. 4. Then these evaluation results are calculated by AHP matrix to obtain each element's priority or global weight. In Delphi Survey, 12 experts (with at least master degree, 8-years of experience in I.40 professional such as digital transformation project leader, transformation director, sensor and cloud expert working in engineering and manufacturing sectors) first prioritized the importance of each pillars among each other by AHP rating 1–9 scale to get each benefit, tool and challenge weight.

Then a combined and innovative survey is designed to get both AHP and QFD evaluations as shown in Fig. 5. The pillar elements are crossed to obtain the relevancies among the intersected relations. For example; *"What benefit can be gained how and what is the challenge?"* Indeed, the survey can also be graphed as a 3-level decision network on which the nodes are importance comparison through AHP, and arcs are the relevancies obtained by QFD rating. QFD model means quality function development that also translated as quality operation development. QFD model is a tool that apply customer's quality demand in different stages of product genesis and service. QFD is a customer-driven tool that is widely used for product planning purposes. It can be

Table 3. List of pillars of analysis - tools of I4.0, benefits and challenges

Tools	Description
RFID	Radio Frequency Identification: Real time inventory. Production level. Transparent value chain
Smart sensor (data collection)	Devices converting physical inputs such as movement,pressure,flow,rate,color,light, humidity,magnetism,noise into digital outputs
Robotization	Machines using for repetitive. Precise and dangerous tasks. Collaborates with human
Production Management Software (MES, APC, ERP)	Computerized systems used in manufacturing. to track and document the transformation of raw materials to finished goods Monitoring all production data real time, getting the instant and last minute data from machinery. Labor and the other entities from the manufacturing environment.
Data analytics	Process of examining data sets to extract meaningful information using specialized mathematical models.
3D printing	The creation of a 3D printed object is achieved using additive processes without using special fixtures.
VR, AR, Remote control	Use of computer technology to create a simulated environment. Integration of dijgital informal Lou with the user's environment in real.
Image processing	Method converting images and videos into digital outputs using algorithms (pattern recognition quality control)
Benifits	Description
Reducing inventories	Raw materials, semi products (WIP), finished products
Increasing process efficiency	Real time & transparent machine monitoring
Reducing man power requirement	Eliminating non-value adding process using leaner methods
Improving quality performance	Quality transparency, real time measurement, early diagnosis of errors, lowering scarps
Improving maintainance performance	Real time visibility of machine parameters: vibration, humidity, temprature and predictive replacement of worn components
Efficienct & Robust production planning	Efficient error-free production planning. on time delivery
Rapid time to market	Shorter R&D, project and sample preparation duration
Energy management	Real time visibility of energy consumption, anomaly detection: leak, over use…etc.
Challenges	Description
Data security concerns	Abuses by unauthorized accesses on confidential data, know- how, cyber attacks
Lack of budget	Financing issues of high cost equipment investments, training and consultancy
Lack of top Mng. involve	Lack of top management support

(*continued*)

Table 3. *(continued)*

Challenges	Description
Lack of expert know-how	Lack of expertise persons in terms of 14.0 technologies
Poor cost-benefit (ROI) ratio	Unclear and poor ROI calculations
Unclear standards	Lack of world- wide approved standards for interoperability and end-to-end seamless communication
Government support & Promotion	Lack of tax and custom exemptions, funding pilot projects

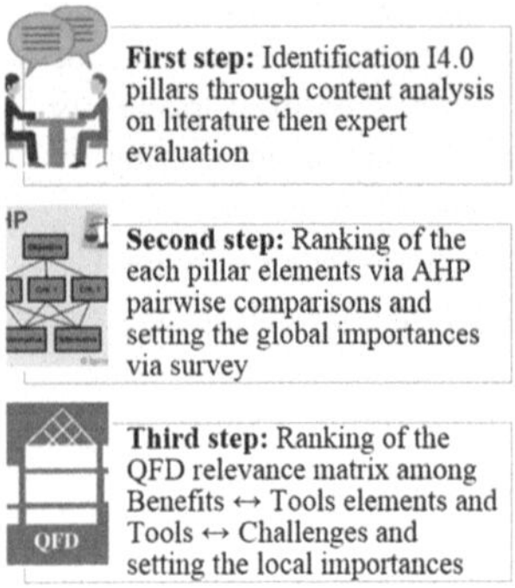

Fig. 2. Construction of the research

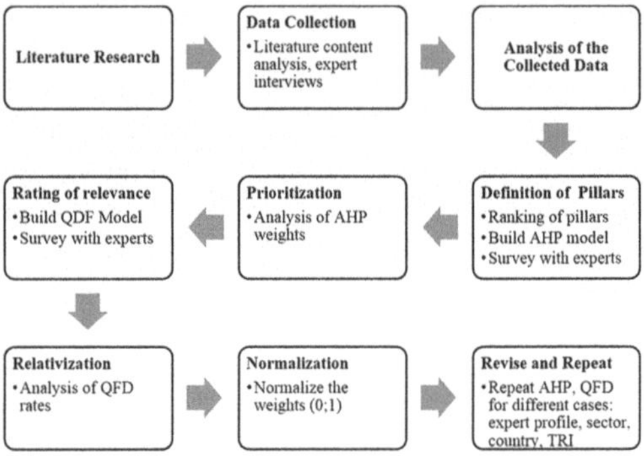

Fig. 3. Process flow of data collection and analysis methodology

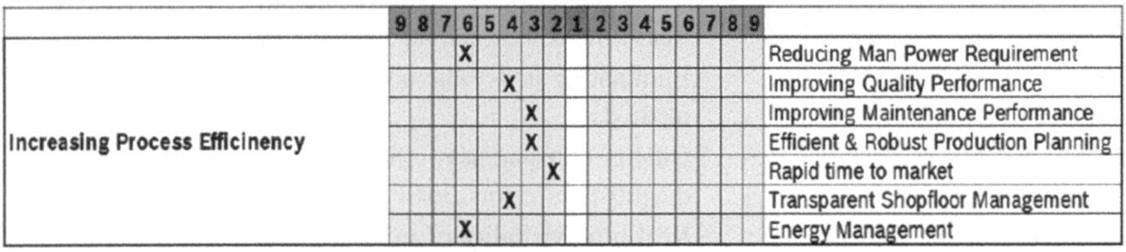

Fig. 4. Example of overall scores of benefits

beneficial to reach higher levels in customer satisfaction. QFD (Quality Function Deployment) is selected as it is needed to choose the preferable strategy among the multiple dimensions and their interactions proposing an intervention matrix. The main I4.0 benefits are defined and are associated with the appropriate tools and challenges in a needs analysis method which is derived from QFD (Herron and Braiden, 2006). By means of combined structured survey, the experts rated the relevancies among the I4.0 elements based on QFD approach in a scale 1 to 4 as shown in Fig. 5. Letters of B, C, T and T' imply the overall scores of the benefits, challenges and tools respectively based on QFD. Those values are normalized to obtain the relevance weights.

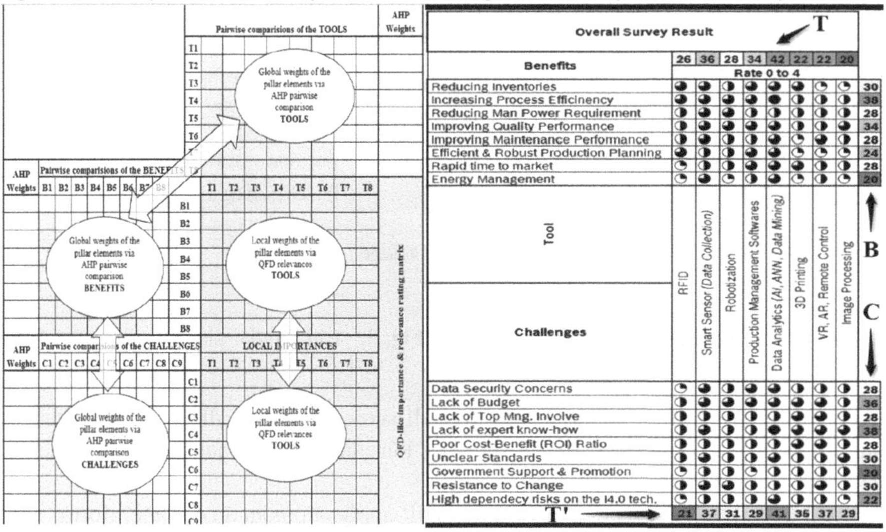

Fig. 5. Example of overall scores of benefits

4 Findings

After all the feedbacks from the respondents, the results for each connection arc (which is the relevance weight among each pillar's elements) are calculated as the geometric mean. Because geometric mean gives more accurate and reliable average results when calculating different feedbacks in AHP (Tomashevskii 2015).

The overall scores of the benefits calculated by the QFD matrix are given in Fig. 5. The need for process efficiency is ranked with the highest score followed by quality performance improvement and reducing inventories respectively. The overall scores of the tools and technologies calculated by the QFD matrix are given in Fig. 6. Data analytics is ranked first with the highest score then followed by smart sensors and

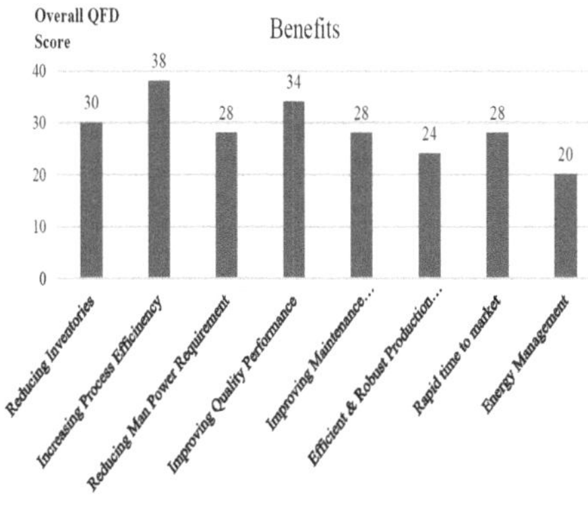

Fig. 6. Overall scores of benefits by QFD

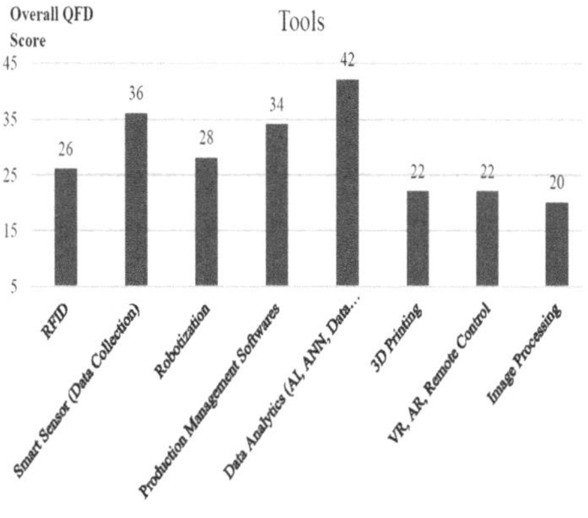

Fig. 7. Overall scores of tools by QFD

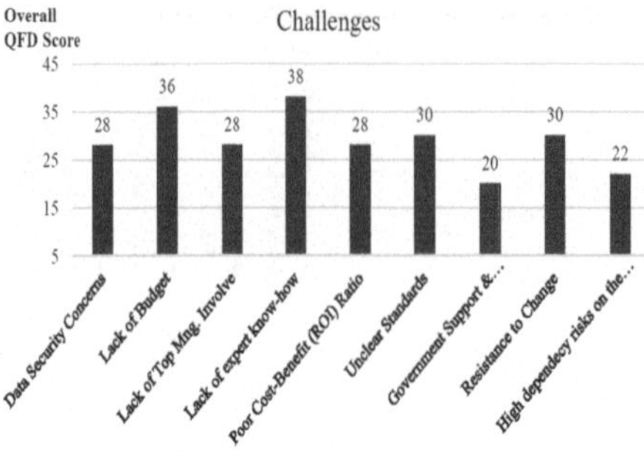

Fig. 8. Overall scores of challenges by QFD

I4.0 Pillar	The top 3 priority with the highest weight
Benefits	Increasing Process Efficiency
	Improving Quality Performance
	Reducing Inventories
Tools	Data Analytics (AI, ANN, Data Mining)
	Smart Sensor (Data Collection)
	Production Management Software (MES, APC, ERP)
Challenges	Lack of expert know-how
	Lack of Budget
	Unclear Standards & Resistance to Change

Fig. 9. Top 3 pillar elements from the QFD

production management software. The overall scores of the challenges calculated by the QFD matrix are given in Fig. 8 showing that the highest challenge is the lack of expert know-how.

As the survey suggests both QFD and AHP applications in a combined form, AHP priorization among the I4.0 pillars as shown in Table 4.

Table 4. AHP weights of the pillars

Benefits	AHP weight	Tools	AHP weight	Challenges	AHP weight
Reducing inventories	12%	RFID	6%	Data Security Concerns	10,2%
Increasing process efficiency	27%	Smart Sensor (Data Collection)	19%	Lack of Budget	19,0%
Reducing man power requirement	10%	Robotization	9%	Lack of Top Mng. Involve	10,1%
Improving quality performance	19%	Production Management Software (MES, APC,ERP)	12%	**Lack of expert know-how**	**22,2%**
Improving maintenance performance	11%	**Data Analytics (AI, ANN, Data Mining)**	**29%**	Pool Cost-Benefit (ROI) Ratio	**10,1%**
Efficient & Robust production planning	6%	3D Printing	10%	Unclear Standards	9,8%
Rapid time to market	12%	VR AR, Remote Control	10%	Government Support & Promotion	3,9%
Energy MANAGEMENT	4%	Image Processing	6%	Resistance to Change	10,1%

(*continued*)

Table 4. (*continued*)

Benefits	AHP weight	Tools	AHP weight	Challenges	AHP weight
				High dependency risks on the I4.0 tech. providers	4.6%
Sum	**100%**		**100%**		**100%**
CR	**0.02**		**0.01**		**0.01**

5 Conclusion

In this study, the primary aim is to discover the current state of the art and state of practice of Industry 4.0 technologies in the manufacturing industry and to provide a strategy map concerning to the potential benefits (WHAT), tools & technologies (HOW) and challenges (DESPITE – AGAINST) from different perspectives. Based on the comprehensive outcome of the research the following conclusions can be drawn:

The main benefits, technologies and challenges of digital transformation in the manufacturing environment are identified via expert interviews, rated in a QFD-like survey, ranked by AHP to suggest a technology selection decision making model for those who target to implement and utilize I4.0 projects in a manufacturing environment. The use of multiple decision making techniques is not new, but there is no combined approach to I4.0 in the literature.

The highest expectation of benefits with regard to I4.0 are explored as process efficiency and quality performance for Turkish automotive industry. The highest potential in terms of benefits perspective of I4.0 is in the areas of increasing process and quality efficiency, reducing inventories. This is compatible with the purposes of the lean manufacturing philosophy. Especially for the SME's in Turkey automotive industry has to improve their productivity to survive under cost reduction pressures from the automobile manufacturers. According to a market report (Tusiad and BCG 2016), the highest potential to be improved via digital technologies are ranked as production efficiency, resource efficiency and product quality that resembles closely to our findings. According to the Global Manufacturing Competitiveness Index (2015), decreasing costs and increasing efficiency as well as improving productivity is vital for Turkish SME's.

Based on the results from AHP in this study, the highest relevance of increasing process efficiency is to data analytics. Manufacturers suffer from low process efficiency, therefore investing on data analytics becomes more critical for those who target to remain competitive among the other rivals. The most critical tools of I4.0 are defined as data analytics and sensor technologies which is not surprising due to high amount of data in the manufacturing environment via sensors. Improvement of quality and being able to produce with zero scraps and rework are another challenge for the manufacturers in the automotive industry that requires uninterrupted supply chain management.

The more machines, products and humans are connected, the more data is generated which is known as connectivity. Data will fuel the digital transformation and data science is already on the rise.

The biggest challenge in the digital transformation is defined as lack of expert know-how and lack of budget with regard to the data analytics and smart sensor technologies. The respondents unexpectedly ranked the lack of expert know-how as higher than lack of budget that implies the need for qualified workforce for sustainability of I4.0 projects is more important than the budget or top management commitment. The essential requirement to overcome the challenges and to create and sustain the development of independent and high value-adding technologies are the investment on the key manufacturing technologies, and revolutionary paradigm changes in the national education system. Both government and private sector should implement short and long term policies to manage the challenges and the change to protect and sustain the competiveness of the organizations and Turkey's position in the manufacturing environment.

References

Aßmann S, Resenhoeft T (2016) At Bosch Industry 4.0 is already a reality

Garbie I (2016) Sustainability in manufacturing enterprises: concepts, analyses and assessments for Industry 4.0. Springer International Publishing, AG Switzerland

Isaacs D (2017) Making factories smarter thorough machine learning

Herron C, Braiden PM (2006) A methodology for developing sustainable quantifiable productivity improvement in manufacturing companies. Int J Prod Econ 104:143–153

Lu Y (2017) Industry 4.0: a survey on technologies, applications and open research issues. J Ind Inf Integr 6:1–10

Luka D (2015) The fourth ICT-based industrial revolution Industry 4.0: HMI and the case of CAE/CAD innovation with EPLAN. In: 23rd Telecommunications Forum (TELFOR), pp 835–838

McKinsey and Company (2015) Industry 4.0: How to navigate digitization of the manufacturing sector

Mongo DB White Paper (2017) Deep Learning and the Artificial Intelligence Revolution

Oesterreich TD, Teuteberg F (2016) Understanding the implications of digitization and automation in the context of Industry 4.0: a triangulation approach and elements of a research agenda for the construction industry. Comput Ind 83:121–139

Pakizehkara H, Sadrabadib MM, Mehrjardic RZ, Eshaghieh AE (2016) The application of integration of Kano's model, AHP technique and QFD matrix in prioritizing the bank's subtractions. Procedia - Soc Behav Sci 230:159–166

RoblekV, Meško M, Krapez A (2016) A complex view of Industry 4.0, SAGE Open 6(2)

Sanders A, Elangeswaran C, Wulfsberg J (2016) Industry 4.0 implies lean manufacturing: research activities in Industry 4.0 function as enablers for lean manufacturing. J Ind Eng Manag 9(3):811–833

Schuh G, Gartzen T (2015) Promoting work-based learning through Industry 4.0. Procedia CIRP 32:83–87

Shafiq SI, Sanin CE, Szczerbicki C (2016) Virtual engineering factory: creating experience base for Industry 4.0. Cybern Syst 47(1–2):32–47

Tomashevskii IL (2015) Geometric mean method for judgement matrices: formulas for errors, Institute of Mathematics, Information and Space Technologies

TUSIAD and BCG (2016) Industry 4.0 in Turkey as an imperative for global competitiveness an emerging market perspective

Zawadzki P, Żywicki K (2016) Smart product design and production control for effective mass customization in the Industry 4.0 concept. Manag Prod Eng Rev 7(3):105–112

The Innovation Performance Under the Shadow of Industry 4.0.

Banu Ozkeser[1(✉)] and Cüneyt Karaarslan[2]

[1] R&D Center/Koluman Otomotiv Endustri, Mersin, Turkey
banuozkeser@gmail.com
[2] Adana Dunyagoz Hospital, Adana, Turkey
cuneyt.karaarslan@dunyagoz.com

Abstract. Today, people are more aware of the importance of Industry 4.0, which increasingly influences the decisions of service or product innovation. Meanwhile, managers take much more care of both this trend and permanence in the market depending on the ability to innovate services and goods that ensure care of sustainable innovation. Because, undoubtedly, innovation must be taken into consideration as a long-term capability. Hence, the organizations have the advantage to enable translation of primary goods/services into higher value-added products. In the direction of this information, this paper was completed in the shadow of Industry 4.0., which is called "big data box". This study, showing the triple interaction, includes various kinds of minor innovation projects, like many branches of a tree. However, every branch in this big data box tree serves to the same target. The research also extracts more value from the inputs used in process, which in turn, improves productivity and increases the capacity of establishment to meet needs of consumers. With the target of proofing the implementation of this big data box was completed in a foundation in health field, "A corporation", having triple collaboration of three main drivers; Industry, university and government. Therefore, this paper has shown that the innovation performance of the "A" corporation keeps going further by means of big data. In addition to this, it should not be forgotten that organizational climate was affected in a positive way, too. Big data box, evaluating the innovation performance, was conceptualized and studied as a phenomenon linked to the activities of research&development done by technical staff in "A" corporation. It's related to form a smart memory, setting light for the future projects. After completing this study, innovation ideas and new collaboration models between Industry and university can easily be formed.

Keywords: Industry 4.0 · Innovation · Big data · Health management Health science

1 Introduction

It's no doubt that there has never been such a period in when data was obtained by different resources from the beginning of the civilization till today. Because, data is generated from both human and machines, intensively. Therefore, outputs, stored in different formats, have been started in utilization whereas there is a traditional way of

N. M. Durakbasa and M. G. Gencyilmaz (Eds.): ISPR 2018, *Proceedings of the International Symposium for Production Research 2018*, pp. 494–502, 2019.
https://doi.org/10.1007/978-3-319-92267-6_42

storage in the previous years. Especially, the world of enterprise computing is now starting to see the in corporation of new technologies that have been developed and adopted by foundations like Instagram, Google, LinkedIn et al. These various kinds of communication technologies include systems which can handle enormously large data sets, cloud computing and virtualization and many more. People are now on the cusp of yet another revolution which will force many industries to change how they process their data and what products and services they can offer. The way of solving customers' demands will be inadequate unless big data is transferred into modern data-processing application. This large both structured and unstructured volume of data creates big data in literature. In briefly, big data, which means complex data sets, has been available in anywhere of the daily life and affects to human directly.

When the components of Industry 4.0. are analysed; cloud computing, simulation, virtual reality, 3D printers, internet of things, the robotics and cyber security can be considered. All of these elements are based on the big data. Thus, Industry 4.0, which is the meeting point of agility, innovation and efficiency, has focused on the big data implementation. In other words, the more performance big data creates, the more potential innovation projects come through. Hence, the innovation performance has the same frequency with the performance of both big data and Industry 4.0. The response of unlimited customer requests defines how big data is used in innovation projects. For instance, internet of things (IoT) makes the big data be used in the application of Androids.

The main drivers of technology collaboration states big data in various storage and processing. In fact, big data can be taken into account like a flexible bridge between human and their demands. Generally used in mobile applications, it can be formed in various shapes but only if the background of big data is based on a strong frame.

2 Literature Review

This study has been focused on three global subjects; big data, innovation and Industry 4.0. Hence, the literature search is based on these fields and done in three parts.

Big Data: Today, effected by Industry 4.0. in different aspects, big data is a popular topic in literature [1]. Though in most of the scenarios data is too massive, it has a significant potential to improve daily activities, like its having various features (technology management, on-line business etc.) [2]. Much of the time, it may be found in unstructured form in weblogs, scientific records, photography or information documents. There are those sorts of data on which researchers should break down into small files and make a decision.

There is another definition explained by Collins [3]. For consumers, big data shows the truth utilizing excessive data from different sources to change into significant data. For instance, an organization can utilize market data to deliver shows custom-made to their groups of onlookers. It's mentioned that a diachronic phenomena by using wide dataset called corpora of ordered language, thus, could identify formerly unknown correlations between usage of language in the paper of Popescu and Strapparava [4]. In addition to this, Rill et al. [5] explains big data demonstrates a frame by implementing political points in Twitter.

Jung and Segev [6] have focused on the methods to seperate how groups change after some time in the reference system diagram without extra outer data and taking into consideration hub and connection forecast and group discovery in their study. A scientific approach by Wua and Tsaib [7] illustrated how an ability to think learning base can be utilized as the establishment to fabricate a bigger dictionary reference by proliferating sentiment values from ideas with known qualities to exhaust ones.

Justo et al. [8] talk about the undertaking of programmed identification of mockery or frightfulness in online written discussion. A methodology for feeling examination in social media situations is reported by Montejo-Raez et al. [9]. As indicated by Flaounas et al. [10] examination of media substance has been focal in social sciences, because of the key part that media plays in shaping public opinion. This sort of investigation ordinarily depends on the preparatory coding of the content being inspected, a stage that includes perusing and expounding it, and that restrains the sizes of the corpora that can be analyzed. Rahnama [11] states that all the more particularly over Social Media, Big-Data pattern has upheld the data-driven frameworks to have ceaseless quick data streams. As of late, real-time analytics on stream data has shaped into another exploration field, which means to answer queries about what-is-occurring now with an unimportant postponement. According to Mukkamala and Hussain [12] developmental approaches to handle social media data is limited to graph database approaches such as social network analysis. Finally, Liu et al. [13] stated that when the dataset is huge, a few algorithms may not scale up well. In this paper, authors intend to assess the versatility of Naïve Bayes classifier (NBC) in substantial datasets rather than utilizing a standard library.

As determined by Waddell [14] an organization requires big data and analytics strategy for three main reasons:

- For creating smarter, leaner organizations: Today, the number of Google queries about lodging and real estate starting with one quarter then onto the next ends up predicting more accurately.
- For equipping organization: As most organizations will agree, it's essentially unrealistic to carry out the conversations which once had with customers. There's a lot of dialog rolling in from various sources.
- For helping your organization get ready for the inevitable future: The primary reversal was in the newspaper business that moved from blasting to near obsolete with the advent of online distributed. This happened inside of a decade.

There are four application domains for Big-Data according to McGuire et al. [15]:

- As organizations create more transactional data.

They can gather more detailed performance information on everything from item inventories to debilitated days and therefore exposes variability and support performance. In fact, some leading companies are utilizing their ability to collect and analyze Big-Data to direct controlled experiments to make better management decisions.

- Big-Data supports narrower segmentation of customers.
- Big-Data analytics can improve prediction, minimize risks.

For instance, firms are using information obtained from machine-sensors embedded in products to create innovative maintenance process. White [16] states that many people view Big-Data as an over-hyped trendy expression.

Innovation: Innnovation, which has the same meaning with the word "new", is one of the most popular subjects in the world literature. Especially, the detailed meaning of "new" can be classified when all the papers are gathered. For instance, the result of the new study may be for the world, Industry, customer or the content of the study can be new product, system or benefits. Therefore, literature is divided into two main groups and the branches of the each group present the details in the Table 1, showing the references of literature review about innovation are listed below.

Industry 4.0: Mostly, the subject "industrial revolution" refers to the changes, taken place after the introduction of steam&water powered manufacturing technics. By means of electricity, the time of second revolution presented major industrial improvements such as the assembly lines and mass production. The time between the second and the third one lasted for only a few decades. Beginning from the 70s, the rapid adoption of electronics and information technologies enabled further automation of manufacturing in factories.

Manufacturing environment with extraordinary changes in the execution of operations is affected by Industry 4.0., began in the 2000s, significantly. Industry 4.0 lets real-time planning of manufacturing plans, along with dynamic self-optimisation unlike conventional forecast based manufacturing planning.

The presentation of knowledge and correspondance systems into industrial network leads to rise in the degree of automation, too. Smart and self-optimising machines in the production line synchronise themselves with the value chain, right from raw-materials from suppliers to delivery of goods to consumers. Simulation of inventory, logistics and usage history of products also allows to effect the manufacturing processes in a positive way. The fourth industrial revolution implementing the principles of cyber-physical systems, Industry 4.0., gathers internet and future-oriented technologies together and may also be thought as an interface with enhanced human-machine interaction. This provides identity and interaction for every entity in the value stream [38].

3 Material and Method

Innovation projects, of which performance is influenced by Industry 4.0, forms the future of every foundation. It's a well-known fact that this development shapes our daily life, too. In that point, creating an innovation pool is a must for especially research and development (R&D) departments or laboraties. However, sustainability should be taken into account with this fact. Therefore, the only way for an innovation approach, going on and on, passes through from big data box (BDB). BDB is like a corporate memory of innovation intelligence. It produces new information by gathering all data set. For instance, the information progressing by the collaboration of Industry and university, should be used in another innovation project by another R&D team. It can only be provided as long as a corporate or smart memory, called big data box. In briefly, BDB includes the unique parts of innovation projects, especially influenced by Industry 4.0.

Table 1. Literature review about innovation

		[17]	[18]	[19]	[20]	[21]	[22]	[23]	[24]	[25]	[26]	[27]	[28]	[29]	[30]	[31]	[32]	[33]	[34]	[35]	[36]	[37]
New to	New to the world							X	X		X		X									
	New to the Industry				X							X				X	X					
	New to scientific community						X					X										
	New to the market (place)	X			X	X			X	X			X			X						
	New to the firm	X		X			X	X				X	X				X			X		
	New to the customer									X			X									
New what	New technology	X		X	X		X	X	X	X		X	X	X			X		X	X	X	X
	New product line	X	X	X		X	X				X	X	X	X		X						
	New product benefits/features	X								X					X	X						
	New product design		X					X	X		X					X					X	
	New process	X		X		X	X		X					X		X						
	New service						X															
	New competition	X		X		X	X							X								
	New customers	X		X			X							X								
	New customer need						X							X								
	New consumption patterns	X	X												X							
	New uses	X													X				X			
	New improvements/changes						X	X	X				X									
	New development skills											X								X		
	New marketing/sales/ distribution skills	X		X			X					X		X		X						
	New managerial skills					X																
	New learning/experience/ knowledge									X	X	X										
	New quality/benefits	X					X															X

The triple main drivers of BDB are Industry, university and government in this implementation. Because, the knowledge framed with the theory in the implementation ensures much more significance. This case-study has been completed in A corporation, a big-scale foundation in health science and having R&D staff including part-time academicians as supervisors. In addition to this, there is one more originality of this study which is based on utilization of theory. Because, the case-study has been taken from real-data set, belonging to the medical/health science. Hence, the method can be repeated in different activity fields due to its flexibility.

Step-1: In this study, gathering uniques of the innovation projects together forms the initial step. (Fig. 1)

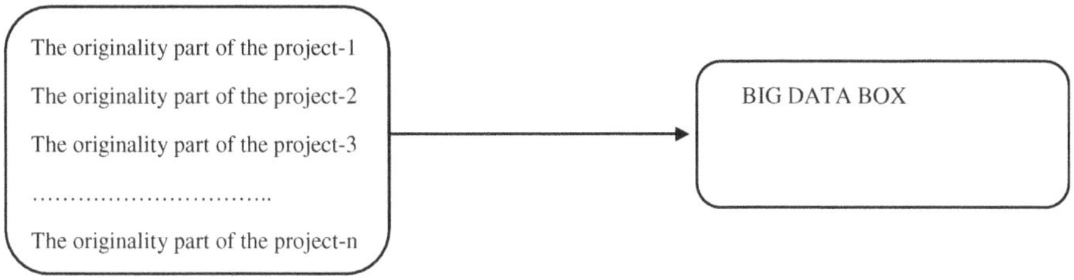

Fig. 1. The form of big data box

The first method used in this approach and in this step is brainstorm. Because every idea has got a deep value and can be evaluated in different aspects. The innovation project ideas are collected from the staff with these brainstorm meetings. So, it also helps the frame of big data box.

Step-2: BDB (Big Data Box) includes the cores of the projects, about the new approaches in product/service field (like a new component/system/product etc.) Hence, it can also be thought as creating of new ideas, which can be turned into financial income. In other words, an innovation pool is formed naturally by means of BDB. This pool is like a sub-division of big data box. Because its content is only based on the uniqueness of the innovation projects. (Fig. 2)

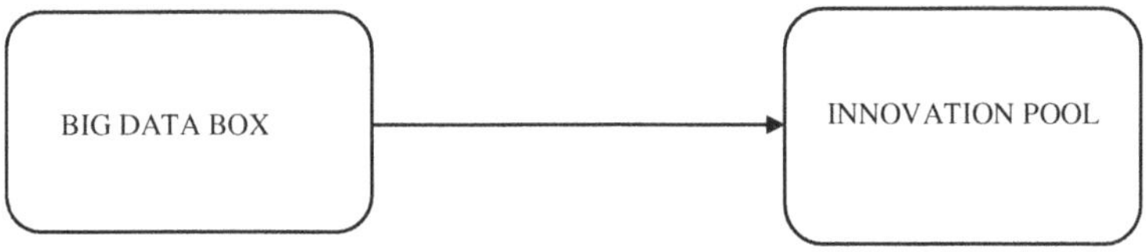

Fig. 2. The form of innovation pool

In this second step, groups of data in big data storage form the innovation pool by classifying. The originalities of the projects are separated from the ordinary data set and have filled the innovation pool.

Step-3: In this step the classification of the information is completed in the form of contents. Then, the priorities are set. By the influence of digitalizm, most of the priorities are concentrated on Industry 4.0, a new era for companies. Because, the people requires a life, which can be carried with themselves. For instance; IoT technologies, smart kitchens, mobile applications can be popular examples about this field. Consequently, Industry 4.0. effects the innovation performance regarding to the trend demands (Fig. 3).

Know-how resulted from the collaborations between universities and industry

The Effects of Industry 4.0. (Digital Life)

The incentives supported by government

Intellectual Properties

The Innovative Ideas from the qualified R&D staff

....

Fig. 3. The contents of innovation pool

In the last step, priorities are taken into account by calculating the average weights and simultaneously, innovation projects in the shadow of Industry 4.0. are provided as the priorities. In this last step, while calculating the average weights, SPSS and AHP have been used, too.

4 Conclusions

This study is prepared so as to show the direct interaction between innovation, Industry 4.0. and big data. Additionally, this may be thought as a guideline of how an innovation map should be used by means of Industry 4.0 and big data collaboration. Should this triple collaboration be taken into account together, the sustainable innovation culture is formed, undoubtedly. Industry 4.0. is most effectively used as long as big data is the base of corporate memory. Additionally, this paper is an example of implementation about Industry 4.0 effect on innovation performance. Thus, these factors must be together in the same implementation.

Nowadays, one of the trend words in our society is innovation, on which a long research done inside of the articles and books. Especially, after the Industry revolution occurs, digital life becomes the first popular sub-field of this innovation world. Thus, key performance indicator of innovation projects should be passed through the Industry 4.0. In that point, the demands of the earth lead the innovation studies to be a part of e-life utilization.

This study shows the interaction between these main drivers. Literature review consists of these studies reflected to each other. Additionally, the implementation done in corporation A is one of the case-studies, which can be one of the examples about this paper. The other advantage of this implementation is that this innovation performance evaluation can be used in both product and service fields. Because, the approach has a flexible frame so as to be used in different field dynamics.

References

1. Jawell D, Barros RD, Diederichs S, Duijvestijn LM, Hammersley M (2014) Performance and Capacity Implications for Big-Data. http://www.redbooks.ibm.com/redpapers/pdfs/redp5070.pdf
2. Srinivas GJ (2012) Big-Data on Microsoft Platform. http://download.microsoft.com/download/A/F/B/AFB1DCA8-7C6F-476B-8159-214500E7A613/Big_Data_On_Microsoft_Platform.pdf
3. Collins E (2014) Big-data in the public cloud. IEEE Cloud Comput 1(2):13–15 ISSN 2325-6095
4. Popescu O, Strapparava C (2014) Time corpora: Epochs, opinions and changes. Knowl-Based Syst 69:3–13. https://doi.org/10.1016/j.knosys.2014.04.029
5. Rill S, Reinela D, Scheidta J, Zicarib RV (2014) PoliTwi: early detection of emerging political topics on twitter and the impact on concept-level sentiment analysis. Knowl-Based Syst 69:23–34. https://doi.org/10.1016/j.knosys.2014.04.022
6. Jung S, Segev A (2014) Analyzing future communities in growing citation networks. Knowl-Based Syst 69:34–44. https://doi.org/10.1016/j.knosys.2014.04.022
7. Wua C, Tsaib RT (2014) Using relation selection to improve value propagation in a ConceptNet-based sentiment dictionary. Knowl-Based Syst 69:100–107. https://doi.org/10.1016/j.knosys.2014.04.043
8. Justo R, Corcoran T, Lukin SM, Walker M, InésTorres M (2014) Extracting relevant knowledge for the detection of sarcasm and nastiness in the social web. Knowl-Based Syst 69:124–133. https://doi.org/10.1016/j.knosys.2014.05.021
9. Montejo-Ráez A, Díaz-Galiano MC, Martínez-Santiago F, Ureña-López LA (2014) Crowd explicit sentiment analysis. Knowl-Based Syst 69:134–139. https://doi.org/10.1016/j.knosys.2014.05.007
10. Flaounas I, Sudhahar S, Lansdall-Welfare T, Hensiger E, Cristianini N (2012) Big-Data Analysis of News and Social Media Content. http://www.see-a-pattern.org/sites/default/files/Big%20Data%20Analysis%2of%20News%20and%20Social%20Media%20Content.pdf
11. Rahnama AHA (2014) Distributed real-time sentiment analysis for big-data social streams. In: 2014 International conference on control, decision and information technologies (CoDIT)
12. Mukkamala RR, Hussain A (2014) Fuzzy-set based sentiment analysis of big social data. In: 2014 IEEE 18th International enterprise distributed object computing conference (EDOC), Ulm. IEEE, pp 71–80. https://doi.org/10.1109/edoc.2014.19. ISSN 1541-7719, INSPEC Accession number 14792906
13. Liu B, Blasch E, Chen Y, Shen D, Chen G (2013) Scalable sentiment classification for big-data analysis using Naïve Bayes classifier. In: 2013 IEEE International conference on big-data, Silicon Valley. IEEE, pp 99–104. https://doi.org/10.1016/bigdata.2013.6691740. INSPEC Accession number 13999322
14. Waddell T (2014) 3 Reasons You Need a Big-Data and Analytics Strategy. http://blogs.adobe.com/Big-Data-analytics-strategy/. Accessed Feb–Mar 2014
15. McGuire T, Manyika J, Chui M (2012) Why Big-Data is the new competitive advantage. http://iveybusinessjournal.com/competitive-advantage/. Accessed Apr 2012

16. White C (2012) What Is Big-Data and Why Do We Need It? http://www.technologytransfer.eu/article/98/2012/1/What_Is_Big_Data_and_Why_Do_We_Need_It_.html
17. Cooper RG (1979) The dimensions of industrial new product success and failure. J Mark 43:93–103
18. Lawton L, Parasuraman A (1980) The impact of the marketing concept on new product planning. J Mark 44:19–25
19. More RA (1982) Risk factors in accepted and rejected new industrial products. Ind Mark Manag 11:9–15
20. Maidique MA, Zirger BJ (1984) A study of success, and failure in product innovation: the case of the US electronics Industry. IEEE Trans Eng Manag EM-31(4):192–203
21. Yoon E, Lilien GL (1985) New industrial product performance: the effect of market characteristics and strategy. J Prod Innov Manag 3:134–144
22. Cooper RG, de Brentani U (1991) New industrial financial services: what distinguishes the winners. J Prod Innov Manag 8:75–90
23. Kleinschmidt EJ, Cooper RG (1991) The impact of product innovativeness on performance. J Prod Innov Manag 8:240–251
24. Lee M (1994) Determinants of technical success in product development when innovative radicalness is considered. J Prod Innov Manag 11:62–68
25. Ali A, Krapfel R, LaBahn D (1995) Product innovativeness and entry strategy: impact on cycle time and break-even time. J Prod Innov Manag 12:54–69
26. Atuahene-Gima K (1995) An exploratory analysis of the impact of market orientation on new product performance: a contingency approach. J Prod Innov Manag 12:275–293
27. Green SG, Gavin MB, Aiman-Smith L (1995) Assessing a multidimensional measure of radical technological innovation. IEEE Trans Eng Manag 42(3):203–214
28. Olson EM, Walker OC, Ruekert RW (1995) Organizing for effective new product development: the moderating role of product innovativeness. J Mark 59:48–62
29. Mishra S, Kim D, Lee DH (1996) Factors affecting new product success: cross country comparisons. J Prod Innov Manag 13:530–550
30. Souder WE, Song MX (1997) Contingent product design, and marketing strategies influencing new product success, and failure in US, and Japanese electronic firms. J Prod Innov Manag 14:21–34
31. Schmidt JB, Calantone RJ (1998) Are really new product development projects harder to shut down? J Prod Innov Manag 15(2):111–123
32. O'Connor C (1998) Market learning and radical innovation: a cross case comparison of eight radical innovation projects. J Prod Innov Manag 15(2):151–166
33. Song MX, Montoya-Weiss MM (1998) Critical development activities for really new versus incremental products. J Prod Innov Manag 15(2):124–135
34. Veryzer RW (1998) Key factors affecting customer evaluation of discontinuous new products. J Prod Innov Manag 15(2):136–150
35. Goldenberg J, Lehmann DR, Mazursky D (1999) The primacy of the idea itself as a predictor of new product success. Marketing Science Institute Working Paper
36. Kessler EH, Chakrabarti AK (1999) Speeding up the pace new product development. J Prod Innov Manag 16:231–247
37. Chandy RK, Tellis GJ (2000) The incumbents curse: incumbency, size, and radical product innovation. J Mark 64:1–17
38. Ganscharn O, Gerlach S, Hämmerle M, Krause T, Schlund S (2013) Produktionsarbeit der Zukunft-Industrie 4.0, pp 50–56. In: Spath D (ed.) Fraunhofer Verlag, Stuttgart

The Role of IoT on Production of Services: A Research on Aviation Industry

Güzide Karakuş(✉), Emre Karşıgil, and Leyla Polat

Necmettin Erbakan University, Konya, Turkey
{gkarakus,ekarsigil,lpolat}@konya.edu.tr

Abstract. At the present time, the demand of air transportation has been increased by global population growth and the rise of middle class in developing countries, economic progress of international trade, and upward trend in sensitivity towards pace and security. This situation consistently prompts to airline businesses on the subjects that are increasing the productivity on their operational activities and developing new business models. While Internet of Things (IoT) which is rapidly spreading around the world at this point is implemented with the dream of that everything would be bigger, faster and cheaper; the reflection of those dreams can affect on aviation industry. Studies of IoT in aviation industry are able to help aviation industry provide benefits namely provide customer satisfaction growth with higher comfort and productivity growth on supply chain activities, reducing the cost with decreasing of human related accidents, developing security activities, decreasing travel time on aviation.

There is a great deal of applications regarding IoT in aviation industry, however, research and academic papers are not adequately conducted in the field. Within this scope, the aim of the study is to examine theoretical aspects of IoT applications in service sectors and analyse IoT tools in process of production of services used by airline and airport operators. Heuristic approach as one of the applied research methods will be used in this study.

In the first part of the study, the theoretical research on the definition, benefits and losses of IOT is given. In the second part, IOT application areas are examined. In the final section, we have included innovations in civil aviation that will serve as examples of IOT and can be regarded as innovations around the world.

Keywords: Internet of Things · IoT · Aviation · Service sector
Production of services · Istanbul new airport

1 What is IoT?

Today, about two billion people around the world use the Internet for a variety of reasons, such as getting information, sending and receiving email, accessing multimedia content and services, playing games, and using social networking applications (Miorandi et al. 2012: 1497). With its powerful communication networks, people are moving beyond geographic boundaries (Gupta and Gupta 2016: 1), and it brings manufacturing and business to the global dimension. Although more and more people are using this kind

N. M. Durakbasa and M. G. Gencyilmaz (Eds.): ISPR 2018, *Proceedings of the International Symposium for Production Research 2018*, pp. 503–511, 2019.
https://doi.org/10.1007/978-3-319-92267-6_43

of information and communication infrastructure, the Internet is making a huge leap in using machines and smart objects as a global platform by allowing them to communicate, dialogue, calculate and coordinate with Internet-of-Things (IOT) (Miorandi et al. 2012: 1497). This technological change process, called Industry 4.0, is seen as a beginning of a new era in real terms.

Internet-of-Things (IOT) envisages a future where digital and physical entities can be connected through appropriate information and communication technologies to provide a new class of applications and services (Miorandi et al. 2012: 1497). The perception provided by IOT technologies has spread to many areas of modern life by providing the means of measuring, understanding and monitoring, from natural sources to industry and urban settlements (Gubbi et al. 2013: 1645). In today's world, people are surrounded by many electronic devices such as automatic devices, smart devices and buildings (Gupta and Gupta 2016: 1). IoT is a platform where devices and processes become smarter and communication is strengthened every day. While IoT is still looking for its own form, the effects have begun to take incredible steps as a universal solution tool (Ray 2016: 1).

In this context, IoT is defined as a network of various nodes belonging to technological, physical and wide socioeconomic environments (Krotov 2017: 834). Information technology has revolutionized products and has become a complex system that combines mechanical and electrical components, hardware, sensors, data storage devices, microprocessors, software, and connectivity in numerous ways (Porter and Heppelmann 2014: 4, Krotov 2017: 834). Five IoT technologies are widely used to implement successful IoT-based products and services (Lee and Lee 2015: 432):

1. Radio frequency identification (RFID),
2. Wireless sensor networks (WSN),
3. Middleware,
4. Cloud computing, and
5. IoT application software.

In the simplest terms, IoT consists of sensors, network communication devices, web storage and data processing applications. Sensors detect and accumulate the necessary data. The network components transmit and store information to the cloud servers. Web storage tools implement data mining techniques to extract the necessary information from a large volume of data. This processed information is then used by applications running on physical devices (Gupta and Gupta 2016: 2). Figure 1 shows the IoT function and its actors.

IoT requires participating organizations to be equipped with device-specific embedded software, sensors and network-supporting components. Device-specific functionality is provided by embedded software. Physical devices sensors detect the presence of other physical entities in the environment and collect the necessary information. The Internet serves as a communication medium between various types of physical devices to provide remote management and access control of these devices (Kortuem et al. 2010: 44).

It is an important necessity to have enough resources to make the right decision whereas one is separated from other creatures by his superior decision-making ability (Bi et al. 2016: 378). The importance of information in strategic decision-making and

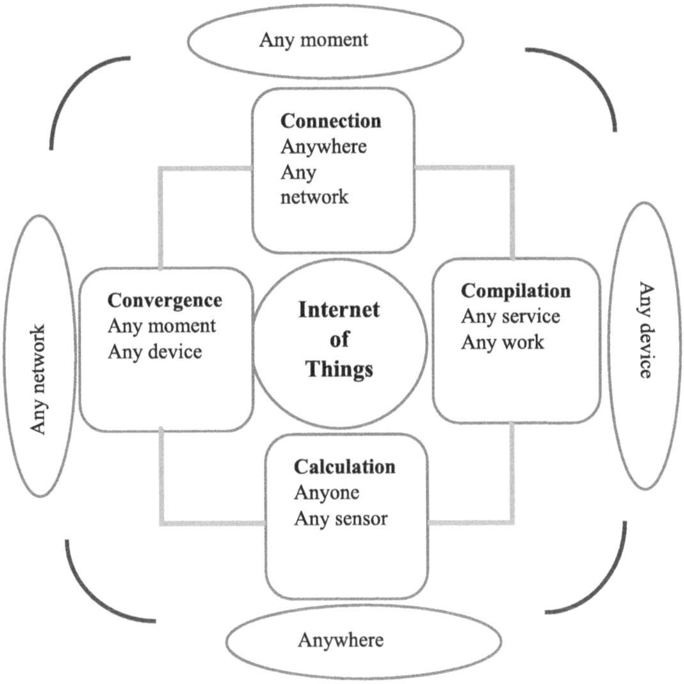

Fig. 1. Scope of IoT Resource: Gupta and Gupta 2016: 2

in achieving competitive advantage is creating concern at the point of information security in IoT technology. While safety in both physical devices and service applications is critical to the success of the IoT (Li and Li 2016: 355), the issue is at the head of criticism raised against system. De Cremer et al. (2017) stated in their study that there are many negative effects besides the benefits of IoT applications. These negativities can be classified as privacy violations, abuse of information, fraud and confusion in customers (De Cremer et al. 2017: 145). The real problem is that the IoT technology and the negativities it can create have not yet been explained with all clarity. In this context, it is important to develop new business models to eliminate potential threats and move the technology forward (Alan et al. 2018: 311).

When it comes to its advantages, IoT offers many advantages for manufacturing enterprises. By redesigning factory workflows with IoT, manufacturers will be able to optimize material and enhance product tracking and optimize distribution costs. This situation will positively affect the efficiency, competitiveness and profitability of the business. In addition to manufacturing enterprises, service providers will be able to offer improved services through IoT, improve their customer satisfaction and increase their incomes and competitiveness (Lee and Lee 2015: 431). As noted by Porter and Heppelmann (2014), IoT should not be regarded only as a technological development, it should be evaluated as an important strategic competitive tool. IoT applications are important tools in increasing customer value for businesses that are in need of chang and transform because of technological, social and competitive pressures (Lee and Lee 2015: 432).

2 Application Areas for IoT

As the Internet does, IoT has the potential to change the world, perhaps even more (Ashton 2009: 97). The "smart, connected products" that emerged with IoT brought a new era of competition (Porter and Heppelmann 2014: 4). Smart products offer exponential opportunities for new functionality, greater reliability, higher product use and features that exceed traditional product limits. The changing nature of products, by developing value chains, forces companies to rethink and remake nearly everything they do in themselves. These new product types expose companies to new competitive opportunities and threats altering the nature of industry structure and competition. The industry is reshaping its borders and creating entirely new industries (Porter and Heppelmann 2014: 4).

For businesses, industrial IoT (IIoT) is being used and it provides benefits in many different categories, such as production management and follow-up, inventory control, security, shipment tracking and energy efficiency as well as customer and supplier relations development (De Cremer et al. 2017: 145). In addition to this, the support of the products produced with a strong service has a significant positive effect on customer satisfaction and loyalty. However, in the near future, service will fall short, and smart services will become a necessity (Allmendinger and Lombreglia 2005: 131).

IoT applications can be classified under the headings such as Smart Wearable Items, Smart Houses, Smart Cities, Smart Initiatives and Smart Environments and the used procuts differ according to their areas (Gupta and Gupta 2016: 5). Thanks to IoT technologies, automation is achieved in almost every sector with the applications at health sector, home automation, service management, traffic management and smart cities (Chan 2015: 552). In Fig. 2, IoT is depicted in the context of application areas and users.

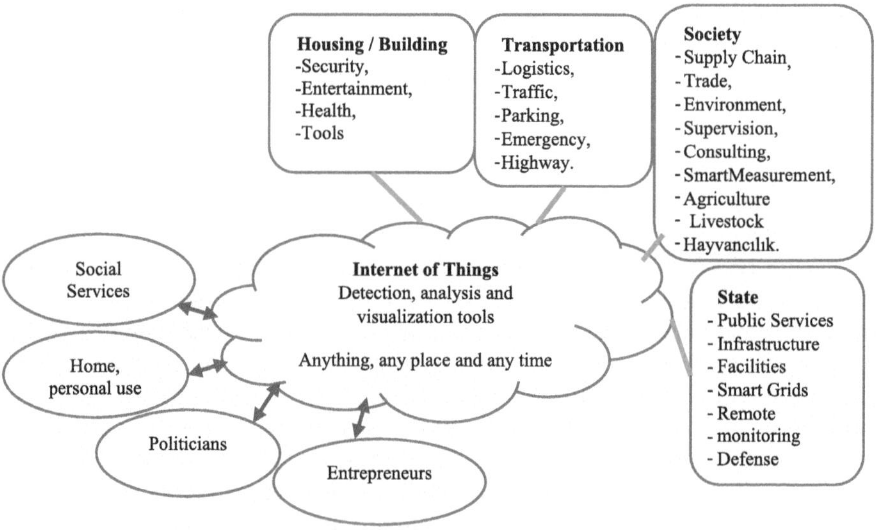

Fig. 2. End users and application areas of IoT Resource: Gubbi et al., 2013: 1647

It is important for entrepreneurs to be able to take new opportunities in order to use existing resources more effectively (Ehret and Wirtz 2017: 115). The IoT is attracted by entrepreneurs at this point, but with some doubt. Entrepreneurs must make and decide the right cost-benefit analysis to use IoT applications to take new business opportunities. For entrepreneurs, IoT offers new possibilities for everything such as smart devices and products, automation, logistics, preventive maintenance, new business/process management, new service and business model development (Atzori et al. 2010: 2787). IoT is not a homogenous concept or paradigm but a whole set of possibilities that can adopt and incorporate a true approach for strategic profits and business needs of each actor (Saarikko et al. 2017: 675).

3 IoT in Civil Aviation Applications

The IoT studies for academic and business life are mainly focused on products, production processes, infrastructure and technology. IoT for service innovation processes has been neglected especially in academic studies. However, the service is in interaction with every business line of work and sector. Even the improvements to be made about service improvement are more understandable by the customers and they give strategic competitive power to the business. In contemporary economic and political debates, it is argued that societies depend on service innovations in terms of living standards, growth, employment, productivity and international competitiveness (Andersson and Mattsson 2015: 85-88). For this reason, in this study, it is aimed to investigate the point reached in IoT applications in civil aviation activities which is one of the service areas.

Growth and technological developments in the aviation sector in the world accelerated the introduction of IoT applications into the aviation sector. Although not as common as other sectors, IoT is also very important for the aviation sector today. While this application is called IoA (Internet of Aviation/Aerospace internet) in the world, this definition has not created its area of use in Turkey yet.

In the field of aviation, where production, maintenance and safety activities are vital, the IoA has been recognized as important and many applications have been carried into effect. For example, IoT is used in the process of controlling and quickly joining aircraft parts that needs much human-power as well as time-consuming. Objects interacting with each other and controlled can be combined in a robust way as desired via RFID commands and specified commands (Bandyopadhyay and Sen 2011: 15). In addition, aircraft maintenance was so much costly and troublesome before, real-time IoA systems now are very helpful to maintenance staff (Matters, Runs and Wind: 5). Another area of use for IoA is the smart airports. Thanks to Internet of Aviation, the airports which became smart at passenger tracking, baggage systems and management systems that these applications provide great convenience and great benefits financially for both passengers and managers (Mukherje 2017). The IoA aims to help passengers not only at the airport, but also in areas where they go after the flight (Matters, Runs and Wind: 6).

Although the concept of Internet of Aviation has made great promises in the field of aviation, it has not yet been widely used. This situation has a much worse structure in Turkey. Although the importance of aviation in our country has been understood, the investment made in this area is profitable, yet the IoA has not yet fully understood in this area and necessary investments have not been realized. While there have been few studies in academic literature on the use of smart systems in different sectors, it has been found that there are no researches in the field of aviation (Gokrem and Bozuklu 2016: 62–63). The idea of working in the direction of this deficiency has been formed and the Internet of Aviation topic has been found worth investigating. Within the scope of this study, it is aimed to investigate the worldwide application areas of the IoT concept (IoA) for air transport which is important logistic support of education, commerce and tourism sectors in our globalizing world.

4 IoT Applications Used in Civil Aviation Services in the World

IoT become an essential tool for every industry in this century. It also plays important role on aviation in order to make the airport future-proof. On the other hand, airlines are expected to adapt IoT developments to enhance passenger experience. According to SITA Aero, 90% of all passengers think that technology helps when traveling. Therefore, technology offers great opportunities for airlines and airports.

IoT application in aviation industry can be described as below:

4.1 Baggage Tracking

SITA Aero states that the industry spends 31 billion US$ on moving bags and 2.3 billion US$ for mishandled bags. Therefore The International Air Transport Association (IATA) started a new programme called "Resolution 753" which becomes effective on 1st June 2018. By launching this programme, it is intended to reduce mishandling by tracking every baggage at the airport. According to that, airlines must track baggage at four key points while baggage are carrying as follows:

- Passenger handover to airline
- Loading to the aircraft
- Deliver to the transfer are
- Return to the passenger

Tracking baggage is crucial to providing real-time information where the location of baggage is and how long it would take to reach the baggage carousel. The baggage tracking saves for airlines and airports are estimated 0.11 US$ per passenger. Mishandling rate has reduced by 70.5% per thousand passenger since 2007. Considering the industry spends, it will help airlines and airports to reduce ground handling costs and charges for passengers as well. The relevant information can be tracked by passenger through the mobile application. SITA Aero indicates that 52% of all passengers require more baggage status information via the smart phones during their journey (www.iata.org, www.sita.aero, A.D: 05.05.2018).

As an example, Delta Airlines introduced Radio Frequency Identification (RFID) baggage tracking technology in 2016, which is a first for U.S. carriers. With the new technology, luggage information will be read through the RFID chip embedded in the luggage tag via scanners which use radio waves to capture accurately relevant data. It will also enable for passengers to see whole luggage journey via push notifications to the mobile app. Delta claims that baggage can be tracked at a 99.9 percent success rate which means much more than consistent baggage handling for passengers (www.delta.com, A.D: 01.05.2018).

Another example is Istanbul New Airport, where will be required to track more than 28,800 bags an hour once the airports opens. Thus, baggage tracking system will play crucial role for the airport to overcome busy traffic in Istanbul (www.avianews.co, A.D: 28.04.2018).

4.2 Biometric Technology

According to SITA, three out of five airports and two out of five airlines are investing in biometric technology to improve security and speed as number of passengers is set to double to 7.8 billion by 2036. Biometric Technologies can be listed as follows:

- Palm vain scanning
- Facial and voice recognition
- Fingerprint scanning
- Digital passports

The SITA report also indicates that three out of four smartphones will have fingerprint sensors by 2020, however only more than half (%57) of passengers states that they are willing to use biometric Technologies at the airport. While low ratio of passenger willingness will be bottleneck for the biometric technology, it will definitely provide faster passenger processing at airports and boosting national security. As a consequence of it, SITA has rolled out 100 kiosks across three Mexican major airports in order to meet the demand of growing international passengers. It is widely accepted that biometric technology will help airports overcoming that challenge and will ensure security for the country. Moreover, the biometric kiosks will improve the pace and accuracy of checks of passengers (www.forbesmiddleeast.com, www.sita.aero, A.D: 25.04.2018).

4.3 Indoor Navigation Systems

Beacons are commonly used for indoor localization. These are small location-based devices that transmit data to mobile devices within a specific range. Recipients' locations are determined via Bluetooth connection using mobile app on their devices.

The use of indoor navigation at airports has some advantages for passengers, airlines, airport operators, stores and restaurants. From passenger perspective, it is essential to take their flight on time, therefore, time is critical for passengers who use bigger airports. As passengers board on aircraft on time, it enables airlines reducing delays and increase their profit. Airports also benefit from decrease of flight delays as airports increase their efficiency especially at peak times. The usage of indoor navigation gives an opportunity for stores and restaurants at airports to attract passenger by offering discounts and sales via mobile app.

4.4 Virtual Reality Tool

In order to enhance passenger pre-flight experience, "seat experience" helps passenger having a chance for looking at every seat on a particular flight. This technology offers passengers the option to get a virtual reality view (360°) of the aircraft cabin before booking the flight. This also enables to gather information about seat dimensions, legroom, material of each seat and exact location. This technology does not require any apps or software and can be used via PCs, tablets, smart phones and virtual reality glasses. While it increases passengers confident and decreases complaints for airlines (www.runwaygirlnetwork.com, A.D: 25.04.2018).

5 Result

Although IoT is put into practice at industrial or service sectors, literature follows from behind. While IoT technology is advancing and a growing number of businesses adopt this technology, it is a necessary to investigate the subject in academic sense in accordance with the themes such as cost, efficiency, quality, process improvement, environmental impact, competitiveness, sustainability etc. Because IoT has a significant investment cost as well as potential benefits including uncertainties. While calculating the opportunities and advantages of IoT for businesses, it is a must that businesses conduct an accurate assessment and a detailed cost-benefit analysis in order to use their resources effectively.

References

Alan AK, Kabadayı ET, Cavdar N (2018) Yeni Nesil "Bağlantı", Yeni Nesil "İletişim": Nesnelerin İnterneti, J Bus Res Turk, 10(1), 294 – 320. https://doi.org/10.20491/isarder.2018.382

İnceleme ÜB, Allmendinger G, Lombreglia R (2005) Four strategies for the age of smart services. Harvard Bus Rev 83(10):131

Andersson P, Mattsson LG (2015) Service innovations enabled by the "Internet of Things". IMP J 9(1):85–106. https://doi.org/10.1108/IMP-01-2015-0002

Ashton K (2009) That 'Internet of Things' thing. RFID J 22(7):97–114

Atzori L, Iera A, Morabito G (2010) The Internet of Things: a survey. Comput Netw 54(15):2787–2805. https://doi.org/10.1016/j.comnet.2010.05.010

Bandyopadhyay D, Sen J (2011) 'Internet of Things-Applications and Challenges in Technology and Standardization'. https://arxiv.org/pdf/1105.1693.pdf. Accessed 19 Jan 2018

Bi Z, Wang G, Xu LD (2016) A visualization platform for internet of things in manufacturing applications. Internet Res 26(2):377–401. https://doi.org/10.1108/IntR-02-2014-0043

Chan HC (2015) Internet of Things business models. J Serv Sci Manag 8(04):552–568. https://doi.org/10.4236/jssm.2015.84056

De Cremer D, Nguyen B, Simkin L (2017) The integrity challenge of the Internet-of-Things (IoT): On understanding its dark side. J Mark Manag 33(1–2):145–158. https://doi.org/10.1080/0267257X.2016.1247517

Ehret M, Wirtz J (2017) Unlocking value from machines: business models and the industrial Internet of Things. J Mark Manag 33(1–2):111–130. https://doi.org/10.1080/0267257X.2016.1248041

Gubbi J, Buyya R, Marusic S, Palaniswami M (2013) Internet of Things (IoT): a vision, architectural elements, and future directions. Future Gener Comput Syst 29(7):1645–1660. https://doi.org/10.1016/j.future.2013.01.010
Gupta R, Gupta R (March 2016). ABC of Internet of Things: advancements, benefits, challenges, enablers and facilities of IoT. In: Symposium on Colossal Data Analysis and Networking (CDAN), IEEE, pp 1–5. https://doi.org/10.1109/CDAN.2016.7570875
Kortuem G, Kawsar F, Fitton D, Sundramoorthy V (2010). Smart objects as building blocks for the internet of things. IEEE Internet Comput 14(1):44–51
Krotov V (2017) The Internet of Things and new business opportunities. Bus Horiz 60(6):831–841. https://doi.org/10.1016/j.bushor.2017.07.009
Lee I, Lee K (2015) The Internet of Things (IoT): applications, investments, and challenges for enterprises. Bus Horiz 58(4):431–440. https://doi.org/10.1016/j.bushor.2015.03.008
Li S, Tryfonas T, Li H (2016) The Internet of Things: a security point of view. Internet Res 26(2): 337–359. https://doi.org/10.1108/IntR-07-2014-0173
Matters, WIT, Runs IT, Wind ON. 'The Internet of Things in Commercial Aviation'. http://www.windriver.com.cn/downloads/whitepaper/1513-IoT-in-Commercial-Aviation-White-Paper.pdf
Miorandi D, Sicari S, De Pellegrini F, Chlamtac I (2012) Internet of Things: vision, applications and research challenges. Ad Hoc Netw 10(7):1497–1516. https://doi.org/10.1016/j.adhoc.2012.02.016
Mukherje JS (2017) IOT: shaping the next epoch of the aviation industry. https://www.hcltech.com/blogs/IoT-shaping-next-epoch-aviation-industry. Accessed 20 Jan 2018
Porter ME, Heppelmann JE (2014) How smart, connected products are transforming competition. Harvard Bus Rev 92(11):64–88
Ray PP (2016) A survey on Internet of Things architectures. J King Saud Univ-Comput Inf Sci. https://doi.org/10.1016/j.jksuci.2016.10.003
Saarikko T, Westergren UH, Blomquist T (2017) The Internet of Things: are you ready for what's coming? Bus Horiz 60(5):667–676. https://doi.org/10.1016/j.bushor.2017.05.010
Gökrem L, Bozuklu M (2016) 'Nesnelerin İnterneti: Yapılan Çalışmalar ve Ülkemizdeki Mevcut Durum. Gaziosmanpaşa J Sci Res 13:47–68
http://www.iata.org/whatwedo/ops-infra/baggage/Pages/baggage-tracking.aspx. Accessed 05 May 2018
https://www.sita.aero/air-transport-it-review/articles/the-tech-powered-airport. Accessed 05 May 2018
https://www.sita.aero/resources/type/videos/bagjourney. Accessed 05 May 2018
https://news.delta.com/delta-introduces-innovative-baggage-tracking-process-0. Accessed 01 May 2018
https://www.avianews.com/wp-content/uploads/2018/04/baggage-report-2018.pdf. Accessed 28 Apr 2018
https://www.forbesmiddleeast.com/en/here-today-gone-tomorrow-crypto-billionaires-net-worth-falls-by-billions/. Accessed 25 Apr 2018
https://www.sita.aero/pressroom/news-releases/mexico-strengthens-border-with-sita-biometrics?utm_source=social&utm_medium=LinkedIn&utm_campaign=press-release. Accessed 25 Apr 2018
https://runwaygirlnetwork.com/2018/04/26/virtual-reality-tool-to-help-you-choose-a-better-aircraft-seat/. Accessed 25 Apr 2018

Lean Production

An Application of Kaizen in a Large-Scale Business

Mahmut Tekin(✉), Murat Arslandere, Mehmet Etlioğlu, and Ertuğrul Tekin

Selçuk University, Konya, Turkey
mahtekins@gmail.com, m_arslandere@hotmail.com,
mehmetetlioglu@gmail.com, ertugrultekin42@gmail.com

Abstract. The ultimate objective of manufacturing industries is to increase productivity with high quality. At present, many manufacturing companies are facing problems such as high quality rejects, high inventories, long lead time of production, high costs of production, and inability to cope with customer orders. They have been trying to reduce the total cost and wastes across their supply chain to remain competitive in the expanding global market. Hence, a collection of reliable tools and techniques are vital for decreasing costs and wastes and for providing effective services for customer demands. One of these techniques is Kaizen. Kaizen which means continuous improvement according to the principle of excellence which is a lean management model that envisages the participation of all managers and employees in businesses. This model predicts that continuous improvement should be made in every aspect of business and social life. Kaizen studies provide great results with improvements in small steps. As a result of Kaizen improvements, innovative results with small but sure steps have been achieved. In this context, Success is achieved by placing process standards and discipline with plan, do, check, take measure cycle that Kaizen in the business management system.

In this study, Kaizen studies was applied in two areas in a large flour factory. First, Conveyor Band System Development Project. Second one is Final Finished Product Loading and Stacking Project. In this context, Conveyor Band System Development Project; reviewing and analyzing the existing situation, proposing and implementing improvement, and controlling and standardization the results. In the Final Finished product and Stacking Workers Project, the examination and analysis of the existing situation, the proposal and implementation of improvement and the checking of the results and the rules of procedure were investigated. As a result of the research, a general evaluation is made about the Kaizen applications in the flour factory and suggestions are presented.

Keywords: Kaizen · Continuous improvement · Lean manufacturing

1 Introduction

The economic, cultural, political and financial globalization significantly enhanced the company business opportunities and the level of competition faced by companies. So, they tried to reduce their structural and production costs through continuous process improvement and waste reduction by using management tools [1]. Businesses have

N. M. Durakbasa and M. G. Gencyilmaz (Eds.): ISPR 2018, *Proceedings of the International Symposium for Production Research 2018*, pp. 515–529, 2019.
https://doi.org/10.1007/978-3-319-92267-6_44

started to operate in the global area depending on technological developments and without borders of the countries [2]. It has become necessary for enterprises to adapt to their environment to change their production systems and methods in order to be able to compete and sustain their existence both nationally and globally. Otherwise, they will not compete and their existence will be in danger. Today, when we look at the leading companies in their sector, it is seen that they are the enterprises that best implement new production approaches [3]. In this context, philosophy of lean which provides many advantages to the users in the field is at the forefront.

Lean manufacturing is a manufacturing concept in which its ultimate goal is to achieve efficient production through comprehensive approach to minimize wastes or muda [4]. It literally means eliminating over production and excess inventory, eliminate wasteful movement of material and human motion, waiting and delays, the need for rework or corrections and wasteful process. It centers on excellence and continuous improvement in manufacturing. It is the primary goal of this production system that all the work done in the production process is the value-added business in which the customers will be willing to pay money. It is also essential for all system components to adopt and internalize this philosophy for lean manufacturing practice.

2 Lean Manufacturing

Businesses are increasingly in competitive to increase their sales since the early years of industry history. While global competitive pressures are spreading throughout the world, the survival of businesses depends on the ability to reduce production costs, continuous improvement of products and the ability to keep up with changes in socio-technological development. For this reason, lean manufacturing forms the basis of being efficient, productive and perfect. Lean is not a new idea. Even though the term "lean" has existed since the early seventies, the techniques can be traced much farther back to ideas developed by Frederick W. Taylor, often called "The Father of Scientific Management." In 1911, he published his methods of applying scientific analysis and testing to manufacturing in a book called The Principles of Scientific Management. Japanese companies used the concepts in this book to rebuild their businesses after World War II. Among these companies, Toyota became particularly noted for its success in applying the techniques, refining them in its automotive manufacturing process and renaming them the Toyota Production System (TPS). Businesses around the world began to adopt these practices in hopes of achieving the same results, and TPS became the model for what would eventually be called lean manufacturing [5].

The production and management system that we call today "lean manufacturing" was developed by engineers Eiji Toyoda and Taiichi Ohno in the Japanese Toyota business in the 1950s. Ohno has developed a production system with flexibility which can be produced in numerous and varied quantities that can meet both low demand and purified from muda in a slow growing economy. Lean production emerged and developed as a whole new understanding of production based on the fundamental of lean philosophy. Lean production system minimizes the elements such as defect, cost, inventory, labor, development process, production area, wastage and customer dissatisfaction which carries unnecessary elements in the structure. In other words, the lean manufacturing system is a system that uses all the production factors flexibly, with zero waste, if possible with the least resources, with the least costly and faultless production which meeting customer demands and expectations [2]. One of the important features of lean manufacturing is significant reduction in manufacturing costs resulting from least inputs [3]. The main aim of lean thought is to reduce costs and ultimately provide excellent value to customers by doing effective and efficiency analyzes of perfect processes that reduced wastage. Businesses get a lot of advantages with lean manufacturing such as improving productivity, improving quality, reducing costs, shortening product cycle time, saving space, eliminating wastes, using energy efficiently, increasing employee skills and motivations, improving customer satisfaction, improving documentation and standardization. There are many techniques used in lean production. Lean production techniques; value flow mapping, kaizen, individual suggestion system, jidoka, Poka-Yoke, standard business, total productive maintenance, 5S, cellular production, heijunka, visual factory/visual management, kanban and pull system, SMED and hoshin kanri [3]. In the lean world, there are six kinds of waste in the manufacturing process [5];

Defect: When a defective part is shipped to the customer, the customer must waste time and labor identifying the defect and returning the defective part to the OEM. The OEM's time and labor have been wasted in manufacturing the part.
Transport: The parts the OEM ships to the customer should only be what the customer needs and should be undamaged at the time of delivery. When these conditions have not been met, time and labor have been wasted for the OEM and the customer.
Timing: When the parts are shipped to the customer before the customer is ready to receive them, time, labor, and other resources are wasted in storing the parts until they are needed. Also, when the customer is ready to receive the part, the specifications for the part may have changed. Again, in this case, the OEM's time and labor have been wasted.
Waiting: This is the opposite of overproduction. When the customer must wait for a parts shipment, the customer must divert manufacturing resources to other processes and the OEM's time and labor are wasted.
Overproduction: When an OEM produces more parts than the customer wants or parts with features that the customer does not need, the OEM's time and labor are wasted.
Motion: To save time, the amount of effort for a worker to perform a task must be minimized. For example, effort is wasted when there is too much distance between workstations, and a line worker must spend more time carrying a part between them.

All lean production techniques try to eliminate wastage by their own methods [6]. Application of lean manufacturing techniques is crucial for businesses which both in phase of establishment and planning to pass for lean production. In the initial phase of newly established businesses, if they adopt lean thoughts, many cost and many possible problems that can be described as possible losses in the future can be prevented from beginning. The adoption and implementation of lean manufacturing techniques within the business is an important factor in achieving the objectives of the operator [7]. To determine the value for a particular product is to identify the value flow of each product, ensure continuous flow of value, enable the customer to pull the value from the manufacturer, and pursue excellence [8];

Value: A product or service that meets the customer's needs and request at a certain price.
Value Flow: Lean manufacturing accepts the system as a whole. It is a fact that processes that do not create value in the system (wastage) are removed from the process.
Flow: Producing as demanded and requested rather than producing more products. It means taking into account the customer's suggestions in shaping the product.
Pull: The product is manufactured when customer requested. The concept of pulling is actually a production control system. It is to establish a balance between demand and production and to provide a synchronized production. For this reason, the idea of eliminating wastes from overproduction has developed in this system [9].
Excellence: Continuous improvement is provided in all areas of the production every time since all faults are eliminated [10].

3 Kaizen

The lean manufacturing system has been accepted worldwide and has been successfully implemented by many companies and still being implemented. The principle of excellence from the five basic principles of this system is an indication that the lean production system is essentially at the top of its reach. The principle of excellence refers to the fact that there is no end as is the beginning of the work done in the lean production system and continuous development. Businesses are forced to adhere to this principle in today's world because intense competition environment and continuous and rapid change. Otherwise, the companies will not be able to keep up with change and competition and will have to stop their activities. Therefore, kaizen which are one of the lean manufacturing techniques that envisage continuous development in simple studies at the crucial stage and it is necessary to give importance to these studies at a high level. Even if it is small, continuous improvement work is the main element that fulfills the principle of excellence in a lean journey. Companies wishing to achieve perfection must continue to do better and adopt continual action. Kaizen application is a lean production technique that feeds the principle of perfection the most because it involves continuous improvement works.

Kaizen (Ky ‘ zen) is a Japanese term that means continuous improvement in Japanese, taken from words ‘Kai’, which means continuous and ‘zen’ which means improvement. Some translate ‘Kai’ to mean change and ‘zen’ to mean good, or for the better [11]. This approach requires that this phenomenon be applied to all aspects of life, both social and business life and advocate for continuous development. Although various authors refer to various key aspects of kaizen, most of them are combined in the following three features.

1. Kaizen is continuous. It includes endless journey through quality and efficiency works.
2. It follows gradual ways due to structure and managers start with organizational changes or technological innovations.
3. It is participant. With the participation of employees, improvement works continue at every stage of the process [12].

All over the world the Kaizen techniques have been particularly distinguished as the best methods of performance improvement within companies since the implementing costs were minimal. It is nowadays more than ever that the relationship between manager and employee is crucial and the Kaizen techniques have a major contribution to the reinforcement of this relationship since the achievements of a company are the result of the mixed efforts of each employee. These methods bring together all the employees of the company ensuring the improvement of the communication process and the reinforcement of the feeling of membership [13]. The Kaizen management originates in the best Japanese management practices and is dedicated to the improvement of productivity, efficiency, quality and, in general, of business excellence. The Kaizen methods are internationally acknowledged as methods of continuous improvement, through small steps, of the economical results of companies. The small improvements applied to key processes will generate the major multiplication of the company's profit while constituting a secure way to obtain the clients' loyalty/fidelity. [14]. Kaizen works seem to be small but it has always produced striking results with its effects. Kaizen is executed with plan, build, control and take measure steps and become internalized as a culture of the company along with permanent standards and discipline [15]. It is necessary to provide 3 basic conditions for Kaizen to take place. These are;

- *Improving human resources:* Human resources are the most valuable asset of an organization. It is people who do everything. Every worker must be made part of continuous improvement activities. Employees should be supported with in-service training to provide problem solving and the results should be standardized to avoid encountering the problem after problem is solved.
- *Finding the current situation inadequate:* There many areas to be developed even in the most functioning system. No system is completely perfect. The system should be developed daily with small and frequent steps.
- *Using problem-solving techniques in a widespread way:* Determined problems need to solve to make improvements. Some tools and techniques need to be used to solve these problems and troubles. In this context, These are effective problem solving techniques which expressed as seven statistical tools used in problem solving. These tools are; histograms, control tables, scatter diagrams, cause result diagrams, graphs, control charts and pareto diagrams.

Kaizen which is a continuous improvement process that empowers people to use their talents and capacities at the point of use and can be used to solve work related, workflow problems or other work related issues. Looking at how employees do their jobs based on quantitative analysis will be a good starting point in the kaizen process. In addition, it will be provided to determine wastage by taking time and work surveys about the tasks by receiving help and information from workers and managers [16]. The general steps to be taken to carry out an Kaizen event are as follows;

1. Choice of problem
2. Examination of current situation
3. Analysis of causes
4. Proposal for improvement
5. Implementation of improvements

6. Checking the results
7. Regulation of working rules
8. Standartization.

The PDCA cycle is a system to ensure the continuation of the kaizen principles. It is a vital part of the process. Plan Do Check Act (PDCA) is a framework that provides a methodical approach to problem solving and continuous improvement. Shewhart and Deming advocated PDCA concept for productivity management, and continuous quality improvement of process and products. PDCA is the “golden cycle for improvement”. PDCA wheel should be considered a never-ending cycle for improvement towards an ideal condition [17].

Plan is to establish objectives and process or countermeasures with expected outcome based on the past performances or future forecasting of work.

Do is to implement the processes or countermeasures planed.

Check is to measure the effectiveness or achievement of processes or countermeasures planed between the actual results and expected results to ascertain any differences.

Act is to analyze the differences to identify the causes of “Gap”, and take necessary action to improve changes.

The hierarchical structure of participating in kaizen studies according to I [18] is given in Table 1 below.

Table 1. Hierarchical structure of participation in kaizen studies (Imai, 2987;7)

Senior management	Middle stage management and staff	Supervisors	Workers
It is determined to launch KAIZEN as a company strategy	It spreads and executes its goals through policy diffusion and inter-functional activities determined by senior management	Use KAIZEN in functional roles	It participates in KAIZEN with its suggestion system and small group activities
It provides support and direction to KAIZEN by providing resources	KAIZEN is used in functional activities	Prepare plans for KAIZEN and guide workers	Apply discipline at work
It creates policy and inter-functional goals for the KAIZEN	It establishes standards, protects and improves standards	It strengthens communication with employees and provides high morale	constantly improves himself in order to solve the problems better
It conducts policy dissemination and audits to achieve its KAIZEN goals	KAİZEN consciousness is given for employees with training programs	Kalite çemberleri gibi küçük grup çalışmalarını ve bireysel öneri sistemlerini destekler	Develops talent and experience with cross-training activities
Establish systems, processes and structures for KAIZEN	Helps employees to develop their skills and problem solving tools	Provides discipline in the workplace	
		It constitutes the KAIZEN recommendations	

When examined Table 1, it can be seen that the Kaizen activities are carried out with the involvement of employees at all levels of the company and that each level of the business has its own responsibilities and duties.

For example; senior management should be committed to launching the company as a corporate strategy and mid-level management will disseminate and manage its goals with policy and inter-functional activities as determined by senior management. While Supervisors use KAIZEN in functional roles, workers participate in kaizen with suggestion system and small group activities. It is nowadays more than ever that the relationship between manager and employee is crucial and the Kaizen technique have a major contribution to the reinforcement of this relationship since the achievements of a company are the results of the mixed efforts of each employee. These methods bring together all the employees of the company ensuring the improvement of the communication process and the reinforcement of the feeling of membership.

4 An Applicatıon of Kaizen

Kaizen constitutes a notion based on the concept of continuous improvement in which no operation/process has taken its final state and can reach even greater perfection. It is kaizen studies that contribute the most to the perspective of lean manufacturing, eliminating wastes and taking continuous steps towards excellence. In this context, it was decided to implement two kaizen projects after the opinions obtained from the employees through the individual recommendation system after the value flow mapping in the flour mill. The kaizen projects are examined by analyzing the current situation, suggesting and applying improvements, checking the results and ruling phases. These kaizen projects are;

- Conveyor Band System Development Project
- Final Finished Product Loading and Stacking Project

4.1 Conveyor Band System Development Project

Following the suggestions from the employees related to the conveyors carried by the wheat in the factory, the details of the project that was developed and applied after the results of the studies and examinations are given below.

Stage 1 - Existing Situation Examination and Analysis

Wheat which transported on conveyor corrodes conveyor's sheet metal over time. It forms deformations and holes through the conveyors. It causes both an ugly image and the spillage of wheat. Besides, the cost of the change of the conveyor's sheet metal together with the deformation occurs. The wheat poured into the conveyor accumulates which in some cases interferes with the operation of the conveyor and stops the system. Figure 1 below shows the existing conveyor screens, and Fig. 2 shows the location images with deformation in the conveyors. As shown in Fig. 1 below, the sheet base is used in the existing conveyor. As shown in Fig. 2, patches are applied to the deformations that occur due to the wear of the sheet metal.

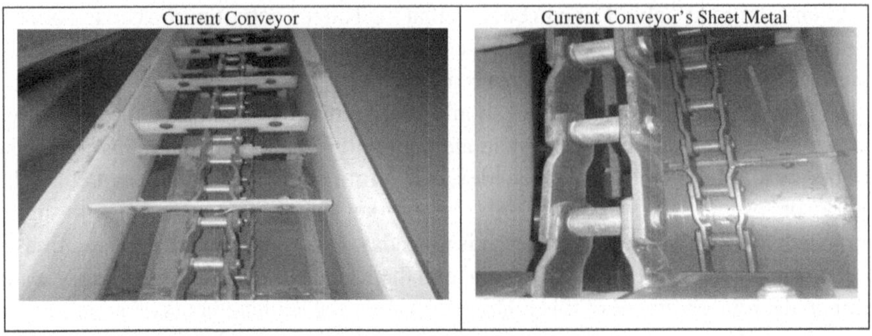

Fig. 1. Current conveyor view

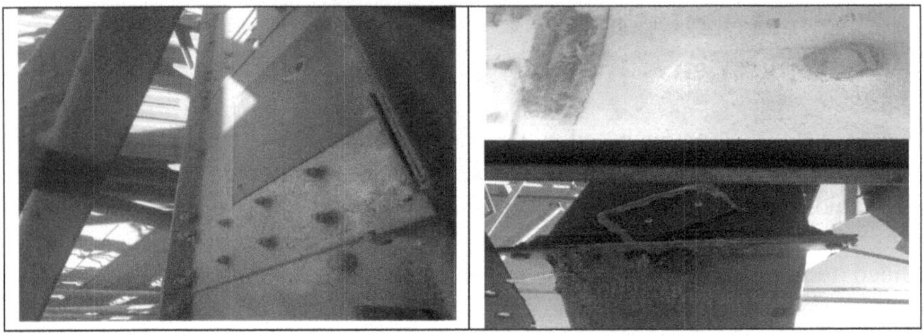

Fig. 2. Deformed location views on current conveyors

The project was carried out by the production managers and the technical team together. Abrasion resistant material was decided instead of the sheet material used in the conveyor. At this point fiber material which is a plastic derivative to be used effectively, was supplied and applied to the system after market research. In addition, small gaps were added to the chains to prevent spilled wheat from clogging and stopping the system. Figure 3 below shows the fiber material and the small reservoirs.

Fig. 3. Fiber material and small reservoir used in new conveyor

With the new application, fiber material is replaced fiber material instead of the sheet metal in the area where the wheat passes and the abrasion of conveyors was prevented, wasting of product and noisy was also eliminated. Along with the newly added reservoir, spilled wheat was prevented from accumulating on the conveyor and the occurrence of fault condition was also prevented. Thanks to these improvements, zero material waste of lean production contributed to zero failure rate targets.

Stage 3 - Checking and Standardization the Results The control of the results of the newly implemented system was carried out in a total of 4 areas. These are;

- Product Quality
- Production capacity
- Discard - Non-Sanitary Wastage Ratio
- Workmanship

When examining the effect of these 4 areas of the new system, it is seen that there is no change in the negative direction of the quality of the product and that the manufacturing of the indirect production capacity is positively affected by the addition of the new small reservoir. In addition, non-sanitary wastage ratio was eliminated after changing fiber material. There has been no change in the workmanship required for the operation of the system. The system has been integrated with the production and technical team and the system has been internalized for the following conveyors.

4.2 Final Finished Product Loading and Stacking Project

In line with the opinions obtained from the employees through the suggestion system, a process of kaizen was carried out at this point by examining the processes of preparing and storing the final products formed after the packaging machine and the packaging machine.

Stage 1 - Current Situation Examination and Analysis

In the current situation, the final products in the production schedule are collected from the package, then reloaded when the vehicle arrives. The packaging machine and example of stacking after packaging are shown in Fig. 4.

Fig. 4. Packing machine and stacking after packaging

The process shown in Fig. 5 below is graphically illustrated.

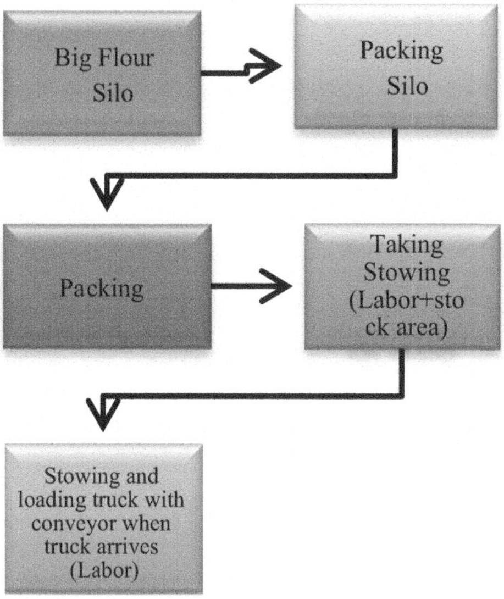

Fig. 5. Packing system current process

As seen in Fig. 5, in the current process, the product comes from the large flour silos to the packaging silos and then the packaging process takes place. Then packed products are stacked and loaded when truck arrives for loading. Furthermore, waste of labor arises for stacking and breaking the stack.

Stage 2 - Suggestion and Application of Improvement
The new process steps shown in Fig. 6 below are seen.

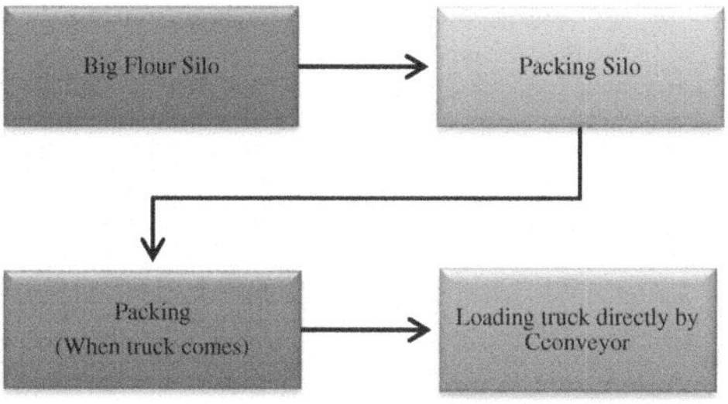

Fig. 6. Recommended process for packaging system

According to the proposed new process (Fig. 6), it will come to the position where the truck can be loaded directly by the conveyor after the package, and loss of labor will be prevented along with waste of waiting time. This recommended process was implemented in the factory and achieved positive results. The project was carried out by the production managers and the technical team together. Figure 7 below shows the new process images.

Fig. 7. New process after packaging

As can be seen in Fig. 7 above, the products are loaded into the truck together with the immediate delivery conveyors and the stacking labor was eliminated. The benefits gained with this new process include;

1. Savings from workers who install packed flour (Zero labor loss/6 staff saving)
2. Saving space by getting rid of packed flour stock (Zero stock/1000 m^2).

Provision of these benefits has contributed to the goal of reducing all elements causing lean production to the highest possible level. Packaging will not work if there is no transportation vehicle. A safety stock is provided to prevent the vehicle from waiting in case of malfunction of the machines. When we talked with the factory official, we have been informed that this situation may not be very frequent at least once in the past. The workforce change that will emerge in the new process is shown in Fig. 8 below.

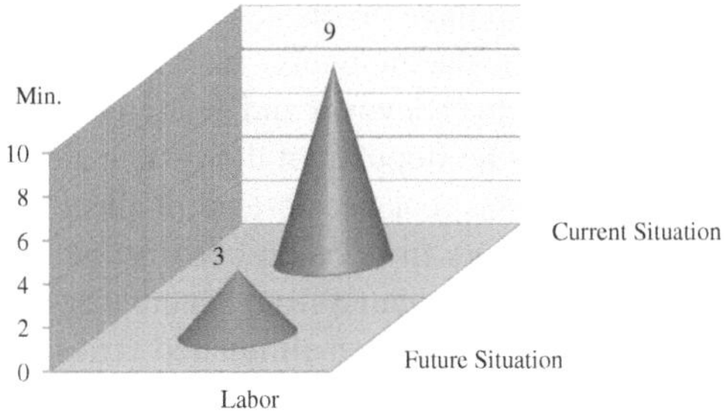

Fig. 8. Current and future situation

As can be seen in Fig. 8, the current labor required for packaging and after packaging is reduced from 9 to 3. Moreover, with the removal of the stacking area, savings of about 1000 m^2 have been achieved, contributing to zero labor costs, zero redundant process, zero stock targets of lean production.

Stage 3 - Checking and Standardization the Results

The control of the results of the newly implemented system has been carried out in 4 areas. These;

- Product Quality
- Production capacity
- Discard - Non-Sanitary Wastage Ratio
- Labor.

When the new process is examined, it has been observed that there is no change in the quality in terms of product quality, production capacity and scrap rate. In the case of labor, 6 personnel were saved in the positive direction. The new application has been internalized to be applied consistently in the subsequent production process by standardizing the production process flow along with the production planning authorities with the staff updates at the required amount of labor.

5 Results and Conclusion

Kaizen is a strategic instrument which is used to achieve and overcome the company's objectives. Kaizen provides a tool to adapt to the global competition by eliminating waste in the process of production, changing corporative culture and encouraging cross-functional links between the managerial staff and production workers, as well as combining between top down and bottom up management.

The kaizen applications were applied in the flour factory and productive results were achieved in our study. In this context, two projects were planned and applied. With the conveyor band system project wheat carried on the conveyor causes the deformation and holes to be formed by the corrugation of the conveyor band over time. This causes both an ugly appearance and the spillage of wheat. As a consequence of this situation, the cost of changing the conveyors, the high noisy working environment and the accumulation of the wheat poured into the conveyor and even in some cases the obstacles to the operation of the conveyor and the stopping of the system are encountered. With the Kaizen operation, fiber material was replaced instead of sheet metal where the wheat passes thus preventing the conveyors from being corrugation. Along with the newly added reservoir, spilled wheat was prevented from accumulating on the conveyor and the occurrence of the fault condition has been eliminated. Thus, conveyor replacement and repair costs (zero material loss), fault condition (zero time loss), scrap (zero material loss) and noisy working environment have been eliminated. With the final product loading and stacking project, applications were carried out and successful results were obtained in the fields of stacking labor and area allocation for stock. With the Kaizen operation, the conveyors after the packing together with the conveying of the packed flour by means of the direct loading of the products into the truck were saved 6 staff savings (zero work loss) and 1000 square meters of area saving (zero stock) by getting rid of the unpacked flour.

With this study, the main objectives of lean production; many advantages such as zero inventory, zero time loss, zero work loss, zero material loss, zero wait, zero unnecessary movement, zero stationery work and expenditures, zero defect production and zero excess process loss are obtained. It is necessary to continue the work in a stable manner in all areas of business from the point of the never-ending improvement works of the principle of excellence. In addition, one of the most important conditions for sustainable lean practices to be able to continue in a stable manner is to give importance to full participation of employees and full support of management. In addition, it is important that lean production practices can be implemented in a planned manner in all areas of the plant.

References

1. Mesquita M, Alliprandini DH (2003) Competências essenciais para melhoria contínua da produção: estudo de caso em empresas da indústria de autopeças. Gestão & Produção 10(1): 17–33
2. Tekin M (2012) Üretim Yönetimi, 8. Baskı. Günay Yayınevi, Konya
3. Tekin M (2017) Üretim Yönetimi, 10. Baskı, Günay Ofset, Konya. 2(8):262
4. Tapping D, Luyster T, Shuker T (2002) Value stream management: eight steps to planning, mapping, and sustaining lean improvements. Productivity Press, New York
5. Fujitsu (2013) Lean manufacturing. Glovia International, Inc. 3. El Segundo

6. Neha S, Singh MG, Simran K (2013) Lean manufacturing tool and techniques in process industry. Int J Sci Res Rev 1(2):54–63
7. Pekin E, Çil İ (2015) Kauçuk Sektörü Poka-Yoke Uygulaması. Sakarya Üniversitesi Fen Bilimleri Enstitüsü Dergisi 19(2):163–170
8. Womack JP, Jones DT (2003) Yalın Düşünce. Sistem Yayıncılık, İstanbul
9. Yingling JC, Detty RB, Sottile J (2000) Lean manufacturing principles and their applicability to the mining industry. Miner Resour Eng 9(2):215–238
10. Arslan S (2006) Yalın Üretim ve MAN Türkiye A.Ş.'de Örnek Bir Yalın Üretim Uygulaması", Gazi Üniversitesi Fen Bilimleri Enstitüsü Endüstri Mühendisliği Anabilim Dalı Yüksek Lisans Tezi
11. Palmer VS (2001) Inventory management Kaizen. In: Proceedings of 2nd international workshop on engineering management for applied technology, Austin, pp. 55–56
12. Brunetand AP, New S (2003) Kaizen in Japan: an empirical study. Int J Oper Prod Manag 23(12):1426–1446
13. Aurel MT, Oprean C, Grecu D (2010). Applying the Kaizen method and the 5S technique in the activity of post-sale services in the knowledge-based organization. In: International multiconference of engineers and computer scientist (IMECS), vol 3, p 17
14. Oprean C, Titu M (2008) Managementul calitatii in economia si organizatia bazate pe cunostinte. Editura Agir, Bucuresti
15. Imai M (2012) Gemba Kaizen: a commonsense approach to a continuous ımprovement strategy. McGraw Hill Professional, New York
16. Ayçın E (2016) Yalın Üretim Uygulamalarında İsrafın Azaltılması İle Performans Ölçütleri Arasındaki İlişkilerin ve Etkileşimin Analizi, Doktora Tezi, Dokuz Eylül Üniversitesi Sosyal Bilimler Enstitüsü, İzmir
17. Basu R (2004) Implementing quality – a practical guide to tools and techniques. Thomson Learning, London
18. Imai M (1986) Kaizen, vol 201. Random House Business Division, New York

An Application of SMED and Jidoka in Lean Production

Mahmut Tekin[1(✉)], Murat Arslandere[1], Mehmet Etlioğlu[2], Özdal Koyuncuoğlu[1], and Ertuğrul Tekin[1]

[1] Selçuk University, Selçuklu, Turkey
mahtekins@gmail.com, m_arslandere@hotmail.com, okoyuncuoglu@konya.edu.tr, ertugrultekin42@gmail.com
[2] Necmettin Erbakan University, Selçuklu, Turkey
mehmetetlioglu@gmail.com

Abstract. Rapid developments in information communication technologies have caused radical changes in all areas of society. This change has led to changes in the lifestyles of individuals and of the ways and modes of doing business of enterprises. It is possible for businesses to be able to survive and compete with the right product, at the right place, the lowest cost and meeting customer demands and expectations. Increased competition, customer orientation resulting from changing customer expectations and demands have also changed the manufacturing systems and methods. Businesses have had to use sustainable flexible production methods and techniques which quality-focused, cost-effective and the least wasteful in order to meet customer demand expectations.

Identification of problems, prevention of mistakes, identification and solving of related problem areas and high preparation times are important problems in manufacturing. Automatically detecting faults and stopping the system makes the process more reliable by reducing the need for physical controllers. Reducing the preparation time can save the waste of waiting time. Many methods and techniques are used to remove wastes which are the basic philosophy of lean manufacturing. Jidoka is the detection of faults in the production process and SMED is shortening of preparation times. Jidoka provides an opportunity to automatically detect the system when an abnormal situation occurs for the machine or operators and to stop the production by detecting the fault. Thus, high-quality products and productivity increase are achieved by producing faultless products, increasing quality, preventing the breakdown of machines, tools and equipments and making processes safe. Large productivity increase can be achieved by avoiding waiting wastage with the SMED. The aim of the work is to create high quality products in production and raise awareness about productivity and to make the manufacturing systems safer by shortening waiting times by stopping and detecting the faults in the production system with jidoka and SMED method as a means of lean management and lean manufacturing. In this context, SMED and Jidoka techniques were applied on called ''vals ball'' in flour factory which operating in Turkey ISO top 500 industrial companies. Successful results were obtained regarding the applied jidoka and SMED techniques.

Keywords: Lean manufacturing · SMED · Jidoka

N. M. Durakbasa and M. G. Gencyilmaz (Eds.): ISPR 2018, *Proceedings of the International Symposium for Production Research 2018*, pp. 530–545, 2019.
https://doi.org/10.1007/978-3-319-92267-6_45

1 Introduction

The production systems have gone through many changes until today. The main reasons for this change can be explained by the changing and developing environmental factors. It is expressed about changing and developing environment that changes in customer demands and requirements, increase in competition with globalization, development of information communication technologies, increasing world economy and environmental sensitivity. Businesses have been affected by these changes and have had to adapt. Businesses have achieved efficiency and efficiency increase especially by changing production and management systems. Otherwise, there will be no competition possibilities. Today, when the leading firms in the sector are examined it is seen that they are the ones that best implement the new production approaches [1].

Businesses are aware of the production techniques that will provide them a competitive advantages in a highly competitive environment. In this context, lean production techniques are at the forefront in terms of the least cost, zero waste, zero stock, zero defect products and quality products for businesses. Lean thinking is everything that does not offer any benefit to the user of the product or service and customer does not agree to pay the extra price. For this reason, all kinds of waste (stocks, waiting time, unnecessary jobs, faults, overproduction etc. must be eliminated. Lean Production aims to minimize the cases that do not add value with its basic structure. All of the tools used to reduce or remove losses in this direction. Jidoka which is one of the lean manufacturing techniques has taken many advantages such as giving the operator the authority to stop the line and solving problems by identifying them, providing that the machines can control the product, gaining ability to stop automatically or give the necessary signals in case of an abnormality, seperation of operator workforce and machine operations and managing multiple machines. Smed is also one of the lean manufacturing techniques which is a big contributor to the realization of JIT (just in time) production. Smed provides flexibility and agility in manufacturing. This technique opens the way to increase production efficiency and to produce with small batches by shortening time spent on model change [2].

2 Lean Manufacturing

Toyota is the first practitioner of lean production. This approach has emerged on a trip to America made by Taiichi Ohno with Toyota executives who form the basis of the Toyota production system. Their aim was to develop production systems to compete with American companies. In those days there were mass production systems developed by Henry Ford in America which still in use today. This system was based on the principle of quality control after production in batches. This system has not been adopted by Taiichi Ohno. Instead, a new production system called the Toyota Production System which started in the 1960s and lasted until the end of the 1980s which emerged in relation with Japan's postwar economic conditions. The so-called lean production system is based on the multiplicity, diversity and flexibility in production which based on competition in global markets instead of product-based strategy. Lean manufacturing basically defines Japanese manufacturing techniques as an alternative to mass production and represents a holistic transformation in manufacturing [3, 4].

Lean production aims to produce with minimal waste by using less energy and less raw material. All elements that do not add value to reduce wastage in lean production need to be removed from production by separating from production. Lean systems need permanent improvements to improve productivity and reduce waste [2]. Lean production system tries to progress on its way by creating a customer-focused flexible structure with zero inventory, zero time loss, zero work loss, zero wait, zero unnecessary movement, zero stationery jobs and expenditures, zero defective production, zero excess process loss and zeroing all other waste items that do not create value.

The concept of lean manufacturing has been used for the first time in the book by Womack, Jones and Ross (1990) "The Machine that Changed the World". They expressed it as a way of thinking style rather than production style and emphasized as a holistic system approach to the creation of a culture of which all of the individuals in the organization should develop their activities continuously [5]. It can be said that the birth point of lean production is the lean thought. Its simplest sense, it refers to the use of available resources in the most efficient way to prevent wasting and to eliminate all unnecessary things. Lean thought is a system that aims to manage efficiently by eliminating unnecessary details with overall viewpoint [6]. In other words, lean production can be defined production with minimum resource, shortest time, cheapest, least waste and faultless product by using flexible manufacturing processes in accordance with customer demands [7].

Although this system was called lean production, Taiichi Ohno expressed it as "Full-Time Production", Shiego Shingo "Non-stock Production System" and Japanese experts and researchers expressed the term "Toyota Production System". In some literature, these terms are still used and are also referred to as "Synchronous Production System". Although the company which called ''Toyota Motor Company'' first developed and implemented this system and later the other companies contributed to the development and dissemination of the system and the system exceeded Toyota's boundaries [8].

3 Scope and Goal of Lean Manufacturing

Lean thought is the idea of reaching a lean production system, a lean company and a lean value chain. The goal of lean thinking is to focus on the production of resources and work that will affect the product and to capture wealth by eliminating wastes rather than changing the center of interest and eliminating the value of the management. It is a production system which carries no unnecessary elements such as fault, cost, stock, labor, development time, production area, wastage, customer dissatisfaction. The main strategy of lean production is to improve the quality, cost and delivery performance by increasing the speed and decreasing the flow time. Lean production is a system that uses time and resources to create value-added activities that shape the material or information that transforms the customer in the direction of customer needs, but distinguishes activities that do not add value to the product in line with customer needs and that do not create added value. It aims to increase customer satisfaction by focusing on the concept of value by eliminating all elements that cause wastage.

The most striking difference between mass production and lean production is their purpose. Serial producers admit themselves a limited target which expressed as "good enough. This means defect products at acceptable level, acceptable stocks at the maximum level and low standardized products. Doing better will cost a lot of money or exceed people's natural abilities according to their ideas. On the other hand, the lean producers are precisely aimed at perfection. Such as constantly reduced costs, zero defective products, zero stock and the endless variety of products. Of course no lean producer has reached this utopia and perhaps it will never reach it. But the endless search for perfection will continue to produce with a surprise change [9]. According to Womack and Jones, lean manufacturing is a process of a number of activities. Purposes of lean production system;

- Accurately determine the value of the product and to increase this value,
- Define the value chain for each product and remove the activities that cause the wastage (zero inventory, zero time lost, zero work lost, zero wait, zero unnecessary move, zero stationery jobs and expenses, zero defective production, zero excess process loss, eliminating all other waste items),
- Ensuring product flow without any interruption,
- Ensuring pulling value customer from the manufacturer,
- Achieving excellence and making it sustainable.

4 Lean Manufacturing Principles

According to [10], we can briefly summarize the principles of the lean production system aimed at preventing waste as five steps. These are; determining a value for a particular product, determining the value flow of each product, providing uninterrupted flow of value, ensuring for customer pulls the value from the manufacturer and pursue excellence.

4.1 Value

Value is concept which critical exit point for lean thinking. Value can only be defined by the last customer and only when expressed form of a particular product (a goods or service and usually both) that meets customer needs at a certain price and certain time. The value is created by the manufacturer. This is the cause of the existence of the producers when viewed from the customer's point of view. It is still very difficult for manufacturers to correctly define the value due to many reasons. So simple thoughts; it must begin with a conscious experiment in order to fully define the value of concrete products that include concrete capabilities which offered at concrete prices through dialogue with concrete customers. The way to do this is to ignore existing assets and technologies and to rethink the on the basis of company's product lines, focused team which have strong company [10]. From a lean manufacturing perspective, the concept of value can be defined solely and exclusively by the customer and is a measure of whether the product can meet customer needs in terms of price and other characteristics. The source of pleasure and appreciation that customers appreciate when buying the finished product is the definition of value they make. Manufacturer who create value from the customer's point of view. For this reason, making production of manufacturers will be more productive according to customer's definition of value.

4.2 Value Flow

The value for the customer is created by means of value flows. Thus, businesses make money through value flows. The main purpose of lean is to focus on value flow processes. As value flow processes are perfected, more value can be created for customers and more can be gained. Value is created in the process of value flow, at the same time wastage is the place where it emerged. By focusing on value flows, waste can be identified and action plans can be developed to eliminate wastage. Value flows must be mapped to determine everything that hampers waste and flow. Recovery efforts must be initiated to eliminate the waste and increase the flow in the value flow [11]. A value flow is a sequence of concrete actions required to achieve a particular product (a product or service, or both) and is achieved through three critical management functions in each business. These; problem solving task process which begin from concept to detailed design and engineering to product and market in process, information management task process from the start of order to detailed delivery through to delivery and the process of physical transformation from raw material to completion of the product to the delivery of the customer. Determining the entire value flow for each product (in some cases for each product group) is the next step in lean thinking. This step which companies rarely test, but brings out large quantities muda.

A value flow can be defined as any value-added and non-value-adding activities needed to deliver a particular product or service to a customer. All products with similar characteristics and qualities are produced in a value flow. It can be said that a value flow consists of all products that have the same route throughout the business. Lean businesses determine value flow so they can organize and manage businesses around them to increase the value they provide to customers. As the value flow for lean businesses becomes the primary organizational necessity, it follows that the income table of the operator is organized in the same way [11].

4.3 Flow

The next step for lean thinking is in order when the value is fully defined, value flow chart for a given product is fully prepared and the steps leading to the wastage are eliminated by the lean business. This is a breathtaking step in the real meaning. After that, it is to provide the flow of the steps that create value. However, this step requires a full restructuring. After World War II, Taiichi Ohno and Shigeo Shingo decided that the real issue was to provide a steady flow of production of dozens or hundreds of small parties rather than millions. This creates the general situation; because the vast majority of human needs reflect these modest rivers, not huge rivers. Ohno and his team have often succeeded in constant flow of small quantities of production by learning to change tools quickly to switch from one product to another without the assembly line and bringing the machines to "right size" (miniaturization). In this way, while the object being manufactured is in continuous flow, different types of processing steps (mold, painting and assembly) are made next to each other in succession.

4.4 Pull

The first visible effect of moving from departments and parties to product teams and product flow is striking drop in process from concept to market, from the sale to delivery and the raw material to reach the customer. When the flow system was passed, the product design period is completed within a month instead of years, taking order can be reduced hours and the usual physical product total transaction time can be reduced minutes or days [10]. The pulling sytem does not produce any products or services in the stages without customer demand at the next stages [12].

4.5 Excellence

Perfection is a simple journey that is not the last point. In this respect, it is necessary to see the concept as "continuous improvement". There is a principle "doing the right job at once" instead of "doing the job right way" in the basis of lean thinking. Perfection is an unlikely goal to be fully attained since wastage can not be totally destroyed. In this case, the real goal should be to ensure the process at the highest performance point and improve value continuously. The concept of zero defect which seen as the key to excellence is an approach that prevents them from happening instead of finding and getting the wrongs. From in this regard, zero defect should not be perceived as merely a defect in products but should be considered as a concept covering all functions of the business. It should not be forgotten that a product was manufactured without defect but was not sold on time could lead to various wastes due to stock cost, depreciation, etc. [12].

5 Lean Manufacturing Techniques

All lean production techniques try to eliminate waste by their own techniques [13]. Applications of lean manufacturing are very important both in the businesses which are to be designed using lean thought principles in the phase of establishment and in the classical type businesses where the lean production transition is planned. When a lean thinking system is created in newly established businesses, future losses, many cost items and many possible problems can be avoided at the very beginning. The adoption and implementation of lean manufacturing techniques within the businesses is an important factor in achieving the objectives of the business [14]. There are many techniques used in lean production. These lean production techniques;

- Value Flow Mapping
- Kaizen
- Individual Suggestion System
- Jidoka

- Poka-Yoke
- Standard Business
- Total Productive Maintenance
- 5S
- Cellular Production
- Heijunka
- Visual Factory/Visual Management
- Kanban And Pulling System
- SMED
- Hoshin Kanri

5.1 Jidoka

The basic idea of the lean manufacturing system is to completely remove waste. One of these ideas is Jidoka [15]. Jidoka is a system that enables the machine or operators to be automatically detected when an abnormal situation occurs and to detect the fault then stop the production [16]. Jidoka which is handled by automation and used as autonomy that aims to stop when there is a fault in production. It tends to increase the efficiency of its equipment with the participation of all of its employees [17]. Autonomy is a mechanism that prevents the flow of faulty parts. Even if a fault occurs during production, each member must be able to obtain the specific information required to keep progressing within the program schedule.

It is a technique which involves the jidoka quality control function in the Toyota system. This is because the jidoka prevents the pass of defective parts from the production line. When a manufacturing fault occured, stopping of production line will prevent to intervene to problems immediately, taking measures and repeating similiar faults.

- The reduction in the number of labor force resulted in a reduction in costs, the design of mechanisms that automatically allow the benchs to stop when a production fault occurs, or when the specified quantity of production is reached, significantly reducing the number of workers following the operation of the looms (zero labor loss).
- The increased ability to adapt to demand changes enables autonomous elimination of excess inventories by virtue of all benches producing only faultless parts and automatically reaching the desired quantity of production (zero inventory).
- It will ensure timely production and rapid adaptation to demand fluctuations (Zero waste of time).
- Incorporation of workers into the problem-solving process has led to the development of a business culture of human respect. Thus, the development of culture of human respect accelerates their intervention and improvement efforts during a problem in the production process (zero labor loss).
- No inspection of damaged parts or machines by other operators (Zero excess process loss, Zero waste of time).
- Reduction of scrap/return ratio (zero false production) [18, 19].

As a result, Jidoka stops the machine prevent from recurrence when there is an fault. It is also an extremely important element in terms of preventing the growth of the problem and showing whether the work has progressed in its normal course. This concept has been adapted to the production lines and workers as well as the machines at Toyota. This means that the worker will stop the band without hesitation when he sees any abnormality. Jidoka allows to identify all the abnormalities that occur in the production band and also prevents faulty production.

5.2 SMED

SMED (Single Minute Exchange of Dies) is one of the lean production techniques which based on performance enhancement by optimizing the necessary times in transitions among products in production together with changing patterns in a single minute [20]. The most important constraint factor on the ability of companies to produce with small parties is the job setting period that occurs during mold change. Rapid mold changing and quick set-up operations make it possible to remove trial production from the ground and make small-batch flexible production. This need is made possible by the fact that the machines work more efficiently in order to react quickly to changing market demands and that the stations achieve high OEE (Overall Equipment Efficiency). Shingo has developed the SMED methodology which a pioneering approach in order to achieve small-batch, flexible production systems. Shingo's technique involves both the necessary theorem for setting operations and the practical applications needed to bring the setting operations to under 10 min. The most important step of Shingo's method is to separate the internal setting operations that can be executed while the machine is turned off from the external setting operations that can also be performed while the machine is running. The SMED technique is applied in three stages;

In stage 1; Almost all adjustment operations are performed after the completion of the previous production, so the activities carried out to correct this situation and reduce the set-up period should be separated as offline and online manufacturing activities. Offline manufacturing activities can be completed before the end of the previous product production. Online manufacturing activities can be carried out after the production of the previous party is completed. Thus, the set-up time is limited only by the in-process time. According to [21] only providing this improves between 30% and 50% of the setting.

In stage 2; It focuses on the activities carried out when a mold is removed or a new mold is attached on the machine being worked on. Efforts focus on converting online activities out of manufacturing, that is, to be done during the time when the production of the previous party is in progress. These efforts can reduce the total settling time by up to 90% with the initial changes.

In stage 3; Arrangements and improvements are made by examining both internal setting activities and external setting activities to the smallest detail. The second and third stages need not be made separately but can be made almost simultaneously. It has been shown separately, analyzed and applied to be seen two separate concepts [21, 22]. Thanks to the SMED technique, the preparation time is shortened and the following benefits are achieved thanks to the rapid production of varieties;

- Ability to produce small batches with reduced mold change times (zero waiting loss, zero stock)
- Flexible production and on-time delivery capability (Zero waste of time)
- Less inventory due to working with smaller parties (Zero stock)
- Needs less operating capital
- Quality production with better mold maintenance (zero defect production)
- More regular stocking area (Zero stock)
- Rapid product variety and labor saving (Zero labor loss)
- Increased production (Zero unnecessary process loss).

6 Advantages of Lean Manufacturing

They are directly beneficial to quality because the customer focuses on the concept of value and is a quality-focused approach. In addition, there is a great contribution to the speed and duration of developing new products at the point of shaping the system based on customer needs. Many benefits are obtained with lean system as follows [23–25];

- Increase in product quality,
- Decrease in production period,
- Production with less work functions,
- Increased on-time deliveries,
- Increase in income,
- Reduced costs,
- Increase in labor productivity,
- Decrease in stocks,
- Increased productivity in production,
- Increase in flexibility,
- Increased efficiency in field use,
- Reduced vehicles and tools requirement,
- Increase in machine productivity,
- Prolonged healthy relationships with suppliers.

7 Methodology

Today businesses need a structure that is flexible and adaptable in an increasingly competitive environment. Businesses need to have the best mix of products in the shortest possible time, at the required quality level, with minimum wastage, zero fault, minimum inventory and cheapest structure to be able to cope with competition and to sustain existence. For this, it is necessary to keep up with the dynamic structure of the age to be open to changes and innovations. Therefore businesses have to adapt new systems, techniques and technologies to their own structures. Otherwise they will be left behind in the race. In this work lean production system was explained and SMED and Jidoka of which lean manufacturing techniques was applied in a flour factory. The aim of the work is to create high quality products in production and raise awareness about productivity and to make the production systems safer by shortening waiting times by stopping and detecting the faults in the production system with jidoka and SMED. In this context, SMED and Jidoka techniques were applied and analyzed in flour factory. In our work, the company was analyzed with a lean manufacturing perspective and detailed interviews were conducted with face to face interviews with the authorities.

7.1 SMED Application

It was observed that the SMED technique will be applied changing of vals ball and the work was performed at this point. In this context vals cylinders are used for grinding wheat. The cleaned and having rested wheat is cracked with the help of cooled steel cylinders called vals. Grinding of the flour does not occur only one vals. This is done step by step every time. it is sifted and the thin parts are separated and thick parts are ground again. The average time for changing the ball is 104 min on average. Figure 1 below shows an example of a vals cylinder. The SMED technique is known in 3 stages. These stages was presented below.

Fig. 1. Vals cylinder

7.1.1 Separation of Internal and External Adjustment Works

In stage 1; almost all adjustment operations are carried out after the completion of the previous production, so the activities carried out to correct this situation and reduce the set-up period should be separated into external (offline) and internal (online) activities. External activities can be completed before the production of the previous product is finished. Internal activities can be carried out after the production of the previous party is completed. As seen in Table 1 below, the work done first by changing vals ball was analyzed with the order and time. As seen in Table 1 below, a total of 19 jobs are performed in turn during roll ball change and all of them start after the production stoppage and the work done is divided into internal and external works and the first stage of SMED technology is realized. As shown in Table 1, 4 actions are external. These tasks can be performed without stopping production, while others are internal which can be performed after production is stopped.

Table 1. Vals ball replacement duration and classification

Order	Action	Time	Internal/External
1	Stopping the factory	–	–
2	Opening the vals safety switches	10 s	Internal
3	Opening the vals cover	5 s	Internal
4	Rescuing vals balls from motor belts	8 min	Internal
5	Disassembling belt pulleys which on the vals	12,5 min	Internal
6	Disassemble the mechanism attached to the cannon	15,5 min	Internal
7	Separation of ball bearings by hydraulics	9 min	Internal
8	Setting up the vals cannon	3 min	External
9	Taking the ball out of the vals	2,5 min	Internal
10	Transfering of old ball to maintenance area	8,5 min	Internal
11	Loading carrier vals ball by lift	2,5 min	External
12	Taking it to elevator	2,5 min	External
13	Taking vals ball from the elevator to the point of change	3,5 min	External
14	Placing the vals's teeth in the correct direction	2 min	Internal
15	Fixing vals ball to vals	11 min	Internal
16	Attaching pulley between vals and motor	5 min	Internal
17	Placing straps to provide movement between ball and motor	12,5 min	Internal
18	Greasing the bearings over the ball with grease	2 min	Internal
19	Ensuring that the motor rotates in the right direction by closing the vals caps and placing safety switches	4,5 min	Internal

7.1.2 Externalization of Internal Adjustment Works

In stage 2; concentrates on the activities carried out when a mold is removed or a new mold is installed on the machine being worked on. The efforts here focus on transforming in manufacturing activities to out of manufacturing that is to be done in time for the production of the previous part. These efforts can reduce the total setting time by up to 90% with the initial changes. All of the work applied to changing the vals ball was already performed as an internal time since it was already done after the factory stopped. In order to convert 4 actions which can be realized as external work in stage 1 into external work that is production can be done without interruption. Action order has been changed as seen in Table 2 below.

As shown in Table 2 below, when the vals cannon car is installed and the new vals ball is loaded with the carrier with lift and the transfer of the vals ball to the vals ball exchange point is carried out in the first 4 processes in turn, the process of going to the former cannon maintenance area it has become a spontaneous external action.

Table 2. New ranking of actions made with Vals ball replacement

Order	Action	Time	Internal/External
1	Set up the Vals cannon	3 min	External
2	Loading carrier vals ball by lift	2,5 min	External
3	Taking it to elevator	2,5 min	External
4	Taking vals ball from the elevator to the point of change	3,5 min	External
5	Stopping the factory	–	–
6	Opening the vals safety switches	10 s	Internal
7	Opening the vals cover	5 s	Internal
8	Rescuing vals balls from motor belts	8 min	Internal
9	Disassembling belt pulleys which on the vals	12,5 min	Internal
10	Disassemble the mechanism attached to the cannon	15,5 min	Internal
11	Separation of ball bearings by hydraulics	9 min	Internal
12	Taking he ball out of the vals	2,5 min	Internal
13	Transfering of old ball to maintenance area	8,5 min	External
14	Placing the vals's teeth in the correct direction	2 min	Internal
15	Fixing vals ball to vals	11 min	Internal
16	Attaching pulley between vals and motor	5 min	Internal
17	Placing straps to provide movement between ball and motor	12,5 min	Internal
18	Greasing the bearings over the ball with grease	2 min	Internal
19	Ensuring that the motor rotates in the right direction by closing the vals caps and placing safety switches	4,5 min	Internal

7.1.3 Final Improvements and Regulations in Interior and Exterior Setting Actions

In the final stage, both the internal setting activities and the external setting activities are examined and refined to the finest detail. With the operations in the middle pulley screws holding the cupboard, on the belts and on the other side of the ball on the pulley and the trigger belt on the other side, the joints in the process, i.e. simultaneous work applications and 16,5 min savings have been achieved. In Fig. 2 below, the vals ball straps and pulleys are visible. The change occurring after the application of the smed technique is shown in Fig. 3 below.

Fig. 2. Vals ball strap and pulleys

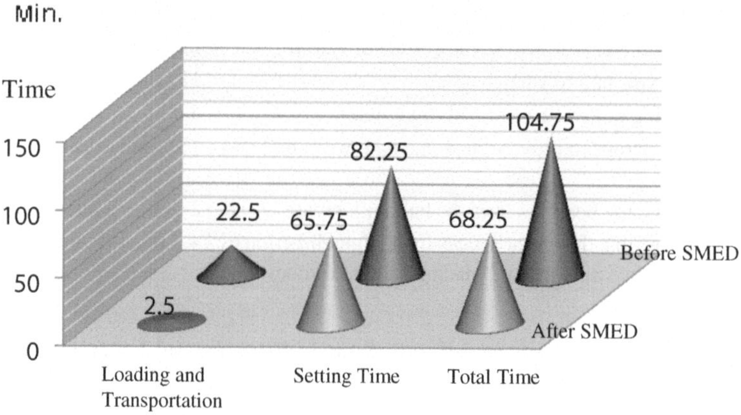

Fig. 3. Vals ball exchange time chart

As can be seen in Fig. 3, the loading and carrying jobs are reduced from 22.5 min to 2.5 min, the setting and assembly jobs reduced from 82.25 min to 65.75 min, the total time from 104.75 min to 68.25 min and zero wait wastage of lean production contributed to zero unnecessary process goals.

7.2 Jidoka Application

The jidoka method which is a system that enables the machine or operators to automatically detect and stop production/machine when an abnormality occurs. Jidoka technique was applied the production line part which called the vals tube. Figure 4 shows the warning signs that are not available in the existing system and integrated into the system through the application process.

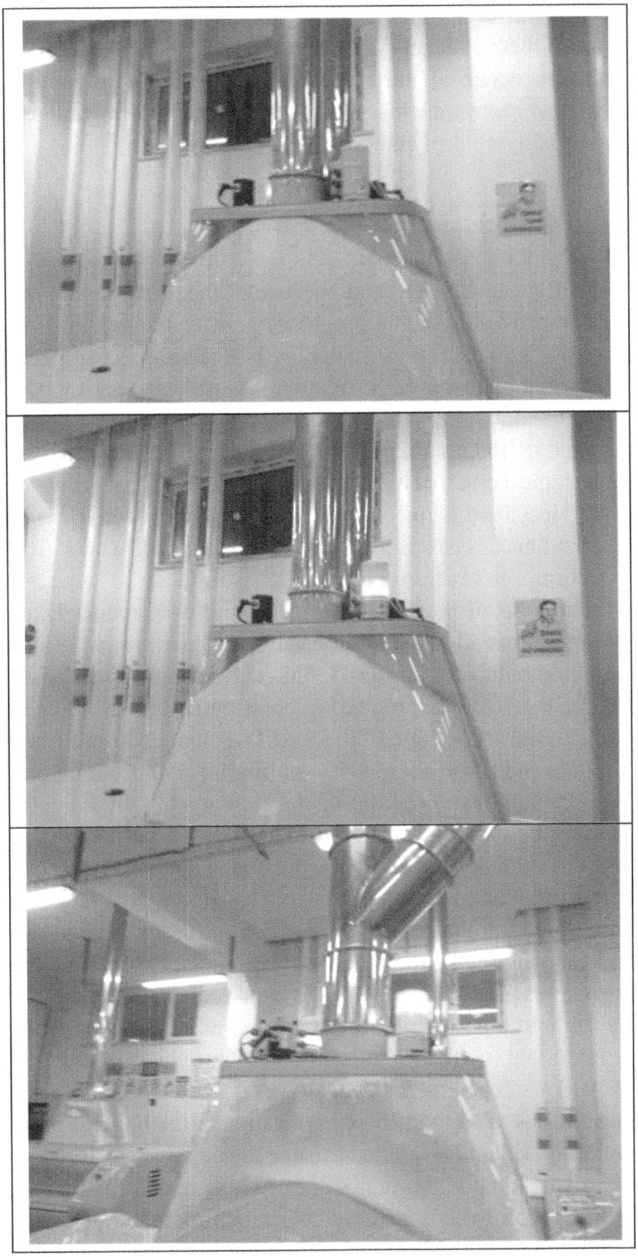

Fig. 4. Vals tube warning ındicators (Color figure online)

As seen in Fig. 4, when the nonstandard value occured on vals tube, the warning lights are activated and the process has the opportunity to intervene by operator. This means that the vals tube is working when the lamp turns green. When it flashes green, the vals is full. when the red flashes, the vals tube stops. when the red light flashes, the vals is off. Thanks to these indicators, the operator has been able to intervene when an abnormal situation has occurred and by preventing the faulty production, contributing to defect production and zero time loss goal of lean production as well as zero labor loss point required for quality control and corrective actions.

8 Results and Conclusion

Lean production will never come to end because of the principle of excellence. There are always better improved systems, processes and techniques. In this context, it is necessary to carry out lean operations including in businesses equipped with high-tech machinery and information technology. In today's world where change has accelerated, both national and global competition has increased and customers have become more conscious and selective, the continuity of their work of leaning and progress towards excellence have become a crucial issue. At this point, the basic logic underlying the lean manufacturing practices is to focus on value and zero stock, zero time loss, zero work loss, zero wait, zero unnecessary movement, zero stationery works and expenditures, zero defect production, zero excess process loss.

In our study, durations of loading and transporting were reduced from 22.5 min to 2.5 min, adjustment and installation duration were reduced from 82.25 min to 65.75 min and total time duration reduced from 104.75 min to 68.25 min with the smed technique. Thus, zero wait wastage of lean production contributed to zero unnecessary process goals. In this context, the application of the SMED technique results in 36.5 min. (Zero labor loss) and more production (Zero waste of time) by contributing production to the earned time. With the application of the jidoka technique, light indicators were added to allow the operator to intervene when any abnormality develops in the process of the vals tupe process. Thus, faulty production is prevented and lean production contributes to zero defect production and zero time loss goal as well as zero labor loss at the point of labor saving required for quality control and corrective actions.

Planning and carrying out of the other lean techniques as well as smed and jidoka techniques will enhance productivity. An individual recommendation system should be set up to ensure full participation of employees and maximum benefit from employee skills. It is very important to incorporate employees into the system with full participation so that lean work can be implemented. However, it is very important and necessary for these administrations to support these applications. It is imperative to ensure discipline at the point of sustainability to standartize of the new system after lean applications have been done. It should be given importance to lean transformation studies in order to strengthen Turkey's global competitiveness in quality and cost-axis and for strong businesses. In this context, training and informing activities about lean techniques should be carried out in cooperation with business, university, industry and trade chambers.

References

1. Kağnıcıoğlu H, Aydın S, Anagün S (2012) Üretim Yönetimi. Anadolu Üniversitesi Yayınları, Eskişehir
2. Tekin M (2012) Üretim Yönetimi (8. Baskı). Günay Yayınevi, Konya
3. Kocabaş F (2004) Endüstri İlişkilerindeki Dönüşüm. Dumlupınar Üniversitesi Sosyal Bilimler Dergisi 10(10):33–53
4. Zerenler M, Iraz R (2006) Japon Yönetim Anlayışı Ve Şirket Ağları (Keiretsu) Analizi. Selcuk Univ Soc Sci Inst J (16):757–776
5. Yüksel H (2010) Üretim/İşlemler Yönetimi (2. Baskı). Nobel Yayınevi, Ankara
6. Gökşen Y (2003) Geleneksel Üretimden Esnek Üretime: Karşılaştırmalı Bir İnceleme. Dokuz Eylül Üniversitesi Sosyal Bilimler Enstitüsü Dergisi 5(4):32–48
7. Özçelik TÖ, Cinoğlu F (2013) Yalın Felsefe Ve Bir Otomotiv Yan Sanayi Uygulaması. İstanbul Ticaret Üniversitesi Fen Bilimleri Dergisi 12(23):79–101
8. Emiliani ML (2006) Origins of lean management in America. J Manag Hist 12(2):167–184
9. Tikici M, Aksoy A, Derin N (2006) Toplam Kalite Yönetiminin Radikal Unsurlarından Biri Olarak Yalın Yönetim. Elektronik Sosyal Bilimler Dergisi 5(15):20–33
10. Womack JP, Jones DT (2003) Yalın Düşünce. Sistem Yayıncılık, İstanbul
11. Özçelİk F, Ertürk H (2010) Yalın Üretim İşletmeleri İçin Değer Akış Yönetim ve Değer Akış Maliyetlemesi. Uludağ Üniversitesi İİBF Dergisi 29(2):51–84
12. Türkan ÖU (2010) Üretimde Yalın Dönüşümün Temel Performans Kriterleri. BAÜ Fen Bil Enst Dergisi 12(2):28–41
13. Neha S, Singh MG, Simran K (2013) Lean manufacturing tool and techniques in process industry. Int J Sci Res Rev 1(2):54–63
14. Pekin E, Çil İ (2015) Kauçuk Sektörü Poka-Yoke Uygulaması. Sakarya Üniversitesi Fen Bilimleri Enstitüsü Dergisi 19(2):163–170
15. Sivaslı E (2006) İşletme Süreçlerinde Yalın Tekniklerin Kullanılması Üzerine Bir Araştırma. Yüksek Lisans Tezi, Dokuz Eylül Üniversitesi Sosyal Bilimler Enstitüsü, İzmir
16. Sugimori Y, Kusunoki K, Cho F, Uchikawa S (1977) toyota production system and kanban system materialization of just-in-time and respect-for-human system. Int J Prod Res 15(6): 553–564
17. Turan H, Çevik H (2016) Üretim İle Yalın Üretimin Karşılaştırılması. 2. Üretim Ekonomisi Kongresi, İstanbul Kültür Üniversitesi. 11–12 Nisan, İstanbul
18. Yazgan H, Sarı R, Seri Ö (1998) V. Toyota Üretim Sisteminin Özellikleri. Sakarya Üniversitesi Fen Bilimleri Enstitüsü Dergisi 2(2):129–134
19. Kılıç A (2014) Yalın Üretimin Gıda Sektöründe Uygulanabilirliği: Gaziantep Ve Kahramanmaş İllerinde Bir Araştırma. Yüksek Lisans Tezi, Kahramanmaraş Sütçü İmam Üniversitesi Sosyal Bilimler Enstitüsü, Kahramanmaraş
20. Jebaraj Benjamin S, Murugaiah U, Srikamaladevi Marathamuthu M (2013) The use of SMED to eliminate small stops in a manufacturing firm. J Manuf Technol Manag 24(5):792–807
21. Dillon AP, Shingo SA (1985) Revolution in manufacturing: the SMED system. CRC Press
22. Tanık M (2010) Kalıp Ayar Sürelerinin Smed Metodolojisi İle İyileştirilmesi: Bir Yalın Altı Sigma Uygulaması. Sosyal Bilimler Ensitüsü Dergisi 25:117–140
23. Pavnaskar SJ, Gershenson JK, Jambekar AB (2003) Classification scheme for lean manufacturing tools. Int J Prod Res 41(13):3075–3090
24. Moore R (2006) Selecting the right manufacturing improvement tools. Elsevier Science & Technology Books
25. Kempton J (2006) Can lean thinking apply to the repair and refurbishment of properties in the registered social landlord sector? Struct Surv 24(3):201–211

Lean Manufacturing Implementations for Process Improvement in a Company Operating in FMCG Sector

Oğuz Emir(✉), Samet Karataş, Eren Ay, Hümra Özker, and Zeynep Gergin

Industrial Engineering Department, Istanbul Kültür University, Istanbul, Turkey
oguzemirr@gmail.com, karatas.samettt@gmail.com, erennay.ea@gmail.com, ozker96@gmail.com, z.gergin@iku.edu.tr

Abstract. In order to survive in the global contest, enterprises have to increase their productivity via better use of their production resources. Hence, companies increasingly implementing lean management principles, which help them generate value and satisfy the requirements of their customers, through the elimination of non-value adding activities and pursuing continuous improvement in the processes.

This study is an example of a process improvement that is applied to one of the fmcg company in Turkey. The project is implemented in packaging lines of Perfetti Van Melle Turkey, which is a global company, engaged in confectionery production and distribution, operating in more than 150 countries. The Turkey plant has been implementing various waste reduction practices for three years, and this project has been initiated to support the design of the packaging zone of the production system, considering the total seven wastes of lean management; namely overproduction, overprocessing, inventory, defects, transportation, motion, and waiting. It is desired to improve the performance of the packaging lines in order to keep up with the growing competition and fulfil the rapidly rising demands.

The study begins by observing processes of packaging lines. The scope is determined by examining utilization rates, general line efficiency (GLE) and operational efficiency (OE). Then, selected lines are analyzed using value stream mapping (VSM) for waste identification, and statistical process control methods are implemented for improvements. The main objective is to increase the line efficiency rates that are calculated based on production quantities and scheduled time parameters. Improvement opportunities are proposed with future state VSMs by utilizing appropriate lean manufacturing techniques. Furthermore, a simulation study using ARENA Software is performed to evaluate proposed enhancements and to compare the alternative implementation scenarios.

Keywords: Lean production · Value stream mapping
Fast moving consumer goods · Simulation · Packaging

N. M. Durakbasa and M. G. Gencyilmaz (Eds.): ISPR 2018, *Proceedings of the International Symposium for Production Research 2018*, pp. 546–561, 2019.
https://doi.org/10.1007/978-3-319-92267-6_46

1 Introduction

Global economic upheaval, competitive price reductions and the increase in the level of quality have made it necessary to focus on time, cost and quality elements. Observed in the economic system, businesses that do not want to lose this competition are faced with the necessity of restructuring their management philosophy. In order to survive in the global contest and be able to move forward, enterprises should aim to increase their productivity by making better use of their production resources. Because of these changes and needs in the field of competition, the notion of lean manufacturing has emerged and its importance has gradually increased.

Lean philosophy is an understanding that emerged in the economic conditions of the Japanese after World War II. It is a production system that is based on Toyota, a Japanese company in the 1950s, and that centers on excellence and continuous improvement in production, which has become widespread all over the world, especially in the western countries since the 1980s. The main goal in lean production is to achieve a continuous value flow to the customer, eliminate all wastes in and around the processes in the system and increase productivity. Waste (muda in Japanese) is defined as every activity that does not add value to the final product and or service, from the customer's viewpoint. For this reason, lean manufacturing systems aim to minimize unnecessary elements in their constitution, for the lowest possible level of wastes such as defect, cost, stock, workforce and production area.

Lean manufacturing becomes one of the most used waste elimination technique in various industries. Many companies accept lean manufacturing improve quality and productivity [1]. Lean is a manufacturing logic that eliminates all types of waste to reduce the lead time between a client order and the delivery of the goods or services. Companies supported by lean are more competitive, responsive, and coordinated in the market through reduction of costs, process durations and non-value added operations [2]. Various objectives for which lean production is applied are – to obtain wide diversity of products with fewer defects [3]. Lean production can be best characterized as a way to deal with conveying the highest value to the client by disposing seven types of waste through process and human style factors. Lean production has turned into a coordinated framework made out of very relevant tools and a wide assortment of administration applications, such as just-in-time (JIT), quality systems, work teams, cell manufacturing [4] Fig. 1 illustrates the seven deadly wastes that the lean tools attack.

This study discusses the improvements achieved in a confectionery company's packaging production lines with lean production implementation. Perfetti Van Melle is one of the strongest players in fast moving consumer goods sector and intends to be the market leader of confectionery segment while improving the performance of its production. The firm has been satisfying diverse customer demands coming from both local and global market through its flexible structure. But, production is affected by tight schedules while increasing demands besides with changing consumer behaviours in recent years. After all, company does not want to encounter the problem of not managing demand. Therefore, it is agreed to cooperate a lean study selecting active bottling packaging production lines. These lines are basically follow same path but

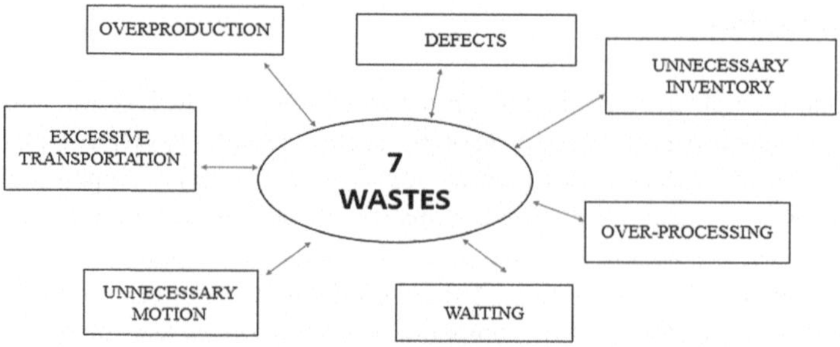

Fig. 1. Seven wastes of lean management

order of the processes can be changed. Mainly, value stream mapping tool is applied to reduce wastes and increase efficiency rates of production lines. In this study, company's packaging lines are monitored and represented by Value Stream Mapping (VSM). VSM is a tool that is used for observing and analyzing a system in its current state and creating a future state for every process that a product goes through [5]. It is a basic and visual process-based instrument which empowers the documentation, representation and cognizance of material and data streams in forms, with a specific end goal to recognize wastes and annihilate them from the system [6]. VSM is an extremely powerful tool used to describe the configuration of value streams [7]. VSM can make the flow of material and information understandable as a product makes its way through the value stream [8]. A VSM is created for both the current and the future states of the system. The current state is a snapshot of the system including material flow, information flow and timeline of activities as it is at present while the future state is an ideal view of how the processes should be. After drawing two maps, it is important to take insights for the improvements and a vision or a target to be achieved. That provides to describe the roadmap of change and improvements in an effective way. VSM can be used in both manufacturing and service industries and can be applied any type of production system (mass production, batch production, continuous production). Literature shows the applicability of lean philosophy for other sectors [9, 10]. Meanwhile, simulation has suggested by many researchers along with VSM in order to eliminate potential shortcomings of VSM in recent years [11, 12]. In this study, the current VSM is reviewed and a set of recommendations is visualised with the future VSM while comparing alternative scenarios with a simulation model developed with ARENA software.

2 Methodology

The methodology flow of this study is as given in Fig. 2. Initially, data and observations are obtained so as to determine the scope of this project. Thereafter, current states of value stream maps are created, and related improvements are analyzed. At last, comparative results are analyzed with the developed simulation model and future state value stream maps are created.

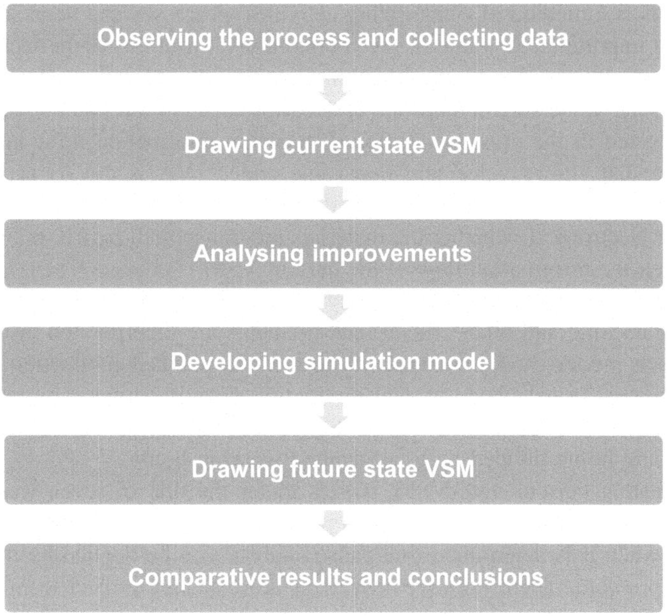

Fig. 2. Methodology flow of the study

The initial step of the methodology is visiting the plant and observing production processes. During that period, there are three critical elements for the proper introduction of the problem: determining the necessary information that is to be collected, selection of the products or product families to be considered, and lastly determination of the scope for an improvement project. In value stream mapping, material and information flows need to be mapped within production flow. Observing all steps involved in forming a packaged product from raw material to finished product is crucial to identify value-added and non-value added activities. Walking down the product's production map through the factory is required to gather information. Therefore, following information is collected;

- How information flow is occurred between customers, suppliers and internal departments
- How material flow is occurred in the system
- General packaging production line process flows
- General downtime details and production efficiency reports
- Planned downtimes such as trainings, breaks or meetings

Selecting production lines can differ in every project. Company can request a VSM study in desired lines or request from project teams to find byself. In such circumstances, focusing product family matrix or performing Pareto & ABC analysis can be used for decision making. Since drawing VSM of all products can be complicated, product family matrix, which is an analytical approach to show the products passing through similar processing steps and over common equipment, can be used. Pareto &

ABC analysis is a method of categorizing activities, processes and items according to their relative importance or value in given conditions. From this point deciding which of the value streams in this case production lines to map becomes increasingly clear.

Current state value stream maps are created for each chosen line by using the data that are collected in the previous section. It contains information flow, material flow and timeline that shows value and non-value added (VA & NVA) activities. The guidance about the time period of orders and forecasts, whether the flow of information is manual or electronical, whether the materials are transported by trucks or freighters, is given when the information flow of the current VSM is created. Then, VA & NVA activities are identified whether they add any value to the customer or not. Process time, cycle time, takt time and other crucial measurements are incorporated inside the data boxes of each processes. GLE considers how much planned production quantity is achieved and that is calculated as real production quantity divided by total planned production quantity. OE considers availability of production lines and that is calculated as real working hours divided by total planned working hours.

After creating current state VSM, wastes under the title of seven wastes in lean manufacturing are analyzed to detect in all processes. Bottleneck in the system is tried to find and when it is determined root cause analysis can be applied by investigating origin of a problem. Then, improvement ideas are listed by project team in order to present and discuss with company's engineers. After organizing workshop with project stakeholders, proposed improvements are implemented into ARENA simulation software to compare results and establish future state VSM.

To develop a simulation model and a future state VSM, approved current state VSM, which is created in previous section, is transferred into the simulation model in ARENA. During the simulation modelling part, it is tried to validate the system of the production lines. After validating processes, proposed improvements are applied to the model and a scenario analysis is implemented. In parallel with simulation modelling, future state VSM is built by using the hypothetical results of the lean tools that is discussed in the improvement section.

At the last step, all obtained results from scenario analysis are evaluated for future state VSM in the direction of eliminating waste and increasing productivity. After determining best scenario for future state VSM, it is compared with current state VSM. Then, contribution of improvements are demonstrated in case of applying in current condition.

3 Implementation and Results

This section begins with visiting facility and making observations about packaging production lines of PVM Turkey. During the visit, it is worth noting that VSM considers a journey of product from raw material form to finished product form through production plant. Hence, it is crucial to understand the information and material flow as a whole. And, another important issue is evaluating activities from the customer's viewpoint whether it adds value or not to achieve continuous process flow.

In Fig. 3, general process flow of bottling production line is represented. The sequence of activities can be changed for different line types. In each bottling line, four employees and one operator work under the coordination of foreman.

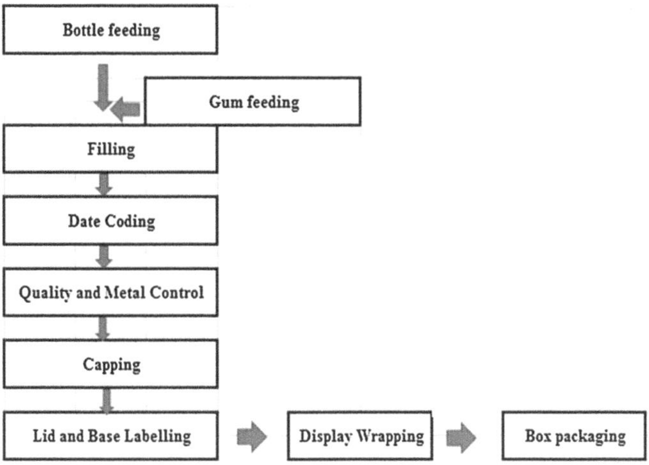

Fig. 3. General process flow of a bottle packaging production line

After collecting essential data, the scope of the project and production lines are determined together with project advisors. It is focused on active bottle lines during project period, which have lower GLE and OE rates and higher product output per time period. Pareto analysis is utilized while deciding production lines. As a result, Bottle 1 line is selected because of the fact that it covers about 20% of produced quantity when compared with all lines and has relatively low GLE and OE rates. Furthermore, since the company has requested a comprehensive lean study, Bottle 5 line is added to project scope. When all these taken into consideration, it is decided to work on Bottle 1 and 5 packaging production lines to identify wastes and increase efficiency rates under the scope of this project. Selected lines basically follow general process flow of a bottle packaging production line. But, bottle 5 has extra processes at the beginning of the flow due to material that is used. Bottle 5 production line uses sleeve for the body of bottles rather than label which is used in Bottle 1 production line. Bottle 5 uses label for only lid of bottles. On the other hand, bottle 1 uses labels for both body and lid of bottles. This difference can give insights to solve unnoticed problems while evaluating alternative scenarios in the next steps. Also, selected lines correspond to almost 40% of bottle production lines. It shows that any improvements achieved in this project, significantly contributed to packaging plant production performance in any case.

After deciding production lines, last year downtime detail report is attained to gain insights for each bottle lines. And these reports are converted to Pareto charts to categorize downtime definitions.

In Fig. 4, it is demonstrated that 60.2% of downtime consists of production organizational and technical categories for Bottle 1 line. When the reasons are analyzed, mostly lack of pack material and quality problem of bottle & lid causes in production organizational downtimes while lid machine and bottle feeding machine breakdown causes in technical downtimes.

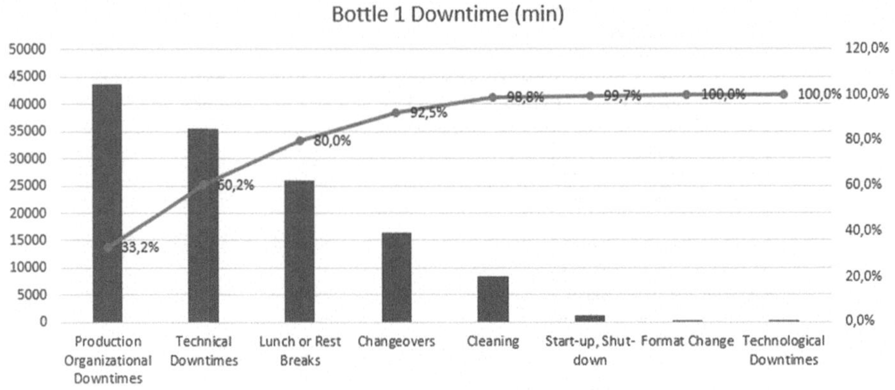

Fig. 4. Downtime report of Bottle 1 in Pareto chart

In Fig. 5, it is seen that Bottle 5 production line downtime, under production organizational category, has lower percentage compared with Bottle 1 line.

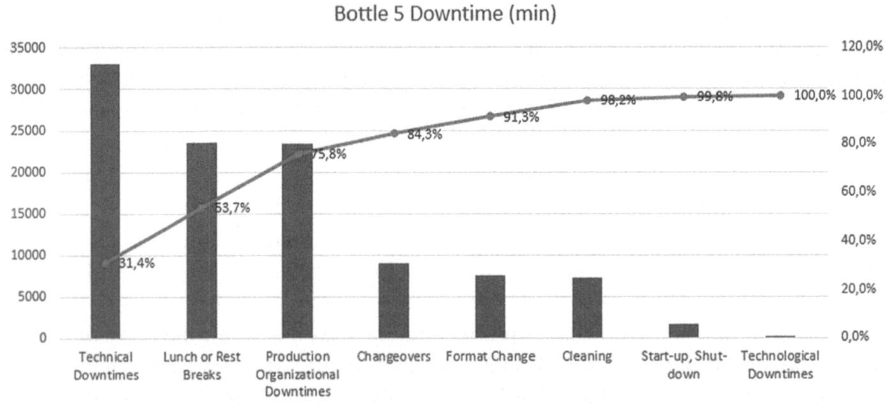

Fig. 5. Downtime report of Bottle 5 in Pareto chart

For all selected lines, a current value stream map (CVSM) is drawn by collecting all necessary data from the previous section. Information flow, process flow, and a timeline are formed respectively in value stream map. Communication between customer, suppliers and internal departments are denoted by wiggle arrows in information flow. Data boxes are filled by measuring process time, cycle time, changeover time, and lead time for each selected bottle line types. Finally, value added (VA) and non-value added (NVA) activities are determined to indicate in the timeline. Measurements are done by visiting the factory for a 3 weeks period, by taking samples from different shifts and days. This is really important to understand the system as a whole. By doing so, it provides sufficient time to analyze wastes and propose improvements. During measurements, 24 bottles are taken as lot size, and work in process inventory is ignored. Because bottle lines have a continuous flow and there is not much considerable amount of accumulated inventory. Kaizen bursts are put onto CVSM by numbering them to explain in the succeeding section. And lastly, takt time, utilization, GLE and OE rates are gathered for each measurement day and demonstrated by taking an average. CVSM of each selected lines are shown in Figs. 6 and 7.

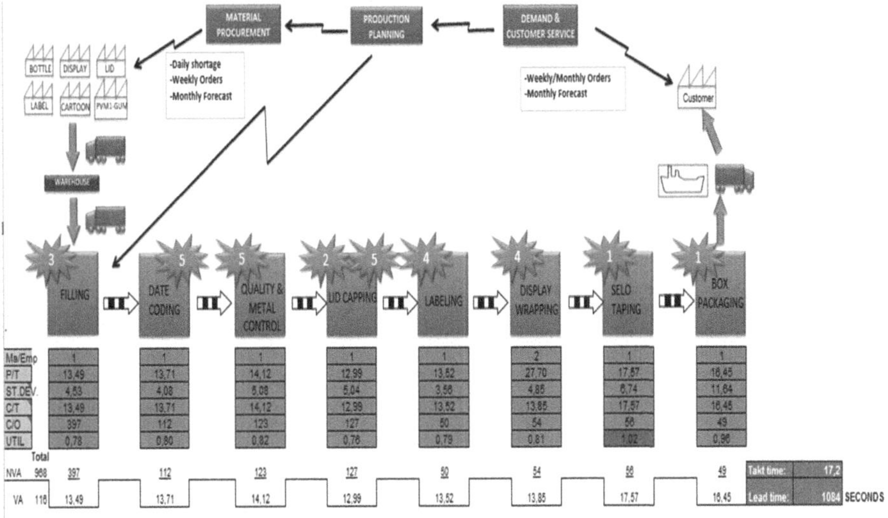

Fig. 6. Current state VSM of Bottle 1 line

After monitoring bottle production lines of PVM Turkey, it can be pointed out that most of the wastes are prevented applying lean practices very well. Therefore, only a few types of lean wastes are observed and summarized in the following paragraphs;

Waiting: It is identified in box packaging process. Process waits for the completion of a previous step work – cycle. The reason for this waste is caused by selo taping process in the observed lines. According to selo taping algorithm, it holds 18 bottles to release 6 bottles to the next process when 24 bottles are collected. That brings longer process time of box packaging due to waiting for the previous section. So, waiting reduces the flow of products to the customers and that resists working on output.

Motion: It is noticed in display wrapping process. Lid labels which are processed in prior flow can be mislabeled. For this reason, operators lose time to adjust labels of lid during wrapping display. They are tried to fix individual labels instead of separating them to process later. Also, it can be specified that the amount of mislabeling is non-negligible according to observations. That brings unnecessary movement of operators

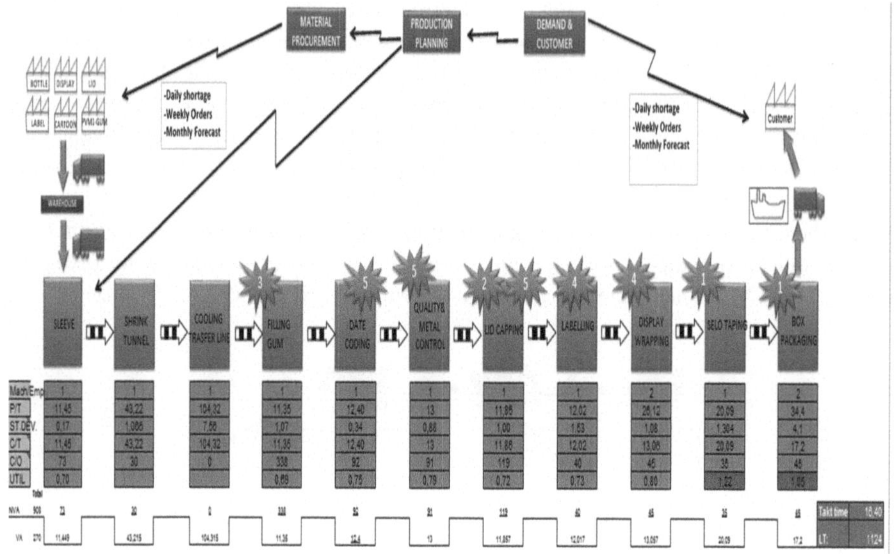

Fig. 7. Current state VSM of Bottle 5 line

to complete their tasks. Another waste of motion occurs during changeover due to operator movement between different controls of machines. Only one operator calibrates filling, date coding, quality & metal control and capping processes. That causes longer changeover time when operator stucks with in some of these processes.

Defects: Some of processes in bottle lines check the bottle units and separate them as rework or scrap that is triggered by sensors. According to observation, production lines have not a considerable defect rate to analyze as a waste.

After analyzing wastes, a workshop is organized with company's engineers to present improvements and discuss whether it is feasible or not. Proposed improvements are listed respectively;

3.1 Reducing Variable Standard Deviation Rates of Bottle 1 Processes

Standard deviation rates of Bottle 1 line's processes are highly variable, and longer when compared with Bottle 5 line's processes. Reducing these rates provides to reach more close to the continuous flow Table 1.

3.2 Reducing Cycle Time of Selo Taping Process in Each Bottle Lines

When it is looked in both Bottle 1 and 5 lines' CVSMs, the cycle time of selo taping process is longer than other processes. Also, it can be seen that selo taping has cycle time of 17.57 and 20.09 s in Bottle 1 and Bottle 5 respectively, that is greater than of Bottle 1 and Bottle 5 lines' takt time (17.2 s for Bottle 1 and 16.4 s for Bottle 5).

Table 1. Mean and standard deviation of each processes for bottle lines

Processes	Bottle 1		Bottle 5	
	Mean (P/T)	Std. Dev	Mean (P/T)	Std. Dev
Filling	13.49	4.53	11.35	1.07
Date Coding	13.71	4.08	12.4	0.34
Quality & Metal Control	14.12	5.08	13	0.88
Lid Capping	12.99	5.03	11.86	1.00
Labelling	13.52	3.56	12.02	1.52
Display Wrapping	27.7	4.85	26.12	1.09
Selo Taping	17.57	6.74	20.09	1.30
Box Packaging	16.45	11.64	34.4	4.10
Sleeve			11.45	0.17

3.3 Reducing Current Takt Time Values for Each Bottle Lines

While analyzing C/T vs T/T graphs for each line given in the figures at the preceeding section, it is observed that almost every process except selo taping works under the target takt time. Thus, reducing takt time can be applied for each bottle lines leaving some time to quality issues within consideration of historical data. In this manner, assigning new T/T rate while reducing the actual rate directly allows to cover more demand (Table 2).

Table 2. Average cycle time, takt time and new assigned takt time values for bottle lines

	Average C/T	Takt time	New T/T can be assigned
Bottle 1	14.46	17.2	15.5
Bottle 5	13.60	16.4	15.0

3.4 Reducing Change Over Times of Bottle 1 and 5

When CVSM are analyzed, it can be seen that Bottle 5 has the lowest change over time value. Therefore, Bottle 1 can be originated in terms of time as Bottle 5 by applying a kaizen study. For instance, replacing molds of filling machine cause to enlarge change over in each bottle lines. This requires a 5S study to arrange filling machine molds in a proper way before a changeover occurs. This can be achieved when it is successfully performed for Bottle 1 and 5. Also since operator has critical tasks that directly increases change over time. And lastly, it is observed that when a congestion or disruption is emerged in any process, workers may be late to take actions because of workload they have. Hence it is proposed hiring a common worker which can support production lines while switching between them. At the same time, this improvement leads also reducing change over time by supporting operator which is mentioned as a waste of motion before. Hence, kaizen burst 5 is put into related processes. But it is also required doing a cost analysis in terms of hiring a worker to amount of saving from one extra worker as well.

3.5 Reducing Lid and Bottle Feeding Machine Breakdown Times

According to downtime report that is obtained during measurement period, the majority of the downtime consists of lid and bottle feeding machine downtime in a technical category. Eliminating this situation possibly improve production and efficiency rate of the packaging lines. So, cost analysis can be done by comparing between an investment for new machine and amount of income loss caused by breakdown. Therefore, kaizen bursts, 2 and 3 are placed for lid and bottle feeding machines.

3.6 Switching Appropriate Label Specifications

As mentioned in the waste analysis, lid labels can be mislabeled and that is induced to lose time while wrapping displays. During the workshop, project advisors of PVM Turkey also confirmed that and they are planning to switch appropriate label specifications to eliminate these condition. Number 4 of kaizen bursts is placed in labelling and display wrapping.

Simulation of bottling lines is a challenging issue due to a large number of entities with a significant number of attributes. Because, it takes time to develop a comprehensive simulation model. Even if it is achieved to model, there can be encountered some certain issues. With the advent of Arena 15, it is easy to model and maintain high-speed processes with a custom template panel called packaging. That allows the user to have more time to increase productivity. Thus, a simulation model is developed for Bottle 1 production line. Cycle per minute process statistics, operator schedules, products, and production plans are assigned to the system. And the developed model is simulated to validate with a real bottle packaging production line. After validating the model, improvements are implemented into Arena model to carry out a scenario analysis. Afterwards, improvements acquired in scenario analysis are transferred to draw the FVSM.

In Fig. 8, the screenshot of the simulation model is shown which is developed for Bottle 1 production line. Cycle per minute statistics are recorded for each process that is obtained from CVSM. For instance, process time of filling which is 13.49 s for 24 bottles is converted to 107 bottles per minute.

Machine modules in Arena packaging custom template allow assigning change-overs according to developed plan and reliability information including expected uptime and time to repair. Then, operators are assigned to specific tasks and production plan is created. It is succeed to validate system according to current state when the developed model is run.

After the development of the model, a scenario analysis is conducted. Proposed improvements are integrated to Arena model in order to create reliable FVSM. With the proposed improvements, it is accomplished to increase GLE rate from 66.9% to 70.5% for Bottle 1. Possibly OE rate also is increased with the reduction of change over and downtime concurrently. It also reveals that crucial progress in efficiency rate can be

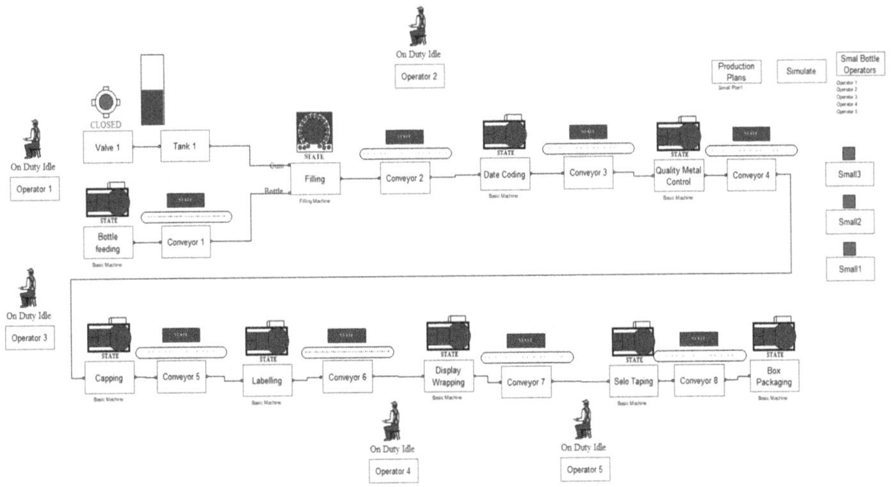

Fig. 8. Developed simulation model for Bottle 1 line

obtained with suggested improvements. It is suggested that, the same adjustments can be applied for other production lines in this way.

Following the simulation model, FVSM is built together with hypothetical and Arena results for each bottling lines. In Table 3, the possible improvements achieved with kaizen studies is shown. According to these improvements, FVSM of each line is also given below in Figs. 9 and 10.

Table 3. Improvements achieved with kaizen suggestions

Kaizen burst	Improvements achieved with kaizen
1	With the change in selo taping process algorithm, process time of selo taping and box packaging is reduced. Thus bottleneck is eliminated
2	With the improvement in capping machine, process time and downtime of the process is reduced
3	With the 5S study in filling process, changeover time is reduced significantly. And also bottle feeding machine is improved to prevent downtime
4	With the change in label specification, display wrapping process is reduced
5	With the new worker which is hired to support common lines by switching them, change over time of some processes are reduced

In Table 4, the comparison of lead time between CVSM and FVSM is shown including NVA and VA times. Lead time improvement rates are shown in the right side of the table. Also, it can be seen that significant improvement has been made in especially NVA time. Since work in process inventory is ignored, only change over times are included in the timeline for NVA.

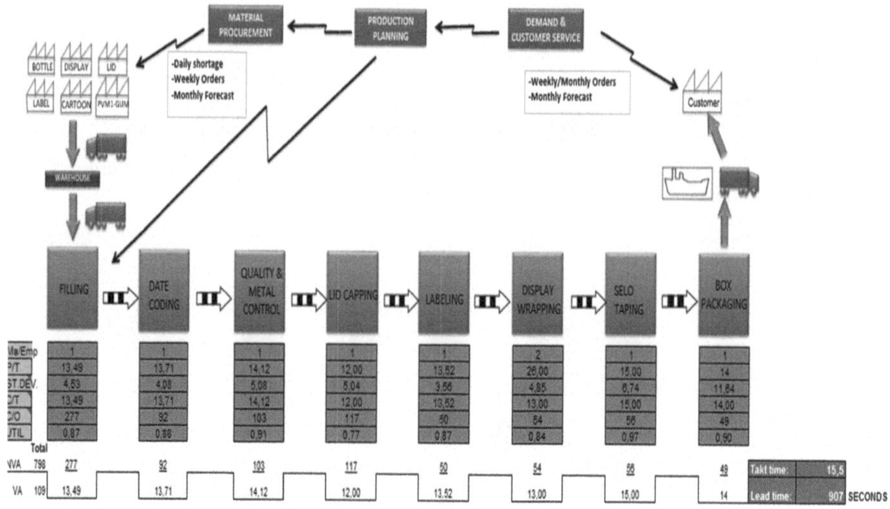

Fig. 9. Future state VSM of Bottle 1

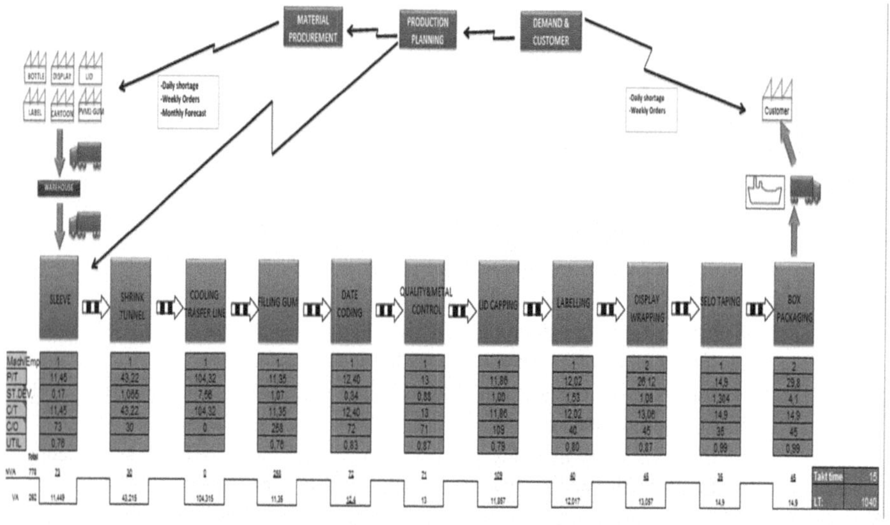

Fig. 10. Future state VSM of Bottle 5

Table 4. Comparative results for lead time improvements

Seconds	CVSM			FVSM			Improvement rate (%)
	NVA	VA	LT	NVA	VA	LT	LT improvement
Bottle 1	968	116	1084	798	109	907	16.33%
Bottle 5	908	270	1178	778	262	1040	11.71%

From the Table 5 below, it can be highlighted that remarkable improvement rates have been achieved which is around 15.5% in change over times for selected packaging lines. Thus, it directly enhances GLE and OE rates with the increase in production quantity and working time.

Table 5. Comparative results for change over time improvement

	CVSM C/O Time (sec)	FVSM C/O Time (sec)	Improvement rate (%)
Bottle 1	968	798	17.56%
Bottle 5	908	778	14.32%

Finally, Table 6 below indicates that increase in production rate can be ensured with the reduction of takt time. That also provides to cover more demand for PVM Turkey in the competitive environment.

Table 6. Comparative results for production rate improvement

	CVSM TT (sec)	FVSM TT (sec)	Increase in production rate (%)
Bottle 1	17.2	15.5	10.97%
Bottle 5	16.4	15.0	9.33%

4 Conclusion

In order to survive in the global contest and be able to move forward, enterprises should aim to increase their productivity by making better use of their production resources. Regarding this goal, companies are increasingly implementing lean manufacturing principles with the objective of reaching superior competitive advantage against other organizations. Thus, lean brings necessity of restructuring management philosophy that centers on excellence to achieve a continuous flow to the customer through the elimination of non-value adding activities and pursuing continuous improvement in the processes.

This study exemplifies the implementation of lean production methodology in a company operating in Turkey. It has been initiated to support the design of the bottling packaging lines to increase efficiency and fulfill the rapidly rising demands by eliminating potential wastes in the system. The study begins by observing processes and collecting data so as to determine the scope of the project. Two bottling production lines called as Bottle 1 and 5 are selected which have lower efficiency rates by utilizing Pareto analysis method. Measurements are conducted attentively by taking samples from different shifts and days. CVSMs are created for each chosen line by illustrating information flow, material flow, and timeline of VA and NVA processes. From seven wastes analysis, motion and waiting have been detected. Also, a bottleneck process is identified that causes to extend production lead time due to the algorithm that it has. Several improvements are proposed and evaluated with company's engineers in the workshop that is organized. Following the workshop, a simulation model is developed in Arena software to one of the selected lines. After validating the model, proposed improvements are implemented into the model to carry out a scenario analysis. Lastly, FVSMs are established with hypothetical and Arena results for each bottling lines. In conclusion, significant improvements are achieved especially in lead time, changeover time, efficiency rate and production rate. It is proved that serious increase in GLE rate from 66.9% to 70.5% can be achieved by applying a simulation study with suggested improvements. Change over times of each bottle packaging lines is reduced at least 14% by implementing a 5S study in filling process and hiring a new worker to assist common lines when required. Also, reduction in change over times and improvements acquired for some processes contribute directly to diminish production lead time of each bottling lines approximately 14%. It is also realized that almost every activity work under the takt time with large differences except the bottleneck. Since it is eliminated from the system, new takt time values for each bottling lines are assigned. Thus, the production rate is increased as 10% in average for chosen lines. That development allows covering more customer demand in the competitive environment. As mentioned previously, cost analysis can be conducted regarding proposed improvements to evaluate whether it is applicable or not.

Future research should further develop and confirm these initial findings by performing a detailed simulation study including scheduled stops, loss, and sensors. Furthermore, a capacity analysis can be investigated which is caused by defects to take advantage in the challenging market.

References

1. Wahab A, Mukthar M, Sulaiman R (2013) A conceptual model of lean manufacturing dimensions. In: The 4th international conference on electrical engineering and informatics (ICEEI 2013), vol 11, pp 1292–1298
2. Alukal G (2003) Create a lean, mean machine. Qual Prog 36(4):29–34
3. Krafcik JF (1988) Triumph of the lean production system. Sloan Manage Rev **30.** Massachusetts Institute of Technology (MIT)
4. Shah R, Ward PT (2003) Lean manufacturing: context, practice bundles, and performance. J Oper Manage 21:129–149
5. Rother M, Shook J (2008) Learning to see: value stream mapping to create value and eliminate muda. The Lean Enterprise Institute Inc, Cambridge
6. Nash MA, Poling SR (2008) Mapping the total value stream: a comprehensive guide for production and transactional processes. CRC Press, Productivity Press, New York
7. Lian YH, Van Landegham H (2007) Analysing the effects of lean manufacturing using a value stream mapping-based simulation generator. Int J Prod Res 45:3037–3058
8. Jones D, Roos D, Womack J (1990) The machine that changed the world. Rawson Associates, New York
9. Lacerda AP, Xambre AR, Alvelos HM (2015) Applying value stream mapping to eliminate waste: a case study of an original equipment manufacturer for the automotive industry. Int J Prod Res 54:1708–1720

10. Abele E, Metternich J, Meudt T (2016) Value stream mappinng 4.0: holistic examination of value stream and information logistics in production. ZWF Zeitschrift für wirtschaftlichen Fabrikbetrieb 111:319–323
11. Akburak D, Gültekin S, Kara B (2017) A value stream mapping study for ground operations processes in atlas global airlines. Undergraduate thesis, İstanbul Kültür University, İstanbul
12. Behnam D, Ayough A, Mirghaderi SH (2017) Value stream mapping approach and analytical network process to identify and prioritize production system's Mudas (case study: natural fibre clothing manufacturing company). J Text Inst 109:64–72

Miscellaneous Topics

A Model Suggestion for Entrepreneurial and Innovative University-Industry Cooperation in Industry 4.0 Context in Turkey

Özdal Koyuncuoğlu[1(✉)] and Mahmut Tekin[2]

[1] Faculty of Applied Sciences, Necmettin Erbakan University, Konya, Turkey
okoyuncuoglu@konya.edu.tr

[2] Faculty of Economic and Administrative Sciences, Selçuk University, Konya, Turkey
mahtekin@selcuk.edu.tr

Abstract. The main philosophy of Industry 4.0 is based on intelligent systems grounded on cyber-physical systems. The more intensive use of automation in production and related functions will lead to intelligent systems and this causes less employment and changes the demographics of the workforce. Digitalization seems to have transformed our everyday lives, as well as our business and educational lives, into a technology-dependent form. Universities, which are the most important building blocks or units of social development, should take place as a driving force that must make an impact in this transformation. As a requirement of competition, the state authorities give importance to the role of the universities in this area as well as offering considerable incentives and support in the field of university entrepreneurship and they have serious expectations related to this issue. It is observed that universities have recently introduced entrepreneurial and innovative university features in their marketing activities. However, what does entrepreneurial and innovative university mean? How can an entrepreneurial and innovative university be measured and evaluated? While looking for an answer to these questions, it is necessary to evaluate the matter in the context of Entrepreneurial and Innovative University-Industry cooperation. For this purpose, based on the data obtained as a result of the interviews conducted with 19 specialist working at universities in 2017 Entrepreneurial and Innovative University Index in Turkey; and considering the impact of digital transformation and the fourth industrial revolution on social and educational processes, an industry-integrated assessment model of Entrepreneurial and Innovative Universities was developed. This model was formed within the categories of Entrepreneurship and Innovation Culture, Potential Activation, Potential Development, Projects and Commercialization and Supervision in the context of Entrepreneurial and Innovative University-Industry cooperation. As a result of the data collection, coding and analysis, the suggestions were given by defining and evaluating the 18 characteristics that the universities with entrepreneurship and innovation cultures should have in the environmental conditions, input and output dimensions.

Keywords: Industry 4.0 · University-Industry cooperation
Evaluation model proposal

N. M. Durakbasa and M. G. Gencyilmaz (Eds.): ISPR 2018, *Proceedings of the International Symposium for Production Research 2018*, pp. 565–580, 2019.
https://doi.org/10.1007/978-3-319-92267-6_47

1 The Importance of Entrepreneurial and Innovative University-Industry Cooperation in Industry 4.0 Context

Entrepreneurial and innovative university-industry has gained importance in the context of Industry 4.0 in the globalizing world. This development has made it necessary for the university system to undergo changes and restructuring. Thus, at the meeting of the Council of Higher Education in the date of 01.03.2018, "From Higher Education to Industry: Qualified Human Power Workshop" was held so as to discuss and reconstruct the current situtation of higher education in Turkey. The chairman of YOK, Prof. Dr. M.A. Yekta Saraç who made a speech at the opening of the workshop underlined the importance of the grauduates from 4-year undergraduate programs and 2-year vocational schools to increase Turkey's economic development and the international competitivenes; and for qualified manpower needed by the private sector and the industry. This situation, together with the digital economy created by "Industry 4.0" revolution, suggests the importance of evaluating the existing university structure in our country according to the wave of "technological unemployment" all over the world. It is also important to determine the future professions in higher education and to bring the students in the skills for these professions [1]. Pointing out that a new strategy should be identified for this purpose, and products and services that will respond to global needs should be produced with this strategy, National Deputy Minister of Education Orhan Erdem stated that Turkey's young population should be in action for innovation and creativity. Ankara Chamber of Industry (ACI) President Nurettin Özdebir pointed out that there is a supply-demand dispute in the labor market and stated that university or vocational college graduates can not find jobs. He also expressed that industrialists complain about not being able to find qualifed employees in accordance with their needs. The main reason for this is that the needs of the industry are rapidly changing in today's global competitive environment and the educational programs of schools fail to follow the developments as it should be. In the workshop, it was emphasized that the sectors with high value added which produce intensive products by means of knowledge and technology need creative individuals who are equipped with basic sciences and mathematics, developed hand skills, and are able to adapt to changing conditions [2].

In the era of digitization, it can be said that providing the qualifications of the ages to the graduates and providing the qualified labor to the sectors is in the authority and responsibility of the universities. With the cooperation of international organizations, governments and industry, the construction of universities that can manage the digital age by developing both global and local strategies is now being debated in the context of higher education [3]. In many countries, it is expected that universities will effectively contribute to regional economic development in their region, as well as the educational and research roles. In this context, information transfer through the education and human capital development of the universities, technology transfer of knowledge produced by the researches carried out within the university, and community development and cultural activities are discussed. In addition, the pursuit of rapid developments in the field of information and communication technologies is a new challenge for universities. So much so that some authors term entrepreneurial and innovative universities as "Entrepreneur-Friendly Universities" and point to the paradigm change there [4, 5]. There must be a search for an answer to the question "which university provides the best conditions for entrepreneurial candidates" rather than the question "which university has most output".

For this reason, universities are a regional actor that provides not only the commercialization of knowledge but also solutions to the problems of society. Accordingly, while considering the entrepreneurial aspects of the universities, the need for the development of an evaluation tool that is appropriate for day-to-day conditions and that makes accurate measurements in the direction of the country's interests and policies has been the starting point of this work. To express with Weberyan terminology, modern practices "appropriate to the spirit of time" (zeitgeist) need to be developed.

This study is intended to create a model that can form the basis for designing a tool that measures the university-industry cooperation and entrepreneurship and innovation at universities in the context of industry 4.0 in Turkey. Face-to-face interviews were held with 19 experts from the universities in four regions of Turkey. The fact that no research has been conducted on the subject based on the data and there is no study for an applicable model in our country has made our study extremely important.

2 Entrepreneurial and Innovative Unıversities in Industry 4.0 Context

One of the most frequent issues in the digitalization of higher education is the transformation of universities from university 4.0 through industry 4.0. It can be said that the research results of the institutues may have impacts in Turkey's success in the international competition and in becoming an industrial country through industry 4.0. In this context, it would not be wrong to say that it is in the authority and responsibility of universities in the digital era to provide qualified labor force to the sectors by bringing the competence of the time for the research centers and the graduates.

According to Gürsoy, industry 4.0 has multi-dimensional effects for socio-technical structure, and at the same time it will be able to change the existing business models and dynamics. In business education, giving the priority to the training of the graduates who can develop innovative business models will give a positive contribution to the transformaiton process. Besides, under industry 4.0, it is evaluated that the development of business graduates' sustainability, decision support systems, managing and adaptation to the change; their training in risk management areas, and developing their innovation and entrepreneurship skills will be important [6]. Dewar (2017) states that university 4.0 represents a new university structure. The characteristic of this structure is that they are the institutions that provide continuous learning opportunities in different channels through the internet, traditional methods or blended methods. These higher education institutions include components such as short-term training and certification programs for the acquisition of a variety of professional qualifications, support for the development of career management and skills for learners, and seamless links and support programs between industry and researchers and learners [7]. According to Lapteva and Efimov (2016), in the era of industry 4.0, it is necessary for the universities to increase the scientific works to transform knowledge into reality, to support the opening of high technology companies in their units, to establish networks and to coordinate between different issues and to be a pioneer for new applications [8].

What should we do today for the adaptation of labor force which is affected by the industry 4.0 transformation process most and to train it according to the needs of new competences? This question must be assessed within the context of high education. The wide range of work to be done in this framework, including not only engineering but also social sciences, medicine and even law, will accelerate the integration process of Industry 4.0 in universities. It is expected that the objectives of academic programs and the change of learning outputs, developed based on competence, will be reflected naturally in educational programs and techniques [6].

The work named "Successful universities towards the improvement of regional competitiveness: Fourth Generation Universities" by Lukovics and Zuti is accepted as another main study in this field [9]. In this work, Lukovics and Zuti construct the functions of the universities on two main pillars. The first of these is the pillar of "Training and Research". Under this pillar are (1) student and research mobility, (2) a rich education program portfolio, (3) innovation, and (4) performance and parameters to become a global university. The second pillar is called "the third mission". The components of the second pillar are (1) information and technology transfer, (2) national and international links, (3) a flexible structure and system that facilitates compliance, and (4) services that support the local economy. In their study, Lukovics and Zuti added the "Fourth Generation Universities" approach to Wissema's comparison of three generations of university and they compared them in the table [10].

In the next-generation university approach to be forward, as shown in Table 1, the Third Generation approach is positioned to a next level in general terms. The region has considerable expectations and needs, and some of these require the cooperation of local governments, central administrations and universities. These collaborations will further enhance the interaction of the universities with the region and make it a strong regional and international actor.

Table 1. The comparison of general lines of generation universities

	First Generation	Second Generation	Third Generation	Fourth Generation
Objective	Education	Training and Research	Training, Research And The Use Of Information	Training, Research, the Use of Information and Proactive Economic Development
Role	Protect the Truth	Knowing the Nature	Create Added Value	Determine Strategic Objectives and Being the Engine of Local Economy
Output	Professionals	Professionals and Scientists	Professionals, Scientists and Entrepreneurs	Professionals, Scientists, Entrepreneurs and Competitive Local Management
Language	Latin	National Language	English	Multilingual (National Language and English)
Administration	Rector	Part-time Scientists	Professional Administration	Professional Administration and Local Experts

Source: Wissema, 2014; Lukovics and Zuti, 2009; Lukovics and Zuti, 2015; Saklı, 2017.

The role attributed to our universities today by the state and society is that the information emerging as a result of research as well as education has to interact with the sector, especially by commercializing in the field of technology, and that they are an important center for the solution of the social problems of the region. After investigating the studies in the literature related to the effects of industry 4.0 on high education, the following explanatory information is given on how the evaluation model was developed about digital age transition of universities.

3 Methodology of the Research

3.1 The Aim of the Research and the Problem Status

The purpose of this study is to put forward a model that could form the basis of university-industry collaboration measuring the entrepreneurship and innovation feature of the univesities in Turkey. The process approach was taken into account when forming a frame for research questions. This process approach is presented below.

As shown in Fig. 1, the process approach consists of environment, input, output and feedback dimensions. It is also seen in the figure that when the entrepreneurial and innovative university is regarded as a system, there are limits of this system. These limits can be seen as a fine line separating the inner and outer circumference of the system. Although the process approach is seen as a system approach according to Tecim, it is possible to investigate "every system that has to operate in a certain environment as open and closed systems in two parts" [12]. In addition, he states that "open systems are the systems where energy, information and materials are exchanged between the system and the external environment in which the system operates" but "closed systems are the systems where there is no such relationship between the system and its surroundings". According to these definitions, it can be said that the concepts of openness or closeness are blurred and each system has to be open at a certain position. For this reason, it is only possible to think of the universe as a closed system properly. Tecim exemplifies this distinction as follows: "While social and biological systems are open systems, mechanical systems can be either closed systems or open systems depending on their relation to the external environment". As shown by the arrow marks in Fig. 1, entrepreneurial and innovative universities are able to use materials, information, energy, etc. from open systems, external environments or other systems; and it processes them, gives them back as outputs to their environment or other systems in various forms of goods and services. In Fig. 1, the system receives input from outside or other systems and transforms it by processing in certain ways with different input components of itself. The inputs subjected to conversions are sent as output to the environment or other systems in the form of goods or services. Although it is mentioned in the open system, it is possible to refer to the limits of the system itself and therefore the inner environment of the system, as seen in Fig. 1. In terms of internal dimension of environment, physical, cultural and similar conditions of the university were dealt with in the scope of the research.

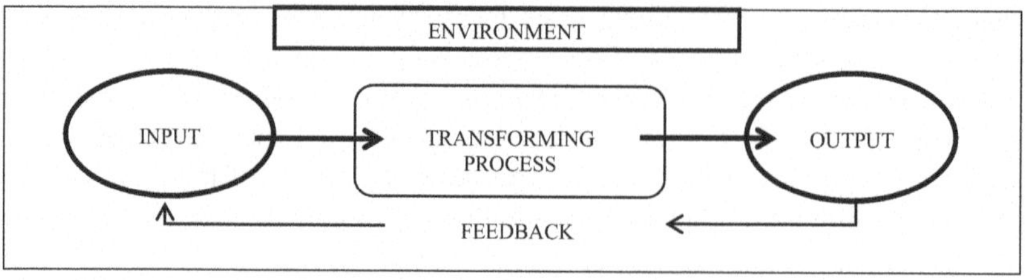

Fig. 1. The process of entrepreneurial and innovative university Source: created by means of Tecim, 2004: 75-100

The input activities of the system at entrepreneurial and innovative university were regarded as the activities with input quality such as consultancy services needed for the conversion of the ideas, training and support activities. The output activities of the system include students who have been trained by the completion of the conversion process, ideas being converted into projects or commercialized products, and so on, and the activities with output quality.

In this study, based on the data after the interviews, it was aimed to search for an answer to the following research questions based on the process approach presented in Fig. 1 so as to establish a theory that will enable to identify and evaluate the components of internal environmental conditions, input and output activities of universities.

1. How can measure how and to what extent entrepreneurial and innovative universities internalize entrepreneurial cultures and how and to what extent they manage to establish them in the system? (Environmental conditions)
2. How can we measure how and to what extent entrepreneurial and innovative universities contact their academists and students and how and to what extent they encourage and support their ideas? (Input)
3. How can we measure how and to what extent the ideas are transformed to projects and enterprises at entrepreneurial and innovative universities? (Output)
4. How can entrepreneurial and innovative universities measure how and to what extent their activities are controlled? (Feedback)

Research questions also show the research framework. After identifying the environmental conditions of the university and how to assess the activities within the scope of the research, answers were sought about how to measure these areas in the second stage. While it is possible to objectively measure the efforts, activities and contributions of the institutions in this area, the determined criteria can still lead to human judgment, and therefore, to faulty measurement.

It can be said that the core competencies of the institutions of the universities, the interest and adaptation to the university entrepreneurship and even their position always show a difference. These differences among universities need to be monitored, measured and evaluated on the basis of objective criteria. This sentence by Işığıçok explains this very well: "if you don't measure one, you can't manage and improve" [12].

Işığıçok defines measurement as "describing an entity or an event that is desired to be measured originally by means of a suitable scale" and states that "evaluation process is a decision process and it difers from measurement". In the evaluation process, the measurement results are taken, compared to the criterion and it is checked whether the results of the measurement meet the condition in the criterion. Therefore, it can be said that the measurement will be meaningful when compared with an ideal value or the reference of success which other performance providers show. This also applies to measurement and evaluation of environmental conditions, input and output activities of entrepreneurial and innovative university system.

In order for the measurement developing the entrepreneurship profiles of universities to be successful, it is first necessary to analyze the environmental conditions, input and output activities and to define these areas. In a sense, the definition of environmental conditions and activities of universities can be regarded as standards or objectives of universities in the field of university entrepreneurship. In this way, the comparison of entrepreneurial profiles of the measured universities with each other will help to determine which university provides the best conditions for students and academists in terms of entrepreneurship and innovation.

Answers were sought discussing with the experts about what can be measured in what way by a measuring instrument that will reveal the entrepreneurial profiles of the universities, and about the problem situation of the study. As a result of these discussions, it was aimed to reveal a theoretical structure that could provide a basis for the design of a measurement and evaluation tool to evaluate the direction of the entrepreneurship of the universities.

3.2 Research Design

This study is intended to develop a model which can be applied in Turkey and which reveals the consept of Entrepreneurial and Innovative Universities in all parts and draws a general framework of domestic ecosystem of Entrepreneurial and Innovative Universities, and it is structured with grounded theory approach from qualitative research designs. It was therefore decided by taking expert opinions on the subject that the research model should be qualitative. The reason for the choice of Grounded Theory research methodology is to deeply examine the internal environment, inputs and outputs of entrepreneurial and innovative universities in our country and to develop a theoretical model explaining the Entrepreneurial and Innovative University based on the transformation process.

The purpose of creating a grounded theory research is to go beyond the description, to make "a unified theoretical interpretation" [13] related to a process or an action and to develop or discover a theory [14]. Strauss and Corbin state that the absence of "a ready-to-use source", and basing and being uncovered within the data provided by the participants are the main idea of developing a theory [13]. Thus, grounded theory creation is a qualitative research design in which the researcher presents an explanatory theory on the basis of data in the context of a large number of participants' views in relation to a process, action or relationship between them [14]. This qualitative pattern is a research strategy that is widely used in social sciences that are quite different and contain unique elements. A grounded theory explains "why" and "how" in a study. The research focuses to develop an inductive theory to explain the phenomenon [15].

3.3 Determination of Working Group

Academic and administrative staffs who are experienced in university entrepreneurship and who have taken place in the first 50 universities, which entered the Entrepreneurial and Innovative University Index which was published by TUBITAK in 2017, constitute the universe of this work.

Criterion sampling technique - a purposeful sampling technique- was used to determine the participants. The basic understanding of the criteria sampling method is that the pre-determined criteria reflect the purpose of the study directly and guide the identification of information-rich situations [16]. Therefore, it was decided that the person to be interviewed for the purpose of the must be in charged in the first 50 universities in the Entrepreneurial and Innovative University Index published in 2017. Participants must be experienced in entrepreneurial entrepreneurship. This means that the participants must have a publishing about University Entrepreneurship, give a lecture for entrepreneurship, organize an activity about entrepreneurship, have a patent, undertake mentoring or counseling duties or they must have features similar to these. Thirdly, to make sure that the participants understand the concept of institutional culture, it was concluded that the participants must work at the same university for at least five years.

Table 2 displays the distribution of people to be interviewed according to universities. The names of the universities in the table are listed according to the ranking in the index order published in 2017. Two persons from an academic or an administrative position were interviewed at Selçuk University, Pamukkale University, Mersin University, Akdeniz University, Bahçeşehir University and Süleyman Demirel University; and one person was interviewed at Gebze Technical University, Bilkent University, Ege University, Gazi University, Kocaeli University and Marmara University.

Table 2. The number of the participants of the universities to be interviewed

University name	University Type	Number of the participants
Gebze Technical University	Public	2
İhsan Doğramacı Bilkent University	Foundation	1
Ege University	Public	2
Selçuk University	Public	2
Gazi University	Public	1
Kocaeli University	Public	1
Yaşar University	Foundation	1
Mersin University	Public	2
Akdeniz University	Public	2
Bahçeşehir University	Foundation	2
Süleyman Demirel University	Public	2
Marmara University	Public	1
Total number of participants		**19**

3.4 Data Collection

Qualitative research methods were used in face to face interviews in the process of gathering qualitative data. In other words, the researcher himself collected the relevant information from the primary source. Semi-structured open-ended questions were asked to the participants through an interview form with the characteristics of a guideline. All interviews were conducted face-to-face by the researcher. Interviews were performed between the 13th of November 2017 and the 23rd of January 2018 in the participants' working rooms or in a public environment (café, cafeteria, etc.) which were deemed appropriate by the participant.

As shown in Table 3, the interviews are ranked according to the date on which they were conducted. The most important point in the data collection process in grounded theory is that data analysis is done simultaneously in this process. As theoretical sampling is done, data collection process should be continued by going to new sample units if necessary, in accordance with the results of the analysis [17]. Interviews were conducted at 2 group time intervals. 13 interviews were conducted in the first group and 6 interviews were conducted in the second group. The areas highlighted in Table 3 refer to interviews conducted at the same time interval.

Table 3. Detailed information between data collection process

Date	University name	University type	Region	Length of interview (minute)	Recording time (minute)	Recording time (second)
13.11.2017	Selçuk University	Public	Central Anatolia	60	32	27
13.11.2017	Selçuk University	Public	Central Anatolia	60	28	2
15.11.2017	Kocaeli University	Public	Mediterranean	120	33	4
17.11.2017	Mersin University	Public	Mediterranean	60	32	4
17.11.2017	Mersin University	Public	Mediterranean	60	No recording	
02.12.2017	Akdeniz University	Public	Mediterranean	60	34	36
02.12.2017	Akdeniz University	Public	Mediterranean	60	26	50
04.12.2017	Süleyman Demirel University	Public	Mediterranean	60	31	3
04.12.2017	Süleyman Demirel University	Public	Mediterranean	60	12	54
08.12.2017	Gazi University	Devlet	Central Anatolia	60	No recording	
15.12.2017	Bahçeşehir University	Foundation	Marmara	60	37	25
15.12.2017	Bahçeşehir University	Foundation	Marmara	60	38	19
15.12.2017	Marmara University	Foundation	Marmara	210	No recording	
16.01.2018	Bilkent University	Foundation	Central Anatolia	60	38	23
19.01.2018	Gebze Technical University	Public	Marmara	90	35	11
19.01.2018	Gebze Technical University	Public	Marmara	90	37	17
23.01.2018	Ege University	Public	Aegean	100	35	55
23.01.2018	Ege University	Public	Aegean	45	36	15
23.01.2018	Yaşar University	Foundation	Aegean	75	32	28
				1450	**516**	**373**
				24 hours 10 minutes	***8 hours 42 minutes 13 second***	

3.5 Data Analysis

Based on the characteristics of the research and collected data, the data analysis plan was developed and applied according to the content analysis approach. Before moving on to the content analysis phase, it was thought that it would be useful to explain some concepts used in content analysis.

Yıldırım and Şimşek show basically a four-step path about how to process qualitative research data [18]: (1) coding the data, (2) finding out the themes, (3) editing of codes and themes, (4) identification and interpretation of findings. In the literature, data analysis phases are mentioned by various denominations related to the coding of data in data analysis of grounded theory studies [16, 19]: (a) Open Coding (b) Free coding (c) Axial coding, (d) Selective coding and (e) Key Point Coding. In this study, five step content analysis method was applied step by step.

The interviews were recorded with the permission of participants. The 3 interviews out of 19 were not recorded due to various reasons and the notes were taken instead. Interviews ranged from 45 min to 210 min and total recorded interview duration was 522 min and 13 s. The recorded interviews were deciphered and converted into electronic text and transferred to text-based Voicedocs data decipher software for data analysis. When the interviews were in progress, deciphering of completed interviews were performed simultaneously. The decipher process was done separately for each participant and was organized according to the word file. For every interview organized in the word file, the data were tried to be separated from the raw data in accordance with the research and research questions. In the second step, the data were combined with the data that had the same meaning in all sections and they were sorted out. Later on, the steps were followed such as: (a) determining the relations between the concepts arising during the processing of the data with the previously prepared themes, in the direction of research questions and process approach (transformation); (b) organizing and defnining the data by theme, category and subcategories. The relationship between the determined themes and the subcategories was explained and interpreted and the results for the purpose of the research were revealed.

3.6 Premises and Restrictions

The data collection tool used in this research (interview form) was assumed to have the characteristics that would determine the environmental conditions in universities, input and output activities. It was also assumed that the answers of university staff participating in the research reflect sincere and genuine opinions.

The working group of the research was limited to the lecturers who were at the first 50 universities in Entrepreneur and Innovative University Index published by TÜBİTAK in 2017, worked at least for 5 years at their current universities and had studies on entrepreneurship felds; and/or to administrative personel in managerial positions in technology transfer offices or entrepreneurship centers.

This research is limited to grounded theory (theory building) out of qualitative research techniques. This study is subject to theoretical sensitivity, which is considered to be an important element in qualitative research, and it is limited to the situations and thoughts of interviewees.

3.7 Validity and Reliability

One of the most important criteria for achieving the purpose of a research is the effectiveness of the data collection tool to be used in the research. The investigator in the measurement process should decide what to measure and which measurement tool will be appropriate. The data collection tool must be in accordance with research questions and this accordance is ensured by validity and reliability.

Validity in qualitative research means that the researcher observes the event as impartial as possible and presents it as is [18]. The fact that the measurement tool decided to be used corresponds to the characteristics to be measured and the obtained data can meet the quality of characteristics to be measured exactly; and the fact that the data are relevant all represent the validity. In other words, the *degree* of the data to give correct information for measurement tool represents effectiveness and usefulness. Analyses applied in the research and the value of the findings is "subject to validity" [20].

The validity and reliability of a qualitative study depends on researcher ethics considerably [16]. In the study, it was taken care of the principles of "intellectual sensitivity, professional integrity and method-scientific competence" which were associated to believability by Patton [21]. Moreover, ethical standards were considered such as not giving the names of the participants and not associating the names of the universities to the expressions.

4 Findings of the Research

4.1 The Findings Obtained as a Result of Open Coding

In the open coding phase, 1,867 reference expressions were reached. The findings of the open coding phase were not included because of the abundance of determined reference numbers and since the expressions did not mean anything singly according to the logic of open coding. For this reason, the presentation of the findings was started with the findings obtained as a result of free coding which was the second stage of data analysis.

4.2 The Findings Obtained as a Result of Free Coding

During the open coding phase, 19 texts were deciphered, the data, important ideas and concepts were marked by the codes consisting of one or several words. Immediately after this stage, free codes were tried to be produced associating the codes in the previous stage with each other, and combining similar expressions and reference expressions refering to the same or near points. The number of free codes obtained at this stage in which the codes were associated with the texts is 86. The code matrix scanner function was used in this process.

4.3 The Results Obtained as a Result of Axis Coding

In axis coding, the third stage of data analysis, codes (categories and properties) were tried to be associated with the others through deductive and inductive thinking [17].

In the axis coding phase, the categorizations were associated with the subcategories by the constant comparison method and the relationship between them was tested. With this method, similarity codes were sometimes combined and sometimes modified so that the differences between the codes were relatively more refined, error-free and more crystallized than the codes in the previous stage. Here, it was possible to link the subcategories to the categories by coding the strategic dimension, environmental conditions and activities. At this stage, the relation scanner function of the codes was used to find the relation of the codes to each other.

4.4 The Results Obtained as a Result of Selective Coding

Selective coding phase is the process of choosing a category as a core category and grouping all other categories into the core category by associating them with this category. Baş and Akturan indicate that the following questions should be answered when creating the core category: "What is the main analytical idea presented in this research? If I should conceptualize my work with a few sentences, how would I express this? What can be the changes between the categories explained by?" It is also stated that the core categories are derived from concepts but should be at a level where the phenomenon is expressed more theoretically [17].

The core category, which will form the basis of the temporay categories of the previous stage and the theoretical structure to be achieved, has been thematized to represent the central (main) case of the study. In doing so, the literature on entrepreneurial and innovative university subjects was also examined and attention was paid to the fact that the core categories developed were partly compatible with the earlier developed theories of research. However, it was not intended to regard the previously mentioned theories as a hypothesis and to test these hypotheses. The main reason for referring to the literature at this stage is to ensure the internal and external consistency of the theory that is tried to be revealed by a single study with a limited working group; and to make the road map to be followed in order to explain the structure of the theory. The theme-category model in the study so far is listed below.

As shown in Table 4; 26 categories generated in the axis coding phase, 5 core categories in this phase were associated with the theme and they were collected under these themes. These themes were named as entrepreneurship and innovation culture, potential activation, potential development, projects and commercialization and audit.

4.5 The Findings Obtained as a Result of Key Point Coding

During the keypoint coding phase, the categories and the themes developed in the selective coding phase were reviewed in the direction of the reference expressions at the free coding stage; and the final categories and themes were developed which would give the final shape of the theoretical structure to be achieved. The number of core category, determined as 5 in the previous step of the data analysis, was combined because both potential activation and potential development themes were input activities. Ultimately, four core categories were created: (1) entrepreneurship orientation, (2) potential activation and development, (3) projects and commercialization, and (4) audit (Table 5).

Table 4. The findings as a result of selective coding

Theme	Category
Entrepreneurship and Innovation Culture	Entrepreneurial Thought
	Structuring
	Entrepreneurship Culture
	Development
Potential Activation	Awareness Activities
	Education Activities
Potential Development	Incentive Activities
	Supporting Activities
Projects and Commercialization	Projects
	Dissertation
	Industrial and Intellectual Property Rights
	Commercialization
	Collaboration and Interaction
	Awards
Audit and Feedback	Information System
	Follow up
	Evaluation
	Feedback

Table 5. The findings as a result of keypoint coding

Research Framework	Research Questions	The Themes	
Environment	Environmental conditions inside University	1	Entrepreneurship Orientation
Input	Reaching and supporting the target group	2	Potential activation and development
Output	Transforming the ideas	3	Projects and commercialization
Feedback	Audit of the activities	4	Audit and Feedback

5 Conclusions

The question "where are the universities, which provide human resources to various sectors with indıstry 4.0, in the world of digitalization and in this fast change?" has gained importance. In response to this question, it is firstly expected from the universities to direct this change, to create changes on national and international scale, and thus to lead the digital age. This study was performed to create a model related to entrepreneurs and innovative university-industry collaboration in the context of industry 4.0 in Turkey. The evaluation model presented within the scope of the research can also be seen as the basis for conceptual mapping. In the direction of the new needs and anticipations of the Industrial 4.0 industrial revolution, the targeted development model was considered to evaluate the input and output dimensions of technology-focused, community-interactive, entrepreneurial and innovative universities equally.

Based on the data obtained from 19 experts working at the universities in 2017 Entrepreneur and Innovative University Index in Turkey, a qualitative and quantitative model was created related to environmental conditions, input and output activities of entrepreneurial and innovative universities. As a result of data collection, coding and analysis, the 18 characteristics which the universities with an entrepreneurship and innovation culture must have were defined in terms of environmental conditions, input and output dimensions. These characteristics were gathered in four categories to develop a model for evaluating the entrepreneurship and innovation culture and the work efforts made in this area.

The developed theoretical model consists of four themes and 18 categories. There are 4 categories under *Entrepreneurship Orientation Theme*. These are; (1) Strategic thinking, (2) Structuring, (3) Entrepreneurship culture and (4) Development. There are 4 categories under *Potential Activation and Development Theme*. These are; (1) Awareness activities, (2) Educational activities, (3) Incentive activities and (4) support activities. There are 6 categories under the theme of *Projects and Commercialization*. These are; (1) Projects, (2) Dissertations, (3) Industrial and Intellectual Property, (4) Commercialization, (5) Cooperation and Interaction, and (6) Awards. There are 4 categories under *Audit and Feedback theme*. These are; (1) Information system, (2) Follow-up, (3) Evaluation and (4) Feedback.

Some writers call entrepreneurial and innovative universities "Entrepreneur-Friendly Universities" and point to the change of paradigm here [4, 5]. It indicates that there must be a search for an answer to the question "which university provides the best conditions for entrepreneurial candidates" rather than the question "which university has most output". The model developed as a result of the study corresponds to this paradigm change.

With the Entrepreneur and Innovative University Index Model, it was aimed to "determine the university giving the most economic contribution" and to increase the competition between the universities in this way. The aim of the model developed within the study was to identify community-oriented universities that provide the best conditions for students in the field of entrepreneurship, to increase the competitiveness of universities at their own discretion, and to internalize entrepreneurial thinking at universities. In assessment and evaluation, it will be aked to determine how and to what extent universities provide entrepreneurial thought. The other dimension to be determined is that how and to what extent the universities reach academicians and students; how and to what extent they encourage and support academicians and students in this regrad. It is also worthwhile to make an evaluation about qualitative and quantitative factors and about how the ideas of university are transformed into initiatives. In order to ensure the sustainability of the activities, it is aimed to learn how and to what extent the university controls the studies.

In order to keep up with the process of the fourth industrial revolution, engineering departments of universities are being developed in line with Industry 4.0 standards and competency needs, and curriculums need to be changed from beginning to end to give priority to the software. The big developments on artificial intelligence show that it requires this. Furthermore, due to the uncertainty in the future labor market, universities need to allow their students to find their own way. For this, unlike traditional university approach, it is possible to give examples of applications like giving start-up skills such as start-up in technoparks, setting up conditions to do a double major, and giving emphasis to general culture education. In the face of the professional uncertainties of the future, the emphasis should be placed on general culture education of the students and the people in social networks should be educated about social sciences as well as software. So, in his article "Engineering plus X" written by the professor Yannis C. Yortsos, the dean of the Faculty of Engineering at the University of Southern California, it was suggested that traditional engineering knowledge plus X is needed to solve the problems that humanity has encountered in the transformation process we are in. X is defined in two ways here. First, engineering students should be equipped with social knowledge as well as technical knowledge. Second, the issues and problems before humanity require engineers to be professionally trained in the field of X [22].

Establishment of Industry 4.0 Application and Research Centers at universities may be advisable. The purpose of these centers: producing intelligent solutions by the instructors and students related to supply chain applications that enable the production systems brought by Industry 4.0 to communicate with each other, sharing the results with the world of science and public opinion, participating in the necessary education, publication and other activities. For this purpose it is expected to operate in the following areas: (a) To conduct research activities related to Industry 4.0, (b) To organize publicity and awareness activities related to Industry 4.0, (c) To prepare and manage projects under Industry 4.0, (d) Transferring knowledge and culture related to Industry 4.0 to industrial establishments, (e) To make collaborations and projects with the sectors where industry 4.0 technologies are used extensively, to use and run University resources in accordance with their aims, (f) Providing scientific and technical consultancy to related organizations, (g) To establish a science center related to the subject within the university, to ensure their operation and to cooperate with other related institutions and organizations in this regard.

In order to provide quality-focused growth and development in our higher education system, the importance of the improvement of university conditions and quality of input activities are inevitable. The future of a country depends on the youth of that country. In this context, universities should focus more on young people who aim at learning, which is the reason for existence, and they must internalize and place integrated approaches in all solution processes of education, research and community problems. Universities are valuable institutions of the community. Universities constitute an important source of welfare and they have an important place in history. They are a culture custodian. For this reason, bringing the universities in compliance with the needs and expectations of the community must be the responsibility of their respective administrations and every individual in society. As a limitation of this study; this research was conducted to improve a model related to Entrepreneurial and Innovative University-Industry Cooperation in the context of industry 4.0 in Turkey and it was carried out with 19 specialists at 12 universities according to an embedded theory which was a qualitative model. Subsequent studies can be done at different universities according to quantitative and qualitative methods.

References

1. YÖK (2018) The YÖK website. http://www.yok.gov.tr/yuksekogretimden-endustriye-nitelikli-insan-gucu-calistayi
2. ASO (2016) The Ankara Chamber of Industry website. http://www.aso.org.tr/wp-content/uploads/2016/09/1-20160927MILLIYET_ANKARA_SF_1.pdf

3. Aybek Y (2017) Üniversite 4.0'a geçiş süreci: kavramsal bir yaklaşım. Açıköğretim Uygulamaları ve Araştırmaları Dergisi 3(2):164–176
4. Tekin M, Geçkil T, Koyuncuoğlu Ö (2017) A model development research: entrepreneurial universities. In: International symposium for production reasearch, Vienna, 13–15 September 2017, pp 707–713
5. Tekin M, Geçkil T, Koyuncuoğlu Ö, Tekin E (2018) Girişimci Dostu Üniversiteler İndeksi ve Bir Model Geliştirilmesi, Selçuk University Social Sciences Institute Journal, vol. 39, April 2018
6. Gürsoy G (2016) 40endustri40.com website. http://40endustri40.com/endustri-4-0-ve-yuksekogretim
7. Dewar J (2017) Call for tertiary sector to gear toward University 4.0. http://www.ceda.com.au/2016/10/call-for-tertiary-sector-to-gear-toward-university-40
8. Lapteva AV, Efimov VS (2016) New Generation of Universities. University 4.0. Journal of Siberian Federal University, Humanit Soc Sci 11(9):2681–2696
9. Lukovics M, Zuti B (2013) Successful universities towards the improvement of regional competitiveness: "Fourth Generation" universities. In: European Regional Science Association (ERSA) 53th Congress Regional Integration: Europe, the Mediterranean and the World economy, At Palermo
10. Wissema JG (2014) Üçüncü Kuşak Üniversitelere Doğru. Özyeğin Üniversitesi Yayıncılık, İstanbul
11. Tecim V (2004) Sistem Yaklaşımı Ve Soft Sistem Düşüncesi. D.E.Ü. İ.İ.B.F. J 19(2):75–100
12. Işığıçok E (2008) Performans Ölçümü, Yönetimi ve İstatiksel Analizi. Ekonometri ve İstatistik J 7(2):1–23
13. Corbin J, Strauss A (2007) Basics of qualitative research: techniques and procedures for developing grounded theory, 3rd edn. Sage, Thousand Oaks
14. Creswell JW (2016) Nitel Araştırma Yöntemleri (Bütün M, Demir SB (eda)). Siyasal Yayıncılık, Ankara
15. Christensen LB, Johnson RB, Turner LA (2015) Araştırma Yöntemleri Desen ve Analiz (Aypay A (ed)). Anı Yayıncılık, Ankara
16. Güler A, Halıcıoğlu MB, Taşğın S (2015) Sosyal Bilimlerde Nitel Araştırma. Seçkin Yayıncılık, Ankara
17. Baş T, Akturan U (2017) Sosyal Bilimlerde Bilgisayar Destekli Nitel Araştırma Yöntemleri. Seçkin Yayıncılık, Ankara
18. Yıldırım A, Şimşek H (2008) Sosyal Bilimlerde Nitel Araştırma Yöntemleri. Seçkin Yayıncılık, Ankara
19. Merriam SB (2013) Nitel Araştırma Desen Ve Uygulama İçin Bir Rehber (Ed.: Selahattin Turan). Nobel Yayıncılık, Ankara
20. Ayazlar RA (2015) Araştırmalarda Güvenirlik ve Geçerlilik. In: Yüksel A, Yanık A, Ayazlar RA (eds) Bilimsel Araştırma Yöntemleri, Seçkin Yayıncılık, Ankara, pp 63–79
21. Patton MQ (2002) Qualitative research and evaluation methods. Sage, Thousand Oaks
22. Coşkunoğlu O (2017) Üniversite Rektörlerinin Endüstri 4.0 Üzerine Görüşleri Farklı. http://www.bthaber.com/yazarlar/universite-rektorlerinin-endustri-4-0-uzerine-gorusleri-farkli/1/21483

An Investigation on Online Purchasing Preferences of Internet Consumers

Emel Celep[1(✉)], Ebru Özer Topaloğlu[2], and H. Serdar Yalçınkaya[3]

[1] Department of Business Administration, Faculty of Economics and Administrative Sciences, Selçuk University, Konya, Turkey
ecelep@selcuk.edu.tr

[2] Department of Management Information Systems, Faculty of Applied Sciences, Necmettin Erbakan University, Konya, Turkey
eozer@konya.edu.tr

[3] Department of Accounting and Tax, Ereğli Kemal Akman Vocational High School, Necmettin Erbakan University, Konya, Turkey
syalcinkaya@konya.edu.tr

Abstract. The main reason why the concept of the internet has begun to take up more space in people's lives in recent years is seen as computer and communication technologies that are constantly evolving and becoming widespread. The beginning of the Internet for business purposes was the beginning of the 1990's when web browsers became available for the first time. The internet has gained importance in terms of electronic commerce activities in recent years, bringing a new trend in the form of shopping over the internet. The fact that this shopping pattern is increasingly preferred by consumers has added new dynamics to the field of business and marketing and this similar increase in internet usage began to be seen in Turkey as well as across the globe. The purpose of this study is to examine the purchasing preferences of internet consumers [1].

Keywords: Internet · Online purchasing · Purchasing preferences
Consumer behaviour · Marketing

1 Introduction

Although there are some similarities between consumers' online and offline decision-making processes, there seem to be some important differences, especially in the shopping environment and marketing communication. According to the traditional consumer decision model, the consumer purchasing decision typically begins with the recognition of need, continues with the collection of information and evaluation of the alternatives and it is finally completed with the decision to purchase and post-purchase behavior [2].

N. M. Durakbasa and M. G. Gencyilmaz (Eds.): ISPR 2018, *Proceedings of the International Symposium for Production Research 2018*, pp. 581–589, 2019.
https://doi.org/10.1007/978-3-319-92267-6_48

2 Electronic Commerce/Online Shopping

With the increase of Internet literacy, it is seen that the rate of online marketing activity increases in general. The fact that millions of people are online most of the time enhances the likelihood that these consumers are potential consumers for the online market. Today, customer shopping behaviors are affected by culture, social class, family, salary level, age, gender and so on; and therefore different customer behaviors are exhibited [3]. The electronic commerce is generally examined in four different ways in terms of application, process, and the format of electronic commerce. These can be listed as follows [4]:

- Business-to-Consumer Electronic Commerce (Business to Customer/B2C)
- Inter-Enterprise Electronic Commerce
- Business to Government Electronic Commerce
- Consumer to Government Electronic Commerce

The way in which electronic commerce is operated as a business to consumer (Business to Customer/B2C) is explaining the purchase over the internet. In other words, B2C form of e-commerce means that consumers can investigate the prices for various goods and services through web sites, compare the goods and services of different web sites in various ways and make use of electronic banking and insurance, electronic payment transactions, consultancy services etc. ([5]; as cited in [1]).

Various definitions of electronic commerce, which is considered to be an important economic activity and which has become increasingly common in information and communication technologies, are as follows [6]:

WTO (World Trade Organisation); The World Trade Organization defines e-commerce as "making the presentation of goods and services, running advertising campaigns, doing sales and marketing activities and ordering through telecommunication networks." [7].

OECD (Organisation for Economic Co-operation and Development); E-commerce is defined by OECD as 'carrying out business activities and organisations generally through transferring the digital data which include text, audio and visual images.' [8].

According to another definition, shopping on the Internet or online shopping is the purchase of goods or services directly from any seller person or corporation on the internet using a web browser [9].

3 The Characteristics of E-Commerce (Shopping Online)

Internet marketing campaigns or online marketing use all aspects of internet advertising for a response to the customers thanks to the wide use of the internet [10].

For the businesses aiming to be successful at e-commerce activities, the web page, the pricing strategies, the service and the other facilities are very effective in motivating and satisfying the online customers. It is important for the businesses to create a website which can respond to the target market and to decide what consumer motives they should focus on such as suitability for the needs and desires of consumers shopping on the internet, the need to acquire information, the desire to have goods and services fast, providing social interaction, retail shopping experience and product variety expectancy [11].

At the center of e-commerce, there is application criteria and online business. At this point, the most basic goal of electronic commerce is to make all of the commercial transactions in electronic environment in a simple, reliable, fast and effective way [12].

Some important features of electronic commerce can be listed as follows ([13] as cited in [6]):

- It is an interactive activity among the parties that are subject to e-commerce.
- Because e-commerce and information, goods and services offered on the internet can reach different parts of the world, implementing businesses need to move away from the locality.
- Due to the suitability for 24/7 operation, limited time on communication and shopping is no longer a problem.
- Thanks to the technological infrastructure of e-commerce, information about purchasing preferences and habits, and demographic characteristics of consumers can be saved and used in private marketing activities in the following periods.

There are many systems that are developed for the secure payment transactions within the scope of electronic commerce activities. Among the different payment methods, the most commonly used payment methods are listed below ([6] as cited in [4]):

- Paying by credit card
- Paying by electronic funds transfer (EFT)
- Paying by electronic money fransfer (e-Money)
- Paying by smart card
- Alternative online payment methods

4 The Advantages and Disadvantages of Online Shopping

The advantages of electronic commerce to the consumer include a comprehensive shopping opportunity, convenient selection, increase in the quality of the shopping service, the opportunity for price reduction and promotion for a certain period, quick response to the requests via internet, being informed about goods and service innovations, saving of time and transportation, being able to have goods and services suitable for consumer demand and expectation etc. [6].

Businesses have had the chance to increase their sales with different types of electronic commerce and have the opportunity to establish relationships directly between themselves and the customer without any person or unit. In this context, electronic commerce has been very influential in the globalization of businesses around the world. Thus, businesses can easily market their products all over the world and are able to provide excellent markets for their products [14].

The evaluation of traditional commerce and e-commerce differences of businesses in terms of customer, profit and cost can be shown as in Table 1. ([15] as cited in [6]):

Table 1. The comparison of traditional commerce and e-commerce

Operation	Traditional commerce	Electronic commerce
Address	The address of the business	web site: www.thenameofthebusiness.com
Customer	Number: Thousands in a month Location: Near the enterprise	Number: hundred of thousands in a month Location: throughout Turkey and abroad
Customer notification	Inside the store, on the phone and while presenting the products	Rapid access to all product options without having to see the product; electronic mail and the presentation of written-visual information
Time	The time spent for each customer is generally high between 09.00–18.00 during working hours	The time spent for customer is low during 7/24 working hours; but easy access to hundred of thousands ones and focused on potential customers
Communication	By telephone, fax etc.	By E-mail, online forms, order boxes, telephon, fax etc.
Profit	Limited profit depending on limited customer	Limitless profit depending on limitless customer
Cost	The rent and the equipment of the store, consumables, telephone, fax and the expenses for the staff	Virtual store and the expenses for image quality of only camera products

Source: ([15] as cited in [6]).

When examined in Table 1; it is seen that there are a number of differences between traditional trade and electronic commerce regarding the number of customers, the tools used to inform customers, the working time, communication tools, the amount of earnings, and the costs to bear. When comparing traditional marketing techniques and shopping on the Internet, it appears that there are some positive and negative distinctions between the two. The advantages of shopping on the internet are the speed advantage and the low cost for the buyers since it allows the comparison of the product, brand, price and the enterprise. Besides, the fact that the consumers cannot contact the products and the factors such as security vulnerabilities that may be occasionally generated in electronic transactions, validity of documents, official validity of electronic contracts, lack of experience of store environment, unauthorized use of personal information, and high cost of delivery are regarded as the disadvantages of e-commerce ([5, 16, 17]; as cited in [1]).

5 Shopping on the Internet (E-Commerce) and the Consumers' Purchase Preferences

Consumer behavior analysis is a study of evaluation in terms of selection, use and disposal in order to meet the goods, services and needs of individuals, groups or organizations; and at the same time, a study to measure the effects of these processes on consumers and society [18].

The model by reference [19], explaining consumer purchasing process on the internet, deals with the steps of consumer purchasing preferences related to searching for information, selection of the altervatives and finally the decision of purchasing. The model acknowledges that online shopping belief is an important determinant of consumers' purchasing preferences related to consumers' behaviors in this regard and the websites that include multi-channel stores. The model which explains the decision making process of consumers is as follows ([19] as cited in [20]) (Fig. 1):

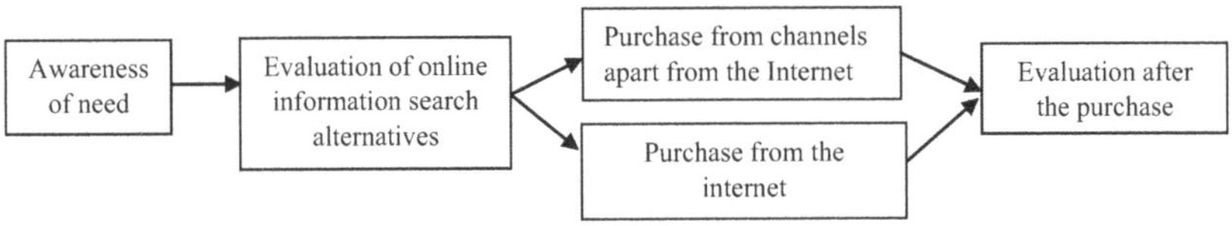

Fig. 1. Online consumer decision making process

Regarding the effects of online shopping on the shopping behaviors of consumers; [21] stated that when the businesses offer the opportunity to shop at a lower price with reduced costs, the consumers purchase more often, they can access stores, goods and services from different parts of the world without any financial expense related to transportation, and they can learn other users' ideas (suggestions, complaints etc.) about goods and services online and they can even buy products they like and join the online auctions in different parts of the world.

6 Studies on Shopping on the Internet and Attitudes to Shopping on the Internet

Among the studies carried out at various periods in order to explain the behaviors of the consumers about the shopping they have done over the Internet, Technology Acceptance Model-TAM [22] and Theory of Planned Behavior-TPB [23] are accepted as two important studies where two theories, which can be regarded as essential in the area, have been experimentally tested [1].

Investigating the results of the study by reference [6] Uluçay entitled 'E-Commerce in the world and in Turkey: An Application on Online Shopping Habits of Consumers'; it was concluded that there were 1,97 times increase in online shopping habits depending on high educational background, so purchasing on the internet and educational status were directly associated with each other. And as a result, this is explained by people's reading habits, research and information equipment, and more courageous behaviors towards newly developed technologies, depending on the increase in the level of education.

According to the results of the study conducted by Ene [11] entitled 'Factors Affecting Consumer Behavior on Shopping on the Internet: An Application on Motivation'; when you look at the behaviors exhibited by the participants during the shopping on the internet, it was seen that 61.5% of them made a decision after checking all web sites in their planned shopping; 11,9% consisted of those who believed in the experience (emotional pleasure, entertainment etc.) they could get on web sites and 9,2% were those who used virtual stores for information gathering only. Besides, the percentage of those who surf on the net and purchased without a plan was 4,6% and it was in the last place.

When investigating the results of the study by Karaçetin [4] entitled ' The Attitudes related to Online Shopping: A Research; about the goods and services prefered by the students', it was found out that the products prefered by the students in a maximum rate were clothing and accessories at the rate of 83,3% (204 people), 59,6% of them (146 people) bought tickets, and books, magazines etc. came in the third place at the rate of 33,1% (81 people), and then computers, electronic products and by-products, mobile devices, digital products were prefered by the students at the same rate.

When evaluating the results of the study by [24] entitled 'Analysis of Factors Affecting the Consumer's Online Shopping Behavior'; the financial risk and the risk of not being delivered under the hypothesis tests were found to have a negative effect on online buying behavior.

It was asserted that the study by [25] entitled 'Discovering Consumer Attitudes and Behavior in Online Hotel Room Reservation' was a starting point to understand the relationship of the factors which affected purchasing behavior and attitudes of consumers in allocating hotel rooms through online travel agencies and especially in terms of Malaysian consumers.

7 Research Method

Questionnaire method was used in the study. The scales related to the subject were examined and a new scale was developed ([6, 11, 20]). The first nine questions consisted of demographic questions. The other 16 questions were designed to measure the purchasing preferences of internet consumers. A scale of 25 questions in total was applied to internet consumers.

8 Results

In the study, the sample was obtained by applying to the students who had education at the department of business administration at a university in Konya and by sharing the data in the social media. A total of 226 people answered the questionnaire but 225 questionnaires were evaluated because there was incomplete information in a questionnaire. According to the Alpha Reliability test, $P > 0.5$ is considered reliable.

According to the reliability test results, the study is considered reliable because P = 0.519. According to the K-S normal distribution test in the questionnaire survey, if P > 0.05, the data are accepted as having normal distribution. However, since the probability values of all variables of the study were less than 0.05, it was assumed that the data in the study did not fit the normal distribution and it was decided to use non-parametric analysis methods in the research.

52.7% of the participants were female, 47.3% were female. 67.7% of the participants were in the age range of 18–25 years. 11,9% of them were in the age range of 26–35, 15% were in the age range of 36–45 and 4,9% were between 46–55. Also, 78,3% were single and 21,7% were married. As for educational background, 73% of them were university graduates. 53,1% of the participants were students. And so, 59,3% of the participants did not have a proffession. The level of income of 65,5% was between 0–2000.

In the study, a chi-square test, a non-parametric analysis that investigated the significant differences between the frequencies of the variables, was applied. According to chi-square test, the results were:

- In both genders, online shopping has mostly begun in the last two years, but this is more related to the age of the participants.
- While women prefer cash on delivery mostly (55,5%), men prefer using a credit card (69,2%).
- The majority of the 18–25 age group has started shopping online in the last two years, others have been doing it for moe than 5 years.
- Young people do shopping mostly with mobile phones, and as the ages of the groups progress, they prefer shopping on the computer more. Tablet using is very low.
- The time at home is generally used for shopping.
- Trust in shopping on the internet is proportional to the age.
- In all age groups, credit card and cash on delivery options have a high rate. However, as the age progresses, paying with a credit card increases as well.
- As the level of education increases, the number of purchases on the internet increases. However, the rate of shopping on the phone is also very high.
- As education level increases, trust decreases.
- Ph.D. graduates use credit cards mostly. The others prefer cash on delivery more.
- Public employees have been using online shopping for a long period.
- The use of telephone and computer is in balance in working groups
- The trust is high in employees but low in non-employed people.
- Public employees use credit card generally, but non-employed people prefer cash on delivery.
- As income goes up, the process of starting shopping is longer
- As income increases, the use of computers for shopping also increases.
- As income increases, the use of credit card increases, too.

Also, according to the results of the study, 38% of the participants stated that they prefer to shop online if the products or the services are at a discount or they are promotional. The most important aspect about the e-commerce site was the trust with the rate of 38%. 25% of the participants prefer clothing and footwear, 15% prefer technology products and 12% prefer movies, music, books and games.

9 Conclusion

The changes and progresses that have taken place in technology in recent years have led to an increase in the number of goods and services that have been operating on the market for businesses in different sectors. The increase in the number of goods and services in the market, in other words, the increase in the number of options that consumers can use to meet their different needs, causes some difficulties in decision making. These difficulties affect the enterprises in terms of attracting attention in a strong competitive environment, and also they affect the consumers in decision making due to the high number of options which can meet the expectations about the product they buy. The rapid increase in the number of goods and services and the fact that the distinguishing criteria for consumers is only perceived as price is causing the real value of the products not to be noticed. Depending on these similar developments, it has become a matter for businesses to use a number of different tools in the form of production of goods and services they produce, and in the way they are presented to the market. One of these tools is the Internet and other technological tools. Especially the development of the internet and the fact that it has become a part of the consumers' daily lives has made it a reason for preference to do shopping fast and effectively for the expectations of the consumers. This study is intended to determine the points that today's consumers are paying attention to in shopping behaviors on the internet and the factors that have caused them to do online shopping. When the data obtained from the research were examined, it was pointed out that especially in the last two years, both genders started shopping on the internet, young people prefered mobile phones for shopping more and their trust in shopping on the internet was proportional to age. At this point, the fact that those who use credit cards during shopping are more likely to be Ph.D. graduates, and the fact that employees trust online shopping more than the non-employed people do shows that the attitude for online shopping has changed positively due to the increase of education level. Considering the changing competition structure, developing technology and the changing consumer demands and expectations in the recent years; the services such as faster delivery of products, faster access to products, after sales returns and product change have made the use of the internet more common for the brands, businesses and the consumers.

References

1. Turan AH (2008) Internet Alışverişi Tüketici Davranışını Belirleyen Etmenler: Geliştirilmiş Teknoloji Kabul Modeli (E-TAM) ile Bir Model Önerisi, Akademik Bilişim 08 Konferansı, Çanakkale Onsekiz Mart Üniversitesi, 30 Ocak, 1 Şubat, 2008, Çanakkale
2. Katawetawaraks C, Wang CL (2011) Online shopper behavior: influences of online shopping decision. Asian J Bus Res 1(2):66–74

3. Lakshmi S (2016) Consumer buying behavior towards online shopping. Int J Res Granthaalayah 4(8):60–65
4. Karaçetin M (2015) Internet Üzerinden Alışverişe Yönelik Tutumlar: Bir Araştırma, Yüksek Lisans Tezi, Mehmet Akif Ersoy Üniversitesi, S.B.E., Burdur
5. Enginkaya E (2006) Elektronik Perakendecilik ve Elektronik Alışveriş. Ege AkademikBakış Dergisi 6(1):10–16
6. Uluçay U (2012) Dünya'da ve Türkiye'de E-Ticaret: Tüketicilerin Internet Üzerinden Alışveriş Alışkanlıkları Üzerine Bir Uygulama, Yüksek Lisans Tezi, Atılım Üniversitesi, S. B.E., Ankara
7. Kırçova İ (1999) İnternette Pazarlama. Beta Basım Yayım Dağıtım, İstanbul
8. Canpolat Ö (2001) E-Ticaret ve Türkiye'deki Gelişmeler. T.C. Sanayi ve Hukuk Müşavirliği
9. Saravanan S, Devi KB (2015) A study on online buying behaviour with special reference to Coimbatore City. IRACST Int J Commer Bus Manage (IJCBM) 4(1):2319–2828
10. Gabriel JK, Kolapo SM (2015) Online marketing and consumer purchase behaviour: a study of Nigerian firms. Br J Mark Stud 3(7):1–14
11. Ene S (2007) Internet Üzerinden Alışverişte Tüketici Davranışlarını Etkileyen Faktörler: Güdülenme Üzerine Bir Uygulama, Doktora Tezi, Marmara Üniversitesi, S.B.E., İstanbul
12. Kieanan B (2002) İşletmeler İçin Çözümler Elektronik Ticaret, Çev. Kaan Öztürk, Okan Cem Çırakoğlu, Serdar Özkaya. Arkadaş Yayınları, Ankara
13. Dolanbay Ç (2000) E-Ticaret Strateji ve Yöntemleri. Sistem Yayınları, Ankara
14. Nazir S, Tayyab A, Sajid A, Rashid H, Javed I (2012) How online shopping is affecting consumers buying behavior in Pakistan? IJCSI Int J Comput Sci Issues 9(3):1
15. Pense F (2008) Küçük Ve Orta Büyüklükteki İşletmelerde, E-Ticaretin Rekabet Şartlarına Etkilerinin Araştırılması. Yayınlanmamış Yüksek Lisans Tezi, Marmara Üniversitesi, Fen Bilimleri Enstitüsü, İstanbul
16. Torlak Ö (2007) Internette Pazarlamada Fiyatlandırma Stratejileri: Kavramsal Bir Çalışma. www.geocities.com/ceteris_tr/o_torlak3.doc. Erişim Tarihi: 13 Aug 2007
17. Aksoy R (2006) Bir Pazarlama Değeri Olarak Güven ve Tüketicilerin Elektronik Pazarlara Yönelik Güven Tutumları. ZKU Sosyal Bilimler Dergisi 2(4):79–90
18. Pahwa B (2015) A review of consumer online buying behaviour. In: International conference on technologies for sustainability-engineering, information technology, management and the environment, SUSTECH 2015. Dav, Institude of Management, India, pp 570–576
19. Kim J (2006) Sensory enabling technology acceptance model (Se-Tam): The usage of sensory enabling technologies for online apparel shopping. Auburn University, Doctor of Philosophy
20. Tümtürk A (2015) İnternet Üzerinden Alışveriş Niyetini Belirleyen Faktörlerin İncelenmesi: Türkiye'de Alışveriş Deneyim Düzeylerinin Farklılığına İlişkin Bir Model Önerisi, Doktora Tezi, Celal Bayar Üniversitesi, Manisa
21. Özbay S, Akyazı S (2004) Elektronik Ticaret. Detay Yayıncılık, Ankara
22. Davis FD (1989) Perceived usefulness, perceived ease of use, and user acceptance of information technology. Manage Inf Syst Q 13(3):319–340. https://doi.org/10.2307/24900
23. Taylor S, Todd PA (1995) Understanding information technology usage: a test of competing models. Inf Syst Res 6(2):144–176
24. Javadi MHM, Dolatabadi HR, Nourbakhsh M, Poursaeedi A, Asadollahi AR (2012) An analysis of factors affecting on online shopping behavior of consumers. Int J Mark Stud 4(5)
25. Li-Ming AK, Wai TB (2013) Exploring consumers' attitudes and behaviours toward online hotel room reservations. Am J Econ 3(5C):6–11

Environmental Risk Assessment of E-waste in Reverse Logistics Systems Using MCDM Methods

Ferhat Duran and İlke Bereketli Zafeirakopoulos(✉)

Department of Industrial Engineering, Galatasaray University, Istanbul, Turkey
ferhattduran@gmail.com, ibereketli@gsu.edu.tr

Abstract. Due to technological developments, industrialization, rapid urbanization, and population growth the impacts of human activities on the environment are increasing all over the world. The expansion of production and consumption activities leads to more intensive use of natural resources. Hence resulting wastes, it also reached a level that threatens the environment and human health. Because of the rapidly diminishing raw material resources, and the pressure on the natural environment, the industry must adapt to more restrictive regulations enforced by governments and required by societies. As result of these environmentally conscious, Reverse Logistics has approved important attention. Reverse logistics includes the return of the products from the consumer market or the warehouse, the remanufacturing, recycling and/or the proper disposal of those products to prevent damage to the environment. E-waste sector is one of the areas where reverse logistics approach is often used in waste management. Nevertheless, there are many environmental risk factors related to the reverse logistics in the e-waste field that must be analyzed. In this study, the assessment of the "environmental risks" such as the risks of global warming due to CO_2 emissions, soil pollution risk etc. is considered as a multi criteria decision making (MCDM) problem. The Analytical Network Process (ANP) based on data collection from the experts, is used to determine the ranking of the criteria by calculating their weights. Complex Proportional Assessment (COPRAS) is applied for evaluating possible precaution strategies against environmental risks. Finally, a case study of fluorescent lamp reverse logistics process is presented to demonstrate the potential of the proposed approach. The major contribution of this work to the literature is that the environmental risk factors will be examined, prioritized and the possible precautions in reverse logistics risk factors will be discussed.

Keywords: Environmental risk assessment · E-waste
Multi criteria decision making · Reverse logistics

1 Introduction

Twenty to thirty years ago, the concept of supply chain was known to be efficient in the logistics of goods from raw material to final consumer. In today's flow system, some changes have occurred due to the environmental sensitivity of the consumers.

N. M. Durakbasa and M. G. Gencyilmaz (Eds.): ISPR 2018, *Proceedings of the International Symposium for Production Research 2018*, pp. 590–603, 2019.
https://doi.org/10.1007/978-3-319-92267-6_49

So, consumer products are now starting to flow back to the origin point. These flows back cover electronic products, textile, pharmaceutical, industrial products, food etc. as well as many other sectors.

Because of increasing world population, technological developments and consumption degree, the economic usage and the recovery of natural resources for the industrialized society become critical for the sustainability of life. As the quantity of the used products increases, the natural sources decline. To get over these two problems, reverse logistics, which is simply the recovery of the used products, becomes more significant [1]. Reverse logistics is part of the concept of sustainability which is the ability to use resources while meeting our present needs without ignoring the future generations' ones. In reverse logistics, firms are known to be able to use product values effectively and efficiently and to reuse them with recovery activities [2]. The common aim in two concepts is leaving a cleaner, more livable world for future generations by focusing on the efficient use of natural resources, avoiding all kinds of activities that might harm the environment.

The waste of electrical and electronic equipment (WEEE) is one of the important materials considered within reverse logistics. The amount of WEEE (e-waste) has been increasing according to the population and the technological developments. Worldwide, the annual quantity of WEEE disposed was about 30–50 million tons in and it is expected to be reach 40–70 million tons by 2015 [3]. The constant increase in the quality of the electrical and electronic devices and the shortening of their use time have accelerated the formation of e-waste [4, 5]. Therefore, the increase in the amount of e-waste causes a significant waste of resources.

Besides, e-waste that cannot be recycled to the desired extent makes it difficult for reverse logistics processes to work effectively. Therefore, the amount of treated e-waste is directly related to the capacity of the reverse logistics systems. The number of products returned by consumers and/or companies for recycle or disposal are affected by this amount of waste.

Nevertheless, there are many environmental risk factors related to the reverse logistics in the e-waste field that must be analyzed. Because, e-waste contains both hazardous and valuable materials that ask special recycling methods to keep off environmental contamination and detrimental effects on human health [6].

Compact Fluorescent Lamps (CFLs) are considered more environmental friendly and more energy efficient than incandescent lamps. Fluorescent lamps are being used more and more in the houses around the world as part of a drive to improve energy efficiency CFLs consume about 75% less energy than incandescent bulbs and last longer. At first glance this seems like a good way to conserve energy and to keep safe the environment. However, there are several serious problems associated with CFL bulbs that need to be examined and corrected. Because the fluorescent lamps, as e-waste, contain mercury, they must be examined in reverse logistics processes.

The aim of this study is twofold: to assess the environmental risks of the reverse logistics of the fluorescent lamps, and to provide precaution strategies for the improvement of the system. As a result of literature review and expert opinions, environmental risk factors have been identified for the reverse logistics of the fluorescent lamps, such as the risks of global warming due to CO_2 emissions, soil pollution risk, public health risk, etc., and analytical techniques have been applied for evaluation of these factors. The studied problem is considered as a multi criteria decision making (MCDM) problem.

Multi criteria decision making (MCDM) is a strong approach, which is extensively used for evaluating problems containing multiple and conflicting criteria. MCDM refers to find the best option from all the feasible alternatives. Priority based, outranking, distance-based and mixed methods could be considered as the primary classes of the current methods [7].

The Analytic Hierarchy Process (AHP) is one of the most prominent MCDM approaches. The approach bases on obtaining the relative weights among the factors and the total values of each alternative based on these weights. Analytical Network Process Method (ANP) is a generalization of the AHP, which considers the dependence between the elements of the hierarchy [8]. ANP determines the criteria weights from subjective judgments of the decision maker group.

First, environmental risk factors obtained from the relevant experts and the literature are evaluated by ANP to find which factors are more likely to impact the environmental risk-causing process. In this study, ANP is chosen because there is a dependence relationship between fuel consumption and CO_2 emissions below the transport criterion. As fuel consumption increases, carbon dioxide emissions also increase.

Secondly, COPRAS is applied to rank and evaluate alternatives in terms of importance and benefit ratings. Criterion values are used to maximize the benefit criterion in evaluating the criterion and to evaluate the most useless criteria for the lowest cost. Compared with other MCDM methods such as AHP and TOPSIS, it is a very simple method to use less calculation time [9]. The method allows evaluation of both qualitative and quantitative criteria.

Finally, the results of the case study are presented for a fluorescent lamp. The application is provided to demonstrate the potential of the proposed approach.

The major contribution of this work to the literature is that the environmental risk factors will be examined, prioritized and the possible precautions in reverse logistics risk factors will be discussed.

2 Reverse Logistics

Reverse logistics is a reverse movement of raw materials, semi-finished products, final products and related information from the point of consumption to the point of origin. Since reverse logistics is a rather new research field, sometimes in the literature; different concepts are encountered, such as reversed logistics, return logistics and retro logistics.

The “repair, reuse, remanufacture, recycling and disposal” options that constitute the reverse logistics process are as shown in Fig. 1 [10]. In this study; transport, storage and disposal processes will be examined.

Two important reasons behind the growing importance of reverse logistics practices are economic and environmental concerns. These aims are not contradictory but overlapping purposes. Adding new value to used products or reusing certain materials brings cost savings and as a result increases profitability rates [11].

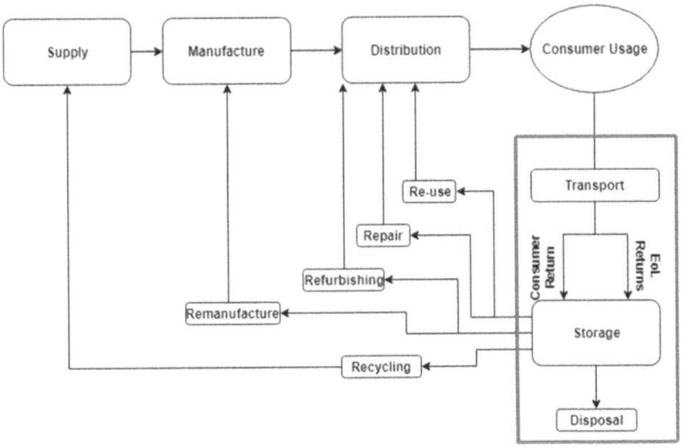

Fig. 1. Reverse logistics process

The significance of the developed supply chain management was begun to understand by many companies [12]. The importance of reverse logistics and closed-loop systems in many industries is accepted as an essential need [13]. The recycling networks needs appropriate logistical systems for the reverse flow of products from users to producers [14].

Literature review shows that past research examines only on a small area of reverse logistics systems, such as network design, production planning or environmental issues. However, there is almost no research on environmental risk management in reverse logistics of e-waste. The present research aims to fill this gap and to explore the opportunities for improved environmental gain.

3 Proposed Methodology

The approach proposed in this study consists of three basic steps:

Step 1. Research about environmental risk management in reverse logistics of fluorescent lamp that gives us the risk factors and the possible precautions strategies.
Step 2. Determining the importance of the evaluation criteria for the risk factors by ANP method.
Step 3. Determining the most appropriate strategy by COPRAS method based on the proposed strategies.

Determining the Criteria Weights by ANP Methodology

AHP is the most widely decision-making model which is developed by Saaty's [22]. AHP is used to determine the priorities among different criteria. ANP is a MCDM tool conceived to be an extension of AHP. While AHP models a decision-making framework using a unidirectional hierarchical relationship between decision levels, the ANP permits for more complex relationships between components and decision levels [23].

To decide in an organized way to generate priorities we need to decompose the decision into the following steps:

Step 1: Identify the decision problem with criteria and sub-criteria.
Step 2: Find the general network of components.
Step 3: Find all dependencies that exist in the decision problem.
Step 4: Make pairwise comparisons on clusters to find the toper matrix.
Step 5: Make consistency analysis of all the pairwise comparisons.
Step 6: Rate the alternatives according all the criteria and sub-criteria.
Step 7: Compute and find the weighted super matrix and the limit super matrix.
Step 8: Determine the final decision to reach the final evaluation of the alternatives.

(1) *Pairwise Comparisons, Eigenvectors and Consistency*

In ANP, relative priorities are constituted as in AHP. Saaty's fundamental scale as given in Table 1.

Table 1. Saaty's fundamental scale

Value of a_{jk}	Explanation
1	*j* and *k* are equally important
3	*j* is slightly more important than *k*
5	*j* is more important than *k*
7	*j* is strongly more important than *k*
9	*j* is absolutely more important than *k*
2, 4, 6, 8	Intermediate values

Using the ratings given in Table 1, the pairwise comparison matrices A = (a_{ij}) are formed as seen below, to calculate the relative priorities of the elements forming these matrices in further steps:

$$A \begin{bmatrix} a_{11} & a_{12} & \cdots & a_{1n} \\ a_{21} & a_{22} & \cdots & a_{2n} \\ \vdots & \vdots & \ddots & \vdots \\ a_{n1} & a_{n2} & \cdots & a_{nn} \end{bmatrix} \text{where } a_{ij} = \frac{1}{a_{ij}} \forall i,j = 1,\ldots,a_{ii} = 1\, \forall i = 1,\ldots,n$$

If the matrix A wouldn't contain errors and the judgments were perfectly consistent, then:

$$a_{kj}a_{kj} = a_{ij} \quad \forall i,j,k = 1,\ldots,n \tag{1}$$

Therefore, all the elements in this matrix could be expressed as follows:

$$a_{ij} = w_i/w_j \quad \forall i,j = 1,\ldots,n \tag{2}$$

And this would yield to the following equality:

$$\begin{bmatrix} w_1/w_1 & w_1/w_2 & \cdots & w_1/w_n \\ w_2/w_1 & w_2/w_2 & \cdots & w_2/w_n \\ \vdots & \vdots & \ddots & \vdots \\ w_n/w_n & w_n/w_2 & \cdots & w_n/w_n \end{bmatrix} \begin{pmatrix} w_1 \\ w_2 \\ \vdots \\ w_n \end{pmatrix} = \mathrm{n} \begin{pmatrix} w_1 \\ w_2 \\ \vdots \\ w_n \end{pmatrix} \tag{3}$$

An easy way to get an approximation of the relative priority vector is to make a column normalization of the matrix A and then take the arithmetic mean of the rows. Hence:

$$w_i = \sum\nolimits_{k=1}^{n} [a_{ij}/\sum\nolimits_{k=1}^{n} a_{ij}]/\mathrm{n} \tag{4}$$

and

$$\lambda_{\max} = \frac{1}{n}\sum\nolimits_{j=1}^{n} \frac{\sum_{j=1}^{n} a_{ij}.w_j}{w_i} \tag{5}$$

It must be underlined that for important application; only the eigenvector derivation procedure has to be used because approximations can lead to a wrong ranking of the alternatives.

The consistency index – CI of a comparison matrix is given by:

$$\mathrm{CI} = \frac{\lambda_{max} - n}{n - 1} \tag{6}$$

And the consistency ratio – CR is obtained by comparing the CI value with the random inconsistency – RI values given in the Table 2. With $\lambda_{\max}$ value, consistency index (CI) and consistency ratio (CR) are found. A value of CR < 0,1 is typically considered acceptable, larger values require the decision maker to reduce inconsistencies in reviewing judgments.

Using these values, the CR value is calculated as follows:

$$\mathrm{CR} = CI/CR \tag{7}$$

$$\lim_{n\to\infty}(W)^{2k+1} \tag{8}$$

A. *Evaluating Reverse Logistics Strategy Alternatives by COPRAS*

The COPRAS method is a method of evaluating alternatives by making step-by-step sequencing of alternatives in terms of importance and utility ratings. The COPRAS

Table 2. The standardized matrix

Criteria	Normalized by cluster	ANP weights
T11	0,62097	0,240784
T12	0,16697	0,064743
T13	0,21206	0,082226
S11	0,16246	0,008432
S12	0,15321	0,007952
S13	0,68432	0,035517
D11	0,46781	0,262136
D12	0,2376	0,133139
D13	0,29459	0,165071

is one of the notable MCDM methods, which select the best alternative among of plausible choices by determining a solution to the best solution to the ratio with the ideal-worst solution [23]. The procedure of the COPRAS method includes the following steps:

Step 1. The decision matrix is formed.
Step 2. Normalize the decision matrix using the following formula

$$x_{ij}^{*} = \frac{x_{ij}}{\sum_{i=1}^{m} x_{ij}} \quad \text{for } (j = 1, 2, \ldots, n) \tag{9}$$

Step 3. Determine the weighted normalized decision matrix

$$d_{ij} = x_{ij}^{*} . w_j \tag{10}$$

Step 4. The sums Si– and Si+ of weighted standardized values are calculated using the following equations for both beneficial and non-beneficial criteria separately:

$$S_{i+} = \sum_{j=1}^{k} d_{ij} \tag{11}$$

$$S_{i-} = \sum_{j=k+1}^{n} d_{ij} \tag{12}$$

Step 5. The Q_i values are relative importance values for each alternative and are calculated using the Eq. (9). The result of the calculations is determined as the best alternative with the highest relative importance value.

$$Q_i = S_{i+} + \frac{\sum_{i=1}^{m} S_{i-}}{S_{i-} * \sum_{i=1}^{m} \frac{1}{S_{i-}}} \tag{13}$$

Step 6. The highest relative priority (Q_{max}) value is found.

Step 7. Calculate the performance index (Pi) of each alternative with this equation:

$$P_i = \left[\frac{Q_i}{Q_{max}}\right] \times 100\% \tag{14}$$

4 Application

Our goal for the study is selecting the best strategy for reverse logistics system. There are five possible alternatives in which to select the strategy: Re-engineering of production system strategy (A1), raising awareness for reverse logistics strategy (A2), increasing operational safety strategy (A3), decreasing environmental impacts of waste management strategy (A4), decreasing environmental impacts of transportation strategy (A5).

The company must take a decision according to the following seven criteria: CO_2 transportation emissions factor per unit of returned product in G/Km (T11), miles per gallon of fuel (T12 release of hazardous chemicals of damaged products (T13), storage and warehouse energy consumption (S11), risks of unexpected situations (fire, natural disasters etc.) (S12), release of hazardous chemicals of damaged products (S13), amount of mercury released to water supply (water pollution), the amount of mercury particles in the air (air pollution) (D12), amount of mercury released to the soil and land (soil pollution) (D13). The five possible alternatives Ai (i = 1, 2, 3, 4, 5) are to be evaluated using the ANP and COPRAS methods.

A. *Evaluation Criteria and Model Components*

There are three main criteria that must be taken into consideration to assess reverse logistics in terms of environmental risks. The goal of the evaluation, the three criteria and their sub-criteria are as in Fig. 2.

- *Criteria 1 - Transport:*

CO_2 Transportation Emissions Factor Per Unit of Returned Product In g/km *(T11):* The simplest for companies to calculate their transport emissions is to record energy and/or fuel use and employ standard emission conversion factors to convert energy or fuel values into CO_2 emissions. Each liter of fuel consumed will result in a certain amount of CO_2 emissions [15].

Miles Per Gallon of Fuel (T12): Gallons per 100 miles. This relates directly to the amount of fuel used and resource depletion [16].

Release of Hazardous Chemicals and Wastes by traffic accident (T13): The mercury in a fluorescent lamp can be released as both dust and vapor if the lamp is broken in an accident. This heavy metal is dangerous to people and animals, and easily migrates through the environment in the air, water, and soil [17].

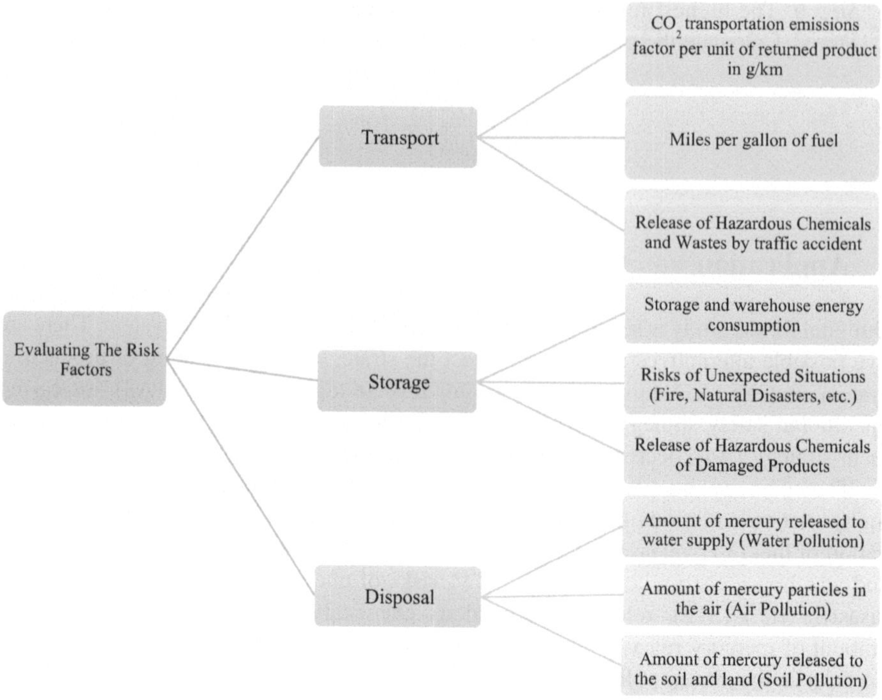

Fig. 2. Evaluating the risk factors

- *Criteria 2 - Storage:*

Storage and Warehouse Energy Consumption (S11): Many products require controlled storage conditions. Therefore, warehouse buildings are equipped with devices that create an appropriate microclimate inside. Thus, these will be cold, air conditioned and heated warehouses. Warehouse operations are clearly linked to energy consumption in its various forms. Among the different forms and varieties of warehouses, cold and heated warehouses are characterized by a relatively high energy demand [17]. As a result, companies need to learn how to save energy while reducing the carbon footprint of the warehouse.

Risks of Unexpected Situations (Fire, Natural Disasters, etc.) (S12): Explosion and/or sudden release of pressure because of unexpected situations such as fire, natural disasters, etc. [18].

Release of Hazardous Chemicals of Damaged Products (S13): Release of hazardous chemicals which poses substantial or potential threats to workers' health in the warehouse and/or the environment [17]. The products can be damaged in the warehouses by several reasons.

- *Criteria 3 - Disposal:*

Amount of mercury released to water supply (Water Pollution) (D11): Release of hazardous water pollutants into the water by landfill [19].

Amount of mercury particles in the air (Air Pollution) (D12): Emission of any kind of air pollutants (mercury etc.) which contain chemicals, particulate matter or biological materials, into the atmosphere, by incineration [20].

Amount of mercury released to the soil and land (Soil Pollution) (D13): Release of hazardous soil pollutants into the soil by landfill [19].

ANP-Stage 1. Evaluation criteria, definitions and hierarchical structures are given above.

ANP-Stage 2. Experts agree on a binary comparison of the criteria underlying the ANP method. A binary comparison matrix has been established for the implementation of ANP.

The decision matrix, which is based on Saaty's nine-point scale, is constructed. The decision maker uses the fundamental 1–9 scale defined by Saaty's to assess the priority score.

ANP-Stage 3. The ANP method was applied to the mathematical model, which was the result of the common opinion of the experts, and the criteria weights were calculated as in Table 2.

ANP-Stage 4. At this level of the application, the consistency ratio is measured according to calculated.

$$\begin{aligned}&Count = 3\\&\lambda_{\max} = 3,095193823\\&\mathrm{CI} = 0,047596912\\&\mathrm{Constant} = 0,58\\&\mathrm{CR} = 0,082063641 < 0,1\end{aligned}$$

B. *Alternative Strategies*

Many strategies can be considered to improve the reverse logistics system. In this case, it is important to choose the strategy that will be most successful and most appropriate for the system. These strategies are as follows:

Re-engineering of production system strategy: Firstly, since mercury can be dangerous to human health, it is important to properly dispose of fluorescent tubes. When collecting used fluorescent light bulbs, it is recommended to store and package them in the ways that minimize lamp breakage. Along with this strategy, organizational arrangements are carried out in order to ensure the systematic planning, coordination and implementation of the reverse logistics system. The implementation of legislative arrangements for the implementation and integration of the reverse logistics system is among the foundations of this strategy [1]. Fluorescent lamps carry potential risks throughout the production process. For this reason; starting from product design, all processes including production technology, packaging process, stocking must be re-engineered with the view of eco-design.

Raising awareness of the reverse logistics strategy: Because of this huge amount of CFLs, the potential danger to the environment increases. In addition, product disposal costs increase with the depletion of landfill and incineration capacity. To prevent dangerous effects on the environment and to realize economic gains, options for recycling and recovery of CFLs are considered. As a result, environmental regulations have been published in most countries of the world such as Taiwan, China and Japan. Similar environmental regulations have been implemented in Europe through the WEEE Directive (2002/96/EC) [1]. The main purpose of this strategy is to increase the awareness of users and practitioners of reverse logistics of fluorescent lamps. So, companies and users will realize that sustainable reverse logistics strategies provide additional benefits that cannot be listed such as eliminating fines from government organization from improper disposal or improving the public perception of the company etc.

Increasing operational safety strategy: This strategy focuses on the regulation of reverse logistics systems in the existing transport. With this strategy, it is aimed to redesign logistics system components for safer reverse logistics processes with lower probability for accidents, and to increase the training of company personnel [21].

Decreasing environmental impacts of waste management strategy: Besides harmful effects on human health, wrong recycling of fluorescent bulbs can also threaten the environment. When CFLs are broken, burned, or buried, metals and other toxic substances can be released into the environment. Mercury in air can be deposited in soil and water bodies, and through microbial activity, be transferred to methylmercury. This harmful element not only contaminates water supplies but also affects fish and those who eat them. This strategy is environment-focused and ensures that all work done is environmentally friendly. With this strategy, it is aimed to develop an environmentally friendly system that will provide energy efficiency [1].

Decreasing environmental impacts of transportation strategy: Solutions aim to reduce emissions in transportation are produced, such as using less fuel-consuming vehicles, switching to electric vehicles, optimized routes etc.

By using the criteria weights found with ANP, five alternative precaution strategies are evaluated with the COPRAS method.

COPRAS-Stage 1. The decision matrix is established.

COPRAS-Stage 2-3. The weighted normalized decision matrix of the alternatives calculated by multiplying the normalized decision matrix and the weights is represented in Table 3. The normalized decision matrix is calculated using Eqs. (9) and (10).

Table 3. The weighted normalized matrix

	T11	T12	T13	S11	S12	S13	D11	D12	D13
A1	0,051	0,016	0,020	0,004	0,002	0,015	0,115	0,059	0,076
A2	0,120	0,038	0,034	0,004	0,004	0,015	0,115	0,059	0,076
A3	0,051	0,016	0,061	0,003	0,006	0,022	0,115	0,059	0,055
A4	0,120	0,027	0,020	0,004	0,002	0,015	0,148	0,075	0,098
A5	0,154	0,038	0,034	0,003	0,002	0,009	0,082	0,042	0,055

COPRAS-Stage 4-5-6-7. The values of Qi, Si+, Si−, Pi were calculated using Eqs. (11)–(14) by COPRAS method. Table 3 shows the results. In view of the proposed model, every option has the preferable values for the maximizing and minimizing indices. At that point, the relative weight and the optimality criterion are figured as appeared in Table 3. According to the value of the optimality criterion, the priority of the alternatives is acquired. Finally, the utility level of every option is measured as displayed in Table 4.

Table 4. The final results

	Si+	Si–	Qi	Pi	Rank
A1	0,000	0,360	0,502	100	1
A2	0,000	0,466	0,388	77,36	4
A3	0,000	0,387	0,468	93,23	2
A4	0,000	0,512	0,354	70,47	5
A5	0,000	0,419	0,432	85,98	3

Ultimately, re-engineering of production system strategy (A1) has become the most desirable strategy among five alternatives with the final performance value of 100; while the strategy increasing operational (A3), the decreasing environmental impacts of transportation strategy (A5), the strategy for rising the awareness reverse logistics (A2), the strategy for decreasing environmental impacts of waste management strategy (A5) have positioned at the second, third, fourth and fifth ranks with 93.23, 85.98, 77.36 and 70.47 as the final performance values, respectively.

5 Conclusions

The recovery of used electronic materials is an essential task. Depending on the importance of the subject, numerous studies have been performed in the recent years. Reverse logistics dealing with the backward flows in a recovery system has become very important.

The purpose of this study is to estimate the strategic risk factors of reverse logistics and to evaluate alternative reverse logistics strategies in terms of environmental risk assessment. This study will show COPRAS method as a MCDM method is effective for strategy evaluation which is a rare method in the literature.

Consequently, at first the basic concepts of reverse logistics are reviewed in terms of environmental risk assessment. The model is built on the criteria and the alternatives for reverse logistics strategy evaluation. As the MCDM approach, the ANP and COPRAS methods are applied. Five alternatives for evaluation of reverse logistics are determined, and the priority order of these five alternatives for environmental risk assessment of reverse logistics is found. The first order in the alternatives is re-engineering of production system strategy (A1). Re-engineering should be the priority for the above-mentioned system. For reverse logistics systems to be successful for the environmental risk assessment, first the existing production system must be analyzed, and the problematic parts concerning the environmental risks should be defined. Then, with a systematical planning re-engineering process must be applied into the current system.

The subject of reverse logistics strategy selection can be advanced in future studies by increasing the number of criteria and the decision makers or using different decision-making methods. Another perspective can be to consider uncertainty using fuzzy approach [24].

Acknowledgment. This study is financially supported by Galatasaray University Research Fund under the project grant number 18.402.012.

References

1. Kilic HS, Cebeci U, Ayhan MB (2015) Reverse logistics system design for the waste of electrical and electronic equipment (WEEE) in Turkey. Resour Conserv Recycl 95:120–132
2. De Brito MP, Dekker R (2002) Reverse logistics-a framework (No. EI 2002-38). Erasmus School of Economics (ESE)
3. Menikpura SNM, Santo A, Hotta Y (2014) Assessing the climate co-benefits from waste electrical and electronic equipment (WEEE) recycling in Japan. J Clean Prod 74:183–190
4. Puckett J, Byster L, Westervelt S, Gutierrez R, Davis S, Hussain A, Dutta M (2002) Exporting harm: the high-tech trashing of Asia. http://ewasteguide.info/biblio/exporting-har. Accessed 28 July 2016
5. Hester RE, Harrison RM (2009) Electronic waste management: design, analysis and applicaiton, vol 27. Royal Society of Chemistry, Cambridge
6. Robinson BH (2009) E-waste: an assessment of global production and environmental impacts. Sci Total Environ 408(2):183–191
7. Pomerol C, Barba Romero S (2000) Multicriterion decision in management: principles and practice, 1st edn. Kluwer Academic Publishers, Norwell
8. Hwang CL, Yoon K (1981) Multiple attribute decision making—methods and applications. Springer, Heidelberg
9. Yardım MS, Akyıldız G (2005) Akıllı Ulaştırma Sistemleri ve Türkiye'deki Uygulamalar, Ulaştırma Kongresi, Istanbul, TMMOB Inşaat Mühendisleri Odası
10. Thierry M, Salomon M, Van Nunen J, Van Wassenhove L (1995) Strategic issues in product recovery management. Calif Manag Rev 37(2):114–136
11. Erol I, Alehan F, Pourbagher MA, Canan O, Yildirim SV (2006) Neuroimaging findings in infantile GM1 gangliosidosis. Eur J Paediatr Neurol 10(5):245–248
12. Kannan G (2009) Fuzzy approach for the selection of third party reverse logistics provider. Asia Pac J Mark Logist 21(3):397–416
13. Govindan K, Soleimani H (2017) A review of reverse logistics and closed-loop supply chains: a journal of cleaner production focus. J Clean Prod 142:371–384
14. Giannotti S, Trombi L, Bottai V, Ghilardi M, D'Alessandro D, Danti S, Dell'Osso G, Guido G, Petrini M (2013) Use of autologous human mesenchymal stromal cell/fibrin clot constructs in upper limb non-unions: long-term assessment. PLoS ONE 8(8):e73893
15. McKinnon A, Piecyk M (2010) Measuring and managing Co_2 emissions. European Chemical Industry Council, Edinburgh
16. Waters MHL, Laker IB (1980) Research on fuel conservation for cars (No. TRRL LR921 Monograph)
17. Taghipour H, Amjad Z, Jafarabadi MA, Gholampour A, Nowrouz P (2014) Determining heavy metals in spent compact fluorescent lamps (CFLs) and their waste management challenges: some strategies for improving current conditions. Waste Manag 34(7):1251–1256

18. Payzan-LeNestour E, Bossaerts P (2011) Risk, unexpected uncertainty, and estimation uncertainty: Bayesian learning in unstable settings. PLoS Comput Biol 7(1):e1001048
19. Lim SR, Kang D, Ogunseitan OA, Schoenung JM (2012) Potential environmental impacts from the metals in incandescent, compact fluorescent lamp (CFL), and light-emitting diode (LED) bulbs. Environ Sci Technol 47(2):1040–1047
20. Nance P, Patterson J, Willis A, Foronda N, Dourson M (2012) Human health risks from mercury exposure from broken compact fluorescent lamps (CFLs). Regul Toxicol Pharmacol 62(3):542–552
21. Murphy PR., Poist RF (2000). Green logistics strategies: an analysis of usage patterns. Transp J 5–16
22. Saaty TL (1980) The Analytic Hierarchy Process. McGraw-Hill, New York
23. Chatterjee NC, Bose GK (2012) A COPRAS-F base multi-criteria group decision making approach for site selection of wind farm. Decis Sci Lett 2(1):1–10
24. Zadeh LA (1965) Fuzzy sets. Inf Control 8(3):338–353

Grey Forecasting Model for CO_2 Emissions of Developed Countries

Asiye Özge Dengiz[1(✉)], Kumru Didem Atalay[1], and Orhan Dengiz[2]

[1] Department of Industrial Engineering, Başkent University, Ankara, Turkey
{aokarahanli,katalay}@baskent.edu.tr
[2] Department of Industrial Engineering,
Nuh Naci Yazgan University, Kayseri, Turkey
orhan.dengiz@gmail.com

Abstract. Global warming endangers our health, jeopardizes our national security, and threatens other basic human needs. Greenhouse gas (GHG) emissions mitigation is a high priority issue for most of the countries in the world. Carbon-dioxide (CO_2) is one of the most important GHG emissions, so prediction of CO_2 emissions is very important issue for the countries. On the purpose of predicting following years' CO_2 emissions of seven developed countries; Australia, China, Italy, Spain, Turkey, United Kingdom and United States, grey forecasting method GM(1,1) which is suitable for solving uncertainty problems with less or lack of information is used in this study. Grey forecasting method is also widely used to forecast carbon emissions in the literature and generally, a few data are considered. The historical data of CO_2 emissions period between 2010 and 2014 is used to forecast the following four-year emissions, up to 2018. For the accuracy of the forecasting model, the post-ratio error (C) indicator is used which is one of the most widely used indicators for similar research. Using this model, some countries that already start to take precautions to decrease emissions could be check if they reach their target and mitigate the effects on climate change in their countries in a long term. This study can also be a counsellor or indicator for the countries that has not improved any environmental policy and chance to take into effect the precautions for their climate change plans. In other words, the methods proposed in this study can be used by countries to review their environmental policies and estimate their outcomes.

Keywords: CO_2 emissions · Grey forecasting · Greenhouse gases
Climate change · Future emissions

1 Introduction

In the last recent years, an ongoing concern is the increasing threat of global warming and climate change worldwide. The Kyoto Protocol, an environmental agreement adopted in 1997 by many of the parties to the United Nations Framework Convention on Climate Change (UNFCCC), is working towards curbing carbon dioxide (CO_2) emissions globally. Reducing the production of greenhouse gases (GHG) at a level that would prevent dangerous anthropogenic interference in the climate is the main

N. M. Durakbasa and M. G. Gencyilmaz (Eds.): ISPR 2018, *Proceedings of the International Symposium for Production Research 2018*, pp. 604–611, 2019.
https://doi.org/10.1007/978-3-319-92267-6_50

objective of these agreements. Carbon dioxide (CO_2), methane (CH_4), nitrous oxide (N_2O), hydrofluorocarbons (HFC_s), perfluorocarbons (PFC_s), hydrofluoride (SF_6) and sulphur hexafluoride (SF_6) are the GHG emissions and among them CO_2 is held responsible for causing climate change with the rate of 58.8% of all GHGs [1]. Carbon dioxide emissions are those stemming from the burning of fossil fuels and the manufacture of cement. They include carbon dioxide produced during consumption of solid, liquid, and gas fuels and gas flaring. Generally, electricity and heat generation, industry, transportation are the main sectors to produce global CO_2 emissions in the world. Therefore, the government of the countries that recognize the Kyoto Protocol are developed the clean energy policies to reduce their emissions. For this purpose, many of them change their energy investment policies to canalize the energy investments to the renewable resources such as solar, wind than fossil fuels, make some regulations in the industry and transportation sectors. No doubt, all these precautions has been affected the carbon emissions of the countries.

In the literature, there are many statistical forecasting methods. However, among these methods, Grey prediction method is one of the most accepted methods because of its ability to provide contented results under uncertain circumstances and predict the future value with only a limited amount of data. Therefore, in the literature, a lot of studies with a variety of Grey Models can be found [2–5] are the studies to predict CO_2 emissions for different countries using grey prediction model. There are also papers that use different forecasting models in the literature [6–11]. Pao et al. [2, 3] used the method to predict CO_2 emissions for Brazil and China respectively. Another study that used grey model to forecast CO_2 of Taiwan is proposed by Lin et al. [4]. Hamzacebi et al. [5] used the same method for Turkey's CO_2 emissions prediction. Chen [6], developed an integrated energy-environment-economy model to generate future energy development and carbon emissions for China throughout year 2050. Liu et al. [7], forecasted the energy consumption, gross carbon emissions and carbon emissions intensity in China from 2013 to 2020 through using system dynamics simulation. Özceylan [8], proposed particle swarm optimization and artificial bee colony techniques to forecast CO_2 emissions in Turkey, based on socio-economic indicators such as energy consumption, population, gross domestic product and number of motor vehicles data. Radojević et al. [9], developed neural network architecture to model, simulate and predict GHG emissions in European countries and the Republic of Serbia. Lotfalipour et al. [10], developed Grey system and ARIMA to predict CO_2 emissions and the result of the study shows that the Grey system forecasting is more accurate than the other forecasting methods. Ge et al. [11] predicted the CO_2 emissions caused by industrial energy consumption of Tianjin from 2005 to 2012 through using regression on population, affluence and technology model, the logistic regression model and GM.

To determine the effect and contribution of the arrangement and make improvement, the analyses and the forecasts for the emissions should be calculated. This study presents a grey forecasting model to predict CO_2 emission of per capita using historical data of some developed and developing countries such as Austria, China, Italy, Spain, Turkey, United Kingdom and United States. This model could be a guide for governments to see their future carbon emissions and provide a reference for future studies on alternative energy technologies to reduce CO_2 emissions for the considered countries in this scope of study. Also, this study can encourage countries that try to decrease their emissions to sustain their clean energy policies and be a notice for the countries that still not take any precaution increasing emissions to develop new, cleaner energy policies.

The remainder of this paper is organized as follows: Sect. 2 describe the development of the theory behind the grey system and is presented the traditional grey forecasting model. Section 3 present the data used and empirical results. Finally, the last section summarizes and concludes the paper.

2 Methodology: Grey System Theory and Grey Forecasting

Grey theory was first proposed by Deng in 1982 [12] to quantify uncertainty and information insufficiency. In grey system theory, a dynamic model with a group of differential equations is developed and it is called grey model (GM). When all the information about system is known, it is called "white"; otherwise it is called "black". The theory of GM is based on understanding of differential and integral calculus and the concept of grey derivatives is introduced so that we can establish models similar to differential equations for sequences of discrete data. Accumulated generating operation (AGO), inverse accumulated generating operation (IAGO) and grey modeling are three basic operations of grey prediction. In grey system, to whitenize a grey process, AGO which is vital characteristic of grey modeling is mostly used. Through AGO, the randomness of data is reduced. In other words; the degree of smoothness of a sequence is increased, which is the main purpose of the operation. In addition, non-negative, smooth, discrete function can also transferred into a series based on an approximate exponential law. In grey theory, this law is called as a grey exponential law and it is used for to establish a suitable foundation in building a differential model. Therefore, tendency of a grey quantity can be obviously detected and hidden special characteristics or laws in the raw data can be adequately disclosed [13–15].

2.1 Model: GM(1,1)

In this section, GM(1,1) model is explained in details. This model has been widely applied in various areas of practical applications of grey system theory. The meaning of the symbol GM(1,1) is given as follows:

G	M	(1,	1)
↓	↓	↓	↓
Grey	Model	First Order	One Variable

Assume that $x^{(0)}$ is a non-negative, initial sequence,

$$X^{(0)} = \left(x^{(0)}(1), x^{(0)}(2), x^{(0)}(i), \ldots, x^{(0)}(n)\right) \tag{1}$$

where $x^{(0)}(i)$ is the time series data at time i. Based on the initial sequence $X^{(0)}$, a new sequence $x^{(1)}$ is generated by the AGO. $X^{(1)} = AGOX^{(0)}$ then the following sequence,

$$X^{(1)} = \left(x^{(1)}(1), x^{(1)}(2), \ldots, x^{(1)}(n)\right) \tag{2}$$

where $x^{(1)}(k)$ is derived as follows,

$$x^{(1)}(1) = x^{(0)}(1) \tag{3}$$

$$x^{(1)}(k) = \sum_{i=1}^{k} x^{(0)}(i) \qquad k = 2, 3, \ldots, n \tag{4}$$

and $Z^{(1)}$ is the mean generated sequence of consecutive neighbors of $X^{(1)}$ given by

$$Z^{(1)} = \left(z^{(1)}(1), z^{(1)}(2), \ldots, z^{(1)}(n)\right) \tag{5}$$

where

$$z^{(1)}(k) = 0.5x^{(1)}(k) + 0.5x^{(1)}(k-1) \qquad k = 1, 2, \ldots, n \tag{6}$$

If $\hat{a} = \begin{bmatrix} a \\ b \end{bmatrix}$ is a sequence of parameters, and

$$x^{(0)}(k) + az^{(1)}(k) = b \qquad k = 1, 2, \ldots, n \tag{7}$$

is a grey differential model, called GM(1, 1) as it includes first order and only one variable. a, b are the coefficients; in Grey System theory, terms a is said to be a developing coefficient and b the grey input, $x^{(0)}(k)$ is a grey derivative which maximizes the information density for a given series to be modeled.

According to the least square method, we have

$$\hat{a} = \begin{bmatrix} a \\ b \end{bmatrix} = \left(B^T B\right)^{-1} B^T y_N \tag{8}$$

there

$$B = \begin{bmatrix} -z^{(1)}(2) & 1 \\ \vdots & \vdots \\ -z^{(1)}(n) & 1 \end{bmatrix}, \qquad y_N = \begin{bmatrix} x^{(0)}(2) \\ \vdots \\ x^{(0)}(n) \end{bmatrix} \tag{9}$$

here B is called a data matrix. By regarding the following equation

$$\frac{dx^{(1)}}{dt} + a\left(x^{(0)}\right) = b \tag{10}$$

as a shadow for $x^{(0)}(k) + az^{(1)}(k) = b$. The response equation for GM(1, 1) are as follows,

$$\hat{x}^{(1)}(k+1) = \left(x^{(0)}(1) - \frac{b}{a}\right)e^{-ak} + \frac{b}{a} \tag{11}$$

The predicted value of $\hat{x}^{(0)}(k+1)$ is calculated by performing IAGO on $\hat{x}^{(1)}(k+1)$.

$$\hat{x}^{(0)}(k+1) = \hat{x}^{(1)}(k+1) - \hat{x}^{(1)}(k) \tag{12}$$

where $\hat{x}^{(1)}(k), \hat{x}^{(0)}(k)$ means calculating values of $x^{(1)}$ and $x^{(0)}$ at point k, respectively [13, 14].

2.2 Test of Grey Forecasting Models

Forecasting is an action based on discussions and studies of the past to provide predictions about the future. Grey forecasting is based on some theoretical treatment of the original data and establishment of Grey models of the data to discover and to control the development laws of the system of interest so that scientific quantitative predictions about the future of the system can be made. For each specific problem, which particular model should be used as the forecasting model depends on a sufficient usage of conclusions of relevant qualitative analysis. The choice of models vary from one case to another. The feasibility and qualification of a model in use need to be checked with various criteria.

To demonstrate the accuracy of the proposed forecasting model, the actual value $X^{(0)}(k)$ and the forecasted value $\widehat{X}^{(0)}(k)$ can be compared. In order to evaluate the forecasting capability of the prediction model, various performance measure are used in the literature. In this study, post-error ratio (C) given in Eq. (13) is used as a performance measurement. This measure is one of the most important indexes to evaluate the model performance.

$$C = \frac{S_2}{S_1} \tag{13}$$

where

$$S_1^2 = \frac{1}{n}\sum_{k=1}^{n}\left[x^{(0)}(k) - \overline{x}^{(0)}\right]^2 \qquad S_2^2 = \frac{1}{n}\sum_{k=1}^{n}\left[\varepsilon^{(0)}(k) - \overline{\varepsilon}^{(0)}\right]^2 \tag{14}$$

$$\bar{x} = \frac{1}{n}\sum_{k=1}^{n} x^{(0)}(k) \qquad \bar{\varepsilon} = \frac{1}{n}\sum_{k=1}^{n} \varepsilon(k) \tag{15}$$

Assume that $X^{(0)}$ is a sequence of actual data, $\widehat{X}^{(0)}$ forecasted value of the model, $\varepsilon^{(0)}$ the error sequence, S_1^2 the variance of $X^{(0)}$, S_2^2 the variance of error Eq. (14), $\bar{x}$ the mean of $X^{(0)}$ and $\bar{\varepsilon}$ the mean error Eq. (15).

3 Grey Forecasting to Predicting the CO_2 Emissions of Selected Countries

In this study, some important countries' CO_2 emissions (metric tons per capita) is used to predict the following years' emissions. For this purpose, as seen in Table 1, CO_2 emissions of Austria, China, Italy, Spain, Turkey, United Kingdom and United States are taken as the leading investor countries to the energy technology. The raw data is obtained from the web site of World Bank (https://data.worldbank.org/indicators). Using the actual values from each countries' data set, forecasted values are calculated according to the formulation described in Sect. 2. The last value of each counties' data is separated as a test value and not added to the calculations with the GM(1,1) model. In Table 1, the values indicated with bold characters show the forecasted values of the related country. The proximity of the forecasted results of GM(1,1) (F_t) and the actual data (A_t) pertaining to seven countries carbon emissions is shown in Fig. 1.

Table 1. Actual and forecasted values of CO_2 emissions (metric tons per capita)

Country	Austria		China		Italy		Spain		Turkey		United Kingdom		United States	
Year	A_t	F_t	A_t	F_t	A_t	F_t	A_t	F_t	A_t	F_t	A_t	F_t	A_t	F_t
2010	8.07	8.07	6.56	6.56	6.84	6.84	5.82	5.82	4.12	4.12	7.86	7.86	17.44	17.44
2011	7.75	7.74	7.24	7.29	6.70	6.71	5.79	5.82	4.37	4.36	7.08	7.31	16.97	16.74
2012	7.39	7.47	7.42	7.39	6.21	6.19	5.66	5.53	4.42	4.38	7.36	7.11	16.30	16.59
2013	7.37	7.21	7.56	7.49	5.73	5.72	5.08	5.24	4.29	4.40	7.15	6.92	16.32	16.45
2014	6.87	6.95	7.54	7.60	5.27	5.28	5.03	4.97	4.49	4.43	6.50	6.74	16.49	16.31
2015		**6.71***		**7.70***		**4.88***		**4.72***		**4.45***		**6.56***		**16.16***
2016		**6.47***		**7.81***		**4.50***		**4.47***		**4.47***		**6.38***		**16.02***
2017		**6.25***		**7.92***		**4.16***		**4.24***		**4.50***		**6.21***		**15.89***
2018		**6.03***		**8.03***		**3.84***		**4.03***		**4.52***		**6.04***		**15.75***

* Forecasted values calculated using previous years data with GM(1,1) model

Finally, to show the accuracy of the forecasting calculations for each country, post error ratio (*C*) is calculated and given in Table 2. Grades of forecasting accuracy for the performance measures are seen in Table 3 [13–15]. As seen in Table 2, forecasted values of Austria, China, Italy and Spain are perfect (in first level), and the rest of the countries such as, Turkey, United Kingdom, United States are well (second level).

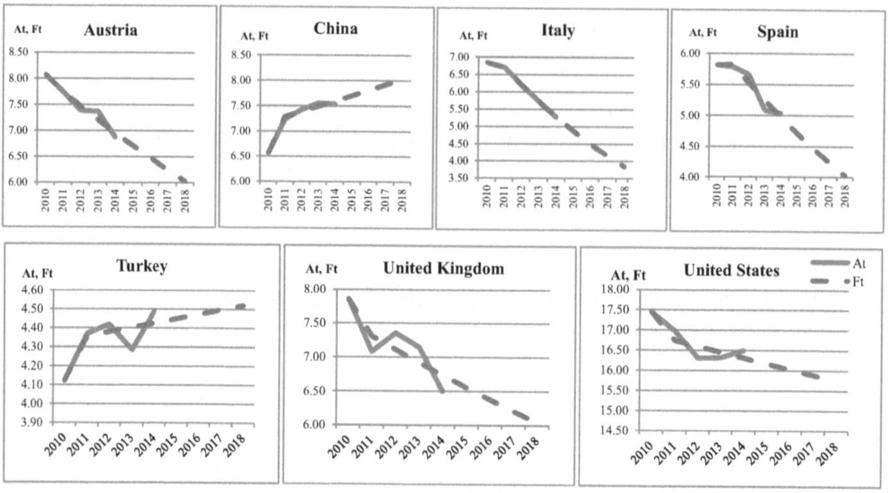

Fig. 1. Actual (solid) and forecasted (dashed) GM(1,1) carbon emissions curves of the countries

Table 2. Performance measures

Country	S_1	S_2	C	Accuracy scale
Austria	0,402	0,089	0.221	Perfect
China	0,370	0,046	0.123	Perfect
Italy	0,588	0,009	0.016	Perfect
Spain	0,345	0,098	0.284	Perfect
Turkey	0,128	0,063	0.494	Well
United Kingdom	0,440	0,209	0.476	Well
United States	0,441	0,196	0.444	Well

Table 3. Grades of forecasting accuracy

Level	C
1st level (Perfect)	<= 0.35
2nd level (Well)	<= 0.50
3rd level (Acceptable)	<= 0.65
4th level (Incapable)	>0.65

4 Conclusion

In recent years, the global warming and climate change has become vital problems across the world. According to the experts, one of the root causes of the global warming is GHGs. Hence, recently GHGs and the global warming have become one of the most significant research areas in science and global politics. CO_2 is held responsible for causing climate change with the rate of 58.8% among all GHGs. The method of Grey system theory is an effective tool for analyzing and forecasting uncertain systems. In this study, a GM(1,1) model is used to forecast CO_2 emissions of seven different developed countries; Australia, China, Italy, Spain, Turkey, United Kingdom and United States. The five-year data from 2010 to 2014 is used for setting up a forecasting model. At the end of the study, these countries' CO_2 emissions for the period between 2015 and 2018 are forecasted. Accuracy of the forecast results are computed by utilizing a widely used performance metric; the post error ratio (C). In general, the obtained forecast accuracy is high and forecasted results are reasonable. The results indicate that the method of GM is a satisfying technique to predict emissions. The data collected over some years after policy implementations can be used to forecast the long-term results of these policies. Therefore, the effectiveness of policies and regulations put into action can be assessed based on the outcomes of Grey forecasting. The results and predictions of this study should be rational in view of optimizing their industrial structure, developing non-fossil energy sources, and improving energy efficiency to promote low-carbon economic development. The research can be used as a guide for planning clean energy investments increasing environmental quality for the future.

References

1. World Bank (2007) The little green data book 2007, The World Bank
2. Pao HT, Tsai CM (2011) Modeling and forecasting the CO_2 emissions, energy consumption, and economic growth in Brazil. Energy 36:2450–2458
3. Pao HT, Fu HC, Tseng CL (2012) Forecasting of CO_2 emissions, energy consumption and economic growth in China using an improved grey model. Energy 40:400–409
4. Lin CS, Liou FM, Huang CP (2011) Grey forecasting model for CO_2 emissions: a Taiwan study. Appl Energy 88:3816–3820
5. Hamzacebi C, Karakurt I (2015) Forecasting the Energy-related CO_2 emissions of Turkey using a grey prediction model. Energy Sources Part A Recov Utilizaiton Environ Effects 3:1023–1031
6. Chen W (2005) The costs of mitigating carbon emissions in China: findings from China MARKAL MACRO modelling. Energy Policy 33(7):885–896
7. Liu X, Mao G, Ren J, Li RYM, Guo J, Zhang L (2015) How might China achieve its 2020 emissions target? A scenario analysis of energy consumption and CO_2 emissions using the system dynamics model. J Clean Prod 103(1):401–410
8. Özceylan E (2015) Forecasting CO_2 emission of Turkey: swarm intelligence approaches. Int J Global Warm 9(3):337–361
9. Radojević D, Pocajt V, Popović I, Perić-Grujić A, Ristić M (2013) Forecasting of greenhouse gas emissions in Serbia using artificial neural networks. Energy Source Part A Recover Util Environ Eff 35(8):733–740
10. Lotfalipour MR, Falahi MA, Bastam M (2013) Prediction of CO_2 emissions in Iran using grey and ARIMA models. Int J Energy Econ Policy 3(3):229–237
11. Ge X, Wang Y, Zhu H, Ding Z (2017) Analysis and forecast of the Tianjin industrial carbon dioxide emissions resulted from energy consumption. Int J Sustain Energy 36(7):1–17
12. Deng J (1982) Control problems of Grey system. Syst Contr Lett 1(5):288–294
13. Liu S, Lin Y (2006) Grey information theory and practical applications. Springer, London
14. Deng J (1989) Introduction to grey system theory. J Grey Syst 1:1–24
15. Hsu C, Wen Y (1998) Improved grey prediction models for the trans-pacific air passenger market. Transp Plann Technol 22(2):87–107

The Examination of Complaints About the Health Sector by Text Mining Analysis

Gamze Yildiz Erduran[1(✉)] and Fatma Lorcu[2]

[1] Department International Trade, Trakya University, Edirne, Turkey
gamzeerduran@trakya.edu.tr

[2] Department of Quantitative Techniques, Trakya University, Edirne, Turkey
fatmalorcu@trakya.edu.tr

Abstract. Since the quality of service is directly connected to human health in the enterprises providing services in the health sector, ensuring satisfaction by the service offered is one of the most important criteria. The complaints arise when the performance of health enterprises does not meet individuals' health-related expectations. The strengthening of the relationship between the businesses and the recipients of the service or bringing them completely to the point of detachment through complaints is linked to the effort of health enterprises to resolve those complaints.

In this research, online complaints of health care with a difficult structure to analyze, written in the natural language (unstructured) in the text form are scrutinized by the clustering method of data mining. Unprocessed (raw) data in online complaint platforms are transformed into information by processing through data and text mining methods. The results of the research are as follows: (1) to enable online complaints of health care to be divided into clusters without having to read them one by one; (2) to identify the most important elements of complaints and to divide them into categories; (3) to offer a new alternative method to health enterprises for assessing complaints vis-à-vis them.

Keywords: Online complaints of health care · Data mining · Text mining Clustering

1 Introduction

Health care is a significant indicator of the level of development of countries. The provision of good, high quality and affordable health services also provides the sustainability of economic and social development along with the ethical and social development of countries. In many OECD countries, the health sector constitutes the most dynamic and most important component of the economy [1].

In OECD countries on average, 9% of GDP is allocated to health expenditures. According to OECD 2016 data, health expenditures have showed an increase in the last five years especially due to outpatient treatment and long-term care services despite economic crises. The support of health expenditures for the economic growth leads the countries to allocate large shares in their budgets for health services. Health expenditures are financed by both private and public institutions. In OECD countries, in

N. M. Durakbasa and M. G. Gencyilmaz (Eds.): ISPR 2018, *Proceedings of the International Symposium for Production Research 2018*, pp. 612–621, 2019.
https://doi.org/10.1007/978-3-319-92267-6_51

general health expenditures are attempted to be shared equally between the public and private sectors. However, in some countries such as Turkey, Korea, Chile and Slovakia health expenditures are still heavily funded by public institutions [2, 3].

It is identified that in the year 2016 health expenditures have increased by 14.5% and total health expenditures comprise 4.6% of GDP. Correspondingly, as stated in OECD data health expenditures are intensively financed by the state budget. In the years 2015 and 2016, 78.5% of health expenditures were covered by the state budget [4].

Health expenditures mainly embrace developments that increase life expectancy and quality of life of individuals and technological and physical investments made in the health sector [5]. The fact that these expenditures made to the health sector are active and efficient provides the economic growth in the countries by developing the health sector and the institutions and organizations that provide services in this sector. However, if health expenditures are inefficient and non-productive, cost increase and low-quality health care services negatively affect economic growth in the country [6].

The effective and efficient development of the health sector, which has a large share in the economies of the countries, and the provision of quality services can only be achieved with an effective feedback and control system. Increasing the quality, efficiency and effectiveness of the enterprises providing services in the health sector is ensured by legal and regular inspections to be made by the government. In addition, in Turkey and in many countries within the scope of the protection of patients' rights, a number of feedback platforms have been set up in order to enable the citizens to convey their health sector related problems to the government. The patient rights complaint forms, the Ministry of Health Communication Centre (SABIM), the Presidential Communication Centre (CIMER), the Prime Ministry Communication Centre (BIMER) system are some of the feedback platforms that enable citizens to transmit their requests, complaints, notices, opinions and suggestions quickly and effectively to the relevant authority.

In this study, individual complaints oriented towards health care in Turkey were examined by text mining methods. The study consists of two phases. At the first stage, complaint texts on the Internet of individuals, who complain about the organizations providing services to the health sector, were collected. In the second stage, the data set consisting of 2380 complaint texts is separated into clusters and word or word groups representing clusters are identified. Correspondingly, in the study the most common complaint topics and the frequency of these topics are explained.

2 The Health Sector and Text Mining

In many countries of the world, there are problems that can cause major crises in health care. The main sources of the problems that cause this crisis are cost increases and the delivery of low quality service. While most of the resources of the countries' public budgets have already been transferred to health services, the growing share allocated within the public budget in the face of cost increases leads many countries to take measures to increase their efficiency in the resource utilization [6].

The most important health parameters used in international comparison for cost-effective and quality health care delivery are the number of beds for patients and the number of physicians and nurses. Considering these parameters, Turkey stays behind other OECD countries despite the positive developments in the economy and in the health sector. Therefore, Turkey holds a place within the countries with low health care cost-efficiency and service quality. Consequently, in the same way as many health organizations operating in Europe, Turkey should make excellence in management and strategic cooperation works. These works are the collaborations with health insurance companies, the encouragement of widespread use of technology in medical establishments, the shortening of the length of stay of patients and the support for outpatient care.

The improvements in information technologies also promote the development of health systems that effectively uses budget resources to improve productivity and reduce costs. For example, the electronic registration system is used to reduce costs and improve service quality. In addition, when information technologies are evaluated from the patient's point of view, they increase the patients' access to information, thus they ensure the patients coming to the health care institutions to become increasingly more conscious. For this reason, patients prefer health institutions that they believe they will benefit the most. This situation causes health institutions to position themselves according to their competitors and to be headed for brand recognition by developing brand image [7].

The positive comments or complaints made by healthcare service recipients regarding the brand value and reputation of healthcare institutions that are directed towards services affect healthcare institutions positively or negatively. It is increasingly easier nowadays to transmit comments and complaints to healthcare providers or healthcare system builders through the opportunities of Internet technology. Therefore, healthcare institutions that aspire to have a higher brand reputation and brand value should assess the comments and complaints and strive to restore their brand images [8].

The complaints inform healthcare providers and healthcare service recipients. At the same time complaints provide opportunities for health institutions to amend their mistakes. The healthcare organizations that embrace complaints as an opportunity are in an effort to increase their brand value and reputation. For this reason, they should utilize all the channels that they can find out complaints. Online platforms as the complaint channels are the most frequently used channels in the health sector as in many sectors in recent years.

The patient complaints are an important source of data that assess the qualities of healthcare institutions. The examination of complaints, also called as negative patient experiences, is essential to be able to realize improvements in healthcare services. Moreover, in this service sector, where human health is at issue, it is vital to quickly assess complaints and intervenes in order to save the reputation of healthcare institutions.

The non-reporting of patient complaints is also important just as the reporting of patient complaints, because in the case of absence of a patient complaint the situation that needs to take precaution is unknown. Therefore, it should be taken into account that the support for the reporting of health care service recipients concerning their dissatisfactions might provide an advantage for health care institutions [9].

Online platforms and new analysis methods are frequently used in the present day to determine patients' most important subjects of complaints and to resolve complaints quickly. One of these analysis methods is the text mining, which is a type of data mining that allows analysing large sets of data and extracting information.

Text mining is defined as the automatic extraction of information from different written sources, the extraction of key items, the creation of associations or hypotheses to form new facts with the extracted information. It can be stated that text mining, which is characterized as distinct from web search, is a variation of data mining. The most significant feature that distinguishes text mining from data mining is the extraction of information from natural language texts rather than structured databases [10].

Text mining, which allows analysing natural language texts, has system architecture similar to data mining in the process of obtaining qualified data and information. Both analyses have similar features, such as pre-processing routines, pattern discovery algorithms, and visualization tools for interpreting the results. All stages, except the stage in which the unstructured text data is converted into the structured data to perform data mining analysis, are the same. In other words, in the discovery of information from a text, text and data mining are used respectively.

The data mining process is composed of five consecutive steps. The first step is the collection of documents that form the data. In this step, in accordance with various programs and the analyst's decision, textual data are collected. The second step is the text pre-processing process in which technical operations such as the cleansing of the data set from the format such as figure, table, picture outside the text, the cleaning of unnecessary words, the marking of words according to orthographic rules or different word ordering probability calculation model, the parsing that allows for giving meaning to a word, the creation of a dictionary, the detection of word roots, are performed. This process is followed by data-dimension reduction in which irrelevant data are removed from the analysis in order to increase the efficiency of the analysis. Subsequently the structured texts are analysed by data mining algorithms and the results are evaluated and interpreted.

In the process of text pre-processing, there are many techniques at present developed for computer systems without language and perception skills, to allow them to analyse text language and meaning. Among these techniques, the most used in text mining analysis are information extraction, linguistic analysis and information discovery. The common point of the three techniques is that computer systems just as the human brain aim to analyse texts. In this regard, computer systems can help people make decisions with text mining techniques without having to read text stacks one by one.

3 Methodology and Results

In this research, within the year 2017 the complaints of 2380 health service recipients concerning the health sector in Turkey were collected from Internet and the data set was created. The data used in the research consist of complaint texts that individuals write with their original thoughts without any restrictions. The purpose of the research is to identify the sources of complaints of individuals from health services, to discover the frequency of complaints, to cluster complaints about health care services, to pinpoint the words that represent health service complaint sets and to prioritize measures to be taken against complaints depending on these clusters.

The complaints appeared in the dataset are unstructured text data. To convert this data into structured data, the data pre-processing process was first applied. The operations used during data pre-processing are the division of word roots, the discovery of words from the same root, the separation of the words between the spaces in the text, optionally setting a letter restriction for words, the conversion of uppercases to lowercases. By applying the data pre-processing process of the text mining to the complaints texts, the data structure was transformed into the appropriate form and the existence, absence and frequency of the word/word groups were calculated. Then, by clustering analysis, health care complaint clusters were established that revealed different problems.

In the course of data pre-processing, the word list in Table 1 was generated by calculating each word as a variable along with its frequency of emergence within the text and the document. According to the word list, the most frequently used words are "hospital, human, insurance, not, physician, medical examination, money, urgent, special".

When the word list is examined, it is likely to make a preliminary assessment of complaints and to make some inferences. For instance, according to the word list the word "hospital" is the most frequent word within the complaints and it comes up in 27% of the complaints. Therefore, it can be said that among health service providers the most prominent topic of complaint is related to hospitals. Following the hospitals, among health service providers insurance service providers come in the second place at the rate of 9% in terms of being the topic of complaints.

When people complain about health services, most of them focus on the matters such as "physician, medical examination, surgery, appointment, time, health insurance, and treatment". Therefore, it can be said that patients have dissatisfactions regarding the issues of physician, surgery, physician's examination and treatment process that they generally have resort to. The presence of complaints about appointments and time can be considered as a sign of the fact that the e-appointment system of hospitals, which started in 2012, still cannot offer a complete solution. The existence of many complaints regarding health insurance can be interpreted in the way that insurance services cannot meet the expectations of people. An important part of complaints about health insurance is connected to policies. It is especially striking that Allianz Insurance takes part immensely among the complaints. The frequent coming out of its name within the complaints is also correlated with the condition that Allianz Insurance is a company with the most widespread customer portfolio [11].

The Ministry of Health frequently comes out among the complaints. Because it is the most competent government department that assesses complaints about health in Turkey. The expectation from this unit to resolve every complaint brings an extra workload to the Ministry of Health. For simple complaints, it might be more meaningful to create intermediary solution units that can provide faster solutions without resorting to the Ministry of Health. The frequent emergence of the word "Istanbul" within the complaints is an expected situation as it is the province of Turkey with the highest population. The appearance of the words "child" and "baby" in the complaint texts pointing out a frequent problem can be considered as a situation that needs to be paid attention by health authorities.

Table 1. The word list

Attribute	Total Value	Document Value
Hospital	928	635
Human	592	504
Insurance	555	225
Despite	500	409
Human health	430	484
Health	442	403
Not	429	350
Physician	450	303
Child	413	289
How	347	275
Medical exam	327	203
Baby	326	186
Money	303	233
Urgent	302	172
He-She/They told	287	242
Special	284	194
Time	257	202
Bad	249	200
Defective	248	150
Patient	245	181
Complaint	236	206
Brand	211	171
Medicine	209	99
Appointment	201	96
Ministry of health	190	180
Surgery	185	89
Client	181	148
Health insurance	163	133
Cancellation	158	94
Food	156	128
Policy	148	80
Allianz insurance	133	72
Victim	129	108
Treatment	129	86
State	127	98
Tooth	124	74
To play with health	116	106
Unhealthy	105	96
İstanbul	102	94
Cheap	102	99

(*continued*)

Table 1. (*continued*)

Attribute	Total Value	Document Value
Customer service	92	74
The date of use	91	75
Attention	84	79
Saw	83	79
Threat	79	71
Cheap as	74	74

Adjectives that have an important place in the complaints are the words "bad, defective and cheap". People often evaluate the object, person or situation that they complain about as being bad or defective. Since the words "cheap" and "cheap as" form an association, they were explored by searching within the complaints. These two words and word groups are often used in the sentences expressed such as "human health should not be so cheap". Therefore, the word "cheap" is actually used to express the value of human health rather than cheap health services. In addition, the pattern of "playing with human's health" is frequently used in the complaints.

The word list generated in the data pre-processing process is a method of providing preliminary evaluation without reading complaints one at a time. However, following the formation of the word list, to make clustering analysis and to generate complaint clusters and words representing clusters can provide more accurate results by strengthening the preliminary evaluation.

Clustering analysis is the process of dividing the data set into similar sub-clusters. In clustering analysis, unlikeness's can also be used to distinguish groups of data as in the same way as the use of similarities. Thus, heterogeneous data sets are transformed into homogeneous data groups. For text clustering, the k-means algorithm as one of the most common non-hierarchical clustering algorithm was employed. By using K-average algorithms, k cluster number- randomly- is tested in various numbers, and 5 clusters form the most meaningful cluster structure. Euclidean distance formula as distance criterion and silhouette width value as performance value was used. The pruning value d as 5–90% and iteration as 10 were determined. The resulting clusters and performance values are given in Table 2.

As a result of the clustering analysis, the word list was created that enables the occurrence of document similarities of clusters. In cluster 1, the words with highest document similarity are "history, health, hospital, as much as it takes, human health,

Table 2. The Results of Clustering Analysis

Cluster 1: 636 items	Average silhouette width: 0,564
Cluster 2: 365 items	Average silhouette width: 0,519
Cluster 3: 264 items	Average silhouette width: 0,567
Cluster 4: 349 items	Average silhouette width: 0,645
Cluster 5: 766 items	Average silhouette width: 0,637
Total number of items: 2380	**Average silhouette width of whole dataset: 0,593**

complaint and healthy". Therefore, Cluster 1 is represented by the complaints directed towards human health and time in the hospitals. The words that represent Cluster 2 best are "the thing it takes, health, time, physician, hospital and human health". The words representing Cluster 3 are also comparable to Cluster 1 and 2. Clusters 1, 2, and 3 are clusters that are formed by very similar complaints. When the cluster neighbourhoods are investigated, it is seen that these three clusters are adjacent to each other. The word "physician" is the most important word that characterizes Cluster 2. It can therefore be said that this cluster is represented by the complaints about physicians in hospitals. The words that represent Cluster 4 best are "ministry of health, physician and complaints". Hence, this cluster is mainly the cluster in which the complaints about the physicians, especially those considered to be complained to the Ministry of Health, are included. The words that represent Cluster 5 best are "hospital, physician, the ministry of health and firm". Cluster 4 and 5 also show similarities.

The significance of the cluster number is determined by the silhouette width value, which is between +1 and −1. This value is calculated by the average distance the clusters in themselves and between each other. As a consequence of the clustering analysis, the mean silhouette width value of the entire data set was found to be 0.593. A silhouette width value greater than 0.5 indicates that an acceptable cluster structure was formed.

Since the number of complaints made is very high, for 10 and 15 clusters silhouette width value is also set to be higher than 0.5 although silhouette width value is formed at the highest in 5 clusters. While the complaint dataset was divided into 10 clusters, in comparison to other cluster structures more heterogeneous clusters were generated. The representative words that separate each cluster if the data set is divided into 10 clusters are given in Table 3.

Table 3. The representative words of clusters

Cluster 1: Physician, Fee, Defective, The Date of Use, Telephone
Cluster 2: Insurance, Firm, Urgent, Defective
Cluster 3: Insurance, Fee, Surgery
Cluster 4: Products, Hospital, Defective, Complaint
Cluster 5: Date, Hospital, Physician, Package
Cluster 6: Medicine, Medical Examination, Appointment
Cluster 7: Insurance, Fee
Cluster 8: Date, Urgent, Physician, Hospital
Cluster 9: The Ministry of Health
Cluster 10: Baby, Child
Total number of items: 2380

When the clusters are evaluated according to Table 3, the division of the data set into 10 clusters produces more significant results than the other cluster numbers. There are three sets related to insurance complaints. The first of these clusters include the matters of insurance along with firm and urgent, the second one contains the topics of insurance together with surgery, and the third one comprises the issues of insurance and fee. The clusters with the most explicit differences are 6, 9 and 10. Cluster 6 is the cluster of complaints linked to medicine, medical examination and appointment, cluster 9 and cluster 10 are the clusters of complaints related to the Ministry of Health and baby and child respectively.

4 Conclusions

The primary elements that compose the health services in Turkey are hospitals; home care services, outpatient care services, retail sales and medical equipment services, community health care, public health services and insurance services. In this research, hospitals and insurance services as the main elements of health services were identified as the most important issues in complaints about health services.

According to the results of this research, the targets of complaints about health services are physicians, hospitals, insurance companies and the Ministry of Health. Despite the presence of many health care personnel and nurses working in health care services, the frequent use of the word "physician" only within the complaints is an important finding to be emphasized. Reader et al. (2014) in their taxonomy review of 59 articles about patient complaints express that 86% of the target of complaints are related to health care personnel whereas 6% of the target of complaints are related to nurses. The targets of complaints about health services in Turkey are oriented towards institutions and physicians. These targets should be also the primary field of study for units seeking solutions to health care related complaints.

The analysis of complaints about health care by clustering analysis provides advantages in transmitting many complains to relevant units by dividing them and providing quick solutions to complaints. In this research when the complaints were divided into ten clusters, significant differences between the clusters of complaints came out.

Complaints related to health care services point out negatives in individual patient experiences not failures in these services. It also reflects conditions that criticize health care providers or personnel, including problems and difficulties with people's health services. It is an important situation that individuals report their complaints and that health care providers take precautions against these complaints. Because health complaints are directly related to human life. In addition, health care related complaints that lead to the emergency cases and the situations that need intervention might damage the reputation of health institutions.

When health care-related complaints are reviewed in total, they indicate potential problems about health services. For this reason, it is necessary for authorized units to find out these complaints and to analyze them meticulously and systemically. Therefore, the causes and prevalence of health care-related complaints will be determined. In addition, the examination of health care- related complaints may help reduce future health problems.

References

1. OECD (2008) China Development Forum 2008: Developing a Health Care System Benefiting All, Pekin, China, 23 March 2008
2. OECD (2017) Health at a Glance 2017: OECD Indicators, OECD Publishing, Paris. http://dx.doi.org/10.1787/health_glance-2017-en
3. OECD (2015) Fiscal Sustainability of Health Systems: Bridging Health and Finance PerspectivesOECD Publishing, Paris. https://doi.org/10.1787/9789264233386-en
4. TUIK (2017) Health Expenditure Statistics 2016, 16 November 2017
5. Akar S (2014) Turkiye'de saglik harcamalari, saglik harcamalarinin nisbi fiyati ve ekonomik buyume arasindaki iliskinin incelenmesi (an investigation of the relationship among health expenditures, relative price of health expenditures and economic growth in Turkey). Yonetim ve Ekonomi (Manag Econ) 21(1):311–322
6. Kilavuz E (2010) Saglik harcamalarindaki artis ve temel bakim hizmetleri (Rising health care cost and primary health care). Sosyal Bilimler Enstitusu Dergisi (J Soc Sci Inst), 29(2): 173–192
7. Deloite (2012) Turkey Health Sector Report, June 2012
8. Guzel FO (2014) Marka itibarini korumada sikayet takibi: Cevrimici seyahat 2.0 bilgi kanallarinda bir uygulama (Complaint chasing in the orientation of protection the brand reputation: a search on the online travel 2.0 information channel), Internet Uygulamalari ve Yonetimi (Internet Appl Manag), 5(1)
9. Reader TW, Gillespie A, Roberts J (2013) Patient complaints in healthcare systems: a systematic review and coding taxonomy. Systematic review. http://qualitysafety.bmj.com/
10. Hearst MA (1999) Untangling text data mining. In: Proceedings of the 37th annual meeting of the association for computational linguistics on computational linguistics, association for computational linguistics, pp 3–10
11. Health insurance. https://www.sigortam.net/saglik-sigortasi/saglik-sigortasi-buyumeye-devam-ediyor/

Operations Research Applications

A Study About Affecting Factors of Development of E-commerce

Mahmut Tekin[1](✉), Haydar İnce[2], Mehmet Etlioğlu[1], Özdal Koyuncuoğlu[3], and Ertuğrul Tekin[1]

[1] Selçuk University, Konya, Turkey
mahtekins@gmail.com, mehmetetlioglu@gmail.com, ertugrultekin42@gmail.com
[2] KSU University, Kahramanmaraş, Turkey
haydarince@ksu.edu.tr
[3] Necmettin Erbakan University, Konya, Turkey
okoyuncuoglu@konya.edu.tr

Abstract. One of the most important reasons of becoming of electronic commerce's popular and widespread in today's global competitive environment is to provide a new communication platform. This new communication platform is the most important feature that distinguishes traditional telephone and fax communication from interactive. This interaction has allowed many things to be done more easily in social and economic life and in particular the ability to access information quickly and cheaply. The use of the Internet and other information technologies has allowed e-commerce to evolve and expand. This study was conducted to determine the reasons why e-commerce has not been developed and the reasons for not doing e-commerce. A questionnaire was applied to 300 enterprises selected by layered sampling method operating in Kahramanmaraş. According to the survey, 118 businesses have e-commerce and 182 businesses have not. The results of the questionnaires were analyzed using SPSS 24 statistical program and the obtained findings were tested by frequency tables, cross tables and Chi-Square method. As a result, despite the fact that e-commerce firms operate, they have found that e-commerce trust, inadequate government support, lack of legal infrastructure, employees' lack of computer and hardware and lack of internet infrastructure are seen as obstacles. The age, educational status and customer relationships subjects are ignored by companies. In this study, it was determined that age, education, and customer relations of business employees affected e-commerce. It has been determined that a majority of non-ecommerce companies's products are not suitable for e-commerce. However, non-ecommerce companies also acknowledge the benefits of e-commerce and acknowledge that they are part of the business culture. If we focus on the above factors that hamper the development of e-commerce and strive to eliminate it, the development will be faster.

Keywords: E-commerce · Business · Marketing

N. M. Durakbasa and M. G. Gencyilmaz (Eds.): ISPR 2018, *Proceedings of the International Symposium for Production Research 2018*, pp. 625–642, 2019.
https://doi.org/10.1007/978-3-319-92267-6_52

1 Introduction

The technological changes that have come to the fore with the influence of globalization, especially the entering of into the human life and commercial has prepared some changes and developments. Businesses using the internet and the commercial area have also gained another dimension of communication with the state and other businesses. However, the greatest development and interaction on this subject has been made with the internet, others have not made such an impact. Basic operations such as promotion of products and services of enterprises, all kinds of commercial transactions, orders, payments, marketing activities, writing works, logistics works and communication are now carried out at a lower cost and at a lower cost through the internet. The widespread use of computers and internet all over the world reveals the concept of "electronic commerce" as a new form of trade with trade crossing the borders of the country. The widespread use of information communication technologies has brought with it the advantages of electronic commerce. Especially the increase of trade capacity, economic efficiency, profitability, communication speed and decrease in costs are the main ones. Now in the new economy; without fully taking advantages of the information and communication technologies it is difficult to develop without spreading the electronic commerce [1].

2 E-commerce

Nowadays potential consumers all over the world can come together on the internet network thanks to the power created by the virtual world. Volume and diversity will promote the companies that want to reach the electronic environment adequately [2]. Electronic marketing is seen as a revolutionary tool that transforms businesses' entire business processes. Electronic marketing has led to a transformation in traditional market structures. The central role in this transformation is the internet [3]. A new order is becoming common with the use of information and communication technologies. However, when the borders were lifted, the markets gained a different dimension and the sellers shifted their commercial activities to web-based systems. This new economic order includes all the economic interactions from one side to the production of science-information and from the other side through processing and distributing, largely through computer networks. There are different definitions of electronic commerce which formed according to the circumstances in such a world. E-commerce is a part of e-business that sets up committed business transactions, agreements and agreements and involves three stages of information exchange, agreement and settlement [4].

3 Economic Effects and Contributions of Electronic Commerce

The national market boundaries which have a continuous and dynamic structure are expanding in favorable areas for electronic trading. Due to the increase in Internet usage rates, the number of consumers and the number of firms increased in these markets, which brought new approaches together. Commercial activities are moving to virtual presence in countries where Internet usage is prevalent with these new approaches to share from e-commerce. As a result, Internet and e-commerce have been in a good position with the potential of people and businesses working together and in a short period of time significant changes have been observed with e-commerce [5].

3.1 The Impact of Electronic Commerce on Cost and Prices

Electronic commerce is one of the most important contributions to the business, in that it reduces the transaction costs and provides savings opportunities. For example, a US company has reduced its purchase costs from $75-250 to $3 because it has reduced paper consumption and order-to-delivery with an electronic ordering system that it has developed [8]. Internet and electronic commerce will change the way of doing business in every sector, especially giving numerical examples about the banking sector. The cost of establishing a Web-based virtual bank in 1997 is $1 million, due to the decline in banking costs in the US via the internet. It emphasizes the inevitability of the virtual banking system with the attractiveness created by this cost-effectiveness [9]. The benefits of electronic commerce can be summarized as follows [10];

- E-commerce provides cost savings during the sale (such as ordering over the Internet, downsizing of the physical infrastructure, acceptance of orders, monitoring of the consumer, number of staff).
- Cost savings are also achieved during the input supply to the e-commerce company (order schedule, paper-based order processing, courier delivery).
- The sale of digital products on the internet, the cost of consulting services is low.
- E-commerce also provides savings in storage costs.
- With e-commerce, operating expenses and additional investments are decreasing.

3.2 The Effect of Competition on Electronic Commerce

Today, competition is focused on technological change. Technological change also shapes the structure and needs of the market. The structure of the market with the developed technologies is also changing [11]. Within the changing market structure, the development of customer loyalty, the differentiation of competitors and the creation of a strategic competitive advantage are other dynamics in the use of e-commerce. Widespread use of electronic commerce has increased competition both nationally and internationally parallel to the development of the Internet. E-commerce activities have facilitated competition in the domestic market and have brought dynamism and action to the economy and company processes with structural changes [12].

3.3 Effects of Electronic Commerce on Market Constructions

The effects of e-commerce on businesses and market structures are explained by the relationship between the two. Its size has changed the position of those who are in the market because the market is virtual. Along with e-commerce, the idea that businesses are a single market consisting of millions of people has changed completely and the market has come to the forefront that millions of markets are composed of individuals [13]. Changing market understanding and structure brought with it some innovations. E-commerce has been shown to increase sales both in the existing market and in the potential market in terms of businesses. In an area study of 34 operations by Wölfle in 2011, 27 operators increased their sales by e-commerce [14].

3.4 The Effect of Electronic Commerce on Efficiency and Productivity

It is important for the management to know the effects and characteristics of e-commerce in its own firm and to carry out activities accordingly in order to be successful on the internet as an enterprise [6]. As e-commerce increases efficiency, it allows more trade with less cost. As the costs fall, the decrease in errors reflects positively on the efficiency and productivity of the enterprises and increases the service quality of the enterprises. For this reason, rapid technological development and electronic commerce practices increase productivity both in the production phase and in the trade phase [1]. The effects that provide changes in the efficiency of e-commerce application can be specified as follows; ease of access to markets, savings in operating budget, saving time and equal opportunity. As e-commerce is adapted to various systems, efficiency and efficiency increase. Especially with the help of e-procurement, electronic media, the productivity increase in companies related to time and money has made progress [16].

3.5 The Effect of Electronic Commerce on Marketing Methods

The widespread use of e-commerce has affected the traditional market structure and the changing of shopping methods. While this effect is compulsive in the sectors where e-commerce is widely used, it takes the form of incentive effect in the non-widely used sectors [1]. The new marketing understanding that has developed along with the internet has brought a different dimension to the commercialization among the companies with the sales oriented to the consumer [17]. In the markets, customers and sellers meet in electronic environment outside of physical boundaries, expand the market and e-commerce creates suitable environments. The potential for e-commerce is unlimited in terms of sales potential, customer orientation, time and cost savings, competitive positioning and the realization of new business models. Internet should be defined as a strategic sales channel, with appropriate marketing objectives defined and target group specific proposals developed in order to be successful using these features. The direct confrontation of electronic commerce with buyers and sellers has also brought some innovations in marketing methods. It is very important for companies to update their technological structure outside the web sites, detailed contents, products, employees and customer relations they prepare to succeed in electronic marketing activities to be able to respond to expectations and to integrate with other systems [15].

3.6 The Effect of Electronic Commerce on Employment

The effect of e-commerce on employment in the narrow sense occurs in two different ways. The first is that e-commerce leads to the emergence of new business areas, and second, the re-definition of tasks on business lines, leaving some of the fields in the middle [18]. Nevertheless, lost jobs can be compensated to a certain extent by the opening of new business fields. The implementation of e-commerce has resulted in a reshaping of the workforce. According to this, physically force is transformed from the workforce which uses the brain power and has a different skills [1]. Employees' skills (especially in the areas of information technology, production and use) must be kept in perspective and human resources must be put into practice in this regard [18].

3.7 The Effect of Electronic Commerce on International Trade

With the use of e-commerce tools, especially the Internet, companies in countries have been able to reach international markets with less cost and faster access, allowing them to compete with international companies. This, in turn, increases the share of the country's economy in international trade. Facilitating trade beyond national borders, e-commerce is an opportunity for this country to grow world trade to a level that can not be compared with previous periods. Exporting companies do not follow their purchases and sales, stocks and customs transactions with paper and pens, check their orders over the internet again and this provides great convenience for exporters. The e-market B2B sites in China, G.Korea and Hong Kong have been successful in the US and EU markets. B2B e-commerce has achieved much larger volumes through fairs; products have been moved to the virtual floor and new markets have been achieved [27].

3.8 The Effect of Electronic Commerce on Growth

The most important effect of e-commerce is that the economy leads to an increase in the total output of the economy with the productivity and the acceleration in growth, thus the cost savings [17]. The cost savings, increased computer and Internet usage, the more advanced interfaces of the web and the reduction of communication costs are creating a growth in e-commerce. In today's economy, e-commerce, an important marketing tool, provides many convenience and new employment opportunities for the firm, while allowing consumers to find goods and services more quickly, easily and cheaply. The advantage of e-commerce is that it saves cost and removes limitations. These advantages provide the growth and development of the enterprises while contributing to the economy of the country [19];

- Stakeholders can access information through Internet network,
- E-commerce tools make transactions fast and efficient,
- Product development with faster sharing of views,
- In virtual shopping malls where businesses are offered in the web environment, virtual catalogs can be created with minimum cost,
- Continuous cash entry with electronic payment tools,
- Reduce inventory costs and ensure customer satisfaction,
- Businesses can reach each other through electronic commerce,
- Uninterrupted communication, continuous shopping opportunities,
- Product information can be delivered to customers,

- Tax reductions in e-commerce and internet services,
- The opportunity to sell without intermediaries,
- Opportunities for branding and personalization in the electronic environment,
- Decrease in mistakes and mistakes,
- The inputs are the most appropriate sources.

4 E-commercial Problems, Obstacles and Sales Increase Efforts

4.1 Security Problem in E-commerce

It is important that people trust the system in electronic commerce as in classical commerce. In a commercial transaction, for information exchange or in any kind of electronic communication, private information about the individual or company should not be open to third parties. Furthermore, in the process of open communication to third parties, the confidentiality of personal information, the integrity of the information communicated and the identity verification must be ensured. Today, companies and institutions prefer applications such as Firewalls, Encryption Methods, Web Security and Internet Security Protocols to overcome the security problem in e-commerce [7]. The security issue in e-commerce is an important issue to focus on. As regards security, matters such as personal privacy, information flow and integrity are important in shopping. For network and transaction security, security protocols and encryption techniques are used to authenticate and grant access privileges. For example it may not be possible to obtain full security as with credit card transactions [20].

1. *Firewalls*

Firewalls are defined as software or hardware that allows protected networks or sites to only access certain features. The user can enter the system using his name, password, IP address [21]. In this context, firewalls control who or what software can log into the server or computers. E-commerce infrastructure provides security of internets open doors of servers using firewalls. E-commerce infrastructure provides security of servers using firewalls.

2. *Encryption and Transaction Security*

It is crucial that any danger be delivered to the buyer securely from life when information flows through networks. Different tools and methods are used for this purpose. Cryptography is a science that produces, develops, and studies decryption methods for secure information communication and hiding. With this process, the information is converted into an unidentifiable numerical value and sent to the receiver. The recipient converts this data into the original by using the password. Some of the encryption and locking methods used today are [21];

Open Key and Secret Key: Open and secret keys are numerical algorithms used in encryption processes.

Approval Authority: These are the organizations that provide the electronic identity document to the persons in exchange of data between the sender and the receiver.

Digital Signature: A "digital signature" is used to ensure that the message is received without modification and the sender's identity.

3. *World Wide Web Security and Internet Protocols*

It is essential that users communicate their identities mutually in e-commerce. In particular, different types of internet security protocols have been developed to provide security for shopping and electronic payment systems on the Internet [21].

Data Security (SSL): A program layer developed by "Netscape" for secure information transfer via webs in web applications. Information security is provided by a program between the application program and the TCP/IP layers. SSL technology is working to encrypt data traffic between content provider servers and clients. In e-commerce systems, it is used to secure the data traffic between the vendor server and the client computer and to verify the identity of the vendor.

Payment Security (SET Protocol): This is a protocol developed by organizations such as VISA, Mastercard and IBM to provide secure information transmission from the Internet in electronic commerce. In card payments, it ensures the confidentiality and security of information.

4.2 Infrastructure Problems and Deficiencies

Difficulties in importing information technologies in a large number of countries make it difficult to purchase and trade computers and information systems for e-commerce. In developing countries, communication costs are high, technological equipment is inadequate and services provided are not appropriate. Especially in countries where there is no competition for infrastructure, costs are doubled and the number of users decreases by 20% compared to other countries [21]. Major infrastructure problems and recommendations can be listed as follows [22];

- The telecommunication infrastructure should be improved and everyone should benefit from these services,
- E-commerce should be done by providing information flows with communication infrastructures,
- The obstacles in front of communication and information technology markets must be removed,
- The technical standards in the field of communication should be determined,
- Service providers should establish a trust environment for users.

4.3 Problems in Electronic Payments

Electronic payment systems are payments made with tools such as electronic money, smart cards, digital money, electronic wallets, smart cards, electronic money. Significant improvement will be achieved if the safe payment problem which is seen as an obstacle to the development and spread of electronic commerce is resolved. Systems known as traditional payment systems such as credit cards and funds transfer are insufficient in terms of security. There are some problems in the development and implementation of electronic payment systems, such as security, confidentiality, identification, and malicious use [21].

4.4 Shopping Habits Problem

At this point, it is necessary to divide shopping habits in the form of shopping habits of the companies and shopping habits of the consumers. Many companies have brought to bankruptcy with which established have great hopes because of having been unchanged of consumer habits [23]. In the survey conducted by TÜİK (Turkish Staistical Enstitute) on "Confidence in E-Commerce", reasons why people do not shop or order online are given buying by seeing product, loyalty and the rate of habits as 81.2%. This is a very serious issue that needs to be considered in terms of e-commerce and is a question to be solved.

4.5 Consumer Protection Problem

In e-commerce, there are separate prescriptions for the protection of the consumer, the physical selection of goods, ordering and payment from the internet and ordering/payment of software and digital products. Therefore, consumer problems related to this process can be solved in similar ways to solve problems in shopping [21].

4.6 Taxation Problem

Manufacturing has become a reality in different countries and the uncertainty of the places of companies selling on the internet brought with it the problem of taxation. The shift of capital and labor from countries with high tax rates to lower rates, the prevalence of global production and being sides who in different countries lead to the problem of taxation [21]. Taxation of electronic commerce is one of the most important considerations in cross-border e-commerce. The most important step taken on the international level regarding taxation of e-commerce is the "Electronic Commerce: The Framework of Taxation Conditions" report [24]. This report is important in that it sets out the general principles that apply to all e-commerce transactions. Countries are shaping the taxation of e-commerce within these principles. These principles are impartiality, effectiveness, certainty, fairness and flexibility.

[24] have found that the Internet has six characteristics that affect the functioning of tax systems. These characteristics can be summarized as follows;

- The possibility to establish public and private global communication systems that are safe and have low operating costs.
- In the process of "brokerage", the internet will significantly reduce the need for intermediation in the sale and distribution of goods and services and in the provision of information.
- Developments in the encryption of information provide the confidentiality of information transmitted over the Internet.

- Business functions such as design and production are intertwined on an increasing scale.
- Internet provides more flexibility to internationally active businesses in determining their structural form.
- Internet has caused fragmentation of economic activity.

4.7 Privacy Issue

It is important that the personal information and confidentiality are secured and the system is made reliable so that people can easily operate on the internet. Easily accessible such information, decryption of confidential information creates human distrust and insecurity for the system. Therefore, it is necessary to regulate the rules for using of information belonging to persons in the electronic environment and to enforce sanctions against those who do not comply [21].

4.8 Problems in Delivery and Customs

Today, when customer satisfaction is the front line, logistics management requires the right product to be in the right place, at the right time and at the right amount; keeping logistics services on the internet makes firm-customer communication fast and healthy [25]. Therefore distribution and delivery of goods and services are important.

5 Employee Demographics, Customer Relationship and the E-commerce

When the literature is examined, there is much work on the relationship between demographic characteristics and e-commerce. The importance of these demographical variables has not been taken seriously and ignored in many e-businesses [26]. It is essential to include such characteristics in analysing. Some characteristics such as age, education level of employee have been commonly employed in the field of marketing and are regarded to be important when understanding technology adoption behaviour as well as the behaviour of technology users [26]. The adopter of a new technology is typically younger, has a good income and appropriate level of education and more reactive to new innovation than the non-adopter. [27] identified seven factors that are linked with e-commerce adoption by the SMEs: decision-maker's age, education level, cosmopolitan outlook, perceived compatibility, cost, customer pressure and the perceived relative advantages of the innovation. Age differences continue to be of great importance in work attitudes as well as behaviour. Despite the general consensus that age differences have effects on the use of technology among both employees and employers, research in this area has so far produced some contradicting results. [28] suggest that there are differences with age in the importance in the adoption and use of e-business technology within firms. They argue that younger employees found attitude toward using a new technology to be more salient than older workers. The level of education is also found to be another important factor that influences employee behaviour and attitudes in technology adoption among organisations. In their study, [28] found that employees that possess a college or graduate degree qualification had high levels of e-business technology than those that did not possess any degree indicate that education promotes a positive attitude than does organisational experience with technology. This means that employees that possess educational qualification or training are more likely to understand how and when to use technology, they find technology use more effectively and fitting in their organisational duties. However, [29] found a negative significant moderating effect of level of education with perceived behavioural control on intention to use technology among Saudis. These results are consistent with those of that found out that the level of educational qualification is insignificant in prompting an individual to adopt and use e-business technology.

CRM is a kind of new business management model that joins organization technology and management tools together in other to increase and maintain the relationship between the firm and its customers. Electronic Commerce Customer Relationship Management or E-CCRM chiefly also relies on Internet or Web-based interaction of companies with their customers. E-CCRM is one of the most developed managerial methods that can be utilized in any organization. It is one of the best approaches that organizations can use to attract and retain their consumers from switching to other companies in order to prevent customer churn [30]. [31] found that companies' ECCRM-capability, as they have modeled it in this set of analyses, proves to correlate positively to corporate success in electronic commerce.

6 Methodology

6.1 The Purpose of the Research and the Importance

There have also been significant changes in the activities of firms especially internet and electronic commerce due to the developments in ICT and they continue to present important contributions to these processes. The technological feature of the Internet has facilitated the use of other means of communication and has provided great advantages to both people and institutions. Firms use these technologies in the promotion of products and services such as ordering, buying and selling transactions, payment transactions, marketing activities, correspondence, logistics activities; they have provided many advantages for country's economies such as saving time and cost.

6.2 Theoretical Model and Hypotheses

The study was carried out in Kahramanmaraş Province with a survey method in order to determine the availability, contributions, the problems and expectations they encountered and effects of e-commerce for 300 companies operating at different scales. Research model is shown in Fig. 1. Questionnaires and interviews was conducted with top and middle level executives in companies. The questionnaire consists of three main parts. The following hypotheses were established to determine the relationship between employees' age, education levels and customer relationships and e-commerce. In this context, the research hypotheses are;

H_1: There is a significant relationship between the level of education of employees of e-commerce companies and sales.
H_2: There is a significant relationship between the level of education of e-commerce employees and e-commerce market area.
H_3: There is a significant relationship between the level of education of e-commerce employees and e-commerce transaction ease.
H_4: There is a significant relationship between age of e-commerce employees and e-commerce growth.
H_5: There is a meaningful relationship between e-commerce and customer relations.
H_6: There is a significant relationship between e-commerce enterprises and export growth.
H_7: There is a meaningful difference between opinions on the advantages of e-commerce and non-e-commerce businesses.
H_8: There is a significant relationship between e-commerce enterprises and their views on competitive advantage.

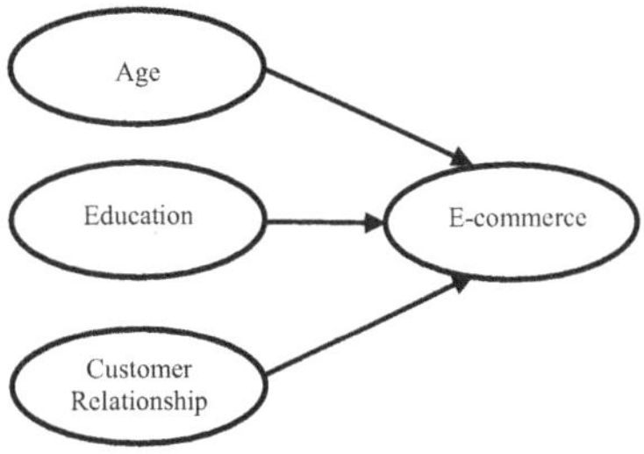

Fig. 1. Research model

6.3 Data Collection Tool and Scales

Survey method was used as data collection tool in this research. Scales used which have been proven validity and reliability in the questionnaire which studied by [9, 32]. Cronbach's Alpha value was found, 91 for business perceptions, attitudes and expectations of e-commerce and these values were found to be well above the acceptable level ($\alpha = 60$).

6.4 Population and Sampling

As of the end of 2016, 4,171 firms are operating at different scales according to the records of Kahramanmaraş Chamber of Commerce. The sample size was limited to 300 companies which selected from these enterprises identified by using stratified sampling method.

6.5 Data Analysis and Findings

The level of education of the employees of the companies participating in the survey is mostly high school graduates (60%); followed by secondary school graduates with 20.7%. In terms of the number of employees, close to half of the firms (47%) employ 1–9 people and the percentage of those employing 10–24 people is 25%. In terms of the average age of the employees, it is seen that 224 employees (75%) of the 300 firms employed employees between the ages of 26–35. 118 of the 300 companies participating in the survey are doing e-commerce; 182 does not do e-commerce. According to this, 39.3% of companies that do e-commerce in K.Maraş and 60.7% of those who do not do e-commerce. Textile (18.3%), food (13%) and steel products (6%) are among the leading companies in the survey. In terms of the duration of the activities, 28% are 1–5 years; 20.7% for 6–10 years; 19.7% for 21 years and over; 17.7% are in operation for 11–15 years and 14% for 16–20 years.

59 of the 118 e-commerce firms respond positively to the question "We have wholesale/retail sales from the Internet". This ratio constitutes 50% of the firms. If the unstable (2.5%) is neglected, 47.4% of the enterprises seem to be responding negatively to this question. So half of the businesses are selling on the internet, not half. This means that other activities related to e-commerce (order, payment, delivery etc.)

92.4% responded positively to the question of using internet in commercial activities. It seems that the internet has been used extensively in commercial activities with dealers. This feature is in line with statistics about the use of the Internet which is spread both in the world and in our country. Businesses believe that foreign trade will increase with the use of the internet. This rate is 88.2%. This situation can also be interpreted as a change in the perspective of foreign firms in Anatolia at the same time. Especially textile and steel ware products in K.Maraş are finding buyers both inside and outside the market and contribute to the increase of export revenues. Similarly, businesses are more easily identified on the internet (97.5%); (93.3%) and supplier relations are easier (87.3%) on the Internet.

6.6 Analysis of Hypotheses

Eight hypotheses was established to identify the reasons for preventing from development of e-commerce and the reasons for non-e-commerce businesses.

H_1: There is a significant relationship between the level of education of employees of e-commerce companies and sales.

It has been tested whether there is a significant relationship between the level of education of e-commerce employees and e-commerce sales activities. I fully agree with those who answered this question positively, 35.3% of them are high school graduates and 23.5% are university graduates; 66.7% of high school graduates and 21.4% of university graduates are employed. The proportion of those who have completely participated in primary school graduation is 11.8%, and the percentage of those who agree is 2.4%. **H_1 hypothesis was accepted** that the value of (P) was 0.000 and $P < 0.05$

according to the constructed Chi-Square test. There is a meaningful relationship between employees' education level and e-commerce sales increase in e-commerce firms.

H_2: There is a significant relationship between the level of education of e-commerce employees and e-commerce market area.

The relationship between the level of education of employees and the expansion of market space is being investigated in e-commerce firms. 65.1% of those who fully participated in this idea are high school graduates, 16.9% are university graduates; 48.4% of the participants are high school students and 19.4% are university graduates. In both cases, the percentage of those who find it positive is high. In primary school graduates, I totally agree with this rate at 1.2%, and I agree at 3.2%. The Chi-square test (P) was found to be 0.037. $P < 0.05$, there is a significant relationship between education level and market area expansion. **H_2 hypothesis was accepted.**

H_3: There is a significant relationship between the level of education of e-commerce employees and e-commerce transaction ease.

It is observed that the majority of high school and university graduates who respond positively question of "e-commerce transactions facilitated" in e-commerce companies. The proportion of high school and university graduates who are totaly agreed to this idea is 66.7% and 15.5%, respectively. The percentage of high school graduates who agreed is 42.4% and the ratio of university graduates is 21.2%. There is a significant relationship between level of education and e-trade and ease of operation since the Chi-square test result (P) is $0.041 < 0.005$. **H_3 hypothesis was accepted.**

H_4: There is a significant relationship between age of e-commerce employees and e-commerce growth.

The employees in companies are divided into five age groups. These are 18–25, 26–35, 36–40, 41–50 and 51–65. In the responses of the companies to the questionnaires, it seems that mostly the 26-35 age group is employed. it has been tested whether there is a significant relationship between age and e-commerce growth in the business. Among the e-commerce firms, those who are totally involved in the question of "foreign trade increases with the use of the Internet" constitute the largest group with 77.9% in the 26-35 age group and 81.5% in the participant. The value of (P) (0.020) is lower than the level of 0.05, it is determined that there is a significant relation between the age group and the increase of foreign trade in the internet, so **H_4 hypothesis was accepted.**

H_5: There is a meaningful relationship between e-commerce and customer relations.

It correlates with e-commerce's increasing customer relationships, again by associating it with the average age. Among the participating firms, 80.8% of those who think e-commerce increases customer relations are in the age group of 26–35. I only agree that this rate is 65.3%. **H_5 hypothesis was accepted** because the value of (P) found according to the Chi-Square test was $0.018 < 0.005$. So it is true that e-commerce increases customer relationships.

H_6: There is a significant relationship between e-commerce enterprises and export growth.

Table 1 shows the distribution of e-commerce firms according to sectors and the ratios of answers given to the question asked about export increase. Participants' distribution of this question is more common in the alternative under the other name and is similar in sectors that maintain the city's economy such as health, food, textile and steel goods. Whether or not there is a meaningful relationship between sectors and e-commerce's export increase is analyzed by Chi-Square test and found to be 0.864 (Table 1). This value was not significant for an output greater than 0.05 and **H_6 hypothesis was not accepted.** In other words, it is not meaningful to link export growth to a specific sector. In other words, export growth can be achieved in every sector depending on e-commerce.

Table 1. E-commerce exports increase by sectors

E-commerce increase export	Health	Food	Textile	Steel	Other	n = 118	
	%	%	%	%	%	χ^2	P
Totally agree	6.2	17.3	16.0	9.9	50.6	4.104	0.864
Agree	8.0	12.0	12.0	12.0	56.0		
Undecided	8.3	8.3	0.0	8.3	75.0		

H_7: There is a meaningful difference between opinions on the advantages of e-commerce and non-e-commerce businesses.

Non-e-commerce companies were asked about e-commerce expectations and benefits. It is seen that the companies that have high school graduates respond positively to this question. The proportion of high school graduates participating in this idea is 69.2% and 61% of those who say they participate. In addition to this, the ratio of those who worked primary school graduate was 0% in total participant and 5.5% in participant. In the analysis performed by Chi-Square test, **H_7 hypothesis was accepted** because (P) value is $0.026 < 0.05$. Therefore, among those who do not do e-commerce, those with high educational level believe in the benefits of e-commerce.

H_8: There is a significant relationship between e-commerce enterprises and their views on competitive advantage.

Numerical values showing that E-Commerce provides competitive advantage to the firm are given in Table 2. 94% of the companies respond positively to this question. In other words, firms believe that it provides competitive advantage. Only 4.2% of these firms are undecided and 1.7% of them did not participate.

Table 2. E-commerce companies and competitive advantage frequency

Features (n = 118)		n	%	Mean	Standard deviation
Current status of E-commerce					
The e-commerce company provides a real competitive advantage	Totally agree	68	57.6	1.50	0.66
	Agree	43	36.4		
	Undecided	5	4.2		
	Not agree	2	1.7		

When the Table 3 examined, high school and university graduates have a higher participation rate. While 72.1% of the totally agree participants were high school students and 17.6% were university graduates, 44.2% of the agree participants and 18.6% of the high schools are university graduates. **H8 hypothesis was accepted** because the value (P) obtained by the Chi-Square test was 0.000 < 0.05. According to this, it can be said that e-commerce provides competitive advantage depending on the education level of employees.

Table 3. E-commerce companies and competitive advantage frequency

The e-commerce company provides a real competitive advantage	Primary school	Secondary school	High school	University	n = 118	
	%	%	%	%	χ^2	P
Totally agree	1.5	8.8	72.1	17.6	27.204	0.000
Agree	2.3	34.9	44.2	18.6		
Undecided	0.0	80.0	20.0	0.0		
Not agree	50.0	0.0	50.0	0.0		

7 Conclusion

39.1% of enterprises participating in the research trade by internet. Businesses participating in the survey know that e-commerce offers advantages in terms of cost, speed, time and productivity. In addition, businesses believe that e-commerce especially in terms of cost savings, opening up to world markets, expanding the market area, providing competitive advantage and increasing trade capacity. Although the internet is widely used all over the world and also in Turkey, it is seen that these figures do not reflect as much in e-commerce as in the survey. The factors that are seen as obstacles to e-commerce are lack of trust, inadequacy of state support, inadequacy of legal infrastructure and lack of computer knowledge and equipment of employees. The rate of internet infrastructure is lower than others. Businesses see e-commerce as their future business model despite the factors seen as an obstacle to e-commerce. They assume that they will become more dependent on e-commerce within 4–5 years, an integral part of business cultures. Even non-ecommerce businesses believe that e-commerce is beneficial. However, the reason for not doing it is that the products are not suitable for internet marketing. Almost all businesses do not see the lack of funding as the reason for not doing e-commerce. Some business executives stated that they could not make a sale on the internet due to the agreement with the dealers in the interviews. Research and development activities is extremely important in terms of the manufacturing, promotion and presentation of new products and services. In particular, the discovery of new technologies or different features of an existing product provides a competitive advantage. Businesses in the research think that the use of the Internet for research and development is necessary. Providing the customer's commitment and loyalty to the relationship with the customer, the product or services; the protection of existing customers is crucial in terms of reaching new customers. In the research findings, businesses in this regard believe that e-commerce has increased customer relations at a high rate. It is accepted by the authorities of the e-commerce business that e-commerce in the research has contributed positively to all activities to increase sales. E-commerce is wholly effective in increasing sales. In this study, especially the activities aimed to increase the sales of e-commerce have been investigated by associating with the education level of the employees and average age. It has been found meaningful that the operations of the enterprises increase their sales activities, such as ease of shopping, retail/wholesale, competitive advantage, low cost, depending on the education level. On the other hand, it is confirmed that the increase in the average age of the firms is correlated with activities such as the age-average exchange, e-commerce, foreign trade, customer relations and the 26–35 age group.

As a result, there are many problems affecting e-commerce such as security, infrastructure problems and deficiencies, electronic payments, customer shopping habits, consumer protection, taxation, privacy issues and delivery and customs. When literature review is examined, it is also seen that there are too many studies related to these problems. However, age, education and customer relations issues that affect e-commerce in the business are ignored by many businesses and are not taken seriously. The age, educational status and customer relationship of the enterprise affects e-commerce. During the research, companies that have started to export by learning the internet marketing have been observed by the young managers who have university education and know the language. Findings obtained in this study overlap with similar studies in the literature. E-commerce is effective in increasing sales activities. It will be possible for e-commerce to become more widespread by removing obstacles and taking seriously. These findings which are the result of the study indicate that e-commerce activities aimed at increasing sales should be carried out nationwide; it can contribute to solving the obstacles in e-commerce and improving e-commerce in our country.

References

1. Akgül B (2004) Elektronik Ticaretin Kalkınma Üzerine Etkileri. Doktora Tezi Selçuk Üniversitesi Sosyal Bilimler Enstitüsü, İktisat Anabilim Dalı, Konya
2. Özbey M (1997) Dijital Dünya, İnternet'te iş kurma ve Geliştirme Fırsatları. Yönetim Geliştirme Merkezi Yayınları, İstanbul
3. Tekin M, Zerenler M (2016) Pazarlama, 2.Baskı. Günay Ofset, Konya
4. Dollmayer T (2001) Characteristika der Internetökonomie unter besonderer Berücksichtigung der Strategien im E-Commerce. Diplomarbeit, Fachhochschule Nürtingen, Sommersemester
5. Bernd W, Krol B (2001) Stand und Entwicklungsperspektiven der Forschung zum Electronic Commerce. Jahrbuch der Absatz-und Verbrauchs- forschung 47(4):332–365
6. Kurth S (2011) E-Commerce-Bedeutung des Absatzkanals für das strategische Management. Bachelorarbeit. Fachhochschule Münster
7. Özgener Ş (2004) KOBİ'lerin E-Ticarette Karşılaştıkları Sorunların Çözümüne Yönelik Alternatif Stratejiler. Erciyes Üniversitesi, Nevşehir İİBF Dergisi 6(22):167–181
8. Mentzel I (2003) Kaufverhalten und Kundenloyalität im E-Commerce –zwei empirische Untersuchungen
9. Kırım A (2009) Yeni Dünya'da Strateji ve Yönetim, 8.Baskı. Sistem Yayıncılık, İstanbul

10. Altınok RA (2008) Elektronik Ticaretin KOBİ'ler Açısından Uygulanabilirliği: Ankara Örneği. Doktora Tezi, Adnan Menderes Üniversitesi Sosyal Bilimler Enstitüsü, İktisat ABD, Aydın
11. Tekin M, Ömürbek N (2013) Küresel Rekabet Ortamında Teknolojik İşbirliği ve Otomotiv Sektörü Uygulamaları, 8.Baskı.Günay Ofset, Konya
12. Gora W, Mann E (2013) Handbuch Electronic Commerce: Kompendium zum elektronischen Handel. Springer
13. Çak M (2002) Dünya'da ve Türkiye'de Elektronik Ticaret ve Vergilendirilmesi. İTO Yayınları, Yayın no:2002-6, İstanbul
14. Wölfle R, Ralf L (2012) E-Commerce-Report 2012. Eine Studie zur Entwicklung des Schweizer E-Commerce, Institut für Wirtschaftsinformatik (IWI)
15. Tekin M, Zerenler M (2011) Pazarlama, 1.Baskı. Günay Ofset, Konya
16. Opuchlik A (2006) E-Commerce-Strategie: Entwicklung und Einführung. BoD-Books on Demand
17. Güler E (2010) İşletmelerin Yabancı Pazara Girişte Elektronik Ticaret ve Endüstriyel Pazar (B2B) Uygulamaları. Yüksek Lisans Tezi, Dokuz Eylül Üniversitesi Sosyal Bilimler Enstitüsü, İşletme ABD, Uluslararası İşletmecilik Proğramı, İzmir
18. Koçak O (2010) Elektronik Ticaretin İstihdama Etkileri. Sosyal Siyaset Konferansları Dergisi, (50):456–478
19. Gökgül M (2014) Türkiye'de Elektronik Ticaret ve İşletmelere, Tüketicilere Sağladığı Avantajlar, Dezavantajlar. Yüksek Lisans Tezi, İstanbul Kültür Üniversitesi Sosyal Bilimler Enstitüsü, İşletme ABD, İstanbul
20. Özbay A (2000) E-Ticaret Rehberi. Hayat Yayıncılık, İletişim, Eğitim Hizmetleri ve Tic.Ltd.Şti, İstanbul
21. Anbar A (2001) E-Ticarette Karşılaşılan Sorunlar ve Çözüm Önerileri. Akdeniz Üniversitesi, İİBF Dergisi (2):18–32
22. Canpolat Ö (2001) E-Ticaret ve Türkiye'deki Gelişmeler. T.C. Sanayi ve Ticaret Bakanlığı, Hukuk Müşavirliği: Ankara
23. Salter Y (2003) Elektronik Ticaret ve İstanbul'daki Bazı KOBİ'ler Üzerine Bir Uygulama. Yüksek Lisans Tezi, Cumhuriyet Üniversitesi, Sosyal Bilimler Enstitüsü, İşletme ABD, Sivas
24. Akçaoğlu SE (2011) Elektronik Ticaretin Vergilendirilmesi, Ulusal ve Uluslararası Perspektiften, Gelir Üzerinden Alınan Vergiler Bakımından Bir İrdeleme. Doktora Tezi, Ankara Üniversitesi Sosyal Bilimler Enstitüsü, Ankara
25. Bakkal M, Demir U (2011) Lojistik Yönetimi ve E-Lojistik, 1.Baskı. Hiperlink Yayınları, İstanbul
26. Hernández B, Jiménez J, Martín MJ (2011) Age, gender and income: do they really moderate online shopping behaviour? Online Inf Rev 35(1):113–133
27. Ching HL, Ellis P (2004) Marketing in cyberspace: what factors drive E-commerce adoption? J Mark Manag 20:409–429
28. Morris MG, Venkatesh V (2000) Age differences in technology adoption decisions: implications for a changing work force. Pers Psychol 53(1):375–403
29. Baker EW, Al-Gahtani SS, Hubona GS (2007) The effects of gender and age on new technology implementation in a developing country: testing the theory of planned behavior (TPB). Inf Technol People 20(4):352–375
30. Wahab S, Mohd Nor NA, Al-Momani K (2010) The relationship between eservice quality and ease of use on electronic customer relationship management (E-CRM) performance: an empirical ınvestigation ın jordan mobile phone services. In: International conference on e-Education, e-Business, e-Management and e-Learning, pp 368–372

31. Schoder D, Madeja N (2004) Is customer relationship management a success factor in electronic commerce? J Electron Commer Res 5(1):38–53
32. Bal V (2003) Elektronik Ticaretin Kobiler Açısından İncelenmesi ve Uygulanması. Yüksek Lisans Tezi, Gaziantep Üniversitesi, Sosyal Bilimler Enstitüsü, G.Antep

Process Management

Analyzing the Delivery Process with TOC

Fatma Serab Onursal[1(✉)], Semra Birgün[2], and Ercan Mızrak[3]

[1] Department of International Logistics, İstanbul Commerce University, Istanbul, Turkey
sonursal@ticaret.edu.tr

[2] Department of Industrial and Systems Engineering, Fenerbahçe University, Istanbul, Turkey
semrabirgun@gmail.com

[3] Sazcılar Composites Company, Bursa, Turkey
ercanmizrak16@gmail.com

Abstract. Today, when transition period to Industry 4.0 is experienced, every business has to increase productivity and reduce costs by establishing intelligent systems, to satisfy customers by increasing quality and enter a strong competition with other businesses. In this competitive environment, businesses need to constantly apply a number of improvements in order to achieve their goals, give importance to service and product improvement and to have products with low cost and high performance to operate properly. In this point, the mould manufacturing industry provides very important services in this respect. While moulds are providing high, easy and continuous production possibility, it also provides automatic production to the factory with minimum work force. In this study, the application of the Theory of Constraints was applied for solving the problems that adversely affect profitability and improving the system. In the study in which Thinking Process is applied, it was determined that the problems experienced by the operator in the process of mould delivery were experienced with "technical problems", "quality", "supply and purchase" and "personnel" problems. The weakest ring of the chain was searched, root problems were detected and evaporated. As a result, the improvement of the delivery period of the business has been ensured with the application of the Theory of Constraints Thinking Processes.

Keywords: Theory of Constraints · Thinking Process · Mould manufacturing Delivery process · Current reality tree · Future reality tree

1 Introduction

From the earliest times of human history, it is known that moulds made of stone are used to shape some kind of tools in periods called bronze and iron age. The tool produced with the desired material whose identical parts are in desired dimensions and in a very short time and obtained to help keeping human power at minimum level is called as "mould", which is obtained in order to produce equal parts, which are equal to each other, in desired dimensions and in a very short time with the desired material, and to help keep human power at minimum level. The person who designs, prepares,

N. M. Durakbasa and M. G. Gencyilmaz (Eds.): ISPR 2018, *Proceedings of the International Symposium for Production Research 2018*, pp. 645–659, 2019.
https://doi.org/10.1007/978-3-319-92267-6_53

asserts and operates this tool is called as "moulder". With the developing and changing technology, moulding has become an unidentifiable situation. The reason for this is that moulding has become a sector that produces parts of many products that have entered our everyday life.

The turnover of the mouldmaking industry is around 75 billion Euros all over the world and the sector shows 6% growth every year. In order to be able to produce multiple parts at sustainable quality at competitive prices, mould is a need for every field. Mould manufacturing is the most important complementary process that transforms innovation and design process into added value. Mould designing and manufacturing started 10 years ago after product design was completed. In today's Industrial Age 4.0, mould designers and product designers work together (co-design) to shape the product. At this point, it is possible to reduce the mould production process by 50% compared to the old one.

This is the common reason why moulds have been treated as a strategic sector in the developed countries since the industrial revolution. Many durable merchandises are produced thanks to the moulds. Inevitably, qualified personnel and high-tech mould makers in this sector make it possible to manufacture many innovative products, from automotive to white goods. The mould sector serves the automotive industry on an average of 70% worldwide. This is also the case for Turkey. One of the most important item of industry and exports for Turkey which is an export-based economy is the automotive industry with 25 billion USD. Mould manufacturing is the most detailed and complicated process in the process for introducing a new vehicle program. Each of the moulds that will manufacture millions of pieces, which generate the vehicles to be produced in millions of units in series, arise as a result of project-based specific design, analysis and manufacturing processes. Each mould must manufacture millions of pieces in series at the same quality. For this reason, every mould has to be done absolutely true in one go by using high-tech design and analysis software, advanced CNC machine tools and qualified workforce. A single mould or mould operation is required to manufacture millions of plastic parts such as bumper and front panel and each of the brattice panels such as engine bonnet or door in sustainable quality at competitive cost.

In this study, it is aimed to eliminate the problems experienced in delivery process in a mould manufacturer company during the delivery process with the help of Theory of Constraints Thinking Process and to improve the delivery process with the solutions suggested. In the following sections, the Theory of Constraints is summarized briefly; the company is introduced, two of the applied Logical Trees of Thinking Process are explained and solutions are developed.

2 Theory of Constraints and Thinking Processes

For many businesses, the aim is to give efficient service and to provide profitability as a result of this service. Since the main goal here is high profits, so the constraints that prevent the system from earning higher profits must be elevated. Every business is a system and the Theory of Constraints is used to reach it to the goodness by improving this system. The Theory of Constraints is a philosophy of management that focuses on the basic processes that prevent the operation of the system and tries to increase the expected benefits from the system in a continuous development cycle. According to the thought of this philosophy, the quality and amount of the utilities expected from the system are limited. Everything that can prevent the system from achieving optimum performance is treated as a constraint.

The theory of constraints put forward by Goldratt has often been described as an intelligence of performance improvement. The development of the Theory of Constraint can be examined in three stages [1].

The first phase is developed between 1975 and 1985. At this stage, scheduling is made by using "*drum-buffer-rope*" concepts. Between 1975–85, it appears that the Theory of Constraints emerged as a inventory management and production flow system. According to Goldratt, there is a process in which the production speed is constantly low in the systems established in the companies. This process is bottleneck or constraint for the system. For this reason, the scheduling system should be realized by considering this constraint or bottleneck. The scheduling should be approached from a rather different and wide-ranging approach, it should be avoided to incline to the sub-optimizations that are integrated into the system and a general optimization in the system should be provided. Goldratt presented a method for general optimization in the emergence of the Theory of Constraints [2].

The second phase is "*Flow*". In his publication, Goldratt [3–5] emphasized the importance of enhancing system output; stated that the thing that should be created in the system is not saving but the enterprise should obtain more goods, in other words more income, than the resources in its hand. After 1990s, "Thinking Process" was developed with the aim of identifying non-physical constraints and producing solutions that are *the third phase* of the theory. The application of Thinking Process consists of *current reality tree, evaporation tree, future reality tree, prerequisite tree* and *transition tree* phases. First, *the current reality tree* is prepared in order to present the problem. In order to achieve the desired result, the problematic and conflicting situations are explained by *evaporation clouds* and the solutions are suggested by making additions. In order to eliminate the problem, a *future reality tree* is created, the condition(s) that prevent reaching to the solution is defined with *the prerequisite tree*, and finally it is made use of the *transition tree* to show how it is passed to the desired situation [6, 7].

According to Goldratt, efficiency should be defined at the enterprise level and should be increased with operational improvements. The system thought is a tool used for performance development thinking and that is very useful in terms of revealing the linking between the factors that are not connected to each other or that are separated from each other and it also helps to identify, understand and prioritize the different perspectives, common goals and problems of the unit. While some constraints prevent the development of performances with high importance, others prevent a number of performances from improving. While some of these constraints have a high impact, some have little impact. Constraints that effect in the development of the system at a high level are defined as "root cause" or "root problem". For all organizations, since high-impact constraints will cause performance to fall behind by the time, focusing on high-impact root problems, identifying and intervening them by the administrators will help them to use the time effectively.

Constraints in organizations are generally divided into two as “physical” and “political”. Physical constraints include machine, material, human, capacity constraints, equipment, demand, etc., while political constraints include procedures, operating policies, external rules affecting the company, and management styles. Organizations often face very few physical constraints. According to Goldratt, 99% of the constraints faced by organizations are political constraints [8]. A constraint without considering its purpose is defined as “Everything that a system that puts forward the purpose of an organization and prevents it from succeeding”.

The most effective and the best method to reveal the optimum performance of a system is seen in five forms as; * Determining the constraints of the system, * Deciding how to turn constraints into benefits, * Subordinating everything else, * Elevating the constraints, * Go back to step 1, but beware of “inertia” [9–11]. These five stages of the Theory of Constraints are used in conjunction with the Logic Trees of Thinking Process. Theory of Constraints Logical Tree of Thinking Process provide a stringent approach to even the most complex problems. In this way, effective procedural support is provided to these five stages. The accurate and effective use of Logical Trees of Thinking Process inevitably leads to root causes and provides effective solutions to root problems [12].

Each Logical Tree structure presents the cause-and-effect relationships in the analysed problem with diagrams and figures.

Dettmer [9] stated that the Current Reality Tree is a logical structure designed to describe the current state of reality. The current reality tree conveys the possible chain of cause and effect relation presented as a set of specific mixed conditions. This structure is looking for the relationship between the indications in the system and the cause that led to this indication and the result. The current reality tree represents the most probable cause-and-effect chain and, when certain fixed conditions are taken into consideration, it is depicted as a systematic structure from top to bottom.

It is also possible to define the current reality tree as a diagnostic part of any problem. When the person who decides to use the Theory of Constraints decides what to change first, then finding a solution to the logic in order to eliminate the main problem in two stages within the process is actually an answer to the question of “to what it will change” [13]. To develop solution for this problem can only be with the “Evaporating Cloud and Future Reality Tree”. The evaporating cloud is composed of a cluster structure with 5 boxes. The person who implements this theory describes two requirements that are contradictory, and opposing to each other and the necessity behind this demand and a quite general goal which he tries to meet this need. Afterwards, the person who implements it reveal the hypothesis immediately behind the connections between the goal and the need and between the need and the desire. This contradiction is exactly the same as that of the “Current Reality Tree”. Goldratt states that the administrators are seeking solutions that compromise while trying to solve such dilemmas (satisfying both requirements at a definite ratio) [9].

At the beginning of the processes of the Evaporating Cloud lies to express the problems clearly. Then, by drawing a cloud diagram that is suitable for the problem, applying a process that is thought to be necessary, revealing a logical conclusion for each arrow, potential solutions are identified and an injection selection is made to fulfil these solutions.

The Future Reality Tree is a very useful tool for previewing and it explains what it will be changed by testing how it will affect the operation. The solution should be examined once more by using cause and effect method by taking into account its effect on the future of the organization. Every step of the Future Reality Tree minimizes the probability of the participants [14].

After the solution defined as integrated solution that the applicant expresses by the evaporating cloud method is revealed, the next application is used to form the future reality tree. By using the cause-effect application, the tree is formed and it should be examined carefully to test the solution [15].

The aim of prerequisite tree is describing the obstacles against the idea that are integrated with the application of evaporating cloud. Prerequisite tree is used to take into account situations that are preventing the achievement of the aim and to provide solutions. All ideas in the current reality tree are applied to create this diagram.

The last tool in the Theory of Constraints and Thinking Process is the Transition Tree [14]. The Transition Tree is planned step by step and the evaluated process shows in detail the steps that can be taken at the end of the decision process to improve the system and in the implementation phase. As it is on a road map, it shows all the details of what to do. The Transition Tree helps to make decisions about how to modify the Transition Tree and the Theory of Constraints.

If it is necessary to present several examples in the historical order without countless studies done by the Theory of Constraints in the literature, [14] have investigated how the Thinking Process influences the politics of businesses in general. [7] investigated how the Theory of Constraints is applied in the manufacturing sector. [16] investigated how the Theory of Constraints is used to identify and solve problems in service enterprises. [17] investigate how the Theory of Constraints affects the performance of the health care system.

In his study, [18] argued that it would be wrong to judge only the quality loss in the process of determining the quality improvement projects; it has been concluded that instead of this situation, more accurate results will be obtained if product mixes are detected by considering the flow and quality loss in the system together. In [19]'s study about order management, it was made use of activity-based costing in the calculation of production costs of the enterprise. The Theory of Constraints provides convenience to the study on the identification and elimination of system constraints in this research. With the use of activity based costing in the calculation of profitability, the calculation of irrelevant costs has been eliminated. [20] examined the impact of the prevention of the current capacity constraint in the process of production after discussing it in production operation under management accounting.

[21] stated that the Theory of Constraints are being used more in terms of businesses since the processes create a status of main way in the improvement studies. He pointed out that the Theory of Constraints provides great facilities for the decision-making stages of management. [22] aimed to improve the points that airline operations could compete by using Thinking Process application which was used as a method of the Theory of Constraints.

[23] investigate the process of establishing inter-organizational systems in large industrial, service and commercial companies with the method of the Theory of Constraints.

In their research about the evaluation of customer complaints with Thinking Process in a call centre by [24], the call centre was examined with the eye of customers, the determined problems were analysed and it was brought the solution offers with the Theory of Constraints and Thinking Process to provide customer satisfaction. In the study carried out by [25] by using the Theory of Constraints, they identified behavioural, managerial and material constraints in a medical imaging unit and lab of a 1000-bed hospital and they described how they solve them.

In this study, it was made use of the Theory of Constraints, which has been applied in many sectors and in different fields with the aim of eliminating the problems experienced during the delivery process of a mould manufacturer.

3 Improvement of the Delivery Performance by Thinking Processes

Here, Theory of Constraints application carried out with the aim of discovering the root causes of the problems and eliminating them in a mould company (KBM Co.) experiencing problems in the delivery process is presented. In the production phase of the products that has entered our everyday life, it is moulding that provides quality, time and measure accuracy, saving from material and providing identity, and at the same time minimizing the cost of labour. All processes from mould design to manufacturing must be carried out in time without sacrificing quality.

The services provided by the company, which has 18 years of business history and which operates in Bursa and making Mould and Machine production, can be collected under 5 main titles.

Mould Design Service,
Mould Analysis Service,
Mould Exercise (MAP) Service,
CNC Outsource Processing Service,
CMM Measurement Service.

KBM Co. is one of the enterprises that started to manufacture moulds in Bursa. The vision of KBM Co. is a company that produces unique technological solutions on the production of the moulds in the frame of total quality management, a supporter of continuous development, has the priority of customer satisfaction, sensitive to society and environment, has national and international leadership and respectability. Its mission is to keep up with the latest technology in mould and renewing; competitiveness, productivity, using the resources effectively and maximizing the satisfaction of the country, customers, stakeholders and the environment with its product and employee qualities.

The process starting with the receipt of the customer order ends at the customer presentation section. After customer received the order, it is required to make preliminary design and to receive approval. Once the preliminary design approval is obtained, the necessary engineering calculations are made according to the type and capacity of the machines and then the approval of the general manager is obtained, the suggestions are prepared and delivered to the customer. After the client approval, the project is detailed, prepared and the presentation of the product is realized.

In this study started after KBM Co. encountered delivery problems after the production of mould, the processes involved in product delivery were analysed with the Theory of Constraints Thinking Process and problems were identified and solution proposals were presented. The Theory of Constraints Thinking Process are implemented step by step in the company, but only two steps will be explained here.

3.1 Current Reality Tree Creation and Analysis

The first step is the creation of the current reality tree. The purpose of the current reality tree is to demonstrate the cause-and-effect relationship between adverse effects and outcomes and reveal the causes of undesirable effects.

KBM Co. receives various complaints from its customers. The main points of these complaints are; failure to observe the contract delivery times and intensive delivery delays; failure to comply with contractual requirements; not changing the materials that used wrong or incomplete; not repairing them; not being able to providing sustainability in the mould; failure to reach the authorities; late reply to e-mail and messages; inadequate technical personnel and failure to deliver project closure files.

As a result of interviews made with customers, information received from managers and technical personnel, and realizations, it was determined that the problems experienced from the drawing process of the mould to the delivery of the customer were collected under four headings. These are "technical constraints", "supply-purchase constraints", "personnel constraints", "poor quality". Figure 1 presents the Current Reality Tree, which shows the root causes, cause-effect relationships of the obstacles to reaching the main goals of the business. That the constraint emerged due to the absence of the Restricted Quality and Productivity Management System and the lack of a competent management and manufacturing engineers is seen in the Current Reality Tree. One of the Root Problems was found as "Frequent Change of the Persons to Give Approval to the Customers". Since the customer cannot be intervened, it is proposed that protective measures be taken in order to ensure that the approvals received from the customers are not in error and the approval mechanism should be increased to higher levels. For this reason, this root problem cannot be eliminated and it remained only advisory. Other root problems were evaporated.

These four basic problems experienced in the delivery process in KBM Co. mould production company are explained under the following headings.

(1) *Technical Constraints*: The reasons for delays in product delivery due to the problems experienced in design and production, which are called technical constraints, are given below.

Delays occur in the planning of designs due to not being delivered on time or due to late sharing of technical specifications. Delays in the preparation of documents describing how the product planned to emerge will be in terms of mechanical, physical, chemical and aesthetic aspects, late feedbacks and incomplete preparation of specifications are among the reasons that cause difficulties in design.

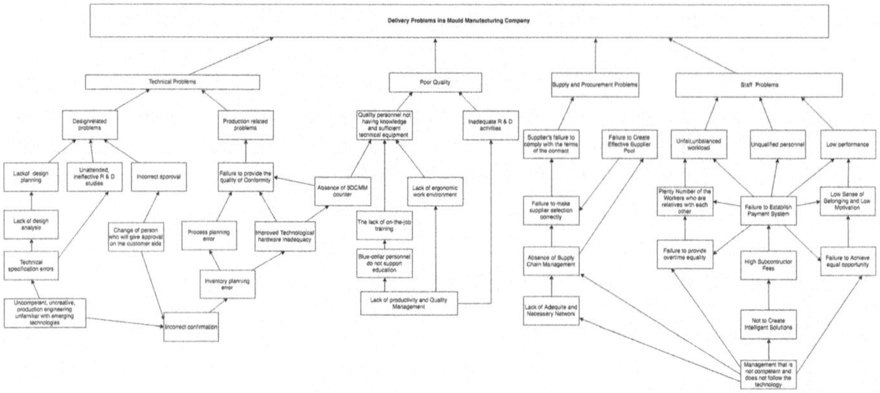

Fig. 1. Current reality tree of KBM Co.

The more efficient a design is; the less problem it has in production progress. Being negatively affected of the efficiency of the design causes the project to enter production late. This situation causes the late delivery of the moulds. The more elaborately the design is created, the more the delays that occur due to design can be avoided. Factors affecting the effectiveness of the design are R&D studies, continuous action, no design analysis, no design planning and constantly changing of the persons who will give approval on behalf of the customer.

The design created by the authorities in line with the customer requests includes problems arising from the authorities not discussing with the stakeholders or sending the chairman's design plan to the customer. Since each client has different demands, there are occasional accumulations in the project period. Because each customer project has its own quality cycle, supplier control systems, the mould designs must be prepared according to the requirements of the factories. In short, it can be expressed as customer's creating his own standards. One of the big mistakes that the firm made is to create a plan and method according to itself not to the customer. This prevents the design from being effective and creates technical problems. The sum of these errors is one of the factors that affects the mould delivery problem.

The more efficient and participatory the R&D work is, the less problems can be encountered in production and delivery. At KBM Co., it seems that R&D work is not carried out effectively and realized in an unattended way.

The mould manufacturing company presents a design despite the problems that arise. After the emerging designs, the process of bringing it the customer's approval comes. Approval meetings on design with customers are held with the participation of authorized specialists. As a result of meetings held with the participation of experts, some actions are being put forward. In order to correct the problems in the actions, a new study is done and after that the approval meeting is held again. Experiencing some situations such as the meetings that are held as a result of the problems are made constantly, being different of the person who gives a new action and the action desired to be made is being the action wanted to be corrected in the previous meeting, etc.

caused not to comply with the design plan. This case also prevents the design from being effective and causes the mould not to be delivered on time.

Production planning and manufacturing issues: The material lists of raw materials to be used in production come late or in part from the R&D section. This leads to a production plan and major manufacturing problems.

Production Engineers: It is the most troublesome and longest process of the mould manufacturing company. The more efficient the techniques used for production, the faster the production process will be completed without any problems.

Failure to make production effectively, not giving due importance to the inventory planning and revisions poses a problem in terms of production. The root cause of the production problems of the company is that it does not provide manufacturing efficiency. Factors affecting manufacturing effectiveness are "production and inventory planning", "design approval", "process planning". Making the manufacturing effective can help solve technical problems.

The production problems of the company include production and inventory planning. Failure to make the necessary inventory plans on time, or making incomplete or incorrectly causes the production to be delayed. That the authorized people from product department do not attend or are not invited to the approval process of the design made by the company for the production of moulds and not receiving advices or opinions for the production of mould designed are the problems encountered during the production process.

Process Planning: That the decisions such as with which method a part of a mould will be produced (will it be cut in wire erosion or with machining in CNC milling) are not made after the project approval of the work is one of the main factors delaying production.

Conformity Quality: Manufacturing of a mould project according to CAD data. Some of the main factors affecting the quality of qualification are "production approval", "technological equipment" and "measurement 3DCMM counter".

Production Approval: The issue of revision in a part of the mould projects and the fact that this revision is made without giving any notification to the R&D department and not reflecting it to the design environment and making production by directly interfering with the production organization are another factors causing production problems. It is necessary to prepare the mould projects finished in serial production according to the serial production conditions, to educate the personnel who can work on serial production conditions on the mould and to be aware of what should be careful about. The main problem that arises in the present situation can be seen as the fact that the authorized persons who need to train the press operator directly do not prepare operating instructions.

(2) *Poor quality*: Due to the quality problems of the produced moulds, the reasons for the delays in product delivery are given below.

R&D Activities: It is seen that the product quality is affected because the R & D activities of the company are not carried out in a sufficient order.

Not being knowledgeable of the quality department and not having sufficient equipment: The lack of necessary equipment and apparatus in the quality section, the lack of an existing quality culture, the lack of environment temperature for the devices, the lack of quality assurance systems, that the training of loss of quality is deemed as a waste of time by the staff and the lack of necessary ergonomic work environment constitute the quality problem.

Lack of 3D CMM Machine: Due to the fact that the coordinate measuring machine is not available at the firm, the part to be extracted from the core is not in the desired quality.

Non-ergonomic Work Environment: It is the failure to provide the required room temperature. In the measurement laboratory, the devices and equipment must be at a certain room temperature in order to function properly. In the present case, precise measurement operations cannot be performed.

Lack of Productivity, Quality Management System and Quality Culture: Since there is not an existing quality policy in the company, there are basic problems related to quality. The lack of on-the-job training, and the fact that blue collar staff do not have the necessary training to educate them prevent the formation of quality cultures.

(3) *Supply and Procurement Constraints*; The reasons for delays in product delivery due to problems experienced in Supply and Procurement departments are given below:

Supply and Purchase Issues: It is one of the root problems of the company. The contract conditions without sanction power, material orders that are not made on time bring about production planning and manufacturing problems. The contract must have a binding feature for the parties. The contracts without sanction power cause many problems, from the delivery of the supplied product to the quality.

Failure to comply with the Terms and Conditions about Time, Quality and Quantity by the Suppliers: The date of termination is not specified when the contract is awarded. Incomplete or incorrect product introduction on term date, not specified of the precise machining places (non-compliance with desired norms and conditions).

Failure to Create Effective Supplier Pool: Absence of Supply Chain Management: Failure to make the supplier selection by the purchasing department and failure to select the supplier correctly causes this pool not to be created.

Lack of Adequate and Necessary Network: Since the supplier selection is made by the chairman of the board and purchasing is not included in the process can cause problems on the purchasing process.

Failure to make supplier selection correctly: Since the supplier selection is made with the rule of the chairman of the board, it prevents the formation of a certain selection criteria.

(4) *Personnel Constraints*: The reasons for the delays in delivery of the products due to personnel problems are given below.

Failure to Establish Payment System: The company's wage policy is very variable. Since job descriptions, job analysis or job specifications are not prepared for the determination of the wage levels, this situation is perceived as a problem for the personnel.

Overtime Fee Policy: Since the company does not have an overtime pay policy, it causes employees not to work overtime.

High Subcontractor Fees: In the present situation, from time to time, moulding service is provided from outside. Since the fees paid for these services are too high, it causes a decline in both the performance of the employees and the quality of the workmanship.

Low Performance: Low motivation, low attendance and lack of equality of opportunity cause the performance of personnel to decrease. In order to get rid of low performance, it is necessary to remove four basic problems.

Low motivation: The reason of the low motivation of the staff is the wages, the overtime and plenty number of relatives working in the firm.

Plenty number of the Workers who are relatives with each other: Being a lot of employees who are relatives with each other causes some problems by other employees. This also prevents a fair and balanced work load distribution.

Failure to Achieve Equal Opportunity: Failure to achieve equal opportunity in the company causes employees to work with lower performance. This situation also forms the basis of personnel problems that are the root cause.

Management Ineffectiveness: Management is the event that brings the company to the future. It is also the responsibility of the management of the persons who will recruit in the company. The fact that the management does not recruit competent personnel or not to operate them in the desired quality and conditions and not to dominate the working and employment legislation create the staff troubles which are our root cause.

Not to create intelligent solutions: It cannot be made use of intelligent factory, intelligent warehouse, intelligent procurement systems, learning robots, intelligent business intelligence systems today when Industry 4.0 is spoken due to inexperienced administrators who are away from technology.

3.2 Future Reality Tree Creation

Within the scope of the project carried out by KBM Co., Evaporating Cloud applications have been implemented to eliminate the problems related to Supply and Purchase, Technic, Quality, Information and Technical Equipment. Due to the implementation of the suggested solutions, it is expected to ensure improvements in Fig. 2. In the future reality tree drawn, the problems of "pattern delivery problems", ideas and its interaction with the future and present values are shown.

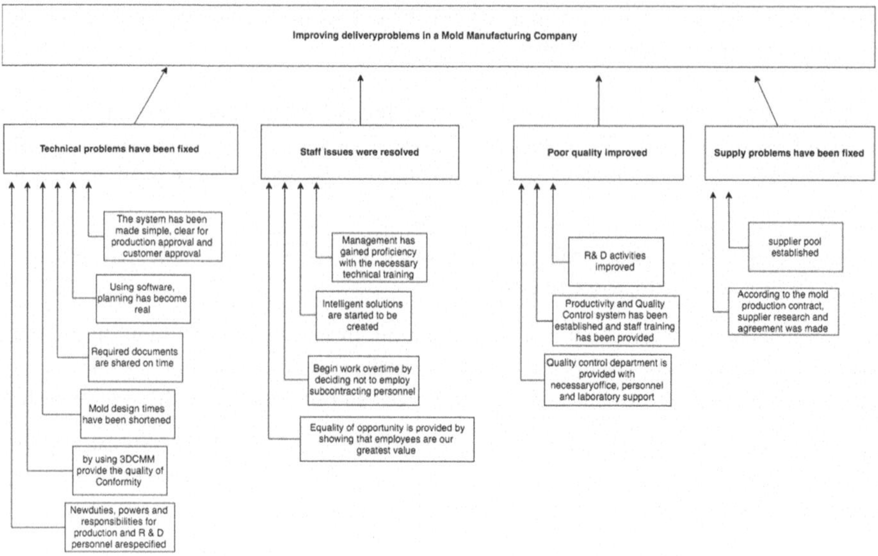

Fig. 2. Future reality tree

The future reality tree determines the relations between cause and effect and the outputs of these elements. At the bottom of this tree, there are ideas and plans, medium-term results in the middle part and the desired conditions in the last part. In order to eliminate the technical problems, receiving approval from the customer before the production and making the production system more plain and understandable can solve some of the delays due to technical problems.

A new job definition for the production and R&D personnel for the same problem and determination of job responsibilities will also be part of the solution. With the establishment of ERP + SAP programs to be used, planning has become more realistic. As a result, mould designing times have been shortened. Full and timely sharing of the documents prepared for production with relevant personnel will ensure to eliminate technical problems.

In order to eliminate supply and procurement problems, it must be strictly adhered to mould production contract and a good research of the supplier enterprises should be carried out. Beginning to investigate the product at the project stage in the supply of raw materials and materials will solve the supply and procurement problems. At the same time, employees should feel that employees are the greatest value for solving employee problems and equal opportunities for employees. Taking the decision not to employ subcontracting personnel, starting the implementation of the overtime concept effectively will result in personnel problems.

In order to solve the problem of poor quality which is another problem which is a constraint to the operation for the delivery problem, the laboratories, offices, personnel and equipment support needed in the quality control section need to be provided. At the same time, it is also necessary for the personnel to gain responsibility and authority.

With the quality department created, a department has been created where the problems are solved by the system, not by the people. By making the quality department independent from the production department, the poor quality has been eliminated and the delivery problem due to the quality problem has been solved.

4 Conclusions

Today, every business, which is mentioned from industry 4.0, has to enter into a strong competition with other businesses to increase productivity by establishing Intelligent Systems, reduce costs, improve quality and satisfy the customer. In this competitive environment, in order to achieve these objectives, enterprises must have low cost and high performance products in order to make continuous improvement, to pay attention to service and product development and to carry out this process.

Moulding ensures to produce many products we use in our everyday life. Like all businesses, mould manufacturing enterprises have targets such as profit and survival. Enterprises may encounter some problems in their activities in order to reach these targets. When these problems are not improved, the business competitiveness breaks down, profit margin falls, economic troubles begin, and at the end the enterprise may have to end its business activities. The Philosophy of the Theory of Constraints includes methods that can determine the weakest point, increase the stability of the enterprise that helps strengthen the weaker ring of chain, and increase the competitiveness of its own by finding the reasons that prevent the success of the company. The Theory of Constraints can also be considered as a problem-solving method. This property allows the Theory of Constraints to use its own tools.

In this study, with the aim of making true intervention to remove the restriction, the problem of delivery in the mould manufacturing operation in KBM Co. has been examined. In the study, first the constraints were determined with the application of the Current Reality Tree. Four basic constraints have been identified according to the Current Reality Tree application. These are "technical constraints, supply-purchase restrictions, personnel constraints and quality". After these four constraints have been identified, in the study in which the thinking processes are applied step by step to identify the weakest ring and improve the process, first of all, the root problems are identified with the application of current reality tree to be resulted from unqualified manufacturing engineers, incompetent administrators and the lack of quality administration system. With the evaporating cloud application, solutions to these problems have been developed and the predicted improvements are explained through the future reality tree application. In order to eliminate problems experienced in delivery, cases such as taking customer requests correctly, designing efficiently, preparing a specification according to received orders, and being a relative of a large number of employees in the company cause not to be managed of the delivery process effectively.

As a result of the study, it is ensured that the delivery period of the operator is improved by applying the Theory of Constraints. From the point of view of mould production and operation, it has been shown that the Theory of Constraint Thinking Process are applicable, and it is seen that the quality of these enterprises can be increased with this theory. The results of such studies can be posted on the board in the company and it is possible to move the research to a different dimension by including employees in the period.

It is expected that this study, which is realized by using the Theory of Constraint Thinking Process will guide the managers in this way and lead to more comprehensive projects.

References

1. Sadıç, Ş, Özdemir D, Gözlü S (2006) Kısıtlar Kuramı Yaklaşımı İle Petrol İthalat ve Ulusallaştırma Sürecinin İyileştirilmesi İstanbul Ticaret Üniversitesi Fen Bilimleri Dergisi, Yıl 5 Sayı 10 Güz 2006/2, pp 99–118
2. Özkaya A (2006) EnZayıfHalka. http://www.biymed.com/pages/makaleler/makale19.htm Accessed 02 May 2017
3. Goldratt EM, Fox RE (1986) The race. Nort River Press, Croton-on-Hudson
4. Goldratt EM (1990) The haystack syndrome. North River Press, Croton-On Hudson
5. Goldratt EM, Cox J (2011) Amaç. Optimist Yayınları, İstanbul
6. Dettmer WH (1995) Quality and the theory of constraints. Qual Prog 28(4):77–81
7. Rahman S (1998) Theory of constraints: a review of the philosophy and its applications. Int J Oper Prod Manag 18(4):336–355
8. Akman G, Karakoç Ç (2005) Yazılım Geliştirme Prosesinde Kısıtlar Teorisinin Düşünce Süreçlerinin Kullanılması. İstanb Ticaret Üniv Bilim Derg 7:103–121
9. Dettmer HW (1997) Goldratt' theory of constraints, a system approach to continuos improvement. ASQC Quality Press, Milwaukee
10. Kerzner H (2013) Project management a systems approach to planning, scheduling and Controlling. Hoboken, Wiley
11. Scheinkopf LJ (1999) Thinking for a change, putting the TOC thinking process to use. St. Luice Press, Boca Raton
12. Mcmullen TB Jr (1998) Introduction to the theory of constraints (TOC) management system. CRC Press LLC, Boca Raton
13. Mabin VJ, Balderstone SJ (1999) The world of the theory of constraints: a review of the international literature. CRC Press, Boca Raton
14. Klein DJ, Debruine M (1995) A thinking process for establishing management policies. Rev Bus 16(3):31–37
15. Gaga O (2009) Süreç Analizi ve Süreç İyileştirme Metodolojisi ve Kısıtlar Teorisi Yöntemiyle Süreç Analizi Uygulaması, Yıldız Teknik Üniversitesi, Fen Bilimleri Enstitüsü, Yüksek Lisans Tezi, İstanbul
16. Siha S (1999) A classified model for applying the theory of constraints to service organizations. Manag Serv Qual 9(4):255–264
17. Womack D, Flowers S (1999) Improving system performance: a case study in the application of theory of constraints. J Health Care Manag 44(5):397–407
18. Köksal G (2004) Selecting quality improvement projects and product mix together in manufacturing: an improvement of a theory of constraints-based approach by incorporating quality loss. Int J Prod Res 42(23):5009–5029
19. Kirche ET, Kadipaşaoğlu SN, Khumawala BM (2005) Mazimizing supply chain profits with effective order management: integration of activity-based costing and theory of constraints with mixed-integer modelling. Int J Prod Res 43(7):1297–1311
20. Ünal EN, Tanış VN, Küçüksavaş N (2005) Kısıtlar Teorisi Ve Bir Üretim İşletmesinde Uygulama. Sos Bilim Enst Derg 14(2):433–448

21. Freeman LN (2006) Theory of Constraints Outlines Path for Improvement. Ophthalmology Times, London
22. Polito T, Watson K, Vokurka RJ (2006) Using the theory of constraints to improve competitiveness: an airline case study. Compet Rev 16(1):44–50
23. Geri N, Ahituv N (2008) A theory of constraints approach to inter organizational systems implementation. Inf Syst e-Bus Manag 6(4):341–360
24. Birgün S, Öztepe T, Şimşit Z (2011) Bir Çağrı Merkezinde Müşteri Şikayetlerinin Düşünce Süreçleri İle Değerlendirilmesi. In: XI. Üretim Araştırmaları Sempozyumu
25. Yükçü S, Yüksel İ (2015) Hastane İşletmelerinde Kısıtlar Teorisi Yaklaşımı ve Örnek Bir Uygulama. İktisadi Bilim Derg **29**(3)

Monitoring of Machining in the Context of Industry 4.0 – Case Study

Wojciech Zębala(✉), Grzegorz Struzikiewicz, and Emilia Franczyk

Production Engineering Institute, Cracow University of Technology, Kraków, Poland
{zebala, struzikiewicz}@mech.pk.edu.pl, franczykemilia@gmail.com

Abstract. The paper presents the algorithm for the construction of a cutting process monitoring system for selected methods of measurement data flow in order to analyze them as a part of the Industry 4.0 idea. The effectiveness of the cutting process information flow was compared directly in the machine tool area (factory zone) and outside of it, e.g. in the office zone. A stand for monitoring of physical phenomena in the cutting zone of formation and flow of chips with the usage of a piezoelectric dynamometer, thermos-vision and quick-cage cameras was presented. The results of the analysis aimed at determining the limitations of the application of the presented measuring techniques in the concept of the idea of Industry 4.0 were presented.

Keywords: Industry 4.0 · Monitoring · Machining · Measuring sensors

1 Introduction

The introduction of various IT and digital techniques into the structure of industrial production is today a response to the quality and cost requirements set by clients. Currently manufactured products should be offered in a wide range, at low production costs and for the needs of a specific customer. To meet the requirements, according to Industry 4.0, it is necessary to combine information and digital techniques at the production stage of a specific product with a production company management system [1–4]. This will be possible only if full communication is provided between machine tools and technological machines, assuming that each of them has an independent system to monitor its status and the process being carried out. This means that one should strive for a state when each of the machine tool components has built-in intelligence and a set of sensors that enables it to participate effectively in the active monitoring of the manufacturing process [4–7]. This is especially important in the broadly understood machining, such as turning, milling and drilling. The dimensional and shape accuracy as well as the quality of the workpiece being processed are influenced by a number of phenomena occurring in the cutting zone, such as vibrations in the OUPN system, the temperature value in the cutting zone or the form and shape of the created chips [4, 8]. For this reason, it is very difficult to achieve the optimal cutting conditions for the workpiece. The use of various forms of monitoring the cutting process parameters, the condition of the tool or the overall machine tool can improve

N. M. Durakbasa and M. G. Gencyilmaz (Eds.): ISPR 2018, *Proceedings of the International Symposium for Production Research 2018*, pp. 660–673, 2019.
https://doi.org/10.1007/978-3-319-92267-6_54

this process [9, 10]. It must be ensured the quick response to the emergency situations and problems occurring in the machining process. Increasing of the productivity and efficiency of machine tool usage, maximizing the permissible feeds, shortening the machining cycle and improving the quality of machining, extending tool life, optimizing cutting parameters and constant machine safety control is the result of using the monitoring and supervising of the cutting processes [11–14]. In addition, downtime is reduced, energy is saved, and the costs of maintaining and servicing the machine tool are reduced. It avoids the necessity of corrections or re-machining of the part and the risk of manufacturing the part beyond the limits of the imposed tolerances [15, 16].

One of the examples of the development of an intelligent industrial plant is Mazak iSMART Factory™ [17]. Production activities are recorded in the form of digital data, which are coordinated with the main computer system and allow for the visualization and analysis of processes in terms of improvements. All machine tools, like peripheral devices (e.g. integrated chip conveyors and automatically guided vehicles), are connected to a network enabling collection of more than 10 million operational data units per day, and as a result - constant monitoring and analyzing the situation. The data comes from all devices and is used to improve the efficiency of the entire machining department. Network communication between different devices, data collection and integrated control follow the open communication industrial protocol MTConnect® and Mazak SMARTBOX™ network interface module, which not only provides cybernetic security of the network, but also acts as a component of "computational fog" for distributed data processing. Another approach in line with the Industry 4.0 strategy is presented by DMG MORI, which offers a unified user interface, the so-called CELOS for all machines that connects them to the superior company structures [18]. This allows the computerization of production and elimination of paper work management and automated machines. Collecting production data in this way and optimizing it allows you to get a product in less time (up to 30%). It is worth noting that similar solutions can be found in China, which coincide with the concept of "Made in China 2025" [19, 20].

The application of monitoring, diagnosing and, as a consequence, monitoring of the machining process can be implemented by means of many techniques, using various sensors and information systems. Obtaining the optimal machining conditions compatible with the Industry 4.0 idea requires, however, determining the limitations of the applied measurement techniques. The authors analyzed the system of monitoring of the physical phenomena in the area of formation and chip flow in the turning process, built with selected measurement techniques. The time of transmission and analysis of measurement results obtained from a piezoelectric dynamometer as well as thermosvision and high speed cameras were analyzed. The aim of the research was to determine the limitations of the measurement techniques presented in the Industry 4.0 concept.

2 The Concept of the Cutting Process Monitoring Structure Within the Framework of the Idea of Industry 4.0

Optimal conditions for conducting the cutting process require the use of active forms to monitor the machine tool parameters, process and tool condition. The machining process is influenced by a number of factors mainly associated with [21, 22]:

- workpiece: type of material, geometry, hardness, surface structure, mass, susceptibility of thin walls, chemical composition, etc.
- machine tool: method of fixing the machining part, kinematic structure of the machine, cooling method, technical condition, spindle rigidity, etc.
- tool: geometry and stereometry, size, type of coating, tool material, rotational speed, feed, cutting edge condition, overhang, etc.
- tool and workpiece holder and rigidity of the clamping system.
- other: programming errors of the process, incorrect machining operations, wrong operator reactions, etc.

Table 1 presents examples of areas where it is possible to use a monitoring system of the manufacturing process, along with the type of information on the course of this process. The most commonly used signal sensors mounted on machine tools are presented in Table 2.

Table 1. Areas of monitoring the manufacturing process along with the type of information on the course of this process

Areas of the monitoring	Information on how the manufacturing process proceeds	
Tool	- Fracture, - Cutting edge wear, - No tool,	- Wrong tool, - Built Up Edge
Cutting process	- Spindle torque, - Cutting resistance, - Vibrations,	- Temperature, - Form of chips, - State of workpiece
Machine tool	- Loads of drives, - Spindle vibrations, - Spindle speed, - Temperature of spindle and drives element, - Coolant expense and pressure, - Certainty of fixing and basing the workpiece,	- Collision detection, - Spindle extension compensation, - Diagnostics of other machine tool assemblies, - Start times

Table 2. Signal sensors mounted in machine tool assemblies, monitored

Machine tool assemblies under monitoring	Sensor
Tool	Force sensor, measuring probe, acoustic emission sensor, laser optical sensor, vibration sensor, touch sensor, vision systems (infrared cameras, fast-cage cameras)
Workpiece	Touch sensor, measuring probe, surface roughness sensor, clamping force sensor, identification sensor
The main drive	Sensor of torque (current), temperature, vibration, force, speed and angular position, power measurement
Feed drives	Position and displacement sensor, speed, torque, end position, temperature
Hydraulic and cooling-lubricating systems	Sensor for pressure, temperature, oil level and cutting fluid
Storage systems	Touch sensor, identification systems, micro-switches
Machine tool	Smoke and flame sensor, infrared camera, video camera

Figure 1 presents an example flow diagram of data obtained from the monitoring areas of the machining process, carried out on a machine tool within the framework of an factory compatible with the idea of Industry 4.0. Three characteristic zones can be distinguished, between which there is a flow of data about the implemented manufacturing process, i.e. office, data transfer and coding, and factory zones.

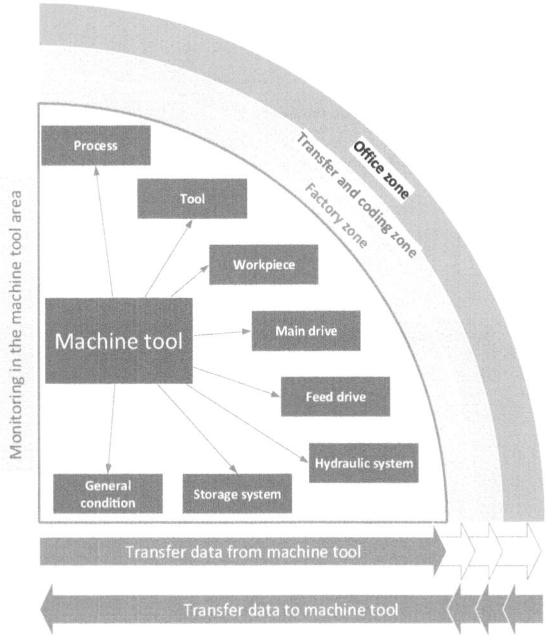

Fig. 1. Diagram showing the flow of data obtained from the monitoring areas of the machining process

In Fig. 2, the authors of the paper presented the proposed algorithm for building the structure of the monitoring system of physical phenomena occurring in the cutting zone, taking into account the flow and analysis of measurement data as part of the Industry 4.0 concept. At the beginning, the methods and types of measurement sensors are selected, possible to build a monitoring system of the cutting process, in particular forming and flow of chips. The available databases and recommendations of measuring apparatus manufacturers are taken into account. The structure of the monitoring system should be compatible with the machine tool's production infrastructure. In the next step, the choice of data transfer method and the place and method of analysis of measurement data obtained from the monitoring system is made (for example, data analysis and inference can be carried out by a human, specialized computer program or artificial intelligence). Next, data transfer times, their analysis and inference are determined, which allows the estimation of limiting values (e.g. the minimum time of data analysis) and the determination of optimization criteria and functions. In the last stage of building the structure of the monitoring system, the selected optimization criterion is checked and the monitoring system is selected or the decision to modify it, e.g. change of the type of sensor used, etc.

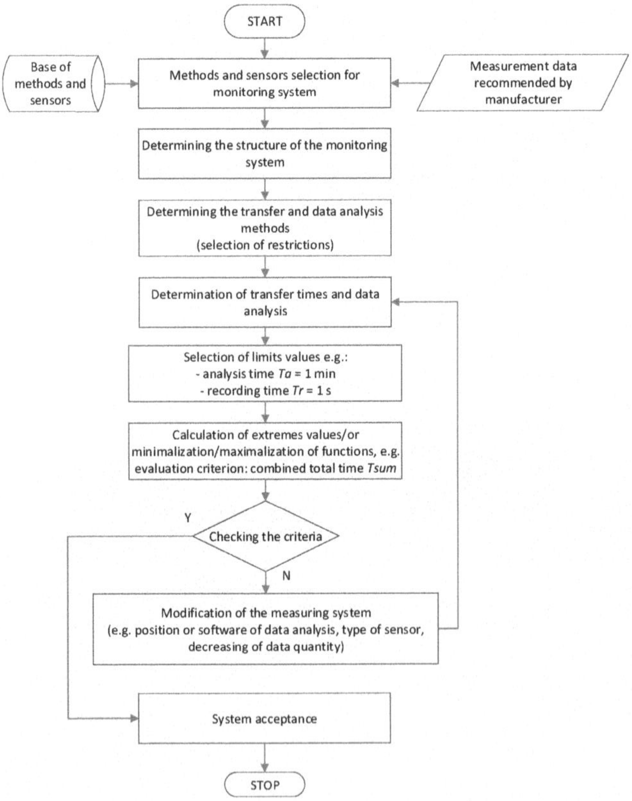

Fig. 2. General algorithm for building the structure of the cutting process monitoring system

According to the Industry 4.0 concept, access and exchange of data collected in the factory network area is possible for users of the office network using various IT techniques. Currently, the number of collected data is huge, but most of the information obtained from the used sensors is easy to analyze, record, interpret and have a small volume, so it can be quickly sent or encoded. However, there is a question about solving the problem of the place and the time needed to conduct data analysis and their interpretation in the case of large volume data. Figure 3 presents examples of different concepts concerning the place where the analysis and interpretation of the results of collected data during the monitoring of the formation and flow of chips in the turning process are presented. This type of system can be based on measurement of the components of the total cutting force, chip temperature and/or in the cutting zone as well as the registration and analysis of fast-changing images. Obtained data from the monitoring process differ significantly in volume. Therefore, the use of different measuring techniques requires high data throughput in the factory network area and in the data transfer and coding zone. The place where the analysis of the results and their interpretation is also important. For example (Fig. 3a) the analysis of the monitoring signals can be carried out in the machine tool area and then the conclusions sent to the office zone for further interpretations or analysis (Fig. 3b). The second approach requires sending all measurement data to the office zone, including the required information coding, from the machine tool area. Therefore, the choice of data analysis place can be treated as a criterion limiting the number of applied methods and measurement sensors. An additional limitation may be the method of analyzing the results and their interpretation. Such analysis can be carried out by a human, a special computer program or it can be made using artificial intelligence. Therefore, the time needed to conduct the analysis and the interpretation of the measurement results can be significantly different.

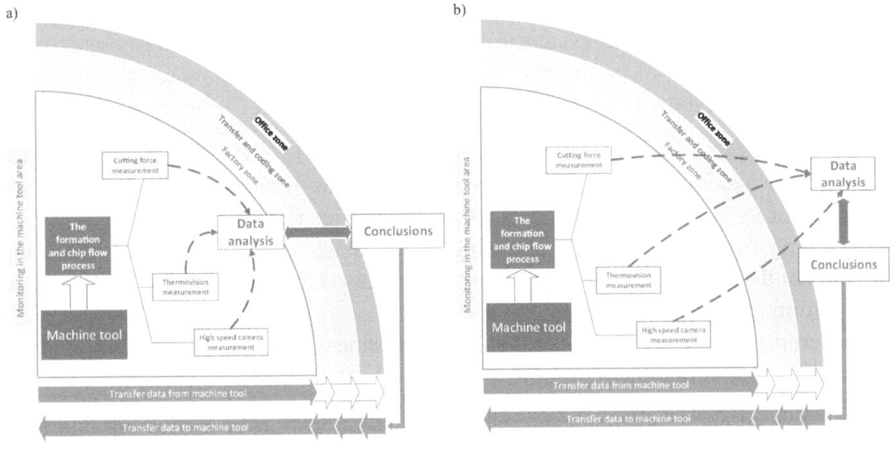

Fig. 3. Diagram of the concept of data analysis and interpretation (a) in the machine tool area, (b) in the office zone

3 Case Study

For registration of phenomena in the chip formation zone, an integrated test stand has been set up for recording fast-changing images, recording cutting forces and measuring the temperature in the cutting zone. The individual measuring systems can be operated individually or run simultaneously. The stand enabled comprehensive and precise observation of the cutting zone, in particular the formation and flow of the formed chips.

The research stand recording the process of chip creation and flow consists of the following components:

High-speed camera, with maximum record speed 148 000 frames/s. Maximum resolution 1152 × 896 piksels is reachable at 1000 frames/s. The recording time was 4 s. Initial registrations of the fast-changing images with speeds from 1000 frames/s to 3000 frames/s were carried out. The analyzes shown that the adopted registration speeds are sufficient to observe the process of forming and breaking chips. To measure the Ff, Fc, Fp components of the total cutting force, a measuring track was built using a Kistler9257B piezoelectric force dynamometer, connected to the Kistler 5070 signal amplifier. The signal amplifier was connected to the computer with a sixteen-channel Pcim-DAS 1602/16 analogue-type digital card. A RS232 type connection was additionally used for the full communication with the DynoWare program.

To measure the components of the total cutting force (Ff, Fc, Fp) a measurement path based on Kistler9257B piezoelectric force meter connected to the Kistler 5070 signal amplifier was used. The signal amplifier was connected to a PC computer with a 16-channel analogue-type Pcim-DAS 1602/16. A RS232 type connection was additionally used for full communication with the DynoWare program.

Temperature measurements in the cutting zone were made using a FLIR SC 620 thermal imager connected via a FireWire serial connection to a PC computer. ThermaCAM Researcher software for acquisition and processing as well as image analysis from the camera was installed on the computer.

The analyzes were aimed at determining the size of files generated during the operation of the monitoring system as well as the time of their analysis and transmission. On the basis of preliminary tests, the following conditions of measurement systems were adopted:

- the frequency of recording the image sequences was f = 1000 frames/second with an image resolution of 1152 × 896 pixels. It was found that these values are sufficient to analyze the method of forming and direction of chips flow.
- sampling frequency f = 1000 Hz using force gauge is sufficient to determine the value of the total cutting force components and also coincides with the frequency of recording high-speed images.
- the temperature range of the thermal imaging camera was from 0 to 500 °C and the number of recorded thermo-grams per second was ir = 30.

For such adopted measurement conditions, cutting tests were carried out based on the calculated average values of the recorded files and their transmission and analysis times. The exemplary results of the analyzes are presented in Tables 3, 4, 5 and 6.

Table 3. Approximate values of file sizes and times for the case of transfer and analysis of measurement data outside the machine (for the time of recording Tr = 0.1 s)

Measuring system of	File size *Wf* [MB] for recording time *Tr* = 0,1 s	Transmission time *Tt* [s] for 100 MB/s	Analysis time [s]			Coding time/data transfer *Tc* [s]	Time of data archiving *Tarch* [s] at recording 120 MB/s	Total time [s]		
			Human *Ta_h*	Program *Ta_p*	AI *Ta_AI*			Human *T_h*	Program *T_p*	AI *T_AI*
Cutting forces	0,004	4,0E-5	10	1	0,5	0,00024	3,3E-5	10,00	1,0003	0,5003
Thermovision images	5	0,05	120	20	5	0,3	0,0417	120,39	20,39	5,39
High-speed camera sequences	200	2	300	180	30	12	1,6667	315,67	195,67	45,67
						Combined total time *Tsum* [s]		446,06	217,06	51,56

Table 4. Approximate values of file sizes and times for the case of transfer and analysis of measurement data outside the machine (for the time of recording Tr = 4.0 s)

Measuring system of	File size *Wf* [MB] for recording time *Tr* = 0,1 s	Transmission time *Tt* [s] for 100 MB/s	Analysis time [s]			Coding time/data transfer *Tc* [s]	Time of data archiving *Tarch* [s] at recording 120 MB/s	Total time [s]		
			Human *Ta_h*	Program *Ta_p*	AI *Ta_AI*			Human *T_h*	Program *T_p*	AI *T_AI*
Cutting forces	0,16	0,0016	10	1	0,5	0,0096	0,0013	10,01	1,01	0,51
Thermovision images	200	2	120	20	5	12	1,6667	135,67	35,67	20,67
High-speed camera sequences	8000	80	300	180	30	480	66,6667	926,67	806,67	656,67
						Combined total time *Tsum* [s]		1072,35	843,35	677,85

Table 5. Approximate values of file sizes and times for the case of transfer and analysis of measurement data at the machine (for the time of recording Tr = 0.1 s)

Measuring system	File size *Wf* [MB] for recording time *Tr* = 0,1 s	Transmission time *Tt* [s] for 100 MB/s	Analysis time [s]			Coding time/data transfer *Tc* [s]	Time of data archiving *Tarch* [s] at recording 120 MB/s	Total time [s]		
			Human *Ta_h*	Program *Ta_p*	AI *Ta_AI*			Human *T_h*	Program *T_p*	AI *T_AI*
Measurement of cutting forces	0,003	3,0E-05	12	1,2	0,6	1,8E-04	2,5E-05	12,0	1,2	0,6
Thermovision measurement	0,004	4,0E-05	144	24	6	2,4E-04	3,3E-05	144,0	24,0	6,0
High-speed camera measurement	0,004	4,0E-05	360	216	36	2,4E-04	3,3E-05	360,0	216,0	36,0
						Combined total time *Tsum* [s]		516,0	241,2	42,6

Table 6. Approximate values of file sizes and times for the case of transfer and analysis of measurement data at the machine (for the time of recording Tr = 4.0 s)

Measuring system	File size *Wf* [MB] for recording time *Tr* = 0,1 s	Transmission time *Tt* [s] for 100 MB/s	Analysis time [s]			Coding time/data transfer *Tc* [s]	Time of data archiving *Tarch* [s] at recording 120 MB/s	Total time [s]		
			Human *Ta_h*	Program *Ta_p*	AI *Ta_AI*			Human *T_h*	Program *T_p*	AI *T_AI*
Measurement of cutting forces	0,12	0,0012	12	1,2	0,6	0,0072	0,0010	12,0	1,21	0,61
Thermovision measurement	0,16	0,0016	144	24	6	0,0096	0,0013	144,0	24,01	6,01
High-speed camera measurement	0,16	0,0016	360	216	36	0,0096	0,0013	360,0	216,01	36,01
						Combined total time *Tsum* [s]		516,03	241,23	42,63

The following designations were adopted:

Tr – time of measurement/recording, which was in the range (0,1–4,0)s,

Wf_i – file size in MB obtained during the usage of an i-th type of measuring sensor for the recorded time Tr,

i – type of measuring sensor, i.e. $i = 1$ (dynamometer, in short “Force”), $i = 2$ (thermo-vision camera, in short “Temp”), $i = 3$ (high-speed camera, in short “Video”),

$Tt_i = Wf_i^* Tt_0$ – transmission time of measurement data, where Tt_0 means transmission time of the 1 MB data for a specific medium (transmission speed was fixed at 100 MB/s, what means that $Tt_0 = 0{,}01$ s), Eq. (1):

$$Tt_i = Tt_0 \cdot \begin{bmatrix} Wf_{Force} \\ Wf_{Temp} \\ Wf_{Video} \end{bmatrix} \tag{1}$$

$Ta_{i,j}$ – time of analyzing the results of the measurement data for the j-th case of the analysis, i.e. $j = 1$ (human, in short “H”), $j = 2$ (computer program, in short “P”), $j = 3$ (artificial intelligence, in short “AI”), Eq. (2):

$$Ta_{i,j} = \begin{bmatrix} Ta_{Force,H} \\ Ta_{Temp,H} \\ Ta_{Video,H} \end{bmatrix}; \quad \begin{bmatrix} Ta_{Force,P} \\ Ta_{Temp,P} \\ Ta_{Video,P} \end{bmatrix}; \quad \begin{bmatrix} Ta_{Force,AI} \\ Ta_{Temp,AI} \\ Ta_{Video,AI} \end{bmatrix} \tag{2}$$

$Tc_i = Wf_i^* Tc_0$ – coding time depends on the method of data coding (it was assumed, that the coding time of the 1 MB data is $Tc_0 = 0{,}06$ s),

$Tarch_i = Wf_i^* Tarch_0$ – time of archiving measurement data (it was assumed, that the recording time of the 1 MB data with the speed 120 MB/s is $Tarch_0$ = 0,012 s)

The total monitoring time for the applied type of sensor $T_{i,j}$ is presented by the Eq. (3):

$$T_{i,j} = Tt_i + Ta_{i,j} + Tc_i + Tarch_i \tag{3}$$

Where, depending on the data analysis method (by human, program or AI):

$$T_{i,j} = \begin{bmatrix} T_{Force,H} \\ T_{Temp,H} \\ T_{Video,H} \end{bmatrix} \text{ or } \begin{bmatrix} T_{Force,P} \\ T_{Temp,P} \\ T_{Video,P} \end{bmatrix} \text{ or } \begin{bmatrix} T_{Force,AI} \\ T_{Temp,AI} \\ T_{Video,AI} \end{bmatrix} \tag{4}$$

The combined total monitoring time $Tsum_{i,j}$ for *j*-th type of the analysis (H, P or AI) and all of the measuring sensors types used are presented in the Eq. (5):

$$Tsum_{i,j} = \sum T_{i,j} \tag{5}$$

so

$$\sum T_{i,j} = \begin{bmatrix} T_{Force,H} + T_{Temp,H} + T_{Video,H} \\ T_{Force,P} + T_{Temp,P} + T_{Video,P} \\ T_{Force,AI} + T_{Temp,AI} + T_{Video,AI} \end{bmatrix} \tag{6}$$

Analyzing the presented results, one can conclude that in order to minimize the total time needed for registration, transmission, analysis, coding and archiving, it is necessary to use specialized computer programs using artificial intelligence modules. In the presented example, it is estimated that the time of analysis of data obtained from recording fast-changing sequences using artificial intelligence may be reduced by 10 times. Due to the size of files obtained during the recording of the fast-changing sequences, and thus the time of their transmission and analysis, it is important to choose the place where they are to be carried out. The file transmission time can be shortened if the data analysis is carried out directly in the machine tool area and only the results of the analysis are sent to subsequent zones of the company management structure (shortening the time from several dozen seconds to milliseconds). In this approach, the coding time of transmitted data is also shortened. A comparison of the analyzed results is presented in Fig. 4.

Figure 5 graphically illustrates the directions of optimization of the cutting zone monitoring due to the total time *Tsum* needed to perform the registration, transfer, coding and archiving of measurement data for different cases of data analysis (human, computer program or AI) and analysis sites (at or outside the machine tool).

Analyzing the obtained results, it can be noticed that with the increase of the recording time *Tr* there are several ways to reduce the total time of monitoring the

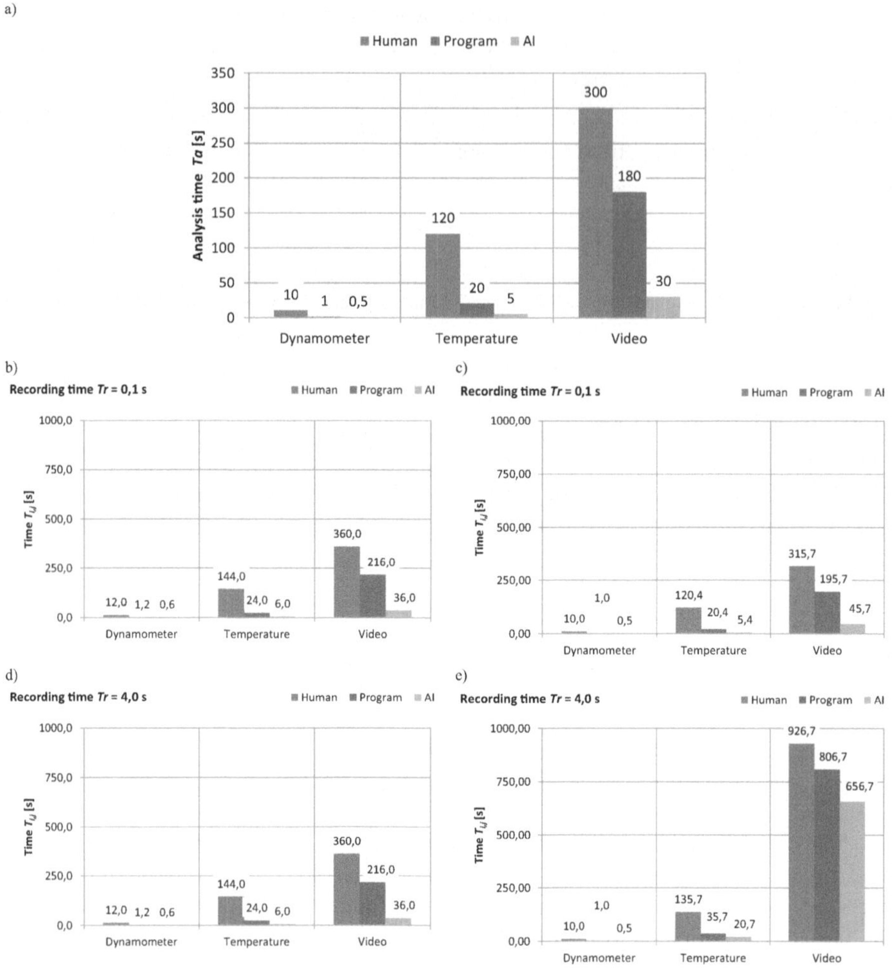

Fig. 4. Comparison of the time of analysis of the results of measurements carried out by human, computer program and artificial intelligence (a) and total times for the case of analysis and interpretation of data in the machine tool area (b, d) and office zone (c, e)

cutting process *Tsum*. One of the ways is to select the method of measuring data analysis. The analysis of the data shows that the use of specialized computer programs can shorten the total monitoring time from about 20 to 50% depending on the recording time *Tr*. The use of programs using artificial intelligence modules can shorten the overall monitoring time by more than 10 times. The second method is the appropriate selection of the place where the measurement data are to be analyzed, which is shown in points A1, B1 and C1 for different cases of analysis of the results, i.e. human, computer program or AI. It can be noticed that in the case of human data analysis, for the time of recording $Tr > 0.6$ s (point A1), the most effective due to the total monitoring time is conducting the analysis directly in the machine tool area. In the case of using a specialized computer program for the analysis of measurement results in the machine tool area, a shorter total monitoring time is obtained for the recording time *Tr* > 0.25 s. Shortening the total time *Tsum* can also be obtained by using specialized computer programs at various stages of monitoring - as shown in points A2, A3 and B2. The most beneficial solution is the use of computer programs using artificial intelligence modules. It should be noted, however, that at the present time there are no many such solutions implemented in the manufacturing industry.

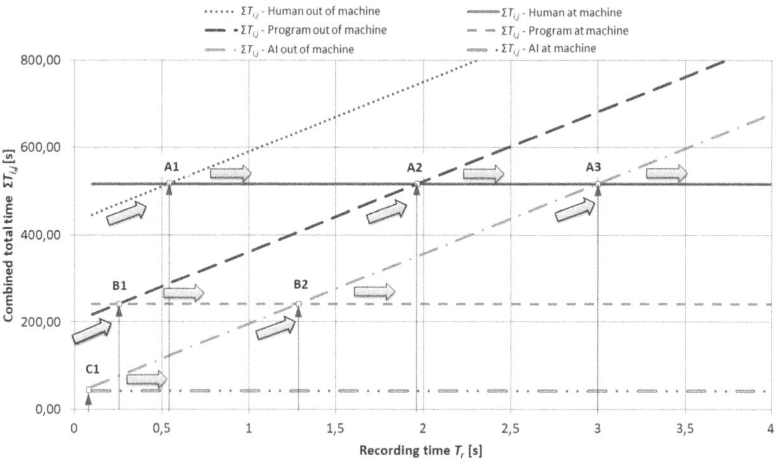

Fig. 5. Combined total time of registration, analysis, coding and archiving of measurement data

4 Conclusions

The following conclusions can be made on the basis of the analyzes:

- the determination of the optimal conditions for monitoring the physical phenomena and the formation and flow of chips in the machining process requires simultaneous application of different measurement techniques (e.g. vision systems and direct measurement). Therefore, there is a need to register, analyze and transmit data of various sizes and formats. Therefore, the actual values of data transmission times and their analysis can be treated as restrictions on the application of selected measurement techniques in the Industry 4.0 concept.
- reducing the time of analysis of measurement data is possible through the use of specialized computer programs or programs using artificial intelligence. It should be noted, however, that this type of solution may have a greater purchase cost.
- due to the total time of data transmission, an important point is the selection of the place where the results of the measurement data will be analyzed. This is especially important when analyzing measurement data (files) of a large size, such as a sequence of high-speed shots. Performing the analysis of results directly in the machine tool area and then sending a ready report to the superior company

management zones can significantly reduce the total time of data transfer and exchange. In addition, such a solution does not require large data volume encoding.
- the general algorithm for selecting the structure of the physical phenomena of the monitoring system in the cutting zone can be useful for the selection and application of measurement methods and sensors in accordance with the Industry 4.0 concept.

References

1. Brecher C, Ozdemir D, Brockmann M (2017) Introduction to integrative production technology. Prod Eng Res Dev 11(2):93–95
2. Wang Y, Ma HS, Yang JH, Wang KS (2017) Industry 4.0: a way from mass customization to mass personalization production. Adv Manuf 5(4):311–320
3. Li Z, Wang Y, Wang KS (2017) Intelligent predictive maintenance for fault diagnosis and prognosis in machine centers: Industry 4.0 scenario. Adv Manuf 5(4):377–387
4. Grzesik W (2017) Advanced machining processes of metallic materials: theory, modelling, and applications, 2nd edn. Elsevier Science Bv, Amsterdam, pp 1–578
5. Melnyk SA, Flynn BB, Awaysheh A (2017) The best of times and the worst of times: empirical operations and supply chain management research. Int J Prod Res 56(1–2):164–192
6. Kusiak A (2017) Smart manufacturing. Int J Prod Res 56(1–2):508–517
7. Strandhagen JW, Alfnes E, Strandhagen JO, Vallandingham LR (2017) The fit of Industry 4.0 applications in manufacturing logistics: a multiple case study. Adv Manuf 5(4):344–358
8. Siddhpura A, Paurobally R (2013) A review of flank wear prediction methods for tool condition monitoring in a turning process. Int J Adv Manuf Technol 65(1–4):371–393
9. Moeuf A, Pellerin R, Lamouri S, Tamayo-Giraldo S, Barbaray R (2018) The industrial management of SMEs in the era of Industry 4.0. Int J Prod Res 56(3):1118–1136
10. Xu X (2017) Machine Tool 4.0 for the new era of manufacturing. Int J Adv Manuf Technol 92(5–8):1893–1900
11. Bombinski S, Blazejak K, Nejman M, Jemielniak K (2016) Sensor signal segmentation for tool condition monitoring. Procedia CIRP 46:155–160
12. Zhou YQ, Xue W (2018) Review of tool condition monitoring methods in milling processes. Int J Adv Manuf Technol 96(5–8):2509–2523
13. Gillich EV, Mocan M, Mituletu IC, Korka ZI (2017) Process monitoring in precision machining using optical sensors. Rom J Acoust Vibr 14(1):50–53
14. Kongchuenjai J, Prombanpong S (2017) An integer programming approach for process planning for mixed-model parts manufacturing on a CNC machining center. Adv Prod Eng Manag 12(3):274–284
15. Malapelle F, Dall'Alba D, Fontana DD, Dall'Alba I, Fiorini P, Muradore R (2017) Cost effective quality assessment in industrial parts manufacturing via optical acquisition. Procedia Manuf 11:1207–1214
16. Abellan-Nebot JV, Subiron FR (2010) A review of machining monitoring systems based on artificial intelligence process models. Int J Adv Manuf Technol 47(1–4):237–257
17. Mazak (2018). https://www.mazakeu.pl/
18. DMG Mori (2018). https://en.dmgmori.com/
19. Li L (2017) China's manufacturing locus in 2015: with a comparison of "Made-in-China 2025"and "Industry 4.0". Technol Forecast Soc Change. http://dx.doi.org/10.1016/j.techfore.2017.05.028

20. Wang L (2016) Comparative research on German "Industry 4.0" and "Made in China 2025". In: 2nd International conference on humanities and social science research, Beijing, pp 27–30
21. Cwikla G (2014) Methods of manufacturing data acquisition for production management - a review. Mod Technol Ind Eng 837:618–623
22. Liang QK, Zhang D, Wu WN, Zou KL (2016) Methods and research for multi-component cutting force sensing devices and approaches in machining. Sensors, 16(11):1926

The Potential Effect of Industry 4.0 on the Literature About Business Processes: A Comparative Before-and-After Evaluation Based on Scientometrics

Güzin Özdağoğlu[1(✉)], Onur Özveri[2], Aşkın Özdağoğlu[1], and Muhammet Damar[3]

[1] Dokuz Eylul University Faculty of Business, İzmir, Turkey
{guzin.kavrukkoca, askin.ozdagoglu}@deu.edu.tr
[2] Dokuz Eylul University Faculty of Economics and Administrative Sciences, İzmir, Turkey
onur.ozveri@deu.edu.tr
[3] Dokuz Eylul University Information Center, İzmir, Turkey
muhammet.damar@deu.edu.tr

Abstract. Industry 4.0 is a collective term that covers many modern automation systems, data exchanges and production technologies representing the fourth revolution in the industry. This revolution allows for the collection and analysis of all the data that can affect production processes, revealing more efficient business models, and creating a structure that will completely change the production-consumption relations. When the bibliometric information of related publications are examined, it is clear that the concept of Industry 4.0 is beginning to be thoroughly debated in 2010 and later (n: 418, 81.48%), and every day of the post-publication period there is an increasing tendency. These developments affecting all areas of the industry can also be considered to affect the studies on the management of the business processes. This study aims to demonstrate how the developments in industry 4.0 are reflected in the studies related to business processes within the scope of the literature. In this context, by taking 2010 breaking point, it is investigated whether a change in the scientometric structure of the research articles related to the business processes after this year and whether the changes are compatible to the concepts conceived together with the concept of industry 4.0. As a necessary data source, the research articles that are focused on business processes in the Web of Science (WoS) database were collected, and the scientometric analyses were conducted on this data for different time intervals. For the comparisons, the similar the data on the concept of industry 4.0 were also collected and analysed. According to the search attempts on WoS on 1st of April, 2018, 6255 research articles were obtained using the phrase "business process" where 3044 of them were published in 2010 or later. Various citation mining techniques were used in the scope of scientometrics to reveal demographics and linkages. Findings show that there are significant differences before and after the year 2010.

Keywords: Business processes · Industry 4.0 · Scientometrics

N. M. Durakbasa and M. G. Gencyilmaz (Eds.): ISPR 2018, *Proceedings of the International Symposium for Production Research 2018*, pp. 674–690, 2019.
https://doi.org/10.1007/978-3-319-92267-6_55

1 Introduction

In the intense global competition, business enterprises have to reduce the cost of doing business, and rapidly develop new services and products for meeting different customer expectations. For reaching this objective, enterprises must continuously take into account the way of doing business and support business processes [1, 2]. Focusing on the business processes is of great significance for an effective management system and thus process management can create a positive effect on all business activities to present right product or service to the market for meeting different customer expectations [3–5].

According to Aguilar-Savén [6], business process is "the combination of a set of activities within an enterprise with a structure describing their logical order and dependence whose objective is to produce a desired result." The author also touches upon the significance of selecting a particular language. In the light of process management approach, modelling the business processes via a standard language is used for the reconstruction of a company. Detailed process models carry the reference document characteristics that construct an effective process management mechanism. Different workflows and scenarios can be performed on these models in order to obtain the most convenient workflow. Planning these kinds of improvement activities in a company by using particular models enables process analysts to try different scenarios without spoiling the active flow [7]. On the other hand, there is no unique solution for constructing process models. Therefore, several frameworks and notations have been developed to represent business processes. Mendling et *al.* [8] discussed the sufficiency of the current frameworks, specially, SEQUAL, that is to say, a well-known framework, and suggested different process modelling guidelines to fulfil the gaps and eliminate the quality issues in existing frameworks.

As of 2000s, digitalization has played a critical role on managing and modelling complex business processes, since it has provided automation systems to carry out the activities and monitor them through complex enterprise systems. In the last decade, sensor and network technologies, internet of things, and big data technologies have been utilized to create autonomous processes where all devices are connected to each other in a network to communicate. This communication generates huge number of data that can be analysed via abovementioned technologies in real time to reveal instant findings for the processes. In 2010, this era was defined as the fourth revolution in industry, i.e., Industry 4.0. Industry 4.0 concept was introduced in Hannover Fair 2011, Germany [9, 10]. With the help of Industry 4.0 concept, production processes can easily react to the requirements of the potential customers [11]. Companies have to adapt the technological developments and produce new models for meeting different customer expectations in an agile manner. Thus, Industry 4.0 related developments provide this agility and can reveal value-added products through the achievements of knowledge management with the extensive integration and use of information [10]. To the date, the concept evolved to a much broader context including smart applications in particular fields, e.g., healthcare, urban and domestic life, construction, logistics, facility management as well as manufacturing [10, 12, 13]. Industry 4.0 has initially included automation and data exchange in facilities [14]. It was asserted that production systems in automotive and information sectors could be improved with the technological innovations.

When the bibliometric information of the publications related to Industry 4.0 are examined, it is clear that the concept is a beginning to be thoroughly debated in 2010 and later (*n:* 418, 81.48%), and every day of the post-publication period, there is an increasing tendency. These developments affecting all areas of the industry, therefore it can also be considered to affect the studies on business processes. This study demonstrates how developments in Industry 4.0 are reflected to the studies related to business processes within the scope of particular literature. In this context, this study aims to visualize how the developments in Industry 4.0 affected the studies related to business processes within the scope of particular literature. In this context, by taking 2010 breaking point, it is investigated whether a change occurred in the scientometric structure of the research articles related to the business processes after this year and whether the changes are compatible to the concepts conceived together with the concept of Industry 4.0.

As a necessary data source, records of the research articles that focused on business processes in the Web of Science (WoS) databases were collected, and the scientometric analyses were conducted on this data for different time intervals, between 1975 and 2010, and later. Scientometrics and bibliometrics are developing approaches, which are used in growing number of research in the last decade. Scientometric techniques provide not only descriptive enriched tables and graphs to reveal the leading papers, journals, countries, keywords, funds, and the other demographics, but also advanced analyses to highlight the relationships between the publications in the literature as densities, networks, and clusters. Techniques of scientometrics have been used for many research areas as well as production and operations management, and business process. Examining the transfer of knowledge across business disciplines [15], performance evaluation of business process management governance [16], depicting a landscape of the scientific literature on the concept of the Smart Factory concept [17], organizational performance evaluation and process management [18], towards mode 2 knowledge production analysis and proposal of a framework for research in business processes [19], business processes of Maneuver's freedom in Latin America [20], mobile supply chain management in the Industry 4.0 [21], economic globalization and its impacts on clustering [22], are among the studies in this context. In parallel to the current trends in production and business process, Liao et *al.* [23] investigated the studies about Industry 4.0 through systematic literature approach, which is another methodology that is close to bibliometrics. In a harmony with these papers in the literature, the main contribution of this paper is that it reveals the changes in the research related to business processes under the impact of Industry 4.0.

2 Method

The objective of this study is to analyse the effect of industry 4.0 concept on business processes. In harmony with this objective, the universe of study covers the research articles that are focused on business processes in the Web of Science (WoS) databases and the data in this universe was divided two parts concerning the rise in Industry 4.0 papers. Bibliometric and scientometric analyses were conducted on these data to find answers for the following questions:

- Is there any trend in the related literature over the years?
- Are there any differences in the metrics of business process management literature between the selected time intervals?
- Is there any difference in the distribution of countries, journals, organizations, and authors?
- Has any alteration happened about the research fields of the studies?
- What kind of changes has revealed in concepts?

This research followed a straightforward methodology. First of all, a bibliometric dataset was retrieved for Industry 4.0. For this purpose, the research articles were searched of which the language is English, the document type is article, time interval is 1975–2017, and the topic includes the keywords such as "Industry 4.0", "industry 3.5", "industrie 3.5", "Industry 4.0", "Industry 4", "Industry four", "industry 4.0", "industry 4", "industry four", "industrie 4.0", "industrie 4", "industrie four", "industry 5.0", "industry five", "industry 5", "fourth industrial revolution". This search resulted in 513 research articles, which were published after the year 2009 (*f*: 418, 80.48%). In this sense, the year 2010 was taken into consideration as a break point, and data retrieval for the papers related to business processes was performed twice regarding this break point through the query keywords "business* process*" searched in the topic field. As a result of 6255 research articles related to business processes, 3211 of them were found to be published before 2010, and remaining 3044 articles in 2010 or later.

Paper distributions over the years were firstly visualised via bar charts. Metrics in the business process management literature were figured out regarding h-index, number of papers, number of citations, impact factors of the journals, thus summary tables were organized to emphasize the top papers, authors, journals, countries, and other demographics, comparatively. Connections and publication densities were visualized via networks, for example, linkages between the keywords were analysed regarding both keyword clusters and keyword changes over years. In summary, various citation mining techniques were used in the scope of scientometrics to reveal demographics and networks for differences between two datasets.

The retrieved content contained the full records and cited references in plain text as a tab-delimited txt file. Some of the tables and graphs were organized according to the reports extracted from Clarivate Analytics (WoS). The data exported was then stored in a relational database designed in Oracle platform through a special coding developed in the Hypertext Preprocessor Programming (PHP) language, with the goal of obtaining queries with Structured Query Language (SQL) and performing customized analyses. For the analytical stages of the methodology, particular software tools were used, namely, VOSviewer [24] for network/density visualizations, and clustering on the networks and MS-Excel for consolidating the queries and descriptive statistics.

3 Findings and Discussion

3.1 Publication Distributions Over Time

Paper distributions over years along with the citation metrics were presented in Fig. 1. The article set had a peak in 2005 for the first dataset (a) then follows a decreasing trend from 2005 to 2009. H-index value was calculated as 112, sum of times cited (STC) as 76484, and average citations per item (*ACPI*) as 23.82 before 2010. An increasing trend from 2015 to 2017 was observed in the second dataset (b) where the most productive year was 2017. H-index value of the second population was naturally lower than the studies before 2010 since there exists direct positive relationship between the year of the study and the total number of citations. Sum of times cited (*STC*) value was 22548 and average citations per item (*ACPI*) 7.41 for the second dataset. All these metrics, in fact, indicated that the number of studies have increased sharply, and it is thought that digitalization and other technological developments related to Industry 4.0 have had at least partial effect on this behaviour. Following in depth network analyses about the keywords generated similar results in parallel with this thought.

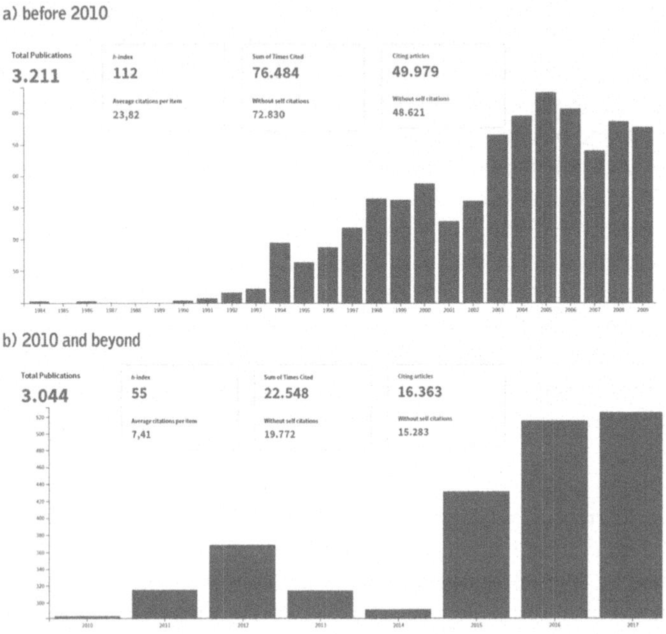

Fig. 1. H-index, sum of times cited, citing articles and total publications over the years

3.2 Authors

The studies before 2010 were written by 5992 different authors. Table 1 exhibits the top 20 authors according to the total number of articles. This list was made concerning the first authors in these studies. The top three authors before 2010 are Van Der Aalst WMP (*f*: 63, *h*: 39), Ter Hofstede AHM (*f*: 19, *h*: 15), and Reijers HA (*f*: 17, *h*: 11). The studies published 2010 and later were written by 7023 different authors. The top three authors for the second set are Mendling J (*f*: 46, *h*: 20), Van Der Aalst WMP (*f*: 36, *h*: 16), and Reijers HA (*f*: 30, h-index: 12) respectively. The most cited authors werw Mendling J (*citations (c)*: 1418), Van Der Aalst WMP (*c*: 839), Reijers HA (*c*: 745).

Table 1. Top 20 authors according to the total number of articles before 2010

Rank	1975–2009						2010–2017					
	Author (3211)	f	H	STC	ACPI	%	Author (3044)	f	h	STC	ACPI	%
1	Van Der Aalst WMP	63	39	5744	91.17	1.96	Mendling J	46	20	1418	30.83	1.51
2	Ter Hofstede AHM	19	15	1279	67.32	0.59	Van Der Aalst WMP	36	16	839	23.31	1.18
3	Reijers HA	17	11	643	37.82	0.52	Reijers HA	30	12	745	24.83	0.98
4	Reichert M	16	11	622	38.88	0.49	Ter Hofstede AHM	19	9	324	17.05	0.62
5	Grefen P	15	9	332	22.13	0.46	La Rosa M	18	10	355	19.72	0.59
6	Weston RH	15	8	171	11.40	0.46	Weber B	18	7	160	8.89	0.59
7	Leymann F	14	12	1229	87.79	0.43	Dumas M	16	8	429	26.81	0.52
8	Dumas M	13	11	827	63.62	0.40	Piattini M	16	6	141	8.81	0.52
9	Grover V	13	11	427	32.85	0.40	Dustdar S	15	8	166	11.07	0.49
10	Teng JTC	13	11	553	42.54	0.40	Poels G	15	6	106	7.07	0.49
11	Brombacher AC	12	6	138	11.50	0.37	Recker J	15	9	287	19.13	0.49
12	Bae H	11	7	181	16.45	0.34	Wang JM	15	6	115	7.67	0.49
13	Dietz JLG	11	8	158	14.36	0.34	Weske M	15	8	260	17.33	0.49
14	Godart C	11	5	93	8.45	0.34	Lee J	14	3	79	5.64	0.46
15	Papazoglou MP	11	9	959	87.18	0.34	Weidlich M	14	8	234	16.71	0.46
16	Yang Y	11	7	174	15.82	0.34	Reichert M	13	6	218	16.77	0.42
17	Dustdar S	10	7	823	82.30	0.31	Dijkman R	12	8	342	28.50	0.39
18	Kim K	10	6	93	9.3	0.31	Grefen P	12	5	109	9.08	0.39
19	Lee S	10	5	109	10.9	0.31	Vanthienen J	12	6	120	10.00	0.39
20	Rinderle S	10	8	175	17.5	0.31	Wen LJ	12	6	110	9.17	0.39

Mendling is in the foreground in 2010 and later, regarding the total number of articles, citations, and *h*-index values.

3.3 Countries and Institutions

Table 2 shows the country-wide review. Top country is United States of America (USA) (*f*: 840, 26.16% before 2010: *f*: 533, 17.51%, 2010 and later) for both article sets. Researchers from 80 different countries wrote the papers before the year 2010. On the other hand, researchers from 95 different countries wrote the papers in the year 2010 and later. One of the remarkable points in Table 2 is that China was in the sixth place before 2010 with 162 articles, however, jumped to the second place after 2010 with 341 articles. There was an increasing trend for Spain, Brazil, Russia, Slovenia, Serbia, and China. Germany, Italy, Austria, Australia, France were almost same positions before and after the year 2010.

Table 2. Top 20 countries about business process management

Rank	1975–2009			2010–2017		
	Country (3211)	*f*	%	Country (3044)	*f*	%
1	USA	840	26.16	USA	533	17.51
2	England	496	15.44	Peoples R China	341	11.20
3	Germany	321	9.99	Germany	297	9.75
4	Netherlands	260	8.09	Australia	230	7.55
5	Australia	210	6.54	England	217	7.12
6	Peoples R China	162	5.04	Netherlands	204	6.70
7	South Korea	127	3.95	Italy	165	5.42
8	Italy	119	3.70	Spain	156	5.12
9	Taiwan	108	3.36	Austria	126	4.13
10	Canada	87	2.70	France	114	3.74
11	Austria	86	2.67	South Korea	107	3.51
12	France	78	2.42	Canada	99	3.25
13	Spain	57	1.77	India	87	2.85
14	Switzerland	56	1.74	Taiwan	86	2.82
15	Scotland	52	1.61	Brazil	76	2.49
16	India	51	1.58	Slovenia	64	2.10
17	Japan	47	1.46	Serbia	62	2.03
18	Singapore	47	1.46	Belgium	59	1.9
19	Finland	44	1.37	Russia	44	1.44
20	Greece	40	1.24	Sweden	44	1.44

Given the density of articles produced over the years the country (Figs. 2 and 3), countries like Slovenia, Estonia, Iran, and Turkey increased their publication performance regarding articles produced about business processes in recent years. It is observed that countries such as Brazil, Peoples R China, Israel, Malaysia, Norway, Singapore, and Denmark were more productive regarding the research in business processes between 2004 and 2006. When the productivity of countries was examined in 2010, and after, the number of countries with a minimum of five documents was 71. In recent years, countries such as Estonia, Tunisia, Morocco, Russia, Qatar, and Turkey have increased the number of contributions in the field of business processes. Russia, Peoples R China, and Turkey increased their related publications in the last decade indicating that there has been a special interest on this field.

The first three universities that made the most contribution to this field in the world rankings for the studies conducted before 2010 were Eindhoven University of Technology (Netherlands, *f*: 128, 3.98%), International Business Machines Corp. (IBM) Company (USA, *f*: 74, 2.30%), Queensland University of Technology (Australia, *f*: 45, 1.40%). 2524 different university researchers contributed to the realization of 3044 articles for the studies conducted in 2010 and later. For this article set, the top three universities that contributed the most in the world rankings were Eindhoven University

Fig. 2. The densities of the countries over the time before 2010

Fig. 3. The densities of the countries over the time in 2010 and later

of Technology (Netherlands, *f*: 109, 3.58%), Queensland University of Technology (Australia, *f*: 63, 2.07%), University of Ljubljana (Slovenia, *f*: 44 1.44%). The status of the universities were found to be compatible with the patterns among the countries. While the Netherlands, the United States, the United Kingdom, and the Australian universities were generally leading ones before 2010, it can be said that important progress was made in the ranking of Slovenian, Estonian and Chinese universities in business process articles after the year 2010 (Table 3).

Table 3. Top ten universities contributing to the field

Rank	1975–2009				2010–2017			
	Organizations	Country	f	%	Organizations	Country	f	%
1	Eindhoven University of Technology	Netherland	128	3.98	Eindhoven University of Technology	Netherland	109	3.58
2	International Business Machines Corp (IBM) Company	USA	74	2.30	Queensland University of Technology	Avustralia	63	2.07
3	Queensland University of Technology	Avustralia	45	1.40	University of Ljubljana	Slovenia	44	1.44
4	Manchester University	England	40	1.24	Tsinghua University	China	37	1.21
5	Polytechnic University of Milan	Italy	33	1.02	Seville University	Spain	30	0.98
6	Texas University	USA	30	0.93	Catholic University of Leuven	Belgium	29	0.95
7	Queensland University	Avustralia	29	0.90	Vienna University of Technology	Austuria	27	0.88
8	National University of Singapore	Singapore	28	0.87	Humboldt University	Germany	26	0.85
9	Tilburg University	Netherland	28	0.87	University of New South Wales	Avustralia	26	0.85
10	Loughborough University	England	27	0.84	Tartu University	Estonia	26	0.85

3.4 Journals and Research Areas

In the related research area, 3211 papers were published in 819 different journals before 2010, while 3044 papers were published in 909 different journals in 2010 and later. The top 10 journals with the highest number of publications are listed in Table 4. It is seen that most of the journals that published articles about business processes before and after 2010 were, mainly from American, German, British and Dutch journals. It seems that the most intensive journals, pre-2010 and later, did not changed significantly in the research domains which accumulated on Computer Science, Operations Research & Management Science, Information Science & Library Science, Business & Economics. When it came to ACPI, Data Knowledge Engineering, Information Management, and International Journal of Production Economics were found to be top three journals. High citation ratios in the journals before 2010 may be attributed to the difference between the numbers of years as the reason for having higher ACPI values.

Table 4. Top-ten journals in this field

Rank	Years	Source titles	Country	FYIF	Research domain	*f*	h-index	STC	ACPI	%
1	1975–2009	Lecture Notes in Computer Science	–	–	–	571	36	6165	10.8	17.78
	2010–2017	Information Systems	England	2.822	Computer Science	75	17	1248	16.64	2.46
2	1975–2009	Computers in Industry	Netherlands	2.731	Computer Science	78	27	2094	26.85	2.42
	2010–2017	Business Process Management Journal	England	–	Business & Economics	64	6	110	1.72	2.10
3	1975–2009	International Journal of Production Economics	Netherlands	4.052	Engineering; Operations Research & Management Science	68	26	2349	34.54	2.11
	2010–2017	Decision Support Systems	Netherlands	4.29	Computer Science; Operations Research & Management Science	53	14	653	12.32	1.74
4	1975–2009	Business Process Management Proceedings	Germany	–	Computer Science	65	16	1203	18.51	2.02
	2010–2017	Expert Systems with Applications	England	3.526	Computer Science; Engineering; Operations Research & Management Science	53	15	593	11.19	1.74
5	1975–2009	International Journal of Operations Production Management	England	4.211	Business & Economics	56	25	1385	24.73	1.74
	2010–2017	Enterprise Information Systems	England	2.859	Computer Science	48	15	564	11.75	1.57
6	1975–2009	Lecture Notes in Artificial Intelligence	–	–	–	56	6	283	5.05	1.74
	2010–2017	Computers in Industry	Netherlands	2.731	Computer Science	47	12	414	8.81	1.54

(continued)

Table 4. (*continued*)

Rank	Years	Source titles	Country	FYIF	Research domain	*f*	h-index	STC	ACPI	%
7	1975–2009	Production Planning Control	England	2.882	Engineering; Operations Research & Management Science	49	13	711	14.51	1.52
	2010–2017	Information and Software Technology	Netherlands	2.924	Computer Science	46	13	569	12.37	1.51
8	1975–2009	Information Management	Netherlands	4.283	Computer Science; Information Science & Library Science; Business & Economics	48	26	1877	39.10	1.49
	2010–2017	IEEE Transactions on Services Computing	USA	4.245	Computer Science	42	12	653	15.55	1.38
9	1975–2009	Industrial Management Data Systems	England	2.343	Computer Science; Engineering	44	18	895	20.34	1.37
	2010–2017	Software and Systems Modeling	Germany	1.869	Computer Science	41	6	161	3.93	1.34
10	1975–2009	Data Knowledge Engineering	Netherlands	2.131	Computer Science	43	22	1735	40.35	1.33
	2010–2017	Journal of Systems and Software	USA	2.619	Computer Science	37	11	343	9.27	1.21

Relevant 3211 articles before 2010 were associated with 77 different research areas as shown in Table 5. Most of the articles were concentrated in the field of Computer Science (*f*: 1960, 61.04), and then other areas can be seen such as Engineering, Business Economics, Operations Research, Management Science, Information Science, Library Science, Telecommunications, Health Care Sciences Services, Automation Control Systems, Informatics. When studies of 2010 and later were examined, 3044 articles were associated with 86 different research areas. The most of the studies were concentrated in the field of Computer Science (*f*: 1557, 51.15%), and other areas were found to be Business Economics, Engineering, Operations Research, Management Science, Information Science, Library Science, Telecommunications, Control Systems, Environmental Sciences, Ecology, Mathematics, Science, and Technology. The fact that the research area of Computer Science was high in both periods was consistent with the research domains of effective journals in the field of interest.

Table 5. Distribution of the research areas used in keyword selection

Rank	1975–2009			2010–2017		
	Research areas (3211)	*f*	%	Research areas (3044)	*f*	%
1	Computer Science	1960	61.04	Computer Science	1557	51.15
2	Engineering	857	26.69	Business Economics	779	25.59
3	Business Economics	768	23.91	Engineering	573	18.82
4	Operations Research Management Science	481	14.98	Operations Research Management Science	292	9.59
5	Information Science Library Science	293	9.12	Information Science Library Science	246	8.08
6	Telecommunications	105	3.27	Telecommunications	65	2.13
7	Health Care Sciences Services	52	1.61	Automation Control Systems	58	1.90
8	Automation Control Systems	48	1.49	Environmental Sciences Ecology	47	1.54
9	Construction Building Technology	41	1.27	Mathematics	47	1.54
10	Medical Informatics	41	1.27	Science Technology Other Topics	46	1.51

3.5 Keywords

In addition to the findings above, keywords were also analysed to reveal the subjects handled in the scope of business processes in two time intervals. Keyword linkages were presented regarding both clusters and change in years. In total, 6020 keywords were used in 3211 articles before 2010. 424 of the total keywords were used at least five times in the dataset. The change of the keywords by years was obtained as shown in Fig. 4 below. 10283 keywords were used in 3044 articles in the year 2010 and later. 749 of the total keywords were used at least five times in the dataset. The change of the keywords by years was obtained as shown in Fig. 5.

In the keyword distributions, information technology related terms draw attention, however, information technology usage can be considered as a necessary issue for process improvement, but it is not adequate individually. All business systems should be taken into consideration for improving the processes. Process improvement will lead to the success of customer relationship management and sustainable market position as well [25, 26]. Besides, in the early 2000 s, the use of business process reengineering, group support systems, expert systems, computer systems and technologies, business process analytics, decision support systems in business processes were shown to have a positive impact on the business processes of the organizations. Mendling et *al.* [27] supported these considerations as well. In the period of 2004-2006, it can be expressed that the studies related to business processes focused on the concepts such as workflow and workflow management, virtual enterprise, e-procurement, electronic commerce, manufacturing systems, information management, and organizational performance. For example, Georgakopoulos et *al.* [1] indicated the importance of workflow

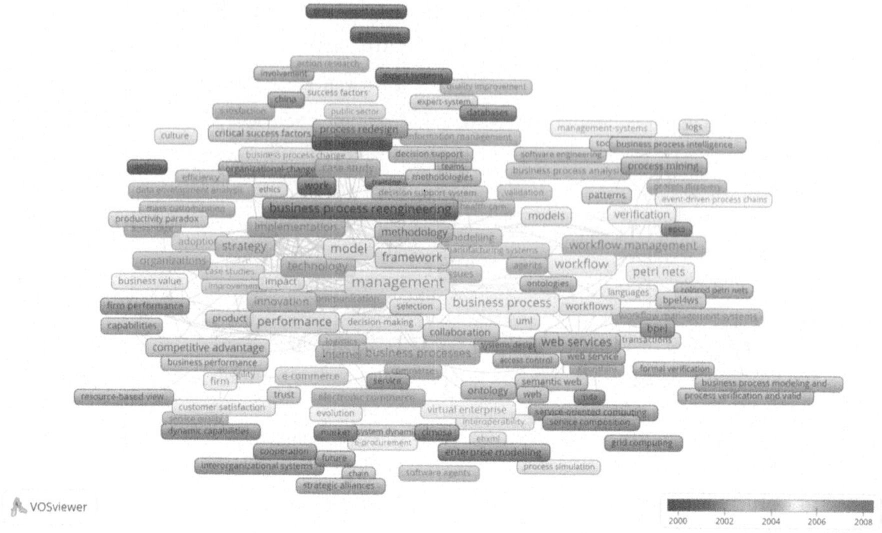

Fig. 4. Keywords over years (1975–2009)

Fig. 5. Keywords over years (2010–2017)

management by emphasizing its role on business process modelling and reengineering, and process automation.

In the period of 2006–2009, it can be seen that China is also in the front line in the keywords. This finding can be interpreted as a supporting issue that China ranked in the second place regarding the number of articles in 2010 and later. In recent years, business process intelligence, business process modelling, process mining, service oriented computing, and grid-computing concepts were covered by many research attempts. The concept of "grid computing" left its place to the concept of cloud and cloud computing after 2010. Another important, interesting finding is that the keywords such as semantic web and formal verification have taken place in recent years providing an infrastructure for the transition to Industry 4.0.

In the period of 2010–2013, UML, electronic data interchange, RFID, agents, web services, service oriented architecture (SOA), SOAP, e-business, virtual enterprises, interoperability and strategic alliances were the most commonly used keywords to indicate the scope of the studies, these concepts are compatible with the philosophy of Industry 4.0 such as collecting data from many points, communication of devices over the internet, and real-time performance monitoring.

The period of 2013–2015 highlighted the concepts of dynamic capabilities, outsourcing, supply chains, China, business process outsourcing, architecture, process mining and process model. When examined regarding keyword content, it can be said that business processes, information, and communication technologies had a direct linkages to each other.

Text clustering was also performed on the keywords to reveal the subjects around particular keyword groups. In the first dataset including the articles published before 2010, ten clusters were obtained over 424 keywords repeated at least five times in the dataset. The cluster structure for this period was presented in Fig. 6. Another ten clusters were obtained from the second dataset over 749 keywords repeated at least five times in the dataset as depicted in Fig. 7. In Fig. 6, intensive clusters indicated particular topics such as work flow management and process analysis; enterprise modelling and reference architectures; web services, service designs, virtual enterprise, and service oriented architecture; business process reengineering and redesign; customer satisfaction, critical success factors, and business value.

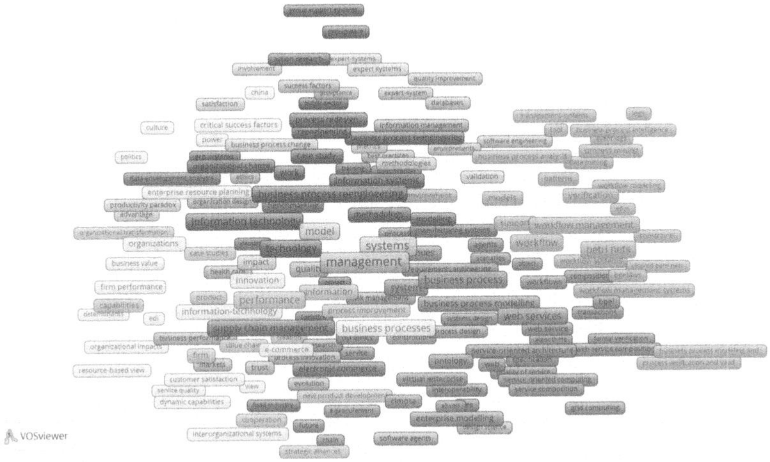

Fig. 6. Keyword clusters (1975–2009)

Fig. 7. Keyword clusters (2010–2017)

In Fig. 7, even if similar cluster structure was obtained, new concepts and systems were included in the clusters such as e-business, electronic data interchange, RFID, internet of things technology adoption, information and communication systems, strategic management and business performance. These findings were found to be compatible with Figs. 4 and 5, besides Figs. 6 and 7 highlights the keyword batches to indicate the which concepts and subjects were used together.

4 Conclusion

This study presented a research to highlight how the developments in Industry 4.0 affected the studies related to business processes within the scope of particular literature. In this context, by taking 2010 breaking point, it is investigated whether a change occurred in the scientometric structure of the research articles related to the business processes after this year and whether the changes are compatible to the concepts conceived together with the concept of Industry 4.0. In this context, the related research articles were investigated in two time intervals, 1975–2009 and 2010–2017. Paper distributions over the years were visualised; metrics in the literature of business process management were figured out regarding h-index, number of papers, number of citations, impact factors of the journals; top papers, authors, journals, countries, and other demographics were summarized comparatively. Furthermore, linkages between the keywords were analysed regarding both keyword clusters and keyword changes over years. In summary, various citation mining techniques were used in the scope of scientometrics to reveal demographics and networks for differences between two datasets. All findings emphasized that the new concepts and technologies by Industry 4 and internet of things have had a positive impact on business process modelling and management. These issues can be seen from the trends in number of publications and keyword analytics. Industry 4.0 have provided multidimensional and large-scale data providing many opportunities to analyse and monitor the processes as well as new approaches to manage the new business processes coming with these new technologies.

This study contributed to the related literature with its findings showing the status of business process modelling and management literature in the scope of Industry 4.0 and the related developments in technology. When all findings are evaluated, it can be said that this study provides valuable information to academicians and practitioners who would like to conduct research on the subject or follow the literature on the subject.

Acknowledgment. The authors declare no conflict of interest.

References

1. Georgakopoulos D, Hornick M, Sheth A (1995) An overview of workflow management: from process modeling to workflow automation infrastructure. Distrib. Parallel Databases 3 (2):119–152. https://doi.org/10.1007/bf01277643
2. Kalpic B, Bernus P (2002) Business process modelling in industry-the powerful tool in enterprise management. Comput Ind 47(3):299–318. https://doi.org/10.1016/s0166-3615(01) 00151-8
3. Turetken O, Van den Hurk H, Karagöz A, Ünal A (2011) Plural Yöntemi ile BPMN Tabanlı Özne Yönelimli Süreç Modelleme: Durum Çalışması. In: 5th national symposium on software engineering, 26–28th September. Middle East Technical University, Ankara
4. Kohlbacher M, Reijers HA (2013) The effects of process-oriented organizational design on firm performance. Bus Process Manag J 19:245–262
5. Jeston J, Nelis J (2013) Business process management, 3rd ed. Routledge
6. Aguilar-Saven RS (2004) Business process modelling: review and framework. Int J Prod Econ 90(2):129–149
7. Dumas M, LaRosa M, Mendling J, Reijers HA (2013) Fundamentals of business process management. Springer. ISBN: 978-3-642-33142-8
8. Mendling J, Reijers HA, van der Aalst VM (2010) Seven process modeling guidelines (7PMG). Inf Softw Technol 52(2):127–136
9. Qin J, Liu Y, Grosvenor R (2016) Changeable, agile, reconfigurable & virtual production a categorical framework of manufacturing for industry 4.0 and beyond. Procedia CIRP 52:173–178. https://doi.org/10.1016/j.procir.2016.08.005
10. Lu Y (2017) Industry 4.0: a survey on technologies, applications and open research issues. J Ind Inf Integr 6:1–10. https://doi.org/10.1016/j.jii.2017.04.005
11. Santos MY, Sa JO, Carina A, Lima FV, Costa E, Costa C, Martinho B, Galvao J (2017) A big data system supporting bosch braga industry 4.0 strategy. Int J Inf Manag 37:750–760. https://doi.org/10.1016/j.ijinfomgt.2017.07.012
12. Oesterreich TD, Frank T (2016) Understanding the implications of digitisation and automation in the context of industry 4.0: a triangulation approach and elements of a research agenda for the construction Industry. Comput Ind 83:121–139. https://doi.org/10.1016/j.compind.2016.09.006
13. Hofmann E, Rüsch M (2017) Industry 4.0 and the current status as well as future prospects on logistics. Comput Ind 89:23–34. https://doi.org/10.1016/j.compind.2017.04.002

14. Peruzzini M, Fabio G, Pellicciari M (2017) Benchmarking of tools for user experience analysis in industry 4.0. In: 27th international conference on flexible automation and intelligent manufacturing, FAIM2017, Modena, Italy, 27–30 June 2017. Procedia Manufacturing, vol 11, pp 806–813. https://doi.org/10.1016/j.promfg.2017.07.182
15. Pratt JA, Hauser K, Sugimoto CR Defining the intellectual structure of information systems and related college of business disciplines: a bibliometric analysis. Scientometrics 93 (2):279–304. https://doi.org/10.1080/08874417.2012.11645610
16. Ensslin L, Ensslin SR, Dutra A, Nunes NA, Reis C (2017) BPM governance: a literature analysis of performance evaluation. Bus Process Manag J 23(1):71–86. https://doi.org/10.1007/s11192-012-0668-y
17. Strozzi F, Colicchia C, Creazza A, Noè C (2017) Literature review on the 'Smart Factory' concept using bibliometric tools. Int J Prod Res 55(22):6572–6591. https://doi.org/10.1080/00207543.2017.1326643
18. Chaves LC, Ensslin L, de Lima MVA, Ensslin SR (2017) Organizational performance evaluation and process management: mapping of the field. Revista Eletronica De Estrategia E Negocios-Reen 10(1):101–140
19. Veit DR, Lacerda DP, Camargo LFR, Kipper LM, Dresch A: towards mode 2 knowledge production analysis and proposal of a framework for research in business processes. Bus Process Manag J. 23(2), 293–328
20. Indriago F, Sarcos H Business processes of maneuver's freedom in Latin America. Revista Cicag 13(2):363–372
21. Barata J, Da Cunha PR, Stal J (2018) Mobile supply chain management in the industry 4.0 era: an annotated bibliography and guide for future research. J Enterp Inf Manag 31(1): 173–193
22. Razminiene K, Tvaronaviciene M (2017) Economic globalization and its impacts on clustering. Terra Econ 15(2):109–121
23. Liao Y, Deschamps F, Loures EFR, Ramos LFP (2017) Past, present and future of industry 4.0 - a systematic literature review and research agenda proposal. Int J Prod Res 55 (12):3609–3629. https://doi.org/10.1080/00207543.2017.1308576
24. Van Eck NJ, Waltman L (2013) Software survey: VOSviewer, a computer program for bibliometric mapping. Scientometrics 84(2):523–538. https://doi.org/10.1007/s11192-009-0146
25. Stojanovic D, Simeunovic B, Tomasevic I, Radovic M (2012) Current state of business process management in serbian industry. Metal Int 17(10):222–226
26. Barnir A, Gallaugher MJ, Auger P (2003) Business process digitization, strategy, and the impact of firm age and size: the case of the magazine publishing industry. J Bus Venturing 18(6):789–814. https://doi.org/10.1016/S0883-9026(03)00030-2
27. Mendling J, Baesens B, Bernstein A, Fellmann M (2017) Challenges of smart business process management: an introduction to the special issue. Decis Support Syst 100:1–5. https://doi.org/10.1016/j.dss.2017.06.009

Production Planning and Control

Production-Integrated Metrology with Modern Coordinate Measuring Machines Using Multisensor and X-Ray Computed Tomography Systems

Joachim Stopp[1(✉)], Raoul Christoph[1], and Ralf Christoph[2]

[1] Technische Universität Dresden, Dresden, Germany
joachim.stopp@mailbox.tu-dresden.de,
raoul.christoph@tu-dresden.de
[2] Werth Messtechnik GmbH, Giessen, Germany
ralf.christoph@werth.de

Abstract. Production related or integrated inspection of workpiece geometry is becoming increasingly relevant in modern production processes. To draw the correct conclusions for production control, the workpiece geometry has to be captured as completely as possible. Coordinate metrology with optical sensors and multisensor systems as well as coordinate metrology with X-ray computed tomography is therefore especially suited for this application. Measurement times of a few seconds can be achieved with modern technology, even with X-ray computed tomography. Thus the technology is able to follow production cycle times.

Keywords: Production metrology · Multisensor systems
X-ray computed tomography · Coordinate metrology · Measuring time

1 Introduction

In order to control and optimize modern production processes, it becomes more important, to capture workpiece geometries as completely as possible. By using innovative design methods and manufacturing processes such as additive manufacturing producing more complex workpiece geometries is possible. Therefore, the metrological system used for measuring has to detect a large number of different and difficult-to-access geometric features. On the other hand, it has to comply to the short cycle times required for close-to-production or production-integrated use. These requirements can be met with coordinate measuring machines that use either a combination of different sensor principles (multisensor systems) [1] or the principle of X-ray computed tomography (X-ray CT) [2]. With multisensor systems, short switching times between the fastest sensors suitable for the respective measuring task enable short total measurement times. The same coordinate system is used for the evaluation of the measured points as well as the calculation of geometric elements and the dimensional evaluation. [3] The principle of X-ray.

N. M. Durakbasa and M. G. Gencyilmaz (Eds.): ISPR 2018, *Proceedings of the International Symposium for Production Research 2018*, pp. 693–701, 2019.
https://doi.org/10.1007/978-3-319-92267-6_56

CT enables fast, complete and non-destructive assessment of all internal and external geometrical characteristics of a workpiece [4].

Figure 1 qualitatively shows the ratio of the number of geometrical characteristics and the required measurement time for various sensors. With multisensors systems, the measurement time is between the values for measuring with conventional probes and only optical image processing sensors. The required measurement time is assumed to be proportional to the number of geometrical characteristics to be measured. Measurement time with X-ray tomography, however, is close to independent to the number of measured geometrical characteristics. Since the entire geometry is detected with one measurement, the evaluation of a great number of geometrical characteristics is possible in almost constant time. However, due to the relatively high measurement time with X-ray CT compared to the measurement time of a single geometric characteristic using a multisensor measuring machine, computed tomography only scales well when many geometric characteristics need to be measured [5].

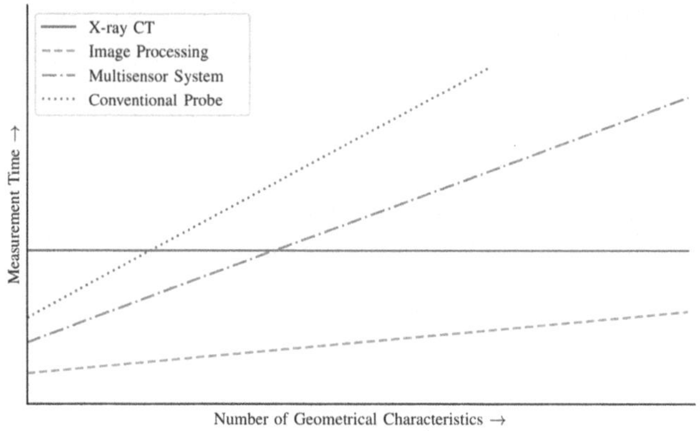

Fig. 1. Required measurement time for different measurement principles in dependence of the number of geometrical characteristics to be captured (qualitative presentation) [4]

In the following, manufacturing-related or manufacturing-integrated measuring with multisensor or computed tomography measuring machines is presented. The focus will be on methods to shorten measurement time.

2 Multisensor Systems

Table 1 lists the suitability of various sensors for different measurement tasks [5]. The image processing with fixed or zoom optics is the fastest method to measure even small geometries without contact [6]. It is also heavily dependent on the optical properties of the surface. Furthermore, three-dimensional contours and undercuts cannot be measured. The topography of a surface of an object can be measured with a sensor based on the principle of focus variation [6]. Measuring with optical image processing sensors (hereinafter IP) is, in most cases, the fastest and most accurate way to measure with a multisensor coordinate machine and they are generally used as a reference for the other sensors [3].

Table 1. Suitability* of sensors for different tasks and characteristics of measured object

	Image Processing Probe	Focus Variation Probe	Laser Sensor	Chromatic Focus Sensor	Interferometric Sensor	Confocal Sensor	Conventional Touch Probe	Fibre Probe	Contour Probe	Computed Tomography
Geometry										
Egdes and Contours	X	(X)	(X)	(X)	(X)	(X)	O	(X)	O	O
3D-Contours and Flatness	O	X	X	X	X	X	X	X	X	X
Surface Topology	O	X	X	X	X	X	X	X	X	X
Micro-Geometries	X	X	X	X	X	0	X	X	(X)	X
Vertical Surfaces/Undercut	O	O	O	O	X	O	X	X	O	X
Inner Geometry	O	O	O	O	O	O	O	O	O	X
Surface										
Sensitive/Deformable	X	X	X	X	X	X	O	X	O	X
High Contrast	X	X	X	X	X	X	X	X	X	X
Low Contrast	O	O	X	X	X	X	X	X	X	X
Reflecting	O	O	(X)	X	X	X	X	X	X	X
Transparent	O	O	O	X	O	X	X	X	X	X
Particularities										
Ultra High Precision	X	(X)	(X)	X	X	X	X	X	X	(X)
Roughness Measuring	O	(X)	(X)	(X)	X	X	O	(X)	X	O

*X suitable
(X) conditionally suitable
O unsuitable

Additionally, non-contact measurement sensors such as confocal sensors, chromatic distance probes and lasers that are based on the principle of triangulation enable the measurement of roughness and of reflective or transparent surfaces [7]. However, since these are point or line sensors, the measurement time for these sensors is higher compared to IP sensors (area sensors).

The devices are complemented by conventional touch probes and contour probes. Tactile measurement is particularly suitable for hard-to-reach geometries (undercuts) where optical methods cannot be used [8].

The fiber probe occupies a special position as an opto-tactile sensor [9]. The probing is done physically with a microscopic stylus sphere (diameter between 20 µm to 100 µm) and thus is independent of optical surface properties. Its deflection is evaluated with a highly accurate image processing sensor. The mechanical transmission of the deflection is not part of the measurement chain. Probing forces are therefore very small and thus also allow the measurement of workpieces where the probing force of conventional probes can cause large measurement deviations or damage [10].

2.1 Particularities for Fast, High Resolution and Production Integrated Measuring

The use of multiple sensors on one device offers many advantages, but also a unique challenge. If the sensors are mounted rigidly relative to each other the measuring volume is narrowed greatly. Additionally, there is an increased risk of collision between sensors which are not involved in the measurement and the workpiece. Mounting the sensors far apart narrows the shared overlapping area of the respective measurement ranges.

Adding additional axes can resolve this by moving the sensors independently. Figure 2 shows a newly developed coordinate measuring machine in bridge construction with two separate rams for the optical sensor and the tactile probe [11].

Fig. 2. Two rams allow for the independent use of optical and tactile sensors and reduce the danger of collision as well as an increase of usable combined measuring volume

Another approach is to switch the sensors via a changing rack. For example, the direct attachment of different sensors in front of the IP sensor is possible. A newly developed changing rack that allows the quick and fully automatic change of the sensors is shown in Fig. 3. [12] It maximizes the shared measuring range. The fiber probe (Fig. 4) occupies a special position in this context, too. For the measurement using a fiber probe, the focal plane of the zoom optics is adjusted to measure the deflection of the probe sphere with the image processing sensor. Due to the shallow depth of focus and low shaft thickness, the stylus shaft disappears in the picture. Underneath this level, the image processing can be used undisturbed. Thus, even without a changing rack and with little or no offset movement, sensors can be switched between the fiber probe and the image processing sensor. This can considerably speed up the measuring process. When a laser sensor is integrated directly in the optical path, it is also possible to change between the laser sensor and the image processing sensor with minimal offset movement [9]. With these methods, the risk of collisions is greatly reduced.

Fig. 3. Changing rack with auxiliary lens, conventional probe, fiber probe and contour probe

Fig. 4. The fiber probe placed in front of the high-precision optics, a sensor change between fiber probe and image processing sensor is possible without or with very small offset movements

Analogous to measuring with scanning probes, non-contact optical sensors can be used with continuously moving machine axes. Precisely synchronized axes enable rapid measuring of many geometric characteristics with optical sensors such as image processing sensors or chromatic distance probes. Using this principle, a distinction is made between raster scanning, contour tracing and *on-the-fly*-measurements. The difference between these methods is explained below using the example of an image processing sensor. When using common raster scanning methods, images are taken at several positions on a grid. The images are composed subsequently. The newly developed Raster Scanning HD superimposes images that are continuously recorded during motion [5]. In both cases, the measurement is made within the resulting composite image. Raster Scanning HD enables measurements with higher accuracy as well as speed. *Contour tracing* controls the motion of the machine axes to follow a contour detected by the IP sensor. The areas needed for the measurement are scanned fully automatically. This allows even large contours to be measured quickly. The benefits of both the Raster Scanning HD and the contour tracing can be obtained by combining them. The measurement *on the fly* does not capture images continuously. The images are taken at defined positions in the same way as in the case of a static measurement with the difference that the recording is done while moving continuously [13]. Thus, the time for positioningand therefore the measurement time is greatly reduced.

3 X-Ray Computed Tomography

In X-ray tomography systems, the measurement object is rotated in the coordinate measuring machine and irradiated from different directions. The radiation remaining behind the workpiece is detected and the information about the local attenuation by the measurement object—contained in the resulting transmission images—is reconstructed into a voxel volume by filtered back projection [14]. A point cloud which describes the complete surface of the workpiece is then extracted from this voxel volume. [4] Sensors based on the principle of X-ray tomography are volume sensors, and accordingly—limited by the structural resolution—all external and internal geometric features are detected with one measurement. Computed tomography enables the measurement of workpieces with undercut geometries that cannot be measured non-destructively with any other sensor that are commonly used in coordinate metrology. Even the measurement of enclosed geometric characteristics, such as those that can be produced with additive manufacturing methods as well as the measurement of assemblies is possible [9]. The measurement time is almost independent of the number of geometric characteristics to be measured. X-ray tomography is thus the ideal method for measuring objects with a large number of geometric characteristics.

3.1 Particularities for Fast, High Resolution and Production Integrated Measuring

In order to be suitable for production related or integrated use, the measurement time must be adjusted to the process cycle. Newly developed X-ray transmission tubes with high tube power and acceleration voltage and at the same time small focal spot size allow measuring with high structural resolution (spatial resolution) and low exposure times [15]. Positioning times and acceleration phases between the individual rotation steps can be eliminated by taking the X-ray images while continuously moving the rotary axis (*on-the-fly tomography*) [16].

If only certain areas of the measurement object are of interest, the measurement time can be reduced by using *Multi-Region-of-Interest CT* (Multi-ROI CT) [4]. For this, a so-called overview tomography of the entire measurement object in low resolution must be performed first. Subsequently, only the areas of interest are scanned with a higher resolution using eccentric tomography. This enables fast high resolution measurements without the necessity to manually reposition the object to be measured.

Time is also saved by reconstructing parallel to the X-ray image acquisition. In this way, the reconstructed volume is available shortly after the projection data has been acquired. Thinning out the generated point cloud within the limits of structural resolution required for the measurement task additionally shortens downstream processes.

In addition to the reduction of the time for the measurement and computational evaluation, an automation of the process, for example, by using the newly developed pallet changer shown in Fig. 5, ensures an optimal integration in the production process.

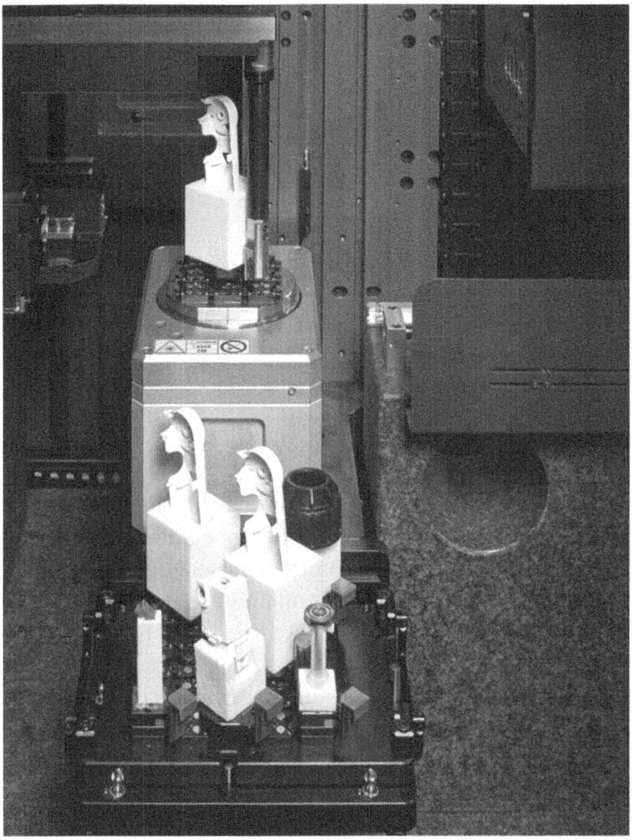

Fig. 5. Automatic part changing garantuees a tight coupling with the production process

To reduce beam hardening artifacts, thin metal filters are inserted between the X-ray source and the measuring object [17, Chap. 2]. Changing these filters can also be automated. The filter weakens and at the same time shifts the emitted spectrum towards higher frequencies. Using a filter changer, as shown in Fig. 6, an adjusted spectrum for the measurement of different parts can be generated fully automatically in combination with adjusting the tube voltage and current.

Fig. 6. In addition to an adjusted tube power and voltage, an automatic filter changer enables the adaption of the emmitted spectrum for different parts

4 Conclusions

Coordinate measuring machines with multisensor technology and X-ray tomography are suitable for production related and production integrated coordinate measurement. They enable the measurement of geometric characteristics of complex workpieces. Thus, they are suitable to form the basis for monitoring and controlling manufacturing processes. Multisensor coordinate machines are best suited for measuring with a low to medium number of geometric characteristics to be measured. As the number of geometric characteristics increases, the measurement time with X-ray tomography decreases when compared to other measurement principles due to the constant measurement time. Additionally, computed tomography is the only technology used in coordinate metrology that is able to completely asses both internal and external geometries nondestructively. For both, measuring machines using multisensory systems as well as computed tomography, methods for reducing the measurement time have been developed. The automation of the measuring process using coordinate measuring machines allows an easy integration into the manufacturing process.

References

1. Weidemeyer B (2017) Ohne Gerätewechsel, Komplettmessung mit Multisensor-Koordinatenmessgeräten. Qualität und Zuverlässigkeit, 12
2. Schmidt I (2015) Vollständig und zerstörungsfrei, Zehn Jahre Computertomografie in der Koordinatenmesstechnik. Qualitiy Engineering, 1

3. Christoph R, Neumann HJ (2006) Multisensor-Koordinatenmesstechnik. Verlag Moderne Industrie, München, vol 248. ISBN: 9783862360505
4. Christoph R, Neumann HJ (2011) Röntgentomografie in der industriellen Messtechnik. Verlag Moderne Industrie, München, vol 352. ISBN: 9783862360321
5. Heidari Bateni S (2018) Multisensorik oder Röntgentomografie?, Die Messaufgabe bestimmt die Wahl des Koordinatenmessgeräts. Qualität und Zuverlässigkeit, 4
6. Fischer M (2018) Licht ins Detail bringen, Bildverarbeitung und Fokusvariation auf unterschiedlichen Oberflächen. Qualität und Zuverlässigkeit, 6
7. Fischer M (2017) Mit hoher Genauigkeit und Geschwindigkeit, Topografie-Scanning mit neuem Chromatic Focus Line Sensor. Qualitiy Engineering, P-01
8. Weidemeyer B (2018) Die Lösung für jede Messaufgabe, Wirtschaftliches Messen mit Multisensor-Koordinatenmessgeräten. Qualität und Zuverlässigkeit, 3
9. Keferstein C, Marxer M, Bach C (2018) Fertigungsmesstechnik, Alles zu Messunsicherheit, konventioneller Messtechnik und Multisensorik. Springer Vieweg, Wiesbaden. ISBN: 978-3-658-17755-3
10. Weidemeyer B (2012) Berührung mit Her(t)z, Fasertaster vermeidet Verformung an weichen Materialien. Qualität und Zuverlässigkeit, 12
11. M. Fischer, "Wenn Flexibilität gefragt ist, Fertigungsbegleitende Messungen mit zwei unabhängigen Sensorachsen", *Qualitiy Engineering*, P-01 2018
12. Fischer M (2016) Flexible Multisensorik, Mit mehreren Sensoren ohne Einschränkungen messen. Qualität und Zuverlässigkeit, 10
13. Ferger D (2016) Messen mit Präzision und Durchblick, Multisensorik überwacht die Qualität in der Fertigung, Kunststoffe, 7
14. Feldkamp L, Davis L, Kress J (1984) Practical cone-beam algorithm. JOSA A 1(6):612–619
15. Christoph R (2018) Das Ganze ist mehr als die Summe seiner Teile, Wie Sie das richtige RöntgentomografieKoordinatenmessgerät wählen. Qualität und Zuverlässigkeit, 2
16. Heidari Bateni S (2017) Messen wie im Flug, Koordinatenmesstechnik mit Computertomografie auf dem Vormarsch. Qualitiy Engineering, 3
17. Buzug TM (2008) Computed tomography: from photon statistics to modern cone-beam CT. Springer. ISBN: 978-3-540-39408-2

Productivity and Performance Management

A Research on Financial Performance Analysis of Informatics Companies in the Scope of Industry 4.0

V. Özlem Akgün[1(✉)] and Ali Akgün[2]

[1] Marketing and Foreign Trade Department, Selcuk University, Konya, Turkey
ozlemakgun@selcuk.edu.tr
[2] Business Department, Selcuk University, Konya, Turkey
aakgun@selcuk.edu.tr

Abstract. The industry sector, which is in the effort to meet market needs quickly and efficiently, has entered into the period of Industry 4.0, which means the Industrial Revolution and is called the new industrial reform, as a result of the rapid development of technology. Compared to the previous three industry periods, Industry 4.0 promises to establish a comprehensive value chain that includes all steps, including product delivery and product recycling, starting from the manufacturing intellectual milieu. This new process, which incorporates a large number of modern automation systems, data exchanges and manufacturing technologies, coupled with physical and digital technology at the same time, offers significant competitive advantages, especially if implemented correctly for developing countries. The aim of this study is to analyze the performance of IT companies operating in Turkey in the Industry 4.0 environment. For this purpose, the performances of the companies in the Borsa Istanbul IT index were analyzed by TOPSİS method and the results of the analysis and the suggestions were taken at the end of the study. R & D expenses have been determined as effective criteria for performance evaluation. BIST Informatics Index includes 6 companies whose R & D expenses are determined. The data is calculated as the ratio because of the scale size is not to affect the analysis. The data as an evaluation factor, 9 financial ratios including R & D expenses were used. As a result, in 2017, the BIST IT Index company with the highest performance was designated as LOGO.

Keywords: Industry 4.0. · Fourth industrial revolution · Digitalization
R & D · TOPSIS

1 Introduction

The first industrialization efforts that started with a mechanical weaving loom, have come to a very different point with the speed of technology which has been at a dizzying pace today, radically changing the traditional production concept. As industry 4.0, this new process of expressing the day-to-day change in the effects of digital conversion on the business and business models of the future has changed to provide a competitive advantage. The basis of the concept; all the related units in industrial

N. M. Durakbasa and M. G. Gencyilmaz (Eds.): ISPR 2018, *Proceedings of the International Symposium for Production Research 2018*, pp. 705–723, 2019.
https://doi.org/10.1007/978-3-319-92267-6_57

production can communicate with each other, all the data can be accessed in real time, and the optimum value added can be provided by these data [1]. Industry 4.0 includes a structure that will completely change production and consumption relationships. On the one hand, production systems that instantly adapt to the changing needs of the consumer, on the other hand, depict the characteristic structure of the new period, which is being entered into automation systems which are in constant communication and coordination with each other [2]. The dizzying developments that have taken place in the Internet and information industries since the year 2000 have laid the foundations of the Industry 4.0 understanding. The innovations that emerged because of the rapid increase in the share of R & D in the information sector have led to improvements in productivity in many sectors, especially manufacturing systems. As a result, Industry 4.0 emerges as a revolutionary name that brings together the informatics industry and factories. Especially, the concept of the internet of objects and the electronic devices started to integrate with each other and their environment by communicating with each other has developed in proportion to the increase of the shares of IT companies in R & D. From this point of view, it turns out that Industry 4.0 has its roots in the development of internet and information companies. The developments in information companies are measured by the investments they make in R & D and the innovations they have emerged. The share allocated to R & D in national income in Turkey in 2016 was realized as 0.98%. The result of research conducted by the company that produces technology that emerged in Turkey are even not aware enough of the industrial innovation. This is a result of the small share of R & D. Industry 4.0 is discussed even as Turkey manpower in a world where the industry began to be spoken by a weighted 5.0 seeks to develop the industry.

The aim of this study is to analyze the performance of IT companies operating in Turkey in the Industry 4.0 environment. In the study, the study of information companies is in a sector that is directly affected by the developments caused by industry 4.0 in a period when the underlying causal industry 4.0 is being intensively discussed. In particular, the development of software development work plays an important role on the basis of almost all of the industry 4.0 components, especially the internet, which are the first values of industry 4.0 that add to our lives. Software development work is done through information companies, so the developments in industry 4.0 will be the most influential, and at the same time the most affected companies will be companies operating in the information sector. For this reason, we aimed to examine the performance of IT companies, which are directly related to industry 4.0. This work consists of two main topics, theoretical and practical. In the theoretical part, the conceptual framework of Industry 4.0 is drawn in general; and related definitions. In the application part, findings, research methods and analysis results related to the performance analysis of the information companies operating in Stock Exchange Istanbul using TOPSİS method are explained. In the conclusion part, comments and suggestions about analysis results are given. The use of the TOPSIS method as a multi-criteria decision-making technique in the study is the subjective determination of the causal factor weights. In the context of Industry 4.0, the most important factor affecting the performance of information companies is the amount invested in R & D. The use of the TOPSIS method was deemed suitable because it was necessary to determine R & D expenditures as the most effective factor in performance analysis.

2 Literature Information

In this part of the study; the development of industrial revolutionaries from this past to the present, Industry 4.0 and Industry 4.0. the descriptions and the information about the technologies used in the literature are mentioned and their places in the literature are mentioned.

2.1 Historical Development of Industrial Revolutions and Industry 4.0

At the end of the 18th century, the founding of the first mechanical looms benefited from water power, the increasing use of steam power and the development of various tools for machines constituted the basic dynamics of the 1st Industrial Revolution. A revolution has been created in manufacturing and production, first emerging in England, first to continental Europe, then to the whole world, replacing instrumental production. In other words; factory type production has been transferred from workshop type production, the use of coal and steam instead of wood has increased the motive power, mechanization and transfer of production to the factory. The start of the use of electricity in the early 20th century, the use of serial production and the first production band to operate electrically are among the most important developmental stages of the 2nd Industrial Revolution. Particularly in the post-World War II period, mass production of Henry Ford's automobile factory was first referred to as Fordism. The mass production band system in the automotive sector and the fact that the factories started to work with electricity have given important impetus to industrialization. In addition, instead of iron, the development of steel production, railway transport and trade has accelerated, innovations such as telephone and radio have also improved communication and communication. Nevertheless, consumer intensification and diversification in consumer preferences and expectations have not fully met customer expectations, single mold production has failed to respond to customer demands, and the system has finally collapsed with the 1973 Oil Crisis. The first programmable machines developed for the first time have assumed a preparatory role in the transition to the 3rd Industrial Revolution. It can be argued that the Third Industrial Revolution started in the 1970s and lasted to the present day. At the beginning of the 1970s, automation of the production processes with electronic and information technology was provided and the automation was carried forward as a result of the development of programmable logic controller PCLs. Flexible production systems that can meet consumer expectations with Ford's fordizmin acquisition have begun to be used in production. The new situation emerged with the formation of information, integration of internet processes into production stages, sensor (data collection), artificial intelligence, robotics technology and interprocess communication networks are called the 4th Industrial Revolution or the 4th Industrial Revolution. It is based on the article that Kagermann wrote in 2011 for the beginning of the revolution. Namely; Kagermann stated that the 4th Industrial Revolution not only includes the development of automation, but also the processes of intelligent observation and decision making [1, 2]. Initially used in Hannover 2011 in Germany, the Industry 4.0 theme is the German government's 2020 high-tech project. In October 2012, the project was submitted to the German Federal Government in a comprehensive report with a report on the fourth industrial revolution proposal prepared by Robert Bosh GmbH and the Kagermann Industry 4.0 working group. On 8 April 2013, the report was completed and the necessary investments were made; science and the business community, the dumping of private sector innovation, the dissemination of leading technologies, the industrialization of R & D activities and the funding of talented entrepreneurs. However, the ten targets for which the concept of Industry 4.0 was first used are stated below [3, 4]:

- Widespread use of alternative renewable biomaterials to oil, Intelligent structuring in energy supply, Effective treatment of diseases with personalized medicines, Quality measures with protective measures and nutrition, Smart cities and buildings, Electric vehicles, Web-based services for businesses, Industry 4.0, Information technologies and security, Neutral, energy efficient climate-friendly cities with no CO2.

Industry 4.0, with the contributions of business and academic world, went beyond Germany and soon became effective in the US and Japan, and eventually became an international concept. In industry 4.0, it is desirable to express the scattered machines and vehicles; to connect intelligent objects with supportive technologies such as internet, cyber physical systems, and to produce intelligent products and intelligent services by integrating these systems [5]. The basis of the concept is mainly to enable all units related to industrial production to communicate with each other, to reach all data in real time and to make it possible to provide optimum added value through these data. The most important feature that distinguishes Industrial 4.0 from other industrial revolutions is; it can be said that the developments in technology are triggered by each other, intertwined, coordinated movement and development of all fields affected together [6]. It is estimated that the Fourth Industrial Revolution will also significantly affect the global economy. With this effect, it can be said that the productivity gains will increase, the growth in the industry will gain speed and the new technologies will be included in the production processes and the work force profiles will change.

2.2 Components of Industry 4.0

Industry 4.0 components can be described as; internet of things, simulation, autonomous robots, additive manufacturing, augmented reality, cloud computing, cyber security, big data and analytics, horizantal and vertical system integration, horizontal and vertical system integration. Briefly, these components can be explained as [1, 3, 6, 7]:

- *Internet of things*; means an intelligent unit that can be measured and controlled by communicating over the internet. The Internet is the common name for software built by means of sensors for the collection, analysis and management of data for many objects at the same time as the internet of objects, people and cloud services connecting to each other over the internet. The internet of objects has many applications in everyday life. Applications such as listening to music and making phone calls while riding a bicycle with a smart bicycle helmet are examples of wearable technologies such as smart coats that allow you to warm up your body via bluetooth over different control centers depending on the temperature of the body and the temperature of the outside. Also, it is possible to flow information through a vehicle, which is a person with a heart in his/her body, an animal carrying a tracking chip, and a vehicle observing the air ratio of the wheels.

- *Simulation;* In the engineering phase, 3D simulations of products, materials and production processes are already in use. However, it is estimated that simulations will be used more extensively in the factory operations in the future. Products that are still in the design process, production processes and 3D simulations of materials can be used, and simulations can be used more effectively in factory production in the future.
- *Autonomous robots;* "Autonomous Robots" can be defined as robotic systems with a certain intelligence rather than robots with automatic job function. Thanks to technological advances, robots are becoming more adaptable and flexible. Developments in sensors have given robots the ability to perceive their surroundings and to be reactive. Beyond the old robot technology, robots can now access information remotely through the cloud, thus enabling them to interact with other robots and network robots.
- *Additive manufacturing;* the use of object technology produced by 3D printers in the industrial dimension is called "Layered Manufacturing". Process briefly; computer, 3D modeling software (CAD), machine equipments and layered material, and then production is done with 3D printer until the ceiling.
- *Augmented reality;* the simplest definition is the enhanced reality of video games played on computers; refers to a live direct or indirect view of a new perception environment created by computer-generated, sensory input, such as sound, video, graphics, or GPS data, combined with the physical, real-world environment of the animated elements.
- *Cloud computing;* With Cloud Computing, users can use applications on the internet via the computers of the service provider instead of keeping the applications for the business in the on-site computer or data center, thus achieving more economical, flexible and agile data management. Incorporation of new product information into the system also increases the performance of cloud technology and shortens the impact response time.
- *Cyber security;* Computer security is also known as Cyber Security or Information Technology security; computer systems, hardware, software or information are harmed or the services they provide are interfered or deflected. In Industry 4.0 environment; it is important that data are only available to authorized persons, and that data sources and integrity can be verified.
- *Big data and* analytics; large data is the way in which electronic data is obtained through machines, devices and services. This method is used by governments, companies and institutions for different purposes. In particular, the state collects many traffic fines belonging to drivers in the case of traffic accompanied by Mobile Electronic System Integration (Mobese) for criminal offenses. Large data; such as logs from web servers, internet statistics, call records from GSM operators, social media publications, RFID tags and information from sensors.

- *Horizontal and vertical system integration;* thanks to the Vertical and Horizontal integration of Industry 4.0, changes to production processes and problems can be responded quickly, customer specific and personalized QA production is facilitated, resource efficiency is improved, and optimization in the global supply chain is achieved. With Industry 4.0, companies, departments, functions and capabilities are becoming more and more compatible because inter-enterprise, universal data integration networks are evolving and enabling fully automated value chains.

2.3 Literature Review

When the literature is examined, it is noteworthy that there are many national and international studies. Table 1 summarizes the work done in Turkey, especially with Topsis method.

Table 1. Literature summary of performance analysis based on the topsis method

Author and Year	Scope and Decision Criteria	Method
Yurdakul and Iç (2003) [15]	Between 1998 and 2001, 5 automotive firms were defined as decision units and 7 financial ratio decision criteria belonging to the firm	Financial performance for 4 different years was examined and TOPSIS performance rankings were obtained and compared with year-end closing prices rankings of firms on the market
Akkaya (2004) [16]	63 selected financial and non-financial performance indicators of a selected airline operation were defined as decision units and a performance-determining model was developed from the criteria	Sixty-three indicators, Gray Relative Analysis and TOPSIS analyzes, have been reduced to 13 indicators under a model that will determine the performance of the end-airlines, and it is proposed to replicate the model in different enterprises
Eleren and Karagül (2008) [17]	The years taken on behalf of Turkey to value the performance of the economy in the years 1986-2006 was determined as per decision criteria and 7 economic indicators	The performance order for the years covered by the TOPSIS method was obtained
Ertuğrul and Karakaşoğlu (2009) [18]	15 cement firm decision units and 18 financial ratios under 5 groups were determined as criteria	The Fuzzy Analytic Hierarchy Process (FAHP) method was used for weighting the decision criteria and the TOPSIS method was applied for performance ranking of firms
Dumanoğlu and Ergül (2010) [19]	11 technology companies listed on the stock exchanges between 2006 and 2009 were defined as decision units and 8 financial ratio decision criteria	The financial performance of the businesses is determined by the TOPSIS method

(*continued*)

Table 1. (*continued*)

Author and Year	Scope and Decision Criteria	Method
Yükçü and Atağan (2010) [20]	Four profitability ratios, which are calculated on the basis of 3 hotel management decision units and 1 year financial tables in 3 different cities belonging to a holding company, were determined as criteria	For the years dealt with by the TOPSIS method, the performance ranking was obtained and the hotel management with the highest financial performance was determined
Akyüz, Bozdoğan and Handekin (2011) [21]	A 10-year period between 1999 and 2008 was defined as a decision-making unit and 19 as a decision criterion	For the years covered by the TOPSIS method, we performed financial performance analysis
Çonkar, Elitaş and Atar (2011) [22]	10 companies in the corporate governance index for 2007 and 2008 were defined as decision units and 8 rate decision criteria	The financial performance of the companies was examined by TOPSIS method over the financial ratios and the performance measures obtained were compared with the Corporate Governance Rating
Özgüven (2011) [23]	Between 2005–2009, 3 firms in the retail sector were determined as decision units and analyzed with 5 decision criteria	For the years covered by the TOPSIS method, we performed financial performance analysis
Soba and Eren (2011) [24]	Between 2007 and 2010, 1 transport company decision unit and 14 financial and non-financial ratio were determined as the decision criterion	For the years covered by the TOPSIS method, performance rankings were obtained and the year in which the operator was most successful was determined
Bülbül and Köse (2011) [25]	Between 2005 and 2008, 19 food companies were defined as decision units and 8 financial ratio criteria	Financial performance rankings according to TOPSIS and ELECTRE methods were obtained
Yılmaz Türkmen and Çağıl (2012) [26]	Between the years of 2007 and 2010, 12 information sector companies listed in Istanbul were determined as the decision unit and 8 financial ratio criteria	3-year financial performance analysis was analyzed by TOPSIS method
Ege, Topaloğlu and Özyamanoğlu (2013) [27]	Between 2009 and 2011, 18 companies in the Corporate Governance Index were defined as decision units and 9 as criteria	For the years covered by the TOPSIS method, performance rankings were obtained and the relationship between corporate governance grades was investigated
Ömürbek and Kınay (2013) [28]	An airline firm operating in BIST and an airline operating in the Frankfurt Stock Exchange as decision units for 2012 and 8 different ratios are defined as criteria	The financial performance comparison was evaluated by the TOPSIS method over 8 different ratios of a single year

(*continued*)

Table 1. (*continued*)

Author and Year	Scope and Decision Criteria	Method
Kazan and Özdemir (2014) [29]	Between 2009 and 2011, 14 holding companies operating in Stock Exchange Istanbul were included in the study as decision units and 19 financial ratios were used as decision criterion	For the years covered by the TOPSIS method, we performed financial performance analysis
Ertuğrul and Özçil (2014) [30]	The highest sales were defined as 8 different A/C model decision units and 7 product characteristic criteria	The criterion weighting was made according to the opinions of the 10 consumers interviewed and the decision was made with TOPSIS and VIKOR methods and the proposals were presented according to the technical features and price of the products
Ergül (2014) [31]	Between 2005 and 2012, 7 companies operating in the BIST tourism sector were defined as decision units and 11 indicator decision criteria	Financial performance ranks were obtained with TOPSIS and ELECTRE methods separately for each year
İşseveroğlu and Sezer (2015) [14]	Between 2008-2012, 16 pension companies were used as decision units and 8 branches were utilized for analysis	For the years covered by the TOPSIS method, we performed financial performance analysis
İç, Tekin, Pamukoğlu and Yıldırım (2015) [32]	Financial performance appraisal software has been developed for 24 different sectors and the relationship with the company value has been analyzed with the analysis made with 13 companies	We have developed a program that evaluates separately from multi-criteria decision making techniques with TOPSIS, VIKOR, GRA and MOORA and shows that when the results obtained from the program are compared with the market values of the companies, the TOPSIS method is the most appropriate model according to the weights determined by the investor group
Akbulut and Rençber (2015) [33]	In the 2010–2012 period, 32 business decision units in the BIST manufacturing sector were identified and used as 11 variable decision criteria	For the years covered by the TOPSIS method, we performed financial performance analysis
Özçelik and Kandemir (2015) [34]	7 tourism enterprises operating in Stock Exchange Istanbul during 2010–2014 period were included in the study as decision units and 8 financial ratios were used as decision criterion	For the years covered by the TOPSIS method, we performed financial performance analysis

(*continued*)

Table 1. (*continued*)

Author and Year	Scope and Decision Criteria	Method
Temizel and Bayçelebi (2016) [35]	Between 2010 and 2014, all businesses except financial institutions included in the BIST 30 index were included in the analysis as decision units and 10 financial ratios were used as the decision criterion	For the years covered by the TOPSIS method, we performed financial performance analysis
Ergüden and Çatlıoğlu (2016) [36]	In 2014, 4 firms were identified as decision units and analyzed using 4 decision criteria	The TOPSIS method performed financial performance analysis for the year covered
Akgün and Soy Temür (2016) [37]	Between 2010 and 2015, 2 firms in the transportation index were determined as decision units and 12 financial ratios were used as decision criterion	For the years covered by the TOPSIS method, we performed financial performance analysis
Orçun and Eren (2017) [12]	13 technology companies that are traded on the stock exchange between 2010 and 2015 have been defined as decision units and 9 indicator decision criteria	For the years covered by the TOPSIS method, we performed financial performance analysis

3 Methodology

In this part of the study, the performance of the companies in the Borsa Istanbul Informatics Index on the performance of the IT companies, which are at the base of the Industry 4.0, is analyzed. For analysis, TOPSIS (Technique for Order Preference by Similarity to Ideal Solution) method is used which is one of the most criterion decision making techniques. Those in the decision-making position in enterprises need to consider multiple alternatives in determining strategy and evaluating business performance. "Multiple Criteria Decision Making Techniques" are used to reach the optimal result in terms of the decision maker in the presence of multiple evaluation criteria. From a literary perspective, it seems that a large number of methods with different approaches are used. Although the methods are developed in-house, the most commonly used methods are generally AHP, AAS, GIA, TOPSIS, VIKOR, ELECTRE, PROMETHEE, Data Envelopment Analysis. It is not the right approach to discuss the superiority of any one to the other, since each of the approaches decision making in a different way. In making a multi-criteria decision, firstly the valuation criteria which will relate the capacity of the system to the targets of the system should be determined. After the criteria are determined, alternatives should be established in order to reach the targets, alternatives should be evaluated in terms of the determined criteria, and one of the alternatives should be considered as optimal. Finally, if the final solution can not be accepted, the most appropriate multi-criteria decision-making method should be applied by collecting new information and repeating the optimization [8].

TOPSIS analysis is known as one of the Multi Criteria Decision Making methods that are used in enterprises to analyze and use the important elements and tools such as profit, cost, production and labor force in an effective way, and especially in business performance analysis [9]. The TOPSIS method developed by Hwang and Yoon (1981) is based on the assumption that the alternative solution point is the shortest distance to the ideal ideal solution and the farthest distance to the negative ideal solution. The positive-ideal solution is the combination of all the best criteria that can be achieved. The negative-ideal solution consists of the worst criterion values that can be achieved. The only assumption in this method is based on the assumption that each measure is either a monotone increasing or monotonously decreasing one-way benefit [10]. The process steps for applying TOPSIS are as follows [11–14]:

Step 1: Creating the Decision Matrix

The decision matrix is a matrix that must be created by the decision maker. The generated matrix will be an mxp-sized matrix. The decision-making rows show the decision points, while the columns contain the factors. This matrix can be shown as follows;

$$A_{ij} = \begin{bmatrix} a_{11} & a_{12} & \cdots & a_{1p} \\ a_{21} & a_{22} & \cdots & a_{2p} \\ \cdot & & & \cdot \\ \cdot & & & \cdot \\ \cdot & & & \cdot \\ a_{m1} & a_{m2} & \cdots & a_{mp} \end{bmatrix}$$

Step 2: Obtaining Normalized Matrix

After the decision matrix is formed, squares of each a_{ij} values (a_{11}, a_{21}, a_{31} ... a_{m1}) are obtained, and column sums composed of the sum of these values are obtained and normalization is performed by dividing each column of the a_{ij} value by the square root thereof. The notation for this process is shown below;

$$N_{ij} = \frac{a_{ij}}{\sqrt{\sum_{i=1}^{m} a_{ij}^2}} \quad (i = 1, \ldots, m \ \ ve \ \ j = 1, \ldots, n)$$

Normalize matrix is obtained as follows;

$$V_{ij} = \begin{bmatrix} w_1 n_{11} & w_2 n_{12} & \cdots & w_n n_{1p} \\ w_1 n_{21} & w_2 n_{22} & \cdots & w_n n_{2p} \\ \cdot & \cdot & & \\ \cdot & \cdot & & \\ \cdot & \cdot & & \\ w_1 n_{m1} & w_2 n_{m2} & \cdots & w_n n_{mp} \end{bmatrix} \Rightarrow V_{ij} = \begin{bmatrix} v_{11} & v_{12} & \cdots & v_{1p} \\ v_{21} & v_{22} & \cdots & v_{2p} \\ \cdot & & & \cdot \\ \cdot & & & \cdot \\ \cdot & & & \cdot \\ v_{m1} & v_{m2} & \cdots & v_{mp} \end{bmatrix}$$

Step 3: Obtaining Weighted Normalized Matrix

Each value of the normalized matrix is weighted with a value such as wij. The weighting process reveals the subjective direction of the TOPSIS method. Because weighting is done according to the importance of the factors. The only subjective parameter of the TOPSIS method is the weight. Note that the sum of wij values is equal to 1. So it will be $\sum_{i=1}^{n} w_i = 1$. The weighted normalized matrix (V matrix) is obtained by multiplying the nij values obtained with the normalized matrix by the wij weights.

$$V_{ij} = \begin{bmatrix} w_1 n_{11} & w_2 n_{12} & \cdots & w_n n_{1p} \\ w_1 n_{21} & w_2 n_{22} & \cdots & w_n n_{2p} \\ . & & & . \\ . & & & . \\ . & & & . \\ w_1 n_{m1} & w_2 n_{m2} & \cdots & w_n n_{mp} \end{bmatrix} \Rightarrow V_{ij} = \begin{bmatrix} v_{11} & v_{12} & \cdots & v_{1p} \\ v_{21} & v_{22} & \cdots & v_{2p} \\ . & & & . \\ . & & & . \\ . & & & . \\ v_{m1} & v_{m2} & \cdots & v_{mp} \end{bmatrix}$$

Step 4: Obtaining Ideal and Negative Ideal Solution Values

Once the weighted normalized matrix (V matrix) is obtained, the maximal values of each column are determined, provided that our goal is maximization, provided that it depends on the structure of the problem. These maximum values are our ideal solution values. Then, minimum values for each column are obtained again. This is the negative ideal solution value. If our goal is minimization, the values obtained will be the exact opposite. The notation for obtaining ideal and negative ideal solution values is shown below;

Ideal solution values:

$$A^* = \left\{ (\max_i v_{ij}\, to\, be \right\}$$
$$\Rightarrow A^* = \left\{ v_1^*, v_2^*, \ldots, v_n^* \right\} \text{max}imum\, values\, for\, each\, column,$$

Negative ideal solution values:

$$A^- = \left\{ (\min_i v_{ij}\, to\, be \right\}$$
$$\Rightarrow A^- = \left\{ v_1^-, v_2^-, \ldots, v_n^- \right\} \text{min}imum\, values\, for\, each\, column,$$

Step 5: Obtaining Ideal and Negative Ideal Distance Values

Euclidean distance is used when calculating ideal and ideal non-point distance values. In the calculation of the distance between the two known x and y coordinates in the coordinate plane, i.e. the calculation of the Euclidean distance the following formula is used.

$$d_{ij} = \sqrt{\sum_{k=1}^{p} (x_{ik} - x_{jk})^2}$$

Here; xik: i. Observe. Variable value xjk: j. Observe. Variable value p: indicates the variable number.

Ideal solving is attempted to determine the nearest Euclidean distance and the negative ideal solving the farthest distance. If this formula is generalized to be able to calculate the ideal and ideal non-point distance, the following calculation method is used;

Ideal distance:

$$S_i^* = \sqrt{\sum_{j=1}^{n} (v_{ij} - v_j^*)^2}$$

Negative ideal distance:

$$S_i^- = \sqrt{\sum_{j=1}^{n} (v_{ij} - v_j^-)^2}$$

Here, the number of decision points will be S_i^* *and* S_i^-.

Step 6: Calculation of Ideal Solving Relative Proximity

Ideal and ideal non-point spacing is used to calculate the ideal resolving relative proximity of each decision point. Ideally, the solution is symbolized by relative proximity $C_i^* = 1$ Here, C_i^* value takes a value in the interval $0 \leq C_i^* \leq 1$, and $C_i^* = 1$ indicates the ideal resolving absolute correspondence of the corresponding decision point, and $C_i^* = 1$ indicates the negative ideal resolving absolute closeness of the respective decision point.

$$C_i^* = \frac{S_i^-}{S_i^- + S_i^*}$$

4 Implementation and Results

The aim of this study is to analyze the performance of IT companies operating in Turkey in the Industry 4.0 environment. Financial ratios for the year 2017 were used for the analysis. When the literature is examined, it is seen that the financial ratios are used predominantly as a valuation factor in the analysis of company performances. Financial ratios can be used as a valuation factor in the performance analysis as well as the financial status of the business. Apart from financial ratios, various physical input and output values are known to be used in performance evaluation. However, the fact that the decision points that make up the subject of study are the information companies and the output that they obtain is a product of knowledge and know-how from the physical product, the financial ratios also need to be used in performance analysis.The reason for using the valuation factors as a ratio is to remove the scale differences between the decision points involved in the analysis. The rates required for the research were obtained from the financial statements submitted to the Public Disclosure Platform (KAP). There are 15 firms in the BIST Informatics Index in the current period. As an evaluation factor in the survey, 6 firms were included in the analysis as the financial ratios and decision points. The decrease in the number of firms included in the analysis is related to the availability of the evaluation factors used in the research on the financial tables. In Tables 2 and 3, research decision points and evaluation factors are given.

Table 2. Decision points and bist codes

BIST codes	Firm names
ALCTL	Alcatel Lucent Teletaş Telekomünikasyon A.Ş.
KAREL	Karel Elektronik Sanayi ve Ticaret A.Ş.
KRONT	Kron Telekomünikasyon Hizmetleri A.Ş.
LINK	Link Bilgisayar Sistemleri Yazilimi Ve Donanimi Sanayi Ve Ticaret A.Ş.
LOGO	Logo Yazilim Sanayi ve Ticaret A.Ş.
NETAŞ	Netaş Telekomünikasyon A.Ş.

The evaluation factors used in the research were determined by taking into consideration the factors used in similar studies after the literature review. It is not expected that there will be a direct relationship between these evaluation factors and industry 4.0. The data identified as the assessment factor in the study are selected bases in order to be able to measure and demonstrate the performance of businesses in an environment where industry 4.0 is alive and discussed. In determining factors, the rating of the industry 4.0 is also taken into account, as well as the ability to measure operating performance. From this point of view, the R & D expenditure ratio of the evaluation factors was included in the survey as the most effective benchmark for the IT companies operating in industry 4.0. Ratio is a source of resources for research and development. It can be said that there is a strong connection between the expenditures of R & D and the performances of information companies. The firms in the BIST Informatics Index, which are not included in the financial statements, were excluded from the survey. The most important factors determining performance after the expense of the R & D are the profitability of the enterprises. For this reason, the ratios related to profitability are determined as the evaluation factors that affect performance the most after R & D expenditures. Profitability is also a result of the activities of information companies in the sector at the same time. It can be said that profitable information companies are more effective than industry 4.0 developments and closely follow the innovations and developments in the sector. In the industry 4.0 environment, the data showing the performance of the businesses and showing the power of paying the operator's debt after the R & D ratio and profitability ratios as the factors with the highest impacts directly affect the performance of the operator. The high payability of the debt indicates that the business operates effectively and efficiently. At the same time, it can be said that in the information sector, where rapid change and development are experienced, enterprises respond positively to these changes and develop high added value. Although payability of debt is not directly related to industry 4.0, industry can be used as a performance measure in a sector directly affected by 4.0.

Table 3. Evaluation factors

Code	Evaluation factors	Formula
CO	Current Ratio	Current Assets/Short Term Liabilities
LO	Acid/Test Ratio	(Current Assets-Inventory)/ Short Term Liabilities
AkDH	Assets Turnover Rate	Net Sales/Total Assets
FinKal	Financial Leverage Ratio	Total Debt/Total Assets
NKO	Net Profit Ratio	Net Profit/Net Sales
AkKar	Return on Assets	Net Ratio/Total Assets
OzKar	Return on Equity	Net Profit/Total Equity
FVOK	Operating Profit Ratio	Operating Profit/Net Sales
ArGe	R & D Expense Ratio	R & D Expense/Total Assets

There is no application in the TOPSIS method to determine the factor weights. It is possible to determine the weights of factors using methods such as Analytic Hierarchy Process, Swara, Entropy, or scoring scale 1–5, 1–9. Scoring system between 1–9 was used in our research. 1 represents the lowest (worst), 9 represents the highest (good) condition. Factor weights were found by dividing the scoring given for each factor by the total score and are shown in Table 4.

Table 4. Factor weights

	CO	LO	AkDH	FinKal	NKO	AkKar	OzKar	FVOK	ArGe
Score	3	4	6	4	8	5	8	8	9
w	0.05454	0.07272	0.10909	0.07272	0.14545	0.09090	0.14545	0.14545	0.16363

The decision matrices to be used in the calculation of the ideal and negative ideal solutions for each criterion were established with the help of relevant decision points and evaluation factors. The decision matrix for 2017 is shown in Table 5 and the TOPSIS method algorithm is processed.

Table 5. Decision matrix

	CO	LO	AkDH	FinKal	NKO	AkKar	OzKar	FVOK	ArGe
ALCTL	2.18	1.89	1.04	0.59	0.04	0.04	0.12	0.06	0.02
KAREL	1.18	0.98	0.81	0.61	0.06	0.05	0.15	0.11	0.04
KRONT	3.61	3.54	0.56	0.2	0.29	0.17	0.21	0.29	0.13
LINK	13.46	13.39	0.38	0.1	0.4	0.15	0.17	0.3	0.05
LOGO	1.4	1.4	0.66	0.45	0.19	0.12	0.23	0.23	0.26
NETAŞ	1.52	1.43	0.72	0.58	0.04	0.03	0.08	0.05	0.002
Square	204.62	200.36	3.15	1.31	0.29	0.07	0.17	0.25	0.09
Square root	14.30450	14.15489	1.77473	1.1441	0.53572	0.26608	0.41133	0.49517	0.29833

Once the decision matrix has been established, it is normalized. Normalization of the decision matrix; by dividing the square of the values in the respective column of each value in the columns by the square of the sum of the values in the corresponding column and reducing it to a single pension. Table 6 shows the normalized decision matrix.

Table 6. Standard (normalized) decision matrix

	CO	LO	AkDH	FinKal	NKO	AkKar	OzKar	FVOK	ArGe
ALCTL	0.15239	0.13352	0.58600	0.51566	0.07466	0.15032	0.29172	0.12116	0.06703
KAREL	0.08249	0.06923	0.45640	0.53314	0.11199	0.18791	0.36466	0.22214	0.13407
KRONT	0.25236	0.25009	0.31553	0.17480	0.54132	0.63889	0.51052	0.58564	0.43575
LINK	0.94096	0.94596	0.21411	0.08740	0.74665	0.56373	0.41328	0.60584	0.16759
LOGO	0.09787	0.09890	0.37188	0.39330	0.35466	0.45098	0.55914	0.46448	0.87150
NETAŞ	0.10626	0.10102	0.40569	0.50692	0.07466	0.11274	0.19448	0.10097	0.00670

Each value in the normalized decision matrix is weighted by the weight value of each factor calculated in Table 4 and the calculated values are shown in Table 7 as the weighted standard decision matrix.

Table 7. Weighted standard decision matrix

	CO	LO	AkDH	FinKal	NKO	AkKar	OzKar	FVOK	ArGe
ALCTL	0.00831	0.00971	0.06392	0.03750	0.01086	0.01366	0.04243	0.01762	0.01096
KAREL	0.00449	0.00503	0.04978	0.03877	0.01629	0.01708	0.05304	0.03231	0.02193
KRONT	0.01376	0.01818	0.03442	0.01271	0.07873	0.05808	0.07425	0.08518	0.07130
LINK	0.05132	0.06879	0.02335	0.00635	0.10860	0.05124	0.06011	0.08812	0.02742
LOGO	0.00533	0.00719	0.04056	0.02860	0.05158	0.04099	0.08133	0.06756	0.14260
NETAŞ	0.00579	0.00734	0.04425	0.03686	0.01086	0.01024	0.02828	0.01468	0.00109

In the weighted decision matrix, ideal ideal solution for each column and negative ideal ideal solution for negative ideal solution are selected to form ideal and negative ideal solution sets. Table 8 shows the ideal and negative ideal solution sets.

Table 8. Ideal and negative ideal solution set

	CO	LO	AkDH	FinKal	NKO	AkKar	OzKar	FVOK	ArGe
A^*	0.05132	0.06879	0.06392	0.03877	0.10860	0.05808	0.08133	0.08812	0.14260
A^-	0.00449	0.00503	0.02335	0.00635	0.01086	0.01024	0.02828	0.01468	0.00109

Table 9 shows the positive ideal solution distance of each decision factor.

In Table 10, the distance of each decision factor from the negative ideal solution was determined.

Table 9. Positive ideal discrimination measures

	CO	LO	AkDH	FinKal	NKO	AkKar	OzKar	FVOK	ArGe
S_1^*	0.18500	0.34912	0	0.00016	0.95538	0.19727	0.15130	0.49700	1.73289
S_2^*	0.21926	0.40656	0.01998	0	0.85217	0.16809	0.08002	0.31148	1.45611
S_3^*	0.14107	0.25612	0.08705	0.06791	0.08919	0	0.00500	0.00086	0.508436616
S_4^*	0	0	0.16458	0.10509	0	0.00466	0.04501	0	1.32674
S_5^*	0.21147	0.37950	0.05456	0.01034	0.32509	0.02918	0	0.04227	0
S_6^*	0.20729	0.37761	0.03869	0.00036	0.95538	0.22879	0.28134	0.53927	2.00257

Table 10. Negative ideal discrepancy measures

	CO	LO	AkDH	FinKal	NKO	AkKar	OzKar	FVOK	ArGe
S_1^-	0.00145	0.0021	0.16458	0.09700	0	0.00116	0.02000	0.00086	0.00974
S_2^-	0	0	0.06986	0.10509	0.00294	0.00466	0.06127	0.03106	0.04344
S_3^-	0.00858	0.01730	0.01224	0.00404	0.46073	0.22879	0.21132	0.49700	0.49291
S_4^-	0.21926	0.40656	0	0	0.95538	0.16809	0.10128	0.53927	0.06931
S_5^-	0.00007	0.00046	0.02962	0.04949	0.16586	0.09455	0.28134	0.27956	2.00257
S_6^-	0.00016	0.00053	0.04367	0.09309	0	0	0	0	0

At the end of the analysis, ideal and ideal nonpolar distances are used to calculate the ideal resolving relative proximity of each decision point. The values calculated with the help of the formula are shown in Table 11. As a result of the analysis, the best company LOGO and the company with the lowest performance were determined as NETAŞ in 2017. The performance ranking of the other companies included in the study was KRONT, LINK, KAREL and ALCTL, respectively, in the worst order. The investment made by LOGO, the company with the best performance in 2017, to R & D is 101,997,313 TL. LOGO, which is the largest investor in all R & D companies in terms of numbers, achieved the return of this investment with the best performance. Looking at the other companies in the line, it seems that there is no linear relationship between investment made in R & D and performance. For example, KAREL, the fourth in terms of performance rankings, is the second largest investor in R & D (25,833,769 TL) among the enterprises included in the analysis. The LINK company (1,224,942 TL), which invested the least amount in R & D, ranked third in performance ranking.

Table 11. Discrimination measures and ranking

	S_i^*		S_i^-		C_i^*		Ranking
S_1^*	2.016964167	S_1^-	0.544996288	C_1^*	0.212726269	ALCTL	5
S_2^*	1.874489354	S_2^-	0.56422186	C_2^*	0.23136067	KAREL	4
S_3^*	1.07502175	S_3^-	1.390298113	C_3^*	0.563942283	KRONT	2
S_4^*	1.28300827	S_4^-	1.568176638	C_4^*	0.550008747	LINK	3
S_5^*	1.025888083	S_5^-	1.703981807	C_5^*	0.624198909	LOGO	1
S_6^*	2.152052637	S_6^-	0.37077112	C_6^*	0.146966715	NETAŞ	6

5 Conclusion and Recommendations

Competition and change of business models in the global sense have made it necessary to establish a new world order and the transition to the fourth industrial revolution. Industry 4.0, which opened the doors of digital fabs as a result of digitalization in production. robotic technology and advanced automation systems allow the workforce to focus on different areas.

The ability of an operator to survive on the market depends on the competitive power and the speed with which it can adapt to technological developments. It is known that information companies are investing at high cost in the sector they are in. However, the return of these investments to the companies takes a long time or takes place under expectations. For this reason, if the investment turnover to R & D expenditures is not sufficient, companies suffer from performance degradation and lag behind their competitors. In this context, R & D investments of IT companies to capture Industry 4.0 are of great importance. In this way, the life span of the companies will be prolonged and at the same time significant gains will be achieved in terms of the country's economy. In this study, the performance of BIST IT companies was analyzed using the financial ratios of 2017. As a result of the analysis conducted by keeping the efficiency level of the R & D expenditures high, it is determined that the company with the best performance for 2017 is LOGO. The company with the lowest performance in 2017 is NETAS. Apart from these, the performance success of the companies included in the analysis was; KRONT, LINK, KAREL, ALCTL.

In future work, the performance of companies in the index can be analyzed before Industry 4.0 and after Industry 4.0. At the same time, the analysis made in this study can be extended to examine the performance of information companies over the years after Industry 4.0. The analysis can be done in more detail by extending the evaluation factors used in performance analysis. In addition to all of these, besides the TOPSİS method used for the analysis, performance analysis can be examined in terms of every method by using other criteria decision making methods.

References

1. Oktay E (2017) Endüstri 4.0. TMMOB EMO Ankara Şubesi, Haber Bülteni, 3. Sayı
2. Sinan A (2016) Üretim için Yeni Bir İzlek: Sanayi: 4.0. J Life Econ 3(2):19–30. http://dx.doi.org/10.15637/jlecon.129
3. Esra K (2016) Endüstri 4.0 ve Dijital Ekonomisi &Dünya ve Türkiye Ekonomisi İçin Fırsatlar Etkiler ve Tehditler. Nobel Yayın Dağıtım, Ankara
4. Bill L (2014) Industry 4.0 Only One-Tenth of Germany's High-Tech Strategy. http://www.automation.com/automation-news/article/industry-40-only-one-tenth-of-germanys-high-tech-strategy. Erişim Tarihi 19 Apr 2018
5. Weyer S, Schmitt M, Ohmer M, Gorecky D (2015) Towards Industry 4.0-standardization as the crucial challenge for highly modular, multi-vendor production systems. IFAC-PapersOnLine 48(3):579–584
6. Schwab K (2016) Dördüncü Sanayi Devrimi, Çev. Zülfü Dicleli, Optimist Yayıncılık, İstanbul

7. Ela B, Taner A (2017) Endüstri 4.0 ve İnovasyon Göstergeleri Kapsamında Türkiye Analizi. ASSAM Uluslararası Hakemli Dergi Sayı:7 Yıl
8. Arif S, İbrahim S (2014) Topsis Yönteminin Finansal Performans Göstergesi Olarak Kullanılabilirliği. Marmara Üniversitesi Öneri Dergisi, 11(41):185–202. 10.14783/öneri.2014414422
9. Kaya A, Gülhan Ü (2010) Küresel Finansal Krizin İşletmelerin Etkinlik ve Performans Düzeylerine Etkileri: 2008 Finansal Kriz Örneği. İstanbul Üniversitesi Ekonometri ve İstatistik Dergisi, 11:61–89
10. Bülbül S, Köse A (2011) Türk Gıda Şirketlerinin Finansal Performansının Çok Amaçlı Karar Verme Yöntemleriyle Değerlendirilmesi. Atatürk Üniversitesi İktisadi ve İdari Bilimler Dergisi, 25:71–95
11. Muhlis Ö, Bolum Adi (2014) TOPSIS, Operasyonel, Yönetsel ve Stratejik Problemlerin Çözümünde Çok Kriterli Karar Verme Yöntemleri, sayfa, pp 133–153, Dora Basım-Yayın Dağıtım, Bursa
12. Orçun Ç, Eren BS (2017) TOPSIS Yöntemi ile Finansal Performans Değerlendirmesi: XUTEK Üzerinde Bir Uygulama. Muhasebe ve Finansman Dergisi (Özel Sayı) 75:39–154
13. Ahmet ŞA, İbrahimH E, Nuri H (2017) BIST'te Ana Metal Sanayi Endeksinde Faaliyet Gösteren İşletmelerin Finansal Performans Ölçümü: 2011–2015 Dönemi, Süleyman Demirel Üniversitesi Vizyoner Dergisi, 8(17):83–91
14. Gülsüm İ, Ozan S (2015) Financial performance of pension companies operating in turkey with TOPSIS analysis method. Int J Acad Res Acc Finan Manag Sci 5(1):137–147
15. Yurdakul M, İç YT (2003) Türk Otomotiv Firmalarının Performans Ölçümü ve Analizine Yönelik TOPSIS Yöntemini Kullanan bir Örnek Çalışma. Gazi Üniversitesi Mühendislik ve Mimarlık Fakültesi Dergisi 18(1):1–18
16. Akkaya C (2004) Finansal Rasyolar Yardı- mıyla Havayolları İşletmelerinin Performansının Değerlendirilmesi. Dokuz Eylül Üniversitesi İ.İ.B.F.Dergisi 19(1):15–29
17. Eleren A, Karagül M (2008) 1986–2006 Türkiye Ekonomisinin Performans Değerlendirmesi. Yönetim ve Ekonomi 15(1):1–14
18. Ertuğrul İ, Karakaşoğlu N (2009) Performance evaluation of turkish cement firms with fuzzy analytic hierarchy process and TOPSIS methods. Expert Syst Appl 36:702–715
19. Dumanoğlu S, Ergül N (2010) İMKB'de İşlem Gören Teknoloji Şirketlerinin Mali Performans Ölçümü. Muhasebe ve Finansman Dergisi 48:101–111
20. Yükçü S, Atağan G (2010) TOPSIS Yöntemine Göre Performans Değerleme. J Acc Finan 45:28–35
21. Akyüz Y, Bozdoğan T, Hantekin E (2011) TOPSIS Yöntemiyle Finansal Performansın Değerlendirilmesi ve bir Uygulama. Afyon Kocatepe Üniversitesi İ.İ.B.F. Dergisi 13(1):73–92
22. Çonkar MK, Elitaş C, Atar G (2011) İMKB Kurumsal Yönetim Endeksi'ndeki (XKURY) Firmaların Finansal Performanslarının TOPSIS Yöntemi ile Ölçümü ve Kurumsal Yönetim Notu ile Analizi. İktisat Fakültesi Mecmuası 61(1):81–115
23. Özgüven N (2011) Kriz Döneminde Küresel Perakendeci Aktörlerin Performanslarının TOPSIS Yöntemi ile Değerlendirilmesi. Atatürk Üniversitesi İİBF Dergisi 25(2):151–162
24. Soba M, Eren K (2011) TOPSIS Yöntemini Kullanarak Finansal ve Finansal Olmayan Oranlara Göre Performans Değerlendirilmesi, Şehirlerarası Otobüs Sektöründe bir Uygulama. Selçuk Üniversitesi Sosyal ve Ekonomik Araştırmalar Dergisi 21:23–40
25. Bülbül S, Köse A (2011) Türk Gıda Şirketlerinin Finansal Performansının Çok Amaçlı Karar Verme Yöntemleriyle Değerlendirilmesi. Atatürk Üniversitesi İ.İ.B.F Dergisi, 10. Ekonometri ve İstatistik Sempozyumu Özel Sayısı, 71–97

26. Yılmaz Türkmen S, Çağıl G (2012) İMKB'ye Kote Bilişim Sektörü Şirketlerinin Finansal Performanslarının TOPSIS Yöntemi ile Değerlendirilmesi. Maliye Finans Yazıları 26 (95):59–78
27. Ege İ, Topaloğlu EE, Özyamanoğlu M (2013) Finansal Performans ile Kurumsal Yönetim Notları Arasındaki İlişki: BİST Üzerine bir Uygulama. Akademik Araştırmalar ve Çalışmalar Dergisi 5(9):100–117
28. Ömürbek V, Kınay ÖGB (2013) Havayolu Taşımacılığı Sektöründe TOPSIS Yöntemiyle Finansal Performans Değerlendirmesi. Süleyman Demirel Üniversitesi İktisadi ve İdari Bilimler Fakültesi Dergisi 18(3):343–363
29. Kazan H, Özdemir Ö (2014) Financial performance assessment of large scale conglomerates via topsis and critic methods. Int J Manag Sustain 3(4):203–224
30. Ertuğrul İ, Özçil A (2014) Çok Kriterli Karar Vermede TOPSIS ve VIKOR Yöntemleriyle Klima Seçimi. Çankırı Karatekin Üniversitesi İktisadi ve İdari Bilimler Fakültesi Dergisi 4 (1):267–282
31. Ergül N (2014) BIST Turizm Sektöründeki Şirketlerin Finansal Performans Analizi. Çankırı Karatekin Üniversitesi İktisadi ve İdari Bilimler Fakültesi Dergisi 4(1):325–340
32. İç YT, Tekin M, Pamukoğlu FZ, Yıldırım SE (2015) Kurumsal Firmalar İçin bir Finansal Performans Karşılaştırma Modelinin Geliştirilmesi. Gazi Üniversitesi Mühendislik-Mimarlık Fakültesi Dergisi 30(1):71–85
33. Akbulut R, Rençber ÖF (2015) BIST'te İmalat Sektöründeki İşletmelerin Finansal Performansları Üzerine Bir Araştırma, Muhasebe ve Finansman Dergisi, 117–136
34. Özçelik H, Kandemir (2015) BIST'de İşlem Gören Turizm İşletmelerinin Topsis Yöntemi ile Finansal Performanslarının Değerlendirilmesi, Balıkesir Üniversitesi Sosyal Bilimler Enstitüsü Dergisi 18(33):97–114
35. Temziel F, Bayçelebi BE (2016) BIST 30 Endeksinde Yer Alan İşletmelerin Finansal Performans Değerlemesinde Topsis Yaklaşımı, TİSK Akademi 11:270–286
36. Ergüden E, Çatlıoğlu E (2016) Sustainability reporting practiceses in energy companies with topsis method. Muhasebe ve Finansman Dergisi, 201–221
37. Akgün M, Soy Temur A (2016) BIST Ulaştırma Endeksine Kayıtlı Şirketlerin Finansal Performanslarının Topsis Yöntemi ile Değerlendirilmesi, Uluslararası Yönetim, İktisat ve İşletme Dergisi, ICAFR Özel Sayısı, 173–186

Analysis of the Relationship Between Enterprise Resource Planning Implementation and Firm Performance: Evidence from Turkish SMEs

Ahmet Safa Kocaaga[1], Beyzanur Cayir Ervural[1,4], Omer F. Demirel[2], and Selim Zaim[3](✉)

[1] Department of Industrial Engineering, Istanbul University, Istanbul, Turkey
Ahmetsafa85@hotmail.com, cayirb@itu.edu.tr
[2] Department of Management, Ibn Haldun University, Istanbul, Turkey
ofahrettin.demirel@ihu.edu.tr
[3] Department of Industrial Engineering, Istanbul Şehir University, Istanbul, Turkey
selimzaim@sehir.edu.tr
[4] Istanbul Technical University, Istanbul, Turkey

Abstract. The principal aim of this study is to determine the critical success factors of Enterprise Resources Planning Implementation and to measure their effect on Firm Performance drawing on a sample of 215 SMEs operating in Turkish textile industry. Firm performance was measured using subjective measures relying on executive's perception of how the firm performed relative to both financial and non-financial performance criteria. The structural equation modelling technique was employed to investigate the relationship between the implementation of Enterprise Resource Planning and Firm Performance. Data analysis reveals that there is a strong positive relationship between Enterprise Resource Planning and organizational performance through Enterprise Resource Planning Performance.

Keywords: Enterprise Resources Planning Implementation · Firm performance Enterprise resources planning performance

1 Introduction

Intensifying global competition and increasing demand for better quality by customers have caused more and more companies to realize that they will have to provide high quality product and/or services in order to compete in the marketplace. These competitions increase the pressure on companies to alter their processes and improve quality and responsiveness causing changes in price, customer service, and delivery. Enterprise Resource Planning (ERP) systems can serve as a technological contributor for companies in fulfilling these needs. To meet the challenge of this global competition, many businesses have invested substantial resources in adapting and implementing ERP systems.

N. M. Durakbasa and M. G. Gencyilmaz (Eds.): ISPR 2018, *Proceedings of the International Symposium for Production Research 2018*, pp. 724–736, 2019.
https://doi.org/10.1007/978-3-319-92267-6_58

ERP systems are integrated software programs, which enable the execution of many internal functions in enterprises through a single system. ERP systems provide a better communication of functions within the enterprise through the efficient use of corporate resources, and facilitate better decision-making when implemented successfully. This property can increase the overall performance and profitability of the enterprise. However, adaptation and integration of ERP projects are not simple and can often be much more difficult or take much longer than expected to implement. If the organizational structure is not compatible with the purchased ERP system, employees' use of this new technology can yield disappointing results. Organizational conformity and training as well as the ERP adaptation are crucial to achieving the gains that are sought. Organizational conformity and organizational resistance of ERP systems can have substantial effects on the success of ERP implementation and the possible ERP System Performance.

Many companies that are going through ERP system implementation process have difficulties in determining critical success factors, which causes ERP implementation problems. Another common problem is passing to ERP implementation stage without making necessary preparations, especially those of non-financial, which leads to ERP implementation failures.

Although there are plenty of studies in the literature regarding ERP system implementation, there is a need for studies focusing on critical success factors and causes of failures. The purpose of this study is to assess the use of ERP in SMEs, to determine critical success factors in ERP implementations, and to find the causes of implementation failures. In this aim, the research uses an examination of the SME financial and non-financial indicators during ERP system implementation. The conclusions will present a comparison of actual to theoretical applications of ERP and some recommendations for SMEs hoping to use ERP systems.

2 Literature Review

In the early 1990's, continuing improvements in technology made it possible to include all operations such as product design, information warehousing, materials planning, capacity planning, communication systems, human resources, finance, and project management to be included into a single system. This led to the development and subsequent growth of ERP systems. Since that time, ERP systems have shown remarkable improvements because of software and hardware developments in computer and software technology. Today, companies are more willing to implement ERP systems in order to take the advantage of information technology to effectively plan, coordinate and monitor all the processes of the company [1].

The description of ERP systems has not changed significantly since inception. According to the APICS (American Production and Inventory Control Society), ERP is a compound-oriented accounting system, which receives, manufactures, delivers and calculates the necessary resources across corporate and enterprises in order to meet customer orders. Comparing the APICS definition to earlier definitions demonstrates that the ultimate goal of ERP systems is to have an overall view, and control the activities throughout the company from a single database.

Kumar et al. [2] defined ERP as a software program which plans and manages all the processes of an organization in order to achieve organizational targets regarding all needs of the business by integrating all the functions of the business. Another definition of ERP is made by Braggs [3] as a proposal package which changes the whole business. The author states that manufacturing, distribution, finance and sales modules are completely different systems from each other but they use a single database, a single application and a single user interface.

ERP provides two major benefits over a non-integrated system. The first is a unified upper level view of the enterprise covering all functions and departments, and the second is the user's ability to record, process, monitor and report all business transactions forming an enterprise database [4]. By 2001, the definition of ERP was becoming more detailed and refined, as the true power and required complexity of these systems were better understood; ERP is a software system, which is designed, in a modular structure that collects all data of an organization in an integrated data system, used for management and planning of business processes. The basic architecture of an ERP consists of database layer, application layer and graphical interface layer [5]. More specifically, an ERP system is composed of many modules such as marketing and sales, field service, industrial facility management, process design and development, product design and development, production and inventory control, purchasing, operations and logistics, production, quality tools, human resources, finance and accounting.

The finance and accounting module includes functions like asset accounting, cash management and forecasting, profitability analysis, standard and period related costing. The human resources module includes functions including payroll, personnel planning and travel expenses and more. The functions of inventory management, materials management, production planning, project management, quality management, routing management are included in the operations and logistics module. The marketing and sales module is composed of functions like order management, pricing, sales management and sales planning.

Successful implementation of an ERP system has become a crucial strategy for a company's competitiveness. Today, instead of a competitive advantage, ERP systems have become a necessary part of basic competitiveness. ERP systems collect all data of an organization and assist in managing and planning issues, but without a successful implementation, they not only waste money but also create complications that reduce competitiveness. The issues surrounding implementation have been studied for some time. Ehie and Madsen [6] identified five main implementation steps for a successful ERP system implementation process.

As the usage of ERP systems increases, research on ERP systems have demonstrated a positive impact on business processes and financial outcomes. Hunton et al. [7] indicated that ERP systems increased the efficiency and the profitability of the businesses in their study. They concluded that the costs and the number of employees decreased for the companies using ERP. Usmanij et al. [8] analyzed human-centred approach in studying the design of ERP system and measuring the user satisfaction on ERP system.

Considering the importance of SMEs to the economic vitality of a country and importance of ERP systems for better business, it is clear that use of ERP systems in SMEs is an important area of research. It is crucial to determine the factors for a successful ERP system implementation in SMEs. Most of the SMEs produce finished products or parts, which are used by large enterprises. An SME that is doing business with a large company as a part of its supply chain is often required to have an information system to ensure prompt response and qualified products. This can require an ERP system to win and keep a contract, but implementation of an ERP system in SMEs can be more difficult than larger organizations.

For SMEs, cost of implementing such a system can be a significant proportion of the company's revenue and the manpower required can be more than the company employs. Another issue is the need for organizational policies and systems to be compatible with the ERP system. Large organizations often redesign processes to support the purchase and implementation of ERP systems, which are then widely distributed throughout the company [9]. Due to the high cost of planning and implementing ERP systems, small and medium-sized businesses are finding it imperative to use ERP systems.

The literature supports the belief that the performance of a company is related to technological improvements that can include ERP systems. Poston and Grabski [10] proposed that companies follow technological improvements closely for improving their performance. The authors pointed out that ERP systems improve financial performance by increasing the number of potential customers, reducing operation times, improving utilization of excess capacity, improving inventory management and building more effective relationships with external partners. It stands to reason that if the company is better able to deliver, has reduced costs and staff, that organizational performance could be improved as well.

Zviran and Erlich [11] defined organizational performance as a measure of the impact of an information system on business performance. Floyd and Zahra [12] concluded that information technology has an impact on organizational performance.

3 Research Model and Hypotheses

Generally, in Turkey there is a lack of awareness, knowledge and resources about the need for SMEs. This research is intended to determine the critical factors for a successful ERP application in small and medium sized businesses in Turkey. In order to measure the effects of ERP on firm performance of the SMEs, a covariance based structural equation model is used. In this study, four hypotheses were used to analyse the relationships between business outcomes and the use of ERP for SMEs in Turkey. The research model and hypotheses are illustrated below in Fig. 1.

H1: ERP system implementation directly affects the firm performance
H2: ERP system implementation directly affects the ERP system performance.
H3: There is a positive and direct relationship between ERP system performance and firm performance
H4: The impact of ERP system implementation on the firm performance increases with a mediating role of ERP system performance.

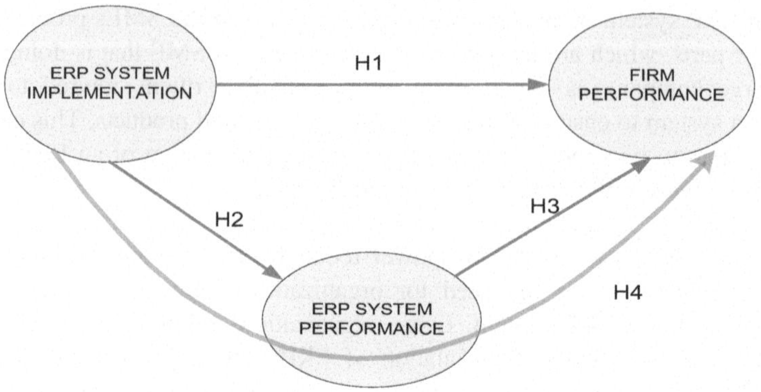

Fig. 1. Hypotheses for ERP implementation in Turkish SMEs

4 Research Methodology

4.1 Survey Instrument

Questionnaire is created to evaluate the success factors of enterprise resource planning. The original version of the questionnaire was in English. This questionnaire was translated into Turkish. The Turkish version was back translated by taking approval of a group of experts that the two versions were comparable. The questionnaire was pre-tested several times to ensure that there is no problem related to the wording, format, and sequencing of questions.

The questionnaire consists of four sections with questions relating to company information, firm performance, ERP system performance and ERP system implementation. Respondents were asked to evaluate ERP system implementation system based on 5-point Likert scale. There are sixteen statements regarding firm performance. These sixteen statements are rated according to a 5-point Likert scale (1: much deteriorated to 5: much recovered). There are 33 statements regarding ERP system implementation. These 33 statements are also rated according to Likert scale (1 = strongly disagree, 2 = Disagree, 3 = Undecided, 4 = Agree, 5 = strongly agree).

The basic success factors that were studied were developed as follows; Role of top management, planning, the role of ERP users, user training, implementation strategy, compliance of ERP systems with business, ERP consulted support, data accuracy, and project success.

4.2 Sample

An SME is identified as one that employs fewer than 100 persons. The minimum of at least 10 employees was also chosen in order to exclude very minor firms that would not be suitable for the purposes of this study. This range is consistent with the definition of an SME adopted by both the Turkish State Institute of Statistics (SIS) and Turkish Small Business Administration and also by a number of European countries such as Norway and Northern Ireland [13, 14].

Data for this study was collected using a self-administered questionnaire that was distributed to 500 SMEs in textile industry in the city of Istanbul in Turkey selected randomly from the database of Turkish Small Business Administration (KOSGEB). The study focused on the textile industry including textile mill products and apparel (SIC codes 22 and 23). Of the 500 questionnaires posted, a total of 215 questionnaires were returned after one follow-up. The overall response rate was thus 43% (215/500), which was considered satisfactory for subsequent analysis.

5 Analysis and Results

The data analysis was completed in three steps.

- Tests of the measurement models for firm performance, ERP System Performance and ERP system implementation by using confirmatory factor analysis (CFA) in order to determine whether a good fit to the data or not. Confirmatory factor analysis was conducted by both first-order model and also second-order models.
- Reliability tests of each construct based on Cronbach alpha (α) values.
- Investigation of the interrelationships among firm performance, ERP System Performance and ERP system implementation dimensions.

5.1 Confirmatory Factor Analysis

Results of CFA for firm performance are shown in Table 1. The standardized regression weights for all variables that are shown in the Table 1 are significant at the 0.001 level. The CFA showed a good fit. The $\chi 2$ statistic was 63.466 (degrees of freedom = 48, $p < 0.05$), with the $\chi 2$/df ratio having a value of 1.32 that is less than 2.0 (it should be between 0 and 3 with lower values indicating a better fit). The goodness of fit index (GFI) was 0.955 and adjusted goodness of fit (AGFI) index was 0.927. These scores are very close to 1.0. The comparative fit index (CFI) was 0.946, Tucker-Lewis coefficient (TLI) was 0.925. All indices are close to a value of 1.0 in CFA indicating that the measurement models provide good support for the factor structure.

To compare the efficacy of first order and second order models the values of the consistent Akaike information criterion (CAIC) are checked as an assessment for improvement over competing models. In general, lower CAIC values reflect the model with better fit [15]. The results indicate that the CAIC of the second-order model (241.898) is less than the CAIC of the first-order model (254.585), suggesting that second order model has better fit.

Results of CFA for ERP system performance are shown in Table 2. The standardized regression weights for all variables that are shown in the Table 2 are significant at the 0.001 level. The CFA showed a good fit. The $\chi 2$ statistic was 19.305 (degrees of freedom = 17, $p < 0.05$), with the $\chi 2$/df ratio having a value of 1.136 that is less than 2.0 (it should be between 0 and 3 with lower values indicating a better fit). The goodness of fit index (GFI) was 0.979 and adjusted goodness of fit (AGFI) index was 0.955. These scores are very close to 1.0 (a value of 1.0 indicates perfect fit). The comparative fit index (CFI) was 0.98, Tucker-Lewis coefficient (TLI) was 0.968. All indices are close to a value of 1.0 in CFA indicating that the measurement models provide good support.

Table 1. CFA for firm performance

Symbol	Variables	Regression weights First order model	Regression weights Second order model
Productivity change			
PC1	Standardization of technologies	0.64*	0.64*
PC2	Regular documentation and reporting	0.52*	0.52*
PC3	Usability of job tracking	0.54*	0.54*
PC4	Costs	0.50*	0.50*
PC5	The length of planning process	0.41*	0.41*
PC6	Production efficiency	0.40*	0.40*
Product availability and delivery change			
PA1	Ratio of on time delivers	0.59*	0.59*
PA2	Raw material and product inventory	0.44*	0.44*
Satisfaction exchange			
SE1	Departmental integration	0.49*	0.50*
SE2	Customer satisfaction rate	0.63*	0.63*
Relationship management			
A41	Supply chain management	0.36*	0.36*
A40	Customer relationship management	0.77*	0.77*

*Significant for $p < 0.01$

Table 2. CFA for ERP system performance

Symbol	Variables	Regression weights First order model	Regression weights Second order model
Access to information			
AC1	Quick access to information	0.50*	0.50*
AC2	Information accuracy	0.52*	0.52*
AC3	Employee communication	0.46*	0.46*
AC4	Better financial control	0.52*	0.52*
Customer satisfaction			
CS1	Closer relationship with customers	0.49*	0.49*
CS2	Less delay in deliveries	0.50*	0.50*
Operational processes			
OP1	Reduction in operational costs	0.33*	0.33*
OP2	Shortening lead times	0.62*	0.62*

*Significant for $p < 0.01$

To compare the efficacy of first order and second order models the values of the consistent Akaike information criterion (CAIC) are checked as an assessment for improvement over competing models. The results indicate that the CAIC of the second-order model (140.348) is equal to CAIC of the first-order model (140.348), suggesting that there is no significant difference between two acceptable models.

Results of CFA for ERP System Implementation are shown in Table 3. The standardized regression weights for all variables that are shown in Table 3 are significant at the 0.001 level. The CFA showed a good fit. The $\chi2$ statistic was 128.549 (degrees of freedom = 125, $p < 0.05$), with the $\chi2/df$ ratio having a value of 1.028 that

Table 3. CFA for ERP system implementation

Symbol	Variables	Regression weights	Regression weights
		First order model	Second order model
User roles			
UR1	User training pre-determined	0.60*	0.60*
UR2	Serious training program for users	0.60*	0.60*
UR3	Proper user interfaces	0.59*	0.59*
UR4	Management care of employees' ERP system concerns	0.54*	0.54*
UR5	Regular monitoring of employees' performances	0.47*	0.47*
Management planning			
MP1	Management is aware of the contributions of ERP system	0.62*	0.62*
MP2	Regular review of ERP qualifications	0.62*	0.62*
MP3	Re-design of ERP system plans for changing conditions	0.54*	0.54*
MP4	Clear definition of ERP objectives	0.53*	0.53*
Application process			
AP1	ERP implementation consultant's ability to speed up the process.	0.35*	0.35*
AP2	Serious time and effort requirement of ERP system	0.64*	0.64*
AP3	Employees' adaptation to the ERP system	0.46*	0.46*
AP4	Data accuracy affects the operations quality	0.49*	0.49*
Administrative support			
AS1	Managements' regular follow of ERP	0.35*	0.35*
AS2	Management's aware for administrative support	0.64*	0.64*
AS3	Compatibility of ERP system	0.52*	0.52*
Worker effect			
WE1	The role of consultants should end gradually	0.53*	0.53*
WE2	All employees are aware of data integrity due to ERP system	0.48*	0.48*

*Significant for $p < 0.01$

is less than 2.0 (it should be between 0 and 3 with lower values indicating a better fit). The goodness of fit index (GFI) was 0.941 and adjusted goodness of fit (AGFI) index was 0.919. These scores are very close to 1.0 (a value of 1.0 indicates perfect fit). The comparative fit index (CFI) was 0.993, Tucker-Lewis coefficient (TLI) was 0.991. All indices are close to a value of 1.0 in CFA indicating that the measurement models provide good support for the factor structure. To compare the efficacy of first order and second order models the values of the consistent Akaike information criterion (CAIC) are checked as an assessment for improvement over competing models. The results indicate that the CAIC of the second-order model (241.898) is less than CAIC of the first-order model (390.218), suggesting that second order model has better fit.

5.2 Reliability Analysis

Reliability analysis includes checking whether the scale used for a construct consistently reflects the construct it is measuring or not. The Cronbach alpha values of reliability for the underlying factors range from 0.67 to 0.80 suggesting satisfactory level of construct reliability [16].

5.3 Path Analysis

A measuring instrument is prepared which is structural equation model analysis consisting of three dimensions which measures different characteristics. The path model is presented in Fig. 2.

The three dimensions included in the path model are: ERP system performance, ERP system implementation and firm performance. For these three dimensions, the indicator variables were the respective factors which are determined by confirmatory factor analyses. As can be seen from Fig. 2, factor analysis results with 5 factors for ERP system implementation, 3 factors for ERP system performance and 4 factors for firm performance, are all reliable and empirically validated.

Structural equations modeling (SEM) using AMOS 20.0 was used to test the relationships in the model as presented in Fig. 2. In order to reveal the degree to which the structural equation model fits the sample data, model fit is used. In the work of Schumacker and Lomax [17], the commonly used model fit criteria are stated as: chi-square ($\chi 2$), goodness of fit index (GFI), comparative fit index (CFI) and root mean square error of approximation (RMSEA). The goodness-of-fit indices for the path model [$\chi 2 = 836.411$ with df = 651; GFI = 0.882; CFI = 0.913; RMSEA = 0.04] are within the generally accepted limits, indicating a good fit to the data.

5.4 Hypothesis Test Results

Figure 2 presents both structural and measurement model results. As we can see from Fig. 2, the standardized regression weight for H1 was found to be 0.09 which is insignificant ($p > 0.1$). This finding does not support H1 that ERP system implementation directly affects firm performance. The first hypothesis was not confirmed since the relationship between the two variables was observed to be meaningless.

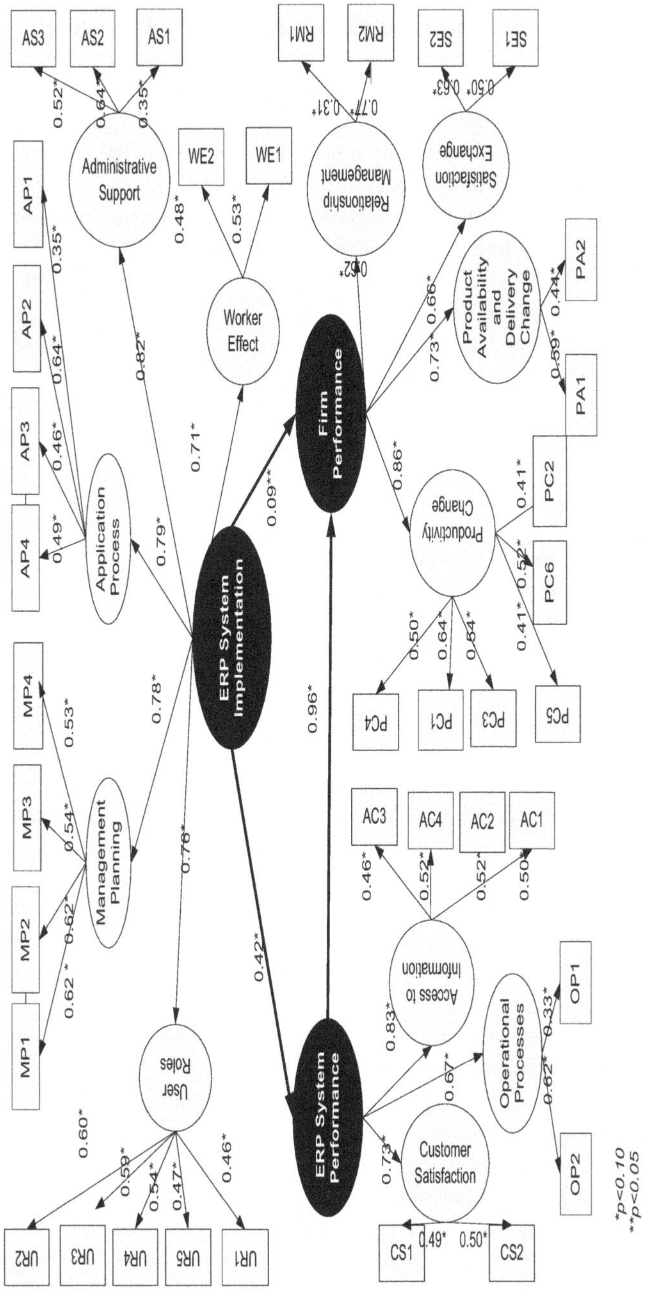

Fig. 2. Structural equation model

Moreover as the figure indicates, the standardized regression weight for H2 was found to be 0.42 which is significant ($p > 0.1$). This finding is supporting H2 that ERP system implementation has a direct and strong impact on ERP system performance of SMEs.

The third hypothesis that ERP system performance affects the firm performance was statistically confirmed since the standardized regression weight for H3 was found to be 0.96. It was revealed that achievement level of the ERP system affects the performance level of companies strongly in a positive way.

The impact of ERP system implementation on firm performance increases with a mediating role of ERP system performance in SMEs, which is shown in Fig. 2. The standardized regression weight was found to be positive and significant ($\beta = 0.403$, $p < 0.1$), providing a good deal of support for H4. Also in order to analyze mediating effect, a helpful approach is conducting Sobel test. Based on calculations, the Sobel test value is found as 2.07 with a p value of 0.019. The result of Sobel test also confirms the fourth hypothesis.

6 Conclusion

The aim of this research is to investigate the direct and indirect effects of Enterprise Resource Planning (ERP) implementation with the mediating effect ERP performance on firm performance. The study is based on survey and data collected from textile companies. Firstly, factors of latent variables were determined, then reliability and validity of these variables were evaluated with confirmatory factor analysis (CFA). In the next step, hypotheses were evaluated with the structural equation modelling method.

Based on the results of the analysis, it was found that there was a statistically positive relationship between ERP system performance and Firm performance. So ERP system performance has a great influence on the firm performance. This result is not surprising since ERP System can be applied to all departments of a company and has increasing effect on their efficiency. For a company to perform well there should be strong emphasis on the ERP system performance. The following three factors respectively affect ERP system performance: access to information, customer satisfaction and operational processes.

However, it is an interesting result that ERP system implementation itself doesn't have a significant relationship with firm performance. This situation can be due to problems with application process or management planning since these factors are found to be the most effective ones on ERP system implementation. On the other hand the effect of ERP system implementation on Firm performance increases with a mediating role of ERP system performance. This result shows that implementing ERP is ineffective without a successful ERP system.

This study presents significant managerial implications for SMEs. In ERP system implementations, the role of the user provides the greatest involvement during the implementation process. Before moving on to the ERP system, companies should emphasize users' needs throughout the stages of the program's use. After passing through the ERP system, the interface of the system should be understandable. System users should be monitored on a regular basis since necessary training affects the ERP system implementations. Consultants accelerate the process during implementation. In ERP applications, management should follow the process of ERP applications and not short cut this process while providing long-term administrative support. The following three factors respectively affect the performance of a company: productivity change, product availability and delivery change, satisfaction exchange and relationship management.

It should also be acknowledged that the study is subject to some limitations. First, it would be highly suggested that the size and nature of the sample must be enhanced to ensure variability and control for possible extraneous variation. While the sample is restricted to only a single region and a single industry, it would be strongly recommended that data should be gathered from various parts of Turkey including both various manufacturing and service industries. Since the data in this study were collected from managers of organizations on the basis of their subjective evaluations, objective performance indicators should also be employed in the analysis.

References

1. Fui-Hoon Nah F, Lee-Shang Lau J, Kuang J (2001) Critical factors for successful implementation of enterprise systems. Bus Proc Manag J 7(3):285–296
2. Kumar V, Maheshwari B, Kumar U (2003) An investigation of critical management issues in ERP implementation: emperical evidence from Canadian organizations. Technovation 23 (10):793–807
3. Braggs S (2005) ERP: the state of the industry. Arc. Insights 12 ECL, New York
4. Angerosa A (1995) The future looks bright for ERP, APICS—The Performance Advantage
5. Wallace TF, Kremzar MH (2001) ERP: making it happen: the implementers' guide to success with enterprise resource planning. Wiley, New York
6. Ehie IC, Madsen M (2005) Identifying critical issues in enterprise resource planning (ERP) implementation. Comput Ind 56(6):545–557
7. Hunton JE, Lippincott B, Reck JL (2003) Enterprise resource planning systems: comparing firm performance of adopters and nonadopters. Int J Acc Inform Syst 4(3):165–184
8. Usmanij PA, Khosla R, Chu M-T (2013) Successful product or successful system? User satisfaction measurement of ERP software. J Intell Manuf 24(6):1131–1144
9. Loh TC, Koh SCL (2004) Critical elements for a successful enterprise resource planning implementation in small-and medium-sized enterprises. Int J Prod Res 42(17):3433–3455
10. Poston R, Grabski S (2001) Financial impacts of enterprise resource planning implementations. Int J Acc Inform Syst 2(4):271–294
11. Zviran M, Erlich Z (2003) Measuring is user satisfaction: review and implications. Commun Assoc Inform Syst 12:81–103
12. Floyd SW, Zahra SA (1990) The effect of fit between competitive strategy and it adoption on organizational performance in small banks. Technol Anal Strateg Manag 2(4):357–372
13. Sun H, Cheng T-K (2002) Comparing reasons, practices and effects of ISO 9000 certification and TQM implementation in Norwegian SMEs and large firms. Int Small Bus J 20(4):421–442
14. Mcadam R (1999) Life after ISO 9000: an analysis of the impact of ISO 9000 and total quality management on small businesses in Northern Ireland. Total Qual Manag Bus Excell 10:229–241

15. Milfont TL, Duckitt J (2004) The structure of environmental attitudes: a first- and second-order confirmatory factor analysis. J Environ Psychol 24(3):289–303
16. Nunnally JC (1978) Psychometric theory. McGraw-Hill, New York
17. Schumacker RE, Lomax RG (2010) A beginner's guide to structural equation modeling, 3rd edn. Routledge, New York

Investigation of System Productivity with Fuzzy Availability Analysis Considering Failure and Repair Times

Berna Dengiz(✉), Merve Uzuner Sahin, and Kumru Didem Atalay

Industrial Engineering Department, Başkent University, Ankara, Turkey
{bdengiz, muzuner, katalay}@baskent.edu.tr

Abstract. Today's more complicated engineering systems make it difficult to analyze system performance accurately due to the increase in the number of components and interconnections within the system. System performance and measures such as failure rate, repair rate and availability are important measures to analyze a system and obtain system productivity. While application of system availability is most widely used for electrical and electronic systems, in recent years it has been started to be used as a performance measure of manufacturing systems. In this study, a novel approach is proposed using both simulation modeling technique and fuzzy availability analysis considering failure and repair times of components to investigate system productivity in a more consistent and logical manner. Simulation model is also used to analyze system behavior and estimate system throughput. Because of insufficient and inaccurate historical data related with component failure and repair times of the considered system, failure and repair data are defined with the fuzzy membership function. This approach provides more detailed information about system characteristics. Thus, more realistic system productivity considering failure and repair times which represents the real behavior of system can be obtained using this approach.

Keywords: Availability analysis · Simulation · Fuzzy sets
Triangular membership function · Triangular fuzzy number

1 Introduction

Today's engineering systems such as manufacturing systems are influenced by changeable customer demands and competitive market. Thus, these systems are getting more complicated than ever before. It is difficult to analyze system behavior and evaluate system performance accurately due to the high technological industrial processes. Because of the stochastic characteristics of these systems, simulation modeling is one of the most widely used approaches to analyze system behavior and obtain system throughput. There are several important measures to evaluate system performance and obtain system productivity such as reliability, availability, system failure and repair rate. Availability is one of the most important measures of component and system performance due to including both system failure and repair information. Availability is the probability that a system or component is operating properly with its

N. M. Durakbasa and M. G. Gencyilmaz (Eds.): ISPR 2018, *Proceedings of the International Symposium for Production Research 2018*, pp. 737–743, 2019.
https://doi.org/10.1007/978-3-319-92267-6_59

required function at a specific point in time or a specific time period, when operated and maintained in a prescribed manner [1].

This paper presents a real case study of a lead-acid battery manufacturing company located in Ankara, Turkey. This system is a complex system as is often the case in manufacturing environments. In this study, a new approach is proposed using the simulation modeling technique and the fuzzy availability analysis together considering failure and repair times of components. Simulation modelling is used to analyze system behavior and estimate system throughput. System failure behavior cannot be considered in the simulation model due to scarce historical data related with component failure and repair times of the considered system. Scarce data are defined with fuzzy membership functions and failure behavior of the system is considered with availability analysis based on fuzzy set theory. Fuzzy Set Theory was introduced by Zadeh [2] to rationalize and model uncertainty and vagueness. It is developed to obtain a more robust and flexible model for incomplete information of real life complex systems [3].

A literature survey is presented in Sect. 2. The novel approach to investigate the system productivity with fuzzy availability analysis is represented in Sect. 3. In Sect. 4, lead-acid battery production line is explained and the proposed approach is illustrated on this system. Conclusions of this work are given in Sect. 5.

2 Literature Review

Simulation is a widely used powerful modelling tool in system design and analysis. In other words, simulation is an effective technique to overcome the complexity of large scale stochastic systems. In the literature, there are various simulation studies that have been used to analyze system design and performance. Moreover, some of relevant studies such as [4–6] are used simulation modeling to improve system performance. However, time consuming computation of simulation for real word application with the failure and repair of the system components is a large drawback. Therefore, there is a very few simulation studies that considered the failure and repairs in the literature. The three examples are [7–9]. Availability analysis has been an important issue in industrial systems and it is used to predict the failure behavior of the system. Although many studies have examined system reliability and availability in electronic fields, studies on reliability and availability of a manufacturing system as a whole are limited in the literature. Loganathan et al. [10] proposed a method to obtain manufacturing system availability with variable failure and repair rate using Semi-Markov model and it is applied in an automative manufacturing system. Gupta et al. [11] presented a method to analyze system reliability and availability in a plastic-pipe manufacturing plant. However, in most of the real life applications, there are limited accurate historical data to estimate precise system parameter values. To overcome the lack of sufficient record data, fuzzy set theory has been used in reliability and availability analysis by several authors [12–14]. For example, Gorkemli and Ulusoy [12] proposed a fuzzy-Bayesian method to analyze system reliability and availability considering exponential failure and repair rates of machines. Knezevic and Odoom [13] developed a quantitative fuzzy Lambda-Tau approach with Petri Nets to compute reliability and availability of an industrial system. Komal [14] analyzed fuzzy reliability of a dual-fuel steam turbine system based on generalized fuzzy lambda-tau technique. In the works cited above, and also in most of the works in the fuzzy system reliability and availability area, the failure and repair rates of system components are obtained from historical data and represented by fuzzy numbers. However in many real world systems, obtaining the precise estimation of these rates is difficult due to the scarcity of historical data. Thus, from the above discussions it is decided to represent failure and repair times by fuzzy numbers that results in more realistic fuzzy system availability.

3 Methodology

Simulation modelling is an alternative method to analytic techniques to overcome the complexities of stochastic characteristics of large-scale systems. Discrete event simulation modelling is one of the dimensions of simulation model that is a system modelling for a specific time period by a representation where the system state variables change at separate points in time. In this study, discrete event simulation is used to represent physical system and failure and repair behaviour is considered with availability analysis. In this study, only the corrective downtime of the system is considered. Thus, availability is computed as steady-state availability. For steady-state availability analysis, failure and repair times are determined for each workstation at which there is only machines for the tasks assigned. Then, mean time between failure (MTBF) and mean time to repair (MTTR) times of all components in the system are computed. MTBF is the expected time between two successive failures (uptime of the system) and MTTR is the expected time that is needed to repair of the component (downtime of the system). Fuzzy Set Theory is used to deal with scarcity of failure and repair times of components. Scarce failure and repair times are defined with fuzzy membership functions for system availability analysis. It is decided to define failure and repair times of the components with triangular membership function due to their features. A bias value is used together with crisp value to obtain the lower and upper bounds of the

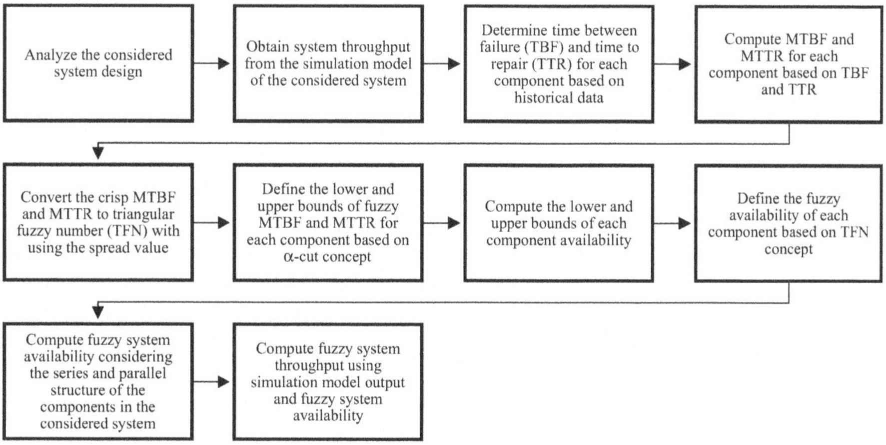

Fig. 1. Flowchart of the proposed approach

triangular fuzzy numbers (TFNs) as Knezevic and Odoom expressed [13]. Then, the simulation output and fuzzy system availability are used to obtain system throughput as mentioned "system throughput for the real situation of the system, that includes system failures, could be obtained with using simulation output and system availability" by Elsayed [15]. The flowchart of the proposed approach to obtain system productivity based on fuzzy system availability analysis is given in Fig. 1.

4 Case Study

The proposed approach is illustrated on a production line of a lead-acid battery manufacturing company located in Ankara, Turkey. The company produces different types of lead-acid batteries and supplies different types of semi-finished products. The flow line of the battery production line is showed in Fig. 2. The battery contains positive and negative grids and the grids are processed in different paste mixing, acid preparing, EXMET and curing workstations. After forwarding, batteries are transported to customers by trucks. There are operators and machines for tasks assigned at workstations. The company works 6 days a week and each day consists of 3 shifts, which are 8 h each. There is no problem associated with the storage area.

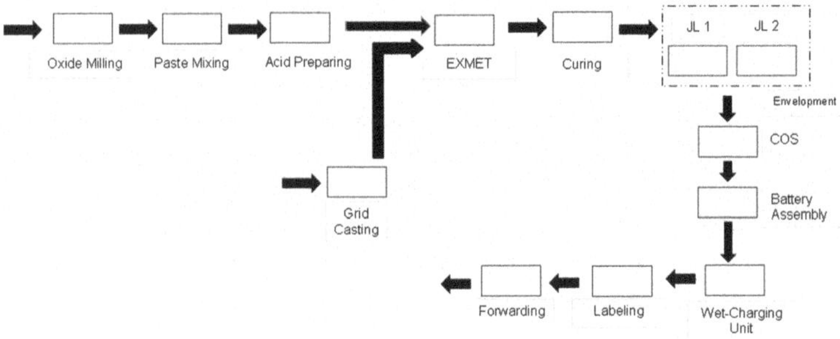

Fig. 2. System flow between workstations

In this study, a simulation model of the system is built using discrete event simulation. The system design is modelled with the ARENA® Simulation Software by Rockwell Automation. In this study, $\pm 15\%$ spread of crisp values was chosen to obtain the lower and upper bounds of TFNs. The proposed approach is applied first with $\pm 15\%$ spread on crisp values and then repeated with $\pm 25\%$ and $\pm 50\%$ spread on crisp values to obtain the lower and upper bounds of TFNs as is expressed in [13]. The production line is analyzed and simulation model is built properly to system flow that is determined considering related workstations as seen in Fig. 2. System throughput is obtained as 7171 batteries per month using simulation model. Time between failure (TBF) and time to repair (TTR) data of all components in the system are determined based on historical failure and repair data in minutes. TBF and TTR data set of the Grid Casting machine is determined and given as an example in Table 1. Then, MTBF and MTTR are computed for each component.

Table 1. TBF and TTR data of the grid casting machine

Grid casting machine		
Failure number	TBF	TTR
1	78000	20
2	460	150
3	12570	120
4	4800	480
5	1440	220
6	500	70
7	7130	60
8	660	300
9	3540	20
10	700	30
11	1410	90
12	5910	105
13	615	30
14	1410	30
15	690	105
16	1815	30
17	690	235
18	2165	120
19	600	200
20	520	45
21	21315	90

Because of the scarce historical data, crisp MTBF and MTTR values are converted to TFN with spread value 15%. For example, MTBF and MTTR representation of Grid Casting Machine are (5947, 6997, 8047) and (103, 121, 139), respectively. Then, with using these bounds, lower and upper bounds of component availability for each component are computed as Ke et al. represented [16]. Lower and upper bounds of Grid Casting machine availability is obtained as $[(1050\alpha + 5947)/(1032\alpha + 6086), (8047 - 1050\alpha)/(8150 - 1032\alpha)]$. Then, fuzzy availability of each component is defined based on TFN concept using these availability bounds. Fuzzy availability of Grid Casting machine is obtained (0.9772, 0.9830, 0.9874). After considering the series and parallel structure of the components in the considered system, fuzzy system availability is computed. Fuzzy system availability is computed as (0.8047, 0.8504, 0.8863). Fuzzy throughput is obtained from simulation output and fuzzy system availability using fuzzy multiplication operation on TFNs and it is computed as (5770, 6098, 6355). It is seen that the lower and upper bounds of fuzzy throughput are 5770 and 6355, respectively and it is impossible to fall below 5770 or exceed 6355. The most possible value for fuzzy throughput is 6098. It is also shown as $[328\alpha + 5770, 6355 - 257\alpha]$ using α-cut concept and the summary of fuzzy throughput for each α level that is ranging from 0 to 1, with increments of 0.1, based on all considered spreads on the crisp value is shown in Fig. 3. Depending upon the value of α, the analysis could predict the capability for the system.

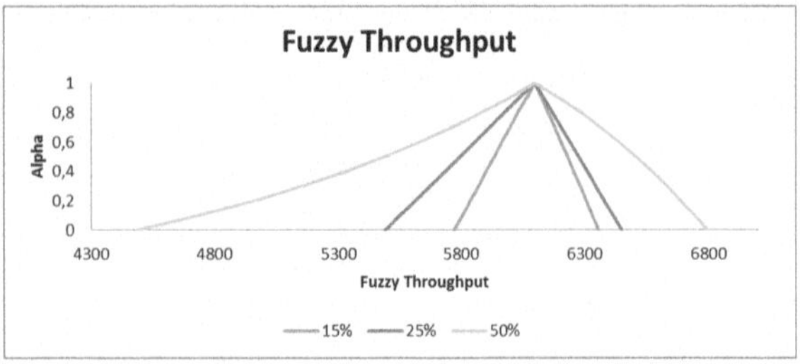

Fig. 3. The membership functions for the fuzzy throughputs of the considered system

The results of fuzzy system availability and fuzzy throughput based on three different spreads on the crisp values are presented in Table 2. As is evident in Table 2, when percentage of spread is increased, upper bounds of both fuzzy system availability and fuzzy throughput are increased and lower bounds are decreased. The most possible values of fuzzy system availability and fuzzy throughput do not change while percentage of spread is changing. It means that the available area of data in both fuzzy system availability and fuzzy throughput are increased and so it can contain an inappropriate value of the system characteristics or operating environment. In that an overestimation of the system parameters may result based on the percentage of spread used.

Table 2. Fuzzy system availabilities and fuzzy throughputs based on TFNs

Percentage of spread	Fuzzy system availability	Fuzzy throughput
15%	(0.8047, 0.8504, 0.8863)	(5770, 6098, 6355)
25%	(0.7656, 0.8504, 0.8993)	(5490, 6098, 6448)
50%	(0.6253, 0.8504, 0.9472)	(4883, 6098, 6792)

5 Conclusions

This study proposes a novel approach based on simulation modelling and fuzzy availability analysis. It provides investigation of system productivity in a more consistent and logical manner for a production line in a lead-acid battery production system. This approach is proposed using both simulation modelling technique and fuzzy availability analysis considering failure and repair times of components. System behavior is analyzed and system throughput is obtained by system simulation model. Because of scarce historical data, failure and repair data are defined with fuzzy membership function in system availability analysis. To the best of our knowledge the proposed approach is the first application in production area and it can be easily applied to any manufacturing system. This approach helps obtain more detailed information about system characteristics even when the system behavior is complex. Thus, using this approach more realistic system productivity which represents the real behavior of system can be obtained.

References

1. Verma AK, Srividya A, Prabhu Gaonkar RS, Rajesh S (2007) Fuzzy-reliability engineering: concepts and applications. Narosa
2. Zadeh LA (1965) Fuzzy sets. Inf Control 8:338–353
3. Lai YJ, Hwang CL (1992) Fuzzy mathematical programming, vol 394. Springer, Heidelberg
4. Dengiz B, Akbay KS (2000) Computer simulation of a PCB production line: metamodeling approach. Int J Prod Econ 63:195–205
5. Dengiz B, Bektas T, Ultanir AE (2006) Simulation optimization based DSS application: a diamond tool production line in industry. Simul Model Pract Theory 14:296–312
6. Padhi SS, Wagner SM, Niranjan TT, Aggarwal V (2013) A simulation-based methodology to analyse production line disruptions. Int J Prod Res 51:1885–1897
7. Liu A, Yang Y, Liang X, Zhu M, Yao H (2010) Dynamic reentrant scheduling simulation for assembly and test production line in semiconductor industry. Adv Mater Res 97–101: 2418–2422
8. Wazed MA, Ahmed S, Nukman Y (2010) Application of Taguchi method to analyze the impacts of commonalities in multistage production under bottleneck and uncertainty. Int J Phys Sci 5:1576–1591
9. Kampa A, Gołda G, Paprocka I (2017) Discrete event simulation method as a tool for improvement of manufacturing systems. Computers 6:1–10
10. Loganathan MK, Kumar G, Gandhi OP (2016) Availability evaluation of manufacturing systems using Semi-Markov model. Int J Comput Integr Manuf 29:720–735
11. Gupta P, Lal AK, Sharma RK, Singh J (2007) Analysis of reliability and availability of serial processes of plastic pipe manufacturing plant: a case study. Int J Qual Reliab Manag 24: 404–419
12. Görkemli L, Ulusoy SK (2010) Fuzzy Bayesian reliability and availability analysis of production systems. Comput Ind Eng 59:690–696
13. Knezevic J, Odoom ER (2001) Reliability modelling of repairable systems using Petri nets and fuzzy Lambda-Tau methodology. Reliab Eng Syst Saf 73:1–17
14. Komal (2018) Fuzzy reliability analysis of DFSMC system in LNG carriers for components with different membership function. Ocean Eng 155:278–294
15. Elsayed EA (1996) Reliability engineering, vol 1. Addison Wesley Longman
16. Ke JC, Huang HI, Lin CH (2006) Fuzzy analysis for steady-state availability: a mathematical programming approach. Eng Optim 38:909–921

Quality Management

Problems of Mathematical Modelling of Rotary Elements

Stanisław Adamczak[1], Dariusz Janecki[2], and Krzysztof Stępień[1(✉)]

[1] Department of Manufacturing and Metrology, Kielce University of Technology, Kielce, Poland
{adamczak,kstepien}@tu.kielce.pl

[2] Centre for Laser Technologies of Metals, Kielce University of Technology, Kielce, Poland
djanecki@tu.kielce.pl

Abstract. Rotary elements constitute a large and significant group of machine parts. Such parts are used in various branches of industry, for example in bearing, automotive, paper, shipping or power industry. Most of rotary machine parts are cylindrical ones but sometimes other types of elements are used, as spheres, cones or barrel-shaped and saddle-shaped elements. Measurements of form deviations of such elements require performing a number of mathematical operations on measurement data, including calculation of reference features. The basis for such operations is to apply an appropriate mathematical model of the surface. Scientific literature and standards refer mainly to measurements of form errors of cylindrical shapes. Apart from the cylindrical surfaces, this paper discusses also the problem of mathematical modelling of cones, spheres and barrel-shaped elements.

Keywords: Measurement · Form deviation · Mathematical model
Rotary elements

1 Introduction

Rotary elements constitute a large and significant group of machine parts. Such parts are used in a number of various branches of industry, for example in bearing, automotive, paper, shipping or power industry [1, 2]. Most of rotary machine parts are cylindrical ones but sometimes other types of elements are used, as spheres, cones or barrel-shaped elements. Usually, such elements are characterised by relatively low value of permitted form deviations [3, 4]. This is why rotary parts are quite often measured with the use of highly accurate special-purpose measuring instruments. Measurements of form deviations require performing a number of mathematical operations on measurement data, including calculation of relevant reference features. The basis for such operations is correct modelling of the surface [5, 6]. Scientific literature and standards refer mainly to the problem of measurements and evaluation of cylindrical parts. Apart from measurements of cylindrical parts, authors of the paper were dealing with measurements of spheres, cones and barrel-shaped parts. This paper is an attempt of synthesis of the information on modelling of surfaces rotary parts used in industry. Contemporary production systems require development of new methods allowing quick and accurate

N. M. Durakbasa and M. G. Gencyilmaz (Eds.): ISPR 2018, *Proceedings of the International Symposium for Production Research 2018*, pp. 747–752, 2019.
https://doi.org/10.1007/978-3-319-92267-6_60

measurement. of rotary elements. Thus it is noticeable that the problem of modelling of rotary surfaces is the matter of great importance for modern industry.

2 Modelling of Cylindrical Surfaces

The easiest case of modelling of rotary elements is to model roundness profiles. It is most convenient to apply polar coordinate system (R,φ) in such task, where R denotes a distance of a given point from the origin of the system and φ is the angle of rotation [7]. Form errors of circular profiles can be easily modelled with the use of trigonometric Fourier series Subsequent Fourier components of the profile have a specific interpretation associated with the form of the profile. For example component denoted by zero is the mean value of the signal. Fourier component denoted by 1 (i.e. cosφ or sinφ) refers to the eccentricity of profile, component no. 2 is associated with the ovality of the profile, component no. 3 with the triangular error of the profile and so on [8].

To illustrate it, let us consider following relationship:

$$R(\varphi) = 5 + 0.005\cos\,\varphi + 0.002\sin 2\varphi + 0,001\cos 10\varphi \tag{1}$$

It is easy to notice that the mean value (zero harmonic component) of this profile is equal to 5. Apart from that the profile is characterized by eccentricity and occurrence of the second and the tenth harmonic component.

It is noticeable that the mean value (zero harmonic component) of this profile is equal to 5. Apart from that the profile is characterized by eccentricity and occurrence of the second and the tenth harmonic component.

Cylindrical surface can be regarded as an expansion of the circular profile, which is two-dimensional case into three dimensions. Therefore if we want to write an equation describing cylindrical surface we should add one more variable to R and φ. The variable to be added is the one that defines location of the point along the vertical axis coinciding the nominal axis of rotation of the cylinder. Usually, this variable is denoted by z. Considering possible errors of cylindrical surfaces it is noteworthy that in the case of real workpieces the axis may not be a straight line. Mathematically we can model it by expressing it as a function of z. Thus, general equation of the cylindrical surface with a constant radius, whose axis is not a straight line can be written as follows:

$$R(\varphi, z) = R + e_x(z)\cos\,\varphi + e_y(z)\sin\,\varphi \tag{2}$$

where: $e_x(z)$ and $e_y(z)$ are the coordinates of the origin of the cross-section of the cylinder in the plane defined by z and R is the nominal radius of the cylinder. The example of the cylinder, whose axis is not a straight line is shown in Fig. 1.

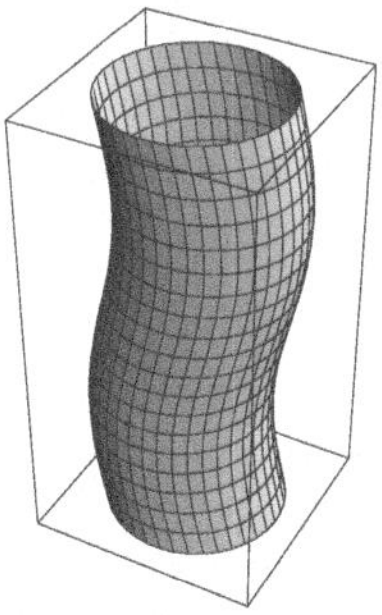

Fig. 1. The roundness profile given by the Eq. (1)

3 Modelling of Barrel-Shaped, Saddle-Shaped and Conical Surfaces

Barrel-shaped, saddle-shaped and conical surfaces can be described in cylindrical coordinates (R,φ,z). It is also noteworthy that if we consider ideal surfaces, then the coordinate φ can be neglected. To start with, let us consider the case of barrel-shaped element. To formulate the equation of the barrel-shaped surface we will use the coordinate system shown in Fig. 2.

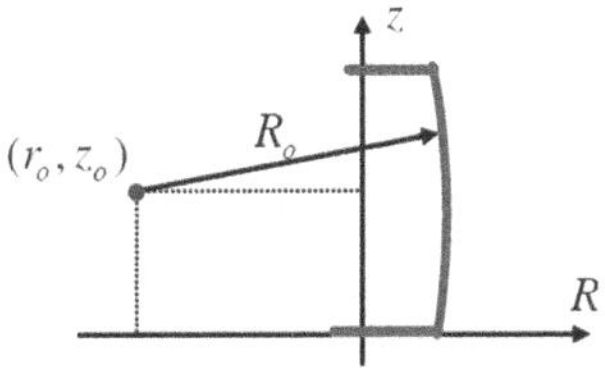

Fig. 2. Coordinate system used to formulate the equation of the barrel-shaped surface [7]

The generatrix of the barrel-shaped surface can be regarded as a fragment of the circle, whose origin is the point defined by (r_0,z_0). The nominal radius of the circle is denoted by R_0. Taking into account the diagram shown in Fig. 2 we can easily formulate the equation of the generatrix of the barrel-shaped surface, which is as follows:

$$R(z) = r_0 + \sqrt{R_0^2 - (z - z_0)^2} \tag{3}$$

It should be noted that the solution for the case of saddle-shaped elements is very similar, as in this case the generatrix is the fragment of the circle, too. The difference is just the location of the origin of the circle. Considering this fact we can write the equation of generatrix of the saddle-shaped element in the following way:

$$R(z) = r_0 - \sqrt{R_0^2 - (z - z_0)^2} \tag{4}$$

Let us now consider the case of conical surfaces. It is noticeable that for such surfaces the generatrix is the linear function of variable z. If we denote a half of the angle of the cone by α, then the equation of the generatrix of the conical surface can be written as follows:

$$R(z) = R_0 - z tg\, \alpha \tag{5}$$

where R_0 is the radius of the surface for the coordinate z equal to zero.

4 Modelling of Spherical Surfaces

Modelling of ideal spherical surfaces can be easily performed with the use of spherical coordinates system (R,θ, φ).

Modelling of an ideal spherical surface in the coordinate system shown in Fig. 3 is not difficult. However, modelling of non-ideal spherical surfaces is one of most complicated tasks in the area of metrology of geometrical quantities. Equations of non-ideal spherical surfaces can be defined with use of so-called spherical surface functions [10]:

$$Y_k^n(\theta, \varphi) = P_k^{|n|}(\cos\theta)e^{in\varphi} \tag{6}$$

where

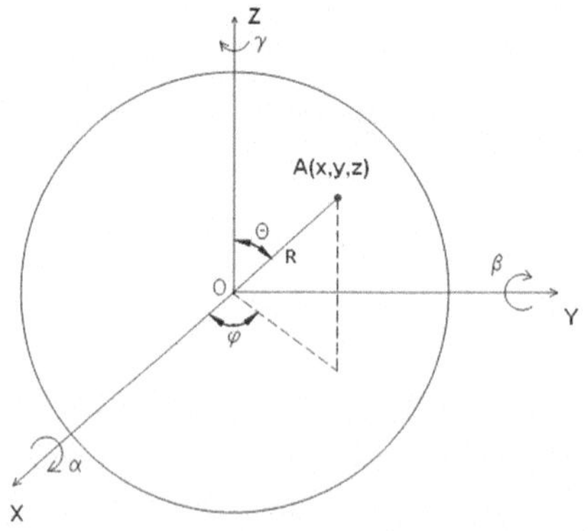

Fig. 3. Coordinate system used to formulate the equation of the spherical surfaces [8]

$P_k^n(\cos\theta)$ - are so-called associated Legendre functions calculated from the formula

$$P_k^n(\cos\theta) = (1 - \cos^2\theta)^{m/2}\frac{d^m}{du^m}P_k(\cos\theta) \tag{7}$$

and $P_k(\cos\theta)$ are Legendre polynomials given by Rodrigues formula:

$$P_k(\cos\theta) = \frac{1}{2^k k!}\cdot\frac{d^k}{dx^k}(\cos^2\theta - 1)^k \tag{8}$$

Equations (6)–(8) allow following description of the spherical surface:

$$R(\theta, \varphi) = R_0 + \psi_s(k, n) + \psi_c(k, n) \tag{9}$$

where

$$\psi_s(k, n) = \sqrt{\frac{2k+1}{2\pi}} \sqrt{\frac{(k-n)!}{(k+n)!}} P_k^n(\cos\theta)\sin(n\varphi) \tag{10}$$

$$\psi_c(k, n) = \sqrt{\frac{2k+1}{2\pi}} \sqrt{\frac{(k-n)!}{(k+n)!}} P_k^n(\cos\theta)\cos(n\varphi) \tag{11}$$

Equations (9)–(11) can be used to model non-ideal spherical surfaces. For example, let us consider equation of the surface given by following equation:

$$R(\theta, \varphi) = 10 + \psi_s(7, 5) + \psi_c(7, 5) \tag{12}$$

The surface describe by this equation is shown in Fig. 4.

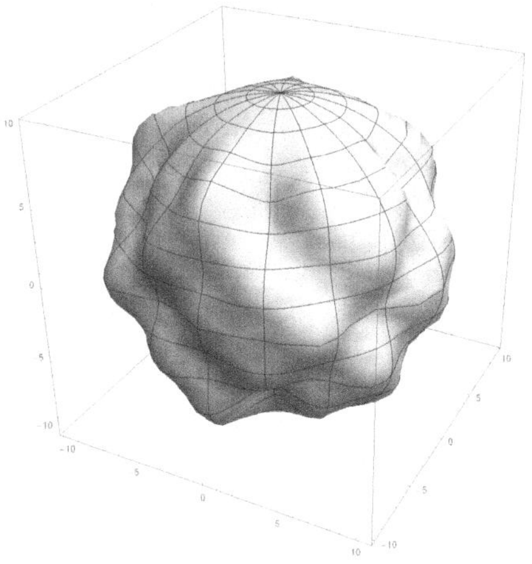

Fig. 4. Model of non-ideal sphere generated with the use of Eq. (12)

5 Conclusions

In order to develop new methods of measurements or analysis of form deviations of various machine parts, firstly it is necessary to conduct simulation-based study. The fundamental of such study is to model the surface to be measured. Therefore the problem of correct modelling of surfaces is a matter of great importance in the field of measurements of form deviations. Considering the problem of modelling of rotary elements it is noticeable that the most complex problem is to model non-ideal spherical surfaces. It requires applying complex mathematical apparatus. It is noteworthy that modelling of the surface is the preliminary stage of the study. After that methods of calculation of reference features should be developed. This problem is well recognized for cylindrical parts. However, there is a lack of references that describe methods of calculation of reference features for barrel-shaped or spherical elements. However, authors have dealt with these problems and the solutions can be found in works [7, 11].

Acknowledgment. The paper has been elaborated within the framework of the research project entitled "Theoretical and experimental problems of integrated 3D measurements of elements' surfaces", reg. no.: 2015/19/B/ST8/02643, ID: 317012, financed by National Science Centre, Poland.

References

1. Jermak CJ, Rucki M (2016) Static characteristics of air gauges applied in the roundness assessment. Metrol Meas Syst 23(1):85–96
2. Zorawski W et al (2015) Microstructure and tribological properties of nanostructured and conventional plasma sprayed alumina-titania coatings. Surf Coat Technol 268:190–197
3. Kundera C, Kozior T (2014) Research of the elastic properties of bellows made in SLS technology. Adv Mater Res 874:77–81
4. Adamczak S, Bochnia J, Kundera C (2012) Stress and strain measurements in static tensile tests. Metrol Meas Syst 19(3):531–540
5. Poniatowska M (2009) Research on spatial interrelations of geometric deviations determined in coordinate measurements of free-form surfaces. Metrol Meas Syst 16(3):501–510
6. Humienny Z, Turek P (2012) Animated visualization of the maximum material requirement. Measurement 45(10):2283–2287
7. Janecki D, Stępień K, Adamczak S (2010) Problems of measurement of barrel- and saddle-shaped elements using the radial method. Measurement 43(5):659–663
8. Janecki D, Stępień K, Adamczak S (2010) Investigating methods of mathematical modelling of measurement and analysis of spherical surfaces. In: Proceedings of the X international symposium on measurement and quality control – ISMQC 2010, 5–9 September 2010, Osaka, Japan
9. Stępień K (2014) In situ measurement of cylindricity—Problems and solutions. Precis Eng 38(3):697–701
10. Kunis S, Potts D (2003) Fast spherical Fourier algorithms. J Comput Appl Math 161:75–98
11. Janecki D, Stępień K, Adamczak S (2016) Sphericity measurements by the radial method: I. Mathematical fundamentals. Meas Sci Technol 27(1):015005

The Effect of Service Quality and Offered Values on Customer Satisfaction and Customer Loyalty: An Implementation on Jewelry Industry

Meltem Diktaş[1,2] and Mahmut Tekin[1,2](✉)

[1] Department of Administration and Organization, Selçuk University, Karapınar Aydoğanlar Vocational School, Konya, Turkey
{mdiktas,mahtekin}@selcuk.edu.tr
[2] Department of Business Administration, Faculty of Economics and Administrative Sciences, Selçuk Unicersity, Konya, Turkey

Abstract. Today, with the development in technology and the increase in rivalry, it is required for businesses to look for the ways to have long- lasting and ever-lasting relations with their customers based on confidence in order to obtain and keep their customers. Businesses, by preserving their present customers and learning more about them, can have long and ever lasting relationships with their customers, gain new customers through the oral advertisements of their present customers and get the edge over rivalry. At this point, relational marketing has its own place in marketing strategies as one of the top priorities in competitive conditions; it also appears as a strategy aiming at having high income as well as ensuring customer satisfaction through the quality of goods and the service and developing long-lasting and permanent relations by offered values. The most important and long term objectives of businesses replacing the service quality as the focus in marketing are to make customers prefer their businesses again and to increase the customer value by creating customer satisfaction and loyalty.

In the implementation part of the study, what is studied is the effect of the service quality as a component of relational marketing strategies being implemented by jewelers; and the satisfaction and the loyalty within the offered values of customers who have bought gold/jewelery from the jewelry stores in Konya.

As it is vital for businesses during implementation to know about the effects of the service quality and relational marketing implementations accepting the offered values factors as a key component in creating customer satisfaction and loyalty, it will be an advantage for them to see their investments made to the relationships not as an owning cost but as an executive strategy that will enable a value of return.

Keywords: Relational marketing · Service quality · Offered values
Customer satisfaction · Customer loyalty

N. M. Durakbasa and M. G. Gencyilmaz (Eds.): ISPR 2018, *Proceedings of the International Symposium for Production Research 2018*, pp. 753–768, 2019.
https://doi.org/10.1007/978-3-319-92267-6_61

1 Introduction

With the globalization today, due to the fact that the development, change and the rivalry in the economical scales of countries have gained an unprecedented speed and content, marketing has gained a new meaning and it has ended with an inevitable final for the changes in the structure of marketing activities for businesses. These changes can be explained as the desire of businesses not to lose their customers and focusing on gaining new customers thanks to their satisfied and loyal customers. Consumers not only buy the products that make them feel satisfied but prefer to buy emotional experiences with the products that make them feel good [1]. And here the material and non-material values offered by businesses play an important role as a key function to keep their customers. The aim not to lose the customers, but to create customer satisfaction and to focus on sharing customers instead of sharing investments have increased the studies for relational marketing [2]. The transition process, from traditional marketing to the sense of relational marketing aiming primarily at getting in contact with customers, strengthening and developing the relations and keeping its stability, has run up as a result of these studies.

Relational marketing is a marketing approach giving competitive advantages for businesses based on creating, sustaining a strong relationship with customers and making it increasing in value to determine the needs of their target groups in a complete and accurate way and formed with the implementations undertaken by businesses within the aim of customer satisfaction and customer loyalty. That's why, businesses, within the scope of today's marketing approach, should analyse the relationship between the consumer and the business well in order to determine, develop and apply the strategies that will bring about brand loyalty [3].

Today, Turkish jewelry industry make its production by making innovational designs carrying the traces of inheritance of the traditional and rich jewel production custom, and by using modern technology and modern technics. In this developing sector, every four tourists visiting Turkey purchase jewelery. About 40% of gold jewels produced annually have been sold to the tourists and to the ones making shuttle trade. Meanwhile, the demand for our traditional jewels, produced with valuable mines, as a present or as an investment tool is in an incontrovertible level in domestic market.

In this framework, the aim of the study is to find out the effects of service quality and offered values components as implementations of relational marketing methods on customer satisfaction and loyalty in jewelry industry which is making progress in production sector.

2 Service Quality of Relational Marketing and Offered Values Components

Relational marketing includes all the marketing activities made in order to form successful changes between the customer and the business, and in order to develop, sustain and strengthen the changing relation sense [4].

The value, formed in relational marketing, based on creating, maintaining relational link between the business and the customer and making it increased in value, is a result of the motive to get profit for both sides. What is required for that is to carry out the value transfer mutually between the business and the customer and to keep the promises as well [5].

Customers', taking place in service quality, being in need of confidence means that the value of the customer will also be high for businesses. And the collaboration formed with confidence and the diminution in the risk for customers will appear within the time [6]. The confidence feeling for the business reduces the worries of customers and provide them with the comfort as they know what to expect from the business [7]. Another advantage for customers of relation marketing in the scope of offered values component is personal privileges such as special prices, discounts, faster service [8], payment advantages, private product design for regular customers. Within the context of offered values, another popular implementation today is the customer clubs where the relationships with the customers are developed and where their relationships with the businesses are increased through customer card privileges [9]. The messages sent to the customers on their special days and the promotions to be applied for discount by means of customer cards attract customers' attention and create pleasure for them.

Customers prefer businesses that they trust on for shopping as it will take time and money to experience other, different offers for a customer who wants to get high quality service and who relies on the business already. After purchasing (shopping), when customers experience a problem that shake their confidence, they regret of their shopping and there appears an abrasion in the relationship based on confidence. Within the context of relation marketing, confidence is defined as the extent of a business relation in which one relies on the promises of the other's. The more confidence they feel for each, the more relations will be permanent and long lasting between the business and the customer [10].

For reliable relations between the customer and the business, making a commitment, like confidence [4, 11], is another important component that affects the sense of customer's service quality and relation quality.

Specialization is defined as the sense of customers towards the seller's technological and commercial competence [12] and the competence includes the knowledge of the seller about marketing, the ability of the seller to give good advice to the customer, the ability to help the customers about their purchase plans and the ability to provide the customers with effective promotion supplies. For this reason, it is an important factor to form the customer's service quality sense.

Another factor for businesses to be superior over the other is communication. The interaction process of relationship marketing will be achieved when the customers feel that the businesses that communicate with them pay real attention to their needs, and take their feedback into consideration [13].

Morgan and Hunt [4] studied the communication as a component of relation development which affects the confidence; and in terms of relationship marketing they defined 'communication' as all the activities -based on communicating continually with valuable customers- carried out during the presentation of the products and services with the aim of providing customers with accurate information at the right time and of intervening in a possible problematic situation.

The more businesses are successful in managing the conflict they have with their customers and the more they avoid the situations that cause negative perceptions for customers, and so loyalty will shape more easily for customers. Another important factor to enhance perceived quality of relationship is the resolution of conflict. The businesses aiming at qualified relationships with their service quality can create customer satisfaction by performing a confident behaviour, showing a real dependence on service and practising their commitments, transfering the information to customers in an effective and correct way, presenting the services effectively and having the superiority over service quality through producing the conflict solution skillfully [14].

According to Bennet and Barkensjo [15], service superiority achieved with relational marketing implementations has been playing an effective role on the tendency of customers to purchase again and more, experience to shop also from other services, have less susceptibility for the price and share their experiences with others [16].

By providing customer satisfaction, complaints about the service or the products of a business decreases, and it is provided that customers purchase the product or have the service again and the sensitivity of customers over the price partially disappear accordingly [17]. It is stated by Hoffman and Bateson [18] that loyal customers tend to pay more when uncertainty is reduced or removed.

3 Service Quality and the Connection Between Offered Values and Customer Satisfaction and Customer Loyalty

After purchasing a product or getting a service, the performance of each meeting the needs or expectations of customers and their reaction of being satisfied with it have been defined as customer satisfaction [19, 20]. Customer Satisfaction is directly related with expectations. Providing customer satisfaction is carried out by meeting customers' needs or expectations [21].

Service quality perceived as a customer satisfaction component [17], is the result of the perception about what customers expect from the service and the performance of the service during its presentation [22]. The perception's being positive will increase customer satisfaction accordingly.

In the study of Noumann et al., in which he searched for the factors that affect customer satisfaction, the quality of the product, transport and delivery process, the design of the product, customer services, price policy and billing are stated as the factors affecting customer satisfaction [23].

According to reference [24], customer satisfaction is also related with the image of a business, the behaviours of employees, the characteristics of customers, the professional manner of a business and the speed of the process. The aim to satisfy the customers forming the basic infrastructure of Total Quality Management is one of the most important indications of the sense of business performance and of total quality [25].

For Oliver, customer loyalty is customer's being persistent and eager to get the goods and services again and again although there exist possible situations and marketing endeavors leading changes in their preferences. A customer's being loyal means that they ignore the pleasure of other brands and that they make their payments regardless of sale price of the brand he has been already depended on [20, 26].

The factors affecting customer loyalty are of vital importance for businesses and they intersect with the topics such as confidence, rewarding [27], price, quality, customer satisfaction values offered to the customers, institutive image, and conflict resolution.

The presentation of qualified service is an important tool for businesses in order to meet customers' needs on the purpose of providing customer satisfaction and creating long term relationships. For businesses, providing satisfaction comes before loyalty and it has an effect on the possibility of a customer to get the product or to use the service again [28].

4 Service Quality and Jewelry Industry Implementation of Offered Values Components

4.1 The Objective and the Importance of the Study

Service Quality and offered values, the main focus implementations in relational marketing, have two objectives: to obtain customers and to keep them stable. To make customers loyal, especially, is a pretty difficult process. Regular customers' being less sensitive to prices, shopping in exorbitant sum and talking out positively to others and coming with new customers with low cost increase the attention to the customer loyalty. In this way, the study oriented whether the implementations in terms of service quality and offered values have changed customers' satisfaction and loyalty in the jewelry industry in Konya.

4.2 Finiteness of the Study

As it does for every kind of study, this study also has some finiteness. This study was carried out between the dates 05.02.2018–27.04.2018, it was expected from participants to answer the questions instantly at one time. This method prevents to get and evaluate the complete sense of a consumer, which is tend to change in time. The changes in the sales of jewelry according to seasons and special days; its getting away from its traditional structure and what is preferred for investment is to get stock certificate, bank accounts or similar investing tools instead of jewelry are also evaluated as other constraints. By the term of the study, the fluctuation in the price of gold/jewelry and having difficulty in reaching its goal in whole because of the pressure of time and cost are stated as the final constraints.

4.3 Model of the Study

By scanning related marketing literature, there existed a survey with four sections. In this framework, secondary data were used first so as to constitute primary data. Included in the relational marketing activities, the expressions used to measure service quality and offered values components; the expressions used in thesis developed by Zhang and Feng [29] themed 'The Effect of Relational Marketing Techniques on Customer Loyalty' were adapted to the jewelry industry

The expressions used to measure customer satisfaction were adapted on the basis of the scales developed by Lam, Shankar, Erramilli and Murthy [30].

The expressions to measure customer loyalty was adapted to the jewelry industry on the basis of scale by Zeithaml and the others [31].

4.4 The Universe and Sampling of the Study

It is aimed to get data within the capacity that helps both to evaluate the sample in the study and the size in conceptive model, and to generalise the findings. In this study, the customers shopping from the jewelery stores in Konya are determined as the centre of the study.

4.5 Methods of the Study

Survey method has been used, one of the primary data acquisition methods, in order to collect necessary data during the implementation process

The survey has been carried out with 475 individuals and the data obtained have been tested through SPSS 21.0 packaged software.

The aforementioned survey consists of four parts. The first part includes the questions about demographic information of the participants, in the second part participants are expected to answer the questions based on measuring the behaviour of jewelry purchasing. In the third part, participants are expected to answer the questions on service quality and offered values of relational marketing implementations, and in the fourth part they are asked for answering about the factors defined in two as customer satisfaction and customer loyalty, which affect the behaviours of customers.

In the survey, quinary likert scale has been used and designed as 1-Totally Agree, 5-Totally Disagree.

Following hypotheses have been developed, taking the aim and the importance of the study into consideration

H_1: Relational marketing implementations affect the customer satisfaction positively in jewelry industry.
H_{1a}: Service quality affects the customer satisfaction positively in jewelry industry.
H_{1b}: Offered values affect the customer satisfaction positively in jewelry industry.
H_2: Relational marketing implementations affect the customer loyalty positively in jewelry industry.
H_{2a}: Service quality affects the customer loyalty positively in jewelry industry.
H_{2b}: Offered values affect the customer loyalty positively in jewelry industry.
H_3: Customer satisfaction affects the customer loyalty positively in jewelry industry.

4.6 The Analysis of Survey Data

The demographic information of the participants evaluating the questionnaire is shown in Table 1.

Out of 475 participants, 232 of them are men and 243 of them are women. In other words, 48,8% of participants are men and 51,1% are women, in the study.

Table 1. Participants demographic information

Gender	Frequency	Percent
Woman	243	51,1
Male	232	48,8
Medeni Durum	Frequency	Percent
Married	331	71,3
Sıngle	133	28,7
Age	Frequency	Percent
15–25	56	11,8
26–34	162	34,2
35–44	158	33,4
45–55	59	12,5
55+	38	8,0
Educational Status	Frequency	Percent
Primary School Graduates	64	13,5
Two-Year Degree Graduates	53	11,2
High School Graduates	115	24,3
Associate Degree Graduates	58	12,3
Bachelor's Degree Graduates	130	27,5
Master's Degree Graduates	53	11,2
Occupational Status	Frequency	Percent
Public Employee	114	24,2
Private Sector Employee	171	36,2
Self-Employment	71	15,0
Housewife	69	14,6
Student	23	4,9
Retired	24	5,1
İncome Status	Frequency	Percent
Less than 2500 TL	156	34,4
2501–5000 TL	207	45,7
5001–7500 TL	64	14,1
7501–10000 TL	18	4,0
100001 TL on	8	1,8

133 of them have stated that they are not married and their rate is 28,7% out of total participants. 331 individuals are in the group of married ones and their rate is 71,3% out of every participant.

Considering age variable, it is possible to say that maximum participation is between the 26–34 age group. This group has 34,2% as a rate and the number of the participants includes 162 individuals. The following age groups, respectively, are 35–44 age group with 158 participants (33,4%), 45–55 age group with 59 participants (13%), 15–25 age group with 56 participants (12%), and participants aged over 55 with 38 participants (8%).

Considering educational status, maximum participation is formed with bachelor's degree graduate and the rate is 27,5% with 130 participants. The following groups respectively are high school graduates with 115 participants (24,3%), primary school graduates with 64 participants (13,5%), and two-year degree graduates with 58 participants (12,3%). At the last row, secondary school graduates and master's degree graduates take their places with 53 participants (11,2%).

Considering occupation of the participants, workers of private sector have the maximum participation with the rate of 36,2% with 171 individuals. The following groups respectively are public officials with 114 participants (24,2%), individuals who have their own businesses with 71 participants (15%), and the other income groups with 69 participants (14,6), 24 participants (5,1%), and 23 participants (4,9%)

Considering monthly income of the participants, individuals earning between 2501–5000 take their place in the first row with 207 participants (45,7%); individuals earning less than 2500, are in the second row with 156 participants (34,4%). The following groups respectively are the ones earning between 5001–7500 with 64 participants (14,1%); 18 participants earn between 7501–10000 (4%), and 8 of them earn more than 10001.

For what purpose do you buy gold/jewelry products? the participants were allowed to mark more than one option, so the frequency and percentage values for the options were realized as shown below in Table 2.

Table 2. For what purpose do you buy gold/jewelry products?

I buy gold/jewelry for beauty and elegance	Frequency	Percent
Positive	161	34,0
Negative	313	66,0
I buy gold/jewelry products for investment purposes	Frequency	Percent
Positive	203	42,8
Negative	271	57,2
Get gold/jewelry for gift purposes	Frequency	Percent
Positive	188	39,7
Negative	286	60,3
I buy because of the gold/jewelry shopping habit	Frequency	Percent
Positive	24	5,1
Negative	450	94,9

Participants state that they spend money on jewelry mostly for investment and the number of these participants is 203 (42,8%); the other 188 participants buy jewelry for present (39,7%); 161 participants want to seem more attractive (34%) and for 24 participants it is just a habitual activity.

450 participants answer the question: 'How often do you buy jewelry?' 253 participants (56,2%) buy more than once a year; 115 of them buy (25,6%) once a year; 43 participants buy once every two years (9,6%). The other 39 participants buy, respectively, (2,7%, 2,4%, 3,6%) once every three, four and five years.

The information given is in Table 3:

Table 3. How often do you buy gold/jewelery products?

How often do you buy gold/jewelery products?	Frequency	Percent
More than once a year	253	56,2
Once a year	115	25,6
Once in two years	43	9,6
Every three years	12	2,7
Every four years	11	2,4
Every five years	16	3,6

(1) ***Factor Analysis***

The suitability of the questionnaire to sample size factor analysis was tested through the SPSS 21.0 package program.

Field (2000) [32] stated that the 0.50 value for the Kaiser-Meyer-Olkin test should be the lower limit and that factor analysis can not be applied if BMO is 0.50. This information should be meaningful in the light of the Barlett Test and greater than 0.50 in the Kaiser-Meyer-Olkin Test.

The results of the Barlett Test and the Kaiser-Meyer-Olkin test on the Relative Marketing Practices scale are given in Table 4.

Table 4. Factors relevant to factor analysis of the relationship marketing scale

KMO and bartlett's test		
Kaiser-Mayer-Olkin (KMO) Örneklem Ölçüm Değer Yeterliliği		0,949
Barlett Testi	Ki-Kare	4851,378
	Serbestlik derecesi	78
	p-değeri	0,000

The factor load is squared to find out how much of a variant describes a variance in a factor.

As seen in Table 4, the value of KMO is obtained as 0.949 and the value obtained confirms that the sample size is "perfect" for factor analysis [33]. In addition, when the results of the Barlett sphericity test are examined, it shows that the value of square (X2 (78) = 4851.378; $p < .05$) is significant.

In order to determine which factors the items are strongly correlated with, a matrix of rotated components has been developed to investigate whether the complexity and factor loadings of the materials meet the acceptance level.

As a result of the confirmatory factor analysis, the articles c1, c2, c3, c4 and c5- in the third part- come under 'Service Quality'; and the articles c14, c15, and c16 come under 'Offered Value' and it is seen in the following: Table 5.

Table 5. Rotated components matrix of the scale of relational marketing implementations

Questions	Factors	
	Service quality	Offered value
c1	,809	
c2	,782	
c3	,775	
c4	,767	
c5	,694	
c14		,801
c15		,839
c16		,700

It is seen that factor load values change between 0,694 and 0,809 for the first factor; and between 0,70 and 0,839 for the second factor.

The results of the Barlett test and the Kaiser-Meyer-Olkin test on customer behavior scales are given in Table 6.

Table 6. Factors relevant to factor analysis of the customer behavior scale

KMO and bartlett's test		
Kaiser-Mayer-Olkin (KMO) Örneklem Ölçüm Değer Yeterliliği		0,961
Barlett Testi	Ki-Kare	5578,654
	Serbestlik Derecesi	45
	p-değeri	0,000

As seen in Table 6, the KMO value was obtained as 0.961, and the value obtained indicates that the sample size is "perfect" for factor analysis [33]. Also, when the results of the Barlett sphericity test are examined, it shows that the value of the square (X2 (45) = 5578.654; $p < .05$) is significant.

The articles d1, d2, d3, d4 and d5 – in the fourth part- come under 'Customer Satisfaction', and the articles d6, d7, d8, and d10 come under customer loyalty as it is seen in the following Table 7:

Table 7. Rotated components matrix of the scale of customer behaviour

Questions	Factors	
	Customer satisfaction	Customer loyalty
d1	,821	
d2	,837	
d3	,718	
d4	,631	
d5	,790	
d6		,653
d7		,726
d8		,718
d9		,795
d10		,857

It is seen that factor load values change between 0,631 and 0,837 for the first factor; and between 0,653 and 0,857 for the second factor.

(2) ***Reliability Analysis***

The Cronbach's Alpha coefficient tests whether the questions in the scale form a whole to explain a homogeneous construction. The Alpha Coefficient is a coefficient that reveals the closeness and similarity of the questions to each other.

SPSS 21 package program was used to calculate the Cronbach's alpha coefficient showing the correlation between the questions.

States that it is enough for Cronbach Alpha value to be over 70 for reliability [34]. As a result of the factor analysis, relational marketing implementation scale has its final shape with 8 items and the reliability analysis of the scale in whole and its sub-dimensions; and reliability analysis on customer behaviours are presented with 10 items in Table 8.

Table 8. Reliability analysis result

Scales and factors	Number of the questions	Cronbach alpha value
Service quality	5	0,931
Offered value	3	0,877
Relational marketing	**8**	**0,954**
Customer satisfaction	5	0,947
Customer loyalty	5	0,949
Customer behaviours	**10**	**0,970**

According to the reliability coefficient data obtained with the SPSS 21 package program, it is seen that all scales related to service quality, presented values, customer satisfaction and customer loyalty have high reliability.

(3) ***Searching Hypotheses with Regression Model***

Whether relational marketing implementations in jewelry industry have effects on customer satisfaction have been searched with regression model.

- 79,4% of the changes in customer satisfaction can be explained with relational marketing implementations (beta = 0,891, t = 42,683 and p = 0<0,01). That is, relational marketing implementations have positive effects on customer satisfaction. R^2 and corrected R^2 values are found as 79,4% and 79,3%.

When the effects of sub-dimensions of relational marketing implementations on customer satisfaction have been studied;

- Service quality (beta = 0,860, t = 36,713, p = 0<0,01) has a positive effect on customer satisfaction.
- Offered values (beta = 0,703, t = 21,449, p = 0<0.01) have positive effect on customer satisfaction

As a result, it has been found out that all the sub-dimensions of relational marketing implementations have positive effects on customer satisfaction.

Whether the measurement of relational marketing in jewelry industry has an effect on customer loyalty has been searched through regression model.

- 78,0% of the changes in customer loyalty can be explained through relational marketing measurement (beta = 0,883, t = 40,949 and p = 0<0,01). That is, relational marketing has positive effects on customer loyalty. R^2 and corrected R^2 values are found as 78,0% and 77,9%.

When the effects of sub-dimensions of relational marketing measurement on customer loyalty have been studied,

- Service quality (beta = 0,820, t = 31,228, p = 0<0,01) has positive effects on customer loyalty
- Offered values (beta = 0,730, t = 23,260, p = 0<0.01) have positive effects on customer loyalty

As a result, it has been found out that all the sub-dimensions of relational marketing implementations have positive effects on customer loyalty.

Whether customer satisfaction has effects on customer loyalty has been studied through regression model.

- 81,8% of the changes in customer loyalty can be explained with customer satisfaction. (beta = 0,904, t = 46,165 and p = 0<0,01). That is, customer satisfaction has positive effects on customer loyalty. R^2 and corrected R^2 values are found as 81,8% and 81,8%.

(4) ***Evaluation of Hypotheses***

Considering the aim and the importance of the study, the results of the hypotheses are as in Table 9.

Table 9. Results of hypotheses

Hypotheses	Results
H_1: Relational marketing implementations in jewelry sector has positive effects on customer satisfaction	Approved
H_{1a}: Service quality in jewelry sector has positive effects on customer satisfaction	Approved
H_{1b}: Offered values in jewelry sector have positive effects on customer satisfaction	Approved
H_2: Relational marketing implementations in jewelry sector has positive effects on customer loyalty	Approved
H_{2a}: Service quality in jewelry sector has positive effects on customer loyalty	Approved
H_{2b}: Offered values in jewelry sector have positive effects on customer loyalty	Approved
H_3: Customer satisfaction in jewelry sector has positive effects on customer loyalty	Approved

5 An Overall Evaluation, Result and Suggestions

Turkey's industrial ranking has risen and has become one of the leader countries due to the fact that the export of the jewelery, in Turkey, made of valuable materials has increased significantly in recent years. In Turkey, jewelry industry has the capacity to process approximately 400 tone gold and 200 tone silver and turn them into jewelery; however, not all the capacity has been used.

On the basis of high demand in domestic and foreign market, the changes and varieties due to today's competetive circumstances in jewelry activities, having an incontrovertible size in production sector, have become the basic topic of the study. The effects of service quality and offered values components included in relational marketing implementations on customer satisfaction and customer loyalty have been studied in the survey.

Preliminary survey has shown that, as the result of the negotiations with jewelers, the service and implementations based on relational marketing in jewelry sector are closely related with satisfaction and loyalty.

Basic goals of the study are (i) to identify the implementations used by the businesses for their relations with their customers, (ii) to identify the satisfaction of customers through relationship marketing (iii) and to identify the loyalty of customers through relationship marketing.

Taking the aim and the importance of the study into consideration, a survey model and hypotheses have been developed. In the study, at first, measuring instruments have been examined in terms of reliability and validity through different kinds of analysis, and it has come up with satisfactory results. To provide casualness and to get the data at the right size that help generalize the findings; study sample, using necessary methods, has been carried out by an analysis with 475 participants chosen from the customers of an institutional business that has branch offices in big shopping malls and in the town centre in Konya and from the customers of other jewelry stores. The sample has a good division in terms of sex, occupation, and age, and includes both educated and middle income individuals.

The survey has resulted that the implementations of jewelry industry, within the context of relational marketing, aiming at service quality and offered values have affected the behaviours of customer satisfaction and loyalty

Within the relational marketing implementations, in jewelry sector, what is evaluated as service quality is to meet customer demands on time, to respond to the complaints fast in the case of any problems with the products, to provide customers with employees who are always eager to understand them, to be stable at the presentation of products or service and to present personal services if required. Businesses caring about these components will create an effect of satisfaction on customers and will have carried out the aim of relational marketing to have long-lasting relationships in this way.

It has been found out that relational marketing implementations applied in jewelry sector have provided customer satisfaction and loyalty. In this regard, another way to have long term relations and to strengthen the relationships for businesses is to provide their customers with customer cards, presents, discounts on special occasions, messages on their special days, and discount offers.

Investments made for service quality and offered values during working with customers are in fact the investments made for relationships set up with the customers. Such an activity, defined as the most correct investment both for economy and value-added for businesses, will help the process of relationship; therefore, longer relationship means higher profit for businesses, and it also means to bring about an additional value to the present brand of a business in long term.

References

1. Morrison S, Crane FG (2007) Building the service brand by creating and managing an emotional brand experience. J Brand Manag 14(5):410–421
2. Sheth JN, Parvatiyar A (2002) Evolving relationship marketing into a discipline. J Relat Mark 1(1):3–16
3. Ballantyne R, Warren A, Nobbs K (2006) The evolution of brand choice. J Brand Manag 13 (4/5):339–352
4. Morgan RM, Hunt SD (1994) The commitment-trust theory of relationship marketing. J Mark 58:20–38
5. Grönroos C (1996) Relationship marketing: strategic and tactical implications. Manag Decis 34(3):5–14
6. Nakıboğlu MAB (2008) Hizmet İşletmelerindeki İlişkisel Pazarlama Uygulamalarının Müşteri Bağlılığı Üzerindeki Etkileri, Doktora Tezi. Çukurova Üniversitesi Sosyal Bilimler Enstitüsü, Adana
7. Berry LL (2002) Relationship marketing of services- perspectives from 1983 and 2000. J Relat Mark 1(1):59–77

8. Bühler A, Nufer G (2010) Relationship marketing in sports. Butterworth Heineman, Elsevier Ltd Yayınları. ISBN: 978-0-7506-8495-8
9. Kılıç N (2010) İlişkisel Pazarlamanın Müşteri Sadakati Yaratmadaki Etkisi: Bir Havayolu Taşıma İşletmesindeki Uygulamanın İncelenmesi. Yüksek Lisans Tezi, Dokuz Eylül Üniversitesi. Sosyal Bilimler Enstitüsü, İzmir
10. Yau OHM, McFetridge PR, Chow RPM, Lee JSY, Sin LYM, Tse ACB (2000) Is relationship marketing for everone? Eur J Mark 34(9/10):1111–1127
11. Palmatier RW, Dant RP, Grewal D, Evans KR (2006) Factors ınfluencing the effectiveness of relationship marketing: a meta-analysis. J Mark 70:136–153
12. Anderson E, Weitz B (1989) Determinants of continuity in conventional ındustrial channels dyads. Mark Sci 8(4):310–332
13. Grönroos C (2000) Service management and marketing: a customer relationship management approach, 2nd edn. Wiley, Hoboken
14. Ndubisi NO, Wah CK (2005) Factorial and discriminant analyses of the underpinnings of relationship marketing and customer satisfaction. Int J Bank Mark 23(7):542–557
15. Bennet R, Barkensjo A (2005) Realitionship quality, relationship marketing, and client perceptions of the levels of service quality of charitable organisations. Int J Serv Ind Manag 16(1):81–106
16. Şendur F (2009) İlişkisel Pazarlama Çerçevesinde Müşteri Değeri Yaratmanın Önemi: Bankacılık Sektöründe Bir Araştırma, Yüksek Lisans Tezi, Balıkesir Üniversitesi Sosyal Bilimler Enstitüsü İşletme Anabilim Dalı, Balıkesir
17. Dökmen T (2003) Havayolu İşletmelerinde Müşteri Tatmini ve İş gören Müşteri Karşılaşmalarının Müşteri Tatmini Üzerindeki Etkisi, Doktora Tezi, Anadolu Üniversitesi Sosyal Bilimler Enstitüsü, Eskişehir
18. Hoffman DK, Bateson JEG (1997) Essential of service marketing. The Dryden Press
19. Kotler P, Armstrong G (2005) Principles of marketing. Prentice Hall, essex
20. Olıver Richard L (2010) Satisfaction: a behavioral perspective on the consumer, 2nd edn. M. E. Sharpe, London
21. Çatı K, Koçoğlu CM (2008) Müşteri İlişkileri Yönetimi. Detay Yayıncılık, Ankara
22. Parasuraman A, Zeithaml VA, Berry LL (1985) A conceptual model of service quality and ıt's ımplications for future. J Mark 49:41–50
23. Noumann E, Jackson DW, Rosenbaum MS (2001) How to implement a customer satisfaction program. Bus Horiz 44:37–46
24. Altunışık R, Özdemir Ş, Torlak Ö (2004) Modern Pazarlama, 3. Baskı. Değişim Yayınları, İstanbul
25. Forza C, Filippini R (1998) TQM Impact on quality conformance and customer satisfaction. Int J Prod Econ 55(1):1–20
26. Keiningham TL, Vavra TG, Aksoy L, Wallard H (2006) Sadakat Söylenceleri. Rota Yayınları, İstanbul
27. Demirbağ E (2004) Sorularla Müşteri İlişkileri Yönetimi. İstanbul Ticaret Odası İşletme Yönetiminde Yeni Eğilimler Dizisi, İstanbul
28. Selvi MS (2007) İlişkisel Pazarlama Stratejiler ve Teknikler. Detay Yayıncılık, Ankara
29. Zhang X, Feng Y (2009) The Impact of Customer Relationship Marketing Tactics On Customer Loyalty- Within Swedish Mobile Telecommunication Industry, Master Thesis. Halmstad University School of Business and Engineering Master Program in International Marketing, Sweden
30. Lam SY, Shankar V, Erramilli MK, Murthy B (2004) Customer value, satisfaction, loyalty, and switching costs: an ıllustration from a businessto-business service context. J Acad Mark Sci 32(3):293–311

31. Zeithaml VA, Berry LL, Parasuraman A (1996) The behavioral consequences of service quality. J Mark 60(2):31–46
32. Field A (2000) Discovering statistics using SPSS for windows. Sage Publications, Thousand Oaks
33. Çokluk Ö, Şekercioğlu G, Büyüköztürk Ş (2012) Sosyal Bilimler İçin Çok Değişkenli İstatistik: SPSS ve Lisrel Uygulamaları. Pegem Akademi Yayıncılık, Ankara
34. Bayram N (2004) Sosyal Bilimlerde SPSS İle Veri Analizi. Ezgi Kitabevi, Bursa

With the Trio of Standards Now Complete, What Does the Future Hold for Integrated Management Systems?

Mihai Dragomir(✉), Călin Neamțu, Sorin Popescu, Daniela Popescu, and Diana Dragomir

Department of Design Engineering and Robotics, Technical University of Cluj-Napoca, Cluj-Napoca, Romania
{mihai.dragomir, calin.neamtu, daniela.popescu, diana.dragomir}@muri.utcluj.ro, posorin@gmail.com

Abstract. The world of standardized management systems is traversing exciting changes with the publishing of the long-awaited ISO standard for occupational health and safety, ISO 45001. It comes to join ISO 9001 and ISO 14001 to form a trio of highly compatible best practice references of interest for many companies that want to become more competitive, especially from the manufacturing sector. For the past at least 20 years there have been many debates and studies on how to best go about using these standards together to set up an integrated management system. Of course, things seem easier now, but together with the benefits, there are also challenges to manage and some are quite significant. This paper discusses the implications of reaching this milestone and presents the results of a focus-group made among production oriented companies, that was aimed at probing the stakeholder community regarding its positioning in the new context. The focus is on the impact of the new regulations upon creating or upgrading an integrated management system, where relationships among components are the key to a successful implementation and to obtaining results.

There are three main parts to the paper, a literature survey meant to catch a glimpse of the changes taking place within the complexity and depth of the topics approached in the field, a theoretical analysis of the main implications of the changes performed within the standards line-up and a qualitative interview asking questions about the benefits and challenges identified by the companies going through the process of upgrading their integrated management systems in the new context.

Keywords: Integrated management systems · Quality management
Environmental management · OHS management

1 Introduction

There is great upheaval in the world of standardized management systems and it has been going on for a few years now. The International Organization for Standardization (ISO) has been developing and publishing new versions of the three most used

N. M. Durakbasa and M. G. Gencyilmaz (Eds.): ISPR 2018, *Proceedings of the International Symposium for Production Research 2018*, pp. 769–778, 2019.
https://doi.org/10.1007/978-3-319-92267-6_62

standards for management systems meant to deal with stakeholder requirements (quality, environment, occupational health and safety - OHS). There is a new approach being promoted across these disciplines and while the first two fields experience their fourth and, respectively third upgrade, health and safety is finally joining them as an ISO referential after being dealt with based on national standards and specifications until now. Companies, consultants, trainers, standardization bodies, registrars and certification bodies, as well as academia and other interested parties are all involved in this long-awaited transition. Beside their respective changes in requirements that prompt changes and reactions from the companies, an important field of discussions and contributions opens up with respect to the topic of integrated management systems (IMS), where all three components mentioned above should function to the best of their abilities and with the lowest resource expenditure possible.

Changes in two fields plus the requirements for the third one in a completely new formulation will bring about considerable amount of work for companies and other parties involved. Many are wondering if this work is worth it and if it will make their life easier when dealing with the challenges of integration, either during implementation or throughout the system's active life. This work attempts to formulate a preliminary answer to this question, based on a combination of literature study, theoretical analysis based on consultancy experience and a limited focus-group interview.

2 Literature Survey

The subject of integrated management systems has been of interest in the stakeholder community for over 20 years, ever since the first edition of ISO 14001 came out in 1996 to join the already successful ISO 9001, which also served as its blueprint. However, the documents diverged over time and that has generated complications during implementation. Companies are always interested in efficiency and having two systems with similar mechanisms and even common requirements seemed like a waste of resources. Moreover, in an ever increasing online and mobile based business environment, organizations seek to optimize their operations and be in constant contact with their stakeholders to be able to address their needs, before the competition can do it. ISO, researchers and practitioners have worked during all this time to find ways to deliver on these goals by achieving integration of the management systems, and the current line-up of standards is another big step in this direction. It is interesting to mention some of the studies performed before the appearance of the 2015/2018 editions (see Table 1).

Also, it is even more compelling to see the changes that occurred since the new duo and then trio of standards came out, in terms of assessment of potential for integration and development of the adequate tools for achieving this goal (see Table 2).

The limited space and scope of this article does not allow for a more in-depth analysis, but this would reveal a bigger picture along the same already mentioned lines. Approaches in literature become more complex over time and the topic of integration gains interest and visibility.

Table 1. Contributions to the field of IMS before 2015

Nr.	Main ideas	Source
1	With 362 Spanish companies reporting as having or working towards and IMS, this complex study performs a grouping of companies based on the integration level and behaviour. The largest such group identified by the authors (87% of their target responses) contains companies not fully integrated, achieving an average of 85% synchronization	Bernardo, Casadesus, Karapetrovic and Heras, 2009 [1]
2	This article finds and describes a correlation in the case of cement factories in Iran between the implementation and maturation of IMS, and the verifiable improvement of safety and productivity indicators such as accident and production rates	Hamidi, Omidvari and Meftahi, 2012 [2]
3	The work identifies 5 main benefits, 3 in terms of reduction of effort (paperwork, cost, certification), and 2 in terms of enhanced management (lower complexity, better continuous improvement)	Zeng, Xie, Tam and Shen, 2011 [3]
4	In this paper, the results of an extensive study on 176 companies are presented. The authors find that the need for IMS is very high (almost 90% of companies employing two or more systems) and that increasing the level of integration over time could be beneficial for the firms	Simon, Karapetrovic and Casadesus, 2012 [4]
5	The authors propose a complex model based on lean management concepts that maximizes the experience of interested parties and sustainability related impacts, while diminishing waste and system implementation overlaps	Rebelo, Santos and Silva, 2014 [5]

Table 2. Recent contributions to the field of IMS

Nr.	Main ideas	Source
1	Domingues, Sampaio & Arezes develop a maturity model specific for IMS that has 6 levels which can be attained through improved performances on three directions: process management, externalities and excellence focused management practices	Domingues, Sampaio and Arezes, 2016 [6]
2	Studying a number of 55 responses from companies in Portugal, these authors conclude that the main drivers of integration are related to internal cost efficiency, while the among the main concerns are listed the lack of qualified personnel on the topic and, interestingly, the continuous appearance of updated versions of the standards	Ribeiro, Santos, Rebelo and Silva, 2017 [7]
3	This work focuses on quality-safety combined systems and arrives at the conclusion, which it validates statistically, that there is positive reinforcement between the two centred upon the effects of process approach	Shevchenko, Pagell, Johnston, Veltri and Robson, 2018 [8]
4	In this study, some of key success enablers of integration are presented, based on a company survey, with two factors becoming apparent: that specific training of top management is needed, and that an integration model should be adopted and understood by the firms	Domingues, Sampaio and Arezes, 2017 [9]
5	A previous proposal of the authors of this paper that uses a colour based algorithm to codify process audit outcomes and combine them to allow specific image processing functions to be used for assessing integration maturity	Dragomir, Popescu, Neamțu, Dragomir and Bodi, 2017 [10]

3 Theoretical Implications

This section of the paper discusses the main changes that happened within the 2015/2018 group of standards by performing a brief analysis of their implications upon creating and maintaining integrated management systems. This is based on the implicit goals of the modifications, the obtained practical results and the consultancy expertise of the authors (see Table 3).

Table 3. Principal changes within the 2015/2018 editions of the standards

Nr.	Modification	Mail implications on IMS
1	High Level Structure (HLS)	Use of the effects of the PDCA cycle for improvement on a system wide scale, allowing for synergies and mutual support among systems Common structure and hierarchy of documentation or information databases that increases operational efficiency and reduces implementation costs
2	Risk based approach	Final synchronization among the systems by adopting the concept of "quality related risks" in tandem with environmental and OHS risks Offers a chance for increased adaptability to the changes on the markets, that grow constantly in dynamics and scope
3	Documented information	Significant simplification and increase in the flexibility of addressing requirements and demonstrating conformity to stakeholders or for certification purposes
4	Common wording	Simplification of trainings and dissemination efforts towards the company's own employees
5	New standard for OHS	Avoidance of differences in vision and formulation of requirements, especially after the publication of the modified 9001 and 14001 standards Simplification of the audits, documentation and certification process and the capability to fully assess integrated components of the system Image benefits for achieving a common certification and possible commercial benefits from the certification bodies (discounts)

Conducting a thorough analysis of the changes that have been incorporated in the updated versions of the standards is an arduous task that is beyond the scope of the present work. However, we must note that developing a long-term plan in the form of Annex SL [11] and applying the detailed ISO process for standards development has proven to be a successful approach. Even the time needed, from 2012 until 2015 and then 2018, has a relatively low value compared to the scale of management standards' updating process that takes around 7 to 8 years.

The main concerns to be discussed here are two-fold. On the one hand, the delayed development and publishing process for the OHS standard and the different structure of the three main families of standards (quality management has a separate and even enhancing guidance standard in the form of ISO 9004, while ISO 14001 and ISO 45001 have integrated guidelines together with the requirements; the same applies for the Terms and definitions sections) makes the trio still appear disjointed to the users which find it difficult to follow the correlation lines. On the other hand, the dual application of the risk approach and process approach within the quality system, and by extension, the IMS is a possible source of confusion and even conflicts in terms actions and resources needed for addressing current needs, correcting previous unsatisfactory performance or preparing for future developments, improvements and innovations.

4 Focus Group Interview

In order to better assess the impact of modifications in real implementation conditions, a focus group was set up with company representatives (focused on medium sized manufacturing companies) that employ integrated management systems with all three components active and are involved in the current transition process. The role of this study is to qualitatively assess the main benefits and challenges perceived by the users and correlate them with the previously documented consultancy and academic experience. The authors believe that such an approach will contribute to generating solutions and best practices for the companies in the area where we are located, Cluj-Napoca, Transylvania, Romania.

The focus group gathered 8 company delegates coming from 5 different industries: metallic structures manufacturing, furniture design and production, plastic products manufacturing, additive manufacturing (3D printing), constructions. The members of the group have been asked to a face to face meeting and a scripted interview was conducted with each of them, based on a questionnaire with 12 items, followed by a common debate/discussion where each of them had a chance to expand or amend their positions. The results of the meeting were documented in the form of single report file that reflects the common ground of all the opinions presented. To ensure a proper idea dialogue, the authors responsible for literature survey and theoretical analysis and those responsible for survey design and the company interviews and meeting did not discuss their findings until their final work was ready.

The structure of the questionnaire is presented in more detail in the table below, without including the full questions and answer options, in order to keep the size of the text manageable (see Table 4). A mix of questions was used, most of them being constrained to a rating scale, to simplify the discussions, while the open questions have been mostly assessed in terms of effort (person-months) or percentage of current workload. During the common meeting, all the previous answers obtained from the participants in the group have been opened up for discussion, with changes being noted down as an agreement and a common vision was reached.

Each company interviewed was investigated first regarding their target markets, products, processes and integrated management system (evolution, results, certification), as such the questions where directly focused on their reaction to the studied changes.

Table 4. Structure of the focus group interview questionnaire

Nr.	Category of question	Quantity/Type	Observation
1	Perceived value of the modifications	5 q., rating scale, ranking, open answer	Rank and analyse each modification, insist on the new OHS standard
2	Estimation of costs and efforts of transition	2 q., rating scale, open answer	Focus on financial costs and time commitments
3	Estimation of reduction in administrative burden	2 q., rating scale, open answer	Focused on diminished needs for documentation and demonstrating conformity
4	Stakeholder value added contribution and feedback	2 q., rating scale, ranking	Identification of top benefits for the top stakeholders
5	Identification of main fears, challenges, and inconsistencies	1 q., multiple choice	Section where participants have been encouraged to address the "worst case scenario"

5 Results and Analysis

This chapter summarizes the main findings of the study performed, while at the same time discussing ideas about the course of action that companies might follow in order to maximize their experience and gains throughout this process. The answers given included both numerical and open answers, especially after the common discussion phase which contributed to the harmonization of opinions among those present. On the plus side, everybody, with no exception, considers that the alignment of the three reference standards is a very good thing in terms of reducing duplicate or triplicate work, managing inconsistencies among the systems, improving training of personnel, and improving the impact and contribution of internal and external audits. The virtual elimination of the quality manual (and by extension of the integrated management manual) as well as granting companies leeway in determining the necessary procedures (i.e. the elimination of the documented or mandatory procedures) are welcomed changes that are perceived to adhere to the principle based management promoted by ISO 9001 since 2000. The firms are more motivated and experience a higher degree of accountability when the burden of finding the best solution to a requirement rests with them. Also, their experience so far with certification or surveillance audits is positive in this direction, with the external auditors encouraging them to streamline quickly their documentation and focus on the other benefits of implementing these systems.

By asking the companies to assess on scale from 1 to 10 the desirability of the changes and the foreseeable workload (the higher the value chosen, the more desirable/difficult the change), it becomes possible to compare the theoretical perspective in Table 4 with the real constraints of the business environment. Implementing the High Level Structure and the ISO 45001 are revealed as bottlenecks (high scores on both scales) where resources should be focused (see Fig. 1). Moreover, these results should be correlated with the perceived value added of the modifications per system (on a similar scale) that shows that the environmental (EMS) and OHS system are seen as more in need of updates to the reference standards, than the quality management system (QMS).

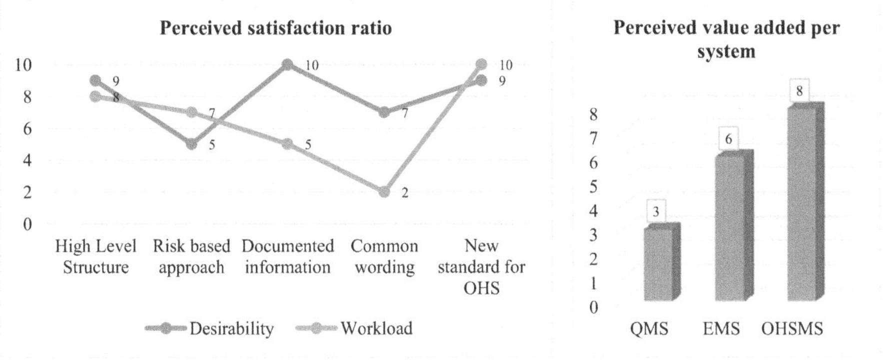

Fig. 1. The perceived value of the modifications (settled rounded scores)

A favourable situation has been revealed in terms of the time commitment considered necessary to make the IMS transition at component level versus the improvement in terms of time savings brought about by the new standards and their respective systems. Half of the contacted companies place the necessary amount of work in the 3–6 person-months ranges, while almost two thirds consider that management systems will save about 3 person-months per year of work to ensure their functioning and the compliance to the requirements coming from the stakeholders and the legislation (see Fig. 2). This means that, on average, the new IMS will pay for itself regarding the time spent in 2 years, which can be considered a very successful result in most companies.

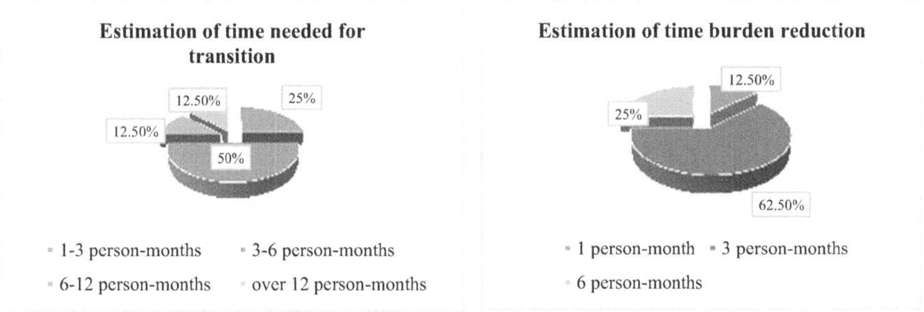

Fig. 2. Time commitment vs. time saved annually

As it became apparent from the interview, companies are not sure how to deal with the new documentation requirements (what should be printed?, what should electronic?, how should it be protected? how it should be used?), they are not clear on how to perform risk management and context analysis especially for quality management and environmental management, and are rather unsatisfied that there is a new reference for occupational health and safety and that it came out with a 2.5 years delay from the others, just when the transition period is getting close to the end. There is a considerable amount of fear regarding the complexity of adopting all the changes, big and small, at the same time and how this will be reflected in the internal consistency of the system (correlation of activities and projects, lines and processes of communication, usefulness of internal audits, etc.). Also, there are uncertainties related to the possible additional costs of certification coupled with a reduced impact of this action when so many companies are obtaining this form of validation. A detailed view of the expressed viewpoints is presented in Fig. 3, below.

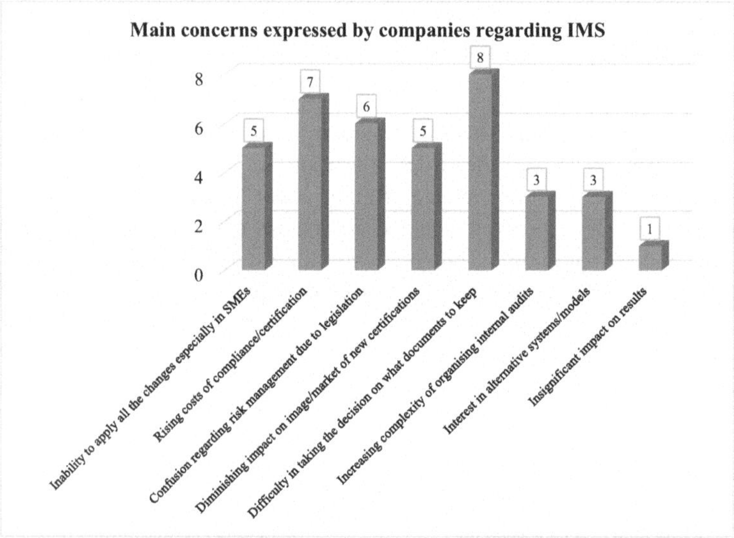

Fig. 3. Possible IMS upgrading issues (unlimited number of choices)

6 Conclusions and Recommendations

In conclusion, although the changes were long awaited and are seen as mostly beneficial, there are serious contention points that require additional work by companies. They might also open an important domain for consultants, as long as they are willing to properly tailor their tools and the requirements of the standards to the situations specific to each company.

Maybe more importantly, it is time to recognize that some of the goals of academic researchers regarding integration are now complete and it is time to refocus the efforts of management systems specialists towards the new needs in the private sector. Manufacturing, as a field of economic activities, would be one of the first beneficiaries of new methods to streamline the bureaucracy associated with these systems as well as one of the most suitable to find the right balance of the risk approach and process approach, due to the nature of its work that relies on well-defined and sequential process chains. Also, due to these factors and due to the clear relationships existing between inputs and outputs, in this area an integrated management system could benefit from the automation of tasks and the introduction of process based solutions such as Six Sigma, Statistical Process Control or Lean management in conjunction with the requirements of the three ISO standards.

In order to tackle some of the challenges revealed by this study, the authors consider that some more innovative approaches to implementing and upgrading the management systems should be deployed at large. Simply updating documents, throwing some away and performing some formal risk analyses will not bring out the spirit of these new standards, nor will it allow companies to transfer some of their benefits towards their stakeholders. Without coming close to a complete approach, there are some obvious opportunities and solutions that are not very used nowadays in integrated management systems, especially in Romania. For example, the IT systems are in many case outdated and amateurish without including fully customized ERP (Enterprise Resource Planning) solutions, cloud based services, decision support systems or distributed computing enabled with sensorics and real-time process control. These elements would prove useful especially for manufacturing to reduce scrap and non-conformities and to diminish environmental aspects.

Also, the managerial processes could benefit from new elements such flat organizational charts to diminish decision complexity, agile management to allow for creativity to flourish and increase timeliness of results, or mobile learning and even online platforms to advance the training of employees. The cost of many of these solutions is now in an acceptable range even for small and medium sized enterprises and implementation times have also been reduced. In many instances, getting in touch with academic partners such as universities, can facilitate this process of adopting and making use of knowledge based solutions. Another important component that can be observed as having a positive impact in Romania is the clusterisation process inside various industries complemented by the connection to international networks. This phenomenon allows for a faster diffusion of best practices in the area of management systems (among others) which are part of the operational excellence, thus facilitating the evolution towards competition based on innovation and new products.

Production companies are subjected to intense competition on the one side and regulatory pressures on the other side, so it is critical for them to have these system transition fast and become effective and auto-piloted in order to contribute to the survival and success of the firm and not generate "death by documentation". The concept and approach of Industry 4.0 could become an excellent blueprint for organizing the above mentioned innovative solutions in a data rich and decentralized decision environment that can enhance the results that the IMS generates towards its stakeholders. In such a context, the goal of the system can move from mere compliance to excellence.

References

1. Bernardo M, Casadesus M, Karapetrovic S, Heras I (2009) How integrated are environmental, quality and other standardized management systems? an empirical study. J Clean Prod 17(8):742–750
2. Hamidi N, Omidvari M, Meftahi M (2012) The effect of integrated management system on safety and productivity indices: case study; Iranian cement industries. Saf Sci 50(5):1180–1189
3. Zeng SX, Xie XM, Tam CM, Shen LY (2011) An empirical examination of benefits from implementing integrated management systems (IMS). Total Qual Manag 22(2):173–186
4. Simon A, Karapetrovic S, Casadesus M (2012) Evolution of integrated management systems in Spanish firms. J Clean Prod 23(1):8–19
5. Rebelo M, Santos G, Silva R (2014) Conception of a flexible integrator and lean model for integrated management systems. Total Qual Manag Bus Excell 25(5–6):683–701
6. Domingues P, Sampaio P, Arezes PM (2016) Integrated management systems assessment: a maturity model proposal. J Clean Prod 124:164–174
7. Ribeiro F, Santos G, Rebelo MF, Silva R (2017) Integrated management systems: trends for Portugal in the 2025 horizon. Procedia Manuf 13:1191–1198
8. Shevchenko A, Pagell M, Johnston D, Veltri A, Robson L (2018) Joint management systems for operations and safety: a routine-based perspective. J Clean Prod 194:635–644
9. Domingues P, Sampaio P, Arezes PM (2017) Management systems integration: survey results. Int J Qual Reliab Manag 34(8):1252–1294
10. Dragomir M, Popescu S, Neamțu C, Dragomir D, Bodi Ș (2017) Seeing the immaterial: a new instrument for evaluating integrated management systems' maturity. Sustainability 9 (9):1643
11. ISO/IEC (2012) Annex SL (normative) - Proposals for management system standards. http://www.kvaliteta.net/files/AnnexSL.pdf

Risk Analysis and Management

A Bayesian Network Analysis for Occupational Accidents of Mining Sector

Fatma Yaşlı[1(✉)] and Bersam Bolat[2]

[1] Transportational Vocational School, Anadolu University, Eskişehir, Turkey
fyasli@anadolu.edu.tr
[2] Management Engineering, Istanbul Technical University, Istanbul, Turkey
bolat@itu.edu.tr

Abstract. The mining sector is one of the most important raw material sources and wealth sources for countries. On the other hand, many work accidents occur during its activities due to the adverse working conditions. Research is being conducted to reduce the security risk factor, which is one of the most critical obstacles to the social sustainability of the mining industry. In this study, under-recorded mine accidents and injuries are handled, rather than the accidents of the roof falling and the explosions which are frequently considered in the literature. In this scope of the study, accidents and incidents that occur during the specified processes (support, face, loading and transportation activities) of an underground chrome mine are investigated. Expert judgments have been used since no past accidents are allowing statistical inferences. BN has been used to find out the issues about the safety risk by addressing the causal relationships between the events. OHS education, OHS inspection, employee attention and, rock and ground structure of the working area have been deduced as the root causes of the accidents which occur mostly during the labor-intensive processes. By using the updating ability of the BN, comprehensive sensitivity analysis has been performed with the new information related to root causes. According to different scenarios associated with the various states of the root causes, the results and the future suggestions are presented.

Keywords: Bayes · Network analysis · Safety risk · Occupational accidents
Underground mining · Expert judgements

1 Introduction

Most injuries and the fatal accidents are experiences in the mining sector. Even though occupational health and safety issues in mining have been profoundly handled in developed countries for the last two decades, accidents and incidents are still being experienced with an unacceptable frequency [1–3]. Due to the severe working conditions, in addition to the recorded accidents resulting in the temporary or permanent injuries, many unrecorded accidents or incidents also occur. For this reason, especially for the mining sector, performing the accident investigation by benefitting from the expert opinions and research on site is an entirely proper approach rather than just analyzing the accidents according to historical data.

N. M. Durakbasa and M. G. Gencyilmaz (Eds.): ISPR 2018, *Proceedings of the International Symposium for Production Research 2018*, pp. 781–799, 2019.
https://doi.org/10.1007/978-3-319-92267-6_63

It is necessary to take into account the decision problems of the mining sector by considering the safety issues related to adverse conditions that the industry has inherently. Therefore, examination the mining accidents and the essential issues related to occupational health and safety is an outstanding research area. It is required that the studies firstly about defining the accidents detailedly and exploring the reasons of the accidents [4] and then the extensive risk analysis methodology especially for the mining sector [5]. Regardless of the frequency or the consequences of the accidents, any undesirable event that may harm employees must be thoroughly examined.

The most important issue addressed in the accident risk analyses is the uncertainty due to the inadequate historical data. For defining the crucial factors and finding out their effects about the accidents, establishing the causal relationships between the events, and the probabilities of the events related to accidents are required. When determining the probabilities of the events without any database, expert judgments are used to remove the data uncertainty. Verbal expressions of uncertainty are that they mean different things to different people and sometimes mean different things to the same person in different contexts. Therefore, they will rarely serve any formal function within elicitation. It needs to be fully clarified the expert opinions by considering them with varying methods of elicitation [6]. Despite the difficulty of the reliable calculations of the probabilities without the historical numerical data, it is possible to get significant and valuable results about the accident risks with the judgments of the field experts. For this reason, there is a crucial triangular which include the elements of "risk, uncertainty and the expert judgments" in the risk investigations without having any statistical data. This situation is mentioned by the reference [7] as, "Studies show that in the future, analysts, engineers, and scientists will need to solve more complex problems with decisions made under conditions of limited resources, thus necessitating increased reliance on the proper treatment of uncertainty and the use of expert opinions".

In this study, the accidents and incident which occur during some specific processes of an underground chrome mine are analyzed using BN. The data gathering method for the analyse, literature, proposed methodology and the analyse with the results are presented respectively.

2 Bayesian Network and Expert Judgments in Risk Analysis

All accidents regardless its sector, occur based upon a number of risky events together in a certain order or simultaneously. BN is a very functional method for examining the relational integrity of the conditions that generate a risky situation and the triggering event that turns this risky situation into an accident. Accident investigation by revealing the events that constitute the risky situation and its relational conditions actually represent a risk analysis depending on the different scenario. To assign a risk score for the events and ranking them according to these scores should not accepted as a sufficient approach to analyze the risk. Since the occurrence of risky events depends on the conditional events that accompany them, risk can not be discussed without considering the conditional circumstances. Therefore, examining the conditional situations of an event that represent its causal relationships is the main focus [8].

BN are an essential tool to perform a risk analysis. BN use conditional probability tables and Bayesian theory to explain the relationships between the variables. Thus, it is possible to understand comprehensively all the events the relationships between them [9]. The Bayes Theorem is presented as follows:

$$P(A|B) = \frac{P(A)P(B|A)}{P(B)}$$

BN represent a graph of the essential determinants of the accidents with the unknown relationship between them [6]. The BN consists of an m-dimensional variable called as a node and a directed acyclic graph. The nodes represent the variables, and arrows indicate the direct causal links between the variables. Acyclic means that the variables are not allowed to influence in a loop [10, 11]. If we denote the variables in the network as $X_i = \{X_1, X_2, X_3, \ldots X_n\}$ and the a_i represents the probabilities of an antecedent variable or called parent variables of the X_i, the joint probability distribution of the variables is presented via the following equation:

$$P(X) = P(X_1, X_2, X_3, \ldots X_n) = \prod_{i=1}^{n} P(X_i|a_i)$$

The Fault Tree Analysis which is also effectively used in the investigation of the accidents and the BN Analysis have a common structure up to a certain level. However, the different modeling features of BN provide a much more effective method to be used in accident risk analysis, because the BN can take into account multiple situations of the variables and explicit conditional dependencies between the variables [12, 13].

The causal relationships between the events in the BN are explained by statistical information or inferential hypotheses [10]. Statistical methods can be used if sufficient and reliable data are available for the past. The presence of the historical data enables the use of learning BN and provides a graphical structure that best describes the conditional dependencies between the events [11]. However, when the historical data are not available or are inadequate, the uncertainty about the reasoning cases is resolved using expert judgments and subjective probability assessments. Subjective probability is based on personal knowledge and the experience and is a numerical measure of a state of knowledge, a degree of belief, a state of confidence [14]. Thus, the BN provides causal models which include the relationships and dependencies of the events by using expert judgments and allows for the quantification of uncertainties about events that reveal the concept of risk [8, 15].

Expert judgments are increasingly seen as a kind of scientific data, and various methods are still being developed to deal with them [16]. For the rational acceptance of the use of expert judgments in scientific research, many appropriate methodological constructs are needed. There are some principles for a rational consensus [17]. To achieve a rational consensus, also known as the classical model, the scientific principles of accountability, reproducibility, empirical control, neutrality, and fairness must be possessed [16]. Experimental control requires the presence of theories and measurements related to the subject matter, but it may not be possible sometimes to measure the considered quantities. When there is not enough data to make a reliable statistical analysis, the calculations of the probabilities are made using expert judgments. The experimental control of subjective probability assessments, which are determined by expert judgment, may only be possible by observing the future.

While eliciting the expert judgments, analysts need to consider the followings:

- Experts think a low likelihood for the events that have never been seen and a higher probability for the events that have been seen.
- There is much less overconfidence in estimating future events than in assessing the likelihood of the past events, as there is no such thing as apparent misleading memory [18].
- Being good at elicitation of the probability is a situation that can be learned and improved. Providing awareness of intuitiveness and bias judgments provides a better probability assessment.
- It is possible to suppress the cognitive bias by informing the experts about the reason why their inferences are needed and by applying different probability elicitation methods to the person [19].

There are many risk assessment studies that have been analyzed through BN constructed or evaluated by the expert judgments. Reference [9] analyzed the ship collision accidents and accident-causing factors with BN. In their study, Van der Gaag et al. established a BN with 70 variables to determine the define the properties of the cancer disease [20]. Reference [21] handled past eight ship accidents with BN, defined the accidents together with their causes and consequences, and reference [22] performed a safety risk analysis on a construction project with expert judgments. Reference [23] also conducted BN analysis through the expert judgments to model the parameters responsible for system error rates in the manufacturing industry.

3 Accident Analysis in Literature

There are many accident analyzes performed in different areas of the literature to prevent accidents by creating appropriate strategies. The common objective of these studies is to provide a comprehensive identification of accidents and to reveal the causes of accidents and to take precautions to eliminate these causes. When the studies that consider the accident investigation in the literature are examined, Reference [9], who have studied ship collisions have determined that the causes of accidents are technical, environmental and human factors. They pointed out that the rate of the human and organizational error is much higher in real life than other factors and that it is difficult to reveal it. Similarly, reference [21] proposed that the human error is four times more important than the technical performance on ship grounding accidents.

The study of reference [24] considered the risk of the roof falling in coal mines where extensive research has been carried out by investigations into mining sector accidents, and they have used expert judgments to determine the probability and the potential consequences of the roof falling, and have presented the significant factors. Reference [25] studied nine past unexpected coal and gas outbursts in Australia from a technical perspective. The investigations on the mining sector accidents often emphasize the elements of the human, education and organizational performance. Reference [1] addressed the role of organization and safety culture in providing the safe working areas in the mining industry. Reference [22] studied fatal accidents related to haul-trucks in the surface mining with fault tree analysis, also they identified the lack of trainings related to the inadequate maintenance and safety precautions, and the insufficient precautions as the root causes of accidents. Reference [26] handled the occupational accidents in the mining but only which were reported to government agency and they revealed the root causes of the accidents. They found the most critical accident cause is the body movement with physical effort or overexertion in the underground mines. Reference [27], in their study related to the past accidents, mentioned that especially the people and the machines have the most significant share in accident factors. Reference [28] examined two underground coal mine past accidents operating in India, to reveal the behavioral factors affecting the mining accident. It is seen that the experience has not a statistical meaning effect for the accidents and the likelihood of accidental exposure increases as increasing the age of the worker, on the other hand, the morale of the worker, is an essential factor for acquiring the accidents and the job dissatisfaction rises the injuries. Also it has been suggested to get rid of the old belief that only engineering control and legal inspections are sufficient for safe environment in the mines. Reference [29] examined the past mining accidents in Spain and they mentioned that the cause of fracture, slip, fall or collapse causing a type of accident of hit or collision with a moving object is the most common in the underground mines. The most prominent root causes of the accidents are electrical problems, slip, falling, loss of machine control, incorrect motion and physical over-exertion. The study of reference [30], 102 well-known and recorded past incidents between 1985 and 2015 of the ten underground and surface mine were reviewed in India. They have extensively addressed the human factor in the accidents and found that the most common errors were attention, decision and technical mistakes and inaccurate state and risk assessments. Just as in the study of reference [30], attention, assessment and decision-making mistakes are also defined as the causes of the accidents in our study. However, it has been concluded that the safety education is the primary factor in the failure of evaluation and decision making mistakes.

In this study, unlike the literature, an underground chrome mine's activities have been investigated on site. According to the working conditions, especially underground mines have significant difficulties inherently. e.g., rock mass tremors, geological conditions which result in many minor or major accidents [24]. Moreover, since the human factor plays an active role at the managerial and operational level, the managerial approach to risk cannot always be maintained to a certain standard. For this reason, OHS management variability and adverse mining conditions together with the issue of the employee's attention are included in the study. The physical exertion or excessive effort, fall, or failure to hit an object are mentioned in the reference [29], and they defined them as root causes of the accidents. But in this study, they are identified as being accidents and are being examined under the natural conditions of the mine, safety management, and the human factor, in this study.

4 Methodology of the Study

An underground chrome mine's accidents and their causes using BN have been considered in this study. Figure 1 shows the methodology of the study. For the analysis, the activities for loading of the loose materials after the drilling and the blasting activities-face, supporting, loading them to transport and the transportation them to surface in underground chromium mine are discussed. There is two information for the configuration of the BN: field experts and statistical data. If the historical data is not sufficient to identify the relationships between the variables in the BN, to construct the network and then determine the necessary parameters, the experts are used. Because, when attempting to determine the statistic information with insufficient data, it causes incorrect evaluations and configurations [11]. For structuring of the BN, analysts and a group of experts need to work together [15]. In the first stage, a three-person expert group consisting of the plant manager, chief engineer and the chief worker of the mining company are established to create the network and determine the parameters of the network. The selected experts are informed about the subject of elicitation and the importance of the research, before the process of the probability elicitation and the awareness of prejudice and biased expressions are provided [32].

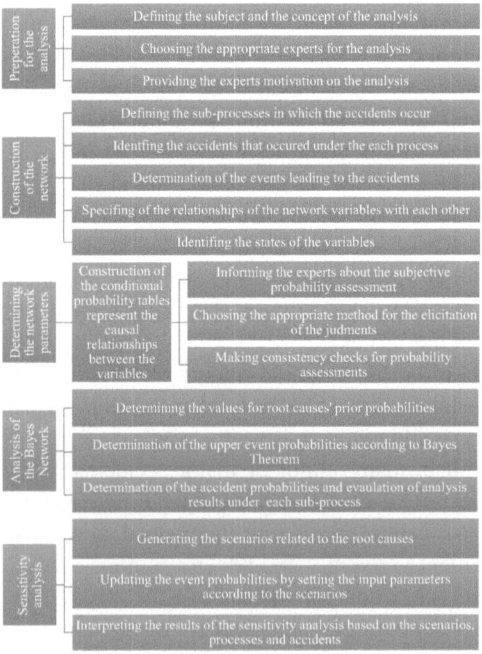

Fig. 1. Methodology of the study

Processes to investigate are determined as support, face, loading and transportation processes in the underground chrome mine. For the study, the limited number of past accident reports of the mining company, literature, and expert opinions are used. Occupational accidents and injuries that occurred during these processes are defined. Establishment of the BN has been ensured by determining the root causes and event composition of the events which cause the accident. The relationships between the variables have been extensively researched, and the links between events have been revealed through repetitive interviews with the experts and the observations on site. The proposed BN can be seen in Fig. 2. The Lucidchart drawing program was used to construct the network.

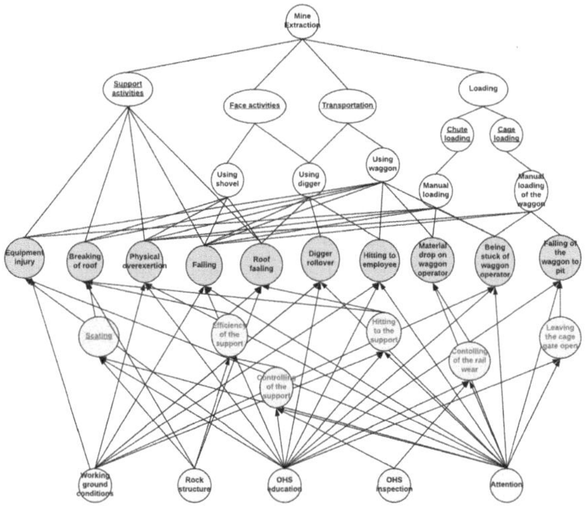

Fig. 2. Bayes network of the underground mining accidents

The most important stage after the establishment of BN is the determination of the conditional probability tables which reflects the relations between events. Possible sources of information for probability calculations are records, literature studies or field experts' experience. While past data and literature are considered as reliable sources, the probability data obtained from expert judgments are being skeptical concerning their scientific relevance. However, it is possible that expert judgments can also be regarded as reliable for probabilistic inferences [6, 33–35]. But it is essential to strengthen the rationality of the elicitated data. Various methods have been developed for the elicitation of the probability from expert judgments, and developments are still needed on the subject. Reference [33] mentioned that the subjective opinions of the experts could be used to reflect uncertainties about the unknown parameters when the historical data are not available or when the quantity of the data is not sufficient. The definition of probabilities for the occurrence of a random phenomenon by experts is elicitation of the subjective probability.

It is difficult to elicitation the probability distribution because experts often do not have the experience of presenting probability judgments and are unfamiliar with the elicitation assessment process [33]. Two methods are used to determine the probability distributions for the elicitation of the expert judgments [36, 37]. These are the determining the probability for the fixed variable value (FV method - P method) and the determining the variable value for the fixed probability (FP method - V method). Spetzler and von Holstein also present a method (PV) in which both methods are used together [37]. These methods are described in reference [36] as querying method, while reference [37] express as encoding methods.

It has been decided to determine the probabilities for occurring the accidents or not in our study. Therefore, the constant variable method is used. In the fixed variable method, the constant value of the variable is used and the associated cumulative probability is queried. The fixed value variable and the probability value believed to correspond are adjusted until it is consistent with the other deduced values. The probability wheel and the probability scale tool methods for elicitations of the probability have been tried for elicitation process. It is thought that there are not enough and appropriate methods to make many and effective inferences. In particular, the probability wheel method fails at very low and very high probability estimates and at the same time requires long working times for multiple probability deductions. The reason why we do not use a probability scale having scales such as ever, possible, low, moderate, high, or the likelihood values on it, is that it has not represented multiple probability options to make the evaluations consistent when making probability assignments. Thus, it has been decided to generate the conditional probability tables according to the direct probability deduction method. In our study, since we keep the variable state numbers to two, we made elicitation of $P(a_i)$ for the probability of the ai event, and elicitation of $P(X_i|a_i)$ for the event of X_i event but in condition with the occurring a_i event. The inferences made throughout the study are compared continuously and tried to verify the consistency of the values. Failure to carefully evaluate the relationships between variables from expert judgments may result in poor quality graphic models [38]. For this reason, maintaining the consistency of the subjective probability values is the most crucial phase of the study.

The probabilities of the root causes (prior probabilities) have been determined following the difficult probabilistic inference to determine the conditional probability table parameters that indicate the relationships between variables. Then the probabilities of these variables are calculated according to Bayes' theorem. The probability values of the underground chrome mine-specific accidents are thus calculated, according to the assigned prior probabilities and causal associations. Various scenarios are generated according to different situations of the root causes, and network sensitivity analysis is made, and the results are presented.

5 Analysis of the Bayes Network

The risk is a variation of every sector, firm, process, even every employee. In organizations where risk elements are present, the performing of the relative assessments of these elements provides essential findings to develop appropriate strategies to mitigate them. In this study, BN analysis is performed for the occupational accidents and injuries occurred in an underground chrome mine.

When the accidents in the underground chrome mine are examined, it is observed that they are mostly involved in falling, equipment damage, falling of objects from the ceilings and walls, using wagons, shovels, and shovels. The answer to the question is "why these accidents happened?" is the direct cause of the event. The answer to the question is "why these accidents were not prevented?" is the fundamental cause of the event [1]. Therefore, an investigation has been conducted to find the answers to the questions about why these accidents, which frequently occur in activities that are subject to continuity, can be carried out and what can be done to prevent them. Although these accidents are not fatal accidents such as roof falling and explosions in the coal mine, no matter what the violence of it, an accident that would cause any employee to be injured is never acceptable. The motivation of this work is to prevent the accidents by understanding the extent to causes of the accidents and distinguishing the effects of the management, geological and the human factors.

5.1 Specifying of the Accidents According to Processes

Processes in the study are "support activities, face activities, loading for transportation and the transportation process" in the underground mine. Detailed information on these processes has been collected according to the review of past accident reports, the examinations made at the mine and the opinions of experts, and the accidents that may occur according to the processes are given in Table 1.

Table 1. Accidents under the processes

Process	Accidents	
Support activities	Equipment injury Breaking of roof/falling of objects from the ceilings and walls Physical overexertion Falling Roof falling	
Face activities	Using shovel Breaking of roof Physical verexertion Falling Roof falling	Using digger Falling Roof falling Digger rollover Hitting to employee
Transportation	Using digger Falling Roof falling Digger rollover Hitting to employee	Using waggon Equipment injury Breaking of roof Physical overexertion Falling Hitting to employee Material drop on waggon operator Being stuck of the waggon operator
Loading for transportation	Chute-Manual loading Equipment injury Physical overexertion Falling Material drop on waggon operator Being stuck of the waggon operator	Cage-Manual loading of the waggon Physical overexertion Falling Being stuck of the waggon operator Falling of the waggon to pit

5.2 Root Causes of the Accidents

The root causes of the underground chrome mine's processes are identified as "working ground conditions, rock structure, OHS education, and OHS inspection and the employee attention" as can be understood from the BN established in Fig. 2 after repetitive interviews and discussions with the mining experts. To the best of our knowledge, there has been no previous study dealing with these accidents together with their processes and these root causes in the literature. Defined root causes can be explained as follows:

5.2.1 Working Ground Conditions

The mines are subject to severe working conditions. The working areas usually have watery, muddy, dark, and most importantly, uneven floors. This conditions can be seen as a root cause for the accidents because when the employees work in the slightest carelessness or not to behave appropriately, they can lose their balance easily and experience a wide variety of injuries. After observations made in the mine and repetitive interviews with experts, it is understood that this root cause is considered to be one of the most natural and normal working conditions of the underground mine. It has been observed that the working ground shows very different characteristics according to the areas and that all employees are firstly trying to stand in the balance due to the uneven ground condition, no matter what they are doing. For this reason, it has been decided that the working ground conditions should be taken under two states, namely straight-rough.

5.2.2 Rock Structure

The mines are generally being operated in faulted rock. The fact that the rock is solid-competent, it eliminates many hazards such as roof falling or falling of objects from the ceilings and walls, and as the opposite, the fact that the rock has a broken-faulted structure triggers the accidents. It also has the effect of reducing the effectiveness of the supports of mine, which has the most responsibility for the underground mine safety and the continuation of the extraction processes. For this reason, this element, which significantly affects the safety of working conditions, is included in the BN analysis with two states: competent rock-faulted rock.

5.2.3 OHS Education

Occupational health and safety education is essential to prevent accidents caused by the theoretical and practical insufficiency of the work that the employee needs. As many researchers in the literature have pointed out, safety education is the reason for most of the accidents happening in the processes which are managed by people. Because employees do not obey the procedures and regulations established to prevent accidents. For example, the worker creates a different option for the way of his work to perform his job easier, practical or comfortable and misjudges the options of these optional ways of the work from the safety side. This situation is defined by reference [39] as a Desirability of options bias. Employees can learn the safety attitude and safety knowledge and can be trained according to the practical side. Employees should be made aware of the dangerous situations, the necessities of using the personal protective equipment, and any accidents they may experience [40]. Accidents can be prevented, and differences between safe and unsafe environments and behaviors can be evident and defined. The conditions and rules of safe conduct are independent of the employee's personality; they are mandatory standards. In addition to all this, the employee must know the difference between being brave and being in safe conditions very well [40]. If the employee does not receive training at the intensity he needs, the management must know the fact that this employee may continue to act in unsafe behaviors. The lack of training of the employee should be eliminated. Therefore, the root cause-OHS education is included in the study, and the variable states of it: adequate and inadequate (less than adequate).

5.2.4 OHS Inspection

While the underground chrome mine accidents are being examined, it is observed that these accidents occurred during the continuous or multiple repetitive activities of the employees. It is determined that the mining engineers at the inspection mission take an active role in controlling the effectiveness of the supporting and the rails in which they are carried. It is seen that these intermediate events are also carried out by the employees, but the mine managers control them as well. It has been accepted that OHS inspection is not a cause for the accidents such as falling or being injured by the equipment or physical exertion and that these accidents are not closely related to the inspection level. Because while inspections are carried out within a specified period, employees continue their activities throughout their shift with the risk of accidents handled in work. Therefore, the states of the variable of the OHS inspection are adequate and less than adequate for the above two intermediate events.

5.2.5 Attention of the Employee

Reference [24] stated that mining accidents are both independent of human activity and partially dependent on the employee, and on the other hand sometimes only reliant on the employee. However, since most of the underground chrome mining operations are labor-intensive activities, it can be said that all of the reasons for the accident are due to people except geological reasons. Planning, inspection, geological conditions, etc. even if all the negative preconceptions about it go away, it is possible for the worker to lose his carelessness and to experience an accident. The basic premise of human performance is that everyone is willing to perform adequately and tries to fulfill his/her expectations. However, mistakes happen, and this can not be avoided [1]. It is observed that the reasons for the unfavorable behavior of employees are either lack of OHS education as mentioned under the related root cause or inevitable causes as also indicated by reference [1]. After interviews with experts and employees, the issue of the inattention is defined as the person is acting erroneously, which can not be avoided due to a momentary unconscious behavior. Thus, unlike the references [40, 41], in our study unsafe behavior may be regarded as a completely random variable. It is independent of safety knowledge, organizational culture, safety climate and managerial issues and considered as just a behavior related to the employee's attention due to a momentary unconsciousness. The management-related factors that play a role in the inaccurate behavior of people are addressed under the OHS education factors. The root cause of the employee's attention that occurs due to instant attention loss, even if all the negative preconceptions about the education, inspection, geological conditions, etc., be removed, this attention issue still causes the workers to make mistakes. Reaching the root cause of the employee inattention behavior under the stated conditions may require the leading to injury a perfect human that does not make a mistake. It is suggested that this factor should be analyzed with the experts in psychology to conduct an unsafe behavior sampling study related to human factors. Because, whatever the case may be, the employee can still be an unsafe act. Also the factors such as age, experience, employment contact, day-hour, experience, working place, psychological status of the employee creates variability in the level of the employee's attentiveness [26]. For this reason, the issue of the "employee attention" which is unmanageable easily is included as a root cause of the network. This root cause is considered in two different states, as well: adequate and less than adequate.

The root causes are defined using the literature review and interviews with experts. Factors such as organizational culture, safety climate, and managerial issues are assumed to be reflected in the root causes of OHS education and OHS inspection. We combined the variables of the OHS education and OHS inspection into "management conditions" to enable more effective interpretation in sensitivity analysis. Because of the quality of the safety education and the inspections depend on the management of the company. Already these activities are known as safety management practices and are directly related to the accident. Also, underground mining working ground conditions and rock structure are also referred to as "geological conditions" in this study. Geological conditions are considered external factors that can not be changed, and they are accepted as nature of the underground mine. The "employee attention" factor, which is independent of the OHS training factor and controls the employee's attention level, is included in the study. It is considered as a "human factor" in sensitivity analysis. The variable of the unsafe behavior in the literature has not been included in the study, because it is accepted to be influenced by the safety attitude, safety knowledge and the supportive environment [40]. Because the "employee attention" and the "OHS education" are thought of as the root causes of the unsafe behaviors factors. Thus, scenarios based on management conditions, geographical conditions and human factors are used to conduct BN sensitivity analysis, which revealed the root causes of accidental probabilities and the causal effects of root causes.

6 Results

Scenarios are established according to management conditions, geological conditions and human factors, and the states of these root causes belonging to each scenario are given in Table 2.

The sensitivity analysis allows for the mine managers to be aware of the potential accidents and obtain information about the risky situations. Each firm can determine its own thresholds for acceptable and unacceptable risk levels for different activities. Thus, where the level of risk is unacceptable, it may take appropriate measures to mitigate the risk.

The prior probabilities of the root causes under each scenario are required to determine the accident probabilities according to the different scenarios. The different

Table 2. Scenarios with root causes

	Geological conditions			Human factor			Management conditions		
Scenarios	Best		Worst	Best		Worst	Best		Worst
Best scenario	√			√			√		
Best-g_worst/Best scenario under the worst geological conditions			√	√			√		
Best-h_worst/Best scenario under the worst condition of the human factor	√					√	√		
Best-m_worst/Best scenario under the worst management conditions	√			√					√
Worst scenario			√			√			√
Worst-g_best/Worst scenario under the best geological conditions	√					√			√
Worst-h_best/Worst scenario under the best human factor's conditions			√	√					√
Worst-m_best/Worst scenario under the best management conditions			√			√	√		
Average case scenario for all factors		√			√			√	
Average-g_best/Average case scenario under the best geological conditions	√				√			√	
Average-g_worst/Average case scenario under the worst geological conditions			√		√			√	
Average-h_best/Average case scenario under the best human factor		√		√				√	
Average-h_worst/Average case scenario under the worst human factor		√				√		√	
Average-m_best/Average case scenario under the best management conditions		√			√		√		
Average-m_worst/Average case scenario under the worst management conditions		√			√				√

worst-best and average values of the root causes according to the scenarios, and the prior probabilities of the root causes are presented in Tables 2 and 3, respectively.

When the probabilities of accidents calculated according to scenarios are examined in Table 4, the probability that is expected for all occupational accidents is 0.01 even under the best scenario where all the conditions are the most positive. Since these accidents are not rare events, it is not possible to reach that probability value easily for underground mines. It appears that there are huge differences between the probability values of the best and the worst scenarios. The accident, which is expected to happen the most, is seen as falling of objects from the ceilings and walls for almost every scenario. This undesired event is already known to be the most common problems of underground mines. The most critical elements of this accident, which may cause a wide variety of injuries, are the fact that the rock is faulted, the efficiency of the mine

Table 3. States of the root causes and the prior probabilities according to scenarios

Scenarios	Working ground conditions	p	Rock structure	p	OHS education	p	OHS inspection	p	Attention	p
Best	Straight	1	Competent rock	1	Adequate	1	Adequate	1	Attentive	1
Best - g_worst	Rough	0	Faulted rock	0	Adequate	1	Adequate	1	Attentive	1
Best - h_worst	Straight	1	Competent rock	1	Adequate	1	Adequate	1	Inattentive	0
Best - m_worst	Straight	1	Competent rock	1	Inadequate	0	Inadequate	0	Attentive	1
Worst	Rough	0	Faulted rock	0	Inadequate	0	Inadequate	0	Inattentive	0
Worst- g_best	Straight	1	Competent rock	1	Inadequate	0	Inadequate	0	Inattentive	0
Worst- h_best	Rough	0	Faulted rock	0	Inadequate	0	Inadequate	0	Attentive	1
Worst- m_best	Rough	0	Faulted rock	0	Adequate	1	Adequate	1	Inattentive	0
Average	Straight	0.5	Competent rock	0.5	Adequate	0.5	Adequate	0.5	Attentive	0.5
Average- g_best	Straight	1	Competent rock	1	Adequate	0.5	Adequate	0.5	Attentive	0.5
Average- g_worst	Rough	0	Faulted rock	0	Adequate	0.5	Adequate	0.5	Attentive	0.5
Average- h_best	Straight	0.5	Competent rock	0.5	Adequate	0.5	Adequate	0.5	Attentive	1
Average- h_worst	Straight	0.5	Competent rock	0.5	Adequate	0.5	Adequate	0.5	Inattentive	0
Average- m_best	Straight	0.5	Competent rock	0.5	Adequate	1	Adequate	1	Attentive	0.5
Average- m_worst	Straight	0.5	Competent rock	0.5	Inadequate	0	Inadequate	0	Attentive	0.5

supports is not ensured, and the scatting process is not performed well enough by the workers. In particular, an employee working in an area with geologically adverse conditions is still at high risk of breaking roof, even if he does not have any unsafe behavior. Even in such areas where the managerial and human factors together have the best conditions, the likelihood of a breaking of the roof (falling of objects from the ceilings and walls) is still high at 35%. It is necessary to focus on the efficiency of the support and scatting activities to prevent this accident. The accident of the roof falling is also an accident where the influence of the support is essential. Since chrome is a metallic mine, it offers more durable structures against collapse, but the mine safety still depends on the rock structure and the supporting efficiencies made by the workers. When the state changes of the root causes are examined, it is possible to see the effect of the factor of "employee attention". Activities in which is sensitive to the attention factor are routine work activities, and these are not easy to prevent with education.

Table 4. Accident probabilities according to scenarios

Scenarios	Equi. injury	Breaking of roof	Physic overex	Falling	Roof falling	Digger rollover	Hitting to emp.	Mat. drop on wagon op.	Being stuck of wagon op.	Falling of wagon to pit
Best	0.01	0.015	0.01	0.01	0.010	0.01	0.01	0.013	0.01	0.012
Best - g_worst	0.05	**0.349**	0.1	0.15	0.041	0.02	0.05	0.013	0.05	0.012
Best - h_worst	0.15	0.024	0.2	0.25	0.011	0.05	0.05	0.14	0.01	0.068
Best - m_worst	0.25	0.051	0.15	0.05	0.011	0.1	0.1	0.097	0.01	0.088
Worst	0.5	0.744	0.5	0.5	0.315	0.3	0.2	0.275	0.2	0.255
Worst-g_best	**0.45**	0.081	**0.4**	0.3	0.012	0.15	**0.15**	**0.275**	0.15	**0.255**
Worst- h_best	0.3	**0.590**	0.2	0.25	**0.240**	0.2	**0.15**	0.097	0.1	0.088
Worst- m_best	0.25	0.491	0.3	**0.4**	0.067	0.1	0.1	0.14	**0.2**	0.068
Average	0.245	0.252	0.233	0.239	0.055	0.116	0.101	0.123	0.091	0.087
Aver- g_best	0.215	0.042	0.19	0.153	0.011	0.078	0.078	0.123	0.045	0.087
Aver- g_worst	0.275	0.541	0.275	0.325	0.159	0.155	0.125	0.123	0.138	0.087
Aver- h_best	0.153	0.214	0.115	0.115	0.046	0.083	0.078	0.054	0.043	0.040
Aver- h_worst	0.338	0.290	0.35	0.363	0.063	0.15	0.125	0.208	0.14	0.147
Aver- m_best	0.115	0.175	0.153	0.203	0.024	0.045	0.053	0.069	0.068	0.035
Aver -m_worst	0.375	0.334	0.313	0.275	0.089	**0.188**	**0.15**	0.178	0.115	0.164

If inattention is combined with the adverse condition of the ground, especially the accidents of the falls and equipment injuries are inevitable. For example, due to unfavorable ground conditions and inattention during an employee's supporting activities. He self-harms with an axe or falls from the chute. These are the unfortunate accident of the underground mines. Despite all the administrative precautions, material dropping, being stuck of the wagon operator and falling off the wagon to pit also depend on the level of having attention when performing the job. If a mining company does not place the necessary emphasis on safety management elements, employees are highly vulnerable to undesired accidents. If employees are not given sufficient OHS education, or if employees do not prefer to act following the rules, the most anticipated accident is equipment damage and physical overexertions. Therefore, if the actions of the employees are different from the safety directives, they should be informed about the accidents via the helpful visuals and animations during the OHS education.

Sensitivity analysis results show that mine work safety cannot be provided solely by management. For safety reason related to the geological conditions, if the carelessness of the employee is also added to this issue, the frequencies of the accidents are observed as quite high. Measures should be taken to reduce the carelessness of the workers. Assuming the underground chrome mine's basic conditions related to this study in an average state, the accidents of the breaking of the roof (falling of objects from the ceilings and walls, equipment injuries and fall accidents expected to occur the most are expected with the 0.25 probability. Already the current situation of the mines is under this scenario. It is possible to assume that these accidents will occur every four shifts and the experts confirm this ratio. Labor-Intensive processes should be maintained as mechanically or using semi-automation technological equipment in the underground mines to mitigate the inattention-human factor element, and the highest administrative safety elements should support the processes.

According to the results of the sensitivity analysis, mine managers should inform the employees about the risk factors according to processes and the accidents. The situation between the best and worst cases of all root cause's conditions is quite striking. It is understood through the BN analysis that employees are most likely to be injured in chrome underground mining, especially during the transport and the face activities.

7 Conclusion

Underground mining operations are generally carried out with labor. The mining processes are carried out by all employees under severe and risky conditions. Managers are under great responsibility both from the social and economic standpoint due to the problems of occupational health and safety arising from these challenging conditions due to the mining industry has one of the highest accident rates. It is essential to carry out comprehensive risk analyzes against any unwanted events. In this study, managerial knowledge and experience are used, and both recorded and unrecorded accidents are considered to increase the efficiency of the analysis.

In the study, an underground chrome mine is investigated by analysis of BN and all accidents which are likely to be harmful to people are defined. For developing the methodology, the literature has been extensively researched. BN is one of the techniques used to model complex systems of uncertainty. In the BN studies conducted with expert opinions, the analyst's influence can never be removed, and it is always necessary to be aware that the results will reflect both the expert and the analyst's knowledge and views [42].

For the construction of the network, and definition of the root causes and intermediate events expert judgments are used. For the judgments especially about the probabilities, repetitive interviews with three experts are benefitted. In contrast to the literature, with the proposed methodology, multiple accidents are addressed in a single BN. When the BN analysis results are examined, ground working conditions, rock structure, OHS education, OHS inspection and employee attention cause the accidents to occur. Findings of the result of the analysis and the effects of root causes are compatible with the literature and the practice.

Due to the geological conditions of the mine cannot be interfered, the OHS education elements of the mine need to be improved. To that end, the critical points for safety risk in each company must be defined, arranged, managed, checked and corrected. In this study, the attention of the employees, which is different from the literature, seems to be one of the main reasons for the accidents. Because sometimes there is no lack of education, but the employee continues of unsafe behavior in the underground mine. This finding is determined by the interviews with the mine experts and the observations at the mine site. There can be many reasons why employees may be in careless behavior, regardless of the company's safety culture. Work stress caused by unfavorable conditions, unfavorable events before work or excessive workload can cause the employee to be careless. For short-term solutions, employees who have the least accident experience caused the issue of attention should be assigned to the activities which most need attention, and paying them more for these sensitive works. All management and employees need to be aware of the hazards and be aware of the necessary precautions or providing the necessary attention. Regulations and the bases should also be arranged following the results of this kind of risk analysis for people working in this sector so that occupational accidents do not cause further damage to them. Most of the processes of underground chrome mining are performed by labor-intensive, the safety of the process depends on the employee's performance on a large scale. As a long-term solution, it is advisable to ensure that processes should be performed as technologically as possible. It is recommended that future studies be done to reveal the effect of the technology related to using equipment has on the accidents.

The managers can use the proposed methodology as a risk investigation tool to see the undesired events closely in a firm. The variables of OHS-education and OHS inspection are not the latent variables of the network. Related standards can be identified for the values of them, and they can be evaluated as a root cause that can be measured by firms or sectors over these standards. It is thus possible that the methodology is used by firms or industries to determine the risk profiles. It is also suggested that the values of these root causes should be defined quantitatively and investigated their variability for the future studies.

Acknowledgement. We would like to offer our thanks to three anonymous field experts for the wealth of the information they freely provided without which this research would not have been possible.

References

1. Komljenovic D, Loiselle G, Kumral M (2017) Organization: a new focus on mine safety improvement in a complex operational and business environment. Int J Min Sci Technol 27(4):617–625
2. Samantra C, Datta S, Mahapatra SS (2016) Analysis of occupational health hazards and associated risks in fuzzy environment: a case research in an Indian underground coal mine. Int J Inj Control Saf Promot 24(3):311–327
3. Boudreau J et al (2014) Social choice violations in rank sum scoring: a formalization of conditions and corrective probability computations. Math Soc Sci 71:20–29
4. Khanzode VV, Maiti J, Ray PK (2011) A methodology for evaluation and monitoring of recurring hazards in underground coal mining. Saf Sci 49(8–9):1172–1179
5. Nawrocki TL, Jonek-Kowalska I (2016) Assessing operational risk in coal mining enterprises–internal, industrial and international perspectives. Resour Policy 48:50–67
6. O'Hagan A et al (2006) Uncertain judgements: eliciting experts' probabilities. Wiley, New York
7. Ayyub BM (2001) Elicitation of expert opinions for uncertainty and risks. CRC Press, Boca Raton
8. Fenton N, Neil M (2012) Risk assessment and decision analysis with Bayesian networks. CRC Press, Boca Raton
9. Hänninen M, Kujala P (2012) Influences of variables on ship collision probability in a Bayesian belief network model. Reliab Eng Syst Saf 102:27–40
10. Morrison RG (2005) Thinking in working memory. In: Holyoak KJ, Morrison RG (eds) Cambridge handbook of thinking and reasoning. Cambridge University Press, New York, pp 457–473
11. Laitila P (2013). Improving the use of ranked nodes in elicitation of conditional probabilities for Bayesian networks. M.Sc. thesis, Aalto University, Espoo, Finland
12. Khakzad N, Khan F, Amyotte P (2011) Safety analysis in process facilities: comparison of fault tree and Bayesian network approaches. Reliab Eng Syst Saf 96(8):925–932
13. Hamza Z, Abdallah T (2015) Mapping fault tree into Bayesian network in safety analysis of process system. In: 2015 4th International conference on electrical engineering (ICEE). IEEE, pp 1–5
14. Kaplan S, Garrick BJ (1981) On the quantitative definition of risk. Risk Anal 1(1):11–27
15. Langseth H, Portinale L (2007) Bayesian networks in reliability. Reliab Eng Syst Saf 92(1):92–108
16. Cooke RM, Goosens LH (2000) Procedures guide for structural expert judgement in accident consequence modelling. Radiat Prot Dosim 90(3):303–309
17. Cooke R (1991) Experts in uncertainty: opinion and subjective probability in science. Oxford University Press, New York
18. Carlson BW (1993) The accuracy of future forecasts and past judgments. Organ Behav Hum Decis Process 54(2):245–276
19. Renooij S (2001) Probability elicitation for belief networks: issues to consider. Knowl Eng Rev 16(3):255–269

20. van der Gaag LC et al (2002) Probabilities for a probabilistic network: a case study in oesophageal cancer. Artif Intell Med 25(2):123–148
21. Veritas N (2003) Risk management in marine-and subsea operations. Det Norske Veritas, Høvik
22. Zhang M, Kecojevic V, Komljenovic D (2014) Investigation of haul truck-related fatal accidents in surface mining using fault tree analysis. Saf Sci 65:106–117
23. Jones B et al (2010) The use of Bayesian network modelling for maintenance planning in a manufacturing industry. Reliab Eng Syst Saf 95(3):267–277
24. Prusek S et al (2017) Assessment of roof fall risk in longwall coal mines. Int J Min Reclam Environ 31(8):558–574
25. Black DJ (2017) Investigations into the identification and control of outburst risk in Australian underground coal mines. Int J Min Sci Technol 27(5):749–753
26. Sanmiquel L et al (2018) Analysis of occupational accidents in underground and surface mining in Spain using data-mining techniques. Int J Environ Res Pub Health 15(3):462
27. Liu Q et al (2016) Accident-causing mechanism in coal mines based on hazards and polarized management. Saf Sci 85:276–281
28. Paul PS, Maiti J (2007) The role of behavioral factors on safety management in underground mines. Saf Sci 45(4):449–471
29. Sanmiquel L, Rossell JM, Vintró C (2015) Study of Spanish mining accidents using data mining techniques. Saf Sci 75:49–55
30. Verma S, Chaudhari S (2017) Safety of workers in Indian mines: study, analysis, and prediction. Saf Health Work 8(3):267–275
31. Clemen RT, Winkler RL (1999) Combining probability distributions from experts in risk analysis. Risk Anal 19(2):187–203
32. Shephard GG, Kirkwood CW (1994) Managing the judgmental probability elicitation process: a case study of analyst/manager interaction. IEEE Trans Eng Manage 41(4): 414–425
33. Morris DE, Oakley JE, Crowe JA (2014) A web-based tool for eliciting probability distributions from experts. Environ Model Softw 52:1–4
34. Lewis MB (1979) A method for simulating fusion reactor radiation damage using triple ion beams. IEEE Trans Nucl Sci 26(1):1320–1322
35. Dewispelare AR, Herren LT, Clemen RT (1995) The use of probability elicitation in the high-level nuclear waste regulation program. Int J Forecast 11(1):5–24
36. Abbas AE et al (2008) A comparison of two probability encoding methods: fixed probability vs. fixed variable values. Decis Anal 5(4):190–202
37. Spetzler CS, Stael von Holstein CAS (1975) Exceptional paper–Probability encoding in decision analysis. Manag Sci 22(3):340–358
38. Wang H (2007) Building Bayesian networks: elicitation, evaluation, and learning. PhD Thesis. University of Pittsburgh, USA
39. Montibeller G, Winterfeldt D (2015) Cognitive and motivational biases in decision and risk analysis. Risk Anal 35(7):1230–1251
40. Mohammadfam I et al (2017) Constructing a Bayesian network model for improving safety behavior of employees at workplaces. Appl Ergon 58:35–47
41. Fleming M, Lardner R (2002) Strategies to promote safe behaviour as part of a health and safety management system. HSE Books, England
42. Hänninen M, Banda OAV, Kujala P (2014) Bayesian network model of maritime safety management. Expert Syst Appl 41(17):7837–7846

Evaluation of Spatial Risks of Nursing Homes by Fuzzy Risk Analysis Method

Semra Boran[1(✉)], Didem Yılmaz[2], Zerrin Funda Ürük[3], and Seda Hatice Gökler[1]

[1] Department of Industrial Engineering, Sakarya University, Serdivan, Turkey
{boran, shgokler}@sakarya.edu.tr

[2] Department of Industrial Engineering, Istanbul Gelişim University, Istanbul, Turkey
dyilmaz@gelisim.edu.tr

[3] Department of Interior Architecture and Environmental Design, Istanbul Gelişim University, Istanbul, Turkey
zfuruk@gelisim.edu.tr

Abstract. The elderly population in the world and Turkey is increasing rapidly. The proportion of elderly people in Turkey that has remained below 5% until the 1990s has reached its highest point in the past 15 years with a significant increase. This leads to an increase in the need for nursing homes. However, for those living in nursing homes, spatial risks can lead to serious hazards such as injury and death. To increase the quality of life for the elderly, these hazards should be assessed, and corrective actions should be taken to eliminate them. The Fine-Kinney risk analysis method is one of the most commonly used methods for risk assessment. In this method, the hazards are evaluated according to the probability, frequency and severity factors. However, the value assignments of these risk factors are often based on expert opinion. Experts often identify risk values in the context of incomplete information and uncertainty. Also, experts tend to use verbal expressions rather than numeric value assignments. In this study, a fuzzy Fine-Kinney risk analysis approach was developed to take advantage of fuzzy logic to remove this shortcoming. This developed approach was applied to the nursing homes located in Istanbul and the areas that need to be improved with high risk were determined first. The order of risk prioritization obtained with the fuzzy Fine-Kinney approach was compared with that of the conventional Fine-Kinney method and it was determined that the fuzzy Fine-Kinney approach gave better results.

Keywords: Fuzzy Fine-Kinney risk analysis approach · Fuzzy logic
Nursing home · Spatial risks · Interior design

1 Introduction

Nowadays the elderly population growth is seen especially in the developed countries. The proportion of elderly population in these countries is expected to rise from 8% to 19% in 2050. Similarly, in Turkey, the proportion of the elderly population, which has remained below 5% until the 1990s, has shown a significant increase in the last 15

N. M. Durakbasa and M. G. Gencyilmaz (Eds.): ISPR 2018, *Proceedings of the International Symposium for Production Research 2018*, pp. 800–810, 2019.
https://doi.org/10.1007/978-3-319-92267-6_64

years and has reached the highest point of Turkey's history. The proportion of population aged 65 and over was 3.9% in the 1935 census, 4.3% in 1990 and reached 8.3% in 2016. It is estimated that in 2023 this ratio will increase rapidly to 10.2% and in the year 2050 it will find 217.5% [1]. This also shows that Turkey will be among the countries with a high elderly population. Increasing elderly population reveals elderly care problems. Demand for nursing homes also increases in proportion to the population. The number of nursing homes in Turkey in September 2016 was nominal 367 and their occupancy rate has reached to 80%. Elderly care homes are residential social service establishments established to protect and care for the elderly people aged 65 and over in a peaceful environment and to meet their social and psychological needs. However, the fact that the spatial characteristics of these institutions are not necessary for the survival of the elderly lead to undesirable hazards such as injury and death. For this reason, it is necessary to analyze the spatial risks and to reduce or eliminate the risks to ensure the quality of life in nursing homes.

Businesses have increased risk assessment studies, particularly with legal obligations. They should apply the risk assessment method that is best suited to them. Several risk analysis methods such as Risk Mapping, Initial Hazard Analysis, Risk Assessment Decision Matrix Methodology (L type and X type Matrix), Fault Tree Analysis, Failure Mode and Effects Analysis (FMEA) and Fine-Kinney Method are in practice.

The Fine-Kinney method based on this study is the most commonly used risk assessment method. With this method, the risks are assessed and the areas to be prioritized according to the evaluation results can be determined. Thus, resources can be used correctly in tasks that need to be done to reduce or eliminate risks. In the method, the risks are evaluated according to three parameters, probability (P), frequency (F) and severity (S), and the risk score of the hazard is obtained by multiplying these three parameters. The risks are divided into 5 classes according to the score of risk: insignificant risk, possible risk, substantial risk, high risk and very high risk. In the absence of sufficient information, the value of the risk factors is determined by the opinion and valuation of experts. A severity factor of risk can be regarded as minor damage by an expert, but it can be considered as significant damage by another expert. This leads to changes in the results and the accuracy of the risk analysis depending on the knowledge, experience and interpretation of the experts involved in the analysis and causes problems. It is caused to ignore significant risks and to use unnecessary effort and resources for insignificant risks. Fuzzy logic must be used to remove the subjectivity. In this study, fuzzy Fine-Kinney approach was developed for this reason.

There are various studies about risk analysis methods based on fuzzy logic in the literature. Cho et al. [2] used the ETA method for the risk assessment of the construction project and tried to eliminate the ambiguity of subjective and probable values by supporting this method with fuzzy logic. Hadjimichael [3] developed the Flight Operations Risk Analysis System (FORAS), a proactive risk identification and decision-making system, using fuzzy set theory in risk analysis in the aviation sector. Mandal and Maiti [4] used fuzzy logic to determine the weights of the probability, severity, and detectability factors used to find risk value in crane operations. A study based on fuzzy logic in the Fine-Kinney method was performed by Oturakçı and Dağsuyu [5]. They aimed to remove this ambiguity by minimizing the differences in expert opinions using the approach they developed to analyze the dangers in construction. In this study, developed fuzzy Fine-Kinney approach has been applied to determine the spatial risks of nursing homes in Istanbul, comprising 30% of nursing homes in Turkey. Thus, it will be possible to bring the quality of life of the people living in the nursing homes to a high level. The risk priority order obtained with this approach was compared with the ranking obtained with the traditional Fine-Kinney method.

2 Methods

2.1 Fuzzy Logic

Zadeh [6] stated that it is insufficient to represent the membership level with only "0" and "1", and that values for the membership level can be used in the range [0, 1]. Fuzzy systems are needed to be able to process the verbal information by means of computers and algorithms. Fuzzy inference system is a calculation system based on fuzzy set theory, fuzzy if-then rules and fuzzy logic concepts. These logic instructions generate output values based on information that is aggregated from all rules [7] entire text should be in Times New Roman. Other font types may be used if needed for special purposes.

2.2 Fine-Kinney Risk Analysis Method

The Fine-Kinney method is a widely used risk analysis method developed by Kinney and Wiruth in 1976 [8]. There are three parameters such as probability (P), frequency (F) and severity (S) in Fine & Kinney method and risk score is obtained multiplying these three parameters. Severity is consequence of an accident. Severity is ranked ranging from 1 to 10 and that is considered catastrophic, highlighted. Frequency is the exposure or frequency of occurrence of a hazard event that could lead to an accident. It is ranked ranging from 1- rarely, once a year, to permanently - continuous 10. Probability is probability of a hazardous event. Probability (P) is ranked ranging from 0.1 - virtually impossible, to 10 – predictable. The risk score defines 5 risk classes, which are insignificant risk, possible risk, significant risk, fundamental risk and unobservable risk.

3 Proposed Fuzzy Fine-Kinney Risk Analysis Method

The Fine-Kinney method is used with fuzzy logic to reduce subjectivity and remove uncertainty in risk assessment. The fuzzy logic method shows high performance in modeling the variables with uncertainty. The steps of the fuzzy Fine -Kinney approach developed in this study are as follows:

- Defining the linguistic variables using Fine-Kinney
- Fuzzification of variables
- Obtaining rules
- Defuzzification of output variables

3.1 Defining the Linguistic Variables Using Fine-Kinney Method

29 nursing homes located in and around the province of Istanbul were taken as a basis. The hazards were identified after examined the spatial plans and places of the nursing homes by experts. In addition to the experts, the opinions of the employees were included in the determination of the hazards. The places with hazards were divided into 5 classes: stairs and elevators, corridors, rooms, WC-bath and social areas. A total of 47 hazards were identified for these places. (Table 5).

3.2 Fuzzification of Variables

A model with three inputs and one output that based on Mamdani type fuzzy inference system was constructed. In the model, input variables are probability, frequency and severity; output variable is risk value. The structure of fuzzy inference system is shown in Fig. 1.

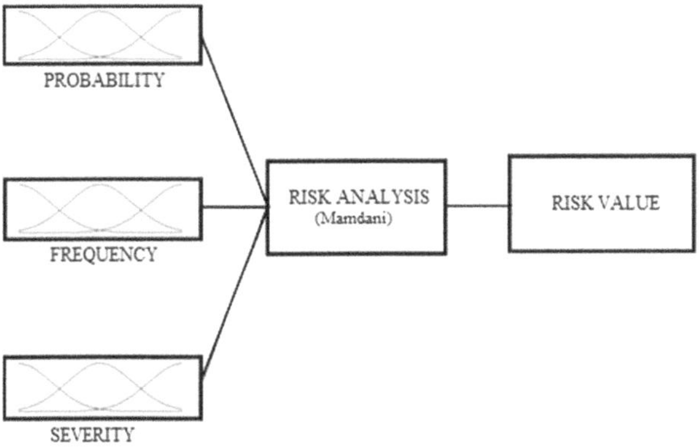

Fig. 1. The structure of fuzzy model

The crisp and fuzzy value of probability input variables is shown in Table 1. In the model; probability variable is evaluated from 0.1 to 10. 7-level triangular membership function is used for probability variable convert into fuzzy number.

Table 1. Crisp and fuzzy value of probability

Probability	Code	Crisp value	Fuzzy value
Virtually impossible	VI	0.1	[0 0.1 0.2]
Practically impossible	PI	0.2	[0.1 0.2 0.5]
Conceivable but very unlikely	CBU	0.5	[0.2 0.5 1]
Only remotely possible	ORP	1	[0.5 1 3]
Unusual but possible	UBP	3	[1 3 6]
Quite possible	QP	6	[3 6 10]
Might well be expected	MWE	10	[6 10 INF]

Triangular membership functions is used for input variables (probability, frequency and severity) and output variable (risk value) according to the user experience. Membership function for probability is shown in Fig. 2.

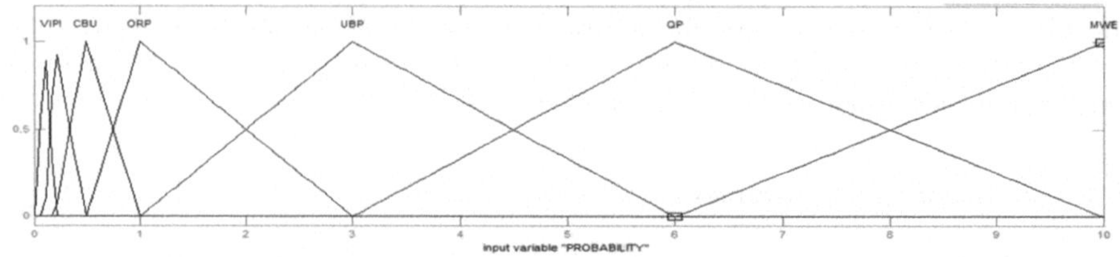

Fig. 2. Triangular membership functions for probability variable

Similarly, the crisp and fuzzy values of the frequency and severity input variables are shown in Tables 2 and 3, respectively.

Table 2. Crisp and fuzzy values of frequency

Frequency	Code	Crisp value	Fuzzy value
Very rare	VR	0,5	[0 0.5 1]
Rare	R	1	[0.5 1 2]
Unusual	U	2	[1 2 3]
Occasional	O	3	[2 3 6]
Frequent	F	6	[3 6 10]
Continuous	C	10	[6 10 INF]

Table 3. Crisp and fuzzy values of severity

Severity	Code	Crisp value	Fuzzy value
Noticeable	N	1	[0 1 3]
Important	I	3	[1 3 7]
Serious	S	7	[3 7 15]
Very serious	VS	15	[7 15 40]
Disaster	D	40	[15 40 100]
Catastrophe	C	100	[40 100 INF]

For the output variable, 5-level triangular membership function is used, and 5 classes is constituted depending on the risk values (Table 4).

The graphical representation of the triangular membership function for frequency, severity and risk is similar to Fig. 2 given for probability. Matlab was used for fuzzy logic.

Table 4. Classic and fuzzy Fine-Kinney value of risk value

Risk value	Code	Crisp value	Fuzzy value	Risk class
Risk	R	R < 20	(0,15,20)	E
Possible risk	PR	20 = < R < 70	(20,30,70)	D
Substantial risk	SB	70 = < R < 200	(65,135,200)	C
High risk	HR	200 = < R < 400	(175,300,400)	B
Very high risk	VHR	400 = < R	(350,450,1000)	A

3.3 Obtaining the Rules

77 rules are constructed to determine the risk classes correctly. For example, rule 1 is as follows.

If (probability is VI) and (frequency is VR) and (severity is N) then (risk is E).

The risk values of the rule base are evaluated and divided into the classes in Table 5. 'A' class hazard requires immediate prevention in the other hand the 'E' class risk value is less risky than the other classification.

Table 5. Risk values found with traditional Fine-Kinney method and fuzzy Fine-Kinney approach

Risk definition	Conventional method risk		Fuzzy method risk	
	Score	Class	Score	Class
WC - Bathroom				
The ground is not made of proper material	10000	A	600	A
Ground is worn	10000	A	600	A
There is a threshold in the transit	2400	A	600	A
Door dimensions do not comply with standards and can not be easily passed	252	B	292	B
Emergency button or phone not available	900	A	600	A
Lightning does not exist correctly and adequately	6000	A	600	A
The holding bars are not mounted according to ergonomic measures	360	B	600	A
No closet and toilet bars	2400	A	600	A
Incorrect adjustment of height measurements of toilet seats	14	E	10	E
Corridor				
The presence of thresholds in transits to rooms and WC-bath spaces or the lack of portability of the existing threshold	540	A	600	A
The floor is not made of the proper material	40000	A	600	A
No bars on the sides of the corridor	63	D	40	D
The holding bars are not mounted according to ergonomic measures	21	D	40	D

(*continued*)

Table 5. (*continued*)

Risk definition	Conventional method risk		Fuzzy method risk	
	Score	Class	Score	Class
Corridor width is not designed to suit wheelchair, stretcher and human pass	70	C	40	D
The orientation plates are not arranged along the corridors	6	E	10	E
No phone or call button to reach the health unit from the corridors	540	A	600	A
Emergency exit, fire escape and fire tubes not available	20	D	600	A
Correct and adequate lighting system not available	2400	A	600	A
Rooms				
The floor is not made of the proper material	2400	A	600	A
The floor is worn	2400	A	600	A
Use of non-ergonomic furniture in rooms	270	B	292	B
The furniture used in the place is made from material that is not suitable for health and the existence of manufacturing defect	270	B	292	B
Lightning is not correct and sufficient	2400	A	600	A
The presence of thresholds in transits to rooms and WC-bath spaces or the lack of portability of the existing threshold	540	A	600	A
No special WC-bath space inside the room	0.6	E	10	E
The shared WC-bathroom space is located away from the room	420	A	600	A
The dimensions of the room doors are not suitable for stretcher and wheelchair measurements	70	C	40	D
Emergency button or phone not available	3000	A	600	A
The emergency button or phone must be of sufficient number and distance in accordance with the size of the room	135	C	133	C
Social Places				
The outer door is unprotected and unlocked	3600	A	600	A
The floor is not made of the proper material	1200	A	600	A
Excessive wear on the floor	1200	A	600	A
Lightning is not correct and sufficient	1200	A	600	A
The in-room use of furniture is not ergonomic	270	B	292	B
The furniture used in the place is made from material that is not suitable for health and the existence of manufacturing defect	270	B	292	B
Emergency button or phone not available	90	C	133	C
The emergency button or phone must be of sufficient number and distance in accordance with the size of the room	14	E	10	E

(*continued*)

Table 5. *(continued)*

Risk definition	Conventional method risk		Fuzzy method risk	
	Score	Class	Score	Class
Stairs and Elevators				
The steps are not made from the proper material	10000	A	600	A
No precautions against skidding on the steps	10000	A	600	A
Stair steps not in correct height and width	6000	A	600	A
There is no holding bar on the edge of the stairs	6000	A	600	A
No lock door at the beginning of the stair	6000	A	600	A
Elevator interior dimensions do not match stretcher and wheelchair size	70	C	40	D
The timing of the elevator doors is not arranged according to elderly	450	A	600	A
No phone in elevator cabin	450	A	600	A
There is no generator to run the elevator	14	E	10	E
Enlightening the stairwell is not sufficient and effective	6000	A	600	A

3.4 Defuzzification of Risk Values

The fuzzy risk value is converted into crisp value by defuzzification method. Centroid of area method of defuzzification is mostly used for estimating crisp output value. Equation 1 represents formula for centroid of area.

$$\frac{\int_i^n \mu_A(x)xdx}{\int_i^n \mu_A(x)dx} \tag{1}$$

where A is a fuzzy set, x is a universe of discourse and μA(x) is aggregated output membership function. Risk values found with conventional Fine-Kinney method and fuzzy Fine-Kinney approach are shown in Table 5.

4 Results

Each risk class was identified using the classification in Table 6. These classes were used in the confusion matrix to compare the results of the traditional Fine-Kinney method and the fuzzy Fine-Kinney approach.

As shown in Table 7, 42 of the 47 hazards were correctly classified and the model worked with 89.3% overall accuracy. Overall accuracy was calculated according to confusion matrix that based on user's accuracy and producer's accuracy. The equation of overall accuracy is as follows (Eq. 2).

$$\text{Overall accuracy} = \frac{total\ number\ of\ correct\ classified}{total\ number\ of\ risks} \tag{2}$$

Table 6. Confusion matrix for risk classes

		Traditional Fine-Kinney Method class					User's accuracy (%)
		A	B	C	D	E	
Fuzzy Fine-Kinney approach class	A	28	1	0	1	0	93.333
	B	0	5	0	0	0	100
	C	0	0	2	0	0	100
	D	0	0	3	2	0	40
	E	0	0	0	0	5	100
Producer's Accuracy (%)		100	83.33	40	66.667	100	

Kappa coefficient (κ) is another measurement to measure similarity the fuzzy approach results with traditional method results (Eq. 3). The kappa value is between −1 and +1. The positive values show high accuracy and the value of zero shows no correlation in the classification [9].

$$\kappa = \frac{n\sum_{i=1}^{p} x_{ii} - \sum_{i=1}^{p}(x_{i+}x_{+i})}{n^{2-}\sum_{i=1}^{p}(x_{i+}x_{+i})} \quad (3)$$

where, n = total number of risk, p = number of classes, $\sum x_{ii}$ = total number elements of confusion matrix, $\sum x_{i+}$ = sum of row i, $\sum x_{+i}$ = sum of column i.

The kappa coefficient is calculated from Eq. 3 as 0.818. This value indicates that the risk classes found in the fuzzy approach are very similar to those found by traditional methods.

5 Conclusions and Recommendations

In this study; the results of the studies on the spatial areas used by the elderly people living in 29 nursing homes belonging to the private institutions in İstanbul were presented with the results of Fine-Kinney and fuzzy Fine-Kinney method and suggestions were introduced.

Within the observed situation, the following results were obtained;

1. The results of the examinations made in the 5 main places of the nursing homes revealed the risks.
2. When the sub-risks of each of the 5 main risk groups are examined, total of 47 sub-risks are identified.
3. The 47 sub-risks were evaluated using the traditional Fine-Kinney method and the fuzzy Fine-Kinney method, and the results were compared.
4. As a result of the evaluation, it was observed that 89.3% of the 47 sub riskie were in the same risk class in both methods and the remaining 10.7% were in different classes. This shows that the values of 42 from 47 sub-risks are preserved, while the other 5 sub-risks show a change.

5. The risk score obtained by the Fuzzy Fine-Kinney method was found to be higher in some cases than in the traditionall method, but sometimes lower (Tables 7 and 8).

Table 7. Risks that score is higher than the traditional method

Risky places	Risks	Score	Score	Class	Class
Corridor	Emergency exit, fire escape and fire tubes not available	20	600	D	A
WC - Bathroom	The holding bars are not mounted according to ergonomic measures	360	600	B	A

Table 8. Risks that score is lower than the traditional method

Risky places	Risks	Score	Score	Class	Class
Corridor	Corridor width is not designed to suit wheelchair, stretcher and human pass	70	40	C	D
Rooms	The dimensions of the room doors are not suitable for stretcher and wheelchair measurements	70	40	C	D
Stairs and Elevators	Elevator interior dimensions do not match stretcher and wheelchair size	70	40	C	D

6. The results obtained by Fuzzy Fine-Kinney approach are similar to those found by these traditional Fine-Kinney methods. The developed method may be an alternative to the conventional method. Particularly in the case of a severity value of 100, the Fine-Kinney method of the risk class has an insignificant risk while the fuzzy Fine-Kinney method has become the very high risk, that is, the highest risk class. It can be said that the fuzzy-based method gives more accurate results.
7. When the 5 different risks that differed were examined in detail, it was found that the difference was caused by the severity risk factor. The developed fuzzy Fine-Kinney method makes the risk more important than the risk determined by the traditional method when the risk score is high.
8. It is suggested that the regulation of the nursing homes should be considered from scratch and that a new regulation should be worked on while adhering to the international norms but sticking to the cultural norms.
9. It is proposed to make improvement according to the new regulations in existing nursing homes and to make spatial designs according to the new regulations in new nursing homes to be constructed.

References

1. Türkiye İstatistik Kurumu Nüfus Projeksiyonları (2014)
2. Cho HN, Choi HH, Kim YB (2002) A risk assessment methodology for incorporating uncertainties using fuzzy concepts. Reliab Eng Syst Saf 78:173–183
3. Hadjimichael MA (2009) Fuzzy expert system for aviation risk assessment, vol 36, pp 6512–6519
4. Mandal S, Maiti J (2014) Risk analysis using FMEA: fuzzy similarity value and possibility theory based approach. Expert Syst Appl 41:3527–3537
5. Oturakçıve M, Dağsuyu C (2017) Risk değerlendirmesinde bulanık Fine-Kinney yöntemi ve uygulaması. Karaelmas Sağlığıve Güvenliği Derg 1:17–25
6. Zadeh LA (1965) Fuzzy Sets. Inf Control 8:338–353
7. Bojadziev G, Bojadziev M (2007) Fuzzy logic for business, finance and management. World Scientific, London
8. Kinney GF, Wiruth AD (1976) Practical risk analysis for safety management (No. NWC-TP-5865). Naval Weapons Center China Lake, CA
9. Taufik A, Syed Ahmad SS (2016) Land cover classification of Landsat 8 satellite data based on fuzzy logic approach. IOP Conf Ser Earth Environ Sci 37:1–7

Simulation and Modelling

Optimisation of Machining in the Context of Industry 4.0 – Case Study

Andrzej Matras(✉) and Wojciech Zębala

Production Engineering Institute,
Cracow University of Technology, Cracow, Poland
{amatras, zebala}@mech.pk.edu.pl

Abstract. This document presents the idea of the machining process optimization in the concept of Industry 4.0. The authors proposed the method which mainly is designed to optimize the machining process of parts made of hard-to-cut materials with complex geometry. The optimization research allowed to gradually increase the efficiency of the machining while ensuring the required surface roughness. The optimization procedure caused also a partial stabilization and reduction of the maximum values of the cutting forces.

Keywords: Optimisation · Simulation · Machining · Industry 4.0

1 Introduction

In modern technology, there is a need to manufacture machine parts from materials with high mechanical properties and characterized by the high surface quality and a complexity of shape. The usage of Inconel 718 alloy for making components of critical aircraft engines such as turbines, bodies and shafts is an example. It is characterized by resistance to high temperatures and corrosion and low creep [1, 2]. Alloy Inconel 718, due to its hardening with high deformation, thermal conductivity and high resistance to high temperatures is included in the group of the hard-to-cut materials [3–5]. When machining the Inconel 718 alloy due to the high temperature and wear of the cutting tool, low values of cutting parameters are used, which results in low process efficiency [5–8]. In order to improve the durability of the cutting edge, tools with blades made of CBN [2, 3, 7, 9, 10] are used. During the machining of the Inconel 718 alloy, MQL lubrication allows achieving the best surface quality [3, 11].

In order to make parts of machines with a complex shape, free surface milling technology is used. This machining is usually carried out on the three or five-axis machine tools using ball milling cutters. The free surface machining technology is a complex and difficult process [12–15]. In order to design it, it is necessary to use CAD/CAM software. The geometric structure of the surface created as a result of milling with spherical cutters is characterized by inequalities arising as a result of mapping the shape of the tool. Their size theoretically depends on the diameter of the cutting tool, the radial depth of cut and the angle of inclination of the machined surface. The quality of surfaces machined using spherical cutters is additionally influenced by

N. M. Durakbasa and M. G. Gencyilmaz (Eds.): ISPR 2018, *Proceedings of the International Symposium for Production Research 2018*, pp. 813–823, 2019.
https://doi.org/10.1007/978-3-319-92267-6_65

many factors, including: cutting tool geometry, cutting data, wear and deflection of the cutting tool [4, 16–21].

Today, many researchers carry out research to improve the machinability of the hard-to-cut materials. They are developed, among others mathematical models that allow prediction of the impact of the applied process parameters on the values of the total cutting force components. Such models can be used to optimize the machining process [8, 22–27]. The idea of the optimization is to control the value of the components of the total cutting force, stresses or temperature in the chip formation zone, which enables the modification of the cutting parameters and improves the machining process [28, 29]. The optimization may shorten the machining time, without deteriorating the quality of the manufactured product or reducing the number of production deficiencies, which results in a reduction of manufacturing costs [12, 27].

In the traditional approach the first step of the optimization process, for example, relies on the defining of the mathematical model, which describes the relationship between the cutting data and the values of cutting forces and the quality of the surface [30]. For this purpose, it is necessary to perform a series of costly and long-term experimental studies or calculations. In the next step, optimization and limitation criteria are defined. In the case of the optimization of the cutting process, it is possible to define criteria in the form of component values of the total cutting force components and limitations in the form of roughness of the obtained surfaces and dimensional and shape accuracy of the machined parts [31]. The next stage of the optimization process consists in the implementation, using the optimized process of the prototype, and then determining the impact of the optimization on the quality and time of manufacturing parts. When the result is satisfactory, serial production of parts takes place, or in the case of unsatisfactory results, optimization criteria are modified or applied to the other mathematical models.

The such approach allows to increase the efficiency of the process or its stabilization without losing the quality of the part and is close to the concept of Industry 3.0. Taking into account the assumptions and technologies included in the concept of Industry 4.0, through their application it is advisable to develop new optimization techniques for the manufacturing process. From the concept of product creation to its implementation and sale, the following steps can be taken in a simplified manner:

- Production planing stage
- Part Machining Stage
- Product assembly stage
- Sale/Use stage

In order to applied the technologies included in the idea of Industry 4.0, a general concept of optimization of the production process was developed. It has been graphically shown in Fig. 1, along with the benefits resulting from its use.

The study proposes the new conception of the manufacturing system structure based on the machining process optimization in the idea of Industry 4.0. In Sect. 1, the literature was analyzed and the subject was characterized. Section 2 discusses the proposed conception. Section 3, via a case study reviews the proposed method. In Sects. 4 and 5 authors discuss the results of the study.

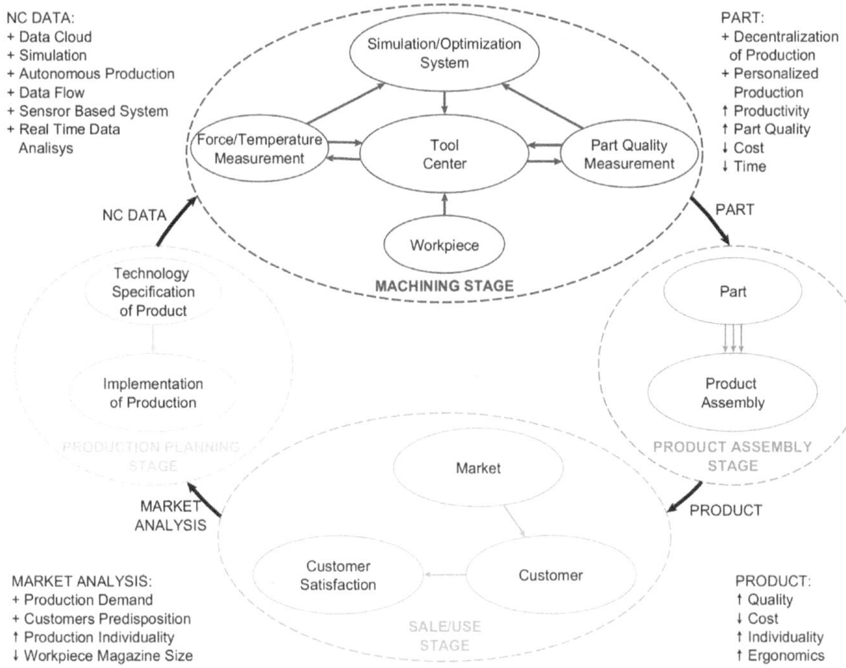

Fig. 1. The general concept of the optimization of the production process

2 Development of a Cutting Process Optimization Method in the Idea of Industry 4.0

Based on the concept shown in Fig. 1, a method for optimizing the feed speed in the machining process has been developed. The method is mainly designed to optimize the machining process of parts made of hard-to-cut materials with complex geometry. Its use is possible both in the case of serial production and single part machining. The developed method is shown in Fig. 2.

The proposed method focuses mainly on activities performed as part of the Part Production/Optimization System. It is possible to use many dispersed Part Production/Optimization System between which also data exchange takes place. Connections between these systems are marked with orange arrows. There are several flows within a single Part Production/Optimization System. The first of these is the data flow of the cutting process (Data Flow). These data contain dependencies between the applied cutting parameters, the quality of the parts and the time and capacity of their manufacture. These data circulate inside the module ensuring their exchange between individual system elements and are updated by measurements and calculations performed as a result of the production. The direction of their flow has been marked in blue in Fig. 2. The next element of the system is Part Flow. It starts from the moment of entry into the blank system, then it goes through the stages of manufacturing parts and assembly of the product and leaves the system as a result of the sale of the product.

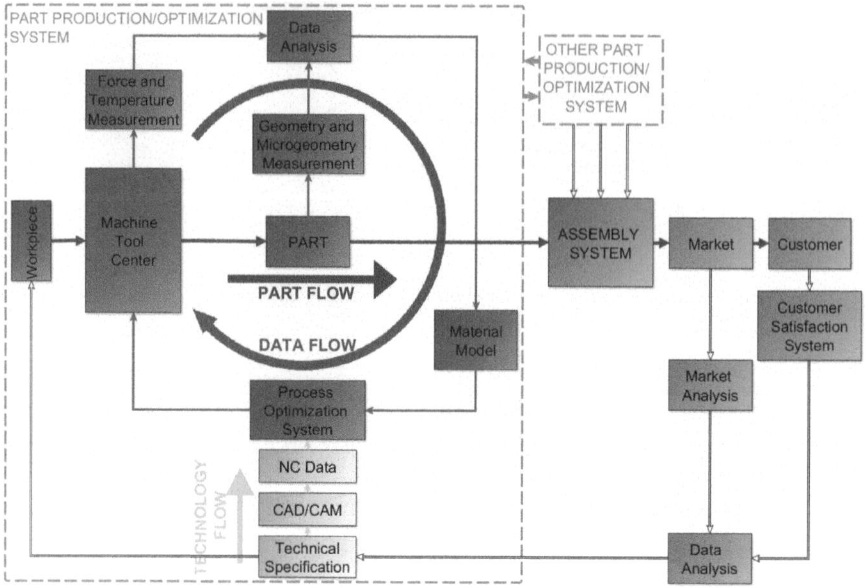

Fig. 2. The proposed method of optimizing the machining process of machine parts (Color figure online)

The third element of the system is Technology Flow. During the process, technological knowledge is supplied to the system, based on which the technological process is carried out. The blocks enter into Technology Flow and connections between them are marked in yellow. The final element of the proposed system is the stage of sale and use of the manufactured product. It has been marked with blocks of green. The use of these modules enables market analysis and customer satisfaction surveys, which allows for making structural and functional changes to the manufactured product and planning the production volume.

The work focuses mainly on the activities and flow of information in the Part Production/Optimization System. The activities performed as part of the module start at the moment of determining the market demand for the product (data from Market Analysis and Customer Satisfaction System blocks) as well as defining its technical specification (Technical Specification block). Next, the technological process is developed and the NC code (CAD/CAM and NC Data blocks) is generated, which makes it possible to make parts on a CNC machine tool and determine the shape and quantity of semi-finished products (Workpiece block). In the next step, optimization of the feed speed (Process Optimization System block) is performed and the part (Machine Tool Center block) and measurements (Force and Temperature Measurement and Geometry and Micro-geometry Measurement blocks) are performed. At this stage, the produced part leaves the Part Production/Optimization System and Stage 1 ends. Then Stage 2 begins. In Stage 2, based on the analysis of the data obtained in the part manufacturing process, the material model equations (Material Model block) and the optimization criterion are updated and a subsequent optimization is performed (Process

Optimization System block). Based on the optimized process, part (Machine Tool Center block) and measurements (Force and Temperature Measuring, and Geometry and Micro-geometry Measurement) are performed again. The part leaves the Part Production/Optimization System module and Stage 2 is terminated and Stage 3, which runs similarly, starts, followed by the next Stage. At any time (based on data from Market Analysis and Customer Satisfaction System blocks) it is possible to update the functionality and technical specification of the product and to introduce it to the system and optimize the parts production process based on data obtained in the Stage previously performed.

3 Case Study

As part of the research, the process of finish milling of the surface made of Inconel 718 alloy with a complex shape was analyzed (Fig. 3). Machining was carried out with a 2 mm diameter spherical cutter equipped with a CBN blade. The machining process was performed with the cutting speed v_c = 40 m/min and feed f_z = 0.03 mm/tooth. Radial variable depth of cut was applied. The inclination angle was adjusted to the surface, the unevenness height resulting from the tool shape mapping was 5 μm. The measurements of the components of the cutting forces were made using a measuring setup based on the Kistler 9257B dynamometer. The surface roughness was measured using a Talysurf Form Intra 50 profilograph in the locations shown in Fig. 3.

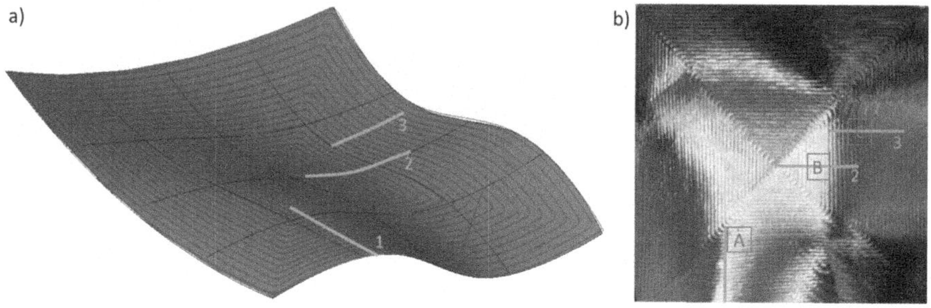

Fig. 3. Outline of the work surface with a simplified view of the tool paths (a) and a view of the obtained surface (b) with the marked places of surface roughness measurements

In order to perform the optimization process in the simulation software, the cutting data, cutting tool, trajectory of its movements, blank and material model of the Inconel 718 alloy were defined. The chemical composition of the Inconel 718 alloy is summarized in Table 1.

Table 1. Chemical composition of Inconel 718

Ni	Cr	Fe	Nb	Mo	Ti	Al	Co	Mn	Si	Cu	C
54,5	17,77	18,08	4,95	2,95	0,93	0,52	0,13	0,08	0,05	0,02	0,018

The material model made possible to calculate the changes of the cutting forces components values as a result of changes of the cutting data values and cross-section of the machined layer. In the first stage (Stage 1), the Inconel 718 alloy material model was implemented in the software. In the subsequent stages (Stage 2 and 3), the material model was updated based on the data obtained during the measurements of the total cutting force component. In this process the feed speed was modified. The optimization criterion was set in the form of the maximum value of the tangential component of the total cutting force F_t= 33 N. The maximum feed speed was also limited to f_{z_max} = 0.05 mm/tooth. A limitation was also assumed in the form of the required surface roughness class No 7 (Ra_{max} = 1.25 μm, Rz_{max} = 6.3 μm).

3.1 Optimization of the Machining Process Carried Out as Part of the Stage 1

On the basis of the material model implemented in the simulation software, calculations of the total cutting forces occurring during machining were made. Then, the feed speed was optimized. The following Fig. 4 shows the changes of the tangential component of the total cutting force for the optimized and non-optimized machining process.

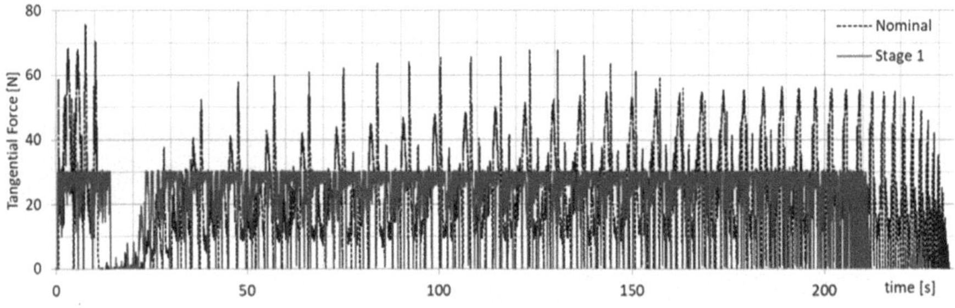

Fig. 4. The course of the tangential component of the total cutting force during the nominal process machining (Interrupted black) and optimized in Stage 1 (Continuous blue) (Color figure online)

As a result of the optimization process, the maximum value of the tangential component of the total cutting force was reduced from 76 N to 33 N. This process resulted also in shortening the cutting time by about 9% and increasing the efficiency of the cutting process, Fig. 5.

The machined surfaces were measured, the average values of Ra = 1.20 μm and Rz = 6.28 μm.

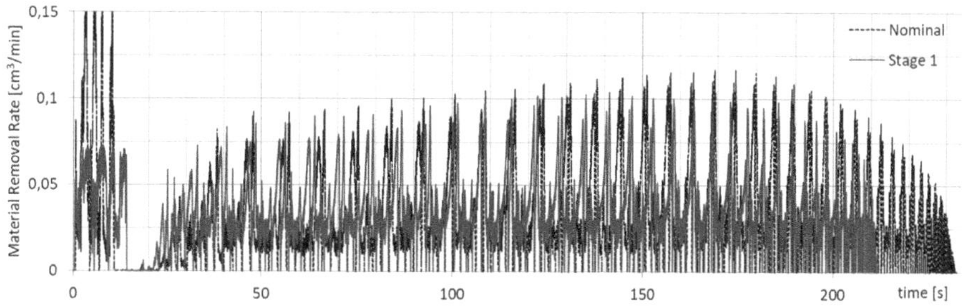

Fig. 5. The course of the material removal rate during the machining in normal stage and Stage 1

3.2 Optimization of the Machining Process Carried Out as Part of the Stage 2 and 3

Simulation tests in Stage 2 and 3 were carried out analogously to the Stage 1. In order to implement them, a modified material model based on the total cutting forces values was used. Figure 6 shows the calculated changes of the tangential component value of the total cutting force occurring during cutting for the optimized machining process in the stage 1, 2 and 3.

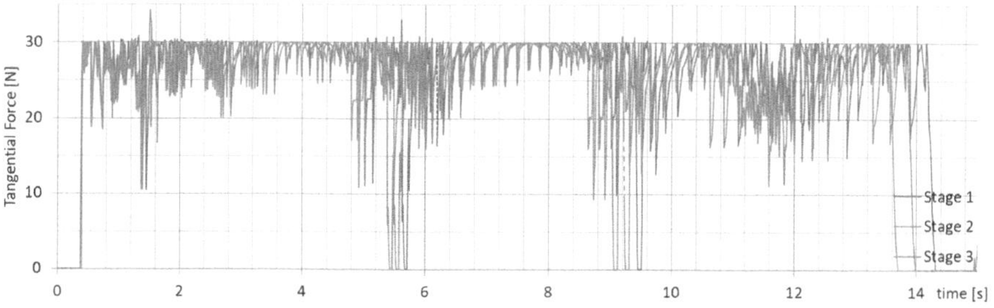

Fig. 6. The course of the tangential component of the total cutting force during the machining in different stages

As a result of the optimization, a reduction (in relation to the non-optimized code) of the maximum value of the tangential component of the total cutting force to 34 N in Stage 2 and 31 N in Stage 3 were achieved. At the same time the optimization process resulted in a further reduction of the cutting time by approximately 10% in Stage 2 and 12% in Stage 3. The surfaces were characterized by average values of Ra = 1.06 μm and Rz = 6.12 μm for surface made in Stage 2 and Ra = 1,14 μm and Rz = 5,96 μm for surface made in Stage 3. The calculated changes of the material removal rate in time are shown in Fig. 7.

Figure 8 shows isometric views of surface roughness measured in places 1–3 (Fig. 3b) for surfaces obtained as a result of machining with NC codes obtained in Stage 1, 2 and 3.

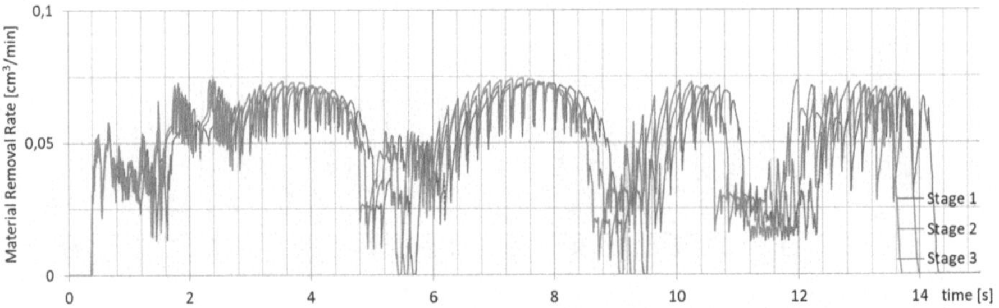

Fig. 7. The course of the material removal rate during the machining in different stages

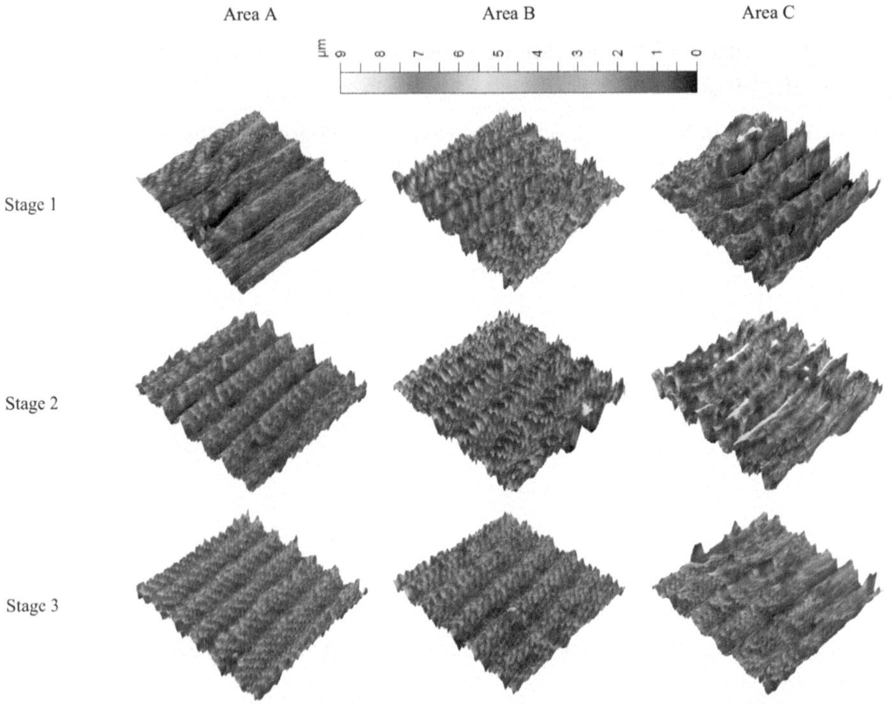

Fig. 8. Isometric views of surface roughness after machining with NC codes obtained in stages Stage 1, 2 and 3

4 Research Analysis

The optimization research allowed to gradually increase the efficiency of the machining while ensuring the required surface roughness. The optimization procedure caused also a partial stabilization and reduction of the maximum values of the cutting forces. Table 2 presents the surface roughness parameters obtained during the tests, as well as the durations and material removal rates using not optimized NC code (Nominal) and optimized code in Stage 1, 2 and 3.

Table 2. Summary of research results

	F_{t_max} [N]	t [s]	Time safe [%]	Average *MRR* [cm^3/min]	Line 1		Line 2		Line 3	
					Ra [μm]	Rz [μm]	Ra [μm]	Rz [μm]	Ra [μm]	Rz [μm]
Nominal	75	233	–	0,0309	–	–	–	–	–	–
Stage 1	33	212	9	0,0330	1.17	6.64	1.06	6.04	1.37	6.17
Stage 2	34	210	10	0,0333	0.98	6.86	0.84	5.99	1.35	5.52
Stage 3	31	206	12	0,0338	1.19	5.99	1.18	5.65	1.05	6.24

5 Conclusions

The method proposed in the work allows to automate the optimization process based on the assumed criteria and limitations. It takes into account all stages of the product life cycle. The analysis of data from the market allows the product to be optimized, which allows to increase its ergonomics and functionality. Using the automated production system, production can be carried out with a small human contribution. The use of computer software allows to automatic analysis of data acquired during the machining process. Due to the progressive optimization calculations (a few steps performed sequentially), the optimization process can be performed without the need to make prototypes.

The work focuses on activities performed in the module *Part Production/ Optimization System*. As a result of the optimization tests, the feed speed was changed. The optimization process resulted in stabilization of the process and reduction of the maximum tangential component of the total cutting force from 75 to 31 N. The machining time was also reduced by 12% and surfaces were achieved with the assumed roughness. Further beneficial results could be obtained at successive stages in which the material model is improved based on the data obtained in previous machining. A larger amount of data may also permits the modification of the optimization criterion which may result in further reduction of the machining time.

References

1. Hadi MA, Ghani JA, CheHaron CH, Kasim MS (2013) Comparison between up-milling and down-milling operations on tool wear in milling Inconel 718. Procedia Eng 68:647–653
2. Sugihara T, Nishimoto Y, Enomoto T (2017) Development of a novel cubic boron nitride cutting tool with a textured flank face for high-speed machining of Inconel 718. Precis Eng 48:75–82
3. Zhou J, Ren J, Yao C (2018) Multi-objective optimization of multi-axis ball-end milling Inconel 718 via grey relational analysis coupled with RBF neural network and PSO algorithm. Measurement 102:271–285
4. Thakur DG, Ramamoorthy B, Vijayaraghavan L (2012) Effect of cutting parameters on the degree of work hardening and tool life during high-speed machining of Inconel 718. Int J Adv Manuf Technol 59:483–489
5. Jafarian F, Amirabadi H, Sadri J (2015) Experimental measurement and optimization of tensile residual stress in turning process of Inconel 718 superalloy. Measurement 63:1–10

6. Liao YS, Lin HM, Wang JH (2008) Behaviors of end milling Inconel 718 superalloy by cemented carbide tools. J Mater Process Technol 201(1–3):460–465
7. Zhang S, Li JF, Wang YW (2012) Tool life and cutting forces in end milling Inconel 718 under dry and minimum quantity cooling lubrication cutting conditions. J Clean Prod 32:81–87
8. Nath Ch, Brooks Z, Kurfess TR (2015) Machinability study and process optimization in face milling of some super alloys with indexable copy face mill inserts. J Manuf Process 20 (1):88–97
9. Costes JP, Guillet Y, Poulachon G, Dessoly M (2007) Tool-life and wear mechanisms of CBN tools in cutting of Inconel 718. Int J Mach Tools Manuf 47:1081–1087
10. Ślusarczyk Ł, Struzikiewicz G (2014) Hardened steel turning by means of modern CBN cutting tools. Key Eng Mater 581:188–193
11. Struzikiewicz G (2016) The application of high-speed camera for analysis of chip creation process during the steel turning. Proc SPIE 10031:100310F
12. He G, Ma W, Yu G, Lang A (2015) Modeling and experimental validation of cutting forces in five-axis ball-end milling based on true tooth trajectory. Int J Adv Manuf Technol 78:189–197
13. Boz Y, Erdim H, Lazoglu I (2011) Modeling cutting forces for five axis milling of sculptured surfaces. Adv Mater Res 223:701–712
14. Bílek O, Žaludek M, Čop J (2016) Cutting tool performance in end milling of glass fiber-reinforced polymer composites. Manuf Technol 16(1):12–16
15. Bilek O, Samek D (2014) Neural networks in modeling of CNC milling of moderate slope surfaces. In: Advances in Intelligent Systems and Computing, vol 285, pp 75–83
16. Sonawane HA, Joshi SS (2015) Modeling of machined surface quality in high-speed ball-end milling of Inconel-718 thin cantilevers. Int J Adv Manuf Technol 78:1751–1768
17. Lee CM, Kim SW, Lee YH, Lee DW (2014) The optimal cutter orientation in ball end milling of cantilever-shaped thin plate. J Mater Process Technol 153–154:900–906
18. Zębala W, Plaza M (2014) Comparative study of 3-and 5-axis CNC centers for free-form machining of difficult-to-cut material. Int J Prod Econ 158:345–358
19. Rokyta L, Bilek O (2012) Design of a casting die in CATIA. Manuf Technol 12:80–82
20. Onozuka H, Utsumi K, Kono I, Hirai J, Numata Y, Obikawa T (2015) High speed milling processes with long oblique cutting edges. J Manuf Process 19:95–101
21. Bilek O, Samek D, Suba O (2014) Investigation of surface roughness while ball milling process, vol 581, pp 335–340
22. Sultan AA, Okafor AC (2016) Effects of geometric parameters of wavy-edge bull-nose helical end-mill on cutting force prediction in end-milling of Inconel 718 under MQL cooling strategy. J Manuf Process 23:102–114
23. Grabowski R, Denkena B, Kohler J (2014) Prediction of process forces and stability of end mills with complex geometries. Proc CIRP 14:119–124
24. Merdol D, Altintas Y (2004) Mechanics and dynamics of serrated cylindrical and tapered endmills. J Manuf Sci Eng 126:317–326
25. Chatelain JF, Zaghbani I (2011) Effect of tool geometry special features on cutting forces of multilayered CFRP laminates. In: 4th international conference on manufacturing engineering, quality and production systems (MEQAPS 2011), Barcelona, pp 85–90
26. Kao YC, Nguyen NT, Chen MS, Su ST (2015) A prediction method of cutting force coefficients with helix angle of flat-end cutter and its application in a virtual three-axis milling simulation system. Int J Adv Manuf Technol 77:1793–1799
27. Zębala W (2010) Milling optimization of difficult to machine alloys. Manag Prod Eng Rev 1 (1):59–70
28. Ślusarczyk Ł (2016) The implementation of a thermal imaging camera for testing the temperature of the cutting zone in turning. Proc SPIE 10031:100310L

29. Li ZZ, Zhang ZH, Zheng L (2004) Feedrate optimization for variant milling process based on cutting force prediction. Int J Adv Manuf Technol 24:541–552
30. Ślusarczyk Ł (2016) The construction of the milling process simulation models. Proc SPIE 10031:100310C
31. Kowalczyk M (2017) Application of Taguchi method to optimization of surface roughness during precise turning of NiTi shape memory alloy. Proc SPIE 10445:104455G

Capstone Projects

A Cargo Vehicle Packing Approach with Delivery Route Considerations

Uğur Eliiyi(✉), Mert Bulan, and Emre Külahlı

Department of Computer Science, Dokuz Eylül University, İzmir, Turkey
ugur.eliiyi@deu.edu.tr, mert.bulan@gmail.com,
emre.kulahli4@gmail.com

Abstract. Our focus in this study is the vehicle dispatching, routing and packing problem of a cargo company branch in Izmir, Turkey. The daily problem involves determining which parcels or packages should be distributed in which vehicle, determining the order in which these packages will be deployed, determining the route through which the vehicles will make the distribution, and finally determining how the packages should be placed in the vehicles. Thus, the whole daily distribution process of the cargo branch is of focus. We develop a web-based decision support tool to handle this hard-to-tackle problem. Within the decision support system (DSS), we handle the problem in stages. By developing an intuitive algorithm, the delivery route-based three-dimensional packing is solved very quickly. The results are shared with the user through a user-friendly interface.

Keywords: Three-dimensional packing · Clustering · Vehicle loading
Decision support system · Optimization · Heuristics

1 Introduction

The focus of this study is to bring a practical solution to the everyday problem of every cargo branch, that is, the distribution of packages from the branch, determining the route of the distribution vehicles, and how the cargo packages should be placed within each vehicle.

The cargo packages should be placed onto the vehicles to maximize capacity utilization in terms of volume. The second stage involves determining the routes of the distribution vehicles. This is important for minimizing the time and travel cost of the fleet, as well as minimizing the effort of loading and unloading the packages within each vehicle. The vehicle route should be taken into consideration when the loading order of the packages onto a vehicle is determined. The third aspect of the daily cargo distribution problem involves three-dimensional (3D) packing. Once the information about the allocation of parcels to vehicles and the order of delivery are obtained, the packages should be loaded onto the distributing vehicles. The consideration here is to minimize the time and the physical effort of the cargo officers' while loading/unloading the vehicles. We handle this problem as a three-dimensional bin packing problem (3D-BPP).

The 3D-BPP is a problem in which packages of different sizes are placed in boxes with similar dimensions into a minimum number of bins, or through occupying the

N. M. Durakbasa and M. G. Gencyilmaz (Eds.): ISPR 2018, *Proceedings of the International Symposium for Production Research 2018*, pp. 827–838, 2019.
https://doi.org/10.1007/978-3-319-92267-6_66

minimum space in a single bin. The size, priority and placement constraints of the packages/parcels/boxes may vary in different versions of the problem. There are many applications of this hard problem in Supply Chain Management, which involve vehicle, container, pallet or load loading, package design, resource distribution, load balancing, scheduling, project management and financial budgeting [1]. It is a problem from the NP-Complete class and resorting to heuristic methods is often necessary in practice to obtain suboptimal solutions.

Scheithauer [2] dealt with packing boxes of fixed length, fixed width and non-fixed height in a way to minimize the bin height in a layered manner. Martello et al. [3] eliminated the layered approach and have developed three algorithms for the problem for selecting a subset of items that can be packed into a single bin, while maximizing the total volume packed. They used the bounds they found to obtain two approximation algorithms and an exact branch-and-bound algorithm. In another heuristic approach [4], a multi-faced buildup technique was used in the packing procedure with no requirement for the packed boxes to form flat layers. The basic algorithm was improved through a look-ahead strategy, yielding an average packing utilization that improved the existing benchmarks significantly.

For dealing with many practical issues in container loading problems, the stability of the packed items or the weight distribution of the cargo are handled as well [5, 6]. In the former study, the authors considered postprocessing approaches, putting forward a new container loading heuristic. The heuristic was evaluated against several existing approaches and was shown to be capable of producing loading arrangements which combined high space utilization with an even weight distribution of the cargo. In another study, a heuristic approach was proposed for tackling problems where the cargo had varying degrees of load bearing strength [7]. For the interested reader, the reviews in [8, 9] present a more general comprehensive typology and a more solution-oriented comparative perspective, respectively.

In this study, for solving the real-life 3D-BPP of the cargo branch, we adapt and extend the approach developed in [10] for solving the 3D-BPP variable bin height. The authors make use of the efficient extreme point-based heuristics mentioned in [11] in their study to determine the locations in which to load the packages within the bin. Before the packages are placed in the box, an index score is calculated using the extreme points of the already packed packages, and a new package is placed at a specific coordinate based on the calculated score.

In [10], the height of the box in which the packages will be placed is assumed to be infinite. However, as we consider the back of a cargo truck, the number of packages that cannot fit in the vehicle after the packing process should be known. The information on how to compress the unpacked packages is adapted from the algorithm in [12]. In this study, the authors developed a global search framework for 3D-BPP with variable carton orientations, through the usage of parallel moves, relative positions and reference point selection. According to their algorithm, the compression of the packages is made in different axis sequences according to the extreme point types.

The rest of the paper is organized as follows. In Sect. 2, we define and propose a mathematical model for the 3D-BPP on hand. In Sect. 3, we present the details of our solution approach for the daily problem of the cargo branch. In Sect. 4, the computational study is presented, along with the developed DSS. The last section includes conclusion and future remarks.

2 The 3D-Bin Packing Problem

The aim in the 3D-BPP of the cargo delivery company is to minimize the time of delivery and the physical effort of the cargo officers' while loading/unloading the vehicles. This means that, when the destination point of a specific package is reached during the day, the cargo officer should not have to take this package from behind or under any other package in the vehicle preventing unnecessary reshuffling time. If the cargo officer is provided with the comfort of picking up the front or topmost package every time the distribution vehicle opens its rear door during distribution, this would be the ideal situation in terms of saving time and effort. For this purpose, the package to be delivered first should be located at the rear of the vehicle and the last package to be delivered should be located towards the front of the vehicle.

For solving this problem, a mathematical model is developed in this study. We present the details of our mathematical model in this section.

2.1 Sets and Parameters

N = total number of packages
p_i = depth of the ith package
q_i = width of the ith package
l_i = height of the ith package
r_i = the order of delivery for i^{th} package
L = depth of the box (container/back of the truck) - Longer edge
W = width of the box - Shorter edge
H = height of the box
F = total number of boxes (cargo trucks or containers) used
M = arbitrary large number

2.2 Decision Variables

(x_{if}, y_{if}, z_{if}) = coordinates of the left-bottom-behind corner of the ith package on truck f.

The binary decision variables:

$$e_{if} = \begin{cases} 1, & \text{if package } i \text{ is placed on truck } f \\ 0, & \text{otherwise} \end{cases}$$

$$s_i = \begin{cases} 1, & \text{if package } i \text{ is placed in parallel to the shorter edge of the back of the truck} \\ 0, & \text{otherwise} \end{cases}$$

$$u_f = \begin{cases} 1, & \text{if truck f is used} \\ 0, & \text{otherwise} \end{cases}$$

For the following decision variables, the point of reference is the viewing angle from the rear of the truck towards the front of the truck. So, for example, the term "*front*" refers to the rear of the truck from where the packages are loaded in and unloaded out:

$$a_{ikf} = \begin{cases} 1, \text{ if package } i \text{ is placed to the left of package } k \text{ on truck } f \\ 0, \text{otherwise} \end{cases}$$

$$b_{ikf} = \begin{cases} 1, \text{ if package } i \text{ is placed to the right of package } k \text{ on truck } f \\ 0, \text{otherwise} \end{cases}$$

$$c_{ikf} = \begin{cases} 1, \text{ if package } i \text{ is placed behind package } k \text{ on truck } f \\ 0, \text{otherwise} \end{cases}$$

$$d_{ikf} = \begin{cases} 1, \text{ if package } i \text{ is placed in front of package } k \text{ on truck } f \\ 0, \text{otherwise} \end{cases}$$

$$g_{ikf} = \begin{cases} 1, \text{ if package } i \text{ is placed under package } k \text{ on truck } f \\ 0, \text{otherwise} \end{cases}$$

$$h_{ikf} = \begin{cases} 1, \text{ if package } i \text{ is placed on top of package } k \text{ on truck } f \\ 0, \text{otherwise} \end{cases}$$

2.3 Mathematical Programming Model

Based on the definitions above, our route-based mixed integer programming model for the 3D-BPP is as follows. The objective function below tries to maximize the number of loaded parcels.

$$max\, z = \sum\nolimits_{i=1}^{N} e_{if}$$

The constraints (1) to (7) are for ensuring that none of the packages loaded onto the same vehicles overlap in the 3-D trailer.

$$x_{if} + p_i s_i + q_i(1 - s_i) \leq x_{kf} + (1 - a_{ikf})M, \qquad \forall i, k, \text{ where } i < k \tag{1}$$

$$x_{kf} + p_k s_k + q_k(1 - s_k) \leq x_{if} + (1 - b_{ikf})M, \qquad \forall i, k, \text{ where } i < k \tag{2}$$

$$y_{if} + q_i s_i + p_i(1 - s_i) \leq y_{kf} + (1 - c_{ikf})M, \qquad \forall i, k, \text{ where } i < k \tag{3}$$

$$y_{kf} + q_k s_k + p_k(1 - s_k) \leq y_{if} + (1 - d_{ikf})M, \qquad \forall i, k, \text{ where } i < k \tag{4}$$

$$z_{if} + l_i \leq z_{kf} + (1 - g_{ikf})M, \qquad \forall i, k, \text{ where } i < k \tag{5}$$

$$z_{kf} + l_k \leq z_{if} + (1 - h_{ikf})M, \qquad \forall i, k, \text{ where } i < k \tag{6}$$

$$a_{ikf} + b_{ikf} + c_{ikf} + d_{ikf} + g_{ikf} + h_{ikf} \geq e_{if} + e_{kf} - 1, \quad \forall f, \text{ and } \forall i, k, \text{ where } i < k \tag{7}$$

Constraints (8) guarantees that a vehicle is used only if a package is loaded onto it.

$$\sum_{i=1}^{N} e_{if} \leq M\left(u_f\right), \qquad \forall f \qquad (8)$$

The volume capacity constraints for all vehicles are ensured by constraints (9) to (11).

$$x_{if} + p_i s_i + q_i(1 - s_i) \leq W, \qquad \forall i,f \qquad (9)$$

$$y_{if} + q_i s_i + p_i(1 - s_i) \leq L, \qquad \forall i,f \qquad (10)$$

$$z_{if} + l_i \leq H, \qquad \forall i,f \qquad (11)$$

The known routes for the vehicles are used in constraints (12) to (15) for ensuring that no reshuffling occurs when the packages are being unloaded.

$$r_i \geq r_k - M\left(1 - c_{ikf}\right) \qquad \forall i,k, \text{ where } i<k \qquad (12)$$

$$r_k \geq r_i - M\left(1 - d_{ikf}\right) \qquad \forall i,k, \text{ where } i<k \qquad (13)$$

$$r_i \geq r_k - M\left(1 - g_{ikf}\right) \qquad \forall i,k, \text{ where } i<k \qquad (14)$$

$$r_i \geq r_k - M\left(1 - h_{ikf}\right) \qquad \forall i,k, \text{ where } i<k \qquad (15)$$

Finally, the below constraints define the binary and continuous decision variables in the model.

$$s_i,\ a_{ikf},\ b_{ikf},\ c_{ikf},\ d_{ikf},\ e_{if},\ g_{if},\ h_{if},\ u_f \text{ binary} \qquad \forall i,k,f \qquad (16)$$

$$x_{if},\ y_{if},\ z_{if} \geq 0, \text{ and } r_i \geq 1 \qquad \forall i,f. \qquad (17)$$

3 Methodology

For obtaining fast and good solutions for the comprehensive problem of the cargo delivery company, that is, the distribution of packages from the branch, determining the route of the distribution vehicles, and the packing of cargo, we utilize a serial solution approach, using a different algorithm at each stage of the solution procedure. Our approach is detailed in this section.

3.1 Clustering Stage

In the first stage of the solution procedure, the packages that the branch should distribute within that day (assuming that all packages are rectangular) are clustered according to the capacity of the distribution vehicles. We assume an identical fleet of vehicles in accordance with the real data, but our approach also handles heterogenous vehicles, as well. During the clustering, the distribution points near each other are grouped. The depth, height and width measurements of all packages are known at the start of the day. In addition, the dimensions of the load area of the distribution vehicles is also known. The weight of the packages and the weight capacity of the vehicle are not considered in this study for obtaining faster solutions, but the developed approach is very flexible, and this constraint can be easily added. The sweep algorithm [13] is used for clustering. In this algorithm, the polar angles of the delivery points are obtained according to the following formula:

$$\sin\alpha = \frac{Y[i,j] - Y_d}{\sqrt{\left((X[i,j] - X_d)^2 + (Y[i,j] - Y_d)^2\right)}}$$

The $Y[i,j]$ in the formula corresponds to the latitude of the delivery point j in polar area i, while $X[i,j]$ to the longitude. The X_d and Y_d represent the longitude and latitude of the cargo branch location, respectively. After the polar angle is obtained for each delivery point, these are ordered in increasing order in a list. Next, starting with the first point on the list and following an anti-clockwise order, the first vehicle is loaded with the parcels until its capacity is filled. Then the next vehicle is opened for distribution for the remaining points in the list. The algorithm continues in this manner until all vehicles are filled, or the list is empty. The algorithm is coded in Python.

3.2 Routing Stage

Once the first stage is completed and the packages are clustered for the vehicles, the second stage involves determining the routes of the distribution vehicles. At this stage, for each vehicle of the cargo branch, a route is generated taking into consideration the delivery addresses. The destination addresses of the packages are known in terms of their respective latitude (lat) and longitude (long) coordinates, as well as the coordinates of the branch. Thus, the routes of the vehicles are computed starting from this single depot point and ending at this point again.

The process of determining the optimal route of each vehicle corresponds to the well-known Traveling Salesman Problem (TSP) [14]. For the solution of the resulting TSP problems after the first stage, a Python library (PyTSP) using the Concorde TSP Solver is used [15, 16]. The solution gives the optimal distance for the real-life instances on hand. This process is repeated for each vehicle.

The routes are then presented to the user visually within the developed decision support system (DSS) on Google Maps environment. The Google Maps interface and the Google Maps Directions application programming interface (API) are used to display route information. Hence, the dispatchers can follow these routes easily for an efficient distribution of the packages.

3.3 Packing Stage

After the routes are determined, the 3D-BPP is solved in the final stage, first by using the model in the previous section. For this purpose, Python language is used with the callable library of IBM ILOG CPLEX 12.8. For 150 packages, the computation time obtained is around 30 min on a PC with 2.6 GHz Core i7 processor and 16 GB RAM. As this number of packages is less than the usual daily delivery amount, an intuitive heuristic algorithm is proposed for the problem to obtain suboptimal solutions.

Based on the fitness evaluation algorithm [10], our heuristic algorithm includes extra three-dimensional packing constraints and route considerations. Our algorithm consists of the following steps: Firstly, the packages are sorted according to their order in the delivery route. Each package is then loaded onto the vehicle. The first package is placed at the (0, 0, 0) coordinates of the vehicle trailer before the next package in order is considered. At this stage of settlement, potential points, extreme points, at which the box can be placed first are calculated. These points represent the left corner (Type 3) of the upper part of the placed packages, the lower left corner (Type 2) of the front part, and the lower right corner (Type 1) of the right part [12].

A penalty score is calculated for each extreme point of the package before the package is placed. While calculating the penalty, the blocked packages in the vehicle are taken into consideration. Then, the extreme point with the least penalty score is selected. If the package does not fit into the respective extreme point, or if it moves out of the vehicle if it is placed at that point, then the extreme point with the next smallest penalty score is selected.

After the box is placed in the extreme point, the vehicle's remaining capacity is determined through a compression procedure, as in [10]. The next packet is then considered for loading. The procedure continues until all packages are loaded. The steps of the algorithm can be summarized as follows:

S1. Packages are ordered by their routes.
S2. The first packet is placed at the (0, 0, 0) coordinates of the vehicle trailer.
S3. After placing the first package, 3 different types of extreme points are calculated, and added to the extreme points array.
S4. For the next package, the penalty score after the package is placed at each extreme point is calculated.
S5. The extreme point with the lowest penalty point is selected.
If the package does not fit into this point, or if it goes out of the box when it is placed on this point, the point with the next lowest penalty point is selected.
S6. After the package is inserted, the point at which the package is placed is deleted from the extreme points array.
S7. The vehicle capacity is compressed according to the extreme point type, and the extreme point set is updated.
S8. If there are remaining packages to be loaded onto the vehicle, return to S4.
S9. After the packing is complete, the total penalty score is calculated.

All packing information is displayed through a web interface to the user, through a Python library (Flask) [17].

4 Implementation

In obtaining data, three branches (Sirinyer, Aliaga and Gaziemir) of a national cargo company in Izmir, Turkey were preferred as the busiest branches, and the locations of the delivery addresses were generated at random around these branches. For a realistic representation of cargo sizes, a set of items in the company web site including dimensions of consumer goods such as refrigerators, televisions and washing machines were used. For test purposes, these three pre-determined branches and 50 delivery points around these branches are taken into consideration.

Figure 1 depicts the screen shot of the developed DSS. At the top of the application screen, the cargo distribution branches and the list of trucks under these distribution branches can be seen. By clicking on the branch, the user can see both the information about how to pack the packages onto the vehicles (left part of the screen), and what route the vehicle should follow during the distribution (right part of the screen). The user can zoom in and out on both sides. The user can also control all packages manually simply by moving the truck around by changing the viewing angle on different axes through the mouse. Moving the packages out of the truck is also possible through drag-and-drop movements via the mouse.

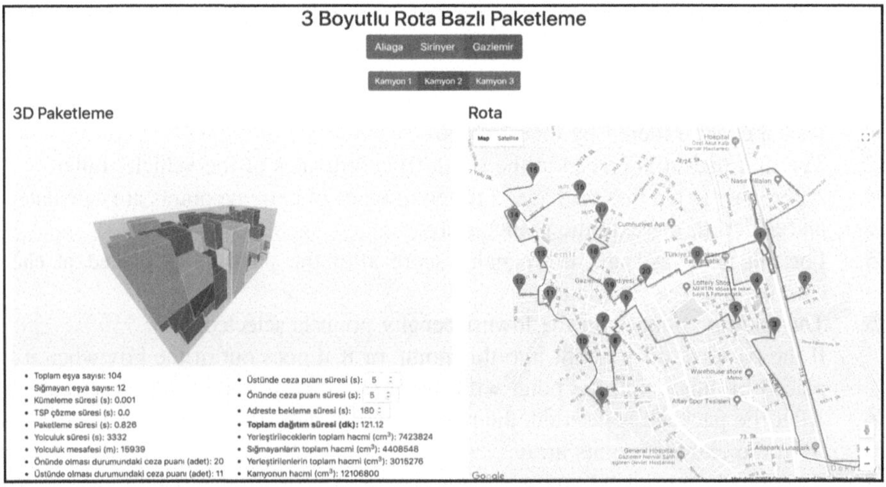

Fig. 1. Screenshot of the developed DSS software application

The screen part under the packing section contains statistics on all stages of the solution. The total number of items represents the number of items that need to be packed. The number of unpacked items is reported. The clustering time represents the duration of the clustering solution through the sweep algorithm. The TSP time represents the duration of the route determination via the Concorde algorithm. The computation time of the packing solution is also reported. The duration of the trip shows the time, in seconds, that the route takes on the actual map, using the real-time information from Google Maps Directions application programming interface (API). The actual travel distances are also obtained via this API and are shown in meters.

The penalty scores are reported as follows. For example, if there are 2 packages to remove before a specific package can be accessed in the truck for unloading, then the penalty point for this package is 2. For each packet, this score is calculated and reported. Penalty scores may differ if another package is on top of, or in front of the needed package, according to the DSS user's priorities and time study parameters of the cargo company. All scores are reported to the user through the web interface.

A testbed was generated to compare the exact and heuristic methods for packing. For this purpose, the packing problem for 29 trucks in 3 different types of branches (80% commercial - 20% residential, 50% commercial - 50% residential, 80% residential - 20% commercial) was solved through the mathematical model and the heuristic, setting a computation time limit of 10 min. The results are presented in the following tables (Tables 1, 2 and 3).

Table 1. Test 1 results

Test 1	Total packages	Packed (Model)	Packed (Heur.)	Front penalty (Heur.)	Top penalty (Heur.)	Computation time (s) (Model)	Computation time (s) (Heur.)	Packing ratio (Model)	Packing ratio (Heur.)
Aliaga Truck 1	136	82	124	36	13	607	2.04	60%	91%
Aliaga Truck 2	109	93	97	79	4	603	1.04	85%	89%
Aliaga Truck 3	145	95	131	157	22	608	2.45	66%	90%
Aliaga Truck 4	40	40	40	3	0	1	0.08	100%	100%
Sirinyer Truck 1	131	87	115	53	13	605	1.71	66%	88%
Sirinyer Truck 2	78	78	77	18	7	375	0.43	100%	99%
Gaziemir Truck 1	136	76	113	96	32	606	1.76	56%	83%
Gaziemir Truck 2	104	79	92	20	11	602	0.86	76%	88%
Gaziemir Truck 3	77	77	75	12	1	219	0.41	100%	97%

The model resulted in zero penalty score for all trucks throughout test instances as a result of including explicit hard constraints. When the number of packages is less than around 80, an average of 5 min is required by the model to pack all items optimally. In cases where the average number of items was 116, the model could place an average of 73% of the packages with zero penalty points within 10 min. The algorithm was able to pack all the packages in 16% of cases in which the model could pack all. However, on

Table 2. Test 2 results

Test 2	Total packages	Packed (Model)	Packed (Heur.)	Front penalty (Heur.)	Top penalty (Heur.)	Computation time (s) (Model)	Computation time (s) (Heur.)	Packing ratio (Model)	Packing ratio (Heur.)
Aliaga Truck 1	130	87	118	77	12	605	1.63	67%	91%
Aliaga Truck 2	109	83	97	30	3	602	0.92	76%	89%
Aliaga Truck 3	132	79	114	17	9	605	1.59	60%	86%
Aliaga Truck 4	74	74	72	15	1	324	0.37	100%	97%
Sirinyer Truck 1	103	86	91	63	11	602	0.84	84%	88%
Sirinyer Truck 2	102	81	92	22	3	602	0.78	79%	90%
Gaziemir Truck 1	113	88	99	92	1	603	1.07	78%	88%
Gaziemir Truck 2	124	83	110	131	9	604	1.46	67%	89%
Gaziemir Truck 3	113	97	103	52	7	603	1.01	86%	91%

Table 3. Test 3 results

Test 3	Total packages	Packed (Model)	Packed (Heur.)	Front penalty (Heur.)	Top penalty (Heur.)	Computation time (s) (Model)	Computation time (s) (Heur.)	Packing ratio (Model)	Packing ratio (Heur.)
Aliaga Truck 1	103	78	87	38	7	605	0.8	76%	84%
Aliaga Truck 2	115	75	99	24	7	603	1.1	65%	86%
Aliaga Truck 3	143	101	129	157	13	607	2.33	71%	90%
Aliaga Truck 4	91	80	83	38	1	605	0.63	88%	91%
Sirinyer Truck 1	90	73	79	70	5	601	0.66	81%	88%
Sirinyer Truck 2	115	78	101	89	7	603	1.05	68%	88%
Sirinyer Truck 3	18	18	17	2	1	601	0	100%	94%
Gaziemir Truck 1	111	93	99	80	6	603	1.09	84%	89%
Gaziemir Truck 2	92	73	80	109	15	601	0.71	79%	87%
Gaziemir Truck 3	103	83	93	103	16	602	0.95	81%	90%
Gaziemir Truck 4	26	26	24	1	0	0	0.02	100%	92%

average, if the number of packages was 116, it could pack 88% of the packages within an average of 1.5 s, at some cost of penalty.

As a result, the mathematical model has packed fewer boxes, but at no penalty cost. The developed procedure was able to pack 15% more items with a penalty, in around 1.5 s. Therefore, we can conclude that the algorithm is preferable to the mathematical model if a solution in less time with acceptable penalty scores is desired in the long run.

5 Conclusion

In this study, the real-life problem of a cargo company in Izmir, Turkey is considered. The daily problem of the company involves the dispatching of the vehicles, as well as routing and packing decisions. Namely, the daily problem of determining which parcels or packages should be distributed in which vehicle, the order in which these packages will be deployed, and the distribution routes of the vehicles is studied. We develop a web-based decision support system for practical usage, tackling this complex problem in a sequential manner for quick and applicable solutions. The developed algorithms and the mathematical model are embedded in the decision support system. The results are presented to the user through a user-friendly interface.

A computational study is carried out with the real-life instances gathered from the company's past data. It indicates that the heuristic approach to packing yields better solutions than the model in much shorter solution times. However, the model dominates the heuristic method in terms of penalties, as it includes hard constraints to avoid reshuffling of the parcels within the truck during the distribution process.

Even for small-sized instances, the run times for optimality seem to be unacceptable for the company for practical purposes, namely for the preparation of the morning loading plans. Hence, the heuristic approach for packing in our overall algorithm can be preferred, since it obtains fast and effective solutions even for large instances for the whole problem.

The developed decision support system can handle the company's daily needs for planning and help save time and effort both for the planners and for the dispatchers. As a future research agenda, our aim is to propose more sophisticated heuristic approaches where the quantities of make-to-stock orders are also determined with the consideration of inventory holding costs.

Acknowledgment. This study is carried out as a senior design project in the Department of Computer Science in Dokuz Eylül University, İzmir. Turkey.

References

1. Eliiyi U, Eliiyi DT (2009) Applications of bin packing models through the supply chain. Int J Bus Manag 1:11–19
2. Scheithauer G (1991) A three-dimensional bin packing algorithm. J Inf Process Cybern 27:263–271

3. Martello S, Pisinger D, Vigo D (2000) The three-dimensional bin packing problem. Oper Res 48:256–267
4. Lim A, Rodrigues B, Wang Y (2003) A multi-faced buildup algorithm for three-dimensional packing problems. Omega 31:471–481
5. Davies AP, Bischoff EE (1999) Weight distribution considerations in container loading. Eur J Oper Res 114:509–527
6. Castro Silva JL, Soma NY, Maculan N (2003) A greedy search for the three-dimensional bin packing problem: the packing static stability case. Int Trans Oper Res 10:141–153
7. Bischoff EE (2006) Three-dimensional packing of items with limited load bearing strength. Eur J Oper Res 168:952–966
8. Wäscher G, Haußner H, Schumann H (2007) An improved typology of cutting and packing problems. Eur J Oper Res 183:1109–1130
9. Zhao X, Bennell JA, Bektaş T, Dowsland K (2016) A comparative review of 3D container loading algorithms. Int Trans Oper Res 23:287–320
10. Wu Y, Li W, Goh M (2010) Three-dimensional bin packing problem with variable bin height. Eur J Oper Res 202:347–355
11. Crainic TG, Perboli G, Tadei R (2008) Extreme point-based heuristics for three-dimensional bin packing. INFORMS J Comput 20:368–384
12. He Y, Wu Y, de Souza R (2012) A global search framework for practical three-dimensional packing with variable carton orientations. Comput Oper Res 39:2395–2414
13. Nurcahyo GW, Alias RA, Shamsuddin SM, Sap MNM (2002) Sweep algorithm in vehicle routing problem for public transport. Asia-Pac J Inf Technol Multimed 2:51–64
14. Lin S, Kernighan BW (1973) An effective heuristic algorithm for the traveling-salesman problem. Oper Res 21:498–516
15. Applegate D, Bixby R, Chvátal V, Cook W (2006) Concorde TSP Solver. http://www.math.uwaterloo.ca/tsp/concorde/
16. GitHub (2018) Python wrapper around the Concorde TSP solver https://github.com/jvkersch/pyconcorde
17. Flask (2018) A microframework for Python. http://flask.pocoo.org

A New Demand Forecasting Framework Based on Reported Customer Forecasts and Historical Data

İlker Mutlu, Doğaç Sancar, Ege Naz Altın, Semih Balaban, Turan Can Cesur, and Önder Bulut(✉)

Industrial Engineering, Yasar University, İzmir, Turkey
ilkermutlu8@gmail.com, egenazaltin@gmail.com, turancancesur@gmail.com, semihbalaban@windowslive.com, dogacsancar@hotmail.com, onder.bulut@yasar.edu.tr

Abstract. The main purpose of the study is to design a forecasting framework through a better understanding of the structure of the customers' own forecasts to minimize the forecast error. Different than the conventional forecasting methods, our framework has an extra input, which is the set of reported customer forecasts; which is mostly based on customers' own production plans and capacities. Furthermore, our framework also takes the advantage of the classical methods: *Winter's Method in its pure form* and *Winter's Method merged with Decomposition Method* are taken into consideration, because these methods are able to handle stability, seasonality and trend in the demand data structures. The general proposed model is designed to embed the customer forecasts to the forecasting framework with a scientific methodology. A new method is created with a logic of a calculation of reliability indices for each customer and product pairs with following some statistical procedures. As a result of the method which can calculate the reliability of each customer, a forecast output is provided. On the other hand, forecast *Combination Method* is adjusted with respect to the nature of the problem to combine the output of *Winter's Method's* and the output of *Customer Forecasts Method.* In order to implement all these methods, the study is practiced in a well-known, international energy company. The demand data are divided into such subgroups as; *Regular, High Variance, New Born, Rare, On Off* and *Inactive* based on the demand structure of the data, to observe how these subgroups would react to the method. In order to achieve all these routines of the forecasting framework, we design a Decision Support System in Excel Visual Basic Applications (VBA).

Keywords: Forecasting · Customer reliability · Combination Method Winter's method · Decomposition · Product classification Decision Support System

1 Introduction

Forecasting is vital for all companies. Based on forecasts, firms develop materials requirement, capacity requirement, master production and production plans. Inaccurate forecasting can create a bullwhip effect. Every modules of supply chain reports the

N. M. Durakbasa and M. G. Gencyilmaz (Eds.): ISPR 2018, *Proceedings of the International Symposium for Production Research 2018*, pp. 839–848, 2019.
https://doi.org/10.1007/978-3-319-92267-6_67

forecast with safety stocks. When a firm overestimates the demand, the firm cannot follow the capacity planning. This leads to the unutilized capacity, inefficient production and excess stock which procures extra costs. Moreover, it could be necessary to transfer workers to other production lines. Additionally, the firm would increase even its market share with accurate forecasting system. The firm can satisfy requirement of new customers, instead of having excess stock.

When a firm underestimates the demand, the plant changes its capacity requirement production plans rapidly in order to satisfy the excess demand of customer. Bottlenecks can occur in production lines. Workers might work overtime. The replenishment of raw material from the supplier can take several months in some companies. It may be necessary to incur additional transportation costs in order to supply raw materials immediately to avoid raw material shortage. The plant might face up with the risk of the halt of the production which is the worst case scenario.

Historical demand data is used for years in forecasting methods (*objective time series models*) as an only input. The demand of next months are forecasted based on this information. An alternative idea is that, customers' preliminary orders (customers' own forecasts) can be also used as another input to the forecasting systems. Many companies have started to request future consumption (demand) estimates of their customers. Therefore, the customers also report forecasts based on their production plans and capacities. This information is very important and beneficial for both companies and customers. However, in order to use this data efficiently, a method is necessary to detect accurate of this information and how it is going to be embed in the forecasting framework.

Even though, the customer forecast is a valuable information, many companies do not know how to interpret and use this data scientifically. Therefore, they could not achieve the potential improvement capacity of their systems. Customer forecasts provide forecasts more accurately. Moreover, this information can be used in other ways in firms. Company could define a different, more efficient business strategy based on this information, with using the reliability of each customer based on the customers' past behaviours. Reliable and unreliable customers can be detected. Business deal can be strengthen with reliable customers.

In literature, not extensive researches are existed about this topic. There is one paper exists which studies on similar topic. "Alan Montgomery, Michael Trick and Nihat Altintas published a research in 2006, named "Using Customers' Reported Forecasts to Predict Future Sales". This study is aimed to adjust the forecasts to provide a better estimate with detecting bias in the forecast and uncertainty in the usage. Our study also adjusts the customer forecasts with a new developed method. However different than the literature, the developed method calculates the reliability index of each customer and use scientifically this data. Additionally, the developed forecasting framework generates forecasts based on historical demand data methods (classical methods) and then combine the output of classical methods and customers' forecasts output based on calculated weights. It is mainly focused on the generation of a new demand forecasting system including historical demand data and customer forecasts. It is aimed, applying a scientific methodology which generates forecasts with acceptable error margin. The paper includes the problem motivation and definition, literature review, model formulation and solution techniques and numerical results and application.

2 Literature Review

Forecasting methods are divided 2 parts as *Subjective* and *Objective* methods. In this project, studies are done based on objective methods since these are derived from an analysis of the past data. Objective methods are consisting of two categories which are *Regression (Casual) models* and *Time Series* models. Regression models construct a casual model that predicts the dependent variable based on change of one or more independent variables. *Time Series Methods* forecasts three types of data such as; stationary, trend and seasonal series. *Moving Average (MA)* and *Exponential Smoothing* are used for the stationary series. Nahmias [1] points that if there is no trend in demand series, these methods could be applied. The only parameter of *MA* is *N (number of recent observations)* and only parameter of exponential smoothing is α (*base value of the series*). Brockwell and Davis [2] points that, *Double Exponential Smoothing (Holt's Method)* can be applied in trend-based series. The method requires specification of two smoothing constants, α and β for the trend. There is no parameter to handle the seasonality.

Decomposition and Winter's Method can handle with the seasonal series. Decomposition deseasonalizes the demand data by removing seasonality. After removing the seasonality on the data, it is need to be implemented a forecast method. In the literature, mostly, *MA* and *Regression Analysis* are applied to generate forecasts and *Double Exponential Smoothing Method* rarely is also used.

According to Kalekar [3], *Winter's Method* as in addition to the double exponential smoothing method's parameters, there is γ factor for the seasonality. *Decomposition* and *Winter's Methods* are explained in detailed in the model formulation part.

In order to combine customer forecasts and output of forecasting methods, the combination weights of forecasts are supposed to be determined. According to [4], since there are two forecasts, called F_1 and F_2. It is assumed that, F_1 has a weight of w and F_2 has a weight of $(1-w)$, w is calculated by dividing the error variance of F_2 by the sum of the error variances of F_1 and F_2. They are under the unbiased with variances, var (F_1) and var (F_2) are supposed uncorrelated.

There are so many ways to measure the performance of the system such as; as it is mentioned in example of a book in [2]; *Mean Absolute Deviation (MAD), Mean Squared Error (MSE), Root Mean Square Error (RMSE), Mean Absolute Percentage Error (MAPE).* Reference [5] is mentioned *Percentage Absolute Error (PAE)*. Since *MAPE* and *PAE* give percentage results, they are taken into considerations to measure the performance of the forecasting system.

3 Model Formulation and Solution Techniques

In first step, historical demand data and reported customer forecasts are defined as the inputs of forecasting system. Meanwhile we benefit from the literature to forecast based on historical demand data. However there is no such a method in the literature which can analyse the customer forecasts with calculation of customer reliability. So, a methodology that can add the customer forecasts by several statistical process and also combine with the historical demand data is improved in this study. So, two main methods are expressed as *M.1. (Forecasting Based on Historical Demand Data)* and *M.2. (Forecasting of Reported Customer Forecast)* step by step in the below expression. Then the outputs of *M.1.* and *M.2.* are combined in the *C (Combination)* step.

M.1. Forecasting Based on Historical Demand Data

M.1.1. Applying only Winter's method in its pure form
M.1.2. Applying Decomposition method merge with Winter's method
M.1.3. Assigning of the optimal sub-method (among outputs of M.1.1 and M.1.2.)

M.2. Forecasting of Reported Customer Forecast

M.2.1. Calculating the reliability of adjusted customer forecast (Adjusted Method)
M.2.2. Calculating the reliability of customer forecasts (Base Method)
M.2.3. Assigning the optimal sub-method (among outputs of M.2.1 and M.2.2.)

C. Combination Method

M.1.1. Winter's Method: Regarding to the problem definition, *Winter's Method* fits to the complex systems, since the method is used for the data that has both seasonality and trend. The method consists three main parameters; α for *base level of time series*, β *for trend* and γ *for the seasonality*. According to Brockwell and Davis [2] Winter's Method consists of two main parts, which are initialization and update. For initialization part, the method needs 36 periods (3 year). First of all, the demand data should be divided into two parts for initialization and update parts. Next step is to calculate the sample means (Vi) for each season. The reason of calculation of Vi is to determine the relations between seasons. Based on Vi values initial values $(G_o, S_o \text{ and } c_t)$ are calculated in the next step. The initial linear trend (G_o) and the value of the series at $t = 0$ (current time) (S_o) are calculated in below formulas.

$$G_o = \frac{(V_m - V_1)}{[(m-1) * N]} \qquad S_o = V_m + G_o * [(N-1)/2]$$

The initial seasonal factors are obtained by dividing each of initial observation by the corresponding point along the line connecting between V_i's. According to Brockwell and Davis [2], the initial seasonal factors (c_t) are computed for each period and then averaged to obtain one set of seasonal factors. This procedure yields exactly n seasonal factors. It needs to be sure that the sum of the seasonal factors in exactly n. Average seasonal factors must be normalized.

The second step of Winter's method is to update S_t, G_o and c_t, based on below equations $(0 \leq \alpha \leq 1, 0 \leq \beta \leq 1, 0 \leq \gamma \leq 1)$.

$$S_t = \alpha\left(\frac{D_t}{c_{t-n}}\right) + (1-\alpha)(S_{t-1} + G_{t-1})$$
$$G_t = \beta(S_t - S_{t-1}) + (1-\beta)(G_{t-1})$$
$$c_t = \gamma\left(\frac{D_t}{S_t}\right) + (1-\gamma)c_{t-n}$$

After the calculation S_t, G_o and c_t the forecast made in period t for any future period t + τ $(F_{t,t+\tau} = (S_t + \tau G_t)c_{t+\tau-n})$.

Optimization: Parameters $(\alpha, \beta, \gamma \text{ and } n)$ are defined separately for each product regarding the different demand structures to increase the performance of the forecasting systems. So, the parameters are optimized in the range of 0 and 1 with the 0.1 increment to minimize the objective function (PAE). n is also optimized among the data set of 1, 3, 4, 6 and 12 which are decided due to the company nature to minimize the objective function. The data set can be easily change for other companies.

$$PAE(\alpha, \beta, \gamma, n) = \frac{\sum_{i=1}^{N} |D_t - F_{t-1,t}(\alpha, \beta, \gamma, n)|}{\sum_{t=1}^{N} D_t} x\, 100$$

And then, the forecast for each product is made with optimized parameters.

M.1.2. Decomposition is not a forecasting method just itself. It prepares the seasonal data by deseasonalazing to generate forecasts. With another saying, the method makes the data smoother. That's why it is needed to use with a forecasting method after it. The method locates moving averages of each season into the centred points of each season. In the next step, seasonal factors are calculated with dividing the demand by the centred moving averages. Then, averages of same season's seasonal factors are calculated and the factors are obtained by normalizing them (c_t). Final step is dividing the demand data by seasonal factors in order to get the deasonalized demand $(D'_t = \mathrm{D_t}/c_t)$.

In order to generate forecasts, *Winter's Method* will be applied on the deasonalized demand. The series still contain all patterns of the original series expect seasonality. Then, forecast output of *Winter's Method* (F'_t) multiplies with seasonal factors (c_t). This procedure provides that forecasts gain seasonality pattern again. F_t symbolizes the output of *Winter's Method Merge* with *Decomposition Method* $(F_t = c_t.F'_t)$.

In the ***M.1.3.*** step, the method which has the lowest error between the outputs are *M.1.1.* and *M.1.2.* are taken and assigned as F_t^2. Then, it goes to the *Combination* step to use in the general forecasting model.

For M.2., After intensive literature search, any appropriate method which can analyse the reported customer forecasts and has an understanding of customer behaviours cannot be found. So, a method is designed with the aim of determining the customer reliability. This can be achieved by calculation of the difference between the customer forecasts and demand and then implies several statistical processes. Customers reports the forecasts monthly in the *specific time (t)* for next 1, 2, …, a months *(i)*.

In the ***Adjusted Customer Forecasting Method (M.2.1.),*** Customer forecasts are adjusted regarding to the forecast's behaviours. The method can cope with the situations that customer forecasts are given with safety stock. Customer reports their forecasts with safety stock to make them sure. First of all, in order to detect customer's tendency, all the $\varepsilon_{t,i}$'s are calculated. The sign of this $\varepsilon_{t,i}$ values gives us a clue about the customer's behaviour *($t = 0$ current month) (n = the size of the data which is examined).*

- $\varepsilon_{t,i} = CF_{t,i} - D_t \qquad t \in \{1-n, -n, \ldots, 0\} \quad i \in \{1, 2, .., a\}$

The *probability of undershooting* of customer forecasts (P_i^1):

- $f_1(\varepsilon_{t,i}) = \begin{cases} 0, \varepsilon_{t,i} \le 0 \\ 1, \varepsilon_{t,i} > 0 \end{cases} \qquad P_i^1 = \sum_{t=1-n}^{t=0} \frac{f_1(\varepsilon_{t,i})}{n}$

The *probability of undershooting* of customer forecasts (P_i^2):

- $f_2(\varepsilon_{t,i}) = \begin{cases} 0, \varepsilon_{t,i} \le 0 \\ 1, \varepsilon_{t,i} > 0 \end{cases} \qquad P_i^2 = \sum_{t=1-n}^{t=0} \frac{f_2(\varepsilon_{t,i})}{n}$

In order to avoid the absorption of errors with opposite signs, the absolute value for $\varepsilon_{t,i}$ is taken.

- $\varepsilon_{mean_i} = \sum_{t=1-n}^{t=0} \frac{|\varepsilon_{t,i}|}{n} \qquad i \in \{1, 2, .., a\}$

Customer forecasts are adjusted with adding the error mean and also using *overshooting and undershooting probabilities* $\left(CF^*_{t,i}\right)$.

- $CF^*_{t,i} = CF_{t,i} - P_i^1 \text{x}\, \varepsilon_{mean_i} + P_i^2 \text{x}\, \varepsilon_{mean_i} \qquad i \in \{1, 2, .., a\}$

Calculation of *Adjusted Forecast Error*;

- $Forecast\ Error(F\varepsilon_{t,i}) = \left|\frac{CF^*_{t,i} - D_t}{D_t}\right| \text{x}\ 100 \qquad i \in \{1, 2, .., a\}$

Forecast Errors are used in order to determine the *forecast accuracy.*

- $Forecast\ Accuracy(FA_{t,i}) = \text{Max}\ (0, 100 - F\varepsilon_{t,i}) \qquad i \in \{1, 2, .., a\}$

Mean of forecast accuracies gives the *reliability index* of each customer for each i. Reliability index is calculated for each customer – product pair.

- $r_i(Reliability\ Index) = \sum_{t=1-n}^{t=0} \frac{|FA_{t,i}|}{n} \qquad i \in \{1, 2, .., a\}$

Customers' future forecasts are adjusted with respect to the previous procedure.

- $F_t^2 = (CF_{t,i} - P_i^1 x\, \varepsilon_{mean_i} + (1 - P_i^2)\, x\, \varepsilon_{mean_i})$

Base Method (M.2.2.) is designed since for cases that adjustment process might make the past customer forecasts and also the final customer forecasts more inaccurate. To prevent these cases, only *forecast error, forecast accuracy* and *the reliability index* are calculated and then customer forecasts are directly taken into the general formula without adjustment.

- $\varepsilon_{t,i} = CF_{t,i} - D_t \quad t \in \{1-n, -n, \ldots, 0\} \qquad i \in \{1, 2, .., a\}$
- $Forecast\ Error(F\varepsilon_{t,i}) = \left|\frac{CF^*_{t,i} - D_t}{D_t}\right| x100 \qquad i \in \{1, 2, .., a\}$
- $Forecast\ Accuracy(FA_{t,i}) = \text{Max}\ (0, 100 - F\varepsilon_{t,i}) \qquad i \in \{1, 2, .., a\}$
- $r_i(Reliability\ Index) = \sum_{t=1-n}^{t=0} \frac{|FA_{t,i}|}{n} \qquad i \in \{1, 2, .., a\}$
- $F_t^2 = CF_{t,i} \qquad i \in \{1, 2, .., a\}$

In the ***Combination (C)***, the outputs of *M.1.* and *M.2.* are combined with respect to reliability conditions.

W_i: Index of weight of the reported customer forecast in the combination $(0 \leq W_i \leq 1)$

- $\sigma_{i,1}$: Variance of error of the reported customer forecast (*M.2.*)
- $\sigma_{i,2}$: Variance error of forecast (*M.1.*)

a, b: threshold values which are designed parametric

General Combination Formula: $F_t^* = (1 - W_i).F_t^1 + W_i.F_t^2 \quad i = t$

(a) ***Customer has a high reliability;*** $r_i \geq a; W_i = 1$

(b) ***Customer has a moderate reliability;*** $b < r_i < a; \quad W_i = \frac{\sigma_{i,1}^2}{\sigma_{i,1}^2 + \sigma_{i,2}^2}$

(c) ***Customer has a low reliability;*** $r_i \leq b; \quad W_i = 0$

4 Numerical Analysis and Applicatıon

To verify *Customer Forecast Method (M.2.);* one sample data is solved with *Winter's Method* and *Customer Forecast Method. CF_i* represents the customer forecasts for after *i* month. Since it is difficult to show the customer forecasts for previous years in the figure, reliability levels (r_i) are defined in the Fig. (1). In Fig. (1), it is shown that *Winter's Method* cannot catch a sudden change in the demand structure. However, the *Customer Forecast Method* is applied to generate forecasts which is closer to the actual data.

For validation of the model, *PAE (Percentage Absolute Error)* (1) is taken into consideration as a performance measurement. When we check the current examples in the literature, MAPE *(Mean Absolute Percentage Error) (2)* is generally used. However, when using MAPE (2) as a criterion, the result will be undefined if the actual value is equal to 0. As an alternative performance measurement PAE (2), unlike MAPE, it is dividing the sum of the difference between demand and forecast by the sum of demands. That's why it has lower possibility that the sum of demand of consecutive periods are all zero also provides error rate in percentage.

$$\text{MAPE} = 100 * \frac{\sum_{i=1}^{N} \frac{|\text{Demand}_i - \text{Forecast}_i|}{\text{Demand}_i}}{n} \tag{1}$$

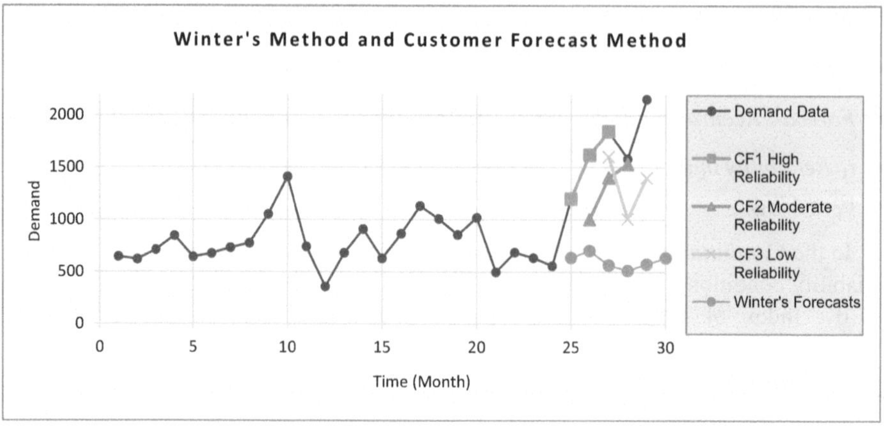

Fig. 1. Verification of customer forecast method and Winter's method

$$PAE = 100 * \frac{\sum_{i=1}^{N} |\text{Demand}_i - \text{Forecast}_i|}{\sum_{i=1}^{N} \text{Demand}_i} \quad (2)$$

This study was conducted in an internationally recognized firm with including 169 products in the current system to implement all these routines which are explained the model formulation. Customers of the firm are reported their estimations for next 3 months, updating following months. All routines based on historical demand data is executed in Excel Visual Basic Applications (VBA) for 169 products and PAE is calculated based on product classification. Since *New Born, On-off and Inactive* classes does not have enough data, the model does not apply on these classes until they have enough data and changes their classes. *Max, Min* and *Average PAE* values for 1, 2 and 3 months for *Regular, High Variance* and *Rare* classes cumulatively are shown in Table 1. As it is expected, average PAE of *High Variance* and *Rare* classes are higher than the *Regular* class. Any method which is based on historical data cannot follow the sudden changes in the demand structures. That's why *Customer Forecast Method (M.2.)* is designed to react to the sudden change in the demand structure. Winter's Method gives better results than Decomposition Method.

Winter's Method is designed optimally for 36 data points. Since the firm has data for 24 months, we run Winter's Method for 24 periods (2 years). However, it would be better to run the method for at least 36 months (3 years). Also the firm is not used to store data reliably. For instance, the demand for some periods are entered by the firm. All these situations make the forecast results less accurate. As another output of the study, the firm could save the data properly on the Decision Support System and the historical demand database will expand 36 months. So, it is expected that PAE values

Table 1. Performance of proposed model

Class name	Winter's merge with decomposition method			Pure Winter's method		
Regular (for 46 products)	PAE for 1 month	PAE for 2 months	PAE for 3 months	PAE for 1 month	PAE for 2 months	PAE for 3 months
Average	48,5%	56,2%	57,6%	44,7%	48,8%	54,3%
Min	3,2%	6,4%	13,7%	3,3%	9,2%	14,7%
Max	367,9%	244,7%	311,3%	347%	223,5%	223,5%
High variance (for 51 products)						
Average	98,1%	91,1%	99,9%	91,5%	82,6%	80,8%
Min	43,5%	36,5%	36,5%	10,4%	14,6%	29,7%
Max	341,3%	220,6%	268, 8%	343,3%	165%	105%
Rare (for 8 products)						
Average	98,6%	99,4%	100%	96,9%	99,7%	102,2%
Min	88,8%	95,8%	96,2%	90%	97,5%	96,6%
Max	100%	100	100	100%	100%	110%

of Winter's Method will be less. Additionally, in order to apply *Customer Forecast Method* and combine with *Winter's Method,* same amount number of data have to exist. As it is mentioned before, the firm has just started to apply collecting forecasts from its customers, that's why there is no 24 months reported customer forecasts in the database of the firm. So, when the reported customer forecasts database enlarges in the future, *Customer Forecasts Method* can combine with *Winter's Method.* Consequently, the PAE especially for *High Variance* and *Rare* classes will decrease.

The improvement in the Table (2) will be higher in the future, when the customer forecasts data pool will expand enough large. Additionally, company gets information during demand realization by calling the customers via phone. This gives not a true analyze of the improvements of the proposed model.

Table 2. Improvement of proposed model

Class name	Products percentage	PAE of pure Winter's method			PAE of Schneider electric			Percentage of improvement		
		PAE for 1 month	PAE for 2 month	PAE for 3 month	PAE for 1 month	PAE for 2 month	PAE for 3 months	PAE for 1 month	PAE for 2 month	PAE for 3 month
Regular	27%	44,7%	48, 8%	54,3%	60,7%	61,7%	84,6%	27,5%	20,9%	35,8%
High variance	25%	91,5%	82,6%	80,8%	70,2%	67,6%	69,3%	–	–	–
Rare	36%	96,9%	99,7%	102,2%	58,5	83,6%	107,6%	–	–	–
New Born	4%	Forecast are not made								
Inactive	2%	Forecast are not made								
on-off	6%	Forecast are not made								

5 Conclusions

The goal of the study was to develop a framework considering both historical demand data and customer forecasts to provide more accurate estimates of future demands. We proposed a new methodology can calculate the reliability of customer forecasts and apply statistical processes within the general forecast framework. More specifically, the proposed model is combining the customer forecasts with the wide spread classical methods *(Winter's and Decomposition Methods).* We believe that this study contributed to the statistical forecasting literature. Therefore, this project contributes to the overall statistical forecasting processes. To implement all these routines, we worked with a well know international company. We calculate the performance of our proposel model and the improvements. Cause they have started to collect the forecast from their customers for a short time ago, reported customer forecasts' pool has not yet reached sufficient size. That's why, we run the methods based on historical demand data. As the pool expands, improvements are expected to yield better results.

Another output of the study is that the companies would have a database with regular, reliable and sustainable use of all the information of the forecasting process via Decision Support System (DSS) Excel Visual Basic Applications (VBA).

Acknowledgment. We thank to Mert Paldrak who is working as research assistant in Industrial Engineering department Yasar University and also Duygu Alan and İsmail Serkan Kabataş who are the students of Industrial Engineering department of Yasar University who provided insight and expertise that greatly assisted the research.

References

1. Nahmias S (2005) Production and operations analysis, 5th edn. McGraw-Hill Irwin, Boston
2. Brockwell PJ, Davis RA (2002) Introduction to time series and forecasting, 2nd edn. Springer, New York
3. Kalekar PS (2004) Time series forecasting using holt-winters exponential smoothing. Kanwal Rekhi School of Information Technology
4. Social Science Computing Cooperative (2018) Homepage. https://www.ssc.wisc.edu/~bhansen/390/2010/390Lecture24.pdf
5. Adalı E, Aktaş, Y, Baykan OM, Güldoğan İ, Koral E (2014-2015) Demand tracking and forecasting system for finished goods. Yasar University Bachelor Thesis Project Book, Izmir

An Application of Permutation Flowshop Scheduling Problem in Quality Control Processes

Göksu Erseven, Gizem Akgün, Aslıhan Karakaş, Gözde Yarıkcan, Özgün Yücel, and Adalet Öner(✉)

Department of Industrial Engineering, Yaşar University, İzmir, Turkey
goksuersvn@hotmail.com, gizemakgunn1510@gmail.com, karakasaslihn@hotmail.com, gozdeyrkcn@hotmail.com, ozgun.ozturk@yasar.edu.tr, adalet.oner@yasar.edu.tr

Abstract. This study involves in a real world application of permutation flowshop scheduling problem (PFSP) in a local company that produces apparel and garments. Various test jobs arrive in the quality control department and a schedule has to be prepared considering the test stations and processing times. A job consists of a series of operations on certain stations. Each station is designed to perform a particular test operation. Therefore, there are jobs waiting to go through a series of quality control operations on various stations. Each job has different processing times at stations. Moreover, different jobs may have different processing times at a particular test station. However, the operation sequences (routes) of all jobs are the same through the stations. It is not allowed to change the order of the jobs on different stations. This problem falls into the realm of *permutation flow shop scheduling problem (PFSP)* in the literature. Various mathematical models are defined in literature for the optimal solution of the problem. One of them was chosen and used to solve small-sized problems. However, as the size of the problem increases, and since the problem is shown to be NP-hard for three or more machines, the solution time increases rapidly and hence it becomes significantly hard to solve the problem in polynomial time. Therefore, a combination of two heuristic methods is employed to solve the problem in reasonable time. First, NEH (Nawaz, Enscore, Ham) method is used to obtain a good initial feasible or a near optimal solution and then iterated local search method (ILS) is engaged to improve the solution. This solution procedure is implemented in a decision support tool in order to develop efficient schedules for quality control jobs in the company. The performance of the procedure is evaluated and verified by comparing the solutions with the optimal solutions of small sized problems. The tool can also be used for educational purposes since it is user friendly and has ability to present outcomes in detail with proper graphics.

Keywords: Permutation Flow Shop Scheduling Problem · Optimization Heuristics · Iterated Local Search · Decision support system

N. M. Durakbasa and M. G. Gencyilmaz (Eds.): ISPR 2018, *Proceedings of the International Symposium for Production Research 2018*, pp. 849–860, 2019.
https://doi.org/10.1007/978-3-319-92267-6_68

1 Introduction

A local company produces apparel and garments for men and women. The competitive attribute of the company is the quality of its products. It exports most of the products. In order to keep the competitive edge and to meet consumer safety standards and regulatory requirements of its destination markets, the company established and currently operates a sophisticated quality control facility equipped with state of art test stations. Among others, some of the quality tests performed in the facility are listed below.

- fiber identification
- test for banned azo colorants
- chemical testing
- dimensional stability: torque, stretch & recovery and shrinkage
- abrasion or piling
- colorfastness test
- flammability and burn test

The facility is busy to check and test the samples coming from different sources. The main source is the regular samples drawn from production lines. The other source is the samples of new designs. Each sample corresponds a job in the facility. A planning tool is desired in order to prepare efficient weekly schedules for the test jobs waiting in the system. In the facility, each job is processed through the same series of tests in a predefined order. That order is the same for all jobs. Each job has different processing times at different stations. Moreover, different jobs may have different processing times at a particular test station. The nature of the problem fits in the *permutation flow shop scheduling problem* in the literature.

Permutation flow shop scheduling problem (PFSP) has been extensively studied in the literature and has important applications in manufacturing and service systems. In the traditional permutation flow shop scheduling problems, n jobs are processed on m machines in the same order. The aim is to find the best sequence of jobs to be processed. The objective function is usually defined as minimizing the makespan. However, there are some other performance criteria employed in literature such as flow time, earliness, lateness, tardiness etc. [1]. Pinedo [2] defines the problem in detail and describe a scheme of classification for extensions and variants of PFSP. Wagner [3] presents initial methods and mathematical models for the solutions of job shop and permutation flowshop scheduling problems. Several other researchers provide mathematical models as in Baker [4], Stafford [5], Wilson [6] and Manne [7].

Rinnooy Kan [8] explains PFSP is a NP-hard problem for three or more machines. Therefore, many heuristics and meta-heuristics are proposed to solve the problem. Constructive heuristics and improvement heuristics are the two main categories of PFSP heuristics. The NEH algorithm is proposed by Nawaz et al. [9] is an example of constructive methods. There are other constructive algorithms presented by Palmer [10] and Campbell et al. [11]. According to Dong et al. [12] and Li et al. [13], NEH heuristic is considered as the best constructive heuristic for solving PFSP.

Meta-heuristic algorithms have also been studied to solve the problem. There are methods based on meta-heuristics such as genetic algorithms, simulated annealing [14], tabu search [15] and ant colony optimization [16]. Also, the iterated local search (ILS) [17] method is a good example for improvement heuristics. ILS is a simple but powerful metaheuristic that have mechanisms to run away from local minima/maxima. Those mechanisms include perturbation process, which is the best-known process to jump to a new restart position.

2 Problem Definition

The scheduling problem in garments and apparel company is a permutation flow shop scheduling problem as described above. The PFSP can be defined simply as follows: There are a set of n jobs and a set of m machines. Each job should pass through a set of m operations which must be done on different machines. All jobs have the same processing order of operations while passing through the machines. There are no precedence constraints among operations of different jobs. Operations cannot be interrupted and each machine can process only one operation at a time. The sequence changes between machines are not allowed. The objective is to find the best job sequence, which minimize the makespan, i.e. the maximum of the completion times of all operations.

The test operations in the company are organized to be done on 7 stations which correspond to the machines in the definition above. Similarly, the test samples correspond to jobs. The nature of the operations and the specifications of the machinery imposes pre-emptive mode of processing. The jobs are ready at the beginning of planning horizon.

There are some mathematical models reported in literature such as Baker [4], Stafford [5], Wilson [6] and Manne [7]. Those models have been investigated and the model proposed by Manne [7] has been chosen to implement for its simplicity. The details of that model are as follows.

$i \in \{1, \ldots, m\}$ stands for machine index

$j \in \{1, \ldots, n\}$ stands for job index.

Decision Variables:

$$D_{jj'} = \begin{cases} 1, & \text{if job j is scheduled before (not necessarily immediately before) job j}' \\ 0, & \text{otherwise} \end{cases}$$

C_{ij} = Completion time of job's j operation on machine i

C_{max} = Finishing time of the last operation on machine m (makespan)

Parameters:

p_{ij} = Processing time of job j on machine i

M = Large number

$$Z = \min C_{max} \quad (1)$$

$$\text{s.t.} \quad C_{1j} \geq p_{1j} \qquad j = 1, \ldots\ldots, n \quad (2)$$

$$C_{ij} - C_{i-1j} \geq p_{ij} \qquad i = 2, \ldots, m;\ j = 1, \ldots, n \quad (3)$$

$$C_{ij} - C_{ij'} + MD_{jj'} \geq p_{ij} \qquad i = 1, \ldots, m;\ j, j' = 1, \ldots., n, j < j' \quad (4)$$

$$C_{ij} - C_{ij'} + M\left(D_{jj'} - 1\right) \leq -p_{ij'} \qquad i = 1, .., m; j, j' = 1, \ldots, n, j < j' \quad (5)$$

$$C_{max} \geq C_{mj} \qquad j = 1, \ldots, n \quad (6)$$

$$C_{max} \geq 0 \quad (7)$$

$$C_{ij} \geq 0 \qquad i = 1, \ldots, m;\ j = 1, \ldots, n \quad (8)$$

$$p_{ij} \geq 0 \qquad i = 1, \ldots, m;\ j = 1, \ldots, n \quad (9)$$

The objective function given in (1) minimizes the makespan. Constraint (2) represents that the completion time of any job j is greater than or equal to the processing time for machine 1. Constraint (3) shows that the completion time difference of any job j between two successive machines $(i-1, i)$ is equal to or greater than the processing time in the ith machine of the same job j. Constraints (4) and (5) include the machine availability. These constraints define the precedence relationship between the jobs j on any machine i. Here, big M corresponds a large number. These constraints insure that job j either precedes job j' or follows job j' in the sequence, but not both. Constraint (6) represents the makespan, which is equal or greater than the maximum completion time of all jobs on the last machine. Constraints (7), (8) and (9) are the non-negativity constraints.

The model has been implemented in order to see the optimal solutions for small sized problems. However, PFSP has been shown to be NP-hard for three or more machines. For this reason, as the size of the problem increases, the solution time increases rapidly and hence it becomes significantly hard to solve the problem in polynomial time. Therefore, heuristic solution methods have been investigated in order to have a way to solve the problem in reasonable time. NEH (Nawaz, Enscore, Ham) algorithm has been chosen as the constructive heuristics since it is simple, easy to implement and proven to be effective. Additionally, iterated local search (ILS) algorithm has been decided to be the improvement algorithm. Therefore, a combination of NEH and ILS algorithms are used to solve the problem. In this setting, NEH algorithm is employed to find a good feasible solution, and then ILS steps in to improve the solution provided by NEH algorithm.

3 NEH and ILS Algorithms

The NEH algorithm is proposed by Nawaz, Enscore and Ham [9] for permutation flow shop scheduling problems that minimizes the makespan. In the NEH algorithm, the jobs are first sorted in descending order depending on the sum of processing times on all machines. Then the first two jobs with highest sum of processing times on the machines are considered for partial scheduling. The best partial schedule of those two jobs (i.e., one that provides lower partial makespan) is determined. This partial sequence is fixed in a sense that the relative order of those two jobs will not change until the end of the procedure. In the next step, the job with the third highest sum of processing times is selected and three possible partial schedules are generated through placing the third chosen job at the beginning, in the middle, and at the end of the fixed partial sequence. These three partial schedules are examined and one that produces minimum partial makespan is chosen. This procedure is repeated until all jobs are fixed and the complete schedule is generated.

Iterated Local Search is a powerful meta-heuristic algorithm proposed by Stützle [17]. The main characteristics of this algorithm is to randomly leap in the determined solution area. With this algorithm, the local optimum solution is obtained. In order not to be confined to a single place, perturbations are made with splashing other places and new local optimum results are found.

The local search starts with some initial sequences and then continually tries to improve the existing sequence with local changes. If a better sequence is found in the neighborhood of the current directory, the current sequence replaces existing sequence and the local search continues. The simplest local search algorithm applies these steps repeatedly until a better sequence is found in the neighborhood and stops at the first local minimum encountered. The pseudo-code of ILS algorithm is shown below in Fig. 1.

π= NEH
π_{best}= π
While "(stopping criterion)" **do**
    π_1= Perturbation (π)
    π_2= Local Search (π_1)
    If f (π_2) < f (π) **then**
        $\pi = \pi_2$
        If f (π) < f (π_{best}) **then**
            $\pi_{best} = \pi$
End while
Return π_{best}
End ILS

Fig. 1. Pseudo-code of ILS

Perturbation is defined as neighborhood for the local search algorithm in the literature for flow type scheduling problems. First, jobs i and $i + 1$ in two adjacent positions are interchanged (binary placement). Then, in the displacement phase, i^{th} and j^{th} job positions are reciprocally exchanged. Finally, the job in position i is removed and inserted in place of the job in position j (placement). The perturbation procedure is shown below (Fig. 2).

```
Do
pt1 = Select a random position of job
pt2 = Select a random position of job
  Loop While (pt1 =pt2)
  ward = Generate a random binary number (0, 1)
  If ward=1 Then
        Temp = p (pt2)
    Do
      p (pt2) = p (pt2 - 1)
      pt2 = pt2 - 1
    Loop While (pt2>pt1)
    p (pt2) = Temp
  Else
    Temp = p (pt1)
    Do
      p (pt1) = p (pt1 + 1)
      pt1 = pt1 + 1
    Loop While (pt1<pt2)
    p (pt1) = Temp
  End If
```

Fig. 2. Perturbation procedure

4 Decision Support System

Manne's mathematical model, NEH and ILS algorithms are implemented in the computer and embedded in a user friendly decision support tool. The interface of the tool is shown in Fig. 3. It provides some functions to help the user for setting and solving the problem quickly. One of the functions is designed for managing the input which includes the job list. A user may list, delete, add or edit the entries in the job list. Therefore, it ensures that the user has the updated version of the list. A part of this function is displayed in Fig. 4.

Once the job list is finalized, the decision tool processes input by incorporating a database, which holds the processing times of each job type, and hence prepares input data for PFSP. In order to solve the problem, there are three options. The first option is the optimal solution by mathematical model. However, the size of the problem is limited for this option since it takes a long time to solve the problem. A commercial

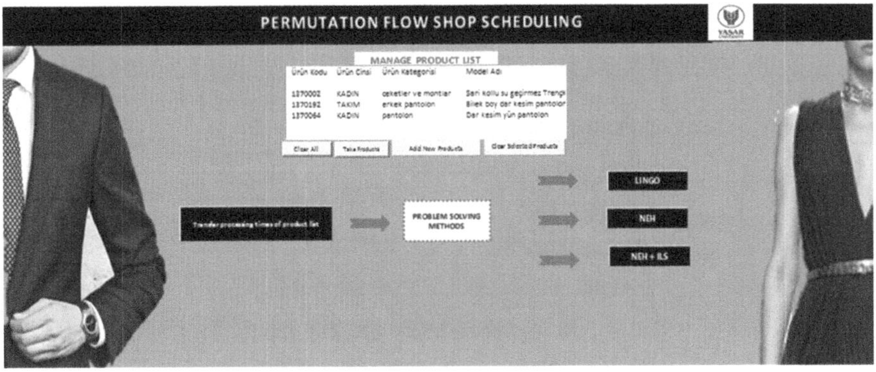

Fig. 3. DSS dashboard

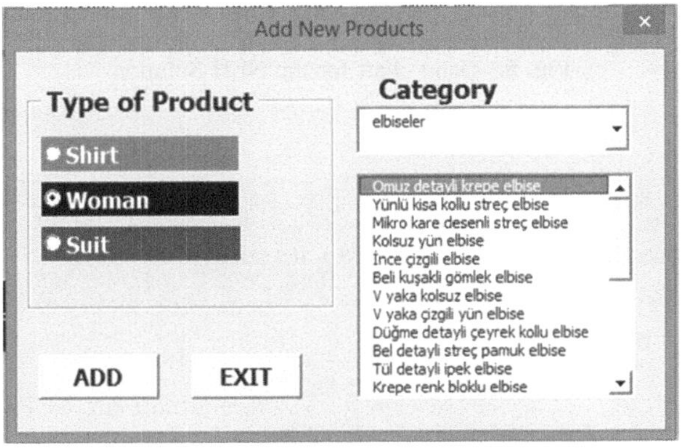

Fig. 4. Managing job list

solver sits behind to take part if mathematical model is desired to solve the problem within allowed size limits.

The second option is the solution by NEH algorithm only, and finally last option is the solution by the combination of NEH and ILS algorithms. The user may choose any option, and corresponding solution is shown in an Excel sheet formatted to present the outcomes in a form that is easy to understand.

The outcomes include the processing times of the jobs, starting and the completion times of the jobs on each station, the order in which the jobs will pass through the stations and the maximum completion time (Cmax). The detailed results may be printed in different formats as desired, and even sent in electronic format via e-mail. Additionally, the tool displays the Gantt chart of the solution. This feature is very useful to present outcomes in graphical format and hence it is easy to track the plan on different test stations. Sample Gantt charts are shown in Figs. 5 and 6.

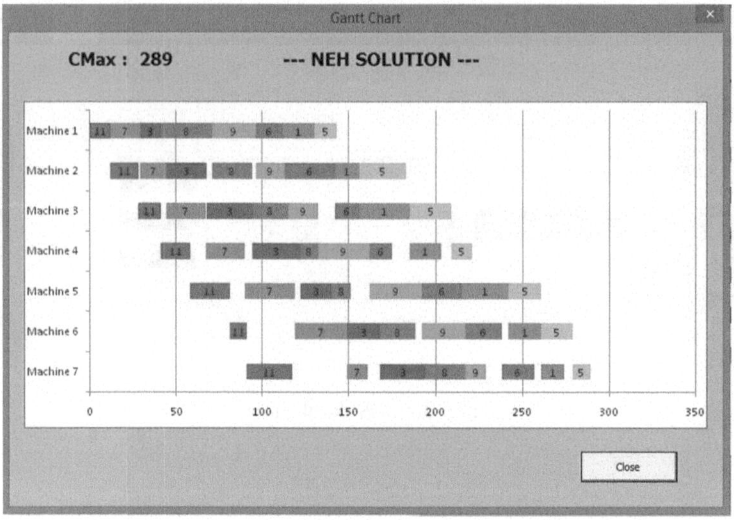

Fig. 5. Gantt chart for the NEH Solution

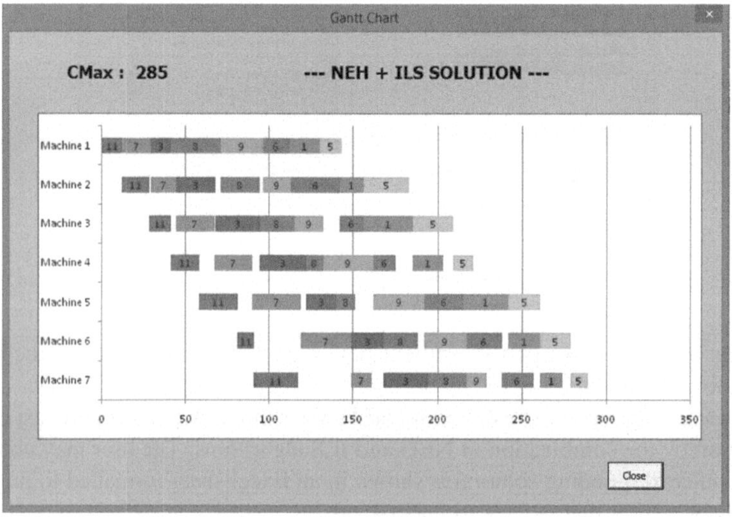

Fig. 6. Gantt chart for the NEH+ILS Solution

5 Computational Experience

This section includes verification of implementation of solution procedures and comparison of performances. All solution methods implemented in this tool has been verified by tracking numerous test problems. Numerical checks are supplemented with visual controls on Gantt charts. The mathematical model is implemented in Lingo 17.0. optimization software. In order to compare the performance of the solution procedures, different sets of test problems have been prepared. The inputs are generated randomly. The following tables show the performance comparisons of NEH, ILS and mathematical model with respect to their objective function values. Initial solution of the ILS algorithm is based on NEH algorithm. In the tables, m represents the number of machines and n represents the number of jobs. In Tables 1 and 2, the outcomes of seven different problem sets are presented. The problem sets are based on different number of jobs and machines. Each set includes 20 instances In Table 1, in all of the instances, except one, ILS algorithm reaches the optimal solution. In Table 2, ILS algorithm solution generates 0.07%, 0.56% and 0.31% average optimality gap in respective problems sets.

Table 1. Results of the Experiments

C_{MAX}	n = 6, m = 2			n = 6, m = 3			n = 8, m = 3			n = 8, m = 5		
Ins #	NEH	ILS	LINGO	NEH	ILS	LINGO	NEH	ILS	LINGO	NEH	ILS	LINGO
1	89	89	89	92	92	92	111	111	111	142	142	142
2	93	93	93	78	78	78	124	124	124	138	138	138
3	70	70	70	96	93	93	139	139	139	149	149	149
4	84	84	84	91	91	91	97	96	96	130	129	129
5	89	89	89	81	81	81	99	99	99	120	118	118
6	62	62	62	94	94	94	68	68	68	137	137	137
7	73	73	73	66	66	66	97	97	97	139	138	138
8	60	60	60	77	74	74	130	130	130	151	147	147
9	71	71	71	95	93	93	105	105	105	133	133	133
10	64	64	64	79	79	79	106	105	103	129	128	128
11	90	90	90	98	98	98	126	126	126	138	136	136
12	88	88	88	95	95	95	119	116	116	132	129	129
13	59	59	59	78	78	78	105	105	105	143	142	142
14	78	78	78	111	111	111	113	113	113	129	129	129
15	76	76	76	81	81	81	136	136	136	153	151	151
16	64	64	64	66	66	66	97	96	96	148	141	141
17	73	73	73	93	93	93	78	77	77	113	113	113
18	89	89	89	92	91	91	94	94	94	131	128	128
19	74	74	74	99	99	99	105	105	105	99	97	97
20	70	70	70	103	101	101	92	92	92	165	160	160

Table 2. Continuation of the Results of the Experiments

C_{MAX}	n = 10, m = 3			n = 10, m = 5			n = 10, m = 6		
Ins #	NEH	ILS	LINGO	NEH	ILS	LINGO	NEH	ILS	LINGO
1	145	145	145	159	152	150	161	159	159
2	139	135	133	189	189	184	152	151	151
3	135	132	132	150	148	148	161	159	159
4	135	132	132	147	142	142	170	169	169
5	120	116	116	145	145	141	171	169	169
6	146	146	146	142	139	139	174	170	170
7	116	116	116	159	159	159	184	177	170
8	136	131	131	155	155	153	196	188	188
9	123	123	123	180	170	169	164	162	162
10	136	136	136	152	149	149	182	179	179
11	135	135	135	154	154	152	170	169	169
12	154	154	154	164	161	160	171	163	163
13	123	123	123	172	172	172	202	197	197
14	116	116	116	146	145	145	158	157	157
15	147	147	147	136	129	129	187	183	183
16	141	141	141	146	145	145	170	163	163
17	143	143	143	179	179	179	194	186	186
18	118	118	118	169	169	169	176	175	171
19	118	117	117	153	146	145	168	168	168
20	126	126	126	171	169	169	158	158	158

In Table 3, relatively large instances are considered and optimal solution cannot obtained within a one-hour time limit. Hence, results of seven different problems set, each having 20 instances, are summarized in Table 3 for both NEH and ILS algorithms. Average gap between NEH and ILS is calculated for all problem sets and the smallest gap is obtained in first problem set (n = 12, m = 3) as 0.21% and the largest gap is obtained in problem set 5 (n = 15, m = 10) as 2.92%.

Table 3. NEH and ILS Comparisons

C_{MAX}	n = 12, m = 3		n = 12, m = 4		n = 12, m = 6		n = 12, m = 8		n = 15, m = 10		n = 20, m = 10		n = 30, m = 15	
Ins #	NEH	ILS	NEH	ILS	NEH	ILS	NEH	ILS	NEH	ILS	NEH	ILS	NEH	ILS
1	144	142	159	158	217	210	231	219	288	278	335	334	547	531
2	155	154	147	147	201	201	237	237	298	277	338	327	513	505
3	129	129	171	170	169	168	235	230	301	297	299	296	553	535
4	167	167	136	134	186	186	208	208	273	271	339	337	512	506
5	159	159	152	152	208	192	252	252	290	284	342	334	531	514
6	153	153	147	145	181	181	203	192	283	271	324	316	534	528
7	164	164	197	193	191	184	195	194	276	266	320	317	530	521
8	159	157	190	190	181	178	241	231	270	264	304	301	500	498
9	123	122	182	182	186	185	233	232	298	295	345	337	532	513
10	147	147	149	143	173	167	238	229	291	276	350	341	515	512
11	168	168	159	158	194	190	246	243	294	268	333	331	521	513
12	163	163	160	154	209	203	214	214	303	289	314	313	504	498
13	134	134	181	179	179	179	243	233	281	276	319	307	548	544
14	132	132	173	173	204	204	236	233	285	279	362	351	529	517
15	179	179	187	184	187	186	209	207	297	286	359	357	474	469
16	136	136	184	181	195	194	200	195	283	280	343	328	545	527
17	156	156	159	159	210	204	233	215	256	256	344	337	543	541
18	157	157	173	173	205	200	217	212	280	279	336	323	552	530
19	132	132	163	159	197	197	229	229	284	275	339	328	530	530
20	126	126	158	158	189	189	225	220	280	275	369	364	518	513

6 Conclusion

This study performs a PFSP in a quality control department of a textile company. As the problem is NP-hard, gathering optimal solutions by a solver (LINGO) is highly expensive. Therefore, NEH and ILS heuristic algorithms are implemented in order to minimize the makespan. Small and large problem sets are generated and small instances reach the optimal solution in a specified time limit. In those instances, NEH algorithm is used as an initial sequence to the ILS algorithm and maximum optimality gap is obtained as 2.76% with respect to ILS algorithm. Relatively large problem instances cannot reach the optimal solution in a given time limit due to the complexity of the problem. However, most of the cases, ILS improves the makespan with respect to NEH heuristic.

A flexible and user friendly tool has been developed for the company to enable the user to prepare quick and efficient schedules. It has been a useful tool for the company since they don't have such a tool before this study. It manages the input of the problem and present the solutions to the user through a user-friendly interface. The solution process and outcomes has been verified and validated. The tool has considerably decreased the amount of scheduling effort since schedules were prepared manually earlier.

The tool can also be used for educational purposes since it is user friendly and has ability to present outcomes in detail with proper graphics. It may be improved further in this field if animation effect can be added.

Acknowledgment. This work cannot be completed without the assistance of Anıl Tekye, Selin Gökkaya and Sinan Maramuroğlu. We are thankful for the contribution of them.

References

1. Chakraborty UK (2009) Computational intelligence in flow shop and job shop scheduling. Springer Science & Business Media, Berlin
2. Pinedo M (2002) Scheduling – theory, algorithms, and systems. Prentice-Hall, Upper Saddle River
3. Wagner HM (1959) An integer linear-programming model for machine scheduling. Nav Res Logist Q 6:131–140
4. Baker KR (1974) Introduction to sequencing and scheduling. Wiley, New York
5. Stafford EF (1988) On the development of a mixed-integer linear programming model for the flowshop sequencing problem. J Oper Res Soc 39:1163–1174
6. Wilson JM (1989) Alternative formulations of a flow-shop scheduling problem. J Oper Res Soc 40:395–399
7. Manne AS (1960) On the job-shop scheduling problem. Oper Res 8:219–223
8. Rinnooy Kan AHG (1976) Machine scheduling problems: classification, complexity, and computations. Nijhoff, The Hague
9. Nawaz M, Enscore EE Jr, Ham I (1983) A heuristic algorithm for the m-machine, n-job flow shop sequencing problem. OMEGA 11(1):91–95
10. Palmer DS (1965) Sequencing jobs through a multistage process in the minimum total time: a quick method of obtaining a near-optimum. Oper Res Q 16:101–107
11. Campbell HG, Dudek RA, Smith ML (1970) A heuristic algorithm for the n job, m machine sequencing problem. Manag Sci 16(10):B630–B637
12. Dong X, Huang H, Chen P (2008) An improved NEH-based heuristic for the permutation flowshop problem. Comput Oper Res 35:3962–3968
13. Li XP, Wang YX, Wu C (2004) Heuristic algorithms for large flowshop scheduling problems. In: Proceedings of the 5th world congress on intelligent control and automation, pp 2999–3003
14. Osman I, Potts C (1989) Simulated annealing for permutation flow shop scheduling. OMEGA 17(6):551–557
15. Grabowski J, Wodecki M (2004) A very fast tabu search algorithm for the permutation flowshop problem with makespan criterion. Comput Oper Res 31(11):1891–1909
16. Rajendran C, Ziegler H (2004) Ant-colony algorithms for permutation flowshop scheduling to minimize makespan/total flowtime of jobs. Eur J Oper Res 155(2):426–438
17. Stützle T (1998) Applying iterated local search to the permutation flowshop problem. Technical report, AIDA-98-04, Intellctics Group, Computer Science Department, Darmstad University of Technology, Darmstad, Germany

Daily Production Planning Problem of an International Energy Management Company

Elif Ercan, Pınar Yunusoğlu, Nilay Yapıcı, Sel Ozcan(✉), and Deniz Türsel Eliiyi

Department of Industrial Engineering, Yasar University, Izmir, Turkey
elifercan95@gmail.com, pyunusoglu@gmail.com, yapici.nilay.1@gmail.com, {sel.ozcan, deniz.eliiyi}@yasar.edu.tr

Abstract. This study is about real-life production and capacity planning problem in an international company which operates in energy management sector in Manisa, Turkey. The company produces different types of circuit breakers and delivers its products to different countries and distribution center, located in France. Within the scope of this problem, the production plan is done for nine products that are manufactured on six production lines. Each product has a unique production line, but some of the products are processed on common production lines. In this study, the production lot amount is determined each day by considering the due date and quantity of the customer orders without exceeding the capacity of the production lines. In the existing system, there are many tardy and early customer orders and the production plan is done manually which causes time loss for the company. A preemptive goal programming model is proposed for solving this problem where the main goal is to minimize total lateness in customer orders and minimizing the number of customer orders that have been split is considered as the secondary objective. The proposed mathematical model is solved optimally for real life instances in IBM ILOG CPLEX Optimization Studio 12.6.3. In addition, a heuristic method is presented in order to decrease the daily production planning duration and the fulfill the company's needs. Moreover, a user-friendly decision support system is developed where both solution techniques are embedded.

Keywords: Production planning · Preemptive goal programming Decision support system · Capacitated lot sizing problem · Optimization

1 Introduction

In recent years, customer satisfaction and high service level performance are accepted as a key factor in the long-term success of the enterprises. It is observed that enterprises try to remain competitive in the fierce business environment by launching the high quality product at low prices and accordingly increase their responsiveness to fluctuating customer demand. In this context, Just-in-time (JIT) manufacturing system becomes more popular in the business environment. JIT production has significant

N. M. Durakbasa and M. G. Gencyilmaz (Eds.): ISPR 2018, *Proceedings of the International Symposium for Production Research 2018*, pp. 861–871, 2019.
https://doi.org/10.1007/978-3-319-92267-6_69

effects on the enterprises by reducing wastes and decreasing inventory level, improving quality of the product and efficiency of the production. In many enterprises, the applicability of JIT manufacturing system has been discussed. The enterprises which have an intention to observe a significant reduction in finished goods inventory and to achieve high service level by the quick response to the customer demand prefer to apply JIT philosophy to their manufacturing system.

JIT philosophy is based on four objectives: zero inventory, zero failures, zero lead time, and zero delays. In order to prevent the delay by satisfying the customer demand on-time, the production capacity becomes more important. Wang investigates the effects of the reduction of earliness and tardiness in the mass production system under the constant capacity and focuses on how to use the JIT philosophy to improve the production planning approach of MRP-II [1]. Sawik focused on production scheduling for single and multi-period orders for master production scheduling in make-to-order manufacturing with various due date related performance measures. In this study, the objective is the minimization of total tardiness, number of tardy orders, maximum tardiness and tardy work ratio. In this sense, it is decided that minimization of tardiness and earliness by considering the capacity restrictions becomes more important to increase customer satisfaction and service level performance of the company [2].

Motivated from a real-life problem of an international company in the electricity and automation sector, this study focuses on providing daily production plans by deciding the production lot amount given the customer order quantities, due dates and weekly capacities of the production lines. The company manufactures circuit breaker products in its manufacturing plant located in Manisa, Turkey. Within the scope of the study, a user-friendly decision support system (DSS) was proposed to the company that creates production plan by minimizing earliness and tardiness in customer orders and total number of orders split. Our DSS proposes two solution techniques for the company, one provides the optimal solution obtained by a global optimizer and the latter yields a near-optimal solution via a heuristic method. Through these options, it is aimed to satisfy company needs by providing an optimal or near optimal solution in a short time.

The rest of the paper is organized as follows: In Sect. 2, we define the problem. Section 3, includes the preemptive goal programming model. In Sect. 4, the heuristic solution methodology is explained in detail. In Sect. 5, the computational study is explained in Sect. 5. Finally, the last section includes our conclusion and future remarks.

2 Problem Definition

Accepted as a global expert in energy management, the company has production facilities in all over the world. One of those is located in Manisa, Turkey and makes production for other plants as well as for the distribution centers. The company has a wide range of products in order to respond to customers' demand. The company gives high importance to continuity of flow, quality of product and distribution of products at the right time which is the general perspective of pull strategy and lean manufacturing. For this purpose, they construct their production system whose production steps are triggered by actual demand. The actual demand consists of two order types:

Make-to-order (MTO) and Make-to-stock (MTS). MTO orders are composed of the products that are produced according to the customer demand, whereas MTS orders are produced for the stock. Each product group has an MRP Controller (MRPC) code so that products are identified among customer orders with these codes.

The company manufactures nine product groups consisting of medium and low voltage circuit breakers. There are six production lines in the manufacturing area. Each product group is manufactured on a predetermined production line and some product groups can use same production line. Each line has a predetermined daily capacity denoting the number of products that can be produced in one day on that line. Production capacities of the lines determined by management in monthly meetings.

In existing production planning system, the production plan is carried out manually by the production planner and there is no systematic approach used. Currently, the planner determines the production date of each customer order by considering the due dates and the predetermined production capacities of the lines. Due to the packaging, shipment and customs procedures, the most suitable date to complete an order is two days before its due date. Since there is a capacity constraint for each line, it is not possible to produce each order two days before its due date. In this situation, the planner shifts the production date of some customer orders either forward or backward. Shifting backward implies an early order yielding to inventory holding costs, whereas forward shift results in tardiness. The company prefers to shift MTS orders backward as they are produced for stock. After the analysis of past data, most of the customer orders are either tardy or early in the existing production planning system. Currently, nearly 8% of the orders are delivered on time. Around 12% of the orders are tardy leading a backlogging cost for the company and almost 78% of the customer orders are early yielding an increase in the inventory cost, as well as customer dissatisfaction.

In this study, a real-life daily production planning problem is examined. One may initially reckon our problem as an application of the well-known capacitated lot sizing problem (CLSP) in the sense that the determination on optimal lot sizes for each customer order is of concern. However, it differs from the classical CLSP, in that the production date of each customer order is also determined according to predetermined due dates and the customer order quantities are decided by considering the problem-specific limitations on the production lines. The main objective is to decide on the production lot size and production date of each customer order to minimize total lateness, whereas the number of orders have been split are aimed to be minimized, as a secondary objective.

3 Mathematical Model

In this section, the proposed preemptive goal programming model is introduced. Firstly, the assumptions of the problem are explained. Then, the preemptive goal programming formulation is presented in detail.

Prior to formulating the problem, we assume that the capacity of each production line (units/day) is known and taken as a parameter for the proposed model. Our formulation only enforces to plan the MTO orders. Thus, we assume that once the MTO orders are planned, the remaining capacities of the production lines will be dedicated for the MTS orders. Moreover, no overtime and outsourcing are considered and there is no shortage of any material and/or equipment before the production starts. Based on these assumptions, the preemptive goal programming model is formulated as follows.

A. *Sets and Parameters*

I	= Set of product groups
T	= Set of days
N	= Set of customer orders
N_i	= Set of customer orders for product group i $N_i \subseteq N$
K	= Set of production lines
L_k	= Set of customer orders that are produced on line k $L_k \subseteq N$
α	= Tardiness penalty coefficient per day
Cap_{kt}	= The capacity of production line k in day t
d_n	= Due date of the order n
q_n	= Quantity of the order n
P_1	= Priority factor of lateness objective
P_2	= Priority factor of order splitting objective
M	: big number

B. *Decision Variables*

X_{nt} = The number of products produced for order n on day t

$$Y_{nt} = \begin{cases} 1, & \text{if } X_{nt} > 0 \\ 0, & \text{otherwise} \end{cases}$$

$$\rho_{nt} = \begin{cases} 1, & \text{if order n is satisfied on day t} \\ 0, & \text{otherwise} \end{cases}$$

$Tard_n$ = Number of days delayed of the tardy order n

$Earl_n$ = Number of early days of early order n

C. *Preemptive Goal Programming Model*

Based on the definitions above, the preemptive goal programming model is as follows.

$$\text{Min } P_1\left(\sum_{n=1}^{N} \alpha Tard_n + \sum_{n=1}^{N} (1-\alpha) Earl_n\right) + P_2\left(\sum_{n=1}^{N}\sum_{t=1}^{T} Y_{nt}\right) \quad (1)$$

The objective function (1) constitutes two different objectives, with different priorities P_1 and P_2. The main objective function of the problem, denoted with priority P_1 minimizes the total weighted tardiness and earliness (total lateness). The secondary objective with priority P_2 minimizes the total number of orders split. Since there is a hierarchy between these two objectives, the problem is handled as a two-stage optimization problem.

In the first stage, the following MILP model is solved.

$$\text{Min } L = \sum_{n=1}^{N} \alpha Tard_n + \sum_{n=1}^{N} (1-\alpha) Earl_n \tag{2}$$

subject to

$$\sum_{n \in L_k} X_{nt} \leqslant Cap_{kt} \quad \forall k, \forall t \tag{3}$$

$$X_{nt} \leqslant M Y_{nt} \quad \forall n, \forall t \tag{4}$$

$$\sum_{t=1}^{T} X_{nt} = q_n \quad \forall n \tag{5}$$

$$Y_{nt} - Y_{n(t+1)} \leqslant M \rho_{nt} \quad \forall n, \forall t \in T - \{t^{last}\} \tag{6}$$

$$Y_{n(t+1)} \leqslant (1 - \rho_{nt}) \quad \forall n, \forall t \in T - \{t^{last}\} \tag{7}$$

$$q_n - \sum_{i=1}^{t} X_{ni} \leqslant M(1 - \rho_{nt}) \quad \forall n, \forall t \in T - \{t^{last}\} \tag{8}$$

$$\sum_{t=1}^{T} \rho_{nt} = 1 \quad \forall n \tag{9}$$

$$Tard_n \geqslant \left(\sum_{t=1}^{T} t \rho_{nt}\right) - d_n \quad \forall n \tag{10}$$

$$Earl_n \geqslant d_n - \left(\sum_{t=1}^{T} t \rho_{nt}\right) \quad \forall n \tag{11}$$

$$\sum_{n \in N_1} X_{nt} \leqslant \frac{2}{5} Cap_{kt} \quad \forall t, n \in N_1, k = 1 \tag{12}$$

$$\sum_{n \in N_4} X_{nt} \leqslant \frac{2}{3} Cap_{kt} \quad \forall t, n \in N_4, k = 3 \tag{13}$$

$$\sum_{n \in N_5 \cup N_6} X_{nt} \leqslant \frac{2}{3} Cap_{kt} \quad \forall t, n \in N_5 \cup N_6, k = 3 \tag{14}$$

$$X_{nt} \in Z^{+} \cup \{0\} \quad \forall n, \forall t \tag{15}$$

$$Tard_n, Earl_n \in Z^+ \cup \{0\} \quad \forall n \tag{16}$$

$$Y_{nt}, \rho_{nt} \in \{0, 1\} \quad \forall n, \forall t \tag{17}$$

The objective function (2) minimizes total weighted tardiness and earliness (total lateness), where α denotes the tardiness coefficient. Constraint set (3) ensures that the total order quantity produced on each line cannot exceed the capacity of that line on that day. Constraint set (4) controls whether the production is done for each customer order that day. Constraint set (5) guarantees demand satisfaction for each customer order. Constraint sets (6–9) control whether the order quantity is split or not and provide the completion date of each customer order when the production quantity is equal to the order quantity. Constraint set (10) calculates the tardiness of the order n which is the difference between completion date of the order and its due date. If the customer order is produced until the due date, tardiness of this order is equal to zero. Constraint set (11) calculates the earliness of the order n that is the difference between the due date of the order and its completion date. Constraint sets (12–14) are problem-specific constraints related with the capacity of two production lines. Constraint set (12) guarantees that daily production amount of a product group with a specific MRPC controller does not exceed the 40% of the production capacity of the corresponding production line. Constraint sets (13) and (14) ensure that the maximum production amount of two product groups should be 2/3 of the production capacity of the corresponding production line, respectively. Constraint sets (15), (16) and (17) define the decision variables.

In the second stage, the mathematical model with the objective of minimizing the total number of orders split is solved. The model defined in (3)–(17) is solved where the objective function is defined (18) and, an additional constraint (19) is introduced.

$$\text{Min} \sum_{n=1}^{N} \sum_{t=1}^{T} Y_{nt} \tag{18}$$

$$\sum_{n=1}^{N} \alpha Tard_n + \sum_{n=1}^{N} (1 - \alpha) Earl_n \leq L^* \tag{19}$$

The objective function (18) minimizes a total number of orders split, as the company prefers. In constraint (19), we limit the total weighted tardiness and earliness cannot exceed the optimal objective function value of the model defined in (1)–(18). With constraint set (19), we guarantee the minimum lateness, denoted as L^* is maintained in the second problem, as well.

4 Heuristic Method

Bitran and Yanasse show that the CLSP is NP-hard even without setup times; no approach is provided for gathering the optimality [3]. Since this problem is NP-hard, we provide an alternative solution approach, embedded into a user-friendly decision support system developed in Excel VBA, so as to obtain fast and effective solutions for large problem sizes. We propose a heuristic method by combining the company's expectations and heuristic method for CLSP that is proposed by Karimi [4]. The decision support system (DSS) first starts with the elimination of the non-value-added activities within production planning process. Our DSS provides a near optimal production plan by minimizing the total lateness of customer orders and by using the capacity of each line in the most efficient way. Flow chart of the heuristic method is given in Fig. 1.

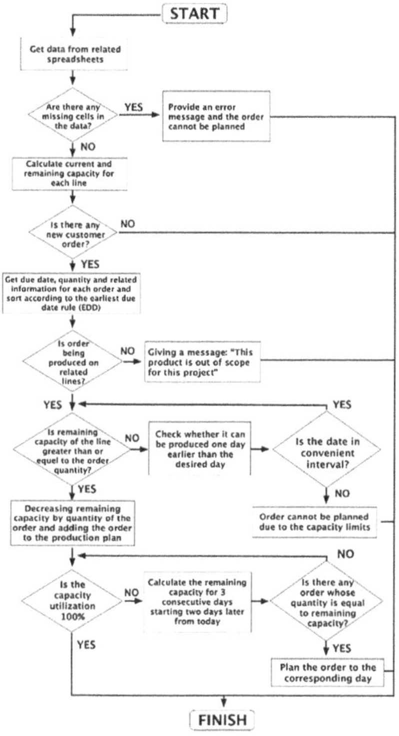

Fig. 1. Flowchart of heuristic method

The inputs of the problem are partial production plan, new customer orders which are taken from SAP and daily production capacities of production lines. In the heuristic method, firstly the related data is obtained from the spreadsheets and validity of the input data is checked. Then, current and remaining capacity is calculated for each line. New orders are sorted according to the EDD (Earliest Due Date) rule and each order's related production line is controlled. If the remaining capacity of the desired date is greater than or equal to the quantity of the order, the order is planned to the related production line. If there is not enough amount of remaining capacity, then the remaining capacity of line is controlled for one day before the desired day and if it fits, the order is planned to that day. This procedure repeats for five days before the desired day. After the fifth day, if the remaining capacity of related line is not enough, the order cannot be planned and in order to inform the user, the system provides a report. Then, the capacity utilization is checked for three (3) consecutive days. If the capacity utilization is 100%, the plan is completed for that day. If it is not, then the customer orders that are proper quantity and produced for stock are shifted to that day in order to use the remaining capacity.

The heuristic method was coded in Excel VBA to reduce the wasted time of current production planning process. A dynamic production planning interface was created so that the company can use for daily production planning activities.

5 Computational Study

The proposed mathematical model is solved using IBM ILOG CPLEX Optimization Studio 12.6.3 on a computer with i7 processor and having 8 GB RAM. As a pilot study, 3 real-life instances of the problem are examined for validation and verification of the proposed mathematical model. Properties of each problem instances and their solution duration and optimality gap are given in Table 1.

Table 1. Properties and solution durations of the test problem instances

Instance #	Number of orders	Number of product types	Number of production lines	CPU (min)	Optimality gap (%)
1	52	8	6	2.04	0
2	144	8	6	5.43	0
3	700	9	6	360	51.68

Note that, the solution times of relatively small instances e.g. instances 1 and 2 are less than a minute. However, as the number of orders increases, the solution time increases, concurrently. For example, the optimal solution cannot be obtained for instance 3 within 6 h and the optimality gap is around 52%.

A sensitivity analysis is conducted to test whether the optimal solution is affected for several α levels. The optimal solution of each test problem, as well as their computational time performances, are reported in Table 2. We can conclude that α does not play an important role in the optimal solution.

Table 2. Sensitivity analysis for various α levels

α	Total earliness (in days)	Total tardiness (in days)	Number of early orders	Number of tardy orders
0.2	3	10	1	4
0.5	3	10	1	4
0.7	3	10	1	4
0.9	3	10	1	4
0.995	3	10	1	4

For three real life instances that are given in Table 1, we compare the computational time performances and solution qualities of the preemptive goal programming model and the heuristic approach in Table 3. In the existing system, the duration of planning 30 (thirty) customer orders takes 90 (ninety) minutes.

Table 3. Comparison of the methodologies

Instance #	# of orders	Methodology	Total earliness (in days)	Total tardiness (in days)	Number of early orders	Number of tardy orders	CPU (min)
1	52	Heuristic method	32	9	7	2	0.92
		Preemptive goal Programming model	0	1	0	1	2.04
2	144	Heuristic method	68	18	23	8	2.31
		Preemptive goal Programming model	3	10	1	4	5.43
3	700	Heuristic method	138	47	82	30	7.2
		Preemptive goal Programming model	Out of memory				360

For all instances, both the preemptive goal programming model and the heuristic method yield better solutions than the current system. In the current system, for instance, 1, total earliness is 127 days, tardiness is 14 days and a total number of early and tardy orders are 45 and 3 respectively. For instance 2, total earliness is 325 days, tardiness is 34 days, number of early orders are 104 and number of tardy orders are 15. There are 549 early and 90 tardy orders in the current system for instance 3 and total earliness and tardiness are 1463 days and 283 days respectively. Thus, the tardiness and earliness in customer orders are reduced with the use of both methodologies and on-time delivery is improved accordingly. We can conclude that the preemptive goal programming model definitely outperforms the heuristic method, however, it is computationally expensive. We report that, for instances 1 and 2, the number of early orders decreased by more than 90% with the preemptive goal programming and number of tardy orders reduced by nearly 70%. However, the preemptive goal programming model does not perform well in terms of computational time. For instance 3, the optimal solution cannot be obtained by the mathematical model, but heuristic method provides a solution for this instance in 7.2 min.

The increase in the solution quality is relatively lower than the preemptive goal programming when the heuristic method is applied. The early and tardy orders are decreased by 80% and 35%, respectively once it is compared with the existing system. Although a clear dominance of the preemptive goal programming model over the heuristic method is observed in terms of solution quality, the daily production planning duration is decreased by 94% with the heuristic method. This implies that the heuristic method provides effective solutions within reasonable computational time.

6 Conclusion

In this study, the real-life daily production and capacity planning problem of a company is studied. In this perspective, the production lot size and production date of each customer order are determined according to due date and the quantity of the customer orders without exceeding the capacity of the production lines. Once the problem-specific constraints are employed, the problem differs from the well-known capacitated lot sizing problem. The problem is formulated as a preemptive goal programming model and solved in two stages. Moreover, a simple heuristic approach is developed to obtain high quality solutions within reasonable time. A user-friendly decision support system is developed in Microsoft Excel Visual Basic for Application. A heuristic approach is also employed into the decision support system, where the inputs of the problem are taken from the company's ERP system and the daily production plans for each production line are constructed thereby. A computational study is carried out with the real-life instances gathered from the company's past data. The computational study indicate that both the preemptive goal programming model and the heuristic method yield better production plans than the existing system. However, preemptive goal programming model definitely dominates the heuristic method in terms of solution quality, whereas relatively good solutions are achieved with the heuristic method in very short computation time. For small-sized instances, the optimal solution is found very quickly by the preemptive goal programming model. However, for larger instances, the run times for optimality seem to be unacceptable. Hence, heuristic method is preferred as it provides fast and effective solutions even for large instances. We believe that our decision support system can handle the company's basic needs on planning and achieve drastic time savings for the planner as well as reduces user errors. As a future research agenda, our aim is to propose more sophisticated heuristic approaches where the quantities of make-to-stock orders are also determined with the consideration of inventory holding costs.

Acknowledgment. We are grateful to the company for sharing their data with us to complete this work. This work cannot be completed without the assistance of Asst. Prof. Dr. Adalet Öner, Asst. Prof. Dr. Canan Pehlivan, Research Assistant Sinem Özkan, and students Alper Uyar, Ece Başar, Fatih Akamca and Irem Amaç. We are thankful for the contribution of them. We also thank The Scientific and Technological Research Council of Turkey (TUBITAK) for funding this study within 2209B-National/International Research Projects Fellowship Programme for Undergraduate Students.

References

1. Wang YM, Parkan C (2007) A preemptive goal programming method for aggregating OWA operator weights in group decision making. Inf Sci 177(2007):1867–1877
2. Sawik T (2003) Integer programming approach to production scheduling for make-to-order manufacturing. Math Comput Model 41(2005):99–118
3. Bitran GR, Yanasse HH (1982) Computational complexity of the capacitated lot size problem. Manag Sci 28(10):1174–1186
4. Karimi B, Fatemi Ghomi SMT, Wilson JM (2003) The capacitated lot sizing problem a review of models and algorithms. Omega 31(2003):365–378

Design of a Decision Support System (DSS) for Housekeeping Operations

Esin Acar, G. Şeyma Demir, Talya Temizçeri, and Levent Kandiller(✉)

Department of Industrial Engineering, Yasar University, İzmir, Turkey
esinnacr@gmail.com, gsymdmr@hotmail.com,
{talya.temizceri,levent.kandiller}@yasar.edu.tr

Abstract. This paper conducts our senior project at Altın Yunus Hotel located in Çeşme, Turkey and it contributes to the improvement of operations management of the hotel, focusing on the front office, reception and housekeeping services. The problem is determined as the housekeeping problem in consideration of other problems described in detail in the following sections. Required data are provided by the hotel management. The aim of the study is to develop a decision support system (DSS) covering and increasing the efficiency in service quality. Through the literature survey, system analysis and the developed mathematical models such as regression, time study, worker assignment, uniform parallel machine scheduling and routing optimization are presented. Finally, achieved time and cost savings are presented.

Keywords: Regression · Time study · Worker assignment
Uniform parallel machine scheduling · Decision support system

1 Introduction

In any hospitality establishment there are three departments particularly concerned withaccommodation: the reception department, whose staff sell and allocate the rooms; thehousekeeping department, whose staff plan, provide and service the rooms; themaintenance department, whose staff provide adequate hot and cold water, sanitation, heating, lighting and ventilation as well as maintaining and repairing individual articlesand area within the rooms operation. This study is among the first studies in which the techniques of production research applied to the above operations at hospitality sector.

The main motivation of this study is our contracted research at Altın Yunus Hotel which is a subsidiary business of Yaşar Holding and one of the largest facilities of Turkey, located on an area of 140.000 m^2 at the Kalem Burnu in the Boyalık District of Çeşme, Izmir, Turkey. It was founded in 1972 and the construction was completed in 18 months and became operative in the 1974 summer season. Altın Yunus is the very first “first class holiday village in Turkey” which could accommodate a thousand people in 465 rooms. At the beginning of our study, a number of symptoms are observed and the problems are identified with the use of related data. One of the most important problems encountered in the hotel is the real-time scenarios created due to system dynamically changing demands. For example, the difficulty of managing the

N. M. Durakbasa and M. G. Gencyilmaz (Eds.): ISPR 2018, *Proceedings of the International Symposium for Production Research 2018*, pp. 872–883, 2019.
https://doi.org/10.1007/978-3-319-92267-6_70

process of emptying rooms and accommodating customers in the cleaned rooms during the day between 12:00 and 14:00. During this hot period, while on one hand attention is paid to customer satisfaction, the teams need to be used efficiently on the other hand. In order to facilitate the solution of similar problems, a study has been conducted in the floor services unit in the hotel reception department. At the system analysis, a demand forecasting model is developed as well as decision models formulated at both the tactical level and the operational level. Our first planning model performs annual personnel levelling. Our second planning model facilitates a daily workforce planning and scheduling including the hot time period mentioned above. The third tool provides a navigation and routing service for both housekeeping and technical services after solving TSP with network distances. Solution methods were developed for these three models and verified by small size problems and the coding of these solution algorithms is completed. An MS Excel-based DSS developed is available for the decision makers.

The remainder of this paper is organized as follows. In Sect. 2, the literature review of studies is presented. In Sect. 3, the problem definition is stated. The problem formulation with observations and input analysis is presented in Sect. 4. In Sect. 5, verification, validation and sensitivity analysis are reported. Computational results, decision support system and output analysis are proposed in the same section. Finally, the conclusions and recommendations for future research are presented in Sect. 6.

2 Literature Review

A standard time is a time required for a work to be defined under specified conditions [1]. A time study can also be defined as a work measurement technique used to record the time and extent of a particular work item under certain conditions and to determine the time required for that work to be performed at a defined work rate (performance) by analysing the collected data. Time standards are used in planning future work and in evaluating past work. The time study also requires the use of concentration techniques such as performance grading so that the working speed can be determined, and the working speed can be correlated with the standard working tempo. When working on the standard work schedule and using the appropriate rest periods, a worker will have reached the standard performance level during workday or shift. The standard time for a job, observing the repetition frequency of components causes all time to occur [2].

'Regression Analysis', 'Time Study', 'Travelling Salesman Problem' 'Parallel Machine Scheduling Problem' and "Scheduling Method" topics are discussed briefly.

The main purpose of the regression analysis, a statistical technique used to relate variables, is to create a mathematical model to relate dependent variables to independent variables [3]. In general, a regression model form defines a single algebraic equation [4]. The multiple linear regression models are used for stating the relation between two or more explicative variables and the response variable by identifying a linear equation between the observed dates. For each value of the independent variable x, it is associated a value of dependent variable y. The individual values of the registered explanatory variables within the linear regression $X_1, X_2, \ldots, X_n$ is defined as: $\mu_y = \beta_0 + \beta_1 X_1 + \beta_2 X_2 + \ldots + \beta_p X_p$.

The parallel machine scheduling problem is one of the important and difficult problems in literature. This problem consists of the scheduling of a set of independent jobs on identical (parallel) machines (processors) with the aim of minimizing maximum job completion time. The rooms to be cleaned by any housekeeper are the jobs and the housekeepers are the machines. Moreover, the goal is to schedule identified jobs and minimize the maximum completion time [5].

The Traveling Salesman Problem (TSP) historically is the core optimization problem studied extensively in the literature. The TSP problem deals with finding the shortest (closed) tour in an n-city situation where each city is visited exactly once [6].

3 Problem Definition

In the current system, the service flow starts at the front desk (reception) when the registration operations are carried out on the system. Transaction entries by the front desk are automatically fed floor services and the check-in operations of customers are initiated. While these operations are taking place, manual preparation of these services causes inability to meet the demand and it leads to some problems. These observed symptoms are unsteady working methods in the housekeeping process and delays in the workflow as a result of those methods. One of the most important reasons for the disruptions in these processes is the difficulty in planning the check-in check-out and cleaning services during hot times (12:00 to 14:00). One of the difficulties observed is that the customer does not comply with the check-out time and continues to stay in the room when he/she is required to leave the room. In the face of such events, floor services get affected by this disruption and cannot complete room cleaning, and the people responsible for meeting customers (receptionist, bellboy etc.) cannot assign the rooms to the customers. This situation puts both the employees and the company in a difficult situation against the customers and the mistakes made by the employees in the operation field interrupt the following processes. In the direction of these symptoms and observations, we can say that there is no systematic arrangement for job assignments in the company, which causes some disruption in the field of operations management. The company needs a decision support system to overcome these problems and to increase productivity.

Along the direction requested by the firm, we are focusing the housekeeping department which the main problems are observed. The housekeeping department is responsible for cleaning all areas, especially the customer rooms. The main issues observed in this department are the enrollment of different employee types according to the seasonal intensity, the identification of the employees who need to be ready for any job, the assignment of the employees in the floor services for the cleaning of the rooms to be prepared and the arrangement of the rooms according to a certain rotation system. As another study, the performance ratings are determined considering the experience of the employees (experienced, inexperienced, intern) and a time study is conducted to collect the preliminary data. After these observations, cleaning processes are recorded and the information about how long each employee cleans an area is obtained according to the time analysis.

According to these observations, reducing the cost, reducing the total time and reducing the total delay (which are our key performance indicators) are the key performance indicators.

4 Problem Formulation

In this section, the four models mentioned above are described in detail together with the solution procedures.

4.1 Demand Forecasting

The timing model used in our first model determines the number of floor personnel, whose estimated capacity will be equivalent. Observing the number of people coming to Altın Yunus and the number of the rooms for the last 4 years, regression analysis is conducted to estimate the year 2018. First, analyzing past years' data, it is noted that the hotel is a seasonal hotel and more customers come on weekends. Thus, a week is divided into two half weeks as weekdays and weekend. It is determined as a half week from Monday to Thursday and a half week from Friday to Sunday, so one year is determined as the independent variable as 104 half-weeks. Our other independent variables are temperature averages, special days, terrorist incidents and their combinations. R^2_{adj} was found as 95%, thus the regression model is good to use. Due to hotel privacy, the specific regression equation is concealed.

4.2 Workforce Planning

This model makes cost-effective long-term team planning. The formulation reduces the total labor cost while meeting the demand for services. This is a simple aggregate service planning (team scheduling) model [6]. The teams that are ready for a certain planning period (half week) in the basic constraint are on the left side of the inequality (full-time employees, interns, part-time employees). On the right side of the equation, there are the requirement values determined by the regression model. As a solution method, CPLEX general solver has been used. Due to the fact that the results of the model are found meaningful by Altın Yunus Executives, proving the validity (Table 1).

The long-term planning formulation is given below:

$$min \sum\nolimits_{w=1}^{n} (C_F F_w + C_P \mathrm{P_w} + C_I I_w) \quad (1)$$

s.t.

$$q_F \sum\nolimits_{t=w-d_F+1mod(n)}^{w} F_w + \mathrm{q_I} \sum\nolimits_{t=w-d_I+1mod(n)}^{w} I_w + \mathrm{q_P} P_w \geq \mathrm{R_w} \quad w = 1, \ldots, n \quad (2)$$

$$\sum\nolimits_{t=w-d_I+1mod(n)}^{w} I_t \leq u_I, P_w \leq u_P, \sum\nolimits_{t=w-d_F+1mod(n)}^{w} F_t \leq u_F \quad w = 1, \ldots, n \quad (3)$$

$$I_W, P_W, F_W \geq 0 \quad \text{and} \quad \text{INT} \quad w = 1, \ldots, n \quad (4)$$

Table 1. Notation for workforce planning model

Parameters: *F: Full timer I: Intern P:Part timer and w half weeks (w =1, ... ,n)*			
C_I	unit cost for intern workers	d_F	working duration of full time
C_P	half week cost for part time workers		
C_F	unit cost for full time workers	d_I	working duration of intern
q_I	max number of rooms cleaned by an intern	u_I	upper bounds set by the decision maker for #interns
q_P	max number of rooms cleaned by a part time	u_P	upper bounds set by the decision maker for #part timers
q_F	max number of rooms cleaned by a full time	u_F	upper bounds set by the decision maker for #full timers
R_w	number of the rooms that will be cleaned		
Decision Variables: *w half weeks (w =1, ... ,n)*			
I_w	number of intern workers starting at half week w and will be available at periods $w, w+1, ..., w+d_I-1 \ (mod\ n)$		
F_w	number of full time workers starting at half week w and will be available at periods $w, w+1, ..., w+d_F-1 \ (mod\ n)$		
P_w	number of part time workers working at w		

The objective function (1) minimizes the total annual cost. The main constraint states that the maximum number of rooms that the existing workers will clean must be more than the number of rooms that are needed to be cleaned during hot times. Constraint set (3) imposes the bounds set by the decision maker. Constraint (4) is the sign restriction.

4.3 Employee Assignment

This model aims to designate rooms to be cleaned at that date to the employees in the most appropriate way (Table 2).

Table 2. Notation for employee assignment model

Parameters: $w = 1, \ldots, W$ and $r = 1, \ldots, R$	
RT_r	Release Time of room r
DD_r	Due Date of room r
P_{rw}	Processing Time of room r by worker w
M	Large number
Decision Variables: $w = 1, \ldots, W$ and $r = 1, \ldots, R$	
L	Maximum lateness
L_r	Lateness
S_r	Starting time of room r
Z_w	If worker w is ever used; 0, otherwise
X_{rw}	1, if room r is assigned to worker w; 0, otherwise
Y_{pqw}	1 , if room q is immediately processed by worker w after room p; 0, otherwise

The employee assignment formulation with three alternative objective functions is given below:

$$Min \sum_{w=1}^{W} Z_w \tag{5}$$

or

$$Min \sum_{l=1}^{R} L_r \tag{6}$$

or

$$\text{Min}\, L \tag{7}$$

s.t.

$$L \geq L_R \quad r = 1, \ldots, R \tag{8}$$

$$\sum_{w=1}^{W} X_{rw} = 1 \quad r = 1, \ldots, R \tag{9}$$

$$\sum_{r=1}^{R} X_{rw} \leq \text{M}Z_w \quad w = 1, \ldots, W \tag{10}$$

$$RT_r \leq S_r \quad r = 1, \ldots, R \tag{11}$$

$$S_r + \sum_{w=1}^{W} P_{rw} X_{rw} - L_r \leq DD_r \quad r = 1, \ldots, R \tag{12}$$

$$Y_{pqw} \leq X_{pw} \quad \forall p, q = 1, \ldots, R, \; p \neq q, \; w = 1, \ldots, W \tag{13}$$

$$Y_{pqw} \leq X_{qw} \quad \forall p, q = 1, \ldots, R, \; p \neq q, \; w = 1, \ldots, W \tag{14}$$

$$S_p + P_{pw} X_{pw} \leq S_q + M\left(1 - Y_{pqw}\right) \quad \forall p, q = 1, \ldots, R, \; p \neq q, \; w = 1, \ldots, W \tag{15}$$

$$S_r, L_r \geq 0 \quad r = 1, \ldots, R \tag{16}$$

$$X_{rw} = 0 \; or \; 1 \quad r = 1, \ldots, R \quad w = 1, \ldots, W \tag{17}$$

$$Y_{pqw} = 0 \; or \; 1 \quad \forall p, q = 1, \ldots, R, \; p \neq q, \; w = 1, \ldots, W \tag{18}$$

$$Z_w = 0 \; or \; 1 \quad w = 1, \ldots, W \tag{19}$$

The objective function (5) minimizes the number of workers used; whereas the objective function (6) minimizes total lateness and the objective function (7) minimizes the maximum lateness. Constraint (8) ensures that the maximum lateness is equal or greater than the lateness of each room. Constraint (9) states that every room should be processed. Constraint (10) marks the usage of worker w. Constraint (11) states that the starting time of a room is later than, its release time. Constraints (12) calculates the lateness. Constraint (13) and (14) relate with x variables to y variables. Constraint (15) ensures that no overlap of rooms by a worker. Constraint (16)–(19) are sign restrictions.

Since the problem is NP-Hard, we develop a heuristic to obtain good solutions in a reasonable time. The algorithm given in Fig. 1 runs as follows: The rooms to be vacated at 12:00 are shown in Group A (ready) and the rooms to be vacated later than 12:00 are shown in Group B (not ready). When the time is beyond ready time, these rooms are moved to the list A. The rooms are assigned to Group A, according to the time of delivery, the nearest employee in terms of distance and the available times of the employees. When the assignment is realized, the dependent time for cleaning the rooms is also taken into consideration and time advances. Our goal here is to assign the employees the most appropriate way by reducing walking and keeping the waiting times to the minimum. A sample problem given in Table 3 is given for illustrative purposes. A sample solution is depicted in Fig. 2.

```
procedure_construct(Romms(RT,DD,NCT),wo
rkforce(rate),Interroom(walking time))

begin
"Initialization"
construct room list A ordered wrt due date
construct room list B ordered wrt ready time
time = 12:00
do while all workforce assigned @12:00
        r=pop(A)
        w=pop(workforce empty 12:00)
        assign room r to worker w between
time and time + CT(r,w),ready(w) = time +
CT(r,w)
end do
"main loop"
do while A ≠ ∅ and B ≠ ∅
        ww=argmin RT(w)
        pop r from B such that RT(r) ≤
RT(ww)
        push r to A with respect to DD(r)
        t time= RT(ww)
        r=ww's last room
        rr=argmin(Interroom(r,rrr))
        pop rr from A
        assign rr to ww from RT(ww) +
        Interroom(r,rr) to RT(ww) +
Interroom(r,rr) + CT(rr,ww)
        RT(ww) = RT(ww) + Interroom(r,rr) +
CT(rr,ww)
        If A = ∅ → Go to
                If A ≠ ∅ or B ≠ 0 → Go
to
end do
"Finish"  " A = ∅"
If B = ∅ then STOP
do while B ≠ ∅
        find rr = argmin(RT(r))
        find available (w: ∀w such that RT(w)
≤ RT(r))
        If available w ≠ ∅ then find ww =
argmin(Interroom(ww's last room,rr))
        else ww = argmin(RT(w) +
Interroom(ww's last room,rr))
        assign room rr to ww
        pop rr from B
end do
```

Fig. 1. Our heuristic algorithm

Table 3. Cleaning time form for rooms

Group A			Group B		
Number of rooms	Ready Time	Due Date	Number of rooms	Ready Time	Due Date
R1	12:00	12:30	R11	12:15	13:30
R14	12:00	12:45	R21	13:30	14:00
R24	12:00	13:15	R2	12:30	13:00
R3	12:00	12:30	R15	12:30	13:15
R5	12:00	13:00	R12	13:00	14:00
R4	12:00	12:45	R23	12:45	13:30
R13	12:00	12:50	R22	13:00	14:00
			R25	12:30	13:00

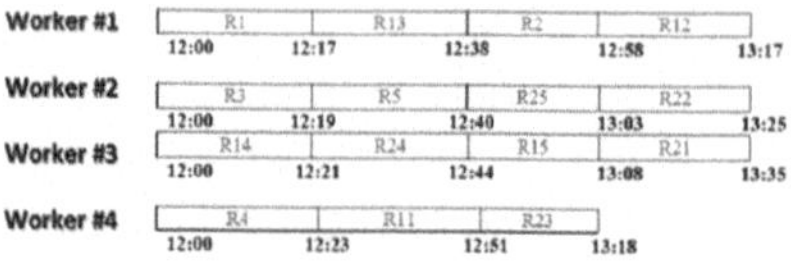

Fig. 2. Distribution of rotation to employees

4.4 Routing Aid

Both housekepping operations and technical maintenance service operations over a large hospitality facility involving more than ten stops needs to decision support for route optimization in terms of time and distance travelled. The underlying decision problem is good old TSP. The formulation of the problem is intensionally omşted since the TSP problem is very well studied in the literature for the last decades. The underlying problem is standard TSP with network distances. In our study, a network distance matrix of more than 500 locations is compiled. As a solution procedure, the nearest neighbor heuristic is used as construction and two-way interchange heuristic is used for greedy improvement. Since we employed standard algorithms, the formulation and algorithmic details are not included here for saving space.

5 Verification Validation and Sensitivity Analysis

The goal of the project is to develop a dynamic decision support system that solves individual problems quickly. The employee planning problem is solved through CPLEX 12.8 Asus on a Core i7,4720 HQ, 2.60 GHz, 16 GB RAM computer. The scheduling problem and the TSP is solved by means of heuristics coded in Netbeans IDE 8.2 and VBA, respectively.

The developed models are run with small-sized problems and it is observed that CPLEX results are significant. By this way, the verification process is completed and the solutions to the real size problem are obtained in the first step for validation. These solutions are shared with the authorized people in Altın Yunus and it is seen that the results are reasonable. After the establishment of DSS at Altın Yunus towards the end of the project, daily performance is measured individually. Comparing these performances with the results suggested by the model, the key performance indicators we determined in the problem definition section become measurable under some circumstances. Sensitivity analysis is performed by enumerating different values to model parameters.

The first model proposes a cost-effective long-term planning. Main restriction of our model is the number of employees. Planning model includes necessary values, the number of rooms needed to be cleaned during hot times at each half week which calculated from the regression equation. The aim is to plan one year of employee planning with 104 half weeks. With the help of CPLEX and Ms Excel VBA codes, the DSS developed in very short time. At opening page of the program shown in Figs. 3 and 4, intern, full-time employee, part-time employee's cost and the total cost are presented. In Table 4, number of rooms that need to be cleaned in hot times is presented. In Table 5 intern, full time and part time working employees, how many half-week works in one year and required time to clean one room in hot and cold times and it can be explained the left of the equation which type of employee will be work in the 104 half weeks and how many employees will work. For the success of the program, the regression analysis results should yield the adjusted R^2 values at least 90%.

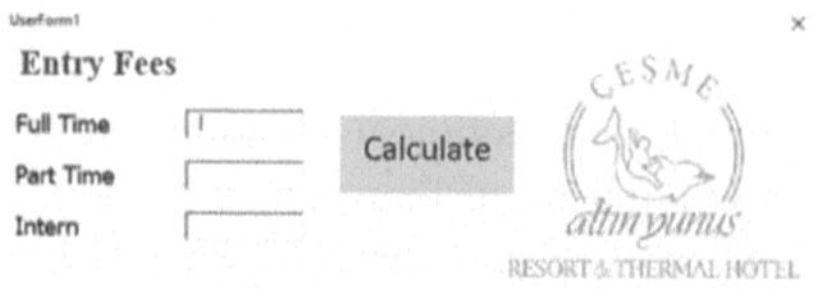

Fig. 3. User form 1

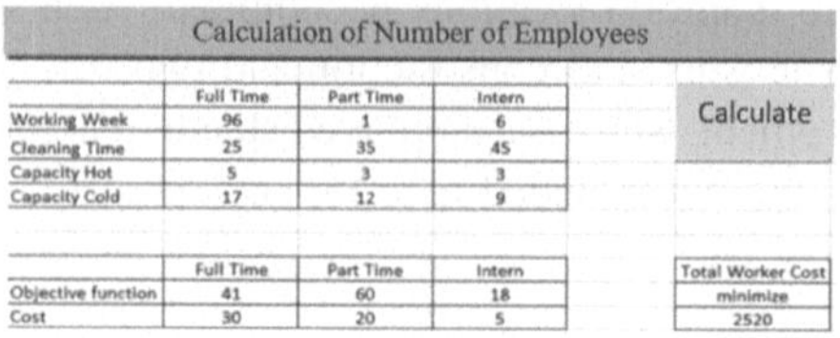

Calculation of Number of Employees			
	Full Time	Part Time	Intern
Working Week	96	1	6
Cleaning Time	25	35	45
Capacity Hot	5	3	3
Capacity Cold	17	12	9

	Full Time	Part Time	Intern
Objective function	41	60	18
Cost	30	20	5

Total Worker Cost
minimize
2520

Fig. 4. Workers calculated

Table 4. Hot time rooms

Year	Month	Room	Weekend?	Hot Time	Cold Time
2018	Jan	718	0	74	9
2018	Jan	243	1	57	13
2018	Jan	177	0	18	2
2018	Jan	252	1	59	13
2018	Jan	176	0	18	2
2018	Jan	269	1	63	14
2018	Jan	703	0	73	8
2018	Jan	620	1	144	33
2018	Jan	720	0	74	9
2018	Feb	421	1	131	28
2018	Feb	26	0	2	0
2018	Feb	89	1	28	6
2018	Feb	466	0	37	0
2018	Feb	148	1	46	10
2018	Feb	22	0	2	0
2018	Feb	65	1	20	4
2018	Feb	10	0	1	0

Table 5. Number of workers per week

	Full Time	Part Time	Intern		R_w
1	21	0	0	≥	74
2	0	0	0	≥	57
3	0	0	0	≥	18
4	0	0	0	≥	59
5	0	0	9	≥	18
6	0	0	0	≥	63
7	0	0	0	≥	73
8	0	7	0	≥	144

The worker assignment model aims to assign employees to the best way to rooms to be cleaned. The model makes the most appropriate assignment, using the workers in the most appropriate way and reducing the delay times of the rooms to minimum values. Walking distances have been added to the processing times, calculated through the time study, so that the workers are assigned to the nearest suitable room and the walking distances are reduced to minimum values. The assignment model is coded in Netbeans IDE 8.2. Our program first selects rooms to be cleaned as shown in Fig. 5. Then, in Fig. 6, the user enters the ready time and due date of the room. In Fig. 7, the number of workers required by the user (the data obtained from the aggregate planning model) is entered. As a result, our program transfers the outputs to Excel, reporting which order the workers will clean the rooms and the delay times of the rooms as illustrated in Fig. 2.

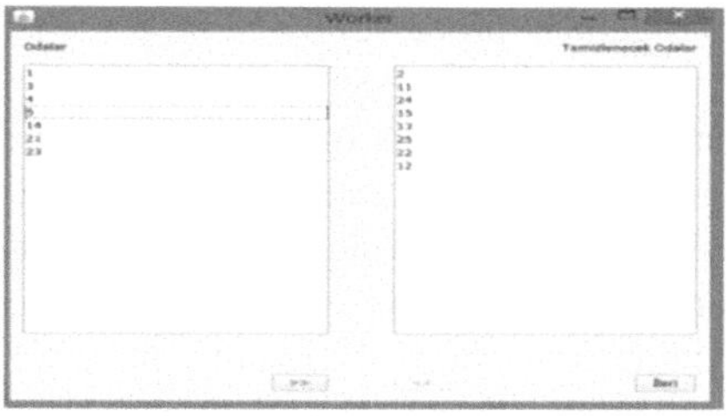

Fig. 5. Entering # of rooms to be cleaned

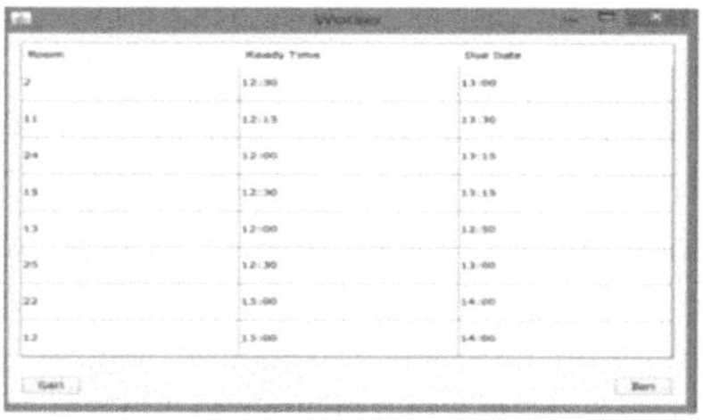

Fig. 6. Entering RT & DD

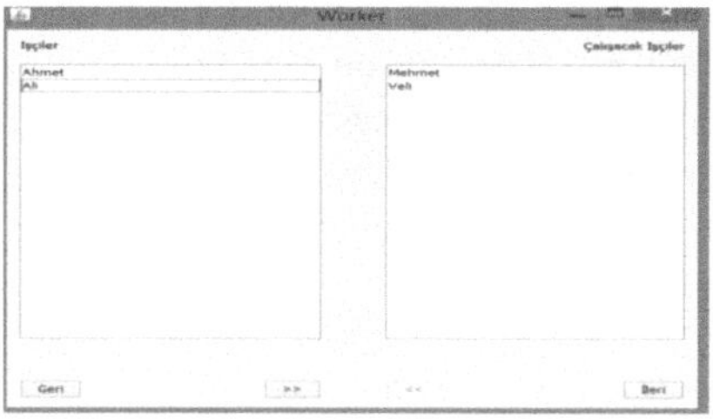

Fig. 7. Entering workers into the system

Another DSS module is set up on the routing system of rooms to be cleaned. For the distance calculations made in this decision support system, an interface is created by means of VBA codes in Excel. When we turn on this DSS, a worksheet shown in Fig. 8 is opened. For this output, the following operations are performed in order; with the "Add Room" button, the room numbers to be cleaned are entered by the user, a small matrix containing the distances of the related rooms is obtained with the "Find Matrix" button and then with the button 'Construct', the room to be entered for the list is made feasible. In this way, cleaning staff and technicians will be able to reach the room with minimum route with minimal effort. By developing the code, the gains obtained by the binary interchange method are calculated for all possibilities. This improved algorithm has been added to the work screen with 'Binary Interchange' button. This myopicim-provement algorithm aims to overcome the disadvantages of the nearest neighbor algorithm, especially the long return path that it uses when completing the tour.

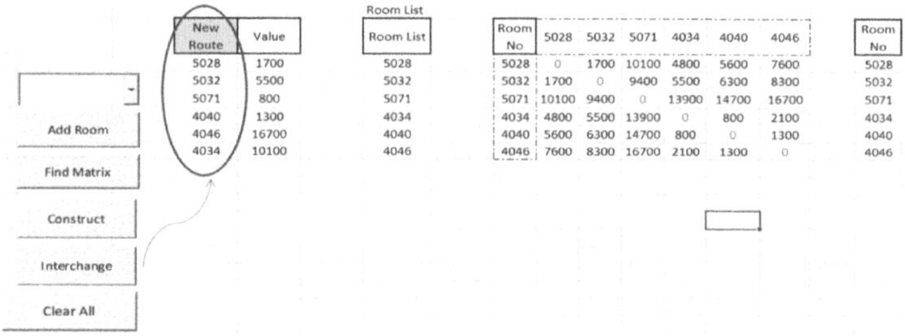

Fig. 8. TSP modelling output

Instead of, manually based estimation and assignment processes based on personal experience, we created a systematic system through the decision support system. Workers were assigned with the most appropriate route, time is minimized, and the minimum delays are provided. At the same time, according to estimation, the occupancy rate and the number of workers to be needed is planned the closest to reality. It is seen that the costs are reduced when the number of employees employed in the hotel is compared with the number of employees providing from the DSS. Finally, the 2018 occupancy rate cannot be reached due to the hotel's privacy policy, so the comparison of 2018 is not made.

As we mentioned in the problem definition, there are three types of employees in the hotel, full time, part time and intern. The most costly of these employees are part-time employees. The hotel was keeping up part-time employees and it caused high costs. By the help of the decision support system, worker assignments are carried out in the most efficient manner, resulting in a 52% cost saving. The costs obtained from the Altın Yunus's manual assignments (left side) and the costs from the decision support system (right side) are given in Table 6.

Table 6. Total costs and percent saving $d_I = 8, d_F = 96, u_I = 10, u_F = \infty, u_p = 25, n = 104$)

AY		2017 Data			Full Time	Part Time	Intern	TOTAL COST
		F	P	I				
Unit Cost	1	3120	65	30	60	58	27	191780
	2	3120	60	30	60	58	27	191490
	3	3120	55	30	60	58	27	191200
	4	3120	50	30	60	58	27	190910
	5	3120	40	30	60	58	27	190330
	6	3120	65	60	60	58	27	192590
	7	3120	60	60	60	58	27	192300
	8	3120	55	60	60	58	27	192010
	9	3120	50	60	60	58	27	191720
	10	3120	40	60	60	58	27	191140

YU		2017 Data			Full Time	Part Time	Intern	TOTAL COST	PERCENT SAVINGS
		F	P	I					
Unit Cost	1	3120	65	30	22	366	92	95190	50,37%
	2	3120	60	30	21	408	92	92760	51,56%
	3	3120	55	30	20	457	103	90625	52,60%
	4	3120	50	30	20	457	103	88340	53,73%
	5	3120	40	30	18	591	108	83040	56,37%
	6	3120	65	60	22	375	78	97695	49,27%
	7	3120	60	60	22	378	75	95820	50,17%
	8	3120	55	60	20	457	103	93715	51,19%
	9	3120	50	60	20	457	103	91430	52,31%
	10	3120	40	60	20	459	100	86760	54,61%

6 Conclusion

This study is among the first studies in which the techniques of production research applied to the housekeeping operations at hospitality sector. The study is motivated from a real life system, Altınyunus Resort and Thermal located in Turkey. A regression model is developed for demand forecasting producing point estimates and prediction intervals given a risk value. Together with a motion and time study conducted, the regression model generates the data required for two planning models at the tactical and operational level. The first model balances the workforce through the year whereas the second model solves worker assignments at daily basis. A third decision problem helps routing within the premises. In this project, a new decision support system has been created, which is going to be live to assist operational managers. Last years' data are used for validating the system and we helped to reduce the annual costs considerably and improve customer satisfaction.

The traditionally turbulent relationship between the front office and housekeeping department is too often the headline act and, as with any sibling rivalry, is based entirely on the quality of communication between the two. Understanding the bottlenecks is important and this has only become possible with the rise of the smart production research that allows for useful software integrations. Dynamic planning, real-time status updates and instant communication can lead to unprecedented cost management for a hotel, let alone reduce heated arguments.

Housekeeping is, arguably, the most inefficient operation in any hotel. Amidst a labyrinth of rooms with enigmatic guests slipping in and out, clipboards and printed reports are too often the only compass housekeepers have to navigate through their daily chores. According to a workflow calculation by productivity management experts, the daily work of a maid consists of 18 tasks. Yet it is estimated that housekeepers spend about 10 to 15% of their time just trying to find the next room to clean. Furthermore, excellent housekeeping is all important for guest satisfaction. According to a survey commissioned by cleaning products brand CLR and conducted by TNS, 86% of hotel guests cited cleanliness as the top criteria they look for when reading online hotel or holiday rental reviews. The survey also revealed that eight out of ten guests would rather give up internet access for the duration of their than stay in a dirty hotel or rental. In short, this means you should be able to run your hotel with less resources and deliver a higher quality experience for guests. The future of hotel housekeeping lies in mobile apps run by a cental decision support system whose model base is pure production research.

Acknowledgment. This paper presents the works of a university-industry collaboration project, under the framework of the capstone design courses (IE 495 Systems Analysis and IE 496 Systems Design) taught in the Industrial Engineering Department of Yasar University. The authors acknowledge the support provided by Necip Atılgan and the project team members; İlber Gündüz, Gamze Urak, Nilay Yıldız, Emre Narin and Merve Kıran as students of Yaşar University, as well as the support of the Department of Industrial Engineering and with Altın Yunus Resort & Thermal Hotel, during the academic year 2017/2018.

References

1. Groover MP (2007) Work systems and the methods, measurements and management of work, University of Lehigh, UK, Chapter 13, vol 352
2. Meyers FE, Stewart JR (2002) Motion and time study for lean manufacturing, 3rd edn. Prentice Hall, New Jersey
3. Anghelache C, SACALÃ C (2016) Multiple linear regression used to analyse the corelation between GDP and some variables. Rom Stat Rev Suppl Rom Stat Rev 64(9):94–99
4. Anghelache C, Manole A, Anghel MG (2015) Analysis of final consumption and gross investment influence on GDP – multiple linear regression model. Theor Appl Econ 22 (3):137–142
5. Hiller FS, Lieberman GJ (2010) Introduction to operations research, University of Stanford, California, Chapter 8, vol 335
6. Taha HA (2007) Operations research: an introduction, 8th edn. Person Education, Harlow. Chapter 9

Efficiency Analysis in Retail Sector: Implementation of Data Envelopment Analysis in a Local Supermarket Chain

Ceren Kahraman, İrem Uluğ, Can Burak Othan,
Yeşim Deniz Özkan-Özen(✉), and Yiğit Kazançoğlu

International Logistics Management Department,
Yaşar University, Izmir, Turkey
cerennkahramann@gmail.com, irem.ulug@hotmail.com,
burakothan15@hotmail.com,
{yesim.ozen,yigit.kazancoglu}@yasar.edu.tr

Abstract. Increased competitiveness leads the importance of efficiency analysis for organizations. There are many tools and techniques for performance measurement. Data Envelopment Analysis is one of the well-known nonparametric, direct programming technique that can deal with multiple inputs and outputs and used for efficiency analysis. In this study, data envelopment analysis is used in retail sector. This study is based on a real life problem and the aim is to measure the store performance of a local retail chain in Turkey. While, size of the store, number of employees, number of deliveries and total cost are used as inputs; number of customers, sales, and store evaluation of customers are used as outputs in this study. Main contribution of this study is using store evaluation of customers as an output by conducting an interview with each stores' customers. Due to the dynamic environment of the retail sector, model that allows variable return to scale while maximizing inputs; input-oriented BCC is used in the study. At the end of the study, numerical results are obtained from a software called Frontier Analyst, and managerial suggestions are presented in the conclusion part.

Keywords: Data envelopment analysis · BCC model · Retail sector
Efficiency analysis · Performance measurement

1 Introduction

Developments in service sector, increase in production amounts, shrinking product life cycles, increase in accessibility of variety of products reveals the importance of gaining competitive advantage in the market place. Therefore, continuous performance monitoring is essential for organizations to be successful. Different tools and techniques are used for measuring performance of organizations including balanced Scorecard [1, 2], EFQM Excellence Model [3] and Key Performance Indicators [4].

As one of a well-known mathematical model based technique, data envelopment analysis (DEA) is also used for measuring performance. DEA is used for assessing productivity and efficiency of decision making units (DMU) and it is a nonparametric

N. M. Durakbasa and M. G. Gencyilmaz (Eds.): ISPR 2018, *Proceedings of the International Symposium for Production Research 2018*, pp. 884–897, 2019.
https://doi.org/10.1007/978-3-319-92267-6_71

linear programming method [5]. Implementation areas of DEA have a high variety including service and manufacturing sectors. Basically, DEA is used to estimate the level of efficiency of similar units of organizations, so called DMU, which utilize the same inputs to produce the same outputs [6].

In this study, DEA is used for the retail sector, which is not new, and in the literature there are many studies. Most of these studies use inputs related to employees, size of the store, economic factor; and outputs related to sales, number of customers etc. However, to the best of our knowledge, using store evaluation of customers as an output while assessing the DMUs has not been conducted. Only [7] used customer satisfaction as an input in residential building management assessment. Therefore, originality of this paper is using store evaluation of customers as an output, in addition to other well-known inputs and outputs for retail sector. This study is conducted based on a real life problem, where an assessment was needed for a close down decision for a store in the retail chain. The objective of this study is to measure performance of the seven stores in a local retail chain in Turkey and to guide the organization about the close down decision according to the numerical results.

This study is divided into six main parts. After introduction, firstly literature review related to general DEA usage and retail sector is presented. Then, methodology, implementation of the study and numerical results are shown. Finally, discussions and conclusion are presented.

2 Literature Review

Data Envelopment Analysis can be defined as nonparametric direct programming strategy for evaluating the productivity and efficiency of Decision Making Unit's (DMU) [8]. DEA is used in many sectors for measuring the performance efficiency including; healthcare, finance, production etc.

To start with the healthcare sector [9–12] can be given as examples where DEA is used for measuring efficiencies. DEA methods varies according to need of implementation environment, even though the sector remains the same. For instance; Input-Oriented CCR Model was used in [10] on the other hand, Output-Oriented CCR Model was used with [11]. Moreover, Dynamic Network and Black Box Model was used by [9] in healthcare sector.

Moreover, education is another sector that DEA is used to measure performance. [13] used Ordinary Least Squares Method and Stochastic Frontier Analysis Model to measure UK universities performance. Output-oriented BCC model was used by [13, 14] to measure efficiencies in the UK and Chinese universities respectively.

In the studies of [15, 16] DEA was used in production sector. In Ref. [15] input-oriented CCR was used for vendor evaluation for a baby food manufacturer. On the other hand, [16] used output-oriented network slack based model to evaluate ten Iranian soft drink manufacturer in e view of green supply chain management.

Finance is another sector that DEA is used frequently. Reference [17] conducted a survey, which focuses on 80 Published DEA applications in 24 different countries. In addition, Regression Model was integrated with DEA and used by [18] to evaluate cost, technical and allocative efficiencies for Brazilian Banks Between 2000–2007.

Moreover, both CCR and BCC Models were used by [19] to measure relative efficiencies of banks. In Table 1, summary of the literature review is given by including sector, focused area and method.

Since this study focuses on performance measurement in retail sector by using DEA, in the following section studies that used DEA in retail sector are investigated.

Table 1. Summary of the literature review

Author(s)	Sector	Focused area	Method
Banker et al. [11]	Healthcare	Comparing about hospital cost and production from 2 estimation models	Translog joint cost function, Output-Oriented CCR Model
Weber [15]	Production	Evaluating the 6 vendor supplying an item to a baby food manufacturer	Input-Oriented CCR
Jemric and Vujcic [20]	Finance	Analyzing bank efficiencies in croatia between 1995 and 2000	Both Output-Oriented BCC and CCR Models
Johnes [21]	Education	Measurement of universities' performance in UK, 1993	Ordinary least squares method, stochastic frontier analysis model, multilevel model
Avkiran [22]	Financial Service	The multi stakeholder perspective on benchmarking rates firms' performance	Input-Oriented and Output-Oriented CCR Model
Johnes [23]	Education	Measuring the content of higher education in England, 2000–2001	Output-Oriented BCC Model
Johnes and Li [14]	Education	Examining the efficiency in the production of 109 Chinese regular universities in 2003 and 2004	Output-Oriented BCC Model
Nayar and Ozcan [10]	Healthcare	Performance measurements of quality in Virginia hospitals	Input-Oriented CCR Model
Staub et al. [18]	Finance	Evaluating cost, technical and allocative efficiencies for Brazilian bank between 2000–2007	Regression model

(*continued*)

Table 1. (*continued*)

Author(s)	Sector	Focused area	Method
Ram Jat and Sebastian [12]	Healthcare	Measuring technical efficiency of 40 public district hospital in India	Input-Oriented BCC Model
Paradi and Zhu [17]	Finance	Surveying 80 published DEA applications in 24 countries that focus on bank branches	Additive and slack-based models (8 of them), BCC and CCR models (72 of them)
Mirhedayatian et al. [16]	Production	Evaluating 10 Iranian soft drinks companies in terms of green supply chain management	Output-oriented network slack based model
Kawaguchi et al. [9]	Healthcare	Evaluating 9000 private and public hospitals in Japan	Dynamic network and black box model
Thomas et al. [24]	Industry	Operational and environmental efficiencies of 47 prefectures in Japan	Input-Oriented CCR
Othman et al. [19]	Finance	Measuring relative efficiency of banks	Both CCR and BCC models

2.1 DEA in Retail Sector

DEA method is a useful technique for performance evaluation in retail sector in terms of measuring multiple inputs and outputs. From retail perspective, DEA is used to ensure better service to customers over time in consequence of input-output analysis to perform new policies in line with the outcomes [24].

Nowadays, there is a perceptible rise and growth in competition in retail sector, therefore, increasing the performance of retail stores to gain competitive advantage and investigating the causes of bottlenecks are essential. In order to do that, measuring current performance is necessary. For efficiency analysis, DEA is very suitable in terms of evaluating multiple inputs and outputs.

According to the needs of research, inputs and outputs that are used in DEA may vary. When the literature is investigated, it has been noticed that value of stock, floor space, and recurrent costs were used as inputs, annual sales, and customer satisfaction were used as outputs with Output-Oriented CCR model by [25]. However, customer satisfaction in [25] refers to the customers that visit the store more than twice and does not consider an evaluation of customers. In addition to that, equivalent number of vehicles used for delivery and total transport cost were used as inputs and number of customers served, number of orders filled and total revenue were used as outputs with Input-Oriented CCR and BCC Model in the study of [26].

Moreover, [27] focused on financial measures and used current ratio, stock turnover and financial lever as inputs, and net profit margin and marketing value as outputs, where Output-Oriented CCR and BCC Models were utilized. On the other hand [28] analysed retail sector from more macro perspective and used quality management applications and systems, internal check, documentation capacity of process and production, company management, design and development opportunities, performance of cost reduction as inputs; Quality Cost, Delivery, Performance of Cost Reduction as outputs. In the same study [28], DEA was integrated with AHP (Table 2).

Table 2. Summary of literature review in terms of retail sector

Author(s)	Methods	Inputs	Outputs
Kamakura et al. [29]	Output-Oriented CCR	Labor, and customer service area	Cash Deposits, Other Deposits Pays (Volume of transit in branch), and MREV (Service fee)
Thomas et al. [24]	Input-Oriented CCR	Labor, experience, location related costs, and internal processes	Sales and profit
Donthu and Yoo [30]	Input-Oriented CCR	Environmental conditions, customers factors, retail firms managerial efforts, and employees factors	Financial or economic outcomes, and behavioral outcomes
Gemici [28]	DEA and AHP	Quality management applications and systems, internal check documentation, capacity of process and production, company management, design and development opportunities, and performance of cost reduction	Quality, cost, delivery, performance of cost reduction
Mishra [25]	Output-Oriented CCR	Value of stock, floor space and recurrent costs	Annual sales, and customer satisfaction
Mostafa [31]	Output-Oriented CCR	Employees, and assets	Revenue, earn share, and market value
Lau [26]	Input-Oriented CCR and BCC	Equivalent number of vehicles used for delivery, and total transport cost	Number of customers served, number of orders filled, and total revenue
Gandhi and Shankar [32]	CCR and BCC	Cost of labor, and capital employed	Profit, and sales
Geyikçi and Bal [27]	Output-Oriented CCR and BCC	Current ratio, stock turnover, and financial lever	Net profit margin, and marketing value

As mentioned before, contribution of this study is using store evaluation of customers as an output, in addition to other well-known inputs and outputs for retail sector. In the following section, methodology of the study is explained.

3 Methodology

DEA is a methodology that evaluates relative efficiency of DMUs with using multiple inputs and outputs. Also, DEA is used for analyzing the managerial performance of productive units. Although, there are different methods under DEA and integration between other methods is possible, CCR (Charnes, Cooper, Rhodes) and BCC (Banker, Charnes, Cooper) are the most well-known DEA methods. CCR model assumes that constant returns to scale whereas BCC model considers that variable returns to scale [33]. Because of this differentiation, CCR and BCC models are separated from each other. In addition, models are categorized as input-oriented and output-oriented models. The aim of the input-oriented model is maximizing the outputs. On the other hand, the goal of the output-oriented model is minimizing the inputs.

Retail sector is a dynamic sector and the changes in inputs or outputs do not affect the other linearly. In other words, changing the values of inputs do not cause a change in outputs in the same amount. Therefore, a model which allows variable return to scale is needed. From this point of view, BCC model is more appropriate for retail sector. Furthermore, maximizing the outputs is the aim of the retail sector. For this reason, input-oriented BCC model was chosen. Linear programs for the input-oriented BCC model (primal) given below;

$$Q_k = \text{Max}\left(\sum\nolimits_{r=1}^{p} \mu_r Y_{rk} - \mu_0\right)$$

Subject to:

$$\sum\nolimits_{i=1}^{m} \omega_i X_{ik} = 1$$

$$\sum\nolimits_{r=1}^{p} \mu_r Y_{rj} - \sum\nolimits_{i=1}^{m} \omega_i X_{ij} - \mu_0 \leq 0$$

$$\mu_r \geq \varepsilon$$

$$\omega_i \geq \varepsilon$$

μ_0: unconstrained

$$j = 1, \ldots, n$$

$$r = 1, \ldots, p$$

$$i = 1, \ldots, m$$

In here;

μ_r: r[th] output is weighted by DMU of k,
ω_i: i[th] input is weighted by DMU of k,
Y_{rk}: r[th] output is produced by DMU of k,
X_{ik}: i[th] input is used by DMU of k,
Y_{rj}: r[th] output is produced by j[th] DMU,
X_{ij}: i[th] input is used by j[th] DMU,
ε: a sufficiently small positive number,
μ_0: variable that is related to the scale according to input direction.

In the following section, details of the implementation of the study are presented by covering company details, inputs and outputs of the model and input oriented BCC model of the problem.

4 Implementation of the Study

Implementation of the study was conducted in a local retail chain in Turkey and mathematical model was structured based on a real life problem. Name of the chain is called as ABC Retail chain in this study. ABC Retail chain currently has 20 stores in different locations. However, in this study 7 stores were selected by company manager for initial performance evaluation. Name of these stores are Narlıdere, İskele, Balçova, Bucakoop, Alaybey, Bayraklı and Karabağlar called DMUs in the mathematical model. As mentioned in the previous section, input oriented BCC model in this study, and in the following subsections firstly inputs and outputs for the study and mathematical model of the problem given respectively.

4.1 Inputs and Outputs of the Model

As mentioned in the literature review part, there are many different inputs and outputs related to retail sector that can be used to evaluate efficiencies of DMU's. In this study 4 inputs and 3 outputs were selected according to the needs of the ABC retail chain. Except store evaluation of customers, all the inputs and outputs are in line with the previous literature. Inputs are modified from [24, 26, 29, 31] and outputs are taken from [24, 32]. Brief explanations of inputs and outputs are given below:

Inputs:

Size of the Land: It refers to the size of the each store in meter square.
Number of Employees: It indicates the number of employees in with different jobs including cashiers, cleaning attendants, security etc.
Number of Deliveries: It refers to number of deliveries that the store received in a month.
Total Cost: It includes all the monthly costs including rent, transportation cost, taxes, staff salaries etc.

Outputs:

Number of Customers: It refers to the number of customer that visited the store for shopping in a month.
Sales: It contains amount of sales for each stores.
Store Evaluation of Customers: It indicates general feelings of customers about the store.

In order to measure store evaluation of customers' output 10 positive statement were prepared and short interviews were conducted with customers separately in each DMUs. Statements are given as below:

- Design of the layout in the store is good, so I can find the products that I want to purchase easily.
- Employees are friendly and helpful.
- Cashiers are experienced and well trained.
- Number of open checkout corner is sufficient, thus, I get served rapidly.
- The store is clean and tidy.
- Store has a good location; it is easy to arrive.
- Price labels and discount tags are accurate on the shelves.
- I can find the products that I want to purchase in store inventory.
- It is easy to return or change the products if it is necessary.
- Number of employees are sufficient.

Customers were asked evaluate the statements with a 5 points scale, where 5 point has the highest rate and refers that customer strongly agree the statement, and 1 point is the lowest and indicates that customer does not agree with the statement. In total, 400 evaluations have been conducted, which were distributed among the stores according to the number of customers of each store.

In Table 3, dataset of inputs and outputs that are going to be used to formulate the mathematical model, is given.

4.2 Mathematical Model of the Problem

According to the identified inputs and outputs, model of the problem was generated. In DEA, separate mathematical models are needed to be formulated for each DMU. However, only objective functions and first constraints are changed according to the DMUs. Other constraints are the same for each DMU. The models are given below;

DMU1: Narlıdere

$$\mathrm{Max}\,\eta_1 = 24748\mu_1 + 621508\mu_2 + 46\mu_3 - \mu_0$$

$$350\omega_1 + 30\omega_2 + 16\omega_3 + 116073\omega_4 = 1$$

Table 3. Input and output data

Name of the store	Inputs				Outputs		
	Land	Number of employee	Number of delivery	Total cost	Number of customer	Sales	Store evaluation of customers
Narlıdere	350	30	16	116073	24748	621508	46
İskele	550	17	12	83487	14496	545600	43
Balçova	350	12	12	50119	17420	397096	40
Bucakoop	300	17	12	74500	21000	435200	44
Alaybey	650	32	24	107000	37768	856000	38
Bayraklı	350	25	24	108768	49036	852000	35
Karabağlar	700	14	12	72411	12272	267792	22

DMU2: İskele (Urla)

$$\text{Max}\,\eta_2 = 14496\mu_1 + 545600\mu_2 + 43\mu_3 - \mu_0$$

$$550\omega_1 + 17\omega_2 + 12\omega_3 + 83487\omega_4 = 1$$

DMU3: Balçova

$$\text{Max}\,\eta_3 = 17420\mu_1 + 397096\mu_2 + 40\mu_3 - \mu_0$$

$$350\omega_1 + 12\omega_2 + 12\omega_3 + 50119\omega_4 = 1$$

DMU4: Bucakoop

$$\text{Max}\,\eta_4 = 21000\mu_1 + 435200\mu_2 + 44\mu_3 - \mu_0$$

$$300\omega_1 + 17\omega_2 + 12\omega_3 + 74500\omega_4 = 1$$

DMU5: Alaybey

$$\text{Max}\,\eta_5 = 37768\mu_1 + 856000\mu_2 + 38\mu_3 - \mu_0$$

$$650\omega_1 + 32\omega_2 + 24\omega_3 + 107000\omega_4 = 1$$

DMU6: Bayraklı

$$\text{Max}\,\eta_6 = 49036\mu_1 + 852000\mu_2 + 35\mu_3 - \mu_0$$

$$350\omega_1 + 25\omega_2 + 24\omega_3 + 108768\omega_4 = 1$$

DMU7: Karabağlar

$$\mathrm{Max}\,\eta_7 = 12272\mu_1 + 267792\mu_2 + 22\mu_3 - \mu_0$$

$$700\omega_1 + 14\omega_2 + 12\omega_3 + 72411\omega_4 = 1$$

$$24748\mu_1 + 621508\mu_2 + 46\mu_3 - 350\omega_1 - 30\omega_2 - 16\omega_3 - 116073\omega_4 - \mu_0 \leq 0$$

$$14496\mu_1 + 545600\mu_2 + 43\mu_3 - 550\omega_1 - 17\omega_2 - 12\omega_3 - 83487\omega_4 - \mu_0 \leq 0$$

$$17420\mu_1 + 397096\mu_2 + 40\mu_3 - 350\omega_1 - 12\omega_2 - 12\omega_3 - 50119\omega_4 - \mu_0 \leq 0$$

$$21000\mu_1 + 435200\mu_2 + 44\mu_3 - 300\omega_1 - 17\omega_2 - 12\omega_3 - 74500\omega_4 - \mu_0 \leq 0$$

$$37768\mu_1 + 856000\mu_2 + 38\mu_3 - 650\omega_1 - 32\omega_2 - 24\omega_3 - 107000\omega_4 - \mu_0 \leq 0$$

$$49036\mu_1 + 852000\mu_2 + 35\mu_3 - 350\omega_1 - 25\omega_2 - 24\omega_3 - 108768\omega_4 - \mu_0 \leq 0$$

$$12272\mu_1 + 267792\mu_2 + 22\mu_3 - 700\omega_1 - 14\omega_2 - 12\omega_3 - 72411\omega_4 - \mu_0 \leq 0$$

$$\mu_r, \omega_i \geq \varepsilon > 0 (r = 1, 2, 3)\,(i = 1, 2, 3, 4)$$

μ_0: variable that is related to the scale according to input direction,
ε: a sufficiently small positive number.

In the following section, numerical results of the given mathematical models are presented by using a software called Frontier Analyst.

5 Numerical Results

After generating the model, the problem was solved with Frontier Analyst which performs an analysis with linear programming to determine the relative efficiency of organizational units that manage. Before entering the data as shown in Fig. 1, maximizing outputs, and varying returns options are selected on the software to receive an input oriented BCC solution.

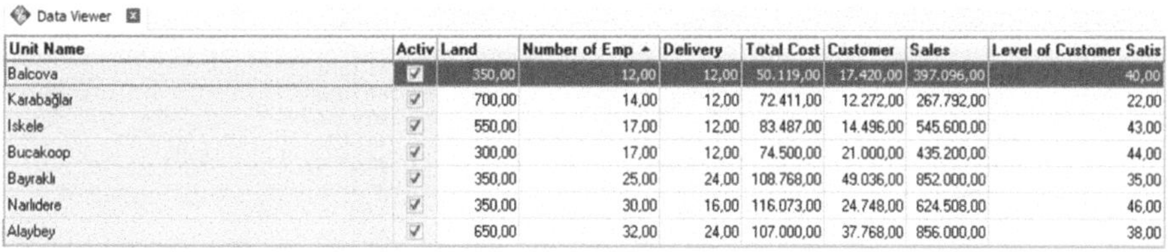
Data Viewer

Unit Name	Activ	Land	Number of Emp	Delivery	Total Cost	Customer	Sales	Level of Customer Satis
Balcova	✓	350,00	12,00	12,00	50.119,00	17.420,00	397.096,00	40,00
Karabağlar	✓	700,00	14,00	12,00	72.411,00	12.272,00	267.792,00	22,00
Iskele	✓	550,00	17,00	12,00	83.487,00	14.496,00	545.600,00	43,00
Bucakoop	✓	300,00	17,00	12,00	74.500,00	21.000,00	435.200,00	44,00
Bayraklı	✓	350,00	25,00	24,00	108.768,00	49.036,00	852.000,00	35,00
Narlıdere	✓	350,00	30,00	16,00	116.073,00	24.748,00	624.508,00	46,00
Alaybey	✓	650,00	32,00	24,00	107.000,00	37.768,00	856.000,00	38,00

Fig. 1. Inputs and outputs

Results in terms of efficient and inefficient DMUs are revealed as shown in Fig. 2. According to the results, only DMU7: Karabağlar, was found as inefficient since its score is less than 1, in other words 100%. Rest of the DMUs revealed as efficient.

Data Viewer | Analysis Options | Efficiency Table

Efficiency scores | Summary graph | Distribution

Units	Comparison 1	
Unit name	Score	Efficient
Alaybey	100,0%	✔
Balcova	100,0%	✔
Bayraklı	100,0%	✔
Bucakoop	100,0%	✔
Iskele	100,0%	✔
Karabağlar	65,1%	
Narlıdere	100,0%	✔

Fig. 2. Efficiency scores of DMUs

Following analysis were conducted for only the inefficient DMU7: Karabağlar Store with a score of 65.1%. Frontier Analysts suggested that DMU3: Balçova and DMU4: Bucakoop as reference to Karabağlar. In Fig. 3, specific results of Karabağlar is presented with actual and target values, potential improvement, and peer contribution results with DMU3: Balçova and DMU4: Bucakoop.

According to the numerical results, shown in Fig. 3, size of the land for DMU7: Karabağlar needs be reduced from 700 m^2 to 330 m^2. Moreover, total cost should be reduced from 72411 TL to 59871 TL. On the other hand, changes in the number of employees, and number of deliveries were not suggested. In addition, when the output results are investigates; number of customers, total sales and level of store evaluation of customers should be increased to reach efficiency target.

65,10% Karabağlar

Peers: 2
References: 0

Potential Improvements

Variable	Actual	Target	Potential Improvement
Customer	12272,00	18852,00	53,62 %
Delivery	12,00	12,00	0,00 %
Land	700,00	330,00	-52,86 %
Level of Customer Satisfaction	22,00	41,60	89,09 %
Number of Employee	14,00	14,00	0,00 %
Sales	267792,00	412337,60	53,98 %
Total Cost	72411,00	59871,40	-17,32 %

Peer Contributions

Balcova	Customer	55,44 %
Balcova	Delivery	60,00 %
Balcova	Land	63,64 %
Balcova	Level of Customer Satisfaction	57,69 %
Balcova	Number of Employee	51,43 %
Balcova	Sales	57,78 %
Balcova	Total Cost	50,23 %
Bucakoop	Customer	44,56 %
Bucakoop	Delivery	40,00 %
Bucakoop	Land	36,36 %
Bucakoop	Level of Customer Satisfaction	42,31 %
Bucakoop	Number of Employee	48,57 %
Bucakoop	Sales	42,22 %
Bucakoop	Total Cost	49,77 %

Fig. 3. Frontier analyst solution of DMU7: Karabağlar

6 Discussions and Conclusions

As mentioned at the beginning of this paper, this study stands on a real life problem. During the initial meeting with the manager of ABC retail chain, their concerns about the DMU7: Karabağlar were revealed in the first place. The numerical results of this study are found in line with their concerns. Therefore, some improvements in Karabağlar store should be conducted in order to avoid close down decision. Firstly, decreasing the size of the store can be suggested due to the results. This could be done by moving the store in a smaller size store alternative or using some parts of the store with different purposes. Moreover, it has been revealed that their customer and sales number is under the target value. In order to increase these numbers, different sales strategies and promotions can be implemented. Furthermore, opinions of customers were found negative according to the results of store evaluation, therefore Karabağlar store needs to focus on customer satisfaction more. Regular service quality surveys can be suggested to the store for improvement, which also helps to increase in number of customers and sales.

In conclusion, this study aimed to make a performance evaluation in a local retail chain by using DEA method. As a contribution to the literature, store evaluation of customers were used as an output. By doing this, not only traditional numerical measures for inputs and outputs were used, but also opinions of customers about the store were considered. In future studies, performance evaluation can be conducted for all the stores of the retail chain and moreover, number of inputs and outputs can be increased.

References

1. Kaplan RS, Norton DP (1996) Using the balanced scorecard as a strategic management system
2. Ittner CD, Larcker DF, Meyer MW (2003) Subjectivity and the weighting of performance measures: evidence from a balanced scorecard. Account Rev 78(3):725–758
3. Wongrassamee S, Simmons JE, Gardiner PD (2003) Performance measurement tools: the balanced scorecard and the EFQM excellence model. Meas Bus Excell 7(1):14–29
4. Chan AP, Chan AP (2004) Key performance indicators for measuring construction success. Benchmarking Int J 11(2):203–221
5. Ji YB, Lee C (2010) Data envelopment analysis. Stata J 10(2):267–280
6. Mardani A, Zavadskas EK, Streimikiene D, Jusoh A, Khoshnoudi M (2017) A comprehensive review of data envelopment analysis (DEA) approach in energy efficiency. Renew Sustain Energy Rev 70:1298–1322
7. Chen WT, Tan PS, Fauzia N, Wang CW (2017) Performance assessment of residential building management utilizing network data envelopment analysis. In: Proceedings of the international symposium on automation and robotics in construction, ISARC, vol 34. Vilnius Gediminas Technical University, Department of Construction Economics & Property
8. Cook WD, Zhu J (2007) Classifying inputs and outputs in data envelopment analysis. Eur J Oper Res 180(2):692–699

9. Kawaguchi H, Tone K, Tsutsui M (2014) Estimation of the efficiency of Japanese hospitals using a dynamic and network data envelopment analysis model. Health Care Manag Sci 17 (2):101–112
10. Nayar P, Ozcan YA (2008) Data envelopment analysis comparison of hospital efficiency and quality. J Med Syst 32(3):193–199
11. Banker RD, Morey RC (1986) The use of categorical variables in data envelopment analysis. Manag Sci 32(12):1613–1627
12. Ram Jat T, San Sebastian M (2013) Technical efficiency of public district hospitals in Madhya Pradesh, India: a data envelopment analysis. Glob Health Action 6(1):21742
13. Johnes J (2006) Data envelopment analysis and its application to the measurement of efficiency in higher education. Econ Educ Rev 25(3):273–288
14. Johnes J, Li YU (2008) Measuring the research performance of Chinese higher education institutions using data envelopment analysis. China Econ Rev 19(4):679–696
15. Weber CA (1996) A data envelopment analysis approach to measuring vendor performance. Supply Chain Manag Int J 1(1):28–39
16. Mirhedayatian SM, Azadi M, Saen RF (2014) A novel network data envelopment analysis model for evaluating green supply chain management. Int J Prod Econ 147:544–554
17. Paradi JC, Zhu H (2013) A survey on bank branch efficiency and performance research with data envelopment analysis. Omega 41(1):61–79
18. Staub RB, e Souza GDS, Tabak BM (2010) Evolution of bank efficiency in Brazil: a DEA approach. Eur J Oper Res 202(1):204–213
19. Othman A, Owen L (2001) Adopting and measuring customer service quality (SQ) in Islamic banks: a case Johnes
20. Jemric I, Vujcic B (2002) Efficiency of banks in Croatia: a DEA approach. Comp Econ Stud 44(2–3):169–193
21. Johnes J (2006) Measuring teaching efficiency in higher education: an application of data envelopment analysis to economics graduates from UK Universities 1993. Eur J Oper Res 174(1):443–456
22. Avkiran NK (2006) Productivity analysis in the service sector with data envelopment analysis
23. Johnes J (2006) Measuring efficiency: a comparison of multilevel modelling and data envelopment analysis in the context of higher education. Bull Econ Res 58(2):75–104
24. Thomas RR, Barr RS, Cron WL, Slocum JW Jr (1998) A process for evaluating retail store efficiency: a restricted DEA approach. Int J Res Mark 15(5):487–503
25. Mishra RK (2009) Benchmarking scheme for retail stores efficiency. Int J Mark Stud 1 (2):131
26. Lau KH (2013) Measuring distribution efficiency of a retail network through data envelopment analysis. Int J Prod Econ 146(2):598–611
27. Geyikçi UB, Bal V (2015) Veri Zarflama Analizi ile Borsa İstanbul AŞ'de Faaliyet Gösteren Toptan ve Perakende Ticaret Sektörü Firmalarının Etkinlik Analizi. Abant İzzet Baysal Üniversitesi Sosyal Bilimler Enstitüsü Dergisi 15(1):21–42
28. Gemici MF (2009) Tedarik Zincirinde Veri Zarflama Analitik Hiyerarşi Prosesi Yöntemiyle Perakende Sektöründe Tedarikçi Performans Değerlendirmesi, Doctoral dissertation, Fen Bilimleri Enstitüsü
29. Kamakura WA, Lenartowicz T, Ratchfrord BT (1996) Productivity assessment of multiple retail outlets. J Retail 72(4):333–356
30. Donthu N, Yoo B (1998) Retail productivity assessment using data envelopment analysis. J Retail 74(1):89–105

31. Mostafa MM (2009) Benchmarking the US specialty retailers and food consumer stores using data envelopment analysis. Int J Retail Distrib Manag 37(8):661–679
32. Gandhi A, Shankar R (2014) Efficiency measurement of Indian retailers using data envelopment analysis. Int J Retail Distrib Manag 42(6):500–520
33. Cooper WW, Seiford LM, Zhu J (2004) Data envelopment analysis. In: Handbook on data envelopment analysis. Springer, Boston, pp 1–39

Model Sequencing and Changeover Time Reduction in Mixed Model Assembly Line

Derya Tataroğlu[1], Renan Dalkıran[1], Sueda Sezen[1], Damla Kesikburun[1(✉)], and Yiğit Kazançoğlu[2]

[1] Department of Industrial Engineering, Yaşar University, İzmir, Turkey
derya_95@hotmail.com, renandalkiran@hotmail.com, sueda_sezen@hotmail.com, damla.kesikburun@yasar.edu.tr
[2] Department of International Logistics Management, Yaşar University, İzmir, Turkey
yigit.kazancoglu@yasar.edu.tr

Abstract. Changeover time is a time between last good product of a specific model, to a first good product of another specific model. During this changeover time, materials, equipment and layout can be changed and this causes non-value added activities. This study aims the reduction of changeover time losses in the sixth line of a worldwide factory which produces combi boilers. Before developing the solution alternatives, causes of time losses were determined. After that, solutions have been searched with process design improvements. A simulation model and decision support system were generated to set the differences between proposed and current situation. While developing solutions, for specific symptoms, various technics were used such as 5S, Lean Manufacturing methods and Quality Function Deployment. Based on these studies, the current situation of the assembly line during changeover process has been improved.

Keywords: Changeover improvement · Process design · Lean manufacturing 5S · Quality Function Deployment

1 Introduction

This study is conducted in a world-wide combi boiler producer factory which is located in Manisa, Turkey. This report presents the solution methodology of model sequencing and changeover time reduction problem in a mixed model assembly line, which is the sixth lean production line of the company. While solving the problem, various methods were used. The steps that have been followed can be listed as the investigation of the literature to find similar problems and their solution approaches, problem definition, presenting solution methodologies such as; Redesign of Material Placement, Mathematical Model, Poke Yoke, 5S, Quality Function Deployment (QFD), verification of results, preparing the Decision Support System (DSS), by integrating Arena Simulation with Excel VBA. As a result, when compared with the current situation of the line, a significant reduction in changeover time was achieved.

N. M. Durakbasa and M. G. Gencyilmaz (Eds.): ISPR 2018, *Proceedings of the International Symposium for Production Research 2018*, pp. 898–913, 2019.
https://doi.org/10.1007/978-3-319-92267-6_72

This project is held in the sixth line of the combi boiler producer factory which uses U type production system as it can be seen in Fig. 1. In this line, 4 models are being produced with 27 operators. These models are Model-1A, Model-1B, Model-2, Model-3, Model-4 and the cycle times are 120, 120, 93, 93 and 80 s, respectively. While these models are being produced, the changeover time losses exist in sixth production line, because of the model transition. According to the production strategy of the combi boiler producer company, model transition occurs from larger cycle time to smaller cycle time. However, this strategy was not always applicable because of the variety of orders.

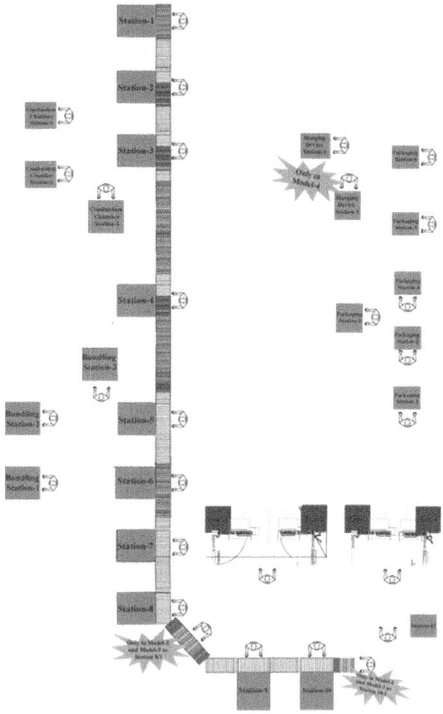

Fig. 1. Line layout with replacement of operators

The production process in the line should be ready when the model transition begins. Before the model transition, the location of materials, that will be used, should be changed according to new model by the logistic operators. In addition, the location of the operators must also be changed in various conditions and the hand tools for production process should be changed by the operators for producing new model in production line. Moreover, during the model transition, BOM list should be controlled, as well. At the end, these changes result in time losses and eventually time losses causes to less amount of production.

2 Literature Review

Changeover process in productive industries is a crucial problem that requires so many progresses to improve as in the combi boilers producer company. It causes huge losses to the company, when it is calculated as daily, monthly and especially annually. There are several studies that are associated with reduction of changeover times. Some common reduction procedure examples are, design for changeover (DFC), simulation and single minute exchange of die (SMED). In this study, the methods of changeover processes are investigated. It can be envisaged that achieving reduction in changeover times affects the total productivity. By supporting this argument, the following literature is included.

In the article of Shingo [1], Single Minute Exchange of Dies (SMED) method is a necessary element for shorter setup times. The target of this approach is to reduce setup times less than 10 min to eliminate waste of time with four phases of SMED. Furthermore, this approach known as Quick Changeover of Tools. The methodology of SMED also tries to make an effort on minimizing the time required to finish all of the tasks that contain changeover.

In the article of Almomani et al. [2] assumed these four phases as; setup process map, classifying activities as external or internal setups, transferring internal to external activities, and simplifying all internal and external activities.

Furthermore, SMED approach is also initiated in the article of Sabadka et al. [3]. A video recording of the machine's manufacturing process is made, based on which the individual operations of the redevelopment are analyzed and included in the individual categories and the specified times of the individual operations.

Thirdly, within the topic of changeover process, according to the article of Mileham et al. [4], DFC improvement can also consist of four main areas, which are, product design, machine design, tooling design and system design and for all these areas, securing, location, adjustment, tool movement, cleaning and access subjects need to be considered. According to Mileham's research to improve changeover times both design and methodology should be considered, since both has more potential time saving and it is more sustainable than only improving the methodology. For methodology improvement there is a list of procedures that consist of three stages, a strategic stage, a preparatory stage and an implementation stage. The understanding and implementation of the rules is relatively simple and not only reduces the transition times considerably, but it also provides the environment in which ecologist production can be done (Mileham et al. 1999). The DFC methods also can be useful for the combi boilers producer company sixth lean line, in logistics design and methodology during changeover and for station designs for tools and equipment.

Also, in the article of Gungor and Evans [5], changeover operations can cause significant burdens through product and time losses, additionally, water and energy use. Cleaning process during the changeover is also important event that cause unignorably losses. According to Gungor and Evans, true cost of a changeover can be calculated as sum of cleaning costs, product losses and time losses (2017). Gungor and Evans (2017) stated that some designs affect the changeover processes such as equipment design, standard operating procedure design and cleaning operations design.

Moreover, McIntosh et al. [6], initiated that DFC is to facilitate rapid dismantling of equipment and subsequent rapid, accurate reassembly of the same equipment for purpose of changeover. It means that it can be used for decreasing losses of changeover processes by facilitating operators' works and usage and maintenance of equipment.

Finally, the topic of changeover process is related with simulation approach. According to Khalid and Kai, in order to be able to increase productivity, the current situation flows should be established (2017) [7]. Solution methods are suggested for problem solving in the current situation. The modeling is carried out using the simulation to show the current state of the flow and the state after the improvement. A case study of the production line of a production facility is used to verify the effectiveness of the proposed solutions and to demonstrate the basic features of the model being developed

3 Problem Definition

In this study, the sixth line of the combi boiler producer company is examined, and the causes of time losses during the changeovers were detected. These causes are material placement, model 1 family transition, movement of operators during changeover, BOM control problem, hand tool and variation of test pads.

A. Material placement is important for producing a new model. When a new model is produced, the location of materials is changed by logistic operators, but sometimes materials are not reachable for operators, therefore, operator change the place of material boxes in order to work in a comfort. Hence, it becomes more difficult to find the materials they need. This leads to loss of time. In addition, observations made on model 1 family suggest that there are further losses at Station-4, Station-5, Station-7, Station-8 and Packing Station-5 due to an improper location of materials during changeover.
B. In model transitions, two of the operators change their places for producing a new model and this situation cause next stations to wait. In model-4, the two operators who are working in the hanging device station which can be seen in Fig. 1, are transferred to the combustion chamber stations when switching to model-1. When the changeover takes place, these two operators are needed in the combustion chamber stations or vice versa. However, since these operators are working in the hanging device station of the rotating model, they continue to work on the hanging device stations until the model-4 device is completely out of production. This place changing problem is also occurring when the models turn to model-2 and model-3. Buffer stock does not have a significant impact on changeover losses. However, buffer stock is not recommended according to the lean production design. For this reason, the company aims to produce as few buffer stock as possible (or without buffer stock).
C. For each product family that will be produced on the model change, the employees check the materials. In the material control; each product is made up of a list of materials prepared for the family and called Bill of Materials (BOM). Each operator, controls the materials in his station based on these lists. However, since

this list includes all the materials of the combi boiler, it takes considerable amount of time for the operator to find and control his own materials from the list.

D. Hand tools change during the model change and operator cannot find required hand tools because of untidy cabinets and variety of hand tools. This causes time loses in the line. There are hand tools that are used for each product family at each station where the assembly production line is operated. Hand tools are used for assembling the parts such as rivet guns and air torque tools. The hand tools are changed at each model based on tolerance ranges of the torque values. For that reason, these hand tools are subject to be changed at each changeover. During the exchange of hand tools, operator must change the tool from material cabinet or from other stations. In the meantime, traveling for the material cabinet and searching other stations for hand tools cause time loss during changeover.
E. Test pads are varied and they are required for testing processes. Unsuitable test pads can damage the product during the test processes. It is time-consuming for the operator to find the appropriate model from various pads.

When the current situation and system are examined, it is observed that the planned time losses are significantly exceeded. This situation is reflected as inefficiency and loss of time when the model conversion is performed in intra-family and inter-family devices in the sixth production line. In this study, it is aimed to reduce time losses and unforeseen losses during model conversion by keeping the number of operators and cycle times constant as they were designed. It is aimed that with the reduction of losses, the productivity of the production line will increase.

4 Problem Formulation and Solution Methodologies

This study aims the reduction of time loss in model conversion. In order to do that, in this section, the following methods have been suggested, which are corresponding to each problem that was defined in the previous section.

4.1 Redesign of Material Placement

In order to solve the logistic operators' material positioning problem, material placement plans for each station were generated. After all the material list analysis have been made, the content of that plan was generated with all materials that were used in the process.

When the list has been created, materials have been grouped as the *Kanban Materials* and *Consumable Materials*. Consumable materials are the most frequently used materials that come in large numbers such as screws, staples, gaskets and etc. Kanban materials are the materials has a critical role in the process that came with certain batches such as boiler covers, display screen and etc. The usage of materials by operators had been considered. After all these processes, material plans are completed and placed to the all stations as operators and logistics operators can see. As a result, the material feeding and usage of them has become easier.

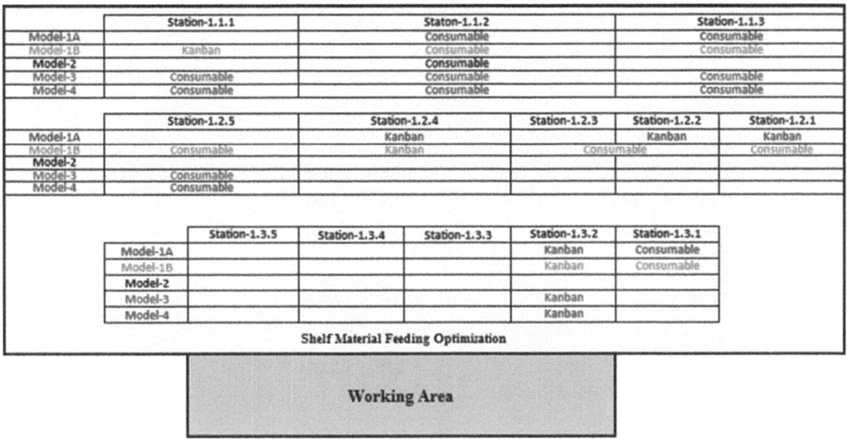

Fig. 2. Material placement plan example for a station (Color figure online)

Observations made on model 1 family suggest that there are further losses at Station-4, Station-5, Station-7, Station-8 and Packing Station-5 due to an improper location of materials during changeover.

The time loss was calculated by the following calculation; "40 s * 6 (station) = 240 s for the line" since 40 s. spend at material changeover when the other stations spend 20 s. As an opportunity cost, this calculated time equals to producing 20 devices. The expectation of the company in the scope of this project study is that this lost time shall be reduced to 120 s for the line. This reduction allows to earn 10 devices. The loss of this time during changeover was tested with the new materials plan and a decrease in time were observed. A list of all materials that were used in model-1 sub models has been compiled and optimization analysis was performed. Only the locations of the kanban materials that are common to the families have been fixed. All other materials, kanban or consumable were returned to the line with hourly material feed. This improved the duration of the model changes compared to the previously set goal, and the reduction in time reflected onto the simulation model.

4.2 Losses Due to Movement of Operators During Changeover

One of the most important losses is, the movement of operators, during changeover for Model 1, 2, 3 and 4. This movement cause time wastes when operators move to their new stations for producing new models. In case of the changeover, switching models from large cycle time to small cycle time generates time wastes.

In Model 4, two operators that work in the hanging device station, move to the combustion chamber stations for switching to Model 1. In Model 2 and 3, two operators that work on the combustion chamber station 1 and 2, move to eighth and tenth stations. These movements retard next stations and the production line. The total time losses are calculated by multiplying number of waiting stations and cycle times of models. Table 1 shows the model change with minimum time waste. For instance, first model change does not waste time until the change from Model 4 to Model 1B as shown in Table 1. This change retards eighteen stations with the cycle time of Model 4.

Table 1. Optimization of model conversion

Changeover (Type to Type)	Calculation of total loss times (sec)
1A-3-2-4-1B	18 * 80 = 1440
1A-4-1B-3-2	18 * 80 = 1440
1A-1B-2-3-4	0*
1A-2-3-1B-4	18 * 93 = 1674
1B-3-2-4-1A	18 * 80 = 1440
1B-4-1A-3-2	18 * 80 = 1440
1B-1A-2-3-4	0*
1B-2-3-1A-4	18 * 93 = 1674
2-3-1A-1B-4	18 * 93 = 1674
2-1B-3-4-1A	(18 * 93) + (18 * 80) = 3114
2-4-1A-1B-3	18 * 80 = 1440*
2-3-4-1A-1B	18 * 80 = 1440*
3-2-1A-1B-4	18 * 93 = 1674
3-1B-2-4-1A	(18 * 93) + (18 * 80) = 3114
3-4-1A-1B-2	18 * 80 = 1440*
3-2-4-1B-1A	(18 * 93) + (18 * 80) = 3114
4-3-2-1A-1B	(11 * 80) + (18 * 93) = 2554
4-1A-1B-3-2	18 * 80 = 1440*
4-1B-3-2-1A	(18 * 93) + (18 * 80) = 3114
4-2-1A-1B-3	(11 * 80) + (18 * 93) = 2554

A mathematical model is formulated for losses due to movement of operators during changeover.

(1) *Parameters Sets and Decision Variables*
This model includes four sets and the definition of these sets are;

$$\begin{array}{ll} n = \textit{Permutation Indices} & n = 1, 2, 3, \ldots, 256 \\ m = \textit{Model Indices} & m = 1, 2, 3, 4 \\ s = \textit{Model Change Indices} & s = 1, 2, 3, 4 \\ r = \textit{Sorting Indices} & r = 1, 2, 3, 4 \end{array}$$

The model parameters are specified as shown below.

DTms = The difference between the number of old and new stations that operators were located m, s
CTm = Cycle Times of Models m
OTm = Order Time m
Tn = Lost time due to movement of operators n

Nmr = The location of model in changeover m, r
Pn(mr) = Permutations of model change n

Decision variables are specified as shown below.

$$\mathrm{Tn} = \begin{cases} 1, & \textit{waste time occurs} \\ 0, & \textit{otherwise} \end{cases}$$

$$\mathrm{OTm} = \begin{cases} 1, & \textit{Nmr} \\ 0, & \textit{otherwise} \end{cases}$$

$$\mathrm{Nmr} = \begin{cases} 1, & \textit{Pn(mr)} \\ 0, & \textit{otherwise} \end{cases}$$

(2) *Mathematical Model*
The mathematical, model that we have formulated is given below.

$$\mathrm{Min} \sum \mathrm{Tn} \sum \mathrm{OTm} \sum \mathrm{Nmr(Pn(mr))} \quad (1)$$

$$\mathrm{DTms} \leq 30 \quad (2)$$

$$\mathrm{Tn} = \mathrm{DTms} * \mathrm{CTm} \quad (3)$$

$$1 \leq \mathrm{Nmr} \leq 4 \quad (4)$$

$$\mathrm{OTm} \geq 0, \mathrm{Tn} \geq 0, \mathrm{CTm} \geq 0, \mathrm{DTms} \geq 0, \mathrm{Pn(mr)} \geq 0 \quad (5)$$

The objective function (1) calculates the weekly production planning according to the orders where there is no time loss in model changes because of the operator changes. The difference between number of stations (2) should be less than or equal to 30, since there are 30 operations in total. Time loss due to movement of operators between models (3) is calculated by multiplying the difference between the number of old and new stations which operators were located and the cycle time of the model. Constraint (4) states that the location of model should be between 1 and 4 during changeover. Model parameters should be greater than or equal to zero (5). Furthermore, Model 1A and Model 1B are defined as Model 1 in common. The reason is that, according to the results of the calculations, the time loss due to the movement of the operator are the same for both Model 1A and 1B.. The mathematical model became more understandable and contributed to have faster results on Excel VBA.

4.3 BOM Control Problem

The Bill of Material lists are consisting all materials used in the line, and the operators were checking the materials, it was too long to control and it was taking approximately 52 s. In order to deal with this problem, Poka-Yoke is used, it is a lean manufacturing technique that aims to prevent, correct, or reduce the errors of misunderstanding, carelessness, or indecisiveness of operators before they happen. After the improvements, the lists become unique to each station and it takes 30 s to control. Also, coloring of materials in the product list has been suggested to make the job even easier. The BOM list is given to the operator of first station by the logistics operator in order to start the work. BOM control is performed by operators who check the codes on the boxes. The signature page at the back is signed and after that BOM control list is fixed with a magnet on the combi boiler. Transition of this list to the next station is ensured and it is expected to be done at every station. Previously, the material codes had to be searched through the whole material lists and these codes were not specified for models and were not in order of importance, for example, before the new list, Kanban and consumable materials was in a mixed order. Thereby, now it became a list that is easily reached and followed by the operator. This study does not only improve production efficiency and quality, but also decreases human error.

4.4 Hand Tool Suggestions

To overcome the problem in hand tool, again Poka-Yoke is used, it is a lean manufacturing technique that aims to prevent, correct, or reduce the errors of misunderstanding, carelessness, or indecisiveness of operators before they happen.

The hand tools used in the production line have specific torque values. As shown in Table 2, combination of torque values was done to minimize the worker error and the time loss while searching for correct hand tools. Some torque values with wide range are narrowed, reducing the number of hand tools to 9.

Table 2. Combination of torque values

Hand tools	Torque values	Number of hand tools	Combination of torque values	Number of common hand tools
Torque Screwdriver	Value -1	1	Value -1	1
	Value -2	1		
Torque wrench	Value -3	2	Value -3	2
	Value -4	2		
Torque device	Value -5	3	Value -6	3
	Value -6	3		
Vertical screwdriver	Value -7	2	Value -8	1
	Value -8	1		
Torque wrench	Value -9	1	Value -10	1
	Value -10	1		
Vertical gun	Value -11	5	Value 12	1
	Value -12	1		

Taking hand tool from cabinet or station is one of the significant loss for the model transformation. By using the 5S technique, the hand tools were examined and classified for each model on a station-by-work basis, based on their torque values. As a result, hand tools will be unique for each station. Figure 2 shows design of hand tool cabinets. Styrofoam was used in the design and it was cut into to the shape of the hand tools. The same hand tools that has different torque values was labeled and colored to be separated from each other (Fig. 3).

Fig. 3. Hand tools cabinet

4.5 Test Pads Design Suggestions

For the solution of the test pad problem, the Quality Function Deployment (QFD) method has been studied and adapted to the problem. With this method, operations were separated and graded according to their importance and priorities. Then, the roof is drawn, and priority order has decided by rating according to relations.

Then, it has been decided to think of the operator as a customer, and to reveal the claims of the operator and to check their feasibility using the quality function deployment method. According to the QFD results, as shown in Fig. 4, ergonomic design found out to be the leading one followed by test subdivisions and new shelf system design, therefore, we have designed a 4-layer test bench rack system to overcome the mess in test pads and sorting of the other bench in the 9th station during the change of the test bench.

The shelf system, shown with the visuals given below, has all the test panels of the models and reduces the time loss by eliminating the need for the operator to put the test pads and order them or place them on the shelves.

Test pads will be painted with different colors for each model and the shelf in the same color will be sorted and we will be able to overcome the problem of mixing any test pad, as shown in Fig. 5.

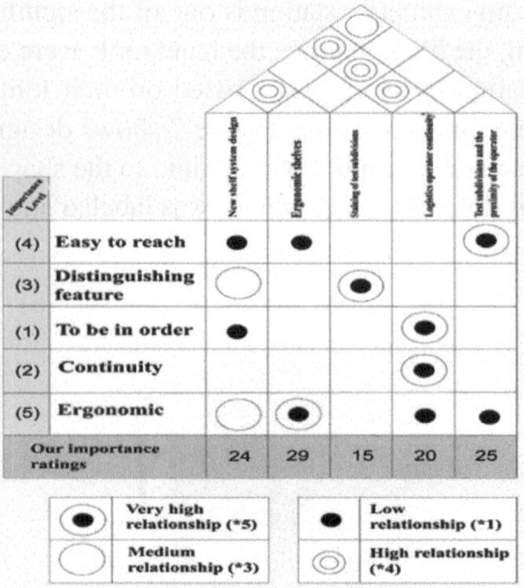

Fig. 4. QFD method usage for test pad design

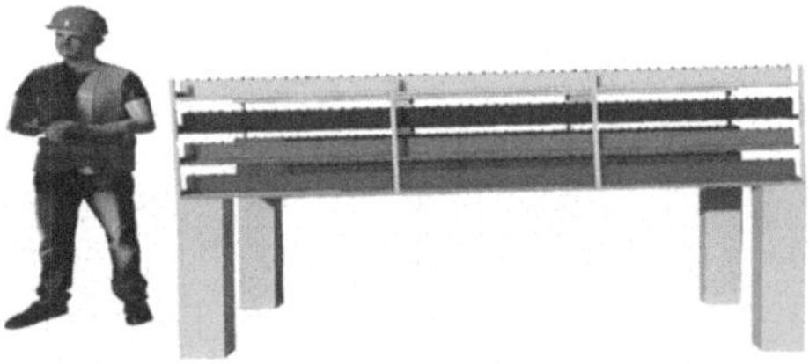

Fig. 5. Test pads design (Color figure online)

Table 3. Comparison of changeover loss times after improvements

Model conversion	First situation in main models (Sec)	Last situation in main models (Sec)
Apparatus exchange	55	45
Material regulation	42	35
BOM reading-checking-signature	52	30
Total	149	110

5 Verification of the Application and Generating a Decision Support System

In this section, the numerical results obtained from the solution of the problem have been analyzed, measured and the verification of the proposed system have been made. The main reason of losses and average times measured can be seen in the Table 3.

These measurements were calculated by the average of the measured times at each station during model conversion.

After all the reductions, the efficiency of the sixth line has been calculated and the detailed calculations has given below:

$$\text{Average changeover in a shift} = 10 \longrightarrow \text{in 3 shifts} : 3 * 10 = 30$$

Initial Situation:
Daily lost time: 30 * 149 s (Maximum Changeover Loss) = 4470 s
Percentage Production for Models

$$\left.\begin{array}{l}\text{Model-1} : \%30\\ \text{Model-2} : \%15\\ \text{Model-3} : \%15\\ \text{Model-4} : \%40\end{array}\right\}(0.3 * 120) + (0.15 * 93) + (0.3 * 93) + (0.4 * 80)$$
$$= 95,9 \text{ Weighted Average Cycle Time}$$

Scheduled run time = 415 s
Production piece in a shift:

$$(415 * 60)/95.9 = 259 \text{ combi boiler} \longrightarrow \text{Daily} : 3 * 259 = 778 \text{ combi boiler}$$

Number of lost devices per day: 4470/95.9 = 46 combi boiler

Final Situation:
Daily lost time: 30 * 110 = 3300
Number of lost devices per day: 3300/95.9 = 34 combi boiler
Gain:
46 – 34 = 12 combi boiler
Annual Productivity Increase: 12/778 = 1.5%

The results are simulated by Arena Simulation Software before generating the DSS. With simulated results, the possible effects were observed. The station-based time losses for all possible models are transferred directly from Decision Support System to Arena Simulation with Read-Write module. The simulation program uses time losses as a variable table in the program itself. Thus, this table reflects non-value-added time in terms of changeover time. The determination of which model should be produced is given by the company. According to this data, simulation runs for one shift and simulates changeovers. These changeovers can be seen from an Andon system which is used by the combi boiler producer company, inside the production line.

In this study Excel VBA is used to make updates based on user input, results of the improvements, comparative graphics showing the initial situation and the final situation. The weekly production plan was arranged to prevent the losses due to the shift of the operator based on the mathematical model.

The home page shown in Fig. 6, provides information to the user about how to use the system and this information is reachable at any time.

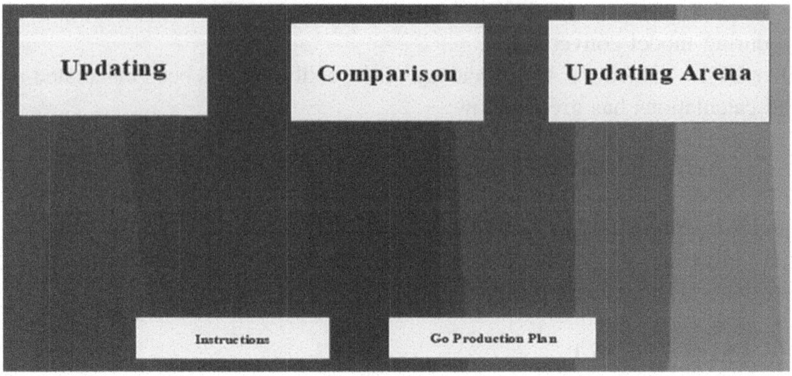

Fig. 6. Home page

In the update section as shown in Fig. 7, the user can update cycle times, changeover times which are based on stations and instructions.

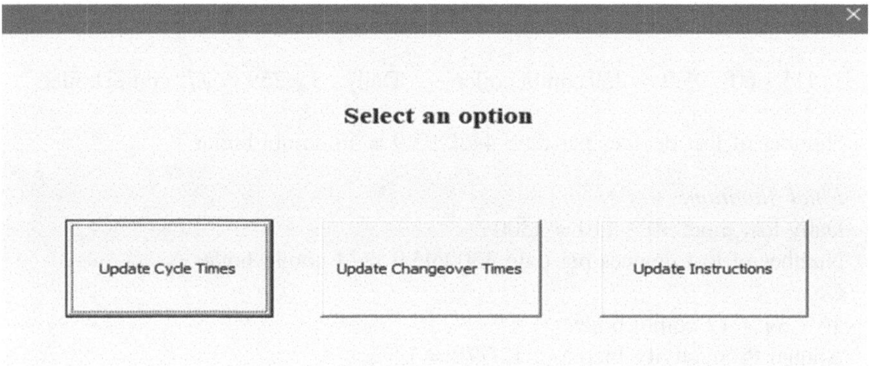

Fig. 7. Update page

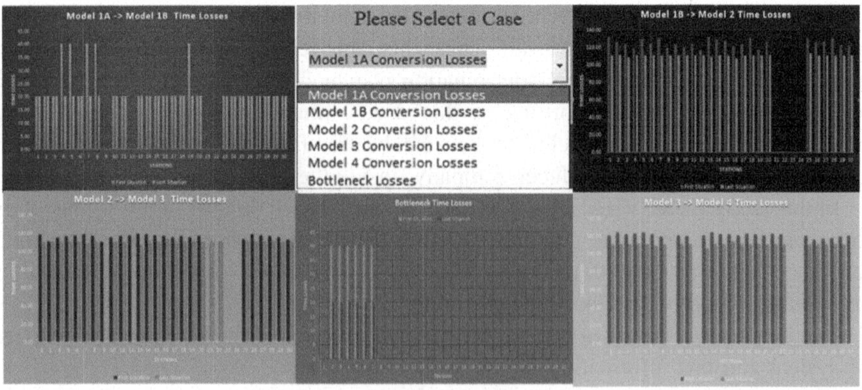

Fig. 8. Comparison dashboard page

In comparison section, user can easily understand the time losses during conversion between models. The bottleneck based time losses of initial situation and improved final situation are shown on Fig. 8 with the comparison graphs.

Arena Update section is aimed to be changed dynamically and the data table that was used in the simulation program, so that Arena and Excel VBA were synchronized.

In production ordering section, as shown in Fig. 9, the models are sorted without any problem and this section creates a weekly plan by using Excel VBA with the mathematical model that is based on decreasing time losses due to movements of operators and expected orders.

Country	Number of Operators	Device	Device Name	Device Specifities	Test Stations	Amount
IT	27	68	Model1	H/L	1	6
IT	27	64	Model1	KD H/L	1	16
IT	27	67	Model1	W H/L	1	6
BE	27	24	Model1	KDW H/L	1	6
IT	27		Model1	KW H/L	1	6
IT	27	69	Model1	W H/L	1	16
BE	27	83	Model1	C 23/21	1	58
IT	27	30	Model1	C 23	1	16
BE	27	84	Model1	23/21	1	20
BE	27	6	Model1	B 23/21	1	6
IT	27	31	Model1	BC 23	1	6
GB	27	5TR	Model4	FSN-GB NG	1	368
GB	27	2TR	Model4	FSN-GB NG	1	312
GB	27	5TR	Model4	FSN-GB NG	1	626
GB	27	2TR	Model4	FSN-GB NG	1	360
GB	27	5TR	Model4	FSN-GB NG	1	354
PL	27	1031	Model2	S5300	1	22
PL	27	1033	Model2	S5300	1	110
PL	27	1032	Model2	S5300	1	30
AT	27	1029	Model2	S1400	1	22
AT	27	1028	Model2	S1400	1	58
AT	27	1030	Model2	S1400	1	18
BE	27	19	Model3	"BE	1	16
BE	27	38	Model3	"BE	1	82
BE	27	21	Model3	"BE	1	6
AT	27	53	Model3	"AT	1	22

Home Page

Fig. 9. Production order page

6 Conclusion and Future Work

As a conclusion, several solution methodologies has been used to decrease the changeover losses in the sixth lean production line at the combi boilers producer company. These solution methodologies are Mathematical Modelling, Poke Yoke, 5S, Quality Function Deployment (QFD), Decision Support System (DSS), by integrating Arena Simulation with Excel VBA. These solution methodologies and approaches have been applied in the factory. The material placement plans have been applied by analyzing material lists based on each station to decrease changeover losses. Model 1 family transition time has been reduced from 240 s to 120 s. Losses due to movement of operators during changeover have been studied based on mathematical model. BOM checking problem has been solved by developing colored product lists for each station. Hand tools has been improved with Poke Yoke and 5S. Test pads have been redesigned based on QFD. After all, DSS has been built as a user-friendly tool to update the data, to compare the initial and the final situation, and to plan the production orders with minimizing time losses. DSS is synchronized with Arena Simulation by changing or updating the changeover times. Finally, changeover losses has been decreased, current situation has been improved significantly, and the productivity of the sixth line has been increased by 1.5%.

Future works may concentrate on deeper analysis of new mechanisms and methods. During this study main focus was on how to reduce the changeover time losses without changing any labor force. However, following suggestions can be analyzed to reduce the changeover loss times in the future:

- In BOM list checking problem, the Industry 4.0 methods can be used and BOM lists can be programmed regarding to warehouse management and can be digitalized with barcode systems. This system can reduce the checking time and it also creates a sustainability since no more papers will be used.
- There are stations that must use the trash box according to the operations performed at the stations in the production line. The stations, where waste containers are used, differ in each model and there is a limited number of waste containers inside the line. It has been observed that these waste containers were moved to stations where they will be used during the changeover. These unnecessary movements also cause time losses.
- Samples can be taken from materials coming from the suppliers these samples can be inspected before they were sent to the production line for quality control purposes. During this process, if there is a defective material among the raw materials and component, there is a risk that it may cause production delay or even production stoppage. Therefore, statistical quality control methods can be applied in the future.
- Products are transported by conveyor system till the 10th station. In the 10th station combi boilers are assembled to the trolleys in order to execute testing and packing operations. During this assembly, trolley pin settings are needed to be changed for each model to fit the combi boilers perfectly. These changes are carried out by workers and cause time losses. There are 17 trolleys on the sixth production line, and during the changeover, the average run time is 5 s. Therefore, more improvement methods can be studied in this situation, as well.

Acknowledgement. We are grateful to the company for sharing their data with us to complete this work. This work cannot be completed without the assistance of Cumhur Kerem Güven, Aygen Aytaç, Emre Akkuzu, and Yasemin Erdem. We are thankful for the contribution of them.

References

1. Shingo S (1985) A revolution in manufacturing: the SMED system. Productivity Press, Cambridge
2. Almomani MA, Aladeemy M, Abdelhadi A, Mumani A (2013) A proposed approach for setup time reduction through integrating conventional SMED method with multiple criteria decision-making techniques. Comput Ind Eng 66(2):461–469
3. Sabadka D, Molnar V, Fedorko G (2017) The use of lean manufacturing techniques – SMED analysis to optimization of the production process. Adv Sci Technol 11(3):187–195
4. Mileham A, Culley S, Owen G, Mclntosh R (1999) Rapid changeover – a pre-requisite for responsive manufacture. Int J Oper Prod Manag 785–796
5. Gungor ZE, Evans S (2017) Understanding the hidden cost and identifying the root causes of changeover impacts. J Clean Prod 167:1138–1147

6. McIntosh RI, Culley SJ, Mileham AR, Owen GW (2000) A critical evaluation of Shingo's SMED (Single Minute Exchange of Die) methodology. Int J Prod Res 38:2377–2395
7. Mustafaa K, Chenga K (2017) Improving production changeovers and the optimization: a simulation based virtual process approach and its application perspectives. In: 27th international conference on flexible automation and intelligent manufacturing, FAIM 2017, 27–30 June 2017, Modena. Procedia Manufacturing, vol 11, pp 2042–2050

Repair Cost Minimization Problem for Containers: A Case Study

Merve Çamlıca, Gülce Çini, Ayşegül Eda Özen, Nilay Çınar, Sel Ozcan(✉), and Deniz Türsel Eliiyi

Department of Industrial Engineering, Yaşar University, Bornova, Turkey
mervecmlca@gmail.com, gulce.cini@gmail.com, ayseguleda.ozen@gmail.com, nilayycinar@gmail.com, {sel.ozcan,deniz.eliiyi}@yasar.edu.tr

Abstract. With the increase of global trade activities and the decrease in transportation costs, the volume of international trade has grown considerably, yielding to an increase in the logistics activities as well as the number of containers used in transport. Hence, containers gradually deteriorate over time and logistics companies start to diminish the repair costs of the containers, as well. This study addresses the repair cost minimization problem for containers where the objective is to minimize the total transportation, repairing and delay time costs. The problem is formulated as a variant of the well-known transportation problem, where we assume that the broken containers transport to repair depots, and the fixed containers are delivered from the repair depots to the customers until the requested due date. In order to monitor the performance of the proposed model, realistic test instances are constructed. Optimal solutions are achieved within 40 min for relatively small and moderate test instances.

Keywords: Logistics · Transshipment problem · Broken container repair
Cost minimization · Delay cost

1 Introduction

Along with the change of the world, the methods of competition of the enterprises also change rapidly. A growing importance of speed, agility and fierce competition in the global supply chain force companies to reconsider using traditional logistics services. As a result of the increase in both market competition and service level expectation of the customers, logistics service providers are forced to re-evaluate and concurrently improve their business services. The main goal in logistics is to reach a high level in customer service, to optimize the use of resources and investments, and to gain competitive advantage in this way. In order to establish the freight transportation in global manner, logistics providers should serve for complex networks with a large number of routing alternatives, which are mainly carried out by different transportation channels, including a collection of truck, rail, barge, air, and ship. Through containerization, all competitors have potentially the same level of access to an efficient and global freight distribution system through ports [1]. Therefore, containerization has become the main driver for global intermodal freight transport, which involves the

N. M. Durakbasa and M. G. Gencyilmaz (Eds.): ISPR 2018, *Proceedings of the International Symposium for Production Research 2018*, pp. 914–921, 2019.
https://doi.org/10.1007/978-3-319-92267-6_73

transportation of freight in containers of standard dimensions with the consideration of high level safety and environmental issues. Many containers are made from materials such as steel, aluminum, and these large boxes gradually deteriorate over time. Considering the damage level and repair alternatives as well as the coordination with the repair centers, the management of the repair operations for containers is a complex issue. Motivated from a real-life problem of a leading logistics company located in Izmir, Turkey, this study is focused on the assignment of broken containers to related repair centers and sending them to the related customers at a minimum cost.

The rest of the paper is organized as follows. In Sect. 2, literature review is proposed. The problem and its characteristics are defined in Sect. 3. The mathematical model is presented in Sect. 4. Section 5 includes the computational study where the solutions of sample problems are listed, tabulated and discussed. Finally, Sect. 6 includes our concluding remarks and future work.

2 Literature Review

Reinhardt et al. studied the drayage problem with the objective of reducing the efficiency of pre- and end-haulage bottlenecks. In this study, the pre- and end-haulages of containers are scheduled using techniques of vehicle routing problems since the problem is based on the movements between the customers and the depots. It has been shown that the model can be easily solved using column enumerating since the number of possible paths in the problem is limited. The effect of side constraints on overall cost has also been analyzed [2].

Häll et al. design vehicle routes and schedules for a dial-a-ride service where some part of each request may be performed by a fixed route service. Passengers can go from one place to another without changing the vehicle, however they can also exchange vehicle. Each request contains one or several passengers and requires a certain capacity in a vehicle, for the persons and any wheelchairs, walkers, luggage etc. They assume that all vehicles have the same capacity and the aim is to minimize the total routing cost [3].

Kiremitci et al. study one of the most important types of vehicle locating problems. The multi-vehicle distribution and collection problem with time windows, where the objective is minimization of total transport costs as well as the number of vehicles required, balancing routes for travel time and vehicle load. The number of variables is reduced by using real values as much as possible. New algorithms approaches are compared with old ones on some problems in the literature and it is observed that the proposed algorithm gives relatively better results [4].

Hlayel and Alia studied the transportation problem with the objective of reducing the transportation cost and time. The most important way of achieving the solution by the best candidate method (BCM) is to start with selecting the most suitable candidates and reduce the solution combinations accordingly. Combinations can be obtained without any means of intersection. By comparing the results obtained, the implementation of BCM in the proposed method achieves the best first applicable solution for a transport problem and achieves faster than current methods with minimal computation time and less complexity [5].

Eliiyi et al. model the transfer problem of a company with a transfer warehouse with many subcontractors and customers in the transfer sector. Taking into account the supply times and the customer's deadlines distinguishes the work from others [6].

3 Problem Definition

As the need for containers increases during transport, the container damages increase accordingly. Logistics companies are trying to repair the broken containers in the shortest time and with the minimum cost and deliver them to the customers. Motivated from the real-life problem of the logistics company, who is serving as a pioneer of the maritime transportation business in Turkey, the decision of where and when to repair the broken containers is a complex issue. The problem is formulated as a transshipment problem, which is a variant of a transportation problem, where the source nodes are the points where the container is broken, and the sink nodes refer to the customer locations. The intermediate nodes between the source and sink nodes, i.e., transshipment points, denote the repair centers. Our problem differs from the classical transshipment problem, since time windows should also be considered. Time window restrictions imply that the customer containers need to arrive within a certain period of time. Moreover, each customer has a due date and additional penalty costs are incurred for each delay. In this direction, a solution method should be developed to minimize the problem of container repair, which is a problem that arise in real life.

4 Mathematical Model

The following section describes our mathematical model. The problem is formulated as a transshipment problem where the objective is to minimize the total transportation, repairing and delay time costs. The mathematical model decides on how to transfer the broken containers from the breakdown points to one of the suitable depots and then to one of the candidate customers. We assume that there are three different vehicle types used for carrying containers from/to depots, breakdown points and customers. The indices and parameters used in the model of transshipment problem are defined as below:

i: breakdown point index
j: depot point index
k: customer point index
S_i: The number of total containers in breakdown point i
D_k: The number of demand of customer k
RT_j: Repairing time of depot j
TT_{ij}: Transportation time from breakdown point i to depot j for truck
TT_{jk}: Transportation time from depot j to customer k for truck
ST_{ij} : Transportation time from breakdown point i to depot j for ship
ST_{jk}: Transportation time from depot j to customer k for ship
RT_{ij}: Transportation time from breakdown point i to depot j for train

RT_{jk}: Transportation time from depot j to customer k for train
WT: Waiting time of container, including the waiting time of the vehicle and in the queue
DD_k: Due date which is given by customer k
t: Delay time of container
tc: Cost of delay
C_{ij} : The transportation and repairing cost for truck from i to j
X_{ij} : The transportation and repairing cost for ship from i to j
Y_{ij} : The transportation and repairing cost for train from i to j
C_{jk} : The transportation cost for truck from j to k
X_{jk} : The transportation cost for ship from j to k
Y_{jk} : The transportation cost for train from j to k
QT_{ij}: The number of broken container which is transported by truck from i to j
QT_{jk}: The number of fixed container which is transported by truck from j to k
QS_{ij}: The number of broken container which is transported by ship from i to j
QS_{jk}: The number of fixed container which is transported by ship from j to k
QR_{ij}: The number of broken container which is transported by train from i to j
QR_{jk}: The number of fixed container which is transported by train from j to k

Decision variables are defined as follows.

$$T_{ij} = \begin{cases} 1, & \text{if the container which is transported by truck from breakdown point } i \text{ to depot } j \\ 0, & \text{otherwise} \end{cases}$$

$$S_{ij} = \begin{cases} 1, & \text{if the container which is transported by ship from breakdown point } i \text{ to depot } j \\ 0, & \text{otherwise} \end{cases}$$

$$R_{ij} = \begin{cases} 1, & \text{if the container which is transported by train from breakdown point } i \text{ to depot } j \\ 0, & \text{otherwise} \end{cases}$$

$$T_{jk} = \begin{cases} 1, & \text{if the container which is transported by truck from depot } j \text{ to customer } k \\ 0, & \text{otherwise} \end{cases}$$

$$S_{jk} = \begin{cases} 1, & \text{if the container which is transported by ship from depot } j \text{ to customer } k \\ 0, & \text{otherwise} \end{cases}$$

$$R_{jk} = \begin{cases} 1, & \text{if the container which is transported by train from depot } j \text{ to customer } k \\ 0, & \text{otherwise} \end{cases}$$

Based on the definitions above, the MILP formulation of the problem is as follows.

$$Min \sum_i \sum_j C_{ij} * QT_{ij} + \sum_j \sum_k C_{jk} * QT_{jk} + \sum_i \sum_j X_{ij} * QS_{ij} + \sum_j \sum_k X_{jk} * QS_{jk} + \sum_i \sum_j Y_{ij} * QR_{ij} + \sum_j \sum_k Y_{jk} * QR_{jk} + t_c * t \quad (1)$$

s.t.

$$T_{ij} + S_{ij} + R_{ij} \leq 1, \quad \forall i,j \quad (2)$$

$$T_{jk} + S_{jk} + R_{jk} \leq 1, \quad \forall j,k \quad (3)$$

$$\sum_j (QT_{ij} + QS_{ij} + QR_{ij}) \leq S_i, \quad \forall i \quad (4)$$

$$\sum_j (QT_{jk} + QS_{jk} + QR_{jk}) \geq D_k, \quad \forall k \quad (5)$$

$$\sum_i S_i = \sum_k D_k \quad (6)$$

$$\sum_i (QT_{ij} + QS_{ij} + QR_{ij}) = \sum_k (QT_{jk} + QS_{jk} + QR_{jk}), \quad \forall j \quad (7)$$

$$QT_{ij} \leq M * T_{ij}, \quad \forall i,j \quad (8)$$

$$QS_{ij} \leq M * S_{ij}, \quad \forall i,j \quad (9)$$

$$QR_{ij} \leq M * R_{ij}, \quad \forall i,j \quad (10)$$

$$QT_{jk} \leq M * T_{jk}, \quad \forall j,k \quad (11)$$

$$QS_{jk} \leq M * S_{jk}, \quad \forall j,k \quad (12)$$

$$QR_{jk} \leq M * R_{jk}, \quad \forall j,k \quad (13)$$

$$RT_j + (TT_{ij} * T_{ij}) + (TS_{ij} * S_{ij}) + (TR_{ij} * R_{ij}) + (TT_{jk} * T_{jk}) + (TS_{jk} * S_{jk}) + (TR_{jk} * R_{jk}) + WT - (DD_k + M * (1 - (T_{ij} + S_{ij} + R_{ij}))) \leq t, \quad \forall i,j,k \quad (14)$$

$$T_{ij}, T_{jk}, S_{ij}, S_{jk}, R_{ij}, R_{jk}\ binary, QT_{ij}, QT_{jk}, QS_{ij}, QS_{jk}, QR_{ij}, QR_{jk}, t \geq 0\ integer, \quad \forall i,j,k \quad (15)$$

The objective function in (1) minimizes the total transportation, repairing and delay time costs. Constraint set (2) enforces that every container must be transported with at most one transportation way (either with land route, shipping way or railway) from breakdown point i to depot point j. Constraint set (3) satisfies that every container must be transported with only one transportation way (land route, shipping way or railway) from depot point j to customer k. Constraint sets (4) and (5) ensure that the broken containers on the breakdown point i are transferred to the relevant customer point k. Constraint set (6) satisfies the supply and demand equality at breakdown and customer points. If a container transports to depot j, it should be send to the customer at depot j. Constraint set (8–13) calculates the total number of containers transferred via each vehicle type. Constraint set (14) ensures that the total repair time for all depots, transshipment times between breakdown points and depots and customers should be smaller or equal to the due date of customers k. Finally, constraint set (15) defines the decision variables.

5 Computational Results

The performance of the proposed mathematical model is tested on several test instances generated by realistic assumptions. The proposed mathematical model is solved using IBM ILOG CPLEX Optimization Studio 12.7.1 on a computer with an i7 processor and 16 GB of RAM. The smallest test instance includes 2 different container breakdown points, 3 different repair depots and 2 different customer points. Then, the instance is extended with 3 and 4 broken containers at each point. Figure 1 visualizes all possible combinations of the smallest instance tested.

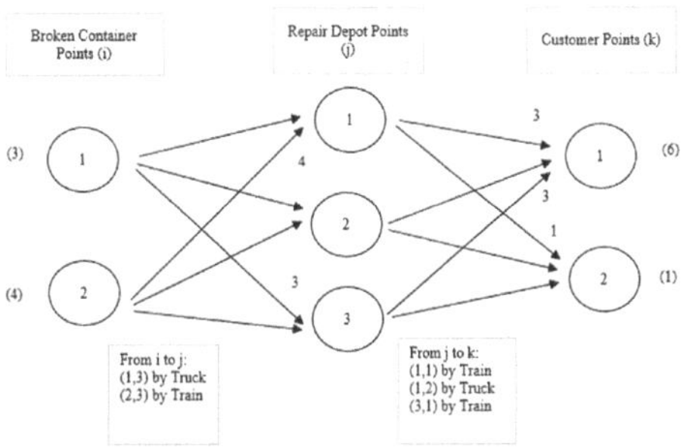

Fig. 1. Network diagram of the toy data solution

The performance of the proposed model is observed in larger test instances. The properties of the large test instances are reported in Table 1. The case of where 10 breakdown, repair and customer points are considered can be denoted as *small,* and the remaining two cases, i.e., with 20 and 30 points in each, are labeled as *moderate* and *large*, respectively. We also expand our computational experiment by employing the number of broken container at each breakdown point as 1, 2 and 3, respectively. All test instances are solved optimally by CPLEX and Tables 2, 3 and 4 report the computational time performance of each problem tested. Note that, the computational time gradually increases as the problem size increases, as expected. The computational time performances of *small* and *moderate* cases are comparable (see Tables 2 and 3). Especially with 30 breakdown points, 30 depots and 30 customer points the solution time increased exponentially as the number of broken containers increased.

Table 1. Characteristics of the large test instances

Number of breakdown points	Number of repair depots	Number of customer points	Number of broken container
10	10	10	1
10	10	10	2
10	10	10	3
20	20	20	1
20	20	20	2
20	20	20	3
30	30	30	1
30	30	30	2
30	30	30	3

Table 2. Computational results for the test instances with 10 points

Number of broken container	Time (in sec)
1	85
2	753
3	1298

Table 3. Computational results for the test instances with 20 points

Number of broken container	Time
1	322
2	545
3	2379

Table 4. Computational results for the test instances with 30 points

Number of broken container	Time
1	2376
2	7510
3	21072

6 Conclusions

In this study, minimization of the transportation and repair costs of containers, which is one of the real-life problems, has been discussed. The problem is formulated as a variant of the transshipment model and solved optimally by IBM ILOG CPLEX Optimization Studio 12.7.1. The test problems with various sizes are constructed so as to reflect the real-life complexity. We can deduce that the model yields similar performances on relatively small and moderate test instances, whereas the computational time performance is low for larger test problems.

Since assignment of the broken containers is an operational level decision making as the future extension, sophisticated heuristic methods can be applied to obtain good quality solutions within reasonable time. Moreover, the formulation can be extended by using the real-time availability on each repair depot.

Acknowledgment. First of all, we would like to thank the Professor Dr. Deniz Türsel Eliiyi for providing this research course opportunity to us. Additionally, we would like to thank to our instructors Sel Ozcan and Hande Öztop for their guidance and enlightenment. We feel enthusiasm and get excited about this special research thanks to them.

References

1. Rodrigue JP, Notteboom T (2015) Looking inside the box: evidence from the containerization of commodities and the cold chain. Marit Policy Manag 42(3):207–227
2. Reinhardt LB, Pisinger D, Spoorendonk S, Sigurd MM (2016) INFOR: information system and operational research: optimization of the drayage problem using exact methods. I N F O R J 54(1):33–51
3. Häll CH, Andersson H, Lundgren JT, Värbrand P (2009) The integrated dial-a-ride problem. Public Transp 1:39–54
4. Kiremitci B, Kiremitci S, Keskintürk T (2015) A real valued genetic algorithm approach for the multiple vehicle pickup and delivery problem with time windows. Istanb Univ J Sch Bus 43:391–403
5. Hlayel AA, Alia MA (2012) Solving transportation problems using the best candidates method. Comput Sci Eng Int J (CSEIJ) 2:23–25
6. Türsel Eliiyi D, Yurtkulu EZ, Yurdakul Şahin D (2010) Supply chain management in apparel industry: a transshipment problem with time constraints

Routing Optimization for Container Dispatching Operations

Hasibe Serap Baş, Ayşe Tolan, Mahmut Ali Gökçe(✉), and Cansu Yurtseven

Industrial Engineering, Yaşar University, Bornova, Turkey
h.serap.bas@gmail.com, ayse.tln@hotmail.com, {ali.gokce, cansu.yurtseven}@yasar.edu.tr

Abstract. An optimal container dispatch planning model for Alkon Logistics is proposed. Alkon Logistics handles container movement between three large ports and a logistics center nearby, connected to major highways and railroad systems. As a solution approach, a novel integer mathematical model has been developed to minimize total cost of operations. Our model is developed to serve three different ports, which are T.C Ege Gübre, Nemport, and APM Terminal. In addition to making daily operational decisions optimally, the proposed model also assists in making critical long-term decisions, such as vehicle investments. Our model is applicable to any transport company, which carries single-commodity due to its flexible structure. Proposed model is solved using IBM ILOG CPLEX Optimization Studio 12.8 program and tested with real-life data from the company. The results show significant improvement over the current situation. To ensure efficient usage of the proposed approach, a decision support system (DSS) is designed and implemented that allows studying different scenarios. Created DSS is dynamic and practical for users and enables making analysis by using mathematical model with different fleet sizes. Results from experimentation show that compared to current situation, number of rented trucks goes down to less than 10 from 17 corresponding to a decrease in renting cost by 65% and a decrease of 25% in total cost. Proposed solution methodology provides a container dispatch planning with improved truck utilization and optimal cost.

Keywords: Routing optimization · Container terminal · Dispatch planning Logistics · Decision Support System

1 Introduction

The amount of container transportation in the world has been increasing at an amazing pace. Starting with 50 million twenty feet equivalent unit (TEU) in 1985, world container turnover has reached more than 350 million TEU in 2004 [1]. That trend continued to increase over the years, without any sign of slowing down. Today, number of cargo ships in the world is approaching 60000 and there are about 20 million containers traveling the oceans every day. For these reasons, container transportation has great importance for logistics companies, which handles the movement of containers between different modes of transportation. An efficient container dispatch planning

N. M. Durakbasa and M. G. Gencyilmaz (Eds.): ISPR 2018, *Proceedings of the International Symposium for Production Research 2018*, pp. 922–936, 2019.
https://doi.org/10.1007/978-3-319-92267-6_74

enables faster and safer dispatching of commodities. In addition, efficient container dispatching operations affects performance of logistics companies positively and, it provides strong relationship with customers.

In container dispatching operations, there are many operational constraints and costs involved. There are also peak times for demand, based on strict timeline of ships' and freight cars' arrivals and departures. These peak times and congestions due to port traffic lead to inefficient trucking operations. This type of situations causes financial loss for logistics firms. Our proposed solution provides an optimal and implementable container-dispatching plan, which can handle all customer demands under strict time constraints. The objective of the problem is minimizing total dispatching cost with increased truck utilization. To provide that, a novel mathematical model is developed on a time-space graph, which discretizes the passing time into manageable short time slots. Proposed model determines optimal routes of trucks and containers on graph.

2 General System Analysis

Alkon Logistics is located in Biçerova, Izmir, Turkey. It manages container dispatching operations between Biçerova Container Terminal and three ports nearby Izmir. All three ports are located in Aliağa, İzmir. Company works in two-shifts; first is between 08:00-16:00 and, second is 16:00-24:00. All trucks must return to Biçerova Container Terminal at the end of each shift. Company operates with its own truck fleet which consists of 15 trucks. Besides, company rents additional trucks at the beginning of each year with an agreement for peak time operations. The agreement is for one year period and currently 17 extra trucks are used.

Containers are carried to the ports via open-bed trucks and they must arrive at given ports before the *cut-off time* of vessels. *Cut-off time* is the latest entrance time for dispatched containers to ports to be able to make it onto the vessel. Not being able to meet this constraint has significant costs associated with it and hence this is taken as a hard constraint. Containers, which are delivered to ports earlier than cut-off times, can be stored at ports for a certain amount of time at no cost. This is called *free storage period* and is determined by port management on a company basis. If containers are stored at ports longer than free storage period, company incurs storage cost that must be paid to port management. Containers are dispatched from Biçerova Container Terminal for inland transportation via freight cars. Departure time of freight cars is 23:00 p.m. every day excluding Sundays. Possible routes of trucks, transportation modes of containers and, locations of Biçerova Container Terminal and three ports are shown in Fig. 1.

As is the case in the world, both 20-feet and 40-feet containers are used and must be transported. Each truck can carry exactly one container between pairs of locations (port-terminal, port-port), whether the container is full or empty. In total, approximately 1000 containers are dispatched within a week, with a tendency of fast growth over the next years.

Fig. 1. Locations of company and three ports

2.1 Observation and Symptoms

There are two main observations made. Firstly, the company has serious difficulty in applying appropriate dispatch planning during peak times. This situation leads inefficient truck operations; meaning more trucks traveling without containers to demand locations for handling customer demands. Second symptom related to the first one is inadequacy of the truck fleet. Number of trucks belonging to company can be insufficient for dispatching operations during peak times. In such a case, company often rents additional trucks unnecessarily. These situations cause waste of money and time, decreasing performance of the company.

3 Problem Identification

There are a number of operational decisions, which have to be made on a daily basis. The main decisions are determining which truck takes which container, at what time and, from which location to which location. In other words, optimal flows of trucks and containers must be done subjects to all constraints. The objective function of the problem is minimizing total operational cost. Total cost includes traveling cost of trucks, renting cost of extra trucks and storage cost, if incurred (Figs. 2 and 3).

There are two strict time related constraints that must be met. Delivery of containers must meet:

- Cut-off time of vessels at ports:

and

- Departure time of freight cars in Biçerova:

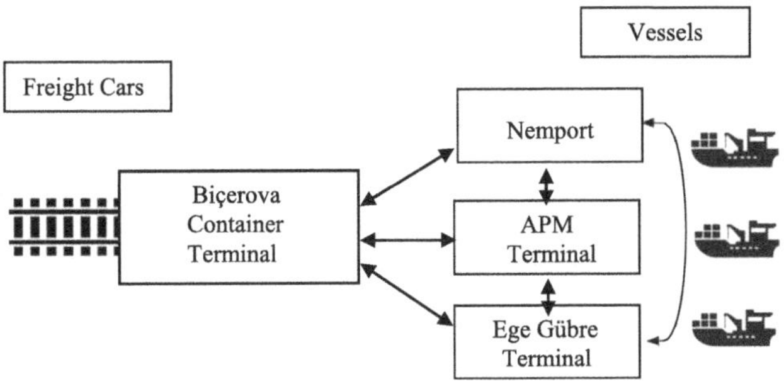

Fig. 2. Operation process

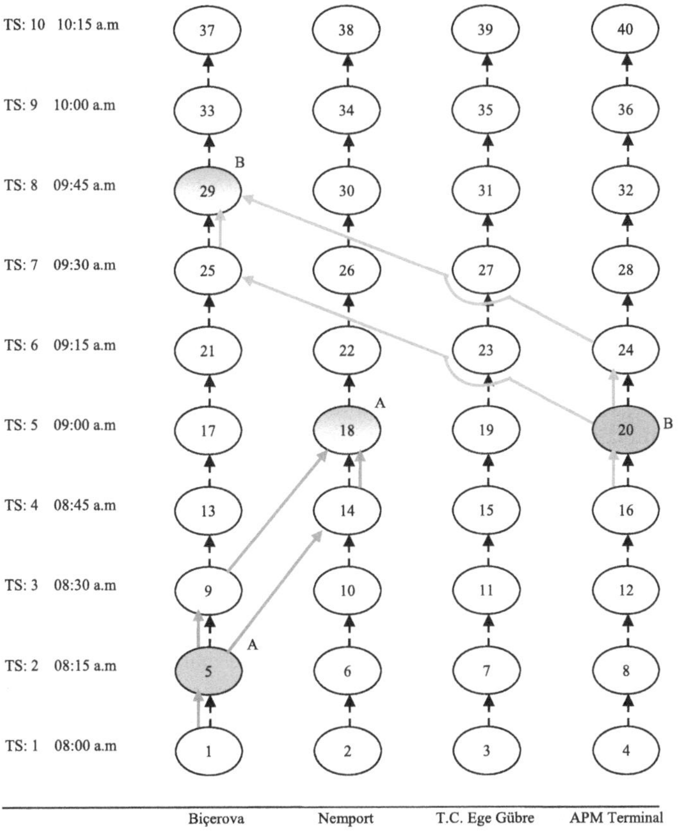

Fig. 3. Example of time-space graph

If containers are dispatched from Biçerova via freight cars, they must arrive in departure time of freight cars in Biçerova.

4 Literature Review

In literature, we find some similar example problems that have been examined, such as inter-terminal transportation, multimodal transportation and, vehicle assignment. To the best knowledge of the authors, no exact model for this problem has been published before. Therefore, this work fills an important void in literature with an exact model that can be adapted to many different situations. However, the most relevant problem is a mathematical model of inter-terminal transportation [2] and our work is based on the idea from this work.

Tierney et al. [2], developed an integer mathematical model for analyzing inter-terminal transportation. Containers can be dispatched between terminals by using different transportation modes such as rail, sea, and road. Inter-terminal transportation leads lateness of container delivery and congestion in peak times. Tierney et al., generated a system which aims minimizing penalized lateness of deliveries.

Chung et al. [3] examine workflow of container transportation and develop mathematical models, which compound various characteristics and classifications of containers and trucks. Objective of this study is to minimize fleet size, total operational cost of vehicles, and total transportation cost. They solved this problem by splitting into three stages. For the first stage, fleet size is tried to minimize by applying the Multiple Traveling Salesman Problem (MTSP) standard formula. For the second and third stages, operational costs are grouped for three different vehicle types and, total operating and transportation cost are optimized by applying insertion heuristics algorithm.

Kim and Nguyen [4] proposed a solution approach for vehicle dispatching at port container terminals. A real-time vehicle dispatching algorithm is developed to assign delivery orders of containers to vehicles considering uncertain travel time of vehicles. This study also can be called as a scheduling problem for vehicles. To study different scenarios and test the dynamic environment, a simulation study is conducted. Their study can be helpful for increasing truck utilization with the developed algorithm.

A mathematical model of inter-terminal transportation [2] was used as a guidance for solving the problem of Alkon Logistics. It is helpful for establishing a solution approach in handling constraints of this problem. The main contributions are allowing of cross-way movements, peak time routing, analysis of capital investment (truck purchasing) under a variety of scenarios and a decision support system that allows running the model.

5 Mathematical Model

Weekly container dispatch planning is modelled on a time-space graph, which includes specially designed constructions. Created model decides routes of trucks during the day; it also allows *cross-way movement*. In *cross-way movement*, a truck travels to port from Biçerova with a container. After delivery, truck travels to another port for carrying another container and return to Biçerova. Cross-way movements are modelled in this work for the first time. These cross-way movements allow better utilization of trucks and lead to much more efficient operation. Besides, model includes components, which can decide how many extra trucks need to be rented. Extra trucks are rented in situations such as inadequate truck fleet for peak operating times.

5.1 Assumptions

First assumption is that capacities of all trucks are same; each truck can carry exactly one container from a location to another location. Second assumption is that, there is no difference in carrying full or empty containers. Third assumption is that daily operating time of two shifts are limited, same and, no overtime.

5.2 Construction of Time-Space Graph

First, a non-temporal base graph is designed. This base graph identifies connections between Biçerova Container Terminal and three ports. Assume that n is the number of nodes. Thus, the base graph can be constructed as $G = (N, A)$, where $N = \{1 \ldots n\}$ is the set of nodes and A is the set of arcs (i, j).

Using the base graph, a time-space graph is designed. This graph is required to model the problem. A time period is needed as a discretized way of representing passing time. Assume that τ is the number of time periods. In this way, the time-space graph can be constructed as $G^T = (N^T, A^T)$, where $N^T = \{1 \ldots \tau . n\}$ is the set of all nodes in the time-space graph and A^T is the set of all arcs (i, j) in the time-space graph.

5.3 Stationary Arcs

Stationary arcs are defined to allow trucks and containers to remain in the same place along the time periods. These arcs connect each node to itself with an arc for the next time periods. Assume that A^S is the set of stationary arcs. A^S can be defined as:

$$A^S = \bigcup_{1 \leq i \leq n.(\tau-1)} \{(i, i+n)\}$$

5.4 Arcs and Node Properties

There are sets of nodes for accessing to each time-space node i with arcs ($i \in N^T$). These sets can be divided into two different types of set. First is the sets of nodes which can enter to each node i with arcs $\{In(i)\}$. Second is the sets of nodes which can leave from each i node with arcs $\{Out(i)\}$.

- $In(i) = \{(j, i) \in A^T \wedge j < i\}$
- $Out(i) = \{(i, j) \in A^T \wedge j > i\}$

At the start of the optimization, each node i is associated with a number of trucks which present at its location ($i \in N^T$). In other words, s_i determines origins of the truck flows. It is expected that all trucks must return to Biçerova at the end of the day. Thus, at the start of the optimization, the only node, which has truck at time step one, is Biçerova Terminal node.

5.5 Demand

Container demands arrive in groups. Each demand group includes one or more containers. This is due to a number of containers for a customer arrive or leave with a single vessel. Assume that D is the number of demand groups $D = \{1 \leq d \leq D\}$. The following parameters are defined for each demand group.

- Origin and destination nodes; $o_d \in N$, $d_d \in N$.
- Release time step; $r_d \in \{1\ldots\tau\}$
- Due time step; $u_d \in \{1\ldots\tau\}$

5.6 Example of Time-Space Graph

Following graph represents an example for 2 h 15 min period. As it was mentioned before, there will be 40 time-space nodes for 10 time step (TS) and 4 specific nodes. All nodes in first column specify Biçerova Container Terminal and the other columns specify Nemport, T.C. Ege Gübre, and APM Terminal, respectively. Dashed lines represent stationary arcs which provide containers and trucks remain in same place through the time periods. Assume that travel time is 30 min between locations and there are two demand groups, which must be dispatched; group A and B.

Node 5, which is shown as solid blue, represents release location of demand group A. Also, node 5 is the member of time step two, this means release time step of demand group A is second time step. At the same time node 18, that is shown as pattern blue, represents delivery location of demand group A. Node 18 is the member of fifth time step, thus due time step of demand group A is fifth time step. In the same way, release and due time steps of demand group B is fifth and eighth time step, respectively. In accordance with these information, some $In(i)$ and $Out(i)$ sets can be defined as:

- $In(5) = 1$
- $In(18) = 9, 14, 11, 12$
- $Out(5) = 9, 14, 15, 16$
- $Out(18) = 25, 22, 27, 28$

$In(5)$ set includes only node 1 as stationary because; time interval does not fit with the travel time for node 2, 3 and 4. $In(18)$ set includes node 14 as stationary and also includes node 9, 11, and 12 for respecting travel time restriction.

5.7 Modeling

Created model routes all trucks on time-space graph while minimizing total cost. Total cost includes loaded and empty travel costs of all trucks, cost of extra rented trucks and storage cost.

Sets and Indices

$N =$ Set of nodes in the base graph, $N = \{1, 2, 3, 4\}$ (More ports can be added, when it is required)
$\tau =$ Number of time periods
$N^T =$ Set of nodes in the time-space graph, $N^T = \{1 \ldots \tau.N\}$
$A^T =$ Set of arcs in the time-space graph
$D =$ Set of demand groups $1 \leq d \leq D$
$W =$ Set of days $1 \leq w \leq W$
$In(i) =$ Set of nodes which can enter to each node i with arcs, $i \in N^T$
$Out(i) =$ Set of nodes which can leave from each i node with arcs, $i \in N^T$

Parameters

$Amount_d =$ Number of containers in demand group d
$o_d =$ Origin node of demand group d, $o_d \in N$
$d_d =$ Destination node of demand group d, $d_d \in N$
$r_d =$ Release time step of demand group d, $r_d \in \{1 \ldots \tau\}$
$u_d =$ Due time step of demand group d, $u_d \in \{1 \ldots \tau\}$
$st_d =$ Time step which gives starting of free storage time for demand group d
$c_{ij} =$ Transportation cost between node i and node j, $i \in N^T, j \in N^T$
$cf_{ij} =$ Cost difference among loaded and empty travel between node i and node j, $i \in N^T, j \in N^T$
$stcost =$ Storage cost per time period
$s_i =$ Number of trucks present at node i at the beginning of the optimization, $i \in N^T$
$z_{ij} =$ A parameter which can take the value of 0 or 1. if arc (i,j) is stationary arc, the parameter will be 0, otherwise it will be 1.
$Hirecost =$ Hiring cost of extra truck

Decision Variables

Decision variable x_{ij}, decides that how many trucks will travel on arc (i,j) and, it routes trucks on time-space graph. Decision variable y_{ijd}, determines number of containers that carried on arc (i,j), it also provide flow of the containers on time-space graph. ($Hired_w$) A decision variable, which determines how many trucks need to be rented from outside, is defined for situations such as inadequate truck fleet. Model determines the value of this variable at time step one, which is the beginning of the optimization. At the beginning of the optimization, all rented trucks must be in Biçerova. Thus, the only node, with any rented trucks at time step one, is terminal node Biçerova. In the following time periods, decision variable h_{ij} routes rented trucks on time-space graph.

$x_{ij} =$ Number of trucks on arc (i,j), $(i,j) \in A^T$
$y_{ijd} =$ Number of container on arc (i,j) for demand group d, $(i,j) \in A^T$
$h_{ij} =$ Number of rented trucks on arc (i,j), $(i,j) \in A^T$
$Hired_w =$ Number of rented trucks at node, $w \in$ W

Objective Function

$$Min\sum_{i,j\in A^T}(x_{ij}+h_{ij}).c_{ij}+\sum_{w\in W}Hired_w.hirecost+\sum_{i,j\in A^T}\sum_{d\in D}y_{ijd}.cf_{ij}+\sum_{i,j\in A^S\wedge j<d_d+(st_d-1).N}\sum_{d\in D}y_{ijd}.stcost \tag{1}$$

Constraints

$$\sum_{d\in D}z_{ij}\cdot y_{ijd}\leq x_{ij}+h_{ij},(i,j)\in A^T \tag{2}$$

$$\sum_{j\in Out(i)}x_{ij}-\sum_{k\in In(i)}x_{ki}\leq s_i,\ i\in N^T \tag{3}$$

$$\sum_{j\in Out(i)}h_{ij}-\sum_{k\in In(i)}h_{ki}\leq Hired_w,\ i=1 \tag{4}$$

$$\sum_{j\in Out(i)}h_{ij}-\sum_{k\in In(i)}h_{ki}\leq 0,\ i\in N^T\backslash\{1\} \tag{5}$$

$$\sum_{k\in In(i)}x_{ki}+\sum_{k\in In(i)}h_{ki}=s_1+Hired_w, w\in \mathrm{W} \tag{6}$$

$$\sum_{j\in Out(i)}y_{ijd}=Amount_d\, d\in D,\in N^T,\ i=o_d+(r_d-1).N \tag{7}$$

$$\sum_{i\in In(i)}y_{ijd}=Amount_d\, d\in D, j\in N^T, j=d_d+(u_d-1).N \tag{8}$$

$$\begin{gathered}\sum_{j\in Out(i)}y_{ijd}-\sum_{k\in In(i)}y_{kid}=0\, d\in D, j\in N^T\\ i\neq o_d+(r_d-1).N\, and\, i\neq d_d+(u_d-1).N\end{gathered} \tag{9}$$

Objective function (1), minimizes the total cost as summation of traveling cost of all trucks, renting cost of extra trucks and, storage cost. Constraint (2) provides that number of containers on an arc cannot be more than number of trucks on same arc. At the same time, if the arc is stationary arc, z_{ij} takes the value 0. In this case, container can remain at same place and no truck needs to carry the container. Constraint (3) is truck flow balance constraint, which allows the trucks to travel on time-space graph. In accordance with constraint, number of trucks leaving a node cannot be more than summation of number of trucks entering the node and number of trucks that start at the node. This constraint also allows trucks to travel without container on time-space graph. Constraint (4) determines the number of rented trucks leaving from Biçerova at the start of the optimization. Constraint (5), is rented truck flow balance constraint which allows rented trucks to travel on time-space graph. Constraint (6) enforce that all trucks return to Biçerova at the end of each shift. Index i takes values, which represents end of each shift through the week in time-space graph. Constraint (7)–(9), flow the containers on time-space graph. Constraint (7) specifies a starting node on time-space graph for all demand group. In other words, this constraint binds the origin node of the demand group. Constraint (8) enables that all demands arrive at the destination node on time. Constraint (9) is balance constraint of containers. It enforces that number of containers leaving a node must be equal to number of containers entering the node excluding origin and destination nodes for all demand groups.

6 Sensitivity Analysis

Real-life data is obtained from the company to test the proposed model with ILOG IBM CPLEX Program. Table 1 shows obtained results from different scenarios based on this data. If results are examined, required number of trucks can be determined according to changing demand information. As shown in Table 1, different scenarios were created by changing number of trucks for 900 containers and 1000 containers. Prepared table includes scenario number, number of containers, number of trucks, and number of rented trucks, storage cost, cost of extra rented trucks, travel cost, total cost and solving time. Number of containers and number of trucks are foreknown information. Total cost and number of rented trucks are determined by the model. Figures 4 and 5 show percentages of all different costs in total.

Table 1. Optimization results

Scenario number	Number of container	Number of trucks	Number of rented trucks	Storage cost (TL)	Cost of renting extra trucks (TL)	Travel cost (TL)	Total cost (TL)	Solving time
1	900	12	4	6.319	2.714	67.184	76.218	00:05:18
2	900	13	3	6.328	2.036	69.890	78.254	00:06:52
3	900	14	2	6.328	1.357	69.890	77.575	00:05:30
4	900	15	1	6.121	678	69.889	76.689	00:07:01
5	900	16	0	6.328	0	69.211	75.539	00:04:11
6	900	17	0	6.308	0	69.211	75.519	00:04:14
7	900	18	0	6.287	0	69.211	75.498	00:04:07
8	1000	12	6	9.414	4.072	79.761	93.247	00:07:00
9	1000	13	5	9.414	3.393	79.760	92.568	00:06:51
10	1000	14	4	9.414	2.714	79.760	91.889	00:06:53
11	1000	15	3	9.413	2.036	79.761	91.210	00:06:38
12	1000	16	2	9.414	1.357,46	79.761	90.532	00:05:18
13	1000	17	1	9.414	678,73	79.760	89.853	00:07:13
14	1000	18	0	9.414	0	79.761	89.175	00:05:01

Based on the results; it is clear that storage cost depends on congestion that arises because of large number of demanded containers. In peak times, model dispatches some containers earlier than free storage time to handle all demands in optimal way. Besides, our model can solve the problem for more than 1000 containers with average 5 min solving time; it shows that our model is reasonable for operations of Alkon Logistics.

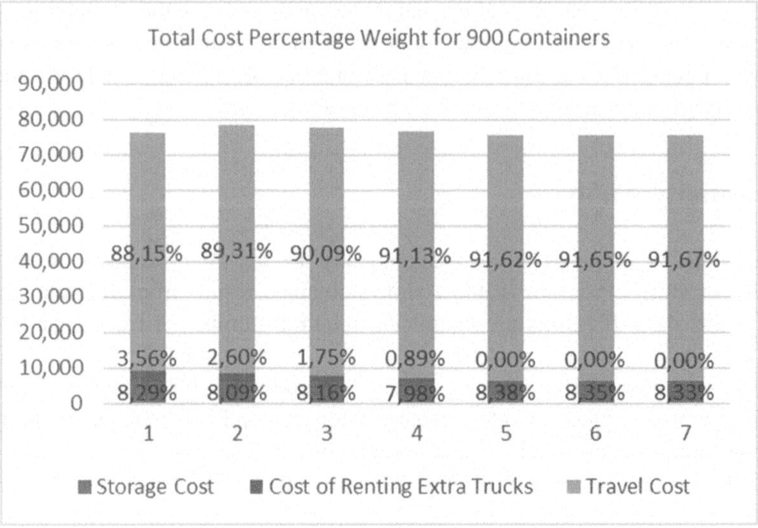

Fig. 4. Cplex result graph for 900 containers

In the current system, company operates with 15 own trucks and 17 additional trucks. It causes crucial financial issues for company in long-term. Our model guarantees solving the problem with minimum number of trucks and total operational cost. Furthermore, 16 trucks can handle the dispatching operations for 900 containers, while 18 trucks are enough for 1000 containers. It should be considered that number of trucks required will change as demand information changes.

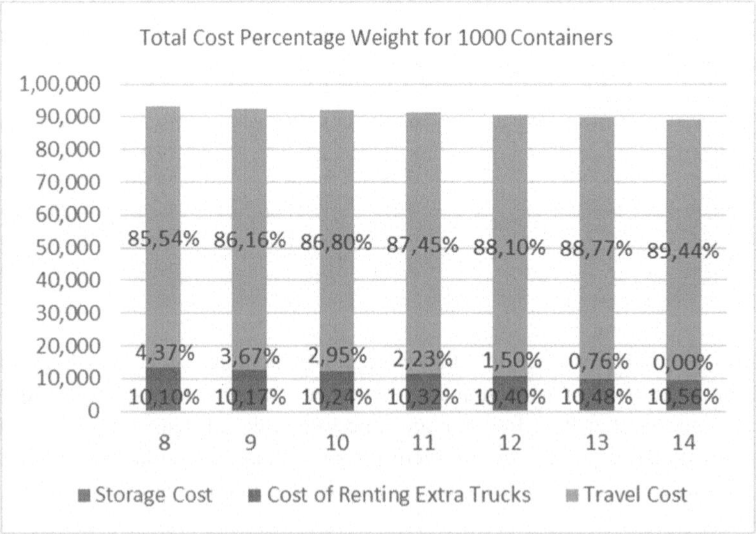

Fig. 5. Cplex result graph for 1000 containers

7 Decision Support System

Decision support system (DSS) provides efficient usage of proposed mathematical model. DSS was created by integrating IBM ILOG CPLEX Optimization Studio 12.8 with Visual Basic for Applications (VBA). DSS allows to respond to all different scenarios that can occur. When Excel file is opened, user is redirected to Excel spreadsheet "Home". "Home" page includes specially designed buttons that is shown in Fig. 6. First, user must select "Change Number of Trucks Belonging to Company" button to decide for the number of trucks to be used in dispatching. This change option is provided to enable that user make changes when number of trucks belonging to company change.

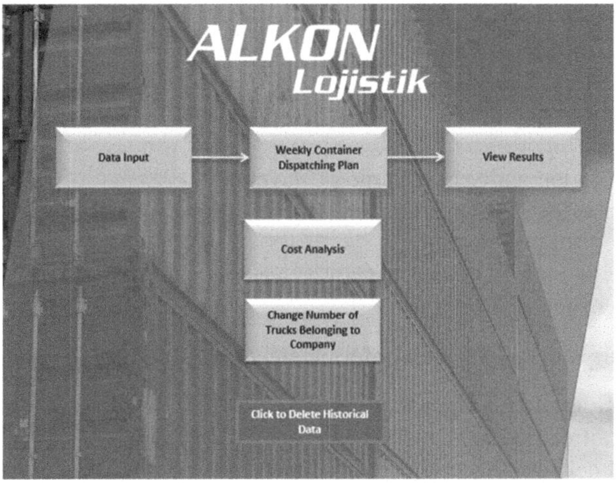

Fig. 6. Excel spreadsheet "Home"

Created DSS is dynamic; this means it can respond to the changes quickly. When user presses "Data Input" button, userform "Data Input" is displayed that is shown in Fig. 7. One of the significant point is that demand information must be recorded as weekly to this form. Critical times are entered for all demand groups such as container release time, container due time and free storage period to"Data Input" form. These time information are converted into time steps on a hidden Excel page. Conversion process is generated in a hidden page for preventing users make any changes in time step information. After entering demand information is completed, data is recorded by pressing "Ok" button.

Weekly dispatch planning can be done after all required demand information are entered. Created mathematical model is resolved by IBM ILOG CPLEX Optimization Studio, when "Weekly Container Dispatching Plan" button is pressed. For displaying results, user should return to "Home" page and press "View Results" button. After, user is directed to Excel sheet called "Dispatching Plan". This page contains dispatching plan for entire week excluding Sunday and it includes different information about trucks and dispatched containers. These information are provided for trucks:

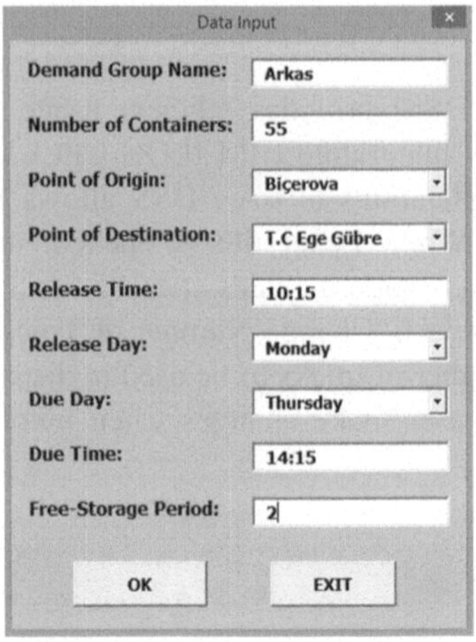

Fig. 7. Userform "Data Input"

(i) Origin location and starting time of traveling
(ii) Destination location and finishing time of travelling
(iii) Number of trucks traveling
(iv) Travel type such as loaded, empty or waiting
(v) Number of rented trucks

If trucks carry containers, these information are provided for dispatched containers:

(i) Name of dispatched demand group
(ii) Number of carried containers
(iii) Demand delivery day and delivery time

After displaying weekly dispatch planning, user can press "Cost Analysis" button to view costs. User is directed to "Cost Analysis" page after button is pressed.

There are two tables on this page. Table 2 shows number of trucks belonging to company, number of rented trucks, number of total dispatched containers, cost of renting extra trucks, traveling cost of trucks, total storage cost and, total cost. If user

Table 2. Cost table on page "cost analysis"

Number of Trucks Belonging to Company	15
Number of Rented Trucks	1
Number of Dispatched Containers	900
Renting Cost of Extra Trucks	678,00 ₺
Traveling Cost of Trucks	69.889,00 ₺
Storage Cost	6.121,00 ₺
Total Cost	76.688,00 ₺

Table 3. Analysis table on page "cost analysis"

Number of Trucks Belonging to Company	Number of Rented Trucks	Number of Dispatched Containers	Renting Cost of Extra Trucks	Traveling Cost of Trucks	Storage Cost	Total Cost
14	2	900	1.357,00 ₺	69.889,00 ₺	6.328,00 ₺	77.574,00 ₺
13	3	900	2.036,00 ₺	69.890,00 ₺	6.328,00 ₺	78.254,00 ₺
15	1	900	678,00 ₺	69.889,00 ₺	6.121,00 ₺	76.688,00 ₺

presses the "Save the Current Data" button on this page, these information will be saved on Table 3. Thus, user can easily perform sensitivity analysis. User can view the cost information by updating number of trucks that is used in dispatching operation and number of dispatched containers. Therefore, user can easily analyse this table, which is composed of different scenarios.

7.1 Implementation Plan for DSS

Decision support system is user friendly and can be used to respond to different scenarios that occur in the system. User can plan weekly container dispatch by updating number of trucks. First weekly container demand must be entered by pressing the "Data Input" button. After data entry is complete, user should press "Weekly Container Dispatching Plan" button. A warning message will be displayed; this message states that Excel file will be saved automatically. After, Excel file will close and model will run. After solution is obtained, user must open Excel file to view results. Finally, user can access information on weekly container dispatch planning easily by using "View Results" and "Cost Analysis" buttons.

8 Conclusion

In this paper, a solution approach is provided for solving the optimal dispatch planning problem for containers. Proposed solution approach includes several constructions that can model real-world elements such as cross-way movements, peak times and congestion. The proposed model and the DSS developed is helpful making long-term critical decisions such as purchasing of new trucks. To ensure this, a novel integer mathematical model was developed to minimize total cost. Proposed model discretizes passing time subject to hard constraints and, it provides optimal dispatch planning for containers. It prevents renting extra trucks unnecessarily by planning ahead of time and considering trade-off between storage, transportation and rental costs. The model is tested with real-life data from the company. The solution from the model provides efficient usage of all trucks under strict time constraints with use of almost half of current number of trucks. Solution from the model is a 65% improvement in rental cost and 25% improvement in total cost. Proposed model has flexible structure; it means that number of shifts and their operational times can be changed. Based on experimentation, higher cost savings are possible by making rental agreements on a weekly basis rather than annual. To efficient usage of mathematical model, also a decision support system (DSS) is generated by integrating IBM ILOG CPLEX Optimization Studio 12.8 with Visual Basic for Application (VBA). Created DSS is dynamic and practical for users. It provides to change number of trucks belonging to company and, to make analysis with provided costs tables. DSS can be used as a guidance for establishing critical long-term investments such as purchasing new trucks.

Acknowledgment. This study was supported by Yaşar University Faculty of Engineering' Department of Industrial Engineering. We also would like to thank Duygu Barış, Duygu Şentürk, Hüseyin Suçsuz and Yasin Özkabak for their input into this work. We especially want to thank to Adalet Öner, as the coordinator of Senior Design Projects this year at Yaşar University Industrial Engineering Department.

References

1. Günther HO, Kim KH (2006) Container terminal and terminal operations. OR Spectr 28:437–445
2. Tierney K, Voß S, Stahlbock R (2013) A mathematical model of inter-terminal transportation. Eur J Oper Res 235:448–460
3. Chung K, Ko C, Shin J, Hwang H, Kim K (2007) Development of mathematical models for the container road transportation in Korean trucking industries. Comput Ind Eng 53:252–262
4. Nguyen VD, Kim KH (2010) Dispatching vehicles considering uncertain handling times at port container terminals, progress in material handling research. In: Proceedings of the 11th international material handling research colloquium, Milwaukee, WI, pp 210–226

The Distributor's Pallet Loading Problem: A Case Study

Selen Burçak Akkaya, Aykut Gül, Zeynep Coşkun, Coşku Karaman, Hande Öztop, and Gizem Mullaoğlu(✉)

Department of Industrial Engineering, Yaşar University, İzmir, Turkey
selenakkaya96@gmail.com, aykutgul45@gmail.com, zeynepcoskun212@gmail.com, coskukaraman@outlook.com, {hande.oztop, gizem.mullaoglu}@yasar.edu.tr

Abstract. The pallet loading problem considers the loading of rectangular-shaped boxes with known dimensions into the pallets. In this paper, a real-life pallet loading problem of a beverage company is studied. In this problem, the aim is to maximize the total loaded box volume on pallets by employing a three-dimensional approach, where overlapping of any two boxes is not allowed and the fragility relationship is regarded. As a solution approach, a heuristic algorithm and a mathematical model have been proposed and verified with a computational experiment. A sensitivity analysis is also performed for the parameters of the problem. Furthermore, a user friendly decision support system, which is integrated with IBM ILOG CPLEX, is developed using Excel VBA interface. The proposed solution approach is embedded in this decision support system. It is estimated that proposed solution method will prevent the field loss due to the improper loading and reduce the financial damage, as it reduces the dependency on operators and the decision-making duration. As this problem can be seen in various companies, the proposed solution approach can also be employed by different sectors and companies.

Keywords: Distributor's pallet loading problem · Three-dimensional loading
Loading models with fragility constraints · Heuristic algorithm
Decision support system

1 Introduction

The pallet loading problem (PLP) aims to determine the best pattern for loading a set of rectangular boxes with known dimensions into a pallet. We consider a real-life pallet loading problem of a beverage company in Izmir, Turkey. The aim is to maximize the total volume of the loaded boxes while ensuring that the boxes do not overlap and fragility constraints are regarded. In this study, a three-dimensional approach is implemented in order to reflect the real world conditions as much as possible.

In the beverage company, products are offered to the customers in various forms of packages such as bottles, cans and barrels. Based on the order lists given by the sales department, different types of products are packed with an associated packaging material and loaded into the pallets by shipping operators. After the loading operation,

N. M. Durakbasa and M. G. Gencyilmaz (Eds.): ISPR 2018, *Proceedings of the International Symposium for Production Research 2018*, pp. 937–948, 2019.
https://doi.org/10.1007/978-3-319-92267-6_75

pallets are loaded into the trucks to be delivered customers using pre-determined routes. The sizes of the packaging materials loaded on the pallets may vary due to product type. However, the dimensions of the pallets are fixed. Furthermore, the dimensions of the trucks and the sections of the trucks, in which the pallets are to be placed, are fixed and known for every truck.

Currently, in the company, the shipping operators act intuitively based on their experience and do not follow any standard loading procedure. This issue leads to time and financial loss for the company. Due to the non-standard pallet loading operation, the desired pallet structure is not always obtained. For this reason, damaged products and unloaded customer orders occur. As there is a fragility relationship between different types of packaging materials, damages due to the improper placement of boxes cause company to suffer from financial loss. The proper placement is crucial for the loading operation, as only one defective product in a pallet can affect the whole pallet. In addition, the extra shipments due to the unloaded demand, especially in summer seasons, cause additional costs to the company. As the shipment operators perform the loading operation based on a trial-and-error method without using any scientific approach, there is a low efficiency in terms of time. Furthermore, in case of an operator resigns, it takes 6 months for a new employee to experience and learn the pallet loading operation. This also creates high dependency on workforce.

The main objective of this study is to develop a user-friendly decision support system for optimizing the pallet structure by deciding which products should be placed on which pallet in which location. In other words, the aim is to standardize and simplify the pallet loading operation by following specific steps and instructions. As mentioned above, it is crucial to determine the most suitable product placements on the pallets, in order to manage the space and use company's resources effectively. Currently, due to the high dependency level on workforce, duration of the decision making process is not same for every operator. Therefore, introducing a scientific approach to pallet loading process is very important for performing the operation efficiently and reducing the loading time. As some of the product packaging materials have a risk of being broken in the wrong pallet placement, consideration of the fragility properties during the organization of pallets is also another significant issue for this study. Consequently, the main performance measures of this study can be summarized as follows: the utilization of pallets, service level of order lists and the duration of the decision process. Therefore, the aim is to improve these performance measures, in other words, to increase the used pallet volume, reduce the amount of unloaded products on the order lists and reduce the duration of the decision making process of shipping operators.

The rest of the paper is organized as follows. A comprehensive literature review on the PLP is presented in Sect. 2. A binary linear programming model is presented in Sect. 3, as well as the assumptions and limitations of the problem. An efficient heuristic solution approach, which is combined with the mathematical model, is presented in Sect. 4. A user-friendly decision support system is also explained in Sect. 4. Section 5 presents the computational results, as well as the sensitivity analysis results for the parameters of the problem. Finally, concluding remarks and future work suggestions are given in Sect. 6.

2 Literature Review

The studied problem can be seen with different names in the literature, such as "bin packing problem", "3D bin packing problem", "cutting stock problem", "container loading problem" and "pallet loading problem". There are two main variants of the pallet loading problem in the literature, namely, the manufacturer's pallet loading problem and the distributor's pallet loading problem. In the manufacturer's pallet loading problem, products are packaged in identical boxes and these boxes are loaded into the identical pallets. Then, the formed pallets are loaded into the trucks that have standard dimensions. The aim of this problem is to choose dimensions of the boxes and pallets that maximize the volume of loaded products in the trucks. Hodgson (1982) defines the distributor's pallet loading problem as follows [1]: The orders of customers are packaged in boxes with varying dimensions. The problem is to load the boxes on a standard pallet, such a way that the volume placed on each pallet is maximized, i.e., the number of pallets used is minimized. Our problem is similar to the distributor's pallet loading problem, since order lists are pre-defined, products are packaged in boxes with varying dimensions and the aim is to maximize the volume placed on each standard pallet.

One of the earlier studies done within this context was published in 1982. In this study, Hodgson [1] studied the two-dimensional pallet loading problem and aimed to improve the transportation operation of US Airforce military equipment. Hodgson's observations show that US Airforce officials usually place the biggest box to origin point and place the rest of the boxes around this box. It is revealed that this intuitive method dramatically decreases the computational time.

In another study, Kang and Park [2] studied the variable size box placement problem and aimed to minimize the total cost of used bins under the assumption that the cost of unit size of each bin does not increase as the bin size increases. This study focuses on variable sized bins and some heuristics were applied for the problem, such as first fit decreasing, best fit decreasing, iterative first fit decreasing and iterative best fit decreasing. In our heuristic algorithm, we inspire from some of these rules.

Later, Lel et al. [3] proposed a heuristic algorithm to pallet packing problem of a beverage manufacturer. The objective is to determine the loading sequence of products with boxes of different sizes and the number of pallets required for the placement. In their algorithm, initially, products with similar cube size are grouped, and full pallet and partial pallet assignments are made in order to decrease the placement combinations. Results of their numerical analysis showed that the proposed algorithm solves the pallet loading problem efficiently within a reasonable computational time.

Junqueira et al. [4] published an article about three-dimensional pallet loading models with consideration of cargo stability and load bearing constraints. This article focuses on box orientation, complete shipment of box groups, box priorities, complexity of arrangement, fragility constraints, and weight distributions. In this study, an optimization model is proposed as a solution methodology for minimizing the empty volume percentage on the pallets and the duration of the loading time of the pallets. Recently, Sheng et al. studied a container loading problem with the consideration of expiration of the products [5]. In this problem, all products on a given order must be placed in one container. The aim is to maximize the utilization of volume on the container. A heuristic algorithm is proposed to standardize the packing process. The literature review is summarized in Table 1. In the table, each article is classified according to studied problem, objective function and proposed solution method.

Table 1. Literature review

Authors	Date	Problem	Objective function	Solution approach
Thom J. Hodgson	1982	Two-dimensional pallet loading problem	Maximize pallet occupancy	Dynamic programming and heuristic approach
Park and Kang	2002	Variable size box placement	Minimize the total cost of boxes used in variable sizes	Greedy algorithm and loading rules (best fit decreasing, first fit decreasing etc.)
Vu et al.	2005	Three-dimensional pallet loading problem	Maximize the volume of products in the pallet	Heuristic approach
Morabito et al.	2010	Three dimensional container loading with load balance, transport and fragility constraints	Maximize the total volume of loaded boxes or minimize remaining free space	Mixed integer programming model and heuristic approach
Sheng et al.	2017	Container loading with multiple constraints	Maximize the volume of loaded boxes	Integer programming model and heuristic approach

The study of Junqueira et al. [4] can be used as a reference as it contains a three-dimensional approach to pallet loading problem and proposes an optimal placement policy. Furthermore, this study integrates fragility issue into the model. However, their model considers the placement of boxes only on a single pallet. Therefore, we extend their model by changing the restrictions on possible location sets and adapting the multiple pallets.

3 Problem Formulation

In this section the mathematical model is presented for the aforementioned problem. Due to the complexity of the problem, below assumptions are made.

- All values are assumed to be integers.
- Boxes can only be placed orthogonally, i.e., either parallel or perpendicular, into the pallet.
- Orientation of the boxes is fixed.

- Each box could be moved down and/or forward and/or to the left, until its bottom, front and left-hand faces are adjacent to other boxes.
- All boxes and the pallets are assumed to be rectangular prisms.

The parameters and decision variables are defined as follows. $J = \{1, \ldots, n\}$ is the set of pallets and $I = \{1, \ldots, m\}$ is the set of product box types. Each product box type $i \in I$: length l_i, width w_i, height h_i, volume v_i and a maximum quantity b_i values are defined. These boxes can be loaded into a standard pallet with length L, width W and maximum allowed height H. To find the possible positions of the boxes, a Cartesian coordinate system is adopted, where the origin point (0, 0, 0) represents the front-left-bottom corner of the pallet. As in Junqueira et al. [4] sets of possible locations (x, y, z) are defined as follows:

$$X = \{0, 1, 2, \ldots, L - min_i(l_i)\},$$

$$Y = \{0, 1, 2, \ldots, W - min_i(w_i)\},$$

$$Z = \{0, 1, 2, \ldots, H - min_i(h_i)\},$$

Two binary decision variables a_{ijxyz} and p_j are defined as follows, to decide which box will be placed on which pallet at which position and which of the pallets will be used, respectively:

$$a_{ijxyz} = \begin{cases} \text{1, if a box } i \text{ is placed on pallet } j \text{ with its} \\ \text{front - left - bottom corner at position(x,y,z)} \\ \text{so that } 0 \leq x \leq L - l_i, 0 \leq y \leq W - w_i, 0 \leq z \leq H - h_i \\ \text{0, otherwise} \end{cases} \forall i \in I, x, x' \in X, y, y' \in Y, z, z' \in Z$$

$$p_j = \begin{cases} \text{1, if pallet } j \text{ is used} \\ \text{0, otherwise} \end{cases}, j \in J$$

The proposed solution approach solves the pallet loading problem with the consideration of the fragility relationship of the boxes. However, this issue is handled within the heuristic algorithm, which is explained in Sect. 4, before the model solution step. Therefore, it is not included in the model as a constraint.

In the pallet loading problem of the beverage company, due to the seasonality of the demand, the objective function changes according to the relationship between customer demand and supply, where supply refers to the capacity of trucks. During the summer seasons, the demand usually increases due to the increase in beverage consumption. As the demand is greater than the supply, the objective function is to maximize the volume used on pallets. On contrary, the demand usually decreases in winter seasons due to the decrease in beverage consumption. As supply exceeds the demand in this case, the objective function is to minimize the number of pallets used.

The optimization model is adapted from the study of Junqueira et al. [4] by introducing a set for pallets, a new binary decision variable for determination of which pallets are used and adding pallet information to the current decision variables. An additional constraint is also defined in order to link these two types of decision variables. The proposed mathematical model is presented below, for the case of demand exceeds the supply amount:

$$max \sum_{j=1}^{n} \sum_{i=1}^{m} \sum_{x \in X} \sum_{y \in Y} \sum_{z \in Z} \sum a_{ijxyz} v_i \tag{1}$$

Subject to

$$\sum_{j=1}^{n} \sum_{x \in X} \sum_{y \in Y} \sum_{z \in Z} a_{ijxyz} \leq b_i \quad \forall i \tag{2}$$

$$\sum_{i=1}^{m} \sum_{x \in X | x' - l_i \leq x \leq x'}, \sum_{y \in Y | y' - w_i \leq y \leq y'}, \sum_{z \in Z | z' - h_i \leq z \leq z'}, a_{ijxyz} \leq 1 \quad \forall j, x' \in X, y' \in Y, z' \in Z \tag{3}$$

$$\sum_{i=1}^{m} \sum_{x \in X} \sum_{y \in Y} \sum_{z \in Z} a_{ijxyz} \leq Mp_j \quad \forall j \tag{4}$$

$$a_{ijxyz} \in \{0, 1\}, \; i \in I, j \in J, x \in X, y \in Y, z \in Z \tag{5}$$

$$p_j \in \{0, 1\}, \; j \in J \tag{6}$$

The objective function (1) aims to maximize the volume used on the pallets, in this case. Constraint (2) restricts the maximum number of packed boxes and constraint (3) prevents overlapping of two boxes. Constraint (4) allows the placement of boxes on a pallet only if that pallet is being used, where M is a large integer. Constraints (5) and (6) define the domain of the decision variables.

For the case of supply exceeds demand, only the objective function and constraint (2) changes as follows:

$$min \sum_{j=1}^{n} p_j \tag{7}$$

Subject to

$$\sum_{j=1}^{n} \sum_{x \in X} \sum_{y \in Y} \sum_{z \in Z} a_{ijxyz} = b_i, \forall i \tag{8}$$

(3)–(6)

The objective function (7) aims to minimize the number of used pallets. Constraint (8) ensures that all products on the order list are fully loaded, as there is enough volume on the pallets in this case.

4 Solution Approach and Implementation

Since the studied problem is NP-hard [6], a heuristic solution approach is employed in the proposed solution procedure. The details of the heuristic algorithm is shown in Fig. 1.

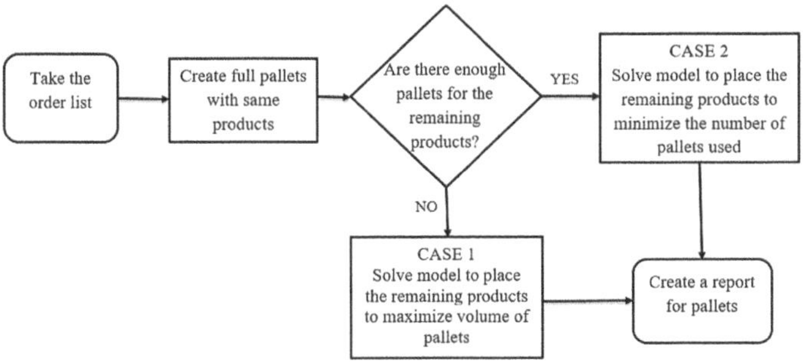

Fig. 1. Algorithm flow chart

Algorithm starts with reading the order list. Firstly, if the order amount of one product type is enough to fulfill a pallet, the same products are placed on the pallet and a full pallet is formed. The full pallets are preferred by the company. In the case of determining the pallet loading structure of the remaining products, a three-dimensional pallet loading optimization model given in Sect. 3 is used.

Before listing the products to be used in the model, case type of products are controlled and placed to the pallets in the form of identical blocks. The main reason of this block system is to satisfy the fragility constraint, as it is not allowed to place other types of products above/under these case type of products. The height of generated blocks is assumed to be equal to the height of the pallets so that any placement of other packaging material types above or under these blocks is avoided. Block system is not used for remaining products as placement above or under them is allowed according to fragility relationships.

After this step, the total volume of the remaining products is calculated and compared with the remaining volume of the pallets that can be used for placement. If the demand exceeds supply, the objective function that maximizes the utilization rate of each pallet is used in the model. In the other case, objective function that minimizes the number of used pallets is used. Finally, the mathematical model is solved under 5 min time limit for on the pre-determined objective by using IBM ILOG CPLEX Optimization Software. A time limit is set for the mathematical model, as this pallet loading operation is performed daily in the company.

Consequently, we develop a user-friendly decision support system (DSS) which uses the above solution methodology. This DSS is developed using the Excel VBA interface and connected with the IBM ILOG CPLEX Optimization Studio, in order to obtain results in short time. At the main screen of the program, there are two main parts namely "Inputs" and "Outputs", as shown in Fig. 2.

In "Inputs" part, users can automatically extract an existing order list in Excel format or create a new order list manually. In "Outputs" part, detailed information of loaded products and loading instructions are reported for each pallet, after running the proposed algorithm.

This application is designed for enabling users to use the program dynamically. After getting the data from the user, the pallet structure is obtained employing the aforementioned algorithm shown in Fig. 1. The usages of each pallet are summarized in a report on the main page. Users are able to add or remove products; create a pallet loading scheme; update the information of pallets or products in database; and report the detailed pallet loading information and loading instructions. Moreover, users are able to print the results and loading instructions, send them by e-mail to other users and export the summary in Excel or PDF format as in Fig. 3. Finally, users can access user's guide so that they can easily use the DSS (Fig. 2).

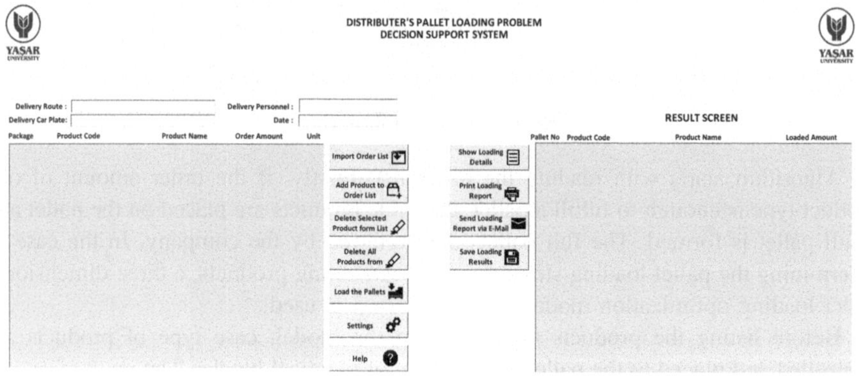

Fig. 2. DSS main screen

LOADING REPORT

Delivery Route : Sample Delivery Route Delivery Personnel : Project Group 11
Delivery Car Plate : 35 YASAR 35 Date : 6.02.2018

Utilizations of Pallets
Pallet 1 : 91,15% Pallet 2 : 69,79% Pallet 3 : 75,00% Pallet 4 : 77,78%
Pallet 5 : 66,67%

Loading Summary

Pallet No	Product Code	Product Name	Loaded Amount	Volume It Covers
1	PRODUCT CODE 1	SAMPLE PRODUCT NAME 1	12	25,00%
1	PRODUCT CODE 14	SAMPLE PRODUCT NAME 14	12	12,50%
1	PRODUCT CODE 1	SAMPLE PRODUCT NAME 1	7	14,58%
1	PRODUCT CODE 9	SAMPLE PRODUCT NAME 9	6	8,33%
1	PRODUCT CODE 18	SAMPLE PRODUCT NAME 18	1	1,56%
1	PRODUCT CODE 14	SAMPLE PRODUCT NAME 14	4	4,17%
1	PRODUCT CODE 9	SAMPLE PRODUCT NAME 9	18	25,00%
2	PRODUCT CODE 18	SAMPLE PRODUCT NAME 18	28	43,75%
2	PRODUCT CODE 10	SAMPLE PRODUCT NAME 10	25	26,04%
3	PRODUCT CODE 9	SAMPLE PRODUCT NAME 9	13	18,06%
3	PRODUCT CODE 11	SAMPLE PRODUCT NAME 11	41	56,94%
4	PRODUCT CODE 11	SAMPLE PRODUCT NAME 11	5	6,94%
4	PRODUCT CODE 41	SAMPLE PRODUCT NAME 41	18	37,50%
4	CRUSHABLE PRODUC	CRUSHABLE PRODUCT	2	33,33%
5	CRUSHABLE PRODUC	CRUSHABLE PRODUCT	4	66,67%

Unloded Products
There are no unloaded products.

Loading Instructions
Pallet 1
PRODUCT CODE 1 is placed in lower left corner of the pallet.
PRODUCT CODE 1 is placed 5 layers to top of previous product.
PRODUCT CODE 1 is placed 6 layers to front of previous product
PRODUCT CODE 14 is placed 6 layers to front of previous product.
PRODUCT CODE 14 is placed 6 layers to right of previous product.
PRODUCT CODE 1 is placed 6 layers to rigth-bottom of previous product.
PRODUCT CODE 1 is placed 1 layers to front of previous product.
PRODUCT CODE 9 is placed 6 layers to top of previous product.
PRODUCT CODE 18 is placed 1 layers to top of previous product
PRODUCT CODE 14 is placed 4 layers to right of previous product
PRODUCT CODE 9 is placed 9 layers to. Rigth-bottom of previous product
Pallet 2
PRODUCT CODE 18 is placed in lower left corner of the pallet.
PRODUCT CODE 18 is placed 5 layers to top of previous product

Fig. 3. DSS output – loading report

5 Computational Results

The computational results of the proposed solution methodology are presented in this section.

5.1 Comparison and Improvements

In order to validate the accuracy of the solution methodology, 15 different order lists for each objective function case are obtained from the company and solved by the suggested method. It is observed that the results are as desired by the company and the model responds quickly and effectively to both objective functions. Furthermore, results of the proposed solution methodology are compared with the results of the current system used by the company.

The computational results are shown in Table 2 for the minimization problem instances, and in Table 3 for the maximization problem instances. For each instance, number of product types and the total number of products are listed. Current method and proposed method are also compared for each instance and improvements are reported in terms of number of used pallets, average pallet utilization and unloaded amount of products.

According to Table 2, improvements are observed in 9 out of 15 instances. On the average, a rate of %10 improvement is obtained in terms of number of pallets used. In addition, results show that the proposed solution method gives better results than the current method in terms of performance measures mentioned in Sect. 1. According to

Table 2. Minimization model comparison results

Minimization problem instances					
Order information			# of used pallets		Improvement (%)
No	# of product type	Total # of prod.	Current method	Proposed method	
1	17	227	5	4	20
2	17	182	4	4	0
3	16	208	3	3	0
4	11	343	7	6	14
5	14	368	8	7	13
6	10	377	7	7	0
7	11	237	6	5	17
8	15	350	7	6	14
9	11	219	5	4	20
10	16	93	2	2	0
11	9	366	6	7	0
12	10	246	7	6	14
13	10	470	8	7	13
14	10	449	8	7	13
15	11	271	5	4	20
Average			**6**	**5**	**10**

Table 3. Maximization model comparison results

Maximization problem instances								
Order information			Avg. pallet utilization			Unloaded amount		
No	# of product type	Total # of prod.	Current method (%)	Proposed method (%)	Imp. (%)	Current method (Amount)	Proposed method (Amount)	Imp. (%)
1	15	577	81	91	13	88	0	100
2	15	554	90	90	0	0	0	0
3	11	645	91	100	9	113	54	52
4	10	697	88	95	8	104	56	46
5	13	600	99	98	0	18	40	0
6	13	689	86	97	12	69	39	43
7	15	740	79	94	18	90	45	50
8	10	767	90	94	4	54	47	13
9	12	780	94	89	0	21	46	0
10	11	620	95	91	0	0	0	0
11	14	819	76	91	20	108	71	34
12	16	815	93	93	0	92	97	0
13	13	758	86	94	10	74	42	43
14	13	748	74	90	22	61	33	46
15	15	394	80	99	23	51	34	33
Average			**87**	**94**	**9**	**63**	**40**	**31**

the observations, the proposed algorithm yields results in maximum 5 min while the current decision duration of the operators is approximately 10 min. In this regard, the decision-making period is improved considerably.

According to the results of the maximization model instances, a significant decrease in the number of unloaded products is observed with the increase in the average utilization rate of pallets. While the average utilization of the pallets is improved by 9%, the improvement rate of the unloaded amount is 31%. Similarly, the decision-making period is also improved considerably by the proposed solution method, as in minimization instances.

5.2 Sensitivity Analysis

A sensitivity analysis is also conducted to observe how the proposed solution method reacts to different scenarios. Two examples of maximization and minimization cases are solved for 8 different settings using real data. In these settings, one of the dimensions of the pallet is increased by 10 or 20 cm, which the company allows it as overflow proportion. The results are shown in Table 4.

Results for minimization cases demonstrate that, in the case of dimensions of the pallets are increased, the number of pallets used decrease for most of the situations, as expected. If the height of the pallet is increased while keeping the floor area constant, the algorithm gives better solutions in minimization cases. Moreover, it is observed that the height and the length values are restrictive for the minimization objective function, whereas the width value does not affect the objective function value for most of the cases.

Table 4. Sensitivity analysis results

No	Minimization			Maximization			
	# of total products	Dimensions (10 cm)	# of used pallets	# of total products	Dimensions (10 cm)	Avg. utilization (%)	Unloaded amount (%)
1	184	12,10,18*	3	739	12,10,18*	97	29
		12,11,18	3		12,11,18	97	31
		12,10,19	3		12,10,19	99	21
		13,10,18	3		13,10,18	98	22
		13,10,19	2		13,10,19	100	20
		12,11,19	3		12,11,19	99	28
		13,11,18	3		13,11,18	97	22
		13,11,19	2		13,11,19	100	20
2	205	12,10,18*	7	620	12,10,18*	91	10
		12,11,18	7		12,11,18	91	10
		12,10,19	6		12,10,20	100	0
		13,10,18	6		13,10,18	91	10
		13,10,19	6		13,10,20	100	0
		12,11,19	6		12,11,20	100	0
		13,11,18	6		13,11,18	91	10
		13,11,19	6		13,11,20	100	0

*Base case

For maximization cases, in the case of dimensions of the pallets are increased by 10 cm, it is concluded that the number of unloaded orders is reduced and the average pallet utilization rates are increased. On the other hand, for the second order list shown in Table 4, no improvement is observed in the number of unloaded products when the pallet dimensions are increased by 10 cm. The main reason of this situation is that adjusted dimensions do not allow any placement of new products, since dimensions of these products are larger than 10 cm. However, in the case of the dimensions are increased by 20 cm, improvements are observed and the order list is fully placed. However, since the altitude limit is legally restricted in national highways, this change is not possible for height value but it may be considered as a recommendation for other dimensions.

6 Conclusions

The aim of this study is to create a pallet loading system that increases the efficiency, removes the dependency on the workforce and reduces the decision-making time and additional costs. We consider a real-life problem of a beverage company, in which operators perform the pallet loading operation based on their experience level, without any scientific method. In this study, a mathematical model, a heuristic algorithm and a convenient decision support system, which makes the pallet placement with a scientific approach, are developed for this problem. The output of the developed user-friendly decision support system is a summary report that shows the locations of the products on the pallets as well as the utilization rates of pallets. Computational results indicate that the proposed solution approach performs much better than the current system, as described in Sect. 5. By applying the proposed solution methodology, the used volume of pallets has been increased in the direction of the observed instances, while the amount of unloaded products, the dependency on the work power and the duration of the decision making period have been reduced. Note that a %10 improvement is obtained in terms of number of pallets used, while the average utilization of the pallets is improved by 9% and the rate of unloaded amount is improved by 31%.

As a future study, the proposed algorithm and model can be extended by allowing rotation of the boxes along *x*-*y* axis. In this manner, additional improvements can be observed in the pallet structures. Furthermore, additional constraints such as load bearing and stability can be adapted into proposed solution methodology.

Acknowledgment. This work cannot be completed without the assistance of Atakan Yurttutan, Serhat Özbıçakçı and Merve Dirik. We are thankful for their contribution. Furthermore, we are grateful to the company for their co-operation.

References

1. Hodgson TJ (1982) A combined approach to the pallet loading problem. IIE Trans 14(3):175–182
2. Kang J, Park S (2003) Algorithms for the variable sized bin packing problem. Eur J Oper Res 147(2): 365–372
3. Lel VT, Creighton D, Nahavandi S (2005) A heuristic algorithm for carton to pallet loading problem. In: 3rd IEEE International Conference on Industrial Informatics, INDIN 2005, pp 593–598
4. Junqueira L, Morabito R, Yamashita DS (2012) Three-dimensional container loading models with cargo stability and load bearing constraints. Comput Oper Res 39(1):74–85
5. Sheng L, Xiuqin S, Changjian C, Hongxia Z, Dayong S, Feiyue W (2017) Heuristic algorithm for the container loading problem with multiple constraints. Comput Ind Eng 108:149–164
6. Garey MR, Johnson DS (1979) Computers and intractability. W.H. Freeman, New York

Three Dimensional Cutting Stock Problem in Mattress Production: A Case Study

Selin Altın, Tezcan Aydilek, Umut Şirvan, Damla Kesikburun, Adalet Öner(✉), and Nejat Kutup

Department of Industrial Engineering, Yaşar University, İzmir, Turkey
selin_altn@hotmail.com, tezcanaydilek@gmail.com, umutsirvan@windowslive.com, {damla.kesikburun, adalet.oner, nejat.kutup}@yasar.edu.tr

Abstract. This study involves in a real-world application of 3-D cutting stock problem in a local bedding company. The company produces various types of mattresses (beds). The bill of materials of a mattress usually includes foam rubber (sponge) components. Each mattress has different foam rubber components in the form of orthogonal rectangular prisms with different dimensions. Those components should be cut from big foam rubber blocks. The properties of current cutting machinery used in the company imposes "guillotine" cuts. The aim is to minimize total waste in cutting operations. There are mathematical models to solve cutting stock problems, however they require "cutting patterns" to be generated in advance. Cutting patterns represent potential combinations of how rectangular prisms are fitted into foam rubber blocks. Column Generation method and a Branch and Bound algorithm are used to generate cutting patterns. Mathematical model then solves the problem provided that the cutting patterns are supplied as input. A user friendly Decision Support System (DSS) has been developed in order to incorporate proposed procedures to solve the problem. It enables the user in the company to prepare efficient cutting plans easily and quickly.

Keywords: Optimization · Cutting stock problem · Guillotine cut
Cutting patterns · Column generation · Branch and bound
Decision Support System

1 Introduction

Foam rubber is used as the cushioning material in mattresses. In the production of mattresses, big foam rubber blocks should be cut into different types of orthogonal rectangular prisms that are the components of a mattress such as *head bars, side bars,* and *comfort layers* as shown in Fig. 1. In this form, the problem is a standard three-dimensional cutting stock problem (3DCSP).

A local bedding company in Izmir produces mattress and suffers a high level of waste in foam rubber cutting operations. Currently, there isn't any systematic or scientific approach for planning of cutting operations. A planning engineer develops cutting plans manually, depending on his/her own judgment and experience. Company

N. M. Durakbasa and M. G. Gencyilmaz (Eds.): ISPR 2018, *Proceedings of the International Symposium for Production Research 2018*, pp. 949–960, 2019.
https://doi.org/10.1007/978-3-319-92267-6_76

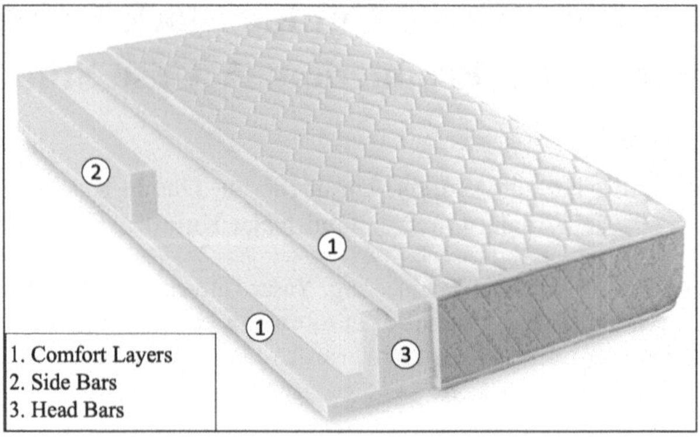

Fig. 1. Foam rubber components of a mattress

management is not happy with the current level of waste. The purpose of this study is to investigate and compile techniques and methods of three-dimensional cutting stock problem and then develop a planning tool that will be used to generate efficient cutting plans. Therefore, the focus of the study is an industrial application of three-dimensional cutting stock problem.

2 Problem Definition

The standard 3-D cutting stock problem (3DCSP) can be defined simply as follows: There is an unlimited quantity of identical big foam rubber block $B = (L, W, H)$ as raw material in producing mattress, where L, W, and H define the length, width and height of the blocks respectively. On the other hand, there is a set of n types of components or items (l, w, h, d) to be cut from big blocks B. The problem is to determine how to cut the smallest possible number of blocks B so as to produce d_i units of each items type i, $i = 1, \ldots, n$. A basic mathematical model can be defined for the standard 3DCSP as follows.

Decision Variables

x_j: Number of Stock material to cut according to pattern j

Cutting patterns represent potential combinations of how components (items) are fitted into foam rubber blocks B.

Parameters

n = Number of items
m = Number of patterns
i = index for *item*, $i = 1, \ldots. n$
j = index for *pattern*, $j = 1, \ldots. m$
p_{ij} = Number of occurances of the ith item in the jth pattern
d_i = Demand for item i

Mathematical Model

$$\min Z = \sum_{j}^{m} x_j \tag{1}$$

Subject to

$$\sum_{j}^{m} p_{ij} x_j \geq d_i \quad \forall i \in \{1, ..n\} \tag{2}$$

$$x_j \in Z^+ \quad \forall j \in \{1, ..m\} \tag{3}$$

The objective function minimizes total number of cutting patterns to be used. Constraint (2) ensures that the items are produced in desired amounts. Constraint (3) indicates that an integer solution is requested. It is obvious that cutting patterns should be determined in advance as the input of the problem. The hard part of the problem is to generate valid patterns.

A cutting pattern shows how many items of each type are cut from stock material. An example of generating cutting patterns are explained below in Figs. 2 and 3. For simplicity, the idea is shown on two dimensions. However, it can be extended easily for three dimensions. Assume that there are four types of items and they should be cut from stock material as shown in Fig. 2.

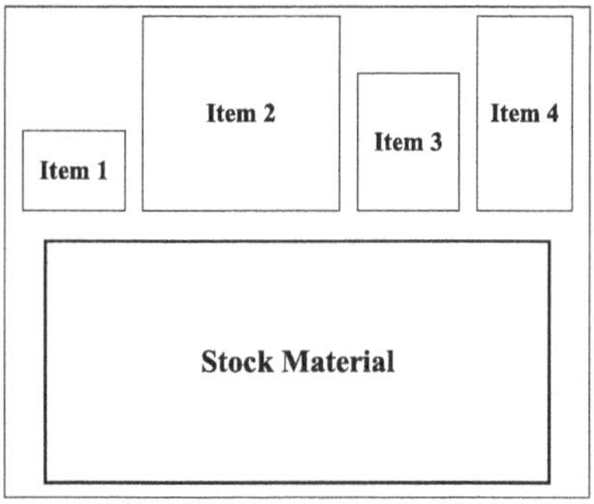

Fig. 2. Example case for generating cutting patterns

The items can be placed on the stock material in different combinations, and each combination represents a pattern. Figure 3 shows that one of the alternative patterns such that five instances of item 1, one instance of item 2 and two instances of item 3 could be placed on the stock material. This placement is just one of many possible combinations. In order to solve the problem optimally, all valid pattern should be determined and listed.

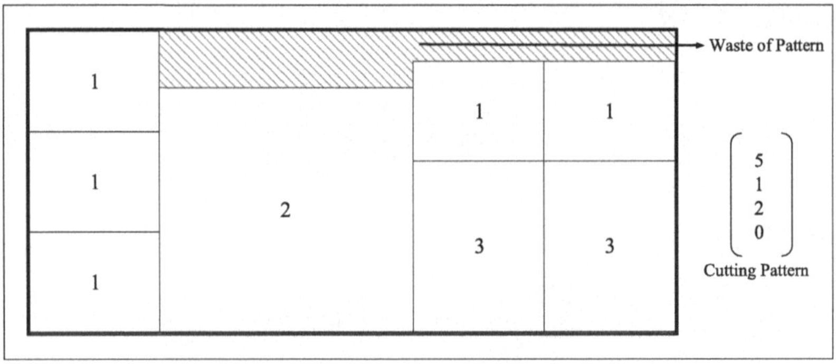

Fig. 3. Generating a cutting pattern

There are many variants and extensions of standard 3DCSP depending on the environment and technological requirements. In this study, we consider the real-world requirements in the bedding company, which can be described as follows:

- Technical specifications and settings of blades of every cutting machine used in the company impose *guillotine* cuts. A guillotine cut is a cut that is parallel to one of the sides of the object and goes from one side to the opposite one.
- There are two types of cutting machines, and each type is set to cut along different Cartesian planes. On one of the machine type, the block B is loaded on the machine and the machine cuts "layers" of foam rubbers parallel to x-y plane that is called "horizontal" cut. The orientation of the block *B* on the machine is fixed and can't be changed. The other type of machine cuts parallel to y-z plane, and it is called "vertical" cut. In this type of machine, the orientation of the objects may be changed as desired. The cutting process is sequential such that any block B is first cut horizontally and generated layers are then cut vertically as many times as required to get components to be used in mattress production. In such a case, cutting process is called *k-staged cutting* in literature [5]. Guillotine cuts are applied in both stages.

The requirements described above leads to a special variant of 3DCSP. The problem turns out to be a 2-staged cutting problem constrained with guillotine cuts. This study focuses on this variant of the problem, and the aim is to develop a solution procedure and then embed it in a decision support tool.

3 Literature Review

In literature, the first known definition and formulation of the cutting stock problem (CSP) were made by Russian economist Kantorovich [1]. Gilmore and Gomory [2] proposed *column generation* method in order to solve single dimension CSP. Generating a column means generating a cutting pattern. Then, they have extended their study to describe how to use column generation method in order to solve multistage CSP of two and more dimensions [3, 4]. However, the proposed method doesn't provide integer solution. Queiroz et al. [5] provide an extensive survey on the 3DCSP.

There are various studies in literature that can be used in order to formulate and solve 2-staged 3DCSP. Among others, some mathematical models are the one-cut concept by Dyckoff [6], the arc-flow concept by Valerio de Carvalho [7], the DP-flow concept by Cambazard and O'Sullivan [8], and the general arc-flow with graph compression by Brandao and Pedroso [9].

There are also exact methods proposed for CSP depending on the *branch & bound* and *branch & cut* algorithms such as Hadjiconstantinou and Christofides [10] and Martello and Toth [11]. In meta-heuristics realm, Kampke [12] and Alvim et al. [13] implemented simulated annealing a tabu search methods respectively.

4 Modelling and Solution Methodology

This study focuses on 2-staged 3DCSP with guillotine cuts as described in "Problem Definition" section. Technological requirements lead us to divide the 3D problem into two consecutive stages. The first stage contains a collection of two-dimensional cutting stock problems (2DCSP), and the second stage contains a one-dimensional cutting stock problem (1DCSP). However, these two stages are not independent of each other, since the output of the first stage (2DCSP) is used as the input in the second stage (1DCSP). Figure 4 shows the relations between the stages.

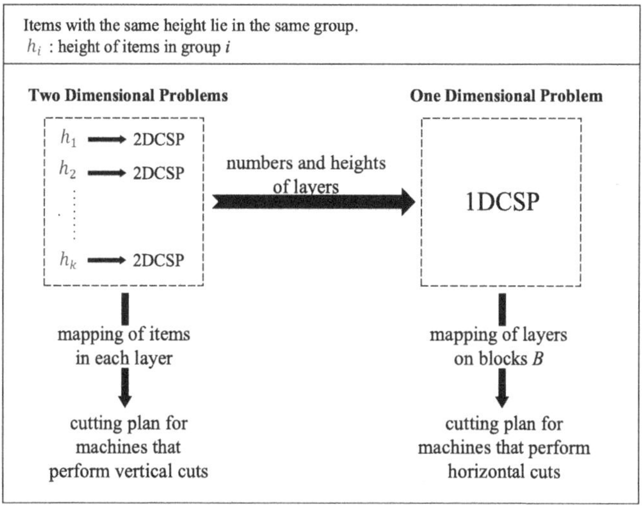

Fig. 4. Problem setting of 2-Staged 3DCSP

In modeling and solution process, the sequence of stages is reversed when compared to the sequence of cutting process in practice. The first stage of the solution procedure corresponds to the second stage of cutting process in practice, which is the vertical cutting stage.

In this problem setting, first, the set of items (components) to be cut are grouped by their heights. The items with the same heights will be in the same group. For each group of items, a 2-D CSP is solved. It leads more than one 2-D problems such that each one corresponds to a particular height. The 2-D problems are independent of each other since they correspond to distinct heights. The solution of each 2-D problem gives the minimum number of "layers" to be cut while ensuring required numbers of items would be produced in a given height. The solutions of those 2-D problems also provide the information of how the items are located on the layers. That information is actually the cutting plan for the machines that performs vertical cuts. (See Fig. 4.)

The output of independent 2-D problems are collected to make a list of layers along with their heights. The layers in that list are to be cut from blocks B. The list of layers is used as the input of the second stage which is a 1DCSP. The solution of the 1-D problem gives the minimum number of blocks B to be used. The solution also provides the information of how the layers are located in block B. That information corresponds to the cutting plan for the machines that performs horizontal cuts.

For the solution procedure, it has been decided to use the model and formulation given in (1)–(3) and it is coded in Lingo 17.0. However, it was required to find a systematic way to determine valid cutting patterns. As mentioned earlier, Gilmore and Gomory [2–4] proposed a column (pattern) generation method.

In column generation method, the problem is divided into two problems: master problem and the auxiliary problem. First, a set of feasible cutting patterns are found and placed in the simplex method as the starting feasible solution. Then a special procedure is applied to find a promising pattern if any. That procedure creates the auxiliary problem which is a knapsack problem whose input is the shadow prices from current basis in the simplex method. The solution of the auxiliary problem is then fed to master problem for the next iteration of simplex method.

Column generation method is easy to implement for 1DCSP but not for multi-dimensional problems. Therefore, other relevant pattern generation methods have been reviewed in the literature. Malaguti et al. [14] present an extensive survey for such methods. Pattern generation methods are usually based on the branch and bound algorithm. Among many others, the methods proposed by Suliman [15] and Rodrigo et al. [16] have been scrutinized, and the latter has been chosen to implement in solution procedure because it is easy to implement and embed into decision support tool. The selected pattern generation methods which are column generation method and the algorithm of Rodrigo et al. [16] are coded in Lingo 17.0 and Excel VBA, respectively. The proposed solution procedure is described in Fig. 5.

A sample problem is prepared below to show that how the proposed solution procedure works. It is assumed that the dimensions of the foam rubber blocks are $200 \times 180 \times 100$ cm that corresponds to length, width and height (L, W, H) respectively. Assume also that production plan is received and it is ordered to produce seven different mattress types. The types and order amounts of mattresses are shown in Table 1.

The BOM data indicates the number and dimensions of the components for each mattress type. Usually, each mattress must have two comfort layers, two head bars, and two side bars (see Fig. 1). Using BOM data, the dimensions and numbers of all the components (items) may be listed easily. For simplicity, comfort layers component were excluded from the list, therefore only head and side bars were considered. The list

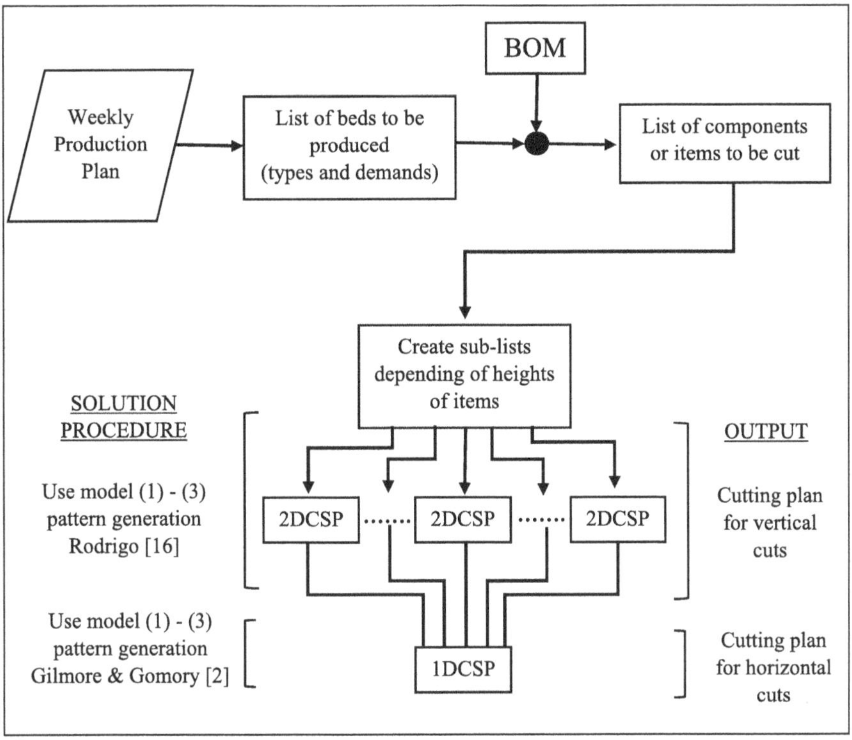

Fig. 5. Proposed solution procedure

Table 1. Sample problem – production orders

Mattress type	Demand
A	15
B	18
C	21
D	14
E	13
F	12
G	20

of the items are then divided into sub-lists depending on their heights. Each sub-list contains the items with the same heights. The production plan and BOM data lead us to have three item sub-lists shown in Tables 2, 3 and 4.

Each sub-list is the input for separate and independent 2DCSPs. The lengths and widths of items are considered in those two-dimensional problems whereas the heights are ignored. The branch and bound method in Rodrigo [16] has been used to generate the cutting patterns and then cutting patterns have been used to find the optimal solution in model defined in (1)–(3). Tables 5, 6 and 7 presents the output of the two-dimensional problems.

Table 2. Item sub-list 1 (All heights = 8 cm)

Mattress type	Component	Length (cm)	Width (cm)	Height (cm)	Demand
A	Head bar	130	16	8	30
	Side bar	170	16	8	30
B	Head bar	150	14	8	36
	Side bar	190	14	8	36
C	Head bar	190	20	8	42
	Side bar	200	20	8	42

Table 3. Item sub-list 2 (All heights = 13 cm)

Mattress type	Component	Length (cm)	Width (cm)	Height (cm)	Demand
D	Head bar	130	16	13	28
	Side bar	170	16	13	28
E	Head bar	190	20	13	26
	Side bar	200	20	13	26

Table 4. Item sub-list 3 (All heights = 17 cm)

Mattress type	Component	Length (cm)	Width (cm)	Height (cm)	Demand
F	Head bar	130	16	17	24
	Side bar	170	16	17	24
G	Head bar	150	14	17	40
	Side bar	190	14	17	40

Table 5. Output of two dimensional problem for sub-list 1 (All heights = 8 cm)

Mattress Type	Component	Cutting patterns selected to be used								
		1	2	3	4	5	6	Demand	Generated	
A	Head bar	0	0	0	0	0	6	30	30	0
A	Side bar	0	0	1	1	7	3	30	30	0
B	Head bar	0	0	0	0	6	6	36	36	0
B	Side bar	0	4	6	3	0	0	36	37	1
C	Head bar	0	0	4	6	0	0	42	42	0
C	Side bar	9	6	0	0	0	0	42	42	0
		Number of								
		4	1	3	5	1	5	= 19	Total number of patterns to be used	

Table 6. Output of two dimensional problem for sub-list 2 (All heights = 13 cm)

Mattress Type	Component	Cutting patterns selected to be used								
		1	2	3	4	5	6	Demand	Generated	Surplus
D	Head Bar	0	0	0	2	4	11	28	28	0
D	Side Bar	0	0	2	10	8	4	28	28	0
E	Head Bar	1	8	7	0	0	0	26	26	0
E	Side Bar	8	1	0	0	0	0	26	26	0
		Number of patterns selected to be used								
		3	2	1	1	1	2	= 10	Total number of patterns to be used	

Table 7. Output of two dimensional problem for sub-list 3 (All heights = 17 cm)

Mattress Type	Component	Cutting patterns selected to be used									
		1	2	3	4	5	6	7	Demand	Generated	Surplus
F	Head Bar	0	0	0	0	1	2	11	24	25	1
F	Side Bar	0	2	3	3	3	3	4	24	25	1
G	Head Bar	0	0	0	12	11	10	0	40	45	5
G	Side Bar	12	10	9	0	0	0	0	40	43	3
		Number of patterns selected to be used									
		2	1	1	2	1	1	2	= 10	Total number of patterns to be used	

Once two-dimensional problems are solved, the output specifies selected cutting patterns. Those patterns have merged into a cutting plan for the machines that perform vertical cuts. The output also provides an information on the total numbers of patterns to be used. Then, the total numbers of patterns along with their respective heights have fed into the problem in the second stage, which is a 1DCSP. Tables 8 and 9 show the input and output of the 1DCSP, respectively.

For the sample problem, the waste level is calculated as 6.25%. The solution procedure has been tested for 5 different problems in local bedding company. As a result of solving these problems, reduction in waste levels are shown in the Table 10. It has been observed that the solution time of test problems vary between 1 and 2 min.

Table 8. Input of one dimensional problem

Height	Demand
8	19
13	10
17	10

Table 9. Output of one dimensional problem

Height	Selected Cutting Patterns				
	1	2	Demand	Generated	Surplus
8	6	1	19	20	1
13	4	0	10	12	2
17	0	5	10	10	0
	Number of blocks required		Total blocks required		
	3	2	5		

Table 10. Reduction in waste level for test problems

Test problem	Reduction in waste level
1	2.79%
2	2.61%
3	4.16%
4	3.28%
5	1.42%

5 Decision Support System

The solution procedure has been implemented in a user friendly decision support tool. The tool enables the user to develop the cutting plans easily and efficiently. The interface of the tool is shown in Fig. 6. The interface incorporates three main sections. The first section is related to managing the production plan which contains the production orders of the mattresses. In this section, it is possible to list, delete, add or edit the entries in the production plan. The order management screen is shown in Fig. 7. It helps the user to have the updated version of the production plan. The second section takes the entries in production plan and process them with BOM data in order to generate item list to be cut and then solves the problem. At this point, the user is

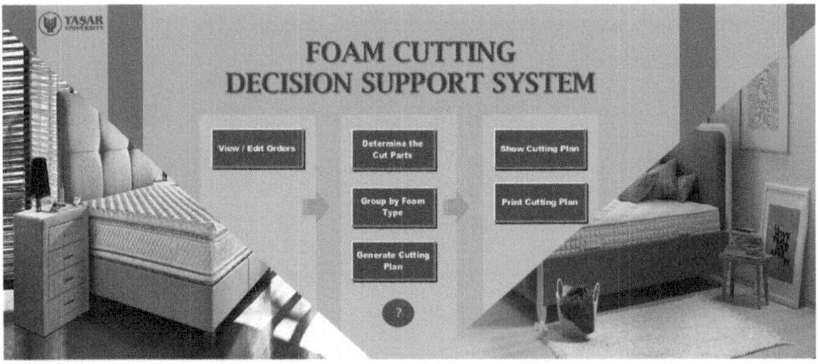

Fig. 6. Main interface of planning tool

informed about the stages of the solution. Finally, the third section allows the user to see and review the cutting plans. The outcome can be displayed in different format and in different detail levels. It is also possible to print the outcome in selected format.

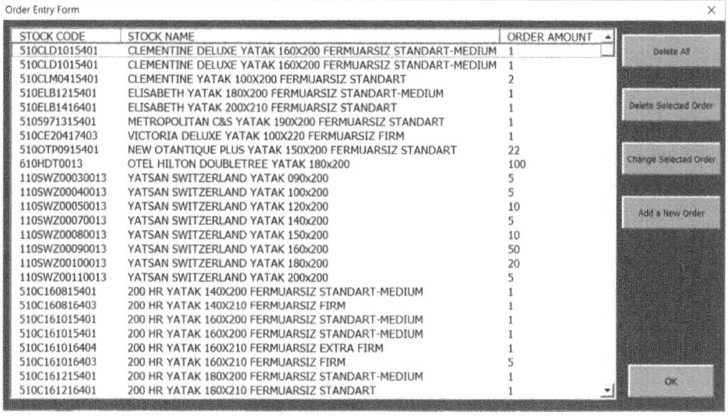
Order Entry Form

STOCK CODE	STOCK NAME	ORDER AMOUNT
510CLD1015401	CLEMENTINE DELUXE YATAK 160X200 FERMUARSIZ STANDART-MEDIUM	1
510CLD1015401	CLEMENTINE DELUXE YATAK 160X200 FERMUARSIZ STANDART-MEDIUM	1
510CLM0415401	CLEMENTINE YATAK 100X200 FERMUARSIZ STANDART	2
510ELB1215401	ELISABETH YATAK 180X200 FERMUARSIZ STANDART-MEDIUM	1
510ELB1416401	ELISABETH YATAK 200X210 FERMUARSIZ STANDART	1
5105971315401	METROPOLITAN C&S YATAK 190X200 FERMUARSIZ STANDART	1
510CE20417403	VICTORIA DELUXE YATAK 100X220 FERMUARSIZ FIRM	1
510OTP0915401	NEW OTANTIQUE PLUS YATAK 150X200 FERMUARSIZ STANDART	22
610HDT0013	OTEL HILTON DOUBLETREE YATAK 180x200	100
110SWZ00030013	YATSAN SWITZERLAND YATAK 090x200	5
110SWZ00040013	YATSAN SWITZERLAND YATAK 100x200	5
110SWZ00050013	YATSAN SWITZERLAND YATAK 120x200	10
110SWZ00070013	YATSAN SWITZERLAND YATAK 140x200	5
110SWZ00080013	YATSAN SWITZERLAND YATAK 150x200	10
110SWZ00090013	YATSAN SWITZERLAND YATAK 160x200	50
110SWZ00100013	YATSAN SWITZERLAND YATAK 180x200	20
110SWZ00110013	YATSAN SWITZERLAND YATAK 200x200	5
510C160815401	200 HR YATAK 140X200 FERMUARSIZ STANDART-MEDIUM	1
510C160816403	200 HR YATAK 140X210 FERMUARSIZ FIRM	1
510C161015401	200 HR YATAK 160X200 FERMUARSIZ STANDART-MEDIUM	1
510C161015401	200 HR YATAK 160X200 FERMUARSIZ STANDART-MEDIUM	1
510C161016404	200 HR YATAK 160X210 FERMUARSIZ EXTRA FIRM	1
510C161016403	200 HR YATAK 160X210 FERMUARSIZ FIRM	5
510C161215401	200 HR YATAK 180X200 FERMUARSIZ STANDART-MEDIUM	1
510C161216401	200 HR YATAK 180X210 FERMUARSIZ STANDART	1

Delete All
Delete Selected Order
Change Selected Order
Add a New Order
OK

Fig. 7. Management of the production plan

6 Conclusion and Future Work

In this study, 3DCSP has been solved in two consecutive stages, since the block orientation is not allowed in the company. The first stage contains a collection of two-dimensional cutting stock problems (2DCSP), and the second stage contains a one-dimensional cutting stock problem (1DCSP). The important point in this study is that the output of the first stage (2DCSP) is used as the input in the second stage (1DCSP). According to the results of 5 sample data, the decrease in waste level is between 1.4% and 4.2%.

A flexible and user friendly tool has been developed for the company to enable the planning engineer to prepare quick and efficient cutting plans. It has been a useful tool for the company since they don't have such a tool before this study. It manages the input of the problem and present the solutions to the user through a user-friendly interface. The solution process and outcomes has been verified and validated. An observable decrease (in between 1.4%–4.2%) in waste level has been reported. On the other hand, the company now have a means of data collection and history of cutting process.

An important point can be studied in the future, if the orientation of the blocks *B* can be changed. Because, more decrease in waste levels may be attained by using some other solution methods, once the orientation of the blocks is allowed within the company.

Acknowledgment. This study is supported by TUBITAK (The Scientific and Technological Research Council of Turkey) in the program of "2209-B Undergraduate Thesis Support Program for Industrial Applications". We would like to thank our colleagues Buğra Akboy, Tevfik Ercan, Simay Karaşahin and Yusuf Rıdvanoğulları for their help and contribution to the study.

References

1. Kantorovich LV (1960) Mathematical methods of organizing and planning production. Manag Sci 6:366–422
2. Gilmore PC, Gomory RE (1961) A linear programming approach to the cutting stock problem. Oper Res 9:849–859
3. Gilmore PC, Gomory RE (1963) A linear Programming approach to the cutting stock problem: Part II. Oper Res 11:863–888
4. Gilmore PC, Gomory RE (1965) Multistage cutting stock problems of two and more dimensions. Oper Res 13:94–120
5. Queiroz TA, Miyazawa FK, Wakabayashi Y, Xavier EC (2012) Algorithms for 3D guillotine cutting problems: unbounded knapsack, cutting stock and strip packing. Comput Oper Res 39:200–212
6. Dyckoff H (1981) A new linear programming approach to the cutting stock problem. Oper Res 29:1092–1104
7. Val´erio de Carvalho JMV (1999) Exact solution of bin packing problems using column generation and branch and bound. Annal Oper Res 86:629–659
8. Cambazard H, O'Sullivan B (2010) Propagating the bin packing constraint using linear programming. In: Principles and practice of constraint programming – CP 2010. LNCS, vol 6308, pp 129–136
9. Brandao F, Pedroso JP (2016) Bin packing and related problems: general arc-flow formulation with graph compression. Comput Oper Res 69:56–67
10. Hadjiconstantinou E, Christofides N (1995) An exact algorithm for general, orthogonal, two-dimensional knapsack problems. Eur J Oper Res 83:39–56
11. Martello S, Toth P (1990) Knapsack problems: algorithms and computer implementations. Wiley, Chichester
12. Kämpke T (1988) Simulated annealing: use of a new tool in bin packing. Annal Oper Res 16:327–332
13. Alvim ACF, Ribeiro CC, Glover F, Aloise DJ (2004) A hybrid improvement heuristic for the one-dimensional bin packing problem. J Heuristics 10:205–229
14. Malaguti E, Durán RM, Toth P (2014) Approaches to real world two-dimensional cutting problems. Int J Manag Sci 47:99–115
15. Suliman MA (2001) Pattern generating procedure for the cutting stock problem. Int J Prod Econ 74:293–301
16. Rodrigo WNP, Daundasekera WB, Perera AAI (2012) Pattern generation for two-dimensional cutting stock problem. Int J Math Trends Technol 3:54–62

Author Index

N. M. Durakbasa and M. G. Gencyilmaz (Eds.): ISPR 2018, *Proceedings of the International Symposium for Production Research 2018*, pp. 961–963, 2019.
https://doi.org/10.1007/978-3-319-92267-6

The manufacturer's authorised representative in the EU is Springer Nature Customer Service Centre GmbH, Europaplatz 3, 69115 Heidelberg, Germany. If you have any concerns regarding our products, please contact ProductSafety@springernature.com

Printed and bound by CPI Group (UK) Ltd, Croydon, CR0 4YY

05/10/2026

02228150-0001